Proceedings of the 75th RILEM Annual Week 2021

RILEM BOOKSERIES

Volume 40

RILEM, The International Union of Laboratories and Experts in Construction Materials, Systems and Structures, founded in 1947, is a non-governmental scientific association whose goal is to contribute to progress in the construction sciences, techniques and industries, essentially by means of the communication it fosters between research and practice. RILEM's focus is on construction materials and their use in building and civil engineering structures, covering all phases of the building process from manufacture to use and recycling of materials. More information on RILEM and its previous publications can be found on www.RILEM.net.

Indexed in SCOPUS, Google Scholar and SpringerLink.

J. Ivan Escalante-Garcia · Pedro Castro Borges ·
Alejandro Duran-Herrera
Editors

Proceedings of the 75th RILEM Annual Week 2021

Advances in Sustainable Construction Materials and Structures

Springer

Editors
J. Ivan Escalante-Garcia
Saltillo Unit
Cinvestav
Ramos Arizpe, Coahuila, Mexico

Pedro Castro Borges
Merida Unit
Cinvestav
Mérida, Mexico

Alejandro Duran-Herrera
Facultad de Ingeniería Civil
Universidad Autónoma de Nuevo León
San Nicolás de los Garza, Nuevo León
Mexico

ISSN 2211-0844 ISSN 2211-0852 (electronic)
RILEM Bookseries
ISBN 978-3-031-21734-0 ISBN 978-3-031-21735-7 (eBook)
https://doi.org/10.1007/978-3-031-21735-7

This Springer imprint is published by the registered company Springer Nature Switzerland AG
The registered company address is: Gewerbestrasse 11, 6330 Cham, Switzerland

Scientific Committee of the Conference by Themes

1. Supplementary Cementitious Materials

Leader

Manu Santhanam IIT Madras, India

Members

Alejandro Durán-Herrera	Universidad Autónoma de Nuevo León (UANL), Mexico
Karen Scrivener	École Polytechnique Fédérale de Lausanne, Switzerland
Ruben Snellings	Flemish Institute for Technological Research, Belgium
Nele De Belie	Ghent University, Belgium
Florence Collet	Université de Rennes 1, France
Said Kenai	Saad Dahlab University, Argelia
Fernando Martirena	Universidad Central de las Villas, Cuba
Douglas Hooton	University of Toronto, Canada
Anya Vollpracht	RWTH Aachen University, Germany
Wolfram Schmidt	The Federal Institute for Materials Research and Testing (BAM), Germany
Maria Juenger	The University of Texas, USA
Arezki Tagnit-Hamou	Université de Sherbrooke, Canada
Mike Otieno	Wits University, South Africa
Pedro Montes-García	CIIDIR-IPN, México
Ricardo X. Magallanes-Rivera	Facultad de Ingeniería, UAdeC, México

2. Durability and Life Cycle Assessment in Urban and Marine Conditions

Segment Scientific Committee

Leader

Daman Panesar University of Toronto, Canada

Members

Pedro Castro-Borges	Cinvestav Unidad Mérida, México
Véronique Baroghel-Bouny	IFSTTAR, France
Esperanza Menendez	IETCC-CSIC, Spain
Alexandra Bertron	Institut National des Sciences Appliquées de Toulouse, France
Arnaud Castel	University of Technology Sydney, Australia
Edgardo F. Irassar	University of the Center of the Buenos Aires Province, Argentina
Roberto J. Torrent	M.A.S. Services Ltd., Argentina
Tezozomoc Pérez-López	Universidad Autónoma de Campeche, Mexico
Gerardo Fajardo-Sanmiguel	Universidad Autónoma de Nuevo León, Mexico
Jose M. Mendoza-Rangel	Universidad Autónoma de Nuevo León, Mexico
José Pacheco	CTL Group USA
Ueli Angst	Institute for Building Materials, Switzerland
Andres Torres-Acosta	Tecnológico de Monterrey, Campus Querétaro, Mexico
Selmo Kuperman	DESEK, Brasil

3. Additive Manufacturing in Concrete Construction

Leader

Viktor Mechtcherine Institute of Construction Materials, Germany

Members

Kamal Khayat	Missouri University of Science and Technology, USA
Dimitri Feys	Missouri University of Science and Technology, USA
Geert De Schutter	Ghent University, Belgium
José Antonio Tenorio	IETCC-CSIC, Spain
Sofiane Amziane	Université Clermont Auvergne, France
Dirk Lowke	The Technical University of Munich, Germany
Mohammed Sonebi	Queen's University Belfast, UK
Dr. Behzad Nematollahi	Swinburne University of Technology, Australia

4. Structural Performance and Design

Leader

Barzin Mobasher Arizona State University, USA

Members

Flavio De Andrade Silva	Pontifical Catholic University of Rio de Janeiro, Brazil
Frank Dehn	Karlsruhe Institute of Technology, Germany
Sergio Cavalaro	Loughborough University, UK
Emmanuel Denarie	École Polytechnique Fédérale de Lausanne, Switzerland
Terje Kanstad	Norwegian University of Science and Technology, Norway
Cristina Zanotti	University of British Columbia, Canada
Hans Beushausen	University of Cape Town, South Africa
Sergio Alcocer	Instituto de Ingeniería UNAM, Mexico
Esteban Astudillo	AG2M Ingeniería, Mexico
Manuel García	Alonso y Asociados, Mexico
Marco Di Prisco	Politecnico di Milano, Italy
Tulio Bittencourt	Universidade de São Paulo, Brazil
Liberato Ferrara	Politecnico de Milano, Italia
Tomoki Shiotani	Kyoto University, Japan

5. Non-Portland Cements and Alkali Activated Cementitious Materials and Eco-concrete

Leader

Susan Bernal University of Leeds, UK

Members

John Provis	The University of Sheffield, UK
Caijun Shi	Hunan University, China
Gregor Gluth	The Federal Institute for Materials Research and Testing, Germany
Francisca Puertas	IETCC-CSIC, Spain
Frank Winnefeld	EMPA, Switzerland
Nuno Cristelo	UTAD, Portugal
Rubí Mejía De Gutiérrez	Universidad del Valle, Colombia
Jordi Paya	Universitat Politècnica de València, Spain
José Monzo	Universitat Politècnica de València, Spain
Oswaldo. Burciaga-Diaz	Instituto Tecnológico de Saltillo, Mexico

Lauren Gómez-Zamorano	Universidad Autónoma de Nuevo León, Mexico
Ismael Flores-Vivian	Universidad Autónoma de Nuevo León, Mexico
Ulises Avila-Lopez	Facultad de Ingeniería, UAdeC, México

6. Cultural Heritage

Leader

Enrico Sassoni University of Bologna, Italy

Members

Arun Menon	Indian Institute of Technology Madras, India
Daniel Oliveira	Universidade do Minho, Portugal
Elia Alonso	UMSNH, Mexico
José Álvarez	Universidad de Navarra, Spain

7. Non-Destructive Testing Techniques

Leader

David Benavente Universidad de Alicante, Spain

Members

Markus Krüger	Graz University of Technology, Austria
José Delgado Rodrigues	National Laboratory of Civil Engineering, Lisbon, Portugal
Elizabeth Marie-Victorie	Laboratoire de Recherche des Monuments Historiques, France
Davide Gulotta	The Getty Conservation Institute, USA
Jason Weiss	Oregon State University, USA

8. Bituminous Materials, Polymers, Timber, Bamboo, Recycling, Masonry, etc.

Leader

Veronique Verges-Belmin Ministère de la culture et de la communication, France

Members

Graça Vasconcelos	Universidade do Minho, Portugal
Jose Norambuena-Contreras	University of Bio-Bio, Chile
Paulo B. Lourenço	Universidad do Minho, Portugal
Chi Sun Poon	The Hong Kong Polytechnic University, Hong Kong
Jorge Branco	Universidade do Minho, Portugal

Preface

One of the main current challenges of mankind is attaining sustainable development. In order to build long-lasting infrastructure and structures, we require improvements and innovation in many research areas, from cementitious materials to construction materials and construction technologies. Improvements in portland cement, alternative cements, concretes of modified properties, alternative construction materials and technologies, new characterization techniques, etc., are among the many areas of opportunity. This book of proceedings of the Conference on Advances in Sustainable Construction Materials and Structures brings more than 100 high-quality peer-reviewed contributions from authors from all continents, covering the eight themes of the conference, including *Supplementary Cementitious Materials*; *Durability and Life Cycle Assessment in Urban and Marine Conditions*; *Additive Manufacturing in Concrete Construction*; *Structural Performance and Design*; *Non-Portland Cements and Alkali Activated Cementitious Materials and Eco-concrete*; *Cultural Heritage*; *Non-Destructive Testing Techniques* and *Bituminous Materials, Polymers, Timber, Bamboo, Recycling, Masonry, etc.*

The conference took place in the middle of a pandemic world contingency and it was a success thanks to the support of RILEM, the logistics teams from the UANL and Cinvestav, our main sponsor Holcim Mexico, and most importantly the outstanding contributions from all delegates, many of which accompanied us in person in Mérida. Finally, we would like to thank all reviewers for their outstanding collaboration during the peer-review process, which was invaluable in making this book a success.

J. Ivan Escalante-Garcia
Pedro Castro Borges
Alejandro Duran-Herrera

Contents

RILEM Publications

The following list is presenting the global offer of RILEM Publications, sorted by series. Each publication is available in printed version and/or in online version.

RILEM Proceedings (PRO)

PRO 1: Durability of High Performance Concrete (ISBN: 2-912143-03-9; e-ISBN: 2-351580-12-5; e-ISBN: 2351580125); *Ed. H. Sommer*

PRO 2: Chloride Penetration into Concrete (ISBN: 2-912143-00-04; e-ISBN: 2912143454); *Eds. L.-O. Nilsson and J.-P. Ollivier*

PRO 3: Evaluation and Strengthening of Existing Masonry Structures (ISBN: 2-912143-02-0; e-ISBN: 2351580141); *Eds. L. Binda and C. Modena*

PRO 4: Concrete: From Material to Structure (ISBN: 2-912143-04-7; e-ISBN: 2351580206); *Eds. J.-P. Bournazel and Y. Malier*

PRO 5: The Role of Admixtures in High Performance Concrete (ISBN: 2-912143-05-5; e-ISBN: 2351580214); *Eds. J. G. Cabrera and R. Rivera-Villarreal*

PRO 6: High Performance Fiber Reinforced Cement Composites - HPFRCC 3 (ISBN: 2-912143-06-3; e-ISBN: 2351580222); *Eds. H. W. Reinhardt and A. E. Naaman*

PRO 7: 1st International RILEM Symposium on Self-Compacting Concrete (ISBN: 2-912143-09-8; e-ISBN: 2912143721); *Eds. Å. Skarendahl and Ö. Petersson*

PRO 8: International RILEM Symposium on Timber Engineering (ISBN: 2-912143-10-1; e-ISBN: 2351580230); *Ed. L. Boström*

PRO 9: 2nd International RILEM Symposium on Adhesion between Polymers and Concrete ISAP '99 (ISBN: 2-912143-11-X; e-ISBN: 2351580249); *Eds. Y. Ohama and M. Puterman*

PRO 10: 3rd International RILEM Symposium on Durability of Building and Construction Sealants (ISBN: 2-912143-13-6; e-ISBN: 2351580257); *Ed. A. T. Wolf*

PRO 11: 4th International RILEM Conference on Reflective Cracking in Pavements (ISBN: 2-912143-14-4; e-ISBN: 2351580265); *Eds. A. O. Abd El Halim, D. A. Taylor and El H. H. Mohamed*

PRO 12: International RILEM Workshop on Historic Mortars: Characteristics and Tests (ISBN: 2-912143-15-2; e-ISBN: 2351580273); *Eds. P. Bartos, C. Groot and J. J. Hughes*

PRO 13: 2nd International RILEM Symposium on Hydration and Setting (ISBN: 2-912143-16-0; e-ISBN: 2351580281); *Ed. A. Nonat*

PRO 14: Integrated Life-Cycle Design of Materials and Structures - ILCDES 2000 (ISBN: 951-758-408-3; e-ISBN: 235158029X); (ISSN: 0356-9403); *Ed. S. Sarja*

PRO 15: Fifth RILEM Symposium on Fibre-Reinforced Concretes (FRC) - BEFIB'2000 (ISBN: 2-912143-18-7; e-ISBN: 291214373X); *Eds. P. Rossi and G. Chanvillard*

PRO 16: Life Prediction and Management of Concrete Structures (ISBN: 2-912143-19-5; e-ISBN: 2351580303); *Ed. D. Naus*

PRO 17: Shrinkage of Concrete – Shrinkage 2000 (ISBN: 2-912143-20-9; e-ISBN: 2351580311); *Eds. V. Baroghel-Bouny and P.-C. Aïtcin*

PRO 18: Measurement and Interpretation of the On-Site Corrosion Rate (ISBN: 2-912143-21-7; e-ISBN: 235158032X); *Eds. C. Andrade, C. Alonso, J. Fullea, J. Polimon and J. Rodriguez*

PRO 19: Testing and Modelling the Chloride Ingress into Concrete (ISBN: 2-912143-22-5; e-ISBN: 2351580338); *Eds. C. Andrade and J. Kropp*

PRO 20: 1st International RILEM Workshop on Microbial Impacts on Building Materials (CD 02) (e-ISBN 978-2-35158-013-4); *Ed. M. Ribas Silva*

PRO 21: International RILEM Symposium on Connections between Steel and Concrete (ISBN: 2-912143-25-X; e-ISBN: 2351580346); *Ed. R. Eligehausen*

PRO 22: International RILEM Symposium on Joints in Timber Structures (ISBN: 2-912143-28-4; e-ISBN: 2351580354); *Eds. S. Aicher and H.-W. Reinhardt*

PRO 23: International RILEM Conference on Early Age Cracking in Cementitious Systems (ISBN: 2-912143-29-2; e-ISBN: 2351580362); *Eds. K. Kovler and A. Bentur*

PRO 24: 2nd International RILEM Workshop on Frost Resistance of Concrete (ISBN: 2-912143-30-6; e-ISBN: 2351580370); *Eds. M. J. Setzer, R. Auberg and H.-J. Keck*

PRO 25: International RILEM Workshop on Frost Damage in Concrete (ISBN: 2-912143-31-4; e-ISBN: 2351580389); *Eds. D. J. Janssen, M. J. Setzer and M. B. Snyder*

PRO 26: International RILEM Workshop on On-Site Control and Evaluation of Masonry Structures (ISBN: 2-912143-34-9; e-ISBN: 2351580141); *Eds. L. Binda and R. C. de Vekey*

PRO 27: International RILEM Symposium on Building Joint Sealants (CD03; e-ISBN: 235158015X); *Ed. A. T. Wolf*

PRO 28: 6th International RILEM Symposium on Performance Testing and Evaluation of Bituminous Materials - PTEBM'03 (ISBN: 2-912143-35-7; e-ISBN: 978-2-912143-77-8); *Ed. M. N. Partl*

PRO 29: 2nd International RILEM Workshop on Life Prediction and Ageing Management of Concrete Structures (ISBN: 2-912143-36-5; e-ISBN: 2912143780); *Ed. D. J. Naus*

PRO 30: 4th International RILEM Workshop on High Performance Fiber Reinforced Cement Composites - HPFRCC 4 (ISBN: 2-912143-37-3; e-ISBN: 2912143799); *Eds. A. E. Naaman and H. W. Reinhardt*

PRO 31: International RILEM Workshop on Test and Design Methods for Steel Fibre Reinforced Concrete: Background and Experiences (ISBN: 2-912143-38-1; e-ISBN: 2351580168); *Eds. B. Schnütgen and L. Vandewalle*

PRO 32: International Conference on Advances in Concrete and Structures 2 vol.(ISBN (set): 2-912143-41-1; e-ISBN: 2351580176); *Eds. Ying-shu Yuan, Surendra P. Shah and Heng-lin Lü*

PRO 33: 3rd International Symposium on Self-Compacting Concrete (ISBN: 2-912143-42-X; e-ISBN: 2912143713); *Eds. Ó. Wallevik and I. Níelsson*

PRO 34: International RILEM Conference on Microbial Impact on Building Materials (ISBN: 2-912143-43-8; e-ISBN: 2351580184); *Ed. M. Ribas Silva*

PRO 35: International RILEM TC 186-ISA on Internal Sulfate Attack and Delayed Ettringite Formation (ISBN: 2-912143-44-6; e-ISBN: 2912143802); *Eds. K. Scrivener and J. Skalny*

PRO 36: International RILEM Symposium on Concrete Science and Engineering – A Tribute to Arnon Bentur (ISBN: 2-912143-46-2; e-ISBN: 2912143586); *Eds. K. Kovler, J. Marchand, S. Mindess and J. Weiss*

PRO 37: 5th International RILEM Conference on Cracking in Pavements – Mitigation, Risk Assessment and Prevention (ISBN: 2-912143-47-0; e-ISBN: 2912143764); *Eds. C. Petit, I. Al-Qadi and A. Millien*

PRO 38: 3rd International RILEM Workshop on Testing and Modelling the Chloride Ingress into Concrete (ISBN: 2-912143-48-9; e-ISBN: 2912143578); *Eds. C. Andrade and J. Kropp*

PRO 39: 6th International RILEM Symposium on Fibre-Reinforced Concretes - BEFIB 2004 (ISBN: 2-912143-51-9; e-ISBN: 2912143748); *Eds. M. Di Prisco, R. Felicetti and G. A. Plizzari*

PRO 40: International RILEM Conference on the Use of Recycled Materials in Buildings and Structures (ISBN: 2-912143-52-7; e-ISBN: 2912143756); *Eds. E. Vázquez, Ch. F. Hendriks and G. M. T. Janssen*

PRO 41: RILEM International Symposium on Environment-Conscious Materials and Systems for Sustainable Development (ISBN: 2-912143-55-1; e-ISBN: 2912143640); *Eds. N. Kashino and Y. Ohama*

PRO 42: SCC'2005 - China: 1st International Symposium on Design, Performance and Use of Self-Consolidating Concrete (ISBN: 2-912143-61-6; e-ISBN: 2912143624); *Eds. Zhiwu Yu, Caijun Shi, Kamal Henri Khayat and Youjun Xie*

PRO 43: International RILEM Workshop on Bonded Concrete Overlays (e-ISBN: 2-912143-83-7); *Eds. J. L. Granju and J. Silfwerbrand*

PRO 44: 2nd International RILEM Workshop on Microbial Impacts on Building Materials (CD11) (e-ISBN: 2-912143-84-5); *Ed. M. Ribas Silva*

PRO 45: 2nd International Symposium on Nanotechnology in Construction, Bilbao (ISBN: 2-912143-87-X; e-ISBN: 2912143888); *Eds. Peter J. M. Bartos, Yolanda de Miguel and Antonio Porro*

PRO 46: ConcreteLife'06 - International RILEM-JCI Seminar on Concrete Durability and Service Life Planning: Curing, Crack Control, Performance in Harsh Environments (ISBN: 2-912143-89-6; e-ISBN: 291214390X); *Ed. K. Kovler*

PRO 47: International RILEM Workshop on Performance Based Evaluation and Indicators for Concrete Durability (ISBN: 978-2-912143-95-2; e-ISBN: 9782912143969); *Eds. V. Baroghel-Bouny, C. Andrade, R. Torrent and K. Scrivener*

PRO 48: 1st International RILEM Symposium on Advances in Concrete through Science and Engineering (e-ISBN: 2-912143-92-6); *Eds. J. Weiss, K. Kovler, J. Marchand, and S. Mindess*

PRO 49: International RILEM Workshop on High Performance Fiber Reinforced Cementitious Composites in Structural Applications (ISBN: 2-912143-93-4; e-ISBN: 2912143942); *Eds. G. Fischer and V. C. Li*

PRO 50: 1st International RILEM Symposium on Textile Reinforced Concrete (ISBN: 2-912143-97-7; e-ISBN: 2351580087); *Eds. Josef Hegger, Wolfgang Brameshuber and Norbert Will*

PRO 51: 2nd International Symposium on Advances in Concrete through Science and Engineering (ISBN: 2-35158-003-6; e-ISBN: 2-35158-002-8); *Eds. J. Marchand, B. Bissonnette, R. Gagné, M. Jolin and F. Paradis*

PRO 52: Volume Changes of Hardening Concrete: Testing and Mitigation (ISBN: 2-35158-004-4; e-ISBN: 2-35158-005-2); *Eds. O. M. Jensen, P. Lura and K. Kovler*

PRO 53: High Performance Fiber Reinforced Cement Composites - HPFRCC5 (ISBN: 978-2-35158-046-2; e-ISBN: 978-2-35158-089-9); *Eds. H. W. Reinhardt and A. E. Naaman*

PRO 54: 5th International RILEM Symposium on Self-Compacting Concrete (ISBN: 978-2-35158-047-9; e-ISBN: 978-2-35158-088-2); *Eds. G. De Schutter and V. Boel*

PRO 55: International RILEM Symposium Photocatalysis, Environment and Construction Materials (ISBN: 978-2-35158-056-1; e-ISBN: 978-2-35158-057-8); *Eds. P. Baglioni and L. Cassar*

PRO 56: International RILEM Workshop on Integral Service Life Modelling of Concrete Structures (ISBN 978-2-35158-058-5; e-ISBN: 978-2-35158-090-5); *Eds. R. M. Ferreira, J. Gulikers and C. Andrade*

PRO 57: RILEM Workshop on Performance of cement-based materials in aggressive aqueous environments (e-ISBN: 978-2-35158-059-2); *Ed. N. De Belie*

PRO 58: International RILEM Symposium on Concrete Modelling - CONMOD'08 (ISBN: 978-2-35158-060-8; e-ISBN: 978-2-35158-076-9); *Eds. E. Schlangen and G. De Schutter*

PRO 59: International RILEM Conference on On Site Assessment of Concrete, Masonry and Timber Structures - SACoMaTiS 2008 (ISBN set: 978-2-35158-061-5; e-ISBN: 978-2-35158-075-2); *Eds. L. Binda, M. di Prisco and R. Felicetti*

PRO 60: Seventh RILEM International Symposium on Fibre Reinforced Concrete: Design and Applications - BEFIB 2008 (ISBN: 978-2-35158-064-6; e-ISBN: 978-2-35158-086-8); *Ed. R. Gettu*

PRO 61: 1st International Conference on Microstructure Related Durability of Cementitious Compo-sites 2 vol., (ISBN: 978-2-35158-065-3; e-ISBN: 978-2-35158-084-4); *Eds. W. Sun, K. van Breugel, C. Miao, G. Ye and H. Chen*

PRO 62: NSF/ RILEM Workshop: In-situ Evaluation of Historic Wood and Masonry Structures (e-ISBN: 978-2-35158-068-4); *Eds. B. Kasal, R. Anthony and M. Drdácký*

PRO 63: Concrete in Aggressive Aqueous Environments: Performance, Testing and Modelling, 2 vol., (ISBN: 978-2-35158-071-4; e-ISBN: 978-2-35158-082-0); *Eds. M. G. Alexander and A. Bertron*

PRO 64: Long Term Performance of Cementitious Barriers and Reinforced Concrete in Nuclear Power Plants and Waste Management - NUCPERF 2009 (ISBN: 978-2-35158-072-1; e-ISBN: 978-2-35158-087-5); *Eds. V. L'Hostis, R. Gens and C. Gallé*

PRO 65: Design Performance and Use of Self-consolidating Concrete - SCC'2009 (ISBN: 978-2-35158-073-8; e-ISBN: 978-2-35158-093-6); *Eds. C. Shi, Z. Yu, K. H. Khayat and P. Yan*

PRO 66: 2nd International RILEM Workshop on Concrete Durability and Service Life Planning - ConcreteLife'09 (ISBN: 978-2-35158-074-5; ISBN: 978-2-35158-074-5); *Ed. K. Kovler*

PRO 67: Repairs Mortars for Historic Masonry (e-ISBN: 978-2-35158-083-7); *Ed. C. Groot*

PRO 68: Proceedings of the 3rd International RILEM Symposium on 'Rheology of Cement Suspensions such as Fresh Concrete (ISBN 978-2-35158-091-2; e-ISBN: 978-2-35158-092-9); *Eds. O. H. Wallevik, S. Kubens and S. Oesterheld*

PRO 69: 3rd International PhD Student Workshop on 'Modelling the Durability of Reinforced Concrete (ISBN: 978-2-35158-095-0); *Eds. R. M. Ferreira, J. Gulikers and C. Andrade*

PRO 70: 2nd International Conference on 'Service Life Design for Infrastructure' (ISBN set: 978-2-35158-096-7, e-ISBN: 978-2-35158-097-4); *Eds. K. van Breugel, G. Ye and Y. Yuan*

PRO 71: Advances in Civil Engineering Materials - The 50-year Teaching Anniversary of Prof. Sun Wei' (ISBN: 978-2-35158-098-1; e-ISBN: 978-2-35158-099-8); *Eds. C. Miao, G. Ye and H. Chen*

PRO 72: First International Conference on 'Advances in Chemically-Activated Materials – CAM'2010' (2010), 264 pp, ISBN: 978-2-35158-101-8; e-ISBN: 978-2-35158-115-5, *Eds. Caijun Shi and Xiaodong Shen*

PRO 73: 2nd International Conference on 'Waste Engineering and Management - ICWEM 2010' (2010), 894 pp, ISBN: 978-2-35158-102-5; e-ISBN: 978-2-35158-103-2, *Eds. J. Zh. Xiao, Y. Zhang, M. S. Cheung and R. Chu*

PRO 74: International RILEM Conference on 'Use of Superabsorbent Polymers and Other New Addditives in Concrete' (2010) 374 pp., ISBN: 978-2-35158-104-9; e-ISBN: 978-2-35158-105-6; *Eds. O.M. Jensen, M.T. Hasholt, and S. Laustsen*

PRO 75: International Conference on 'Material Science - 2nd ICTRC - Textile Reinforced Concrete - Theme 1' (2010) 436 pp., ISBN: 978-2-35158-106-3; e-ISBN: 978-2-35158-107-0; *Ed. W. Brameshuber*

PRO 76: International Conference on 'Material Science - HetMat - Modelling of Heterogeneous Ma-terials - Theme 2' (2010) 255 pp., ISBN: 978-2-35158-108-7; e-ISBN: 978-2-35158-109-4; *Ed. W. Brameshuber*

PRO 77: International Conference on 'Material Science - AdIPoC - Additions Improving Properties of Concrete - Theme 3' (2010) 459 pp., ISBN: 978-2-35158-110-0; e-ISBN: 978-2-35158-111-7; *Ed. W. Brameshuber*

PRO 78: 2nd Historic Mortars Conference and RILEM TC 203-RHM Final Workshop – HMC2010 (2010) 1416 pp., e-ISBN: 978-2-35158-112-4; *Eds. J. Válek, C. Groot and J. J. Hughes*

PRO 79: International RILEM Conference on Advances in Construction Materials Through Science and Engineering (2011) 213 pp., ISBN: 978-2-35158-116-2, e-ISBN: 978-2-35158-117-9; *Eds. Christopher Leung and K.T. Wan*

PRO 80: 2nd International RILEM Conference on Concrete Spalling due to Fire Exposure (2011) 453 pp., ISBN: 978-2-35158-118-6, e-ISBN: 978-2-35158-119-3; *Eds. E.A.B. Koenders and F. Dehn*

PRO 81: 2nd International RILEM Conference on Strain Hardening Cementitious Composites (SHCC2-Rio) (2011) 451 pp., ISBN: 978-2-35158-120-9, e-ISBN: 978-2-35158-121-6; *Eds. R.D. Toledo Filho, F.A. Silva, E.A.B. Koenders and E.M.R. Fairbairn*

PRO 82: 2nd International RILEM Conference on Progress of Recycling in the Built Environment (2011) 507 pp., e-ISBN: 978-2-35158-122-3; *Eds. V.M. John, E. Vazquez, S.C. Angulo and C. Ulsen*

PRO 83: 2nd International Conference on Microstructural-related Durability of Cementitious Composites (2012) 250 pp., ISBN: 978-2-35158-129-2; e-ISBN: 978-2-35158-123-0; *Eds. G. Ye, K. van Breugel, W. Sun and C. Miao*

PRO 84: CONSEC13 - Seventh International Conference on Concrete under Severe Conditions – Environment and Loading (2013) 1930 pp., ISBN: 978-2-35158-124-7; e-ISBN: 978-2- 35158-134-6; *Eds. Z.J. Li, W. Sun, C.W. Miao, K. Sakai, O.E. Gjorv and N. Banthia*

PRO 85: RILEM-JCI International Workshop on Crack Control of Mass Concrete and Related issues concerning Early-Age of Concrete Structures – ConCrack 3 – Control of Cracking in Concrete Structures 3 (2012) 237 pp., ISBN: 978-2-35158-125-4; e-ISBN: 978-2-35158-126-1; *Eds. F. Toutlemonde and J.-M. Torrenti*

PRO 86: International Symposium on Life Cycle Assessment and Construction (2012) 414 pp., ISBN: 978-2-35158-127-8, e-ISBN: 978-2-35158-128-5; *Eds. A. Ventura and C. de la Roche*

PRO 87: UHPFRC 2013 – RILEM-fib-AFGC International Symposium on Ultra-High Performance Fibre-Reinforced Concrete (2013), ISBN: 978-2-35158-130-8, e-ISBN: 978-2-35158-131-5; *Eds. F. Toutlemonde*

PRO 88: 8th RILEM International Symposium on Fibre Reinforced Concrete (2012) 344 pp., ISBN: 978-2-35158-132-2, e-ISBN: 978-2-35158-133-9; *Eds. Joaquim A.O. Barros*

PRO 89: RILEM International workshop on performance-based specification and control of concrete durability (2014) 678 pp, ISBN: 978-2-35158-135-3, e-ISBN: 978-2-35158-136-0; *Eds. D. Bjegović, H. Beushausen and M. Serdar*

PRO 90: 7th RILEM International Conference on Self-Compacting Concrete and of the 1st RILEM International Conference on Rheology and Processing of Construction Materials (2013) 396 pp, ISBN: 978-2-35158-137-7, e-ISBN: 978-2-35158-138-4; *Eds. Nicolas Roussel and Hela Bessaies-Bey*

PRO 91: CONMOD 2014 - RILEM International Symposium on Concrete Modelling (2014), ISBN: 978-2-35158-139-1; e-ISBN: 978-2-35158-140-7; *Eds. Kefei Li, Peiyu Yan and Rongwei Yang*

PRO 92: CAM 2014 - 2nd International Conference on advances in chemically-activated materials (2014) 392 pp., ISBN: 978-2-35158-141-4; e-ISBN: 978-2-35158-142-1; *Eds. Caijun Shi and Xiadong Shen*

PRO 93: SCC 2014 - 3rd International Symposium on Design, Performance and Use of Self-Consolidating Concrete (2014) 438 pp., ISBN: 978-2-35158-143-8; e-ISBN: 978-2-35158-144-5; *Eds. Caijun Shi, Zhihua Ou and Kamal H. Khayat*

PRO 94 (online version): HPFRCC-7 - 7th RILEM conference on High performance fiber reinforced cement compo-sites (2015), e-ISBN: 978-2-35158-146-9; *Eds. H.W. Reinhardt, G.J. Parra-Montesinos and H. Garrecht*

PRO 95: International RILEM Conference on Application of superabsorbent polymers and other new admixtures in concrete construction (2014), ISBN: 978-2-35158-147-6; e-ISBN: 978-2-35158-148-3; *Eds. Viktor Mechtcherine and Christof Schroefl*

PRO 96 (online version): XIII DBMC: XIII International Conference on Durability of Building Materials and Components(2015), e-ISBN: 978-2-35158-149-0; *Eds. M. Quattrone and V.M. John*

PRO 97: SHCC3 – 3rd International RILEM Conference on Strain Hardening Cementitious Compo-sites (2014), ISBN: 978-2-35158-150-6; e-ISBN: 978-2-35158-151-3; *Eds. E. Schlangen, M.G. Sierra Beltran, M. Lukovic and G. Ye*

PRO 98: FERRO-11 – 11th International Symposium on Ferrocement and 3rd ICTRC - International Conference on Textile Reinforced Concrete (2015), ISBN: 978-2-35158-152-0; e-ISBN: 978-2-35158-153-7; *Ed. W. Brameshuber*

PRO 99 (online version): ICBBM 2015 - 1st International Conference on Bio-Based Building Materials (2015), e-ISBN: 978-2-35158-154-4; *Eds. S. Amziane and M. Sonebi*

PRO 100: SCC16 - RILEM Self-Consolidating Concrete Conference (2016), ISBN: 978-2-35158-156-8; e-ISBN: 978-2-35158-157-5; *Ed. Kamal H. Kayat*

PRO 101 (online version): III Progress of Recycling in the Built Environment (2015), e-ISBN: 978-2-35158-158-2; *Eds I. Martins, C. Ulsen and S. C. Angulo*

PRO 102 (online version): RILEM Conference on Microorganisms-Cementitious Materials Interactions (2016), e-ISBN: 978-2-35158-160-5; *Eds. Alexandra Bertron, Henk Jonkers and Virginie Wiktor*

PRO 103 (online version): ACESC'16 - Advances in Civil Engineering and Sustainable Construction (2016), e-ISBN: 978-2-35158-161-2; *Eds. T.Ch. Madhavi, G. Prabhakar, Santhosh Ram and P.M. Rameshwaran*

PRO 104 (online version): SSCS'2015 - Numerical Modeling - Strategies for Sustainable Concrete Structures (2015), e-ISBN: 978-2-35158-162-9

PRO 105: 1st International Conference on UHPC Materials and Structures (2016), ISBN: 978-2-35158-164-3, e-ISBN: 978-2-35158-165-0

PRO 106: AFGC-ACI-fib-RILEM International Conference on Ultra-High-Performance Fibre-Reinforced Concrete – UHPFRC 2017 (2017), ISBN: 978-2-35158-166-7, e-ISBN: 978-2-35158-167-4; *Eds. François Toutlemonde and Jacques Resplendino*

PRO 107 (online version): XIV DBMC – 14th International Conference on Durability of Building Materials and Components (2017), e-ISBN: 978-2-35158-159-9; *Eds. Geert De Schutter, Nele De Belie, Arnold Janssens and Nathan Van Den Bossche*

PRO 108: MSSCE 2016 - Innovation of Teaching in Materials and Structures (2016), ISBN: 978-2-35158-178-0, e-ISBN: 978-2-35158-179-7; *Ed. Per Goltermann*

PRO 109 (2 volumes): MSSCE 2016 - Service Life of Cement-Based Materials and Structures (2016), ISBN Vol. 1: 978-2-35158-170-4, Vol. 2: 978-2-35158-171-4, Set Vol. 1&2: 978-2-35158-172-8, e-ISBN : 978-2-35158-173-5; *Eds. Miguel Azenha, Ivan Gabrijel, Dirk Schlicke, Terje Kanstad and Ole Mejlhede Jensen*

PRO 110: MSSCE 2016 - Historical Masonry (2016), ISBN: 978-2-35158-178-0, e-ISBN: 978-2-35158-179-7; *Eds. Inge Rörig-Dalgaard and Ioannis Ioannou*

PRO 111: MSSCE 2016 - Electrochemistry in Civil Engineering (2016), ISBN: 978-2-35158-176-6, e-ISBN: 978-2-35158-177-3; *Ed. Lisbeth M. Ottosen*

PRO 112: MSSCE 2016 - Moisture in Materials and Structures (2016), ISBN: 978-2-35158-178-0, e-ISBN: 978-2-35158-179-7; *Eds. Kurt Kielsgaard Hansen, Carsten Rode and Lars-Olof Nilsson*

PRO 113: MSSCE 2016 - Concrete with Supplementary Cementitious Materials (2016), ISBN: 978-2-35158-178-0, e-ISBN: 978-2-35158-179-7; *Eds. Ole Mejlhede Jensen, Konstantin Kovler and Nele De Belie*

PRO 114: MSSCE 2016 - Frost Action in Concrete (2016), ISBN: 978-2-35158-182-7, e-ISBN: 978-2-35158-183-4; *Eds. Marianne Tange Hasholt, Katja Fridh and R. Doug Hooton*

PRO 115: MSSCE 2016 - Fresh Concrete (2016), ISBN: 978-2-35158-184-1, e-ISBN: 978-2-35158-185-8; *Eds. Lars N. Thrane, Claus Pade, Oldrich Svec and Nicolas Roussel*

PRO 116: BEFIB 2016 – 9th RILEM International Symposium on Fiber Reinforced Concrete (2016), ISBN: 978-2-35158-187-2, e-ISBN: 978-2-35158-186-5; *Eds. N. Banthia, M. di Prisco and S. Soleimani-Dashtaki*

PRO 117: 3rd International RILEM Conference on Microstructure Related Durability of Cementitious Composites (2016), ISBN: 978-2-35158-188-9, e-ISBN: 978-2-35158-189-6; *Eds. Changwen Miao, Wei Sun, Jiaping Liu, Huisu Chen, Guang Ye and Klaas van Breugel*

PRO 118 (4 volumes): International Conference on Advances in Construction Materials and Systems (2017), ISBN Set: 978-2-35158-190-2, Vol. 1: 978-2-35158-193-3,

Vol. 2: 978-2-35158-194-0, Vol. 3: ISBN:978-2-35158-195-7, Vol. 4: ISBN:978-2-35158-196-4, e-ISBN: 978-2-35158-191-9; *Eds. Manu Santhanam, Ravindra Gettu, Radhakrishna G. Pillai and Sunitha K. Nayar*

PRO 119 (online version): ICBBM 2017 - Second International RILEM Conference on Bio-based Building Materials, (2017), e-ISBN: 978-2-35158-192-6; *Ed. Sofiane Amziane*

PRO 120 (2 volumes): EAC-02 - 2nd International RILEM/COST Conference on Early Age Cracking and Serviceability in Cement-based Materials and Structures, (2017), Vol. 1: 978-2-35158-199-5, Vol. 2: 978-2-35158-200-8, Set: 978-2-35158-197-1, e-ISBN: 978-2-35158-198-8; *Eds. Stéphanie Staquet and Dimitrios Aggelis*

PRO 121 (2 volumes): SynerCrete18: Interdisciplinary Approaches for Cement-based Materials and Structural Concrete: Synergizing Expertise and Bridging Scales of Space and Time, (2018), Set: 978-2-35158-202-2, Vol.1: 978-2-35158-211-4, Vol.2: 978-2-35158-212-1, e-ISBN: 978-2-35158-203-9; *Eds. Miguel Azenha, Dirk Schlicke, Farid Benboudjema, Agnieszka Knoppik*

PRO 122: SCC'2018 China - Fourth International Symposium on Design, Performance and Use of Self-Consolidating Concrete, (2018), ISBN: 978-2-35158-204-6, e-ISBN: 978-2-35158-205-3; *Eds. C. Shi, Z. Zhang, K. H. Khayat*

PRO 123: Final Conference of RILEM TC 253-MCI: Microorganisms-Cementitious Materials Interactions (2018), Set: 978-2-35158-207-7, Vol.1: 978-2-35158-209-1, Vol.2: 978-2-35158-210-7, e-ISBN: 978-2-35158-206-0; *Ed. Alexandra Bertron*

PRO 124 (online version): Fourth International Conference Progress of Recycling in the Built Environment (2018), e-ISBN: 978-2-35158-208-4; *Eds. Isabel M. Martins, Carina Ulsen, Yury Villagran*

PRO 125 (online version): SLD4 - 4th International Conference on Service Life Design for Infrastructures (2018), e-ISBN: 978-2-35158-213-8; *Eds. Guang Ye, Yong Yuan, Claudia Romero Rodriguez, Hongzhi Zhang, Branko Savija*

PRO 126: Workshop on Concrete Modelling and Material Behaviour in honor of Professor Klaas van Breugel (2018), ISBN: 978-2-35158-214-5, e-ISBN: 978-2-35158-215-2; *Ed. Guang Ye*

PRO 127 (online version): CONMOD2018 - Symposium on Concrete Modelling (2018), e-ISBN: 978-2-35158-216-9; *Eds. Erik Schlangen, Geert de Schutter, Branko Savija, Hongzhi Zhang, Claudia Romero Rodriguez*

PRO 128: SMSS2019 - International Conference on Sustainable Materials, Systems and Structures (2019), ISBN: 978-2-35158-217-6, e-ISBN: 978-2-35158-218-3

PRO 129: 2nd International Conference on UHPC Materials and Structures (UHPC2018-China), ISBN: 978-2-35158-219-0, e-ISBN: 978-2-35158-220-6

PRO 130: 5th Historic Mortars Conference (2019), ISBN: 978-2-35158-221-3, e-ISBN: 978-2-35158-222-0; *Eds. José Ignacio Álvarez, José María Fernández, Íñigo Navarro, Adrián Durán, Rafael Sirera*

PRO 131 (online version): 3rd International Conference on Bio-Based Building Materials (ICBBM2019), e-ISBN: 978-2-35158-229-9; *Eds. Mohammed Sonebi, Sofiane Amziane, Jonathan Page*

PRO 132: IRWRMC'18 - International RILEM Workshop on Rheological Measurements of Cement-based Materials (2018), ISBN: 978-2-35158-230-5, e-ISBN: 978-2-35158-231-2; *Eds. Chafika Djelal, Yannick Vanhove*

PRO 133 (online version): CO2STO2019 - International Workshop CO2 Storage in Concrete (2019), e-ISBN: 978-2-35158-232-9; *Eds. Assia Djerbi, Othman Omikrine-Metalssi, Teddy Fen-Chong*

PRO 134: 3rd ACF/HNU International Conference on UHPC Materials and Structures - UHPC'2020, ISBN: 978-2-35158-233-6, e-ISBN: 978-2-35158-234-3; *Eds. Caijun Shi & Jiaping Liu*

PRO 135: Fourth International Conference on Chemically Activated Materials (CAM2021), ISBN: 978-2-35158-235-0, e-ISBN: 978-2-35158-236-7; *Eds. Caijun Shi & Xiang Hu*

RILEM Reports (REP)

Report 19: Considerations for Use in Managing the Aging of Nuclear Power Plant Concrete Structures (ISBN: 2-912143-07-1); *Ed. D. J. Naus*

Report 20: Engineering and Transport Properties of the Interfacial Transition Zone in Cementitious Compo-sites (ISBN: 2-912143-08-X); *Eds. M. G. Alexander, G. Arliguie, G. Ballivy, A. Bentur and J. Marchand*

Report 21: Durability of Building Sealants (ISBN: 2-912143-12-8); *Ed. A. T. Wolf*

Report 22: Sustainable Raw Materials - Construction and Demolition Waste (ISBN: 2-912143-17-9); *Eds. C. F. Hendriks and H. S. Pietersen*

Report 23: Self-Compacting Concrete state-of-the-art report (ISBN: 2-912143-23-3); *Eds. Å. Skarendahl and Ö. Petersson*

Report 24: Workability and Rheology of Fresh Concrete: Compendium of Tests (ISBN: 2-912143-32-2); *Eds. P. J. M. Bartos, M. Sonebi and A. K. Tamimi*

Report 25: Early Age Cracking in Cementitious Systems (ISBN: 2-912143-33-0); *Ed. A. Bentur*

Report 26: Towards Sustainable Roofing (Joint Committee CIB/RILEM) (CD 07)(e-ISBN 978-2-912143-65-5); *Eds. Thomas W. Hutchinson and Keith Roberts*

Report 27: Condition Assessment of Roofs (Joint Committee CIB/RILEM) (CD 08)(e-ISBN 978-2-912143-66-2); *Ed. CIB W 83/RILEM TC166-RMS*

Report 28: Final report of RILEM TC 167-COM 'Characterisation of Old Mortars with Respect to Their Repair (ISBN: 978-2-912143-56-3); *Eds. C. Groot, G. Ashall and J. Hughes*

Report 29: Pavement Performance Prediction and Evaluation (PPPE): Interlaboratory Tests (e-ISBN: 2-912143-68-3); *Eds. M. Partl and H. Piber*

Report 30: Final Report of RILEM TC 198-URM 'Use of Recycled Materials' (ISBN: 2-912143-82-9; e-ISBN: 2-912143-69-1); *Eds. Ch. F. Hendriks, G. M. T. Janssen and E. Vázquez*

Report 31: Final Report of RILEM TC 185-ATC 'Advanced testing of cement-based materials during setting and hardening' (ISBN: 2-912143-81-0; e-ISBN: 2-912143-70-5); *Eds. H. W. Reinhardt and C. U. Grosse*

Report 32: Probabilistic Assessment of Existing Structures. A JCSS publication (ISBN 2-912143-24-1); *Ed. D. Diamantidis*

Report 33: State-of-the-Art Report of RILEM Technical Committee TC 184-IFE 'Industrial Floors' (ISBN 2-35158-006-0); *Ed. P. Seidler*

Report 34: Report of RILEM Technical Committee TC 147-FMB 'Fracture mechanics applications to an-chorage and bond' Tension of Reinforced Concrete Prisms – Round Robin Analysis and Tests on Bond (e-ISBN 2-912143-91-8); *Eds. L. Elfgren and K. Noghabai*

Report 35: Final Report of RILEM Technical Committee TC 188-CSC 'Casting of Self Compacting Con-crete' (ISBN 2-35158-001-X; e-ISBN: 2-912143-98-5); *Eds. Å. Skarendahl and P. Billberg*

Report 36: State-of-the-Art Report of RILEM Technical Committee TC 201-TRC 'Textile Reinforced Con-crete' (ISBN 2-912143-99-3); *Ed. W. Brameshuber*

Report 37: State-of-the-Art Report of RILEM Technical Committee TC 192-ECM 'Environment-conscious construction materials and systems' (ISBN: 978-2-35158-053-0); *Eds. N. Kashino, D. Van Gemert and K. Imamoto*

Report 38: State-of-the-Art Report of RILEM Technical Committee TC 205-DSC 'Durability of Self-Compacting Concrete' (ISBN: 978-2-35158-048-6); *Eds. G. De Schutter and K. Audenaert*

Report 39: Final Report of RILEM Technical Committee TC 187-SOC 'Experimental determination of the stress-crack opening curve for concrete in tension' (ISBN 978-2-35158-049-3); *Ed. J. Planas*

Report 40: State-of-the-Art Report of RILEM Technical Committee TC 189-NEC 'Non-Destructive Evalua-tion of the Penetrability and Thickness of the Concrete Cover' (ISBN 978-2-35158-054-7); *Eds. R. Torrent and L. Fernández Luco*

Report 41: State-of-the-Art Report of RILEM Technical Committee TC 196-ICC 'Internal Curing of Con-crete' (ISBN 978-2-35158-009-7); ; *Eds. K. Kovler and O. M. Jensen*

Report 42: 'Acoustic Emission and Related Non-destructive Evaluation Techniques for Crack Detection and Damage Evaluation in Concrete' - Final Report of RILEM Technical Committee 212-ACD (e-ISBN: 978-2-35158-100-1); *Ed. M. Ohtsu*

Report 45: Repair Mortars for Historic Masonry - State-of-the-Art Report of RILEM Technical Committee TC 203-RHM (e-ISBN: 978-2-35158-163-6); *Eds. Paul Maurenbrecher and Caspar Groot*

Report 46: Surface delamination of concrete industrial floors and other durability related aspects guide - Report of RILEM Technical Committee TC 268-SIF (e-ISBN: 978-2-35158-201-5); *Ed. Valerie Pollet*

Supplementary Cementitious Materials

Freeze Thaw Resistance of Non-ferrous Slag Concrete

Pithchai Pandian Sivakumar[1,2,3](✉), Nele De Belie[1], Stijn Matthys[1], and Elke Gruyaert[2]

[1] Magnel-Vandepitte Laboratory for Structural Engineering and Building Materials, Ghent University, Technologiepark Zwijnaarde 60, 9052 Ghent, Belgium
pithchaipandian.sivakumar@ugent.de
[2] Department of Civil Engineering, Materials and Constructions, Ghent Technology Campus, KU Leuven, Gebroeders De Smetstraat 1, 9000 Ghent, Belgium
[3] SIM Vzw, Technologiepark 48, 9052 Zwijnaarde, Belgium

Abstract. The objective of this work is to study the freeze thaw resistance of supplementary cementitious materials (SCM) based concrete made from non-ferrous slag (NFS) benchmarked with CEM I 52.5 N and CEM III 42.5 B concrete. NFS is synthesized during the production of Cu metal from Cu scraps. The freeze thaw resistance of NFS concrete containing 70% CEM I 52.5 R and 30% NFS (w/b = 0.45) as binder, as well as of CEM I 52.5 N and CEM III 42.5 B concrete was tested following CEN TR 15177 (2006). The analysis was based on a calculation of the relative dynamic elastic modulus determined by ultrasonic measurements and a determination of the water absorption by mass in function of the number of freeze thaw cycles. Furthermore the relative tensile strength loss after 56 cycles was considered and a microstructural analysis was performed. All concrete mixes showed a relative tensile strength after 56 freeze thaw cycles lower than 100% of the initial value, whereas the CEM III 42.5 B concrete showed the highest strength loss of around 15% followed by 11% for NFS concrete. NFS concrete also showed highest water uptake of around 4% whereas CEM I 52.5 N and CEM III 42.5 B concrete showed values of 1.2% and 2.2% respectively.

Keywords: Non-ferrous slag · Freeze thaw resistance · Ultrasonic measurement · Sustainability & circular economy

1 Introduction

Based on the expected evolution of the population and demand, the IEA CSI cement technology roadmap predicts that the global cement production is set to grow by 12–25% by 2050 [1, 2]. However, by considering the usage of supplementary cementitious material (SCM) based cement in high volume, the huge increase in the cement production units can be reduced. SCM based cements are used widely in concrete application for implementing the principle of sustainability and circular economy. A commonly used SCM material is blast furnace slag (BFS), a by-product from ferrous metallurgy and used for the synthesis of BFS cements. In contrast non-ferrous slag (NFS) is only used

J. I. Escalante-Garcia et al. (Eds.): RW 2021, RILEM Bookseries 40, pp. 3–12, 2023.
https://doi.org/10.1007/978-3-031-21735-7_1

in limited volumes due to the presence of higher iron contents and low calcium contents in the slag chemistry. Moreover, the availability is lower compared to ferrous slags such as BFS.

In the work of Hallet et al. [3], the impact of slag fineness on the reactivity of blended cements with high volume NFS (replacing Portland cement (PC) with 30, 50 and 70 wt% NFS) was investigated. The authors stated that NFS with increased fineness showed similar reactivity as siliceous fly ash. In addition, Feng et al. [4] studied the pozzolanic activity of granulated copper (Cu) slag with the help of calorimetry, thermogravimetric analysis (TGA) and scanning electron microscopy (SEM). The final findings stated that the blended cements with 30% granulated Cu slag with CaO addition showed increased reactivity after 7 days compared to the 100% PC system. Moreover, in the recent work of Sivakumar et al. [5], reactivity of one kind of NFS was evaluated by the novel R^3 method. The assessment showed that the NFS acts as a reactive pozzolanic material with an acceptable performance in heat release, bound water content and calcium hydroxide (CH) consumption.

Studies with respect to the freeze thaw resistance of BFS concrete are abundantly available while no studies corresponding to the freeze thaw resistance of NFS as SCM binder could be found. The laboratory studies reported in [6–9] proved that the replacement of cement by BFS showed poor freeze thaw resistance (with/without de-icing salts). However, certain optimizations can be carried out in the concrete composition to improve the freeze thaw resistance. Deja et al. [9] studied the freezing and de-icing salt resistance of concrete containing 57% BFS in presence of air entraining agents and microfibers. The final finding stated that the concrete with an air content above 5% improved the freeze thaw scaling resistance. Moreover, the age of the BFS concrete is crucial. Chidiac et al. [10, 11] investigated the scaling resistance of the BFS concrete as a function of age. One of the interesting findings stated that the BFS concrete cured for a period of 2 years showed improved scaling resistance compared to the sample cured for 28 days. In addition to the age and air content, carbonation shrinkage and interfacial transition zone also play a vital role in the scaling resistance of BFS concrete [12].

The principal aim of this research was to investigate the freeze thaw resistance following CEN TR 15177 (2006) of the novel binder with NFS as SCM and compare the performance with that of traditional binders as PC and blast furnace slag cement.

2 Materials and Methodology

2.1 Materials

Patented NFS (designated as Modified Ferro Silicate slag) from Metallo Belgium was used as the SCM to prepare concrete together with CEM I 52.5 R. Traditional cement used in the study was a Portland cement type CEM I 52.5 N and a blast-furnace slag cement type CEM III 42.5 B. Natural sand 0/1 and 0/8 mm together with gravel 4/16 and 4/32 mm were used as the aggregates. Polycarboxylic ether based superplasticizer was used to improve the workability of the concrete.

2.2 Characterization of Raw Materials

Wavelength-dispersive X-ray fluorescence spectrometry was used to analyse the chemical composition of the starting materials. X-ray diffraction with 10% internal standard (crystalline ZnO) was used to investigate the mineralogy of the NFS slag and PC. Rietveld analysis was used to quantify the diffractogram.

2.3 Concrete Composition, Slump and Air Content

Reference concrete was designed with a strength class of C35/45 fulfilling the requirements for an environmental class EE4 (humid interior environment or outer environment with frost and de-icing salts) according to the standard NBN EN 206 + NBN B 15-001. Concrete containing 30% NFS and 70% CEM I 52.5 R was bench marked against CEM III 42.5 B and CEM I 52.5 N. Table 1 shows the three concrete compositions. After casting, the samples were stored at a temperature and relative humidity of 20 °C and 95% respectively. Demolding was carried out after 24 h. Slump and air content were determined 15 min after mixing with respect to the norm NBN EN 12350-2 (2009) and NBN EN 12350-7 (2009) respectively.

Table 1. Concrete mix design (kg/m^3)

Component	CEM I 52.5 R + NFS	CEM III 42.5 B	CEM I 52.5 N
Sand 0/1	88	84	84
Sand 0/8	719	681	683
Gravel 4/16	170	161	161
Gravel 4/32	930	880	883
CEM I 52.5 R	273	0	0
CEM I 52.5 N	0	0	390
CEM III 42.5 B	0	390	0
NFS	117	0	0
Superplasticizer	1.287	1.287	1.560
Water	175	175	175

2.4 Freeze Thaw Experiment

Ten prisms with dimensions 400 × 400 × 100 mm^3 were made per mixture to investigate the freeze thaw internal damage resistance. The freeze thaw test was carried out on the basis of the standard CEN/TR 15177 (2006). The standard prescribes the test pieces to be stored in plastic foil for 6 days after demolding. The plastic wrap was removed when the test specimens were 7 days old. Subsequently, the test specimens were stored under water for 21 days. After this, the freeze thaw test was carried out in a freeze-thaw cabinet for 56 cycles. According to the standard NBN B15-100 (2016), 5 test specimens per concrete type must be subjected to a splitting test before exposure to the freeze-thaw cycles. After 56 cycles, another 5 exposed test specimens must be subjected to a splitting test. The splitting test was performed according to the standard NBN EN 12390-6 (2010). Relative strength loss (Δf) was calculated by (tensile strength before frost-thaw—tensile strength after frost-thaw)/tensile strength before frost-thaw. CEN/TR 15177 (2006) states that the degradation of test pieces must be monitored by determining the relative dynamic elastic modulus (RDM). The measurements were taken using the Proceq Pundit Lab Ultrasonic Instrument. The degradation as a function of the number of freeze-thaw cycles (0, 8, 14, 28, 42, 56) was determined on the basis of the ultrasonic pulse velocity. Equation (2) defines the calculation of RDM, where V_n and V_0 are the longitudinal velocity (m/s) after n and 0 freeze-thaw cycles respectively.

$$RDM\,(\%) = V_n^2/V_0^2 \cdot 100 \tag{1}$$

The water absorption is expressed in % with respect to the mass after 35 days and as a function of the number of freeze thaw cycles. The final value of water uptake is then calculated as the average after 56 freeze thaw cycles per sample. The water absorption was calculated as per the Eq. 2, where m_{56F} and m_{35d} are the mass of the specimen after 56 freeze thaw cycles and mass of the specimen after 35 days in grams.

$$Water\ absorption(\%) = \left(m_{56f} - m_{35d}\right)/m_{35d} \cdot 100 \tag{2}$$

2.5 Fluorescence Microscopic Analysis

Fluorescence microscopic analysis was carried out only on the freeze thaw exposed NFS slag concrete. 50 mm diameter cores were vacuum impregnated with fluorescent epoxy. After hardening, the samples were sawn in longitudinal direction. The sawn samples were then glued onto 30 mm × 50 mm glass plates, cut to a thickness of 10 mm and plane ground. Later, the samples were cleaned and placed under vacuum for 2 h after which they were impregnated with fluorescent epoxy. Afterwards, the impregnated concrete samples were ground to remove excess epoxy. Subsequently, the ground surface was glued onto an object glass and further ground to a thickness of 20 μm. Afterwards, a cover glass was glued onto the sample with UV hardening glue.

3 Results and Discussion

3.1 Characterization of Raw Materials

Table 2 shows the chemical composition of the raw materials. The main constituents of NFS used in this work are Fe_2O_3 and SiO_2. XRD analysis showed the presence of around 92 wt% amorphous phase (=glass) and 8 wt% crystalline phase, mainly spinel. NFS was milled in the similar range of particle size with respect to the PC.

Table 2. Chemical composition of the starting materials in wt%

Component	CEM I 52.5 N	CEM I 52.5 R	CEM III 42.5 B	NFS
SiO_2	18.3	18.0	25.8	32.3
Fe_2O_3	4.0	4.1	2.6	40.9
Al_2O_3	5.2	5.2	8.2	11.0
CaO	64.3	64.1	52	3.9
Others	8.2	8.6	11.4	11.9

3.2 Slump, Air Content and Compressive Strength

Table 4 shows the consistency of the three different mixes. All three mixes showed a desirable slump class S4. CEM I 52.5 N showed an air content of 1.5% whereas the CEM III 42.5 B and CEM I 52.5 R + NFS showed an air content of 1.3% and 1.2% respectively. Table 4 also shows the 28 days compressive strength of the synthesized concretes.

Table 4. Slump and air content of the concrete

Binder	Slump (mm)	Air content (%)	Air to paste (%)	Compressive strength (28 d)
CEM I 52.5 N	180	1.5	5.1	58.6
CEM III 42.5 B	190	1.3	4.4	51.5
CEM I 52.5 R + NFS	200	1.2	4.1	52.1

3.3 Freeze Thaw Resistance

Relative Tensile Strength Loss

Figure 1a) shows the splitting tensile strength of the CEM I 52.5 N, CEM III 42.5 B and CEM I 52.5 R + NFS concrete before/after freeze thaw exposure. All samples showed a decreased splitting tensile strength compared to the initial value. The decrease in strength is explained by internal damage due to freeze thaw cycles, as further discussed in Sect. 3.4. Figure 1b) shows the relative strength loss of the concrete caused by the freeze thaw mechanism and is calculated based on the mean of 5 values per mix before/after freeze-thaw attack. A relative strength loss of 15% for CEM III 42.5 B was obtained whereas the CEM I 52.5 N and CEM I 52.5 R + NFS showed a relative strength loss of around 11%.

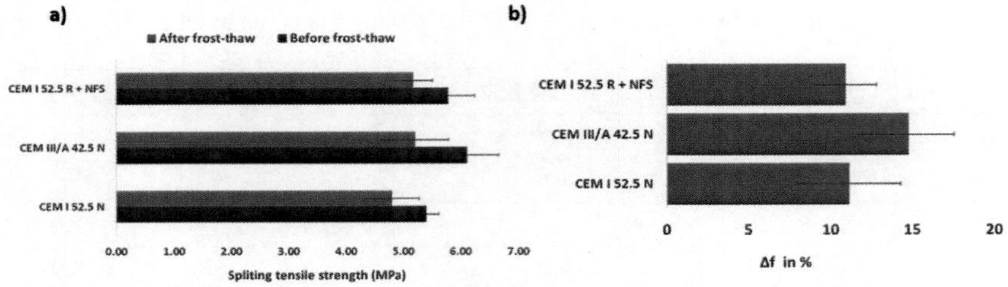

Fig. 1. Splitting tensile strength a) splitting tensile strength before/after freeze thaw b) relative strength loss

Water Absorption

Figure 2 shows the water uptake calculated according to formula 2. NFS slag concrete showed an increase in mass of around 4% after 56 freeze thaw cycles whereas CEM I 52.5 N and CEM III 42.5 B concrete showed values of 1.2% and 2.2% respectively.

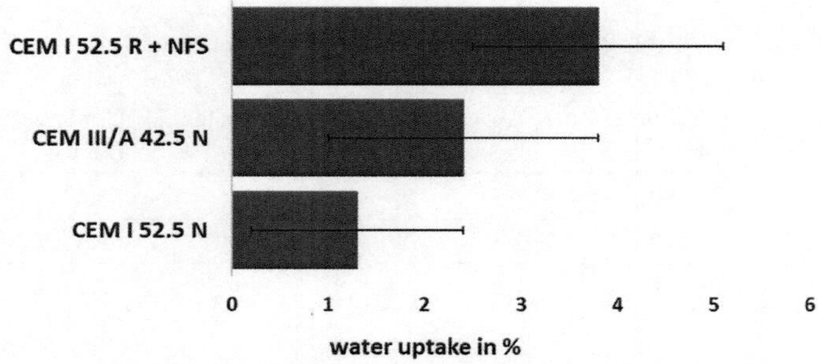

Fig. 2. Mass change (water uptake) after 56 cycles

3.3.1 Relative Dynamic Elastic Modulus

Figure 3 shows the mean RDM values determined on the basis of the ultrasonic measurements after 0, 8, 14, 28, 42 and 56 freeze thaw cycles. CEM I 52.5 N concrete samples showed around 4% reduction in their mean RDM after exposure to 8 freeze thaw cycles whereas CEM III 42.5 B and CEM I 52.5 R + NFS concrete samples showed around 7% and 6% reduction in mean RDM to its initial value. CEM III 42.5 B and CEM I 52.5 R + NFS showed a decrease in the RDM from 93% to 90% and 94% to 92% between 8 and 14 cycles whereas CEM I 52.5 N showed only a slight decrease from 96% to 95%. After 14 cycles, CEM I 52.5 N showed a constant decrease losing only an additional 1% after 28 and 42 cycles. CEM III 42.5 B showed a decrease of 13% and 15% whereas CEM I 52.5 R + NFS showed a decrease of 10% and 12.5% after 28 and 42 cycles with respect to its initial value. After 56 cycles, CEM III 42.5 B showed the highest loss of

18% followed by 15% for CEM I 52.5 R + NFS. CEM I 52.5 N showed only 8.5% loss of RDM outperforming both CEM III 42.5 B and CEM I 52.5 R + NFS.

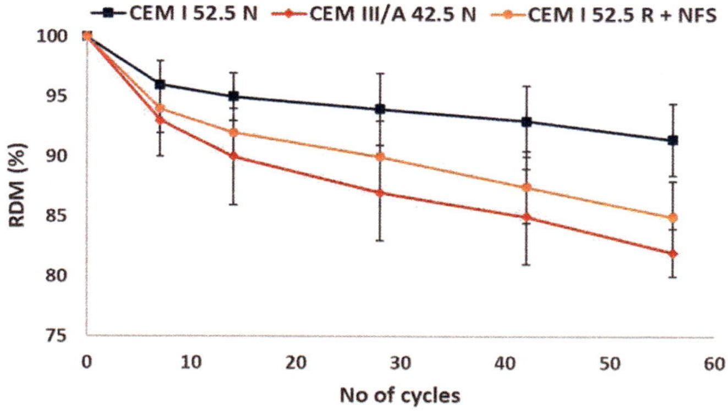

Fig. 3. Relative dynamic elastic modulus calculated by ultrasonic measurement

3.4 Relative Dynamic Elastic Modulus Versus Relative Tensile Strength Loss

While comparing the results between the RDM (Fig. 3) and relative tensile strength (Fig. 1b) of the different mixes before and after frost thaw attack, contradictory behaviour can be found. Relative tensile strength loss measured by the tensile strength difference for CEM III 42.5 B concrete showed a decrease of 15% whereas the RDM results showed a higher decrease of 18%. Moreover, similar co-relation can also be seen in the CEM I 52.5 R + NFS where RDM results showed higher decrease value (15%) compared to the relative tensile strength loss of 11%. However, CEM I 52.5 N concrete showed lower decrease of RDM value (8.5%) compared to relative strength loss of 11.1%.

3.5 Fluorescence Microscopic Analysis—General Observations

Figure 4 shows the binder matrix (yellow region) and aggregate (black region) of the CEM I 52.5 R + NFS concrete exposed to the freeze thaw cycles. Figure 4a) shows the porous structure of the ITZ whereas the Fig. 4b) shows the presence of cracks which are mainly propagating via the ITZ. The ITZ tends to become wider, weaker and it can be considered as the weakest link in the concrete [13]. The possible cause (wide & weak ITZ) could lead to an increase in liquid transport rate creating larger pores while freeze thaw cycles are causing loss in tensile strength. This behaviour could be observed in the Fig. 4a and 4b as well as in Fig. 1 (relative tensile strength loss). Figure 5 mainly shows the binder matrix of CEM I 52.5 R + NFS concrete after freeze thaw cycles. In Fig. 5a), cracks can be seen clearly in the binder matrix. Figure 5b, c show the presence of voids/pores in the binder matrix. As one of the interesting findings, Fig. 5c) shows the cracks generating from the pores. This could be possibly explained due to the thermal expansion coefficient difference between ice and concrete. Since, these voids are filled with water and resulting in the formation of ice crystals during freeze cycles, the expansion and shrinkage of ice exerts tensile stress resulting in cracks.

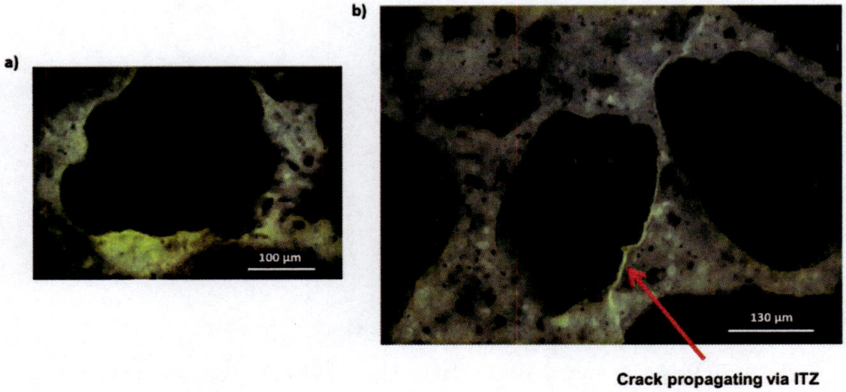

Fig. 4. Fluorescence microscopic analysis a) porous ITZ b) crack in ITZ

4 Conclusion

In this study, concrete containing 30% NFS and 70% CEM I 52.5 R was prepared and bench marked against CEM III 42.5 B and CEM I 52.5 N to investigate the freeze thaw resistance according to CEN TR 15177 (2006).

Following can be stated as the important findings:

1) All three tests showed that CEM I 52.5 N performed best
2) Performance of CEM III 42.5 is in all tests worse in comparison to the CEM I 52.5 N. Although based on the splitting tensile strength tests the performance of CEM I 52.5R + NFS is comparable to that of CEM I 52.5N; the other tests show a an increased water uptake and increased loss in RDM
3) Fluorescence microstructural analysis showed cracks along the ITZ suggesting that the ITZ could be the weakest link in concrete.

As further investigation, NFS concrete with air entraining agent should be produced to examine the influence of entrained air voids against freeze thaw resistance. Moreover, curing regime and period are important factors influencing freeze-thaw resistance, Freeze thaw resistance of NFS concrete with different curing period should also be investigated as a further study.

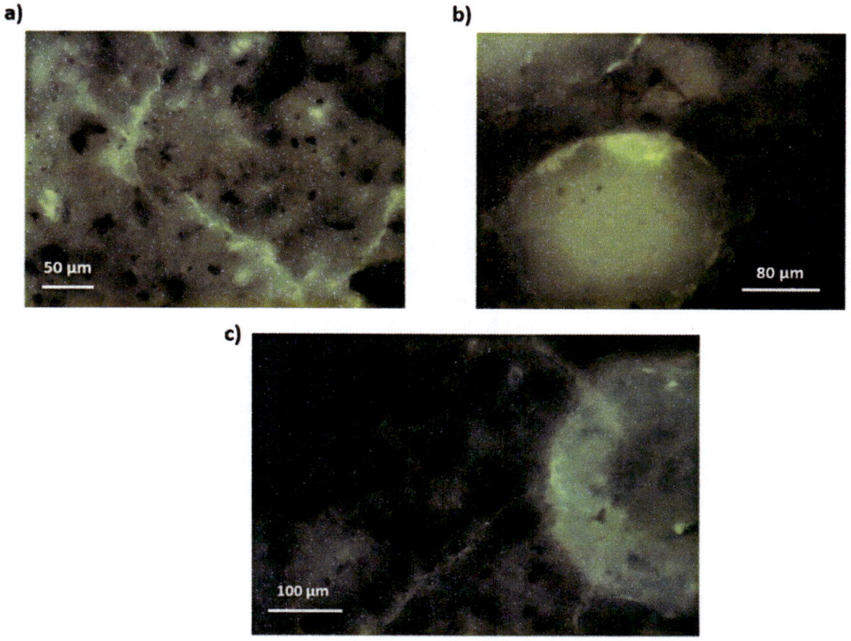

Fig. 5. Fluorescence microscopic analysis a) binder matrix b) voids in binder matrix c) cracks originating from voids

Acknowledgments. The work has been financed by the SIM MARES program, grant number HBC.2017.0607. The author would like to thank the industrial partner Van Pelt for producing concrete. The authors would also like to thank Hanne Vanoutrive (KU Leuven) for assisting with the freeze thaw measurements.

References

1. IEA: Technology Roadmap Low-Carbon Transition in the Cement Industry (2018)
2. Herbst, A., Fleiter, T., Rehfeldt, M.: Scenario analysis of a low-carbon transition of the EU industry by 2050: Extending the scope of mitigation options. In: Eceee Ind. Summer Study Proc. (2018)
3. Hallet, V., De Belie, N., Pontikes, Y.: The impact of slag fineness on the reactivity of blended cements with high-volume non-ferrous metallurgy slag. Constr. Build. Mater. **257** (2020). https://doi.org/10.1016/j.conbuildmat.2020.119400
4. Feng, Y., Yang, Q., Chen, Q., Kero, J., Andersson, A., Ahmed, H., Engström, F., Samuelsson, C.: Characterization and evaluation of the pozzolanic activity of granulated copper slag modified with CaO. J. Clean. Prod. **232** (2019). https://doi.org/10.1016/j.jclepro.2019.06.062
5. Sivakumar, P.P., Matthys, S., De Belie, N., Gruyaert, E.: Reactivity assessment of modified ferro silicate slag by R^3 method. Appl. Sci. **11** (2021). https://doi.org/10.3390/app11010366
6. Deicing salt scaling resistance of concrete incorporating supplementary cementing materials: CANMET research. Free. Durab. Concr. (2020). https://doi.org/10.1201/9781482271553-16

7. Nicula, L.M., Corbu, O., Iliescu, M., Dumitraş, D.G.: Using the blast furnace slag as alternative source in mixtures for the road concrete for a more sustainable and a cleaner environment. Rev. Rom. Mater. Rom. J. Mater. **50** (2020)
8. Nicula, L.M., Corbu, O., Iliescu, M.: Influence of blast furnace slag on the durability characteristic of road concrete such as freeze-thaw resistance. Procedia Manuf. (2020). https://doi.org/10.1016/j.promfg.2020.03.029
9. Deja, J.: Freezing and de-icing salt resistance of blast furnace slag concretes. Cem. Concr. Compos. **25** (2003). https://doi.org/10.1016/S0958-9465(02)00052-5
10. Chidiac, S.E., Panesar, D.K.: Evolution of mechanical properties of concrete containing ground granulated blast furnace slag and effects on the scaling resistance test at 28 days. Cem. Concr. Compos. **30** (2008). https://doi.org/10.1016/j.cemconcomp.2007.09.003
11. Panesar, D.K., Chidiac, S.E.: Capillary suction model for characterizing salt scaling resistance of concrete containing GGBFS. Cem. Concr. Compos. **31** (2009). https://doi.org/10.1016/j.cemconcomp.2009.01.004
12. Stark, J., Ludwig, H.M.: Freeze-thaw and freeze-deicing salt resistance of concretes containing cement rich in granulated blast furnace slag. ACI Mater. J. **94** (1997). https://doi.org/10.14359/284
13. Çopuroğlu, O., Schlangen, E.: Modeling of frost salt scaling. Cem. Concr. Res. **38** (2008). https://doi.org/10.1016/j.cemconres.2007.09.003

Investigating Supplementary Cementing Materials for Alkalis and Pore Fluid Properties

Mahipal Kasaniya(✉) and Michael D. A. Thomas

University of New Brunswick, Fredericton, NB 3B 1Y1, Canada
mkasaniy@unb.ca

Abstract. Dwindling supplies of quality fly ash in many parts of the world necessitates an expeditious search for supplementary cementing materials (SCMs) to conform to the growing demand for sustainable materials in partially substituting for Portland cement in concrete. In recent years finely-ground waste glass, landfilled/ponded coal ashes and natural pozzolans have (re)drawn the attention of researchers and the industry to become potential SCMs. In this study a wide range of such SCMs were characterized for water-soluble and available alkalis using ASTM C114 and ASTM C311. Inductively coupled plasma emission spectroscopy (ICP-ES) was employed to determine the amounts of alkalis. Cementitious pastes containing Portland cement and SCM were prepared at a 0.5 water-to-cementing materials ratio and stored in airtight containers at 23 °C. The pore solution of these pastes at 28 and 91 days was analyzed for hydroxyl ion (OH^-) concentration, and strengths of alkali ions (Na^+ and K^+) using ICP-ES. The results indicate that all alkali oxides of SCMs are generally neither water-soluble nor available. Except for the high-alkali ground glass and high-calcium fly ash, the use of all other reactive SCMs considerably decrease the presence of the hydroxyl and alkali ions in the pore solution in the long-term.

Keywords: Supplementary cementing materials · Alkalis · Pore solution · Alkalis-silica reaction

1 Introduction

It has been well-established that partially replacing portland cement (PC) with supplementary cementitious materials (SCMs) such as fly ash and silica is a feasible way to improve the longevity of concrete containing blended cements [1]. The long-term performance of concrete can be evaluated in several aggressive conditions, including internal and external attacks on hydration products and aggregates [2, 3]. One of the most prevalent internal attacks on concrete in North America is the alkali-silica reaction (ASR) causing deleterious expansion. The chemical reaction between alkali hydroxides (Na^+, K^+ and OH^-) present in the pore solution of concrete and the amorphous silica available in aggregates is defined as ASR. The use of appropriate SCMs in adequate proportions to replace PC is one of the effective means to drop alkali concentration in the pore solution, thereby mitigating ASR [4]. However, the availability of quality fly

J. I. Escalante-Garcia et al. (Eds.): RW 2021, RILEM Bookseries 40, pp. 13–21, 2023.
https://doi.org/10.1007/978-3-031-21735-7_2

ash, the most widely used SCM, is forecast to decline due to a continuous reduction in fly ash production as coal-based power generating stations have being phased out because of environmental concerns [5]. Therefore, there is an immediate need to fill the growing gap created by fly ash shortages to meet the increasing demand for SCMs.

In the past years, researchers have examined several types of SCMs (e.g., ground coal bottom ash or ground glass) to determine whether these materials have a satisfactory reactivity and the potential to improve certain durability properties of concrete or mortar. Oruji et al. [6] found that coal bottom ash shows a reasonable reactivity and efficacy to mitigate ASR if ground to a high fineness, approximately 1100 m^2/kg specific surface area. Similarly, finely-ground waste glass (10-μm) can reduce the expansion due to ASR expansion when used at a 20% PC replacement level with a low-CaO fly ash addition [7]. Stanton [8] showed that natural pozzolans can reduce ASR expansion when blended with a high-alkali PC.

Similar to low-CaO fly ashes, most other SCMs are generally pozzolanic materials and chemically combine with portlandite produced in PC hydration to form additional cementitious hydrates with a low Ca/Si atomic ratio and enhanced capability to encapsulate alkalis (Na^+ and K^+) [9]. However, workers argue that some, if not all, of alkalis encapsulated in hydrates are released into the pore solution over time [10], and a portion of leached alkalis are available to participate in further ASR. On the other hand, a portion of the total alkalis of SCMs may be tied up in crystalline phases, which generally do not dissolve and provide alkalis to entrap in low Ca/Si hydrates and play a role in ASR [11].

This study was aimed at quantifying the alkali contribution of SCMs using different techniques. The water-soluble and available alkalis of SCMs were determined using the ASTM C114 and C311 test methods. Pastes with blended cements were developed to investigate the effects of SCMs on the chemical composition of the pore solution.

2 Materials and Methods

The cementing materials used in this study are one high-alkali ASTM C150 Type I/II PC, two fly ashes having low- and high-CaO contents, one ground coal bottom ash, eight natural pozzolans (pumices, perlite, lassenite and metakaolin), one silica fume, three ground glasses of low- and high-alkali contents, and two blended pozzolans. In addition, two inert fillers (ground quartz and limestone) were also used. The blended pozzolans were a 60/20/20 fly ash/lassenite/limestone and a 50/50 fly ash/pumice mixtures. The chemical and physical properties of these materials are presented in Table 1. A laboratory-grade portlandite ($Ca(OH)_2$) was used to prepare paste samples for available alkalis in ASTM C311.

2.1 Water-Soluble and Available Alkalis

The water-soluble alkali (Na_2O_e) of PC and SCMs were determined in accordance with ASTM C114. In this test, 25 g of cementing material was added to 250 mL of distilled water in a flask with a screw cap. The suspension was vigorously mixed by shaking the flask for 10 min followed by filtrating using a quantitative-grade filter paper. The filtrate was analyzed for alkalis using inductively coupled plasma emission spectroscopy

(ICP-ES). An internal standard made up of known amounts of sodium chloride, potassium chloride and distilled water was used to calibrate the spectrometer for the alkali determination.

Table 1. Chemical composition (mass %) and physical properties of materials.

Material		SiO_2	Al_2O_3	Fe_2O_3	CaO	MgO	SO_3	TiO_2	Na_2O_e	LOI[a]	SG[b]	D_{50} (μm)
Portland cement	PC	18.8	5.7	2.5	61.3	3.0	4.2	–	**1.0**	2.7	3.1	–
Ground glasses	GE	53.6	13.7	0.4	21.9	1.1	–	0.1	0.9	0.6	2.6	11.0
	GS-A	71.5	1.3	<0.1	10.1	1.4	0.2	<0.1	**13.1**	0.2	2.5	8.0
	GS-B	71.0	0.8	0.3	9.4	3.1	0.1	<0.1	**13.3**	1.1	2.5	7.4
Coal ashes	CFA	35.6	17.9	5.8	**27.4**	6.8	1.6	1.4	1.6	0	2.6	9.8
	FFA	55.6	20.1	5.7	4.4	–	1.7	–	1.6	–	2.5	6.5
	GBA	52.0	19.4	9.1	8.8	1.5	0.4	1.1	0.9	5.9	2.8	2.4
Blended pozzolans	FAB-A	66.4	14.3	5.4	1.7	0.8	1.0	0.6	2.8	5.7	2.4	11.8
	FAB-B	65.6	19.2	3.8	2.3	0.8	0.2	–	3.5	2.5	2.3	9.7
Pumices	PM-A	75.2	11.4	1.4	0.5	–	–	–	1.3	3.4	2.3	6.4
	PM-B	75.2	11.4	1.4	0.5	–	–	–	1.3	3.4	2.4	2.8
	PM-C	73.0	12.2	1.1	1.0	0.2	<0.1	0.1	5.6	4.8	2.4	17.6
	PM-D	72.5	13.2	0.9	1.3	–	0.1	–	1.5	4.1	2.4	4.6
	PM-E	71.4	11.2	1.3	1.5	0.6	0.1	–	5.7	5.3	2.4	6.1
Perlite	PR	73.9	12.7	1.1	0.8	0.1	<0.1	<0.1	6.0	3.6	2.4	19.8
Lassenite	LSN	64.2	13.7	5.7	1.6	0.8	1.0	0.5	2.7	8.8	2.5	4.9
Metakaolin	ME	51.5	44.1	1.4	0.1	0.2	–	1.9	0.2	0.4	2.6	4.0
Silica fume	SF	94.4	0.2	0.1	0.6	0.1	–	–	0.4	4.0	2.2	1.2
Limestone	LMS	3.1	0.1	0.1	52.8	1.3	<0.1	<0.1	0.2	42.1	2.3	4.0
Ground quartz	GQ	98.5	0.5	<0.1	<0.1	<0.1	<0.1	<0.1	0.2	0.5	2.7	3.7

Paste samples were prepared to quantify the available alkalis of PC and SCMs using the procedure in ASTM C311 at ages of 28 and 91 days. In this test, 5 g cementitious material was blended with 2 g portlandite and 10 mL distilled water in a 20 mL plastic vial that, after filling up the materials, was closed with a screw cap and vigorously shaken for 2 min to homogenize the paste mixture. The vials were then stored over water in large airtight containers at 38 ± 2 °C in an oven. At 28 and 91 days, the vials were opened and cut with a knife to retrieve the contents (i.e., hydrated pastes) that were ground with a pestle to produce a uniform slurry in distilled water. The slurry was filtered through a medium-textured filter paper which was washed thoroughly 8 to 10 times with hot water. The filtrate was eventually analyzed for alkalis using ICP-ES.

2.2 Pore Solution Expression and Analysis

Cementitious pastes of the control (100% PC) and blended mixes with 25% SCM at a 0.50 water-to-cementing materials (w/cm) ratio were tested. The materials were transferred into a high-shear mixer and mixed for four minutes to produce homogenous pastes. The pastes were poured into 44-mm dia. × 89-mm tall airtight containers and then stored over water in large airtight containers at 23 ± 2 °C. One specimen for each mix at each age, 28 and 91 days, was used for pore solution expression under cyclic loading ranging up to 600 MPa. Subsequently, the pore solutions were titrated against hydraulic chloric acid for hydroxyl ion (OH^-) concentration and analyzed for alkalis (Na^+ and K^+) using ICP-ES.

3 Results

3.1 Water-Soluble and Available Alkalis

The water-soluble (ASTM C114) and available (ASTM C311) alkali results for cementing materials are presented in Fig. 1. The data shown in Fig. 2 are the percent of the total alkalis ($Na_2O_e = Na_2O + 0.658 \times K_2O$) of cementing materials determined using x-ray fluorescence. In all cases, less than 50% of the total alkalis of a material (PC or SCM) is water-soluble. The PC and high-CaO fly ash (CFA) have the highest proportion (40–45%) of their total alkalis as water-soluble. Natural pozzolans, ground bottom ash and ground glasses, including high-alkali glasses (~13% Na_2O_e) contain less than one-tenth of their total alkalis as water-soluble. Similarly, a portion of the total alkalis is found available in lime-SCM pastes in ASTM C311. The available alkalis of a material at both ages, 28 and 91 days, are higher than their water-soluble alkalis. The PC has almost all alkalis available at 91 days; whereas SCMs show a wide range in available alkalis – from 9% for lassenite (LSN) to about 70% for high-CaO fly ash and pumice PM-D. Compared to 28 days, more alkalis were available at 91 days for all materials except for ground coal ash (GBA), lassenite (LSN), silica fume (SF), and metakaolin (MK).

3.2 Pore Solution

The alkalis (Na^+ and K^+) and hydroxyl ion (OH^-) concentration in the pore solution of the control mix (100% PC) and blended mixes with 75% PC + 25% SCM are presented in Figs. 3 and 4, respectively. The alkali concentration of the control increases slightly from about 800 mmol/L at 28 days to 866 mmol/L at 91 days. The blended samples with high-alkali ground glasses (~13% Na_2O_e) or high-CaO fly ash (27.4% CaO) demonstrate a similar trend and have slightly higher alkalis present in the pore solution than the control. On the other hand, the use of other SCMs (and inert materials) reduces the amount of alkalis at the later age of 91 days; however, the degree of reduction in alkalis is highly variable, depending on the SCM used to partially replace PC. For example, the silica fume (SF) mix has the lowest alkalis concertation of 59 mmol/L, whereas the mix with perlite (PR) contains the highest alkalis, 530 mmol/L at 91 days. With respect to low-CaO fly ash, low-alkali ground glass, natural pozzolans, ground coal bottom ash and blended pozzolans having low-CaO fly ash and natural pozzolan exhibit comparable or decreased alkali concentration in the long-term. Moreover, the hydroxyl ion concentration of the control and blended mixes (see Fig. 4) shows trends similar to those found in alkali concentration.

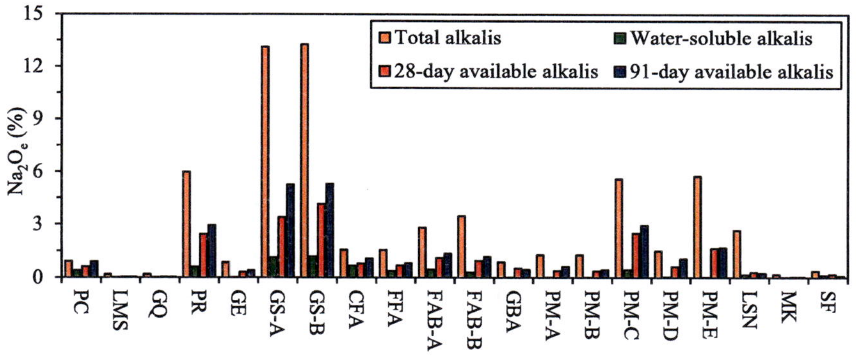

Fig. 1. Water-soluble and available alkalis of cementing materials.

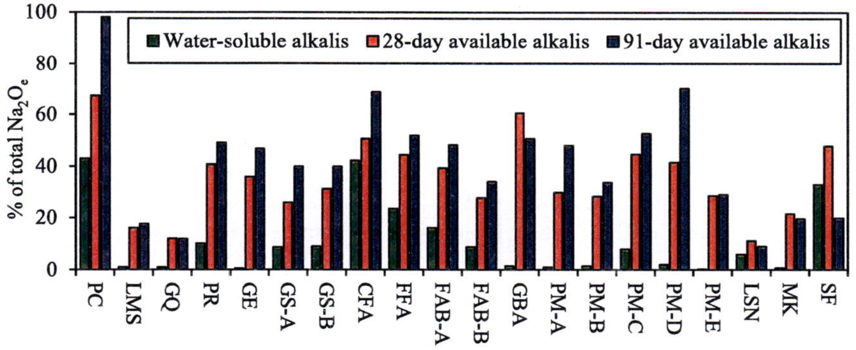

Fig. 2. Water-soluble and available alkalis of cementing materials as percent of total alkalis.

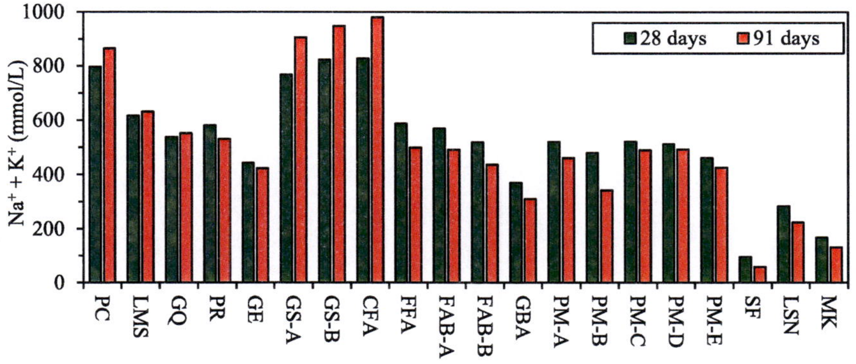

Fig. 3. Alkali concentration in the pore solution of high-alkali pastes at 25% level.

4 Discussion

The use of SCMs in partially replacing PC in concrete generally improves durability performance, including reduced expansion due to damaging alkali-silica reaction (ASR) if amorphous silica is present in the aggregate used [7]. Shehata and Thomas [12] found that the degree of ASR mitigation is related to the type of SCM incorporated and its PC replacement level. For example, the authors have shown that high-CaO fly ash (27.4% CaO) mitigates ASR to a considerably less extent than low-CaO fly ash (4.4% CaO) used at the same replacement level [13]. This is mainly a result of differences in the available alkalis of SCMs that influence the pore solution chemistry of blended cements (PC + SCM) with regards to alkalis (Na^+ and K^+) and hydroxyl (OH^-) concentrations.

The water-soluble alkalis of SCMs in C114 are determined after SCMs are immersed in water for a very short duration, 10 min. The results indicate that a portion (1 to 42%) of the total alkalis present in SCMs are rapidly water-soluble. For Portland cement approximately 50% of the alkalis are immediately water soluble but for some SCM, particularly natural pozzolans the portion of alkalis that are water soluble are very small, less than 1% in some cases. In addition, researchers have shown that the alkalis which are rapidly water-soluble do not necessarily come from alkali oxides (Na_2O and K_2O) but alkali salts in the forms of sulfate and chloride [10]. The reactivity and reactive nature (pozzolanic, hydraulic, or both) of SCMs do not influence the water-soluble alkali content of SCMs since the C114 test is rapid (SCMs are immersed in water for 10-min) and does not provide lime for SCMs to react and form cementitious hydrates. Whereas SCMs in the C311 available alkalis test chemically react with lime at 38 °C for at least 28 days. SCMs that react pozzolanically (e.g., natural pozzolans, ground glasses) form cementitious hydrates with a low Ca/Si ratio [9]. Such hydrates have an increased capability to encapsulate available alkalis [14]. The results indicate that a majority of SCMs demonstrate a considerable increase in available alkalis between 28 and 91 days. Such behavior of SCMs requires an extended test duration beyond 91 days until the available alkali values reach plateaus to determine their available alkali contents accurately.

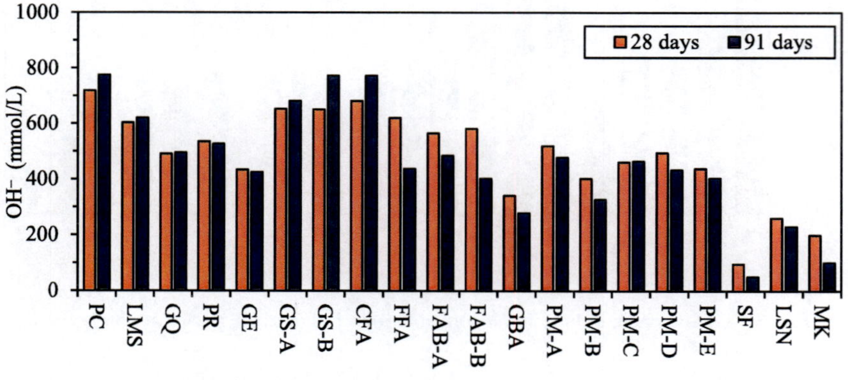

Fig. 4. Hydroxyl ion concentration in the pore solution of high-alkali pastes at 25% level.

A portion of the total alkalis of SCMs may also be tied up in crystalline phases, which generally do not dissolve and contribute to available alkalis [11, 15]. Although the total alkali composition appears to provide a reliable indication of the available alkalis for some SCMs (see Fig. 5a), there are natural pozzolans that have approximately $\leq$ 3% total alkali content but with a small portion ($\leq$one-third) of these alkalis being available because most of the alkalis are bound in inert crystalline phases [15], as depicted by Fig. 5a. Note that inert materials (ground quartz and limestone) are excluded from the correlations presented in this paper. Workers argue that the C311 test overestimates the available alkalis that ultimately participate in ASR because the hydrated lime-SCM pastes are ground to fine in a slurry and then washed using hot water 10 times [10]. This process potentially results in the leaching of some of the alkalis entrapped in hydration products into the test solutions that are analyzed for alkali concentration.

Alkalis that release from PC and SCM in the long-term but are not bound in the cementitious hydrates formed by SCMs eventually end up in the pore solution of the hydrated cementitious matrix, thereby increasing the risk of deleterious ASR [10, 16]. The lower the available alkalis of SCMs in C311, the lower the ionic strength of alkalis in the pore solution of PC-SCM pastes (see Fig. 5b). It is evident that the degree of alkali reduction in PC-SCM samples by the inclusion of SCMs is beyond the level of mere dilution brought by limestone (LMS) inert filler (see Fig. 3). This suggests that natural pozzolans, ground coal bottom ash and low-alkali ground glass tested are reactive and combine a considerable portion of (their own and PC's) alkalis in the hydrates. Conversely, high-alkali ground glasses (~13% Na_2O_e) and high-CaO fly ash (27.4% CaO) are reactive but do not reduce the alkali (Na^+ and K^+) concentration compared to the control at 91 days. In a previous publication the authors have shown that the same high-alkali glasses and high-CaO fly ash have more or less no impact on ASR mitigation when used at a 25% level [13]. However, a mix with high-volume (40%) high-alkali glass exhibited improved efficacy in decreasing ASR expansion.

The expansion of Pyrex mortar prisms, in ASTM C441, containing the same binary blends (75% PC + 25% SCM) due to ASR at 38 °C is published elsewhere [13] but included in this paper to compare with available alkalis and pore solution data generated in this study. The alkali concentration in the pore solution of PC-SCM pastes seems to be better predicting the extent of ASR mitigation than the available alkalis of SCMs do (see Fig. 5c, d). This is predominately because the availability of alkalis in the pore solution is dictated by the available alkalis of SCMs and the capability of SCMs to bind a portion of PC's alkalis (that are almost 100% available). Thus, as determined by the C311 test procedure, available alkalis do not necessarily capture the alkalis that are truly "available" to participate in ASR. Improvements to the existing C311 available alkalis test could include: 1) the use of PC rather than lime with SCMs, 2) curing at 38 °C for an extended period (e.g., 91 days), and 3) leaching crushed samples, at the end of the test, with a solution that better represents the pore of concrete having a pH of 13.0–13.2 rather than distilled water [11].

5 Conclusion

Based on the results presented in this paper, the following conclusions can be drawn:

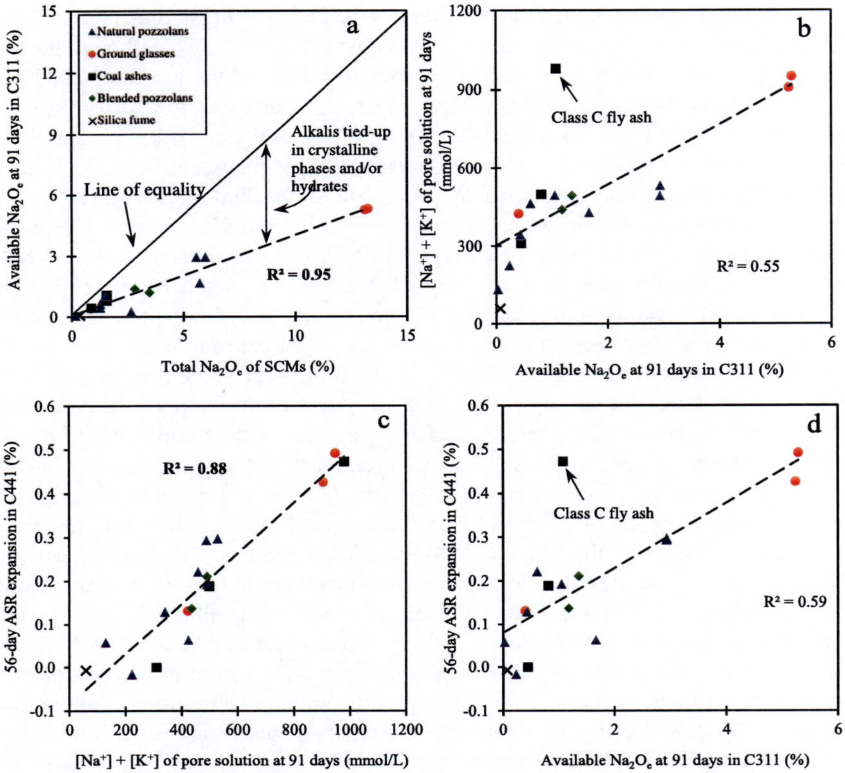

Fig. 5. Correlations between available alkalis, pore solution composition and ASR expansion.

1. For all SCMs tested, the water-soluble alkalis represent less than 50% of the total alkalis and for some, particularly natural pozzolans, less than 2% of the total alkalis were found to be water soluble by the C114 test method.
2. The available alkali contents of SCMs as determined by the C311 test procedure at 28 and 91 days are significantly higher than the water-soluble alkali content tested using C114 and the values vary significantly depending on the SCM tested. For instance, for materials within the class "natural pozzolans" the portion of the total alkalis that are "available" by C311 ranges from 10 to 70% of the total alkali content of the SCM.
3. To a comparable or higher degree than low-CaO fly ash, natural pozzolans, low-alkali ground glass and ground bottom ash at the same 25% replacement level lower the alkali ion concentration in the pore solution of high-alkali PC mixes by encapsulating available alkalis in hydration products.
4. Compare to the available alkalis of SCMs, the pore solution composition of the PC-SCM blends with regards to alkalis (Na^+ and K^+) correlates well with the ASR expansion in C441. The available alkalis in C311 increase considerably for most

SCMs until 91 days and appear to overlook the alkali binding capacity of SCMs when blended with PC.

References

1. Moffatt, E.G., Thomas, M.D.A.: Performance of 25-year-old silica fume and fly ash lightweight concrete blocks in a harsh marine environment. Cem. Concr. Res. **113**, 65–73 (2018)
2. Dunstan, E.R.: The effect of fly ash on concrete alkali–aggregate reaction. ASTM Cem. Concr. Aggr. **3**(2), 101–104 (1981)
3. Lee, S.T., Moon, H.Y., Swamy, R.N.: Sulfate attack and role of silica fume in resisting strength loss. Cem. Concr. Compos. **27**(1), 65–76 (2005)
4. Ramlochan, T., Thomas, M.D.A.: The effect of metakaolin on external sulphate attack. In: Proceedings of the 4th ACI/CANMET International Conference on the Durability of Concrete. Malhotra, V.M. (ed.), ACI SP-192, vol. 1, pp. 239–251 (2000)
5. McCarthy, M.J., Rob, T., Csetenyi, L.J.: Recovery, recovery, processing, and usage of wet-stored fly ash. In: Coal Combustion Products (CCPs): Characteristics, Utilization and Beneficiation, pp. 343–367. Elsevier (2017)
6. Oruji, S., et al.: Mitigation of ASR expansion in concrete using ultra-fine coal bottom ash. Constr. Build. Mater. **202**, 814–824 (2019)
7. Schwarz, N., Cam, H., Neithalath, N.: Influence of a fine glass powder on the durability characteristics of concrete and its comparison to fly ash. Cem. Concr. Compos. **30**(6), 486–496 (2008)
8. Stanton, T.E.: Studies of use of pozzolans for counteracting excessive concrete expansion resulting from reaction between aggregates and the alkalis in cement. In: Stanton, T., Blanks, R. (eds) Symposium on Use of Pozzolanic Materials in Mortars and Concretes, pp. 178–203. ASTM STP99, ASTM, Philadelphia, USA (1950)
9. Glasser, F.P., Marr, J.: The alkali binding potential of OPC and blended cements. Cemento **82**, 85–94 (1985)
10. Duchesne, J., Bérubé, M.: Available Alkalies from supplementary cementing materials. ACI Mater. Jour. **91**(3), 289–299 (1994)
11. Shehata, M.H., Thomas, M.D.A.: Alkali release characteristics of blended cements. Cem. Concr. Res. **35**, 1166–1175 (2006)
12. Shehata, M.H., Thomas, M.D.A.: The effect of fly ash composition on the expansion of concrete due to alkali silica reaction. Cem. Concr. Res. **30**, 1063–1072 (2000)
13. Kasaniya, M., Thomas, M.D.A, Moffatt, E.G.: Efficiency of natural pozzolans, ground glasses and coal bottom ashes in mitigating sulfate attack and alkali-silica reaction. Cem. Concr. Res. (under review) (2021)
14. Chappex, T., Scrivener, K.: Alkali fixation of C–S–H in blended cement pastes and its relation to alkali silica reaction. Cem. Concr. Res. **42**, 1049–1054 (2012)
15. Kalina, R.D., Al-Shmaisani, S., Seraj, S., Cano, R. Ferron, R.D., Juenger, M.C.G.: Role of alkalis in natural pozzolans on alkali-silica reaction. ACI Mater. J. **118**(3), 83–89 (2021)
16. Shehata, M.H., Thomas, M.D.A., Bleszynski, R.F.: The effects of fly ash composition on the chemistry of pore solution in hydrated cement pastes. Cem. Concr. Res. **29**, 1915–1920 (1999)

Calcination and Reactivity of Marlstones: A Comparison Between Palygorskite and Smectite Bearing Marlstones

Victor Poussardin[1,2,3](✉), Michael Paris[2], William Wilson[3], Arezki Tagnit-Hamou[3], and Dimitri Deneele[1,2]

[1] GERS-GIE, Univ Gustave Eiffel, IFSTTAR, 44344 Bouguenais, France
victor.poussardin@usherbrooke.ca
[2] Institut Des Matériaux Jean Rouxel, CNRS, Université de Nantes, IMN, 44000 Nantes, France
[3] Université de Sherbrooke, Sherbrooke, QC, Canada

Abstract. In the race to reduce greenhouse gas emissions, the cement industry appears to be a bad pupil and is responsible for non-negligible pollution, particularly due to clinkerization. Calcined clays are among the most promising Supplementary Cementitious Materials (SCMs) and allow a significant reduction in the environmental footprint of cement manufacturing. Many secondary resources, which are considered as waste by the mining industry, contain clay phases, and are therefore potentially recoverable as SCMs once calcined. This study focuses on the calcination and reactivity of two marlstones (MS1 and MS2) from a phosphate mine. The evolution of the phases as well as the physico-chemical modifications induced by calcination and hydration were characterised by X-ray diffraction (XRD) and Solide State Nuclear Magnetic Resonance (MAS NMR). The two materials are relatively similar and mainly composed of dolomite, associated with clay phases (smectite and biotite), quartz and other minority phases. The main difference between the two materials is the occurrence of palygorskite – a particular TOT clay mineral - in the MS1 sample. The calcination of these two materials resulted in a significant dehydroxylation of the clay phases, a dissociation of the dolomite into lime and periclase and the formation of dicalcium silicate. The self-reactivity of the samples in the presence of water was then tested in order to determine the evolution of the reactive phases (lime, periclase, dicalcium silicate) and to test the pozzolanic reactivity of the calcined clay phases. The results show a strong reactivity of the MS1 sample (mainly pozzolanic) while MS2 exhibits a lower reactivity. The presence of palygorskite in the MS1 material improves its reactivity once calcined, it thus seems that this unconventional clay has a promising pozzolanic reactivity, which opens new ways of valorization for this type of secondary resources.

Keywords: Supplementary cementitious materials · Palygorskite · ^{29}Si MAS NMR · ^{27}Al MAS NMR · Calcined clay · Pozzolanic reactivity · Marlstones

J. I. Escalante-Garcia et al. (Eds.): RW 2021, RILEM Bookseries 40, pp. 22–29, 2023.
https://doi.org/10.1007/978-3-031-21735-7_3

1 Introduction

It is reported that the cement industry account for 5 to 8% of global anthropogenic CO_2 emissions [1]. In this context, companies and academics are developing new technological innovations to reduce this number. The use of calcined clays as a substitute for clinker is one of the solutions to reduce the environmental footprint of cement. Among the clays used, those containing kaolinite seem to be the most reactive. However, there is a certain lack of interest from the scientific community in studying other types of clays. It is also important to remember that clays are often associated with other minerals such as carbonates in nature. These mixtures (carbonate-clay), also called marlstones, are complex multi-phase systems that require special studies before they can be used as supplementary cementitious materials (SCMs). The objective of this study is to investigate the calcination and reactivity of two marlstones considered as waste by the mining industry in order to be used as SCMs.

2 Methods and Materials

2.1 X-ray Diffraction Analysis

The diffractograms were acquired using a Bruker D8 Advance diffractometer (Bragg-Brentano geometry, Cu anode source tube (40 kV/40 mA), 8 keV, $\lambda = 1.5418$ Å). X-ray diffraction analysis were recorded between 4° and 60° with 0.017° 2θ step size and a 1s measurement time per step.

2.2 Solid State Nuclear Magnetic Resonance

The ^{27}Al MAS NMR analysis were performed on a Bruker Avance III 500 MHz spectrometer using MAS probe of 2.5 mm. Rotation speed of the rotor was 30 kHz, $\pi/13$ pulse, 11 kHz radio frequency field. A repetition time of 1s was chosen. All spectra were performed with ^{1}H decoupling and referenced against an aqueous solution of $Al(NO_3)_3$.

The ^{29}Si MAS NMR analysis were performed on a Bruker NEO 300 MHz spectrometer using MAS probe of 7 mm. Rotation speed of the rotor was 5 kHz, $\pi/2$ pulse duration. Repetition time of 10s was chosen. All spectra were performed with ^{1}H decoupling and referenced against TMS (Tetramethylsilane).

2.3 Materials

Two marlstones were studied (MS1 and MS2) which are mining wastes. Table 1 shows the quantitative phase analysis using the Rietveld method for each material.

Both samples are marlstones mainly composed of dolomite associated with clay phases (smectite and biotite), quartz and hydroxylapatite. The main difference between the two materials is the occurrence of palygorskite in MS1.

Table 1. Wt.% of the crystalline phases of the MS1 and MS2 materials

	Phase	Dolomite	Palygorskite	Smectite	Biotite	Hydroxylapatite	Quartz
MS1	wt.%	54	17	16	2	3	8
MS2	wt.%	59		27	1	2	11

2.4 Calcination and Hydration

The materials were calcined with a laboratory furnace in alumina crucibles from room temperature to 800 °C with a heating rate of 300 °C/h and 1h at maximum temperature. For hydration 1 g of sample was manually mixed with 0.8 g of water until obtaining a homogeneous paste and then left to react for 7, 14 and 180 days closed. Samples were then frozen and freeze-dried in order to stop hydration.

3 Calcination Results

3.1 X-ray Diffraction Analysis

Figure 1 shows the evolution of the diffractograms of MS1 and MS2 before and after calcination at 800 °C.

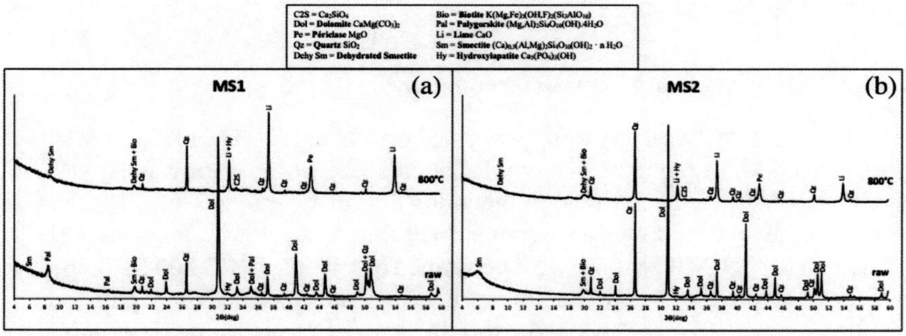

Fig. 1. Evolution of the X-ray diffractograms of MS1 (A) and MS2 (B) before and after calcination at 800 °C.

Concerning MS1, the characteristic peaks of palygorskite have disappeared after calcination, which indicates a structural loss of this clay during calcination at 800 °C. On the other hand, smectite and biotite remain detectable, although we observe a shift of the 001 peak of smectite from 6.2 to 8.75° which is due to its dehydration. The other important information is the dissociation of dolomite into lime and periclase after calcination at 800 °C. At 800 °C a characteristic C_2S (dicalcium silicate) peak appears, however its weak intensity does not allow to determine the type of polymorph. The MS2 material responds in the same way to heat treatment, with the only difference being the absence of palygorskite.

3.2 ^{27}Al MAS NMR Analysis

Figure 2 shows the evolution of the ^{27}Al MAS NMR spectra of MS1 (Fig. 2a) and MS2 (Fig. 2b) before and after calcination at 800 °C.

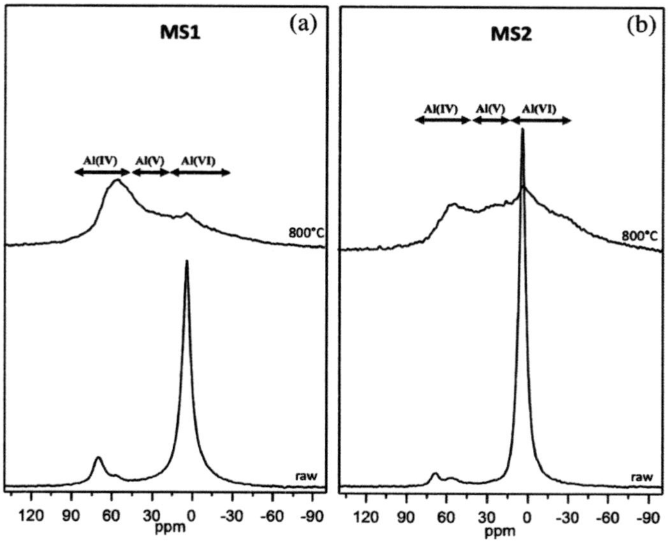

Fig. 2. ^{27}Al MAS NMR spectra of MS1 (A) and MS2 (B) before and after calcination at 800 °C.

The raw spectra of MS1 and MS2 are similar with two main resonances at 3 and 70 ppm which can be associated with 6-fold and 4-fold aluminium respectively. 6-fold aluminium refers to the octahedral aluminum of the clay phases and 4-fold aluminium refers to the tetrahedral aluminum of the clay phases [2, 3].

During calcination both materials behave in a similar way. The resonance intensity associated with 6-fold aluminium decreases significantly and 4- and 5-fold aluminium are formed. This phenomenon is explained by the dehydroxylation of the clay phases due to calcination. MS1 appears to react differently to heat treatment than MS2. Indeed, the change from 6-fold to 5- and 4-fold aluminium after calcination is more pronounced in MS1 than in MS2. Knowing that the pozzolanic activity of a calcined clay closely linked to the appearance of 5-fold and 4-fold aluminium [4] we can hypothesis that MS1 will be more reactive than MS2. The main difference between the two materials is the presence of palygorskite in MS1, this clay is therefore more sensitive to heat treatment than smectite and biotite.

3.3 ^{29}Si MAS NMR Analysis

Figure 3 shows the evolution of the ^{29}Si MAS NMR spectra of MS1 (Fig. 3a) and MS2 (Fig. 3b) before and after calcination at 800 °C.

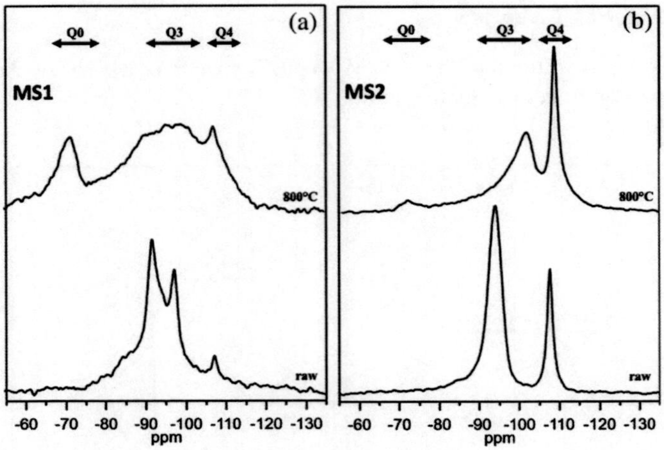

Fig. 3. ^{29}Si MAS NMR spectra of MS1 (A) and MS2 (B) before and after calcination at 800 °C.

The two raw spectra of MS1 and MS2 have a resonance at −108 ppm which corresponds to silicon in the Q4 configuration of quartz [5]. Concerning the clay phases, for MS2 the total signal is composed of Q3 silicon of smectite at −93 ppm [6] and Q2(1Al) silicon of biotite at −86 ppm [7].

For MS1, the total signal is composed of Q3 silicon of smectite, Q3(1Al) silicon of biotite and the two Q3 silicon sites of palygorskite at −92 and −98 ppm which correspond respectively to silicon situated in the center and edges of palygorskite tetrahedra [8].

After calcination, a broadening of the resonances associated with the clay phases is observed in MS1 and MS2. This is explained by a distribution of the chemical environments of silicon and reflects a loss of structure [9]. There is the appearance of a new resonance at −71 ppm in MS1 and MS2. This resonance corresponds to silicon in the Q0 configuration and can be associated with C_2S [10] already observed during the XRD analysis. The intensity of the resonance associated with C_2S is much higher in MS1 than in MS2, so it seems that palygorskite facilitates the formation of this C_2S during calcination.

The study of calcination in these two materials therefore revealed the formation of several reactive phases (lime, periclase, C_2S) and a dehydroxylation phenomenon of the clay phases. It was therefore decided to characterize the reactivity of these calcined samples by carrying out hydration tests. By hydrating the samples the hydraulic reactivity of the neoformed C_2S can be tested as well as the pozzolanic reactivity of the calcined clay phases since the lime is transformed into portlandite and becomes available to react.

4 Reactivity Results

4.1 X-ray Diffraction Analysis

Figure 4 shows the evolution of the X-ray diffractograms of MS1 (Fig. 4a) and MS2 (Fig. 4b) calcined at 800 °C and hydrated during 7, 28 and 180 days. In 800 °C calcined MS1, hydrotalcite, portlandite and calcite and/or CSH are formed after 7 days of

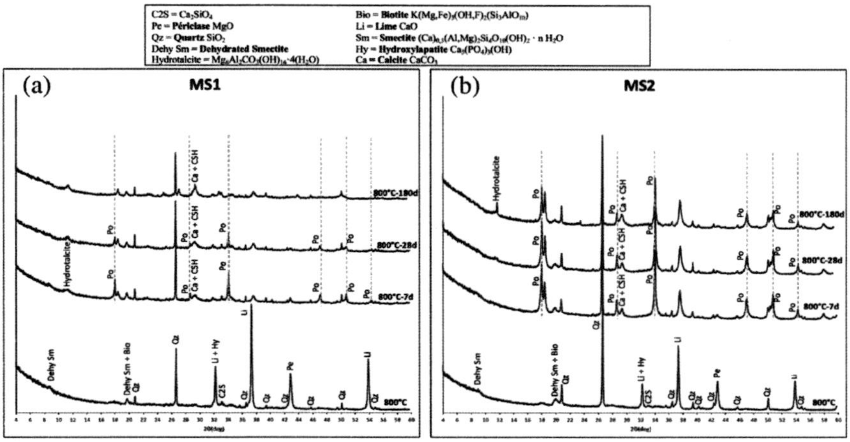

Fig. 4. Evolution of the X-ray diffractograms of MS1 (A) and MS2 (B) calcined at 800 °C and hydrated during 7, 28 and 180 days.

hydration. It seems that the magnesium from the periclase is incorporated into hydrotalcite rather than in brucite. Portlandite comes from the lime hydration, CSH from the pozzolanic reaction and calcite from carbonation phenomena.

Hydration of 800 °C calcined MS2 leads to the formation of the same phases as MS1 except that hydrotalcite is only detectable after 14 days of hydration. Subsequently, with increasing hydration time, a decrease in the intensity of the peaks associated with portlandite is observed for both materials, indicating portlandite consumption. However, the consumption of portlandite is much higher for MS1 than for MS2 since after 180 days of hydration no detectable portlandite remains for MS1.

4.2 ^{29}Si MAS NMR Analysis

Figure 5 shows the evolution of the ^{29}Si MAS NMR spectra of MS1 (Fig. 5a) and MS2 (Fig. 5b) calcined at 800 °C and hydrated during 7, 28 and 180 days.

After 7 days of hydration for MS1 and MS2 we observe a decrease in the intensity of the resonances associated with the calcined clay phases and the formation of two new resonances at −78 and −85 ppm. The resonance at −78 ppm corresponds to terminal Q1 [SiO_4] tetrahedral groups of C–S–H and the resonance at −85 ppm to tetrahedral Q2 [SiO_4] intermediates and bridging groups of C–S–H [11].

After 7 days, the Q0 associated with the neoformed C2S disappeared, thus indicating a reactivity of the C2S from 7 days of hydration for both MS1 and MS2.

With increasing hydration time there is a progressive consumption of Q3 associated with calcined clay phases and the formation of Q1 and Q2 characteristic of CSH. However, the relative amount of Q1 and Q2 silicon (to total silicon) formed is much higher for MS1 than for MS2.

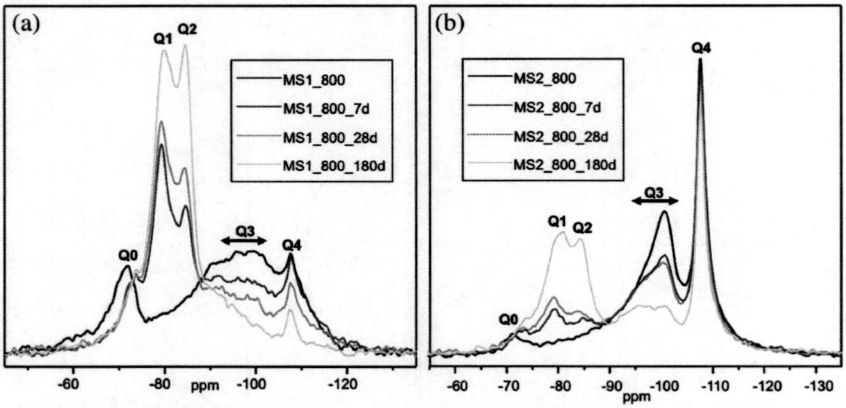

Fig. 5. ^{29}Si MAS NMR spectra of MS1 (A) and MS2 (B) calcined at 800 °C and hydrated during 7, 28 and 180 days.

It seems that the pozzolanic reactivity of MS1 is much higher than for MS2, which explains the low consumption of portlandite during the hydration of MS2. The main difference between the two materials is the presence of palygorskite in MS1. This calcined clay therefore has a higher pozzolanic activity than calcined smectite and biotite.

5 Conclusions

In this paper, a comparative study of the calcination and reactivity of two marlstones was made. The following conclusions can be drawn:

- The calcination at 800 °C resulted in a dehydroxylation phenomenon of the clay phases for both materials studied. However, this phenomenon is much more pronounced for MS1 which contains palygorskite.
- The calcination of both samples at 800 °C resulted in C2S formation.
- The hydration tests indicated a self-reactivity of the two calcined materials (pozzolanic reactivity of the calcined clay phases and reactivity of the neoformed C2S).
- Calcined palygorskite-bearing marlstone shows a much higher reactivity than the calcined smectite-bearing marlstone, consequently calcined palygorskite appears to have significant pozzolanic activity compared to calcined smectite.

These types of complex multi-phase marlstones have the potential to be recovered as SCMs, in particular those containing palygorskite. Further experiments, as mechanical performance tests on cements incorporating these calcined materials, will be interesting to carry out in order to confirm this reactivity.

References

1. Huntzinger, D.N., Eatmon, T.D.: A life-cycle assessment of Portland cement manufacturing: comparing the traditional process with alternative technologies. J. Clean. Prod. **17**(7), 668–675 (2009). https://doi.org/10.1016/j.jclepro.2008.04.007
2. Maia, A.Á.B., Angélica, R. S., de Freitas Neves, R., Pöllmann, H., Straub, C., Saalwächter, K.: Use of 29Si and 27Al MAS NMR to study thermal activation of kaolinites from Brazilian Amazon kaolin wastes. Appl. Clay Sci. **87**, 189–196 (2014). https://doi.org/10.1016/j.clay.2013.10.028
3. Sanz, J., Serratosa, J.M.: Silicon-29 and aluminum-27 high-resolution MAS-NMR spectra of phyllosilicates. J. Am. Chem. Soc. **106**(17), 4790–4793 (1984). https://doi.org/10.1021/ja00329a024
4. Fernandez, R., Martirena, F., Scrivener, K.L.: The origin of the pozzolanic activity of calcined clay minerals: A comparison between kaolinite, illite and montmorillonite. Cem. Concr. Res. **41**(1), 113–122 (2011). https://doi.org/10.1016/j.cemconres.2010.09.013
5. Lippmaa, E., Maegi, M., Samoson, A., Engelhardt, G., Grimmer, A.R.: Structural studies of silicates by solid-state high-resolution silicon-29 NMR. J. Am. Chem. Soc. **102**(15), 4889–4893 (1980). https://doi.org/10.1021/ja00535a008
6. Brown, I.W., MacKenzie, K.J.D., Meinhold, R.H.: The thermal reactions of montmorillonite studied by high-resolution solide-state 29Si and 27Al NMR. J. Mater. Sci. **22** (1987)
7. Mackenzie, K.J.D., Brown, I.W.M., Cardile, C.M., Meinhold, R.H.: The thermal reactions of muscovite studied by high-resolution solid-state 29-Si and 27-Al NMR. J. Mater. Sci. **22**(7), 2645–2654 (1987). https://doi.org/10.1007/BF01082158
8. Barron, P.F., Frost, R.L., Qlil, N.: Solid state 29Si NMR examination of the 2:1 ribbon magnesium silicates, sepiolite and palygorskite. Am. Mineral. **70**, 758–766 (1985)
9. MacKenzie, K.J.D., Smith, M.E.: Multinuclear Solid-State NMR of Inorganic Materials. Pergamon Materials Series
10. Skibsted, J., Jakobsen, H. J., Hall, C.: Quantification of calcium silicate phases in Portland cements by 29Si MAS NMR spectroscopy. J. Chem. Soc. Faraday Trans. **91**(24), 4423 (1995). https://doi.org/10.1039/ft9959104423
11. Andersen, M.D., Jakobsen, H.J., Skibsted, J.: Characterization of white Portland cement hydration and the C-S-H structure in the presence of sodium aluminate by 27Al and 29Si MAS NMR spectroscopy. Cem. Concr. Res. **34**(5), 857–868 (2004). https://doi.org/10.1016/j.cemconres.2003.10.009

Effect of the Formulation on the Hydric and Mechanical Properties of Flax Concrete: Comparison with Hemp Concrete

Ferhat Benmahiddine[1,2](✉), Rafik Belarbi[1,2], and Abdelkader Tahakourt[3]

[1] LaSIE, UMR 7356, CNRS, La Rochelle University, Avenue Michel Crépeau, 17042 La Rochelle cedex1, France
ferhat.benmahiddine1@univ-lr.fr
[2] 4evLab, LaSIE, CNRS, EDF R&D, La Rochelle University, Avenue Michel Crépeau, 17042 La Rochelle cedex1, France
[3] Laboratoire de Génie de La Construction Et Architecture (LGCA), Faculté de Technologie, Université de Bejaia, 06000 Bejaia, Algérie

Abstract. The incorporation of waste plant particles in a cementitious matrix has led to obtain high-performance materials with good economic and ecological benefits. In this paper, a characterization of the main hydric and mechanical properties of a new material composed by flax shives was performed. The aim is to study the impact of the flax shives content on its hydric and mechanical behavior. Moreover, a comparison between the behavior of this material and those of hemp concrete was carried out. This allows, on the one hand, to apprehend better the hydric behavior of the different applications of this material, and on the other hand, to promote new economic perspectives for these products. The results showed that this material has a good hydric behavior. The formulation with a high flax shives content showed a moisture buffer value (MBV index) of 2.80, which allowed it to be classified as an excellent moisture regulator. The comparison of the properties of flax concrete with those of standardized hemp concrete showed that both materials are relatively identical with advantageous and satisfactory properties.

Keywords: Bio-based materials · Flax concrete · Microstructure · Moisture buffer value

1 Introduction

The use of biobased building materials for construction is perfectly in line with a sustainable development approach and helps to meet the current societal challenges surrounding the achievement of a zero-carbon territory in 2040. Indeed, materials with a low environmental footprint have a negative carbon balance [1] and constitute an essential solution for reducing energy and environmental impacts. However, the behavior of these materials remains little discussed in the literature, which limits their use by the industry professionals [2]. Among these vegetative granulates that can be used for the formulation of

J. I. Escalante-Garcia et al. (Eds.): RW 2021, RILEM Bookseries 40, pp. 30–37, 2023.
https://doi.org/10.1007/978-3-031-21735-7_4

the bio-based materials, we can find flax shives. Flax shives have a low thermal conductivity [3–5] and a high water absorption capacity of about 2 to 3 times their weight in 48 h. This is due to the nature of their microstructure (cellulose and hemicellulose) and their pore size distribution organized in a parallel capillary form [6, 7]. For this purpose, we proposed an experimental investigation of the characterization of flax concrete by highlighting the effect of the vegetative particles content on its functional properties. The concerned properties are water porosity, adsorption and desorption isotherms, moisture storage capacity, moisture buffering capacity (MBV) and compressive strength. In addition, microstructure analysis of the vegetative particles was also performed using a scanning electron microscope in environmental mode (SEM). This allows us, on the one hand, to better understand the hydric and mechanical behavior of the material and to evaluate its various applications in the construction field, and on the other hand, to promote new economic perspectives for these products. These formulations were defined with reference to hemp concrete, which is part of a normative framework. A comparison of the properties of these two materials was also carried out in order to be able to classify and situate this new biobased material.

2 Materials and Methods

With reference to hemp concrete (standardized material), we have proposed 03 formulations to be studied. The objective is to highlight the effect of the vegetative particles content on the hydric and mechanical properties of the flax concrete. For this purpose, we proposed three mass percentages of flax shives content, namely: 11.5%; 14.5% and 17.5%. The binder used in this study is the Tradical® PF70 lime. This binder is composed of 75% air lime, 15% hydraulic binder, and 10% pozzolanic binder. The aim is to meet the different applications of flax concrete, which depend on the formulation of the material. These mass percentages of flax concrete were defined based on the standard formulation of hemp concrete [8]. In addition, three corresponding water/binder ratios (W/B) were used: 1.35; 1.45 and 1.55. The formulations are summarized in Table 1.

Table 1. Studied flax and hemp concrete formulations.

Materials	Formulation abbreviations	Percentage Mass			G/L	W/L
		Vegetative aggregates G (%)	Binder (Tradical® PF70) L (%)	Water W (%)		
Flax concrete	F 1	11.50	37.66	50.84	0.31	1.35
	F 2	14.5	34.90	50.60	0.42	1.45
	F 3	17.50	32.35	50.15	0.54	1.55
Hemp concrete	HC	16.00	34.00	50.00	0.47	1.47

3 Résults and Discussion

3.1 SEM Analysis of Flax and Hemp Shives

Before starting the characterization of the materials, we first performed Scanning Electron Microscope observations in Environmental mode on the flax and hemp shives using an FEI Quanta 200 microscope. Figure 1 shows the observations performed on the cross section of the studied vegetative aggregates at different resolutions. First, it can be clearly noticed that the microstructure of the flax shives is different to that of the hemp shives, especially in terms of their thicknesses. Despite the small thickness of the flax shives (about 300 μm), its cross section shows a very complex porous structure with pores of different categories and sizes. The walls forming these porous cells are mainly composed of cellulose (up to 70%), hemicellulose, lignin and pectin [9, 10]. This porous and complex structure allows to dampen the transmission of heat in the vegetative particles, which confers a low thermal conductivity (between 0.037 and 0.042 w/(m.K)). The incorporation of these vegetative aggregates in an air binder allows not only to give good thermal properties to the material, but also to improve its hydric properties, especially the moisture storage capacity and the moisture buffer value (MBV).

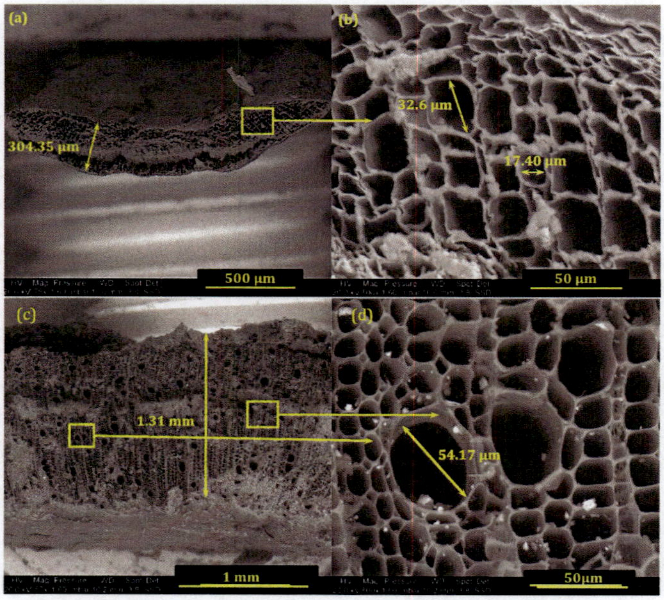

Fig. 1. SEM observations of cross section of flax (a, b) and hemp (c, d) shives.

3.2 Total Water Porosity

Knowing the total porosity of our materials can be useful for possible explanations and interpretations of our results. The water porosity of the studied materials was measured on three samples of each formulation in order to ensure the repeatability of the measurements. The characterization was performed following the AFPC-AFREM procedure [11]. Results presented in Fig. 2 are therefore the average of the different measurements performed. The average porosity of the flax concretes studied varies between 72 and 76%. These values are in agreement with the porosity of hemp concrete, which is around 71%. In general, we can observe a slight increase in the total porosity of the flax concrete as a function of the flax shives content. This increase can be clearly seen between flax concretes containing 11.5% flax shives and those containing 17.5% flax shives. This is due to the high porosity of the flax particles (as shown in the SEM images of Fig. 1) and to the increase of the intergranular voids.

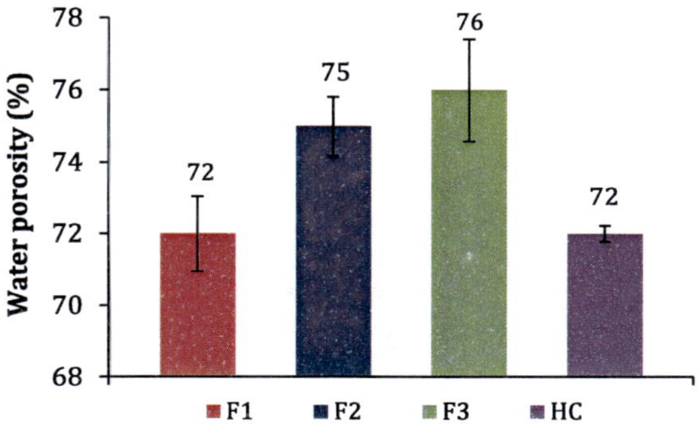

Fig. 2. Total porosity of flax and hemp concretes.

3.3 Sorption Isotherms

The measurement of sorption isotherms was performed based on a gravimetric method using the ProUmid device. Before starting the characterization, the 1 cm cubic samples were dried in an oven at 40 °C for 24 h and then degassed under vacuum to complete the drying.

In Fig. 3, we can observe that the flax shives content has a significant influence on the maximal adsorbed water contents. For the same relative humidity, the higher the G/L ratio, the higher the equilibrium water content. This observation is clearly visible at high humidity levels (RH > 90%). This is due to the high absorption capacity of flax shives, which can absorb up to 3 times their weight in 48 h. That being said, the absorption capacity of flax concrete is highly dependent on the amount of flax shives used. Furthermore, we can see that both materials (hemp concrete and flax concrete) show almost the same behavior but with a different maximum water content at high humidity values (RH > 90%). The maximum water content absorbed by hemp concrete is higher than that of flax concrete. This is due to the high absorption capacity of hemp, which is about 400% of its weight in 48 h.

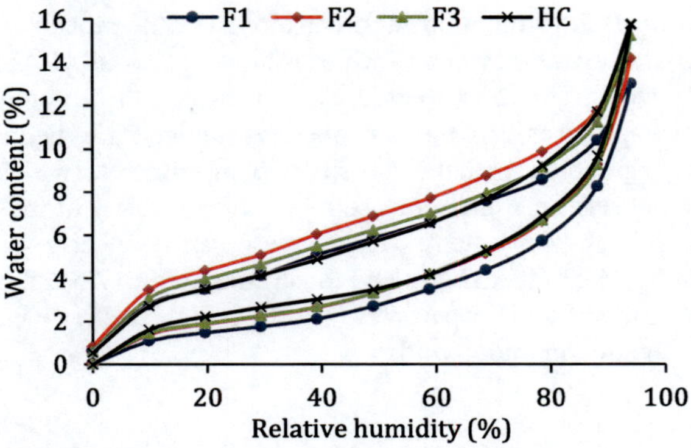

Fig. 3. Adsorption/desorption isotherms of flax and hemp concrete.

3.4 Moisture Storage Capacity

The measurement of sorption isotherms allow us to determine the moisture storage capacity of the studied materials by calculating the slope of the isotherms. The obtained results are presented in Fig. 4. It can be seen that when the relative humidity is below 80% (low relative humidity), the flax shives content doesn't have any influence on the moisture storage capacity. However, at high humidity (RH > 80%), the moisture storage capacity increases with the increase of the flax shives in the material. On the other hand, based on the results of formulations F1 and F2, we can clearly observe that the moisture storage capacity of flax concrete is almost identical to that of hemp concrete over the whole range of relative humidity.

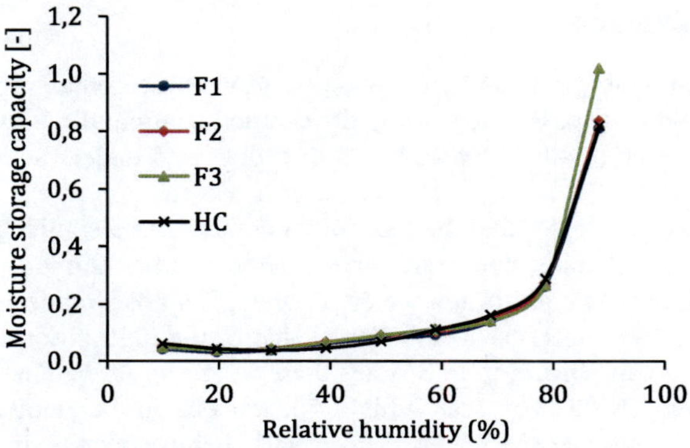

Fig. 4. Moisture storage capacity of flax and hemp concretes.

3.5 Moisture Buffer Value (MBV)

The moisture buffering capacity characterizes the capacity of the material to moderate the relative humidity variations of the ambient air. This index (MBV) can therefore inform us about the quality of the material with respect to the natural regulation of relative humidity in the living environment. This parameter was measured with reference to the Nordtest project, which proposes to condition the samples in a climatic chamber for 8 h at 75% relative humidity, followed by 16 h at 33%, maintaining a constant temperature of 23 °C. The characterization was carried out in a climatic chamber that allows to regulate with precision the temperature and relative humidity. Based on the results presented in Fig. 5, we can observe that the moisture buffering capacity of flax concrete increases with the vegetative particles content in the material. This is due to the high regulation capacity of the vegetative particles by adsorption and restitution of relative humidity in case of surplus or lack, respectively. Therefore, increasing the G/L ratio leads to a higher MBV value. Compared to hemp concrete, flax concrete is also an excellent moisture regulator since its MBV value is higher than 2.

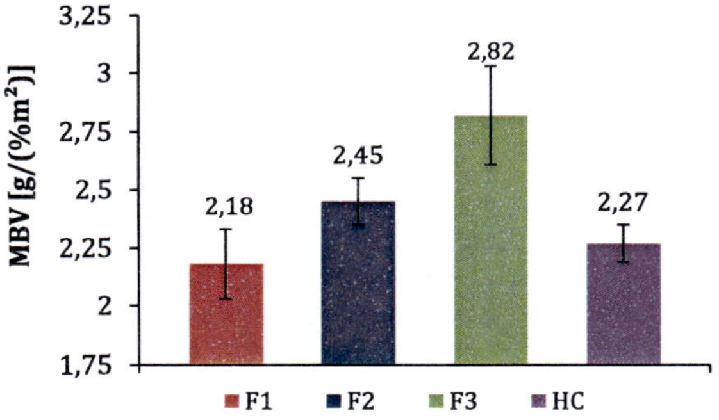

Fig. 5. Moisture buffer values of flax and hemp concretes.

3.6 Compressive Strength

The mechanical characterization of flax concrete was performed with reference to the literature [12–14] and the EN 826 [15] & ASTM C109 [16] standards. The test was carried out on 5 cm square cubic specimens with a load speed of 5 mm/min using a ZwickRoell measuring device. The obtained results are presented in Fig. 6.

Based on these results, we can clearly notice that the compressive strength of flax concrete is closely related to the vegetative particle content. As the flax shives content increases, the compressive strength decreases. This is mainly due to the dependence between the intergranular porosity and the vegetative particles content in the material. Indeed, the increase of the volume of voids in the material causes the embrittlement of the mechanical behavior of the material and consequently, its compressive strength decreases. In comparison with hemp concrete, and taking into account the proportions of vegetative aggregates incorporated in the material, the mechanical behavior of flax

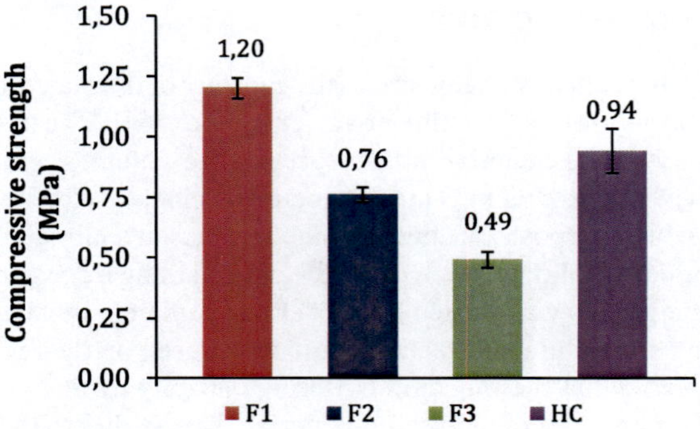

Fig. 6. Compressive strength of flax and hemp concretes.

concrete is almost identical to that of hemp concrete. Both materials can therefore be used in the construction industry as insulation and not as structural materials.

4 Conclusion

In this study, an experimental investigation of the hydric and mechanical behavior of flax concrete was presented. The aim is to highlight the effect of the vegetative particles content on the functional properties of biobased materials. For this purpose, three formulations were characterized with different mass percentages of flax shives. The results were then compared with those of the hemp concrete which was included in a normative framework. Scanning electron microscope (SEM) observations showed that flax and hemp shives have a very complex and porous microstructure. The incorporation of these vegetative particles in a lime-based binder allowed to obtain a lightweight material with good hydric properties. We have shown that an increase in the amount of these vegetative particles leads, on the one hand, to an increase in the total porosity, moisture buffering capacity and maximum water content absorbed, and on the other hand, to a reduction in the compressive strength of the materials. In fact, flax concrete can be classified as an excellent moisture regulator with a maximum MBV index of about 2.8. In general, the results showed that this material has a good hydric behavior, which allows to address perfectly the energetic and environmental issues. Moreover, the behavior of flax concrete is almost identical to that of hemp concrete. Based on the obtained results, both materials can be used as insulation in the construction field for renovation and rehabilitation.

Acknowledgements. The Region and the European Union support the project <CPER-FEDER Bâtiment durable Axis 2 MADUR Project: High-performance building materials with low environmental impact, sustainable and resilient > within the framework of the Operational Program FEDER/FSE 2015–2020 and Energy saving certificate program of the Ministry of Ecological and Solidarity Transition "SmartReno support" 2019–2021. The authors thank the electron microscopy platform of the LaSIE laboratory of La Rochelle University for performing the SEM observations.

References

1. Balaras, C.A., et al.: European residential buildings and empirical assessment of the Hellenic building stock, energy consumption, emissions and potential energy savings. Build. Environ. **42**, 1298–1314 (2007)
2. Benmahiddine, F., et al.: Experimental investigation on the influence of immersion/drying cycles on the hygrothermal and mechanical properties of hemp concrete. J. Build. Eng. **32**, 101758 (2020)
3. Garikapati, K.P., Sadeghian, P.: Mechanical behavior of flax-lime concrete blocks made of waste flax shives and lime binder reinforced with jute fabric. J. Build. Eng. 101187 (2020)
4. Li, X., Tabil, L.G., Oguocha, I.N., Panigrahi, S.: Thermal diffusivity, thermal conductivity, and specific heat of flax fiber-HDPE biocomposites at processing temperatures. Compos. Sci. Technol. **68**, 1753–1758 (2008)
5. Dubois, V., Leblanc, A., Carpentier, O., Alhaik, G., Wirquin, E.: Performances of flax shive-based lightweight composites with rapid hardening. Constr. Build. Mater. **165**, 17–27 (2018)
6. Rahim, M., et al.: Characterization of flax lime and hemp lime concretes: hygric properties and moisture buffer capacity. Energy Build. **88**, 91–99 (2014)
7. Khazma, M., Goullieux, A., Dheilly, R.M., Rougier, A., Quéneudec, M.: Optimization of flax shive-cementitious composites: impact of different aggregate treatments using linseed oil. Ind. Crops Prod. **61**, 442–452 (2014)
8. Bennai, F., Issaadi, N., Abahri, K., Belarbi, R., Tahakourt, A.: Experimental characterization of thermal and hygric properties of hemp concrete with consideration of the material age evolution. Heat Mass Transf. **54**(4), 1189–1197 (2017). https://doi.org/10.1007/s00231-017-2221-2
9. Baley, C., Gomina, M., Breard, J., Bourmaud, A., Davies, P.: Variability of mechanical properties of flax fibres for composite reinforcement. A review. Ind. Crops Prod. 111984 (2019)
10. Sathish, S., et al.: Experimental investigation of mechanical and morphological properties of flax fiber reinforced epoxy composites incorporating SiC and Al_2O_3. Mater. Today Proc. (2019)
11. AFPC-AFREM : Méthodes recommandées pour la mesure des grandeurs associées à la durabilité, Compte rendu des journées techniques AFPC-AFREM, Toulouse, 11–12 (décembre 1997)
12. Walker, R., Pavia, S., Mitchell, R.: Mechanical properties and durability of hemp-lime concretes. Constr. Build. Mater. **61**, 340–348 (2014)
13. Glass, S., Sonebi, M.: Investigation of mix composition hemp concrete on transport properties, mechanical properties and durability. In: 2nd International Conference on Bio-based Building Materials & 1st Conference on ECOlogicalvalorisation of GRAnular and FIbrous materials (2017)
14. Sassoni, E., Manzi, S., Motori, A., Montecchi, M., Canti, M.: Experimental study on the physical-mechanical durability of innovative hemp-based composites for the building industry. Energy Build. **104**, 316–322 (2015)
15. EN 826: Thermal Insulating Products for Building Applications—Determination of Compression Behavior (1998)
16. Committee, A.S.T.M.: ASTM C109/C109M-02 standard test method for compressive strength of hydraulic cement mortars. Annu. B. ASTM Stand. **04**, 1–6 (2002)

Hydration Characteristics of Cementitious Paste with Low Grade Limestone

B. Asha(✉) and Manu Santhanam

Indian Institute of Technology Madras, Chennai 600036, India
bahulayanasha@gmail.com

Abstract. In limestone quarries, a major portion of limestone is rejected because of the presence of large amounts of impurities that makes cement production difficult. Such limestones have a significant potential to be used for replacing clinker in cement. The use of ground limestone powder as cement replacement has proved to be very promising and has led to several developments in terms of binary and ternary binders. Addition of fine limestone accelerates the initial hydration of cement by providing extra sites for nucleation, in addition to forming new products of hydration like carboaluminates, when there is a reactive alumina source present in the binder. This paper reports a study that was conducted to follow the impact of low-grade limestone on the hydration behaviour of cementitious systems by isothermal calorimetry on paste samples and by compressive strength test of mortar specimens The study was conducted for various replacement levels of low-grade limestone (from two different mines in India) and also for different particle sizes of limestone. Addition of low-grade limestone was found to accelerate hydration in cementitious systems. Acceleration in hydration increased with fineness of limestone and was reflected in compressive strength of mortar cubes as well.

Keywords: Blended cement · Limestone · Hydration · Calorimetry · Filler · Compressive strength

1 Introduction

Supplementary cementitious materials have been gaining application worldwide in the last few decades. finely ground limestone is one such material. Both binary binders as well as ternary binders formed by the partial replacement of cement by finely ground limestone have been studied in detail by various researchers across the world. Studies have shown that finely ground limestone has a chemical effect on the hydration of cement in addition to the filler effect [1]. This chemical effect of limestone powder is found to be more prominent if alumina is available in plenty in the system. This adds more scope for ternary cement systems with limestone and alumina rich supplementary cementitious materials.

In limestone quarries a major portion of limestone is being rejected due to the presence of impurities like silica, dolomite etc. which makes it unsuitable for cement production. These low-grade limestones also may have a potential to be used as a partial

J. I. Escalante-Garcia et al. (Eds.): RW 2021, RILEM Bookseries 40, pp. 38–43, 2023.
https://doi.org/10.1007/978-3-031-21735-7_5

replacement in cements. So far, very few studies have been found in literature regarding the utilisation of these limestones. Binary blends as well as ternary blends with this low-grade limestone in finely ground form are potentially expected to produce cements of required strength. If this limestone can be utilised in the production of cement, the advantages are twofold: it can lead to sustainable construction through the production of cement of low carbon footprint and the environmental issues associated with the dumping of this limestone reject will be solved.

This study discusses the results of experiments done to understand the hydration behaviour of finely ground low-grade limestone blended ordinary Portland cement. Isothermal calorimetry was employed to study the reaction kinetics. Compressive strength tests on cement mortar cubes were done to study the variation in strength due to the addition of finely ground low-grade limestone. The effect of particle size of ground limestone on cement hydration also was studied.

2 Materials and Methods

Ordinary Portland cement 53 grade conforming to IS 12269:2013 was used in the study. Two low-grade limestone samples were used for replacement of OPC. Limestone 1 (LS1) was collected from Tadipatri mines in Andhra Pradesh and limestone 2 (LS2) from Ariyalur mines in Tamil Nadu. Both the limestone samples were ground to two different fineness, one finer than cement and the second one coarser than cement. Thermogravimetric analysis of the limestone samples showed that both the limestone samples had $CaCO_3$ content below 75%. Hence, they do not qualify for cement production or for use even as a performance improver (up to 5%) as per the existing Indian standards. Replacement levels of 15 and 35% by weight were considered in the present study. The properties of materials used in the study are listed in Table 1.

Table 1. Properties of materials used in the study

Property	Cement	LS 1		LS 2	
		Coarse	Fine	Coarse	Fine
Specific gravity	3.15	2.78		2.82	
Residue on 45 μm sieve (%)	15	50	10	47.5	12.5
$CaCO_3$ Content (%)	–		62		51

Isothermal calorimetry was used to study the reaction kinetics of the cementitious systems containing ordinary Portland cement and finely ground low-grade limestone. The experiment was conducted using a Cal metrix I-Cal HPC 4 channel calorimeter. 20 g of cement was used for the preparation of cement paste. A water/binder ratio of 0.45 was used in the preparation of the samples. Mixing with deionized water was done in a high shear mixer at 1600 rpm for two minutes. Water was preconditioned at 23 °C for 24 hours before mixing. Data was collected for the first 72 hours.

Cubes of size 50 × 50 × 50 mm were prepared with a cement: sand ratio of 1:3 using standard Ennore sand. Water/binder ratio used was 0.45. Cubes were prepared at a temperature of 25 °C and relative humidity 65% and kept under the same environment for 24 hours. Later the cubes were demoulded and kept in the curing room under 25 °C and 90% relative humidity until testing. Cubes were tested at 1 day and 7 days of curing using Controls 250 kN compression testing machine at a loading rate of 35 N/mm^2/min as per IS 4031(Part 6):2005.

3 Results and Discussions

The isothermal calorimetry curves obtained for fine (LS1F) and coarse (LS1C) powders of limestone sample LS1 normalized by the weight of binder and weight of cement are shown in Figs. 1 and 2 respectively. Results of isothermal calorimetry of fine (LS2F) and coarse powders (LS2C) of limestone sample LS2 are given in Figs. 3 and 4, normalized by weight of binder and weight of cement respectively.

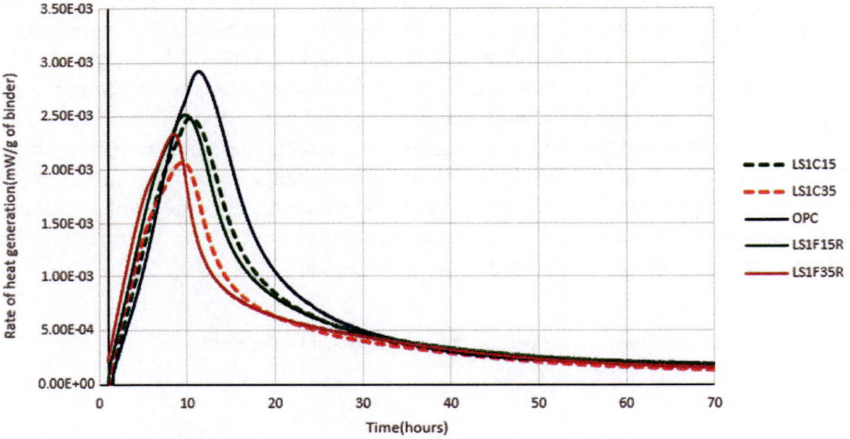

Fig. 1. Isothermal calorimetry curves of LS1F and LS1C normalized by the weight of binder

Both the low-grade limestone samples in finely ground form when mixed with OPC at 15 and 35% replacement levels showed acceleration of hydration. The peak heat rate was increased and time to peak was reduced. Higher replacement level showed higher acceleration. Similar acceleration was reported in studies when pure limestone in fine powder form was used for replacement of OPC [2–4]. However, fine samples of both LS1 and LS2 were found to exhibit a higher acceleration rate compared to the corresponding coarse limestone powder. In the case of pure limestone also higher initial acceleration was reported when limestone powder was of higher fineness [5].

Compressive strengths of cement mortar cubes with 0, 15 and 35% replacement of OPC using fine and coarse powders of limestone powders LS1 and LS2 are shown in Figs. 5 and 6 respectively.

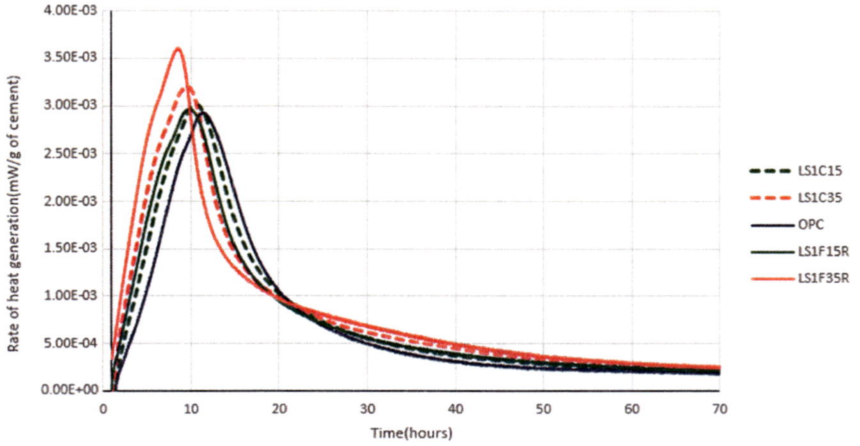

Fig. 2. Isothermal calorimetry curves of LS1F and LS1C normalized by the weight of cement

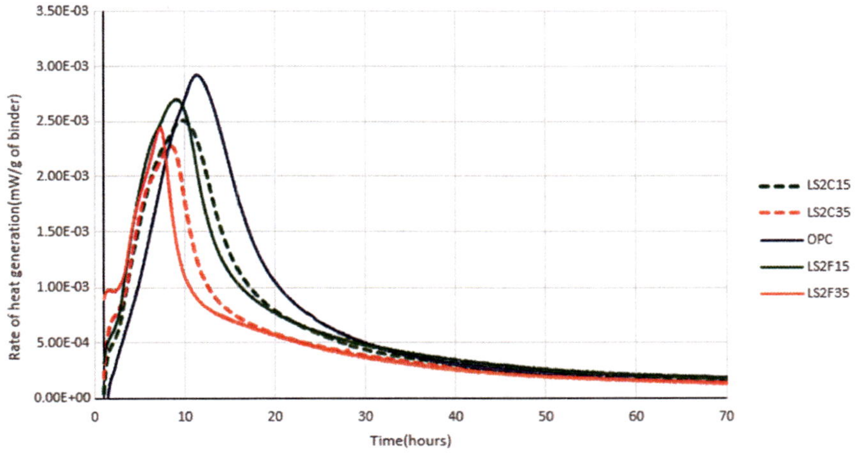

Fig. 3. Isothermal calorimetry curves of LS2F and LS2C normalized by the weight of binder

The results show a reduction in the compressive strength of mortar cubes containing low-grade limestone powder with increasing percentage replacement of cement irrespective of the limestone powder used. The compressive strength results are consistent with the observations made by calorimetry. There is a decrease in the heat rate released when the size of limestone increases and/or when the rate of replacement of cement by limestone increases, inducing a decrease in short-term compressive strengths. Mortar cubes with fine limestone powder had higher compressive strength compared to the corresponding cubes with coarse powder. A reduction in compressive strength beyond a limestone replacement level of 10% is reported in literature [6]. However, almost 75%

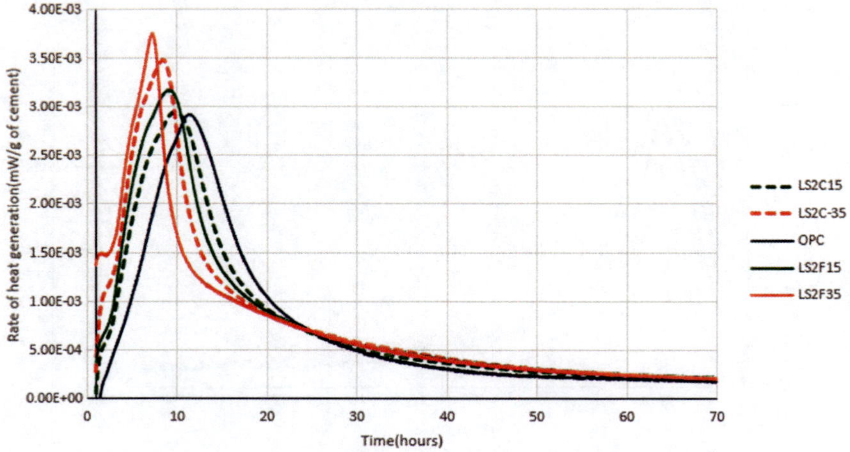

Fig. 4. Isothermal calorimetry curves of LS2F and LS2C normalized by the weight of cement

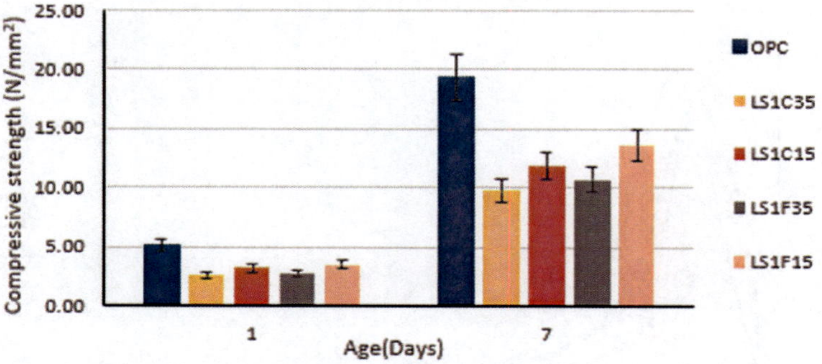

Fig. 5. Compressive strength of mortar cubes using LS1

of the strength of OPC was achieved by a replacement of 15% of finely ground limestone powder LS2. The present study indicates that particle size of low-grade limestone powder has a major influence on the compressive strength at 1 day and 7-day strength of mortar cubes.

4 Conclusions

The major conclusions drawn from the study are as follows.

- The use of low-grade limestone powder as partial cement replacement led to the acceleration of the hydration rates, as seen from isothermal calorimetry.
- Both fine and coarse powders of low-grade limestone samples were found to exhibit acceleration. Fine powders were found to enhance the cement hydration at a higher rate compared to coarse powders of both the limestone samples.

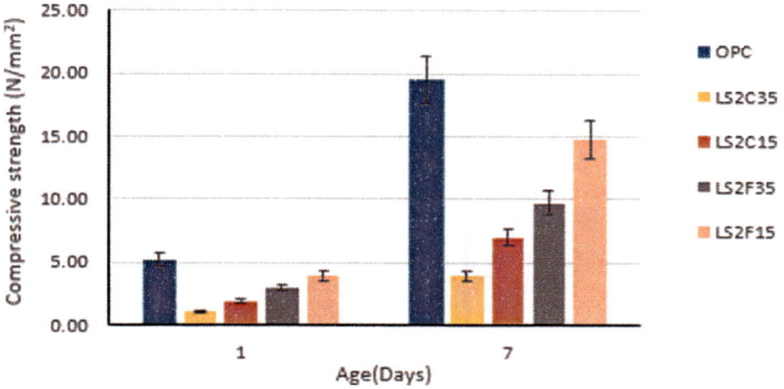

Fig. 6. Compressive strength of mortar cubes using LS2

- Compressive strength of cement mortar cubes reduces with increase in replacement level of low-grade limestone powder. However, a replacement of 15% with LS2 was found to give 75% strength of OPC at 1 day and 7 days.
- Particle size of limestone was found to influence the compressive strength. Mortar cubes with coarser particles were found to give very low strengths compared to those with the corresponding fine limestone powders.

References

1. Soroka, I., Setter, N.: The effect of fillers on strength of cement mortars. Cem. Concr. Res. **7**(4), 449–4569 (1977)
2. Péra, J., Husson, S., Guilhot, B.: Influence of finely ground limestone on cement hydration. Cem. Concr. Compos. **21**(2), 99–105 (1999)
3. Lothenbach, B., Le Saout, G., Gallucci, E., Scrivener, K.: Influence of limestone on the hydration of Portland cements. Cem. Concr. Res. **38**(6), 848–860 (2008)
4. Oey, T., Kumar, A., Bullard, J.W., Neithalath, N., Sant, G.: The filler effect: the influence of filler content and surface area on cementitious reaction rates. J. Am. Ceram. Soc. **96**(6), 1978–1990 (2013)
5. Briki, Y., Zajac, M., Haha, M.B., Scrivener, K.: Impact of limestone fineness on cement hydration at early age. Cem. Concr. Res. **147**, 106515 (2021)
6. Tsivilis, S., Chaniotakis, E., Badogiannis, E., Pahoulas, G., Ilias, A: A study on the parameters affecting the properties of Portland limestone cements. Cem. Concr. Compos. **21**(2), 107–116 (1999)

The Resistance of Composite Cement Paste with Calcined Diatomite to Chloride Attack

Muttaqin Hasan(✉), Taufiq Saidi, Rudiansyah Putra, and Nurmasyitah

Department of Civil Engineering, Universitas Syiah Kuala, Banda Aceh, Indonesia
muttaqin@unsyiah.ac.id

Abstract. An experimental study was carried out to understand the resistance of composite cement paste containing different levels of calcined diatomite powder (CDP) to chloride attack. The amount of water used was determined so that the composite cement paste has a normal consistency. The cube specimens with 50 mm size were cast and cured in water for 28 days after which the specimens were immersed in 10% NaCl solution for 1 month and 2 months. The loss in mass and the degradation of compressive strength after 1 and 2 month immersion time were examined. The results show that the CDP cement paste had less surface deterioration compared to Ordinary Portland Cement (OPC) paste. Based on visual inspection, mass loss, and degradation of compressive strength, the composite cement paste with 20% CDP has better resistance to chloride attack.

Keywords: Composite cement · Calcined diatomite · Chloride · Mass · Compressive strength

1 Introduction

With the rapid development of technology, especially in the construction sector, the development of infrastructure is not only carried out in general residential areas that do not require special requirements. However, at this time many constructions have been built in aggressive environments to meet the needs in areas with high chloride and sulfate content, such as in swampy areas, offshore areas, industrial areas, and mining areas. According to Putra [1], aggressive environments such as groundwater and seawater usually contain 3.6–4% dissolved salts consisting of 75% sodium chloride (NaCl), 10% magnesium chloride ($MgCl_2$), and the remaining 10% sodium sulfate (Na_2SO_4), magnesium sulfate ($MgSO_4$), gypsum (Ca_2SO_4), potassium sulfate (K_2SO_4). Besides, underground structures are also prone to exposure to sulfate and chloride salts [2, 3]. The high chloride content in the environment will degrade concrete performance and cause the corrosion of the reinforcing steel bars which eventually results in the concrete cover cracking [4–6].

The use of materials containing high silica such as fly ash, silica fume, and blast furnace slag can reduce the penetration of chloride ions into the concrete which ultimately increases the performance of the concrete against chloride ion attack [7]. Diatomite as one of the materials containing high silica is widely available in Aceh Besar District,

J. I. Escalante-Garcia et al. (Eds.): RW 2021, RILEM Bookseries 40, pp. 44–50, 2023.
https://doi.org/10.1007/978-3-031-21735-7_6

Indonesia. Previous studies show that calcined diatomite powder (CDP) from Aceh Besar District contains 78.73% reactive silica so it can be used as cementing materials [8–14]. The mechanical and microstructural properties of the composite cement paste using CDP have also been studied [15, 16]. To study the durability of concrete with the addition of CDP against chloride attack, it can be done by studying the resistance of cement paste with the addition of CDP to NaCl solution. This paper, therefore, presents the resistance of composite cement paste with CDP against NaCl attack.

2 Materials and Method

Composite cement in this study was produced in the laboratory by replacing Ordinary Portland Cement (OPC) with Calcined Diatomite Powder (CDP) at a level of 10, 20, and 30% by weight. OPC and CDP were mixed and stirred evenly. The OPC used in this study has a specific gravity of 3.16 dan specific surface area of 539.8 m^2/kg. The CDP was made by grinding diatomite chunk collected from Beureunuet Village, Aceh Besar District, Indonesia, and sieving under #200 mesh. The diatomite powder was then dried in the oven at the temperature of 100 °C for 24 h and calcined in the laboratory furnace at the temperature of 650 °C for 5 h. The CDP has a specific gravity of 1.89 and a specific surface area of 675.6 m^2/kg. The chemical composition of CDP is 78.73% SiO_2, 16.34% CaO, 2.89% Fe_2O_3, 1.11% MgO, 0.39% Al_2O_3 and 0.39% SO_3.

Table 1. The amount of water and the flow of cement paste

CDP content (%)	Amount of water (%)	Flow of cement paste (%)
0	24.63	107.49
10	26.22	103.38
20	29.32	101.34
30	32.43	97.86

The cementitious paste was made by mixing the composite cement with water. The amount of water was determined by adjusting the cement paste to have normal consistency as specified in EN 196-3 and ASTM C191-01a standards [17, 18] and shown in Table 1. The OPC paste was also prepared for comparison purposes. The flow of cement paste was tested following ASTM C230/C230M-08 standard [19] and shown in Table 1. Nine cube specimens with 50 mm size were prepared for each mixture. The specimens were cured in water for 28 days. After 28 days the specimens were removed from the water, dried, and weighed. A compression test was conducted on 3 specimens for each mixture while the other 6 specimens were immersed in 10% NaCl solution for 1 and 2 months. After 1 month immersion time, 3 specimens of each mixture were removed from the bath and their appearances were observed as well as their mass and compressive strength were measured. the same procedures were conducted for the other 3 specimens after the immersion time of 2 months.

3 Results and Discussion

Before being immersed in NaCl solution, OPC cement paste was dark gray, while composite cement paste with CDP was slightly yellow. After 1 month of immersion, there was a color change in both OPC cement paste and composite cement paste where the color of both cement pastes changed to light gray. Meanwhile, after being soaked for 2 months, all the tested cement pastes changed their color to whiter. The color changes of all the tested cement paste specimens are shown in Figs. 1, 2, 3 and 4. The figures also show that after 1 month of immersion, the OPC cement paste had little surface damage, while the composite cement paste had no surface damage. After 2 months of immersion, the OPC cement paste and composite cement paste with 30% CDP had surface damage, while the composite cement paste with 10% and 20% CDP had no damage at all. The surface damage as shown in Figs. 1, 2, 3 and 4 is in the form of the appearance of pores on the surface of the specimen due to the peeling of a small amount of cement paste and peeling off the edges of the cement paste at the corners of the specimens.

Fig. 1. The visual appearance of OPC paste specimens: **a** before immersion; **b** after 1-month immersion; and **c** after 2-month immersion

Fig. 2. The visual appearance of composite paste specimens with 10% CDP: **a** before immersion; **b** after 1-month immersion; and **c** after 2-month immersion

The peeling of the surface of the specimen causes the cement paste to lose its mass. The amount of mass loss of cement paste expressed in percent is shown in Fig. 5. The figure shows that up to 2 months immersion in a 10% NaCl solution, composite cement paste did not experience any mass loss at all. Meanwhile, OPC cement paste has lost mass after 2 months of immersion, although at 1 month of immersion it has not lost mass. From these results, it can be said that according to the condition used, the CDP

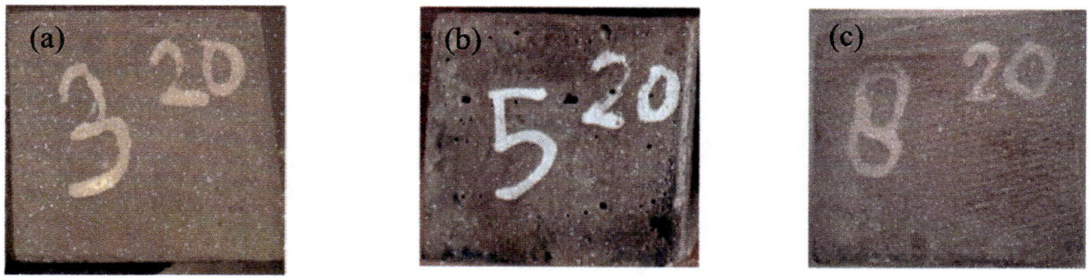

Fig. 3. The visual appearance of composite paste specimens with 20% CDP: **a** before immersion; **b** after 1-month immersion; and **c** after 2-months immersion

Fig. 4. The visual appearance of composite paste specimens with 30% CDP: **a** before immersion; **b** after 1-month immersion; and **c** after 2-month immersion

specimens had less surface deterioration and mass loss to chloride solution than OPC neat paste.

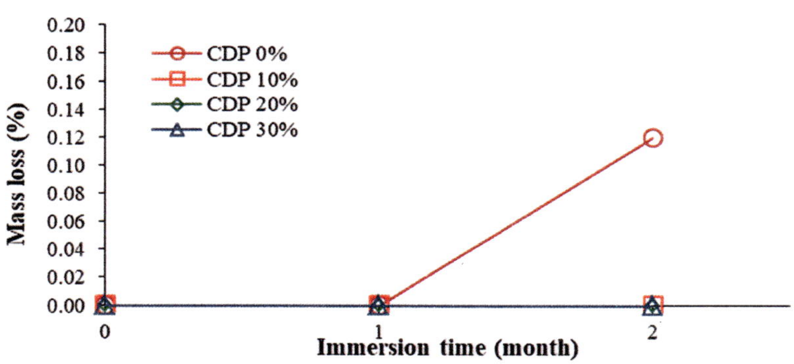

Fig. 5. Mass loss of OPC and composite cement paste after immersion in 10% NaCl solution

Cement paste exposed to a chloride-containing environment also causes internal cracks. Those internal cracks cause degradation of the compressive strength of the cement paste. The compressive strength of OPC cement paste and composite cement paste tested in this study before and after immersion in 10% NaCl solution is shown in Fig. 6. Figure 7 shows the comparison of the compressive strength of cement paste after immersion in NaCl solution (f'_{cc}) to the compressive strength of cement paste before immersion (f'_{co}). Figure 6 shows that the compressive strength of composite cement paste before being immersed in NaCl solution is lower than the compressive strength of OPC cement paste where the compressive strength decreases with increasing CDP content. However, the composite cement paste with 20% CDP had the lowest compressive strength degradation after being immersed in NaCl solution for 1 month and 2 months. This, along with the visual inspection and mass loss, shows that the composite cement paste with 20% CDP has better resistance to chloride attack. However, the results of this study only presented the durability of cement paste against chloride attack during the immersion time of up to 2 months where the compressive strength degradation of all cement pastes had not reached 10%. Therefore, further studies are needed by soaking cement paste for up to one year and will be carried out in future studies.

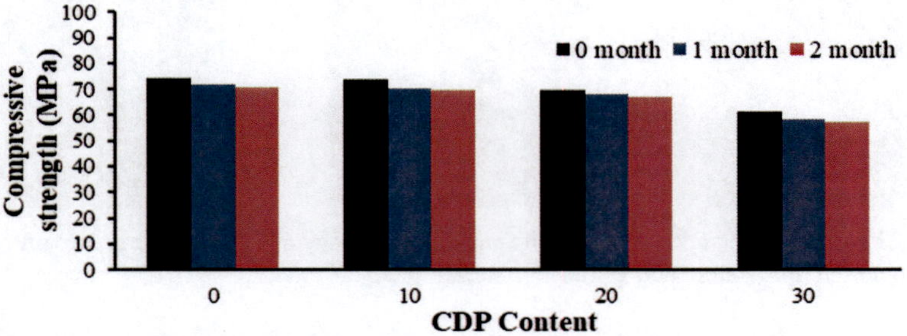

Fig. 6. Compressive strength of OPC and composite cement pastes before and after immersion in NaCl solution

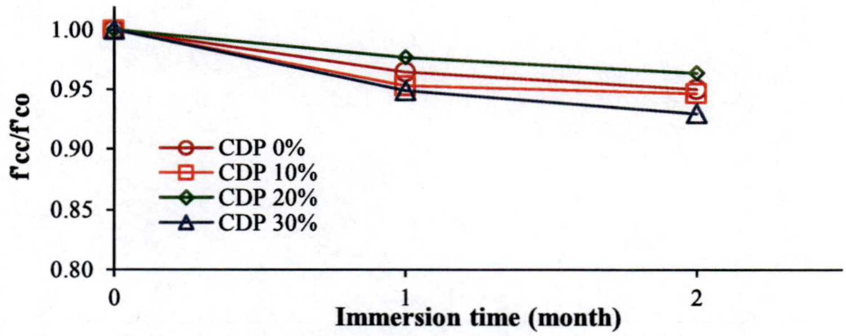

Fig. 7. Degradation of compressive strength of cement pastes due to NaCl attack

4 Summary

The resistance of composite cement pastes with calcined diatomite powder (CDP) was studied by immersing the composite cement paste with 10, 20, and 30% CDP content in 10% NaCl solution for 1 month and 2 months and measuring its mass loss and compressive strength loss. The same treatment was also carried out on OPC cement paste. The results showed that the composite cement pastes had not lost mass until the age of 2 months of immersion, while the OPC cement paste had experienced a mass loss of 0.12%. The smallest compressive strength degradation was found in the composite cement paste with 20% CDP.

Acknowledgment. This research was supported by a grant provided by Syiah Kuala University Research and Community Service Center (LPPM) under contract number of 172/UN11/SPK/PNBP/2021.

References

1. Putra, D.: Penambahan abu sekam pada beton dalam mengantisipasi kerusakan akibat magnesium sulfat pada air laut. Jurnal Ilmiah Teknik Sipil **10**(2), 195–203 (2006)
2. Leemann, A., Loser, R.: 2011 Analysis of concrete in a vertical ventilation shaft exposed to sulfate-containing groundwater for 45 years. Cem. Concr. Compos. **33**(1), 74–83 (2011)
3. Hasan, M., Husin, S., Nursaniah, C.: Mechanical properties of concrete in compression exposed to sulfuric acid. Key Eng. Mater. **711**, 302–309 (2016)
4. Shi, X., Xie, N., Fortune, K., Gong, J.: Durability of steel reinforced concrete in chloride environments: an overview. Constr. Build. Mater. **30**, 125–138 (2011)
5. Broomfield, J.: Corrosion of Steel in Concrete: Understanding. Investigation and Repair. CRC Press, London (2006)
6. Sarja, A.: Durability design of concrete structures—committee report 130-CSL. Mater. Struct. **33**, 14 (2000)
7. Ellis, W.E.Jr., Rigg, E.H., Butter, W.: Comperative results of utilization of fly ash, silica fume and GGBFS in reducing the chloride permeability of concrete. Durability Concrete, SP-126, American Concrete. Institute, Detroit, MI 443–458 (1991)
8. Hasan, M., Saidi, T., Muyasir. A., Alkhaly, Y.R., Muslimsyah, M.: Characteristic of calcined diatomaceous earth from Aceh Besar District—Indonesia as cementitious binder. In: IOP Conference Series: Material Science and Engineering, vol. 933, p. 012008 (2020)
9. Saidi, T., Hasan, M.: The effect of partial replacement of cement with diatomaceous earth (DE) on the compressive strength and absorption of mortar. J. King Saud Univ. Eng. Sci. **34**, 250–259 (2020). https://doi.org/10.1016/j.jksues.2020.10.003
10. Hasan, M., Saidi, T., Afiffuddin, M.: Mechanical properties and absorption of lightweight concrete using lightweight aggregate from diatomaceous earth. Constr. Build. Mater. **277**, 122324 (2021)
11. Saidi, T., Hasan, M., Riski, A.D.D., Ayunizar, R.R., Mubarak, A.: Mix design and properties of reactive powder concrete with diatomaceous earth as cement replacement. IOP Conf. Ser. Mater. Sci. Eng. **933**, 012007 (2020)
12. Hasan, M., Saidi, T., Jamil, M., Rhina, Z.: Characteristics of lightweight bricks composed of clay and diatomite. In: Akhyar (ed.) Proceeding of the 2nd International Conference on Experimental and Computational Mechanics in Engineering, Lecture Notes in Mechanical Engineering, pp. 9–19. Springer Nature, Singapore (2021)
13. Hasan, M., Muyasir, A., Saidi, T., Husaini., Azzahra, R.: Properties of high strength concrete with calcined diatomaceous earth as cement replacement under compression. Defect Diffus. Forum **402**, 7–13 (2020)
14. Hasan, M., Riski, A.D.D., Saidi, T., Husaini., Rahman, P.N.: Flexural and splitting tensile strength of high strength concrete with diatomite micro particles as mineral additive. Defect Diffus. Forum **402**, 50–55 (2020)

15. Hasan, M., Saidi, T.: Properties of blended cement paste with diatomite from Aceh Province Indonesia Properties of blended cement paste with diatomite from Aceh Province Indonesia. IOP Conf. Ser. Mater. Sci. Eng. **796**, 012034 (2020)
16. Hasan, M., Saidi, T., Husaini: Properties and microstructure of composite cement paste with diatomaceous earth powder (DEP) from Aceh Besar district—Indonesia. Asia-Pac. J. Sci. Technol. **27**(01), 1–11 (2021)
17. EN 196-3.: Methods of testing cement—Part 1: determination of setting time and soundness. European Committee for Standardization (2018)
18. ASTM C191-01a.: Standard test method for time of setting of hydraulic cement by Vicat needle. ASTM International (2001)
19. ASTM C230/C230M-08.: Standard specification for flow table for use in tests of hydraulic cement. ASTM International (2008)

Co-activation of Blended Blast Furnace Slag and Fly Ash with Sodium Sulfate and Hydroxide

Marcello Mutti(✉), Shiju Joseph, and Özlem Cizer

Department of Civil Engineering, KU Leuven, Leuven, Belgium
marcello.mutti@kuleuven.be

Abstract. The addition of sodium sulfate (Na_2SO_4) to alkali-activated blast furnace slag (BFS), specifically when sodium hydroxide (NaOH) is used as an activator, allows to reduce the amount of hydroxide necessary to achieve desirable mechanical properties. This kind of activation is suitable for BFS that does not require very high pH to promote its hydration. However, it might fall short in case of hybrid binders such as mix of BFS and fly ash (FA). The objective of this study is to elucidate whether the co-activation with Na_2SO_4 and NaOH of these blended mixes leads to advantages compared to just NaOH. The effectiveness of such activation on blended systems is evaluated by means of isothermal calorimetry, products assemblage and compressive strength. The FA replacement significantly affects the induction period length, especially when low NaOH concentration is used, leading to poor early age properties. Nevertheless, the obtained results show that, at later ages, it is still possible to obtain mechanical properties suitable for practical application.

Keywords: Alkali-activated materials · Sodium hydroxide · Sodium sulfate · reaction kinetics · Compressive strength

1 Introduction

Alkali-activated materials (AAMs) provide a promising route in dealing with the environmental impacts of Portland-cement-based building materials. Despite some controversies about the actual carbon emission of alkali-activated binders, it is well accepted that these materials have a significantly lower carbon foot print, compared to Portland cement, due to the different sourcing of the binders that are commonly obtained as industrial by-products [1, 2]. One big disadvantage of these binders is their low or even nil reactivity in water. This drawback is overcome by mixing those binders with high alkaline solutions, made from sodium hydroxide and sodium silicate, leading to materials with high mechanical properties, even superior to those of Portland cement. However, a significant carbon emission is associated to the production of both NaOH and waterglass [3]. A possible alternative to those activators is represented by the near-neutral salts, e.g. carbonates and sulfate [4]. The use of Na_2SO_4, obtained from natural sources (e.g. thernardite and mirabilite) and as industrial by-products, offers unquestionable advantages in terms of environmental impact, cost, ease of use and reduced shrinkage. On the

J. I. Escalante-Garcia et al. (Eds.): RW 2021, RILEM Bookseries 40, pp. 51–59, 2023.
https://doi.org/10.1007/978-3-031-21735-7_7

other hand, sulfate activation often results in a low reactivity of the system, leading to weak mechanical properties [5]. Moreover, the neutral nature of this salt also hinders the possibility to use fly ash (FA) and metakaolin whose activation is strongly pH dependent [6]. These low-calcium materials require pH much higher than 14 to reach acceptable reaction rate. Many studies investigated how to enhance the reactivity of sodium sulfate activated blast furnace slag (BFS) based on BFS composition, fineness and curing temperature [4, 7, 8]. Compressive strength in the order of 45 MPa was obtained from very fine sulfate activated BFS cured at 40 °C while strength of 41 MPa was achieved at room temperature by using wet-grinded BFS with a particle size of 3.87 μm [7]. Another approach to enhance the reactivity of the system consists of providing an initial boost of hydroxyl ions through the addition of limited amount of NaOH, several times lower than what it is generally used [9]. The carbon emissions associated to these systems can be further reduced by partially replacing BFS with FA. A recent study investigated the sodium sulfate activation of ultra-fine BFS (D50 of 3.1 μm) blended with FA [10]. A compressive strength of 60.8 MPa was achieved, at 90 days, with 20% FA and 4% Na_2O, even though in the same a strength of 77.3 MPa was achieved with pure slag. The pH values between 12 and 13 are reported for solutions leached from sulfate activated slag and slag/fly ash [8, 10]. In this pH interval the reactivity of FA is very limited and presumably mainly acts as a filler [11]. In this study the NaOH-Na_2SO_4 approach is followed. The addition of both sodium hydroxide and sulfate shows a synergistic effect on the activation of BFS and leads to appreciable reaction kinetics and mechanical properties. Compressive strength of around 20 and 30 MPa, at 7 and 28 days respectively, was obtained by using 1M $NaOH_{(aq)}$ in combination with Na_2SO_4 while the same BFS activated with higher [NaOH] (e.g. 2M) reached around 17 and 25MPa at the same ages [9]. The presence of a certain concentration of OH^- ions in the activating solution can also promote the hydration of the low reactive FA and be beneficial for the later ages mechanical properties. The aim of this study is to investigated the possibility associated to the activation of blended BFS/FA binders with relatively low alkaline solutions; compared to high alkaline and near-neutral AAMs' compositions these kind of systems might offer a favorable compromise between mechanical properties, low environmental impact, cost and ease of use. The reaction kinetics, products assemblage and the mechanical properties development of different BFS/FA ratios and NaOH concentration are investigated to understand whether or not the hydroxide-sulfate co-activation is able to sufficiently promote the hydration of the binders and lead to materials suitable for practical applications.

2 Materials and Method

The raw materials used in this study are BFS, with a D50 of 9μm, and FA F10, purchased respectively from Ecocem and Bauminerals. Their chemical compositions, retrieved from XRF, are reported in Table 1. While the XRD patterns are reported in Fig. 1. The activating solutions were made by dissolving the appropriate amounts of sodium hydroxide (≥97%) and sodium sulfate anhydrous (≥98%) in distilled water.

BFS and FA were dry mixed in two different ratios: 80% BFS-20% FA and 60% BFS-40% FA. Each blend was activated with three different solutions where the Na_2SO_4

concentration was kept constant at 0.6M while NaOH was added to achieve concentrations of 0.5, 1 and 2M. All the solutions were prepared in advance and let cool down to room temperature (20 °C) before being used. The solution to binder ratio of all the investigated compositions was fixed at 0.5. All the details concerning the mix design are reported in Table 2.

Table 1. Chemical composition of BFS and FA retrieved by XRF

Oxides wt.%	BFS	FA
CaO	40.80	4.33
SiO_2	33.30	56.70
Al_2O_3	12.30	23.50
MgO	7.84	1.43
SO_3	2.30	1.16
TiO_2	1.29	1.23
K_2O	0.67	2.65
Na_2O	0.44	0.91
Fe_2O_3	0.39	5.92
MnO	0.36	–
BaO	0.11	0.21
P_2O_5	–	1.49
SrO	–	0.15

Table 2. Details of the mix design

	BFS %	FA %	Solution/binder	[Na_2SO_4]	[NaOH]	Na_2O-E %	Water/binder
A0.5	80	20	0.5	0.6	0.5	2.42	0.45
A1	80	20	0.5	0.6	1	3.22	0.44
A2	80	20	0.5	0.6	2	4.33	0.43
B0.5	60	40	0.5	0.6	0.5	2.42	0.45
B1	60	40	0.5	0.6	1	3.22	0.44
B2	60	40	0.5	0.6	2	4.33	0.43

The hydration kinetics was investigated by means of isothermal calorimetry, up to 72 h, using a TAM Air isothermal calorimeter kept at 20 °C. The signals are normalized by the mass of binder in the measured sample.

The compressive strength was determined on mortar specimens prepared following the standard EN 196-1 with a sand/binder ratio of 3. The mortar was casted in steel molds, wrapped in plastic foils and cured in a room kept at 20 °C and 60% RH. At the

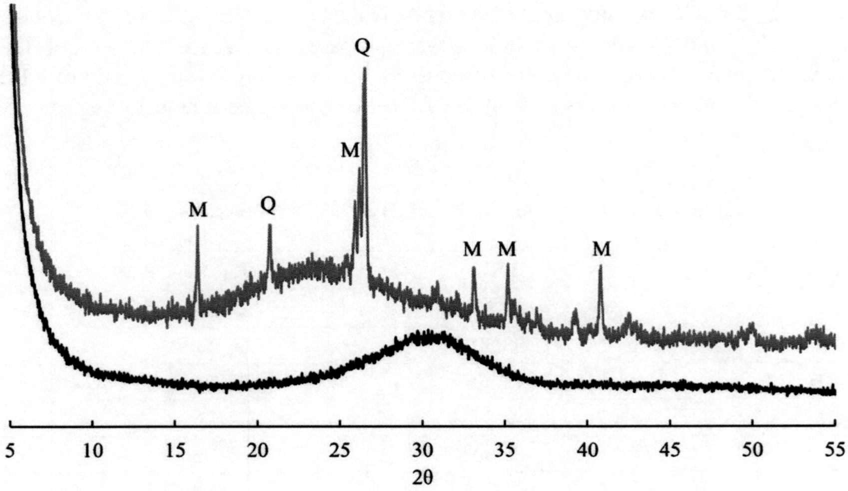

Fig. 1. XRD of the raw materials (Q = Quartz, M = Mullite)

first day of testing the specimens were removed from the molds and cured in a room at 20 °C and ≥85% RH. The compressive strength was tested at 3, 7 and 28 days on pieces of around 40 × 40 mm obtained by splitting the prism into two halves. The composition B0.5 was not tested at 3 days due to its prolonged setting.

The reaction products assemblage was investigated by XRD. A thin slice was cut from paste specimens cured at 20 °C in sealed condition and immediately placed on the sample holder of a Bruker D2 Phaser (Cu Kα radiation). The XRD patterns were recorded from 5 to 55° 2θ with a step width of 0.02 2θ and counting time of 0.3s per step.

3 Results and Discussion

3.1 Reaction Kinetics and Reaction Products Assemblage

The heat flow and total heat curves of the investigated compositions are reported in Fig. 2a, b respectively. It can be noted that the FA replacement has less influence on the reactivity than the concentration of OH^- ions in the activating solution. Nevertheless, the difference between different blends activated with the same solution increases as [NaOH] is reduced. This behavior is expected since it is known that high alkalinity is required to promote FA dissolution and thus its hydration [12]. Due to relatively low [NaOH] investigated in this study, and the short period over which the heat release is evaluated, it is plausible that, in this initial stage, FA mostly behaves as filler. Despite the different intensity there is not a much shift in the hydration peak for samples activated with 2M NaOH, estimated in a couple of hours. If we compare instead A1 and B1, we observe that the time occurring to reach the maximum heat flow in B1 is twice that of

A1. The same comparison cannot be done for A0.5 and B0.5 since the latter does not reach its maximum in the investigated time frame.

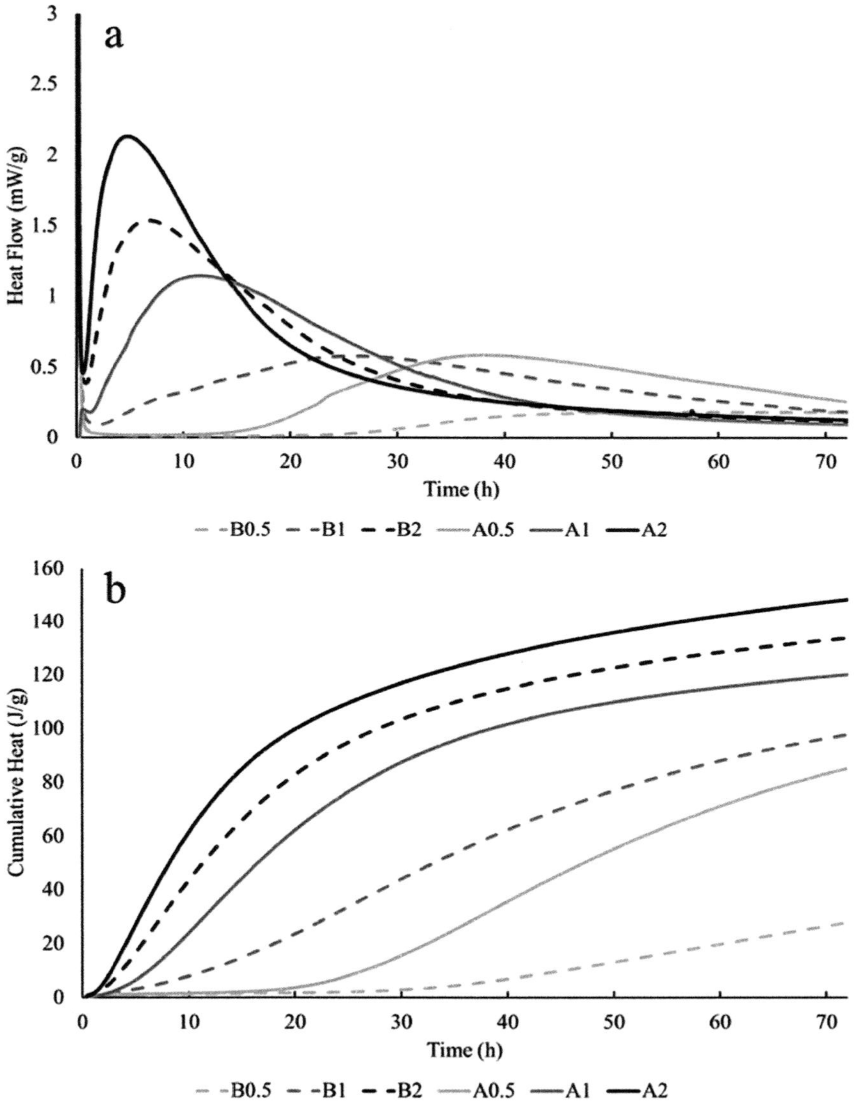

Fig. 2. Heat flow (a) and cumulative heat release (b) evaluated up to 72 h

Among the studied compositions only A0.5 and B0.5 exhibit a proper induction period while the peak's shift of the others is mainly due to the broadening of the peak itself. The cumulative heat release in Fig. 2b confirms the observation made on the heat flow. After 3 days A2 has the highest total heat, 148 J/g, followed by B2 that reaches 133 J/g. The gap evaluated between compositions with different FA replacement and same activating solution increases to 23 and 57 by reducing [NaOH] to 1 and 0.5M respectively.

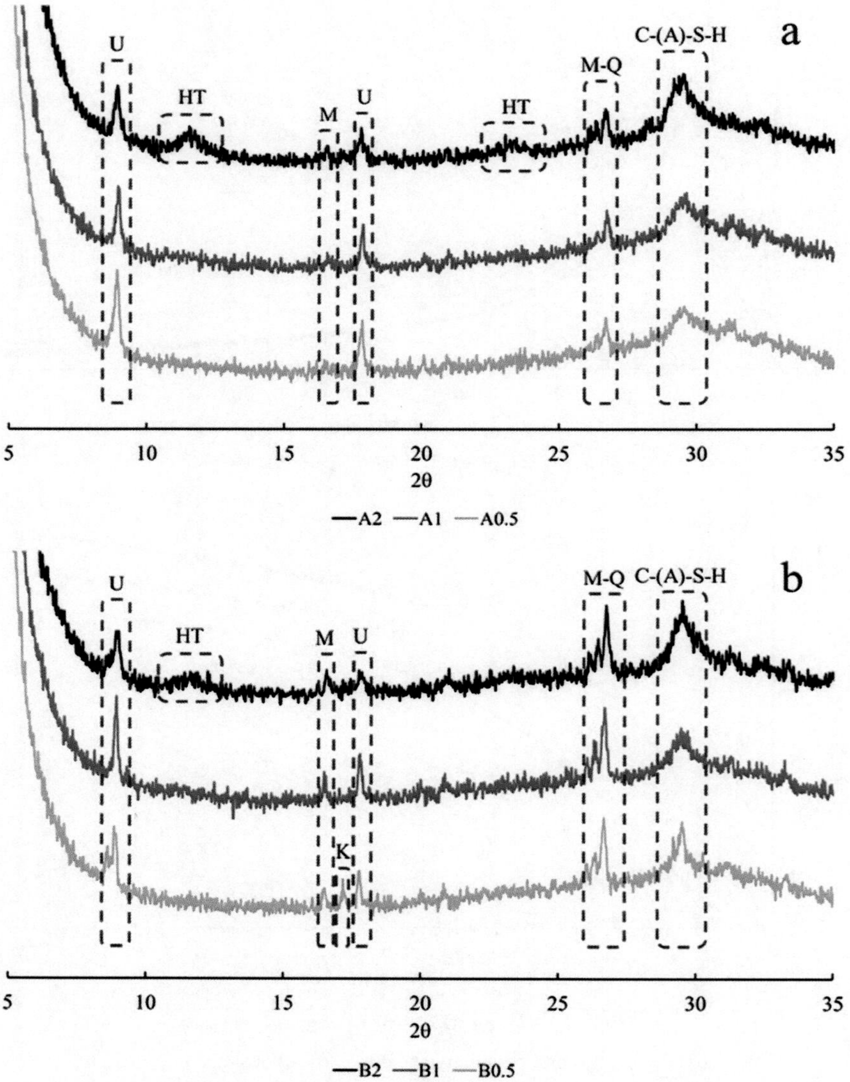

Fig. 3. XRD of A (a) and B (b) series recorded at 7 days without stopping the hydration

The diffraction patterns are reported in Fig. 3a, b for A and B series respectively. The same phases are detected in all the patterns with the exception of hydrotalcite (HT), formed only when 2M NaOH was used in the activating solution, and katoite (K), whose peak is present only in B0.5. The C-(A)-S-H broad peak and two very sharp peaks associated to the U-phase (U), a sodium substituted AFm phase with composition $4CaO.0.9Al_2O_3.1.1SO_3.0.5Na_2O.16H_2O$ [13], are clearly visible in all the patterns. The identified C-(A)-S-H peak, growing on top of BFS' amorphous bump, indicates a partial crystallinity of the product. As reported by many authors, the C-(A)-S-H originated from BFS hydration is characterized by higher crystallinity compared to that from OPC hydration [14–16].The detected mullite (M) and quartz (Q) are due to the presence of FA and the intensity of their peak is proportional to the percentage of FA replacement. Despite the significant amounts of sulfate added through the activating solution, ettringite was not detected and the only sulfate-phase in the investigated compositions is the U-phase. The latter is reported to be stable only at high pH; conditions in which ettringite is unable to precipitate [17, 18].

3.2 Compressive Strength

The results obtained from compressive strength tests are reported in Fig. 4. Each composition was tested at 3, 7, and 28 days except for B0.5 that was still not set at 3 days. At low FA replacement (A series), a substantial difference is present only in the strength developed at 3 days when A0.5 reaches only 3 MPa while A2 has around 19 MPa of compressive strength.

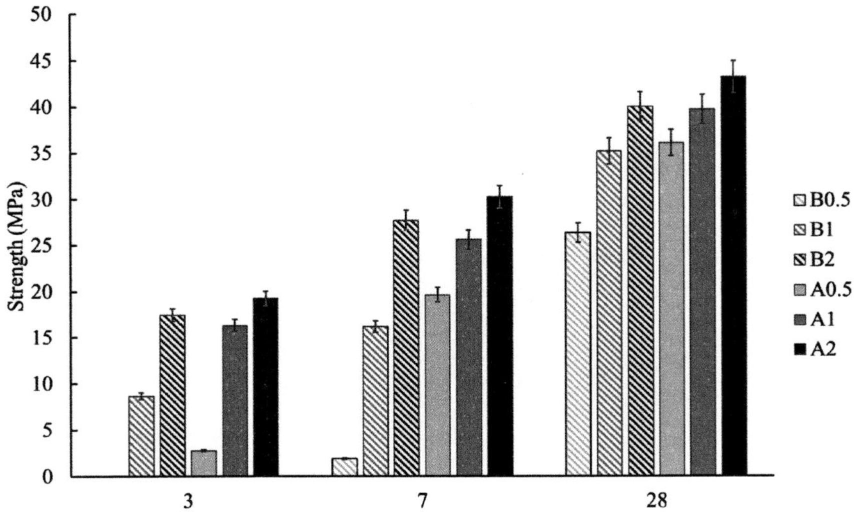

Fig. 4. Compressive strength developed at 3, 7, and 28 days

This difference is substantially reduced at later ages when a contribution from FA is expected. At 28 days A0.5 reaches a notable 36 MPa while A2 reaches 43 MPa. In contrast to the A series, in the B series a significant difference persists between the three compositions also at 28 days, even though this is drastically reduced compared to 3 and 7 days. The 28-day strengths achieved in this study is comparable to those obtained from Zhang et al. [10] on ultra-fine BFS/FA blends activated with sodium sulfate. While they observed a reduction of the mechanical properties due to the incorporation of FA, compared to 100% BFS, the mechanical properties of these blended systems are higher than in pure ones when sulfate and hydroxide solutions are used together. The 28-day compressive strength obtained with 100% BFS and the same activating solution used for B1 and A1(Na_2SO_4 0.6M and NaOH 1M) was significantly lower compared to what is achieved with blended binders within this study [9]. This further corroborates the idea that, in absence of an initial alkalinity boost, the pH reached by the pore solution is not sufficient to promote the reaction of FA even at later ages. In contrast, if a certain $[OH^-]$ is provided, even as low as in A/B0.5 and thus not able to efficiently promote FA reaction within the first days, as shown by calorimetric results, it seems that the alkalinity achieved by the pore solution leads to a FA contribution after prolonged curing.

4 Conclusions

In this study the activation of blended BFS/FA with sodium sulfate and hydroxide was studied in terms of reaction kinetics, products assemblage and compressive strength. Both the isothermal calorimetry and the compressive strength development show that the FA replacement influence on the reaction kinetics and mechanical properties increase as the alkalinity of the activating solution decreases. The reported results show that it is possible to efficiently activate BFS/FA blended binders with a combination of sodium sulfate and hydroxide even when solutions with a low sodium content (i.e. 2.42%) are used. The addition of FA induces a significant increase in the induction period and subsequent delay in the mechanical properties development. However, a minimum of around 25 MPa can still be achieved at 28 days at high FA percentage (40%) and with a limited addition of NaOH (0.5M). It seems that a certain OH^- concentration should be initially provided through the activating solution, even without exceeding a pH of 14, to observe contributions due to the hydration of FA thus it might not be possible to properly activate blended AAMs by using only near-neutral activators such as sodium sulfate.

Acknowledgements. This work is supported by the EOS-programme (The Excellence of Science) through research project 30439691.

References

1. Habert, G., D'Espinose De Lacaillerie, J.B., Roussel, N.: An environmental evaluation of geopolymer based concrete production: reviewing current research trends. J. Clean. Prod. **19**, 1229–1238 (2011). https://doi.org/10.1016/j.jclepro.2011.03.012

2. Turner, L.K., Collins, F.G.: Carbon dioxide equivalent (CO_2-e) emissions: a comparison between geopolymer and OPC cement concrete. Constr. Build. Mater. **43**, 125–130 (2013). https://doi.org/10.1016/j.conbuildmat.2013.01.023
3. Ouellet-Plamondon, C., Habert, G.: Life cycle assessment (LCA) of alkali-activated cements and concretes. Woodhead Publishing Limited (2015). https://doi.org/10.1533/9781782422884.5.663
4. Bernal, S.A.: Advances in nearneutral salts activation of blast furnace slags. RILEM Tech. Lett. **1**, 3944 (2016)
5. Mobasher, N., Bernal, S.A., Provis, J.L.: Structural evolution of an alkali sulfate activated slag cement. J. Nucl. Mater. **468**, 97–104 (2016). https://doi.org/10.1016/j.jnucmat.2015.11.016
6. Provis, J.L., van Deventer, J.S.J.: Alkali Activated Materials: State-of-the-Art Report, RILEM TC 2. RILEM/Springer (2014)
7. Tan, H., et al.: Compressive strength and hydration process of wet-grinded granulated blast-furnace slag activated by sodium sulfate and sodium carbonate. Cem. Concr. Compos. **97**, 387–398 (2019). https://doi.org/10.1016/j.cemconcomp.2019.01.012
8. Rashad, A.M., Bai, Y., Basheer, P.A.M., Milestone, N.B., Collier, N.C.: Hydration and properties of sodium sulfate activated slag. Cem. Concr. Compos. **37**, 20–29 (2013). https://doi.org/10.1016/j.cemconcomp.2012.12.010
9. Mutti, M., Joseph, S., Cizer, Ö.: Sodium sulfate and sodium hydroxide co-activation of blast furnace slag. In: 74th RILEM Annual WEEK 40TH Cement and Concrete Science Conference (2020)
10. Zhang, J., Tan, H., Bao, M., Liu, X., Luo, Z., Wang, P.: Low carbon cementitious materials: sodium sulfate activated ultra-fine slag/fly ash blends at ambient temperature. J. Clean. Prod. **280** (2021). https://doi.org/10.1016/j.jclepro.2020.124363
11. Deschner, F., et al.: Hydration of Portland cement with high replacement by siliceous fly ash. Cem. Concr. Res. **42**, 1389–1400 (2012). https://doi.org/10.1016/j.cemconres.2012.06.009
12. Khale, D., Chaudhary, R.: Mechanism of geopolymerization and factors influencing its development: a review. J. Mater. Sci. **42**(3), 729–746 (2007). https://doi.org/10.1007/s10853-006-0401-4
13. Li, G., Le Bescop, P., Moranville-Regourd, M.: Synthesis of the U phase (4CaO . 0.9Al2O3 . 1.1SO3 . 0.5Na2O . 16H2O), Cem. Concr. Res. **27**:7–13 (1997)
14. Gruskovnjak, A., Lothenbach, B., Holzer, L., Figi, R., Winnefeld, F.: Hydration of alkali-activated slag: comparison with ordinary Portland cement. EMPA Act **34** (2006)
15. Wang, S.D., Scrivener, K.L.: 29Si and 27Al NMR study of alkali-activated slag. Cem. Concr. Res. **33**, 769–774 (2003). https://doi.org/10.1016/S0008-8846(02)01044-X
16. Schneider, J., Cincotto, M.A., Panepucci, H.: 29Si and 27Al high-resolution NMR characterization of calcium silicate hydrate phases in activated blast-furnace slag pastes. Cem. Concr. Res. **31**, 993–1001 (2001). https://doi.org/10.1016/S0008-8846(01)00530-0
17. Tambara, L.U.D., Cheriaf, M., Rocha, J.C., Palomo, A., Fernández-Jiménez, A.: Effect of alkalis content on calcium sulfoaluminate (CSA) cement hydration. Cem. Concr. Res. **128**, 105953 (2020). https://doi.org/10.1016/j.cemconres.2019.105953
18. Dosch, W., Zur Strassen, H.: An alkali-containing calcium aluminate sulfate hydrate. Zement-Kalk-Gips **20**, 392–401 (n.d.)

Going Below 50% Clinker Factor in Limestone Calcined Clay Cements (LC^3): A Comparison with Pozzolanic Cements from the South American Market

Franco Zunino(✉) and Karen L. Scrivener

Laboratory of construction materials, EPFL STI IMX LMC, Station 12, CH-1015 Lausanne, Switzerland

franco.zunino@epfl.ch

Abstract. Limestone calcined clays are a promising technology as they offer similar performance to OPC from 7 days onwards, while enabling a reduction of the clinker content of 50%. In some regions of the world like South America, pozzolanic cements (i.e., blended cements that combine clinker with natural pozzolans) have been used for decades as an effective solution to reduce the clinker content. Their clinker factors range from 80 down to 60% on average. However, their mechanical properties are in general lower as compared to OPC/CEM I. In this study, we show that by using LC^3-type formulations, cements with the same performance as commercial pozzolanic cements can be produced with clinker contents significantly below 50%. This is explained by the high reactivity of calcined clays and the synergetic reaction of metakaolin and limestone that allows offsetting the clinker reduction.

Keywords: Metakaolin · Reactivity · Clinker factor · Pozzolans · Sustainability

1 Introduction

Reducing the embodied carbon associated with cement production is driving the industry to find solutions that are, simultaneously, technically and economically feasible. The most effective strategy to tackle this challenge in a global scale is to reduce the clinker factor [1]. This has promoted the adoption of blended cements, which incorporate supplementary cementitious materials (SCMs) replacing part of the clinker [2, 3]. The International Energy Association (IEA) projects in their sustainability roadmap that by 2050 the average clinker factor worldwide will be around 72% [4]. This projection above the reduction in clinker required to substantially mitigate the carbon footprint associated with cement. In the last decade, a slowdown in clinker reduction trends has been observed, mainly due to the unavailability of traditionally used SCMs such as fly ash and slags [1].

Limestone calcined clay cements (LC^3) are a family of blended cements that incorporate limestone and calcined clays replacing part of the clinker Limestone and calcined

J. I. Escalante-Garcia et al. (Eds.): RW 2021, RILEM Bookseries 40, pp. 60–64, 2023.
https://doi.org/10.1007/978-3-031-21735-7_8

clay are the only SCMs available in the quantities required for a reduction of the clinker factor on a global-scale [1]. They allow reducing the clinker factor down to 50%, while retaining the same performance as OPC from 7 days onwards [5, 6]. The clinker reduction allows to save about 400 Mt of CO_2 per year [7, 8].

In regions where volcanic ash deposits are available, like in Chile and the rest of the pacific coast of South America, pozzolanic cements have been used since the 1960s. Pozzolanic cements are mixtures of clinker, natural pozzolans (volcanic ashes) and gypsum. The clinker factors of pozzolanic cements usually range from 80 to 60% despite standards allowing lower clinker factors [9]. The reduction in clinker contend is limited to meet the strength requirements of the standards. Natural pozzolans are in general less reactive than calcined clays [10]. Consequently, the strength achieved by pozzolanic cements is in most cases lower than OPC, particularly between 1 and 28 days.

This project explores the feasibility of producing LC^3-type cements that meet the same strength requirements as commercial pozzolanic cements available in the Chilean market (i.e., that meet the current standards in the countries where they are used) with clinker contents lower that 50%.

2 Materials and Methods

2.1 Raw Materials

A commercial ordinary Portland cement (OPC) conforming to EN 197–1 as CEM I 42.5N was used in this study as a reference to benchmark the commercial pozzolanic cements against a classic European OPC. Two Chilean commercial pozzolanic cements were used. First, a high-strength Portland pozzolanic cement conforming to Chilean standard NCh 148 with a clinker factor of about 80%, determined by X-ray diffraction. Second, a general use pozzolanic cement conforming to NCh 148 with a clinker factor of approximately 60%. In both pozzolanic cements, the only SCM incorporated is natural pozzolans. In general, high-strength Portland pozzolanic cements are ground finer at the cement factory to increase their early-age strength and fulfill the requirements of the standard. To account for this factor, a blend of 50% CEM I 42.5N and 50% of the same natural pozzolans as in the commercial pozzolanic cements was prepared, to provide a benchmark comparison of natural pozzolans against a conventional LC3–50 (50% clinker) formulation.

The LC^3 systems were prepared using the OPC and a commercial high purity calcined clay with 95% metakaolin content. Limestone from Omya (Betoflow D) was used. The clinker factors selected for this study were 35% and 20%, and a clay-to-limestone ratio of 2-to-1 was selected. Consequently, in LC^3–35 systems, calcined clay accounts for 40% of the mass and limestone 20%. In LC^3–20 systems, calcined clay corresponds to 53% while limestone to 27%, considering in all systems an addition of 5% gypsum. In the calcined clay fraction, the high-purity calcined clay was mixed with limestone at calcined-to-limestone ratio of 0.8 to achieve total metakaolin contents of 63% (namely MK63). In all cases, the sulfate content was adjusted by means of isothermal calorimetry [11].

The w/b ratio was fixed at 0.4. Paste samples for isothermal calorimetry were prepared using a high-shear vertical mixer at 1600 rpm for 2 min. Mortar samples were prepared with each cement confirming to EN 196 standard. Samples were seal-cured until the time of testing in an environmentally controlled chamber.

2.2 Experimental Methods

Strength measurements were conducted at 1, 2, 3, 7, 28 and 90 days of hydration, using a 300 kN load cell with a 100 N resolution and a loading rate of 2.4 kN/s.

3 Results and Discussion

3.1 Strength Development

Figure 1 shows the measured values of compressive strength of mortar specimens of LC^3–35 and LC^3–20 with 63% metakaolin content in thew calcined clay fraction, along the two commercial pozzolanic cements and the 50/50 blend of CEM I with natural pozzolans.

As observed, the strength at early-age of all systems is lower compared to the OPC reference mortar. From 7 days onwards, the LC^3–35 exhibits equivalent strength to the OPC system. This can be explained by the contribution of the reaction of metakaolin and limestone that leads to the formation of carboaluminates, effectively offsetting the effect of the reduction in OPC content on porosity and, consequently, strength [12].

On the other hand, LC^3–20 is unable to achieve the same degree of mechanical performance. At later ages, the relative difference between LC^3–35 and LC^3–20 suggests that another factor, in addition to clinker content reduction, plays a role in the observed results. Among the possible explanations, an early portlandite depletion due to the low amount of PC in the system appears as a feasible hypothesis.

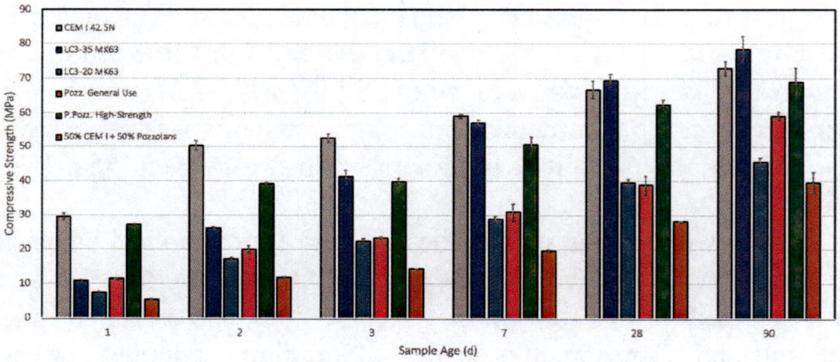

Fig. 1. Compressive strength of mortar specimens of low-clinker (35 and 20%) LC^3, commercial pozzolanic cements and a 50/50 CEM I + natural pozzolans blend.

By comparing the low-clinker LC^3 systems with the commercial pozzolanic cements, it can be noted that LC^3–35 has a similar strength development than the Portland pozzolanic high-strength cement, while LC^3–20 has a similar strength development as the pozzolanic general use cement. This indicates that low-clinker LC^3 formulations are feasible solutions to achieve sustainable cements that comply with the current requirements of the standard, while allowing additional reductions of 45 (80 to 35) and 40 (60 to 20) percent of clinker respectively. Further experiments are required to validate these findings under different experimental conditions (w/b, admixture compatibility, etc.).

The 50/50 blend system of CEM I and pozzolans exhibits a lower strength than the general use commercial cement at all ages. The performance is also significantly lower than LC^3–50 (i.e., an LC^3 formulation with the same clinker factor) cements with different grades of clay that use the same OPC as in this study [12]. This further highlights the importance of using reactive SCMs and proper blend designs (particularly in terms of sulfate balance) to enable substantial clinker reductions and CO_2 savings, while ensuring good performance of the resulting cements.

Table 1 summarizes the strength requirements for pozzolanic cements contained in the Chilean standard NCh 148. As expected, both commercial pozzolanic cements meet the compressive strength requirements of their corresponding strength classes, but also do the LC^3–35 and LC^3–20 studied for high-strength and general use applications, respectively. Consequently, we show here that LC^3 can satisfy the requirements of this particular market, saving an additional 40–45% of clinker in the process.

Table 1. Strength classes defined for pozzolanic cements in Chilean standard NCh 148 [9]

	Min. Compressive strength (MPa)		Min. Flexural strength (MPa)	
Strength class	*7 days*	*28 days*	*7 days*	*28 days*
General use	18	25	3.5	4.5
High strength	25	35	4.5	5.5

4 Conclusions

This study highlighted the potential to use limestone and calcined clays to develop low-clinker factor blended cements that meet the requirements of current standards, while saving an additional 40% of clinker in the process.

Based on the results of this study, the following conclusions are drawn:

1. LC^3-type formulations enable to produce a blended cement with equivalent performance to a commercial pozzolanic high-strength and general use cements, but with an additional saving of around 40% of the clinker.
2. The additional reduction of clinker content does not lead to an additional sulfate demand in low-clinker formulations as compared to LC^3–50, which is explained by the similarities of the alite hydration rates among these systems.
3. The reaction between metakaolin and limestone is enhanced in low-clinker LC^3 formulations, leading to more carboaluminates precipitated which effectively offset the effect of the reduction of clinker content in strength.

4. The low reactivity of natural pozzolans compared to calcined clays explain the results presented in this study in terms of strength development.

Acknowledgements. The authors would like to acknowledge financial support by the Swiss Agency of Development and Cooperation (SDC) grant 81026665.

References

1. Scrivener, K.L., John, V., Gartner, E.M.: Eco-efficient cements: potential, economically viable solutions for a low-CO2, cement-based materials industry. United Nations Environmental Programme (UNEP) (2016)
2. Lothenbach, B., Scrivener, K., Hooton, R.D.: Supplementary cementitious materials. Cem. Concr. Res. **41**, 1244–1256 (2011). https://doi.org/10.1016/j.cemconres.2010.12.001
3. Schneider, M., Romer, M., Tschudin, M., Bolio, H.: Sustainable cement production—present and future. Cem. Concr. Res. **41**, 642–650 (2011). https://doi.org/10.1016/j.cemconres.2011.03.019
4. IEA, Cement Technology Roadmap: Carbon Emissions Reductions up to 2050, OECD Publishing, Paris, n.d. https://doi.org/10.1787/9789264088061-en
5. Scrivener, K., et al.: Impacting factors and properties of limestone calcined clay cements (LC ^{3}). Green Mater. **7**, 3–14 (2019). https://doi.org/10.1680/jgrma.18.00029
6. Zunino, F.: Limestone calcined clay cements (LC3): raw material processing, sulfate balance and hydration kinetics, EPFL Thesis 8173 (2020)
7. Pillai, R.G., et al.: Service life and life cycle assessment of reinforced concrete systems with limestone calcined clay cement (LC3). Cem. Concr. Res. **118**, 111–119 (2019). https://doi.org/10.1016/j.cemconres.2018.11.019
8. Zunino, F., Martirena, F., Scrivener, K.: Limestone calcined clay cements (LC3), ACI Mater. J. 118 (2021) 49–60. https://doi.org/10.14359/51730422
9. I.N.N., NCh148.Of68 : Cemento—Terminología , clasificaci6n y especificaciones generales (1968)
10. Li, X., et al.: Reactivity tests for supplementary cementitious materials: RILEM TC 267-TRM phase 1. Mater. Struct. **51**(6), 1–14 (2018). https://doi.org/10.1617/s11527-018-1269-x
11. Zunino, F., Scrivener, K.: The influence of the filler effect on the sulfate requirement of blended cements. Cem. Concr. Res. **126**, 105918 (2019). https://doi.org/10.1016/j.cemconres.2019.105918
12. Zunino, F., Scrivener, K.: The reaction between metakaolin and limestone and its effect in porosity refinement and mechanical properties, Cem. Concr. Res. 140 (2021). https://doi.org/10.1016/j.cemconres.2020.106307

Microstructure and Durability Properties of Concretes Based on Oyster Shell Co-products

Alexandra Bourdot[1](✉), Camille Martin-Cavaillé[1], Marc Vacher[1], Tulio Honorio[1], Nassim Sebaibi[2], and Rachid Bennacer[1]

[1] LMT - Laboratoire de Mécanique Et Technologie, Université Paris-Saclay, ENS Paris-Saclay, CNRS, 91190 Gif-Sur-Yvette, France
alexandra.bourdot@ens-paris-saclay.fr

[2] Laboratoire de Recherche ESITC Caen, COMUE NU, 1 Rue Pierre Et Marie Curie, 14610 Epron, France

Abstract. One of the solutions to reduce the environmental impact of the construction sector is the use of alternative materials, more sustainable and with tailored or new properties. This paper proposes the valorization of oyster shell co-products within cementitious materials. These co-products from oyster farm could be used to produce composites with interesting properties. Shell co-products are used in this study to produce load-bearing concretes. These concretes show good durability properties with a lower chloride diffusion coefficient and a higher resistivity compared to references concretes. The optimum mix is composed of CEM III and 50%vol of oyster shell as aggregate. A good cohesion between the particles of oyster shell and the binding matrix is observed, which explains the feasibility to get concretes of mechanical resistance such as the references.

Keywords: Oyster shell · Cementitious materials · SEM · Microstructure · Durability · Porosity

1 Introduction

Oyster shells are part of aquaculture by-products that generate large amounts of waste. Each year more than 17.3 million tons of shelled molluscs are produced, representing 56.2% of the production of marine and coastal aquaculture in 2018 [1]. The shells of bivalves, such as mussels and oysters, can be turned into calcium carbonate or calcium oxide, two highly versatile chemical compounds with wide industrial applications, but are poorly valued and are a global concern for coastal countries. Oyster shells represent about 33% of the major molluscan species groups. In France, the production of bivalves represents 144 800 tons, the country is thus part of the top ten countries in the World for aquaculture production.

The material characteristics suggest that similar to limestone, seashell waste could be a reactive material. Seashell waste would be expected to present some reactivity in cement systems, contributing to the formation of monocarboaluminate [2]. It is shown in past research that while seashell waste has been used as a replacement for both cement

J. I. Escalante-Garcia et al. (Eds.): RW 2021, RILEM Bookseries 40, pp. 65–73, 2023.
https://doi.org/10.1007/978-3-031-21735-7_9

and aggregate, there is still a lack of investigations concerning its durability [3], with a lack of consensus on the influence of these materials [4].

This paper studies the durability of concretes using oyster shell co-products as aggregates in substitution of natural aggregates up to 50% in volume. Two types of binders are used: a CEM I and a CEM III. The concrete formulations studied have a similar mechanical strength at 28 days and a comparable workability. These materials are characterized in terms of durability by chloride diffusion tests, electrical resistivity and open porosity to water. Microscopic observations allow characterizing the shells and concretes in terms of microstructure.

2 Materials

2.1 Raw Materials

Oyster shell co-products have been substituted to natural gravel in concrete. The natural and shell aggregates have a grain size of 4/10 mm. The oyster shell gravel has a bulk density of 809 kg/m^3 and an absolute density of 2512 kg/m^3 close to that of the natural gravel. Oyster shells have an absorption coefficient of 6.2% higher than that of natural aggregate. The sand used for the concrete formulation is a 0/2 mm with few fines. All the data are available in Table 1.

Table 1. Physical characteristics of the raw materials.

	Properties				
Aggregate	size (mm)	Bulk density (kg/m^3)	Absolute density (kg/m^3)	Water absorption at 24h (%)	Sand equivalent (%)
Sand	0/2	1464	2625	1.79	92
Natural aggregate	4/10	1531	2677	2.85	–
Oyster shell aggregate	4/10	809	2512	6.2	–

2.2 Concrete Formulations

Two types of concrete are studied to determine the impact of oyster shells on durability and microstructure: one with CEM I 52,5 N as reference and the other with CEM III/C 32.5 N, as low carbon footprint cement. These concretes with oyster shells substituting 50% of natural aggregate by volume, with a CEM I and CEM III are noted BI_50 and BIII_50, respectively. These formulations will be compared to references noted BI_0 and BIII_0 (0%vol of oyster shell), with CEM I and CEM III, respectively. The proportions of the four formulations are presented in Table 2. The samples were cured under sealed conditions at 22 ± 1 °C, 50 ± 5%RH. The concrete formulations studied have similar mechanical strength at 28 days and comparable workability. The addition of oyster shell co-products leads to an increase in air content.

Table 2. Concrete mixes (in kg/m^3).

Mix notation	BI_0	BI_50	BIII_0	BIII_50
Sand	718.9	701.5	687.4	661.5
Aggregate	1012.8	494.1	949.2	465.9
Oyster shell aggregate	–	463.7	–	437.2
Binder	350.1	366.0	454.3	475.0
Water	250.1	271.6	247.7	270.2
$W_{efficent}/B$	0.60	0.60	0.46	0.46
Slump (cm)	9.0	10.1	10.5	9.5
Air content (%)	2.9	4.1	2.8	3.8
Compressive strength at 28 days (MPa)	33.8	31.8	28.5	31.0

3 Methods

3.1 Durability Study

Porosity measurement. The porosity accessible by water of the specimens are measured according to NF P 18 459 standard [5]. The samples are cut at 50 mm thickness from cylinders of ø11×22 cm, and then are placed in a desiccator for vacuum treatment for 4 h. Next, the container is filled by demineralized water under vacuum for 44 h. The test is repeated three times.

Chloride diffusion. Chloride diffusion measurement protocol is based on the French standard NF P 18–462 [6]. Cylindrical samples are cut into 3 cm thick slices. All samples are then coated with epoxy resin to ensure an easy sawing. Samples are soaked with NaOH in a vacuum desiccator during 48 h, then are placed in testing cells. The cells are connected to the testing apparatus, set to 20 V and 6 h of testing. At the end of the test, samples are broken in two parts sprayed with $AgNO_3$. The penetration front can be measured and the apparent diffusion coefficient can be calculated using Eq. (1) and (2).

$$D_{app} = \frac{RTL}{ZFE} \cdot \frac{X_d - \alpha X_d^{0.5}}{t} \quad (1)$$

where: L is the thickness of the sample; R is the gas constant; T is the temperature (294 K); Z is the valence of the chloride ion; $F = 9{:}64 \times 10^4$ J/(mol.V) is the Faraday constant; E is the mean potential difference between the faces of the sample (20V); X_d

is the mean measure of the penetration front [m]; *t* is the test duration (t = 6h); and

$$\alpha = 2.\left(\frac{RTL}{ZFE}\right)^{0.5}.erfc^{-1}\left(1 - \frac{2C_d}{C_0}\right) \quad (2)$$

where: C_d, the chloride concentration for which the colorimetric test changes color; C_0 is the initial concentration in the solution, and $erfc^{-1}$ is the inverse of the complementary error function.

Electrical resistivity. The protocol for measuring the electrical resistivity is based on the European project ChlorTest [7] for the preparation of the specimen. Three samples are tested for each mix. The samples follow the same vacuum treatment as for the porosity measurement. The measure consists in placing the saturated sample of ø5×22 cm between two electrodes of copper. The contact between the sample and the electrodes is provided by two sponges wetted with tap water.

The electrical resistivity is obtained from the resistances of the device Re and of the sponges with the electrodes *Rs*, the section of the concrete sample and its thickness, using Eq. (3).

$$\rho = \frac{S}{L}.(Re - Rs) = \frac{\pi\left(\frac{D}{2}\right)^2}{L}.\left(\frac{U}{Ie} - \frac{U}{Is}\right) \quad (3)$$

where: ρ is the resistivity (Ω.m), *S* is the section, *D* is the diameter of the sample (m), *L* the thickness of the sample (m), *U* the imposed tension (V), *Ie* the current with the sample and the sponges, *Is* the current with the sponges.

3.2 Microstructure Characterization

The oyster shell aggregates are observed by optical and electronic microscope.

The optical microscope Keyence VH-Z100R was used of oyster shell characterization.

The SEM apparatus used in this study is a Hitachi S-3400N. The concrete samples for the microstructural investigation are prepared with polishing. The observations are carried out after 28 days.

4 Results and Discussion

4.1 Durability Properties

The durability results, in terms of porosity accessible to water, resistivity and chloride diffusion coefficient, of the 4 concrete formulations are presented in Fig. 1.

The CEM III concretes exhibits a lower porosity accessible to water when compared to CEM I concretes which would be explained by a denser microstructure. This reduction is 41% in the presence of oyster shells and 42% for the reference concretes. The use of shell co-products induces a slight increase in the porosity accessible to water. This increase is of the order of 0.44% of porosity, i.e. a relative increase of 5.8% with CEM I and 0.36% of porosity with CEM III, i.e. a relative increase of 8.1%. This could be explained by the porosity of the shell as suggested by the water absorption rate at 24 h or by additional porosity introduced in the mixing process due to the oblate shape of shell aggregates [8]. Indeed, the oblate shape could influence the properties of concrete according to De Larrard et al. [9].

Nevertheless, this porosity does not induce a negative effect on the transport and durability properties. The chloride coefficient D_{app} is reduced with the use of shell co-products and even more with the use of CEM III compared with CEM I. The reduction is 18% between the diffusion coefficients of BI_0 and BI_50, and 32% between the diffusion coefficients of BIII_0 and BIII_50. The use of CEM III with 50%vol of oyster shells, BIII_50, allows a 41% reduction of D_{app} compared to BI_50 concrete. For the studied concretes, the diffusion coefficient remains nevertheless, a relatively high with a rather weak resistance against chlorides according to the classification proposed by [10–12] except for the BIII_50 concrete which shows a moderate resistance against chloride ingress.

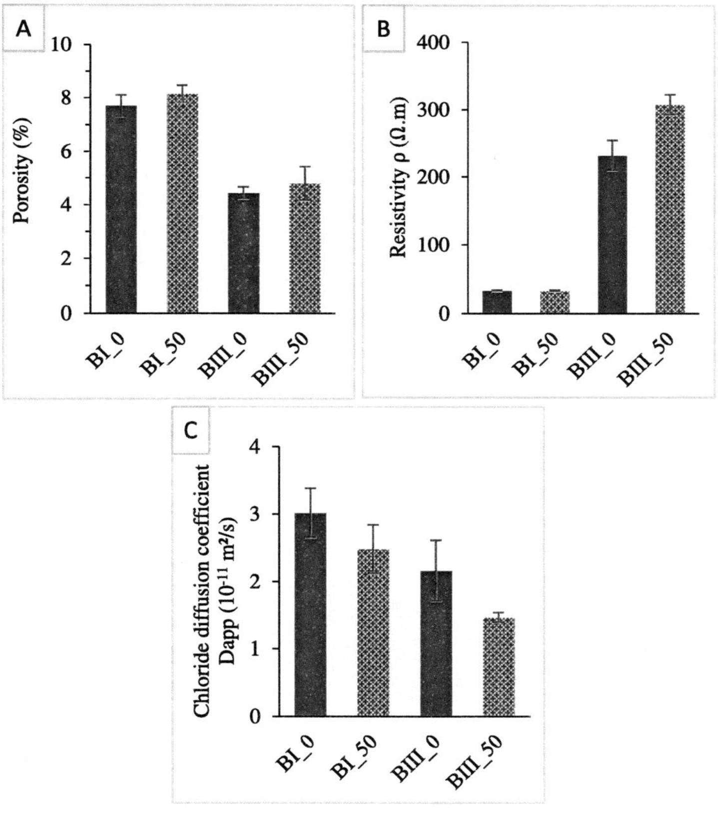

Fig. 1. Porosity (A), Resistivity (B) and Chloride diffusion coefficient (C) of concretes based on oyster shell compared to references at 28 days.

The electrical resistivity is more dependent on the cement used with an increase of 100 times for CEM III based concretes compared to CEM I based concretes, i.e. more than 800% relative increase. A difference in the electrical response is expected when different cements (or SCMs) are used since different ions are released into the pore according to the cement type and ion dynamics [13] and concentration is directly related to the electrical conductivity [14]. The use of shell co-products allows a significant improvement of 32% with CEM III. The CEM I based concretes would have a very low potential durability while the shell concrete with CEM III, BIII_50, allows a high potential durability higher than its reference BIII_0 at medium value according to [15]. Oyster shell co-product, finely graded as well as the surfaces of the aggregates could be a source of calcareous filler and a zone that favors the growth of the hydrates, respectively. Indeed, calcareous fillers have already shown their influence on the development of C-S-H [16]. This impact of oyster shell powders has been demonstrated by Ez-zaki et al. [17] with better transport properties. The addition of calcareous material is also recognized as allowing the development and stabilization of monocarboaluminate and the acceleration of slag hydration [18, 19] that could explain better durability properties with oyster shell.

4.2 Microstructure Observations

Oyster shell characterization. The oyster shells were observed by optical microscopy, the images are shown in Fig. 2. The oyster shells are presented as a layered structure (Fig. 2a), characterized by variable microstructures. Two types of microstructures were highlighted (Fig. 2b). These two microstructures were observed more precisely by SEM. One microstructure appears rather dense with very few micropores (Fig. 3a). The second microstructure is a little less present but is very porous as can be seen in Fig. 3b. Indeed, it has been highlighted by some authors and summarized by Bonnard [20], calcitic and foliated structures, consistent with the observations of this study. This last microstructure can explain an increase in the water-accessible porosity of concrete made of these materials.

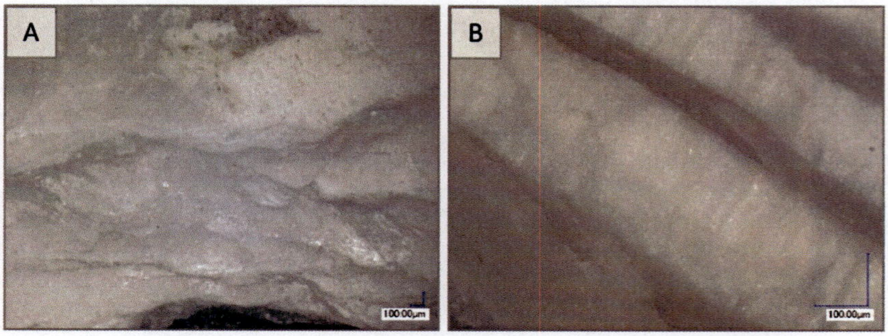

Fig. 2. Optical observations of oyster shell aggregates: foil stacking (A) and alternance of two microstructures (B).

SEM observations of concretes. The microstructure of the shell-based concretes shows a good cohesion between the cementitious pastes and the aggregates (Fig. 4). We can notice a good binding between the porous microstructure of the oyster shell and the matrix (Fig. 4d). This matrix does not integrate the microporosity of the oyster shell allowing to keep this interesting property with regard to the hygrothermal properties possible with the use of oyster shell co-products.

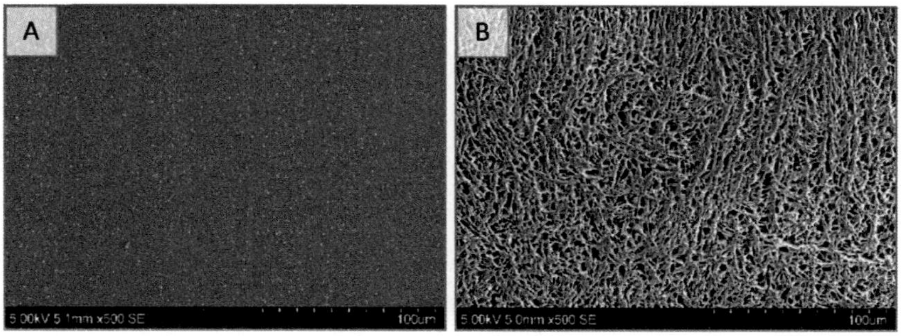

Fig. 3. SEM observations of oyster shell aggregates: dense (A) and porous (B) microstructures.

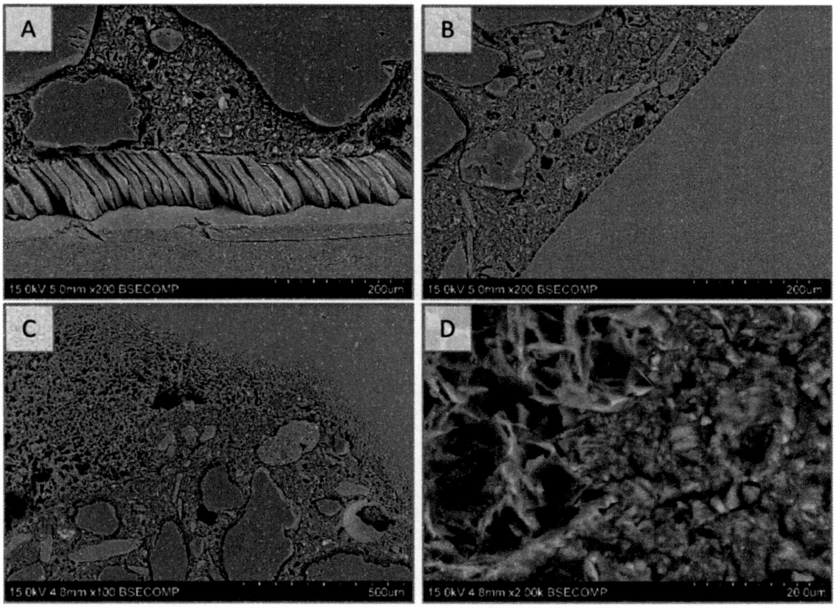

Fig. 4. SEM observations of concretes BI_50 based on CEMI and oyster shell aggregate.

5 Conclusion

The purpose of this study was the characterization of concretes valorizing oyster shell as aggregate through durability properties and microscopic observations.

The substitution of natural aggregates by oyster shell up to 50%vol induces an increase of the porosity, explained by one of the microstructures of oyster shell. Nevertheless, this porosity does not induce a negative effect on the transport and durability properties. The chloride coefficient is reduced with the use of shells co-products and even more with the use of CEM III. According to the results of the resistivity tests, the concretes based on CEM I would have a very low potential durability while the concrete based on oyster shells with CEM III, allows a high potential durability higher than its reference at medium value.

The microstructure of the concretes based on shell co-products show a good cohesion between the cementitious pastes and the aggregates. This matrix does not interpenetrate the microporosity of the oyster shell allowing to keep this interesting property with regard to the possible hygrothermal properties with the use of oyster shell co-products.

References

1. FAO: The State of World Fisheries and Aquaculture 2020. Sustainability in action, Rome (2020)
2. Lothenbach, B., Le Saout, G., Gallucci, E., Scrivener, K.: Influence of limestone on the hydration of Portland cements. Cem. Concr. Res. **38**, 848–860 (2008). https://doi.org/10.1016/j.cemconres.2008.01.002
3. Mo, K.H., Alengaram, U.J., Jumaat, M.Z., Lee, S.C., Goh, W.I., Yuen, C.W.: Recycling of seashell waste in concrete: A review. Constr. Build. Mater. **162**, 751–764 (2018). https://doi.org/10.1016/j.conbuildmat.2017.12.009
4. Eziefula, U.G., Ezeh, J.C., Eziefula, B.I.: Properties of seashell aggregate concrete: A review. Constr. Build. Mater. **192**, 287–300 (2018). https://doi.org/10.1016/j.conbuildmat.2018.10.096
5. AFNOR NF P18–459: Essai pour béton durci - Essai de porosité et de masse volumique (Tests for determining porosity and density for hard concrete) (2010)
6. AFNOR NF P18–462: Essai sur béton durci: Essai accéléré de migration des ions chlorure en régime non-stationnaire, Détermination du coefficient de diffusion apparent des ions chlorure (2012)
7. Tang, L.: Guideline for practical use of methods for testing the resistance of concrete to chloride ingress, CHLORTEST - EU funded research project "Resistance of concrete to chloride ingress - from laboratory tests to in-field performance" G6RD-CT-2002–00855 (2005)
8. Mizuta, D.D., Wikfors, G.H.: Seeking the perfect oyster shell: a brief review of current knowledge. Rev. Aquac. **11**, 586–602 (2019). https://doi.org/10.1111/raq.12247
9. De Larrard, T., Bary, B., Adam, E., Kloss, F.: Influence of aggregate shapes on drying and carbonation phenomena in 3D concrete numerical samples. Comput. Mater. Sci. **72**, 1–14 (2013). https://doi.org/10.1016/j.commatsci.2013.01.039
10. Jepsen, M.T., Mathiesen, D., Munch-Petersen, C., Bager, D.: Durability of resource saving "green" types of concrete. Proc. - fib-Symposium Concr. Environ. **2001**, 41–42 (2001)
11. Tang L.: Chloride transport in concrete – measurement and prediction (2010)
12. Jóźwiak-Niedźwiedzka, D.: Effect of fluidized bed combustion fly ash on the chloride resistance and scaling resistance of concrete. In: RILEM TC 211-PAE Final Conference on Concrete in Aggressive Aqueous Environments. Performance, Testing and Modeling, vol. 2, pp. 556–563. Toulouse, France (2009)
13. Vollpracht, A., Lothenbach, B., Snellings, R., Haufe, J.: The pore solution of blended cements: a review. Mater. Struct. **49**(8), 3341–3367 (2015). https://doi.org/10.1617/s11527-015-0724-1

14. Honorio, T., Carasek, H., Cascudo, O.: May self-diffusion of ions computed from molecular dynamics explain the electrical conductivity of pore solutions in cement-based materials? Mater. Struct. **53**(3), 1–13 (2020). https://doi.org/10.1617/s11527-020-01507-7
15. Baroghel-bouny, V., Henry, D.: Vieillissement des bétons en milieu naturel : une expérimentation pour le XXI e siècle III – Propriétés de durabilité des bétons mesurées sur éprouvettes conservées en laboratoire. 13–59 (2002)
16. Berodier, E., Scrivener, K.: Understanding the filler effect on the nucleation and growth of C-S-H. J. Am. Ceram. Soc. **97**, 3764–3773 (2014). https://doi.org/10.1111/jace.13177
17. Ez-zaki, H., Diouri, A., Kamali-Bernard, S.: Transport properties of blended cement based on dredged sediment and shells. Adv. Mater. Lett. **8**, 481–485 (2017). https://doi.org/10.5185/amlett.2017.1434
18. Adu-Amankwah, S., Zajac, M., Stabler, C., Lothenbach, B., Black, L.: Influence of limestone on the hydration of ternary slag cements. Cem. Concr. Res. **100**, 96–109 (2017). https://doi.org/10.1016/j.cemconres.2017.05.013
19. Deboucha, W., Sebaibi, N., Mendili, Y. El, Fabien, A., Alengaram, U.J., Leklou, N., Hamdadou, M.N., Bourdot, A., Gascoin, S.: Reactivity effect of calcium carbonate on the formation of carboaluminate phases in ground granulated blast furnace slag blended cements. Sustain **13** (2021). https://doi.org/10.3390/su13116504
20. Bonnard, M.: Identification of valuable compounds from the shell of the edible oyster Crassostrea gigas to cite this version : HAL Id : tel-03282617 DE L ' UNIVERSITÉ DE MONTPELLIER En Chimie Séparative , Matériaux et Procédés École doctorale Sciences Chimiques Balard (2020)

Correlation Between the Reactivity of Supplementary Cementitious Materials and Their Efficacy to Prevent Alkali-Silica Reaction

Krishna Siva Teja Chopperla(✉), Anuj Parashar, and Jason H. Ideker

Oregon State University, Corvallis, OR 97331, U.S.
chopperk@oregonstate.edu

Abstract. Supplementary cementitious materials (SCMs) have been extensively used to partly replace ordinary portland cement to reduce concrete carbon footprint and improve concrete durability. However, SCMs with varying reactivities affect the hardened properties and durability of concrete. Current prescriptive specifications to prevent alkali-silica reaction (ASR) are based only on the chemical composition of SCMs and known past performance test results. SCMs with similar chemical compositions but different reactivities will affect their efficacy to prevent ASR. In this study, SCM reactivities were correlated to their ASR prevention efficacy in mortar and concrete mixtures. SCM reactivities were determined based on the calcium hydroxide consumed by SCM at 50 °C in a high pH environment. The accelerated mortar bar test and the miniature concrete prism test were used to monitor the expansion and evaluate the efficacy of SCMs in terms of ASR prevention. A very highly reactive fine aggregate and several SCMs including three fly ashes, a slag, a silica fume, and a natural pozzolan with varying compositions were used. The results showed that the reactivity of SCM in the mixture and their mass fraction in the cementitious material blend was correlated well to the expansion of the mortar and concrete mixtures. The findings showed that the higher the SCM reactivity, the lower the expansion due to ASR. A possible new approach to determine the SCM efficacy in preventing ASR was discussed.

Keywords: Supplementary cementitious materials · Reactivity · Alkali-silica reaction · Accelerated laboratory tests · Expansion

1 Introduction

Alkali-silica reaction (ASR) is a chemical reaction between alkali hydroxides in the pore solution of concrete and reactive silica of aggregates in concrete. ASR products can absorb water and cause expansion, and eventually crack the concrete. One of the most used ASR prevention strategies is the usage of supplementary cementitious materials by replacing required amount of portland cement. As portland cement is one of the major sources of alkalis in concrete, replacing a certain amount of portland cement with SCMs can reduce the alkalis in concrete due to the dilution [1]. In addition, hydration products

J. I. Escalante-Garcia et al. (Eds.): RW 2021, RILEM Bookseries 40, pp. 74–82, 2023.
https://doi.org/10.1007/978-3-031-21735-7_10

of SCMs adsorb alkalis from the pore solution, thus contributing towards ASR control [1–3]. However, a few SCMs may contribute more alkalis to pore solution than they bind. SCMs can also improve the pore structure due to both filler and chemical effects thus reducing the ingress of moisture and external alkalis [4]. SCMs rich in alumina are known to lower the ASR expansion probably due to reduction in silica dissolution from aggregates when aluminum ions are present in the pore solution [5, 6].

The amount of SCM required to prevent ASR generally needs to be increased for SCMs with higher calcium and alkali content and lower alumina and silica content [1]. Several studies [7–13] correlated the mortar or concrete expansion due to ASR to the chemical oxide composition of cements and SCM blends to develop empirical models. The empirical models were developed to determine the efficacy of SCM to prevent ASR [7–13]. Also, ASR prevention guides such as ASTM C1778 recommend a minimum amount of SCM(s) to prevent ASR based on their chemical composition and not their reactivities. However, considering only chemical oxide composition may not explain how reactive they are because SCMs with similar chemical composition may have different amounts of reactive alumina and silica and hence different reactivities. Also, SCMs with similar chemical compositions may have different fineness, which can affect their reactivity [14]. Therefore, there is a need to study the effect of SCMs reactivity on ASR prevention and develop empirical models that can help determining the SCMs efficacy to prevent ASR. This study aims to determine and evaluate the correlation between the SCM reactivity and the expansion due to ASR.

2 Materials and Methods

2.1 Materials

In this study, an ASTM Type II/V cement was used. Several SCMs including fly ashes, slag, silica fume, and natural pozzolan were considered. The chemical compositions of the ordinary portland cement (OPC) and SCMs are listed in Table 1.

A very highly reactive fine aggregate (according to ASTM C1778) was used in the study. The 14-d ASTM C1260 expansion of the mixture with the reactive fine aggregate is 0.54%. A non-reactive high purity limestone rock was used as a coarse aggregate in this study. The SCM combinations used in the study were 1) 15/25% FA1 (15/25FA1), 2) 25/30% FA2 (25/30FA2), 3) 25% FA3 (25FA3) 4) 25% FA4 (25FA4) 5) 20% FA1 + 5% SF (20FA1-5SF), 6)) 30/50% SL (30/50SL), 7) 25% FA1 + 25% SL (25FA1-25SL), and 8) 25% NP (25NP). The percent replacement of cements with SCMs was done on a mass basis. Table 2 shows the mixtures and the expansion tests considered for this study.

2.2 Methods

SCM reactivity test

The reactivities of the SCMs were determined by following the approach proposed by Suraneni and Weiss [15]. SCM and reagent grade calcium hydroxide (CH) were mixed at 1 to 3 mass ratio, and then the powder blend was mixed with 0.5 M potassium hydroxide

Table 1. The chemical oxide composition of the OPC and SCMs used in the study (n.a. – not available)

Constituents (%)	Fly Ash 1 (FA1)	Fly Ash 2 (FA2)	Fly Ash 3 (FA3)	Fly Ash 4 (FA4)	Slag (SL)	Silica Fume (SF)	Natural Pozzolan (NP)	OPC
SiO_2	51.86	47.15	41.61	48.55	31.58	95.88	66.42	20.54
Al_2O_3 (%)	21.70	16.57	21.23	26.28	12.13	0.69	11.98	4.05
Fe_2O_3 (%)	5.04	5.88	20.91	13.89	0.55	0.12	0.86	3.62
CaO (%)	8.61	12.54	6.03	2.67	41.34	0.70	4.06	61.72
MgO (%)	2.58	4.80	0.71	0.84	6.97	0.26	0.18	2.52
SO_3 (%)	0.78	0.60	1.14	0.16	3.75	0.15	0.19	1.80
Na_2O (%)	2.58	3.65	0.59	0.32	0.24	0.16	3.57	0.17
K_2O (%)	1.45	1.72	1.50	2.30	0.28	0.49	4.35	0.69
TiO_2 (%)	1.19	1.17	1.01	1.40	0.47	0.01	0.09	–
P_2O_5 (%)	0.23	0.24	0.07	0.31	0.00	0.05	0.00	–
ZnO (%)	0.02	0.01	0.01	0.03	0.00	0.06	0.00	–
Mn_2O_3 (%)	0.03	0.09	0.03	0.05	0.19	0.04	0.07	–
Cl (%)	0.01	0.00	0.01	0.01	0.00	0.01	0.02	–
LOI (%)	1.42	n.a	5.12	3.19	n.a	4.30	4.09	1.96

Table 2. Experimental matrix (AMBT: accelerated mortar test; MCPT: miniature concrete prism test; X represents the mixtures that were cast)

Mixture	AMBT	MCPT
25FA1	X	X
25FA2	X	–
25FA3	X	–
25FA4	X	–
20FA1-5SF	X	X
50SL	X	X
25FA1-25SL	X	X
25NP	X	–
15FA1	–	X
30FA2	–	X
30SL	–	X

solution by maintaining a liquid-to-solid ratio of 0.9. All three reactants were mixed using a spatula in a plastic container for 4 min. After mixing, the paste was placed in a sealed glass vial, which was then placed in a 50 °C chamber for 240 h. The test conditions are expected to react the SCM entirely in most cases [16]. Approximately 40 mg of the sample was used for thermogravimetric analysis (TGA) to determine the CH left in the system. The sample was heated from 23 °C to 1000 °C at 20 °C per minute in a nitrogen-purged atmosphere in the instrument. From the initial and final amounts of CH, the amount of SCM reacted with CH was determined. The average CH consumed of two samples per SCM was reported.

Accelerated laboratory ASR test methods

The accelerated mortar bar test (AMBT – ASTM C1567) and the miniature concrete prism test (MCPT - AASHTO T 380) were used to monitor the expansion due to ASR.

For the AMBT, mortar bars of dimensions 25 mm × 25 mm × 285 mm were cast. The bars were cured and placed in 1M NaOH solution for 14 d according to ASTM C1567 to monitor their expansions. Three bars were cast per mixture, and their average expansion and standard deviation were reported. The expansion criteria for determining SCM's efficiency to prevent ASR is 0.1% average expansion at 14 d.

For the MCPT, concrete bars of dimensions 50 mm × 50 mm × 285 mm were cast and cured according to AASHTO T 380. The concrete bars were boosted to 1.25% Na_2O_e of the cement mass by adding 50%w/w NaOH solution to the fresh concrete mixtures. The bars were later immersed in 1M NaOH solution as per the standard for up to 56 d to monitor expansions. Three concrete bars were cast per mixture, and their average expansion and standard deviation were reported. The expansion criteria for determining SCM's efficiency to prevent ASR is 0.025% average expansion at 56 d, according to AASHTO T 380.

3 Results and Discussion

3.1 SCM Reactivity

The amounts of calcium hydroxide consumed by the SCM(s) in the reactivity test are listed in Table 3. Table 3 also lists the mixtures and their respective [CH consumed by SCM]*[SCM mass fraction]. The parameter: [CH consumed]*[SCM fraction] represents both the SCM reactivity and the amount of SCM in the mixture. It was observed that among the fly ashes tested, FA1 and FA4 have higher reactivity and FA2 and FA3 have lower reactivity. SF has the highest reactivity among all the SCMs or SCM combinations tested. It was observed that the blends, 20FA1 + 5SF and 25FA1 + 25SL, have the highest [CH consumed]*[SCM fraction] among all the mixtures.

3.2 ASR Expansion

Figures 1 and 2 show 14-d AMBT expansions and 56-d MCPT expansions, respectively. It was observed that FA2 mixture expanded the most in the AMBT, and this could be probably due to its higher alkali content of 4.78% Na_2O_e and relatively lower reactivity.

Table 3. CH consumed by SCMs in the reactivity test (example: for 25% SCM in the mixture, SCM fraction = 0.25)

SCM	CH consumed (g/100g SCM)	Mixtures	[CH consumed] *[SCM fraction] (g/100g SCM)
FA1	83.7	15FA1	12.6
FA2	72.8	25FA1	20.9
FA3	70.0	25FA2	18.2
FA4	85.3	30FA2	21.8
SL	38.6	25FA3	17.5
SF	187.4	25FA4	21.3
NP	87.0	20FA1-5SF	26.1
4:1 FA1-SF (for 20FA1-5SF)	104.4	30SL	11.6
1:1 FA1-SL (for 25FA1-25SL)	61.2	50SL	19.3
		25FA1-25SL	30.6
		25NP	21.8

Fly ashes with greater than 4% Na_2O_e may contribute alkalis to pore solution than they bind, thus increasing the expansion due to ASR [12]. In addition to 25FA2, 25FA3 and 50SL mixtures failed in the AMBT; all three mixtures have CH consumed*SCM fraction less than 20 g/100gSCM. All the mixtures with [CH consumed]*[SCM fraction] above 20 g/100gSCM were effective in preventing ASR as they passed the test, suggesting a possibility of proposing a threshold value to differentiate SCM combinations that could prevent ASR. However, this might not always be valid for the blends with Slag, as the heat of hydration or bound water were found to be better parameters than CH consumed to determine the Slag blends reactivity [14].

In the MCPT, except 25FA1 and 25FA1-25SL, the rest of the mixtures tested failed by expanding more than 0.025% at 56 d. Similar to the AMBT, FA2 mixture failed in the MCPT probably due to high alkali contribution and relatively lower reactivity. It was interesting to find that 20FA1-5SF failed in the MCPT despite 25FA1 passing the test. This could probably be due to improper dispersion of SF in the concrete mixture that could lead to agglomeration of SF and could act as ASR sites as shown previously [17].

3.3 Correlation Between SCM Reactivity and Expansion

Figures 3 and 4 show the results of the correlation between CH consumed*SCM fraction and 14-d AMBT expansions and 56-d MCPT expansions, respectively. The SCM fraction implies the fraction of SCM in the cementitious material blend by mass. In both figures, FA2 mixtures were excluded from the correlation as FA2 contributes higher alkalis than it binds, thus expanding more.

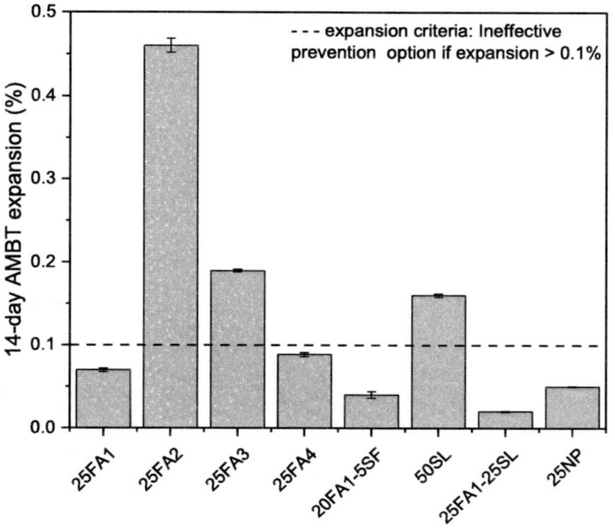

Fig. 1. 14-d AMBT expansions

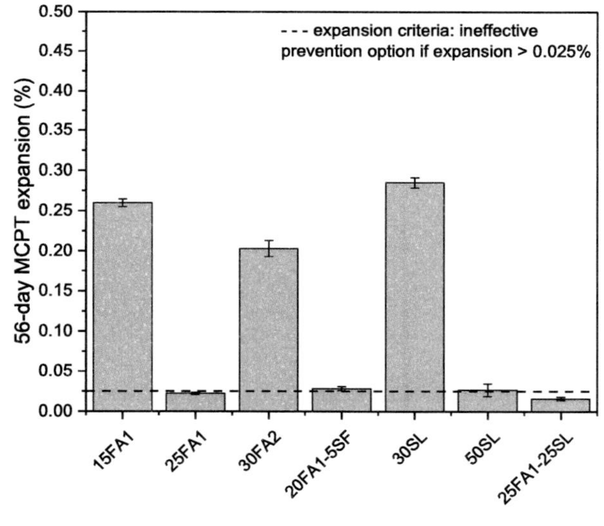

Fig. 2. 56-d MCPT expansions

From Figs. 3 and 4, it was observed that the expansions and [CH consumed by SCM in the reactivity test]*[SCM fraction] correlated well. The SCM combinations with higher SCM amount and SCM reactivity showed higher efficiency in lowering the expansion due to ASR. This is probably due to higher SCM amount contributing towards alkali dilution and higher SCM reactivity contributing towards higher amounts of hydration products

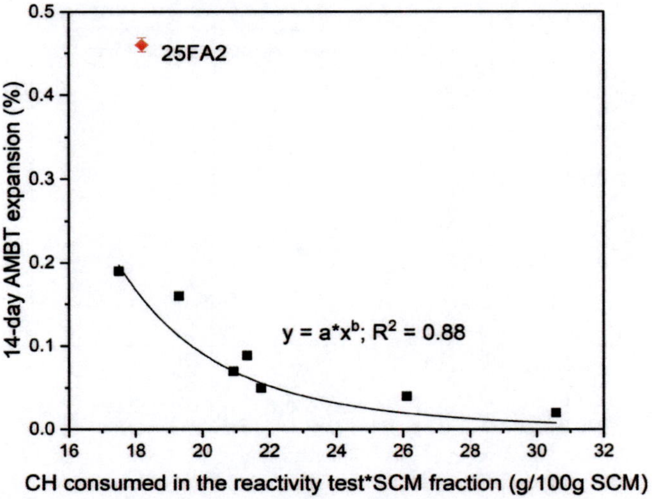

Fig. 3. The correlation between the SCM reactivity*SCM fraction and 14-d AMBT expansion (example: for 25% SCM, SCM fraction = 0.25; FA2 mixture was excluded from the fit)

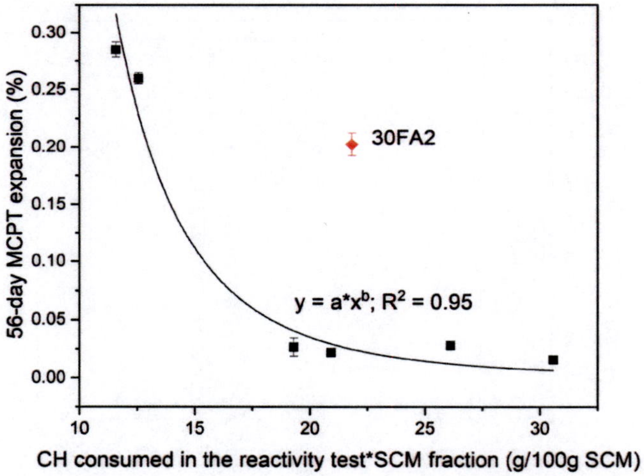

Fig. 4. The correlation between the SCM reactivity*SCM fraction and 56-d MCPT expansion (example: for 25% SCM, SCM fraction = 0.25; FA2 mixture was excluded from the fit)

and improved alkali binding. Thus, the approach of evaluating SCM combinations using the reactivity test to determine their efficacy in preventing ASR is showing promising results. However, the proposed approach might not accurately evaluate the SCMs that may contribute higher alkalis to pore solution as can be seen here in the case with FA2 mixtures. More SCM and reactive aggregate combinations need to be evaluated to be able to recommend the proposed approach to evaluate SCMs' efficacy in preventing ASR.

It should be noted that the AMBT test is aggressive as it is performed at a very high temperature and needs significant aggregate processing. The MCPT test conditions are comparatively realistic as it needs minimal aggregate processing, and it is performed at a relatively lower temperature than the AMBT. In addition, the MCPT was recently shown as a promising test method that correlates better to the field concrete in terms of determining the efficient SCM combinations to prevent ASR [18]. Therefore, the MCPT method could be used in future research to develop empirical models to determine SCM efficacy in preventing ASR.

4 Conclusions and Future Work

In this study, several SCMs and their combinations' reactivities and their mass fraction in the cementitious blend were correlated to the expansions of the mortar and concrete mixtures with very highly reactive fine aggregate (according to ASTM C1778). The findings from the study are summarized as follows:

- The mixtures with highly reactive SCMs and higher SCM mass fractions had lower ASR expansions.
- CH consumed by SCM in the reactivity test and SCM mass fraction correlated well with both AMBT and MCPT expansions; thus, showing the possibility of evaluating SCMs efficacy to prevent ASR using this approach.
- SCMs that contribute higher alkali content to pore solution than they bind might not be evaluated with the proposed approach.

Future work includes evaluating the amount of SCM in the mixture on its reactivity and including the effect of alumina content of SCMs on aggregate dissolution or ASR expansion could improve the proposed approach to evaluate SCM efficacy to prevent ASR. In addition, the MCPT could be used in future work instead of the AMBT to develop empirical models developed using the SCM reactivity to determine their efficacy to prevent ASR.

References

1. Thomas, M.: The effect of supplementary cementing materials on alkali-silica reaction: a review. Cem. Concr. Res. **41**, 1224–1231 (2011)
2. Glasser, F.P., Hong, S.-Y.: Alkali binding in cement pastes Part I. The C-S-H phase. Cem. Concr. Res. **29**, 1893–1903 (1999)
3. Bhatty, M.S.Y., Greening, N.R.: Interaction of alkalis with hydrating and hydrated calcium silicates. In: The Fourth International Conference on the Effects of Alkalis in Cement and Concrete, Purdue, pp. 87–112 (1978)
4. Lothenbach, B., Scrivener, K., Hooton, R.D.: Supplementary cementitious materials. Cem. Concr. Res. **41**, 1244–1256 (2011)
5. Chappex, T., Scrivener, K.L.: The influence of aluminum on the dissolution of amorphous silica and its relation to alkali-silica reaction. Cem. Concr. Res. **42**, 1645–1649 (2012)

6. Shi, Z., Ma, B., Lothenbach, B.: Effect of Al on the formation and structure of alkali-silica reaction products. Cem. Concr. Res. **140**, 106311 (2021)
7. Deschenes, R., Jones, C., Giannini, E., Hale, M.: A modified chemical index to predict fly ash dosage for mitigating alkali-silica reaction. Adv. Civil Eng. Mater. **8**, 20190191 (2019)
8. Kawabata, Y., Hamada, H., Sagawa, Y., Miyake, J., Ikeda, T.: Fly ash characterization related to mitigation of expansion due to ASR. In: 13th International Conference on Alkali-Aggregate Reaction in Concrete, Norway, pp. 184–191 (2008)
9. Mahyar, M., Erdoğan, S.T., Tokyay, M.: Extension of the chemical index model for estimating AlkaliSilica reaction mitigation efficiency to slags and natural pozzolans. Constr. Build. Mater. **179**, 587–597 (2018)
10. Malvar, L.J., Lenke, L.R.: Efficiency of fly ash in mitigating alkali-silica reaction based on chemical composition. ACI Mater. J. **103**, 319–326 (2006)
11. Schumacher, K.A., Ideker, J.H.: New considerations in predicting mitigation of alkali silica reaction based on fly ash chemistry. J. Mater. Civ. Eng. **27**, 04014144 (2015)
12. Thomas, M., Shehata, M.: Use of blended cements to control expansion of concrete due to alkali silica reaction. In: Gupta, P. (Ed.) Eighth CANMET/ACI International Conference on Fly ash, Silica Fume, Slag and Natural Pozzolans in Concrete, Las Vegas, U.S.A., pp. 591–607 (2004)
13. Vayghan, A.G., Wright, J.R., Rajabipour, F.: An extended chemical index model to predict the fly ash dosage necessary for mitigating alkali-silica reaction in concrete. Cem. Concr. Res. **82**, 1–10 (2016)
14. Parashar, A., Bishnoi, S.: A comparison of test methods to assess the strength potential of plain and blended supplementary cementitious materials. Constr. Build. Mater. **256**, 119292 (2020)
15. Suraneni, P., Weiss, J.: Examining the pozzolanicity of supplementary cementitious materials using isothermal calorimetry and thermogravimetric analysis. Cement Concr. Compos. **83**, 273–278 (2017)
16. Choudhary, A., Bharadwaj, K., Ghantous, R., Isgor, B., Weiss, J.: Examining factors that influence the measurement of pozzolanic reactivity of supplementary cementitious materials. ACI Mater. J. (Under Review)
17. Chopperla, K.S.T., et al.: Combining reliable performance testing and binder properties to determine preventive measures for alkali-silica reaction. Cem. Concr. Res. **151**, 106641 (2022)
18. Tanesi, J., Drimalas, T., Chopperla, K.S.T., Beyene, M., Ideker, J.H., Kim, H., Montanari, L, Ardani, A.: Divergence between performance in the field and laboratory test results for alkali-silica reaction. Transp. Res. Rec. **2674**, 120–134 (2020)

Water Absorption by Capillarity, Physical and Mechanical Properties of Concrete Containing Recycled Concrete Aggregate with Partial Cement Replacement by Metakaolin

Juliano F. Dutra[1](✉), Lucas C. Menegatti[1], Mayara Amario[1,2], Caroline S. Rangel[1], Marco Pepe[3,4], and Romildo D. Toledo Filho[1]

[1] Department of Civil Engineering, COPPE, Federal University of Rio de Janeiro, Rio de Janeiro, Brazil
juliano.dutra@coc.ufrj.br
[2] Department of Civil Construction, POLI, Federal University of Rio de Janeiro, Rio de Janeiro, Brazil
[3] Department of Civil Engineering, University of Salerno, Fisciano, Salerno, Italy
[4] TESIS srl, Fisciano, Salerno, Italy

Abstract. An alternative way to mitigate the impact generated by the Construction and Demolition Waste (CDW) on the environment is recycling this material in order to utilize it as Recycled Concrete Aggregates (RCA) in concrete. Studies carried out demonstrated that while using a scientific mix-design, it is possible to have a Recycled Aggregate Concrete (RAC) with mechanical properties similar the ones in conventional concrete. However, the durability of the recycled concrete is still a matter that should be explored further. The use of Supplementary Cementitious Materials (SCMs), such as metakaolin, has been shown to be efficient when partially replacing cement. It has the ability to improve the packaging of concrete by decreasing the volume of the pores of the material and, consequently, reducing the water absorption capacity of RAC, since RCA presents greater water absorption compared to natural aggregates. Therefore, the aim of this study is to evaluate the durability and mechanical behavior of RAC with 60 MPa of compressive strength by totally replacing the natural coarse aggregate by RCA. Also, part of the cement was replaced by metakaolin. The coarse aggregate considered in this research has 19 mm as nominal maximum size. Total absorption, capillarity absorption, compressive strength test and tensile strength test were carried out for all mixtures. The research results showed that the use of metakaolin offset the increase in capillary absorption in recycled concrete caused by the use of RCA, reaching a value close to concrete with only natural aggregate and without metakaolin for this durability parameter.

Keywords: Recycled Concrete Aggregate (RCA) · Recycled Aggregate Concrete (RAC) · Metakaolin · Durability · Mechanical properties

J. I. Escalante-Garcia et al. (Eds.): RW 2021, RILEM Bookseries 40, pp. 83–91, 2023.
https://doi.org/10.1007/978-3-031-21735-7_11

1 Introduction

Over the last few decades, environmental issues have become a concern in the construction sector. Several studies on recycling of Construction and Demolition Waste (CDW) have been conducted with the purpose of replacing Natural Aggregate (NA) by Recycled Concrete Aggregate (RCA) in the production of new concrete [1, 2]. In addition to reducing irregular waste disposal, this replacement aims to preserve natural resources, which are widely exploited to produce natural aggregates [1].

Aggregate is a fundamental component for the production of concrete and its quality exerts high influence on the composite properties [3]. RCAs differ from NAs due to the presence of an Attached Mortar (AM) layer on the aggregate surface, which consequently leads to the formation of an Interfacial Transition Zone (ITZ) between AM and the original natural aggregate. Due to these characteristics, RCA presents high water absorption capacity, which directly impacts physical, mechanical and durability properties of the Recycled Aggregate Concrete (RAC) [1, 4].

However, there are two main concerns when using recycled concretes. The first is the lack of a specific mix design method for concrete with recycled aggregates. The second is the composite high porosity, which consequently negatively affects its performance.

Regarding the mix design, some researchers [4–6] report that the Compressible Packing Method (CPM), developed by De Larrard [7] for conventional concrete, presented satisfactory results for RAC, since this method considers the intrinsic properties of the materials used. To reduce RAC porosity, studies [8–12] indicate that the partial replacement of cement by Supplementary Cementitious Materials (SCM), such as metakaolin (MK), can provide improvement in the mechanical and durability properties. However, due to the metakaolin higher absorption of free water, mixtures with metakaolin tend to present reduced workability [12].

As for the mechanical properties, studies [12–14] showed that the use of metakaolin benefits the mechanical behavior of recycled concrete, especially at early ages, offsetting the loss of strength caused by the use of RCA. Muduli and Mukharjee [12] explain that the increase in mechanical strength of concrete with partial metakaolin replacement is associated with pozzolanic effect. The pozzolanic reaction of metakaolin with portlandite forms an additional C-S-H, which contributes to the pore refinement and to concrete strength improvement.

For RAC with metakaolin, Muduli and Mukharjee [12] report a decrease in the water absorption capacity. The authors explain that this effect occurs due to the presence of insoluble hydration products resulting from the pozzolanic reaction, which block the concrete pores and prevent the entry of water.

Therefore, the present experimental study aims to analyze the combined effects of the incorporation of RCA and the partial replacement of cement by metakaolin (15% by weight) in high-strength concretes (60 MPa), designed with CPM method, evaluating its physical, mechanical and durability behavior. As a durability parameter, in this research, the absorption of water by capillary action was analyzed, since it can allow the entry of ions and aggressive agents into the concrete.

2 Materials and Methods

2.1 Materials

In this research, fine and coarse aggregates were used (Fig. 1). As natural fine aggregate, quartz sand with nominal diameter smaller than 4.75 mm was used. Also, two fractions of natural coarse aggregates from granite rocks were used: Nat_C0, with nominal diameter between 4.75 and 9.5 mm; and Nat_C1, with nominal diameter between 9.5 and 19 mm.

The RCA was derived from a demolition waste recycling plant and processed by Rangel et al. [6]. The CDW was submitted to four steps: crushing, drying, sieving and homogenization. The recycled material of nominal diameter between 9.5 and 19 mm was classified as coarse aggregate 1, named "RCA_C1". As CWD is composed of different demolition concretes, its properties are unknown. Therefore, it is important to characterize the RCA to use them as aggregates in concrete. Table 1 reports the properties of all aggregates used in this study. Specific gravity and water absorption were performed according to NBR 16916 [15] and NBR 16917 [16] for fine and coarse aggregates, respectively. The Attached Mortar content was determined by thermal shock (more details in Rangel et al. [6]).

Fig. 1. Natural and recycled aggregates used.

Table 1. Properties of natural and recycled aggregates.

Properties	Aggregates			
	Sand	Nat_C0	Nat_C1	RCA_C1
Maximum grain size (mm)	4.75	9.5	19	19
Specific gravity (kg/m^3)	2574	2618	2645	2287
Water absorption rate (%)	0.4	1.7	1.1	5.8
Attached mortar content (%)	–	–	–	35.1

The cement used was a "high initial strength Portland cement" (as established by NBR 16697 [17]), which reached 32.3 MPa of compressive strength at 28 days, in test performed according to NBR 7215 [18]. Cement and metakaolin presented specific gravity equal to 3067 and 2655 kg/m^3, respectively. Although the literature [12] indicates that metakaolin presents filler effect due to the smaller particle size in relation to cement grains, the metakaolin used in the research did not present this behavior. The d_{90} for the metakaolin and cement used were, respectively, 54.3 and 35.5 μm. In order to control workability, a polycarboxilate superplasticizer (labeled "Glenium 51") with specific gravity of 1087 kg/m^3 and solid concentration content of 30% was used.

2.2 Mixture Proportioning

The scientific mix-design of the concrete mixtures was determined according to the CPM [7]. This method takes into account the individual properties of each material and allows the effectively prediction of certain properties of the concrete, such as its compressive strength.

Four mixtures of high strength concrete (60MPa) were designed: a concrete with only natural aggregates (C60-NAC); one with incorporation of RCA, by replacing 100% of NAT_C1 by RCA_C1 (C60-RAC); another one with natural aggregates and 15% replacement of cement weight by metakaolin (C60-NAC-MK); and the last one by replacing 100% of NAT_C1 by RCA_C1 and replacing 15% of cement by metakaolin (C60-RAC-MK). Table 2 presents the composition of the concrete mixtures.

Table 2. Mixture proportioning of the concrete mixtures.

Mixture	Materials (kg/m^3)							
	Nat_C1	RCA_C1	Nat_C0	Sand	Cement	MK	Total water	Effective water
C60-NAC	411.5	–	415.6	896.9	471.4	–	152.2	150.0
C60-RAC	–	349.8	413.0	891.3	496.8	–	161.9	151.0
C60-NAC-MK	410.2	–	414.3	894.1	365.5	64.5	161.0	140.0
C60-RAC-MK	–	350.0	413.2	891.9	391.9	69.2	166.8	145.0

The mix-design method adopted considers that part of the total water is absorbed by the aggregates during the mixing process, while the remaining water, called effective water (w_{eff}), remains available to the cement hydration reactions and to ensure the concrete workability. Studies [5, 19] report that during the concrete mixing process, RCAs absorb only half of their total water absorption capacity. Therefore, for the mix-design, only 50% of the water absorption of the RCA was considered.

For mixtures with and without metakaolin, respectively, the amount of superplasticizer used was 0.60 and 0.35% of solids of SP based on the cementitious materials (cement and metakaolin) weight.

2.3 Mixing Procedure

The mixing procedure used consists of an adaptation made by Amario et al. [5] of the Two-Stage Mixing Approach (TSMA), developed and improved for recycled concretes by Tam et al. [20]. This method considers the addition of water in two steps due to the high absorption capacity of recycled aggregates.

The mixing process was carried out according to the following steps: fine and coarse aggregates were placed in dry condition into the mixer and homogenized for 1 min; then, 50% of the total water was added and the materials were mixed for another 1 min; then, the cement was poured and mixed for 1 min (for the mixtures with metakaolin, MK was previously homogenized with the cement); finally, the superplasticizer and the second half of water were add and all materials were mixed for 8 min.

2.4 Test Methods

Slump tests [21] were performed to determine the workability on fresh state of concrete. Compressive strength [22], elastic modulus [23] and tensile splitting strength [24] were determined on cylindrical specimens with 75 mm of diameter and 150 mm of height at 28 days. Total water absorption, void index and density [25] were performed at 28 days on cylindrical specimens with 100 mm of diameter and 200mm of height. For capillary absorption [26], cylindrical specimens with 75 mm of diameter and 150 mm of height were tested.

3 Results and Analysis

3.1 Workability and Physical Properties of RAC

Table 3 summarizes the results obtained for workability of fresh concrete (slump) and for physical properties of hardened concrete at 28 days (water absorption, void index, and specific gravity).

Table 3. Workability of fresh concrete and physical properties of hardened concrete.

Mixture	Slump (mm)	Water absorption (%)	Void index (%)	Specific gravity (kg/m^3)
C60-NAC	200	0.74 (±2.9%)	1.74 (±4.0%)	2361 (±1.1%)
C60-RAC	190	1.32 (±3.2%)	2.95 (±2.9%)	2235 (±0.3%)
C60-NAC-MK	200	0.93 (±11.0%)	2.20 (±10.6%)	2364 (±0.4%)
C60-RAC-MK	190	0.76 (±1.1%)	1.76 (±2.2%)	2308 (±1.1%)

The mixtures produced only with natural aggregates (C60-NAC and C60-NAC-MK) presented slump values equal to 200 mm, while the mixtures with incorporation of RCA (C60-RAC and C60-RAC-MK) reached slump values equal to 190 mm. The use of recycled aggregate did not significantly affect the workability of the recycled concretes produced in this research. Regarding the mixtures with metakaolin, it was observed that the workability was maintained when compared to the reference concretes. These satisfactory workability results for all the mixtures were achieved due to the testing step carried out during the mix-design process.

Regarding physical properties, the use of RCA caused more significant changes. The total water absorption of the C60-RAC was 1.32% and for the C60-NAC it was 0.74%, representing an increase of 79%. The partial replacement of cement by MK provided improvement for concretes that contained RCA. The water absorption value of C60-RAC-MK mixture was 0.76%, close to the value obtained for the C60-NAC mixture. This result shows that the use of metakaolin offset the high absorption caused by the incorporation of RCA in the concrete mixtures. However, the use of Metakaolin did not presented the same behavior for the mixture that contained only natural aggregates (C60-NAC-MK). This result is justified by the fact that C60-NAC presented better packaging than C60-NAC-MK and, as consequence, lower values of void index and water absorption.

The void index is directly related to water absorption. The higher the void index, the higher the absorption values. Therefore, RACs presented higher values compared to natural concretes, except when metakaolin was added in the mixture with RCA.

Recycled concretes had lower specific gravity values when compared to reference concretes. This occurred due to the presence of RCA in C60-RAC and C60-RAC-MK mixtures, since recycled aggregates presents lower specific gravity than natural ones (see Table 1).

3.2 Mechanical Behavior of RAC

Table 4 shows the means values and the coefficient of variations of mechanical properties analyzed at 28 days: compressive strength ($f_{c,28}$), strain at maximum stress ($\varepsilon_{c,28}$), elastic modulus ($E_{c,28}$), splitting tensile strength ($f_{t,28}$) and the relation between 28-day splitting tensile strength and 28-day compressive strength ($f_{t,28}/f_{c,28}$).

Table 4. Mechanical properties of hardened concrete.

Mixture	$f_{c,28}$ (MPa)	$\varepsilon_{c,28}$ ($\mu\varepsilon$)	$E_{c,28}$ (GPa)	$f_{t,28}$ (MPa)	$f_{t,28}/f_{c,28}$ (%)
C60-NAC	65.0 (±0.7%)	2647 (±1.2%)	34.6 (±3.9%)	4.7 (±1.9%)	7.2
C60-RAC	59.5 (±0.2%)	2650 (±4.0%)	31.7 (±3.9%)	4.3 (±7.8%)	7.2
C60-NAC-MK	61.0 (±3.2%)	2690 (±4.8%)	31.0 (±4.8%)	4.2 (±0.8%)	6.9
C60-RAC-MK	59.0 (±4.0%)	2409 (±1.4%)	31.7 (±1.4%)	4.2 (±9.2%)	7.1

The 28-day compressive strength values for mixtures without metakaolin were 65.0 MPa and 59.5 MPa for C60-NAC and C60-RAC, respectively. The value obtained for C60-NAC presented a variation of 8.3% in relation to the expected strength. Regarding the mixtures with partial replacement of cement by metakaolin, the strength values achieved at 28 days were 61.0 MPa for C60-NAC-MK and 59.0 MPa for C60-RAC-MK, with lower variation in relation to the expected strength (1.7% for both mixtures).

The elastic modulus results of mixtures without metakaolin were 34.6 GPa and 31.7 GPa for C60-NAC and C60-RAC, respectively. When using metakaolin, the results obtained were 31.0 GPa for C60-NAC-MK and 31.7 GPa for C60-RAC-MK.

Analyzing the relationship between tensile and compression strength at 28 days for the mixtures without metakaolin, the value of 7.2% was obtained for both mixtures (C60-NAC and C60-RAC). When using metakaolin, it was observed that the mixture with RCA (C60-RAC-MK) presented better tensile behavior (7.1%) when compared to C60-NAC-MK (6.9%).

3.3 Water Absorption by Capillarity

The evaluation of water absorption by capillarity was performed over 72 h and the results are summarized in Fig. 2. It can be highlighted that, as expected, the mixture with the highest absorption by capillary was C60-RAC (0.219 g/cm^3), as shown in Fig. 2.b. This is explained since it was the mixture with the highest water absorption and voids index among the mixtures analyzed. An improvement was observed for this mixture when metakaolin was used (C60-RAC-MK), reaching a capillary absorption of 0.155 g/cm^3. The same behavior was not observed for the mixtures produced only with natural aggregates (Fig. 2.a). Although C60-NAC-MK achieved slightly higher value for absorption by capillary than C60-NAC (0.168 and 0.154 g/cm^3, respectively), it is possible to affirm that the use of metakaolin did not cause significant impacts in these properties for concrete mixtures without RCA.

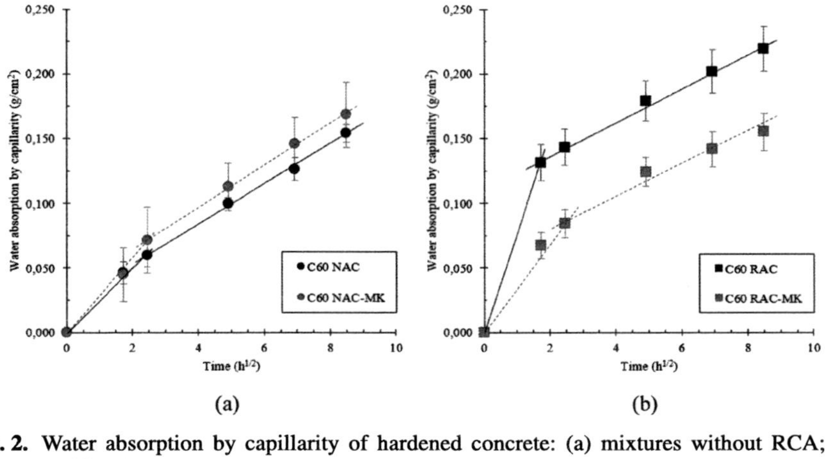

Fig. 2. Water absorption by capillarity of hardened concrete: (a) mixtures without RCA; (b) mixtures with RCA

4 Conclusion

This research evaluated the combined effect of using recycled concrete aggregates and a supplementary cementitious material (metakaolin) on mechanical and physical properties and on water absorption by capillarity as a durability indicator. According to the obtained results, the following considerations can be highlighted:

- The use of recycled concrete aggregate and metakaolin did not significantly affect compressive strength of the mixtures when compared to the reference concretes. Therefore, it can be ensured that CPM is suitable for the mix-design of recycled concretes with metakaolin.
- Elastic modulus and tensile strength were also not affected by the use of RCA and metakaolin. These properties are strongly related to the compressive strength of concrete and all mixtures showed similar mechanical behavior.
- Physical properties were influenced by the use of recycled concrete aggregate. Recycled concrete presented higher void index and, consequently, higher water absorption. This fact is due to the presence of attached mortar on RCA particles, which is responsible for the higher porosity of this material in relation to natural aggregates.
- Metakaolin also influenced physical properties of recycled concrete due its filler effect. Since RAC presents higher porosity, metakaolin particles fills part of the existing pores, obtaining lower void index and, consequently, lower water absorption for the recycled concrete with metakaolin.
- Results also showed that the mixture with incorporation of RCA (C60-RAC) reached the highest result of water absorption by capillarity. By partially replacing cement by metakaolin (C60-RAC-MK), lower absorption values were obtained when compared to its reference, showing that it is possible to obtain satisfactory results when combining the use of recycled aggregate and metakaolin in concrete mixtures.

Finally, this study indicates that with an adequate mix-design method, it is possible to produce recycled concrete with RCA with physical and mechanical performance equivalents to conventional concretes. Furthermore, the use of metakaolin can reduce recycled concrete porosity, improving its durability.

References

1. Verian, K.P., Ashraf, W., Cao, Y.: Properties of recycled concrete aggregate and their influence in new concrete production. Resour. Conserv. Recycl. **133**, 30–49 (2018)
2. Salesa, A., et al.: Physico-mechanical properties of multi- recycled concrete from precast concrete industry. J. Clean. Prod. **141**, 248–255 (2017)
3. Koper, A., Koper, W., Koper, M.: Influence of raw concrete material quality on selected properties of recycled concrete aggregates. Procedia Eng. **172**, 536–543 (2017)
4. Ouyang, K., Shi, C., Chu, H., Guo, H., Song, B., Ding, Y., Guan, X., Zhu, J., Zhang, H., Wang, Y., Zheng, J.: An overview on the efficiency of different pretreatment techniques for recycled concrete aggregate. J. Clean. Prod. **263** (2020)
5. Amario, M., Rangel, C.S., Pepe, M., Toledo Filho, R.D.: Optimization of normal and high strength recycled aggregate concrete mixtures by using packing model. Cement Concr. Compos. **84**, 83–92 (2017)

6. Rangel, C.S., Toledo Filho, R.D., Amario, M., Pepe, M., de Castro Polisseni, G., de Andrade, G.P.: Generalized quality control parameter for heterogenous recycled concrete aggregates: a pilot scale case study. J. Clean. Prod. **208**, 589–601 (2019)
7. De Larrard, F., Ferraris, C.F., Sedran, T.: Fresh concrete: a Herschel-Bulkley material. Mater. Struct. **31**(7), 494–498 (1998)
8. Kisku, N., Joshi, H., Ansari, M., Panda, S.K., Nayak, S., Dutta, S.C.: A critical review and assessment for usage of recycled aggregate as sustainable construction material. Constr. Build. Mater. **131**, 721–740 (2017)
9. Kou, S., Poon, C.: Long-term mechanical and durability properties of recycled aggregate concrete prepared with the incorporation of fly ash. Cement Concr. Compos. **37**, 12–19 (2013)
10. Kurad, R., Silvestre, J.D., De Brito, J., Ahmed, H.: Effect of incorporation of high volume of recycled concrete aggregates and fly ash on the strength and global warming potential of concrete. J. Clean. Prod. **166**, 485–502 (2017)
11. Muduli, R., Mukharjee, B.B.: Influence of recycled coarse aggregate and metakaolin on properties of concrete. Int. J. Eng. Technol. Sci. Res. **5**, 649–655 (2018)
12. Muduli, R., Mukharjee, B.B.: Effect of incorporation of metakaolin and recycled coarse aggregate on properties of concrete. J. Clean. Prod. **209**, 398–414 (2019)
13. Bui, N.K., Satomi, T., Takahashi, H.: Effect of mineral admixtures on properties of recycled aggregate concrete at high temperature. Constr. Build. Mater. **184**, 361–373 (2018)
14. Younis, K.H., Amin, A.A., Ahmed, H.G., Maruf, S.M.: Recycled aggregate concrete including various contents of metakaolin: mechanical behavior. Adv. Mater. Sci. Eng. **2020**, 17 (2020)
15. NBR 16916: Fine aggregate—determination of density and water absorption. ABNT (2021)
16. NBR 16917: Coarse aggregate—determination of density and water absorption. ABNT (2021)
17. NBR 16697: Portland cement—requirements. ABNT (2018)
18. NBR 7215: Portland cement—determination of compressive strength of cylindrical test specimens. ABNT (2019)
19. Pepe, M., Toledo Filho, R.D., Koenders, E.A.B., Martinelli, E.: A novel mix design methodology for recycled aggregate concrete. Constr. Build. Mater. **122**, 362–372 (2016)
20. Tam, V.W., Gao, X.F., Tam, C.M.: Microstructural analysis of recycled aggregate concrete produced from two-stage mixing approach. Cem. Concr. Res. **35**, 1195–1203 (2005)
21. NBR 16889: Fresh concrete—slump test. ABNT (2020)
22. NBR 5739: Concrete—compression test of cylindrical specimens. ABNT (2018)
23. NBR 8522: Concrete—determination of static modulus of elasticity and deformation by compression. ABNT (2017)
24. NBR 7222: Concrete and mortar—determination of the tension strength by diametrical compression of cylindrical test specimens. ABNT (2011)
25. NBR 9778: Hardened mortar and concrete—determination of absorption, voids and specific gravity. ABNT (2005)
26. NBR 9779: Mortar and hardened concrete—determination of water absorption by capillarity. ABNT (2012)

Check for updates

Evaluation of the Potential of Single-Wall Carbon Nanotubes in Improving the Properties of Cement-Composites

J. A. Gonzalez-Calderon[1], Isidro Montes Zavala[1], Eldho Choorackal[2(✉)], and Diego A. Santamaria Razo[2]

[1] Instituto de Física, Universidad Autónoma de San Luis Potosí, Zona Universitaria, Av. Manuel Nava 6, 78290 San Luis Potosí, México

[2] Innovation Centre for Sustainable Construction, CRH Nederland BV, Amsterdam, Netherlands
echooracka1@crh.com

Abstract. The experimental investigation described in this paper deals with the characterization, functionalization, dispersion and use of single-wall carbon nanotubes (SWCNT) in cement-composites. The dispersion of SWCNT was achieved using sonication and modification with a surfactant. The rheological modelling of the SWCNT-surfactant system was carried out, which is shown to reduce the viscosity of the system. In order to better understand the effect of SWCNT in cementitious systems, detailed characterization of the SWCNT and sedimentation tests in water were carried out. To assess the effect on mechanical properties, different loading percentages of the SWCNT were added to the cement mortar, which lead to significant improvement in compressive strength compared to the reference mortar. A threshold dosage range of SWCNT was identified for the optimum mechanical performance of the cement mortar. These results demonstrate the potential of using SWCNT to produce better performing cement-composites.

Keywords: Low clinker cement concrete · Single-wall carbon nanotubes · Nano-modification

1 Introduction

The cement industry is responsible for a significant part of the atmospheric carbon dioxide (CO_2) emission. To reduce this carbon footprint, reduction of clinker content from the cement-concrete mixtures and carbon capture, utilization and storage (CCUS) are two main strategies adopted by the cement-concrete industry [1]. The CCUS is still largely under research and needs to solve several other technical challenges before its large-scale implementation. Other measures like improving the energy efficiency of cement plants, reducing the clinker content from cement and reducing cement content from the concrete mix designs are currently being adopted [2]. Use of alternate binders like geopolymer concrete and use of supplementary cementitious materials (SCM) like silica fume, fly ash and slags in the concrete mixes, as a partial replacement to cement are also being practiced [3, 4]. However, these SCM and alternate binders alone cannot

J. I. Escalante-Garcia et al. (Eds.): RW 2021, RILEM Bookseries 40, pp. 92–99, 2023.
https://doi.org/10.1007/978-3-031-21735-7_12

achieve the clinker reduction goals. Moreover, achieving sufficient early-age strength for the low-clinker concrete products and identification of suitable alternative binders instead of clinker are remaining as a challenge for the concrete industry. The application of nano materials in cement-concrete is hence interesting for the researchers, as there is a growing need of identifying materials that can contribute to the clinker reduction objective.

Materials in nano metric size were of interest because their material properties differ from the same bulk material. Higher surface area and small particle size of nano materials are the main reasons for this peculiar behavior. Different nano materials exhibit different properties depending upon their morphological features and chemical composition. Selection of suitable nano material based on the target application and their uniform dispersion are the two key factors that govern the success in nano modification of materials.

1.1 Nanomaterials in Cement Concrete

The use of nano materials as an additive in the concrete mix designs modifies the properties of cement concrete in several ways. They increase the viscosity of the liquid phase and thereby improve segregation resistance and workability. They fill the voids between cement grains, act as centers of crystallization of the cement hydrates, which also accelerates the rate of hydration [5]. The interfacial transition zone between the aggregates and cement paste also gets improved in presence of nano materials.

The benefits of nano modification vary with the type and dosage of nanomaterial used. Nano SiO_2 participates in the pozzolanic reactions, resulting in the consumption of $Ca(OH)_2$ and the formation of an additional calcium silicate hydrate which would contribute to the strength development [6, 7]. Nano Al_2O_3 and nano TiO_2 improve the compressive strength, latter also enhances the Photocatalytic properties [8, 9]. Other materials like nano $CaCO_3$, nano Carbon black, and graphite nano particles are all found to be improving the hydration kinetics and compressive strength [10]. Other benefits like crack arrest, improvement in the toughness were also observed with nano modification. Thus, proper application of nano materials improves the fresh, hardened and durability properties of the cementitious materials.

CNTs are the type nano particles, which increase both compressive and tensile properties of the cement composites at their optimum dosage [11]. Due to the special morphology of the CNT, proper dosing and uniform mixing of them in cement concrete is important to observe the beneficial effects. Improper dosing and insufficient mixing could lead to adverse effects in the strength development of the composite. Extensive studies on the surface modification of SWCNT and their modeling were performed to identify the optimum usage of surfactants [12].

Challenges and Way forward. Despite all this laboratory-scale research, the transfer of technology from the laboratory to industrial-scale production was not smooth due to several reasons. Firstly, the uniform dispersion of nanomaterials in the cementitious matrix was difficult to achieve during the mass production of concrete. Secondly, the high cost and energy required for the production of nano-materials made this economically not viable for commercial cement concrete producers.

To address some of the challenges mentioned above, the following specific objectives were identified for the experimental investigation described in this paper.

- Development of a specific dispersion protocol for the SWCNT in the cementitious matrix, which can also be used in industrial-scale cement concrete production.
- Quantification of the effect of SWCNT modification in the mechanical properties of cementitious matrix.

2 Experimental Investigation

2.1 Materials and Methods

Characterization. The SWCNT used in all the tests were supplied by InnTeG and the surfactant Sodium dodecylsulfate (SDS) supplied by Sigma Aldrich. TEM analysis was performed to observe the morphology of the nanoparticles. The powder samples were dispersed in an aqueous medium, making serial dilutions until the optimum concentration of particles was achieved. A drop was placed on a copper grid and observed in a JEOL equipment (model JEM 1230 microscope), with a resolution of 0.4 nm and an acceleration voltage of 100 kV.

The Z potential measurements were performed using a Delsa Nano C Particle Analyzer (Beckman Coulter). In order to monitor the behavior of the different nanomaterials for Z potential tests, the pH was varied from 3 to 11. The carbon nanotubes were analyzed in a range of 800–4000 cm^{-1}, using a diamond ATR device for solid samples.

Dispersion of carbon nanotubes and sedimentation tests. First, tests were carried out to monitor the sedimentation of carbon nanotubes in an aqueous medium. This is to ensure that proper dispersion of SWCNT will be achieved when it is incorporated into the cementitious matrix. For the dispersion of the carbon nanotubes, SDS was used to improve the colloidal behavior. 30 g of deionized water was added to 0.6% by weight of SWCNT (0.18g) with two SWCNT: SDS ratios (1:0, 1:1) (labeled as T).

The carbon nanotube suspensions were subjected to sonication tests and subsequently accelerated sedimentation tests. The sonication process was carried out for 5 min and the behavior of the suspensions was visualized for 40 min. For the accelerated sedimentation process, special equipment manufactured in the laboratory was used, in which the suspensions were rotated at an angular velocity of $\omega = 10.68$ rad/s and lateral acceleration of $a = 29.66 m/s^2$. If it is considered that $1g = 9.81\ m/s^2$, the equipment causes an acceleration of approximately 3g. Each suspension was again subjected to ultrasonic vibration for 5 min before the accelerated sedimentation process.

Preparation of cementitious nanocomposites. The cement mortar was prepared using a water to cement ratio of 0.75 and a Cement to sand ratio of 1:3. A laboratory mortar mixer was used for the preparation of the mortar. The dosage of carbon nanotubes were varied in steps as 0.1, 0.05, 0.01 and 0.005% with respect to the weight of the cement. The mortar samples were tested at 2, 7 and 28 days of age for their compressive strength.

3 Results and Discussions

3.1 Characterization of the Nanoparticles

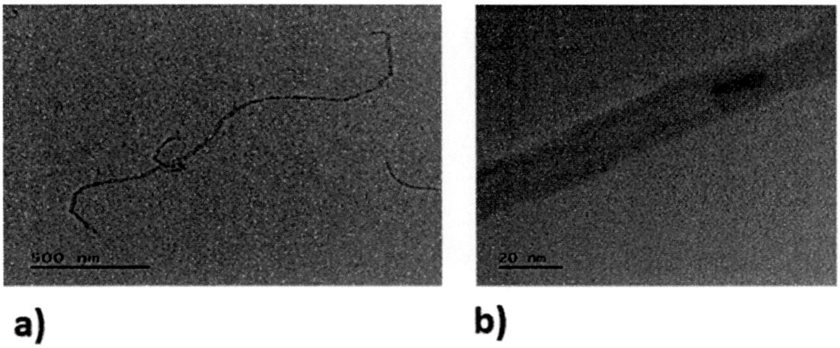

Fig. 1. TEM images of carbon nanotubes. (a) 500 nm and (b) 20 nm.

The transmission electron microscopy images are shown in Fig. 1. Carbon nanotubes of approximately 2 microns are visualized in Fig. 1a. While the diameter of the nanotubes reaches 20 nm (see Fig. 1b).

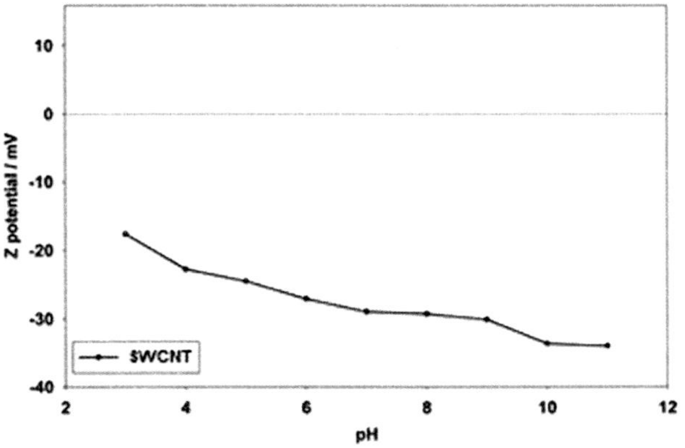

Fig. 2. Z potential of carbon nanotubes.

In Fig. 2, it was observed that the carbon nanotubes maintain a good dispersion in the pH range of 3–11. This indicates that the incorporation of carbon nanotubes will have a positive influence on the cementitious matrix, promoting an improvement in their mechanical properties.

The sample was analyzed in a range of 800–4000 cm^{-1} (see Fig. 3), using a diamond ATR device for solid samples. The FTIR pattern does not show signs of organic groups

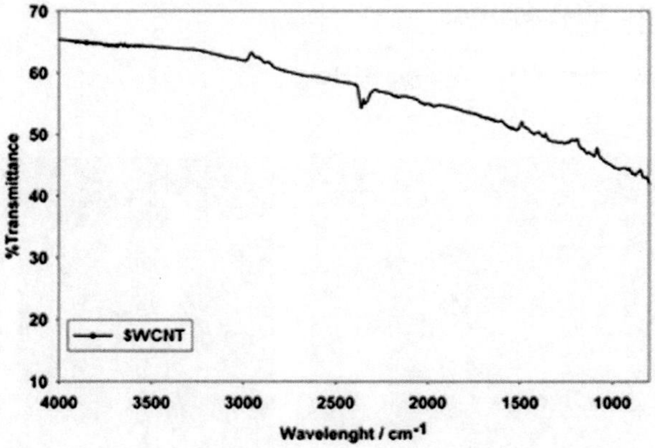

Fig. 3. FTIR of carbon nanotubes.

other than carbon, the signal obtained by the equipment is typical of non-oxidized carbon materials (Fig. 3).

3.2 Sonication of SWCNT/Surfactant Systems

The suspensions of SWCNT/surfactant systems were subjected to ultrasonic vibration for 5 min. In Figs. 4 and 5, the suspensions obtained, before and after sonication are shown. It is observed that there is no visually significant difference when a surfactant is used during sonication.

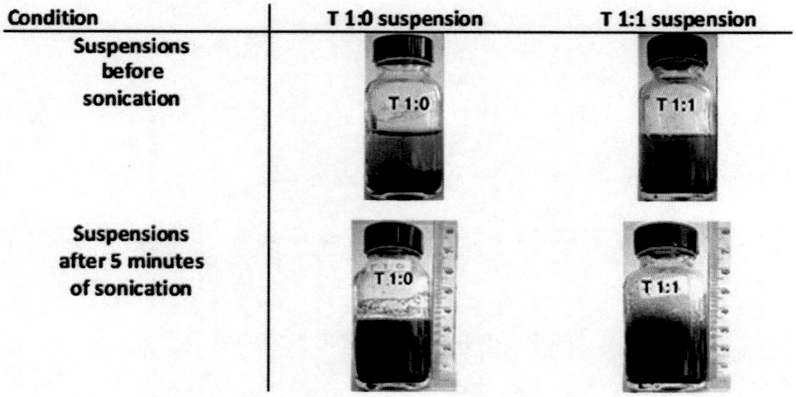

Fig. 4. Carbon nanotubes in suspension.

Sedimentation in suspensions after ultrasonic vibration. After sonication, the samples were visually monitored to observe the sedimentation of particles. There was no significant difference between the suspensions until 30 min from the end of the sonication in the sedimentation of particles. Unfortunately, the foam produced does not disappear in the suspension containing SDS. Therefore, the suspension without surfactant is emerging as a good candidate, if the time available for dosing SWCNT suspension into the cementitious mix is less than 30 min.

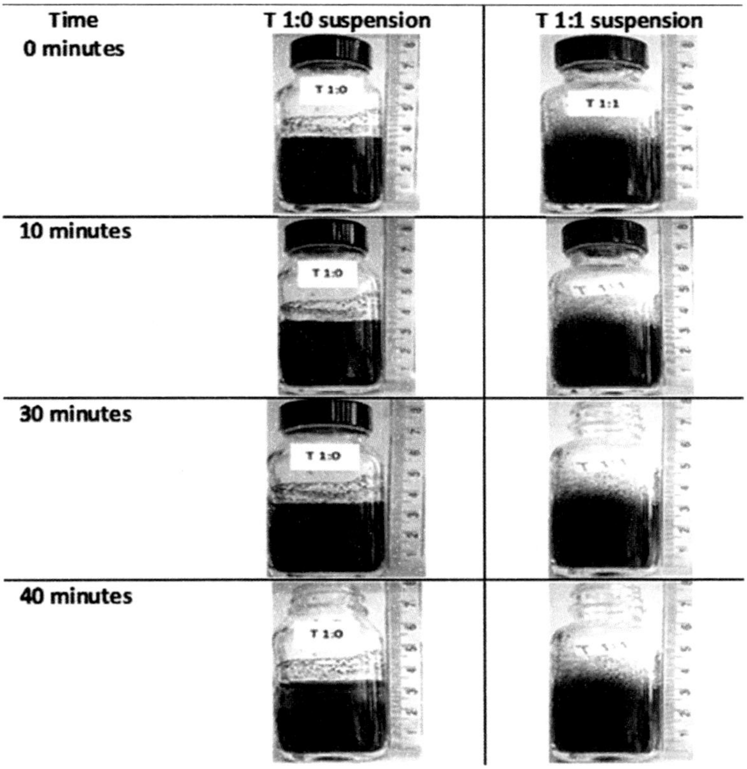

Fig. 5. Carbon nanotubes in suspension under sonication conditions.

3.3 Mechanical Testing of Cementite Compounds with Carbon Nanotubes

The compressive strength data obtained from the testing of mortar samples at the ages of 2, 7 and 28 days is plotted in Fig. 6 with the SWNCT dosage. It can be seen that at low concentrations of SWCNT (0.005 and 0.01%), the compressive strength is significantly improved with respect to the reference mortar. At higher dosages, a negative trend was observed at all ages. This indicates an optimum range of SWCNT dosage varying from 0.005 to 0.010% by weight of the cement to achieve better mechanical properties. The substantial improvement in mechanical properties obtained from the use of SWCNT can be further translated to a reduced cement content to achieve target properties.

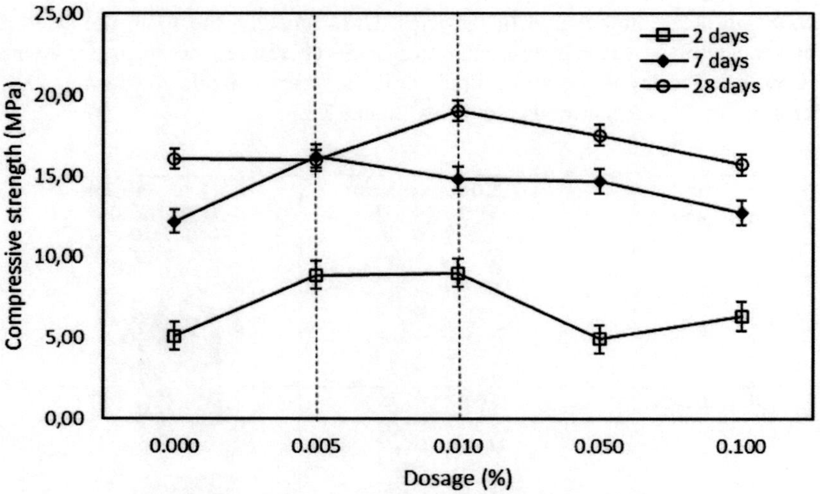

Fig. 6. Compressive strength of the mortar at different dosages of SWCNT.

4 Conclusions

This paper demonstrated the use of SWCNT in the cementitious matrix to enhance their mechanical properties. Expertise gained from this study, in the surface modification and dispersion of the SWCNT can be translated to engineering the properties of the cement concrete as desired.

- A specific dispersion protocol for the SWCNT in the cementitious matrix was developed in the study, which can also be adapted to the requirements of industrial-scale cement concrete production.
- The effect of SWCNT modification in the mechanical properties of the cementitious matrix was quantified and an optimum dosage range was identified for the target application in cement mortar.

The developed knowledge on nano modification can be applied to the production of high-performance concrete, multi-functional concrete and low clinker concrete. Specifically, with the production of cheaper nanomaterials like nano $CaCO_3$, this technology is worth pursuing for reducing the carbon footprint of the cement-concrete industry.

References

1. Benhelal, E., Shamsaei, E., Rashid, M.I.: Challenges against CO2 abatement strategies in cement industry: a review. J. Environ. Sci. **104**, 84–101 (2021)
2. Favier, A., De Wolf, C., Scrivener, K., Habert, G.: A sustainable future for the European cement and concrete industry: technology assessment for full decarbonisation of the industry by 2050. ETH Zurich (2018)

3. Juenger, M.C., Siddique, R.: Recent advances in understanding the role of supplementary cementitious materials in concrete. Cement Concrete Res. **78**, 71–80 (2015)
4. Assi, L., Carter, K., Deaver, E.E., Anay, R., Ziehl, P.: Sustainable concrete: building a greener future. J. Clean. Prod. **198**, 1641–1651 (2018)
5. Sanchez, F., Sobolev, K.: Nanotechnology in concrete–a review. Constr. Build. Mater. **24**(11), 2060–2071 (2010)
6. Abhilash, P.P., Nayak, D.K., Sangoju, B., Kumar, R., Kumar, V.: Effect of nano-silica in concrete; a review. Constr. Build. Mater. **278**, 122347 (2021)
7. Yu, R., Spiesz, P., Brouwers, H.J.H.: Effect of nano-silica on the hydration and microstructure development of Ultra-High Performance Concrete (UHPC) with a low binder amount. Constr. Build. Mater. **65**, 140–150 (2014)
8. Joshaghani, A., Balapour, M., Mashhadian, M., Ozbakkaloglu, T.: Effects of nano-TiO2, nano-Al2O3, and nano-Fe2O3 on rheology, mechanical and durability properties of self-consolidating concrete (SCC): An experimental study. Constr. Build. Mater. **245**, 118444 (2020)
9. Yousefi, A., Allahverdi, A., Hejazi, P.: Effective dispersion of nano-TiO_2 powder for enhancement of photocatalytic properties in cement mixes. Constr. Build. Mater. **41**, 224–230 (2013)
10. Zhang, L., Ding, S., Sun, S., Han, B., Yu, X., Ou, J.: Nano-scale behavior and nano-modification of cement and concrete materials. Advanced Research on Nanotechnology for Civil Engineering Applications, pp. 28–79. IGI Global (2016)
11. Kumar, S., Kolay, P., Malla, S., Mishra, S.: Effect of multiwalled carbon nanotubes on mechanical strength of cement paste. J. Mater. Civ. Eng. **24**(1), 84–91 (2012)
12. Montes-Zavala, I., Castrejón-González, E.O., Rico-Ramírez, V., Pérez, E., Santamaría-Razo, D.A., González-Calderón, J.A.: Which is better? Experimental and simulation analyses of the chemical modification of carbon nanotubes to improve their dispersion in water. J. Dispers. Sci. Technol. 1–12 (2020)

Simulating the Hydration Process of Low Water to Cement (w/c) Ternary Pastes Incorporating Superabsorbent Polymers (SAP)

Judy Kheir[1,2](✉), Benoît Hilloulin[1], Ahmed Loukili[1], and Nele De Belie[2]

[1] Institut de Recherche en Génie Civil Et Mécanique (GeM), UMR-CNRS 6183, Ecole Centrale de Nantes, 1 Rue de La Noë, 44321 Nantes, France
judy.kheir@ugent.be, benoit.hilloulin@ec-antes.fr, ahmed.loukili@ec-nantes.fr

[2] Magnel-Vandepitte Laboratory for Structural Engineering and Building Materials, Ghent University, Technologiepark Zwijnaarde 60, B-9052 Ghent, Belgium
nele.debelie@ugent.be

Abstract. Early age cracking is a high risk for low water to cement (w/c) concrete known as the ultra-high performance concrete (UHPC). The lack of water to fully hydrate cement particles at the early age, provides a rapid increase of self-desiccation and autogenous shrinkage which might lead to cracking. However, internal curing agents such as superabsorbent polymers (SAP) can reduce/eliminate crack development by providing water to the mixture. In this study, a numerical model called CemPP was presented, using a modified version of CEMHYD3D v3 by creating a dedicated pass for the simulation of the hydration process and the microstructure development of low w/c ternary pastes (with silica fume and filler) containing SAPs. Both the filler effect and pozzolanic reaction of silica fume were considered in the numerical simulation, as well as the effect of the water release of SAPs on the hydration process of cement at the early age. The evolution of small pores was lower for SAP pastes than for the reference ones as seen in the experimental results, which indirectly implies lower shrinkage for SAP pastes. The model has been verified to successfully simulate SAP behaviour in cement pastes, and give a good idea about the shrinkage evolution.

Keywords: Superabsorbent polymers · Ultra high performance concrete · Early autogenous shrinkage · CemPP

1 Introduction

Ultra High Performance Concrete (UHPC) are known by their low water to cement (w/c) ratio (typically between 0.2 and 0.3 or even lower), this type of concrete is very susceptible to autogenous shrinkage (AS) because of its low w/c ratio. [1]. Early AS can enhance the risk of crack development. The consumption of water during the hydration process causes a lack of free water which results in a drop of the (and when no external water source is present) internal relative humidity (RH $<$ 100%) implying self-desiccation

J. I. Escalante-Garcia et al. (Eds.): RW 2021, RILEM Bookseries 40, pp. 100–107, 2023.
https://doi.org/10.1007/978-3-031-21735-7_13

[2, 3]. A solution to that problem, would be to add an internal curing (IC) agent to the mixture so that extra water will be present to counter the self-desiccation process. In this study, Superabsorbent Polymers (SAPs) were used as IC. SAPs are able to absorb up to 500 times their own weight and gradually release their water to maintain a high RH [4]. When the hydration process in concrete or cement pastes start, water-filled pores are emptied quickly in low w/c pastes and biggest pores are emptied first. As the pores in the paste decrease in size, the capillary pressure that would develop upon their emptying increases. The formation of the liquid-vapor menisci in the pores leads to shrinkage of the 3D microstructure, and the smaller the meniscus radius the higher the capillary pressure resulting in higher shrinkage. Thus, the presence of smaller pores are the main reason behind high early autogenous shrinkage in low w/c pastes [5]. However, the capillary pressure that starts to develop is the driving force that pulls water out of SAPs. Therefore, the higher the volume fraction of SAPs inside a mixture the higher the release of internal curing water, the lower the shrinkage.

Numerical models could help design and optimize the formulation of concrete. Over the last decades, several interfaces were used to model the evolution of microstructure of concrete, mortar and cement paste over time starting from hydration to degradation. CEMHYD3D is one of the model that simulates concrete hydration [6], it can also lead to good estimations of mechanical properties [7]. Very few models were developed to account for the SAP role in the cement hydration process. Therefore, in this study CemPP model (modified version of CEMHYD3D), developed by Hilloulin et al. [8], was used in order to take into consideration the role of SAPs in cement pastes and to monitor the water release from SAPs at pre-defined time periods. Hence, a dedicated pass for SAPs has been added in the model. The model was calibrated using experimental calorimetry test values. At first, the experimental autogenous shrinkage results were represented for both UHPC mixtures with and without SAPs. Then the model was briefly represented for the use of SAPs in low w/c pastes. Lastly, a comparison between a reference and a SAP group including filler and silica fume was considered in the numerical simulation to try and understand the role of each mineral addition on the evolution of small pore volume, as well as the effect of the water release of SAPs on the hydration process of cement at the early age.

2 Materials and Methods

2.1 Materials

Portland cement CEM I 52.5 R was used in the experimental and numercial implementations, it has a density of 3.15 g/cm^3 and a Bogue's composition as follows: 66% of C_3S, 11% of C_2S, 7% of C_3A and 10% of C_4AF. The mineral additives added in the mixture were filler and silica fume with densities equal to 2.7 g/cm^3 and 2.2 g/cm^3 and with the following proportions 23% and 19% by weight of cement respectively. A superplasticizer with a density of 1.08 kg/liter and 40% of active materials was used with a dosage of 1.1% by weight of cement. In addition, basalt 4/8 mm were used as the coarse aggregates and sand 0/4 mm as fine aggregate. A commercially available SAP with a dry diameter of 100 μm was added in the UHPC mixture. The composition of the experimental concrete is represented in Table 1.

Table 1. Experimental UHPC composition without SAPs (REF) and with SAPS (SAP)

Materials	REF (kg/m^3)	SAP (kg/m^3)
Basalt 4/8	649.3	649.3
Sand 0/4	401.9	401.9
CEM I 52.5 R	778.2	778.2
Filler	185.5	185.5
Silica fume	153.8	153.8
Water	186.4	186.4
Superplasticizer	8.56	8.56
SAP	–	2.17
Extra water needed for SAPs	–	58.63

2.2 Experimental Autogenous Shrinkage

Concrete was poured into a $7 \times 7 \times 28$ mm^3 horizontal mold and completely covered for autogenous shrinkage deformations measurements. Early deformations were monitered by laser sensors from approximately 1 h after water to cement contact up until 24 h. After 24 h, AS measurements were monitored through LVDT sensors placed on top of the fully covered vertical concrete specimens, see Fig. 1a and b. The horizontal molds are essential in order to represent the AS results from the moment of the final setting time. Measurements were followed up to 1 week as the study deals with early age deformations.

2.3 Model of SAP Influence on Autogenous Shrinkage Development in Cement Paste

In order to simulate the behavior of a SAP in UHPC and how it affects the autogenous shrinkage, CemPP was used. CemPP is an interface developed to model the hydration reactions of a given concrete and generates its microstructures. It is a modified version of CEMHYD3D introducing 'box transformation passes'. These transformation passes are coded beforehand as C++ classes and have acces to the entire simulation state to make it easier for the user to run the desired change on the microstrcuture. The functionality of the 'box transformation passes' can be either to replace some chemical elements in a given region by other elements (for example removing phases to create a crack and simulate self-healing) or, as in this study, simulating a SAP and performing some analysis and getting a report on the chemical structure by measuring the size of pores in the structure (gives an idea about the shrinkage development). These transformation passes were developed to be enabled/disabled at specific time periods given in hours, and at a certain frequency i.e. number of simulation cycles between two executions of the same pass.

In this study, two new box passes were created in order to account for SAP inclusion and to monitor the water release from SAPs at pre-defined time periods. 3D

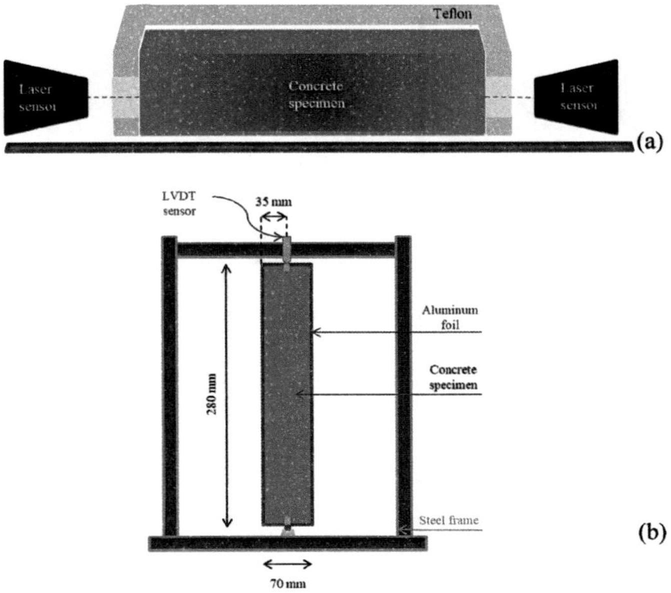

Fig. 1. Schematic representation of the autogenous shrinkage measurement setups for fresh concrete (a) and hardened concrete (b)

microstructures, developed with the same experimental UHPC paste were simulated. The microstructures for a ternary cement paste having a water-to-binder ratio of 0.2 containing SAPs with a volume size of 40 μm^3 discretized by 1 μm^3 were generated using CemPP code until 3 days, see Fig. 2. SAPs appear as grey cubes in this figure. The volume fraction of SAP used was 2%, two other volume fractions (5 and 8%) were also used to further validate the model. Water release time was accounted to start at 16 h after water to cement contact, and to take end at 72 h as it was shown by Lefever et al. [9] that SAPs activity starts from few hours after final setting time, between 12 to 16 h after w/c contact, and can lasts up until 3 days.

3 Results and Discussion

3.1 Experimental Autogenous Shrinkage

Autogenous shrinkage evolution results are represented in Fig. 3 for an ultra high performance reference concrete denoted as REF and a UHPC containing SAPs (SAP) over a period of one week. Shrinkage curves start from the moment of final setting measured from the FreshCon test. REF concrete has a shrinkage curve that starts with a very high rate at the beginning for few hours and then it becomes slower as the skeleton becomes more rigid to resist volume changes induced by autogenous shrinkage. After 8 days, the shrinkage value exhibited by the reference specimen accounted for around 500 μm/m.

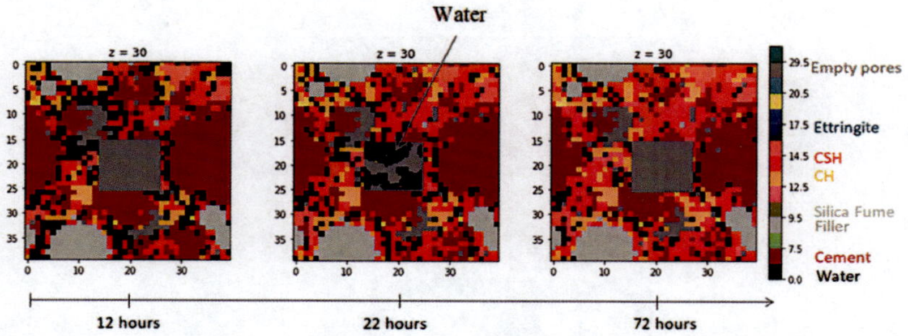

Fig. 2. 2D sections of a simulated 3D microstructure of a ternary cement paste incorporating a SAP generated from CemPP

A shrinkage reduction of almost 80% is shown in the SAP deformation curve a constant value of 100 μm/m against 500 μm/m at 8 days. It can be clearly seen how the presence of polymers reduce/mitigate the shrinkage as SAPs can gradually release their water at the appropriate moment during the hydration process maintaining the RH and minimizing the self-desiccation, thus reducing autogenous shrinkage [2]. In Fig. 3, a shrinkage increase is seen in the SAP curves for the first few hours after final setting time to reach a peak at around 110 μm/m followed by a decrease and a stabilizitaion at around 92.5 μm/m. The decrease in the curve represents the moment of the water release from SAPs in order to keep the RH above 95%, and reduces the development of small pores that are responsible for AS. Therefore, the role of SAPs at early age has been confirmed to have a major effect on the AS reduction.

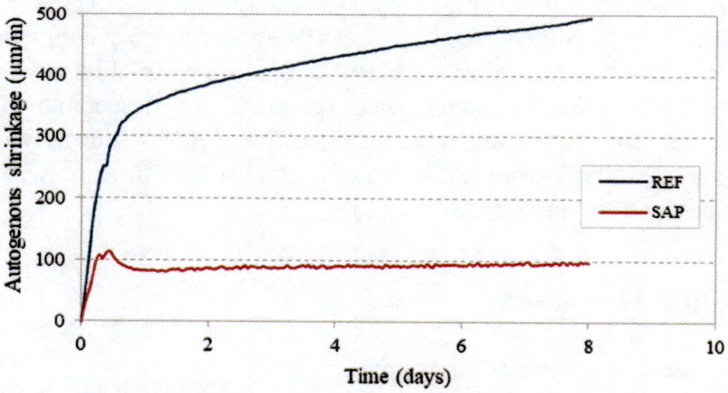

Fig. 3. Autogenous shrinkage evolution of a reference UHPC (REF) and a UHPC containing superabsorbent polymers (SAP)

3.2 Model of SAP Influence on Autogenous Shrinkage Development in Cement Paste

First, in order to verify the model capability, different volume fractions of SAPs were introduced the 2, 5 and the 8%. In Fig. 4, the evolution of small pore volume of size 1 μm has been represented in function of time at different SAP volume fraction, the water release period time has been assigned from 16 h until 72 h as described above in section II.c). It can be shown that the higher the volume fraction of SAP, the lower the formation of the small pores which indicates low autogenous shrinkage. Hence, the model can be considered as a valid model that can successfully simulate the hydration of cementitious materials in the presence of SAPs.

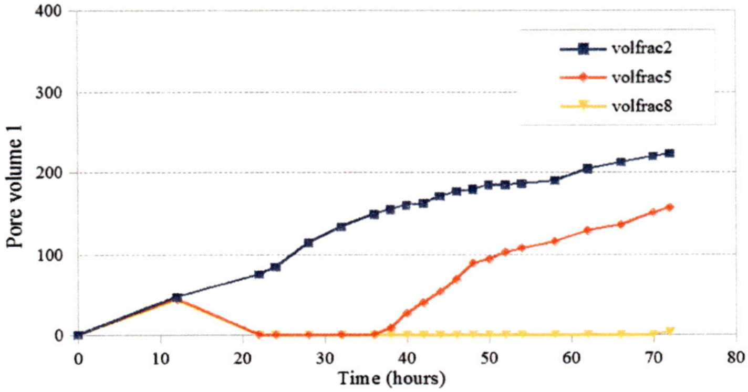

Fig. 4. Evolution of small pores of 1 μm size for a low w/c reference cement paste containing superabsorbent polymer (SAP) at different SAP volume fractions of 2, 5 and 8%

In Fig. 5, the evolution of the small pore volume has been represented for two groups: the reference group and the SAP group with a volume fraction of 2%. Each group has a reference paste and three other pastes with mineral additions as follows: 2 binary pastes, 1 containing cement and silica fume and the other one cement and filler, and a ternary paste with the combination of cement, silica fume and filler. These pastes combinations can give an idea about the effect of each mineral addition on the shrinkage evolution. The evolution of small pores for both pastes incorporating silica fume (SF) was the highest at 3 days although the filler content was higher than the SF content (which implies that normally a lower amount of cement was present in the pastes containing filler). But, filler is not a reactive product in the hydration reaction, it decreases the porosity by increasing the packing density due to its fine grains compared to cement. Even though SF is known to reduce chemical shrinkage [10], it increases the autogenous shrinkage [11]. This justifies the lowest slope of the SF curves at the early age, because it takes a little bit of time to participate in the hydration process. In both cases of, the ternary paste has the lowest autogenous shrinkage, because in the ternary pastes the advantages of both mineral additions take place, bonding between the particles is enhanced in the cement paste and fewer voids will be present in the microstructure skeleton. The shape curves in the reference curves have a slight resemblance to the experimental autogenous shrinkage curve in Fig. 3. Finally, the group with SAP has the lowest curves than the reference group, which represents once more the effectiveness of these internal curing agents toward reducing the autogenous shrinkage.

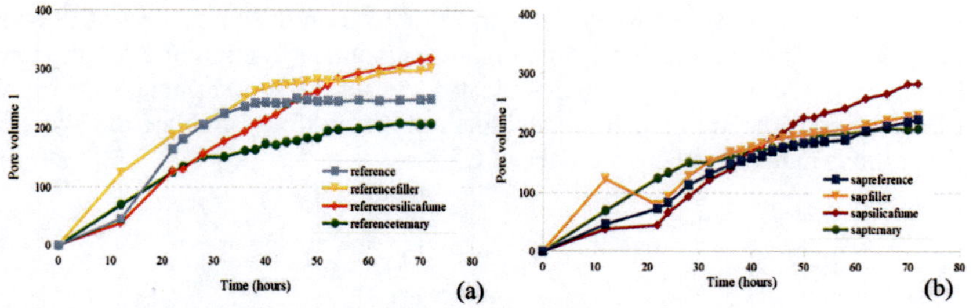

Fig. 5. Comparison of the evolution of small pore volume of 1 µm in function of hydration time for a low w/c reference cement paste (a) and a SAP paste of volume fraction 2% (b)

4 Conclusion

A numerical model CemPP was presented in this study, developed from a modified version of CEMHYD3D by creating a dedicated transformation pass for the simulation of the hydration process and the microstructure development of low w/c ternary pastes (with silica fume and filler) containing superabsorbent polymers (SAPs). The filler effect and pozzolanic reaction of silica fume were considered in the numerical simulation, as well as the effect of the water release of SAPs on the hydration process of cement at the early age. Experimental results were presented as well to show the SAP behaviour towards autogenous shrinkage. It can be concluded that:

- From the experimental results, SAPs highly reduce early age autogenous shrinkage in low w/c concrete, by gradually releasing their water few hours after final setting to reduce high capillary pressure that appears in smaller pores with the hydration process.
- The incorporation of mineral addition such as silica fume (SF) and filler (F) has been presented in the model to see the effect of each mineral addition on the evolution of AS. It turned out that SF and F tend to increase the number of pores which was expected as these mineral additions are known in litterature to increases the autogenous shrinkage, whereas a combination of both decreases the number of small pores developed during hydration.
- SAPs behaviour has been successfully represented by the model. The evolution of small pore volume was the lowest for the SAP group than for the reference one. Thus, the model has been verified to successfully simulate SAP behaviour in cement pastes.

To conclude the model simulates successfully the SAP behaviour in cement pastes, and can give a good idea about the shrinkage evolution, but it further needs some adjustments to account for the real shrinkage shape behaviour of SAP pastes.

References

1. Bentz, D.P., Jensen, O.M., Hansen, K.K., Olesen, J.F., Stang, H., Haecker, C.: Influence of cement particle-size distribution on early age autogenous strains and stresses in cement-based materials, vol. 35, no. 188803, pp. 129–135 (2001)
2. Snoeck, D., Jensen, O.M., De Belie, N.: The influence of superabsorbent polymers on the autogenous shrinkage properties of cement pastes with supplementary cementitious materials. Cem. Concr. Res. **74**, 59–67 (2015). https://doi.org/10.1016/j.cemconres.2015.03.020
3. Kovler, K., Zhutovsky, S.: Overview and future trends of shrinkage research. Mater. Struct. Constr. **39**(293), 827–847 (2006). https://doi.org/10.1617/s11527-006-9114-z
4. Mechtcherine, V., et al.: Effect of internal curing by using superabsorbent polymers (SAP) on autogenous shrinkage and other properties of a high-performance fine-grained concrete: results of a RILEM round-robin test. Mater. Struct. **47**(3), 541–562 (2013). https://doi.org/10.1617/s11527-013-0078-5
5. Henkensiefken, R., Bentz, D., Nantung, T., Weiss, J.: Volume change and cracking in internally cured mixtures made with saturated lightweight aggregate under sealed and unsealed conditions. Cem. Concr. Compos. **31**(7), 427–437 (2009). https://doi.org/10.1016/j.cemconcomp.2009.04.003
6. Bentz, D.P.: Three-dimensional computer simulation of portland cement hydration and microstructure development. J. Am. Ceram. Soc. **80**(1), 3–21 (1997). https://doi.org/10.1111/j.1151-2916.1997.tb02785.x
7. Bernard, F., Kamali-Bernard, S., Prince, W.: 3D multi-scale modelling of mechanical behaviour of sound and leached mortar. Cem. Concr. Res. **38**(4), 449–458 (2008). https://doi.org/10.1016/j.cemconres.2007.11.015
8. Hilloulin, B., Hilloulin, D., Grondin, F., Loukili, A., De Belie, N.: Mechanical regains due to self-healing in cementitious materials: experimental measurements and micro-mechanical model. Cem. Concr. Res. **80**, 21–32 (2016). https://doi.org/10.1016/j.cemconres.2015.11.005
9. Lefever, G., Snoeck, D., De Belie, N., Van Hemelrijck, D., Aggelis, D.G.: Elastic wave monitoring of cementitious mixtures including internal curing mechanisms. Sensors **21**(7), 1–14 (2021). https://doi.org/10.3390/s21072463
10. Kheir, J., Hilloulin, B., Loukili, A., De Belie, N.: Chemical shrinkage of low water to cement (w/c) ratio CEM I and CEM III cement pastes incorporating silica fume and filler. Materials (Basel) 1–17 (2021)
11. Zhang, M.H., Tam, C.T., Leow, M.P.: Effect of water-to-cementitious materials ratio and silica fume on the autogenous shrinkage of concrete. Cem. Concr. Res. **33**(10), 1687–1694 (2003). https://doi.org/10.1016/S0008-8846(03)00149-2

Study of the Pozzolanic Reactivity of Amazonian Metakaolin for Use as a Supplementary Cementitious Material

Ana Paula de Lima Mendes[1], Daniela Oliveira de Lima[1](✉), Luciane Farias Ribas[2], and João de Almeida Melo Filho[3]

[1] University of São Paulo, São Paulo, Brazil
daniela.engcivil@usp.br
[2] Centro Universitário Fametro, Manaus, Brazil
[3] Federal University of Amazonas, Manaus, Brazil

Abstract. The application of supplementary cementitious materials (SCMs) as a partial replacement for Portland cement (PC) can significantly reduce carbon dioxide emissions, making the mortar or concrete more sustainable and environmentally friendly. Among the supplementary materials, metakaolin (MK) has received considerable attention, as it provides improvements in mechanical properties and increased durability of the material. The Amazon region presents large deposits of kaolin and therefore this study investigated its use as SCMs. In the study, three metakaolin were produced and characterized using X-ray diffraction (XRD), X-ray fluorescence spectroscopy (XFR), differential thermal analysis (DTA), and laser diffraction techniques. The pozzolanic reactivity study was evaluated by the indirect electrical conductivity method, known as the Luxán Method. In addition, cement pastes were produced, where PC was replaced in the fractions of 40 and 50%, the samples were evaluated by thermogravimetric analysis (TGA) at 28 days of curing. This procedure was adopted to confirm the pozzolanic of MK, by monitoring the reduction of calcium hydroxide. The results show that the wet processing method of kaolin and its calcination at 750 °C for 4 h showed satisfactory results, being sufficiently able to eliminate all the calcium hydroxide at 28 days, confirming the pozzolanic test by the Luxán method.

Keywords: Metakaolin · Pozzolanic reactivity · Electric conductivity · Luxán method

1 Introduction

Portland cement is one of the most widely used materials in the construction industry. However, its production harms the environment, as it is responsible for approximately 6–8% of CO_2 emissions in the world [1, 2]. To mitigate this problem, several strategies are being studied, one of them is to replace a proportion of cement with SCMs [3]. Pozzolanic materials are one of the main substitutes for cement, among which we highlight fly ash, rice husk ash, silica fume, and blast furnace slag, and calcined clays,

J. I. Escalante-Garcia et al. (Eds.): RW 2021, RILEM Bookseries 40, pp. 108–117, 2023.
https://doi.org/10.1007/978-3-031-21735-7_14

such as metakaolin [4]. The use of supplementary materials allows the cement matrix (paste, mortar and concrete) to improve properties related to durability, mechanical performance, and waterproofing [5]. These benefits are the results of the hydrated products due to the pozzolanic reaction. In addition, these pozzolans can be included in cementitious matrices to aid in the conversion of $Ca(OH)_2$, which is a less desirable hydration product, into the more desirable calcium silicate hydrate (C-S-H) [6].

In recent years, the use of calcined clay in the form of metakaolin as supplementary material for mortar and concrete has received considerable attention due to its high pozzolanic properties. Metakaolin ($Al_2Si_2O_7$) is a type of clay resulting from the calcination of kaolinite clay [7, 8]. It is a manufactured product and not a by-product, which has smaller particles than cement, but not as fine as silica fume [9]. When kaolin is heated to a temperature between 500 and 800 °C there is dehydroxylation of the crystalline structure of kaolinite, forming metakaolin [10, 11].

However, to use MK it is necessary to previously evaluate its reactivity. For this, there is a wide variety of methods, classified as direct and indirect. The direct method monitors the pozzolanic reaction by the variation of reduction of $Ca(OH)_2$ over time, using XRD techniques, TGA, or chemical methods. The direct method, using thermogravimetric analysis, determines pozzolanic activity by the reaction of the cement with the supplementary cementitious material. The technique is based on determining the type and amount of hydrated phases at different ages. As the $Ca(OH)_2$ concentration decreases, the amount of C-S-H phases and some alumina-hydrated phases increase, evidencing the pozzolanic reaction [12].

The indirect method measures a physical property of the sample that indicates the extent of pozzolanic activity and can be evaluated by compressive strength tests, electrical conductivity, or heat release by calorimetry [13]. The indirect method of electrical conductivity was designed by Raask & Bhaskar [14] to measure the amount of silica dissolved in a hydrofluoric acid solution, in which the active material is dispersed, from the pozzolanic index is calculated by measuring the electrical conductivity. Based on this study, Luxán et al. [15] created a fast and efficient method to evaluate the pozzolanic reactivity of natural products. The method consists of the compensated measurement of a saturated solution of calcium hydroxide, to which the material is added. This technique has as its main characteristic the rapid obtaining of results, from a minimum time of two minutes.

This research aims to characterize and monitor the pozzolanic reactions of metakaolins produced from Amazonian kaolin, with the potential to be used as SCM. For this, the pozzolanic was evaluated by electrical conductivity, and thermogravimetry and compression analyses were performed in addition on the MK with higher reactivity.

2 Materials and Methods

2.1 Materials

The kaolin in nature (K) was collected from a deposit located on BR 174, near the Igarapé Cabeça Branca (Km 44), Manaus—AM, Latitude: 2°35′13.28″ S and Longitude: 60°01′58.53″ W. From kaolinite clay, three MK were produced, adopting different methodologies for the production process. First, the MKI was produced, in which, initially, the K was taken to the oven for 24 h at a temperature of 100 °C ± 5 °C, to remove excess moisture. After this period, the K was placed in a mortar and crushed manually with a pistol. Next, the K was passed through a 100 μm sieve. For calcination, samples were distributed in porcelain crucibles and taken to the Linn Elektro Therm muffle furnace at a temperature of 750 °C for 2 h. After calcination and cooling, the sample was sieved on a 200 μm sieve.

The metakaolins MKII and MKIII were produced wet. Thus, after collection, the K was deposited in a container with water for 24 h (see Fig. 1a). It was then passed through a 200 μm wet sieve (see Fig. 1b). Subsequently, the material was allowed to decant for 24 h (see Fig. 1c). After this procedure, the excess water was removed and the K was placed in trays and kept in an oven until it was completely dry, at a temperature of 100 ± 5 °C (see Fig. 1d). Soon after, the K was placed in a mortar and manually crushed with a pistol. Finally, to perform the calcination, the K was distributed in porcelain containers and taken to the muffle furnace for 4 h (see Fig. 1e). However, the calcination temperature of the samples was varied, thus producing metakaolin calcined at 650 and 750 °C, called MK II and MK III, respectively.

(a) **(b)** **(c)** **(d)** **(e)**

Fig. 1. Kaolin softening (a), kaolin washing (b), kaolin decantation (c), kaolin after oven drying (d), kaolin in the porcelain crucible (e).

2.2 Methods

Instrumental Techniques

The chemical and mineralogical compositions of the materials were characterized with the instrumental techniques described below.

The particle size distribution of the MK was determined by the analytical technique of laser beam diffraction, wet method, with a CILAS 1180—Particle Size Analyzer, with detection capability between 0.00004 and 2.5 mm.

The chemical composition of the elements was determined by X-Ray Fluorescence spectrometry (XRF), using PANalytical equipment, model EPSILON 3-XL, with a maximum voltage of 50 kV, performed with a maximum current of 3 mA and helium gas with a pressure of 10 kgf/cm^2.

The mineralogical analysis of the clay in nature and the metakaolins was determined by X-ray diffraction (XRD), adopting the powder method. The samples were analyzed in a Bruker D2 Phaser diffractometer, with copper radiation (Cu-kα, $\lambda = 1.5406$ Å), operating with 30 kV voltage generator and 10 mA current setting, at a step of 0.02° per second and measurement range between Bragg angles (2θ) from 10° to 70°. To verify the purity and crystallinity of the minerals, the information generated in the diffractometry was treated from the X'Pert HighScore Plus software, using the Powder Diffraction File (PDF 2).

Thermogravimetry (TG) and differential thermal analysis (DTA) were used to characterize the Amazon kaolin. TA Instrument, model SDT Q600, was used to perform the analyses, operating with a heating rate of 10 °C/min, from room temperature to 1000 °C, in a nitrogen atmosphere and a flow rate of 30 ml/min.

Pozzolanicity Test

To perform the analysis, 200 ml of a saturated solution of $Ca(OH)_2$ at 40 °C ± 1 °C was used, with the addition of 5 mg of MK, submitted to constant stirring with a magnetic stirrer. After the addition of the pozzolan, the variation of conductivity was measured before and after 120 s of the addition of the sample. The pozzolanic was classified according to the conductivity variation proposed by Luxán et al. [15], as presented in Table 1.

Table 1. Classification of pozzolanic activity by conductivity measurement.

Classification	No activity	Moderate activity	Good activity
Conductivity variation (mS/cm)	<0.4	0.4 a 1.2	>1.2

After the electrical conductivity test, the most reactive MK was chosen to evaluate its behavior through thermogravimetry and axial compression. The TG was performed on pastes produced with Portland cement type CP II E32 containing granulated blast furnace slag addition, MK, and adopted a water/binder factor of 0.4%. A reference paste (M0) was produced and used for comparison with the pastes containing 40 and 50% substitution, (M40 and M50), respectively. The samples were prepared and analyzed at 28 days. The equipment and technical parameters are the same as previously described, except the temperature was 900 °C. The pozzolanic was evaluated by the reduction of the $Ca(OH)_2$ content over time.

For the compression tests, mortars were produced using a 1:1 (cement:sand) mix with partial replacement of cement by MK, with 50% content (M1) and a reference matrix (M0). The natural sand came from a river with a modulus of 1.71, water/binder factor of 0.4%, and addition of 0.8% by weight of superplasticizer. Cylindrical specimens with dimensions of 100 × 50 mm (height × diameter) were produced according to NRB 7215 [16], molded four specimens for each matrix, broken at 28 days. The specimens were cured in water until the test was performed. The mechanical test was performed in a Universal EQUILAM machine, model EQRB 100, with 100 kN load cell, at a constant speed of 0.5 mm/min, until the rupture of the specimen.

3 Results and Discussion

3.1 Starting Material Characterization

The particle size distribution of MK is presented in Fig. 2. It can be seen that all MK curves are similar, with particle sizes between 0.1 and 70 µm. For comparison, we added the size distribution curve of Portland cement used in cement pastes, which can prove that the MK has smaller particles than the cement, which has particles between 0.1 and 100 µm. According to Sahmaran et al. [17], the MK should have particles smaller than the cement, so it acts as filler by increasing the densification of the matrix, filling the voids, improving the physical effects, and, consequently, provided greater reactivity.

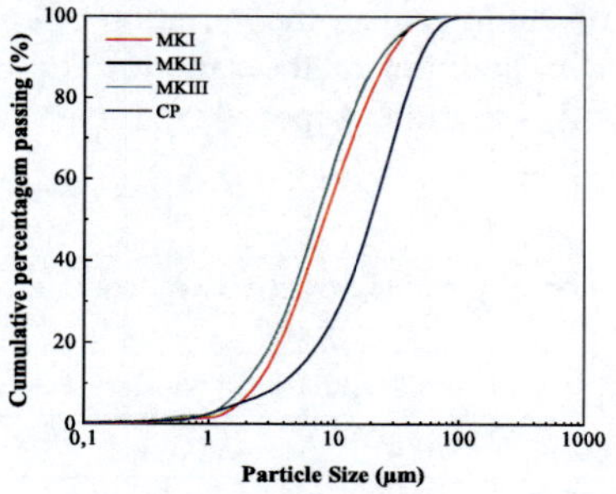

Fig. 2. Particle size curve of metakaolin and cement.

Table 2 shows the chemical composition in the form of more stable oxide identified by XRF. It is observed that the K the MKs are composed of Al_2O_3 and SiO_2, added together total approximately 97% of the composition of MK and 98% of the composition of K. It is noted that there was no significant change in the chemical composition due to calcination of K, and that the content of SiO_2+Al_2O_3+Fe_2O_3 is greater than 70%, meeting the requirements of ASTM C618 [18].

Table 2. Chemical composition (% by mass) of kaolin and metakaolin.

	SiO_2	Al_2O_3	Fe_2O_3	Ti	P_2O_5	K_2O
K	55.15	43.04	0.58	0.37	0.42	0.26
MK I	53.68	44.70	0.54	0.34	0.35	0.24
MK II	55.94	41.21	0.92	0.48	0.77	0.38
MK III	55.86	41.19	0.97	0.49	0.78	0.39

The X-ray diffractograms of K and MK are shown in Fig. 3. It was found that the K consists mainly of the mineral kaolinite and presents as a secondary constituent of the quartz in less intensity. It is noted that despite the different production methodologies for MKI and MKII the diffractogram of both is very similar. While the diffractogram of MKIII after calcination showed an increase in the background between the range of 2θ = 20 and 30°, indicating the dehydroxylation of the crystalline structure of kaolinite, resulting in the total conversion of kaolinite into a pozzolanic additive [19].

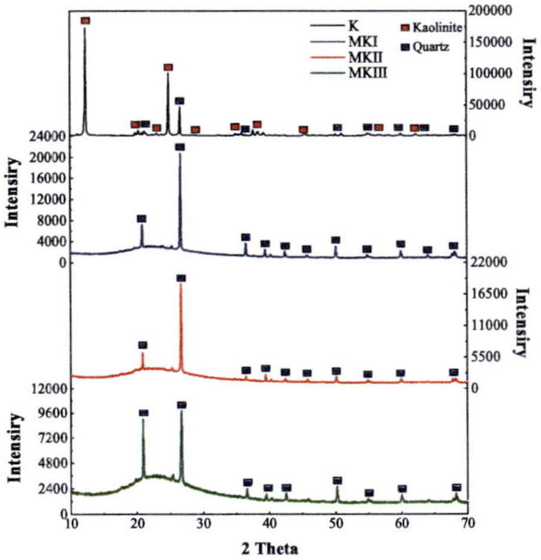

Fig. 3. Diffractogram of kaolin and metakaolin.

Figure 4a shows the TG and DTA curves obtained for the fresh kaolinite. Initially, at temperatures <100 °C occurs the release of water absorbed at low temperature in the pores, surfaces, among others. Between 100–400 °C is related to weight loss, which can be correlated with a pre-dehydration process, which occurs first at the OH of the surface [20].

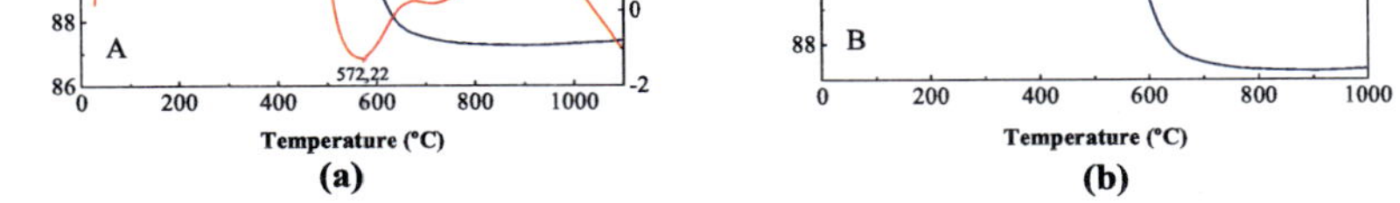

Fig. 4. TG and DTA curves of kaolin (a), TG curves of kaolin in natural and calcined (b).

The DTA curve shows a mass loss at the temperature of 572 °C, the mass loss in this region represents the dehydroxylation of kaolinite into metakaolin, which corresponds to the transition from SiO_2. $2Al_2O_3$. $2H_2O$ to SiO_2. $2Al_2O_3$, which occurs between 500–600 °C [21], the clay loses mass equivalent to 11.3%. In addition, the DTA peak at the temperature of 995 °C is related to the formation of the spinel (Al-Si) phase [22]. In Fig. 4b the TG curves for the calcined samples used are presented, and the efficiency of the calcination of the material can be observed.

3.2 Evaluation of Pozzolanic Reactivity

Figure 5 shows the values found for the electrical conductivity for metakaolin, where it is observed that the values fell sharply in the first 2 min, followed by some oscillations until 30 min, a behavior similar to other studies with calcined clays in [23, 24]. According to [23], this behavior is related to the high consumption of Ca_2+OH ions and the fixation of dissolved $Ca(OH)_2$ by metakaolin particles. For [24], this phenomenon indicates the high reactivity of the calcined clay with $Ca(OH)_2$, therefore being a measurement of the pozzolanic activity. Analyzing the reaction time of 30 min, the reactivity increases for all MK. It is observed that the MK sample showed a greater drop initially, and during the course obtained several variations due to temperature fluctuation during the test. In the end, the highest reactivity recorded was MKIII, which also showed greater amorphous, as shown in XRD.

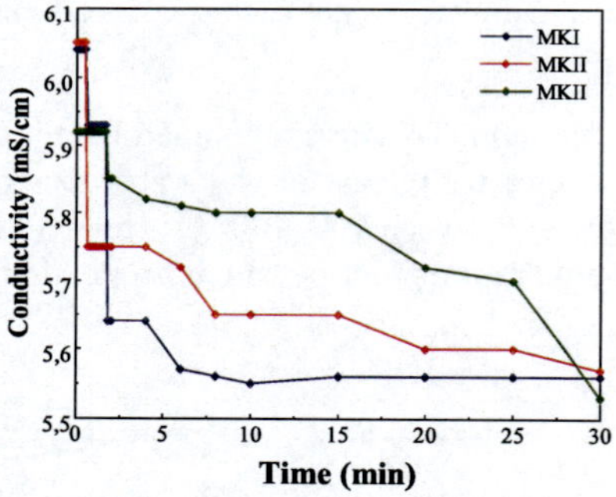

Fig. 5. Electrical conductivity of MK samples as a function of time.

The results of the thermogravimetric analysis performed on the pastes are presented in Fig. 6. The results were obtained through the methodology adopted by Dweck et al. [25]. The identification of the decomposition phases of the cement pastes was determined using the peak temperatures of the derived thermogravimetry (DTG) curves. The first mass loss corresponds to the decomposition of ettringite. The second and third peaks correspond to the decomposition of calcium silicate hydrate (C-S-H) and calcium silicoaluminate hydrate (CASH), respectively. The decomposition of calcium aluminate hydrate (CAH) was observed as the fourth peak, being found in all pastes [26]. The calcium hydroxide of the M0 sample can be observed in the DTG curve between the temperatures of 416.43–482.90 °C. For the M40 paste, $Ca(OH)_2$ can be found between

429.74 and 457.85 °C, and the M50 paste did not present a peak referring to calcium hydroxide. This behavior indicates that the partial replacement of Portland cement by 50% metakaolin ensured the total consumption of Portlandite, allowing a matrix-free of calcium hydroxide to be obtained. Finally, the decomposition of calcium carbonate can be observed.

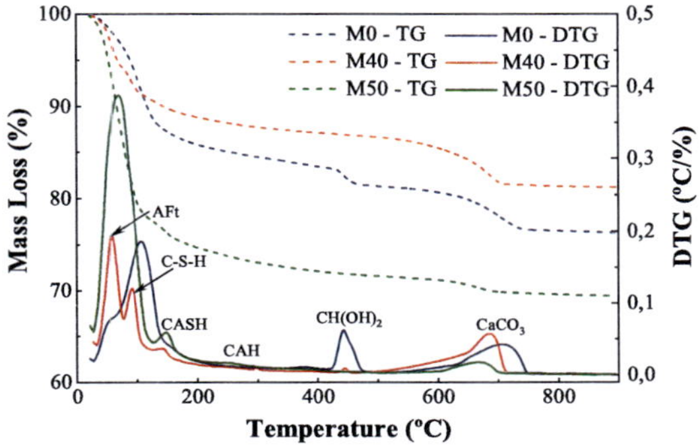

Fig. 6. TG and DTG analysis of the pastes at 28 days.

Table 3 presents the calcium hydroxide content present in the cement pastes. The values found were quite divergent from previous studies [26, 27]. The calcium hydroxide content decreased from 7.73% in the M0 paste and 0.69% in the M40 paste. As can be observed, the substitution completely consumed the Ca(OH)2, proving the reactivity of MKIII from Amazonian kaolin.

Table 3. Calcium hydroxide content of the pastes.

	M0	M40	M50
Calcium hydroxide content (%)	7.40	0.69	0

Figure 7 the data obtained from the compressive strength tests are presented, with the respective standard deviation value, for a curing time of 28 days, performed on the reference matrix (M0) and with metakaolin substitution (M1). It can be observed that the use of metakaolin caused an increase in the average strength value of 14.22% compared to the mortar without substitution. This increase in strength results from the formation of more hydration products due to pozzolanic reactions and the decrease in porosity.

4 Conclusion

Based on the results obtained in this study, the following conclusions can be drawn from the present work:

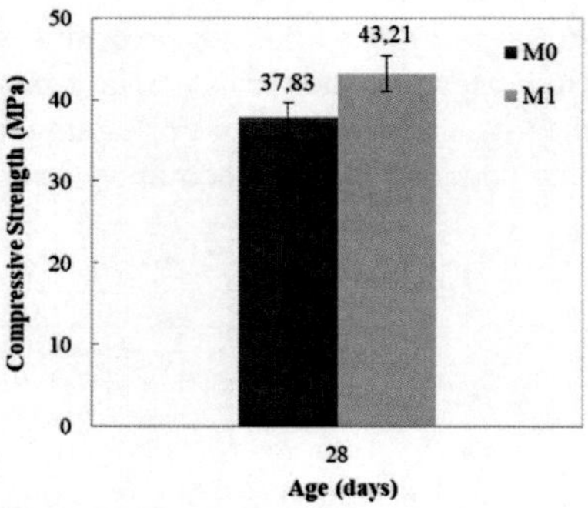

Fig. 7. Average values of compressive strength.

The MKIII produced presented a good potential as a pozzolan. The X-ray diffraction technique pointed out the significant transformation of the crystalline structure, no longer being identified the peaks of kaolinite, transforming it into amorphous material.

The pozzolanic activity, studied by the method of electrical conductivity, can be confirmed by thermogravimetric and axial compression tests, proving the effectiveness of MKIII. In which, the partial replacement of cement by 50% of MK was efficient to obtain a matrix-free of calcium hydroxide at 28 days and helped to increase the mechanical strength of the cement matrix.

References

1. Andrew, R.M.: Global CO_2 emissions from cement prodution. Earth Syst. Sci. Data **10**, 195–217 (2018)
2. Villar-Cociña, E., Rodier, L., Savastano, H., et al.: A comparative study on the pozzolanic activity between bamboo leaves ash and sílica fuma: kinetic parameters. Waste Biomass **11**, 1627–1634 (2020)
3. Juenger, M.C.G., Snellings, R., Bernal, S.A.: Supplementary cementitious materials: new sources, characterization, and performance insights. Cem. Concr. Res. **122**, 257–273 (2019)
4. Asadollahfardi, G., Mohsenzadeh, P., Saghravani, S.F., Mohamadzadeh, N.: The effects of using metakaolin and micro-nanobubble water on concrete properties. J. Build. Eng. **25**, 100781 (2019)
5. El-Diadamony, H., Amer, A.A., Sokkary, T.M., El-Hoseny, S.: Hydration and characteristics of metakaolin pozzolanic cement pastes. HBRC J. **14**(2), 150–158 (2018)
6. Ramli, R.H., Alonge, O.R.: Characterization of metakaolin and study on early age mechanical strength of hybrid cementitious composites. Constr. Build. Mater. **121**, 599–611 (2016)
7. Abdelli, K., Tahlaiti, M., Belarbi, R., Oudjit, M.N.: Influence of the origin of metakaolin on pozzolanic reactivity of mortars. Energy Procedia **139**, 230–235 (2017)
8. Khatib, J.M., Baalbaki, O., Elkordi, A., Metakaolin, A.: In.: Waste and Supplementary Cementitious Materials in Concrete: Characterisation, Properties and Applications, pp. 493–511 (2018)
9. Thankam, G.L., Renganathan, N.T.: Ideal supplementary cementing material—Metakaolin: a review. Int. Rev. Appl. Sci. Eng. **11**, 58–65 (2020)

10. Joshaghani, A., Moeini, M.A., Balapour, M.: Evaluation of incorporating metakaolin to evaluate durability and mechanical properties of concrete. Adv. Concr. Constr. **5**(3), 241–255 (2017)
11. Rakhimova, N.R., Rakhimov, R.Z.: Reaction products, structure and properties of alkali-activated metakaolin cements incorporated with supplementary materials—a review. J. Market. Res. **8**, 1522–1531 (2019)
12. Tironi, A., Trezza, M.A., Scian, A.N., Irassar, E.F.: Thermal analysis to assess pozzolanic activity of calcined kaolinitic clays. J. Therm. Anal. Calorim. **117**(2), 547–556 (2014). https://doi.org/10.1007/s10973-014-3816-1
13. Donatello, S., Tyrer, M., Cheeseman, C.R.: Comparison of test methods to assess pozzolanic activity. Cem. Concr. Compos. **32**, 121–127 (2010)
14. Raask, E., Bhaskar, M.C.: Pozzolanic activity of pulverized fuel ash. Cem. Concr. Res. **5**(4), 363–375 (1975)
15. Luxán, M.P., Madruga, F., Saavedra, J.: Rapid evaluation of pozzolanic activity of natural products by conductivity measurement. Cem. Concr. Res. **5**, 63–68 (1989)
16. Associação Brasileira de Normas Técnicas (ABNT). NBR 7215: Cimento Portland – Determinação da resistência à compressão. Rio de Janeiro (2019) [in Portuguese]
17. Sahmaran, M., Christianto, H.A., Yaman, I.O.: The effect of chemical admixtures and mineral additives on the properties of self-compacting mortars. Cement Concr. Compos. **28**(5), 432–440 (2006)
18. American Society For Testing And Materials. ASTM C618: Standard Specification for Coal Fly Ash and or Calcined Natural Pozzolan for Use in Concrete. USA (2019)
19. Loureiro, A.M.S., Paz, S.P.A., Veiga, M. do R., Angélica, R.S.: Assessment of compatibility between historic mortars and lime-METAKAOLIN restoration mortars made from amazon industrial waste. Appl. Clay Sci. **198**, 105843 (2020)
20. Kakali, G., Perraki, T., Tsivilis, S., Badogiannis, E.: Thermal treatment of kaolin: the effect of mineralogy on the pozzolanic activity. Appl. Clay Sci. **20**(1–2), 73–80 (2001)
21. Rashad, A.M.: Alkali-activated metakaolin: a short guide for civil engineer—an overview. Constr. Build. Mater. **41**, 751–765 (2013)
22. Ptáček, P., Šoukal, F., Opravil, T., Havlica, J., Brandštetr, J.: Crystallization of spinel phase from metakaoline: the nonisothermal thermodilatometric CRH study. Powder Technol. **243**, 40–45 (2013)
23. Roy, B. Chatteriee, A. Aiyer, H.: Effect of mineralogy and temperature treatment on the pozzolanic activity of Indian clays. Mater. Res. Innov. (2018)
24. Tironi, A., Cravero, F., Scian, A.N., Irassara, E.F.: Pozzolanic activity of calcined halloysite-rich kaolinitic clays. Appl. Clay Sci. **147**, 11–18 (2017)
25. Dweck, J., Cherem da Cunha, A.L., Pinto, C.A., Pereira Gonçalves, J., Büchler, P.M.: Thermogravimetry on calcined mass basis—hydrated cement phases and pozzolanic activity quantitative analysis. J. Therm. Anal. Calorim. **97**(1), 85–89 (2009)
26. Cavalcante, D.G., Marques, M.G.S., Melo filho, J.A.: Influence of the levels of replacement of portland cement by metakaolin and silica extracted from rice husk ash in the physical and mechanical characteristics of cement pastes. Cem. Concr. Compos. **94**, 296–306 (2018)
27. Melo Filho, J.A., Silva, F.A., Toledo Filho, R.D.: Degradation kinetics and aging mechanisms on sisal fiber cement composite systems. Cem. Concr. Compos. **40**, 30–39 (2013)

Evaluation of the Mechanical and Chemical Performance of Mortar Calcined Sludge at 600 °C and Nitrophosphogypsum for Its Application in Construction

Nemesio Daza[1](✉), Andrés Guzmán[1], and Yoleimy Avila[2]

[1] Universidad del Norte, km 5 Vía Puerto Colombia, 081007 Barranquilla, Colombia
nemesiod@uninorte.edu.co

[2] Universidade de Lisboa, Lisbon, Portugal

Abstract. In recent years, the reuse of industrial waste and by-products have gained relevance within the construction sector. Thereby generating an alternative solution for reducing CO_2 emissions originated from Portland cement manufacturing and incorporation processes, finding the optimal dosage that does not affect the mortar properties. This research investigates the effects on fresh and hardened properties in mortar mixtures incorporating two residues as supplementary cementitious materials: nitrophosphogypsum in conjunction with drinking water treatment plant sludge calcined at 600 °C; characterization of sludge and NPG was carried out. Fresh and hardened properties of seven (7) types of mortar mixtures were evaluated (including a control mixture). Results obtained show the incorporation of these agro-industrial by-products, in conjunction, is recommended as supplementary cementitious material for the preparation of mortar mixtures in dosages from 5 to 10 wt%.

Keywords: Sludge ash · Nitrophosphogypsum · Supplementary cementitious material · Mortars

1 Introduction

In recent years, several researchers have focused their efforts on implementing supplementary cementitious materials (SCM) being capable of a partial or total replacement for construction materials (i.e. mortars, concrete) [1]. The interest in replacing cement within the construction sector lies in the high CO_2 emissions produced at the time of its manufacture, thus generating high pollution levels in the environment. On average, ordinary Portland cement (OPC) production generates 7% of total CO_2 emissions; for the year 2050, an increase in consumption of 100% of this material is expected [2].

Drinking water treatment plant sludge (WTPS) is a waste from the water purification process. This material has been extensively studied for its implementation in the construction sector [3–8]. This waste is disposed of in bodies of water which generate a negative impact on the environment due to its physical and chemical composition

J. I. Escalante-Garcia et al. (Eds.): RW 2021, RILEM Bookseries 40, pp. 118–130, 2023.
https://doi.org/10.1007/978-3-031-21735-7_15

(high content of sand, clay, among others) [9–12]. Worldwide, approximately 19.7 × 10^9 m^3/year of WTPS are produced [13, 14]. The correct deposit of this by-product is a common struggle associated with its cost [15].

Secondly, a fertilizer manufacturing industry generates an average of 62 000 tons/year of nitrophosphogypsum (NPG) in Barranquilla, Colombia. This material is originated from the nitrophosphate route, where it starts from phosphoric rock ($Ca_{10}F_2P_6O_{24}$, fluorapatite) as a source of phosphorus and nitric acid (HNO_3) as a source of nitrogen. It is estimated that for every ton of phosphate rock, 1.5 tons of NPG are generated [16]. This material contains impurities and toxic elements, which, for the most part, are deposited in quarries for their natural degradation. At present, the NPG does not have the chemical and mechanical characterization to evaluate its performance within the construction sector. However, this by-product has properties very similar to phosphogypsum (PG); the latter is originated from the interaction between phosphoric rock and phosphoric acid (H_3PO_4). PG has been widely studied for its incorporation into the construction sector (i.e. mortar and concrete) [17–20], generating an alternative solution to reduce its negative impact on the environment.

The purpose of this research is to characterize the mechanical performance of mixture mortars made from the use of WTPS jointly with NPG as a partial replacement for OPC by 10%. The following tests have been performed and discussed: characterization of sand [21–23], microstructure-chemical analysis of samples (XRF, SEM and pH) [24], consistency [25], setting time [26], workability [27], compressive strength [28], absorption of water by capillarity [29] and mortar density in a fresh and hardened state [30, 31].

2 Materials and Experimental Methods

The current section describes the materials and standards followed in the experimental design for the research.

2.1 Mixture Constituents

OPC, WTPS and NPG were used as cemented materials in this study. The local water treatment company donated 15 L of sludges; the material was collected before its disposal in the Magdalena River. A local fertilizer company donated 50 kg of NPG.

Just one type of natural sand was used and characterized according to the ASTM C136 [23] and ASTM C128 [22], complying with ASTM C778 [21]. Tap water was used, complying with the Colombian regulations for physical, chemical and microbiological characteristics for drinking water [32]; a temperature of 23 ± 1.7 °C was guaranteed according to ASTM C511 [33].

2.2 Sample Preparation and Test Methods

The total sample of sludges from WTP (23.75 kg of sludges with a moisture content of 63.84%) was first sun-dried for 48 h with an average temperature of 35 °C (Fig. 1a), getting a weight of 8.90 kg, achieving a mass yield of 36.16% on a dry base (Fig. 1b).

Then the resulting sample was ground in a Mortar mill model PM200, calcined later at a temperature of 600 °C for 2 h (SC600) at a heating rate of 10 °C per minute. This temperature is sufficient to remove the organic matter from the dry sludge and improve its interaction [5]. After the calcination process was completed, the mass obtained was 2.18 kg with a yield of 89.47%. Finally, the resulting sample was ground again in Mortar mill, and the total mass was passed through sieve No. 200 not to lose properties by discarded material (Fig. 1c).

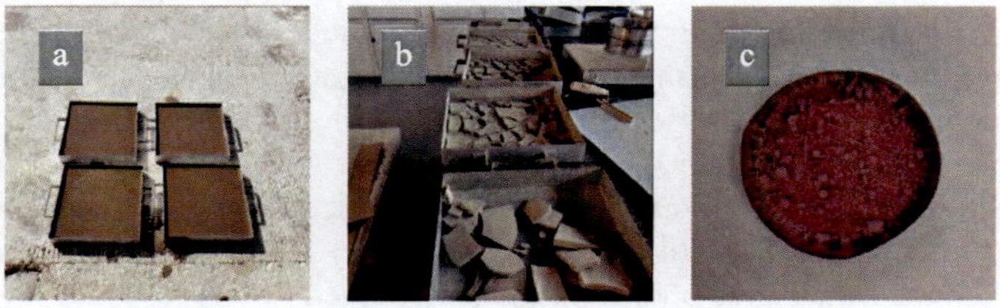

Fig. 1. Sludge from the drinking water treatment plant. (a) Wet sample, (b) Sun-dried sample, (c) Sample calcined at 600 °C and sieved.

Two (2) kg of NPG samples were dried at 105 °C for 24 h (NPG105), getting a mass yield of 81.84%. The resulting sample of this treatment was ground and passed through sieve No. 16 (NPG-1). Simultaneously, another 2 kg of NPG samples (Fig. 2a) were calcined at 600 °C (Fig. 2b) for 2 h (NPG600), the mass yield was 72.26%. Then the resulting sample was ground in Mortar mill, and the total mass was divided into three equal parts to be passed through the sieves No. 16, 70 and 200, respectively (NPG-2, NPG-3 and NPG-4) (Fig. 2c). The heat treatment of NPG converts it into β-hemihydrate and, in turn, reduces its impurities, mainly caused by the nature of phosphate rock [34, 35].

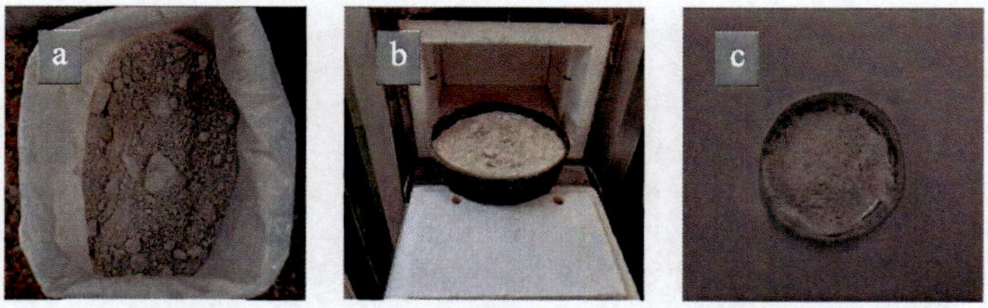

Fig. 2. NPG. (a) Wet sample, (b) Calcined sample at 600 °C, (c) Sample calcined at 600 °C and sieved.

After the treatments for each material, the chemical composition of samples was determined with X-ray fluorescence (XRF) analysis on an Epsilon 3Lp Panalytical device. The pH was determined following the standard ASTM D4972 [24], allowing quantifying the degree of alkalinity or acidity of solids suspended in water. Scanning electron microscopy (SEM) was tested on a JEOL JSM-5600 device, allowing morphological visualization. The density of the materials was determined following the standard ASTM C188 [36].

2.3 Mix Proportions

Mortar mixtures design was carried out based on the following concepts: treatment temperature and maximum particle size. NPG samples were treated at two different temperatures: 105 °C at 24 h and 600 °C at 2 h. Three particle sizes were controlled using sieves No. 16, No. 70 and No. 200 (maximum size of 1.18 mm, 0.212 mm and 0.75 mm, respectively). Moreover, WTPS were calcined at 600 °C with particle size not greater than 0.75 mm; a single treatment for sludge was considered [14, 37].

Two experimental campaigns were developed to study the mechanical behavior of mixtures made only with NPG and NPG together with SC600; the replacement percentages were indicated in weight by weight (%wtc, percentage weight to cement). During the first experimental campaign, NPG was used as a replacement of 10% of the cement weight by NPG products. The combinations obtained from the first experimental design were the following: M0 (control), M1 (mix with NPG-1), M2 (mix with NPG-2), M3 (mix with NPG-3) and M4 (mix with NPG-4). Additionally, for the second experimental campaign, these mixes were elaborated: M0 (control), M5 (SC600 10%wtc replacement), and M6 (SC600 and NPG-4 5%wtc for each one, 10%wtc in total).

The preparation of mortar mixes followed the standard ASTM C109 [28], water-cement (w/c) ratio of 0.485 and a setting of 0.78% by absorption aggregate. Mix proportions of the mortar samples are listed in Table 1. Curing conditions of samples were at 23 ± 1.7 °C and 63 ± 2% RH.

Table 1. Mortar material proportions (kg/m^3) w/c ratio of 0.485

Mixture type	Sand	Cement	NPG	Sludge
M0	1833.28	666.65	0.00	0.00
M1	1833.28	599.98	66.66	0.00
M2	1833.28	599.98	66.66	0.00
M3	1833.28	599.98	66.66	0.00
M4	1833.28	599.98	66.66	0.00
M5	1833.28	599.98	0.00	66.66
M6	1833.28	599.98	33.33	33.33

2.4 Performance of Cement Pastes and Fresh State Mortars

The setting time and water requirement of normal consistency of cement pastes were evaluated following the standard ASTM C191 [26] and ASTM C305 [25], respectively.

Table 2 presents the mix proportions for paste samples. The water requirement for M0 was adjusted until normal consistency as defined standard ASTM C187 [38]. The other mixture types were elaborated for the same w/c ratio to analyze the change of consistency.

The bulk density in the fresh state was calculated according to BS EN 1015-6:1999 [31]. The workability of fresh mortars was carried out following the standard ASTM 1437 [27].

Table 2. Cement paste material proportions (g)

Mixture type	Cement	NPG	Sludge	Water
M0	650.00	0.00	0.00	169.00
M1	585.00	65.00	0.00	169.00
M2	585.00	65.00	0.00	169.00
M3	585.00	65.00	0.00	169.00
M4	585.00	65.00	0.00	169.00
M5	585.00	0.00	65.00	169.00

2.5 Performance of Hardened Mortar

The compressive strength of mortars (fm) was measured according to standard ASTM C109 [27] at 7 days, 14 days and 28 days; each measuring age got 3 replicas. The rate of water absorption of mortars was performed following ASTM C1403 [29], being affected by the type of aggregate, cementitious materials and water-cement ratio. The bulk density of hardened mortar was calculated following the BS EN 1015-10:1999 [30]. The pH of mortar mixtures was determined following the ASTM D4972 [24]; it is worth mentioning that this regulation does not present a defined methodology for mortars. It was adapted to approximate the impact of incorporating NPG and SC600 in the degree of alkalinity of samples compared to the control test.

3 Analysis and Discussion of Results

This section presents the results and further analysis of each variable considered.

3.1 Characterization of Sand

In the present research, one type of natural sand was used of national origin, complying with ASTM C778 [21], with a specific gravity of 2.61, fineness modulus of 1.98 and water absorption of 0.78%; no chemical admixtures were used. The particle size distribution is shown in Fig. 3.

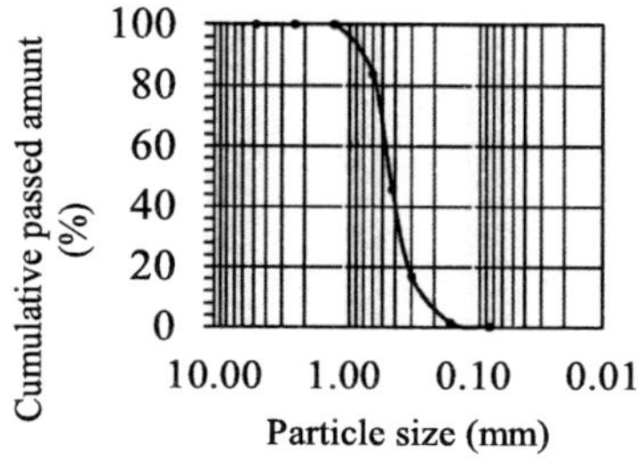

Fig. 3. Granulometry curve for fine aggregate

3.2 Characterization of NPG and SC600

Table 3 gives the chemical composition of the cement materials. XRF analysis does not yield logical results NPG; thus, it is not presented herein. For SC600, three major oxides were identified: silicon, aluminum, and iron.

Table 3. Chemical composition of SCM (%)

Oxide	Na_2O	MgO	Al_2O_3	SiO_2	P_2O_5	SO_3	K_2O	CaO	TiO_2	MnO	Fe_2O_3	SrO
OPC	0.12	0.98	0.85	28.24	0.42	3.7	0.93	59.19	0.39	1.09	3.96	0.1
SC600	0.72	1.44	21.36	61.59	0.25	0.18	1.91	1.39	0.82	0.22	10.09	0.03

Results of pH for the cementitious materials (Fig. 4) shows acidity for the NPG-1 sample, probably due to the presence of residual acids and impurities that are not removed at the temperature considered. Similar results were obtained by Ennaciri et al. [35]. NPG samples calcined at 600 °C (NPG-2, NPG-3 and NPG-4) presented at least two units higher on the pH scale than NPG dried at 105 °C; the calcination process decreased the degree of acidity (slightly acidic). Furthermore, SC600 can be classified as slightly acidic on the pH scale.

SEM images obtained from NPG samples (Fig. 5) confirmed that the particle size control improves the homogeneity of the structure. NPG-4 presented greater homogeneity in its particle distribution and fewer pores. Geraldo et al. [34] and da Silva et al. [17] reported similar results. NPG samples calcined at 600 °C showed lower impurities in their microstructure [39]; in general, NPG samples presented a slatted structure. SC600 particles were round and porous, evidencing a homogeneous morphological structure. Husillos et al. [8] and Liu et al. [4] reported similar results.

3.3 Cement Pastes and Fresh Mortar

The results of consistency and settling time for pastes in campaigns 1 and 2 are shown in Tables 4 and 5, respectively, together with Fig. 6. The mixtures made with the incorporation of NPG presented a reduction in the initial consistency of the paste. M4 presented

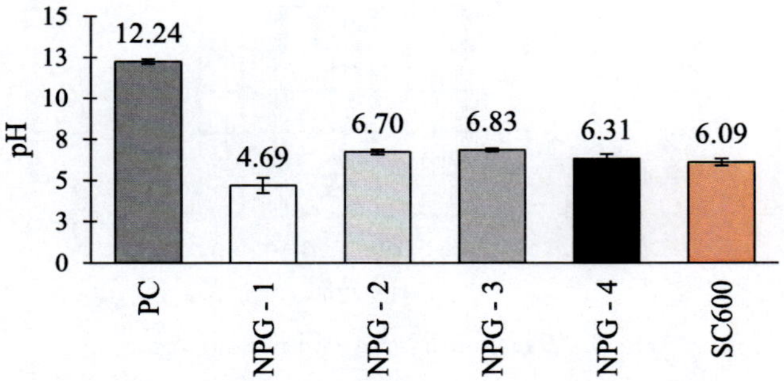

Fig. 4. Results of pH measurements of samples. Average temperature of 24.38 ± 1.04 °C

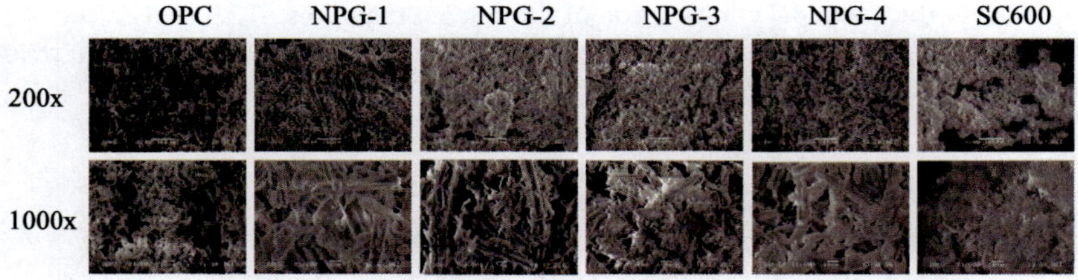

Fig. 5. SEM images of samples

a better consistency contrasted with M2 and M3 because of the smaller particle size of implemented NPG (NPG-4) related to a small absorption area of the sample. Absorption allows the formation of cement hydration products first, thus avoiding the formation of NPG products (β-hemihydrate and dihemihydrate). Similar results were reported by Liu et al. [39]. The heat treatment on the NPG transforms its gypsum phase to β-hemihydrate and anhydrate, which improves its hydration properties. B-Hemihydrate forms at temperatures greater than 100 °C, while anhydrate forms at temperatures greater than 500 °C.

Furthermore, incorporating SC600 with NPG improved the final settling time compared to only incorporating NPG-4 in the mixtures (results of M4).

According to these results, M4 has a longer final setting time than M0, which allows greater durability in the plastic state of the mix (Fig. 6a). On the other hand, M6 began to set faster; however, it had a longer time in the plastic state (Fig. 6b).

The workability of mortars mixtures for experimental campaigns 1 and 2 are showed in Tables 7 and 8, respectively. The particle size distribution of NPG was essential to the workability of M4. The incorporation of SC600 significantly reduces the workability of mortar mixes because of the high aluminosilicate content leading to the early formation of C-S-H and C-A-S-H. These properties accelerate the setting time and the workability of the mix [1]. The bulk density of fresh mortar did not present a relevant behavior associated with the other properties evaluated (Table 6).

Table 4. Results of consistency and setting time of pastes—experimental campaign 1

Mixture type	Consistency[a] (mm)	Setting time/min[b]	
		Initial	Final
M0	9	119	150
M1	3	30	135
M2	4	114	165
M3	4	100	165
M4	6	111	195

[a] ASTM C305 [25]
[b] ASTM C191 [26]

Table 5. Results of consistency and setting time of pastes—experimental campaign 2

Mixture Type	Consistency[a] (mm)	Setting time/min[b]	
		Initial	Final
M0	10	119	150
M5	5	52	120
M6	5	93	165

[a] ASTM C305 [25]
[b] ASTM C191[26]

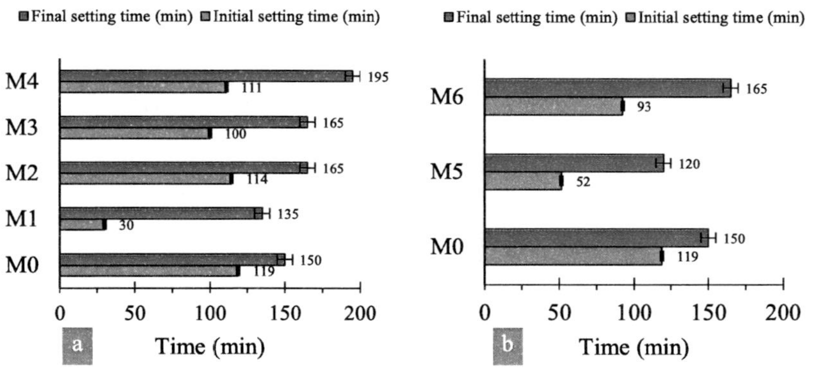

Fig. 6. Results of setting time per C191 [26] (a) experimental campaign 1 (b) experimental campaign 2.

3.4 Performance of Hardened Mortar

Figure 7 shows the compressive strength for mortars at 7 days, 14 days and 28 days for experimental campaigns 1 and 2. Overall, mixtures made with NPG showed a loss

Table 6. Results of workability and bulk density of fresh mortar experimental campaign 1

Mixture type	Workability[a] (mm)	ρ fresh mortar[b] (kg/m^3)
M0	135.60 ± 2.22	2134.79 ± 4.36
M1	110.70 ± 0.24	2125.43 ± 5.83
M2	118.15 ± 2.21	2143.64 ± 7.71
M3	121.05 ± 2.01	2120.52 ± 10.39
M4	127.37 ± 1.25	2143.35 ± 12.86

[a] ASTM 1437 [27]
[b] BS EN 1015–6:1999 [31]

Table 7. Results of workability and bulk density of fresh mortar experimental campaign 2.

Mixture type	Workability (mm)	ρ fresh mortar (kg/m^3)
M0	131.90 ± 1.28	2125.39 ± 7.87
M5	117.51 ± 1.14	2141.89 ± 0.40
M6	115.48 ± 1.41	2124.47 ± 6.07

[a] ASTM 1437 [27]
[b] BS EN 1015–6:1999 [31]

of compressive strength between 52.52 to 64.34% with only 10% wtc. This behavior reveals that NPG does not favor compression strength because this material has a higher water absorption than OPC, thus avoiding the formation of cement hydration products [18]. On the other hand, for experimental campaign 2, M4 obtained a loss in compressive strength of 7.92%, while M5 got a loss of 15.15% at 28 days.

Results of water absorption by capillary uptake, pH, bulk density of hardened mortars, and compressive strength (MPa) at 28 d of mixes for experimental campaigns 1 and 2 are shown in Tables 8 and 9. The use of NPG in mortar mixes increases the water absorption capacity. SC600 improves the absorption properties of mortar mixtures because of the low Ca/Si ratio, which allows a rapid formation of C-S-H and C-A-S-H, improving the mechanical properties and, thus, durability [1]. The bulk density of hardened mortar did not present a relevant behavior associated with the other properties evaluated. Incorporating SC600 together with NPG-4 in mortar mixtures (M4) improves the properties of compressive strength (63.33%) and water absorption (7.35%) compared to only NPG as SMC.

4 Conclusions

From the results obtained, it can be concluded that NPG is not recommended solely as SCM. Incorporating SC600 with NPG in dosages of 5% each (10% wtc in total) is

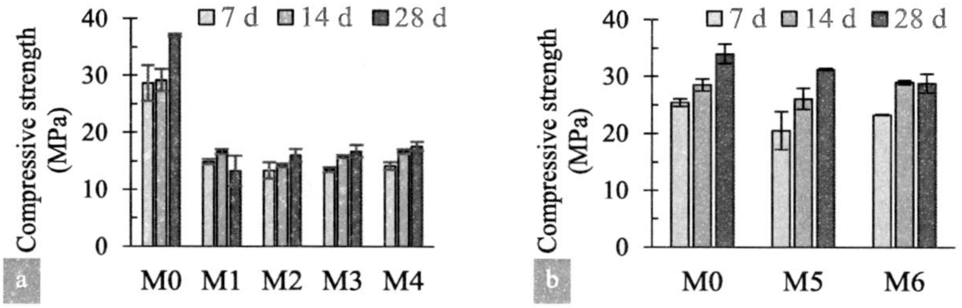

Fig. 7. Results of compressive strength per ASTM C109 [27] (a) experimental campaign 1 (b) experimental campaign 2.

Table 8. Results of water absorption by capillary uptake, pH, bulk density of hardened and compressive strength at 28 d of mortars—experimental campaign 1

Mixture type	As[a] (kg/m^2)	pH[1,b]	ρ hardened mortar[c] (kg/m^3)	fm (MPa)
M0	9.53 ± 0.07	12.01 ± 0.02	2018.05 ± 4.17	37.13 ± 0.22
M1	11.69 ± 0.44	11.40 ± 0.16	1982.36 ± 9.50	13.25 ± 2.71
M2	11.53 ± 0.05	11.73 ± 0.09	1962.81 ± 1.94	16.00 ± 1.11
M3	10.17 ± 0.28	11.81 ± 0.04	2020.61 ± 17.12	16.75 ± 1.07
M4	10.18 ± 0.13	11.74 ± 0.09	1982.93 ± 23.54	17.67 ± 0.76

[1] Average temperature of 23.14 ± 0.40 °C
[a] ASTM C1403 [29]
[b] ASTM D4972 [24]
[c] BS EN 1015-10:1999 [30]

Table 9. Results of water absorption by capillary uptake, pH, bulk density of hardened and compressive strength at 28 d of mortars—experimental campaign 1

Mixture type	As[a] (kg/m^2)	pH	ρ hardened mortar[c] (kg/m^3)	fm (MPa)
M0	9.12 ± 0.02	12.00 ± 0.04	2009.64 ± 3.12	34.02 ± 1.70
M5	8.92 ± 0.07	11.81 ± 0.02	2009.76 ± 8.06	31.32 ± 0.17
M6	9.43± 0.07	11.94 ± 0.07	1995.44 ± 9.73	28.86 ± 1.63

[1] Average temperature of 23.17 ± 0.33 °C
[a] ASTM C1403 [29]
[b] ASTM D4972 [24]
[c] BS EN 1015-10:1999 [30]

recommended. The sludge produced in Barranquilla, Colombia, can be classified as an SCM. The compressive strength loss can be adjusted with an increase in the w/c ratio, which in comparison with the environmental benefit, can be an efficient alternative for the approximate reduction of 220 tons of CO_2 emissions to the environment per year due to manufacturing of the OPC.

References

1. Lothenbach, B., Scrivener, K., Hooton, R.D.: Supplementary cementitious materials. Cem. Concr. Res. **41**(12), 1244–1256 (2011). https://doi.org/10.1016/j.cemconres.2010.12.001
2. Pinto, S.R., Angulski da Luz, C., Munhoz, G.S., Medeiros-Junior, R.A.: Resistance of phosphogypsum-based supersulfated cement to carbonation and chloride ingress. Constr. Build. Mater. **263**, 120640 (2020). https://doi.org/10.1016/j.conbuildmat.2020.120640
3. Chang, Z., Long, G., Zhou, J.L., Ma, C.: Valorization of sewage sludge in the fabrication of construction and building materials: a review. Resour. Conserv. Recycl. **154**, 104606 (2020). https://doi.org/10.1016/j.resconrec.2019.104606
4. Liu, Y., et al.: Utilization of drinking water treatment sludge in concrete paving blocks: microstructural analysis, durability and leaching properties. J. Environ. Manag. **262**, 110352 (2020). https://doi.org/10.1016/j.jenvman.2020.110352
5. de Oliveira Andrade, J.J., Wenzel, M.C., da Rocha, G.H., da Silva, S.R.: Performance of rendering mortars containing sludge from water treatment plants as fine recycled aggregate. J. Clean. Prod. **192**, 159–168 (2018). https://doi.org/10.1016/j.jclepro.2018.04.246
6. Vouk, D., Nakic, D., Stirmer, N., Cheeseman, C.R.: Use of sewage of sludge ash in cementitious materials. Rev. Adv. Mater. Sci. **49**(2), 158–170 (2017) [En línea]. http://www.ipme.ru/e-journals/RAMS/no_24917/05_24917_nakic.pdf
7. Kappel, A., Ottosen, L.M., Kirkelund, G.M.: Colour, compressive strength and workability of mortars with an iron rich sewage sludge ash. Constr. Build. Mater. **157**, 1199–1205 (2017). https://doi.org/10.1016/j.conbuildmat.2017.09.157
8. Rodríguez, N.H., et al.: Re-use of drinking water treatment plant (DWTP) sludge: characterization and technological behaviour of cement mortars with atomized sludge additions. Cem. Concr. Res. **40**(5), 778–786 (2010). https://doi.org/10.1016/j.cemconres.2009.11.012
9. Rocha, B., Zavala, F., Peralta, M.: Revisión global de los contaminantes emergentes PBDE y el caso particular de México, p. 10 (2015)
10. Perilla, J.: Estrategias para el reúso de los lodos en la zona dos de la empresa de acueducto y alcantarillado de Bogotá y plantas de tratamiento de aguas potables y residuales. Universidad Militar Nueva Granada, Bogotá, Colombia (2015)
11. Sandoval, L., Montellano, L., Martín, A., Sánchez, L., Santana, M., Morán, M.: Tratabilidad de los lodos producidos en la potabilización del agua, p. 10 (2005)
12. Gallo, J.A., Uribe, J.C.: Reutilización de lodos de planta de potabilización en el tratamiento de aguas residuales. Universidad Nacional de Colombia, Manizales, Colombia (2003)
13. Świerczek, L., Cieślik, B.M., Konieczka, P.: The potential of raw sewage sludge in construction industry—a review. J. Clean. Prod. **200**, 342–356 (2018). https://doi.org/10.1016/j.jclepro.2018.07.188
14. Ávila, Y., Restrepo, S., Jiménez, J., Castillo, M., Parody, A.: Análisis comparativo de la concentración de óxidos presentes en el cemento Portland y lodos de plantas de tratamiento de agua potable. Afinidad **74**(577) (2016) [En línea]. https://www.raco.cat/index.php/afinidad/article/view/320832. Accessed 27 June 2018
15. Ahmad, T., Ahmad, K., Alam, M.: Sustainable management of water treatment sludge through 3'R' concept. J. Clean. Prod. **124**, 1–13 (2016). https://doi.org/10.1016/j.jclepro.2016.02.073
16. Palma, A., Sierra Serje, K.: Beneficio de nitrofosfoyeso proveniente de la fabricación de fertilizante compuestos para su uso en la industria de la construcción. Thesis (2011)

17. da Silva Calderón, B., García, A., Pineda, P., García, R.: Valorization of phosphogypsum in cement-based materials: limits and potential in eco-efficient construction. J. Build. Eng. 102506 (2021). https://doi.org/10.1016/j.jobe.2021.102506
18. Pang, M., et al.: Influence of phosphorus slag on physical and mechanical properties of cement mortars. Materials **13**(10), Art. n.º 10 (2020). https://doi.org/10.3390/ma13102390
19. Flores, V., Rodríguez, M., Romero, I., Esquivias, L.: Caracterización de morteros mixtos de cal obtenida del reciclado de fosfoyeso. Bol. Soc. Esp. Cerámica Vidr. **59**(3), 129–136 (2020). https://doi.org/10.1016/j.bsecv.2019.09.002
20. Rashad, A.M.: Phosphogypsum as a construction material. J. Clean. Prod. **166**, 732–743 (2017). https://doi.org/10.1016/j.jclepro.2017.08.049
21. ASTM International: ASTM C778-17: Standard Chemical Standard Specification for Standard Sand. ASTM International, West Conshohocken, PA (2017) [En línea]. https://www.astm.org/Standards/C778.htm
22. ASTM International: ASTM C128-15: Standard Test Method for Relative Density (Specific Gravity) and Absorption of Fine Aggregate. ASTM International, West Conshohocken, PA (2015). https://doi.org/10.1520/C0128-15
23. ASTM International: ASTM C136/C136M-14: Standard Test Method for Sieve Analysis of Fine and Coarse Aggregates. ASTM International, West Conshohocken, PA (2014). https://doi.org/10.1520/C0136_C0136M-14
24. ASTM International: ASTM D4972–19: Standard Test Methods for pH of Soils. ASTM International (2019). https://doi.org/10.1520/D4972-19
25. ASTM International: ASTM C305-20: Standard Practice for Mechanical Mixing of Hydraulic Cement Pastes and Mortars of Plastic Consistency. ASTM International, West Conshohocken, PA (2020) [En línea]. https://doi.org/10.1520/C0305-20
26. ASTM International: ASTM C191-19: Standard Test Methods for Time of Setting of Hydraulic Cement by Vicat Needle. ASTM International, West Conshohocken, PA, USA (2019). https://doi.org/10.1520/C1611_C1611M-18
27. ASTM International: ASTM C1437-20: Standard Test Method for Flow of Hydraulic Cement Mortar. ASTM International, West Conshohocken, PA (2020) [En línea]. https://www.astm.org/Standards/C1437. Accessed 05 May 2020
28. ASTM International: ASTM C109/C109M-20b: Standard Test Method for Compressive Strength of Hydraulic Cement Mortars (Using 2-in. or [50-mm] Cube Specimens). ASTM International, West Conshohocken, PA (2020). https://doi.org/10.1520/C0109_C0109M-16A
29. ASTM International: ASTM C1403-15: Standard Test Method for Rate of Water Absorption of Masonry Mortars. ASTM International, West Conshohocken, PA (2015) [En línea]. https://compass.astm.org/EDIT/html_annot.cgi?C1403+15
30. BSI: BS EN 1015-10:1999: Methods of Test for Mortar for Masonry. Determination of Dry Bulk Density of Hardened Mortar. British Standard Institution, London (1999) [En línea]. https://doi.org/10.3403/01905430. Accessed 05 May 2020
31. BSI: BS EN 1015-6:1999: Methods of Test for Mortar for Masonry. Part 6: Determination of Bulk Density of Fresh Mortar. British Standard Institution, London (1999) [En línea]. https://shop.bsigroup.com/ProductDetail?pid=000000000030151085. Accessed 05 May 2020
32. Ministerio de la protección social y Ministerio de ambiente, vivienda y desarrollo territorial: Resolucion 2115 de 2007 Calidad del agua potable.pdf, p. 23 (2007) [En línea]. http://www.confia.com.co/normatividad/Resolucion%202115%20de%202007%20Calidad%20del%20agua%20potable.pdf. Accessed 29 June 2018
33. ASTM International: ASTM C511-19: Standard Specification for Mixing Rooms, Moist Cabinets, Moist Rooms, and Water Storage Tanks Used in the Testing of Hydraulic Cements and Concretes. ASTM International, West Conshohocken, PA (2019) [En línea]. https://www.astm.org/Standards/C511. Accessed 05 May 2020

34. Geraldo, R.H., et al.: Calcination parameters on phosphogypsum waste recycling. Constr. Build. Mater. **256**, 119406 (2020). https://doi.org/10.1016/j.conbuildmat.2020.119406
35. Ennaciri, Y., Zdah, I., Alaoui-Belghiti, H.E., Bettach, M.: Characterization and purification of waste phosphogypsum to make it suitable for use in the plaster and the cement industry. Chem. Eng. Commun. **207**(3), 382–392 (2020). https://doi.org/10.1080/00986445.2019.1599865
36. ASTM International: ASTM C188-17: Standard Test Method for Density of Hydraulic Cement. ASTM International, West Conshohocken, PA (2017). https://doi.org/10.1520/C0188-17
37. Bohórquez González, K., Pacheco, E., Guzmán, A., Avila Pereira, Y., Cano Cuadro, H., Valencia, J.A.F.: Use of sludge ash from drinking water treatment plant in hydraulic mortars. Mater. Today Commun. **23**, 100930 (2020). https://doi.org/10.1016/j.mtcomm.2020.100930
38. ASTM International: ASTM C187-16: Standard Test Method for Amount of Water Required for Normal Consistency of Hydraulic Cement Paste. ASTM International, West Conshohocken, PA (2016) [En línea]. https://www.astm.org/Standards/C187.htm. Accessed 05 May 2020
39. Liu, S., Ouyang, J., Ren, J.: Mechanism of calcination modification of phosphogypsum and its effect on the hydration properties of phosphogypsum-based supersulfated cement. Constr. Build. Mater. **243**, 118226 (2020). https://doi.org/10.1016/j.conbuildmat.2020.118226

Supplementary Cementitious Materials Reactivity: From Model Systems to Concrete

Prannoy Suraneni(✉)

University of Miami, Coral Gables, FL 33135, USA
suranenip@miami.edu

Abstract. Newly developed reactivity tests for supplementary cementitious materials have shown the ability to quantify reactivity and to differentiate reactive and inert materials. Recent results using the modified R^3 test are summarized here. The test uses a model system of supplementary cementitious materials mixed with calcium hydroxide in a simulated pore solution at 50 °C. Heat release and calcium hydroxide consumption of the supplementary cementitious materials are measured. The reactivity of a large number of conventional and alternative materials has been measured using this test. By using both heat release and calcium hydroxide consumption, two measures of reactivity are obtained, and the test is able to differentiate pozzolanic, latent hydraulic, and inert materials. While the measured reactivity correlates with the properties of cementitious pastes, other factors including the supplementary cementitious material replacement level and the age of testing also influence these predictions. Thoughts on the use of reactivity testing to predict concrete durability are presented.

Keywords: Supplementary cementitious materials · Reactivity · Durability

1 Introduction to Reactivity Tests and the Modified R^3 Test

Supplementary cementitious materials (SCMs) are critical for the production of sustainable and durable concrete. Shortfalls in the supply of conventional SCMs, such as fly ash, have necessitated development of alternative SCMs for use in concrete. Detailed concrete durability testing is needed before alternative SCMs can be widely used in concrete. However, before such complex and time-consuming testing is carried out, it is important to quantify the reactivity of the SCM. If the SCM is inert, it is unlikely to improve concrete durability. Standard tests, such as the strength activity test, which is used in fly ash and natural pozzolan specifications are severely flawed and cannot accurately screen inert materials from reactive ones [1]. Therefore, if alternative SCMs are to be widely used in concrete, development of accurate SCM reactivity tests is critical.

The last decade has seen important advances in the development of SCM reactivity tests; a detailed discussion is presented in [2]. The R^3 suite of tests involve the measurement of heat release and/or bound water in a SCM-calcium hydroxide model system in a simulated pore solution at 40 °C [3]. The heat release is obtained using an isothermal calorimeter and the bound water is obtained using an oven and/or furnace. Heat release

J. I. Escalante-Garcia et al. (Eds.): RW 2021, RILEM Bookseries 40, pp. 131–140, 2023.
https://doi.org/10.1007/978-3-031-21735-7_16

and bound water measured in these tests show excellent correlation with strength of mortars containing 30% SCM replacement. The R^3 tests using heat release and bound water were recently standardized as ASTM C1897. While measuring the heat release and/or bound water is an important measure of reactivity, it can be argued that is an overall measure of reactivity. Or, in other words, a heat release measurement is unable to differentiate between pozzolanic and latent hydraulic materials. Indeed, one could obtain very similar values of heat release from a slag and a calcined clay. However, at the same replacement level, these two materials would have very different alkali-silica reaction mitigation abilities in concrete. Since pozzolanic reactivity is important for concrete durability, measurement of heat release and calcium hydroxide consumption in model systems should be considered. These dual measurements are the principle behind the modified R^3 test [4]. In this test, the heat release and calcium hydroxide consumption of SCM-calcium hydroxide model systems in a simulated pore solution at 40 °C are measured using an isothermal calorimeter and thermogravimetric analysis, respectively. The temperature is higher in the modified R^3 test, and, unlike the R^3 tests, it does not use sulfates and carbonates added to the pore solution. The R^3 test is typically carried out for 7 days while the modified R^3 test is typically carried out for 10 days.

2 Modified R^3 Test—Results

Over 100 SCMs have been tested using the modified R^3 test. Figure 1 shows representative results for several conventional and alternative SCMs. While the boundaries are subjective and need to be systematically validated, the test is able to quantify reactivity (reactivity increases as the points move further away from origin) and differentiate pozzolanic and latent hydraulic materials (latent hydraulic materials have a higher slope, or heat release/calcium hydroxide consumption ratio than pozzolanic materials).

By drawing horizontal or vertical lines on Fig. 1, it is apparent that very different materials could have similar heat release *or* calcium hydroxide consumption values, but it is unlikely that they would have similar heat release *and* calcium hydroxide consumption values. Figure 2 shows the effects of the (bulk) chemical composition on the modified R^3 test results. High silica materials, aluminosilicate materials, and high calcium materials all show different slopes of the calcium hydroxide consumption-heat release curve. Based on this information, it could be argued that—once the chemical composition is known—the heat release can be predicted from the calcium hydroxide consumption, and vice versa. Therefore, the measurement could be simplified so that only one of the two reactivity parameters needs to be measured. However, there is obvious scatter in Fig. 2 which means there would be some error in the predictions. More importantly, the bulk composition includes (typically) unreactive crystalline phases, so the relationships in Fig. 2 are approximate at best. If only one measurement is to be made, it is of interest to consider whether this would be heat release or calcium hydroxide consumption. Both use complex and expensive laboratory equipment, so the choice could ultimately be based on equipment availability. However, the bound water correlates with the heat release [3], meaning that that an oven/furnace could be used to obtain this information. Testing is ongoing to evaluate whether adding a further heating step in the furnace could be used to estimate calcium hydroxide consumption. Having the ability to measure bound water and calcium hydroxide consumption in a furnace will substantially reduce the cost of testing and increasing the ability of labs to run such tests. The ability to cheaply run these tests cheaply is also critical if they are to replace the strength activity index test in specifications such as in ASTM C618.

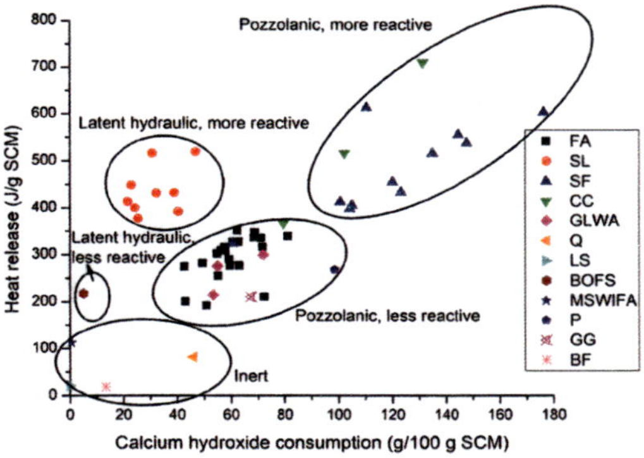

Fig. 1. Modified R^3 test results for conventional and alternative SCMs [5].

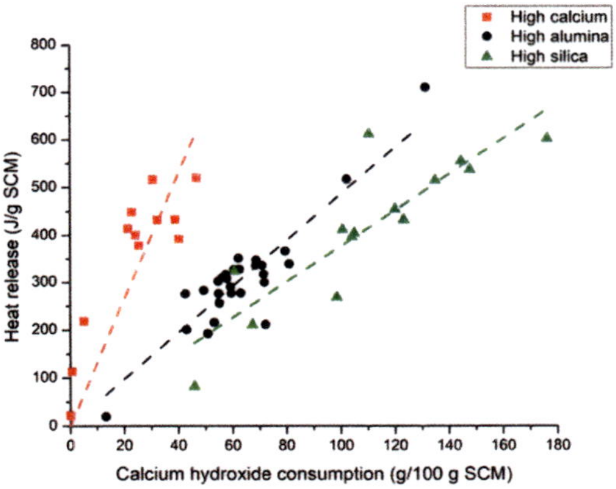

Fig. 2. Chemical composition effects on modified R^3 test results [5].

The influences of fineness and amorphous content on SCM reactivity have been studied using the modified R^3 test [6, 7]. The advantage of using the modified R^3 test is that direct measures of reactivity are obtained. When using indirect tests, the influence of fineness on reactivity are confounded by its impacts on the filler effect. Figure 3 shows the linear effect of fineness on the heat release for pumice and ground glass [6]. For these materials, the bulk chemical composition and amorphous content are approximately the same, meaning the fineness is the only variable affecting the reactivity. As only three fineness levels were tested, it is unclear if the linearity holds at other fineness levels. Nevertheless, this is an important observation, especially in light of the lack of influence of changing fineness on limestone reactivity. It is noted that for the coarsest limestone and pumice, the heat release values (around 100 J/g SCM) would qualify them as inert on the basis of Fig. 1. Since fineness strongly influences reactivity, its effects must be accounted for in reactivity testing. Materials could all be tested at the same fineness levels (which is impractical), or tested at the fineness levels they would be used in practice, or tested at two different fineness levels. Figure 3 suggests that inert or completely crystalline materials are not affected by changing fineness, whereas reactive materials are. How the fineness is changed, either through size fractionation or through grinding will also affect reactivity. Figure 4 shows that amorphous content affects reactivity in a non-linear manner. There is a threshold level of amorphous content over which the reactivity increases linearly. The dataset in Fig. 4 is obtained for mine tailings which are ground to increase their reactivity. For these low amorphous content materials, as the grinding duration increases, both fineness and amorphous content increase. As both parameters are changing, it is not trivial to separate out the effects of amorphous content on the reactivity. Better controlled experiments with systematic variation of both fineness and amorphous content are needed. Ideally these would involve more materials and larger ranges of fineness and amorphous content to confirm or extend the findings in [6, 7].

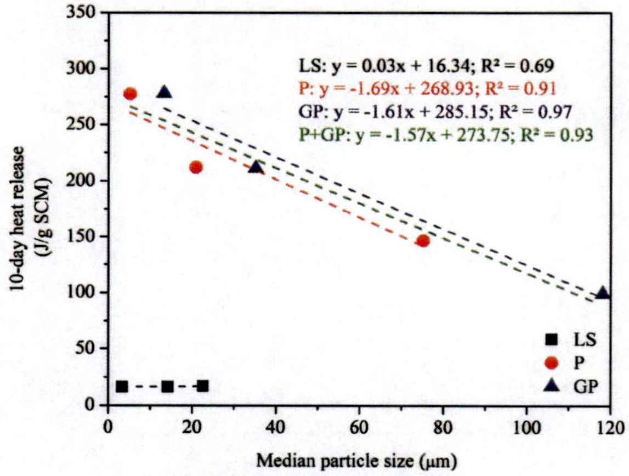

Fig. 3. Fineness effect on reactivity [6].

All results presented so far have been at the end of the 10-day period of testing. However, the kinetics of SCM reaction can also be obtained from the modified R^3 test. Figure 5 shows the importance of SCM kinetics in the test. At 10-days, the Class F fly ash and Class C fly ash in Fig. 5 have similar heat release (~320 J/g), but the values at 1-day are very different (~50 J/g for the Class F fly ash and ~150 J/g for the Class C fly ash). It is a reasonable hypothesis that early-age results in the test are reflective of early-age cement paste properties, while later-age results in the test are reflective of later-age cement paste properties. This hypothesis is consistent with the differences between the behaviors of Class F and Class C fly ashes in cement paste.

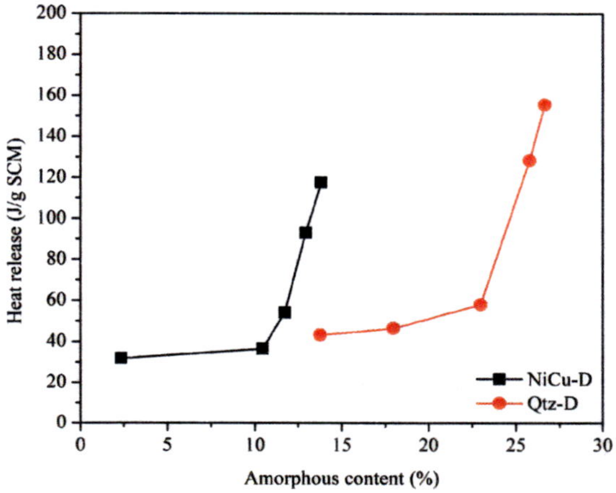

Fig. 4. Amorphous content effects on reactivity [7].

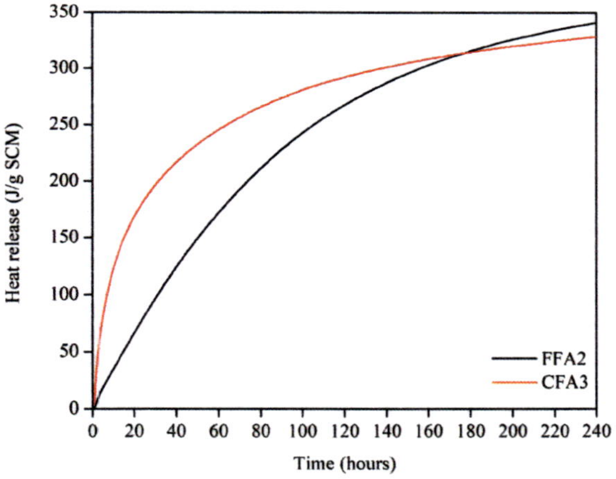

Fig. 5. Differences in SCM reaction kinetics for Class F and Class C fly ash [8].

Figure 6 shows that the 10-day and 3-day heat release values are well correlated. The Class F fly ashes, which react slower than most other materials may be considered outliers on the fit. Considering the excellent relationship between these two parameters, the testing duration could be reduced to 3 days which would be advantageous in saving equipment time.

3 Modified R^3 Test, Paste, and Concrete Properties

Correlating the results of the modified R^3 tests with cement paste, mortar, or concrete properties is not trivial. If low SCM replacements are used, unless the SCM is highly

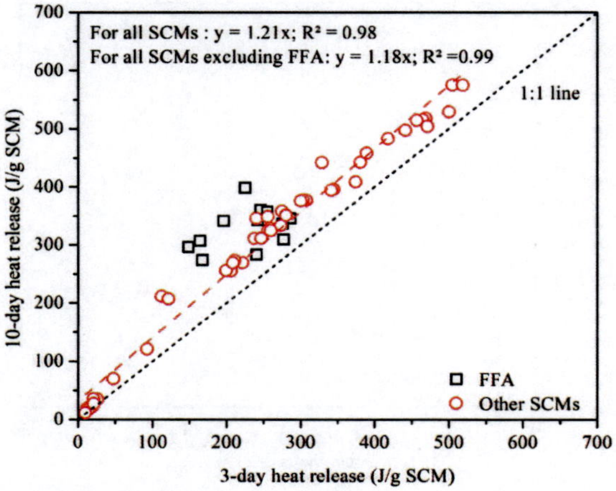

Fig. 6. Correlation between 10-day and 3-day heat release values [8].

reactive, filler effects dominate, and the effects of the SCM on cementitious pastes are relatively minor. Under such conditions, it may not be possible to differentiate reactive and inert materials, when using properties such as strength. Therefore, a correlation with reactivity would not be possible, or even make sense. This lack of distinction between inert and reactive materials is especially true at low replacements *and* early ages. As the age increases, the effects of SCMs become more observable, hence correlating later-age properties of cementitious pastes with reactivity test outputs is more logical. As replacement level increases, SCM degree of reaction actually decreases. However, the effects of SCMs are more obvious at higher replacement levels, at least up to a certain limit. Therefore, cement paste properties and reactivity test outputs are likely to be correlated if using higher replacement levels and later ages. This has been shown to be the case, and paste calcium hydroxide content, compressive strength, bulk resistivity have all been correlated to heat release and calcium hydroxide consumption in the modified R^3 test. In this case, the paste was made with 30% SCM and the correlations were obtained using 56-day properties. An example using compressive strength is shown in Fig. 7.

Much evidence indicates that reactivity test outputs can be used to predict strength. However, it can be argued that the ability to predict strength is not particularly useful, considering that achieving strength is typically not a major concern for most real-life projects. SCMs are largely used due to their improvements to concrete durability. While SCMs do reduce early-age strength, reactivity tests cannot predict such reductions, due to low reactivity at early ages. While screening reactive and inert materials is indeed useful, it can be argued that for reactivity tests to be really useful, they should be able to provide some indication regarding concrete durability. Since reactive materials typically improve concrete durability and inert materials do not, it is plausible that reactivity tests will be able to predict some aspects of concrete durability. The correlation of heat release and calcium hydroxide consumption with bulk resistivity is important in this

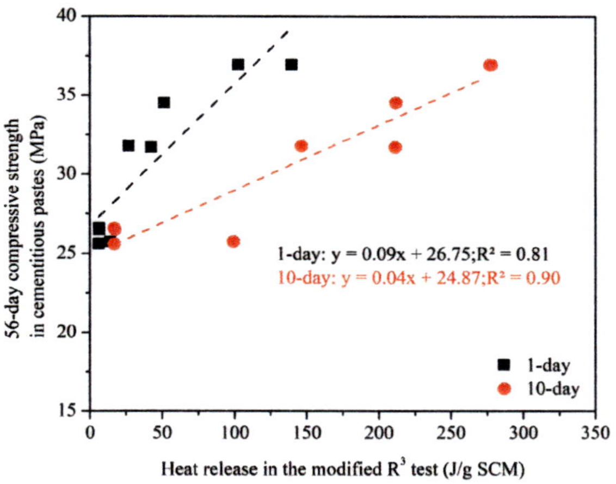

Fig. 7. Correlations between 56-day compressive strength in cement paste and heat release in the modified R^3 test [6].

context because the bulk resistivity is a common indicator of certain concrete durability properties. Our data suggests that this relationship is non-linear and mortar bulk resistivity increases drastically once calcium hydroxide consumption increases beyond a certain value. This finding is possibly explained by the presence of a certain calcium hydroxide consumption beyond which pores depercolate, but this hypothesis needs to be validated. Since in cement pastes, the calcium hydroxide consumption leads to further hydration products and pore refinement, the calcium hydroxide consumption is in one sense a more fundamental parameter to measure than the heat release. Therefore, we can tentatively suggest that the calcium hydroxide consumption in the model system could be used to predict to concrete quality or concrete resistance to chloride ingress due to the relationships between these parameters are the bulk resistivity.

Since the calcium hydroxide consumption in the modified R^3 test is linked to (later-age) paste calcium hydroxide content, it could also be linked to calcium oxychloride contents measured in cement pastes. Figure 8 shows this to be the case, and the calcium hydroxide contents measured for several SCMs in the modified R^3 test show a strong linear correlation with calcium oxychloride contents in pastes made with 30% of these SCMs. Because calcium oxychloride amounts are reduced by both pozzolanic reaction and dilution, the obtained relationship depends on replacement level. A threshold level of calcium oxychloride is known to limit calcium oxychloride damage, therefore, the modified R^3 test could directly be used to predict the replacement level at which materials would have to be used to successfully mitigate calcium oxychloride damage. For inert materials, these replacement levels typically turn out to greater than values that can be used in concrete without causing unacceptable losses in mechanical properties.

It is known that the pozzolanic reaction is critical in the ability of SCMs ability to control alkali-silica reaction. Therefore, the modified R^3 test could predict this ability. Figure 9 shows the correlation between alkali silica reaction expansion (%, as measured using the accelerated mortar bar test) and the product of calcium hydroxide consumption in the modified R^3 test (g/100 g SCM) and SCM replacement (%). Similar to calcium oxychloride consumption, both pozzolanic reaction and dilution affect the durability measure, hence replacement level has to be taken into account when considering the role of reaction. Figure 9 suggests that this can be done simply by multiplying contributions from reactivity and dilution, although further data points are needed to confirm this finding. Using higher replacement levels or more pozzolanic SCMs (higher calcium hydroxide consumption) will mitigate alkali-silica reaction expansion. However, as with calcium oxychloride damage, the replacement levels needed for inert materials to mitigate alkali-silica reaction are not feasible in practice. Alkali-silica reaction and its associated damage is complex, and when expansion from the accelerated mortar bar test is within certain ranges, further study is needed.

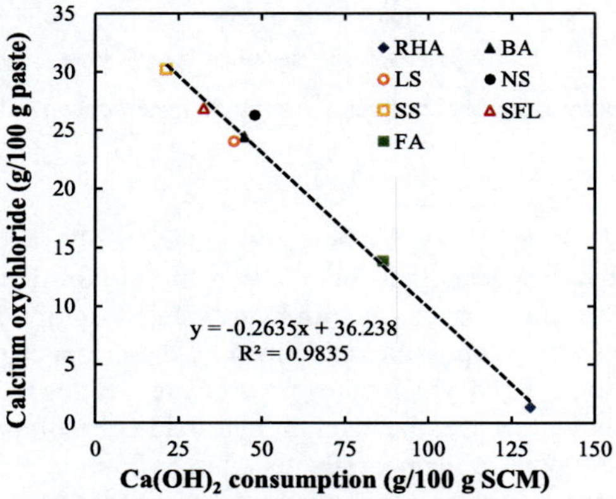

Fig. 8. Correlations between paste calcium oxychloride content and calcium hydroxide consumption in the modified R^3 test [9].

The modified R^3 test clearly provides valuable information about concrete durability. However, its results should not be over interpreted. As an example, two SCMs could have very similar calcium hydroxide consumption but one could have much higher free alkali than the other. These SCMs, while having similar reactivity, would have dissimilar abilities to mitigate alkali silica reaction, although their ability to mitigate calcium oxychloride damage would likely be similar. Such complex durability behavior cannot be predicted only by using one or two measures of reactivity. However, as we have demonstrated, reactivity tests can certainly be used for material screening—including for durability behavior.

4 Conclusions

The heat release and calcium hydroxide consumption of SCM-calcium hydroxide model systems in a simulated pore solution at 40 °C are measured in the modified R^3 test. Numerous SCMs have been studied using this test and it is able to quantify reactivity and

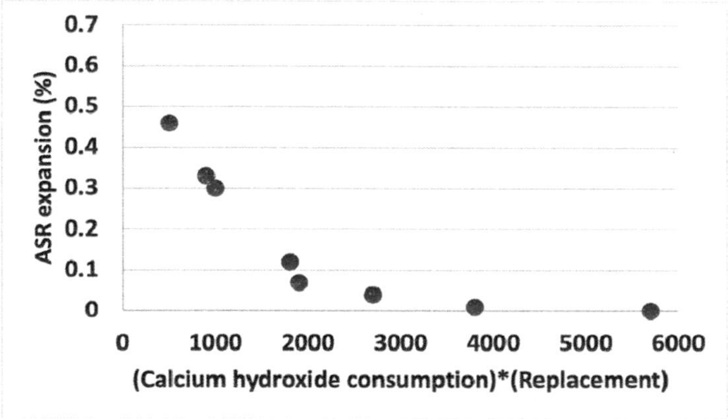

Fig. 9. Correlation between accelerated mortar bar test alkali-silica reaction expansion and calcium hydroxide consumption in the modified R^3 test multiplied by the SCM replacement level.

show differences between pozzolanic, latent hydraulic, and inert materials. The chemical composition of the SCMs affects their reactivity. Reactivity increases with fineness and amorphous content, although further data is needed to confirm whether the relationship is linear or not. Paste properties can be predicted by the modified R^3 test outputs, especially at later ages and when using higher SCM replacement levels. Reactivity can be linked to durability behavior through bulk resistivity, calcium oxychloride content, accelerated mortar bar expansion. The effects of replacement should be considered when making such durability predictions. We conclude that the modified R^3 test can be used to screen SCMs, including for concrete durability.

References

1. Kalina, R.D., Al-Shmaisani, S., Ferron, R.D., Juenger, M.C.: False positives in ASTM C618 specifications for natural pozzolans. ACI Mater. J. **116**(1), 165–172 (2019)
2. Snellings, R., Scrivener, K.L.: Rapid screening tests for supplementary cementitious materials: past and future. Mater. Struct. **49**(8), 3265–3279 (2016)
3. Li, X., et al.: Reactivity tests for supplementary cementitious materials: RILEM TC 267-TRM phase 1. Mater. Struct. **51**(6), 1–14 (2018). https://doi.org/10.1617/s11527-018-1269-x
4. Suraneni, P., Weiss, J.: Examining the pozzolanicity of supplementary cementitious materials using isothermal calorimetry and thermogravimetric analysis. Cement Concr. Compos. **83**, 273–278 (2017)
5. Suraneni, P., Hajibabaee, A., Ramanathan, S., Wang, Y., Weiss, J.: New insights from reactivity testing of supplementary cementitious materials. Cement Concr. Compos. **103**, 331–338 (2019)
6. Ramanathan, S., Tuen, M., Suraneni, P.: Influence of supplementary cementitious material and filler fineness on their reactivity in model systems and cementitious pastes. Mater. Struct. **55**, 1–25 (2022)
7. Ramanathan, S., Perumal, P., Illikainen, M., Suraneni, P.: Mechanically activated mine tailings for use as supplementary cementitious materials. RILEM Tech. Lett. **6**, 61–69 (2021)

8. Ramanathan, S.: Reactivity of Supplementary Cementitious Materials in Model Systems and Cementitious Pastes. Ph.D. Dissertation, University of Miami, Coral Gables (2021)
9. Jones, C., Ramanathan, S., Suraneni, P., Hale, W.M.: Mitigating calcium oxychloride formation in cementitious paste using alternative supplementary cementitious materials. Constr. Build. Mater. (2022) (under review)

Experimental Studies on the Mechanical Properties of Concrete Made up with Processed GGBS

Sanjay R. Salla[1](✉), Chetankumar Modhera[2], and U. Raghu Babu[1,2]

[1] Civil Engineering Department, S V National Institute of Technology, Surat, Gujarat, India
er.sallasanjay@gmail.com

[2] Civil Engineering Department, Srinivasa Ramanujan Institute of Technology, Anantapuramu, Andhra Pradesh, India

Abstract. The current research article presents the utilization and recycling of industrial waste material to help the environment. The production of cement alone contributes 8% of CO_2 emissions of global pollution. The article has also enlightened the effectiveness of processed granulated ground blast furnace slag (GGBS) as a supplementary cementitious material (SCM). Processed GGBS replaced the cement with 0–50%, at 5% interval in concrete as a binder. This study also carries the effect of processed GGBS on the fresh and mechanical properties of concrete. The optimization done by the analytical hierarchy process (AHP) shows that the 20% replacement is more effective. This effective utilization of processed GGBS increases the split tensile strength by 15%, flexural strength by 28%, and shear strength by 48%. The overall cost of concrete can be reduced by 8.62% with 20% of processed GGBS.

Keywords: Concrete · Processed GGBS · AHP · Cost analysis · Optimization

1 Theoretical Background

The utilization and recycling of industrial waste material may help protect the environment. Cement production alone contributes 8% of CO_2 emissions of global pollution [1]. Numerous researchers are also working on alternative fuels in cement production, changing binder chemistry, and working on industrial by-products. The utilization of SCMs is one of the ways to reduce the consumption of cement. The use of SCMs in cement production is widely recognized worldwide either by addition during manufacturing or by replacing cement [2]. The present study also enlightened the effectiveness of processed GGBS (PG) as a supplementary cementitious material (SCM).

Processed GGBS is a finer granulated powder of granulated ground blast furnace slag known as alccofine [3]. Processed GGBS help to reduce water content and increase the compressive strength, and it is also observed that processed GGBS is nearer to silica fume. The ultrafine GGBS has a more specific surface area; it may benefit from better hydration, high compressive strength, and durability [4]. The incorporation of alccofine increases fly-ash-based geopolymer concrete's fresh and hardened properties [5]. The

J. I. Escalante-Garcia et al. (Eds.): RW 2021, RILEM Bookseries 40, pp. 141–150, 2023.
https://doi.org/10.1007/978-3-031-21735-7_17

ultrafine GGBS is beneficial for fly ash-based geopolymer and self-compacting concrete [6].

Furthermore, according to this limited literature, it reduces concrete workability and water absorption while increasing the compressive strength by 18% [3]. Using alccofine with GGBS has also been stated to have a negative effect on the ternary blend. It means that the combination of GGBS and alccofine as a cement substitute might not help prepare ternary blend concrete [7]. Although the utilization of proceeds GGBS has been reported, the optimal dosage of it has not yet been confirmed. Hence, the current work investigates its properties and finds the optimal dosage to replace cement in concrete.

2 Material

In the current experimental work, cement is used, confirming IS 269: 2015. Processed GGBS having the fineness 851.28 m^2/kg, specific gravity 2.84 was used in the present study [8]. The sieve analysis of locally available river sand (RS) and manufactured sand (MS) shows that both confirmed Zone II [9]. The physical properties of RS were fineness modulus 2.56, specific gravity 2.75, water absorption and free moisture 1.76 and 3.85%, respectively, and bulk density 1600 kg/m^3 [10, 11]. Similarly, the properties of MS were fineness modulus 2.69, specific gravity 2.72, water absorption and free moisture 3.28 and 5.83%, respectively, and bulk density 1640 kg/m^3 [10, 11]. The physical properties of coarse aggregate (20 mm) were fineness modulus 7.06, specific gravity 2.87, water absorption 1.76%, and bulk density 1530 kg/m^3, aggregate flakiness index, and elongation index 13.14 and 14.30%, and aggregate impact value 7.55% [10–12]. Similarly, the properties of coarse aggregate (10mm) were fineness modulus 7.05, specific gravity 2.80, water absorption 1.35%, and bulk density 1575 kg/m^3, aggregate flakiness index, and elongation index 12.25 and 12.98% [10–12]. The admixture used low polycarboxylate based on confirming IS 9103: 1999 [14], with a specific gravity of 1.145. The pozzolanic reactivity of processed GGBS was confirmed as IS 1727-1967, and the results show that up to 40% utilization is effective [8]. A concrete mix for M25 grade concrete with medium workability was designed based on material properties and Indian standards. During the laboratory mixing, mild environmental conditions and a high level of quality control were maintained [13].

3 Methodology

3.1 Workability

The slump test is widely accepted for the easiness and simplicity of its test procedure. The slump cone is used to measure the slump, and the value of the slump is expressed in millimeter (mm).

3.2 Mechanical Properties

The concrete cube specimens with size 150 × 150 × 150 mm size were cast for compressive strength and split tensile strength. The compressive strength test was carried out as per IS 516: 1959, and the split tensile strength test was done as per IS 5816: 1999 with the compressive testing machine [14, 15]. The beam specimens with 100 × 100 × 500 mm were cast for flexural strength, tested by applying two points loading using a universal testing machine as per IS 516: 1959 [14]. The shear strength test was also carried out [16]. The test setup for the shear test is as shown in Fig. 1.

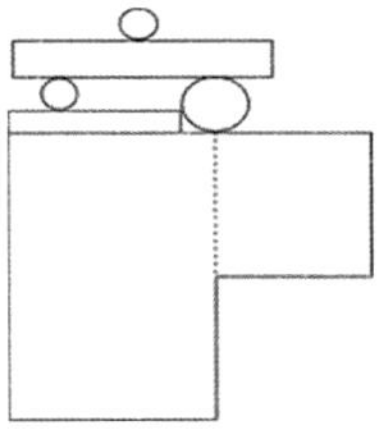

Fig. 1. Test setup with specimen for shear strength

4 Analytical Approach

Multi-criteria decision-making (MCDM) is most beneficial to find an optimum solution that fulfills the various properties checked as per the experimental program. There are many methods available for MCDM, but the Analytical Hierarchy Process (AHP) was adopted to find out the optimum solution or best alternative among the all prepared mixes [17] (Table 1). In the present study, the weights of attributes were calculated for the entropy method.

Table 1. Selected attributes and their weights.

Workability	Density	Compressive strength	Split tensile strength	Flexural strength	Shear strength
0.17	0.17	0.17	0.17	0.17	0.17

AHP is a mathematical tool that helps make a decision and also helps to determine the relative importance of selected attributes. The following steps are used to determine the relative importance of different attributes [18].

Step-1: Construct the pairwise comparison matrix using a linker scale.

$$B_{m\times m} = \begin{matrix} B_1 \\ B_2 \\ B_3 \\ - \\ - \\ B_m \end{matrix} \begin{bmatrix} 1 & b_{12} & b_{13} & - & - & b_{1m} \\ b_{12} & 1 & b_{23} & - & - & b_{2m} \\ b_{31} & b_{32} & 1 & - & - & b_{3m} \\ - & - & - & - & - & - \\ - & - & - & - & - & - \\ b_{m1} & b_{m2} & b_{m3} & - & - & 1 \end{bmatrix} \quad (1)$$

Step-2: Find the relative normalized weight for each attribute by the geometric mean of each raw.

$$GM_j = \left[\prod_{j=1}^{m} b_{ij}\right]^{\frac{1}{m}} \tag{2}$$

$$w_j = \frac{GM_j}{\sum_{j=1}^{m} GM_j} \tag{3}$$

Step-3: Determine the maximum Eigenvalue λ_{max},

$$\lambda_{max} = \sum_{j=1}^{i} a_{ij} \frac{w_j}{w_i} \tag{4}$$

Step-4: Calculate the consistency index CI and consistency ratio CR,

$$CI = \frac{\lambda_{max} - m}{m - 1} \tag{5}$$

$$CR = \frac{CI}{RI} \tag{6}$$

The value of CR is 0.1 or less acceptable [18]; more than that shows that inconsistency in pairwise comparison [17] and must be discarded.

Step-5: Determination of priority weights for each criterion and calculates each criterion's global weights.

5 Experimental Results and Discussion

5.1 Workability, Water Absorption, and Density

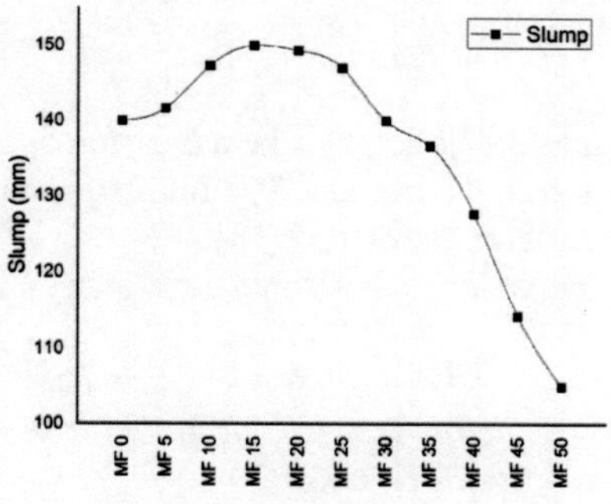

Fig. 2. Workability

Figure 2 shows the graphical representation of results of workability. The test results show that the increment of processed GGBS decrement in the workability. The test results of water absorption and density of hardened concrete not much more affected by incorporating the processed GGBS. Figure 3 depicts the density of hardened concrete.

The effect of GGBS on density depends on its use, like paste, mortar, or concrete. In GGBS, It id found that slag decreases concrete density by around 0.5–1% [19].

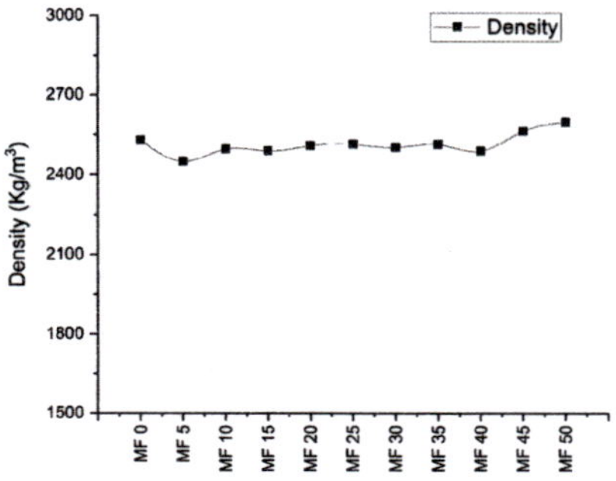

Fig. 3. Density (hardened concrete)

5.2 Compressive Strength

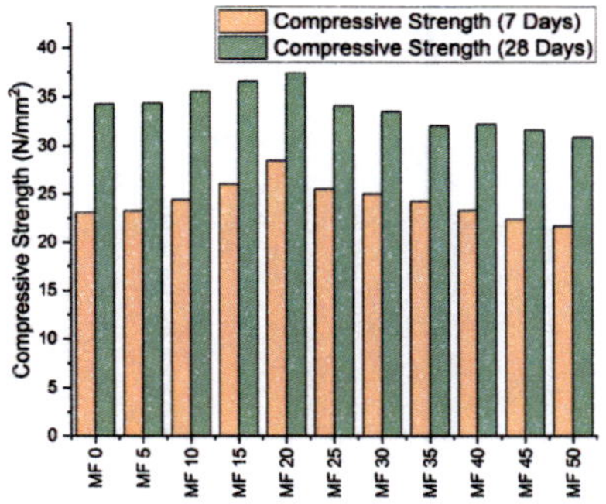

Fig. 4. Compressive strength

Figure 4 depicts the compressive strength test findings after seven and twenty-eight days. The findings show that incorporating processed GGBS improves compressive strength by up to 20%; however, more than 20% substitution reduces strength but meets the lower grade concrete requirements. The compressive strength increases by 10% when 20% of the cement is replaced with processed GGBS [20]. The utilization of GGBS $\geq$ 50% in paste and mortar reduces the compressive strength [7, 21]. The literature shows that the effective utilization of GGBS is around 45–50% for concrete [19, 22, 23].

5.3 Split Tensile Strength, Flexural Strength, and Shear Strength

The split tensile strength and flexural strength of concrete are tested, as shown in Fig. 5. The use of processed GGBS instead of cement improves the split tensile strength of concrete by up to 18%.

The results of the experiments indicate that the incorporation of processed GGBS improves flexural strength. The utilization of processed GGBS instead of cement improves the flexural strength of concrete by up to 28% [21].

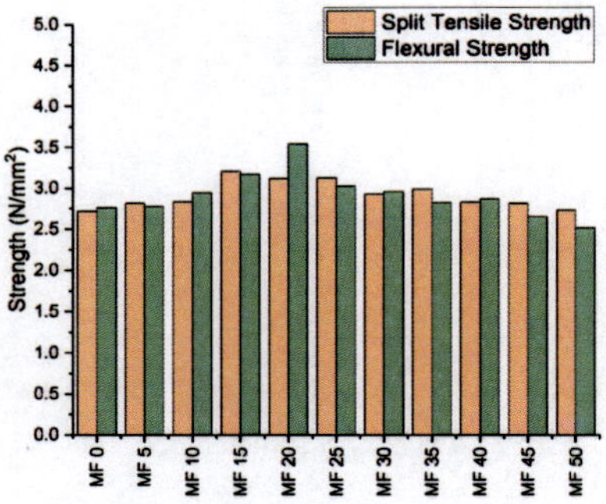

Fig. 5. Results of split tensile and flexural strength

Figure 6 depicts the concrete shear strength test findings. The results of the experiments indicate that incorporation of processed GGBS improves shear strength as well. The use of processed GGBS instead of cement improves the shear strength of the concrete by up to 50%.

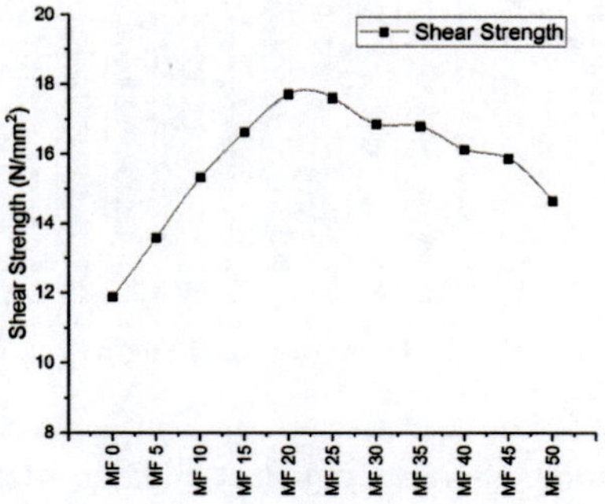

Fig. 6. Results of shear strength

6 Analytical Hierarchy Process

A hierarchical structural was established by setting up effective utilization of processed GGBS in concrete at the top level with six different attributes: workability, density, compressive strength, split tensile strength, flexural strength, and shear strength for each mix. Besides, this pairwise comparison was carried out for every five attributes for each mix, calculation done based on Eqs. 1 and 2. Figure 7 shows the graphical representation of pairwise comparisons of all six attributes for all different mixes. Besides this, the priority matrix and normalization use maximum Eigenvalue for each attribute and calculated based on Eq. 4, as shown in Table 2 and Fig. 7. The consistency ratio of 0.1 was maintained by using Eqs. 5 and 6. The AHP technique results show that the effective utilization of processed GGBS is 20%. Ultrafine grinding of GGBS is expected to increase performance by increased surface area with grinding and may not result in a significant difference in cost. However, this should be verified by further study on the performance and cost analysis of concrete with and without GGBS grinding.

Table 2. Weighted normalized priority vectors for all different mixes and their summation

Samples	Slump	Density	Compressive strength	Split tensile strength	Shear strength	Flexural strength	Raw sum
MF 0	0.0156	0.0154	0.0154	0.0141	0.0113	0.0143	0.0861
MF 5	0.0158	0.0149	0.0154	0.0146	0.0130	0.0144	0.0880
MF 10	0.0165	0.0152	0.0160	0.0147	0.0146	0.0152	0.0921
MF 15	0.0168	0.0151	0.0164	0.0166	0.0159	0.0164	0.0972
MF 20	0.0167	0.0152	0.0168	0.0162	0.0169	0.0184	0.1001
MF 25	0.0164	0.0153	0.0153	0.0162	0.0168	0.0157	0.0957
MF 30	0.0156	0.0152	0.0150	0.0152	0.0161	0.0153	0.0924
MF 35	0.0153	0.0153	0.0144	0.0155	0.0160	0.0146	0.0910
MF 40	0.0143	0.0151	0.0145	0.0147	0.0154	0.0149	0.0887
MF 45	0.0128	0.0156	0.0142	0.0146	0.0151	0.0137	0.0860
MF 50	0.0117	0.0158	0.0138	0.0141	0.0140	0.0130	0.0825

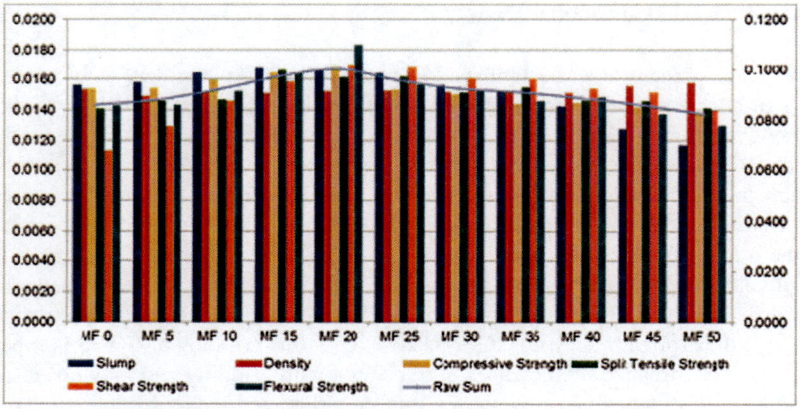

Fig. 7. Priority vectors for all different mixes

7 Cost Analysis

The processed GGBS as a binder requires just grinding, so it is available at a lower cost. It gives economic and sustainable solutions compare to control concrete. For example, Table 3 shows the comparative cost of concrete, control concrete, and prepared with concrete made with 20% processed GGBS. Costs of all the materials are taken based on the available local market. The overall cost of concrete was reduced by 8.62%, incorporating 20% processed GGBS in the concrete.

Table 3. Cost analysis for 1 m^3 concrete

Sr. No.	Material	Quantity (kg/m^3)		Unit cost	Cost (Rs.)	
		Control mix	MF 20		Control mix	MF 20
1	Cement	335	268	6.00	2010.00	1608.00
2	PG	0	61	1.50	0.00	91.50
3	FA	614	614	0.75	460.50	460.50
4	MS	205	205	0.35	71.25	71.25
5	CA	1143	1143	0.85	971.55	971.55
6	Water	168	168	0.50	84.00	84.00
Overall cost					≈3597	≈3287

PG: Processed GGBS, FA: River Sand, MS; Manufactured Sand, CA: Coarse Aggregate

8 Conclusions

This study aimed to learn more about the effects of using processed GGBS as a concrete binder. The following conclusions can be drawn from the mechanical properties of the various mixes investigated in this study:

With increasing the amount of processed GGBS as a binder,

1. The workability of different mixes is decreased.
2. The density of concrete was not much more affected.
3. The mechanical properties like compressive strength, split tensile strength, flexural strength and shear strength increases by incorporating processed GGBS.
4. The result of AHP shows that the effective utilization of processed GGBS is 20%, which increases the split tensile strength by 15%, flexural strength by 28%, and shear strength by 48%.
5. The total expense can be cut by 8.62% per meter cube by 20% processed GGBS.

References

1. Rodgers, L.: Climate change: the massive CO_2 emitter you may not know about. BBC News (2018). https://www.bbc.com/news/science-environment-46455844
2. Siddique, R., Khan, M.I.: Supplementary Cementing Materials. Springer (2011)
3. Kumar, M.P., Mini, K.M., Rangarajan, M.: Ultrafine GGBS and calcium nitrate as concrete admixtures for improved mechanical properties and corrosion resistance. Constr. Build. Mater. **182**, 249–257 (2018). https://doi.org/10.1016/j.conbuildmat.2018.06.096
4. Teng, S., Lim, T.Y.D., Divsholi, B.S.: Durability and mechanical properties of high strength concrete incorporating ultra fine ground granulated blast-furnace slag. Constr. Build. Mater. **40**, 875–881 (2013). https://doi.org/10.1016/j.conbuildmat.2012.11.052
5. Saloni, Singh, A., Sandhu, V., Jatin, Parveen: Effects of alccofine and curing conditions on properties of low calcium fly ash-based geopolymer concrete. Mat. Today: Proc. (2020). https://doi.org/10.1016/j.matpr.2020.02.763
6. Mohan, A., Mini, K.M.: Strength and durability studies of SCC incorporating silica fume and ultra fine GGBS. Constr. Build. Mater. **171**, 919–928 (2018). https://doi.org/10.1016/j.conbuildmat.2018.03.186
7. Narender Reddy, A., Meena, T.: A study on compressive behavior of ternary blended concrete incorporating alccofine. Mater. Today: Proc. **5**, 11356–11363 (2018). https://doi.org/10.1016/j.matpr.2018.02.102
8. Salla, S.R., Modhera, C.D., Babu, U.R.: An experimental study on various industrial wastes. Concr. Sustain. Constr. **19**, 133–148 (2021). https://doi.org/10.3151/jact.19.133
9. IS 383-2016, Coarse and fine aggregate for concrete—specification (2016)
10. IS 2386 (Part-I)-1963, Methods of test for aggregates for concrete (2016)
11. IS 2386 (Part-III)-1963, Methods of test for aggregates for concrete (2016)
12. IS 2863 (Part-IV)-1963, Methods of test for aggregates for concrete (2016)
13. IS 10262-2019, Concrete mix proportioning—guidelines (2019)
14. IS 516-1959, Methods of tests for strength of concrete (2018)
15. IS 5816-1999, Splitting tensile strength of concrete—method of test (2013)
16. Bairagi, N.K., Modhera, C.D.: Shear strength reinforced concrete. ICI J. 47–52 (2001)

17. Rashid, K., Hameed, R., Ahmad, H.A., Razzaq, A., Ahmad, M., Mahmood, A.: Analytical framework for value added utilization of glass in concrete: mechanical and environmental performance. Waste Manag. 312–323 (2018). https://doi.org/10.1016/j.wasman.2018.07.052
18. Venkata Rao, R.: Decision Making in the Manufacturing Environment: Using Graph Theory and Fuzzy Multiple Attribute Decision Making Methods. Springer Science & Business Media (2007)
19. Rashad, A.M.: An overview on rheology, mechanical properties and durability of high volume slag used as cement replacement in paste, mortar and concrete. Constr. Build. Mater. **187**, 89–117 (2018)
20. Reddy, A.N., Meena, T.: A study on compressive behavior of ternary blended concrete incorporating alccofine. Mater. Today: Proc. 11356–11363 (2018). www.sciencedirect.comwww.materialstoday.com/proceedings
21. Lim, D.T.Y., Xu, D., Divsholi, B.S., Teng, S.: Effect of ultra fine slag replacement on durability and mechanical properties of high strength concrete. In: 36th Conference on Our World in Concrete & Structures, 14–16 August 2011, Singapore, pp. 14–16 (2011). http://cipremier.com/100036039www.cipremier.com
22. Yi, H., Xu, G., Cheng, H., Wang, J., Wan, Y., Chen, H.: An overview of utilization of steel slag. Procedia Environ. Sci. 791–801 (2012). https://doi.org/10.1016/j.proenv.2012.10.108
23. Jiang, Y., Ling, T.-C., Shi, C., Pan, S.-Y.: Characteristics of steel slag and their use in cement and concrete—a review. Resour. Conserv. Recycl. **136**, 187–197 (2018)

Effect of Mineral Addition LC2 on Shrinkage Produced in Concrete

María B. Díaz García(✉), Yosvany Díaz Cárdenas, and José F. Martirena Hernández

Center for Research and Development of Structures and Materials (CIDEM), Havana, Cuba
mbetania@uclv.cu

Abstract. Together with the additives, the use of Supplementary Cementitious Materials (SMCs) is one of the most sustainable solutions worked on to achieve high performance concretes, more resistant to aggressive environments and as a measure to mitigate different phenomena, such as shrinkage cracks. Mineral additions are one of the most widespread, as they not only favour the partial replacement of Portland cement, but also provide certain chemical and physical properties that make the concrete more durable in different environments. The development of technologies that allow these new materials to appear requires studies to determine their influence on construction. This is why this research is focus on evaluating the effect of the active mineral addition of calcined clay, limestone and gypsum (LC2), on the shrinkage that occurs in fluid concretes and their behaviour in the paste-concrete ratio. Samples P35 (Cement P-35 and Sika Plast additive 0.65%) and LC65 (Cement P-35, addition LC2 30% and Sika Plast additive 1%) are defined. They are evaluate by means of chemical shrinkage in cement pastes and then a study is made of the shrinkage in the concretes by means of ASTM c-157 and the shrinkage channel. This research demonstrates the shrinkage reducing effect of the mineral addition LC2 and achieves an important step forward in the sustainable development of new materials for the Construction Industry in Cuba.

Keywords: Pozzolans · Cement hydration · Volume changes

1 Introduction

The use of pozzolans in Portland cement has an effect on reducing chemical shrinkage, drying shrinkage and hydration heat. Concretes with pozzolanic additions reduce the consequences of the alkaline-aggregate reaction, such as cracking and reduced strength. It incorporates very fine particles into the concrete, helping to improve the workability of the mix, and even reducing the water demand for a given consistency. The calcined clays provide the concrete with an increase in impermeability, in resistance to sulphate attack, as well as an increase in mechanical resistance at advanced ages [2].

The reduction of chemical and autogenous shrinkage is also contribute by the incorporation of pozzolan in the concrete. In some case, pozzolanic cements can develop in

J. I. Escalante-Garcia et al. (Eds.): RW 2021, RILEM Bookseries 40, pp. 151–158, 2023.
https://doi.org/10.1007/978-3-031-21735-7_18

the long term shrinkage of comparatively equal or less length than that achieved by Portland cement. This may be due to the fact that the pozzolanic addition forms compounds with a similar structure to the tobermorite only in the medium and long term [3, 4].

These cracks can be explained as the result of a physical-chemical process that the cement paste undergoes, after being mixed with the aggregates, and which consists mainly of a variation in volume, which may depend on many factors, including the type of cement, the aggregates, the water-cement ratio, and fundamentally the humidity and temperature conditions during curing [2, 5].

In this context, it is of capital importance to identify and develop new sources of pozzolanic materials that allow the production of binders with high volumes of clinker substitution at an affordable cost from an economic, energy and environmental point of view.

In view of the above, the Centre for Research and Development of Structures and Materials (CIDEM) proposes a study on the evaluation of shrinkage in pastes and concretes of Portland P 35 cement with 30% active mineral additions (LC2); which is no more than a product of calcined clay, limestone and gypsum in suitable proportions that substitute cement in the manufacture of concretes. All this done to know the influence of the hydration process on the volume changes of LC2 as an active mineral addition and relation with shrinkage in concrete.

2 Methodology Design

The use of additions requires a greater demand for water than the concretes produced with ordinary Portland cement, therefore they tend to suffer a greater contraction and therefore the use of plasticizing additives is recommended for those structural elements susceptible to this type of fissures. For this reason, a study of concretes with the active mineral addition LC2 is necessary to define the effect that the addition of calcined clay, limestone and gypsum (LC2) has on the fissures produced by volume changes in concretes, with the action of the superplasticizer additive.

For the research, two samples are analysed to compare the action of the LC2 addition on hydration and its relationship with the changes in volume that occur in the concrete, the first with the P35 cement without addition and the second with 30%. This study takes into account the chemical shrinkage in the cement pastes and the deformations that occur in the concrete through the shrinkage channel and ASTMc-157.

2.1 Materials and Method

The raw materials used in this research are P-35 cement from the Karl Marx factory in the province of Cienfuegos, Sika Plast additive. The aggregates come from the El Purio quarry in Encrucijada and the active mineral addition LC2 composed of, calcined clay, limestone and gypsum from the industrial test 2 that was carried out in 2018 at the Siguaney factory, Sancti Spríritus.

The active mineral addition is formulate in the Faculty of Construction where the thermal activation of the clay is develop by calcination at 800 °C, then proceeds to grinding in a ball mill MB 800, and mixes the clay with limestone in a 2:1 ratio with

7% gypsum total to grind. The cement to be used is produced at the Karl Marx cement factory in the province of Cienfuegos, according to NC: 95, 2011 Portland Cement—Specifications, classified as Portland P-35 cement (Table 1).

Table 1. Composition of the active mineral addition LC2

Materials	Quantity (Kg)	%
Clay	19.8	62
Limestone	9.6	31
Gypsum	0.6	7

To carry out the analysis of the microstructure of the cement paste, 6 samples with cement P 35 are prepared as a standard whit the water/cement ratio = 0.40 and the incorporation of superplasticizer additive. Then, under the same conditions, the paste samples are elaborate with cement P 35 and the mineral addition LC2 at 15 and 30%. For the development of the research, a dosage is define for the tests on cement pastes, where the water/cement ratio is varied (Table 2).

Table 2. Samples

Samples	Composition
P35	P 35, Ra/c = 0.40, 0.5% SP
LC65	70%P 35 + 30% LC2, Ra/c = 0.40, 0.5%SP

In the study of cement pastes, calorimetry techniques are used to study the hydration of cement with mineral addition LC2 according to NC:525, 2014 Hydraulic Cement. Test methods. Determination of the heat of hydration and chemical contraction to measure volume changes that occur between 0 and 7 days, for this method the rule NC: 504, 2013 Cement. Volume stability is consulte.

The investigation has a procedure that begins with the definition of the dosages to use in the elaboration of concretes for the taking of samples of each mixture. In the case of concrete, we study the deformation at early ages through the shrinkage channel and at late ages according to the ASTM c-157 Standard Test Method for Length Change of Hardened Hydraulic-Cement Mortar and Concrete (Table 3).

In concrete work with 1% additive, because it is the dosage that meets the fluid consistency required in this type of concrete.

3 Discussion of Results

To define the entire research process, as reflected in the stages, tests are performed on Portland cement pastes and cement with the active mineral addition LC^2, to evaluate the effect in the hydration kinetic in pastes and relation with shrinkage concrete.

Table 3 Dosage in concrete

Material	Cement (kg)	LC2 (kg)	Sand (kg)	Coarse (kg)	Water (Lts)	Adittive SP (Lts)
P35	8.4	0	18.7	17.82	3.42	0.055
LC65	5.88	2.52	18.74	17.82	3.41	0.084

3.1 Influence of the LC2 Addition on Volume Change in Pastes

The hydration process of Portland cement with 30% LC2 mineral addition is slow compare to pure Portland cement. Previous studies have shown that these cements have greater chemical stability and a denser microstructure. The aim of this research is to evaluate the LC2 influence on the hydration of cement pastes with a 30% substitution.

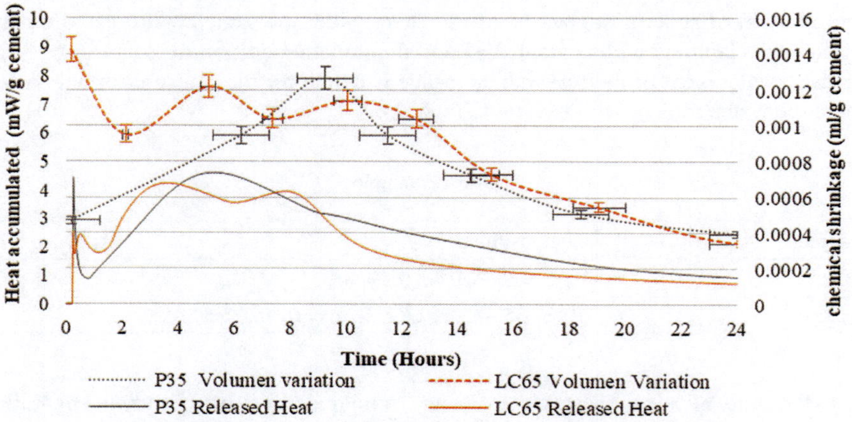

Fig. 1. Relation between chemical contraction and calorimetry

Figure 1 shows the results of the chemical contraction and defines the maximum values of volume change that occur in the pastes with the different binders. The samples with LC2 have the greatest volume changes at ages of 6–10 h which coincide with the maximum heat releases.

It is observe that the samples with LC2 have two peaks of greater change, the first when the C-S-H gel is formed and the second with the formation of the ettringite. It is define that in the binder with LC2 each of the phases forms more rapidly, but with less variation in volume than Portland cement.

3.2 Influence of the LC2 Addition on Volume Change in Concrete

For the evaluation of volume changes in the concrete, the analysis of the deformations in the early ages is of great importance to determine the influence of each of the stages of cement hydration and the effect that the LC2 addition has. At early ages the chemical shrinkage is evaluate where the formation of the phases is influenced and at later ages the shrinkage by drying is evaluated, which varies according to the factor being analysed, curing, water-cement ratio, surface moisture. For this study, the temperature and humidity were kept constant to ensure that both concretes were kept in the same conditions in order to evaluate only the effect of cement hydration and the relationship between this and the deformations in the concretes with LC2.

Early Age

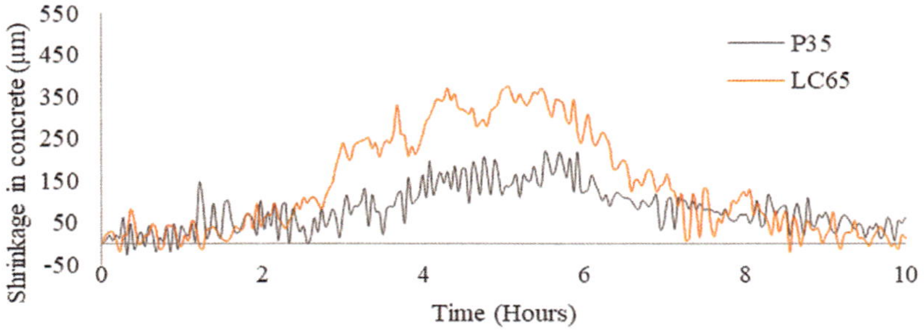

Fig. 2. Variation in concrete volume at early ages

The graph shows the variations in volume of the concrete in the early ages. The sample with LC2 increases the volume in less time although the variations do not exceed 50 µm in 3 min, while for the concrete with P-35 it can exceed 100 µm in three minutes. In the first 3 h, it can be seen that concretes with LC2 increase in volume and then begin to decrease until the variation remains constant. In the case of samples with P-35, an increase in volume is define after 6 h, which, although the variations are greater, expands and contracts to the same extent. The samples with LC2 expand in the first ages to a greater extent and after 5 h they contract and maintain constant with less variation. In the case of P-35 cement, it expands and contracts throughout the hydration process in which it reaches greater deformations than in LC2 (Fig. 2).

Late Age

In the case of concrete evaluated at a later age, the action of the LC2 is observe over time and the decrease in volume with respect to the P-35. In graph (a) the maximum variations of concretes with LC2 are between 5 and 10 µm in one day, while concretes with P-35 are between 20 and 25 µm in half a day. In graph (b) the accumulated deformations over the 28 days, where the LC2 sample with the least shrinkage is defined. This decrease in shrinkage by drying over time is because the greatest changes in volume of the concretes with LC2 occur in the first 24 h, as the clay absorbs the greatest amount of water in the system and the hydration compounds are formed more quickly. In the case of concretes with P-35, the chemical shrinkage occurs more slowly than when they enter the drying stage, and therefore the volume changes due to shrinkage by drying are greater (Fig. 3).

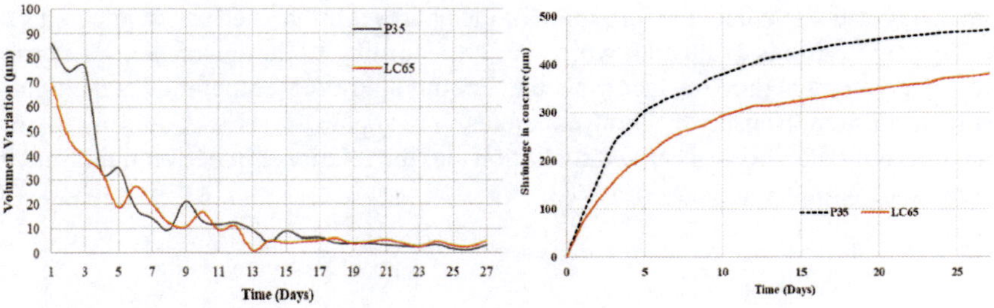

a) b)

Fig. 3. Change in volume at late ages. a) Change in volume b) Shrinkage by drying

3.3 Relationship Between Deformations in Pastes and Concretes

In a cement with pozzolanic addition, the calcium hydroxide, product of the reaction with the silicates, combines with the pozzolanic material producing more tobermorite gel, which is mainly responsible for hardening and gaining mechanical resistance. This is why the relationship between the formation of the phases of the hydrated cement and the properties of the concrete, all related to changes in volume, is being investigate (Fig. 4).

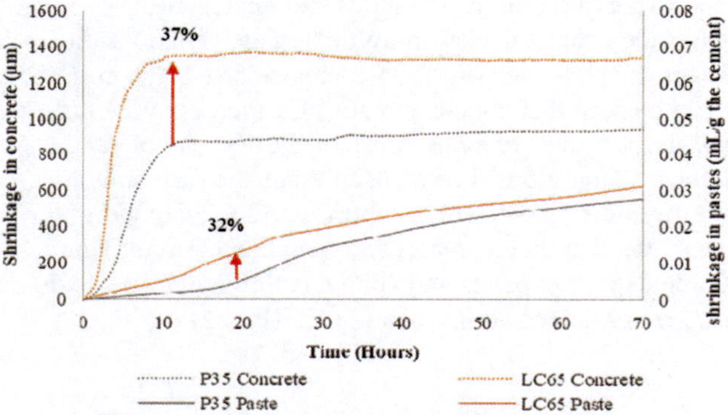

Fig. 4. Comparison between shrinkage in concrete and cement paste

When a relation is made between chemical contraction in pastes and deformation in concretes, it is observed that for both cases the samples with LC2 increase in relation to 37%. The maximum values are observe before 10 h, which coincides with the formation of the gel and the ettringite, and in the case of the concrete, it is observed that they occur. It should be noted that these changes in volume that occur at an early age do not cause cracking, but rather achieve greater compaction in the mass and greater filling of the pores in the paste, as they occur in the fresh state.

Figure 5 shows the deformations that occur in the concrete and the heat that is given off in the cement pastes in the 24-h period for samples with P-35 cement and the LC2

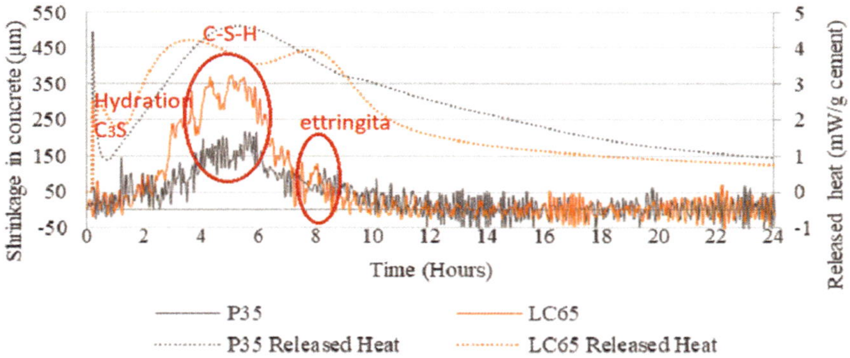

Fig. 5. Correlation between shrinkage in concrete and hydration in paste

addition. It can be seen that from 3 h onwards, concretes with LC2 have a greater variation in volume, they increase in volume together with the greater release of heat. This occurs due to the pozzolanic reaction, since in this process more C-S-H gel and Portlandite are formed, which in this period of time occupy the pores and solidify, which makes the porosity of the system decrease as a consequence of the deposition of the hydrates. It can be seen in the sample with the addition of LC2, a slight increase in volume after 8 h which coincides with the second peak in hydration which is the formation of ettringite. No major changes are defined at this point, as hemicarboaluminates (C3A.CH0, 5$C\overline{C}$ 0.5.H12) and monocarboaluminates (C3A.$C\overline{C}$.H12) from pozzolan and clinker are formed. This type of phases is stabilize in the presence of high quantities of reactive alumina and CO32-(ac) ion source materials, such as finely ground limestone, as is the case with LC2. The SO42-(ac) ions released in the formation of this type of phase, contribute to stabilizing the ettringite in the cementitious matrix so that the major volume changes do not occur in this stage (Sergito).

For the P-35 cement it is observed that the biggest changes occur in the same stage, which defines that for both cases, the first 12 h of the concrete elaboration is the stage of biggest volume changes. These variations in the first ages do not cause cracking in the concrete, because they occur in a fresh mass and their effect is given in providing some compaction and filling of the pores of the cement paste that favors the compactness and resistance of the concrete.

4 Conclusions

1. The addition of clay + limestone + gypsum called LC2 can be used as a mitigation measure of the fissures produced by shrinkage, since these volume changes occur in the cement paste and with the substitution of 30%, this phenomenon is reduced.
2. The maximum values of deformation that occur in the concretes with the addition of LC2 are the first ages of elaboration, mainly in the stage of average hydration and to a lesser extent in the reaction of the aluminates, where the mass is still in fresh state and the accommodation of the components of the concrete occurs.

3. Concretes with the LC2 addition show less variation in volume at later ages, a characteristic which makes them more resistant to damage during the 28-day period, such as poor curing, a determining factor in achieving adequate quality.

References

1. Habert, G., Choupay, N., Montel, J.M., Guillaume, D., Escadeillas, G.: Effects of the secondary minerals of the natural pozzolans on their pozzolanic activity. Cem. Concr. Res. **38**(7), 963–975 (2008)
2. Monteagudo, S.: Estudio microestructural y de los procesos de hidratación de cementos con adiciones. Tesis Doctoral (2014)
3. Grassl, P., Wong, H.S., Buenfeld, N.R.: Influence of aggregate size and volume fraction on shrinkage induced micro-cracking of concrete and mortar. Cem. Concr. Res. [Internet] **40**(1), 85–93 (2010). https://doi.org/10.1016/j.cemconres.(2009.09.012)
4. Idiart, A., Bisschop, J., Caballero, A., Lura, P.: A numerical and experimental study of aggregate-induced shrinkage cracking in cementitious composites. Cem. Concr. Res. [Internet] **42**(2), 272–281 (2012). https://doi.org/10.1016/j.cemconres.(2011.09.013)
5. Scrivener, K., Snellings, R., Lothenbach, B.: A Practical Guide to Microstructural Analysis of Cementitious Materials. Escuela Politécnica Federal de Lausanne, Suiza (2015)

Formulation of Expansive Mortars Based on Calcareous-Based Additives and Ternary Cement with Calcined Clay and Limestone

Dania Betancourt-Cura[1](✉) and Fernando Martirena-Hernandez[2]

[1] Department of Civil Engineering, School of Construction, Central University of Las Villas, Santa Clara, Cuba
daniab@uclv.edu.cu
[2] CIDEM, School of Construction, Central University of Las Villas, Santa Clara, Cuba
f.martirena@enet.cu

Abstract. The present work deals with the evaluation of the addition of a calcareous-based expansion additive in formulations of type III mortars, consisting of Portland cement, LC2 and stone powder as aggregate, to obtain expansive mortars. The design of the experiment was carried out in order to evaluate the influence of the calcareous-based expansion additive on masonry mortars with mineral addition LC2, following the criteria established in NC 175: 2018 "Masonry mortars - specifications".

The work included the evaluation of additions of 2, 4 and 7% of the weight of the binder in additive on the compressive strength and expansion of the formulated mortars. The results of the tests carried out showed that the compressive strength of the samples at 28 days is higher than the 5.2 MPa required by NC 175: 2018. As the addition of the calcareous-based expander additive increases, the compressive strength of the mortar decreases, indicating an influence on the hydration of the cement and the impact of the increase in the water/cement ratio.

Mortars without expansion additive show the typical shrinkage expected from the use of cement. Minimal additions, in this case of 2% of the binder weight, already cause deformations in the same order as the reference ones (obtained from international references), over 0.12%. The addition of 4% by weight of binder causes deformations in the order of 0.18%, which in this case are considered the desired ones.

The sample that yielded the best results in the Compressive Strength and Retraction Channel tests is M-2 for a 4% addition of calcareous-based expander additive, reaching a Compressive Strength of 6.75 MPa at 28 days and the highest expansion value at 24 h. This is, therefore, the recommended dose in the formulation of the expansive mortar for channel joints.

1 Background Information

The National Institute of Hydraulic Resources (INRH) currently has a network of channels throughout the country, which guarantee the irrigation of agricultural products, mainly rice. Said channels are today in frank deterioration, especially the expansion

J. I. Escalante-Garcia et al. (Eds.): RW 2021, RILEM Bookseries 40, pp. 159–167, 2023.
https://doi.org/10.1007/978-3-031-21735-7_19

joints. To repair these and specifically said joints, expansive mortars are needed, for this the country would have to import a whole range of products that today are produced and marketed in the world for these purposes.

The economic situation of the country urgently demands the substitution of imports and border production in this case of expansive mortars with national raw materials that allow the repair of said channels.

This research work evaluated in formulations of type III mortars, made up of Portland cement, LC2 and stone powder as aggregate, the addition of a calcareous-based expansive additive to obtain expansive mortars.

2 Objective

Evaluate in formulations of type III mortars, made up of Portland cement, LC2 and stone dust as aggregate, the addition of calcium oxide to obtain expansive mortars.

3 Materials and Methods

The design of the experiment was carried out in order to evaluate the influence of the addition of CaO to masonry mortars with mineral addition LC2, following the criteria established in NC 175: 2018.

(1) Carrying out characterization tests on raw materials: aggregate, Portland cement, LC2 and calcium oxide.
(2) Manufacture of type III masonry mortars, fixing the fluidity, attending to the dosages and the parameters established in NC 175: 2018 Masonry mortars.
(3) Elaboration of 36 specimens of 40 × 40 × 160 mm, to carry out the following test:

    (a) Mechanical resistance to bending-compression at 3.7 and 28 days, (3 specimens per mortar sample).

(4) Preparation of the mortar to carry out the test of the retraction channel (Table 1).

Table 1. Proportions of the samples for the tests of consistency, resistance to flexo-compression and retraction channel.

Samples	P-35	LC2	Aggregate	% Additive
MP-1-P35+LC2	0.5	1	4	0
M-1-P35+LC2	0.5	1	4	2
M-2-P35+LC2	0.5	1	4	4
M-3-P35+LC2	0.5	1	4	7

4 Results

4.1 Consistency

Table 2 shows the results of the consistency test on the samples tested.

Table 2. .

Nomenclature	Cement P-35+LC2 (g)	Additive (g)	Water (ml)	W/c ratio
MP-1-P35+LC2	228	0	210	0.92
M-1-P35+LC2	228	4.56	220	0.96
M-2-P35+LC2	228	9.12	225	0.98
M-3-P35+LC2	228	15.56	235	1.03

The consistency to be achieved is checked on the shaking table, where a minimum diameter of 190 mm ± 5 mm must be guaranteed following the protocol of the "NC 170: 2002 standard" Fresh mortar.

The water-cement ratio for all the tested samples is high and as the addition of the calcareous-based expander additive increases, the amount of water to be added to the mortar increases to achieve the fluidity established by the NC 170: 2002 standard" Fresh mortar. This occurs due to the ability of calcined clays to retain water, which cannot be used to guarantee flow [7].

4.2 Compressive Strength Tests

Next, Table 3 shows the results of the Compressive Strength test of the samples at the ages of 3, 7 and 28 days (Fig. 1).

Despite the high water/cement ratio at which the desired consistency is achieved, the compressive strength values of mortars are very high from an early age, and after 7 days the mortar reaches more than 80% of the minimum resistance requested in the NC 175: 2018 standard. Masonry mortars - specifications at 28 days (5.2 MPa). Increasing the addition of the calcareous-based expander additive decreases the compressive strength. The literature explains this phenomenon by the impact of the increase in the concentration of Ca^{++} ions in the pore solution, which inhibit the hydration of the anhydrous phases of clinker [4, 8, 10, 11]. This problem is solved by increasing the sulfate concentration in the system.

4.3 Expansion in the Shrinkage Drain

Figure 2 presents the results of the volume change produced with the addition of the calcareous-based expander additive. The results of similar mortars obtained from the international literature are shown for reference. All the tests were carried out with the pure additive, that is, with more than 90% CaO; hence its maximum expandability.

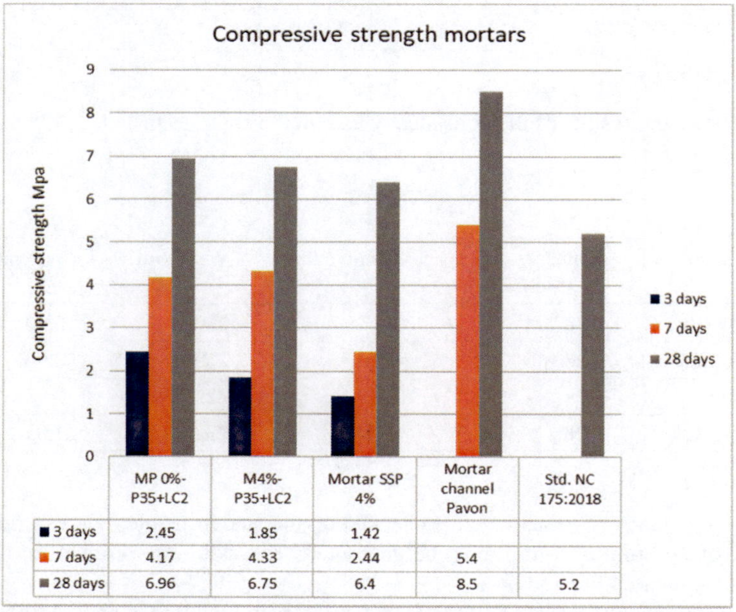

Fig. 1. Compressive strength results.

Table 3. .

Nomenclature	Compressive strength (MPa)		
	3d	7d	28d
MP 0%-P35+LC2	2.45	4.17	6.96
M 2%-P35+LC2	1.96	4.86	6.32
M4%-P35+LC2	1.85	4.33	6.75
M7%-P35+LC2	1.91	3.12	5.43

Mortars without expansion additive show the typical shrinkage expected from the use of cement. Minimal additions, in this case of 2% of the binder weight, already cause deformations in the same order as those expected (obtained from international references), over 0.12%. The addition of 4% by weight of binder causes deformations in the order of 0.18%, which in this case are considered the desired ones. The tests carried out with the dose of 7% by weight of binder caused a breakdown of the specimen, and for this reason they are discarded.

It is still pending to evaluate the impact that the expansion of the mortar can generate in a confined environment, and the compression stresses that they generate at very early ages, and if the mortar is capable of resisting them without failing.

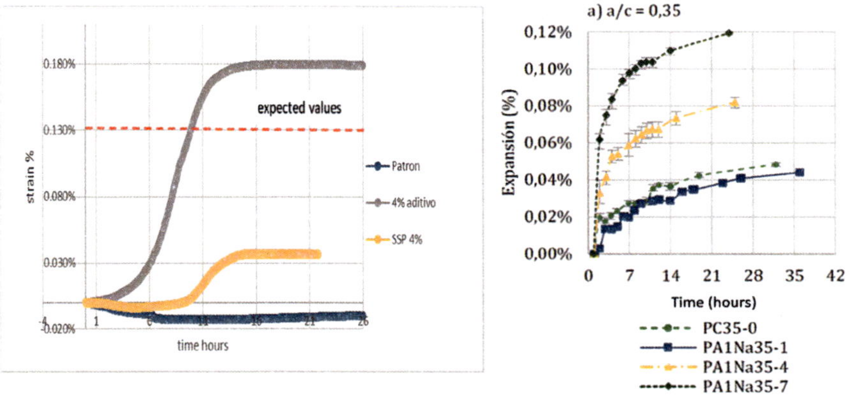

Fig. 2. Results of the retraction channel (left) Experimental results; (right) Results obtained from the literature.

4.4 Hydration Kinetics

The curves of heat released normalized to the clinker content in the paste and to the paste content, indicative of the hydration kinetics, are shown in Figs. 3 and 4 respectively. The accumulated total heat values at different ages, normalized to the clinker content in the paste and to the paste content, are shown in Figs. 5 and 6. In all cases, an additional reference series, called P-35, was included to complement the analysis and consisting of the same P-35 cement used in the mix designs, prepared at a water/cement ratio of 0.40.

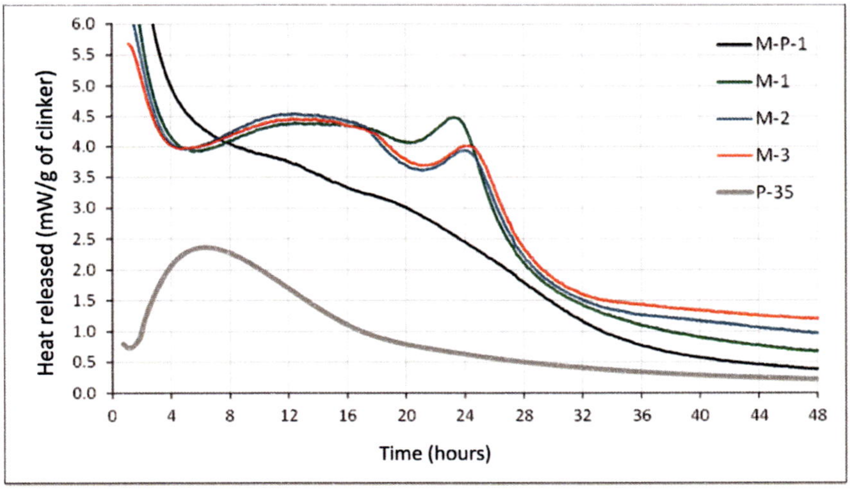

Fig. 3. Heat released curves normalized to clinker content.

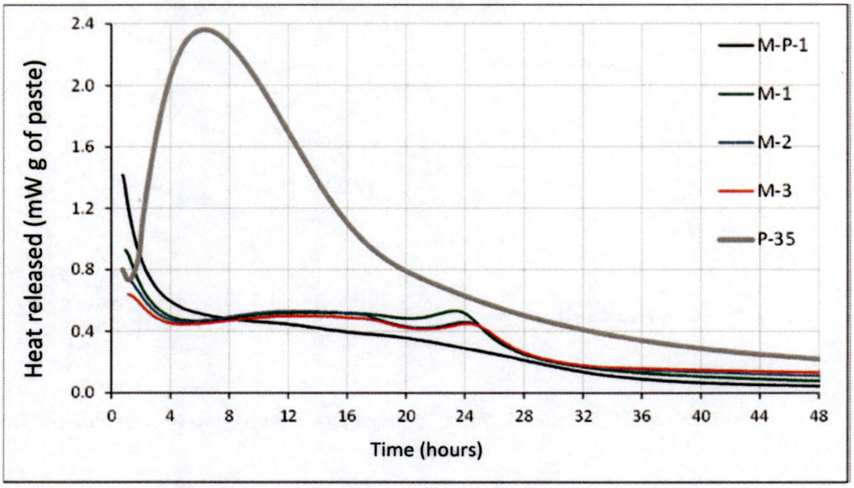

Fig. 4. Curves of heat released normalized to paste content.

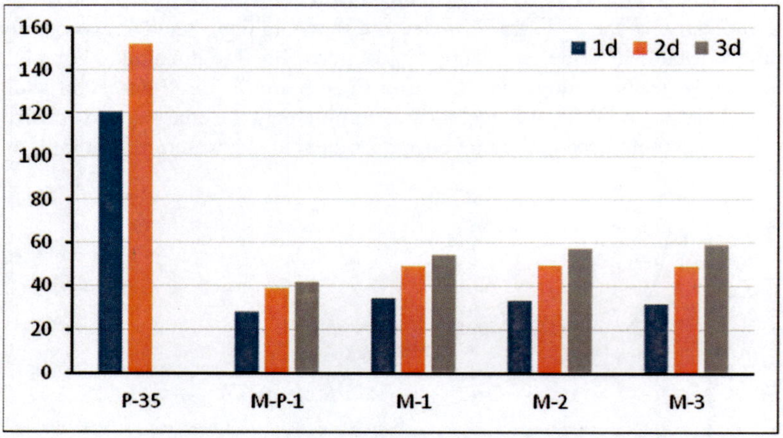

Fig. 5. Total heat accumulated at different ages normalized to clinker content.

The control series M-P-1, without additions of CaO, presents an abnormal behavior, with a prolonged period of dissolution and induction. Nor can the main maxima corresponding to the hydration of the cement be clearly distinguished. This behavior is attributed to the low clinker content in the system and may also be associated with insufficient sulphation.

The series with the incorporation of CaO dosages show better behavior, with a clear distinction between the maximums corresponding to the hydration of the cement, mainly the one corresponding to the secondary formation of ettringite. In all cases, the combined effect of the addition of CaO and the increase in clinker dilution due to the

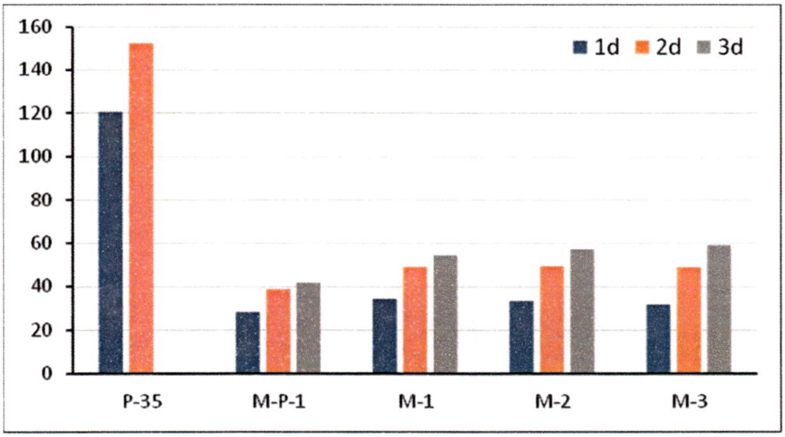

Fig. 6. Total heat accumulated at different ages normalized to pasta content.

increase in the water/solid ratio in the paste increases the filler effect as can be deduced from the comparison of the heat released curves and the values of total accumulated heat normalized to the clinker content in the paste. The development of hydration products is slow in time, much more than for the reference series P-35, and appreciable values of the degree of hydration development are only reached after 36 h. This phenomenon is referred to in the literature.

This behavior compromises the development of resistance at an early age. The increase in the addition of CaO and the increase in the water/solid ratio in the paste only seems to have an appreciable impact on the development of the hydration products after three days, as observed in Figs. 5 and 6.

The volume change analysis (Fig. 7) clearly shows a notable decrease in chemical shrinkage in pastes with CaO incorporation, very possibly due to the expansive effect of the addition of CaO, as referred to in the literature [4, 8, 10, 11]. However, this effect does not show a behavior that is directly proportional to the CaO content since the chemical shrinkage decreases according to the sequence M-2 (4% CaO) > M-1 (2% CaO) > M-3 (7% CaO). In all cases, it is evident that M-2 provides the best results. This decrease in chemical shrinkage tends to decrease appreciably after 48 h, and in this way the controlled nature of the expansion caused is ratified.

5 Conclusions

- The lime-based additive developed to cause the mortars to expand has proven to work as expected. The mortars produced show a controlled expansion, which occurs after 12 h and ends after 48 h. This will solve the problems of waterproofing channel joints.
- The addition in proportions of 2, 4 and 7% of the weight of the binder produces an increase in the demand for water; As the addition of the calcareous-based expansion additive increases, the amount of water to be added to the mortar increases to achieve the fluidity established by NC 170: 2002.

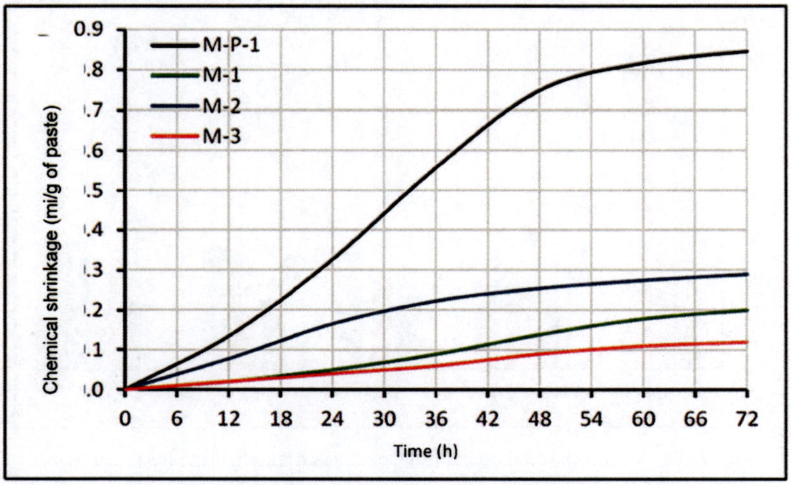

Fig. 7. Chemical shrinkage normalized by paste content.

- The increase in water demand does not affect the compressive strength results of the samples, which at 3 days already reach the values required by the standards at 28 days.
- Increases in the expander additive have a negative impact on the resistance at early ages, which is explained by the inhibition of the hydration of the cement reaction products due to the excess of calcium in the system.
- Mortars without expansion additive show the typical shrinkage expected from the use of cement. Minimal additions, in this case of 2% of the binder weight, already cause deformations in the same order as those expected (obtained from international references), over 0.12%. The addition of 4% by weight of binder causes deformations in the order of 0.18%, which in this case are considered the desired ones.
- The sample that yielded the best results in the Compressive Strength and Retraction Channel tests is M-2 for 4% addition of calcareous-based expander additive, reaching a compressive strength of 6.75 MPa at 28 days and the highest expansion value at 24 h. This is, therefore, the recommended dose in the formulation of the expanding mortar for channel joints.
- Studies of the hydration kinetics of the binder used in pastes confirm the results described in the literature: (i) the impact of the increase in Ca^{++} concentration on the decrease in hydration at 24 h, and (ii) the controlled character of the expansion caused by the addition of CaO, which is considered finished after 48 h.

Reference Literatura

1. Forster, A.: Hot-lime mortars: a current perspective. J. Archit. Conserv. 37–41 (2014). https://doi.org/10.1080/13556207.2004.10784923
2. Moropoulou, A., Bakolas, A., Aggelakopoulou, E.: The effects of limestone characteristics and calcination temperature to the reactivity of the quicklime. Cem. Concr. Res. **31**, 633–639 (2001)
3. Shahraki, B.K., Mehrabi, B., Gholizadeh, K., Mohammadinasab: Thermal behaviour of calcite as expansive agent. J. Min. Metall. **47**, 89–97 (2011). https://doi.org/10.2298/JMMB1101089S
4. Rubio, C.F.: Development of materials with compensated shrinkage from calcium oxide (CaO). Universitat Politècnica de Catalunya (2018)
5. Sisomphon, K., Copuroglu, O.: Self healing mortars by using different cementitious materials (2010)
6. Collepardi, M.: Method of producing expansive and high strength cementitious pastes, mortars and concretes, US4046583A (2020)
7. Collepardi, M., Troli, R., Bressan, M., Liberatore, F., Sforza, G.: Crack-free concrete for outside industrial floors in the absence of wet curing and contraction joints. Cem. Concr. Compos. **30**, 887–891 (2008). https://doi.org/10.1016/j.cemconcomp.2008.07.002
8. Hou, P., et al.: Mechanisms dominating thixotropy in limestone calcined clay cement (LC3). Cem. Concr. Res. **140**, 106316 (2021). https://doi.org/10.1016/j.cemconres.2020.106316
9. Polat, R., Demirboga, R., Karagöl, F.: Mechanical and physical behavior of cement paste and mortar incorporating nano-CaO. Wiley Fib. 11–17 (2018). https://doi.org/10.1002/suco.201800132
10. Khushefati, W.H., Demirbog, R.: Effects of nano and micro size of CaO and MgO, nano-clay and expanded perlite aggregate on the autogenous shrinkage of mortar. Constr. Build. Mater. J. **81**, 268–275 (2015). https://doi.org/10.1016/j.conbuildmat.2015.02.032
11. Cavalaro, S.H.P.: Development of materials with compensated shrinkage from calcium oxide (CaO) (2017)
12. Antiohos, S.K., Papageorgiou, A., Papadakis, V.G., Tsimas, S.: Influence of quicklime addition on the mechanical properties and hydration degree of blended cements containing different fly ashes. Constr. Build. Mater. **22**, 1191–1200 (2008). https://doi.org/10.1016/j.conbuildmat.2007.02.001

Data Mining Strategies to Handle State of Art Knowledge on Self-healing Capacity of Cementitious Materials

Shashank Gupta[1(✉)], Salam Al-Obaidi[2,3], and Liberato Ferrara[2]

[1] Princeton University, Princeton, NJ, USA
[2] Politecnico di Milano, 20133 Milan, MI, Italy
liberato.ferrara@polimi.it
[3] University of Al-Qadisiyah, Al Diwaniyah, Iraq

Abstract. Concrete and cement-based materials inherently possess an autogenous self-healing capacity, which is even high in High and Ultra-High-Performance Concretes because of the high amount of cement and supplementary cementitious materials (SCM) and low water/cementitious material (w/c) ratio. Despite the huge amount of literature on the topic self-healing concepts still fail to consistently enter design strategies to effectively quantify their benefits on the structural performance. In this study, quantitative relationships have been developed through design charts and artificial neural network (ANN) models. The employed approaches aimed at establishing a correlation between the mix proportions, exposure time, the width of the initial crack, and volume of fibers against suitably defined self-healing indices (SHI), quantifying the recovery of material performances which can be of interest for intended applications. Therefore, this study provides, for the first time in the literature to the authors' knowledge, a holistic investigation on the autogenous self-healing capacity of cement-based materials based on extensive articles focused on the literature data mining. The design charts are developed to pave the way towards consistent incorporation of self-healing concepts into durability-based design approaches for reinforced concrete structures, aimed at quantifying, with reliable confidence, the benefits in terms of slower degradation of the structural performance and extension of the service lifespan. Finally, through ANN, a straightforward input-output model is developed to quickly predict and evaluate the self-healing efficiency of cement-based materials which can significantly reduce, in the design stage, the time, money, and efforts of laboratory investigation.

Keywords: Self-healing concrete · Durability-based design · Artificial Neural Network

1 Introduction

Autogenous healing is the natural process of total or partial self-closure of cracks in cementitious materials. The durability performance and initial strength of concrete are partially or completely regained following the closure of cracks. Autogenous self-healing

J. I. Escalante-Garcia et al. (Eds.): RW 2021, RILEM Bookseries 40, pp. 168–182, 2023.
https://doi.org/10.1007/978-3-031-21735-7_20

is not a recent discovery, in contrast, this phenomenon was identified by the French Academy of Science as early as 1836 [1]. In earlier permeability experiments of saturated concrete, it was observed that the permeability coefficient was noticeably deviating from the one calculated utilizing Darcy's law of fluid flow through porous solids [2]. It was observed that the permeability coefficient was reducing with time violating Darcy's assumption of the linear relationship between the hydraulic gradient and water flow velocity.

The mechanisms of autogenous healing are continuing hydration, dissolution & crystallization, particle clogging, and carbonation. Based on the mechanism of autogenous self-healing of cementitious material, several parameters have been determined that influence the rate of self-healing. Crack width, mix design, fiber content, together with treatment duration, and environmental exposure conditions are the most significant variables that affect the self-healing of concrete [1].

Numerous rigorous studies have been performed to determine the relationship between recovery rate and crack width for different types of concrete [3–6]. A logarithmic relation was obtained between the normalized flow rate and time of exposure for high-performance concrete with fly ash, micro-silica, and water-reducing agent. It was observed that for larger cracks, the permeability was increasing at a higher rate. The experimental results were also in close agreement with the law of Hagen–Poiseuille [6].

The addition of SCM and crystalline admixtures (CA) can additionally improve the rate and efficiency of self-healing as a "stimulated-autogenous healing" approach [7–9]. It has been observed that both compressive strength and electrical resistivity can be recovered up to 94% by partial replacement of cement with fly ash and/or the addition of crystalline admixture [10]. Similarly, in a separate study, the crack healing rate of around 81% to 93% was reported for concrete containing CA compared to 83% value for concrete specimens without CA [11].

The addition of fiber has a substantial effect on the self-healing of concrete as fibers control the opening of the cracks that make self-healing products easily attach to the surface of the crack and fill them [12, 13]. The microstructural characterization of the concrete containing different synthetic fibers, like polypropylene (PP), ethylene vinyl alcohol (EVA), and poly-vinyl acrylate (PVA) has concluded that the precipitation of calcium carbonate potentially increases with the inclusion of synthetic fibers [14].

The existing literature shows that the self-healing of concrete has been extensively studied over the last few decades. However, the parameters affecting autogenous healing have been well investigated qualitatively, not quantitatively. Moreover, there is a lack of self-healing concepts into durability-based design approaches for reinforced concrete structures. This study takes a preliminary step toward the standardization and codification of the existing findings of self-healing of cementitious materials through the aid of design charts and neural network models.

2 Methodology

2.1 Data Retrieval

The relevant data has been collected from multiple peer-reviewed articles. In the preliminary survey, 236 articles associated with the self-healing capacity of cementitious materials were obtained from "Scopus". However, following the scrutiny of the data, numerous papers were removed based on the parameters investigated in this study. The data points used to prepare both design charts and neural network modeling are finally collected from 100 publications. It should be noted that most of the data have been obtained from recent publications. The articles published between 2018 and 2021 comprises 50% of the total sources.

The input data retrieved from each publication was comprised of the dosage of cement (kg/m^3), the dosage of SCM (kg/m^3), fiber content (%Vol), w/c ratio, crack width (μm), and period of healing (days). It has been further observed that in a large proportion of studies cracked concrete specimens have been subjected to complete immersion at ambient temperature as a healing environment. Therefore, the data for such specimens have been used in this study.

The output parameters collected are appropriate results of the self-healing characterization test and geometrical closure of the crack. The relevant results are then converted into "Self-Healing Indices" (SHI) to reflect the healing capacity of cementitious material. The values of SHI vary from 0 to 1, signifying no healing to full healing of crack, respectively. The formulas of SHI for several tests are presented in Table 1. Moreover, the number of articles used to study the effect of crack size, healing period, w/c ratio, fiber content, and SCM are 42, 37, 14, 7, and 14.

Table 1. SHI formulas for different tests.

S. no.	Experiment	Parameters	Formula
1	Crack size	w_i = initial width & w_f = final width	$(w_i - w_f)/w_i$
2	Permeability test	Q_i = initial flow & Q_f = final flow	$(Q_i - Q_f)/Q_i$
3	Ultrasonic Pulse Velocity (UPV) Test	U_h = UPV of healed sample & U_c = UPV of cracked sample	$(U_h - U_c)/U_h$
4	Mechanical strength	M_h = Strength of healed sample & M_c = Strength of cracked sample	$(M_h - M_c)/M_h$

2.2 Development of Charts

Based on the collected data, optimization charts are developed for optimized design concerning the self-healing capacity of cementitious material. The Delaunay triangulation has been implemented to evaluate the interpolated values of SHI to generate the contours lines. Delaunay triangulation is used as an optimal method to generate a linear interpolation function that can be used to reconstruct the functional approximation in 3-dimensional space. The Delaunay triangulation is applied to the pair of chosen parameters (like cement content, crack width, etc.) and linear interpolation functions are constructed based on the value of response (SHI).

2.3 ANN Analysis

The total number of data points obtained from the collected literature was around 2800. For ANN analysis, the data corresponding to cases where crack size and age of treatment had not been provided were eliminated. The 5 input variables are taken into consideration, i.e., cement dosage (kg/m^3), SCM dosage (kg/m^3), fiber content (%vol), w/c ratio, crack width (μm), and age of treatment (days).

For this preliminary study, a single-layer feed-forward neural network has been developed. This model has been optimized in terms of several parameters viz. Training algorithms, adaptation learning functions, number of neurons, and activation function. In this study, the ratio of training to validation to testing subset has been chosen as 0.7:0.15:0.15 [15]. The weights are calculated repetitively until they become constant for each cycle.

3 Results

3.1 Optimization Chart

The optimization charts of some of the design parameters are shown in Figs. 1, 2, 3, 4 and 5. The SHI contours for the w/c ratio and the dosage of cement (kg/m^3) in concrete are presented in Fig. 1. The results obtained depend on the composition of concrete, generally having a high w/c ratio for low-cement concrete (low-quality concrete) and vice-versa (high-performance concrete). The SHI also gets significantly influenced by other factors like the period of healing and width of cracks. It is notable from the chart that SHI increases with cement content even with low w/c ratios, which is reasonable because this will correspond to the higher availability of un-hydrated cement for further hydration reactions.

The effect of SCM on the SHI is presented in Fig. 2 using cement content and the amount of total cementitious material (TCM). It has been observed that the most common SCM used in studies were fly ash, ground granulated blast-furnace slag (GGBS), and silica fumes. As anticipated, the SHI increases with the amount of total cementitious material in concrete. However, it can be noted that there is a range for SCM content for the given amount of total cementitious material for optimum SHI.

Following the peak, the increase in cement content in total cementitious constituent is preferred for improved SHI. This can be attributed to the thermo-chemistry of hydration

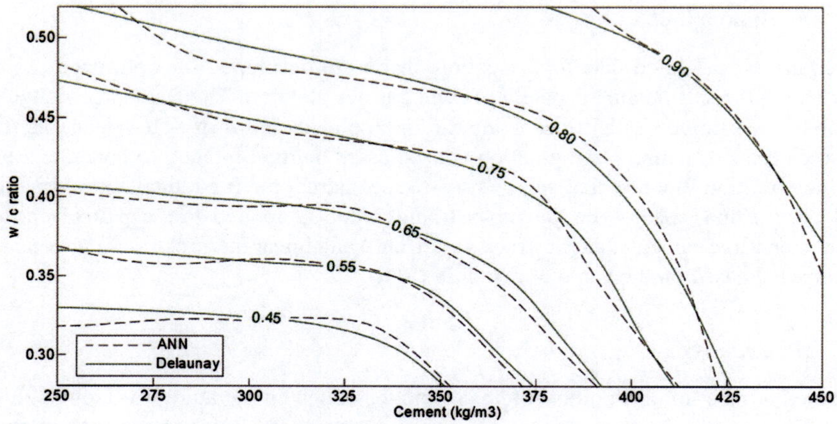

Fig. 1. Optimization chart for cement content and w/c ratio.

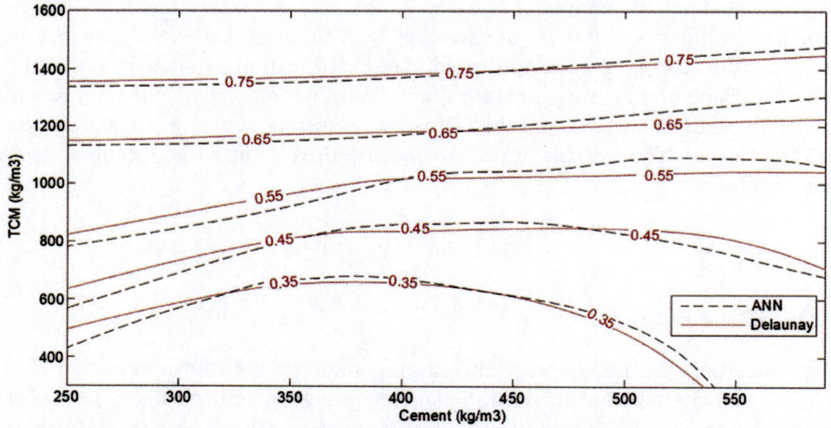

Fig. 2. Optimization chart for cement content and total cementitious material (TCM).

reaction of the cement-fly ash blend in concrete. The hydration reaction of fly ash utilizes the hydration products of cement, namely the calcium hydroxide to produce new calcium silicate hydrates [16]. Hence, the low amount of cement in the blend results in a lack of activation of the hydration reaction of fly ash. This can be overcome with the addition of activators like sodium hydroxide and sodium silicate in the mix design [17].

The influence of crack width on the self-healing efficiency is presented in Fig. 3. It is evident from the chart that small cracks (50–100μ) have high SHI, around 0.75–1, even in concrete with low cement content. The SHI index drops significantly with higher crack size, going as low as 0.15 for concrete with cracks of 400 microns. The effect of cement content is distinct in the smaller cracks, where cracks heal with a higher healing index in high-cement concrete.

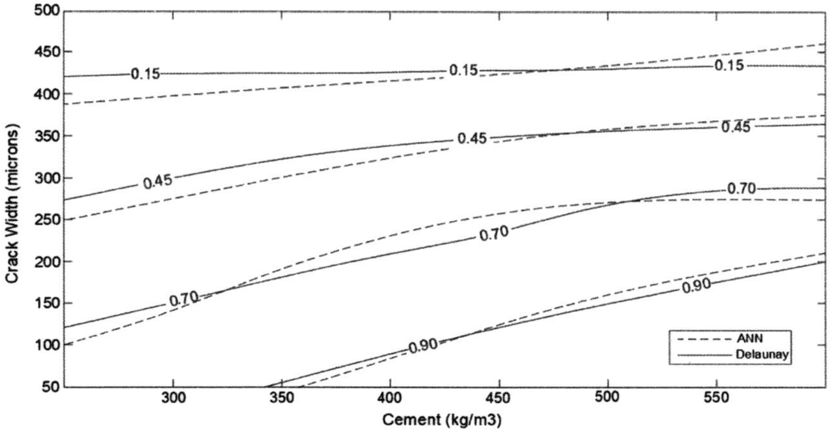

Fig. 3. Optimization chart for cement content and crack width.

The contour graph of SHI, mapping the effect of the period of healing as well as cement content, is presented in Fig. 4. For a longer period of healing, the effect of cement becomes more pronounced on healing capacity as the hydration reaction is completed to a greater degree following the initial 28 days of the interaction of cement and water [18]. Therefore, for a longer period of exposure, the SHI depends on the time water takes to reach the available un-hydrated cement in numerous pores.

The influence of fiber content on the self-healing capacity of concrete is shown in Fig. 5. The increase in the amount of fiber causes a slight improvement in the SHI of concrete. It is visible from the chart that for low-cement concrete, the increase in SHI due to the addition of fiber is slightly more pronounced than high-cement concrete. A peculiar behavior has been shown by the chart in which the healing indices seem to decrease with cement for high-cement concrete for a particular volume of fiber. This can be due to the influence of other significant parameters like large crack size or low treatment period.

3.2 ANN Analysis

The stepwise optimization of the ANN model has been performed in MATLAB. Firstly, the root means square error (RMSE) values are calculated for the optimization of the training algorithm and the number of neurons. The minimum training RMSE value was obtained for Levenberg-Marquardt (LM) algorithm. In the LM algorithm, the lowest RMSE of the training set and validation set such that RSME of validation set is slightly higher than training set is for 18 number of neurons in the hidden layer. Therefore, the LM algorithm and 18 neurons were adopted for further investigation.

The optimization corresponding to adaptation learning functions and activation functions were also performed w.r.t. correlation coefficient, R. The two adaptation learning functions were gradient descent (learngd) and gradient descent with momentum (learngdm), while two activation functions are hyperbolic tangent sigmoid (tansig) and

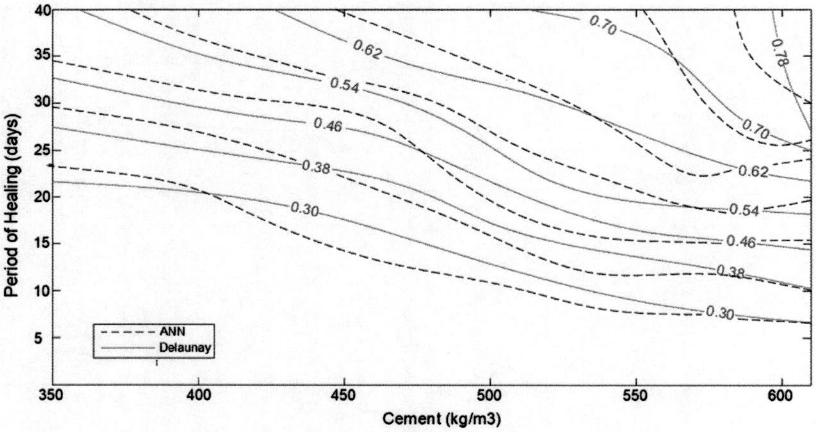

Fig. 4. Optimization chart for cement content and period of healing.

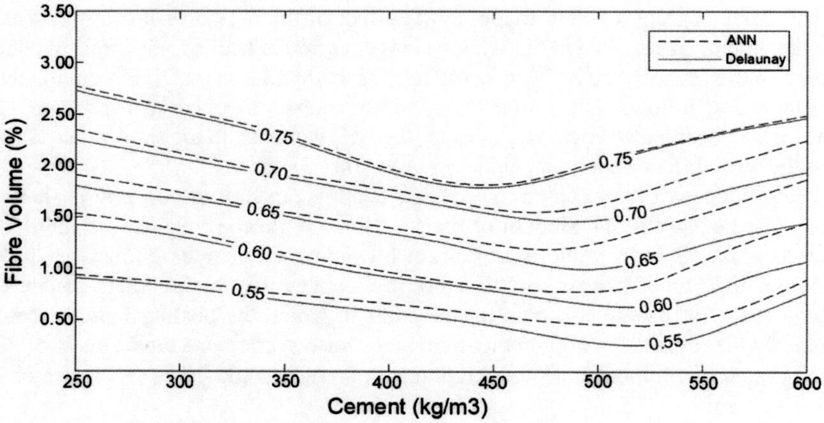

Fig. 5. Optimization chart for cement content and fiber content.

hyperbolic tangent (tanh) function. The four combinations based on two parameters are learngd-tansig, learngd-tanh, learngdm-tansig, and learngdm-tanh. The best regression model was obtained for the learngdm-tansig combination (Fig. 6). The design charts are generated using the optimized model and have been compared with the charts obtained from Delaunay's approximation (Figs. 1, 2, 3, 4 and 5). The output is the predicted SHI value of the neural network model and the target is the inputted SHI value. This further investigation yields a robust mechanism that would provide a more accurate approximation of SHI based on all the input parameters involved.

The SHI contour plots based on crack size, TCM, and w/c ratio (Figs. 1, 2 and 3) are in good agreement with the theoretical interpolation chart with small errors. It can be observed that the neural network model is showing the dependence of healing capacity on

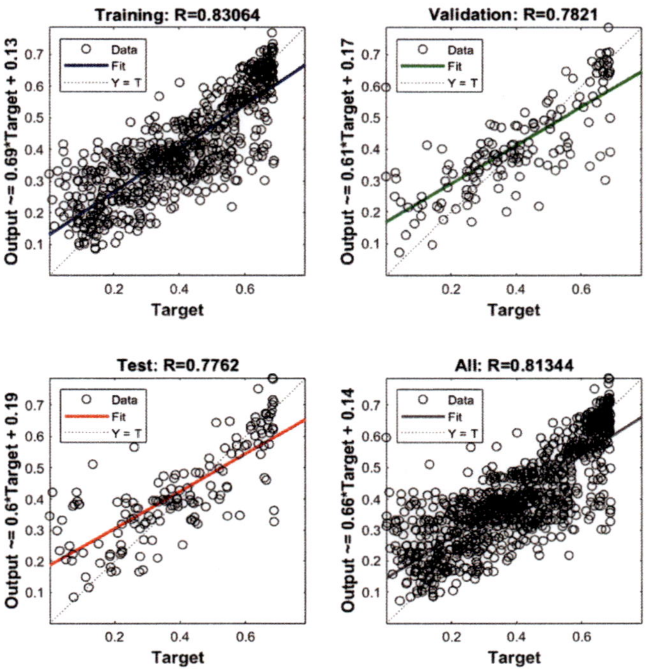

Fig. 6. Regression model for gradient descent (momentum) and hyperbolic tangent sigmoid.

the amount of cement in high-cement concrete contrary to interpolation approximation. This is because neural networks consist of several nodes that provide a better estimate than Delaunay triangulation interpolation.

4 Conclusion

A quantitative review of more than 100 papers has been carried out to study the effect of several factors on the self-healing capacity of cementitious systems. The numerical values of these factors and corresponding SHI (calculated) are obtained to carry on quantitative analysis. The summary of important findings corresponding to meta-analysis, design charts, and neural network model is presented below:

1. The design chart of w/c ratio and cement content shows that SHI increases with cement content even with low w/c ratios.
2. The analysis of SCM shows that there is a peak range for SCM content for the given amount of total cementitious material for optimum SHI.
3. Design chart investigating the effect of the period of healing shows rapid recovery at the beginning of the treatment process. For a longer period of healing, the effect of cement becomes more pronounced on healing capacity as the hydration reaction

is completed to a greater degree following the initial 28 days of the interaction of cement and water.

4. The increase in the volume of fibers causes a slight improvement in the SHI of concrete.
5. The results obtained were in good agreement with the existing literature. However, more range of parameters should be included, and more relevant papers should be documented to enable meta-analysis to cover a wider range of applicability.
6. A single-layer back-propagation optimized ANN algorithm has been developed to predict the self-healing efficiency of cement-based materials. The predictions were a good agreement for SHI corresponding to crack size, total cementitious material, and w/c ratio. A moderate error was observed for fiber content and period of healing. This can be attributed to the lack of demarcation of data points based on more parameters.
7. The results obtained were in good agreement with the existing literature. However, more range of parameters, like treatment conditions, type of admixtures (including chemical admixtures), types of fibers, etc. should be included and more relevant papers should be documented to enable this analysis to cover a wider range of applicability.

References

1. De Belie, N., et al.: A review of self-healing concrete for damage management of structures. Adv. Mater. Interf. **5**(17), 1800074 (2018)
2. Hearn, N.: Recording permeameter for measuring time-sensitive permeability of concrete. Ceram. Trans. **16**, 463 (1991)
3. Ferrara, L., Krelani, V., Moretti, F.: Autogenous healing on the recovery of mechanical performance of high-performance fibre reinforced cementitious composites (HPFRCCs): part 2–correlation between healing of mechanical performance and crack sealing. Cement Concr. Compos. **73**, 299–315 (2016)
4. Ma, H., Qian, S., Zhang, Z.: Effect of self-healing on water permeability and mechanical property of medium-early-strength engineered cementitious composites. Constr. Build. Mater. **68**, 92–101 (2014)
5. Li, V.C., Yang, E.H.: Self-healing in concrete materials. In Self-healing Materials, pp. 161–193. Springer, Dordrecht (2007)
6. Reinhardt, H.W., Jooss, M.: Permeability and self-healing of cracked concrete as a function of temperature and crack width. Cem. Concr. Res. **33**(7), 981–985 (2003)
7. Ferrara, L., Krelani, V., Moretti, F.: On the use of crystalline admixtures in cement-based construction materials: from porosity reducers to promoters of self-healing. Smart Mater. Struct. **25**(8), 084002 (2016)
8. Cuenca, E., Mezzena, A., Ferrara, L.: Synergy between crystalline admixtures and nano-constituents in enhancing autogenous healing capacity of cementitious composites under cracking and healing cycles in aggressive waters. Constr. Build. Mater. **266**, 121447 (2021)
9. Monte, F.L., Ferrara, L.: Self-healing characterization of UHPFRCC with crystalline admixture: experimental assessment via multi-test/multi-parameter approach. Constr. Build. Mater. **283**, 122579 (2021)
10. Nasim, M., Dewangan, U.K., Deo, S.V.: Effect of crystalline admixture, fly ash, and PVA fiber on self-healing capacity of concrete. Mater. Today: Proc. **32**, 844–849 (2020)

11. Roig-Flores, M., Moscato, S., Serna, P., Ferrara, L.: Self-healing capability of concrete with crystalline admixtures in different environments. Construct. Build. Mater. **86**, 1–11 (2015)
12. Monte, F.L., Ferrara, L.: Tensile behaviour identification in ultra-high-performance fibre reinforced cementitious composites: indirect tension tests and back analysis of flexural test results. Mater. Struct. **53**(6), 1–12 (2020)
13. Ferrara, L., Krelani, V., Moretti, F., Flores, M.R., Ros, P.S.: Effects of autogenous healing on the recovery of mechanical performance of high-performance fibre reinforced cementitious composites (HPFRCCs): part 1. Cem. Concr. Compos. **83**, 76–100 (2017)
14. Nishikawa, T., Yoshida, J., Sugiyama, T., Fujino, Y.: Concrete crack detection by multiple sequential image filtering. Comput.-Aided Civil Infrastruct. Eng. **27**(1), 29–47 (2012)
15. Shahmansouri, A.A., Yazdani, M., Ghanbari, S., Bengar, H.A., Jafari, A., Ghatte, H.F.: Artificial neural network model to predict the compressive strength of eco-friendly geopolymer concrete incorporating silica fume and natural zeolite. J. Clean. Prod. **279**, 123697 (2021)
16. Moghaddam, F., Sirivivatnanon, V., Vessalas, K.: The effect of fly ash fineness on heat of hydration, microstructure, flow, and compressive strength of blended cement pastes. Case Stud. Construct. Mater. **10**, 00218 (2019)
17. Ramagiri, K.K., Chauhan, D.R., Gupta, S., Kar, A., Adak, D., Mukherjee, A.: High-temperature performance of ambient-cured alkali-activated binder concrete. Innov. Infrastruct. Solut. **6**(2), 1–11 (2021). https://doi.org/10.1007/s41062-020-00448-y
18. Termkhajornkit, P., Nawa, T., Yamashiro, Y., Saito, T.: Self-healing ability of fly ash–cement systems. Cem. Concr. Compos. **31**(3), 195–203 (2009)

Database: List of Sources

The following 100 articles are used to carry out the ANN analysis.

19. Abd_Elmoaty, A.M.: Self-healing of polymer modified concrete. Alex. Eng. J. **50**(2), 171–178 (2011)
20. Alghamri, R., Al-Tabbaa, A.: Self-healing of cracks in mortars using novel PVA-coated pellets of different expansive agents. Construct. Build. Mater. **254**, 119254 (2020)
21. Araújo, M., Van Tittelboom, K., Dubruel, P., Van Vlierberghe, S., De Belie, N.: Acrylate-endcapped polymer precursors: effect of chemical composition on the healing efficiency of active concrete cracks. Smart Mater. Struct. **26**, 055031 (2017)
22. Azarsa, P., Gupta, R., Biparva, A.: Assessment of self-healing and durability parameters of concretes incorporating crystalline admixtures and Portland limestone cement. Cem. Concr. Compos. **99**, 17–31 (2019)
23. Chindasiriphan, P., Yokota, H., Pimpakan, P.: Effect of fly ash and superabsorbent polymer on concrete self-healing ability. Constr. Build. Mater. **233**, 116975 (2020)
24. Darquennes, A., Olivier, K., Benboudjema, F., Gagné, R.: Self-healing at early-age, a way to improve the chloride resistance of blast-furnace slag cementitious materials. Constr. Build. Mater. **113**, 1017–1028 (2016)
25. Deng, H., Liao, G.: Assessment of influence of self-healing behavior on water permeability and mechanical performance of ECC incorporating superabsorbent polymer (SAP) particles. Constr. Build. Mater. **170**, 455–465 (2018)
26. Du, W., Yu, J., Gu, Y., Li, Y., Han, X., Liu, Q.: Preparation and Application of microcapsules containing toluene-di-isocyanate for self-healing of concrete. Constr. Build. Mater. **202**, 762–769 (2019)
27. Feng, J., Su, Y., Qian, C.: Coupled effect of PP fiber, PVA fiber and bacteria on self-healing efficiency of early-age cracks in concrete. Constr. Build. Mater. **228**, 116810 (2019)

28. Gao, S., Jin, J., Hu, G., Qi, L.: Experimental investigation of the interface bond properties between SHCC and concrete under sulfate attack. Constr. Build. Mater. **217**, 651–663 (2019)
29. Granger, S., Loukili, A., Pijaudier-Cabot, G., Chanvillard, G.: Experimental characterization of the self-healing of cracks in an ultra high performance cementitious material: mechanical tests and acoustic emission analysis. Cem. Concrete Res. **37**, 519–527 (2007)
30. Gray, R.J.: Autogenous healing of fibre/matrix interfacial bond in fibre-reinforced mortar. Cem. Concr. Res. **14**, 315–317 (1984)
31. Gupta, S., Dai Pang, S., Kua, H.W.: Autonomous healing in concrete by bio-based healing agents—a review. Constr. Build. Mater. **146**, 419–428 (2017)
32. Guzlena, S., Sakale, G.: Self-healing of glass fibre reinforced concrete (GRC) and polymer glass fibre reinforced concrete (PGRC) using crystalline admixtures. Constr. Build. Mater. **267**, 120963 (2021)
33. He, H., Zhu, Y., Zhou, A.: Electrochemical impedance spectroscopy (EIS) used to evaluate influence of different external pressures, curing ages and self-healing environments on the self-healing behavior of engineered cementitious composites (ECC). Constr. Build. Mater. **188**, 153–160 (2018)
34. Homma, D., Mihashi, H., Nishiwaki, T.: Self-healing capability of fibre reinforced cementitious composites. J. Adv. Concr. Technol. **7**, 217–228 (2009)
35. Hong, G., Choi, S.: Rapid self-sealing of cracks in cementitious materials incorporating superabsorbent polymers. Constr. Build. Mater. **143**, 366–375 (2017)
36. Hung, C.-C., Su, Y.-F.: Medium-term self-healing evaluation of engineered cementitious composites with varying amounts of fly ash and exposure durations. Constr. Build. Mater. **118**, 194–203 (2016)
37. Hung, C.-C., Su, Y.-F., Hung, H.-H.: Impact of natural weathering on medium-term self-healing performance of fiber reinforced cementitious composites with intrinsic crack-width control capability. Cement Concr. Compos. **80**, 200–209 (2017)
38. In, C.-W., Holland, R.B., Kim, J.-Y., Kurtis, K.E., Kahn, L.F., Jacobs, L.J.: Monitoring and evaluation of self-healing in concrete using diffuse ultrasound. NDT and E Int. **57**, 36–44 (2013)
39. Jiang, Z., Li, W., Yuan, Z.: Influence of mineral additives and environmental conditions on the self-healing capabilities of cementitious materials. Cement Concr. Compos. **57**, 116–127 (2015)
40. Jiang, Z., Yuan, Z., Li, W.: Acoustic emission analysis of characteristics of healing products in steam-cured cementitious materials with mineral additives. Constr. Build. Mater. **201**, 807–817 (2019)
41. Kan, L., Shi, H.: Investigation of self-healing behavior of engineered cementitious composites (ECC) materials. Constr. Build. Mater. **29**, 348–356 (2012)
42. Khaliq, W., Ehsan, M.B.: Crack healing in concrete using various bio influenced self-healing techniques. Constr. Build. Mater. **102**, 349–357 (2016)
43. Kim, H.G., Qudoos, A., Ryou, J.-S.: Self-healing performance of GGBFS based cementitious mortar with granulated activators exposed to a seawater environment. Constr. Build. Mater. **188**, 569–582 (2018)
44. Kim, S., Yoo, D.-Y., Kim, M.-J., Banthia, N.: Self-healing capability of ultra-high-performance fiber-reinforced concrete after exposure to cryogenic temperature. Cem. Concr. Compos. **104**, 103335 (2019)
45. Kobayashi, K., Suzuki, M., Rokugo, K.: The effects of PE and PVA fiber and water cement ratio on chloride penetration and rebar corrosion protection performance of cracked SHCC. Constr. Build. Mater. **178**, 372–383 (2018)
46. Kwon, S., Nishiwaki, T., Kikuta, T., Mihashi, H.: Experimental study on self-healing capability of cracked ultra-high-performance hybrid-fiber-reinforced cementitious composites.

In: Proceedings of the 3rd International Conference on Sustainable Construction Materials and Technologies, Kyoto, Japan (2013)
47. Lee, H.X.D., Wong, H.S., Buenfeld, N.R.: Self-sealing of cracks in concrete using superabsorbent polymers. Cem. Concr. Res. **79**, 194–208 (2016)
48. Lee, H.X.D., Wong, H.S., Buenfeld, N.R.: Potential of superabsorbent polymer for self-sealing cracks in concrete. Adv. Appl. Ceram. **109**(5), 296–302 (2010)
49. Li, G., Liu, S., Niu, M., Liu, Q., Yang, X., Deng, M.: Effect of granulated blast furnace slag on the self-healing capability of mortar incorporating crystalline admixture. Constr. Build. Mater. **239**, 117818 (2020)
50. Litina, C., Al-Tabbaa, A.: First generation microcapsule-based self-healing cementitious construction repair materials. Construct. Build. Mater. **255**, 119389 (2020)
51. Liu, H.: Effects of external multi-ions and wet-dry cycles in a marine environment on autogenous self-healing of cracks in cement paste. Cem. Concrete Res. **9** (2019)
52. Liu, H.: Promotion on self-healing of cracked cement paste by triethanolamine in a marine environment. Construct. Build. Mater. **10** (2020)
53. Liu, H.: Self-healing of microcracks in engineered cementitious composites under sulfate and chloride environment. Construct. Build. Mater. **9** (2017)
54. Ma, H.: Effect of Self-healing on water permeability and mechanical property of medium-early-strength engineered cementitious composites. Construct. Build. Mater. **10** (2014)
55. Maes, M.: Chloride penetration in cracked mortar and the influence of autogenous crack healing. Construct. Build. Mater. **11** (2016)
56. Medjigbodo, S., Bendimerad, A.Z., Rozière, E., Loukili, A.: How do recycled concrete aggregates modify the shrinkage and self-healing properties? Cem. Concr. Compos. **86**, 72–86 (2018)
57. Mehdipour, I.: Feasibility of using near-field microwave reflectometry for monitoring autogenous crack healing in cementitious materials. Cem. Concrete Comp. **13** (2018)
58. Mircea, A.C., Szilagyi, H., Hegyi, A., Baera, C.: Study of self-healing engineered cementitious composites for durable and sustainable infrastructure. Procedia Manuf. **46**, 871–878 (2020). https://doi.org/10.1016/j.promfg.2020.05.002
59. Mu, S., Schutter, G.D., Ma, B.: Non-steady state chloride diffusion in concrete with different crack densities. Mater. Struct. **11** (2013)
60. Nasim, M.: Effect of crystalline admixture, fly ash, and PVA fiber on self-healing capacity of concrete. Mater. Today **6** (2020)
61. Nguy, H.H.: Self-healing properties of cement-based and Alkali-activated slag-based fiber-reinforced composites. Construct. Build. Mater. **11** (2018)
62. Nguy˜ên, H.H.: Autogenous healing of high strength engineered cementitious composites (ECC) using calcium-containing binders. Construct. Build. Mater. **16** (2020)
63. Nishiwaki, T., Koda, M.: Experimental study on self-healing capability of FRCC using different types of synthetic fibers, **10**, 13 (2012)
64. Palin, D., Jonkers, H.M., Wiktor, V.: Autogenous healing of sea-water exposed mortar: quantification through a simple and rapid permeability test. Cem. Concrete Res. **7** (2016)
65. Park, B.: Quantitative evaluation of crack self-healing in cement-based materials by absorption test. Construct. Build. Mater. **10** (2018)
66. Park, B.: Self-healing capability of cementitious materials with crystalline admixtures and super absorbent polymers (SAPs). Construct. Build. Mater. **13** (2018)
67. Qian, S., Zhou, J.: Self-healing behavior of strain hardening cementitious composites incorporating local waste materials, **9** (2009)
68. Qiu, J., He, S., Yang, E.-H.: Polymeric fiber and hydraulic cement matrix. Cem. Concrete Res. **9** (2019)
69. Qiu, J.: Autogenous healing of fiber-reinforced reactive magnesia-based tensile strain-hardening composites. Cem. Concrete Res. **13** (2019)

70. Qiu, J.: Coupled effects of crack width, slag content, and conditioning alkalinity on autogenous healing of engineered cementitious composites. Cem. Concrete Compos. **10** (2016)
71. Rajasegar, M.: Hybrid effect of poly vinyl alcohol, expansive minerals, nano-silica and rice husk ash on the self-healing ability of concrete. Mater. Today **9** (2021)
72. Rauf, M.: Comparative Performance of different bacteria immobilized in natural fibers for self-healing in concrete. Construct. Build. Mater. **13** (2020)
73. Reddy, C.M.K.: Effect of crystalline admixtures, polymers and fibers on self healing concrete—a review. Mater. Today **8** (2020)
74. Reddy, T.C.S.: Macro mechanical properties of self healing concrete with crystalline admixture under different environments. Ain Shams Eng. J. **10** (2019)
75. Reinhardt, H.-W., Jooss, M.: Permeability and self-healing of cracked concrete as a function of temperature and crack width. Cem. Concr. Res. **33**, 981–985 (2003)
76. Roig-Flores, M.: Effect of crystalline admixtures on the self-healing capability of early-age concrete studied by means of permeability and crack closing tests. Construct. Build. Mater. **11** (2016)
77. Roig-Flores, M.: Self-healing capability of concrete with crystalline admixtures in different environments. Construct. Build. Mater. **11** (2015)
78. Ruan, S.: Influence of crack width on the stiffness recovery and self-healing of reactive magnesia-based binders under CO_2-H_2O conditioning. Construct. Build. Mater. **13** (2021)
79. Beshr, B.S.S.: Feasibility assessment on self-healing ability of cementitious composites with MgO. J. Build. Eng. **8** (2021)
80. Shao-feng, Z., Chun-hua, L., Rong-gui, L.: Experimental determination of chloride penetration in cracked concrete beams. Procedia Eng. **24**, 380–384 (2011)
81. Siad, H.: Influence of limestone powder on mechanical, physical and self-healing behavior of engineered cementitious composites. Construct. Build. Mater. **10** (2015)
82. Sidiq, A.: Microstructural analysis of healing efficiency in highly durable concrete. Construct. Build. Mater. **15** (2019)
83. Singh, H.: Influence of cellulose fiber addition on self-healing and water permeability of concrete. Case Stud. Construct. Mater. **15** (2020)
84. Sisomphon, K.: Self-healing of surface cracks in mortars with expansive additive and crystalline additive, **9** (2012)
85. Snoeck, D., Van Tittelboom, K., Steuperaert, S., Dubruel, P., De Belie, N.: Self-healing cementitious materials by the combination of microfibres and superabsorbent polymers. J. Intell. Mater. Syst. Struct. **25**, 13–24 (2014)
86. Snoeck, D., De Belie, N.: Mechanical and self-healing properties of cementitious composites reinforced with flax and cottonised flax, and compared with polyvinyl alcohol fibres. Biosys. Eng. **111**(4), 325–335 (2012)
87. Snoeck, D., Smetryns, P.A., De Belie, N.: Improved multiple cracking and autogenous healing in cementitious materials by means of chemically-treated natural fibres. Biosys. Eng. **139**, 87–99 (2015)
88. Song, X.F., Wei, J.F., He, T.S.: A method to repair concrete leakage through cracks by synthesizing super-absorbent resin in situ. Construct. Build. Mater. **6** (2009)
89. Su, Y.-F.: Autogenous healing performance of internal curing agent-based self-healing cementitious composite. Cem. Concrete Compos. **8** (2020)
90. Suleiman, A.R.: Effect of environmental exposure on autogenous self-healing of cracked cement-based materials. Cem. Concrete Res. **12** (2018)
91. Tomczak, K., Jakubowski, J.: The effects of age, cement content, and healing time on the self-healing ability of high-strength concrete. Constr. Build. Mater. **187**, 149–159 (2018). https://doi.org/10.1016/j.conbuildmat.2018.07.176

92. Tziviloglou, E., Wiktor, V., Jonkers, H.M., Schlangen, E.: Bacteria-based self-healing concrete to increase liquid tightness of cracks. Constr. Build. Mater. **122**, 118–125 (2016). https://doi.org/10.1016/j.conbuildmat.2016.06.080
93. Van Tittelboom, K., De Belie, N., Van Loo, D., Jacobs, P.: Self-healing efficiency of cementitious materials containing tubular capsules filled with healing agent. Cem. Concr. Compos. **33**, 497–505 (2011). https://doi.org/10.1016/j.cemconcomp.2011.01.004
94. Van Tittelboom, K., Wang, J., Araújo, M., Snoeck, D., Gruyaert, E., Debbaut, B., Derluyn, H., Cnudde, V., Tsangouri, E., Van Hemelrijck, D., et al.: Comparison of different approaches for self-healing concrete in a large-scale lab test. Construct. Build. Mater. **107**, 125–137 (2016). https://doi.org/10.1016/j.conbuildmat.2015.12.186
95. Wang, R., Yu, J., Gu, S., He, P., Han, X., Liu, Q.: Investigation of self-healing capability on surface and internal cracks of cement mortar with ion chelator. Constr. Build. Mater. **236**, 117598 (2020). https://doi.org/10.1016/j.conbuildmat.2019.117598
96. Wang, W., Liu, J., Agostini, F., Davy, C.A., Skoczylas, F., Corvez, D.: Durability of an ultra high performance fiber reinforced concrete (UHPFRC) under progressive aging. Cem. Concr. Res. **55**, 1–13 (2014). https://doi.org/10.1016/j.cemconres.2013.09.008
97. Wang, X., Huang, Y., Huang, Y., Zhang, J., Fang, C., Yu, K., Chen, Q., Li, T., Han, R., Yang, Z., et al.: Laboratory and field study on the performance of microcapsule-based self-healing concrete in tunnel engineering. Construct. Build. Mater. **220**, 90–101 (2019). https://doi.org/10.1016/j.conbuildmat.2019.06.017
98. Wiktor, V., Jonkers, H.M.: Quantification of crack-healing in novel bacteria-based self-healing concrete. Cem. Concr. Compos. **33**, 763–770 (2011). https://doi.org/10.1016/j.cemconcomp.2011.03.012
99. Win, P.P., Watanabe, M., Machida, A.: Penetration profile of chloride ion in cracked reinforced concrete. Cem. Concr. Res. **34**, 1073–1079 (2004). https://doi.org/10.1016/j.cemconres.2003.11.020
100. Wu, M., Hu, X., Zhang, Q., Cheng, W., Xue, D., Zhao, Y.: Application of bacterial spores coated by a green inorganic cementitious material for the self-healing of concrete cracks. Cem. Concr. Compos. **113**, 103718 (2020). https://doi.org/10.1016/j.cemconcomp.2020.103718
101. Xu, J., Wang, X.: Self-healing of concrete cracks by use of bacteria-containing low alkali cementitious material. Constr. Build. Mater. **167**, 1–14 (2018). https://doi.org/10.1016/j.conbuildmat.2018.02.020
102. Xue, C., Li, W., Qu, F., Sun, Z., Shah, S.P.: Self-healing efficiency and crack closure of smart cementitious composite with crystalline admixture and structural polyurethane. Constr. Build. Mater. **260**, 119955 (2020). https://doi.org/10.1016/j.conbuildmat.2020.119955
103. Xue, C., Li, W., Luo, Z., Wang, K., Castel, A.: Effect of chloride ingress on self-healing recovery of smart cementitious composite incorporating crystalline admixture and MgO expansive agent. Cem. Concr. Res. **139**, 106252 (2021). https://doi.org/10.1016/j.cemconres.2020.106252
104. Yang, Y., Lepech, M.D., Yang, E.-H., Li, V.C.: Autogenous healing of engineered cementitious composites under wet-dry cycles. Cem. Concr. Res. **39**, 382–390 (2009). https://doi.org/10.1016/j.cemconres.2009.01.013
105. Yang, Y., Yang, E.-H., Li, V.C.: Autogenous healing of engineered cementitious composites at early age. Cem. Concr. Res. **41**, 176–183 (2011). https://doi.org/10.1016/j.cemconres.2010.11.002
106. Yang, Z., Hollar, J., He, X., Shi, X.: Laboratory assessment of a self-healing cementitious composite. Transp. Res. Rec. **2142**(1), 9–17 (2010)
107. Ye, H., Jin, N., Jin, X., Fu, C.: Model of chloride penetration into cracked concrete subject to drying-wetting cycles. Constr. Build. Mater. **36**, 259–269 (2012). https://doi.org/10.1016/j.conbuildmat.2012.05.027

108. Yıldırım, G., Khiavi, A.H., Yeşilmen, S., Şahmaran, M.: Self-healing performance of aged cementitious composites. Cem. Concr. Compos. **87**, 172–186 (2018). https://doi.org/10.1016/j.cemconcomp.2018.01.004
109. Yildirim, G., Aras, G.H., Banyhussan, Q.S., Şahmaran, M., Lachemi, M.: Estimating the self-healing capability of cementitious composites through non-destructive electrical-based monitoring. NDT & E Int. **76**, 26–37 (2015). https://doi.org/10.1016/j.ndteint.2015.08.005
110. Yildirim, G., Sahmaran, M., Balcikanli, M., Ozbay, E., Lachemi, M.: Influence of cracking and healing on the gas permeability of cementitious composites. Constr. Build. Mater. **85**, 217–226 (2015). https://doi.org/10.1016/j.conbuildmat.2015.02.095
111. Zha, Y., Yu, J., Wang, R., He, P., Cao, Z.: Effect of ion chelating agent on self-healing performance of cement-based materials. Constr. Build. Mater. **190**, 308–316 (2018). https://doi.org/10.1016/j.conbuildmat.2018.09.115
112. Zhang, L.V., Suleiman, A.R., Nehdi, M.L.: Self-healing in fiber-reinforced alkali-activated slag composites incorporating different additives. Constr. Build. Mater. **262**, 120059 (2020). https://doi.org/10.1016/j.conbuildmat.2020.120059
113. Zhang, P., Dai, Y., Ding, X., Zhou, C., Xue, X., Zhao, T.: Self-healing behaviour of multiple microcracks of strain hardening cementitious composites (SHCC). Constr. Build. Mater. **169**, 705–715 (2018). https://doi.org/10.1016/j.conbuildmat.2018.03.032
114. Zhang, Z., Zhang, Q.: Self-healing ability of engineered cementitious composites (ECC) under different exposure environments. Constr. Build. Mater. **156**, 142–151 (2017). https://doi.org/10.1016/j.conbuildmat.2017.08.166
115. Zhang, Z., Qian, S., Ma, H.: Investigating mechanical properties and self-healing behavior of micro-cracked ECC with different volume of fly ash. Constr. Build. Mater. **52**, 17–23 (2014). https://doi.org/10.1016/j.conbuildmat.2013.11.001
116. Zhao, G., Li, J., Shi, M., Cui, J., Xie, F.: Degradation of cast-in-situ concrete subjected to sulphate-chloride combined attack. Constr. Build. Mater. **241**, 117995 (2020). https://doi.org/10.1016/j.conbuildmat.2019.117995
117. Zhong, W., Yao, W.: Influence of damage degree on self-healing of concrete. Constr. Build. Mater. **22**, 1137–1142 (2008). https://doi.org/10.1016/j.conbuildmat.2007.02.006
118. Zhu, H., Zhang, D., Wang, T., Wu, H., Li, V.C.: Mechanical and self-healing behavior of low carbon engineered cementitious composites reinforced with PP-fibers. Constr. Build. Mater. **259**, 119805 (2020). https://doi.org/10.1016/j.conbuildmat.2020.119805

Carbonation Results in Diverse Cement Pastes with 10% Substitution of Binder by Bentonite

Carmen Andrade[1](✉), Ana Martinez[2], Miguel Ángel Sanjuan[3], and José A. Tenorio[3]

[1] International Center of Numerical Methods in Engineering, CIMNE (UPC), Barcelona, Spain
andrade@cimne.upc.edu
[2] Institute of Construction Sciences "Eduardo Torroja", CSIC, Madrid, Spain
[3] Institute of Cement and its Applications, IECA, Madrid, Spain

Abstract. The greenhouse effect is accelerating to reduce environmental footprint of concrete. In addition to the traditional mineral additions which are blast furnace slags, pozzolans and fly ashes, worldwide are tested new products in order to have a wider variety of substances capable of maintaining the performance and good durability of the concrete. Among these substance, calcined clays are proposed due to their availability in numerous places. However, the pre-calcination ads waste of energy and therefore the use of non-calcined clays appears as a possibility very little explored, as few literature references are found. In present work non calcined bentonite has been tested by substituting 10% of different Portland cements. Paste specimens have been submitted during 3 and 6 months to natural carbonation in three different exposures: indoors and outdoors sheltered and unsheltered from rain. The results show that the presence of bentonite either improves the carbonation resistance or does not modify it. With respect to the resistance to chloride ingress, the tests have been in mortar in accelerated conditions through an electrical field. The results indicate that the presence of bentonite also resulted beneficial.

Keywords: Cement · Bentonite · Clays · Clinker · Carbonation

1 Introduction

In the case of cement production, greenhouse gases are accelerating the need to reduce the CO_2 emissions by, among others, reducing the proportion of clinker. The need to reduce clinker proportion is not new, as the petrol crisis in the 1970s generalized the practice, already made by several countries, of adding mineral compounds to substitute the clinker as they are blast-furnace slag, coal fly ash, silica fume and limestone with no adverse effect on concrete mechanical performance or durability [1]. However, at present, the need is more crucial because some of these additions are disappearing precisely by the same reason of reduction of CO_2 emissions [2]. One of the alternatives suggested is the use of pre-calcined clays [3, 4].

The use of calcined clay as a mineral addition to cement is not new [5, 6]. Its application was not adopted in the past mainly because it resulted more expensive than

J. I. Escalante-Garcia et al. (Eds.): RW 2021, RILEM Bookseries 40, pp. 183–192, 2023.
https://doi.org/10.1007/978-3-031-21735-7_21

the other additions, due to the heating pretreatment. The clays are very adequate as additions because they have similar chemical composition, although without cementing properties, than the clinker as they are used as raw material for clinkerization. In the heating pretreatment, the water in the clay is removed and the resulting mineral is very active to reincorporate the water [7]. During hydration of cement phases, the clays act as pozzolanic material by combination with the calcium ions leading to the same cement hydrated phases than with other pozzolanic additions [5, 6, 8]. However, as said, the pre-treatment is intensive in energy and then the need to find additions with lower carbon footprint remains as one of the main objectives in the cement industry.

In present study is reported the use of non-calcined clays, in particular a bentonite [9] added as substitute of clinker in 10% proportion and its carbonation resistance, because higher carbonation rate is one of the weakness noticed when calcined clays are used [4]. The reason of thinking in non-calcined clay was based in that it is widely available in numerous places, and precisely in the cement plants as it is used for obtaining the clinker and then, it seems that the same clay as raw material might be incorporated in the grinding step, if it results compatible and does not modify the mechanical and durability properties of the concrete. The results presented are part of a wider study [10] where it has been shown that until 10% of substitution by a commercial bentonite did not alter the mechanical strength. In the tests made, carbonation resistance either remained essentially unchanged or improved in a set of different cements with and without traditional mineral additions.

2 Experimental

The specimens used for the carbonation test were 10 × 10 × 60 mm in paste at a water/cement ratio of 0.5. They were cured in a climatic chamber at >95% relative humidity, first in the moulds for 24 h and after removal for 28 d prior to be submitted to the carbonation exposure. The cements used are shown in Table 1 were in the carbonation tests.

A commercial bentonite (Mapeiproof Seal), distributed by Mapei for purposes other than studied here, was used to ensure consistent composition and particle size distribution throughout. According to the vendor's specifications sheet, the material contained over 95 wt% montmorillonite.

The particle size distribution of the bentonite powder showed that 90% of the particles were <86 μm, whilst most lay within the 10–15 μm range (see the volume density curve). Such greater fineness than observed for the cement was initially deemed suitable, although optimisable. Addition fineness plays a significant role in the strength and rheology of composite cements, for the distribution curves for those materials complement the curve for the cement itself. This parameter must consequently be analysed in depth, in the future.

Another factor of particular interest in bentonite materials is water demand, affected not only by fineness but also by its cation charge and especially its sodium content. This property was not a variable in present study.

The specimens once cured the 28 days in humid chamber were exposed to carbonation in an indoor laboratory environment and two outdoor environments at the natural

Table 1. Composition (%) of the cements used in the tests

Cement	K	V	L/LL	S	P	Remaining
CEM I 42,5R	95					5
CEM I/ 42,5 SR	95					5
CEM II/A-P 42,5R	83				11	6
CEM II/A-P 42,5R	80				16	4
CEM II/A-S 42,5N	83			12		5
CEM II/A-V 42,5 R	80		15			5
CEM III/A 42,5N	59			39		2
CEM IV/A(V) 42,5 R-SR	74	23				3
IV/A(P) 42,5 R/MR	87				13	0
BL II/B-LL 42,5R	74		26		0	0

atmospheric CO_2 pressure prevailing in the city of Madrid, one protected and the other unprotected from rainfall, i.e., environments with varying relative humidities and temperatures (Fig. 1). Carbonation depth, as tested through phenolphthalein indicator, was recorded after 3 and 6 months of exposure.

Fig. 1. Exposure to natural carbonation_environments

3 Results

Table 2 depicts the phenolphthalein printing in the specimens from which carbonation depth was deduced. In general, the lowest values of carbonation were observed in the laboratory, intermediate penetration under outdoor sheltered conditions and the deepest in the specimens exposed to rainfall. That order of environmental aggressiveness is diametrically opposed to what is found in other previous experiments comparing the same environments. According to those data, penetration was deepest in specimens exposed to indoor environments or sheltered outdoor conditions, whilst carbonation was least intense in those exposed to rainfall, due to their higher or nearly optimal moisture content. The present findings were considered correct as the values at 3 months were repeated in 6 month specimens. It has to be mentioned however that in this study the specimens exposed 6 months were tested also during the summer under high temperature, low RH conditions. In other seasons with higher RH and more rain the order might have differed. That is scantly relevant, however, for inasmuch the purpose of the test was to compare the behaviour in the various cements.

Figure 2 plots the 3 month carbonation depths in the specimens bearing 10% bentonite with respect to the values in the reference (without the substitution) and Fig. 3 the same parameters in the 6 month samples. After 3 months carbonation was less intense in a larger number of 10% bentonite than of reference specimens. In the 6 months values the better performance when the samples have the 10% substitution with bentonite is more evident in outdoor unsheltered conditions.

Figures 4, 5, 6 and 7 show the values of the flexural and compressive strength of the carbonated samples in order to compare the possible effect of carbonation as well in the mechanical properties. As a rule, the flexural strengths were rather similar in the reference and substituted samples. The compressive strength values were even closer in the two types of specimens. Inasmuch as the experiment was designed for purposes of comparison, the inference drawn from these findings is that adding 10% bentonite to the cement mix had no consistent effect on the behaviour.

4 Discussion

From the figures it is possible to deduce that as an average, the substitution of 10% of clinker by bentonite, is not producing a noticeable effect. It has to be emphasized as well that the cements used are having already other types of additions and then, they are like ternary cements. In the case of the two CEM I cement used, the substitution of clinker by 10% bentonite even lower the carbonation depth. Of the other cements, the ones bearing natural pozzolans appeared to perform better than those carrying fly ash or slag. In neither case did the use of bentonite induce clearly poorer performance than already observed in those cements. Such behaviour cannot be attributed to a reaction between the clay and carbon dioxide, because being the bentonite an anionic clay, it cannot accommodate anions in its interlayers. The mechanism by which the carbonation is not modified or improved remains to be deeper studied.

This feature, the reduction or at least non-alteration of carbonation depth, is very relevant finding because their lower carbonation resistance, is one of the shortcomings identified of mineral additions in general. Bentonite could consequently be used to advantage instead of the 5% of inert matter or the up to 10% of limestone routinely added to clinker, it may improve one or several properties of the end product. Confirmation of the foregoing will nonetheless call for much more testing, in particular to detect possible adverse effects on shrinkage or creep.

It has to eb remarked that a new mineral addition to cement has to be studied comprehensively in its possible impact in several properties. Thus, to be compatible with

Table 2. .

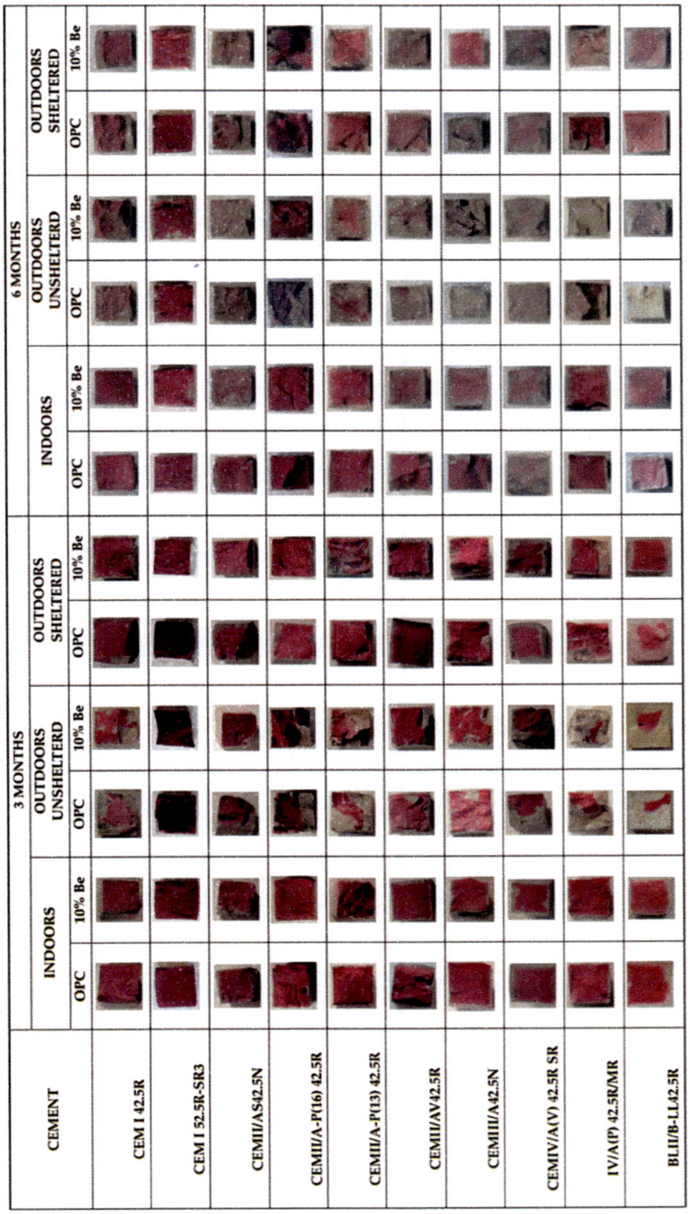
CEMENT
3 MONTHS
6 MONTHS
INDOORS
OUTDOORS UNSHELTERD
OUTDOORS SHELTERED
OPC
10% Be
CEM I 42.5R
CEM I 52.5R-SR3
CEMII/AS42.5N
CEMII/A-P(16) 42.5R
CEMII/A-P(13) 42.5R
CEMII/AV42.5R
CEMIII/A42.5N
CEMIV/A(V) 42.5R SR
IV/A(P) 42.5R/MR
BLII/B-LL42.5R

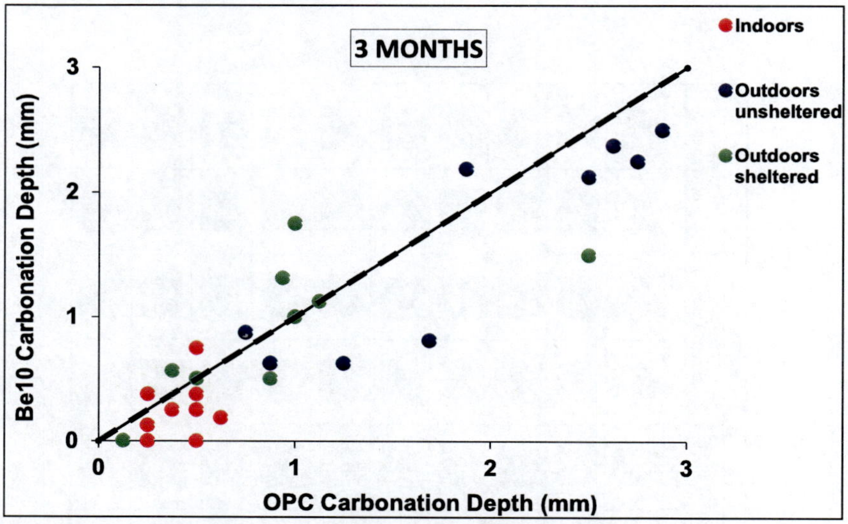

Fig. 2. Comparison of Carbonation depth in 3 months specimens between reference and with a substitution of 10% bentonite

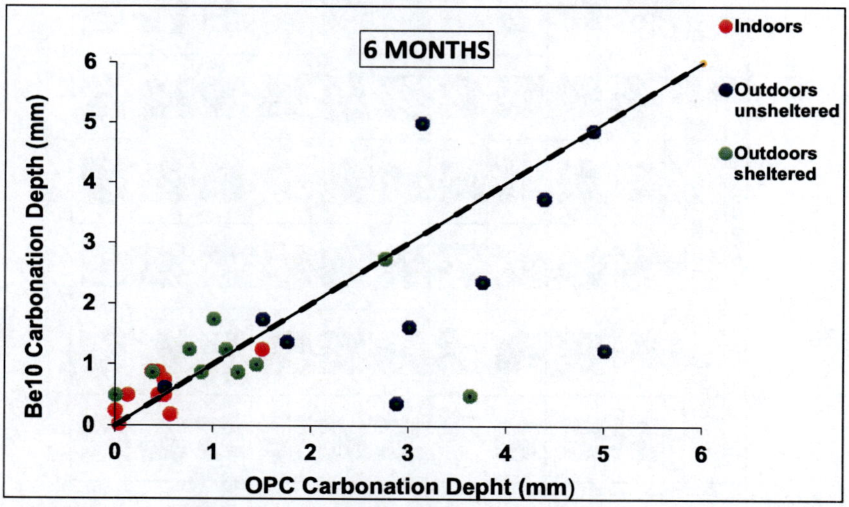

Fig. 3. Comparison of Carbonation depth in 6 months specimens between reference and with a substitution of 10% of bentonite

cement and usable in concrete, additions must meet a series of short- and long-term requisites, summarised below.

– They must be inert or at least not induce expansion or degenerative reactions.

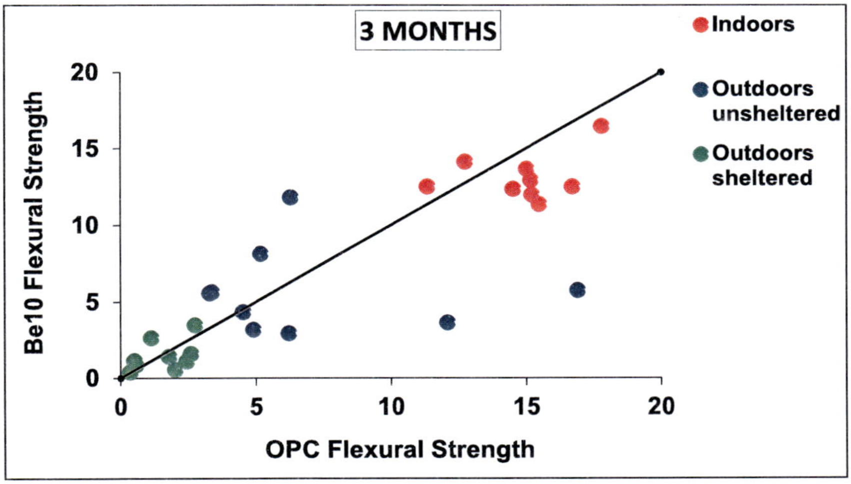

Fig. 4. Comparison of flexural strength in 3 months specimens

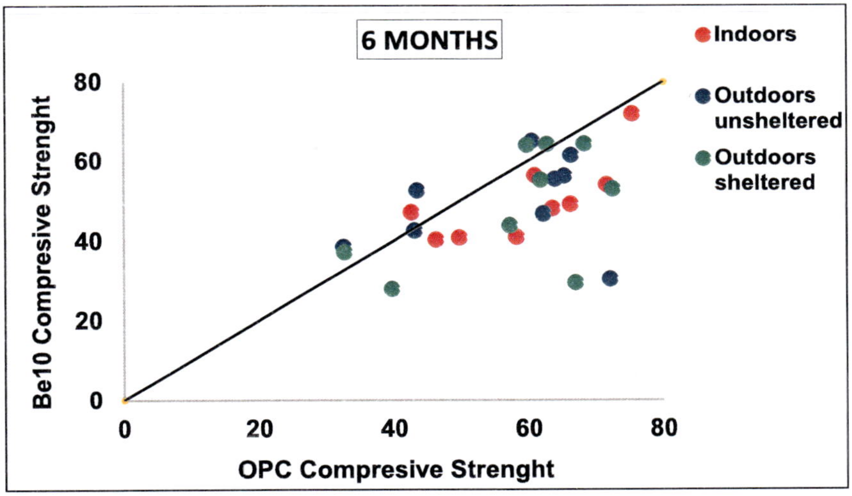

Fig. 5. Comparison of compressive strength in 3 months specimens

- They must improve or at least not alter volumetric stability of concrete (in terms of shrinkage and creep especially).
- They must improve or at least not alter mechanical performance.
- They must lengthen or at least not shorten concrete or steel durability.

In light of its vast diversity and fairly widespread geographic availability, clay is just one more local raw material deployed in cement manufacture. The same clays used

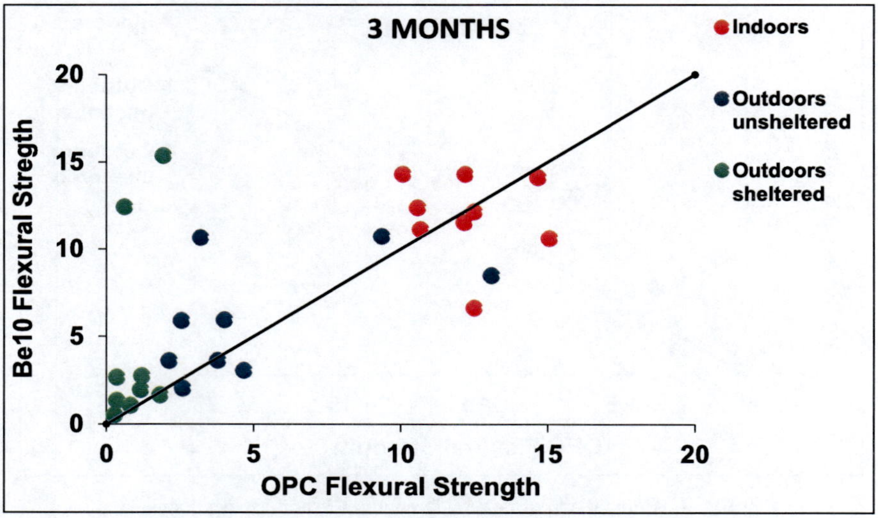

Fig. 6. Flexural strength in 6 months specimens

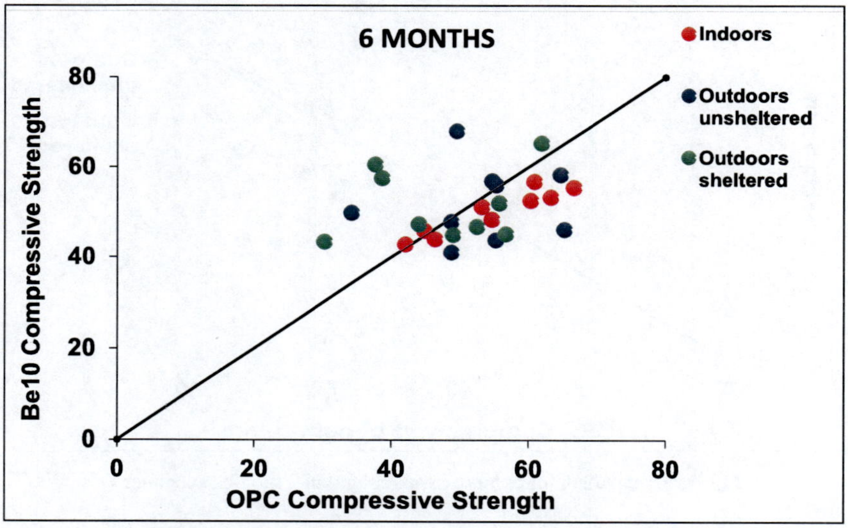

Fig. 7. Comparison of compressive strength in 6 months specimens

in clinker kilns might, in certain proportions, constitute compatible additions to lower milling-related CO_2 emissions (since milling clay is less energy-intensive than grinding clinker).

Surprisingly, not only the studies on non-calcined clays are very scarce in the literature [11], but even the presence of clays in the aggregates is forbidden in many codes.

No research on its non-pre-calcined use has been found in the literature, except when it is used in foundations for stabilizing the soil [12], however, for that approach has consistently posed rheological problems. For those reasons, many building codes limit the presence of clay materials in aggregates. However, today's admixtures afford fresh concrete properties [13] impossible to attain in the past and modern laboratory techniques now in place can substantially shorten the time needed to design an optimal mix.

A review of the literature on the long-term performance of bentonite showed that it has exhibited excellent durability in underground works, where it has been used profusely as permanent formwork in concrete foundations often built long ago [14]. It is likewise used in conjunction with concrete to generate impermeable slurry walls in highly radioactive waste storage facilities (designed to last for thousands of years) [15], where the caverns holding the radioactive waste are shotcreted and the encapsulated waste itself lies on a bed of bentonite. Such studies have verified the interaction between cement alkalinity and bentonite phases or equivalently, phase reactivity with clinker hydrated phases, with initially promising results from the standpoint of their role as a clinker replacement because of the slow reactivity. This reactivity is due to the high alkalinity of the pore solution which solubilize the silico-aluminates of the bentonite.

5 Conclusions

From the results it can be drawn up that the substitution of cement by 10 m% of bentonite is not detrimental with respect to the carbonation depth of the paste made without such substitution, because the substitution reduces or maintains the carbonation depth observed in the reference material, deemed to be a very promising development.

In addition it has been verified that the substitution is neither affecting the compressive or flexural strength with respect to the reference cements.

References

1. Taylor, H.F.W.: Cement Chemistry, 2nd edn. Thomas Telford Ltd; N.° 2 edition (1997)
2. Gartner, E., Hirao, H.: A review of alternative approaches to the reduction of CO_2 emissions associated with the manufacture of the binder phase in concrete. Cem. Concr. Res. **78**, 126–142 (2015)
3. Fernandez, R.: Martirena, F.: Scrivener, K.L.B.: The origin of the pozzolanic activity of calcined clay minerals: a comparison between kaolinite, illite and montmorillonite. Cem. Concr. Res. **41**, 113–122 (2011)
4. Pillai, R.G., et al.: Service life and life cycle assessment of reinforced concrete systems with limestone calcined clay cement (LC3). Cem. Concr. Res. **118**, 111–119 (2019)
5. Medina C., Sanchez de Rojas M.I., Frias M.: Valorizacion del árido ceramino de sanitarios como árido reciclado
6. Sánchez de Rojas M.I, Frias M., Rivera J., Marin F.P.: Aprovechamiento del cascote cerámico como material puzolánico - Cemento Hormigón, pp. 32–41, ISSN 0008-8919, N°. 915 (2008)
7. Frias, M., Sabador, E., Ferreiro, S., Sanchez de Rojas, M.I., Vigil, M.R.: Influencia de la activación de un residuo arcilloso de la industria papelera en el comportamiento de matrices de cemento - Materiales de construcción. ISSN 0465-2746, N°. 292, pp. 67–80 (2008)

8. Courard, L., Darimont, A., Schouterden, M., Ferauche, F., Willem, X., Degeimbre, R.: Durability of mortars modified with metakaolin. Cem. Concr. Res. **33**, 1473–1479 (2003)
9. Lima Souza, P.S., Dal Molin, D.C.C.: Viability of using calcined clays from industrial by-products. As pozzolans of high reactivity. Cem. Concr. Res. **35**, 1993–1998 (2005)
10. Memon, S.A., Arsalan, R., Khan, S., Lo, T.Y.: Utilization of Pakistani bentonite as partial replacement of cement in concrete. Constr. Build. Mater. **30**, 237–242 (2012)
11. Andrade C., Martínez-Serrano A., Sanjuán M.A., Tenorio J.A.: Reduced carbonation, sulfate and chloride ingress due to the substitution of cement by 10% non-precalcined bentonite. Materials **14**, 1300 (2021). https://doi.org/10.3390/ma14051300
12. Wei, J., Gencturk, B.: Hydration of ternary Portland cement blends containing metakaolin and sodium bentonite. Cem. Concr. Res. **123**, 105772 (2019)
13. Liu, M., et al.: Influence of various bentonites on the mechanical properties and impermeability of cement mortars. Cem. Concr. Res. **241**, 118015 (2020)
14. Lei, L., Plank, J.: A study on the impact of different clay minerals on the dispersing force of conventional and modified vinyl ether based polycarboxylate superplasticizers. Cem. Concr. Res. **60**, 1–10 (2014)
15. Ata, A.A., Salem, T.N., Elkhawas, N.M.: Properties of soil bentonite-cement bypass mixture for cutoff walls. Constr. Build. Mater. **93**, 950–956 (2015)
16. Trivedi, D.P., Holmes, R.G.G., Brown, D.: Monitoring the in-situ performance of a cement/bentonite cut-off wall at a low level waste disposal site. Cem. Concr. Res. **22**, 339–349 (1992)

Durability and Life Cycle Assessment in Urban and Marine Conditions

Expanded Clay as Bacteria Cells Protector in Manufacturing Self-healing Mortar: The Possibility of Adding Less Nutrient to Maintain the Mechanical Properties of Mortar

Puput Risdanareni[1,2], Jianyun Wang[3], and Nele De Belie[1](✉)

[1] Magnel-Vandepitte Laboratory for Structural Engineering and Building Materials, Ghent University, Tech Lane Ghent Science Park, Campus A, Technologiepark Zwijnaarde 60, B-9052 Ghent, Belgium
{puput.risdanareni,nele.debelie}@ugent.be
[2] Department of Civil Engineering, Faculty of Engineering, State University of Malang, Semarang Street 5, 65145 Malang, Indonesia
[3] Department of Civil Engineering, Xi'an Jiaotong University, Xianning West Road 28, Xi'an 710049, China

Abstract. Bacteria-based self-healing is a promising method due to its effectiveness to completely heal the crack. Expanded clay was chosen as a bacteria protector as it can retain a high amount of bacteria liquid in its pores and it is compatible with the mortar matrix. Most researchers employed bacteria in the form of spores to produce self-healing concrete due to its survivability in a harsh environment. However, to activate the spores, yeast extract, which has a negative effect on mortar strength, needs to be added. In this research, *Bacillus sphaericus* (BS) vegetative cells were used as a healing agent to determine whether these cells survive in the mortar and could precipitate $CaCO_3$. The variation in the amount of yeast extract as a bacteria nutrition was also investigated. The mechanical properties of fresh and hardened mortar, such as flow, compressive strength, and bulk density, were determined. The healing efficiency was observed through crack closure observation. In the end, the morphology of $CaCO_3$ precipitation was observed under a scanning electron microscope (SEM). The result shows that it is possible to remove yeast from the mixture when producing self-healing concrete with vegetative cells as a healing agent. The resulting mortar's mechanical properties are similar to the reference, and the healing efficiency was still acceptable.

Keywords: Self-healing concrete · Vegetative cells · Expanded clay LWA · Yeast extract · Crack closure · Compressive strength

1 Introduction

Bacteria based self-healing concrete was reported to deliver excellent healing performance [1, 2]. Several types of bacterial strains which can survive in alkaline environment, such as *Bacillus sphaericus*, *Sporosarcina pasteurii*, *Bacillus subtillis* and *Bacillus alkalinitrilicus* are used as healing agent [3–6]. In this research, a strain of *Bacillus*

J. I. Escalante-Garcia et al. (Eds.): RW 2021, RILEM Bookseries 40, pp. 195–204, 2023.
https://doi.org/10.1007/978-3-031-21735-7_22

sphaericus (BS) was selected as healing agent as it has been reported to deliver satisfying healing performance when encapsulated into various carrier [2, 7–9]. To protect the bacteria from harsh environment, cells need to encapsulated into carriers such as hydrogel, porous material, microcapsules etc. [2, 4, 8]. In this research, porous expanded clay lightweight aggregate (EC LWA) was selected as the bacterial carrier as it is compatible with the mortar matrix, and has been reported to have ability to heal large crack widths up to 0.46 mm [4, 10–12]. As expanded clay is an open carrier with quite big pore size, bacteria could leak out which could decrease the healing ability. A study on finding suitable encapsulation methods with expanded clay as carrier suggested the application of vacuum and pressure treatment to push the bacteria deep into the pores of EC LWA [13].

Researchers are mostly working with the spore form of bacteria, as it is less sensitive to the harsh environment of fresh concrete [1, 2, 5]. However, spores need to germinate into vegetative cells first before they could actively heal the crack. Unfortunately, yeast extract that is usually employed as a germination agent negatively affects the resulting concrete or mortar [14–16]. Thus directly applying vegetative cells as a healing agent that will not require yeast for germinating processes could be an option to maintain the mechanical properties of the self-healing concrete matrix. Several articles on employing vegetative cells as healing agent reported that encapsulated vegetative cells are still viable inside the concrete matrix and are capable to heal the cracks [7, 14, 17].

Summarizing, the aim of this study is to evaluate the possibility of omitting the yeast extract from the self-healing bacteria based matrix. By using the vegetative cells form of *B. sphaericus*, encapsulated into expanded clay LWA, sufficient healing ability was expected even though the yeast was remove. On the other hand it was investigated whether the properties of fresh and hardened mortar assessed by flowability, compressive strength and bulk density could be maintained with omitting the yeast from the mixture.

2 Materials and Methods

2.1 Materials

Vegetative cells of *Bacillus sphaericus* LMG 22257 (Belgian Co-ordinated Collections of Micro-organisms, Ghent) were used as a healing agent. The cells were cultivated in a sterile liquid medium containing 20 g/l yeast extract and 20 g/l urea on a shaking table (120 rpm) for 24 h in a 28 °C room. The vegetative cells were then harvested by separating the bacteria pellets from the liquids by means of centrifuging (7000 rpm for 7 min). Finally, the cells pellet were resuspended in yeast extract 5 g/l. The final concentration of the vegetative cells used was 2×10^9 cells/ml.

The initial properties of LWA needed for mortar mix design is presented at Table 1. Yeast extract, calcium nitrate tetrahydrate, and urea (Carl Roth Belgium) with a purity of 99% used as bioagents in the mortar. In this study, two variations of yeast extract concentration of 5 g/l and 0 g/l were applied while the concentration of urea and calcium nitrate were kept constant (Table 2).

A commercial expanded clay lightweight aggregate (EC LWA) with a 2–4 mm fraction from Argex Nv was used as a bacterial protector. A cement type 1 52.5N from Holcim was employed as a binder, while river sand with the fraction size of 0–2 mm was obtained by sieving 0/4 sand. The whole volume of sand with the size of 2–4 mm was replaced with EC LWA (Table 2).

The cells of BS were encapsulated into EC LWA through vacuum and pressure treatment. First, the EC LWA were put in sealed tubes. Vacuum pressure (−0.8 bar) was applied for 30 min, followed by adding bacteria solution in a certain amount based on the water absorption of the EC LWA, and the pressure was maintained for another 30 min. Finally, the pressure of +1 bar was introduced and maintained for 24 h to push the bacteria deep into the pores of EC LWA.

Table 1. Properties of expanded clay lightweight aggregate (EC LWA)

Test	EC LWA
Apparent particle density (AD) (kg/m^3)	1.25 ± 0.01
Oven dried density (OD) (kg/m^3)	0.99 ± 0.01
SSD particle density (SSD) (kg/m^3)	1.19 ± 0.01
Water absorption 24 h (WA_{24h}) (%)	20.85 ± 0.09
Open porosity (OP) (%)	21.14 ± 0.3

Table 2. Mix design of mortar

Type	LWA OD 2/4 (g)	Entrained water (ml)	Bacteria (ml)	Sand (g)	Cement (g)	Water (g)	Yeast (g)	Urea (g)	Calcium nitrate (g)
Ref	82	17	0	1251	450	225	0	0	0
NY5	82	0	17	1251	450	214	3.83	18	36
NY0	82	0	17	1251	450	214	0	18	36

2.2 Methods

The flowability of fresh mortar was tested according to European standard NBN EN 1015-3 (1999). The fresh mortar was cast into two different mortar moulds. The samples with the dimension of 40 × 40 × 160 mm were used to test the compressive strength and the bulk density (3 replicates), while samples with the dimension of 30 × 30 × 360 mm and reinforced with a 6 mm steel bar were used to fabricate multiple cracks for crack closure observation. After being cast, the mortar samples were covered with plastic film and were cured in a curing chamber, which has a temperature of 20 ± 2 °C and humidity of 95 ± 5% for 24 h. Finally, the mortars were demoulded, covered with plastic film, and cured in the same chamber until reaching the testing age of 28 days.

The compressive strength and bulk density of 28 days old mortar were determined according to EN 196 standard. The multiple cracks were fabricated by subjecting the reinforced prism to a tensile load (Instron Universal Testing Machine). The crack width was measured by monitoring the displacement recorded. The load was stopped after each crack had approximately 0.4 mm width in average.

After cracks were fabricated, the initial crack width was measured under the microscope (Leica 2900). The samples were then subjected to a wet dry cycle curing regime with 4 h dry and 4 h wet in tap water. The cracks closure was monitored after 10 days and 20 days of curing through microscopic observation (Leica 2900). Finally, the healing ratio (HR) was calculated by Eq. 1.

$$HR = \left(C_{wi} - C_{wf}\right)/C_{wi} \times 100\% \quad (1)$$

where C_{Wi} is the initial crack width, and C_{wf} is the final crack width measured after 10 days or 20 days.

For SEM analysis, a piece of mortar containing a whitish healing product was collected from the crack mouth. The samples were patched on the sample holder, dried in 40 °C oven for 24 h, coated with a platinum coating, and were kept in the vacuum chamber until the testing day. Finally, the morphology of the healing products was observed with Jeol JSM-7600F Field Emission Scanning Electron Microscope (FESEM) equipped with AZtecEnergy software from Oxford Instruments for EDX mapping. The EDS mapping was performed for calcium, oxygen, and carbon elements.

3 Results and Discussion

3.1 Workability of Fresh Mortar

Increasing the yeast concentration in the mortar mixture tends to increase the workability of fresh mortar determined by the flow value (Fig. 1). Larger effects on workability have been reported for samples with similar yeast concentrations without any additional bioagent such as calcium nitrate and urea [15]. On the other hand, the sample NY0 with no yeast but containing the same amount of calcium nitrate had lower flow value than the reference sample. The addition of calcium in the mixture has been reported to deliver a flash setting, as the hydration processes that require calcium could be accelerated when the amount of calcium in the mixture increased [18]. It seems that adding yeast into the mix could improve the workability of fresh mortar that tends to decrease due to the presence of calcium.

3.2 Bulk Density and Compressive Strength of Mortar

A significant decrease in bulk density and compressive strength was observed in samples NY5 compared to the Ref samples (Figs. 2 and 3). The compressive strength of NY5 samples decreased by almost 45% compared to the reference samples. On the other hand, the strength reduction in NY0 samples was less than 5% compared to the reference samples. A similar reduction in bulk density was also observed. The yeast extract creates nitrogen gas during its autolysis that later on induces the formation of pores in the hardened mortar. This process could lead to the significant decrease of bulk density and compressive strength of NY5 samples [16]. The low flowability of the NY0 sample did

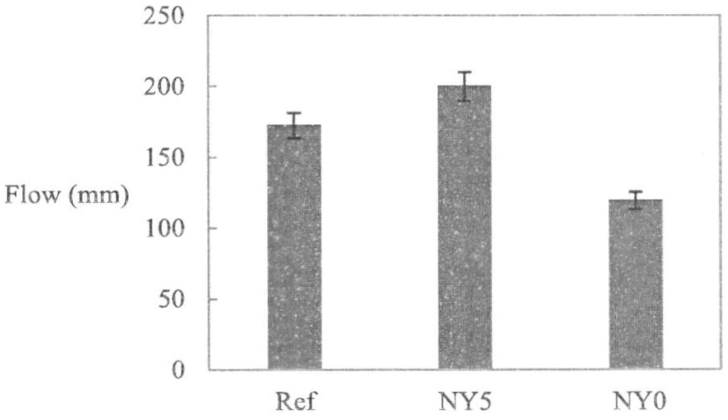

Fig. 1. The effect of yeast extract concentration on the workability of fresh mortar

not induce a decrease of bulk density and compressive strength in its hardened state. Urea that was reported to decrease the hydration heat could help to mitigate the flash setting caused by the excess of calcium in the mixture [19]. Thus, even though the fresh mortar was quite stiff, no issue was found while casting that could decrease its mechanical properties.

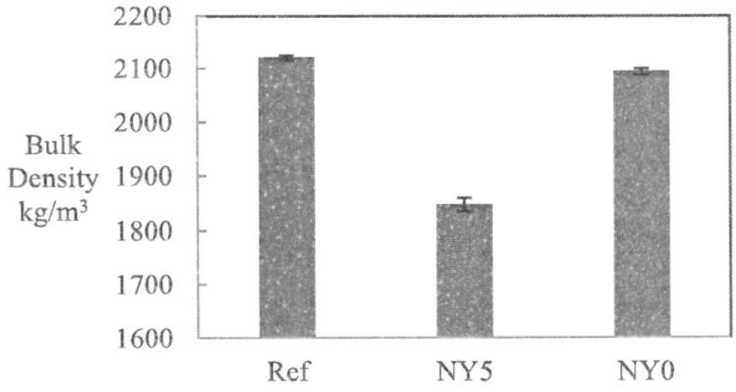

Fig. 2. The effect of yeast extract concentration on the bulk density of mortar

3.3 Crack Closure Observation

The healing ratio after curing for 10 and 20 days was plotted as a function of initial crack width was shown in Fig. 4. It can be seen that in NY5 and NY0 samples, some cracks with a width larger than 0.3 mm could completely heal. However, the amount of these sealed cracks is less in NY0 samples than in NY5.

In the reference samples, cracks larger than 0.3 mm could not completely sealed. However, some complete healing of cracks in the range of 0.2–0.25 mm was observed. Similar results were reported in the control samples used in a study on mortar with expanded clay as a carrier of spores from *B. alkalinitrilicus* [20]. After 20 days immersion in tap water, some cracks with the width of 0.18 mm were completely healed. The role

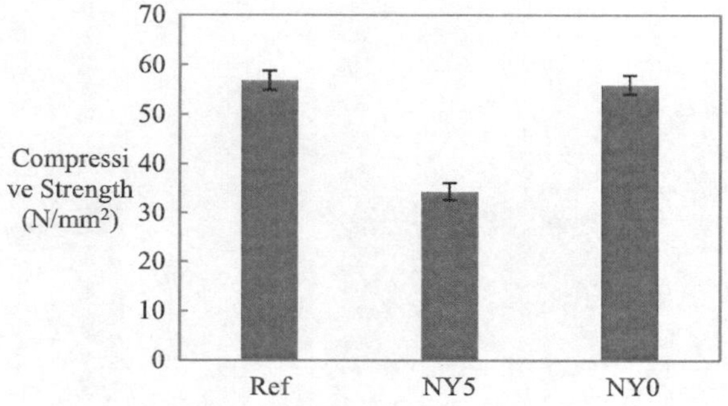

Fig. 3. The effect of yeast extract concentration on the compressive strength of mortar

of EC LWA as an internal curing agent could contribute to this healing. The EC LWA can retain water in their pores, which could react with un-hydrated cement and seal the small cracks [21].

Indeed the healing capacity of a no-yeast sample was decreased compared to the sample with yeast. In the NY5 sample, crack widths up to 0.4 mm were entirely healed after subjected to wet-dry cycles for 20 days. While in the NY0, given the similar crack width and curing duration, partial healing of approximately 80% was observed. The number of cells that could survive in the mortar without yeast might decrease or some may turn into spores. In the sample with yeast, the spore forms of bacteria still could germinate again by consuming yeast in the matrix and actively healed the crack, but this could not happen in the no-yeast sample. However, it is still possible to improve the performance of a no-yeast sample by increasing the number of cells encapsulated into EC LWA and prolong the curing period. It was reported that the threshold curing time of *B. alkalinitrilicus's* spores encapsulated into EC LWA was 40 days to completely heal the crack width of 0.4 mm with replacing rate of EC LWA of 24% by weight [20].

3.4 Morphology of the Healing Products

The cube shape of calcium carbonate could be observed in the healing products of the reference and NY0 sample (Fig. 5). However, the healing product's shape of a sample with bacteria is rather sharp-edged and triangular, while a cube-like form was observed in the reference sample. A similar cube-like shape of the healing product was also observed in the literature and was confirmed as $CaCO_3$ [22]. It has been reported that the addition of calcium sources combined with bacterial activity contributed to the different morphology of $CaCO_3$ precipitation [4, 23].

Even though all samples were confirmed to have $CaCO_3$ healing products, through EDS mapping analysis, it was observed that the sample containing bacteria has higher oxygen content compared to the reference sample (Fig. 6 and Table 3). Higher oxygen content was also observed previously in the bacterial precipitation product in an alkaline environment (pH 10) [24].

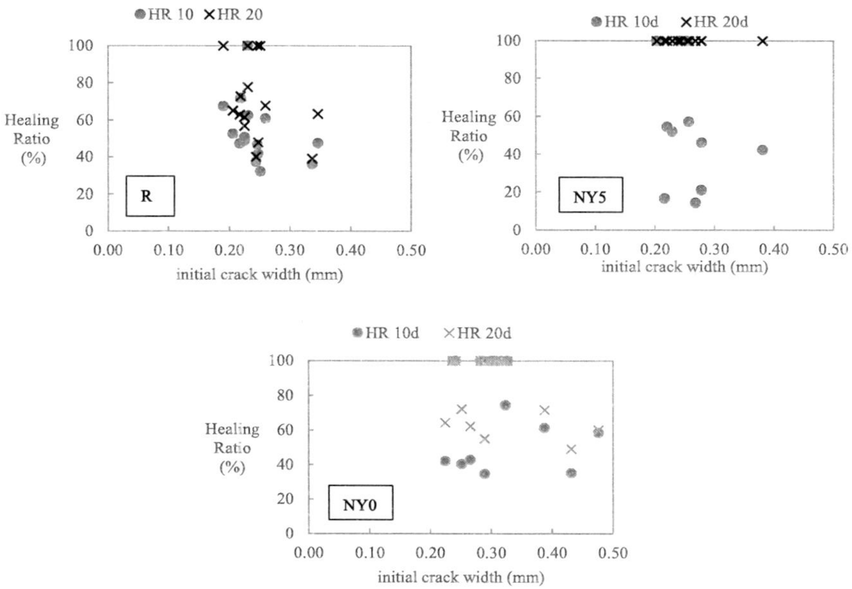

Fig. 4. The initial crack width as a function of healing ratio (%) of samples subjected to wet dry cycles for 10 days (HR 10d) and 20 days (HR 20d)

4 Conclusions

The lower flow value of a no-yeast sample did not affect the mechanical properties of its hardened mortar. A similar compressive strength and bulk density compared to the reference sample was observed in no-yeast samples. Omitting the yeast from the matrix decreased the healing capacity compared to the sample with yeast. Crack widths of 0.4 mm could be healed for 80% in this series, while completely healing was obtained in the samples with yeast. Through SEM EDS the healing product of the bacteria sample was confirmed to be $CaCO_3$. In conclusion, omitting the yeast from the mortar matrix is possible. A slight decrease in healing capacity was observed while the mechanical properties of the resulting mortar are maintained.

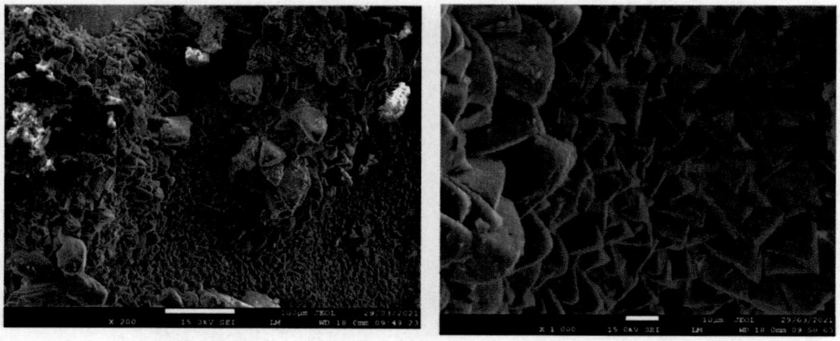

(a)

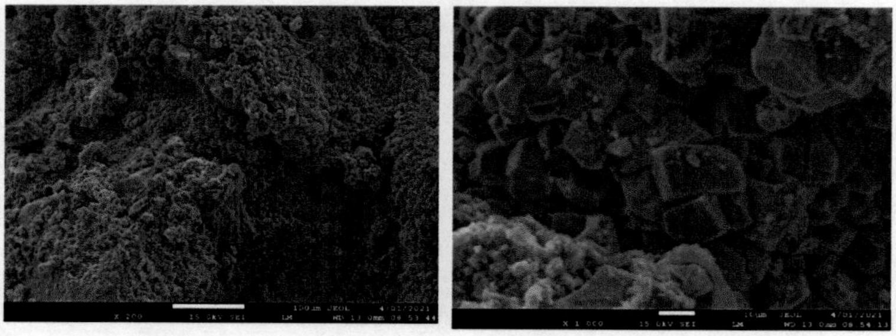

(b)

Fig. 5. The Morphology of Healing Product Observed at (a) NY0 and (b) Ref Sample with the Magnification of 200x (left) and 1000x (right)

(a) (b)

Fig. 6. EDS layered image mapping on (a) Ref sample and (b) NY0 sample. The magnification was set to 1000x. In both images, the red colour is representing calcium. In the Ref sample oxygen is presented in purple colour, and green in NY0 sample

Table 3. EDS mapping analysis for Ca, O and C elements

Sample	Ca	O	C	Ca/O
Ref	56.7	27.8	15.3	2.04
NY0	35.9	50.2	13.7	0.72

References

1. Jonkers HM, Schlangen E (2007) Crack repair by concrete-immobilized bacteria. Civ. Eng. 1–7
2. Wang, J., Mignon, A., Trenson, G., et al.: A chitosan based pH-responsive hydrogel for encapsulation of bacteria for self-sealing concrete. Cement Concr. Compos. **93**, 309–322 (2018). https://doi.org/10.1016/j.cemconcomp.2018.08.007
3. Wang, J., Jonkers, H.M., Boon, N., De Belie, N.: Bacillus sphaericus LMG 22257 is physiologically suitable for self-healing concrete. Appl. Microbiol. Biotechnol. **101**(12), 5101–5114 (2017). https://doi.org/10.1007/s00253-017-8260-2
4. Jonkers HM, Schlangen E (2009) A two component bacteria-based self-healing concrete. In: Concrete Repair, Rehabilitation and Retrofitting II - Proceedings of the 2nd International Conference on Concrete Repair, Rehabilitation and Retrofitting, ICCRRR, pp. 119–120. https://doi.org/10.1201/9781439828403.ch27
5. Jadhav, U.U., Lahoti, M., Chen, Z., et al.: Viability of bacterial spores and crack healing in bacteria-containing geopolymer. Constr. Build. Mater. **169**, 716–723 (2018). https://doi.org/10.1016/j.conbuildmat.2018.03.039
6. Feng, J., Chen, B., Sun, W., Wang, Y.: Microbial induced calcium carbonate precipitation study using Bacillus subtilis with application to self-healing concrete preparation and characterization. Constr. Build. Mater. **280**, 122460 (2021). https://doi.org/10.1016/j.conbuildmat.2021.122460
7. Wang, J.Y., De Belie, N., Verstraete, W.: Diatomaceous earth as a protective vehicle for bacteria applied for self-healing concrete. J. Ind. Microbiol. Biotechnol. **39**, 567–577 (2012). https://doi.org/10.1007/s10295-011-1037-1
8. Wang, J.Y., Soens, H., Verstraete, W., De, B.N.: Self-healing concrete by use of microencapsulated bacterial spores. Cem. Concr. Res. **56**, 139–152 (2014). https://doi.org/10.1016/j.cemconres.2013.11.009
9. Wang, J., Vandevyvere, B., Vanhessche, S., et al.: Microbial carbonate precipitation for the improvement of quality of recycled aggregates. J. Clean. Prod. **156**, 355–366 (2017). https://doi.org/10.1016/j.jclepro.2017.04.051
10. Zhang, J., Liu, Y., Feng, T., et al.: Immobilizing bacteria in expanded perlite for the crack self-healing in concrete. Constr. Build. Mater. **148**, 610–617 (2017). https://doi.org/10.1016/j.conbuildmat.2017.05.021
11. Alazhari, M., Sharma, T., Heath, A., et al.: Application of expanded perlite encapsulated bacteria and growth media for self-healing concrete. Constr. Build. Mater. **160**, 610–619 (2018). https://doi.org/10.1016/j.conbuildmat.2017.11.086
12. Lucas, S.S., Moxham, C., Tziviloglou, E., Jonkers, H.: Study of self-healing properties in concrete with bacteria encapsulated in expanded clay. Sci. Technol. Mater. **30**, 93–98 (2018). https://doi.org/10.1016/j.stmat.2018.11.006
13. Ersan, Y., Bravo, F., Silva, D., et al.: Screening of bacteria and concrete compatible protection materials. Constr. Build. Mater. **88**, 196–203 (2015). https://doi.org/10.1016/j.conbuildmat.2015.04.027

14. Bundur, Z.B., Kirisits, M.J., Ferron, R.D.: Biomineralized cement-based materials: impact of inoculating vegetative bacterial cells on hydration and strength. Cem. Concr. Res. **67**, 237–245 (2015). https://doi.org/10.1016/j.cemconres.2014.10.002
15. Vandervoort M (2019) Impact of bioagents on the cementitious matrix. Ghent University
16. Bolobova, A.V., Kondrashchenko, V.I.: Use of yeast fermentation waste as a biomodifier of concrete (Review). Appl. Biochem. Microbiol. **36**, 205–214 (2000)
17. Bundur, Z.B., Kirisits, M.J., Ferron, R.D.: Use of pre-wetted lightweight fine expanded shale aggregates as internal nutrient reservoirs for microorganisms in bio-mineralized mortar. Cement Concr. Compos. **84**, 167–174 (2017). https://doi.org/10.1016/j.cemconcomp.2017.09.003
18. Ogunbode EB, Hassan IO (2011) Effect of addition of calcium nitrate on selected properties of concrete containing volcanic ash. Leonardo Electron. J. Pract. Technol. 29–38
19. Ying, H., Zhang, Y., Cheng, J.: Dynamic urea bond for the design of reversible and self-healing polymers. Nat. Commun. (2014). https://doi.org/10.1038/ncomms4218
20. Wiktor, V., Jonkers, H.M.: Quantification of crack-healing in novel bacteria-based self-healing concrete. Cement Concr. Compos. **33**, 763–770 (2011). https://doi.org/10.1016/j.cemconcomp.2011.03.012
21. Liu, C., Yang, L., Wang, F., Hu, S.: Enhance the durability of heat-cured mortars by internal curing and pozzolanic activity of lightweight fine aggregates. Constr. Build. Mater. 121439 (2020). https://doi.org/10.1016/j.conbuildmat.2020.121439
22. Li, W., Liu, L., Chen, W., et al.: Calcium carbonate precipitation and crystal morphology induced by microbial carbonic anhydrase and other biological factors. Process Biochem. **45**, 1017–1021 (2010). https://doi.org/10.1016/j.procbio.2010.03.004
23. De Muynck, W., Verbeken, K., De Belie, N., Verstraete, W.: Influence of urea and calcium dosage on the effectiveness of bacterially induced carbonate precipitation on limestone. Ecol. Eng. **36**, 99–111 (2010). https://doi.org/10.1016/j.ecoleng.2009.03.025
24. Zheng, T., Qian, C.: Influencing factors and formation mechanism of $CaCO_3$ precipitation induced by microbial carbonic anhydrase. Process Biochem. **91**, 271–281 (2020). https://doi.org/10.1016/j.procbio.2019.12.018

Comparative Life Cycle Assessment of Concrete with Coarse Natural and Recycled Concrete Aggregates for Household Construction, Case Study in the Hamburg Area

Lionel Reichart(✉), Juan Cassiani, and Sylvia Kessler

Helmut-Schmidt-University/University of the Federal Armed Forces Hamburg, Hamburg, Germany
lionel.reichart@hsu-hh.de

Abstract. Society's awareness of sustainability has been increasing in the last years, hence the use of recycled materials is becoming crucial in order to avoid negative impacts of human activities on the planet. Concrete is one of the most important building material in the construction industry. For the sake of concrete sustainability, the introduction of recycled concrete aggregate (RCA) in new concrete represents a possibility of avoiding construction waste and saving resources. The aim of this study is to evaluate the environmental impacts of three different sources of coarse aggregates for concrete production through a comparative Life Cycle Assessment (LCA). Two aggregates are produced with recycled construction and demolition waste, also known as RCA, and the third is produced from natural gravel also referred as natural concrete aggregate (NCA). The analyzed products are several concrete mix designs, covering different strength classes which are commonly used for household construction. The level of RCA replacement for the mix designs were established to a German guideline. The processes and materials evaluated comprehend the concrete production in a ready-mix plant located in Hamburg, Germany. Based on a data collection from the literature and different datasets out of the databases Ecoinvent version 3.7 and ELCD version 3.2, a LCA-Model has been developed. The results are presented using five midpoint indicators calculated by the CML-baseline and ReCiPe Method. During the evaluation of the LCA-results, it was found that the inclusion of RCA does not inevitably reduce the environmental impacts of concrete. Depending on the considered impact categories, it is possible that additional strains are eliminating the emerging benefits of RCA contribution. Supplementary transportations, changes in the mix design, compressive strength, and general demands for recycling old demolition waste are in focus of the discussion. In order to improve the sustainable performance of RCA in concrete some factors need to be considered.

Keywords: Concrete · Sustainability · Recycled concrete aggregates · Life cycle assessment

J. I. Escalante-Garcia et al. (Eds.): RW 2021, RILEM Bookseries 40, pp. 205–213, 2023.
https://doi.org/10.1007/978-3-031-21735-7_23

1 Introduction

Adequate housing has been acknowledged as a fundamental human right by the United Nations [1]. The household provides basic elements such as shelter, security, water, electricity and sanitary services, which are necessary to keep a decent life. In 2018 the German government launched an initiative to promote the construction of 1.5 Million flats during four years. Just the budget of social housing construction nationwide is assigned with 5 billion Euros [2]. This social program is aimed to make the household economically affordable for the growing population, while at the same time promoting sustainable practices in terms of resources and energy consumption.

In Germany, the construction industry is responsible for about 228.1 billion tons of waste per year, corresponding to more than 60% of the total waste generated in the country [3]. In addition, the sector produces a significant amount of harmful emission to the air with 10929 Giga tons of CO_2 and 2970 tons of fine dust particles every year [4]. Due to the significant environmental burdens of the construction industry, the European Union has established goals to improve the sustainability and resource efficiency of the sector [5]. Since concrete is the most used material in construction work, it is imperative to reduce its environmental impacts in order to achieve the desired sustainability in the construction sector. The replacement of natural concrete aggregate (NCA) with recycled concrete aggregate (RCA) has been recognized as a way of reducing the environmental impact of concrete [6, 7]. The RCA creates the possibility of minimizing the extraction of natural aggregates. Furthermore, using demolition waste material can reduce the requirement of landfill areas thus reducing the deforestation and the transportation required for that.

Including RCA in new concretes is particularly attractive in cities and metropolitan regions. Given the increase of new homes construction in urbanized areas, the demand of natural aggregates for producing concrete will also increase. Meaning the further depletion of the current quarries or in the worst case the need for new mining sites. At the same time, the construction waste generation also increases as product of old structures demolition. The aim of this study is to evaluate the environmental impacts of the concrete production with two types of RCA and with NCA, through a comparative life cycle assessment (LCA) according to DIN EN ISO 14040 [8] and DIN EN ISO 14044 [9]. The study consists of four steps; the goal and scope definition, the inventory analysis, the impact assessment and discussion section. To investigate the product system a LCA-model is created using the software SimaPro [10].

2 Life Cycle Assessment

2.1 Goal and Scope Definition

The Goal of this study is to determine the environmental impacts of the production of different concretes with RCA and NCA. By comparing the obtained results of the concretes produced with RCA and with NCA, the environmental implications of using RCA in new concrete will be evaluated. The selected concrete mixes are commonly used for household construction according to the German regulations. The inclusion of different strength classes allows for a wide range of applications, which improves the study.

2.1.1 Functional Unit

The functional unit is defined as 1 m^3 of concrete with a compressive strength at 28 days given by its strength classes. The strength class are C8/10, C16/20, C25/30, C30/37 according to the European regulation [11]. The first and second number represents the characteristic compressive strength evaluated with cylinders and cubes, respectively. The concretes are designed to have a plastic consistency (class F2). The mass amounts in the concrete mix designs are explained in the following chapter.

2.1.2 Concrete Mix Designs

The concrete mixes are designed without superplasticizer or other concrete admixtures, hence the constituents for the analysed system are cement, natural aggregates (fine and coarse), RCA type 1, RCA type 2 and water. The used cement is a CEM III/A 32,5 also known as blast furnace slag cement. The RCA1, which is called "concrete chippings", is composed with 90% of concrete waste products and 10% of bricks and blocks. The RCA2, which also is called as "building chippings", is composed with 70% of concrete waste products and 30% of bricks and blocks. In this study only the replacement of natural coarse aggregates by RCA is addressed. The RCA replacement is set as the maximum replaceable allowed by the country standards [11]. In the classes C8/10 and C16/20 this proportion of RCA1 and RCA2 are 45 and 35% in weight, respectively. For the classes C25/30 and C30/37 the share are 35 and 25% in weight for RCA1 and RCA, respectively.

The water amount of concrete mixtures with RCA is increased by 3.8% in order to compensate water absorption by the RCA based on the findings of [12]. The cement content of those mixes was also adjusted in order to preserve the water-cement ratio. Hence assuring that those mixes will achieve the required compressive strength. The minimum cement content and water-cement ratio of the mixes were checked according to their corresponding exposure classes stated in DIN 1045-2 and DIN EN 206-1 [13]. Table 1 shows the mass of each constituent for the studied strength classes.

2.1.3 System Description and Boundaries

The analysed product system represents the production of concrete in a ready-mix plant located in the west of Hamburg, Germany. The processes considered for the concrete production are shown in Fig. 1, as well as the transportation distances. All processes are analysed using cradle-to-gate approach. The studied system includes all processes upcoming from the extraction of raw materials ending with the mixed concrete at the gate of the ready-mix plant. The concrete transport from the plant to the construction site is not included. Since the RCA mix design is modified to have the same flow properties as NCA concrete, the construction process for RCA and NCA concretes is basically the same. Thus, it is assumed that there is no significant difference between the environmental impacts in the construction phase. Furthermore, the concrete is intended to be used for household applications, hence there are no aggressive exposure conditions which can affect the durability of the reinforced concrete elements. Therefore, the construction and service phases are also not considered in this study.

For the sake of this study feasibility, the following assumptions are made:

Table 1. Overview about concrete components, strength-classes and mix-designs

	Building component	Foundation C8/10	Interior wall C16/20	Exterior wall C25/30	Basement wall (outside) C30/37
Amounts for concrete with NCA	Water (kg/m^3)	180	180	180	180
	Cement (kg/m^3)	217	300	310	419
	Sand [kg/m^3]	1067	1028	1023	972
	Gravel (kg/m^3)	854	823	819	778
Amounts for concrete with RCA	Water (kg/m^3)	187	187	187	187
	Cement (kg/m^3)	225	311	322	435
	Sand (kg/m^3)	1053	1012	1007	954
(Type 1)	Gravel (kg/m^3)	464	446	524	497
	RCA1 (kg/m^3)	315	303	234	222
(Type 2)	Gravel (kg/m^3)	548	527	605	573
	RCA2 (kg/m^3)	245	236	167	159

- The demolition waste necessary to manufacture both RCA was assumed to be burden free, hence only environmental impacts of transporting and processing are considered.
- Demolition waste with high shares of concrete is used of the production of RCA1.
- Demolition waste with less shares of concrete is used of the production of RCA2.
- The demolition waste is first sorted in a classification plant, then crushed and processed in another plant.
- The materials transportation is expected with 22t lorry.

2.2 Life Cycle Inventory

In this phase of the LCA study all the inputs and outputs of the system processes are collected, to compute the overall environmental impact of the analysed system. Datasets from other LCA studies were used to gather information about the emerging environmental impacts of the production of cement, water, sand and gravel, and those of transportation and concrete manufacture. The data of cement, water, electricity and transportation

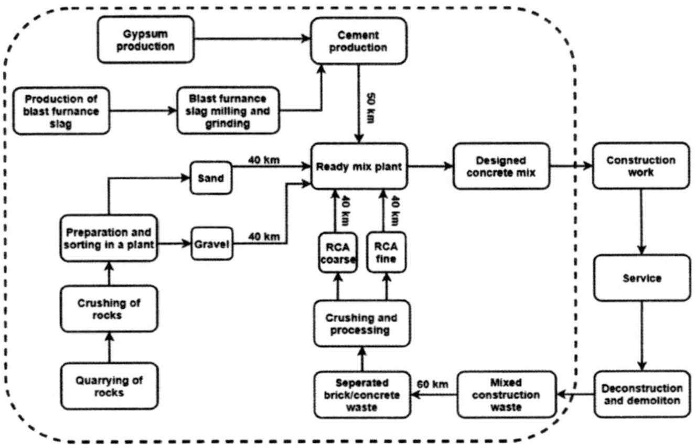

Fig. 1. System under investigation with boundary

were taken from the Ecoinvent database version 3.7 (2020) [14]. The data for sand and gravel production were found in the ELCD database version 3.2 [15]. Due to the lack of information regarding land occupation for the natural aggregates in the ELCD database, this data was complemented with the values from the Ecoinvent database. The inputs and outputs of the RCA sorting and manufacture were taken from the literature [16] and [17], the detailed procedure can be found in [18].

2.3 Life Cycle Impact Assessment

The six chosen impact category indicators are abiotic depletion (fossil fuels), global warming potential, ozone layer depletion, acidification, eutrophication and land use. The calculation for the first five impact category indicators is done with the "CML-IA Baseline Method", version 3.06. The calculation of the impact category "land use" is made with the "ReCiPe 2016 v1.1 midpoint method, Egalitarian version". Both are predefined in the software SimaPro.

Figure 2 shows the results of the impact categories for all the concrete mixed analysed. The inclusion of the recycled aggregates affected negatively the environmental performance of the concrete in the indicators abiotic depletion, global warming potential, acidification and eutrophication. In the category abiotic depletion, the usage of RCA leads to an increase of the results between 2.5 and 3.0% with respect to the mix with NCA. Those differences were presented in the mixes C30/37 and C16/20 respectively. There is no difference on the indicator values between the RCA1 and RCA2, except for the class C25/30 where the difference is 0.4%. The global warming potential also presented an increment when using RCA. This increase ranges between 2.9 and 3.5%, corresponding to the mix C16/20 for RCA1 and RCA2 respectively. Similarly, regarding the abiotic depletion indicator, only in the C25/30 class there is significant difference between the RCA1 and the RCA2 values. In the acidification indicator the inclusion of RCA presented mixed results. Although there is a reduction of −0.3% in C8/10 class, there is an increase in all the other strength classes with a maximum value of 18.8% in the class C30/37. Finally, the eutrophication indicator value is increased between 2.7 and 10.6%, corresponding to the classes C25/30 and C16/20. Here in this last strength class the RCA2 had a more unfavourable impact than the RCA1.

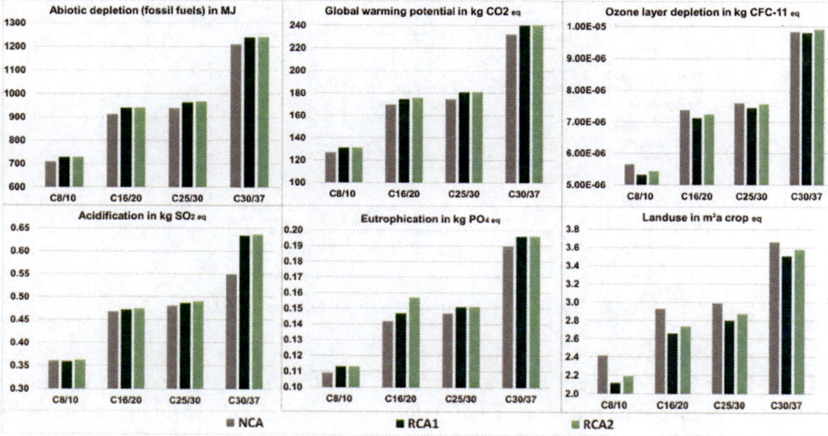

Fig. 2. Results of the impact category indicators

On the other hand, the usage of recycled aggregates had positive impacts on the ozone layer depletion and land use indicators. The RCA1 reduced the ozone layer depletion indicator between −6.0 and −0.3%, corresponding to the C8/10 and C30/37 classes. In contrast the RCA2 reductions were slightly smaller with a maximum value of −3.9% in class C8/10, however presented a slight increment of 0.7% in class C30/37. Here it should be noted that in this impact category the benefit from RCA decreases as the strength class increases. The land use indicator had the same trends as the observed in the ozone layer depletion. Firstly, the RCA1 presented the major reductions ranging between −12.4 and −4.1%, while in the RCA2 the reductions were lower with values between −9.2 and −2.3%. These limits also correspond to the strength classes C8/10 and C30/37, respectively. Secondly, the benefits of both RCA decrease with increasing strength class. An interesting finding is that the RCA1 provides better environmental savings than RCA2 in both impact indicators.

3 Discussion

The results of the analysed case study show that the usage of RCA is not improving the environmental performance of concrete in 3 of the 5 chosen impact categories. In order to understand the reason for those results, it is necessary to identify the key factors that are affecting the environmental impacts of concrete. For that purpose, the contribution of each material and unit process on each impact indicator was computed. Given the space limitation of this document, only the results for the class C25/30 are plotted in Fig. 3. In the graph it is noticeable that the cement production is the main contributor of the values of all category indicators. In the C25/30 strength class the cement contribution ranges between 68.3 and 92.8%, corresponding to the indicators land use and global warming potential respectively. The main reason for the high cement contribution is that the clinker production process demands a high amount of fossil fuels and minerals.

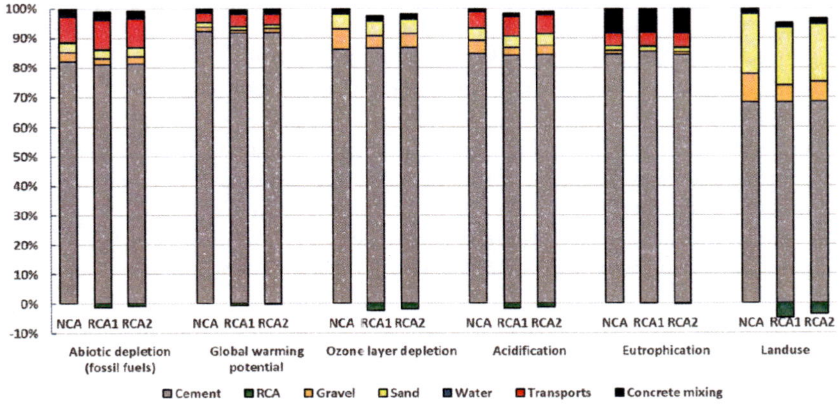

Fig. 3. Impact category composition for strength class C25/30

Due to the water absorption of RCA the water content of all mixes was increased by 3.8% in mass, leading to a rise of the cement amount to meet the compressive strength specification. The increment of the cement content in mass started at 8.2% for classes C8/10, and increase with the upper strength classes up to 15.9% (C30/37 class). Given that the cement is the main contributor on all impact category results, the increased cement content in the RCA mixes affects the environmental results of concrete negatively. In particular the impact categories global warming potential, acidification and eutrophication. Here the higher the strength class of the concrete, the bigger the negative impacts in those indicators.

Another factor affecting the environmental results of the mixes with RCA is the extra transport required for manufacturing the RCA. The transportation of the concrete waste material to the RCA producing plant and then to the concrete mixing plant consumes fossil fuels, thus generating emissions of harmful gases and dust particles. Therefore, producing additional negative environmental impacts in the categories abiotic depletion (fossil fuels), global warming potential, acidification and eutrophication. In the ozone layer depletion, the transportation has a small influence so the changes in this category are negligible. Although the RCA inclusion lead to environmental saving in those indicators, the additional transportation creates more environmental burdens than the RCA avoids.

The environmental benefits provided by the RCA in the land use and ozone layer depletion indicators are due to two reasons. Firstly, the natural gravel contribution on those indicator values is significant with values around 10 and 7% for the land use and ozone layer depletion, respectively. Therefore, reducing the natural gravel amount in the mixes results in a decrease of the indicators values. Secondly, for the land use and ozone layer depletion the RCA provides negative impacts which are in other words environmental savings. Hence, adding them to the concrete mixes will reduce more the final value of the indicators. Finally, there is a difference in the environmental performance of the analysed recycled aggregates. The RCA1 is providing bigger environmental benefits and less negative impacts than the RCA2. The main reason for this is the difference in the maximum replacement allowed by the German guideline [11]. Due to the higher quality of the RCA1 the maximum replacement is 10% higher than the allowed for RCA2 in all strength classes. Hence, the introduction of the RCA1 leads to more reduction of the NCA in the mix which results in a better environmental performance.

The results of this study show that the usage of RCA presents an interesting opportunity to reduce the environmental impacts of concrete. The re-use of construction waste in new concrete reduces the requirement of landfill area, which can help to reduce deforestation. In addition, the RCA usage reduces the consumption of natural gravel resources hence lowering the depletion of non-renewable resources. Nevertheless, the RCA is generating negative impacts in 4 of the 6 category indicators due to the increased cement content and extra transportation distances. Therefore, the partial replacement of cement with supplementary cementitious materials should be considered in order to mitigate the impacts of the increased cement content [6]. On the other hand, the planning of the supply chain for the RCA manufacture must also be addressed. Here, it is necessary to determine the maximum distances so that the environmental impacts of the RCA production and concrete are kept at a minimum.

Finally, the inclusion of RCA in the mixes also reduces the costs of the concrete production. For the analysed mixes, the economic savings are 1.5% for the highest strength class (C30/37) and reaches up to 4.6% for the lowest strength class (C8/10). Taking into account that the mixes between those strength classes are around the 82% of the total annual concrete production in Germany (49 Mio m^3 a) [19], those costs savings will definitely have an impact in the construction market.

4 Conclusion

The following conclusions can be drawn from this comparative LCA case study of the inclusion of coarse RCA and NCA in different concrete strength classes.

- The use of RCA in new concrete had positive impacts on the ozone layer depletion and land use indicators. The RCA helps to save landfill space and reduce the exhaustion of natural gravel deposits.
- The RCA deteriorated the results of the impact categories abiotic depletion, global warming potential, acidification and eutrophication. Those negative impacts are due to the extra cement and additional transportation distances.
- The cement is the major contributor in the environmental impacts of concrete production. The increase of the cement content due to the usage of RCA should be managed to a minimum or mitigated by alternative cementitious materials.
- The transportation distances for the RCA manufacture must be kept as minimal as possible to optimize the environmental performance of concrete with RCA.

References

1. UN-Habitat: The Right to Adequate Housing: Fact Sheet No. 21/Rev. 1, Fact Sheet No. 21/Rev. 1, 2014
2. Bundesministerium des Innern, für Bau und Heimat (BMI): Die Wohnraumoffensive und ihr Umsetzungsstand: Bezahlbares Wohnen und Bauen – Bilanz der Wohnraumoffensive am 23. Februar 2021, Feb. 2021
3. Statistisches Bundesamt (Destatis): Umwelt Abfallbilanz 2018: Abfallaufkommen/-verbleib, Abfallintensität, Abfallaufkommen nach Wirtschaftszweigen, article nr: 5321001187004, May 2021. Accessed 4 May 2021
4. Statistisches Bundesamt (Destatis): Umweltnutzung und Wirtschaft 2019: Tabellen zu den Umweltökonomischen GesamtrechnungenTeil 3: Anthropogene Luftemissionen, Article nr: 5850007197006, Sep. 2019. Accessed 27 May 2021
5. European Commission: The European construction sector—a global partner, Jan. 2016 [Online]. https://ec.europa.eu/growth/content/european-construction-sector-global-partner-0_en
6. Kurda, R., Silvestre, J.D., de Brito, J.: Life cycle assessment of concrete made with high volume of recycled concrete aggregates and fly ash. Resour. Conserv. Recycl. **139**, 407–417 (2018). https://doi.org/10.1016/j.resconrec.2018.07.004
7. Reza, F., Wilde, W.J., Izevbekhai, B.: Sustainability of using recycled concrete aggregates as coarse aggregate in concrete pavements. Transp. Res. Rec. **2672**(27), 99–108 (2018). https://doi.org/10.1177/0361198118801357
8. DIN Deutsches Institut für Normung e. V.: Umweltmanagement—Ökobilanz—Grundsätze und Rahmenbedingungen (ISO 14040:2006) 14040, Nov. 2009
9. DIN Deutsches Institut für Normung e. V.: Umweltmanagement—Ökobilanz—Anforderungen und Anleitungen (ISO 14044:2006 + Amd 1:2017) 14044, May 2018
10. SimaPro Version 9.1.0.7: PRé Sustainability B.V., 2020 [Online]. https://simapro.com/
11. Deutscher Ausschuss für Stahlbeton: Beton nach DIN EN 206-1 und DIN 1045-2 mit rezyklierten Gesteinskörnungen nach DIN EN 12620: Teil 1: Anforderungen an den Beton für die Bemessung nach DIN EN 1992-1-1, Berlin, Sep. 2010
12. Braga, A.M., Silvestre, J.D., de Brito, J.: Compared environmental and economic impact from cradle to gate of concrete with natural and recycled coarse aggregates. J. Clean. Prod. **162**, 529–543 (2017). https://doi.org/10.1016/j.jclepro.2017.06.057
13. DIN Deutsches Institut für Normung e. V.: Tragwerke aus Beton, Stahlbeton und Spannbeton—Teil 2: Beton—Festlegung, Eigenschaften, Herstellung und Konformität—Anwendungsregeln zu DIN EN 206-1. 1045-2, Aug. 2008
14. Ecoinvent, The Ecoinvent database: version 3.7 (2020) [Online]. https://www.ecoinvent.org/database/database.html. Accessed 30 Nov 2020
15. European Commission: European platform on life cycle assessment: database [Online]. https://eplca.jrc.ec.europa.eu/ELCD3/
16. Heyn, S., Mettk, A.: Ökologische Prozessbetrachtungen—RC-Beton (Stofffluss, Energieaufwand, Emissionen). Forschungebericht, Brandenburgische Technische Universität Cottbus, Cottbus (2010)
17. Holcim (Schweiz) AG, Ökobilanzen rezyklierter Gesteinskörnung für Beton: Ein Forschungsbericht der Holcim (Schweiz) AG, Zürich [Online]. www.holcim.ch
18. Reichart, L.: Towards a circular economy in construction: environmental assessment of the recycled concrete aggregate usage in new concrete. Bachelorthesis, Helmut-Schmidt-Universität/Universität der Bundeswehr Hamburg, Hamburg (2020)
19. Bundesverband der DeutschenTransportbetonindustrie e. V. (BTB), GUTES KLIMA - Jahresbericht 2020, Aug. 2020 [Online]. www.transportbeton.org

Deformations in Pastes During Capillary Imbibition and Their Implications

Natalia Alderete(✉), Yury Villagrán Zaccardi, and Nele De Belie

Department of Structural Engineering and Building Materials, Ghent University, Technologiepark Zwijnaarde 60, Campus Ardoyen, B-9052 Gent, Belgium
NataliaMariel.Alderete@ugent.be

Abstract. In capillary imbibition (or absorption) experiments performed on cementitious materials, a non-linear evolution of water uptake with the square root of time is frequently reported. This anomaly is derived from analytical methods that assume that the pore structure remains invariable during water ingress. A more comprehensive understanding of the transport phenomena should consider calcium silicate hydrate (C-S-H) swelling during water ingress. Such swelling may cause changes in the pore structure and interferes with water uptake. To analyse that process, the external deformations occurring in cement paste samples during capillary imbibition tests were measured. Pastes were prepared using Portland cement (CEM I 52.5 N) and water/binder = 0.4 and 0.6. After 7 days of curing, strain gauges were glued on the samples to register the external strain. Differences in the deformation results show that the restriction of the system plays an important role. Results indicate that the anomalous water ingress during capillary imbibition coincides with the measured external deformations. In this sense, calculations based on the model of Thomas and Jennings showed that pores shrink. Such pore changes are synchronous with the capillary rise. Hence, an improved description of the delaying process in the transport mechanism is here presented considering the effects of the changing pore structure. The extent of this physical effect is related to the porosity and amount of non-deforming and deforming phases (C-S-H) of each system.

Keywords: Capillary imbibition · Deformations · C-S-H · Internal restrain · Swelling

1 Introduction

Traditionally, capillary imbibition in cementitious materials is described and modelled considering a rigid pore structure during the transport process [1]. However, in a previous research deformations during imbibition were measured in mortar and concrete [2]. Such changing pore structure was linked to the ability of C-S-H to retain water during water ingress with a consequent swelling. This is likely to hinder the water flow causing the anomalous behavior of cementitious materials: lack of linearity with the square root of time. In fact, this anomaly is manifested by a water ingress rate that decreases faster than the square root of time. This delaying effect, finds a suitable model in the evolution with

J. I. Escalante-Garcia et al. (Eds.): RW 2021, RILEM Bookseries 40, pp. 214–220, 2023.
https://doi.org/10.1007/978-3-031-21735-7_24

the fourth root of time considering a variable hydraulic diffusivity [3]. Then, considering a rigid pore structure during imbibition does not seem to describe the whole transport process. The idea of a dynamic porosity has been the focus of many researchers lately [4, 5] and it is highly relevant to better describe water ingress in cementitious materials.

The cause of the changing pore structure seems to be the restricted C-S-H swelling, and it is translated into volumetric variations. The factors which influence these variations still need further research. Based on the model proposed by Thomas and Jennings [6], the cementitious matrix can be considered as composed by 3 phases: deformable phase (namely C-S-H), non-deformable phase (rest of the paste) and porosity. The aim of this research is to quantify these 3 phases and their influence on the deformations during capillary imbibition in cement pastes with different water-to-cement ratios. Porosity was assessed by water absorption under vacuum and C-S-H was determined by XRD coupled with Rietveld. The results of C-S-H and porosity in relation to the deformations suffered by the paste are presented. The analysis brings light to the interaction that occurs during capillary imbibition and the consequent deformations.

2 Methodology

Paste prisms of 10 mm × 15 mm × 40 mm were prepared using CEM I 52.5 N as binder with water-to-binder ratio of 0.4 and 0.6 (denoted as P4 and P6 respectively). After 48 h of casting, specimens were unmolded and kept in a conditioned room at 95% relative humidity and (20 ± 1) °C for 1 week and then they were preconditioned for testing.

To monitor the external volumetric stability of the specimens during capillary imbibition, two strain gauges were attached as indicated in Fig. 1. The cables were connected to a data logger that continuously recorded the deformations for 28 days. After all connections were made, samples were laterally covered with epoxy paint leaving the top and bottom of the sample uncovered. This allowed one-directional flow and protected the strain gauges connections from direct contact with liquid water.

For the capillary imbibition test, five specimens were preconditioned by first placing them under immersion for 72 h and then drying in an oven at 40 °C. The drying was stopped when the weight loss was lower than 0.1% in a 24 h period. This took around 10 days. Then samples were wrapped in double plastic bags and left 1 week for moisture homogenization. After that, the bottom of the specimens (10 m × 15 mm) were placed in contact with water with a 3 mm immersion of their bottom face.

Significant challenges were faced when determining C-S-H content because of its amorphous structure and indefinite composition [7]. To measure the C-S-H in the cement pastes, XRD coupled with Rietveld + PONKCS [8] was performed, using zincite as internal standard. Specimens hydration was stopped following the RILEM recommendations [9]. A prismatic specimen per mix was crushed with hammer and the interior freshly crushed pieces were collected. Such particles were then ground with mortar and pestle to pass a 1 mm sieve. The whole crushing and sieving procedure was performed as fast as possible (<10 min) to minimize the exposure to air of the sample. Immediately after, the obtained particles were immersed in isopropanol. The suspension was filtered using a Buchner filter and flask (filter paper with pore size smaller than 2 μm). The filter paper was placed on a watch glass and dried for 8 min in a ventilated oven at 40 °C. After that,

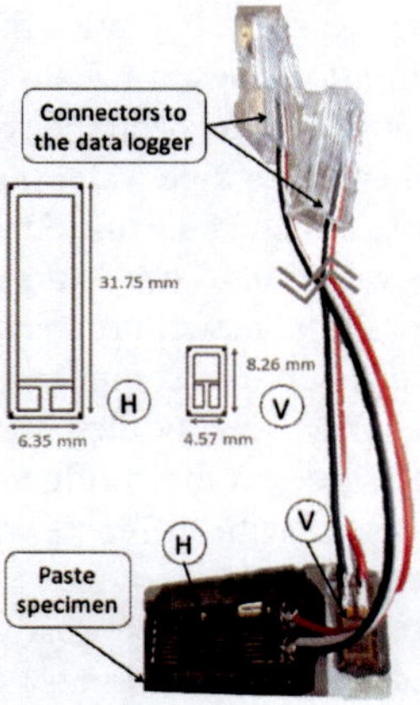

Fig. 1. Paste specimen with the attached strain gauges and cable connections (H = horizontal strain gauge, V = vertical strain gauge).

the powder was ground, homogeneously mixed with 10% zincite. For the XRD measurements, side loading was chosen to prevent preferential orientation of some phases. All samples were tested within 48 h after hydration stoppage.

To determine the open porosity under vacuum, 3 prismatic samples were cut into slices of (10 ± 2) mm, so 9 specimens were tested for each cement paste mix. Such specimens were put in a vacuum chamber (<0.1 bar) for 2 h. Afterwards water was drawn into the vacuum chamber until the sample became fully immersed. Samples were kept under immersion for 24 h, and then the saturated mass (msat) and immersed mass (mim) were registered. The dry mass (md) was determined by putting the samples in an oven at 105 °C and until stable mass (dm/dt <0.1 wt.%/24 h). The open porosity was determined as (msat-md)/(msat-mim).

3 Results and Discussions

To make an easy visualization of the results, C-S-H content determined by XRD coupled with Rietveld + PONKCS and porosity measured by water absorption under vacuum are shown together in Fig. 2. The C-S-H amount (rounded markers) of P4 is slightly higher than P6, but the difference is rather small. For the case of porosity (triangular markers), differences are clear, driven by the difference is the water-to-binder ratio [10].

Results of capillary imbibition are shown in Fig. 3. Due to the larger porosity of P6 in comparison to P4, the CIR of P6 is noticeably higher than P4. The water ingress in P6 specimens starts to decrease faster in samples (around after 1 week). This could be because the water front has reached the top of the samples faster in P6 specimens than in P4 specimens. As P6 has a higher porosity, the water flow advances quicker and the capillary potential is equalized sooner. Neutron radiography experiments are planned to visualize the water ingress front and evaluate the water ingress rate.

To monitor the deformations during imbibition, continuous measurements were performed for 28 days. Results shown in Fig. 4 indicate that there are almost immediate

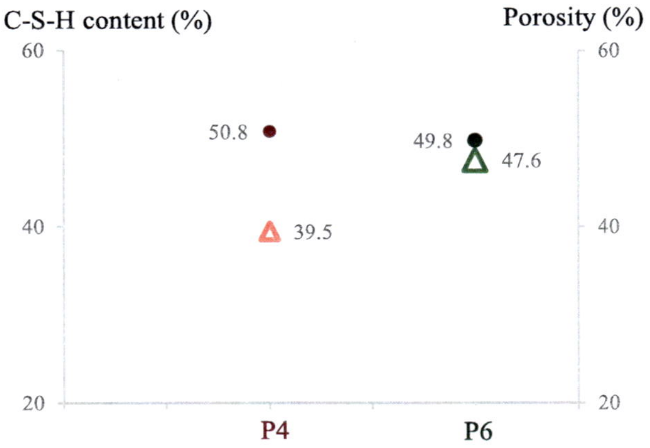

Fig. 2. Results of P4 and P6 pastes of: C-S-H content -XRD coupled with Rietveld- (rounded markers, left axis); and accessible porosity -water absorption under vacuum- (triangular markers, right axis).

deformations during imbibition. These are mainly positive, i.e., expansion or swelling. A small remark should be made for the deformations registered with the vertical strain gauges: the expansion produced only at the bottom of the sample causes negative recordings in the vertical direction. However, as water advances through the samples there is a progressive expansion along the height of the sample and hence vertical positive deformations are registered.

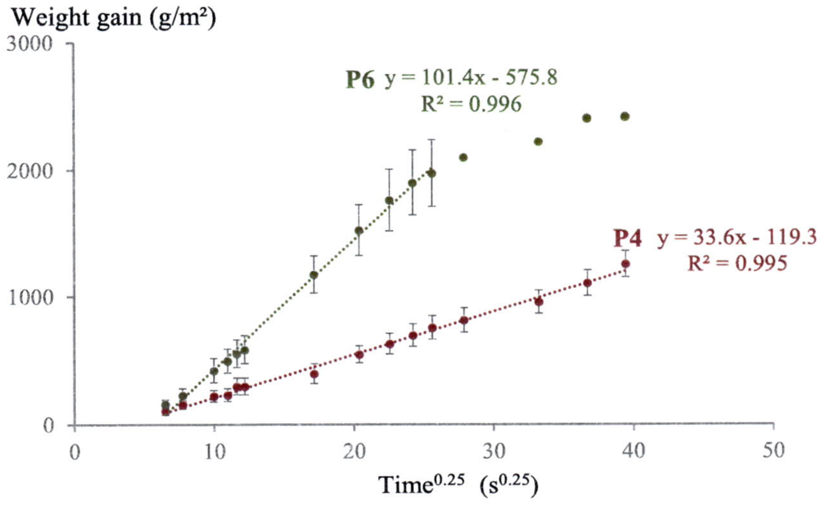

Fig. 3. Results of capillary imbibition during 28 days of P4 and P6 pastes.

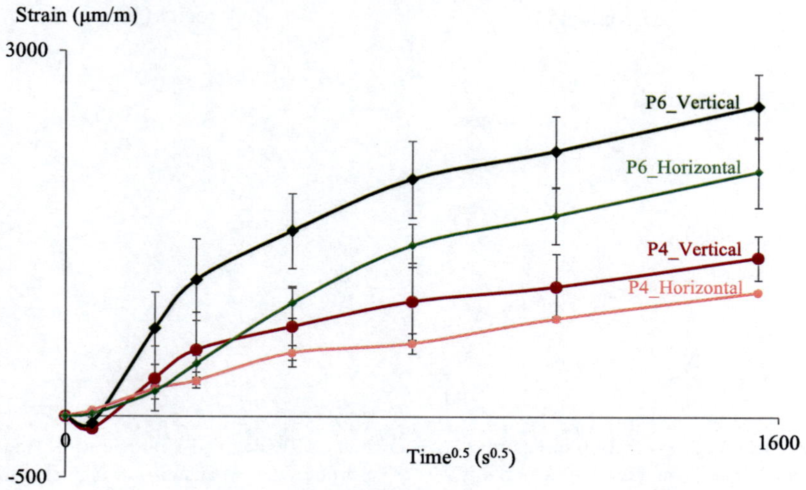

Fig. 4. Deformations measured during capillary imbibition of pastes P4 (round markers) and P6 (rhomboidal markers) for a period of 28 days. Error bars indicate the standard deviation (n = 5).

The expansion experienced in P6 samples is much larger than in P4 samples, for both horizontal and vertical strain gauges. These differences in the external deformation are probably linked to the relation between the deformable phase, the non-deformable phase and the porosity of each system. As described by Thomas and Jennings [6], the degree of restriction of the system is given by the relative amount of porosity and deformable phases. For the case of P6, there is a higher porosity with a similar amount of deformable phase (C-S-H content) in comparison to P4. This would indicate that there are less restrictions in the system. As seen in the deformations obtained, same exposition time to water led to higher deformations in P6 than P4. This is a relevant point to highlight, because it clearly means that the amount of the deformable phase is not directly linked to the magnitude of the volumetric deformations. In fact, the relative amount of porosity, non-deformable phases and C-S-H content seem all to play a relevant role.

After 28 days of continuous capillary imbibition, volumetric deformations continue to increase and even values of P6 are almost double as those of P4. From the results it could be observed that (1) expansion lasts for a prolonged time, (2) with similar C-S-H content a relative small increase of porosity of P6 (20.5%) in comparison to P4 leads to a much larger increase (96%) in deformations, (3) vertical deformations are larger than horizontal deformations. For the case of (1), it is interesting to see that deformations continue to increase even after 28 days of imbibition. These prolonged volumetric deformations are important to describe the long-term behavior and should be taken into account in modelling to describe the transport process in a comprehensive manner. When considering point (2), even though only 2 pastes were tested, it seems that a certain change in porosity (with similar C-S-H content) leads to a larger impact in the final deformations. This can be attributed to the restrictions of each mix, because with more porosity, the P6 paste is less restricted and hence deformations can occur more easily. Finally, point (3) provides insight into the anisotropic behavior of the sample. It seems that vertical deformations are always larger than horizontal deformations (also in mortar and concrete [2]). This might be related to the geometry of the sample, but further research is needed to verify this.

4 Conclusions

Volumetric changes during capillary imbibition in cement pastes (P4 and P6) were registered for 28 days. These deformations are related to the affinity of C-S-H with water. However, not directly proportional to its amount. The relative amount of porosity and C-S-H content plays a vital role, as seen in the results of P6 and P4. Here, one of the most relevant points is that the amount of deformable phase (C-S-H) is actually not directly related to the deformations. Measurements of C-S-H and porosity in cement pastes indicated that the relative amount of porosity and C-S-H has a larger effect on the deformations.

The registered deformations change with time suggest a changing porosity during capillary imbibition. As a result, the analysis of water movement during imbibition must consider the cementitious materials as a non-rigid media, including potential changes in the pore structure of the system. Hence, further research might contribute to connect the measured deformations with the changing porosity as an explanation of the anomalous behaviour of cementitious materials.

Acknowledgments. Natalia Alderete and Yury Villagrán Zaccardi are postdoctoral fellows of the Research Foundation—Flanders (FWO) and acknowledge its financial support (projects 12ZG820N and 12ZZD21N, respectively).

References

1. Villagrán Zaccardi, Y.A., Alderete, N.M., De Belie, N.: Lucas-Washburn vs Richards equation for the modelling of water absorption in cementitious materials. Matec Web Conf. **02019**, 4 (2018)
2. Alderete, N.M., Villagrán Zaccardi, Y.A., De Belie, N.: Physical evidence of swelling as the cause of anomalous capillary water uptake by cementitious materials. Cem. Concr. Res. **120**, 256–266 (2019)
3. Villagrán Zaccardi, Y.A., Alderete, N.M., De Belie, N.: Improved model for capillary absorption in cementitious materials: progress over the fourth root of time. Cem. Concr. Res. **100**, 153–165 (2017)
4. Mcdonald, P.J., Istok, O., Janota, M., Agata, M., Faux, D.A.: Sorption, anomalous water transport and dynamic porosity in cement paste: a spatially localised 1 H NMR relaxation study and a proposed mechanism. Cem. Concr. Res. **133**, 106045 (2020)
5. Hall, C.: Capillary imbibition in cement-based materials with time-dependent permeability. Cem. Concr. Res. **124** (2019)
6. Thomas, J.J., Jennings, H.M.: Changes in the size of pores during shrinkage (or expansion) of cement paste and concrete. Cem. Concr. Res. **33**, 1897–1900 (2003)

7. Olson, R.A., Jennings, H.M.: Estimation of C-S-H content in a blended cement paste using water adsorption. Cem. Concr. Res. **31**, 351–356 (2001)
8. Scarlett, N.V.Y., Madsen, I.C.: Quantification of phases with partial or no known crystal structures. Powder Diffr. **21**, 278–284 (2006)
9. Snellings, R., et al.: Report of TC 238-SCM: hydration stoppage methods for phase assemblage studies of blended cements—results of a round robin test. Mater. Struct. **51**(4), 1–12 (2018). https://doi.org/10.1617/s11527-018-1237-5
10. Aïtcin P.C.: The influence of the water/cement ratio on the sustainability of concrete. Lea's Chem. Cem. Concr. 807–826 (2019)

Repair of Concrete in Environments with Chlorides or Subjected to Freeze-Thaw Scaling

Vanessa Giaretton Cappellesso[1](✉), Tim Van Mullem[1], Elke Gruyaert[2], Kim Van Tittelboom[1], and Nele De Belie[1]

[1] Ghent University, Gent, Belgium
vanessa.cappellesso@ugent.be
[2] KU Leuven, Gent, Belgium

Abstract. Crack formation further decreases the durability of structures when chloride ions associated with freezing temperatures are present. Therefore, preventing the entry of aggressiveness is imperative to guarantee the service life. Repair actions might recover the liquid-tightness when cracks occur. A water repellent agent (WRA) and a sodium silicate (SS) solution were applied to self-repair cracks in the current research. The repair occurred by manual injection of cracks to obtain a proof-of-concept for the possible self-healing efficiency. Two extreme conditions have been assessed after the healing period, the first referring to continuous immersion in a chloride solution, and the second applying freeze-thaw conditions with de-icing salts. Chloride ingress was evaluated through the colour change boundary test. In addition, optical microscopy analysis was used to measure the crack width and to observe differences before and after exposure. SS prevented the chloride ingress through the crack in both conditions. However, the method used to verify chloride ingress did not give consistent results for the WRA due to its hydrophobicity. Microscopic analysis showed that both agents could avoid chloride ingress in the cracks. For the samples exposed to freeze-thaw cycles, only chloride ingress measurement could indicate the healing performance as the scaling destroyed the surface.

Keywords: Concrete · Repair · Chlorides · Freeze-thaw · Water repellent agent · Sodium Silicate

1 Introduction

Seawater or de-icing salts are an aggressive environment for concrete structures. Damage associated with extreme temperatures can increase degradation, since ice formation in the pores creates internal tension generating more cracks. Moreover, the presence of chlorides within reinforced concrete causes corrosion and results in spalling and delamination. The moisture content is an essential factor contributing to chloride penetration and can increases aggressiveness [1]. The transport mechanism switches in saturated conditions. Three different ways of transport can occur, such as sorption, diffusion and

J. I. Escalante-Garcia et al. (Eds.): RW 2021, RILEM Bookseries 40, pp. 221–230, 2023.
https://doi.org/10.1007/978-3-031-21735-7_25

penetration. Diffusion occurs in a completely saturated condition, e.g. structures submerged in seawater or covered by a salt solution layer, for instance, pavements covered by ice when de-icing salts contribute to thawing.

Preventing chloride ingress allows to increase the service life of the concrete and reduces the cost of maintenance. Hence, efforts in finding solutions to guarantee the durability of concrete structures exposed to chlorides are widespread in the literature [2, 3]. Many alternatives are available, from surface treatments to self-healing technologies. In practice, surface treatments based on hydrophobic features or silica-based compounds are used to block the aggressiveness of the environment. However, cracks generally occur, allowing the ingress of aggressive substances via destroyed surface protection, necessitating new actions.

Self-healing technologies are the solution to guarantee protection when damage occurs. In addition, it can help to ensure repair by quick response, not depending on maintenance activities. Self-healing agents such as hydrophobic or silica-based materials can be introduced in cementitious materials using macrocapsules [4] or microcapsules [5, 6]. Nevertheless, most research has focused on analysing these hydrophobic products as surface protection, not as a potential self-healing material [7]. The literature reports several materials to prevent water or chloride ingress, such as water repellents with silane/siloxane/acrylate bases, sodium acetate, sodium silicate, and fluoropolymers. However, their efficiency as a self-healing mechanism for cracked concrete to prevent chloride ingress and water ingress at ambient and freezing temperatures is unknown [4–6, 8].

Water repellent agent (WRA) has the aim to protect surfaces by hydrophobic impregnation. Three different active components can promote the hydrophobic ability, such as silane-based [2, 4, 7–10], silane-siloxane based [4, 8], and siloxane-acrylate-based [4, 8]. Van Tittelboom et al. [4] have described the advantages of using a water repellent as healing agent applied in macrocapsules is that they remain functional after reloading. Besides, permanent sealing of the cracks avoids ingress of water and aggressive agents [4, 8]. However, the WRA cannot block carbon dioxide penetration [4]. Al-Kheetan et al. [2] present that the moisture content during the application has an important role in preventing water and chloride ingress. The water repellent penetrates through the capillary pores. Therefore, impregnation of specimens in dry condition increases the benefits of the product due to higher penetration. Water repellents as a surface treatment also showed higher resistance and less mass loss under freezing conditions [11].

Sodium silicate (SS) has been studied as cargo in a microencapsulation system to promote the self-healing ability in cementitious matrices [6]. The advantage is the suitable interaction with cement-based materials due to its composition based on silica. Furthermore, sodium silicate reacts with calcium hydroxide (CH) in the presence of water to form a calcium silicate hydrate (C-S-H) rich in silica that is the main product of cement hydration [5, 6, 10]. Sodium silicate reduces the water uptake due to healing the cracks [5, 6]. In contrast with WRA, the SS needs water to trigger the reaction. Since the cracks are physically closed, a higher strength is expected, together with a significant potential to block chloride ingress even when considering freezing temperatures.

The significance of this study starts from the lack of knowledge related to the performance of a potential autonomous healing agent. Repairs consume time and are expensive, even more, if they are delayed, and damage increases with time. In this way, self-repair provided by encapsulated materials can promote the recovery in time, decreasing extra costs and improving the performance. An autonomous healing agent can be a great solution as cargo in encapsulation technologies. However, first, the possible products need to be tested under extreme conditions to prove the efficiency before investing time and budget in developing a methodology to encapsulate them.

This research presents the assessment of the efficacy of two protective materials (SS and WRA) that can be used for self-repair. The potential of those materials is investigated by manual injection in cracked concretes and by evaluating the resistance against chloride ingress and freeze-thaw scaling. It aims to assess the performance of a water repellent agent with solvent-based silane and a sodium silicate solution. Their details are in Sect. 25.3. Two severe exposure classes were assumed following the European standard [1]. The class XS2 corresponds with permanent submersion in a chloride environment. The class XF4 relates to conditions of high water saturation with the presence of de-icing salts under freezing and thawing cycles.

The main objective is to verify the product's performance to prevent chloride ingress considering a suitable self-repair in cracked concretes under those environmental conditions. After their efficiency is proven, further studies evaluating encapsulation systems for concrete exposed to extreme conditions will be conducted.

2 Materials and Methods

2.1 Specimens Preparation and Concrete Composition

The geometry of the cast cylindrical specimens was 100 mm in diameter and 50 mm in height. After casting, the specimens were stored in an air-conditioned room (at 20 ± 2 °C and at least 95% relative humidity) for 24 h. Subsequently, samples were demoulded and placed in the same air-conditioned room until the cracks were created at 56 days. Three series of samples were considered: (1) cracked, (2) cracked and repaired with WRA, and (3) cracked and repaired with SS. The samples were subjected to two different environment conditions (described in detail in Sect. 25.4): Three samples were cast for each series for chloride immersion and four samples for freeze-thaw conditions.

For the concrete composition, some thresholds were considered, such as a water/cement ratio (w/c) of 0.45, a minimum strength class C35/45, and minimum cement content of 340 kg/m^3. Those limitations are related to the requirements of the exposure class, XS2 and XF4, in EN 206 [1]. Table 1 gives the concrete composition.

Table 2 shows the fresh and hardened concrete properties. Compressive strength and air voids analysis were done 28 days after casting. The compressive strength was tested on three cubes of 150 mm, according to NBN EN 12390–3. The mean compressive strength was equal to 80.3 MPa. The air void analysis is an automated test for analysing the spacing factor of hardened concrete. Four slices of $100 \times 100 \times 20$mm were extracted from two cubes of 100 mm. The study requires polishing the concrete surface and executing a contrast enhancement of the surface, as described in ASTM C 457 [12]. The spacing of the air voids is considered a good indicator for freeze-thaw durability. It is the distance from any point in the paste to the wall of the nearest air void. Pigeon et al. [13] indicated a critical spacing factor for different types of concrete, i.e., if the value is

Table. 1. Concrete composition.

Materials	Quantity (kg/m^3)
CEM I 52.5 N	341
Limestone filler	182
Sand 0–4 mm	672
Gravel 2–8 mm	766
Water	153
Superplasticizer (Glenium 27 BASF)	8.98

larger than the critical value, the deterioration of the concrete under freeze-thaw will be higher. For a high-performance concrete, the critical spacing factor is 600 μm. A regular concrete with a w/c ratio of 0.5 has a critical value of 500 μm, and a silica fume concrete has 250 μm as critical spacing factor [13]. The measured spacing factor will be compared with the limit of 600 μm considering the compressive strength result in this study. The results presented in Table 2 show that the protection offered by these air voids against the harmful effects of freezing and thawing cycles is probably not sufficient. In this way, extra protection could help to improve this resistance, and the repairs can promote it around the crack wall compared to concrete without any treatment.

Table. 2. Concrete features in the fresh and hardened stage.

Concrete features				
Fresh properties	Density (kg/m^3)	2344	(NBN EN 12350–6:2019)	
	Air content (%)	3.2	(NBN EN 12350–7:2019)	
	Slump flow (cm)	53	(NBN EN 12350–8:2019)	
		μ	σ	CoV (%)
Hardened properties	Compressive strength (MPa)	80.3	0.9	1.1
	Spacing factor	639	57.4	9.0

2.2 Crack Creation

The Brazilian splitting test was executed to crack the cylindrical specimens. First, the samples were opened entirely into halves. Next, a silicon sheet with a thickness of 100 μm was placed on the edges of the crack to obtain a specific crack width. Finally, two steel clamps tied the halves together with the silicone sheets in between to guarantee the crack width [14, 15].

The crack width of the realistic cracks was measured at maximum six selected zones and at least four along the crack length at the top side of the specimens, i.e., the surface in contact with salt solutions, with five measurements performed in each zone. Crack widths were measured at the same places immediately after crack creation, after curing of the repair agents, and after exposure to the harsh condition. Table 3 describes the data collected for each series immediately after crack creation. The measurements were executed with a stereomicroscope, Leica S8 APO, with DFC 295 camera.

Table 3. Data related to the crack width for each series immediately after crack creation.

	Reference (REF)	Sodium silicate (SS)	Water repellent (WRA)
Minimum (μm)	124	102	105
Maximum (μm)	160	149	170
μ (μm)	143	119	135
σ (μm)	16	14	26
CoV (%)	11%	11%	19%

2.3 Repair Products

Two commercial products generally used as a protective surface treatment were employed. First, the products were injected into the crack by means of a syringe. The syringe is placed in a hole with a diameter of 3 mm and a depth of 30 mm made in one side of the sample, indicated by the red arrow in Fig. 1a. Next, an aluminium butyl tape was applied over the crack to protect the area near the crack mouth (Fig. 1b). Finally, the product was injected until the release of the agent was noticed on the other side of the crack on the same surface. This procedure guarantees an appropriate and complete distribution of the product inside the crack. The syringe was kept in place for around one minute to assure proper absorption by the matrix. It is important to mention that the specimens already had the sides and bottom sealed by a waterproofing resin beforehand.

Two different procedures were selected considering different saturation and curing time before exposure to the harsh conditions to guarantee an adequate performance of the repair products used. The WRA has better efficacy when applied in dry conditions. Hence, the specimens were maintained in the oven for three days at 40 °C to guarantee lower moisture content, increasing the product's penetration in the concrete matrix. For the curing period, no water is required. Thus, after being injected with WRA, the specimens were stored in an air-conditioned room (at 20 ± 2 °C and 60% relative humidity) for two days. Subsequently, they were placed in the exposure conditions. In contrast, sodium silicate performs better in wet environments. In this case, the injection was made in the surface dry condition, i.e., the specimens were stored in an air-conditioned room (at 20 ± 2 °C and 60% relative humidity) for one day before the injection. The curing time was 14 days, and the specimens were placed with the crack facing downwards in

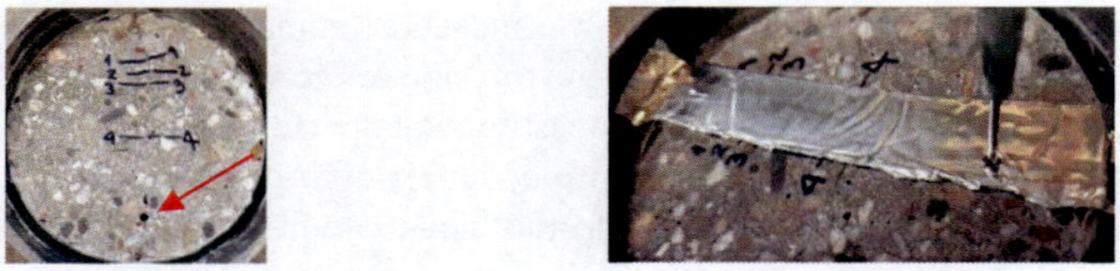

(a) (b)

Fig. 1. Repair execution: (a) hole made at the edge of the crack indicated by the red arrow, and (b) syringe placed in the hole while the crack is covered by aluminium butyl tape.

contact with water (water level 5 mm above the bottom of the specimens) until they were subjected to the exposure condition.

Water repellent agent. A commercially available agent named Sikagard®-705L produced by the company Sika was applied. It is a one-component, solvent-free agent with low viscosity (1.9 mPa.s at 25 °C) based on silane technology, and a hydrophobic coating is formed by covalent bonds. The density at 20 °C is 0.88 g/cm^3, and it has the appearance of a water-like liquid.

Sodium silicate. A commercial product was used, with a density of 1.35 g/cm^3, viscosity of 130 mPa.s at 20 °C, and pH between 11 to 11.5. The appearance is a colourless liquid product.

2.4 Exposure Conditions

Two chloride environments were considered as exposure conditions. They are described in detail in the following sections. In both environments, there is the penetration of chlorides by diffusion as the samples were maintained saturated during the complete experiment.

Chloride immersion. The chloride attack was carried out by immersion in a 33 g NaCl /liter solution at 20 °C for 30 days. The specimens were waterproofed with an epoxy resin at the bottom, and on the sides, only the top surface had contact with the salt solution. The samples were placed with the top facing upwards. The salt solution was 10 mm above the surface.

Freeze-thaw with de-icing salts. The experiment is based on the standard CEN/TS 12390–9 [16]. The chamber is automatically programmed to perform one cycle per day, changing the temperature from –20 °C to 20 °C. The specimens were covered by a 3 mm layer of 3% NaCl salt solution. In total, 56 cycles were conducted.

2.5 Method to Assess the Repair Efficiency

Chloride ingress. The chloride penetration was measured perpendicular to the crack after splitting the specimens, after 30 days in chloride immersion and after 56 days in freeze-thaw cycles. Once the surface was dry, a 0.1 mol/L silver nitrate solution was sprayed on the fractured cylinder halves to visualise the chloride ingress. The chloride-contaminated zone appeared lighter. A marker was used to trace the penetration front.

The location of the colour change boundary was measured from the top for chloride immersion condition as the surface remained intact during the exposure (Fig. 2a). In contrast, the chloride ingress was calculated from the bottom to the top for the specimens subjected to freeze-thaw cycles since each sample's height was reduced due to the surface deterioration by freezing. Figure 2b illustrates the measurements using different colour arrows with a picture on a greyscale image. The red arrow represents the measurement for freeze-thaw specimens, the blue arrow the height of the specimen, and the green the measurement of the chloride immersion specimens. The first 20 mm from the crack surface measurements were taken each 5 mm, and then each 10 mm (Fig. 2c). The measurements were made using a digital calliper gauge.

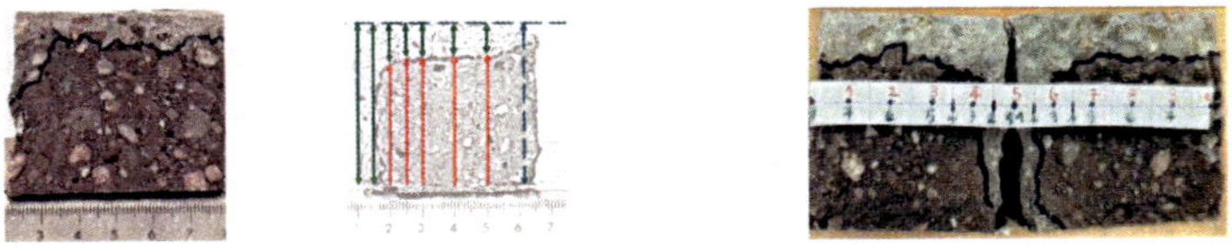

(a) (b) (c)

Fig. 2. Chloride ingress in cracked specimens: (a) example of a sample after chloride immersion, (b) example of a sample after exposure to freeze-thaw cycles on a greyscale image with a focus for the arrows that indicate the measurement method, and (c) example of both sides of the sample with the detail of the measurements.

3 Results and Discussions

3.1 Chloride Ingress

Figure 3 presents the chloride ingress after 30 days of submersion in the chloride solution for each test series. Three samples were done for each series, resulting in six analysed surfaces for each series. The markers represent the mean value and the error bar the standard deviation. The Y-axis represents the crack wall, i.e., if the chloride ingress reaches 50 mm, it means that chlorides cross the specimens completely. Along the X-axis is the half diameter of the specimen equivalent to 50 mm. The sodium silicate was efficient to prevent the chloride ingress through the crack according to the colour boundary difference.

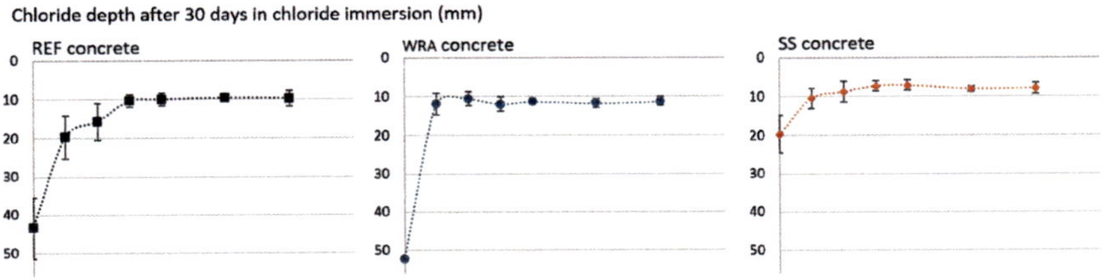

Fig. 3. Chloride ingress after 30 days in chloride immersion (33 g NaCl/l) for each series.

The water repellent agent demonstrated a possible inconsistent result when spraying silver nitrate due to its hydrophobic characteristic. Hydrophobic material increases the contact angle of the surface, repelling any liquid. Therefore, the silver nitrate was repelled in the areas where the water repellent was impregnated. Hence, it did not show a colour difference in those areas preventing a proper verification. In this way, further analysis to measure chloride ingress quantitatively by extracting powder in different layers needs to be performed to confirm the obtained results.

Figure 4 shows the results obtained for chloride ingress after 56 cycles of freezing and thawing with de-icing salts for each test series. The graphs follow the same parameters as described in Fig. 3. Again, sodium silicate was able to prevent the chloride ingress, even after freezing and thawing cycles. In this way, it can be supposed that the formed healing products are able to survive the freezing temperatures. However, the water repellent agent, like described previously, did not allow the penetration of the silver nitrate due to its hydrophobicity to verify the colour change boundary in the treated area. Hence, it is possible that the presented results do not give a reliable estimation of the chloride penetration. Therefore, additional chemical analysis can be used to prove the efficiency of WRA to prevent chloride ingress.

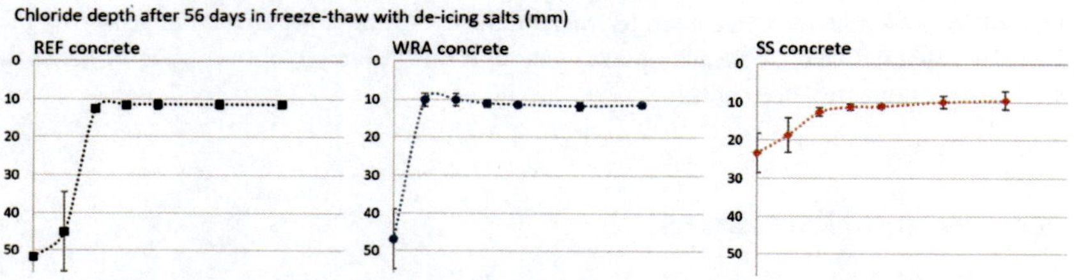

Fig. 4. Chloride ingress after 56 cycles of freezing and thawing with de-icing salts for each series.

3.2 Crack Closure

However, in freeze-thaw cycles, only the chloride ingress method allows evaluating the performance. The scaling destroyed the surface, and microscopic images could not be taken after exposure.

Table 4 shows one example for each series of images collected at the crack creation day, before exposure to chloride immersion, i.e., after injection and curing, and after exposure to chloride immersion for each series. The water repellent agent appears colourless after injection only with a smooth brightness due to its coating ability. However, after chloride immersion, the crack was closed entirely, probably combining autogenous healing products with the water repellent or it could be related to the WRA itself. Further analysis needs to be done to understand better the healing effects formed.

Sodium silicate demonstrates healing capacity after injection and curing. Studies with similar sodium silicate solution used as cargo in microcapsules showed by elemental analysis that the formed products are mainly ettringite and calcium-silicate-hydrate (C-S-H) [6]. After chloride immersion, the healing effects became more visually consistent, also changing the colour. However, in freeze-thaw cycles, only the chloride ingress method allows evaluating the performance. The scaling destroyed the surface, and microscopic images could not be taken after exposure.

Table 4. Microscopic images for concretes subjected to chloride immersion condition, (a) after crack creation, (b) before exposure, and (c) after exposure. The scale bar represents 500 μm.

(a) Crack creation day	(b) Before exposure	(c) After exposure
Reference (REF)		
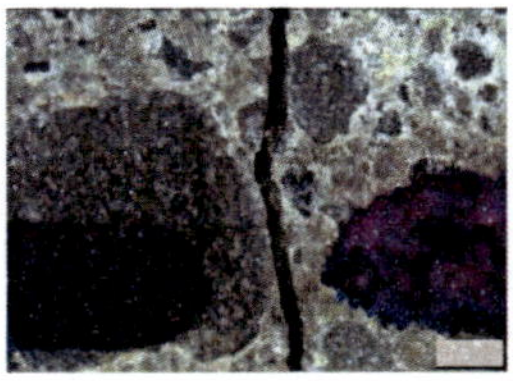		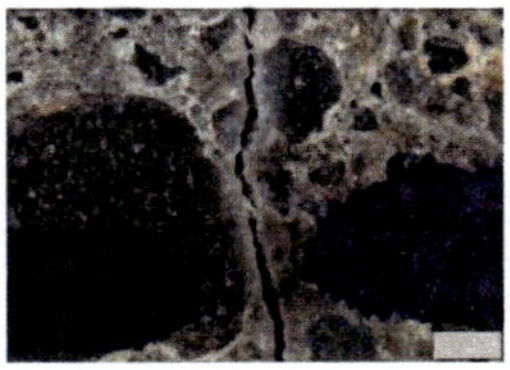
Water repellent agent (WRA)		
	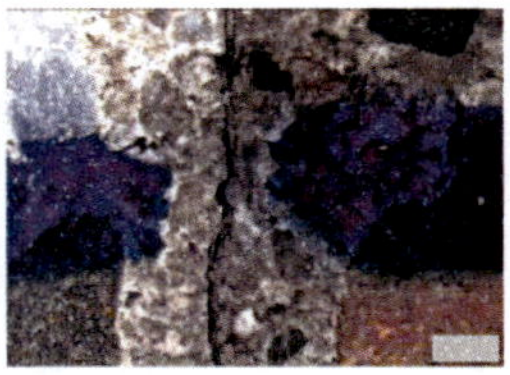	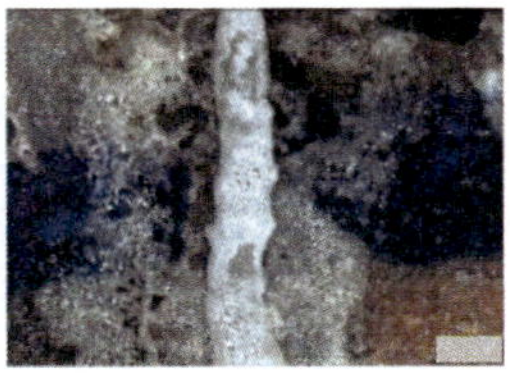
Sodium silicate (SS)		
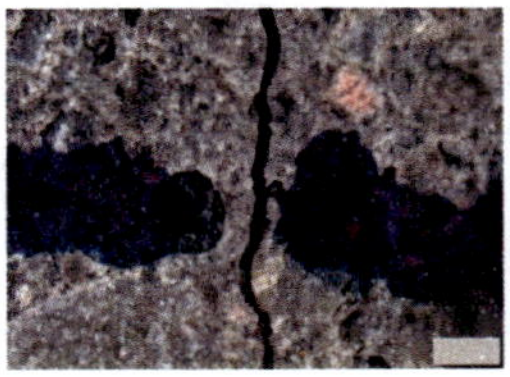	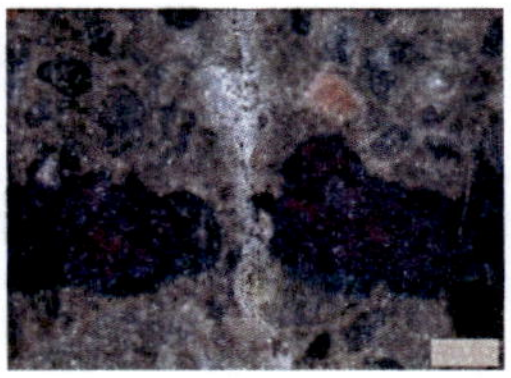	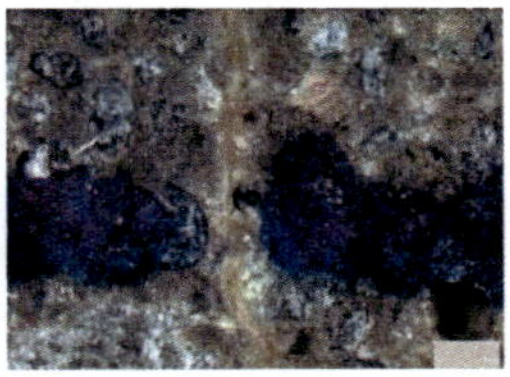

4 Conclusion

The sodium silicate prevented the chloride ingress through the crack during chloride immersion and freeze-thaw cycles with de-icing salts. Likewise, the healing products formed by sodium silicate could survive the extreme conditions assessed. However, the method used to verify chloride ingress by spraying silver nitrate does not give consistent results for the water repellent agent due to its hydrophobicity. Therefore, further chemical analysis to confirm the chloride content must be done to verify the results obtained and prove the ability of WRA to prevent chloride ingress. In contrast, the microscopic analysis showed that both WRA and SS were able to close the cracks under chloride immersion condition.

Overall, both products used for repair showed good efficiency and reduced chloride ingress compared to the reference concrete. Their ability enables them to be used in encapsulation systems and produce self-healing concrete for application in extreme conditions.

Acknowledgement. This project has received funding from the European Union's Horizon 2020 research and innovation programme under the Marie Skłodowska-Curie grant agreement No 860006.T. Van Mullem and N. De Belie were supported by a grant (21SCIP-C158977–02) from the Construction Technology Research Program funded by the Ministry of Land, Infrastructure and Transport of the Korean government.

References

1. NBN EN 206:2013+A1:2016 Concrete. Specification, performance, production and conformity. Belgium- Brussels (2016)
2. Al-Kheetan, M.J., Rahman, M.M., Chamberlain, D.A.: Fundamental interaction of hydrophobic materials in concrete with different moisture contents in saline environment. Constr. Build. Mater. **207**, 122–135 (2019)
3. Xue, X., Li, Y., Yang, Z., et al.: A systematic investigation of the waterproofing performance and chloride resistance of a self-developed waterborne silane-based hydrophobic agent for mortar and concrete. Constr. Build. Mater. **4**(155), 939–946 (2017)
4. Van Tittelboom, K., Gruyaert, E., De Backer, P., et al.: Self-repair of thermal cracks in concrete sandwich panels. Struct. Concr. **16**, 273–288 (2015)
5. Giannaros P., Kanellopoulos A., Al-Tabbaa A. Sealing of cracks in cement using microencapsulated sodium silicate. Smart Materials and Structures (25), 084005 (2016)
6. Kanellopoulos, A., Giannaros, P., Al-Tabbaa, A.: The effect of varying volume fraction of microcapsules on fresh, mechanical and self-healing properties of mortars. Constr. Build. Mater. **122**, 577–593 (2016)
7. Bao, J., Li, S., Zhang, P., et al.: Influence of exposure environments and moisture content on water repellency of surface impregnation of cement-based materials. J. Market. Res. **9**, 12115–12125 (2020)
8. Gruyaert E., Van Tittelboom K., De Backer P, et al.: Self-healing of thermal cracks in sandwich panels. ICSHM2013. p. 196–200 (2013)
9. Zhang, P., Wittmann, F.H., Zhao, T., et al.: Neutron imaging of water penetration into cracked steel reinforced concrete. Phys. B **405**, 1866–1871 (2010)
10. Van Belleghem B.: Effect of capsule-based self-healing on chloride induced corrosion of reinforced concrete. Ph.D Thesis. Ghent University, Belgium (2018)
11. Song Z., Lu Z., Lai Z.: Influence of hydrophobic coating on freeze-thaw cycle resistance of cement mortar. Adv. Mater. Sci. Eng., 1–12. (2019)
12. ASTM International. ASTM C457/C457M-16, standard test method for microscopical determination of parameters of the air-void system in hardened concrete. West Conshohocken (2016)
13. Pigeon, M., Marchand, J., Pleau, R.: Frost resistant concrete. Constr. Build. Mater. **10**, 339–348 (1996)
14. Shin, K.J., Bae, W., Choi, S.-W., et al.: Parameters influencing water permeability coefficient of cracked concrete specimens. Constr. Build. Mater. **151**, 907–915 (2017)
15. Van Mullem T. Development of standard testing methods to evaluate the self-healing efficiency of concrete Ghent University, PhD Thesis (2021)
16. NBN EN 12390–9. Testing hardened concrete Part 9: Freeze-thaw resistance with de-icing salts - Scaling (2016)

Influence of Self-healing via Embedded Macrocapsules Filled with Polyurethane on Carbonation of High-Volume Fly Ash Mortar

Tim Van Mullem, Arne Sintobin, Philip Van den Heede, Laurence De Meyst, Robby Caspeele, and Nele De Belie(✉)

Laboratory Magnel-Vandepitte, Department of Structural Engineering and Building Materials, Faculty of Engineering and Architecture, Ghent University, Technologiepark Zwijnaarde 60, B-9052 Gent, Belgium
Nele.DeBelie@UGent.be

Abstract. During carbonation of concrete, calcium hydroxide is consumed which results in a drop of the pH which can cause steel corrosion. In uncracked cementitious materials it can take several years before the carbonation front reaches the level of the reinforcement. However, most structures are cracked allowing for a fast ingress of carbon dioxide at the cracks. Self-healing cementitious materials are able to close cracks without a manual repair intervention. Most of this research has focused on a regain in liquid tightness or strength, yet results on the influence of carbonation are scarcely reported. In the current study, High-Volume Fly Ash mortar samples were made with pre-placed glass macrocapsules filled with polyurethane. Three different series were studied: reference specimens, specimens with non-pressurized capsules and specimens with pressurized capsules. After cracking of the specimens, the self-healing efficiency was assessed via a capillary water absorption test, prior to placing the specimens in a carbonation chamber and studying the carbon dioxide ingress at the location of the crack. The specimens with non-pressurized capsules showed a sealing efficiency of more than 80% compared to only 25% for specimens with pressurized capsules. The capsules were pressurized by mixing the polyurethane with benzoyl peroxide. This most likely resulted in a premature polymerization of the polyurethane inside some capsules, preventing a good crack filling. Nevertheless, the carbonation front was less deep at the location of the crack compared to the reference series without capsules. The specimens with non-pressurized capsules had a good performance; on average the carbonation depth at the location of the crack was only half of that of reference specimens.

Keywords: Carbonation · Self-Healing · Cracks · Polyurethane

1 Introduction

Carbonation of concrete is a process in which (atmospheric) carbon dioxide (CO_2) reacts with the cementitious matrix. As CO_2 diffuses into concrete, it reacts with water (H_2O) forming carbonic acid (H_2CO_3), see Eq. (1). The formed carbonic acid reacts with

J. I. Escalante-Garcia et al. (Eds.): RW 2021, RILEM Bookseries 40, pp. 231–240, 2023.
https://doi.org/10.1007/978-3-031-21735-7_26

portlandite ($Ca(OH)_2$), which is present in the cementitious matrix, to form calcium carbonate ($CaCO_3$), see Eq. (2).

$$CO_2 + H_2O \rightarrow H_2CO_3 \tag{1}$$

$$H_2CO_3 + Ca(OH)_2 \rightarrow CaCO_3 + 2\,H_2O \tag{2}$$

The consumption of portlandite results in a drop in pH in the highly alkaline cementitious environment from 13 to below 9 [1]. When the carbonation front reaches the reinforcement, this pH drop destroys the passivation layer of the steel, initiating corrosion.

Most concrete structures have cracks throughout their entire service life. However, the influence of cracks on carbonation has been scarcely studied in literature. The studies which have been executed all indicate that cracks act as a highway via which CO_2 can quickly ingress into the concrete [2–6]. As such, carbonation-induced corrosion will start sooner than predicted from service-life models.

In the past decade, significant improvements have been made on the development of self-healing concrete. Many different types of self-healing concrete exist, e.g. via the addition of mineral additions, crystalline admixtures, superabsorbent polymers, microcapsules, macrocapsules, vascular networks, bacteria, etc. [7]. In contrast with conventional concrete, in which cracks need to be detected and subsequently manually repaired, self-healing concrete is able to heal its own cracks without the need for manual repair. As such, the influence of cracks is greatly reduced in self-healing concrete, and in some cases the durability might even be restored to an uncracked state. One way to make concrete self-healing is via the addition of brittle macrocapsules filled with a healing agent. The moment that the concrete cracks, the macrocapsules rupture and the healing agent flows into the crack filling it up. Macrocapsules are usually tubular in shape and different capsule materials have already been used, e.g. glass [8], ceramics [8], polymer-modified cement paste [9], etc. As a healing agent, one-component or two-component polyurethane (PU) has often been used [8, 9]. Often not all PU flows from the capsules. To remediate this, research has been done on pressurizing capsules filled with PU to increase the outflow [10].

Up until now, the influence of self-healing of cracks on carbonation has scarcely been investigated. A recent study did investigate High Volume Fly Ash (HVFA) mortar specimens, as well as self-healing HFVA mortar specimens with macrocapsules [11]. A HVFA mortar was used because the replacement of cement by fly ash allows for a faster CO_2 ingress. The study showed very promising results, although it should be remarked that the exposure time at an elevated CO_2 concentration was only 14 days, and cracks were created artificially by pulling out thin metal plates.

The significance of this study is that it aims to expand upon the limited literature which is available about carbonation of self-healing concrete by studying HVFA mortar prisms cracked in a three-point bending test instead of cracked artificially, because a previous study found that the use of artificial cracks (by pulling out thin metal plates) is not necessarily representative of the carbonation around realistic cracks [6]. Similar to the previous research, a HVFA mix was used in order to have a quicker development of the carbonation front; however, the current specimens were only exposed to an elevated

CO_2 concentration after 3.5 months (allowing for a more mature cementitious matrix) and they were stored in the carbonation chamber for a longer period (up to 68 days). In the current study, specimens with and without pre-placed macrocapsules filled with polyurethane were prepared. Three different series were studied: reference specimens, specimens with non-pressurized capsules and specimens with pressurized capsules. Prior to exposure in the carbonation chamber, the specimens were tested in a capillary water absorption test, allowing to link the sealing efficiency to the carbonation results.

2 Materials and Methods

2.1 Mortar Mix Composition

All specimens had the same mortar mix design. The mortar was made with a sand-to-binder ratio of 3. The sand had a grading 0–5 mm. In order to allow for a fast development of the carbonation front half of the cement (CEM I 52.5 N) was replaced by fly ash (class F). The water-to-binder ratio was 0.5.

2.2 Macrocapsules

All capsules had a length of 49 mm and were cut from long glass tubes with an external diameter of 3.35 mm and a wall thickness of 0.18 mm. Two different types of macrocapsules were prepared: pressurized and non-pressurized capsules.

The non-pressurized capsules were prepared by first sealing one end of the capsule with methylmetacrylate, subsequently injecting the capsule with a single-component polyurethane (PU), and finally sealing the other end of the capsule with methylmetacrylate.

The pressurized capsules were also filled with PU, but the PU was first mixed with benzoyl peroxide (BPO) in powder form for 15 min via the use of a magnetic stirrer. BPO decomposes at elevated temperatures and releases CO_2 (with the aim to pressurize the capsules and allow for a better release of the PU inside the crack when the capsules rupture [10]). In order to have a stronger sealing, these capsules were sealed with a two-component epoxy instead of methylmetacrylate. After the epoxy had cured, the capsules were placed for 3 h in an oven at 80 °C to activate the BPO.

2.3 Prismatic Mortar Specimens

Prismatic specimens were cast with a dimension of 60 x 60 x 220 mm. The moulds were provided with two smooth reinforcement bars with a diameter of 3 mm, which were positioned at 12 mm from the bottom of the specimens. Three different series were studied: Reference specimens, specimens with non-pressurized capsules and specimens with pressurized capsules. All specimens were cast from the same mix design as described in Sect. 26.1. To cast the specimens with capsules, five capsules were preplaced in the moulds just above the reinforcement by gluing them on a thin iron wire. A schematic representation of the cross-section is given in Fig. 1a. Note that the placement of the capsules is similar to the one used in an inter-laboratory study which assessed test methods for concrete with macrocapsules [12]. In total 21 specimens were prepared for this study: 3 reference specimens which were cracked (REF), 3 reference specimens which remained initially uncracked (UNCR), 9 specimens with non-pressurized capsules (N-PRES), and 6 specimens with pressurized capsules (PRES).

After casting, the specimens were placed in a curing room (20 °C, > 95% RH). After 3 days they were demoulded, wrapped in plastic foil and stored again in the curing room.

The specimens were cracked only at an age of 66 and 67 days. Cracking was done in a displacement-controlled three-point bending setup, similar to the crack creation described in the previously mentioned inter-laboratory study [12]. Prior to cracking, the specimens were provided with a notch with a depth of 4 ± 1 mm. The target crack width of the specimens after unloading was 300 μm. To obtain this crack width, the specimens were loaded until a Linear Variable Differential Transformer (LVDT; mounted on the bottom side of the specimens perpendicular to the notch, see Fig. 1b) measured a value of 485 μm.

After cracking, the average crack width of each specimen was determined by taking four microscopic images inside the notch. For each image five measurements of the crack width were performed. The crack width of a sample was taken as the average of the measurements of all four images. It is noted that for some specimens it was not possible to find four locations to measure the crack, as large parts were filled with PU. The mean crack width of the three series corresponded well with the target crack width of 300 μm (reference series: 300 μm; non-pressurized capsules series: 305 μm, pressurized capsules series: 310 μm).

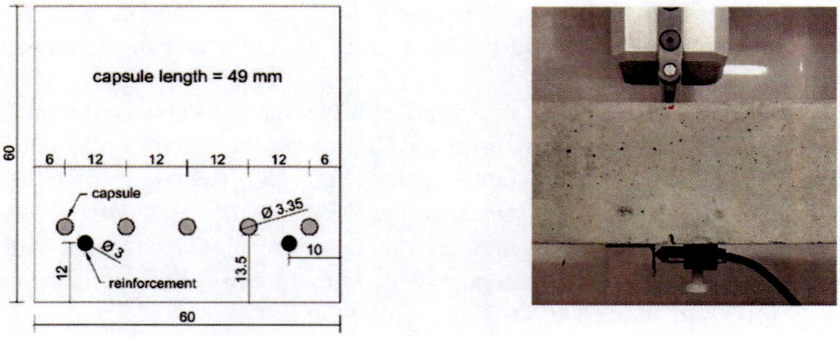

(a) (b)

Fig. 1. **(a)** Cross-section of prismatic mortar specimens with macrocapsules (dimensions are in mm) [12]. **(b)** Position of LVDT during cracking.

2.4 Capillary Water Absorption Test

After the crack width was measured, the specimens were dried in an oven at 40 °C for 12 days. After the oven drying, the specimens were acclimatized for 24 h in an air-conditioned room (20 °C and 60% RH). During this acclimatization time the specimens were waterproofed with aluminum tape, except for a zone of 14 mm centered around the crack on the bottom and side faces. After the capillary water absorption test, the specimens had to be placed in a carbonation chamber, for which the aluminum tape had to be removed. Since this is a very time-consuming process, pieces of paper were placed on the specimens prior to waterproofing them, to allow for a quicker removal of the tape. Figure 2 shows a picture of the waterproofing of one of the specimens at the location of the crack.

For the capillary water absorption test the specimens were placed in contact with water, and during a period of 24 h they were taken out at regular time intervals to determine the mass gain. The sorption coefficient *SC* could be determined as the slope of the data plotted in a mass versus square root of time graph. The same approach was followed as in the previously mentioned inter-laboratory study [12], meaning that aside from the sorption coefficient from 10 min of water contact to 24 h of water contact ($SC_{0\text{–}24}$), also a sorption coefficient from 10 min of water contact to 6 h of water contact ($SC_{0\text{–}6}$) and from 6 h of water contact to 24 h of water contact ($SC_{6\text{–}24}$) were determined. For each set of sorption coefficients (expressed in (g/$\sqrt{\text{h}}$)) the sealing efficiency *SE* could be determined according to Eq. (3)

$$SE = (SC_{REF} - SC_{HEALED})/(SC_{REF} - SC_{UNCR}) \quad (3)$$

with: SC_{REF} the average sorption coefficient of the cracked reference specimens, SC_{HEALED} the average sorption coefficient of the cracked self-healing specimens (PRES or N-PRES), and SC_{UNCR} the average sorption coefficient of the uncracked reference specimens.

Fig. 2. Waterproofing of specimens for capillary water absorption test.

2.5 Carbonation Test

After the capillary water absorption test, the aluminum tape was removed from the specimens and they were dried for 26 days in an air-conditioned room (20 ± 2 °C and RH 60% ± 5%). Prior to placing the specimens in the carbonation chamber, they were coated with a double layer of epoxy, leaving only the bottom surface exposed.

All specimens were placed in a carbonation chamber with a CO_2 concentration of approximately 0.7% (20 ± 2 °C and RH 60% ± 5%) at an age of 110 days. Due to a failure of the CO_2 sensor used to control the CO_2 concentration throughout the test, the accurateness of the CO_2 concentration cannot be guaranteed. Nevertheless, all specimens described in the current study were placed in the same carbonation chamber at exactly the same time. Therefore, the comparison between the results of the different series is valid. However, specific care needs to be taken when comparing the results of this paper to results in literature.

At specific time intervals specimens were taken out of the carbonation chamber and were sawn in half perpendicular to their crack. Subsequently, their carbonation front was visualized using a 1% phenolphthalein solution.

During the capillary water absorption test, there were six reference specimens: Three uncracked and three cracked specimens. After the capillary water absorption test, the three uncracked specimens were also cracked to a nominal crack width of 300 μm, so that six cracked reference specimens were obtained.

3 Results and Discussion

3.1 Mortar Characterization

Before casting the carbonation specimens, the workability of the mortar was determined as 19.5 cm via flow table testing (NBN EN 1015–3 [13]).

The compressive strength of the mortar was determined at an age of 66 days (similar to the cracking age of the carbonation specimens) on control prisms (40 x 40 160 mm) according to NBN EN 196–1 [14] and was equal to 49.1 MPa (standard deviation 1.3 MPa). The density as derived from the hardened specimens was equal to 2238 kg/m^3 (standard deviation 7 kg/m^3).

3.2 Capillary Water Absorption

For some specimens an unexpected increase was noticed in the mass versus square root of time graphs, which was attributed to the water front reaching the piece of paper, after which not only the mortar, but also the paper started absorbing water. From the moment this was expected, the specimen was excluded from the analysis. Table 1 gives the sorption coefficient from 10 min of water contact to 6 h of water contact ($SC_{0\text{–}6}$), as well as the sorption coefficient from 10 min of water contact to 24 h of water contact ($SC_{0\text{–}24}$).

Equation (3) allows to calculate the sealing efficiency. Using $SC_{0\text{–}6}$ a sealing efficiency of 89% and 36% can be found for the N-PRES, respectively PRES series. When using $SC_{0\text{–}24}$ similar sealing efficiencies can be found: 84% for the N-PRES series and 25% for the PRES series. $SC_{6\text{–}24}$ is not reported here, as it leads to sealing efficiencies in the same order of magnitude.

The series with specimens with pressurized capsules (PRES) behaved significantly worse than the series with non-pressurized capsules (N-PRES). The capsules were pressurized via the addition of BPO to facilitate the outflow of the healing agent into the crack. However, BPO is an unstable compound and the smallest amount of water can result in the premature hardening of the healing agent in the capsules [10]. As such, the healing agent cannot flow out of the capsules at the moment of rupture.

Table 1. Overview of average sorption coefficient, with n the number of specimens taken into account, μ the mean and SD the standard deviation.

	$SC_{0\text{–}6}$ (g/$\sqrt{}$h)				$SC_{0\text{–}24}$ (g/$\sqrt{}$h)		
	n	μ	SD		n	μ	SD
REF	3	8.09	0.55		3	7.02	0.58
UNCR	3	2.72	0.25		3	2.92	0.25
N-PRES	7	3.29	0.64		4	3.59	0.66
PRES	6	6.15	2.30		5	5.99	1.60

3.3 Carbonation

At an age of 19, 40 and 68 days three specimens from the N-PRES series were removed from the carbonation chamber to determine the CO_2 ingress. At each of these exposure times also one, two and respectively three reference specimens were removed from the carbonation chamber. Originally, it was intended to remove three PRES specimens after 19 days and three after 68 days, but due to the poor crack healing as well as the relatively high variation of the capillary water absorption results (see Sect. 26.2), it was decided to remove them all after 68 days.

Figure 3 shows the carbonation depth of the N-PRES specimens after the different exposure times in the carbonation chamber. The crack healing with PU clearly delays the CO_2 ingress at the location of the crack. In the uncracked zone, the carbonation depth is comparable, which is to be expected since the addition of the capsules did not change the cementitious matrix. After 68 days of storage in the carbonation chamber the carbonation depth in the N-PRES specimens is, on average, only half of the carbonation ingress in the reference specimens (19.1 ± 7.7 mm versus 40.6 ± 9.4 mm).

After 68 days, specimens of all three series (reference, N-PRES and PRES) were removed from the carbonation chamber. The determined carbonation depth is compared in Fig. 4. The performance of the N-PRES specimens was better than that of the PRES specimens. The mean carbonation depth in the crack was 19.1 ± 7.7 mm for the N-PRES specimens (determined from three specimens) compared to 29.9 ± 13.3 mm for the PRES specimens (determined from six specimens). The influence of the crack was also much more pronounced for the PRES series. In addition, there was a high amount of variation on the carbonation front of the PRES specimens. These results might be explained by premature hardening of the healing agent inside the PRES capsules of certain specimens resulting in a bad crack healing. The bad results of the PRES specimens is in contrast with a preliminary study by Van den Heede et al. [11] which showed very promising results for PU mixed with BPO. The difference might be explained by the cracking age which was much earlier (7 days) than in the current study (66 days), resulting in a lower risk of premature hardening of the healing agent.

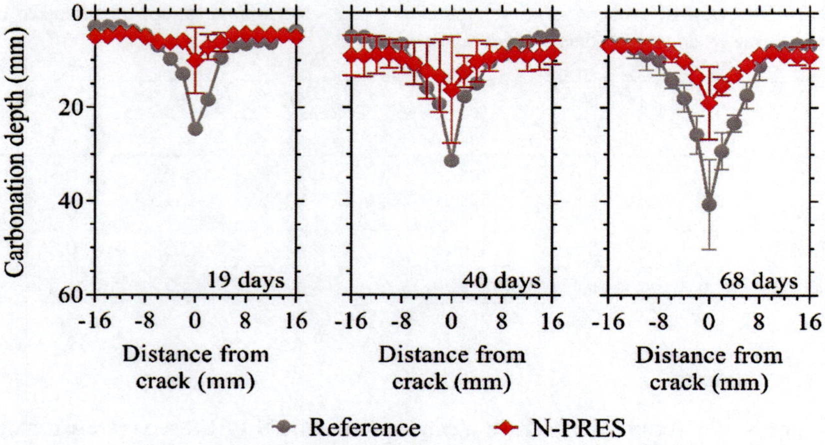

Fig. 3. Carbonation depth of the reference specimens and specimens with non-pressurized capsules (N-PRES) after 19, 40 and 68 days of storage in a carbonation chamber. Error bars represent the standard deviation.

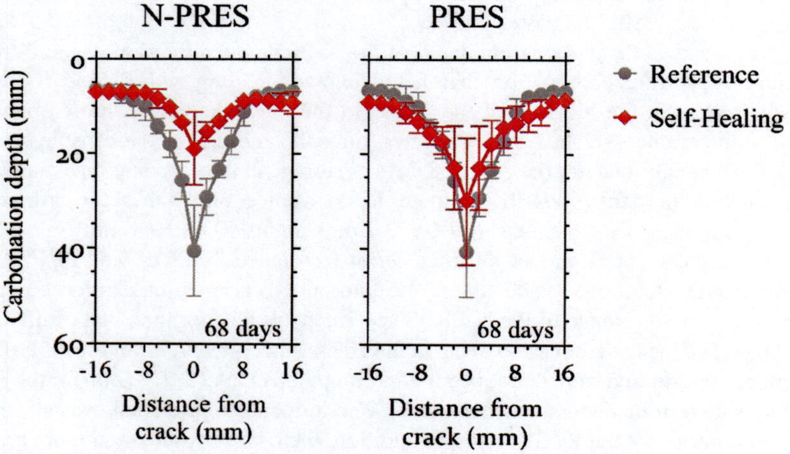

Fig. 4. Comparison of carbonation depth of PRES and N-PRES specimens after 68 days of storage in a carbonation chamber. Error bars represent the standard deviation.

4 Conclusion

The current study investigated the carbonation of HVFA mortar prisms with and without pre-placed macrocapsules filled with polyurethane. Three different series were studied: reference specimens, specimens with non-pressurized capsules and specimens with pressurized capsules.

Prior to testing the carbonation of the different series, the efficiency of the specimens with macrocapsules was determined in a capillary water absorption test. The two series with macrocapsules performed better than the reference series, although it should be noted that the sealing efficiency of the specimens with non-pressurized capsules was significantly higher ($> 80\%$) than the one of the specimens with pressurized capsules ($< 40\%$).

The results of the carbonation study were consistent with the capillary water absorption test: The two series with macrocapsules performed better than the reference series, but the series with non-pressurized capsules clearly performed best. After 68 days of storage in a carbonation chamber, the crack filling in the specimens with non-pressurized capsules resulted in a carbonation depth in the crack which was only half of the carbonation depth in the reference specimens.

The unsatisfactory results obtained from the specimens with pressurized capsules is most likely a consequence of the addition of benzoyl peroxide. This makes the capsules more unstable which, together with the late cracking age of the specimens, will have resulted in a hardening of the polyurethane in many of the capsules before the specimens were cracked.

Acknowledgements. This work is supported by the Korean Agency for Infrastructure Technology Advancement (KAIA) by a grant funded by the Ministry of Land, Infrastructure and Transport (Grant 22SCIP-C158977–03). Philip Van den Heede is a postdoctoral researcher of the Research Foundation—Flanders (FWO) (project No. 3E013917 & project No. G062720N). The financial support of FWO is gratefully acknowledged.

This project has received funding from the European Union's Horizon 2020 research and innovation programme under the Marie Skłodowska-Curie grant agreement No 860006.

References

1. Papadakis, V.G., Vayenas, C.G., Fardis, M.N.: Fundamental modeling and experimental Investigation of. ACI materials journal **88**(4) (1991)
2. Alahmad, S., Toumi, A., Verdier, J., François, R.: Effect of crack opening on carbon dioxide penetration in cracked mortar samples. Mater. Struct. **42**(5), 559–566 (2009)
3. Ann, K.Y., Pack, S.W., Hwang, J.P., Song, H.W., Kim, S.H.: Service life prediction of a concrete bridge structure subjected to carbonation. Constr. Build. Mater. **24**(8), 1494–1501 (2010). https://doi.org/10.1016/j.conbuildmat.2010.01.023
4. De Schutter, G.: Quantification of the influence of cracks in concrete structures on carbonation and chloride penetration. Mag. Concr. Res. **51**(6), 427–435 (1999). https://doi.org/10.1680/macr.1999.51.6.427
5. Song, H.-W., Kwon, S.-J., Byun, K.-J., Park, C.-K.: Predicting carbonation in early-aged cracked concrete. Cem. Concr. Res. **36**(5), 979–989 (2006). https://doi.org/10.1016/j.cemconres.2005.12.019
6. Van Mullem, T., De Meyst, L., Handoyo, J.P., Caspeele, R., De Belie, N., Van den Heede, P.: Influence of crack geometry and crack width on carbonation of High-Volume fly Ash

(HVFA) mortar. Paper presented at the RILEM Spring Convention 2020, University of Minho, Guimarães, Portugal, (2020)
7. De Belie, N., et al.: A review of self-healing concrete for damage management of structures. Adv. Mater. Interfaces **5**(17), 1800074 (2018). https://doi.org/10.1002/admi.201800074
8. Van Tittelboom, K., De Belie, N., Van Loo, D., Jacobs, P.: Self-healing efficiency of cementitious materials containing tubular capsules filled with healing agent. Cement Concr. Compos. **33**(4), 497–505 (2011)
9. Anglani, G., et al.: Sealing efficiency of cement-based materials containing extruded cementitious capsules. Constr. Build. Mater. **251**, 119039 (2020). https://doi.org/10.1016/j.conbuildmat.2020.119039
10. Feiteira J: Self-Healing concrete–Encapsulated polymer precursors as healing agents for active cracks. Ph.D thesis, Ghent University, Fac. Eng. Arch., Ghent, Belgium, (2017)
11. Van den Heede, P., Van,: Belleghem B, Araújo MA, Feiteira J, De Belie N Screening encapsulated polymeric healing agents for carbonation-exposed self-healing concrete, service life extension, and environmental benefit. In: Int. Congr. Polym. Concr., Washington DC, USA, pp. 83–89. Springer (2018)
12. Van Mullem, T., et al.: Addressing the need for standardization of test methods for self-healing concrete: an inter-laboratory study on concrete with macrocapsules. Sci. Technol. Adv. Mater. **21**(1), 661–682 (2020). https://doi.org/10.1080/14686996.2020.1814117
13. NBN EN 1015-3 Methods of test for mortar for masonry - Part 3: Determination of consistence of fresh mortar (by flow table). Brussels, Belgium (2007)
14. NBN EN 196–1 Methods of testing cement—Part 1: Determination of strength (2016)

Reliability Analysis for Service Life of Concrete Structures Subject to Various Carbon Dioxide Concentrations

Gokul Dev Vasudevan(✉) and David Trejo

Oregon State University, Corvallis, OR 97330, USA
vasudevg@oregonstate.edu

Abstract. Carbonation is one of the factors that can reduce the service life of reinforced concrete structures, especially in urban environments where carbon dioxide (CO_2) concentrations are elevated. While concrete is generally considered to be resistant to carbonation under normal atmospheric conditions, in polluted urban environments and accelerated laboratory conditions, carbonation can occur at accelerated rates. Concentration of carbon dioxide (CO_2) can have a significant impact on the rate at which carbonation occurs. There are no current guidelines for determining the carbonation rate in reinforced concrete as a function of CO_2 concentrations. This work presents a risk-based framework that can be used to quantify carbonation depths and rates for various levels of CO_2. This paper also identifies the key variables that affect the service life of reinforced concrete structures, from the point of view of carbonation and the resulting corrosion. These variables include the quantities of the various hydrated/anhydrous phases and the effective diffusion coefficient of the cementitious system. Monte-Carlo simulations are also performed to identify the change in time to carbonation with change in both CO_2 concentration exposure and the concentrations of the hydrated and anhydrous phases of concrete.

Keywords: Carbonation · Corrosion · CO_2 concentration · Monte-Carlo simulation

1 Introduction

Corrosion of the reinforcement embedded in concrete structures is responsible for around 95% of the durability related issues of concrete. The steel reinforcement bar is passive in nature due to the high pH environment created by the concrete pore solution. Corrosion is considered to be initiated when the passive layer on the surface of the steel reinforcement bar is disrupted due to the reduction of pH of the concrete pore solution. The penetration of atmospheric carbon dioxide (CO_2) and its subsequent reactions with hydrated and anhydrous phases of cement contributes to this reduction of pH. This process is termed as carbonation and it is quantified using the RILEM CPC-18 method *(Measurement of hardened concrete carbonation depth)* [1].

J. I. Escalante-Garcia et al. (Eds.): RW 2021, RILEM Bookseries 40, pp. 241–250, 2023.
https://doi.org/10.1007/978-3-031-21735-7_27

The three environmental factors that have a significant impact on concrete carbonation are: (1) relative humidity, (2) ambient temperature, and (3) CO_2 concentration. The effect of relative humidity and temperature have been thoroughly discussed in past literature [2]. There is consensus, among these studies, that 40–90% RH and higher temperatures are conducive to concrete carbonation [3–5]. Although, plenty of literature capture the effect of CO_2 concentration on concrete carbonation, there is no decision-making model available to the decision makers that quantify the time and rate of carbonation based on the ambient CO_2 levels. Such a tool will be valuable for decision makers to compute the service life of the structure, especially when carbonation-induced corrosion is of utmost concern. The average atmospheric concentration of CO_2, 390 parts per million by volume (ppmv), is still relatively low from the perspective of carbonation of concrete [6]. However, local concentrations of CO_2 sampled in urban environments can be as high as 700 ppmv [2]. Figure 1 illustrates the CO_2 concentration of the various states in the US based on the information provided by the US Energy Information Administration (EIA) [7]. Note here that CO_2 concentration values specified for each state is an average value for the state and on-site measurements are recommended to obtain more accurate values. Ambient CO_2 concentration influences the carbonation process in two ways: – (1) higher concentration gradient of CO_2 increases the diffusion of CO_2 into the concrete and (2) higher CO_2 concentration leads to a faster rate of reaction.

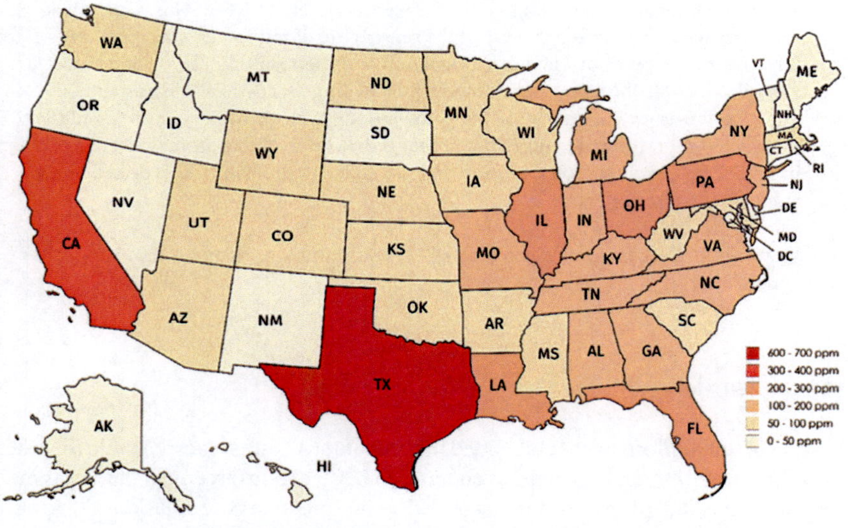

Fig. 1. CO_2 levels of the states in the US based on the EIA data (2019)

1.1 Carbonation Depth Modeling

Most of the models developed to predict the carbonation depth and rate have been either empirical or statistical in nature [8, 9]. However, equations based on the theoretical science governing the process and the stoichiometry of the reactions involved are generally more accurate than the empirical models. The following equation was determined by Papadakis et al. (1991) in a seminal paper on fundamental modeling of concrete carbonation [10].

$$x_c = \sqrt{\frac{2D^c_{e,CO_2}[CO_2]_0}{[Ca(OH)_2]_0 + 3[C{-}S{-}H]_0 + 3[C_3S]_0 + 2[C_2S]_0}t} \tag{1}$$

where x_c is the carbonation depth in m, $D_{e,CO2}$ is the effective diffusivity of CO_2 through the concrete (m^2/s), and $[CO_2]_0$, $[Ca(OH)_2]_0$, $[C\text{-}S\text{-}H]_0$, $[C_3S]_0$, and $[C_2S]_0$ are the concentrations of the respective compounds at the end of moist curing and the onset of CO_2 exposure. The molar concentration of the hydrated and anhydrous phases in Eq. (1) are represented as moles per unit volume of concrete (mol/m^3). It can be noted that Eq. 1 has the same form as the common empirical function that is generally used to model the carbonation front [10].

$$x_c = \sqrt{At} \tag{2}$$

To identify the time to carbonation, Eq. 1 can be rewritten as:

$$t = \frac{[Ca(OH)_2]_0 + 3[C - S - H]_0 + 3[C_3S]_0 + 2[C_2S]_0}{2D^c_{e,CO_2}[CO_2]_0}x_c^2 \tag{3}$$

Here t is the time to carbonate the depth of concrete x_c and it can be seen in Eq. (3) that it depends upon the CO_2 concentration, diffusivity of CO_2, and the concentration of the hydrated and unhydrated phases. However, this equation can be simplified further. The concentration of the unhydrated phases (C_3S, C_2S) becomes small when compared to the concentration of the hydrated phases ($Ca(OH)_2$, CSH) as the hydration reaction proceeds over time. So, as the degree of hydration (DoH) approaches 100%, the concentrations of the unhydrated phases are negligible when compared to the concentrations of the hydrated phases [11]. This is especially valid for the timeframe that is observed in the outcome of this study. So, Eq. (3) simplifies to Eq. (4) as follows:

$$t = \frac{[Ca(OH)_2]_0 + 3[C{-}S{-}H]_0}{2D^c_{e,CO_2}[CO_2]_0}x_c^2 \tag{4}$$

It can be observed in this equation that $[C\text{-}S\text{-}H]_0$ and $[Ca(OH)_2]_0$ have stoichiometric coefficients of 3 and 1 respectively. This implies that the time to carbonation is thrice as sensitive to CSH concentration when compared to the $Ca(OH)_2$ concentration. The authors would like to note that this could be a potential drawback of the model as the concrete carbonation process primarily involves the reaction of $Ca(OH)_2$ with atmospheric CO_2. Table 1 shows a typical sample concrete mixture that will be used to assess the time to carbonation.

Table 2 shows the parameters, based on the mixture proportions in Table 1, which can be used to compute the time to carbonation. The molar concentrations of the hydrated and anhydrous phases were calculated, based on the mixture design, using an online calculator developed by the Swiss Federal Laboratories for Materials Science and Technology

Table 1. Mixture design chosen for the study

Mixture proportions	
Ingredient	kg/m^3 (lb/yd^3)
Cement	426 (718)
Water	213 (359)
Coarse aggregate	1041 (1755)
Fine aggregate	1041 (1755)

Table 2. Time to carbonation parameters based on the chosen mixture design

Time to carbonation		
Parameters	Quantity	Reference
x_c	38 mm (1.5 inch)	Assumed
CH	986.88 mol/m^3	[12]
$C_{1.75}A_{0.05}SH_{4.3}$	1057.42 mol/m^3	[12]
$D_{e,CO2}$	5e−8 m^2/s	[13]

[12]. This calculator is based on the principles of thermodynamic modeling and serves as a powerful tool to estimate the concentrations of hydrated phases of a cementitious system.

Using Eq. (4), a sensitivity plot for time to carbonation is created and is shown in Fig. 2. The values shown in Table 2 were used to calculate the baseline value for time to carbonation. Percentage change in the variables is represented in the abscissa and the percent changes in the time to carbonation are shown on the ordinate. The slope of the lines indicates the sensitivity of time to carbonation with respect to each parameter. The steeper the slope, the higher the sensitivity associated with that parameter.

It can be observed in Fig. 2 that $D_{e,CO2}$ and CO_2 are inversely related to the time to carbonation. For a 10% increase in these two parameters, the time to carbonation decreases by approximately 9%. It can also be seen that concrete cover is the most effective way to extend the time to carbonation. A 20% increase in cover depth can lead to almost a 44% increase in the time to carbonation (Assuming the increase in cover does not result in cracking). Another item of note in this figure is the sensitivity due to changes in $Ca(OH)_2$. It can be seen that for a 100% decrease in the value of $Ca(OH)_2$, the time to carbonation decreases by only 23.7%. Research shows that time to carbonation is very likely more sensitive to $Ca(OH)_2$ content than what is indicated in this figure [14].

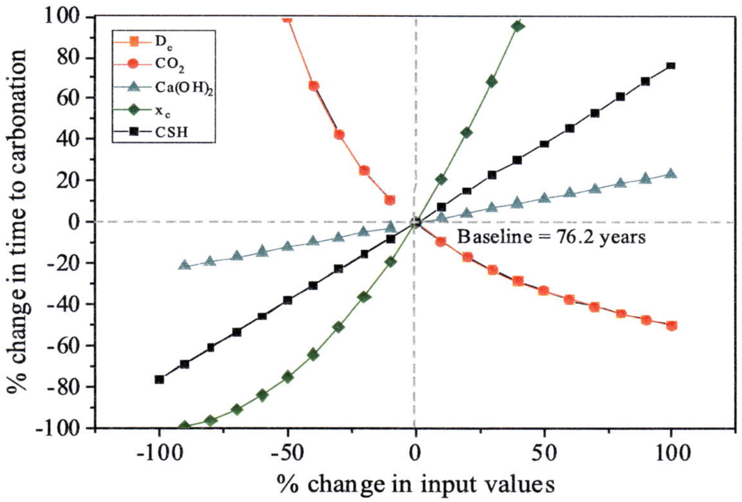

Fig. 2. Sensitivity plot showing the effect of parameters on time to carbonation.

2 Methodology

2.1 Monte-Carlo Simulation

The outcome of the simulations shown in this paper can assist users and decision makers to better understand the effects of ambient CO_2 concentration on the time to carbonation of reinforced concrete structures and the associated corrosion risks. This model, as noted earlier, depends primarily on the concrete cover depth, effective diffusivity, and concentrations of hydrated and anhydrous phases. Similar works on chloride-induced corrosion are available, but at the time of this writing no analogous works are available for carbonation-induced corrosion [15]. To generate a dataset large enough to develop such a model, Monte-Carlo simulation was used [16]. This simulation method uses random sampling and statistical modeling to estimate mathematical functions and can simulate the operations of complex systems. This method relies on probability density functions (PDFs) and random number generators (RNGs) to introduce the element of randomness, to what would be otherwise, a deterministic approach. In this work, a large dataset is generated by identifying the probabilistic distributions for all the parameters shown in Eq. 4, except for the CO_2 concentration. A normal distribution was assumed for the parameters for the sake of simplicity and ease of interpretation. In this work, CO_2 concentration will be treated as a deterministic parameter with values ranging from 300 to 800 ppm incrementally. Using CO_2 concentration as a deterministic parameter enables the understanding of its effect on the time to carbonation on a case-by-case basis.

2.2 Kaplan Meier Reliability Analysis

The Kaplan-Meier method is a non-parametric survival analysis technique that is used to estimate survival and hazard functions for a given application [17]. Equation 4 can be used to carry out reliability analysis by constructing survival and hazard curves. These plots can help inform decision-makers about the service life of the structure based on the ambient carbon dioxide concentration. In this scenario, where it is intended to compute the time to carbonation of a reinforced concrete structure, the survival curve can be established using the following Eq. (5):

$$S(t/CO_2) = 1 - P(t \leq t_i/CO_2) = 1 - F(t/CO_2) \tag{5}$$

In Eq. (5), S is the survival function at for a duration of time t, in years and CO_2 is the ambient CO_2 concentration in ppm. F is the cumulative density function. The survival function *S(t)* gives the probability that the structure does not undergo complete concrete cover carbonation up to a specified time, t. In other words, *S(t)* gives the probability that the random variable T exceeds the specified time, t. The survival function is a crucial aspect of the survival analysis because it provides survival probabilities of structures for different values of t.

The hazard function, denoted by *H(t)*, gives the instantaneous potential per unit time for the event to occur, i.e., that the structure has undergone complete concrete cover carbonation at time t. The hazard function is computed by calculating the ratio of the probability density function to the survival function. This function is then summed up until the required duration to determine the cumulative hazard function. In this study, cumulative hazard curves were constructed using the following Eq. (6):

$$H(t/CO_2) = \int_0^t \frac{f(t/CO_2)}{S(t/CO_2)} = \int_0^t \frac{f(t/CO_2)}{1 - F(t/CO_2)} = 1 - F(t/CO_2) \tag{6}$$

These curves are widely used because they can be easily interpreted. These plots can be utilized by stakeholders and decision-makers to determine the effects of CO_2 concentration on the time to carbonation, and hence the service life of the structure.

3 Results and Discussion

The results from the Monte-Carlo simulations are summarized in Fig. 4. Ten thousand simulations were performed and 6 cases are illustrated, one for each ambient CO_2 concentration. For example, Fig. 3a shows the histogram of time to carbonation at a CO_2 ambient concentration of 300 ppm. A normal distribution was superimposed on the histogram and it can be seen that the mean time to carbonation is 169 years. This indicates that, at low CO_2 concentrations, such as 300 ppm, the structure will be safe from carbonation-induced corrosion over its service life. Figure 3b–f illustrate the time to carbonation histograms at CO_2 concentrations of 400 ppm, 500 ppm, 600 ppm, 700 ppm, and 800 ppm respectively. Overall, it can be observed that, as the CO_2 concentrations increase, the spread of the histograms are reduced and the mean time to carbonation decreases. For the highest CO_2 concentration investigated in this study (800 ppm), the mean time to carbonation is 63 years. This indicates that a CO_2 concentration of 800 ppm poses a challenge to structures that were intended to have a service life longer than 63 years.

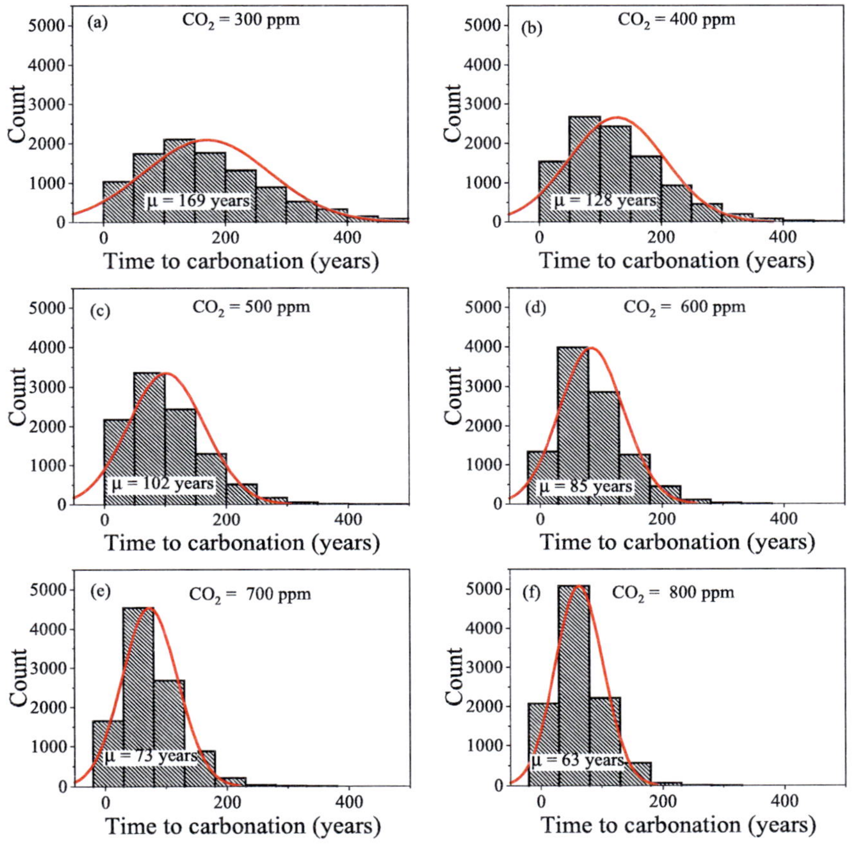

Fig. 3. Monte-Carlo simulation results illustrating the time to carbonation at various ambient CO_2 concentrations.

Figures 4 and 5 illustrate the survival and hazard curves for various ambient CO_2 concentrations. In Fig. 4, the survival duration is presented in the abscissa and the cumulative survival probability is presented in the ordinate. Overall, it can be seen that as CO_2 concentrations increase the probability of survival up to a specified time decreases. For example, at a duration of 200 years, the cumulative survival probability at 300 ppm is around 0.45. At the same duration, the cumulative survival probability is around zero at 800 ppm. A closer inspection of the 800-ppm curve clearly indicates that the survival probability is approximately zero beyond 100 years of service life. This corroborates with the sensitivity plot shown in Fig. 2. Another important piece of information that can be gathered from the survival plots is the risk associated with each case. For example, the 700-ppm case curve reaches a cumulative survival probability of 0.5 around 82 years. This implies that there is a 50% chance of survival up to that duration and hence, 50% risk associated with it.

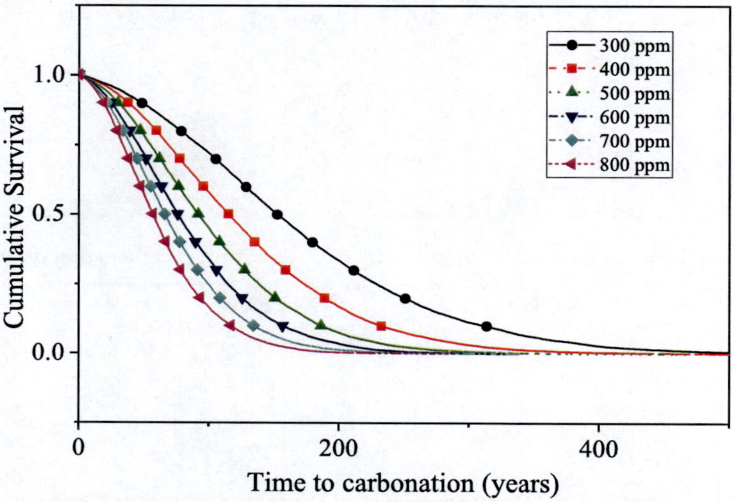

Fig. 4. Cumulative survival curves for various ambient CO_2 concentrations.

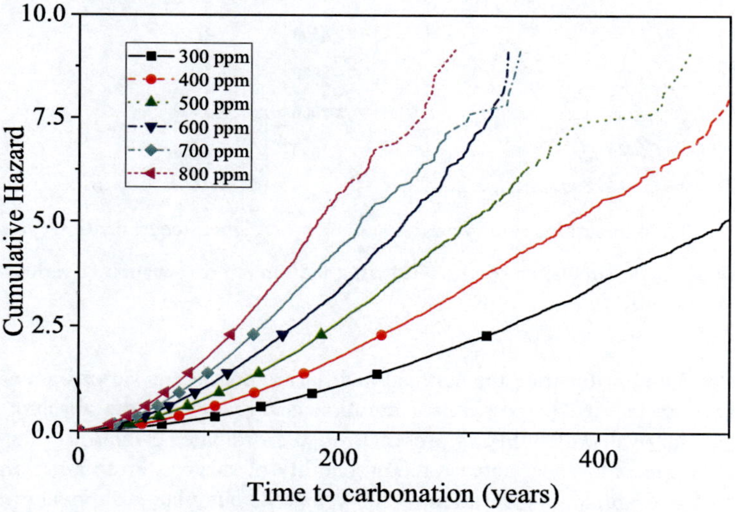

Fig. 5. Cumulative hazard curves for various ambient CO_2 concentrations.

Figure 5 shows the cumulative hazard plots. The cumulative hazard plot consists of a plot of the cumulative hazard *H(t)* versus time *t* of the ith failure. In the context of this study, failure refers to the complete carbonation of concrete cover. For a structure exposed to 300 ppm of CO_2, the probability of failure is approximately zero for the first 25 years, implying no hazards in that time frame. For the same case, at 300 years, the cumulative hazard is slightly higher than 2. This indicates that 2 out of 100 structures will be subjected to the hazard (complete carbonation) within that duration. At any duration, it can be seen that, low CO_2 concentrations have the least cumulative hazard. It can also be noticed that the slopes are steeper for higher ambient CO_2 concentrations indicating a higher rate of failure with respect to time.

3.1 Case Study

In this section, an example is presented that synthesizes all the findings presented in the work. In Fig. 1, it was shown earlier that California and Texas are the top-two states in the US for ambient CO_2 concentrations. Texas has an average CO_2 concentration of 657 ppm and California has an average CO_2 concentration of 363 ppm. The information presented in Fig. 3, along with linear interpolation, can be used to predict the time to carbonation for both cases. Therefore, it can be estimated that reinforced concrete structures at Texas, which were built with the sample mixture proportions shown in Table 1, are expected to have a time to carbonation of 78 years. Similarly, structures in California are expected to have a time to carbonation of around 148 years. At a duration of 100 years, the cumulative survival probabilities for structures in Texas and California are around 0.4 and 0.8 respectively. If the design service life is set at 100 years, it can be seen that most of the structures in the US are relatively immune to carbonation-induced corrosion caused by ambient CO_2. However, in cases of elevated local CO_2 concentrations, there would be a higher risk of carbonation-induced corrosion. Industrial chimneys are a structure that could be at risk of higher rates of carbonation due to continuous elevated CO_2 levels.

4 Conclusion

Time to carbonation was calculated in this work, based on a fundamental model developed by Papadakis et al. [10], using statistical techniques. Results from the sensitivity analysis showed the sensitivity and influence of ambient CO_2 concentration, effective diffusion coefficient, concentration of hydrated and unhydrated phases, and concrete cover depth. The outcomes from the Monte-Carlo simulation showed that mean time to carbonation reduces as the ambient CO_2 concentration increases. At concentrations below 600-ppm, the mean values of time to carbonation were greater than 100 years. This indicates that carbonation-induced corrosion is likely not a major factor affecting the service life of reinforced concrete structures when lower ambient CO_2 concentrations are present.

The survival and hazard plots that were created as part of the reliability analysis can be used by decision makers to better understand the risk associated with the service life of their structure. The results from the reliability analysis showed that ambient CO_2 concentrations are capable of significantly affecting the risk associated with the survival of the structure. Beyond 200 years, the cumulative probability of survival is approximately zero for CO_2 concentrations of 600, 700, and 800 ppm. This clearly indicates that, at elevated CO_2 concentrations, carbonation-induced corrosion must be considered while estimating the expected service life of the structure.

References

1. De La Rilem Recommendations: CPC-18. Measurement of hardened concrete carbonation depth. Mater. Struct. **21**, 453–455 (1988)
2. Richardson, M.G.: Fundamentals of Durable Reinforced Concrete, 273 pp. CRC Press (2002)
3. Bunte, D.: Zum Karbonatisierungsbedingten Verlust der Dauerhaftigkeit vonAußenbauteilen aus Stahlbeton. Ph.D. thesis, TU Braunschweig, 346 pp (1994)
4. Wierig, H.J.: Longtime studies on the carbonation on concrete under normal outdoor exposure. RILEM seminar durability of concrete structures under normal outdoor exposure, pp. 239–249 (1984)
5. Miller, L.B., Witt, J.C.: Solubility of calcium hydroxide. J. Phys. Chem. **33**(2), 285–289 (2002)
6. Dyer, T.: Concrete Durability, 420 pp. CRC Press (2014)
7. US EPA webpage. https://www.eia.gov/environment/emissions/state/analysis/. Accessed 20 June 2021
8. Silva, A., Neves, R., De Brito, J.: Statistical modelling of carbonation in reinforced concrete. Cement Concr. Compos. **50**, 73–81 (2014)
9. Te Liang, M., Qu, W., Liang, C.H.: Mathematical modeling and prediction method of concrete carbonation and its applications. J. Mar. Sci. Technol. **10**(2), 128–135 (2002)
10. Papadakis, G., Costas, G., Michael, N.: Experimental investigation and mathematical modeling of the concrete carbonation problem. Chem. Eng. Sci. **46**, 1333–1338 (1991)
11. Scrivener, K., Snellings, R., Lothenbach, B.: A Practical Guide to Microstructural Analysis of Cementitious Materials. Crc Press (2018)
12. EMPA webpage. https://www.empa.ch/web/s308/ternary-diagram. Accessed 20 June 2021
13. Jung, S.H., Lee, M.K., Oh, B.H.: Measurement device and characteristics of diffusion coefficient of carbon dioxide in concrete. ACI Mater. J. **108**(6), 589 (2011)
14. Sisomphon, K., Franke, L.: Evaluation of calcium hydroxide contents in pozzolanic cement pastes by a chemical extraction method. Constr. Build. Mater. **25**(1), 190–194 (2011)
15. Shakouri, M., Trejo, D., Gardoni, P.: A risk-based model for determining allowable admixed chloride limits in concrete. In: International RILEM Conference on Materials, Systems and Structures in Civil Engineering, pp. 631–640 (2016)
16. Mooney, C.Z.: Monte Carlo Simulation, no. 116. Sage (1997)
17. Kleinbaum, D.G., Klein, M.: Survival Analysis. Springer (2010)

Durability Analysis of Ultra High-Performance Fiber Reinforced Concrete Structures in Cracked Serviceability Limit State

Salam Al-Obaidi[1,2](✉), Fabio De Sandre[1], and Liberato Ferrara[1]

[1] Department of Civil and Environmental Engineering, Politecnico Di Milano, Milan, Italy
salammaytham.alobaidi@polimi.it

[2] Roads and Transportations Engineering Department, University of Al-Qadisiyah, Diwaniyah, Iraq

Abstract. Ultra High Performance (Fibre Reinforced) Concrete and Cementitious Composites (UHPFRCCs) have not only the potential to control the crack width but, even better, the potential ability to spread a damage state into a set of multiple tiny and tightly spaced cracks in the strain hardening regime, which would otherwise localize into a single crack. This is extremely advantageous for durability since the governing parameter is the opening of the single crack. Within the framework of the H2020 ReSHEALience project, one of the main objectives is to upgrade the concept of Ultra High Performance Concrete (UHPC) structures to Ultra High Durability Concrete (UHDC) structures, by implementing micro- and nano-scale constituents, which stimulate the autogenous self-healing capacity of concrete. From 4-point bending tests on UHDC 30 mm thick beams, it was observed a sequential cracking behaviour with evenly distributed spacing, which also resulted into a deflection and strain-hardening behaviour. In this paper, based on material identification test data, a sectional analysis of UHDC structural elements has been performed taking into account the sequential cracking behaviour and, from the sectional analysis, the member displacement behaviour was evaluated. Degradation mechanism governing laws, referring to penetration of chlorides and sulphates, have then been incorporated into the aforementioned structural design algorithms, in order to predict the long-term serviceability flexural behaviour of UHDC structures under structural service scenarios characterized by extremely aggressive conditions. The results are being validated against experimental monitoring of a real scale basin for collecting water from a geothermal power plant made with selected UHDC mixes.

Keywords: UHDC · Durability based design · Self-healing concrete · Service life

1 Introduction

It has been proved by many researchers that the Ultra High Performance Fibre Reinforced Concrete (UHPFRC) is an excellent candidate material for "durability-challenged" structures such as marine structures and industrial chemical facilities due to its dense matrix

J. I. Escalante-Garcia et al. (Eds.): RW 2021, RILEM Bookseries 40, pp. 251–261, 2023.
https://doi.org/10.1007/978-3-031-21735-7_28

composition and compact micro structure [1]. On the other hand, the presence of steel fibres in the UHPFRC makes it possible for the material to undergo multiple small cracks at a certain characteristic length spacing. The combination of these two features make the UHPFRC superior to the conventional concrete in terms of durability performance. Nonetheless, implementing the durability concept with structural design methodology of UHPFRC structures is still under investigation particularly in structures that are entirely built with the UHPFRC.

Several studies are available in the literature dealing with analytical and computational nonlinear analysis of the Fibre Reinforced Concrete (FRC) and UHPC structures. Soranakom and Mobasher [2] presented a closed-form solution for flexural response of FRC beams which has a good agreement with experimental result data. The same researches expanded the FRC closed form solution so it can capture the strain hardening behaviour of UHPC structural members as well [3]. Yao et al. [4] developed an analytical approach for the design of flexural elements made of reinforced ultra-high performance concrete, where they incorporated the previous closed form solutions of UHPC to produce linearized moment-curvature diagrams, from which the load deflection response can be predicted. Since structural and non-structural cracks are unavoidable in the concrete, which makes it prone to the chemical ingress that can easily penetrate to the internal steel reinforcement or steel fibres and lead to corrosion, leaching degradations or combination of both and eventually impact the service life of the structures, the durability of UHPC is expected to be superior to that of the conventional concrete, thanks to the dense matrix, discontinuous microstructure and well- dispersed micro cracking [5]. Within the framework of the H2020 ReSHEALience project [6], one of the main objectives is to upgrade the concept of Ultra High Performance Concrete (UHPC) structures to Ultra High Durability Concrete (UHDC) structures, by implementing micro- and nano-scale constituents, which stimulate the autogenous self-healing capacity of concrete. Beside increasing the service life of the concrete element (durability aspect), the superior mechanical performance of UHPC/UHDC makes it also possible to decrease the volume of the used concrete in structures and infrastructures application, thus contributing to enhance the overall sustainability signature of the UHPC material and structural concept and applications. However, due to the lack of standardizations and user-friendly tools, the extensive use of this material is still misperceived as a challenge. Therefore, in this study and through a combination between the available degradation laws and nonlinear structural analysis procedures, the long term performance evaluation of UHDC materials and structures has been investigated. The investigation included different structural elements assumed as case studies (60 mm thick wall and 30 mm thick precast slabs stiffened with columns) for the same functionality and applied loads in attempt to reduce and optimize the use of the materials and provide more confidence to the designers. The results are being validated against experimental monitoring of a real scale basin for collecting water from a geothermal power plant made with selected UHDC mixes, whose experimentally measured mechanical properties have been assumed as reference for the design, within the framework of the H2020 ReSHEALience project.

2 Structural Analysis Method

The first phase of the modelling consists of performing a nonlinear analysis to evaluate the moment- curvature response of the UHDC section described in Fig. 1. In this phase, instead of evaluating the axial and flexural rigidity by finding the neutral axis corresponding for each strain increment which is rather a time consuming approach, the secant axial and flexural rigidity can be evaluated as $S_a = \frac{P}{\varepsilon_a}$, and $S_b = \frac{M}{\phi}$ to calculate S_a and S_b corresponding to the bending moment, M and axial load, P, the deformation vector $\left[\, \varepsilon_o \; \phi \,\right]^t$ must be obtained first. Sectional tangent stiffness has been used to link an infinitesimal load vector $\left[\, dP \; dM \,\right]^t$ and an infinitesimal deformation vector $\left[\, d\varepsilon \; d\phi \,\right]^t$ as follows.

$$dP = \frac{\partial P}{\partial \varepsilon} d\varepsilon + \frac{\partial P}{\partial \phi} d\phi \tag{1}$$

$$dM = \frac{\partial M}{\partial \varepsilon} d\varepsilon + \frac{\partial M}{\partial \phi} d \tag{2}$$

$$\begin{bmatrix} dP \\ dM \end{bmatrix} = [K_t] \begin{bmatrix} d\varepsilon \\ d\phi \end{bmatrix} \tag{3}$$

Using this approach, and by dividing the concrete section into sufficient number of small strips, strains in the middle of the strips can be iteratively imposed to find the solution. Newton-Raphson method has been used to find the convergence of the solution with respect to the applied axial load. The flow chart described in Fig. 2 illustrates the algorithm of the proposed sectional nonlinear analysis.

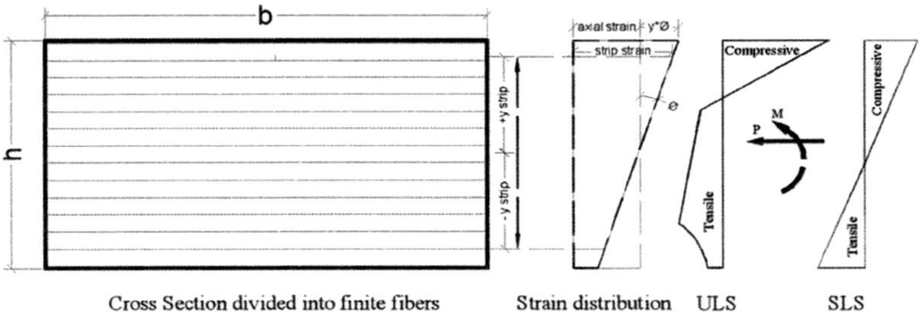

Fig. 1. Details of the section, strain and stress distribution at Ultimate Limit State (ULS), and Serviceability Limit State (SLS).

3 Constitutive Laws of the Materials

All the mechanical properties have been measured after 90 days of curing to allow the cementitious materials such as blast furnace slag to be hydrated. The compressive strength and elastic modulus have been determined to be 100 MPa and 41.7 GPa respectively. On the other hand, the tensile strength was identified by means of inverse analysis based on indirect tensile tests as shown in Fig. 3 [7].

(1) Reading the input parameters

(2) Loop of increment the curvature $\phi_{next} = \phi_{current} + d_{\phi}$

(3) Loop of increment the strain $\varepsilon_{onext} = \varepsilon_{ocurrent} + d_{\varepsilon o}$

(4) $\varepsilon_{strip} = \varepsilon_o + \phi * y_{strip}$

(5) Subroutine to build the section based on the desired strip size

(6) Using the constitutive laws of the compressive and tensile each strip will have corresponding stress based on its position

(7) calculating the resulting force $P_{\varepsilon o(n)} = \sum_{w=1}^{n.strip} f_{cw} A_{cw}$

(8) $\mathrm{d}P = P_{applied} - P_{\varepsilon o(n)}$

(9) If dP is greater than specified tolerance apply Newton Raphson method for convergence

$$\varepsilon o_n = \varepsilon o_{(n-1)} + \frac{\mathrm{d}P}{\frac{\partial P}{\partial \varepsilon_o}},\ \frac{\partial P}{\partial \varepsilon_o} = \frac{P_{\varepsilon o(n)} - P_{\varepsilon o(n-1)}}{\varepsilon_{o(n)} - \varepsilon o_{(n-1)}}$$

(10) If dP is less or equal to the specified tolerance, compute $M = \sum_{w=1}^{nconcrete} f_{cw} A_{cw} y_{cw}$

(11) By using eq.1 the corresponding strain increment is evaluated as $dP = \frac{\partial P}{\partial \varepsilon} d\varepsilon + \frac{\partial P}{\partial \phi} d\phi$

$$d\varepsilon o = \frac{dp - \frac{\partial P}{\partial \phi} d\phi}{\frac{\partial P}{\partial \varepsilon}},\ \mathrm{d}P = P_{applied} - P,\ \frac{\partial P}{\partial \varepsilon} = \frac{P_{current} - P_{previous}}{\varepsilon o_{current} - \varepsilon_{oprevious}},\ \frac{\partial P}{\partial \emptyset} = \frac{P_{current} - P_{previous}}{\emptyset_{current} - \emptyset_{previous}}$$

(12) Compute the displacement response $\Delta_{(i)} = \sum_{i=1}^{curveture\ points} \left(\frac{\emptyset_i * x_i + \emptyset_{i+1} * x_{i+1}}{2}\right) * \Delta x$

(13) Plot the required responses

Fig. 2. Flow chart explaining the algorithm of the nonlinear analysis of the section -first phase

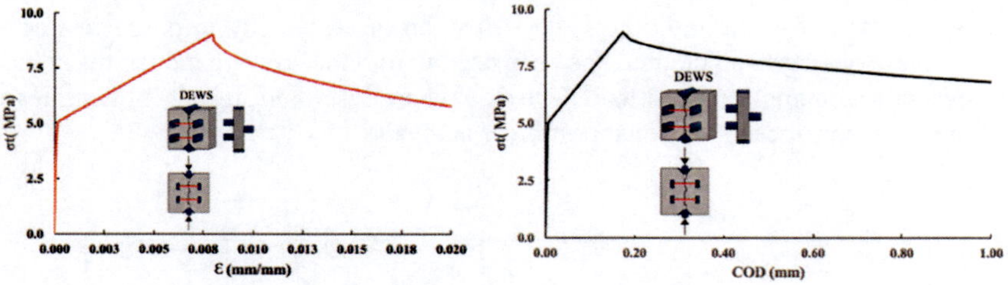

Fig. 3. Tensile constitutive laws derived from DEWS (Double Edge Wedge Splitting) test.

4 Analysis of the UHDC Elements Implemented in the Pilot

The structure under the investigation is a concrete basin to contain geothermal water which has the dimension of 7.5 × 21 m and divided into three cells as shown in Fig. 4. Different structural element and material types have been characterized in this structure to investigate the performance and durability of the Ultra-High Performance Concrete. The first cell was constructed with 100 mm ordinary reinforced concrete walls, the second cell has been made with only 60 mm Ultra-High Performance/Durability fibre reinforced Concrete (UHPC/UHDC) (sec. A-A), and the third cell was made with 30 mm precast UHPC/UHDC slabs and supported by 200 × 200 mm UHPC/UHDC columns (sec. B-B). By implementing the concrete tensile stress-strain relationship described in the previous section, the analysis procedure described previously was adapted for the second and the third basin where the section consists of only UHPC material to perform the first phase analysis (structural analysis) which are represented by the moment-curvature, moment-crack width, and moment-displacement. On the other hands, the second phase (durability analysis phase) is inserted as a function of the exposure time, where the degradation laws in terms of chloride induced corrosion, sulphate leaching and erosion were embedded in the structural analysis (the first phase) to account for the loss of the structural performance (bending moment, crack width and displacement). The same degradation models and life prediction procedures presented in Al-Obaidi et al. 2020 [8] have been used in this analysis. However, giving the advantage of the incremental increase of the curvature and strain presented in the non-linear analysis of the section and observing the sequential and multiple cracking behaviour of the UHDC elements during the experimental tests, the current study takes in consideration the evolution of the degradation with respect to the cracking regime along the wall as schematically shown in Fig. 4 where the cracking is mostly concentrated at the base of the wall and gradually decreases approaching the tip of the wall where less induced stress is expected.

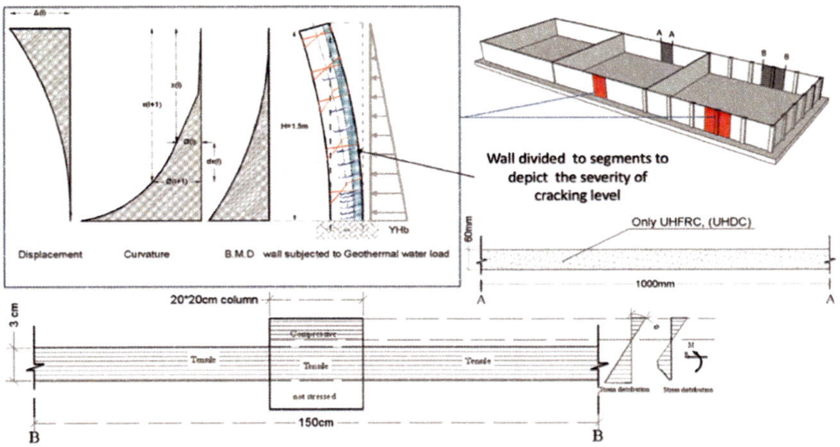

Fig. 4. Description of the pilot wall sections and the adopted analysis.

5 Degradation Mechanisms Implementation

In order to adapt the design process to the use of UHDC, there are some durability indicators to consider. In presence of moisture, the processes of interaction between concrete and the environment takes place due to the penetration of the aggressive elements into the matrix. These substances then interact with the pore solution, the hydrated cement phases, and the steel bars or fibres modifying their properties and eventually effecting the overall behaviour of the concrete. In this case, considering the environment in which the structure operates, the focus was put on chloride-induced corrosion, acid and sulphate attack and surface erosion.

The walls of the structure are analysed using a cantilever beam model subjected to a triangular load given by the water contained in the basin. The cantilever is discretized into 15 elements along its high and for each element, knowing the moment acting on it, it is possible to correlate a curvature or a width of the cracks, using the results of the sectional analysis. The curvature is then integrated twice with a trapezoidal method to obtain the displacement, necessary to evaluate the serviceability of the structure, while the crack width is used to obtain the corresponding value of the degradation coefficients, Chloride diffusion coefficient D_{app}, and Erosion coefficient k_e. At this point, it is possible to evaluate the amount of the section loss due to degradation processes, X_{crit} by adding the contributions of chloride corrosion and acid attack. Then the new thinner sections undergo a new sectional analysis and the process is repeated until a certain displacement limit is reached or until the resisting bending moment is no longer able to sustain the given load as illustrated in Fig. 5.

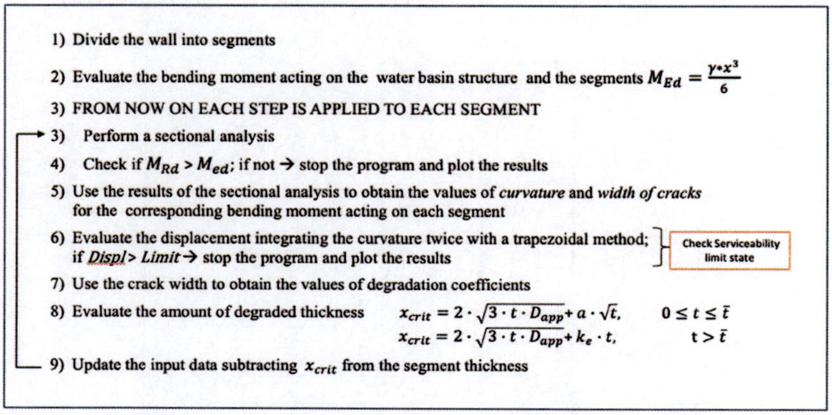

Fig. 5. Flow chart explaining the algorithm of the degradation analysis -second analysis phase

6 Evolution of Crack Width

Regarding the cracking behaviour of the UHPC, propagation of cracks in the matrix at service limit state (before the initiation of steel fibres debonding or elongation) is still not thoroughly evaluated in the codes or standards. Usually at zero strain and at a very low strain level, the behaviour of UHPC element under direct tensile load is homogenous, which means that the concrete matrix and steel fibres works as a one homogenous material and the overall behaviour is linear elastic. When tensile strain increases, stress in the matrix increases further which will trigger micro cracks within the matrix, when the strain increases, these micro cracks will evolve into macro cracks, which indicates the cracking strength of the concrete matrix. After this point, possibly three tensile behaviours of the concrete, depending on the amount and alignment of steel fibres, can be achieved. When no steel fibres is available in the mix, no residual strength in the concrete is obtained. However, for concrete having steel fibres, the steel fibres will carry the extra loads and the post cracking regime is depending on the amount of the steel fibres presented in the active cracks which varies from strain softening to strain hardening behaviour. For strain hardening and pseudo strain hardening propagation, which usually characterizes by the multiple crack regime, the energy to initiate new crack is less than the required energy to pull the steel fibres or elongate it. Therefore, new matrix cracks will initiate until the applied tensile stress in the section is larger than the fibre to matrix bonding stress. At this stage, the debonding of fibre/fibres will occur at the one of the formed cracks and localized cracking regime starts Fig. 6. Many researchers have done several attempts to predict the cracking behaviour of UHPFRC [9]. Some models are mainly based on the models of FRC, then experimental test results of UHPFRC with probabilistic approach can be incorporated to account for the multiple cracking behaviour [10]. Other researches established the sequential cracking model to account for the multiple cracking regime then the load-deflection/ crack width was rationally predicted [11]. In this study, the crack with evolution was modelled depending on the experimental behaviour of the Double Edge Wedge Splitting, DEWS tests shown in Fig. 3. The crack width therefore, depends on the level of the applied stress and on the corresponding phase. Referring to Fig. 6 described below, three phases can be highlighted: zero crack width before strain level of 0.00016, multiple cracking regime phase, when the strain is limited between 0.00016 and 0.006, and localized cracking phase when the strain level exceeds 0.006. The crack width of the multiple and localized phases can be evaluated as shown in Eq. (4) and Eq. (5) respectively where Lc is the space between the cracks as explained in Fig. 6-c and was observed during the experimental test to be about the same of the steel fibres length 2 cm. As reported by *Djerbi, S. et al.* [12], apparent diffusion coefficient D_{app} increases with the increasing of crack width and is almost constant when the crack width reaches approximately 80 μm or more. This has been implemented in the current model, creating a function that increases linearly up to a crack width of 80 μm and becomes constant after that value. Moreover, the layer of the structure which is penetrated by the chloride is considered to be no more able to give any structural contribution and it is then subtracted by the sectional thickness. The same procedure is applied for acid leaching and erosion, but since there are no studies related to the relation between the coefficients a and k_e which are related to leaching and erosion respectively and the crack opening, for each coefficient the minimum and maximum values have been taken respectively as uncracked and cracked coefficients with k_e varying from 0.00011 to 0.0011(mm/day) respectively. It has been assumed that the two values are interpolated by means of an exponential function up to a crack opening of 80 μm and are constant for greater values.

$$w_M = \frac{(\varepsilon_{total} - 0.00016)}{number\ of\ segments} Lc \tag{4}$$

$$w_L = (\varepsilon_{total} - 0.006) Lc \tag{5}$$

7 Results and Discussion

The results of the analysis described above is illustrated in Figs. 7 and 8 for section A-A (60 mm cast in place UHDC wall), and section B-B (30 mm precast UHDC panel supported by 200 × 200 mm columns) respectively.

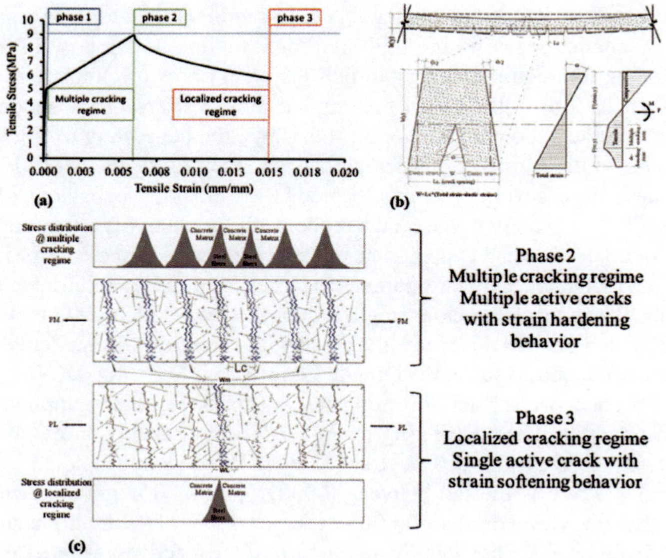

Fig. 6 Sequential cracking behaviour (a) experimental observed strain-stress relationship, (b) crack modelling in the current study, and (c) cracking phases of UHPC.

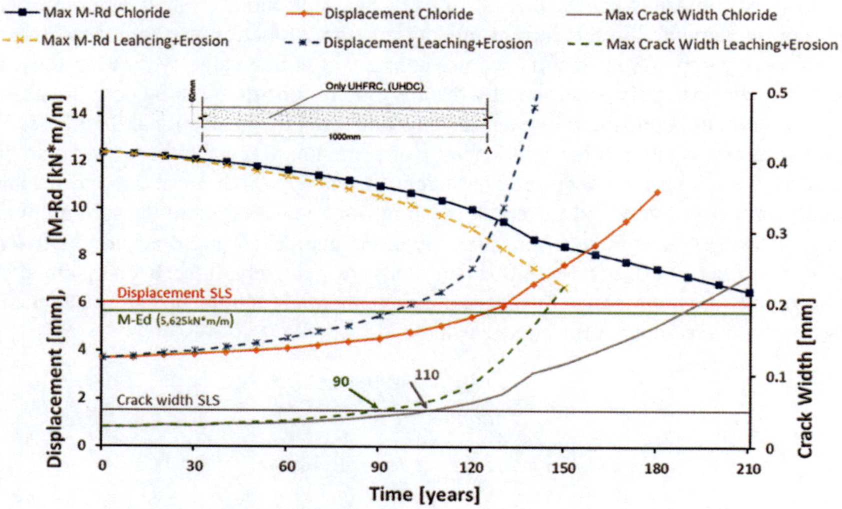

Fig. 7. Performance of section A-A compared to ULS and SLS.

The figures show the performance of the section under the structural loading (hydrostatic pressure or retained geothermal water) and chemical aggressive environment exposures (chloride penetration, and Acid leaching combined with erosion), the performance

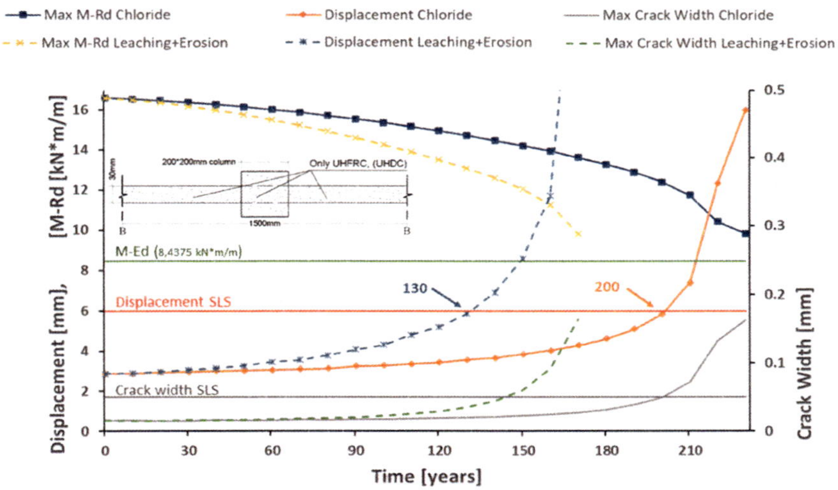

Fig. 8. Performance of section B-B compared to ULS and SLS.

is compared to the ultimate limit state, the maximum required bending moment to resist the hydrostatic pressure (5.6 kN.m/m for section A-A, and 8.4 kN.m/m for section B-B), while for serviceability limit states, the deformation limited to (L/250) as described in EC-2 chapter 7.4.1 [13], and crack width limited according to EN1992 part3 [14] where the crack width is recommended as 0.05 mm for effective sealing of cracks and to 0.2 mm max crack width if some leakage is permitted. Comparing these limit states to the performance of the two basins, the service life can be predicted as 90 years for section A-A with no leakage and 130 years for the section B-B, this indicates the solution of the section B-B is performing better than section A-A where the volume of consumed UHDC materials has reduced to almost half of the amount in section A-A.

8 Conclusions

The study incorporated durability parameters of UHPC into long term mechanical performance, where different aggressive environments have been applied. Based on the obtained results, the following points can be withdrawn.

1. Serviceability limit state (displacement and crack width) is the dominant criterion to define the service life of UHPC thin sections.
2. The structural detailing of UHPC has a pronounced effect on the durability performance, where 3.0 cm slab supported by columns performed better than 6.0 cm wall in terms of serviceability limit state.
3. The sever exposure condition affecting the thin UHPC element appeared to be the leaching combined by the Erosion. Nonetheless, the thinnest UHPC section (3.0 cm thickness) provided at least 90 years as service life.

4. The overall project framework presented in this paper has to be intended as a pioneer study in moving towards performance based design approach for UHPC structures also pushing forward the possibility for a widespread use of this category of materials also in new structural concepts conceived to exploit at the best their performance even in extremely aggressive scenarios.

Acknowledgements. The research activity reported in this paper has been performed in the framework of the ReSHEALience project (Rethinking coastal defence and Green-energy Service infrastructures through enHancEd-durAbiLity high-performance cement-based materials) which has received funding from the European Union's Horizon 2020 research and innovation program under grant agreement No 760824. The information and views set out in this publication do not necessarily reflect the official opinion of the European Commission.

References

1. Shi, C., Chen, B. (eds.) Proceedings of the 2nd International Conference on UHPC Materials and Structures (UHPC2018-China), p. 832. RILEM Pubs., Fuzhou, China (2018)
2. Soranakom, C., Mobasher, B.: Closed-form solutions for flexural response of fiber-reinforced concrete beams. J. Eng. Mech. **133**, 933–941 (2007). https://doi.org/10.1061/(ASCE)0733-9399(2007)133:8(933)
3. Soranakom, C., Mobasher, B.: Correlation of tensile and flexural responses of strain softening and strain hardening cement composites. Cem. Concr. Compos. **30**, 465–477 (2008). https://doi.org/10.1016/J.CEMCONCOMP.2008.01.007
4. Yao, Y., Mobasher, B., Wang, J., Xu, Q.: Analytical approach for the design of flexural elements made of reinforced ultra-high performance concrete. Struct. Concr. **22**, 298–317 (2021). https://doi.org/10.1002/SUCO.201900404
5. Cuenca, E., D'Ambrosio, L., Lizunov, D., Tretjakov, A., Volobujeva, O., Ferrara, L.: Mechanical properties and self-healing capacity of Ultra High Performance Fibre Reinforced Concrete with alumina nano-fibres: Tailoring Ultra High Durability Concrete for aggressive exposure scenarios. Cem. Concr. Compos. **118**, 103956 (2021). https://doi.org/10.1016/J.CEMCONCOMP.2021.103956
6. Home - ReSHEALience Available online: https://uhdc.eu/ (accessed on Sep 21, 2020)
7. Lo Monte, F., Ferrara, L.: Tensile behaviour identification in Ultra-High Performance Fibre Reinforced Cementitious Composites: indirect tension tests and back analysis of flexural test results. Mater. Struct. Constr. **53**, 1–12 (2020). https://doi.org/10.1617/s11527-020-01576-8
8. Al-Obaidi, S., Bamonte, P., Ferrara, L., Luchini, M., Mazzantini, I.: Durability-based design of structures made with ultra-high-performance/ultra-high-durability concrete in extremely aggressive scenarios: Application to a geothermal water basin case study. Infrastructures **5**, 1–44 (2020). https://doi.org/10.3390/infrastructures5110102
9. Lee, S.C., Oh, J.H., Cho, J.Y.: Fiber efficiency in SFRC members subjected to uniaxial tension. Constr. Build. Mater. **113**, 479–487 (2016). https://doi.org/10.1016/J.CONBUILDMAT.2016.03.076
10. Choi, S.W., Choi, J., Lee, S.C.: Probabilistic analysis for strain-hardening behavior of high-performance fiber-reinforced concrete. Materials **12**, 2399. https://doi.org/10.3390/MA12152399
11. Shin, K.J., Jang, K.H., Choi, Y.C., Lee, S.C.: Flexural behavior of HPFRCC members with inhomogeneous material properties. Materials **8**, 1934–1950 (2015). https://doi.org/10.3390/MA8041934

12. Djerbi, A., Bonnet, S., Khelidj, A., Baroghel-bouny, V.: Influence of traversing crack on chloride diffusion into concrete. Cem. Concr. Res. **38**, 877–883 (2008). https://doi.org/10.1016/j.cemconres.2007.10.007
13. EN 1992-1-1: Eurocode 2: Design of concrete structures-Part 1-1: General rules and rules for buildings (2004)
14. Europeenne, N., Norm, E., Version, E.: EUROPEAN STANDARD Eurocode 2-Design of concrete structures-Part 3: Liquid retaining and containment structures, vol. 4 (1992)

Investigating the Use of Off-Specification Ashes to Prevent Alkali-Silica Reaction

Anuj Parashar(✉), Krishna Siva Teja Chopperla, and Jason H. Ideker

Oregon State University, Corvallis, OR 97331, USA
anuj.parashar@oregonstate.edu

Abstract. Two fly ashes and one milled bottom ash that do not meet standard specification requirements per ASTM C618 (commonly termed as off-spec) were investigated for their use in preventing alkali-silica reaction (ASR). The ashes were characterized for their chemical composition, pozzolanic reactivity, and particle size distribution. A very-highly reactive fine aggregate was used in this study. The accelerated mortar bar test (AMBT) was done at two different replacement levels of 25% and 50% of ordinary portland cement (OPC) using the off-spec ashes. The AMBT results showed that only one mixture was able to prevent deleterious ASR. Also, the mixture with 25% ash had a high sulfate content showed higher expansion than the control (100% OPC) mixture. The miniature concrete prism test (MCPT) was also used to evaluate this mixture for up to 56 days. Further, the mixtures were investigated to differentiate ASR expansions and the potential for other deleterious chemical reactions from alkalis and sulfates in the ashes. The results of this study highlight the need for supplemental test methods to standard ASR accelerated test methods to evaluate the efficacy of off-spec ashes for ASR prevention.

Keywords: Off – specification fly ash · Reactivity · Alkali-silica reaction · Accelerated laboratory tests

1 Introduction

The demand for sustainable construction material is pushing the cement and concrete producers to adopt various alternatives. The utilization of supplementary cementitious materials (SCMs), alternative fuels, alternative cements, etc., are well-known and are already in practice [1–3]. Challenges occur when the waste materials are burned for ash and the incorporated impurities in them affect the properties of cement and concrete. One such example is utilizing alternative fuels (such as municipal waste, biomass, etc.) mixed with coal at the thermal power plants that alter ash's physical and chemical properties and make the fly ash (FA) unsuitable as per the standard requirements [4]. It also increases the variability in the FA behavior and makes it difficult to use with cement and concrete. Such alternatives for sustainable solutions also demand alteration in the standard requirement for utilizing by-products for cement and concrete production [4, 5].

J. I. Escalante-Garcia et al. (Eds.): RW 2021, RILEM Bookseries 40, pp. 262–268, 2023.
https://doi.org/10.1007/978-3-031-21735-7_29

In recent decades, the availability of traditionally used SCMs has been reduced significantly [3]. The demand for good quality FA is forcing processed pond ash/low reactive and off-specification (off-spec) ashes for use in cement and concrete. Although many studies have ensured the reactive behavior and suitability of such processed ashes for cement clinker replacement, and some study focused on concrete durability [4]. Some studies focused on the utilization of such off-spec ashes in soil stabilization, cement clinker replacement, etc., but the behavior of such ashes on alkali-silica reaction (ASR) is concerning and has not been explored in much depth [4, 6, 7]. ASR is commonly defined as the reaction between the reactive silica present in aggregate with the alkalis present in pore solution that generates ASR gel and cause expansion to concrete. In general, ASR is very slow and all the laboratory tests tries to accelerate the reaction by boosting additional alkalis or increasing the temperature [8, 9]. The presence of soluble alkalis in SCMs also could influence the rate of reaction [10]. Many studies tried to look at the behavior of different SCMs on ASR expansion but a major disconnect has also been reported between the accelerated laboratory testing and the expansion of concrete in the field [8, 11]. It could be due to the lack of testing protocol or conditions for measuring the potential of ASR mitigation especially with alternative SCMs. The problem becomes more challenging when SCMs contains impurities such as high sulfate content, high available alkali content, and other impurities [6, 10].

This study focuses on the utilization of off-spec fly ashes in mitigating ASR and the effect of impurities in such ashes on their expansion behavior in accelerated laboratory testing. Three different off-spec ashes were procured and tested for their reactivity and potential of ASR prevention. The off-spec ashes were checked to prevent ASR expansion in mixtures with one very highly reactive (R3 reactive as per ASTM 1778 [12]) fine aggregate. The effect of the chemical impurities in the ashes on the accelerated laboratory testing is focused.

2 Materials and Methods

Three different off specification ashes including two fly ashes and one milled bottom ash were procured for the study. One commercially available Type II OPC as per ASTM C150/C150M was used for the study. The fly ashes and the OPC were characterized for their oxide phase composition using X-ray fluorescence (XRF) and their oxide phase compositions are presented in Table 1. The particle size distribution of all the fly ashes measured using laser diffractometry is also presented in Table 1. The results show that the FA1 and FA2 contains a significant amount of SO_3 and their LOI was also high.

The reactivities of the ashes were characterized using isothermal calorimetry and calcium hydroxide (CH) consumption [13]. The paste samples were prepared by mixing CH and FA in 3:1 ratio, and 0.5M potassium hydroxide solution was used for mixing the powders at a water to powder ratio of 0.9. The cumulative heat of hydration and CH consumption in the CH-SCMs paste samples were measured as indicative of reactivity.

One very highly reactive (R3 – reactive) fine aggregate was selected to check the influence of the off-spec ashes on the ASR. The expansion of the control mortar sample with OPC was checked using the accelerated mortar bar test (AMBT) in accordance with ASTM C 1260 [14]. The feasibility of off-spec ashes against ASR expansion was

Table 1. The chemical oxide composition of the SCMs used in the study.

Constituents (weight %)		Portland cement (OPC)	Fly Ash 1 (FA1)	Fly Ash 2 (FA2)	Fly Ash 3 (FA3)
SiO_2		20.88	21.88	28.91	42.31
Al_2O_3		3.62	17.42	14.82	16.82
Fe_2O_3		3.15	5.7	4.67	6.97
CaO		61.62	29.21	27.33	21.21
MgO		3.19	7	3.78	4.2
SO_3		2.13	5.03	11.9	0.44
Na_2O		0.17	1.67	1.09	1.11
K_2O		0.52	0.47	0.54	0.43
TiO_2		0.28	1.26	1.07	1.33
P_2O_5		0.11	2.63	0.71	0.54
ZnO		0.08	0.02	0.02	0.01
Mn_2O_3		0.08	0.11	0.03	0.04
LOI		0.52	3.03	3.22	2.49
$SiO_2 + Al_2O_3$		27.63	45	48.4	66.1
Na_2Oe		0.51	1.98	1.44	1.39
PSD (μm)	$d_v(10)$	3.8	2.2	2.5	4.2
	$d_v(50)$	11.4	9.4	12.8	13.8
	$d_v(90)$	38.9	42.9	43.3	34.1

investigated using AMBT followed by ASTM C1567 [15]. As the ASTM C1567 specifies that the 14-day expansion of the mortar bars including SCMs should be less than 0.1%. All the off-spec ashes were used for 25% replacement (FA1-25, FA2-25 and FA3-25) and at 50% replacement for FA2 and FA3 (FA2-50 and FA3-50).

As the pore solution alkalinity affect the expansion due to ASR, the ions concentrations in the pore solution of the mortar bars were evaluated. The AMBT specimens after the 14 days measurements were used. The pore solution was extracted using a mechanical pore press and the solution was filtered to remove any solid particles. Later, the filtered solution was used for determining its ionic concentrations of Na, K, $\bar{S}$ using XRF [16].

The ASR prevention potential of 25% FA2 (FA2-25) was also tested using the miniature concrete prism test (MCPT - AASHTO T 380) using the same aggregate used for AMBT [17]. The MCPT testing was considered to differentiate the influence of test methods on the expansion behavior of reactive aggregate in the presence of off-spec ash. Since the AMBT uses the 25 mm × 25 mm × 285 mm mortar bars and after the demolding of bars, the specimens are cured at 80 °C underwater for one day for the zero reading and under 1M NaOH solution for the next 14 days [14]. Whereas the MCPT uses the 50 mm × 50 mm × 285 mm concrete bars. The concrete bars are boosted by adding 50/50 wt.% NaOH to the mixing water to reach 1.25% Na_2O_{eq} of the cement mass. The concrete bars after demolding are cured at 60 °C under 1M NaOH solution until 56 days [17]. As in the case of AMBT, the expansion of mortar bars including SCMs should be below 0.1%, and for the MCPT, it should be below 0.025% for the optimum dosage of an SCM to prevent ASR.

3 Results and Discussion

3.1 Reactivity

The cumulative heat release and the CH consumption in the samples hydrated at 50 °C for ten days are reported in Table 2. The results showed that FA2 released maximum cumulative heat, but its CH consumption was lower than FA1. The FA3 was the least reactive but the ash can be considered as pozzolanic reactive [13]. It is to be noted that due to the presence of impurities in these ashes, the cumulative heat of hydration results alone can be misleading to conclude about their behaviors in the cement system. The higher heat release and lower CH consumption by FA2 in comparison with FA1 indicate that the heat release can be mostly due to the presence of reactive SO_3 phases in the ash that might support the formation of ettringite/monosulphate or other hydration products. Overall, the reactivity results do not just highlight the reactive behavior of such ashes, but also highlight the presence of impurities in the ash and their effect on reactivity behavior.

Table 2. Cumulative heat release and CH consumed by SCMs in the reactivity test.

SCM	Cumulative heat release (J/g SCM)	CH consumed (g/100g SCM)
FA1	223.5	59.2
FA2	396.9	48.8
FA3	177.4	27.8

3.2 ASR Expansion

The results for 14 days AMBT expansion are shown in Fig. 1. The control specimen showed the 14 days expansion higher than 0.45% and thus fell in the category of very highly reactive aggregate as per ASTM C1260 [14]. It can be noted that none of the off-spec ashes at 25% replacement mitigate the expansion below 0.1%. Interestingly, the expansion for FA2-25 mix was found to be higher than the control specimens. Even the 50% replacement of OPC with FA2 was not able to reduce the expansion below 0.1%. Only for FA3-50 the expansion was 0.04% and helps in mitigating the ASR. The additional increase in the expansion in samples containing FA2 was suspected due to the presence of the significant amount of SO_3 in the ash.

To ensure the solubility of SO_3 in FA2 during the AMBT testing, the pore solution in the AMBT bars were extracted for FA2-25 and FA3-25 samples. The ions concentration for Na, K and $\bar{S}$ ions in mmol/l measured using XRF are reported in Fig. 2. The results show that the pH of FA2-25 was lower than FA3-25 due to its lower Na_2O_{eq}. It was clear from the results that the FA2 had soluble $\bar{S}$ ions. The reason for lower pH and Na+K ions can also be due to the alkali binding in the formation of sulfate hydration products that incorporate alkalis and produce expansive hydration products such as ettringite.

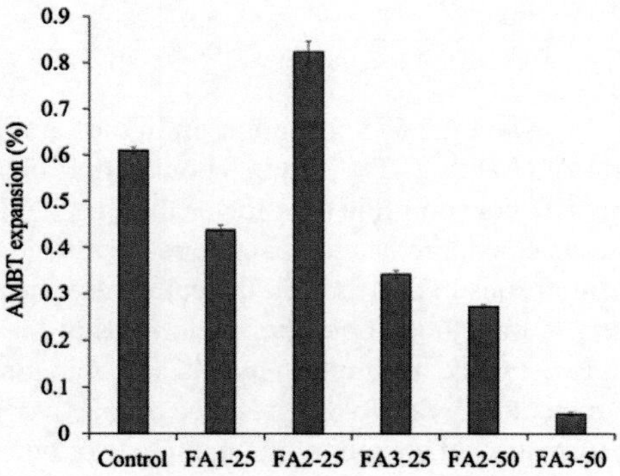

Fig. 1. The AMBT expansion results at 14 days.

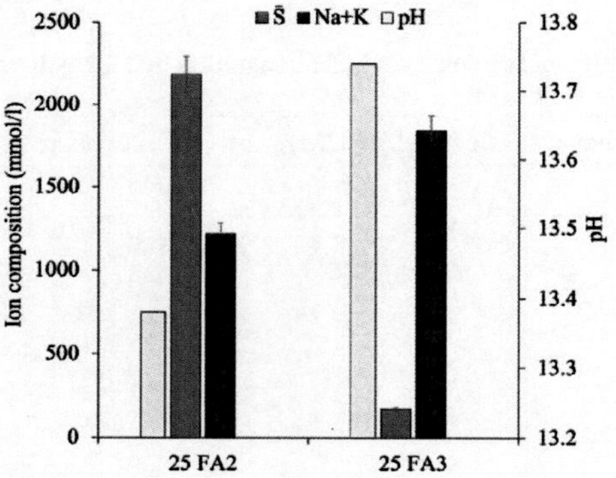

Fig. 2. Pore solution analysis of AMBT bars tested at 14 days.

The MCPT results for FA2-25 with a very highly reactive aggregate are shown in Fig. 3. Contrary to the AMBT results, the FA2-25 data shows lower expansion than the control mix. MCPT data confirms that the off-spec ashes do have the potential to mitigate the ASR up to a certain extent depending on their pozzolanic reactivity and replacement levels. The results not only show the potential of utilizing off-spec ashes in ASR prevention but also the need for an alternative methodology for studying the behavior of such off-spec ashes for ASR.

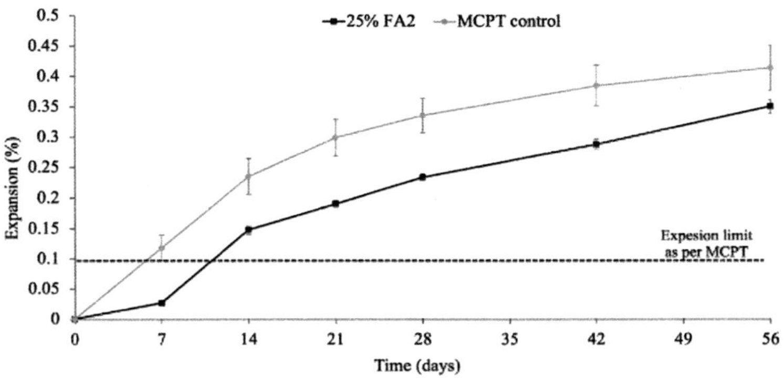

Fig. 3. The MCPT expansion results up to 56 days.

4 Conclusions

Following conclusions can be made from the results:

- The results indicate that the reactivity of off-spec ashes cannot be considered for evaluation their behavior concrete. The influence of impurities such as SO_3 or alkalis can highly hinder their behavior. The solubility of SO_3 and alkali was confirmed from the pore solution analysis.
- The AMBT test performed at 80 °C was found unsuitable for the evaluation of off-spec fly ashes in mitigating ASR. The MCPT test showed a better potential of investigating the ASR mitigation with such ashes. Further investigation is necessary to understand the effect of testing conditions on the behavior of such ashes.
- The presence of impurities such as SO_3 and soluble alkalis were found to be reactive in the AMBT test and can be suspected of the additional expansion than control (100% OPC) mixture.
- The results indicate that the using off-specification ashes may result in poor durability of concrete. The study also points out that the current standard specifications cannot be imposed directly on the off-spec ashes. A detailed investigation on the effect of type and amount of impurities and their impact on the testing approach is required.

References

1. Chatterjee, A., Sui, T.: Alternative fuels–effects on clinker process and properties. Cem. Concr. Res. **123**, 105777 (2019). https://doi.org/10.1016/j.cemconres.2019.105777
2. Parashar, A., Bishnoi, S.: A comparison of test methods to assess the strength potential of plain and blended supplementary cementitious materials. Constr. Build. Mater., 256 (2020)https://doi.org/10.1016/j.conbuildmat.2020.119292
3. Juenger, M.C.G., Snellings, R., Bernal, S.A.: Supplementary cementitious materials: new sources, characterization, and performance insights. Cem. Concr. Res. **122**, 257–273 (2019). https://doi.org/10.1016/j.cemconres.2019.05.008

4. Wirth, X., Benkeser, D., Nortey Yeboah, N.N., et al.: Evaluation of alternative fly ashes as supplementary cementitious materials. ACI Mater. J. **116**, 69–77 (2019). https://doi.org/10.14359/51716712
5. ASTM C618 Standard specification for coal fly Ash and raw or calcined natural pozzolan for use (2022)
6. Rajabipour, F., Zahedi, M., Kaladharan, G.: Evaluating the performance and feasibility of using recovered fly ash and fluidized bed combustion (FBC) fly ash as concrete Pozzolan (2020)
7. Rosa, M.G., Cetin, B., Edil, T.B., Benson, C.H.: Freeze–thaw performance of fly ash–stabilized materials and recycled pavement materials. J. Mater. Civ. Eng. **29**, 04017015 (2017). https://doi.org/10.1061/(asce)mt.1943-5533.0001844
8. Thomas, M., Fournier, B., Folliard, K., et al.: Test methods for evaluating preventive measures for controlling expansion due to alkali-silica reaction in concrete. Cem. Concr. Res. **36**, 1842–1856 (2006). https://doi.org/10.1016/j.cemconres.2006.01.014
9. Rajabipour, F., Giannini, E., Dunant, C., et al.: Alkali-silica reaction: current understanding of the reaction mechanisms and the knowledge gaps. Cem. Concr. Res. **76**, 130–146 (2015). https://doi.org/10.1016/j.cemconres.2015.05.024
10. Thomas, M.D.A., Blackwell, B.Q., Nixon, P.J.: Estimating the alkali contribution from fly ash to expansion due to alkali-aggregate reaction in concrete. Mag. Concr. Res. **48**, 251–264 (1996). https://doi.org/10.1680/macr.1996.48.177.251
11. Ideker, J.H., East, B.L., Folliard, K.J., et al.: The current state of the accelerated concrete prism test. Cem. Concr. Res. **40**, 550–555 (2010). https://doi.org/10.1016/j.cemconres.2009.08.030
12. ASTM C 1778-16: Standard Guide for Reducing the Risk of Deleterious Alkali-Aggregate Reaction. ASTM International (2014)
13. Suraneni, P., Weiss, J.: Examining the pozzolanicity of supplementary cementitious materials using isothermal calorimetry and thermogravimetric analysis. Cem. Concr. Compos. **83**, 273–278 (2017)
14. ASTM C1260-14: Standard Test Method for Potential Alkali Reactivity of Cement-Aggregate Combinations (Mortar-Bar Method). ASTM International (2003)
15. ASTM C1567-04: Standard test method for determining the potential alkali-silica reactivity of combinations of cementitious materials and aggregate (Accelerated Mortar-Bar Method). ASTM International. Annu B ASTM Stand 04.02:774–778 (2005)https://doi.org/10.1520/C1567-13.2
16. Tsui-Chang, M., Suraneni, P., Montanari, L., et al.: Determination of chemical composition and electrical resistivity of expressed cementitious pore solutions using X-ray fluorescence. ACI Mater. J. **116**, 155–164 (2019). https://doi.org/10.14359/51712242
17. AASTHO R 80-17: Standard Practice for Determining the Reactivity of Concrete Aggregates and Selecting Appropriate Measures for Preventing Deleterious Expansion in New Concrete, Construction, American Association of State Highway and Transportation Officials (2017)

Check for updates

A Review on Durability Performance of Calcined Clay Binders for Adoption in the Construction Industry

Yuvaraj Dhandapani(✉) and Susan A. Bernal

School of Civil Engineering, University of Leeds, Leeds, UK
{y.dhandapani,s.a.bernallopez}@leeds.ac.uk

Abstract. Calcined clays are considered one of the most promising sources of supplementary cementitious materials (SCMs) for substituting Portland clinker in the production of concrete. Natural clays have been extensively studied for optimizing pre-conditioning treatments to increase their reactivity, and consequently the hydration, and mechanical properties of low energy binders produced with them. Studies on durability performance are essential for providing the confidence required for the widespread adoption of calcined clays as SCMs. This paper analyse the suitability of calcined clay containing binders in different exposure environments to facilitate its adoption, and presents a brief overview of durability performance of cementitious materials produced with kaolinitic or non-kaolinitic calcined clays. Also, the research needs that will facilitate the adoption of calcined clays in construction projects, as well as the physico-chemical alterations introduced by calcined clays in the hydrated cement matrix is discussed.

Keywords: Calcined clays · Durability · Carbonation · Chlorides · Performance

1 Introduction

Activated clays are a widely available resource and have been extensively used as Portland cement substitutes in the production of concretes. Typically, the reactivity of the clays is increased after activation by thermal (calcination) [1] or mechanical (grinding) [2] treatments. Several sources of clays are available globally with variations in mineralogy and purity [3–5]. These differences make treated clays highly versatile materials to produce SCMs that enables the production of cementitious systems with bespoke performance. Kaolinitic clays have been reported to be more reactive than 2:1 clays, such as illite, montmorillonite, when thermally activated [1]; however, it is well known that optimized reactivity can be achieved if appropriate calcination conditions are selected considering the clay mineralogy [6, 7]. Kaolinite clay has been studied extensively in the context of construction materials [4]. Several other alternative clays sources have also been identified, such as marine dredges [8, 9], clay waste from the paper manufacturing industry [5], and mixed clays or clayed excavated soils [10]. The present overview will only centre on durability aspects of materials containing calcined natural clays.

J. I. Escalante-Garcia et al. (Eds.): RW 2021, RILEM Bookseries 40, pp. 269–279, 2023.
https://doi.org/10.1007/978-3-031-21735-7_30

The aluminosilicates reactive phases in different clays induces distinct variations in the reaction chemistry and microstructure evolution when blended with Portland cements [11]. Upon hydration, aluminates from the calcined clay sources react with carbonate added to the cementitious system such as limestone, dolomite or marble dust to produce additional reaction products such as carboaluminates [12–14]. The reaction from aluminates in calcined clay and calcite present in limestone sources leads to the formation of mono- and hemi- carboaluminates, i.e., CO_3-AFm (Alumino Ferrite- mono) phases. These hydration products have better space-filling characteristics due to lower density states compared to calcium silicates hydrate (C-S-H) and portlandite (CH) [15]. The combination of calcined clay with limestone induces a significantly improved strength development at an early age than the results observed when using calcined clays alone [11, 16–18]. When using lower grade clay seems to enable higher replacement levels of Portland clinker to those achievable with pure metakaolin (below 20%) [19].

The physicochemical characteristics of the blended calcined clay-Portland cement binders are governed by several factors, including (but not limited to) the clay types and activation treatment, purity, mineralogy, substitution level, among others. Such a wide range of factors consequently induces differences in the durability performance of concrete produced with calcined clay containing binders. Depending on the exposure environment, such as those rich in chlorides, sulphates or CO_2, the performance of blended concretes varies compared to concretes produced with plain Portland cement (PC). Based on the environmental exposure, the response of the material is also altered in terms of chloride/CO_2 binding and/ or chemical interaction with aggressive solutions, which can impact the long-term performance of concretes produced with calcined clays.

This paper presents a brief overview of the performance of calcined clay binders in different durability exposure environments. The need for more systematic studies, particularly on the carbonation of calcined clay containing binders is emphasized.

2 Durability Performance of Calcined Clay Containing Binders

2.1 Chloride Ingress

Ingress of chlorides is one of the primary causes for deterioration of steel reinforced concretes. There is a good consensus that concrete chloride resistance is improved significantly with the addition of calcined kaolinite clay due to the combined effect of refined pore structure [20, 21], and improved chloride binding capacity due to the reactive aluminates from calcined clays [21, 22] which favors the formation of Friedel's salt. The apparent chloride diffusion (and migration) coefficient was found to be reduced by nearly an order of magnitude in mortar or concrete containing calcined clay [16, 20, 21, 23, 24]. Chloride ingress is strongly influenced by the kaolinite content of the clay, i.e., the higher the kaolinite content, the lower the apparent chloride diffusion coefficient [21], as shown in Fig. 1a. However, this trend was evident only at lower kaolinite content (10 to 50%). At high kaolinite content (above 50%), the enhancement in chloride resistance was minimal due to limited pore refinement for calcined clay with kaolinite content above 60% [11]. A comparison of binary cementitious systems with calcined kaolinite clay and ternary combination with calcined clay-limestone have shown a minimal variation in chloride ingress rate in concretes produced with them [17].

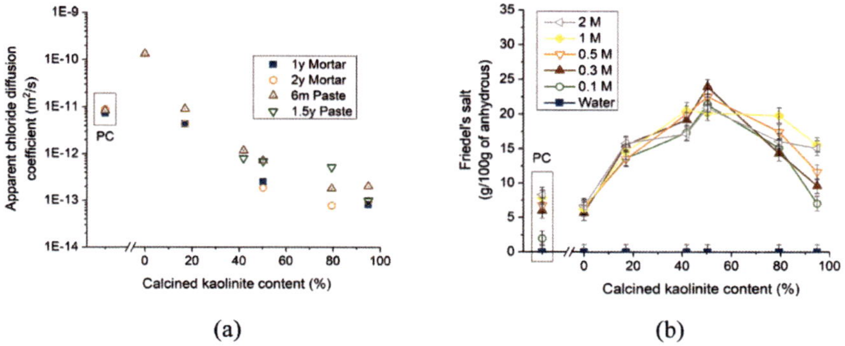

Fig. 1. Influence of kaolinite content on (a) chloride diffusion coefficient and (b) chloride binding, from [21].

Although chloride binding has been demonstrated to increase with the blending of calcined clays with Portland cement [21, 22], it seems that a higher binding capacity is achievable when using calcined clay with 40–50% kaolinite clay than what could be obtained with high purity clays. Specifically, in calcined clay-limestone, more carboaluminate form with lower grade calcined clays which facilitate more chloride binding, as shown in Fig. 1b. Furthermore, it has been reported that there is an impact on the chloride binding capacity identified in these materials can be impacted by the pH of the chloride rich solution. A reduction in pH of the chloride exposure is likely to increase the chloride binding capacity of calcined clay containing cements [25, 26]. On the contrary, internal pH reduction associated with CH consumption due to the pozzolanic reaction of calcined clay, has been shown to reduce the chloride threshold value [24]. Herein, the reduced alkalinity could affect the corrosion resistance of reinforced concrete structures containing calcined clays, which needs to be assessed in field exposure studies.

The role of calcined kaolinite clays regarding chloride ingress in concrete are well understood. However, studies on calcined clays with varying clay mineralogy are spare. In addition, it remains largely unclear what is the role of microstructure, in terms of pore structure, tortuosity and pore solution conductivity in the chloride resistance of calcined clay containing binders. In [20], the improved chloride ingress in calcined clay containing concretes was attributed to the highly tortuous pore network formed in these materials (see Fig. 2); while other studies have attributed the chloride ingress to be a function of the pore solution conductivity [23, 27]. Nevertheless, the use of calcined clays in the production of concrete seems to be beneficial to produce durable concrete exposed to chloride rich environments [24].

2.2 Carbonation Resistance

Replacement of Portland cement with SCMs reduces the calcium buffering capacity of the binder, and makes the cementitious systems more prone to carbonation [28]. However, the refined and more tortuous pore structure typically identified in blended Portland cement, compared with plain systems, could offset the lower Ca^{2+} buffering

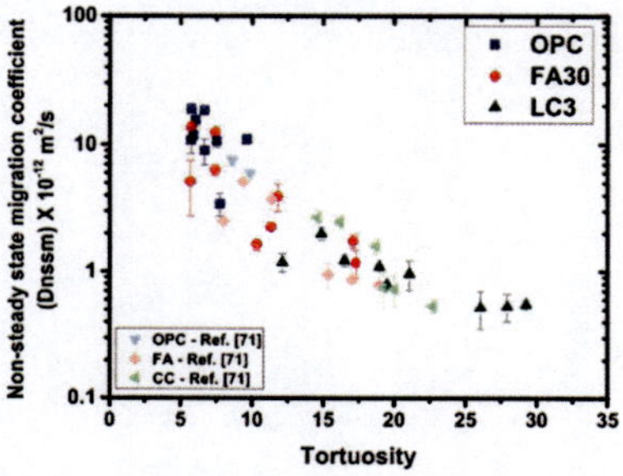

Fig. 2. Improved chloride resistance in calcined clay and calcined clay-limestone concrete systems as a function of (electrical) tortuosity from [20]. *Note* FA – fly ash, CC- calcined clay, and LC3 is limestone calcined clay systems

capacity to some extent. This being strongly influenced by the SCM replacement level, its reactivity, the water-binder ratio, among other factors [28].

In a recent review of Bakera et al. [29], it is reported an improved carbonation performance with metakaolin (MK) addition at lower replacement levels <20%, as shown in Fig. 3 (A). However, high volume replacement of calcined clay and calcined clay-limestone significantly affect carbonation performance [30, 31] (see Fig. 3 B to E). It is worth noting that the majority of the studies evaluating carbonation performance of metakaolin containing concretes have been conducted inducing accelerated carbonation by exposing specimens to high CO_2 concentration for example, from 50% CO_{2eq} [32] to 5% CO_{2eq} [33]. These CO_2 concentrations values are far above what is recommended in the EN 12390-12:2020 (3% CO_{2eq}) [34], which suggest that historical accelerated carbonation results might not be very representative of what could be expected in terms of carbonation of these material under in-service conditions.

Natural carbonation depths of MK blends with 15, 20 and 25% replacement with CEM I and CEM II were compared with commercial CEM I, CEM II (limestone filler), CEM III (Slag) and CEM V (fly ash + Slag) in [35]. Carbonation depth after 365 days and 730 days were used for assessing performance. MK blended with CEM II exhibited reduced carbonation compared to commercial cements, while MK blended with CEM I was found to have average carbonation depths marginally above the upper limits of commercial binders. The natural carbonation coefficient of ternary cements combining calcined clays, limestone and PC was reported for concrete with varying water-to-binder (w/b) ratios in [36]. The carbonation coefficient were found to be 0.42, 0.37, 0.34 mm/d$^{0.5}$ (recalculated from [36]) for concretes produced with 0.50, 0.45 and 0.40 w/b, respectively. Higher carbonation coefficients, determined via accelerated carbonation testing, have been reported for calcined clay-limestone combinations in numerous studies [30, 37, 38], compared with systems produced without limestone addition as show in Fig. 3, and when compared to the values determined after natural carbonation exposure mentioned above.

From the consulted literature, the significant differences in carbonation coefficients are expected considering the differences in the type of samples studied (concrete vs mortars whose pore structure is different), varying accelerated carbonation exposure conditions adopted (temperature, relative humidity and CO_2 concentration) and curing/preconditioning duration as well as environment conditions (See Fig. 3, C, D and E).

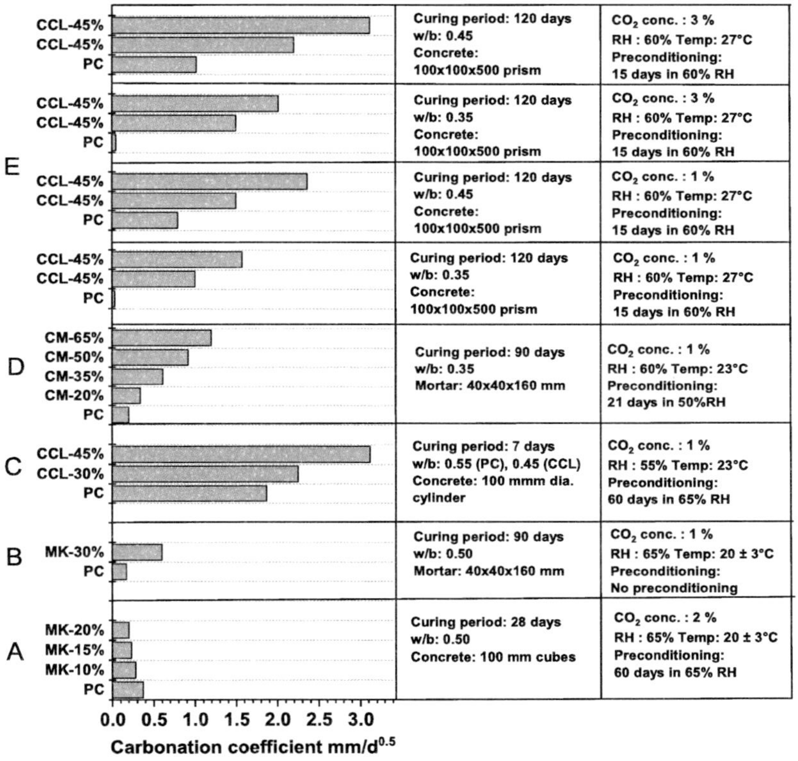

Fig. 3. Summary of carbonation coefficient from literature with metakaolin (MK), calcined clay-limestone (CCL) and calcined marl (CM). Data in A from [29], B from [31], C from [30], D from [37], E from [38]. *Note* carbonation coefficients were calculated from carbonation depth data or converted to similar units for comparison.

Unlike chloride ingress, it remains unclear what is the role of kaolinite content or varying clay types, as well as the impact of exposure conditions in the carbonation resistance of concretes containing calcined clays. Consequently, there is an urgent need to determine what are the major differences in microstructure and CO_2 interaction mechanism within calcined clay cementitious systems, compared to other SCMs containing materials. For practical adoption, provision of higher cover depth seems to be required for metakaolin concrete to produce similar service life to those expected in plain PC concretes, due to the higher carbonation coefficient [35].

Figure 4a shows the pH profile after carbonation of limestone calcined kaolinitic clay cement with 30% and 45% PC replacement by clay, based on the results reported in [30]. At higher the replacement levels of limestone-calcined clay, the pH profile after accelerated carbonation was lower, with a more significant difference between non-carbonation and carbonation pH identified in the carbonation zone. In [31], it was reported that a lower amount of reacted CO_2 could lead to decalcification of the low Ca/Si (C/S) ratio aluminium substituted calcium silicate hydrate (C-A-S-H) typically formed in metakaolin systems compared with plain PC (Fig. 4b).

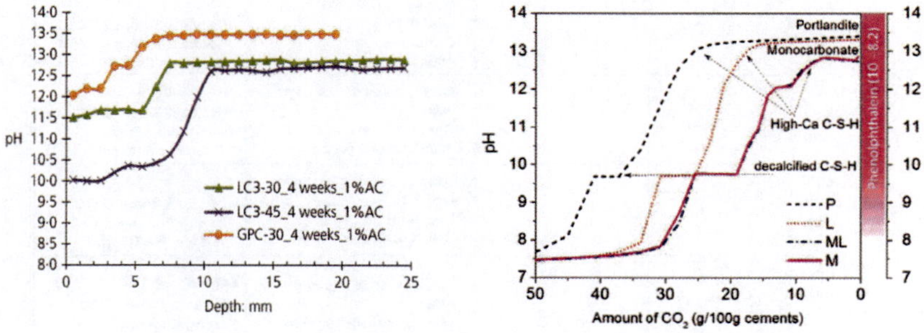

(a) (b)

Fig. 4. Carbonation performance in terms of (a) pH profile in calcined clay limestone systems from [30], and pH changes linked to simulated phase assemblage alterations as a function of CO_2 exposure concentration [31]

To the best of the authors' knowledge, there are no studies reporting on the specific role of clay mineralogical or kaolinite content on carbonation performance of blended Portland cement systems, from a mechanistic perspective. There is a lack of understanding about any potential differences in the carbonation mechanism of impure calcined clay containing cement, particularly on the microstructure changes that the exposure to CO_2 might induce.

A systematic study on the carbonation mechanism in terms of decalcification of low C/S C-A-S-H type gel and carbonation of carboaluminate phase was not found, with a very limited number of studies on carbonation of C-A-S-H type gels [39–41]. This has been highlighted in a recent review by the RILEM TC-281 CCC [28] as one of the main research gaps for understanding the carbonation mechanism of SCMs containing concretes.

Thermodynamic predictions of phase assemblage have shown the potential carbonation sequence of different hydrate phases formed in calcined clay-limestone (see Fig. 4b). However, the difference in phase assemblage in natural and accelerated conditions for the AFm rich phase assemblage formed in ternary binders with calcined clay is not well established.

Two promising strategies to control carbonation and carbonation induced corrosion in calcined clay containing systems are the provision of increased curing duration and higher cover depth, as presented in [35]. More studies have to focus on tailoring the performance of calcined clay systems for carbonation exposure from a fundamental perspective by considering microstructure-performance relationships. It is essential to revisit the existing carbonation relationship with respect to exposure relative humidity and temperature on calcined clay and calcined clay-limestone systems due to the complex microstructure formed in these systems, which might be susceptible to alterations under standardized accelerated carbonation conditions.

2.3 Sulphate Ingress

Sulphate resistance has been shown to improve with the addition of metakaolin. Shi et al. 2019 [42] studied expansion for metakaolin and metakaolin-limestone combination studied at 20 and 5 °C in sulphate exposure for 700 days. They reported no mass change and expansion at both 5 and 20 °C for both binary and ternary systems containing metakaolin. Lower-grade kaolinite clay has similarly improved performance in sulphate resistance with no expansion up to 90 weeks of exposure to sulphate solution in [43]. The addition of metakaolin was found to reduce the tendency for thaumasite attack caused in limestone systems at low temperatures [44, 45]. Rossetti et al. [46] studied two different illite calcined clay (calcined at 950 °C) with limestone addition. The study reported expansion and mass loss up to 720 days and showed that illite clay with higher amorphous showed no expansion at 30% replacement level. The study also found that an increase in replacement level from 15 to 30% calcined illite improved sulphate resistance.

3 Discussion

Based on the recent studies, the major observation on durability performance of calcined clays containing cementitious materials are summarized in Table 1. The role of physical change, in terms of pore structure, and chemical interactions with the aggressive environment, on the durability performance in different exposure conditions is summarized. In general, there is a good understanding on the performance of calcined clay containing concrete when exposed to a chloride rich environment. However, there is limited understanding of their potential carbonation performance. The reaction of calcined clay and calcined clay-limestone is expected to consume most of CH at an early age, and this could facilitate faster carbonation in concretes produced with these binders. This behavior is typically reported in SCMs containing concretes, so it is not exclusive to calcined clays; however, the limited number of studies available on accelerated carbonation of concretes under varying exposure conditions, makes it really difficult to draw any reliable conclusions about their potential performance. Thus, carbonation of calcined clays containing concretes needs to be studied in more detail and more extensively to provide the evidence needed for the confident use of these promising binders in structural concrete applications. Another research gap identified in this overview is the need to focus on assessment of the different clay mineralogy and its potential impact on various durability performances. This should be attempted with the aim to elucidate what potential sources of clay deposits will be more effective SCMs from a durability viewpoint.

4 Conclusion and Perspectives

This paper presents a brief overview of the durability performance of cementitious systems containing calcined clay when exposed to different environments. Chloride resistance and sulphate resistance are expected to be improved with the addition of calcined

Table 1. Summary of factors controlling durability performance of calcined clay containing systems

Durability exposure	Physical	Chemical	Role on durability performance
Chloride resistance	+	+	• Improved pore structure increase chloride resistance • Aluminate from calcined clay promotes chloride binding • CO_3-AFm phase promotes chloride binding
Carbonation resistance	+	−	• Significant reduction in CH due to high reactivity of calcined clay reduce carbonation resistance • Low Ca/Si ratio C-A-S-H formed to become more prone to decalcification • The difference in carbonation tendencies of calcined clay with dominant CASH phase and calcined clay-limestone with CO_3-AFm phase is not clear
Sulphate resistance	+	+	• Lower ingress due to improved pore structure • Limited gypsum formation due to lower CH • Lesser tendency for CO_3-AFm to Ettringite conversion compared to monosulphate

Note: '+' and '−' implies positive and negative impact on performance respectively

kaolinitic clay, and the mechanism taking place when these materials are exposed to those environments have been widely studied, and relatively well understood. However, there are limited studies on durability performance of systems containing low content kaolinite, or non-kaolinitic clay sources. Given the increasing interest in utilizing calcined clays as SCMs from different regions, it is then important that durability studies of concretes produced with this new generation of SCMs are conducted.

Carbonation performance of calcined clays containing systems, determined by accelerated carbonation, varies significantly from study to study. This is attributed to differences in testing condition, curing, and exposure conditions (relative humidity, CO_2 concentration or temperature), among other factors. The carbonation coefficient reported for these materials also varies significantly at different replacement levels of calcined clay. More systematic studies on carbonation focusing on establishing mechanism leading to microstructure alterations, so that performance of these concretes can be tailored as needed. This will facilitate the adoption of calcined clay systems in carbonation prone exposure environments.

Acknowledgements. This study is part of the RENACEM project sponsored by the National Science Foundation (NSF) award 1903457, and the UK Engineering Physical Science Research Council (EPSRC) grant EP/R001642/1, via the NSF-EPSRC lead agency programme.

References

1. Fernandez, R., Martirena, F., Scrivener, K.L.: The origin of the pozzolanic activity of calcined clay minerals: a comparison between kaolinite, illite and montmorillonite. Cem. Concr. Res. **41**, 113–122 (2011)
2. Tole, I., Habermehl-Cwirzen, K., Cwirzen, A.: Mechanochemical activation of natural clay minerals: an alternative to produce sustainable cementitious binders–review. Mineral. Petrol. **113**(4), 449–462 (2019). https://doi.org/10.1007/s00710-019-00666-y
3. Kaminskas, R., Kubiliute, R., Prialgauskaite, B.: Smectite clay waste as an additive for Portland cement. Cem. Concr. Compos. **113**, 103710 (2020)
4. Sabir, B., Wild, S., Bai, J.: Metakaolin and calcined clays as pozzolans for concrete: a review. Cem. Concr. Compos. **23**, 441–454 (2001)
5. Frías, M., Rodríguez, O., Vegas, I., Vigil, R.: Properties of calcined clay waste and its influence on blended cement behavior. J. Am. Ceram. Soc. **91**, 1226–1230 (2008)
6. He, C., Osbaeck, B., Makovicky, E.: Pozzolanic reactions of six principal clay minerals: activation, reactivity assessments and technological effects. Cem. Concr. Res. **25**, 1691–1702 (1995)
7. Msinjili, N.S., Gluth, G.J.G., Sturm, P., Vogler, N., Kühne, H.-C.: Comparison of calcined illitic clays (brick clays) and low-grade kaolinitic clays as supplementary cementitious materials. Mater. Struct. **52**(5), 1–14 (2019). https://doi.org/10.1617/s11527-019-1393-2
8. Dixit, A., Du, H., Pang, S.D.: Marine clay in ultra-high performance concrete for filler substitution. Constr. Build. Mater. **263**, 120250 (2020)
9. Snellings, R., Cizer, Ö., Horckmans, L., et al.: Properties and pozzolanic reactivity of flash calcined dredging sediments. Appl. Clay Sci. **129**, 35–39 (2016)
10. Zhou, D., Wang, R., Tyrer, M., et al.: Sustainable infrastructure development through use of calcined excavated waste clay as a supplementary cementitious material. J. Clean. Prod. **168**, 1180–1192 (2017)
11. Avet, F., Scrivener, K.: Investigation of the calcined kaolinite content on the hydration of Limestone Calcined Clay Cement (LC 3). Cem. Concr. Res. **107**, 124–135 (2018)
12. Antoni, M., Rossen, J., Martirena, F., Scrivener, K.: Cement substitution by a combination of metakaolin and limestone. Cem. Concr. Res. **42**, 1579–1589 (2012)
13. Dhandapani, Y., Santhanam, M., Kaladharan, G., Ramanathan, S.: Towards ternary binders involving limestone additions—a review. Cem. Concr. Res. **143**, 106396 (2021)
14. Krishnan, S., Bishnoi, S.: Understanding the hydration of dolomite in cementitious systems with reactive aluminosilicates such as calcined clay. Cem. Concr. Res. **108**, 116–128 (2018)
15. Matschei, T., Lothenbach, B., Glasser, F.P.: The AFm phase in Portland cement. Cem. Concr. Res. **37**, 551–558 (2007)
16. Dhandapani, Y., Sakthivel, T., Santhanam, M., et al.: Mechanical properties and durability performance of concretes with Limestone Calcined Clay Cement (LC3). Cem. Concr. Res. **107**, 136–151 (2018)
17. Dhandapani, Y., Santhanam, M., Gettu, R., Pillai, R.G.: Perspectives on blended cementitious systems with calcined clay- limestone combination for sustainable low carbon cement transition. Indian Concr. J. **94**, 25–38 (2020)
18. Tironi, A., Scian, A.N., Irassar, E.F., Eng, C.: Blended cements with limestone filler and kaolinitic calcined clay: filler and pozzolanic effects. J. Mater. Civ. Eng. **29**, 1–8 (2017)

19. Krishnan, S., Gopala Rao, D., Bishnoi, S.: Why low-grade calcined clays are the ideal for the production of limestone calcined clay cement (LC3) BT-calcined clays for sustainable concrete. In: Bishnoi, S. (ed.) Springer Singapore, Singapore, pp 125–130 (2020)
20. Dhandapani, Y., Santhanam, M.: Investigation on the microstructure-related characteristics to elucidate performance of composite cement with limestone-calcined clay combination. Cem. Concr. Res. **129**, 105959 (2020)
21. Maraghechi, H., Avet, F., Wong, H., et al.: Performance of limestone calcined clay cement (LC3) with various kaolinite contents with respect to chloride transport. Mater. Struct. **51**, 125 (2018)
22. Shi, Z., Geiker, M.R., Lothenbach, B., et al.: Friedel's salt profiles from thermogravimetric analysis and thermodynamic modelling of Portland cement-based mortars exposed to sodium chloride solution. Cem. Concr. Compos. **78**, 73–83 (2017)
23. Wilson, W., Georget, F., Scrivener, K.: Unravelling chloride transport/microstructure relationships for blended-cement pastes with the mini-migration method. Cem. Concr. Res. **140**, 106264 (2021)
24. Pillai, R.G., Gettu, R., Santhanam, M., et al.: Service life and life cycle assessment of reinforced concrete systems with limestone calcined clay cement (LC3). Cem. Concr. Res. **118**, 111–119 (2019)
25. Avet, F., Scrivener, K.: Influence of pH on the chloride binding capacity of Limestone Calcined Clay Cements (LC3). Cem. Concr. Res. **131**, 106031 (2020)
26. Ipavec, A., Vuk, T., Gabrovšek, R., Kaučič, V.: Chloride binding into hydrated blended cements: The influence of limestone and alkalinity. Cem. Concr. Res. **48**, 74–85 (2013)
27. Sui, S., Georget, F., Maraghechi, H., et al.: Towards a generic approach to durability: Factors affecting chloride transport in binary and ternary cementitious materials. Cem. Concr. Res. **124**, 105783 (2019)
28. von Greve-Dierfeld, S., et al.: Understanding the carbonation of concrete with supplementary cementitious materials: a critical review by RILEM TC 281-CCC. Mater. Struct. **53**(6), 1–34 (2020). https://doi.org/10.1617/s11527-020-01558-w
29. Bakera, A.T., Alexander, M.G.: Use of metakaolin as a supplementary cementitious material in concrete, with a focus on durability properties. RILEM Tech. Lett. **4**, 89–102 (2019)
30. Khan MSH, Nguyen QD, Castel A (2019) Performance of limestone calcined clay blended cement-based concrete against carbonation. Adv Cem Res Feb:1–36
31. Shi, Z., Lothenbach, B., Geiker, M.R., et al.: Experimental studies and thermodynamic modeling of the carbonation of Portland cement, metakaolin and limestone mortars. Cem. Concr. Res. **88**, 60–72 (2016)
32. Graybeal, B.: Tanesi J (2008) A cementitious long-life wearing course to reduce frequency of maintenance works on high-traffic roads. Transp Res Arena Eur **1561**, 454–461 (2008)
33. San Nicolas, R., Cyr, M., Escadeillas, G.: Performance-based approach to durability of concrete containing flash-calcined metakaolin as cement replacement. Constr Build Mater **55**, 313–322 (2014)
34. BS EN 12350–8:2019 (2019) BSI Standards Publication Testing hardened concrete. Br Stand 18
35. Bucher, R., Diederich, P., Escadeillas, G., Cyr, M.: Service life of metakaolin-based concrete exposed to carbonation: Comparison with blended cement containing fly ash, blast furnace slag and limestone filler. Cem. Concr. Res. **99**, 18–29 (2017)
36. Gettu, R., Pillai, R.G., Santhanam, M., Basavaraj, A.S., Rathnarajan, S., Dhanya, B.S.: Sustainability-based decision support framework for choosing concrete mixture proportions. Mater. Struct. **51**(6), 1–16 (2018). https://doi.org/10.1617/s11527-018-1291-z
37. Østnor, T.A., Justnes, H.: Durability of mortar with calcined marl as supplementary cementing material. Adv Cem Res **26**, 344–352 (2014)

38. Shah, V., Bishnoi, S.: Carbonation resistance of cements containing supplementary cementitious materials and its relation to various parameters of concrete. Constr Build Mater **178**, 219–232 (2018)
39. Ortaboy, S., Li, J., Geng, G., et al.: Effects of CO2 and temperature on the structure and chemistry of C-(A-)S-H investigated by Raman spectroscopy. RSC Adv **7**, 48925–48933 (2017)
40. Li J, Yu Q, Huang H, Yin S (2019) Effects of Ca/Si ratio, aluminum and magnesium on the carbonation behavior of calcium silicate hydrate. Materials (Basel) 12:
41. Irbe L (2019) The role of aluminium in C-A-S-H during chemical attack on concrete, Dissertation TU München, Germany
42. Shi, Z., Ferreiro, S., Lothenbach, B., et al.: Sulfate resistance of calcined clay – Limestone – Portland cements. Cem. Concr. Res. **116**, 238–251 (2019)
43. Santhanam M, Suma F, Dhandapani Y (2017) Sulphate Resistance of Limestone Calcined Clay Cement. In: The 2nd ACF Symposium 2017 - Innovations for Sustainable Concrete Infrastructures. pp 1–5
44. Skaropoulou, A., Tsivilis, S., Kakali, G., et al.: Thaumasite form of sulfate attack in limestone cement mortars: A study on long term efficiency of mineral admixtures. Constr Build Mater **23**, 2338–2345 (2009)
45. Skaropoulou, A., Sotiriadis, K., Kakali, G., Tsivilis, S.: Use of mineral admixtures to improve the resistance of limestone cement concrete against thaumasite form of sulfate attack. Cem Concr Compos **37**, 267–275 (2013)
46. Rossetti, A., Ikumi, T., Segura, I., Irassar, E.F.: Sulfate performance of blended cements (limestone and illite calcined clay) exposed to aggressive environment after casting. Cem. Concr. Res. **147**, 106495 (2021)

Estimating Service Life of Prestressed Concrete Systems Exposed to Chlorides

Dyana Joseline(✉) and Radhakrishna G. Pillai

Indian Institute of Technology Madras, Chennai, India
dyanajoseline@gmail.com, pillai@civil.iitm.ac.in

Abstract. In general, pretensioned concrete (PTC) structures are being designed with a target service life of about 100 years. However, cases of premature/catastrophic corrosion-induced strand failures have been reported. This necessitates a re-examination of the current practices for service life-based design (SLD) as well as corrosion assessment of new and existing PTC structures respectively, especially the choice of limit states and the parameters considered. The work presented in this paper is a component of the research carried out to develop tools to assist SLD and assessment of PTC structures, and focusses on the estimation of time required for Passive-to-Active (P-to-A) transition, which is the first step of chloride-induced corrosion of prestressed steel strands. For this, ordinary Portland cement-based reinforced mortar specimens (both unstressed and stressed categories) were prepared using prestressing (PS) steel king wires extracted from 7-wire strands. Totally, 10 specimens (5 + 5) were cast, cured for 28 days and then subjected to dry-wet (5-day drying followed by 2-day wetting) cycles in simulated concrete pore solution containing sodium chloride. The P-to-A transition was detected by continuous monitoring of the open circuit potential and impedance measurements at the end of every wet exposure period. The specimens were then autopsied and the concentration of chlorides in the mortar at the level of steel was determined and defined as Cl_{th}. It was found that the Cl_{th} of PS steel can reduce by half in OPC system when prestress is applied. A case study was performed to understand the implications of using overestimated values of Cl_{th} on the service life. An overestimation in service life of about 40% was observed when the Cl_{th} of unstressed PS steel was used - emphasizing the importance of determining the Cl_{th} of stressed steel and using it for service life estimation.

Keywords: Prestressed concrete · Corrosion · Service life-based design · Corrosion assessment

1 Introduction

A significant proportion of the world's bridges consist of pretensioned concrete (PTC) members. Usually, these important (and expensive) PTC bridges are expected to be serviceable for a 'deemed-to-satisfy' life of more than 100 years. However, many of them have started showing signs of corrosion-related distress in even less than 40 years [1]. Corrosion induced failures in PTC bridges are catastrophic in nature [2]. Though the

J. I. Escalante-Garcia et al. (Eds.): RW 2021, RILEM Bookseries 40, pp. 280–288, 2023.
https://doi.org/10.1007/978-3-031-21735-7_31

number of reported failures is low at present, it is still a point of concern as most of the existing PTC structures are relatively young (couple of decades old) and may not have sufficient build-up of chlorides yet. But if they start corroding, structural performance can be affected adversely and more drastically due to corrosion of prestressing (PS) steel at a given rate, as opposed to that of conventional reinforcement since the PS strands are under a tensile stress. Engineers will also find it difficult to cope up with the repair and rehabilitation work that this would necessitate. Ensuring an improved durability of these structures will also indirectly contribute to sustainability, as usage of large quantities of materials for restoring or replacing the corroded members will be avoided. In addition to adversely affecting the economy and sustainability, corrosion of PTC structures can result in safety issues. If a single wire of a strand breaks due to corrosion, the carried load is redistributed to the remaining wires, increasing the probability of failure of the entire strand [3]. A PTC element with corroded reinforcement may thus fail in a brittle manner, as demonstrated in the cases of failed PTC bridge in North Carolina [4] and failed girder in Pennsylvania [5] and in many other bridges. This possibility of brittle failure should be considered in the service life design (SLD) and corrosion assessment of PTC structures; and was the focus of a doctoral research work at IIT Madras. This paper intends to give an overarching view on the key findings from the doctoral research work reported in Joseline et al. [6, 7], Joseline [8]

The initiation of corrosion (phenomenon of Passive-to-Active (P-to-A) transition) is the first step of degradation of steel due to chloride-induced corrosion. Assuming that the transport of Cl^- is primarily by diffusion mechanism, the time to corrosion initiation (t_i) can be estimated based on Fick's second law if the concentration of chloride at the concrete surface (Cl_s), the chloride diffusion coefficient (D_{Cl}) and the chloride threshold (Cl_{th}), which is the minimum concentration of Cl^- necessary to initiate corrosion, are known. Very limited information on the Cl_{th} of unstressed PS steel is available with inadequate evaluation of the effect of stress [9]. It is reported that the application of tensile stress, even well below yield limit, can degrade the passivity of carbon steel. Presence of tensile stress is reported to cause retardation in passive film formation, and result in a film structure with more defects and low resistance to breakdown [10]. However, in field structures, the complete formation of passive film occurs in cementitious environment prior to chloride buildup at the level of steel. This must be adequately simulated in the laboratory studies. The in-service stress level may vary across countries, but is almost close to 75% of the ultimate tensile strength (f_{pu}) of the PS strand—which is very high. In this paper, an attempt was made to understand the implications of using the Cl_{th} of unstressed PS steel for estimation of t_i of PTC members of bridges such as girders.

2 Research Significance

The realistic quantification of t_i has two-fold importance in the civil engineering field. Firstly, the quantified t_i for various steel-cementitious systems can be compared with the target service life and the most suitable steel-cementitious system can be chosen for a particular project. Secondly, the correct quantification of residual service life (RSL, number of years left until corrosion initiation) for a PTC structural element (using D_{Cl} and Cl_s obtained from concrete core samples) can help in assessing the corrosion state of the PTC element. This is dependent on the accuracy of the Cl_{th} input. The methodology presented in this paper can help in a more realistic estimation of Cl_{th}.

3 Methodology

Figure 1a–b)shows the schematic of the test setups used to estimate the Cl_{th} of unstressed and stressed PS steel wire embedded in mortar. Central (king) wires of 5.28 mm diameter, extracted from 7 wire PS strands of nominal diameter 15.2 mm, were used. The mortar was made with ordinary Portland cement, and a water: binder: fine aggregate ratio of 0.5:1:2.75 was adopted. The fine aggregate consisted of a 50:50 mix of standard sand [11] of particle sizes 1–0.5 mm and 0.5–0.09 mm, to ensure adequate packing. Distilled water was used to prepare the corrosion test specimens and test solutions. The chemical composition of the PS steel and cement used in the study are presented in Tables 1 and 2. Self-reacting frames to stress the king wires to the desired prestress level were fabricated and the prestress used in Indian practice [12], i.e. 0.76 f_{pu} was selected for the study. Unstressed PS steel samples were also prepared for comparison.

Table 1. Elemental composition of PS steel (in weight %)

C	Si	Mn	S	Cr	Ni	Cu	Al	Pb	Nb	Ti	W	N	Re
0.89	0.19	0.72	0.01	0.56	0.09	0.03	0.03	0.09	0.05	0.06	0.11	0.01	96.6

*Re-remaining including Fe and other trace elements

Table 2. Chemical composition of ordinary Portland cement (in weight%)

CaO	SiO_2	Al_2O_3	MgO	Fe_2O_3	Na_2O	K_2O	TiO_2	SO_3	LOI
64.59	19.01	4.17	0.88	3.89	0.16	0.59	0.23	1.70	1.40

LOI – Loss on ignition

After 28 days of curing, the unstressed specimens (OPC_0 f_{pu}) were exposed to dry-wet cycles (5 and 2 days, respectively) in simulated pore solution, SPS, (with 0.3 g $Ca(OH)_2$ + 10.4 g NaOH + 23.2 KOH and 967 g H_2O per litre of solution) containing 3.5% NaCl. Because of a previous experience with very early initiation in stressed specimens resulting in less data points for analysis, the concentration of chloride used in the exposure solution was reduced to 2% for the stressed specimens (OPC_0.76 f_{pu}). Circumferentially placed nichrome mesh was used as the counter electrode (CE), saturated calomel electrode was the reference electrode (RE), and the king wire was the working electrode (WE). Open circuit potential (E_{corr}) and electrochemical impedance spectroscopy (EIS) techniques were used to detect corrosion initiation. For the EIS test, an AC signal of 10 mV amplitude was applied over a frequency range of 0.01–10^5 Hz. Upon corrosion initiation, the specimens were split at the level of steel and the mortar

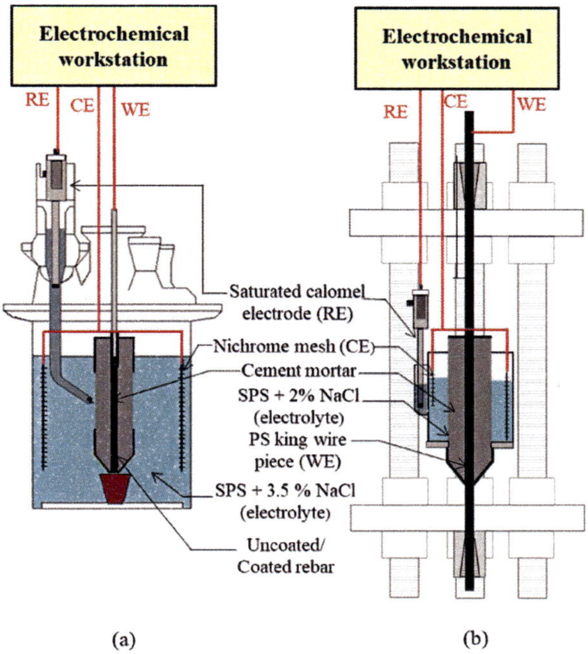

Fig. 1. Experimental setup for testing chloride threshold of prestressing steel embedded in cement mortar: (**a**) Unstressed wire and **b** Stressed wire

adjacent to the steel was powdered and collected. The chloride content (% bwob - percent by weight of binder) was determined as per the guidelines available in SHRP S 330 [13] and defined as the Cl_{th} of PS steel in OPC mortar.

A case study was then performed to understand the implications of using the Cl_{th} of unstressed PS steel to determine the t_i of a typical PTC bridge girder. For this, an in-house developed Matlab® program for service life estimation (SL-Chlor [14]) was used. The probabilistic estimation of T_i was done by means of 1000 random realizations of the cover depth, D_{Cl} and Cl_{th}. During this, the concentration of chloride across the depth was calculated for each year and compared with the Cl_{th}. The probability of corrosion initiation was calculated based on the number of instances of the chloride concentration at the level of steel exceeding $Cl_{th,}$ and a cumulative distribution function (CDF) was generated. Refer Joseline (2021) [8] for more details.

4 Results and Discussions

4.1 Effect of Stress on the Chloride Threshold

The E_{corr} measured during the test period is shown in Fig. 2(a). It can be observed that the potentials transitioned from around −100 mV Vs, SCE to very negative values as the duration of exposure to chlorides increased. The time required for this transition was found to be less in the case of stressed specimens although the exposure solution was less aggressive (2% NaCl instead of 3.5% NaCl; reason is presented in Sect. 3). This indicates that a lesser amount of chlorides can lead to active corrosion of PS steel under in-service stress conditions.

The interpretation from the impedance spectra obtained during the test period was in agreement with that from the E_{corr} results. At the same cycle when a drastic drop in E_{corr} occurred, the low frequency impedance of the system reduced drastically (by about an order). A representative evolution of impedance spectra with exposure to chloride ions is presented in Fig. 2(b). The pattern was observed to be similar for the unstressed and stressed wires embedded in similar mortar, the only difference being in the test cycle at which the reduction was observed – indicating a difference in the Cl_{th}. Corrosion was initiated much earlier in the case of stressed (OPC_0.76 f_{pu}) specimens than unstressed (OPC_0 f_{pu}) specimens.

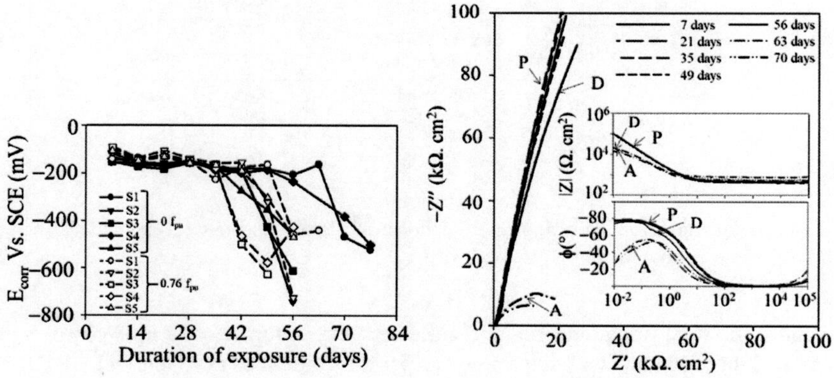

(a) E_{corr} (b) Typical evolution of EIS response

Fig. 2. Variation in electrochemical responses during the P-to-A transition

(P-Passive; D-Depassivated; A-Active). The occurrence of a drastic reduction in the Ecorr and EIS responses during P-to-A transition is attributed to the passive film cracking indicated by scanning electron micrographs of the surface of PS steel after P-to-A transition Fig. 3(a)). This phenomenon is believed to have occurred either instead of or along with localized pits due to chloride attack. This is expected to be due to various metallurgical factors- the key factor being the presence of tensile stress (residual/applied in the unstressed and stressed case respectively) at the time of passivation. Because of this tensile stress, the inner passive film layers are expected to be defective, and the expansive stress of the corrosion products is believed to cause the cracking in both directions as illustrated in Fig. 3(b) and explained in detail in Joseline et al. [7].

Upon chloride analysis, it was observed that the Cl_{th} of unstressed PS steel was around 0.4% bwob while that of stressed PS steel was around 0.2% bwob (a 50% reduction due to stress). Also, the variation in the values obtained for the stressed specimens was about 40% as opposed to 15% in case of unstressed PS steel - necessitating the testing of a

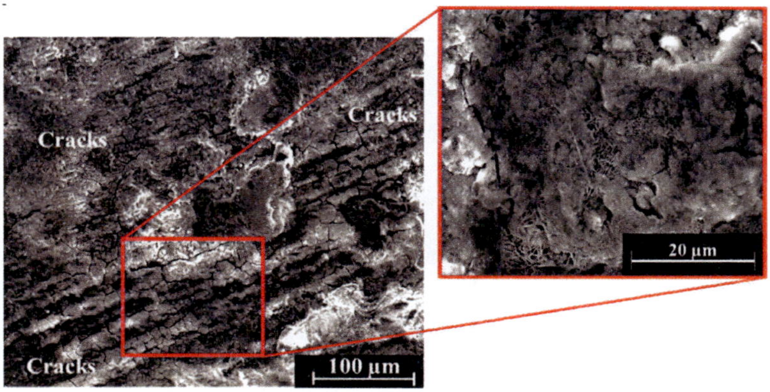

(a) Micrographs showing cracks on the passive film after P-to-A transition

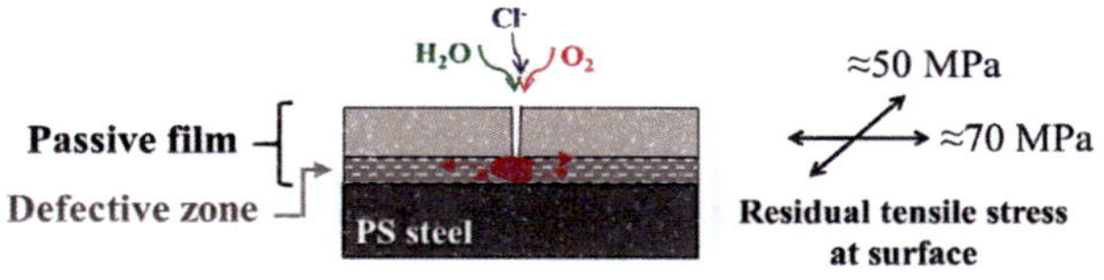

(b) Mechanism proposed for passive film cracking (adapted from [7])

Fig. 3. Cracking of passive film – Micrographs and proposed mechanisms

larger number of specimens to arrive at a good estimate of Cl_{th} for a stressed PS steel. Considering the mean values, it can be said that the Cl_{th} reduced by about 50% when service level prestress was applied.

4.2 Implications of Using Overestimated Cl_{th} for Estimation of T_i

A case study was performed to understand the consequence of using Cl_{th} of unstressed PS steel on the estimated T_i. A typical PTC bridge girder in a highway bridge was selected for the study. The geometric details of the girder are shown in Fig. 4(a).

SL-Chlor (a Matlab® program developed in-house) was used to estimate the t_i. The concrete was assumed to have a cement content of 350 kg/m^3. A maximum surface chloride concentration of 0.6% by weight of concrete was assumed and the time required to for its built up was assumed to be 15 years in all the cases. The value of the ageing factor, m, which captures the reduction in the chloride diffusion coefficient (D_{Cl}) as a function of time, was assumed to be 0.17 [15]. Typical magnitude of D_{Cl} of OPC concretes is in the order of 10^{-11} m^2/s [6]. However, OPC concretes can be designed to have lower values by adjusting the mix design. Hence, since the existing PTC bridges may be from different era of construction resulting in a possibility of D_{Cl} in the order of 10^{-11} to 10^{-12} m^2/s, three cases, namely (a) poor (b) moderate and (c) good quality cover concretes were considered with differences in the assumed D_{Cl}. The obtained CDFs for these 3 cases each of unstressed and stressed PS steels considered are presented in Fig. 4(b).

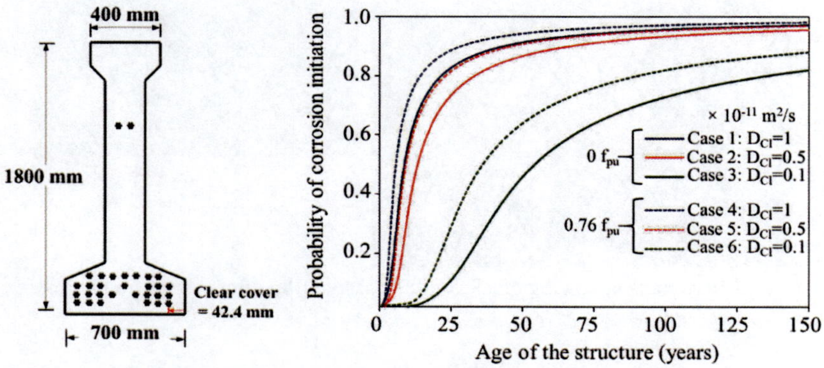

(a) Cross section of a typical PTC girder

(b) Probability of corrosion initiation

Fig. 4. Probability of corrosion initiation for various stress-DCl combinations

The effect of Cl_{th} on the estimated t_i can be understood from Fig. 4(b). Usage of Cl_{th} determined for OPC_0 f_{pu} for estimation of T_i resulted in about 40% overestimation. Also, for the given inputs, the T_i was very less for the cases with high D_{Cl}. This means that the usage of correct Cl_{th} value is more critical for concretes which have poor resistance to chloride ingress. The concretes in the middle-aged and old bridge structures are also expected to fall in this category as the usage of supplementary cementitious materials and low water-binder ratios, which can produce concretes with compact microstructure and high chloride resistance, was not practiced at that time. This emphasizes the need for usage of Cl_{th} of stressed PS steel (instead of unstressed PS steel) for estimation of t_i of PTC systems. A test methodology (setup, test parameters, statistically valid estimates) was developed to address this need [8]; and is kept outside the scope of this paper.

5 Summary and Conclusions

Literature provides limited information on the Cl_{th} of prestressing steel. Moreover, the reported values are predominantly for unstressed case. A methodology to determine the Cl_{th} of stressed steels (at in-service stress levels) is presented in the paper. For the OPC system considered in this study, it was observed that in-service prestress of 0.76 f_{pu} can reduce the Cl_{th} by about 50%. Although a higher scatter was observed in the Cl_{th} of stressed specimens (COV = 40%) as opposed to unstressed PS steel (COV = 15%), it can be concluded that the applied stress has a significant effect on the Cl_{th}. From a case study, it was inferred that a 50% overestimation of Cl_{th} (because of the usage of Cl_{th} of unstressed PS steel as input) can lead to a 40% overestimation of time for corrosion initiation, t_i (and hence the residual service life as well). The overestimation can be critical in the cases where the cover concrete had higher rate of chloride ingress. The findings presented in this paper emphasize the need of using Cl_{th} obtained from stressed steel as input for achieving a more realistic estimation of t_i of PTC systems.

Acknowledgments. The authors are grateful for the guidance provided by Dr. Lakshman Neelakantan, Prof. Manu Santhanam and Prof. Ravindra Gettu at various stages. The financial support from the Department of Science and Technology (Sanction No. EMR/2017/004687) and the Ministry of Education, Government of India, and the Department of Civil Engineering, Indian Institute of Technology Madras is also acknowledged. The support from laboratory staff in the Construction Materials Research Laboratory, the Civil Engineering Department Workshop, and other laboratories within IITM, and from the research associates is highly appreciated.

References

1. Rogers, R.A., Al-Ani, M., Ingham, J.M.: Assessing pre-tensioned reinforcement corrosion within the New Zealand concrete bridge stock. New Zealand Transport Agency Research Report 502 (2013)
2. Lau, K., Lasa, I.: Corrosion of prestress and post-tension reinforced-concrete bridges. In: Corrosion of Steel in Concrete Structures, vol. 2, pp. 37–57, Elsevier Ltd. (2016)
3. Darmawan, M.S., Stewart, M.G.: Spatial time-dependent reliability analysis of corroding pretensioned prestressed concrete bridge girders. Struct. Saf. **29**(1), 16–31 (2007). https://doi.org/10.1016/j.strusafe.2005.11.002
4. Sly, C.: An initial look at the Lowe's motor speedway pedestrian bridge collapse. J. Fail. Anal. Prev. **1**(2), 7–9 (2001). https://doi.org/10.1361/152981501770352969
5. Naito, C., Sause, R., Hodgson, I., Pessiki, S., Macioce, T.: Forensic examination of a noncomposite adjacent precast prestressed concrete box beam bridge. J. Bridg. Eng. **15**(4), 408–418 (2010). https://doi.org/10.1061/(asce)be.1943-5592.0000110
6. Joseline, D., Pillai, R.G.: Enhancing service life of prestressed concrete structures by using fly ash and corrosion inhibitors. Indian Concr. J. **94**(11), 54–67 (2020)
7. Joseline, D., Pillai, R.G., Neelakantan, L.: Initiation of stress corrosion cracking in cold-drawn prestressing steel in hardened cement mortar exposed to chlorides. Corrosion **77**(8), 906–922 (2021). https://doi.org/10.5006/3730
8. Joseline, D.: Electrochemical and chloride-induced corrosion parameters for service life design and assessment of prestressed concrete. Ph.D. Thesis, Department of Civil Engineering, Indian Institute of Technology Madras, Chennai, India (2021). https://civil.iitm.ac.in/faculty/pillai/people.php
9. Virmani, P.Y., Ghasemi, H.: Literature review of chloride threshold values for grouted post-tensioned tendons. FHWA Publication No. FHWA-HRT-12-067 (2012)
10. Lu, B.T., Zeng, Y., Pang, X., Luo, J.L.: Effects of hydrogen and tensile stress on passivity of carbon steel. Corros. Eng. Sci. Technol. **50**(3), 186–190 (2015). https://doi.org/10.1179/1743278215Y.0000000020
11. IS 650: Specification for Standard Sand for Testing of Cement, Bureau of Indian Standards (1991)
12. IS 1343: Prestressed Concrete-Code of Practice, Bureau of Indian Standards (2012)
13. SHRP-S-330: Standard Test Method for Chloride Content in Concrete Using the Specific Ion Probe, pp. 1–118. Strategic Highway Research Program, National Research Council, Washington, DC (1993)

14. Rengaraju, S.: Electrochemical response and chloride threshold of steel in highly resistive concrete systems. Ph.D. Thesis submitted to the Indian Institute of Technology Madras (2019)
15. Pillai, R.G. Gettu, R., Santhanam, M., Rengaraju, S., Dhandapani, Y., Rathnarajan, S., Basavaraj, A.S.: Service life and life cycle assessment of reinforced concrete systems with limestone calcined clay cement (LC3). Cem. Concr. Res. **118**(February), 111–119 (2019). https://doi.org/10.1016/j.cemconres.2018.11.019

Relationship Between the Vertical Profile of Marine Aerosol Salinity and the Chloride Accumulation into Concrete

Gibson Meira[1,2](✉) and Pablo Ferreira[1,2]

[1] Federal Institute of Education, Science and Technology of Paraíba, João Pessoa, PB 58036-860, Brazil
gibson.meira@ifpb.edu.br

[2] Federal University of Paraíba, João Pessoa, PB 58051-900, Brazil

Abstract. This paper analyses the influence of height from the ground on the vertical distribution of marine aerosol salinity. It considers altitudes between 1.5 and 10 m from the ground in a region under the major influence of surf produced aerosol and its impact on chloride accumulation into concrete along time. Results show that the rate of chloride deposition on wet candle markedly decreases when the altitude increases and this relationship can be represented by a power decay function. The impact of the salinity decrease with altitude on chloride accumulation into concrete is significant and can surpass the level of 50% after 50 years of exposure. As a consequence, this behaviour should be considered when analysing the performance of vertical concrete structures placed in marine atmosphere zone.

Keywords: Chloride · Concrete · Corrosion · Marine Aerosol

1 Introduction

Chloride ions present in the marine aerosol can induce reinforcement corrosion in concrete structures placed in the marine atmosphere zone. These ions are transported inland with the aerosol and, when in contact with concrete surfaces, they are partially captured and transported into concrete. As a consequence, when chloride concentration exceeds a critical content on the reinforcement surface corrosion can start [1].

Marine aerosol is mainly produced in the breaking waves zone, which characterises the surf produced aerosol [2, 3]. This aerosol is more concentrated due to the production of larger and more numerous salt particles [4, 5] and it is responsible for the significant concentration of salts in marine atmosphere zone. As a result, buildings placed along the coastal cities, in the marine atmosphere zone, are subjected to the influence of this kind of aerosol.

The influence of the distance from sea [6–8], as well as the influence of wind characteristics on marine aerosol salt concentration were previously studied [9, 10]. However, the vertical distribution of marine aerosol salinity is a topic that was not fully addressed. Some of the previous studies focused on the height influence were based on measurements taken over the sea, which environment is significantly different from inland sites

J. I. Escalante-Garcia et al. (Eds.): RW 2021, RILEM Bookseries 40, pp. 289–295, 2023.
https://doi.org/10.1007/978-3-031-21735-7_32

[11–13]. Furthermore, the impact of the height from the ground on chloride transport into concrete has been even less studied. This work focuses on this gap of knowledge and its main contribution is the analysis of the influence of the variation on the vertical chloride concentration on the chloride accumulation into concrete along time taking into account long-term exposure simulations.

2 Experimental Work

Marine aerosol salinity was measured along one year through the wet candle method [14]. Wet candle apparatus were placed at four monitoring stations located 100 m from the sea and at 1.5, 4, 7 and 10 m of height from the ground level. This exposure site was located in a coastal region of João Pessoa city, Brazil (Fig. 1). Temperature, relative humidity and wind characteristics were also monitored along the same period in a weather station close to the exposure site. Climatic data were collected following recommended standards for each kind of measurement.

Taking the average results obtained along the monitoring period at each height from the ground as input parameters, simulations on chloride accumulation into concrete were done considering a relationship previously published by the authors [15]. This experimental part is completely detailed elsewhere [15] and can be fully accessed in that publication. There, the experimental work was composed by a first part focused on the environmental characterisation and a second one concentrated on studying chloride concentrations in concrete along time. As a result, a relationship between chlorides from atmosphere and those accumulated into concrete was obtained, which was adopted in the present work. Here, the concrete details are repeated (Table 1) to make easier the simulations analysis.

Fig. 1. Exposure site location.

Table 1. Concrete mixtures and properties

Concrete	C1	C2	C3
Cement (kg/m^3)	320	356	406
Sand (kg/m^3)	840	812	769
Gravel (kg/m^3)	947	947	947
Platicizer (kg/m^3)	–	1.06	1.22
w/c ratio	0.65	0.57	0.50
Slump (mm)	90	80	80
Compressive strength (28d)—MPa	21	27	32
Total porosity (% vol.)	15.5	13.8	12.8

3 Results and Discussion

3.1 Climatic Parameters and Marine Aerosol Salinity

Regarding climatic parameters, daily average temperature remained between 20 and 31 °C, with little variation along the year. Relative humidity ranged between 65 and 95%, which means that this parameter remained at high levels along the year. The region alternated periods of scarce rainfall (between September and March) and of heavy rainfall (between April and August). This behaviour is similar to 30-year average historical data for the same region [16].

Wind speed ranged between 1.5 and 4 m/s, with low variability within each month. Predominant winds were form from south (S), southeast (SE) and East (E), with a preponderance of SE winds.

Figure 2 shows the behaviour of marine aerosol salinity at the different levels from the ground. Results show that there is a decrease in salinity when the level from the ground increases, as previously observed [12, 13] and that this decrease follows a power decay function (Fig. 2), following Eq. (1), where *Dep* is the average chloride deposition on wet candle and *h* is the height from the ground.

$$Dep = 203.09\ h^{-0.956} \quad (1)$$

This way, as lower salt concentrations are observed at higher altitudes it is expected that chloride accumulation into concrete structures take place at lower rates at higher levels from the ground as well. This aspect is analysed in the next section.

3.2 Simulations of Chloride Penetration into Concrete

Considering a previous published work by the authors [15], simulations on chloride transport into concrete were carried out, which were based on the relationship described in Eq. (2), where C_{tot} is the average total chlorides accumulated into concrete, C_O is the initial concentration of chlorides in concrete, *kd* is a fitting coefficient that depends on material characteristics and D_{ac} is the accumulated chloride deposition on wet candle

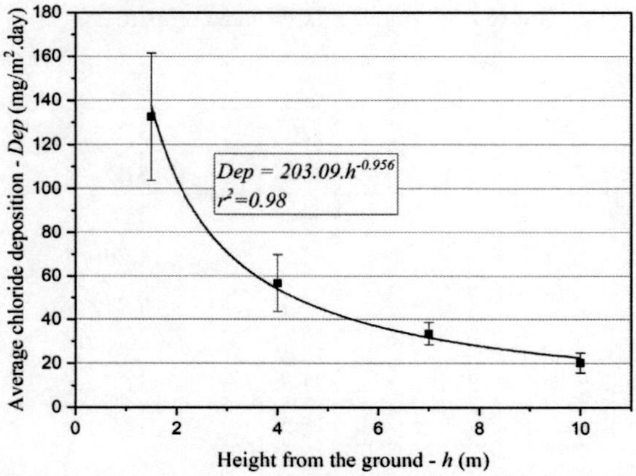

Fig. 2. Relationship between the average chloride deposition on wet candle and height from the ground.

along the years, which means the sum month-to-month of chloride deposition on wet candle. About C_{tot}, it is important to clarify that it does not represent a chloride concentration at an specific depth, but an integral of the whole area below the chloride profile divided by the maximum depth of this profile.

$$C_{tot} = C_0 + k_d (D_{ac})^{0.5} \tag{2}$$

Considering the values of C_0 and k_d previously determined by the authors [15], simulations of C_{tot} values were carried out considering the maximum exposure period of 50 years. Results of these simulations are presented in Fig. 3 for concretes C1 (Fig. 3a), C2 (Fig. 3b) and C3 (Fig. 3c).

Results show that the increase of the height from the ground can lead to decreases in C_{tot} values ranging between 27% and 47% after 10 years and between 31% and 54% after 50 years of exposure time for the less porous concrete (C3). For the more porous concrete (C1), the decreases in C_{tot} values ranged between 29 and 51% after 10 years and between 32% and 56% after 50 years of exposure time. These decreases remained between 28% and 48% after 10 years and between 31% and 54% after 50 years of exposure time for concrete C2. Although the internal distribution of chlorides in concrete is not considered in Eq. (2), which certainly would lead to zones closer to the surface with higher chloride concentrations and deeper zones with lower chloride concentrations, this behaviour shows that the decrease in C_{tot} values along time is significant in the first 10 m above the ground level. As a consequence, the lower parts of concrete structures in marine atmosphere zone are subjected to a more aggressive microenvironment than the upper ones. Therefore, this needs to be taken into in reinforced concrete structures with a significant vertical projection.

Another way to see the influence of the height from the ground on chloride accumulation into concrete is thorough comparing the curves of each concrete at a pre-defined exposure time. Figure 4 shows these curves after 50 years of exposure. It can be observed a clear reduction in *Ctot*, as the height from the ground increases from 1.5 to 10 m, and

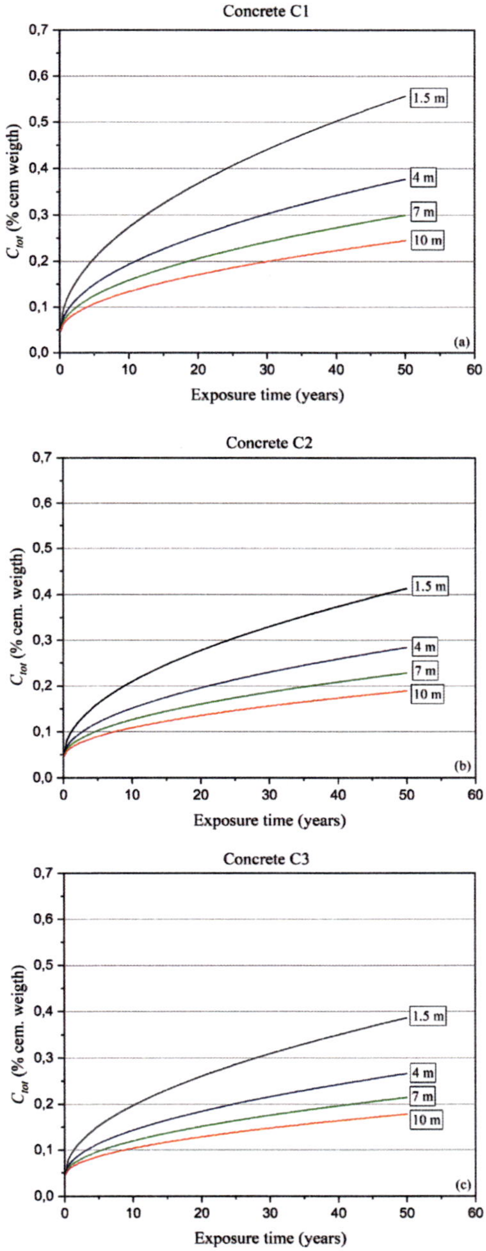

Fig. 3. Simulations of chloride accumulation into concrete considering mixtures C1 (**a**), C2 (**b**) and C3 (**c**)

this reduction is higher than 50% when considering the difference between 1.5 and 10 m from the ground.

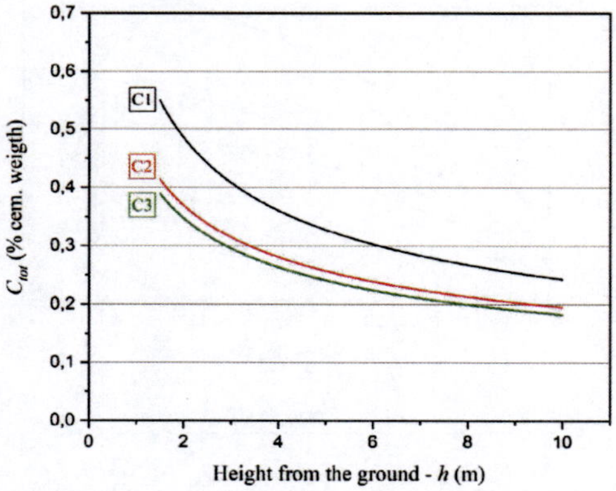

Fig. 4. Simulations of chloride accumulation into concretes C1, C2 and C3 after 50 years of exposure considering the influence of the height from the ground.

4 Concluding Remarks

This study shows that salt concentration in marine atmosphere zone at heights lower than 10 m from the ground strongly decreases with altitude and this relationship follows a power decay tendency. The equation $Dep = a \cdot h^{-b}$ well represents this behaviour.

The impact of differences on vertical distribution of chlorides in atmosphere on chlorides accumulation into concrete is significant and can lead to differences of more than 50% in the average total amount of chlorides accumulated into concrete along a 50-year-exposure time when comparing heights of 1.5 and 10 m from the ground. This shows the relevance of taking into account this reduction of aggressiveness when analysing the performance of vertical reinforced concrete structures.

References

1. Meira, G.R., Andrade, C., Vilar, E.O., Nery, K.D.: Analysis of chloride threshold from laboratory and field experiments in marine atmosphere zone. Constr. Build. Mater. **55**, 289–298 (2014)
2. McKay, W.A., Garland, J.A., Livesley, D., Halliwell, C.M., Walker, M.I.: The characteristics of the shore-line sea spray aerosol and the landward transfer of radionuclides discharged to coastal sea water. Atmos. Environ. **28**, 3299–3309 (1994)
3. Cole, I.S., Paterson, D.A., Ganther W.D.: Holistic model for atmospheric corrosion. Part 1—theoretical framework for production, transportation and deposition of marine salts. Corrosion Eng. Sci. Technol. **38**, 129–134 (2003)
4. Fitzgerald, J.W.: Marine aerosols: a review. Atmos. Environ. **25**A, 533–545 (1991)

5. O'Dowd, C.D., Smith, M.H., Consterdine, I.A., Lowe, J.A.: Marine aerosol, sea-salt, and the marine sulphur cycle: a short review. Atmos. Environ. **31**, 73–78 (1997)
6. Mustafa, M.A., Yusof, K.M.: Atmospheric chloride penetration into concrete in semi-tropical marine environment. Cem. Concr. Res. **24**, 661–670 (1994)
7. Feliu, S., Morcillo, M., Chico, B.: Effect of distance from sea on atmospheric corrosion rate. Corrosoin Eng. **55**, 883–891 (1999)
8. Meira, G.R., Padaratz, I.J., Alonso, C., Andrade, C.: Effect of distance from sea on chloride aggressiveness in concrete structures in Brazilian coastal site. Materilaes de Construcción **53**, 179–188 (2003)
9. Morcillo, M., Chico, B., Mariaca, L., Otero, E.: Salinity in marine atmospheric corrosion: its dependence on the wind regime existing in the site. Corros. Sci. **42**, 91–104 (2000)
10. Meira, G.R., Andrade, C., Alonso, C., Padaratz, I.J., Borba, J.C., Jr.: Salinity of marine aerosols in a Brazilian coastal area—influence of wind regime. Atmos. Environ. **41**, 8431–8441 (2007)
11. Piazzola, J., Despiau, S.: Contribution of marine aerosols in the particle size distributions observed in Mediterranean coastal zone. Atmos. Environ. **18**, 2991–3009 (1997)
12. Blanchard, D.C., Cipriano, R.J.: The vertical distribution of the concentration Osea salt in the marine atmosphere near Hawaii. Tellus **36B**, 118–125 (1984)
13. Ohba, R., Okabayashi, K., Yamamoto, M., Tsuru, M.: A method for predicting the content of sea salt particles in the atmosphere. Atmos. Environ. **24A**(4), 925–935 (1990)
14. American Society for Testing and Materials, ASTM G140–02: Standard Test Method for Determining Atmospheric Chloride Deposition Rate by Wet Candle Method. Annual Book of ASTM Standards, Philadelphia (2008)
15. Meira, G.R., Andrade, C., Alonso, C., Padaradtz, I.J., Borba, J.C., Jr.: Chloride penetration into concrete structures in the marine atmosphere zone—relationship between deposition of chlorides on the wet candle and chlorides accumulated into concrete. Cement Concr. Compos. **29**, 667–676 (2007)
16. Meira, G.R., Andrade, M.C., Alonso, M.C., Padaratz, I.J., Borba, J.C., Jr.: Measurements and modelling of marine salt transportation and deposition in a tropical region in Brazil. Atmos. Environ. **40**, 5596–5607 (2006)

Effect of Size and Preconditioning of Concrete Cores Against Sulfate Attack (Test)

Lautaro R. Santillán[1,2](✉), Yury A. Villagrán-Zaccardi[3], Edgardo F. Irassar[4], and Claudio J. Zega[2]

[1] IITCI, CONICET, Neuquén, Argentina
lrsantillan@conicet.gov.ar
[2] LEMIT, CONICET, La Plata, Argentina
[3] Magnel-Vandepitte Laboratory for Structural Engineering and Building Materials, Ghent University, Zwijnaarde, Belgium
[4] CIFICEN, UNCPBA-CICPBA-CONICET, Olavarría, Argentina

Abstract. Some of the proposed accelerated tests for evaluating the sulfate resistance of concrete address only the Portland cement, as the hydrated cement paste contains unstable compounds that may react with sulfate ions, while others are based on concrete performance. In search of comprehensive and practical assessments, some researchers proposed the evaluation of core samples as performance tests. The kinetics of attack is accelerated by further exposing the inner matrix of the concrete to the aggressive solution. Another accelerating strategy reported in the literature is to precondition the samples by vacuum saturation with sulfate solution, which speeds up the penetration of sulfate ions into concrete at the start of the test.

In this work, both strategies were applied to develop an accelerated performance test for concrete. Core specimens of two sizes were drilled off from two concrete types and exposed to a highly concentrated sulfate solution. Two preconditioning procedures were applied to the samples prior to exposure. The results show a large reduction in failure time when a vacuum step is used as preconditioning (one-sixth). This methodology could be useful in recycled concrete production to evaluate the potential reactivity of source concrete against sulfates.

Keywords: Sulfate test · Preconditioning · Concrete cores · Accelerated test

1 Introduction

In previous studies, core specimens were used as performance tests to evaluate the sulfate resistance of concrete [1–5]. This technique was even incorporated in a Swiss standard as accelerated performance test [6]. The direct exposure of the interfacial transition zone at the surface (drilled area) opens larger pores for rapid sulfate ingress. This approach might make the evolution of the test more dependent on the chemical quality of the matrix, i.e., the aluminate content in the hardened cement paste and the portlandite/C-S-H ratio.

Like other durability issues, External Sulfate Attack (ESA) has induction and propagation periods. The size of core specimens has a known effect on sulfate attack kinetics,

J. I. Escalante-Garcia et al. (Eds.): RW 2021, RILEM Bookseries 40, pp. 296–305, 2023.
https://doi.org/10.1007/978-3-031-21735-7_33

specifically the volume/surface ratio [2]. For a high volume/surface ratio, the induction period is longer. The diffusion rate of sulfate ions into the core specimen is driven by the concentration gradient and occurs through the microporosity of concrete [7, 8]. Afterward, expansion tests enter into the propagation phase when the ESA reaches a sufficient proportion of the cross-section, leading to a certain incidence of slenderness [9].

Preconditioning of concrete samples with vacuum saturation is an alternative procedure that accelerates the kinetics of sulfate resistance tests, as it allows quick ingress of sulfates into the whole section. Some accelerated evaluation methods proposed in the literature for testing concrete and mortar specimens use this strategy [10, 11]. The high concentration of sulfate (and its associated cation) in the pore solution leads to massive ettringite formation across the whole section. This situation completely cancels the transport kinetics restraint of the attack and makes the test fully thermodynamically controlled. Unfortunately, few results on this acceleration by preconditioning core specimens can be found in the literature.

This work originates from the search of a quick test for evaluating source concrete for recycled aggregates production. The initial objective was to evaluate the chemical quality of a source concrete and its reactivity to a sulfates aggression. The core extractions allow evaluation of existing concrete (i.e., residual concrete); meanwhile, the preconditioning by vacuum saturation may accelerate the results. In this work, both strategies were used. Two different sizes of core specimens were taken to evaluate their applicability to this test. Then, twin samples of one size were preconditioned by two different protocols. The results of this work are a new benchmark in testing core specimens as a performance test of concrete resistance against sulfate attack.

2 Experimental

Two source concretes were prepared, evaluated, and used to produce the recycled aggregates afterward. The design criterion was to produce two source concretes with similar physical and mechanical properties but with different chemical quality in their cement paste phase.

Source concrete A (SC-A) was made with an ordinary low C_3A Portland cement (cA) similar to CEM I 52.5 SR-5 according to EN 197-1. Source concrete B (SC-B) was made with high C_3A limestone Portland cement (cB), similar to CEM II/A-L 52.5 according to EN 197-1. Table 1 summarizes the cements properties: (a) the chemical composition of both cement types obtained by XRF, (b) the mineralogical composition obtained using XRD and Rietveld analysis, and (c) physical properties. Although both cements have a high strength level, cement B has a lower sulfate resistance than cement A due to its higher C_3A and limestone filler content [12]. Regarding physical properties, cement A showed a higher mechanical strength. However, both cements showed a high compressive strength (>50 MPa at 28 days) and similar fineness.

Table 2 shows the mixture proportions of the source concretes. Coarse aggregate was crushed granite with a maximum size of 19 mm, and the fine aggregate was a 70:30 w/w mixture of river siliceous sand and crushed granite 0–6 mm. As both cement types have different densities, the content of cement and sand was corrected to obtain the same coarse aggregate/mortar volume ratio in both concretes (0.59). A superplasticizer

Table 1. Characterization of cements

Chemical composition, [%]								
Cement	CaO	SiO_2	Al_2O_3	Fe_2O_3	SO_3	MgO	Na_2O	K_2O
A	62.5	20.2	3.6	3.3	2.3	1.7	0.5	1.0
B	65.7	19.4	3.7	0.2	2.6	1.3	0.1	0.2

Mineralogical composition [%]						
Cement	C_3S	C_2S	C_3A	C_4AF	Gypsum	Calcite
A	59.6	13.1	4.0	10.1	5.5	4.1
B	49.6	18.3	9.4	0.5	3.7	17.2

Physical properties						
Cement	Density	SSA Blaine [m^2/kg]	f'c 28d, [MPa]	Water demand, [%]	Loss on ignition, [%]	Insoluble residue, [%]
A	3.05	431	62.2	29.4	2.6	0.85
B	3.02	416	52.3	33.5	8.8	1.30

(dosage 0.3% by mass of cement) was used to obtain a slump of 80 ± 20 mm (following ASTM C 143 [13]). Table 2 also presents the fresh state properties and compressive strength at 28 days for both concretes.

Table 2. Proportions and properties of source concretes.

Material, kg/m^3	SC-A	SC-B
Water	158	158
Cement	351	350
River siliceous sand	628	617
Crushed granite 0–6 mm	268	266
Crushed Granite 4–19 mm	970	970
Superplasticizer	1.05	1.05
Air content, % [ASTM C 231]	3.2	3.5
Slump, mm [ASTM C143]	70	60
Compressive strength at 28 days, MPa [ASTM C 39]	48.7	45.4

Prismatic (7 × 10 × 43 cm^3) specimens were cast, demoulded after 24 h, and stored in a curing room (RH > 95%, Temp: 23 ± 2 °C) for up to 28 days. After that, specimens were kept for ~1 month in the laboratory environment (RH: 60%; Temp: 20 ± 2 °C). Then, core specimens were drilled from the prisms: 8 having 25 mm in diameter and 3 having 75 mm in diameter for each concrete. Core specimens were covered in their extremes with waterproofing paint, and stainless steel pins were glued with an epoxy resin to both ends. Figure 1 shows the visual aspect of the cores before the preconditioned stage.

For the normal test, the preconditioning stage consisted of immersing the cores (the three with 75 mm and four with 25 mm) in a saturated solution of calcium hydroxide for 72 h at laboratory temperature (21 ± 2 °C).

Fig. 1. Visual aspect of core specimens before the preconditioned stage.

For the accelerated test, the preconditioning stage consisted of vacuum-saturating the samples (the four with 25 mm) with sulfate solution, as follows: (a) cores were stored in a vacuum chamber with a pressure of −75 cm Hg for 4 h; (b) a 50 g/L sodium sulfate solution was carefully introduced into chamber while keeping the vacuum until saturating all cores with it; (c) when submersion was complete vacuum pressure was released, and cores were kept submerged in the same sulfate solution for another 20 h, at atmospheric pressure, before testing.

After the preconditioning stage, the initial length and weight of samples were measured. Then, samples were immersed in plastic containers with 50 g/L sodium sulfate solution at 21 ± 2 °C, with a solution-to-core volume ratio of 1:3. The pH of the solution was corrected once a month by the addition of solutions of sulfuric acid (2 N) and sodium sulfate (0.35 M). Progressive measurements were made once a week during months 0–2 of exposure, once a month between months 2–6, and every three months from month 6 onwards. In the normal tests, measurements were made up to 2 years of exposure. In the accelerated test, measurements were made for up to 6 months, as expansion and general deterioration of samples in that period were in the same order as normal test samples.

3 Results and Discussion

3.1 Normal Test

Figure 2 presents the average values and standard deviation for the expansion of 75 mm and 25 mm cores up to 2 years of exposure. As expected, the slenderness factor of cylinders had a critical effect on the kinetics of sulfate attack. In both cases, an initial expansion was observed after immersion in the solution, and then the measurements stabilized. Only 25 mm cores of SC-B showed propagation stage of ESA starting after around 400 days of exposure.

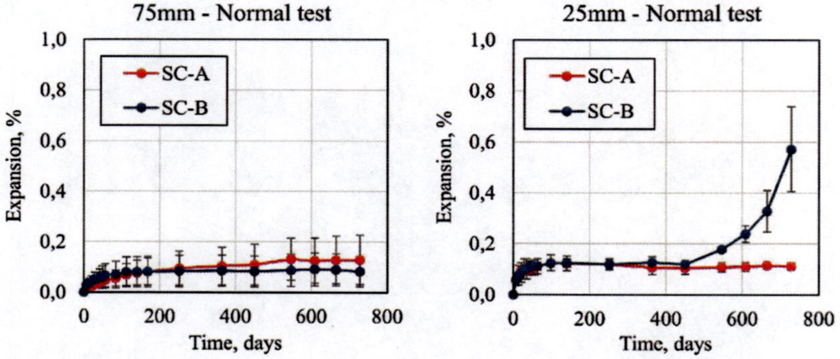

Fig. 2. Expansion for both core sizes in normal test up to 2 years of exposure.

The initial expansion is attributed to additional water absorption of core specimens. Although samples were saturated before testing, the time may be insufficient to reach a full saturation degree (72 h). It is noted that 25 mm cores had a higher initial relative expansion than 75 mm cores.

For 75 mm cores, a slightly higher absolute expansion for SC-A is observed, but the difference is lower than the standard deviation. For 25 mm cores, samples of SC-B reached the propagation stage of ESA. After 2 years of exposure, the average expansion for SC-B and SC-A was 0.58% and 0.16%, respectively, showing the poor sulfate resistance of cement B in comparison with cement A. The higher expansion of SC-B is attributed to the high C_3A content of its cement, and its limestone filler content in a second order.

Figure 3 presents the average values and standard deviation for the weight variation of 75 mm and 25 mm cores up to 2 years of exposure. Results are in concordance with expansion results. An initial weight increment is observed for all samples, then weight trends stabilize. For 25 mm cores, a consistent weight increase of SC-B samples is observed after 270 days of exposure.

The initial weight variation was lower in 25 mm cores than in 75 mm cores. This is attributed to a different degree of saturation. In the case of 75 mm cores, the time to fully saturate samples should be longer than for 25 mm cores, but all samples were submerged simultaneously in lime water (72 h). Then, in 75 mm cores, higher water absorption was observed for SC-A. The small deviation in these curves makes this difference consistent.

For 25 mm cylinders, the SC-B weight variation associated with ESA started after 200 days of exposure. The average weight variation of SC-B and SC-A at 2 years was 0.96% and 0.26%, respectively. The induction period in weight variation curves was shorter than the same in expansion curves. It is consistent as sample expansion starts after ESA covers most of the cross-section [9].

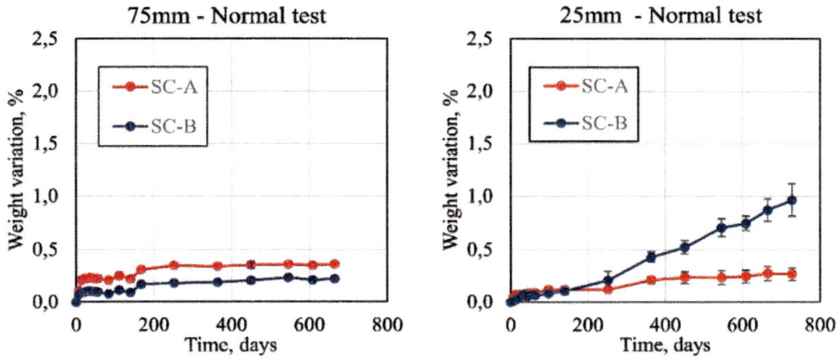

Fig. 3. Weight variation of cores in normal test up to 2 years of exposure.

Figure 4 presents the correlation between expansion and weigh variation of all series in the normal test. Different test stages can be identified, with different relations between both parameters depending on sample size and concrete type.

For 25 mm cores, a first stage of weight variation and fast expansion is observed. This stage is associated with additional water absorption. The correlation of both parameters is similar for both concrete types. Then is observed a second stage of weight increment without expansion. This stage is associated with ESA product precipitation during the induction period. Only for SC-B samples this weight increment reaches high values and then the third stage of the test is reached. In this last stage, the expansion to weight increment ratio suddenly increases. This stage is associated with the propagation period of ESA, and both effects are related to ettringite and gypsum precipitation in the cement paste of samples.

For 75 mm cores, higher water absorption of SC-A with respect to SC-B is observed in the first stage of the test. Two reasons could explain this: (1) a difference in saturation degree of both concretes (may SC-A had a lower water absorption rate and needed more time to be fully saturated in the preparation step); (2) a higher relative content of portlandite in its matrix, which is a very soluble phase (higher relative content of portlandite is expected in SC-A matrix due to the higher C_3S/C_2S ratio of cement. This may opens microporosity in cement paste to an additional water imbibition. This hypothesis is based on the fact that clear evidence of lixiviation was observed at the beginning of the tests. The first measurements observed a high pH level in the solution (~12). Each time, the pH level was corrected by adding a buffer solution. In the following measurements (after one month of exposure) the pH tended to stabilize. Indeed, progressive smaller quantities of buffer solution were necessary to reach a neutral pH in the solution. This suggests that samples suffer a lixiviation of soluble compounds (mainly portlandite), and small changes in their microporosity might occur. There is evidence of the physical swelling of concrete even after saturation through water migration in its microporosity [14]. Finally, this initial expansion (and weight variation) due to additional absorption is not connected with sulfate attack, so it is essential to consider and correct the absolute expansion and weight variation in the test reports.

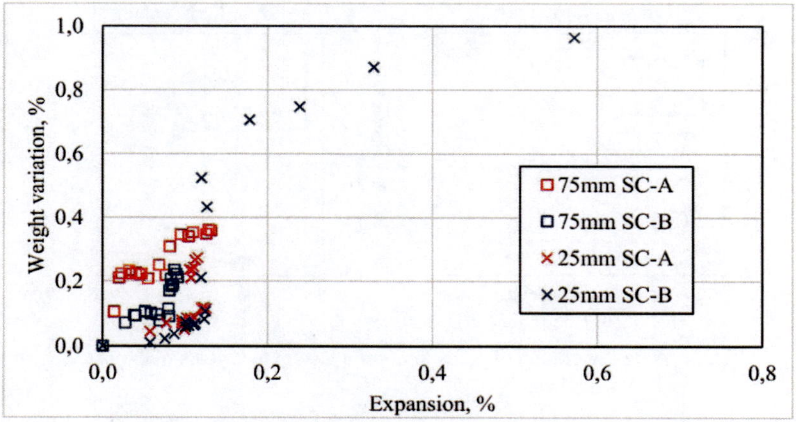

Fig. 4. Expansion vs. weight variation for 25 and 75 mm-diameter cores.

3.2 Accelerated Test

Figure 5 presents average expansion and weight variation, and their respective standard deviations, for 25 mm-diameter cores registered under the accelerate test up to 6 months of exposure. As in the normal test, SC-A cores showed lower expansion and weight increment, indicating their better sulfate performance, while the SC-B cores, judged by these same parameters, showed a poor sulfate performance.

An initial expansion and weigh increment are observed in the same manner as for normal test samples. This is associated with additional water absorption, but in this case, the saturation degree of samples was the same since all samples were saturated under vacuum pressure. This supports the hypothesis of a secondary imbibition period of samples favored by the initial lixiviation until equilibrium. In this case, a progressive stabilization of the pH of the solution in each measurement was also observed.

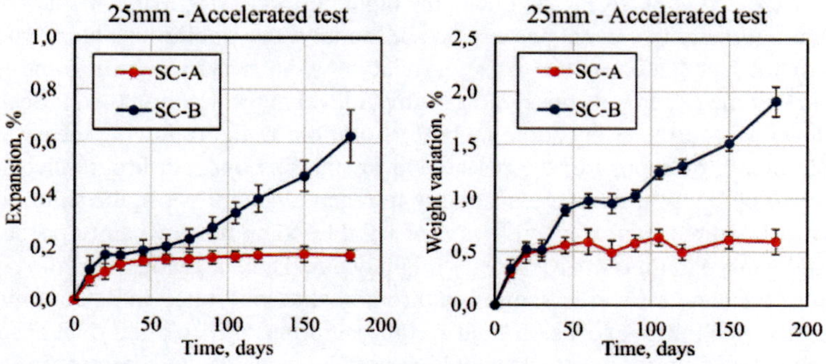

Fig. 5. Expansion and weight variation of 25 mm diameter cores in accelerated test up to 6 months.

Then, after the plateau, the weight variation of SC-B deviates from the weight variation of SC-Aafter 25 days of exposure. Moreover, the expansion rate for SC-B accelerates after 60 days of exposure (2 months). At 6 months of exposure, the average expansions for SC-A and SC-B were 0.17% and 0.62%, respectively, and the average weight variations were 0.60% and 1.91%, respectively.

Comparing the normal and accelerated test results, they confirm the effectiveness of the preconditioning with vacuum saturation with sulfate solution to shorten the exposure time and detect the non-sulfate resistant concrete early. For 25 mm cores, the preconditioning of samples with sulfate solution significantly reduces induction period of the ESA: from approximately 400 days in the normal test to 60 days in the accelerated test. Regarding the average expansion, the threshold of 0.3% was reached at 650 days using the normal test and at 100 days of exposure using the accelerated test (one-sixth). This reduction in the test duration makes it suitable for a performance control test (for example, to evaluate the chemical quality of a source concrete to be used to produce recycled concrete aggregate).

Figure 6 compares the expansion and the weight variation for 25 mm cores subjected to both tests. A different correlation between both parameters is observed depending on the preconditioning type, especially in the propagation stage (third stage of the test). Thereby accelerated test samples showed a higher weight variation than normal test samples for the same expansion level. This is attributed to a different attack mechanism since the massive precipitation of ettringite occurred in the whole section already at the beginning of the test. This probably leads to a different failure mechanism of samples in the accelerated test compared to the normal test. To corroborate the test reliability, it is necessary to increase the number of results, including additional mixes and materials.

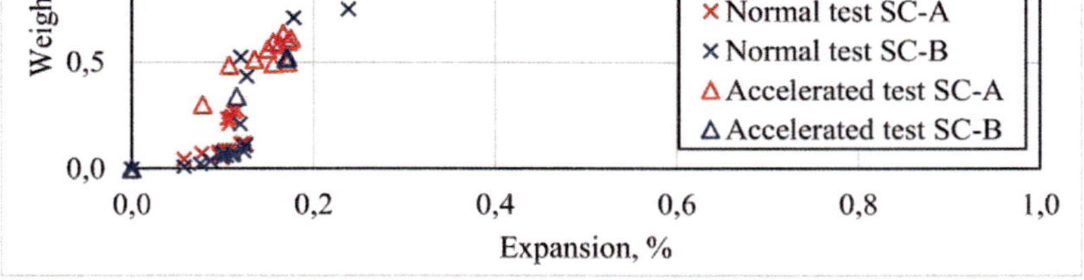

Fig. 6. Expansion vs. weight variation of 25 mm-diameter cores under both normal and accelerated test.

4 Conclusions

This research studied the effects of size and preconditioning of cores obtained from hardened concretes on the expansion and weight increase when exposed to sulfate solution. Results indicated a good correlation between the normal and the proposed accelerated tests and a reduction in the induction period for the latter. For example, with an expansion of 0.30% as limit criterion, the reduction in time from normal to accelerated test is around 6 times.

A strong effect of the core slenderness ratio was observed in the duration of the induction period. 75-mm samples showed a time to failure incompatible with the requirements for a control test. Initial weight gain unconnected with ESA was observed at the start of all tests. For evaluating the proper sulfate expansion, it is important to discount the early expansion and weight gain from the later absolute expansion and weight variation to assess the sulfate resistance of concrete individually.

The time reduction in the accelerated test portrays this procedure as an excellent alternative for the routine characterization of source concrete regarding potential sulfate reactivity before producing recycled concrete aggregates. However, further research is necessary to increase the number of results, assess the proposed procedure's reliability, and define limits within the expansion and mass gain tests.

Acknowledgment. Financial support from FONCyT through PICT 2015–3339 and PICT 2017–0091 Prest BID for this research is greatly appreciated.

References

1. Girardi, F., Vaona, W., Di Maggio, R.: Resistance of different types of concretes to cyclic sulfuric acid and sodium sulfate attack. Cem. Concr. Compos. **8**, 595–602 (2010). https://doi.org/10.1016/j.cemconcomp.2010.07.002
2. Brunetaud, X., Khelifa, M.R., Al-Mukhtar, M.: Size effect of concrete samples on the kinetics of external sulfate attack. Cem. Concr. Compos. **34**(3), 370–376 (2012). https://doi.org/10.1016/j.cemconcomp.2011.08.014
3. Loser, R., Leemann, A.: An accelerated sulfate resistance test for concrete. Mater. Struct. **49**(8), 3445–3457 (2015). https://doi.org/10.1617/s11527-015-0731-2
4. Menéndez, E., García-Rovés, R., Aldea, B., Ruíz, S., Baroghel-Bouny, V.: Combination of immersion and semi-immersion tests to evaluate concretes manufactured with sulfate-resisting cements. J. Sustain. Cem. Mater. **8**(6), 337–352 (2019). https://doi.org/10.1080/21650373.2019.1624659
5. Achour, R., Zentar, R., Abriak, N.E., Rivard, P., Gregoire, P.: Durability study of concrete incorporating dredged sediments. Case Stud. Constr. Mater. **11**, e00244 (2019). https://doi.org/10.1016/j.cscm.2019.e00244
6. Standard S. SIA 262/1, Appendix D
7. Alexander, M., Bertron, A., De Belie, N. (eds.): Performance of cement-based materials in aggressive aqueous environments. RSR, vol. 10. Springer, Dordrecht (2013). https://doi.org/10.1007/978-94-007-5413-3
8. Whittaker, M., Black, L.: Current knowledge of external sulfate attack. Adv. Cem. Res. **27**(9), 532–545 (2015). https://doi.org/10.1680/adcr.14.00089

9. Suma, M.F., Santhanam, M.: Influence of specimen size on the expansion of portland cement mortars immersed in sodium sulphate solution. In: Menéndez, E., Baroghel-Bouny, V. (eds.) External Sulphate Attack – Field Aspects and Lab Tests, RILEM Final Workshop of TC 251-SRT (Madrid - SPAIN, 2018), RILEM Bookseries, RB 21, p. 81. Springer, Cham (2020). https://doi.org/10.1007/978-3-030-20331-3_7
10. Messad, S., Carcasses, M., Linger, L.: Design of an accelerated test method for external sulfate attack. In: Alexander, M.G., Bertron, B. (eds.) Concrete in aggressive aqueous environments – Perform., Testing and Modeling, RILEM Book, pp. 307–317. Springer (2009)
11. Aguayo, F.M.: External sulfate attack of concrete : an accelerated test method. mechanisms and mitigation techniques. Dissertation, The University of Texas at Austin (2016). https://repositories.lib.utexas.edu/handle/2152/39727
12. Irassar, E.F., González, M., Rahhal, V.: Sulphate resistance of type V cements with limestone filler and natural pozzolana. Cem. Concr. Compos. **22**(5), 361–368 (2000). https://doi.org/10.1016/S0958-9465(00)00019-6
13. ASTM International C143-03: Standard test method for slump of hydraulic-cement concrete (2004)
14. Alderete, N.M., Villagrán-Zaccardi, Y.A., De Belie, N.: Physical evidence of swelling as the cause of anomalous capillary water uptake by cementitious materials. Cem. Concr. Res. **120**, 256–266 (2019). https://doi.org/10.1016/j.cemconres.2019.04.001

CFRP Reinforcement for Historical Buildings at Mexico City: Concrete Frame Structures Around 1940 to 1960

Juan Manuel García Garduño(✉)

Universidad Nacional Autónoma de México, Coyoacán, CDMX, México
manuel.garduno1914@gmail.com

Abstract. Around 1940 the introduction of skyscrapers improved the development of reinforced concrete systems capable of adapting to seismic, soil, and subsoil conditions. Nowadays, most of them still operative as iconic historical constructions. Nevertheless, they exhibit differential subsidence considering their age, natural aging, and seismic activity; then, they require an analysis to deem them safe structures according to codes demands. Furthermore, because of the historical value of those buildings, it is necessary to reach the structural adaptation without disturbing its integrity. To accomplish these purposes, the use of FRP materials, focus on carbon fiber was considered. This paper aims to present the results of the fiber influence to control plastic flow and its effect on the ductility of structural members by bending tests to study the distribution of the internal forces.

Keywords: Composites · Moment-curvature · Plastic flown · Shear · Concrete

1 Introduction

The constructive systems previous to 1940 utilized concrete in foundations design because the characteristics of the material helped confront the complex soil configuration [1]; the subsoil in this area presents high water content: plasticity ratio is more significant than 300%, and the compression ratio (Cc) is between 7–10 [2, 3]. The other important variable for construction at the Mexican territory is its seismic activity, which is related to its localization over North American plate mostly, and Rivera and Cocos plates that are subsiding under the first [4–6]. The *Lotería Nacional* skyscraper (1934–1946) was one of the most representative constructions in the use of reinforced concrete [7, 8]. Its understructure exhibited a complex concrete design over T beams invested [9, 10]. Those T beams are intended to grant freedom of movement for the foundation to adjust ground reverse movements under a flexible layer [11, 12]. This kind of design and application of concrete renovated the vision for architecture and structural systems [13]. It was an example that impulse concrete frame buildings at the city for the next twenty years, skyscrapers most of them [14–16].

Many of those buildings are nearby to be obsolete. They can not accomplish current regulation codes [17] because its steel quantity is less than the established so they are

J. I. Escalante-Garcia et al. (Eds.): RW 2021, RILEM Bookseries 40, pp. 306–315, 2023.
https://doi.org/10.1007/978-3-031-21735-7_34

considered non ductile [18]. Also, natural aging [19] could be accelerated due to soil transformations and cyclic seismic stresses [20, 21]: plastic flown deformations could cause a disbalance between the load/stress distribution on structural members [22–24]. The actual research is developed to analyze the possibility of improving FRP reinforcement to control the displacement and deformations in old concrete members to delay its aging [25].

2 Manufacture of Specimens

2.1 Method

This research FRP refers to a composite material made with carbon fiber (CF) impregnated with epoxy resin, attached to concrete by wet lay-up procedure [26]. The selection of CF is because its semi-crystalline microstructure exhibit a tensile strength of over 12,330–22,940 kg/cm^2 [27], and its significant advantages for in situ installation wherefore CF sheet can adapt to complex geometries [28, 29].

Materials and Manufacture of Specimens

It is selected unidirectional carbon tow sheet; thus, one direction cloth has stronger tensile resistance [30]. It was used two parts of epoxy adhesive To attach the fiber. Table 1 shows the properties of CF. Two components form epoxy adhesive: Both shall mix in a 100g:34.5gr relation, demonstrating its properties in Table 2. The campaign used twelve concrete beams to compare the variance utilizing a different number of CF plates, as shown in Table 3.

Table 1. Mechanical properties of carbon fiber.

Material	Ultimate tensile strength	Young modulus	Weigh
SikaWrap®-530 C©	7,268 kg/cm^2	500,500 kg/cm^2	1.8 g/m^3

Table 2. Mechanical properties of epoxy resin.

Material	Flexural modulus	Tensile strength	Tensile modulus
Sikadur®300	28,500 kg/cm^2	460 kg/cm^2	35,690 kg/cm^2

The geometry of specimens was 15 cm × 15 cm × 60 cm using 9.5 mm steel bars as reinforcement. Concrete mix for all items were design with a compression resistance target of 250 kg/cm^2 [31, 32], with a slump ratio of 16–18 cm. The moisture of the beams after casting was controlled into a curing chamber for 20 days [33, 34].

On day 21 from the cast, carbon fiber was applied. A thin concrete layer was removed with an average thickness from 0.5 mm to 2 mm, and then the surface was cleaned off dust [26] to increase phase adherence before using it on the beams. After attaching each

Table 3. Test specimen designation.

Name	Characteristic	Amount
Control	Reinforced concrete	3 beams
RC + 1 FRP	Reinforced concrete plus one plate of CF	3 beams
RC + 2 FRP	Reinforced concrete plus two plates of CF	3 beams
RC + 3 FRP	Reinforced concrete plus three plates of CF	3 beams

CF plate, it was an epoxy curing time of 12 h, previous to bind another. Before pasting other plates, the beam reinforced side was pulled with fine ultra grain sandpaper then cleaned again off dust. The entire lot was ending after three days. For flexural tests, a waiting period of four days was considered, so the specimens reach a period of 28 days after cast.

3 Flexural Test

The specimens were tested according to the three-point flexural test parameters [35, 36]. All beams were tested into Laboratory of Structures, at Engineering Institute of UNAM, by putting them in a SATEC Systems Inc. Universal testing machine, screw-type model 500WHVL with capacity for 220 tons, as shown in Fig. 1a. Specimens were instrumented with two high sensitive displacement transducers model CDP-50/200 $\times$ $10^{-6/mm}$/BBD03240; as shown on Fig. 1b.

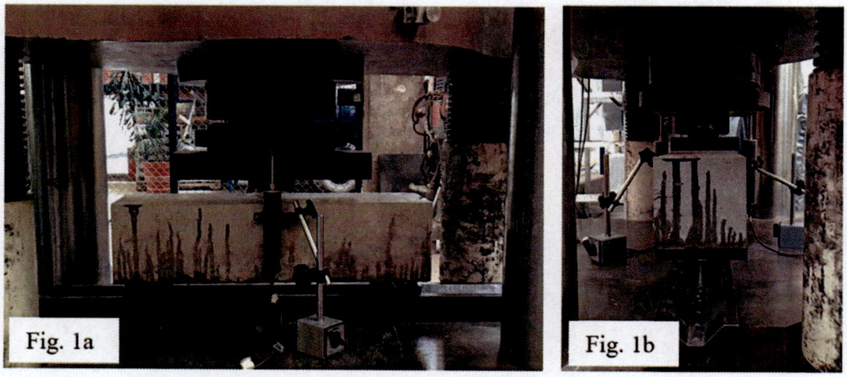

Fig. 1. Concrete beam specimen mounted into the universal testing machine and instrumented with transducers at Engineering Institute, UNAM.

3.1 Results

The flexural deformation for FRP beams was not reached. The average results are shown in Table 4. All the tested specimens exhibited a compression shear failure, which is shown in Fig. 2a. The main reason for that trend was the excessive distance between stirrups, 15 cm; about the beam length, 60 cm, shown in Fig. 2b.

Table 4. Average results of beams after flexural test

Name	Ultimate load	Maximal bending moment	Ultimate shear capacity	Rigidity
Control	6.56 T	816 T-mm	3.26 T	2.64 T/mm
RC + 1 FRP	9.06 T	1,132 T-mm	4.53 T	5.68 T/mm
RC + 2 FRP	9.11 T	1,139 T-mm	4.56 T	4.41 T/mm
RC + 3 FRP	11.51 T	1,438 T-mm	5.76 T	5.78 T/mm

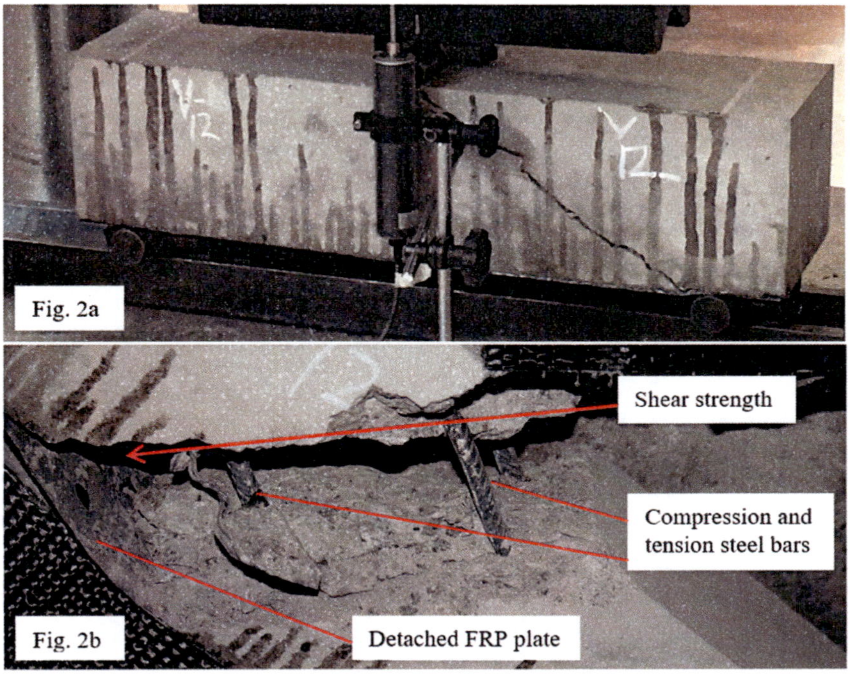

Fig. 2. Concrete beam reinforced with carbon fiber plates. It showed shear failure.

Analysis

In correspondence with results in Table 4, carbon fiber layers increase the rigidity of concrete specimens, which means the FRP reinforcement induces a fragile failure trend. On the other hand, CF contributes to extend load capacity and restrict displacement. Nevertheless, the results of RC + 2FRP specimens exhibit a lower value than expected.

The behavior of the load-displacement curve of two CF plates lot shown a low performance since the beginning, and it has less rigidity than the one-carbon plate specimens. Figure 3 shows the average load-displacement curves of all samples.

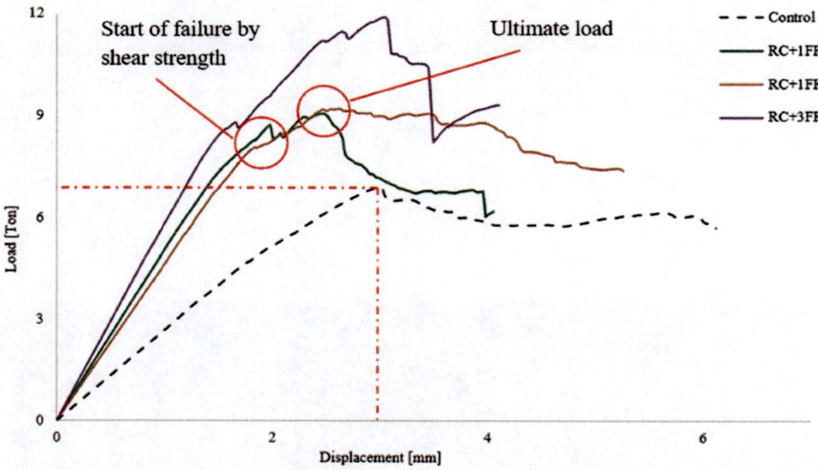

Fig. 3. Average load-displacement curves of tested specimens. The average values were taken considering 40% of the load. FRP beams are in contrast with Control specimens (red guides).

On Fig. 3, it observes the curve of FRP reinforced items have primary fluence before they cracked. Carbon fiber reduces the strain while increasing load capacity: they showed a decreasing peak before they reach two millimeters of displacement, indicating where shear failure started. Contrasting FRP values with Control specimens suggests the carbon fiber layers restrict the expansion at the bottom of the beam while the load was applied. Therefore, it is assumed when the load was applied, the insufficient vertical steel can not dissipate shear energy, and when it achieved the textile reinforcement, transmit the stress directly into the interphase concrete-epoxy, which pulled off the plate as shown in Fig. 4.

Physical exploration of tested specimens exhibited that carbon fiber plates were detached by shear stress, based on: there wasn't delamination within carbon fiber layers; over the surface of detached plates, there are torn concrete residues; it also had sequences of array voids with length from 10–50 mm and 3–12 mm wide, these are weak zones that interrupt stress transmission and its irregular shape absorb energy at the perimeter generating micro tangential tensions that perk stress to detach the plates; it also exhibited a flat white coloration whose means the polymeric matrix has been under tension stress. Over the beam surface, it observes a shading tone over the beam surface too, which is a polymer residue, confirming the epoxy matrix was pulled apart by shear strength. The predominance of those tones over the plate and beam surfaces suggests a low adherence between phases, as shown in Fig. 5.

Fig. 4. Failure by compression shear: transmission of shear stress through the beam to carbon fiber plate: detaching of reinforcement plates.

Fig. 5. Surfaces of beam and carbon fiber plate after the flexural test.

To measuring carbon fiber influence in ductility, it used moment-curvature (M−φ), considering that CF layers work as reinforcement bars. It used the Eqs. (1)–(3) [26] to calculate the theoretical FRP bending moments whose value was added to the calculations for M−φ.

$$\varphi Mn = \varphi(Mns + \Psi\ Mnf) \tag{1}$$

$$Mns = Asfs[d(\beta 1kd/2)] \tag{2}$$

$$\Psi Mnf:\ 0.85\{Mnf = Af\ ffe[h(\beta 1kd/2)]\} \tag{3}$$

Considering the experimental results of Table 4, the moment-curvature values were evaluated considering only 25% of FRP bending moment because of shear failure. The average results of all specimens are shown in Table 5.

Table 5. M−φ Average results

	Control	RC + 1 FRP	RC + 2 FRP	RC + 3 FRP
Mcr	27.6 T-cm	87.5 T-cm	73.8 T-cm	99.4 T-cm
φMcr	0.000092 rad/cm	0.000180 rad/cm	0.000132 rad/cm	0.000118 rad/cm
My	77.5 T-cm	109.2 T-cm	109.7 T-cm	139.6 T-cm
φy	0.000676 rad/cm	0.000677 rad/cm	0.000657 rad/cm	0.000658 rad/cm
Mu	81.9 T-cm	113.4 T-cm	113.7 T-cm	143.6 T-cm
φu	0.001996 rad/cm	0.001997 rad/cm	0.001741 rad/cm	0.001741 rad/cm

According to the values in Table 5, the moment after cracking (Mcr) in FRP beams with one and two CF plates has a rotation with an effective aperture than Control, over 45–95%; in the case of specimens with three CF plates aperture is around 30%. These research moments had similar behavior about rotation, but with higher load performance on FRP beams; when the reinforced sections cracked, steel rebars started to yield considering the contribution of the carbon fiber. The control and CR + 1FRP sections exhibited a similar and major aperture angle at failure level than CR + 2FRP and CR + 3FRP, as presented in Fig. 6.

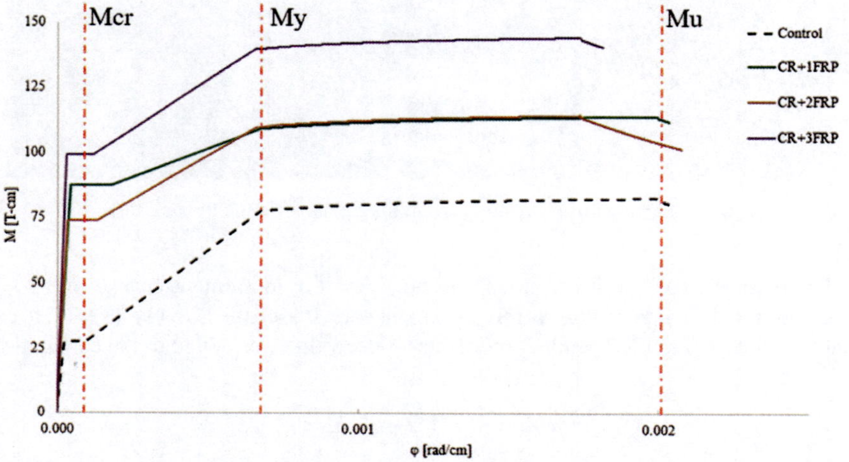

Fig. 6. Moment-curvature diagrams. Moments and rotation of FRP beams are in contrast with Control specimens (red guides).

In Fig. 6, the curve of the RC + 2FRP lot presented an uncommon development in correspondence with the FRP trend: load capacity is lower than beams with one CF plate, and it also fails before RC + 1FRP. This condition showed a manufacturing deficiency.

4 Conclusions

Carbon fiber reinforcement exhibited an accurate performance for load capacity. Its tensile strength made the concrete beam absorb three times the energy stress before it cracked, then it fluent simultaneously with steel bars until the CF plate is detached. However, under that assumption on these tests, carbon fiber induces a fragile behavior until concrete crack adapts to steel flown so FRP only provides ductility after concrete fail.

Given the results that RC + 2FRP lot presented, an inaccurate strain development can be concluded that they can not consider such representative. These values are contrasted only because the trend after concrete failure corresponds with CF contribution's previous assumption.

The consistent and similitudes behaviour that all models showed confirms the importance of the quantity and separation of horizontal steel rebar about the shear failure. However, even the CF plates were designed for flexural strength. Furthermore, the results demonstrate an apport from the fiber to absorb and dispel shear forces, which was unexpected because most of the literature refers to FRP reinforcement for these phenomena separately. According to the previous, it can be stablished the following preliminary conclusions:

- Carbon fiber could control plastic flow by restricting the concrete displacement but reducing the ductility.
- The FRP reinforcement was considered for flexural deformations, but it exhibited an increased shear resistance ratio of about 40%.
- The moment-curvature of FRP beams had a similar development as Control specimens cracked: then carbon fiber increased ductility of reinforcement steel bars.
- The angle of aperture at cracking in FRP specimens depends on the number of carbon fiber plates. Contrasting Control items with one CF plate, the beams have a rotation angle over 50%, indicating the shear failure was fragile. On the other hand, the three CF plate specimens reach an aperture of 30%, which supposed a better energy absorption. A significant displacement restriction is given by increasing the CF plates, although rigidity is improved simultaneously.

In order to reach bending deformation in FRP items, its planned to perform another experimental campaign considering the gained experience into manufacture. Finally, following the main objective, it pretends to construct a virtual model based on a non-destructive probe of an iconic concrete frame building to contrast the collected data.

References

1. Brothers, M.: Steel, Iron and others Products for Building and Bridges. Milliken Brothers. Estados Unidos, New York (1905)

2. Anza, A.M.: Los Procedimiento de Cimentación en el Valle de México. El Arte y la Cien-cia, pp. 117–119 (1900)
3. Marsal Raúl, J., Mazari, M.: El Subsuelo de la Ciudad de México, Ciudad de México: Instituto de Ingeniería UNAM (2016)
4. Martínez López, M.R., Mendoza, C.: Acoplamiento sismogénico en la zona de subducción de Michoacán-Colima-Jalisco, México, vol. 68, no. 2, pp. 199–214 (2016)
5. Moiseevich Zobin, V.: Los terremotos y sus peligros: ¿cómo sobrevivir a ellos?, Ciudad de México: Universidad de Colíma (2004)
6. SSN: Zona de subducción mexicana y su potencial para un sismo mayor (2019). http://www.ssn.unam.mx/jsp/reportesEspeciales/sismoMayor.pdf
7. Cuevas, J.A.: Datos relativos al proyecto para el nuevo edificio de la Lotería Nacional. Biblioteca de la Lotería Nacional, Ciudad de México (1933)
8. Groenenwold Ramos Arizpe, K.: Anexo N. 7, Biblioteca de la Lotería Nacional, Ciudad de México (1944)
9. Cuevas, J.A.: The floating foundation of the new building of National Lottery of Mexico: an actual size study of the deforma-tion of a flocculent-structured deep soil. In: First International Conference on Soil Mechanics and Foundation Engineering, Massachusetts, Estados Unidos (1936)
10. Cuevas, J.A.: Especificaciones relativas a las obras de concreto y concreto armado para el nuevo edificio de la Lotería Nacional. Biblioteca de la Lotería Nacional, Ciudad de México (1934)
11. Geotecnia: José Antonio Cuevas, impulsor de la mecánica de suelos, Geotecnia, no 239, pp. 7–8, Marzo-Mayo (2016)
12. Shiordia C.J.: Historia de la construcción del edificio "Moro". Biblioteca de la Lotería Nacional, Ciudad de México (1996)
13. ACI, ACI: A century of progress, p. 127. American Concrete Institute (2004)
14. Martínez Lizárraga, A.E.: Historia de la Arquitectura Latinoamericana: la Integración Plástica como Modelo de Identidad Arquitectónica Mexicanista Moderna. Heurística (18), 214–225 (2015)
15. Arquitectura México: La Ciudad Universitaria de México. Arquitectura México, no. 32, pp. 99–107, Octubre 1950
16. Chávez, D.C.M.: La Ciudad Universitaria en la arquitectura mexicana contemporánea. La revista de la Universidad de México, vol. Número extraordinario, no. Número extraordinar-io, pp. 12–19 (1994)
17. Moehle, J.P.: State of research on seismic retrofit of concrete building structures in the US (1997). http://www.ironwarrior.org/ARE/Lateral_Forces/us
18. Arana, J.L., González, J.J.: Mecánica de fractura, País Vasco: Universidad del País Vasco (2003)
19. Sanjuan, B.M.Á., Castro, B.P.: Acción de los agentes químicos y físicos sobre el concre-to. CDMX: IMCyC (2001)
20. Juárez Luna, G., Tenorio Montero, E.: Formulación de elementos finitos viga-columna con discontinuidades embebidas para modelar daño en elementos prismáticos de concreto refor-zado. Ingenieria Sísmica, no. 96, Enero-Junio (2017)
21. Metha, P.K., Monteiro, J.P.: Concreto: Estructura, propiedades y materiales, Ciudad de Mexico: Instituto Mexicano del Cemento y el Concreto, A.C. (1998)
22. Park, R., Paulay, T.: Estructuras de Concreto Reforzado, Ciudad de México: Editorial Limusa. S.A. de C.V (1988)
23. López López, A., Porras Navarro González, D., Rosado Tamariz, E.: Metodología para el Análisis de Fatiga de Estructuras de Concreto Reforzado. Sociedad Méxicana de Ingenie-ría Estructural, Ciudad de México (2010)

24. Zanuy Sánchez, C.: Análisis seccional de elementos de hormigón armado sometidos a fatiga, incluyendo secciones entre fisuras. Universidad Politécnica de Medrid (2008)
25. Al-Mahaidi, R., Kalfat, R.: Rehabilitation of Concrete Structures with Fiber-Reinforced Polymer. Butterworth-Heinemann (2018)
26. ACI: Guide for the Design and Construction of Externally Bonded FRP Systems for Strengthening Concrete Structures, vols. %1 de %2ACI 440.2R-08, American Concrete Institute (2008)
27. Hull, D.: Materiales Compuestos. Editorial Reverté, España (2010)
28. Park, S.-J.: Carbon Fibers. Springer, Korea (2018)
29. Teng, J., Chen, J., Smith, S., Lam, L.: FRP Strengthened RC Structures. Wiley, England (2002)
30. Chanda, M.: Plastics Technology Handbook, Florida. CRC Press, Taylor & Francis Group, USA (2018)
31. CEMEX: Manuel del Constructor. CEMEX concretos
32. IMCyC. http://imcyc.com, http://imcyc.com/ct2009/guiadeproductosjun09/concreto/holcim apasco.htm
33. Organismo Nacional de Normalización y Certificación de la Construcción y la Edificación, S.C.: Industria de la Construcción -Concreto- Elaboración y Curado de Especímenes en el Laboratorio. ONNCCE, Ciudad de México (2016)
34. American Concrete Institute: C-192/C-192M 02 Making and Curing Concrete Test Specimens in the Laboratory. ASTM (2002)
35. Organismo Nacional de Normalización y Cetificación de la Construcción y la Edificación: Industria de la Construcción-Concteto-Determinación de la resistencia a la flexión del concreto usando una viga simple con carga en los tercios del claro. ONNCCE, Ciudad de México (2015)
36. American Concrete Institute: C-293-02 Flexural Strength of Concrete (Using Simple Beam With Center-Point Loading). ASTM (2002)

Role of Oxygen and Humidity in the Reinforcement Corrosion

Carmen Andrade(✉)

International Center for Numerical Methods in Engineering (CIMNE), Barcelona, Spain
candrade@cimne.upc.edu

Abstract. Corrosion of reinforcement is produced when the carbonation front or the chloride threshold reaches the surface of the steel bars. The carbonation progresses when the concrete pores are not filled with water being prevented when these pores are highly water saturated. In the case of chlorides, reversely, the penetration is slower if the pores are not saturated, being negligible if the relative humidity is low. Degree of concrete saturation plays then a crucial role in the advance rate of the aggressive fronts. In addition to water, oxygen plays as well an important role in the steel depassivation stage. In carbonation case, oxygen is available as the pores are not filled with water, while in the case of the chlorides is the opposite and the amount of oxygen is that dissolved in the pore solution. In the stage of corrosion propagation, the oxygen is much less critical due to the autocatalytic nature of the corrosion, being however the humidity an essential element. These aspects will be commented and illustrated with results. The correct interpretation of the role of humidity and oxygen is an essential element in the assessment of existing structures.

Keywords: Reinforcement · Corrosion · Propagation · Oxygen · Humidity

1 Introduction

Corrosion of reinforcement in recent years is mentioned to be the motivation of numerous papers because concrete durability is becoming of primary interest once the concrete strength is easily provided. Lack of durability and shortening of performance has economic consequences and should be an essential component of the sustainability and the lowering of carbon footprint of cementitious materials because carbon footprint in terms of the resources spent should be calculated taking into account the whole life cycle and not only considering the CO_2 released during the material fabrication. As longer is the performance in service, smaller would be the impact in the amount of raw materials used in the construction sector.

This lecture is related to some fundamentals of the corrosion of reinforcement that need continuous updating in order not misinterpreting the reasons behind a lack of durability due to steel corrosion. The updating is necessary providing the fundamentals were established many years ago [1–8] and recent researchers do not consult the pioneering work assuming it has been overcome by new research. However, the fundamentals

J. I. Escalante-Garcia et al. (Eds.): RW 2021, RILEM Bookseries 40, pp. 316–325, 2023.
https://doi.org/10.1007/978-3-031-21735-7_35

of steel corrosion are still valid and old papers are sometimes much clear and free of errors while some more recent ones have re-interpreted the meaning of several facts and propagate errors [9, 10] that should be avoided for future generations.

In this lecture after the description of the corrosion cell and the role of microcells in the process, it will be commented the role of oxygen in the corrosion reaction and the role of water contained in the concrete pores.

1.1 Anodic and Cathodic Reaction of the Corrosion Process

The corrosion of metals is named "dry" if it happens at elevated temperature where water is not stable and the metal is forming the oxides by direct reaction with the oxygen, or is named "aqueous corrosion" when water is present in the environment and the process is of electrochemical nature with simultaneous involvement of electron exchanges. It is known that in the case of aqueous corrosion there are formed two regions "anode and cathode" on the metal surface, where the metal dissolves in the anodic zones to form ions, which liberate electrons which circulates through the metal to be captured by a substance which is "reduced" in the cathodic zones by forming a new compound. This compound interacts with the metal ions and the water molecules to form oxides or new products which can or not precipitate.

The anodic reaction is always the formation of metal ions as indicated in Eq. 1.

$$Fe^0 \rightarrow Fe^{+2} + 2\,e^- \tag{1}$$

The cathodic reactions are more complex as it can be.

- the reduction of the protons (H^*) in neutral and acid media to form hydrogen gas:

$$2H^+ + 2e^- \rightarrow H_2 \uparrow \text{(gas)} \tag{2}$$

- the reduction of the oxygen dissolved in the water in neutral and basic media to form hydroxides:

$$O_2 + 2H_2O + 4e^- \rightarrow 4\,OH^- \tag{3}$$

However, depending on the potential, the cathodic reaction can be also the decomposition of the water to give also hydroxides:

$$2H_2O + 4e^- \rightarrow 2OH^- + H_2 \uparrow \text{(gas)} \tag{4}$$

The basic electrochemical features of the corrosion were established by Evans [1] and experimentally shown through his famous "water drop experiment" which was more widely commented in [11].

An essential part of the matter is that it is electrically neutral and then, the anodic current equals the cathodic one and it is equal what is named the "corrosion current" or "corrosion rate". If divided by the corresponding areas or its addition, it can be formulated as current density:

$$i_a = i_c = i_{corr} \tag{5}$$

This basic process evolves with time and the local conditions change. This evolution is an aspect usually neglected in the case of reinforcement corrosion. It has a key step well described decades ago by [2–4] among other. This key aspect is that the anodic zone acidifies due to the immediate reaction of the Fe^{+2} ions with the water molecules following the (simplified) reactions:

$$Fe^{+2} + H_2O \rightarrow FeOH^+ + H^+ \tag{6}$$

$$FeOH^+ + H_2O \rightarrow Fe(OH)_2 + H^+ \tag{7}$$

These key steps are crucial in concrete where the protons are neutralized by the hydroxides of the high pH value of the pore solution. The corrosion in concrete does not develops further if the pore solution alkalinity neutralizes (by forming water molecules) the acidity generated during the first corrosion phases. In this neutralization is based the known "buffering capacity" of the portlandite. Thus, the portlandite is in equilibrium with its ions;

$$Ca(OH)_2 \leftrightarrow Ca^{2+} + OH^- \tag{8}$$

Thus, the hydroxides neutralize (to give water molecules) the H^+ formed from the reaction of the Fe^{*2} ion with the water and further new portlandite dissolves following its solubility product until all portlandite is dissolved. When the portlandite is exhausted, it is the C-S-H gel which starts to be dissolved. The surrounding of the anodic zones are steadily acidified until an equilibrium is reached and the corrosion zone does not extends more.

The main consequence of this acidity is that the cathodic reaction can be the reduction of the H^+ or of the Fe^{*3} ions generated by further oxidation of the Fe^{+2}. From that moment, the oxygen does not play a key role anymore. The corrosion progresses by its own products and that is why it is said it is an "autocatalytic" process.

That is, another feature of the corrosion process which is critical is that the ion Fe^{+2} is oxidized to Fe^{+3} following a redox equilibrium between both ions which can be as well the reversal [12]:

$$\begin{aligned} &2Fe^{+2} + 1/2O_2 + H_2O \leftrightarrow 2Fe^{+3} + 2OH^- + 2e^- \\ &Fe^{+2} \leftrightarrow Fe^{+3} + e^- \end{aligned} \tag{9}$$

Then, if Fe^{+3} ions are formed, their reduction can be also feeding the overall cathodic reaction [13].

In summary, the corrosion process is dynamic and involves several simultaneous chemical changes that are related to the buffering capacity of the cement paste phases, and depending on the corrosion rate, the corrosion stops or develops with no more need of the presence of oxygen as reactive product in the corrosion process. These simultaneous side reactions are usually neglected in many of the present papers of reinforcement corrosion while they are critical when trying to model the basic corrosion process.

1.2 Micro and Macrocells

Anode and cathode formation is the basic of the aqueous corrosion, however the anodic zones are not usually pure anodes but in the corroding zones there are micro-anodes and micro-cathodes as Fig. 1 shows. That is, if the corrosion is homogenous, as is in the carbonated zones, the corrosion is due to the action of multitude microcells but if pits are formed, as in chloride contaminated concrete, inside the pits (the corroding zones) also micro-anodes and micro-cathodes (microcells) are formed. This means that the pit can progress alone from any contribution of the surrounding passive (cathodic) surface. The connection with this cathodic area will increase the corrosion rate, but the contribution of this cathodic zone to the overall corrosion rate would depend on the pH inside the pit and the ratio of anodic/cathodic areas [14].

This fact has several consequences among which can be summarized the following:

- The galvanic current between a corroding zone and a passive one is not the corrosion current, but only a fraction of it. The precise fraction would depend on the particular geometrical arrangement and of the particularities of the electrolyte (amount of chlorides, pH, etc.).
- Then a galvanic sensor which measures the current between two metal pieces is not informing of the corrosion rate but only of the galvanic current. The galvanic sensors can detect corrosion development but cannot give accurate values of the corrosion rates.

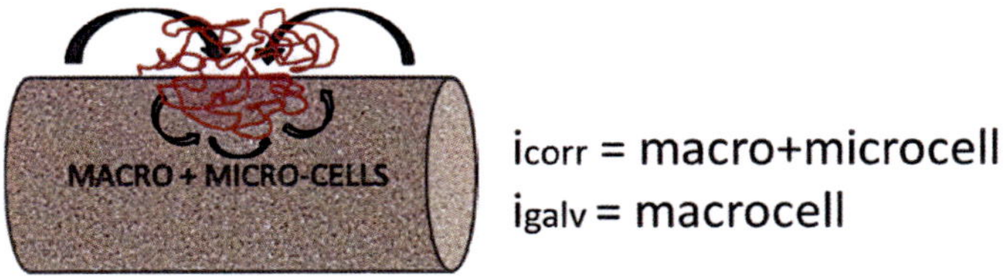

Fig. 1. There are microcells inside a pit. The external passive regions contribute to the cathodic current as a fraction of the whole corrosion rate (Eq. 5).

2 Role of Oxygen in the Corrosion of Steel in Concrete

Oxygen is a gas which is dissolved in water in a quantity of around 8 ppm. The fishes can breathe thanks to this content and therefore where there are fishes or plants is oxygen in enough quantity, in spite how deep into the water the place is. If the water contains electrolytes, the concentration of oxygen may decrease. In chloride solutions for instance the maximum amount of oxygen is observed for the chloride concentration of the sea water (around 30 g/L of NaCl). As a general rule as more concentrated is the solution, the water can afford less amount of oxygen.

Which is the role of the oxygen in the reinforcement corrosion? It is critical in the corrosion initiation while its importance decay or is nil during corrosion propagation.

The oxygen is crucial in the corrosion initiation because it is the responsible of the oxide layer exiting in the steel surface when in contact to the atmosphere or embedded in concrete (passive layer).

In solutions simulating the pore solution the oxygen content is between 4–8 ppm depending on its concentration. This is enough to passivate the steel in solution and when embedded in concrete. This is noticed by the corrosion potential which evolves to positive values when passivation is progressing with age (as shown in Fig. 2). If the oxygen decreases, the potential evolves to negative values and then, the corrosion potential in passivated steel is a clear indicator of the oxygen presence in the pore solution. That is the oxygen content fixes the level of the corrosion potential in absence of depassivator agents (chlorides).

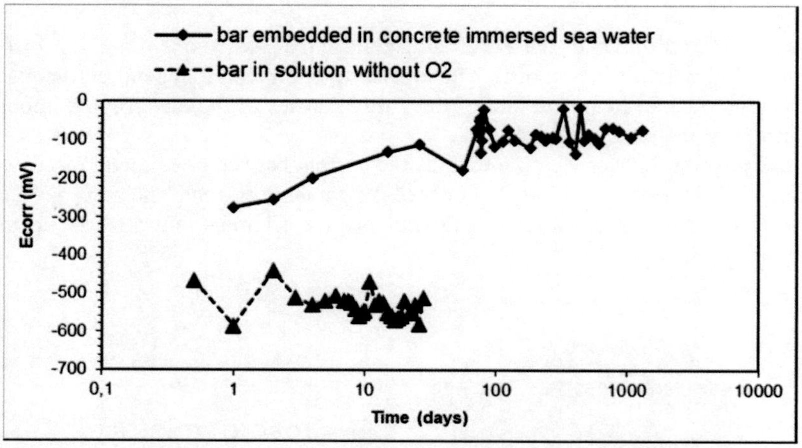

Fig. 2. Corrosion potential of a bar embedded in mortar submerged in saline water (diamond symbol) showing a trend to positive values of the potential (not corroding and not lack of oxygen). Other bar (triangle symbol) submerged in a non-aerated solution showing, on the opposite, much cathodic values of the corrosion potential due to the low oxygen concentration.

As a consequence of the local acidification, in the metals suffering pitting corrosion, the corrosion potential shifts to cathodic values and it is not anymore a good indicator of the oxygen content, because the potential measured is the result of the new anodic and the previous cathodic zones, and the anodic zones now have very negative potential as shown experimentally by Pourbaix [3].

The situation is somehow different when the concrete carbonates. As the pH drops due to the carbonation, the first acidification is produced by that carbonation which prevents the further neutralization by the portlandite (unless not all is carbonated). The evolution of the potential is as indicated in Fig. 3 [15] because it follows Nernst law of potential variation with pH, that, is it shifts 60 mV each pH unit. At the neutral pH the passivating film has disappeared and the corrosion has started with an equal amount of protons and hydroxides in the pore solution. The corrosion results homogenous with multiple microcells and acidifies more as the carbonate/bicarbonate equilibrium fixes the pH value of the pore solution. Existing protons in the pore solution and being the oxygen not necessary for maintaining the passive state, its role is less critical for the corrosion to progress.

Several tests were made by the author to study these phenomena on which here are presented some summary. From all the experiments performed [16, 17] we could conclude that:

- Oxygen and "aging" of the passivating film are responsible of the corrosion potential when the steel is passivated in alkaline solutions.
- The corrosion potential shifts to more cathodic values when the oxygen is reduced by passing nitrogen flow in an alkaline solution with passivated bars immersed.
- When the corrosion starts by adding chlorides, the afterwards removal of oxygen does not stop the corrosion or reduces the corrosion rate.
- In carbonated solution where the CO_2 gas is flowed continuously, there is not oxygen but the corrosion is very active.

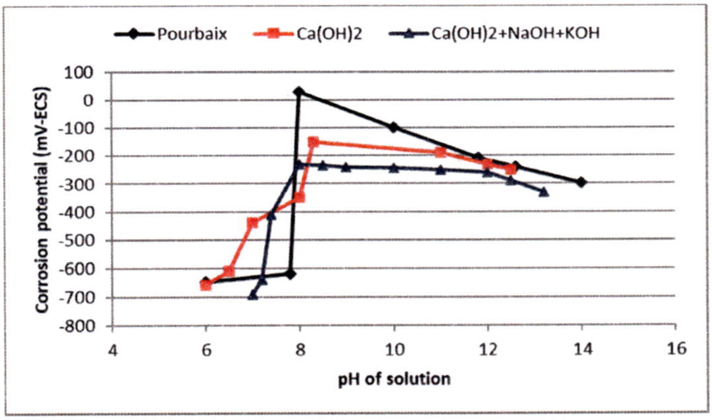

Fig. 3. The corrosion potential follows Nernst equation when carbonating the alkaline solution. The potential moves towards more anodic values every lowering of pH unit, but falls abruptly when depassivation occurs around pH = 8 [15].

All these evidences confirm that oxygen only is important for the corrosion initiation (as for cooperating in forming and maintaining the passive film which is noticed by the level of the corrosion potential and to oxidize the Fe^{+2} ions at the initial steps of corrosion) but not for the corrosion propagation. That is, the corrosion rate is not directly dependent on the amount of oxygen in the pore solution. The anodic current is sustained by other cathodic reaction with the addition of that of oxygen. The presence of oxygen can increase the corrosion rate, but there is not a direct relation to it. When modelling, to make the cathodic current dependent only of the oxygen flow through the cover depth is an incorrect assumption, as the corrosion is not directly dependent on the cover thickness.

3 Role of Moisture, Mechanisms of Corrosion Control in Concrete

In the Introduction it was mentioned the two kind of corrosion types: dry and aqueous: Then water, as enabling the formation of electrolytes, is an essential substance to produce corrosion. If there is not water, corrosion is stopped. This general principle is particularly important in porous materials or in atmospheric conditions where the degree of saturation can be variable. That is in solution, water is always covering the metallic surface, however in the atmosphere the covering will depend on the relative humidity of the air. In the porous materials the external relative humidity and absorption of water will control the degree of saturation.

In concrete, it has been experimentally verified decades ago that the degree of water saturation is the controlling parameter of the corrosion rate [18]. As indicated in the Fig. 4, a higher degree of saturation means a higher corrosion rate. This performance was confirmed by C. Page a bit later [19]. Further the confirmation has been made also in specimens held outdoor exposed to rain or sheltered form it [20].

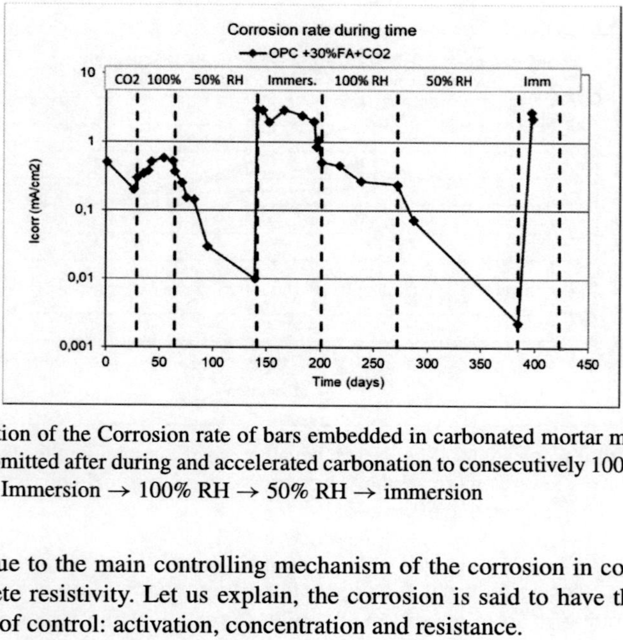

Fig. 4. Evolution of the Corrosion rate of bars embedded in carbonated mortar made with OPC + 20% FA submitted after during and accelerated carbonation to consecutively 100% RH → 50% RH → Partial Immersion → 100% RH → 50% RH → immersion

This is due to the main controlling mechanism of the corrosion in concrete is due to the concrete resistivity. Let us explain, the corrosion is said to have three possible mechanisms of control: activation, concentration and resistance.

- Activation control is related to the "activation energy" (energy barrier for the process) needed for the metal to become the metallic ion. This energy barrier is specific to each metal/system. It depends on the type of metal being those more noble or passivated which have a higher activation energy for starting to corrode. In the case of concrete, the activation energy is depending on the amount of chlorides and on the pH value (carbonation).

- Concentration control happens when the amount of ions and corrosion products difficult the generation of new corrosion products. In concrete having micropores, it might happen either in carbonated specimens or containing chlorides, but in fully saturated concrete the ions formed and the oxides can diffuse, and it seems that concentration is not so a critical controlling mechanism as are the activation energy and the electrical resistance.
- Resistance control seems the most important one in concrete once the corrosion is active. It would depend on the amount of water in the pores that the corrosion rate is higher or smaller. The degree of saturation is detected through the measurement of the electrical resistance (or resistivity) of the concrete: as much dry is the concrete higher is the electrical resistance and smaller is the corrosion rate. This is why there is so direct relation between the corrosion rate and the concrete electrical resistance (or resistivity) (Fig. 5) as shown also many years ago in [21]. This relation has been always confirmed [23] and recently for example by [22].

In the case of concrete, then it can be deduced that the Activation control is important when the steel is passive, but as soon as corrosion starts, the control is by Resistance. Perhaps Resistance plus Concentration, because when the concrete is dry, the oxides cannot diffuse because the lack of water. However, the primary mechanism of control seems to be the resistance.

The relation between I_{corr} and the resistivity enabled us to formulate the following equation which is the base of a model deduced from the Fig. 5 [23] for predicting the corrosion evolution and a manner to check the validity of corrosion rate measurements made on site.

$$I_{corr}\left(\mu A/cm^2\right) = \frac{26\,(mV)}{\rho(K\Omega \cdot cm)} \tag{11}$$

Please notice the similitude between this expression and that given in [18] for the relation between the Polarization Resistance and the Corrosion rate. It is out of the scope of present work to explain the reasons of this similitude [24] (but it reinforces the idea that the corrosion rate it primarily controlled by the electrical resistance).

Then, the water is the key parameter in the corrosion process and the relation between Resistivity and Corrosion rate makes this last parameter also directly dependent on the degree of water saturation or, in volumetric terms (as is expressed the resistivity), of the volumetric fraction of water (f_v). Then, substituting the Resistivity in the Eq. 11 by the expression found for the f_v [25] it results:

$$I_{corr}\left(\mu A/cm^2\right) = \frac{26\,(mV)}{\rho_0 \cdot f_v^{-\tau}} \tag{12}$$

where tau (τ) is the tortuosity factor with a value around 2 in the concretes [25] and ρ_0 is the resistivity at 100% of water saturation.

4 Comments as Conclusions

In the reinforcement corrosion, oxygen and water content of concrete play a key role, being the most important the water or degree of saturation. It is necessary a water

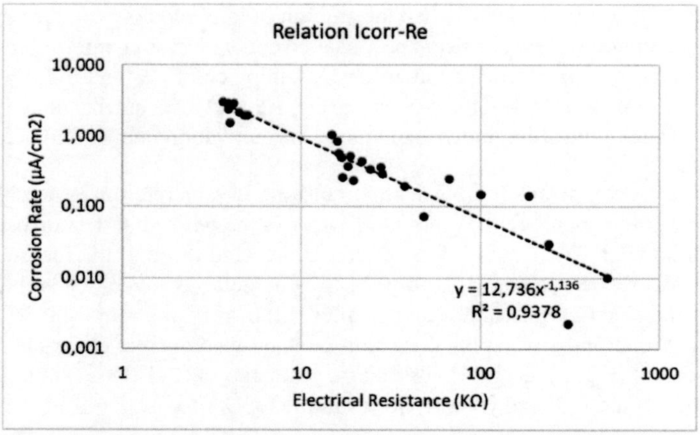

Fig. 5. Relation between the corrosion rate and their electrical resistance of the specimens represented in Fig. 4.

saturation threshold in each concrete to produce corrosion which could be significant in quantitative terms. The threshold is not unique because each concrete has a different pore size distribution, which is also unique in each concrete. The water saturation is well characterized by measuring the resistivity and that is why the corrosion rate is inversely related to the resistivity and directly dependent to the water saturation degree.

Regarding the role of the oxygen, it has to be reminded that all water contains around 8 ppm of that gas and therefore it is in the mixing water and then, in the pore solution. This is the amount of oxygen that sustains corrosion in solutions. Being the corrosion potential of the steel, when passivated, dependent on its content, experiments have shown that even with very low amount of oxygen, the corrosion can develop and progress. Oxygen is important for the corrosion initiation, but not for the corrosion propagation.

References

1. Evans, Ulick Richardson.Metallic corrosion, passivity and protection Longmans Green and Co., New York (1948)
2. Hoar T.P.: Electrode Processes. Ed. Butterworths, London (1961)
3. Pourbaix, M.: Lectures in Electrochemical Corrosion. Plenum Press, New York and London (1973)
4. Galvele, J.: Pitting Corrosion-Treatise on Materials Science and Technology, vol. 23, pp. 1–57 (1983)
5. Page, C.L., Treadaway, K.W.J.: Aspects of the electrochemistry of steel in concrete. Nature **297**(5862), 109–115 (1982)
6. Gouda, V.K.: Anodic polarization measurements of corrosion and corrosion inhibition of steel in concrete. Br. Corros. J. 136 (1966)
7. Tuutti, K.: Corrosion of steel in concrete. Swedish Cement and Concrete Institute (CBI) no. 4–82. Stockholm (1982)
8. Andrade, C., Rodriguez-Maribona, I.: Macrocell versus microcell corrosion of reinforcement placed in parallel. Paper no 194, Corrosion- NACE (1992)

9. Liu, T., Weyers, R.: Modeling the time-to-corrosion cracking in chloride contaminated concrete structures. ACI Mater. J. **95**(6) (1998)
10. Dangla, P., Dridi, W.: Rebar corrosion in carbonated concrete exposed to variable humidity conditions. Interpretation of Tuutti's curve. Corros. Sci. **51**, 1747–1756 (2009)
11. Andrade, C.: Propagation of reinforcement corrosion: principles, testing and modelling. Mater. Struct. (2018)
12. Stratmann, M., Bohnenkamp, K., Engell, H.J.: Electrochemical study of phase-transitions in rust layers. Corros. Sci. **23**(9), 969–985 (1983)
13. Alonso, C., Andrade, C., Izquierdo, M., Novoa, X.R., Pérez, M.C.: Effect of protective oxide scales in the macrogalvanic behaviour of concrete reinforcements. Corros. Sci. **40**(8) 1379–1389 (1998)
14. Andrade, C., Rodriguez-Maribona, I., Feliu, S., Gonzalez, J.A., Feliu, S. Jr.: The effect of macrocells between active and passive areas of steel reinforcements. Corros. Sci. **33**(2), 237–249 (1992)
15. Andrade, C., Alonso, C., Santos, P., Macias, A.: Corrosion behaviour of steel during accelerated carbonation of solutions which simulate pore concrete solution. In: 8th International Congress of Cement chemistry, Rio de Janeiro, Brazil Paper 4.3 (1986)
16. Andrade, C., Alonso, C., García, A.M.: Oxygen availability in the corrosion of reinforcements. Adv. Cement Res. 3(1), 127–132 (1990)
17. Toro, L., Andrade, C., Fullea, J., Martinez, I., Rebolledo, N.: Steel corrosion process in a non oxygenated media. In: RILEM 4th International PhD RILEM Workshop, November, Madrid, Spain (2010)
18. González, J.A., Andrade, C.: Effect of carbonation, chlorides and relative ambient humidity on the corrosion of galvanized rebars embedded in concrete. Br. Corros. J. 17(1), 21–28 (1982)
19. Page, C., Glass, G.K., Page, C.L., Short, N.R.: Factors affecting the corrosion rate of steel in carbonated mortars. Corros. Sci. **32**(12), 1283–1294 (1991)
20. Sarria, J., Andrade, C., Castillo, A.: Water content of concrete in natural atmospheres and its impact in the corrosion parameters. In: International RILEM Conference on Material Science, MATSCI, Aachen, Germany, vol. II, pp. 43–51. Springer Edition (2010)
21. Alonso, C., Andrade, C., González, J.A.: Relation between concrete resistivity and corrosion rate of the reinforcements in carbonated mortar made with several cement types, Cement Concrete Res. **18**, 687–698 (1988)
22. Hornbostel, K., Larsen, C.K., Geiker, M.R.: Relationship between concrete resistivity and corrosion rate—a literature review. Cement Concrete Compos. **39**, 60–72 (2013)
23. Andrade, C., Fullea J., Alonso, C.: The use of the graph corrosion rate-resistivity in the measurement of the corrosion current. In: Rilem Proceedings no. 18: Measurement and Interpretation of the on-Site Corrosion Rate, pp. 157–165 (1999)
24. Andrade, C.: From diffusivity to corrosion rate through concrete resistivity values. In: Workshop on Corrosion Control of Concrete structures C^3S- Rilem Week, Madras, Chennai, India (2017)
25. Andrade, C., Rebolledo, N., Menéndez, E., Perez, R., Baz, M.: Characterization through mercury porosimetry of tortuosity factors of concretes containing mineral additions. In: XIII DCBM- International Conference on Durability of Building Materials and Components, San Paulo, Brazil, September (2014)

Experimental Investigation on the Influence of Partial Immersion and Drying Cycles on Hemp Concrete Properties

M. Maaroufi(✉), A. Bourdot, M. El Assaad, and K. Abahri

LMT - Laboratoire de Mécanique et Technologie, Université Paris-Saclay, ENS Paris-Saclay, CNRS, 91190 Gif-sur-Yvette, France

maroua.maaroufi@ens-paris-saclay.fr

Abstract. The construction sector has a significant environmental impact, since it is the largest energy consumer and is responsible for 25% of greenhouse gas emissions. Vegetal concretes represent a good alternative to reduce the environmental footprint of building materials, because of their thermal insulation properties and the use of vegetal waste. In addition, they have the outstanding environmental quality of being carbon negative. Hemp lime concrete is the most used bio-based material in building sector and its potential as an environmental-friendly material was quickly realized. It is a relatively new building material and its use is limited by the lack of data concerning its durability and the evolution of its properties over time.

This paper focuses on the influence of an accelerated aging of hemp lime concrete on its hydric and thermal properties. Samples will undergo several cycles of partial immersion and drying and will be characterized before and after the accelerated aging. Partial immersion was preferred to complete immersion, as it is more representative of the real climatic conditions. The thermal conductivity and the capillary absorption coefficient will be determined. Microscopic observations of the surface topology will also be carried on, in order to evaluate the microstructure alteration.

Keywords: Hemp lime concrete · Durability · Partial immersion and drying · Hygrothermal properties

1 Introduction

Construction sector is a huge energy consumer and is responsible for a large amount of CO_2 emissions, thus becoming a source of concern. This is mainly due to the poor thermal insulation of buildings that requires higher energy demands and therefore generates more greenhouse gas emissions. Bio-based materials have proven to be more efficient in terms of thermal and hygroscopic properties, compared to usual insulation materials. Indeed, they offer many advantages concerning energy needs, thermal comfort and environmental impact as they also allow reducing the carbon footprint [1–3].

J. I. Escalante-Garcia et al. (Eds.): RW 2021, RILEM Bookseries 40, pp. 326–334, 2023.
https://doi.org/10.1007/978-3-031-21735-7_36

Bio-based materials are increasingly used in the construction sector, as they have interesting hygrothermal and acoustic properties [4, 5]. It is generally composed of vegetal aggregates, called shiv particles, coated by a mineral binder, usually based on lime in order to maintain the hygroscopic properties of the resulting material.

Hemp concrete is one of the most used bio-based materials in the construction sector. It has interesting thermal and phonic insulation performances and moisture regulation properties that works as a moisture buffer [6–8].

The evolution of the properties of hemp lime concretes is highly dependent on the climatic conditions to which they are exposed. Indeed, bio-based materials are highly affected by relative humidity, temperature and liquid water, as they can cause swelling, chemical reactions and microstructural variations [9]. Recently, studies were interested in durability of hemp concrete, through accelerated aging tests. These tests consist in varying the relative humidity, the temperature and water immersion [10–12]. Results show that variations in functional properties are observed, such as an increase of thermal conductivity and water vapor sorption capacity and a decrease of acoustic performance [13], but also a decrease in compressive strength [12]. These aging tests are highly aggressive, as they are based on immerging the samples and/or putting them at a very high or very low temperature. They do not represent the real conditions in continental climate of buildings and one could wonder if it corresponds to a real aging of construction materials.

In this work, a different aging protocol is carried on to characterize the evolution of the hygrothermal properties of bio-based materials. In order to be more representative of the real aging of building materials, the aging tests will be based on partial immersion and drying at a relatively low temperature. Different samples of hemp lime concrete will be tested and their properties will be determined before and after the laboratory aging. The thermal conductivity, the capillary absorption coefficient and the moisture buffer capacity will be determined.

2 Materials and Methods

2.1 Materials

The material of interest in this work is hemp lime concrete. Hemp lime concrete is constituted of hemp shives, a lime-based binder and water. The mix proportions are presented in Table 1 and the water/binder ratio was adapted to the shives, with consideration of their water absorption coefficient. The used mixer is of type Eirich and the mixing lasts approximately 10 min.

Table 1. Mix proportions of hemp lime concrete

Material	Shives	Binder	Water
Hemp concrete	7 kg	35 kg	28 kg

The used hemp shives are provided by BIOFIB (from France) and they result from the defibration of hemp fibers, with no use of water or solvent. Their hygrothermal and physical properties were determined following the RILEM recommendations [14]. Their measured thermal conductivity is 64 ± 7 mW/m.K and their measured density is 107 ± 4 kg/m^3.

The water absorption of hemp shives was also determined following RILEM recommendations [14] and the results are presented in Fig. 1. The evolution of water content as a logarithmic function of time as proposed in literature [15] and two specific parameters were determined using Eq. (1).

$$W = IRA + K_1 \times ln(t) \quad (1)$$

W is the water absorption, IRA is the initial rate of absorption and K_1 is a parameter indicating the diffusion rate of water that is linked to the intrinsic porosity.

The obtained IRA after one minute is 231% and the parameter K_1 is 18 and these values are consistent with literature for other bio-aggregates, including different types of hemp shives [16]. The logarithmic evolution of the water absorption as a function of time is in good agreement with literature [15, 16], as it follows a linear evolution with a coefficient of determination R2 = 0,99.

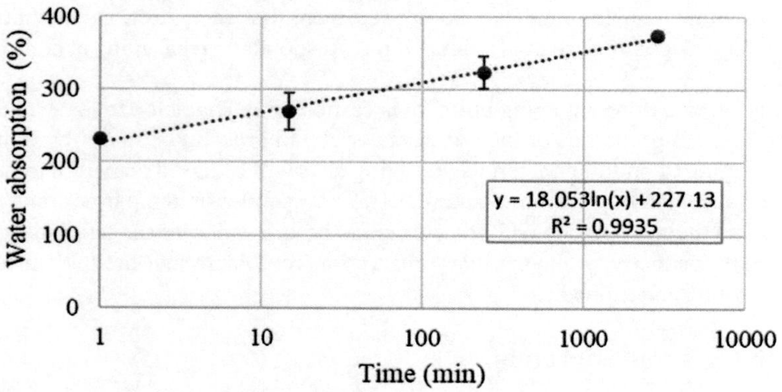

Fig. 1. Water absorption of hemp shives

Hemp lime concrete mixes were then put into different molds corresponding to the characterization tests. They were compacted on a vibrating table, to ensure homogenization and to eliminate the trapped air bubbles. After manufacturing, the samples were stored at 21 °C and 30% of relative humidity for 28 days. The reference samples are tested at 28 days and the aged samples are tested after the aging protocol.

2.2 Aging Protocol

The accelerated aging protocol is based on successive cycles of immersion and drying. The partial immersion was at room temperature (25 °C) and lasted 2 days, while the water immersion level was maintained at 8 mm (Fig. 2). The sides of the samples were covered to ensure a unidirectional flow, as shown in Fig. 2. The drying was carried on in a ventilated oven for 3 to 5 days at 40 °C. Samples were exposed to 8 cycles of immersion/drying before the determination of their hygrothermal properties.

Fig. 2. Accelerated aging setup (Partial immersion)

2.3 Methods

Thermal conductivity. The thermal conductivity was measured using the hot wire method [19] (Fig. 3a). Samples were conditioned at 21 °C and 30% until mass equilibrium before characterizing them. Three 70 × 70 × 280 mm^3 parallelepiped samples of each material were tested before and after the aging. The thermal conductivity measurements were carried on at a temperature of 22 °C and a relative humidity of 60%.

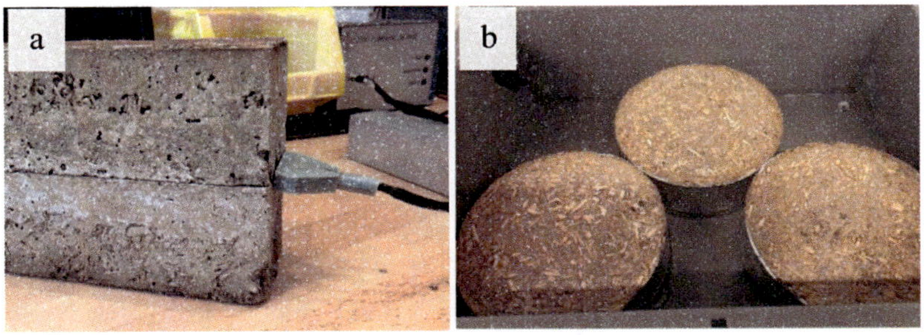

Fig. 3. (a) Thermal conductivity measurement; (b) Water absorption setup

Water absorption coefficient. The water absorption coefficient was determined following the standard given by AFPC-AFREM [17]. The test was carried on three cylindrical samples of hemp lime concrete and their dimensions were 150 mm of diameter and 50 mm of thickness. Preconditioning was based on drying the materials in a ventilated oven at 80 °C until mass variation is less than 0,1% over 24 h. Samples were covered on their sides to ensure a unidirectional water flow, as can be seen in Fig. 3b. The test was performed in a room controlled in temperature (22 °C) and relative humidity (55%).

The water absorption coefficient is calculated following the Eq. (2) and is expressed in kg/m^2.

$$C_a = \frac{M_x - M_0}{A} \tag{2}$$

M_x (kg) is the sample mass at a certain time, M_0 (kg) is the initial sample mass and A (m^2) is the exposed sample area.

Moisture Buffer Value. The moisture buffer value is an indicator of the materials capacity to moderate the variations of indoor humidity in buildings. It was determined following the NORDTEST method [18] that proposed a standardized quantity to characterize the moisture buffering capacity of a material. The test was performed on three cylindrical samples of 110 mm of diameter and 40 mm of thickness, as can be seen in Fig. 4. It consists in submitting the samples to multiple cycles of 8 h of high humidity (75%) followed intermittently by 16 h of low humidity (33%). Temperature is kept constant at 23 °C. Specimens were sealed on all but one surface and weighed or intermittently during the test. The test is over when the weight amplitude do not vary by more than 5% from day to day.

The moisture buffer value is obtained using the Eq. (3) and it is expressed in g/m^2.%HR)

$$MBV = \frac{\Delta m}{A * (RH_{High} - RH_{Low})} \tag{3}$$

Δm (g) is the mass loss/gain during a cycle, A (m^2) is the sample exposed area and RH_{High} (resp. RH_{Low}) is the high (resp. Low) relative humidity value (%).

Fig. 4. Hemp concrete samples in a climatic chamber for MBV measurements

3 Results

3.1 Thermal Conductivity

Thermal conductivity of hemp lime concrete was measured at the reference state after 28 days of manufacturing and after the accelerated aging. The obtained results are displayed in Fig. 5 and show a decrease of the thermal conductivity after the aging.

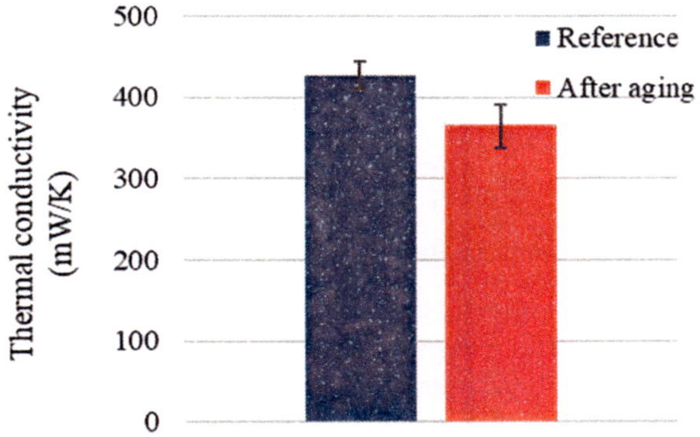

Fig. 5. Thermal conductivity of hemp lime concrete before and after aging

Indeed, thermal conductivity of hemp lime concrete at the reference state is 426 mW/m.K and it reaches only 364 mW/m.K after the accelerated aging, which corresponds to a decrease of 15%. This is due to the immersion and drying cycles that alter the microstructure of the material, thus creating more porosity and a lower thermal conductivity. This is consistent with previous works [12]. Their results show a decrease of 6,6% in thermal conductivity and link it to an increase in porosity because of the modification of the microstructure.

3.2 Water Absorption Coefficient

Water absorption of hemp lime concrete was determined after 0.5, 1, 2, 4, 8 and 24 h. Samples were tested both at the initial state and after the accelerated aging. Figure 6 presents the evolution of the capillary absorption over time. Results show that water absorption of hemp lime concrete decreases after the laboratory aging and the decrease is higher with time. The water absorption coefficient is defined as the water absorption after 24 h of immersion. It is equal to 4,26 kg/m^2 at the reference state and 3.02 kg/m^2 after aging, which corresponds to a decrease of 30%. This could be linked to a change in the materials microstructure that would induce less capillary flow.

3.3 Moisture Buffer Value

Moisture buffer value of hemp lime concrete was determined before and after the accelerated aging. The mass variation of the different samples during the test is presented in

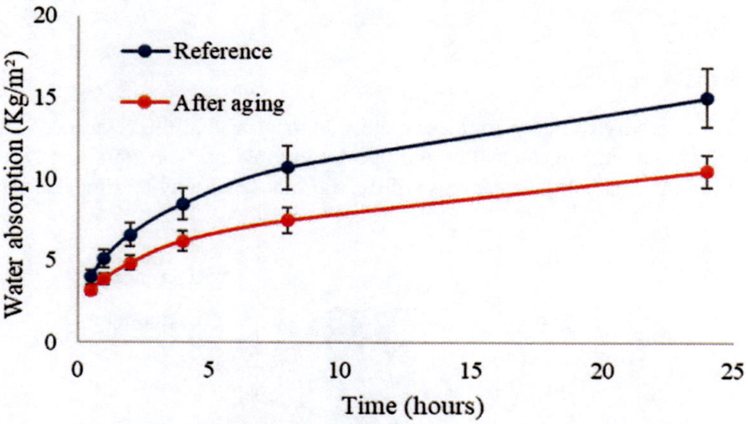

Fig. 6. Water absorption of hemp lime concrete before and after aging

Fig. 7, but only the last 5 cycles were presented, from 120 h. Results show that the mass loss/gain is slightly higher for the reference samples.

The moisture buffer capacity of hemp lime concrete is 1,7 g/m^2.%RH and is equal to 1,4 g/m^2.%RH after aging, which corresponds to a decrease of 16%. The results indicate a good moisture buffer capacity of hemp lime concrete, as it is mentioned in the classification proposed by the NORDTEST project [18], but the results are slightly lower than those found in previous works [7, 12]. The environmental aging decreases the moisture buffer value of the materials and thus alters their moisture regulation capacity. This may be due to the aging of the hemp shives that causes a degradation of their capacity to adsorb and restore moisture.

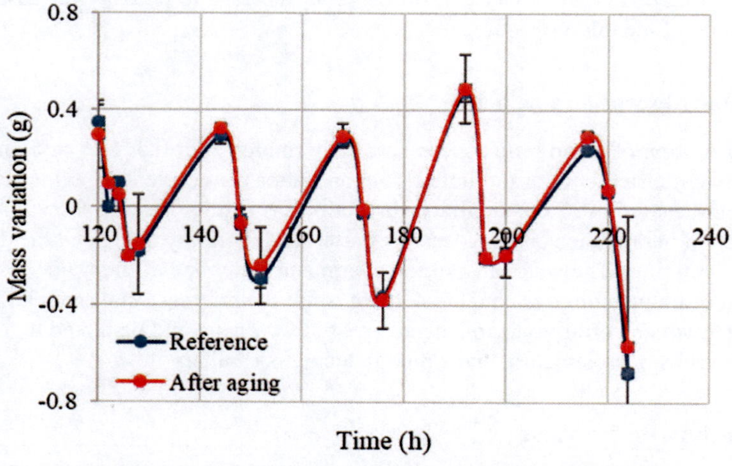

Fig. 7. Mass variation for MBV measurements before and after aging

4 Conclusion

In this work, durability of hemp lime concrete was investigated through a laboratory accelerated aging. A new aging protocol was followed, in order to get closer to real aging of building materials. The aging protocol was based on partial immersion only and drying at a moderate temperature. Multiple samples were put in 8 mm height of water during 2 days and then dried at 40 °C in a ventilated oven during 3 to 5 days. Their thermal conductivity, water absorption and moisture buffer capacity were determined before and after the accelerated aging, in order to evaluate its influence on materials durability. Results showed that the accelerated aging of hemp lime concrete decreased its thermal conductivity, its water absorption and its moisture buffer capacity.

Further work needs to be carried on, regarding the microscopic variations of the materials structure, but also its mechanical performance. Different compositions of hemp concrete but also flax concrete are going to be tested, in order to assess the influence of the used shives and/or binder on the hygrothermal and mechanical properties of the materials.

References

1. Osanyintola, O.F., Simonson, C.J.: Moisture buffering capacity of hygroscopic building materials: experimental facilities and energy impact. Energy Build. **38**, 1270–1282 (2006)
2. Peñaloza, D., Erlandsson, M., Falk, A.: Exploring the climate impact effects of increased use of bio-based materials in buildings. Constr. Build. Mater. **125**, 219–226 (2016)
3. Holopainen, R., Tuomaala, P., Hernandez, P., Hakkinen, T., Piira, K., Piippo, J.: Comfort assessment in the context of sustainable buildings. Build. Environ., 60–70 (2013)
4. Amziane, S., Collet, F. (eds.): Bio-aggregates Based Building Materials. RSR, vol. 23. Springer, Dordrecht (2017). https://doi.org/10.1007/978-94-024-1031-0
5. Rahim, M., et al.: Characterization of flax lime and hemp lime concretes: hygric properties and moisture buffer capacity. Energy Build. **88**, 91–99 (2015)
6. Limam, A., Zerizer, A., Quenard, D., Sallee, H., Chenak, A.: Experimental thermal characterization of bio-based materials (Aleppo Pine wood, cork and their composites) for building insulation. Energy Build. **116**, 89–95 (2016)
7. Asli, M., Brachelet, F., Sassine, E., Antczak, E.: Thermal and hygroscopic study of hemp concrete in real ambient conditions. J. Build. Eng., 44 (2021)
8. Palumbo, M., Lacasta, A.M., Holcroft, N., Shea, A., Walker, P.: Determination of hygrothermal parameters of experimental and commercial bio-based insulation materials. Constr. Build. Mater. **124**, 269–275 (2016)
9. Nozahic, V., Amziane, S., Torrent, G., Saïdi, K.D., Baynast, H.: Design of green concrete made of plant-derived aggregates and a pumice–lime binder. Cement Concr. Compos. **34**, 231–241 (2012)
10. Arizzi, A., Viles, H., Martín-Sanchez, I., Cultrone, G.: Predicting the long-term durability of hemp–lime renders in inland and coastal areas using Mediterranean, Tropical and Semi-arid climatic simulations. Sci. Total Environ., 542 (2016)
11. Marceau, S., et al.: Influence of accelerated aging on the properties of hemp concretes. Constr. Build. Mater. **139**, 524–530 (2017)
12. Benmahiddine, F., Bennai, F., Cherif, R., Belarbi, R., Tahakourt, A., Abahri, K.: Experimental investigation on the influence of immersion/drying cycles on the hygrothermal and mechanical properties of hemp concrete. J. Build. Eng., 32 (2020)

13. Delannoy, G., Marceau, S., Glé, P., Gourlay, E., Guéguen-Minerbe, M., Amziane, S., Farcas, F.: Durability of hemp concretes exposed to accelerated environmental aging. Constr. Build. Mater., 252 (2020)
14. Amziane, S., Collet, F., Lawrence, M., Magniont, C., Picandet, V., Sonebi, M.: Recommendation of the RILEM TC 236-BBM: characterization testing of hemp shiv to determine the initial water content, water absorption, dry density, particle size distribution and thermal conductivity. Mater. Struct. **50**, 167 (2017)
15. Nozahic, V. : Vers une nouvelle démarche de conception des bétons de végétaux lignocellulosiques basée sur la compréhension et l'amélioration de l'interface liant/végétal. PhD thesis (2012)
16. Bourdot, A., Magniont, C., Lagouin, M., Niyigena, C., Evon, P., Amziane, S.: Impact of bio-aggregates properties on the chemical ineractions with mineral binder, application to vegetal concrete. J. Adv. Concrete Technol., 17 (2019)
17. AFPC-AFREM, Méthodes recommandées pour la mesure des grandeurs associées à la durabilité, Compte rendu des journées techniques AFPC-AFREM (1997)
18. Rode, C., Peuhkuri, R., Time, B., Svennberg, K., Ojanen, T.: Moisture buffer value of building materials. J. ASTM Int. **4**(5) (2007)
19. Alvarado, S., Marin, E., Juarez, A.G., Calderon, A., Ivanov, R.: A hot-wire method based thermal conductivity measurement apparatus for teaching purposes. Europ. J. Phys., 33 (2012)

Sustainability and Economic Viability of Self-healing Concrete Containing Super Absorbent Polymers

Davide di Summa[1,3](✉), Didier Snoeck[1], José Roberto Tenório Filho[1,4], Philip Van den Heede[1], Sandra Van Vlierberghe[2], Nele De Belie[1], and Liberato Ferrara[3]

[1] Department of Structural Engineering and Building Materials, Ghent University, Magnel-Vandepitte Laboratory, Tech Lane Ghent Science Park, Campus A, Technologiepark Zwijnaarde 60, 9052 Ghent, Belgium
davide.disumma@UGent.be
[2] Department of Organic and Macromolecular Chemistry, Centre of Macromolecular Chemistry, Polymer Chemistry & Biomaterials Group, Ghent University, Krijgslaan 281, S4, 9000 Ghent, Belgium
[3] Department of Civil and Environmental Engineering, Politecnico di Milano, Piazza Leonardo da Vinci 32, 20133 Milan, Italy
[4] SIM zw, Technologiepark Zwijnaarde 48, B-9052 Ghent, Belgium

Abstract. Recent studies highlighted the need to investigate the sustainability of innovative cement-based composites. In this regard, some works focused their attention on the use of Super Absorbent Polymers (SAPs) blended into the concrete matrix also employed to promote the autogenous healing, which can result into extended durability. In this study the Life Cycle Assessment (LCA) methodology takes into account the impacts associated to the whole service life of a structure. Thus, the eco-profile of a wall made up of concrete containing SAPs, was compared to the one of a reference wall without those additions. Four scenarios were considered to estimate the frequency of the repairing activities needed because of the chloride induced corrosion. Two corrosion models were adopted: a uniform one for scenarios 1 and 2, with a service life of 50 and 100 years respectively and the hemispherical pit model, for scenarios 3 and 4 with 50 and 100 years of service life as well. Additionally, a Life Cycle Cost (LCC) analysis was developed to investigate the overall costs. The results highlight the better performances for SAP-containing concrete with a reduction up to 11% for the overall costs and up to 55% for the environmental burdens.

Keywords: Self-healing · Concrete · LCA · LCC · SAPs · Sustainability

1 Introduction

Environmental and cost sustainability of concrete are critical issues that must be addressed in view of the fact that concrete is the most used construction material worldwide. In this framework, not only the environmental impacts associated to the cement

J. I. Escalante-Garcia et al. (Eds.): RW 2021, RILEM Bookseries 40, pp. 335–343, 2023.
https://doi.org/10.1007/978-3-031-21735-7_37

production itself (800 kg of CO_2 per ton of cement and use of non-renewable raw materials) have to be highlighted [1, 2] but also the implications related first of all to the structural use of the material, aimed to achieve a target structural performance, as well as to the deterioration of concrete structures [3]. As a matter of fact, concrete durability has environmental, economic and social consequences for the society due to the fact that the crack susceptibility, increasing the frequency of various repairing activities, implies the use of further raw materials, energy and workmanship to re-establish the functionality of a given structure. In this regard, recent studies have developed innovative cementitious materials able to ensure both the shrinkage reduction, and the related cracking potential, and the self-repair of cracks upon occurrence, acting, indirectly, on the overall durability of the structure [4–8]. Superabsorbent polymers (SAPs) are natural or synthetic water-insoluble 3D networks of polymeric chains cross-linked by chemical or physical bonding that, blended into the concrete matrix, could give substantial advantages. Being characterized by the possibility to take up an amount of fluid up to 1500 times their own weight [9] they can act as a water reservoirs for the system, keeping the level of internal relative humidity high and preventing the shrinkage cracking due to self-desiccation [5, 10, 11]. Additionally, the absorbed water can stimulate the hydration of unhydrated binder materials favouring the autogenous healing of the cracks [12]. This study employs Life Cycle Assessment (LCA) and Life Cycle Cost (LCC) analysis to address the potential environmental sustainability and economic viability of concrete containing SAPs exposed to a real structural service scenario. Therefore, the used methodological approach integrates the structural performances of the case study into LCA and LCC analysis to better predict the repetitiveness of the maintenance activities throughout the entire service life. To this purpose, the work starts taking as a reference two mock-up walls casted in Bruges (Belgium) resembling a tunnel element. The structures have been cast in the framework of the iSAP project that investigated the use of SAPs to produce smart concrete mixtures able to reduce water permeability through cracks of up to 250 μm in width and increasing, at the same time, the resistance to salt-scaling under frost attack [13, 14].

2 Case Studies

The analysed earth-retaining walls are 14 m long, 2.75 m high with a 0.80 m deep cross section. The wall henceforth indicated with Wall_Ref has been cast employing concrete C35/45, and reinforced with horizontal ϕ16 mm steel bars spaced at 96 mm and vertical ϕ12 mm bars spaced at 140 mm on either side. The second wall, indicated as Wall_SAP has been built with concrete C30/37 with a reinforcement that, compared to the one of Wall_Ref, simply differs for the longitudinal ϕ16 mm steel bars spaced at 107 mm instead of 96 mm. Both walls have been designed to allow a maximum crack width of 300 μm under service loads. As a matter of fact, the internal curing promoted by the SAPs could counteract the shrinkage-cracking, thus resulting into the reinforcement reduction as above for Wall_SAP in comparison to Wall_Ref. Both walls have been built on two foundation slabs cast at least three months in advance with concrete C35/45. Table 1 details the two adopted mix designs.

Table 1. Wall_Ref and Wall_SAP mix design per 1 m3

	Wall_Ref [kg]	Wall_SAP [kg]
CEM III-B 42.5N	359.78	359.82
Limestone 2/20	1124.44	1084.55
Marine sand 0/4	782.22	744.50
Sika ViscoFlow-26 Superplasticizer	1.56	1.55
Tixo - Superplasticizer	2.42	2.42
Water	144.40	137.45
Commercial SAP	–	1.36

3 Concrete Repair Techniques

Several methods are nowadays available to refurbish damaged concrete, using different materials and technologies according to the environmental conditions to which the structures are exposed. For the scope of this research the concrete cover replacement with the substitution of the damaged reinforcement steel bars has been taken into account.

3.1 Concrete Cover and Reinforcement Substitution

Several experimental studies have been developed to investigate the radial expansion of corroded reinforcement after corrosion initiation by chlorides or carbonation. [15–18]. The substitution of the damaged reinforcement bars is usually adopted, followed by the casting of a new concrete layer after a prior treatment with a primer to ensure a better adhesion to the substrate. In this regard, a focus of this research has been to determine when the reduced structural stability, due to the rebar corrosion, necessitates the execution of these maintenance activities. In this case the lateral earth pressure using the Coulomb theory was calculated using following equation.

$$S = \frac{\gamma_t}{2} \times H^2 \times \tan^2\left(\frac{90 - \varphi}{2}\right) \tag{1}$$

where H corresponds to the height of the structure while γ_t and φ are the specific weight and the internal friction coefficient of the earth piled against the walls, assumed equal to 18 kN/m^3 and 35° respectively. A cantilever model was employed to re-design the structure and to calculate the bending moment action due to the earth thrust which has to be resisted through the moment resisting capacity of the wall base cross-section. Therefore, using Eq. 1, for a 1 m wide strip of the wall, a total earth thrust equal to 18.44 kN/m was calculated, to be applied at H/3, resulting into a bending moment (M_{ED}) at the base cross section equal to 16.90 kNm/m. The calculated bending moment value has then been amplified by a coefficient γ_G of 1.35 as per EN 1990:2002. Equation 2 was then adopted to estimate the resistent moment M_{RD} equal to 212.9 kNm.

$$M_{rd} = 0.9 \times f_{yd} \times A_S \times d \tag{2}$$

Additionally, since it must always be verified that $M_{ED} < M_{RD}$ it can easily be calculated that the A_s value, which represents the area of the tension reinforcement bars per unit width of the wall, must be higher than 8,443 mm^2. This means that, as in one meter of the wall 7 ϕ 12 steel bars are located, assuming a uniform corrosion along the entire wall, the reinforcement bar cannot loose more than 89% of its cross section. Higher ratios will imply that the safety of the structure can no longer be guaranteed.

3.2 Corrosion Initiation

To calculate the time the chloride ions need to penetrate the concrete matrix, the solution to Fick's 2nd law of diffusion presented in Eq. 3 was generally employed.

$$x_{crit} = 2\sqrt{3(t - t_0) \times D_{app}} \times \left[1 - \sqrt{\frac{C_{crit} - C_I}{C_s - C_i}}\right] \quad (3)$$

where x_{crit} is the critical chloride depth assumed equal to the concrete layer, C_i is the initial chloride content of concrete equal to 0.2% by weight of cement according to EN 206 [19]; C_s is the chloride concentration at surface being 0.92% by weight of soil assumed for this case study [20] and C_{crit} is the critical chloride concentration equal to 0.44% by weight of cement according to Alonso et al. [21]. With regard to the apparent chloride diffusion coefficient D_{app}, a value of 10^{-10} m^2/s can be found in literature for the case of cracked concrete, while for the uncracked one with CEM I, a value of 10^{-11} m^2/s can be adopted [3, 22–24]. Moreover, a theoretical correction factor of 0.2 has been applied to the latter for CEM III/B according to [25] with a resulting value in the order of 10^{-12} m^2/s. This, since currently no details are available regarding the effect of SAP addition on D_{app}, has been taken for both uncracked Wall_Ref and Wall_SAP. With these values an "instantaneous" initiation time (0 days) was calculated for cracked Wall_Re (corresponding to the wall in its current state, since several cracks have already been observed) using 10^{-10} m^2/s. This is also in line with what noticed in an earlier research with regard to the corrosion initiation of reinforcement in cracked concrete after only three weeks of exposure to chloride solution [26]. With regard to Wall_SAP and Wall_Ref after its first maintenance activity, since no cracks have been observed up to 9 months after casting for the first one and no cracks are supposed for the second one, a corrosion initiation time of 13 years in total has been calculated using a value for Dapp equal to 10^{-12} m^2/s.

3.3 Corrosion Propagation

Two different models have been adopted to estimate the corrosion propagation for the scope of this research. According to what is indicated in [27], one model consisted of adopting a uniform corrosion rate expressed in μm/year that, for the case of a concrete contaminated by chloride and subjected to 80–90% of relative humidity, the closest assumption to the case studies of this research is equal to 50 μm/year [27]. The second model considered a loss of steel concentrated in a small zone (pit corrosion) that increases in size as the corrosion propagates in time and generates a volumetric mass loss of steel.

Van Belleghem [26] carried out a research on reinforced fly ash containing concrete specimens (water to binder ratio 0.41 and fly ash-to-binder ratio of 15%). From the experimental campaign developed on pre-cracked specimens exposed to a simulated chloride containing environment (equal to an exposure class XS2) a progressive loss of cross section of 13.36 mm^3 was registered for the initial period of 20 weeks and 0.58 mm^3 per week from that onwards. These values have been used to calculate the progressive loss of area of the cross section of the transversal ϕ 12 steel bars. Moreover, despite the structure stability is compromised when the tension bars lose 89% of their cross-section, for the scope of this research it has been reasonably considered that the repairing activities must be carried out when the reinforcement bars lose 20% of their cross-section adopting the two corrosion propagation methods above described.

4 Life Cycle Assessment

The LCA has been developed according to ISO 14040–14-44 using a cradle-to-gate system boundary and adopting Ecoinvent 3.6 as data source for all the Life Cycle Inventories (LCIs) of the raw materials. This was possible for most of the common constituents but needed further investigation for the missing ones such as the commercially available products like SAPs and superplasticizers.

4.1 Impact and Cost Analysis

To develop the analysis, a functional unit (FU) corresponding to the walls themselves has been selected for both Wall_Ref and Wall_SAP with a service life ranging from 50 years up to 100 years according to the different adopted propagation models, as done in similar works [28]. This results into 4 scenarios summarized in the following Table 2.

While scenarios 1 and 3 are limited to 50 years of service life, scenarios 2 and 4 are extended up to 100 years in total. The aforementioned estimations for both the initiation and propagation time were used to calculate the frequency of the repairing activities. This resulted, for the case of scenarios 1 and 2 into 38 years in total for Wall_SAP while for Wall Ref it has been considered an initially time of 25 years until the first repairing activity (since it is already cracked and the correspondent initiation time is 0 years) and 38 years starting from that onwards (with an initiation time of 13 years after the casting of the new uncracked concrete layer). With regard to scenarios 3 and 4, adopting the same initiation periods of scenarios 1 and 2, 3.8 years have been taken into account for both Wall_SAP and Wall_Ref. The environmental footprint of the FU has been calculated using the CML-IA impact method. It provides a general overview of 10 impact indicators in total: global warming; acidification; ozone depletion; photochemical oxidation; eutrophication; abiotic depletion potential; human toxicity potential; freshwater aquatic ecotoxicity potential; marine aquatic ecotoxicity and terrestrial ecotoxicity. Figure 1 shows the obtained results indicating the percentage decrease between Wall_Ref and Wall_SAP.

Wall_SAP always performs better in comparison to Wall_Ref in terms of material production related environmental burdens reaching a reduction higher than 50% and for

Table 2. Summary of the adopted scenarios

	Wall_SAP				Wall_ Ref		
	Adopted corrosion model	Initiation time	Penetration time	Number of maintenance activities	Initiation time	Penetration time	Number of maintenance activities
Scenario 1 (50 years of SL)	50 μm/year	13 years	25 years	**1**	0 for cracked cover, 13 years for the uncracked one	25 years	**1**
Scenario 2 (100 years of SL)	50 μm/year	13 years	25 years	**2**	As above	25 years	**3**
Scenario 3 (50 years of SL)	Hemispherical pit model	13 years	3.8 years	**2**	As Above	3.8 years	**3**
Scenario 4 (100 years of SL)	Hemispherical pit model	13 years	3.8 years	**5**	As above	3.8 years	**6**

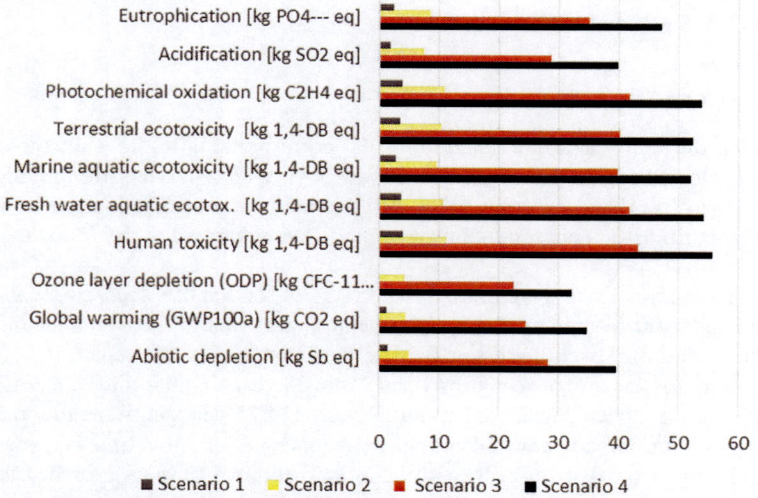

Fig. 1. Percentage reduction of Wall_SAP impacts in comparison to Wall_Ref.

the case of photochemical oxidation, Terrestrial ecotoxicity, marine aquatic ecotoxicity, fresh water aquatic ecotoxicity and human toxicity for the case of scenario 4. Furthermore, also the economic viability has been assessed. In this regard, after a market survey to collect all the construction cost items, for the activities to be carried out in the future, each rate has been recalculated according to the discount factor reported in [29] and in Eq. 4.

$$(T) = 1/(1 + r)^{T} \tag{4}$$

where r is the annual real discount rate and T is the number of future years. For the case of r, an annual real discount rate equal to 3% has been adopted. Figure 2 outlines the obtained results. While for scenario 1 and 3 the differences between Wall_Ref and Wall_SAP are almost neglectable, a more pronounced reduction can be observed for the Scenario 4 with an overall costs reduction of around 10% for Wall_SAP.

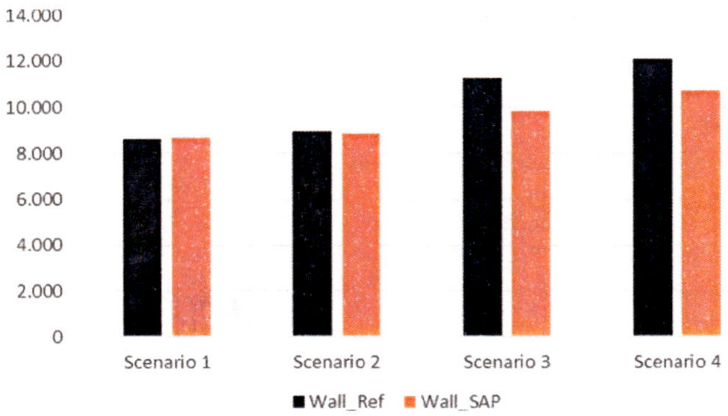

Fig. 2. Costs comparison between Wall_Ref and Wall_SAP.

5 Conclusions

In this paper LCA and LCC methodologies have been employed as part of a holistic structural design approach to address the ecological and economic profiles of a given structure and to support the potential advantages in the use of SAP-containing concrete in comparison to traditional reinforced solutions. The obtained results highlight that the structure made with concrete containing SAP, especially with a service life extended up to 100 years, not only performs better in terms of environmental impacts, as compared to the reference one, but could also represent the most cost-saving solution mainly due to the delay and the frequency reduction of the maintenance activities along the entire service life. This paves the way to highlight the importance of LCA and LCC to favour the design of more sustainable reinforced concrete solutions and to stimulate at the same time, the use of advanced materials, products and technologies, including self-healing concrete, for the concrete construction market.

Acknowledgements. This project has received funding from the European Union's Horizon 2020 research and innovation programme under the Marie Skłodowska-Curie grant agreement No 860006.

The work within the ICON project iSAP (Innovative SuperAbsorbent Polymers for crack mitigation and increased service life of concrete structures) has been financed by the SIM program SHE (Engineered Self-Healing Materials). Philip Van den Heede is a postdoctoral researcher of the Research Foundation - Flanders (FWO) (project No. G062720N). The financial support of FWO is gratefully acknowledged.

As a Postdoctoral Research Fellow of the Research Foundation-Flanders (FWO-Vlaanderen), D. Snoeck would like to thank the foundation for the financial support (12J3620N).

References

1. Meyer, C.: The greening of the concrete industry. Cem. Concr. Compos. **31**, 601–605 (2009). https://doi.org/10.1016/j.cemconcomp.2008.12.010
2. Federico, L.M., Chidiac, S.E.: Waste glass as a supplementary cementitious material in concrete - critical review of treatment methods. Cem. Concr. Compos. **31**, 606–610 (2009). https://doi.org/10.1016/j.cemconcomp.2009.02.001
3. Al-Obaidi, S., Bamonte, P., Luchini, M., Mazzantini, I., Ferrara, L.: Durability-based design of structures made with UHP/UHDC in extremely aggressive scenarios: application to a geothermal water basin case study. MDPI Infrastructures **5**(11), 1–44 (2020). https://doi.org/10.3390/infrastructures5110102
4. Liu, J., Farzadnia, N., Khayat, K.H., Shi, C.: Effects of SAP characteristics on internal curing of UHPC matrix. Constr. Build. Mater., 280 (2021). https://doi.org/10.1016/j.conbuildmat.2021.122530
5. Craeye, B., Geirnaert, M., Schutter, G.D.: Super absorbing polymers as an internal curing agent for mitigation of early-age cracking of high-performance concrete bridge decks. Constr. Build. Mater. **25**, 1–13 (2011). https://doi.org/10.1016/j.conbuildmat.2010.06.063
6. Feiteira, J., Gruyaert, E., De Belie, N.: Self-healing of moving cracks in concrete by means of encapsulated polymer precursors. Constr. Build. Mater. **102**, 671–678 (2016). https://doi.org/10.1016/j.conbuildmat.2015.10.192
7. Gruyaert, E., Debbaut, B., Snoeck, D., Díaz, P., Arizo, A., Tziviloglou, E., Schlangen, E., De Belie, N.: Self-healing mortar with pH-sensitive superabsorbent polymers: Testing of the sealing efficiency by water flow tests. Smart Mater. Struct., 25 (2016). https://doi.org/10.1088/0964-1726/25/8/084007
8. Pelto, J., Leivo, M., Gruyaert, E., Debbaut, B., Snoeck, D., De Belie, N.: Application of encapsulated superabsorbent polymers in cementitious materials for stimulated autogenous healing, Smart Mater. Struct., 26 (2017). https://doi.org/10.1088/1361-665X/aa8497
9. Mechtcherine, V., Reinhardt, H.W.: Application of Super Absorbent Polymers (SAP) in Concrete Construction, in State-of-the-Art Report Prepared by Technical Committee 225-SAP., RILEM. p. 165 (2012)
10. Piérard, J., Pollet, V., Cauberg, N.: Mitigating autogenous shrinkage in HPC by internal curing using superabsorbent polymers. In: International RILEM Conference on Volume Changes of Hardening Concrete: Testing and Mitigation. (2006). Lyngby, Denmark: RILEM Publications SARL
11. Snoeck, D., Jensen, O.M., De Belie, N.: The influence of superabsorbent polymers on the autogenous shrinkage properties of cement pastes with supplementary cementitious materials. Cem. Concr. Res. **74**, 59–67 (2015). https://doi.org/10.1016/j.cemconres.2015.03.020
12. Snoeck, D., Van Tittelboom, K., Steuperaert, S., Dubruel, P. and De Belie, N.: online (2012).Self-helaing cementitious materials by the combination of microfibres and superabsorbent polymers. J. Intell. Mater. Syst. Struct. **25**(1), 13–24 (2014)
13. Tenório Filho, J.R., Mannekens, E., Van Tittelboom, K., Van Vlierberghe, S., De Belie, N., Snoeck, D.: Innovative superabsorbent polymers (iSAPs) to construct crack-free reinforced concrete walls: an in-field large-scale testing campaign. J. Build. Eng. **43**, 102639 (2021). https://doi.org/10.1016/j.jobe.2021.102639

14. Tenório Filho, J.R., Mannekens, E., Van Tittelboom, K., Snoeck, D., De Belie, N.: Assessment of the potential of superabsorbent polymers as internal curing agents in concrete by means of optical fiber sensors. Constr. Build. Mater., 238 (2020). https://doi.org/10.1016/j.conbuildmat.2019.117751
15. Andrade, C., Alonso, C., Rodriguez, J.: Cover cracking and amount of rebar corrosion: importance of the current applied accelerated tests C. Sjostrom (Ed.), Durability of Building Materials and Components, 7, E&FN Spon, London, pp. 263–273 (1996)
16. Liu, Y., Weyers, R.E.: Modelling the time-to-corrosion cracking in chloride contaminated reinforced concrete structures ACI Mater. J. **95**(6), 675–681 (1998)
17. Clark, L.A., Saifullah, M.: Effect of corrosion on reinforcement bond strength. In: Forde, M. (Ed.), Proceedings of 5th International Conference on Structural Faults and Repairs, Edinburgh: Engineering Technical Press, vol. 3, pp. 113–119 (1993)
18. Al-Sulaimani, G.J., Kaleemullah, M., Basunbul, L.A.: Influence of corrosion and cracking on bond behaviour and strength of reinforced concrete members ACI Struct. J. **87**(2), 220–231 (1990)
19. Cen, UNI EN 206:2016. Concrete - Specification, performance, production and conformity (2016)
20. Regione Emilia Romagna. Carta della salinita‘ dei suoli della pianura emiliano-romagnola strato 50–100 cm (2011)
21. Alonso, M.C., Sanchez, M.: Analysis of the variability of chloride threshold values in the literature. Mater. Corros. **60**, 631–637 (2009). https://doi.org/10.1002/maco.200905296
22. Shafikhani, M., Chidiac, S.E.: Quantification of concrete chloride diffusion coefficient – a critical review. Cem. Concr. Compos. **99**, 225–250 (2019). https://doi.org/10.1016/j.cemconcomp.2019.03.011
23. Torres-Acosta, A.A., Presuel-Moreno, F.J., Andrade, C.: Electrical resistivity as durability index for concrete structures. ACI Mater. J. **116**, 245–253 (2019). https://doi.org/10.14359/51718057
24. Liu, Y., Presuel-Moreno, F.J., Paredes, M.A.: Determination of chloride diffusion coefficients in concrete by electrical resistivity method. ACI Mater. J. **112**, 631–640 (2015). https://doi.org/10.14359/51687777
25. Coppola, L.: Concretum, MC-Graw Hill Education (2007)
26. Van Belleghem, B.: Effect of Capsule-Based Self-Healing on Chloride Induced Corrosionof Reinforced Concrete, Ghent University (2018)
27. Bertolini, L., Elsener, B., Pedeferri, P.: Corrosion of Steel in Concrete (2004)
28. di Summa, D., Tenório Filho, J.R., Snoeck, D., Van den Heede, P., Van Vlierberghe, S., Ferrara, L., De Belie, N.: Environmental and economic sustainability of crack mitigation in reinforced concrete with SuperAbsorbent polymers (SAPs). J. Cleaner Product. **358**, 131998 (2022), ISSN: 0959-6526. https://doi.org/10.1016/j.jclepro.2022.131998
29. UNI, UNI EN ISO 16627:2015 - Sustainability of construction works - Assessment of economic performance of buildings - Calculation methods (2015)

Self-healing Capabilities of Ultra-High Performance Fiber Reinforced Concrete with Recycled Aggregates

Niranjan Prabhu Kannikachalam[1,2(✉)], Ruben Paul Borg[3], Estefania Cuenca[1], Nele De Belie[2], and Liberato Ferrara[1]

[1] Department of Civil and Environmental Engineering, Politecnico di Milano, Piazza Leonardo da Vinci 32, 20133 Milan, Italy
niranjanprabhu.kannikachalam@polimi.it

[2] Department of Structural Engineering and Building Materials, Ghent University, Magnel-Vandepitte Laboratory, Tech Lane Ghent Science Park, Campus A, Technologiepark Zwijnaarde 60, 9052 Ghent, Belgium

[3] Faculty for the Built Environment, University of Malta, Msida 2080, MSD, Malta

Abstract. This study examines the effect of recycled aggregates on the stimulated autogenous self-healing of Ultra-High Performance Fiber Reinforced Concrete (UHPFRC) when exposed to wet and dry conditions. Recycled aggregates have been produced by crushing four months old UHPFRC specimens with an average compressive strength of 150 MPa. Two different percentages (50% & 100% by weight) of recycled aggregate (0–2 mm) have been used as a substitute for the natural aggregate in a reference UHPFRC mix to produce recycled UHPFRC. After an assessment of the overall mechanical properties of the recycled UHPFRC mixes, one year old notched beam specimens were pre-cracked to the width of 150 μm through a three-point flexural test. The self-healing capacity of recycled UHPFRCs has been investigated in terms of water absorption tests, regain in flexural strength and microscopic crack healing, at scheduled times after precracking (0 days, 1 month, 3 months and 6 months). Constant wet/dry healing conditions were maintained throughout the experiment. The specimens with recycled UHPC aggregate showed better and longer self-healing than the specimens with natural aggregate, which provides additional value to the overall environmental sustainability of the investigated category of materials.

Keywords: Self-healing · Ultra-High Performance Fiber Reinforced Concrete · Recycled aggregate · Crystalline Admixture · UHPFRC

1 Introduction

The construction industry is one of the world's largest consumers of natural resources and one of the world's top waste producers. Construction and demolition waste (CDW) accounts for 30–40% of total solid waste [1, 2]. In the European Union, CDW accounted for 36% of total solid waste generated in 2016 [3] and 67% in the USA [4]. However,

J. I. Escalante-Garcia et al. (Eds.): RW 2021, RILEM Bookseries 40, pp. 344–352, 2023.
https://doi.org/10.1007/978-3-031-21735-7_38

a large volume of construction waste is disposed in landfill due to the inefficient waste management process even though there is increasing interest in recycling and reuse. In addition, the disposal to landfill has a negative effect on the environment, causing greenhouse gas emissions, land degradation, and water pollution. Globally, 20–30% of construction and demolition waste is currently recovered [5]. Therefore, there is a growing concern about resource consumption and waste production in construction and demolition. Under the Waste Framework Directive 2008/98/EC (WFD), the EU established as a goal 70% recovery and recycling of CDW by 2020, and 15 member states have already achieved this target; however, there are still 11 member states that have yet to meet the target. The recovery rate in the United States stands at 70% [6].

The construction industry actively promotes recycled aggregates (RA) made from crushed CDW as a partial or complete replacement for natural aggregates, which can help decrease waste disposal and demand on natural resources. In addition, research has shown that recycled aggregate concrete (RAC) has sufficient mechanical strength for usage as structural concrete, even in Ultra-High Performance Concrete [7]. However, the properties of RAC are influenced by the recycled aggregate replacement ratio and the performance of the parent concrete of RA. On the contrary, RA use in concrete structures is currently limited due to concerns about its long-term durability. The main difference between recycled and natural aggregate is the presence of adhered cement paste and the inherent weakness of the old interfacial transition zone (ITZ), due to cracks formed during the crushing process in RA production. CO_2 treatment of RAs and the use of additives in RAC production improves the adhered paste and ITZ and the durability of the concrete.

Recently, Borg et al. [7] have investigated the possibility of recycling UHPC/UHPFRC, showing that, likely because of the high strength and high amount of still unreacted binder which may remain attached to the recycled aggregate particles, the recycled UHPC/UHPFRC can feature no decrease in both mechanical and durability performance, as compared to the parent reference mix from which the same recycled aggregates were produced.

UHPC structures have a longer structural service life and a limited production volume, resulting in a smaller supply of recyclable materials. However, from an engineering standpoint, UHPC structures that have reached a particular point in their lifespan may no longer meet the requirements of owners and stakeholders. Urban and infrastructure design concepts must shift from a "additional" model to a "circular" one, which must imply that when the need for a new building/structure/infrastructure facility arises, the old one is demolished and a new, better-suited one is constructed in its place, rather than the original structure being added to or incorporated into the new one. Cement-based construction materials and structures would need to incorporate recyclability into their concept and design.

There are not many studies available in the literature for the autogenous self-healing capacity of RAC, which especially in the case of UHPC and UHPFRC may benefit from the amount of unhydrated binder material which remains attached to the recycled aggregate particles. In the aforementioned perspective, this paper discusses the autogenous healing performance of Recycled Ultra High-Performance Fiber Reinforced Concrete (R-UHPFRC), where recycled aggregates produced from UHPFRC have been employed in partial and total replacement of the natural sand in the parent mix-design. The sorptivity test was used to determine the specimens durability, and the specimens recurrent healing behavior was also examined by re-cracking them at scheduled healing times, respectively equal to 1, 3 and 6 months, under wet and dry conditions.

2 Experimental Program

2.1 Mix Design

Table 1 shows the composition of the UHPFRC mix used as a reference in this research, which was developed and validated at the University of Malta as part of the Horizon 2020 ReSHEALience Project [8]. Four-month-old UHPFRC specimens with an average compressive strength of 150 MPa were crushed to less than 2 mm particle size to produce recycled aggregates. A magnetic separator was used to separate the steel fibers from the cement matrix. The particle size distribution of recycled aggregates was then sorted out and blended to be as close as possible to the natural aggregates particle distribution—more details about production and properties of recycled aggregates being available in [7]. Two more concrete mixes (Recycled-UHPFRC) were created by substituting about 50% and 100% of natural aggregates with the recycled aggregates obtained as above. Two specimens from the reference mix and three specimens from each of the recycled mixes were utilized for this research.

Table 1. UHPFRC mix design

Constituents	Reference concrete (kg/m^3)
Cement 52.5 R	700
Silica Fume	400
Superplasticizer (ACE 442)	64
Water	231
Natural Aggregate	817
Steel fibers ($l_f = 22$ mm, $d_f = 0.2$ mm)	160
Crystalline admixture (Penetron Admix®)	5.6

2.2 Sorptivity Test

The experiment was conducted on one year-old UHPFRC and R-UHPFRC specimens with a dimension of 160 × 40 × 40 mm^3. At the mid-span, a 6 ± 1 mm deep notch was cut to localize the fracture in the beam's center. Specimens were then pre-cracked up to 0.15 mm of crack width under a three-point flexural test at a constant Crack Mouth Opening Displacement (CMOD) rate of 0.5 μm/s. The CMOD was measured using a clip gauge position on the knife-edge plates across the notch.

After pre-cracking, a silicon layer was applied to the bottom and lateral surfaces of the specimen, up to a height of 20 mm, except for a 20 mm wide area at the fracture zone (mid-span) to perform the sorptivity tests and monitor capillary water absorption only through the cracked zone. Each time before the sorptivity test, the silicon layer was renovated, and the specimens were oven-dried at 40 °C for 24 h, followed by 24 h in a dry environment. The specimens were mounted on two supports so that the water level remained 7 mm above the depth of the notch, illustration of the sorptivity test was shown in Fig. 1. Water uptake was measured at 15 min, 30 min, 1 h, and every hour afterward until 8 h and then a final 24 h reading was taken. The sorptivity tests were performed immediately after per-cracking and at the end of the 1st, 3rd, and 6th month of healing. The healing environment adopted for this study is four days wet (submerged in water), and three days dry condition maintained throughout the experimental program and every month the water was replaced with fresh water. At each scheduled time digital microscopy crack imaging was also performed to monitor the crack sealing. One specimen from each R-UHPFRC was re-cracked to 25 μm at the end of 1st and 3rd month of healing. The purpose of selecting a fraction of the pre-crack width for re-cracking is to check the repeatability of self-healing without further crack propagation.

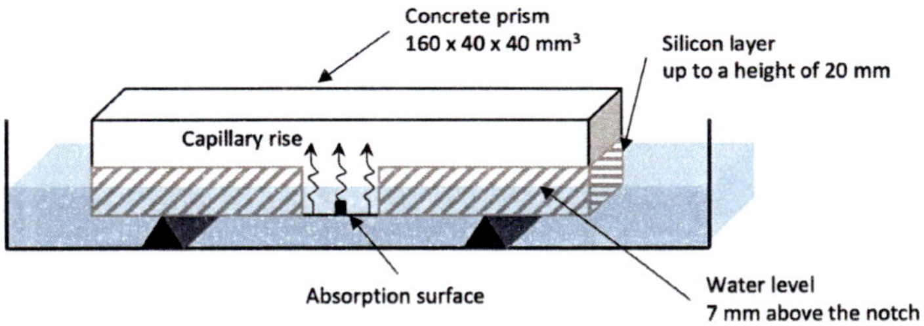

Fig. 1. Illustration of sorptivity test.

3 Results and Discussions

The capillary water absorption test indicated consistent sealing of cracks throughout the healing period (see Fig. 2). According to Fig. 2a, f, reference specimens demonstrated a 31 percent sorptivity recovery after one month under wet and dry cycles, followed by a further 14 percent recovery over the second and third month. The following final three months of healing resulted in only 6 percent recovery. Based on these results, the effective self-healing duration for the reference specimen can be assessed as equal to three months. In comparison, specimens with 50% and 100% recycled aggregates showed significant healing even after three months, with a healing rate per month slightly higher than 5%, which is about three times as much as the one featured by the reference specimens. This can be likely attributed to the presence of unhydrated binder which remains attached to the recycled aggregate particles and which represents an additional source of reactive material, which, upon contact with water penetrating through the cracks, can undergo

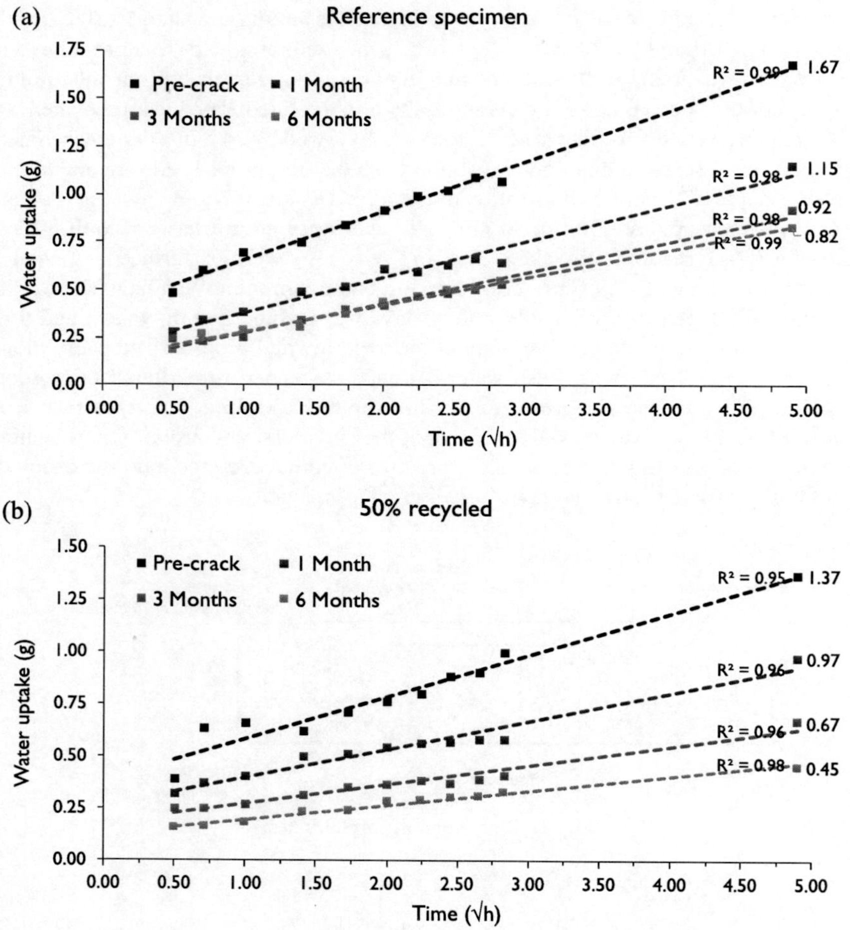

Fig. 2. Sorptivity chart in water uptake vs square root of time; **a** Reference specimens, **b** 50% Recycled-UHPFRC, **c** 100% Recycled-UHPFRC, **d** Re-cracked −50% recycled specimens, **e** Re-cracked −100% recycled specimens, and **f** % reduction in water uptake.

further delayed hydration and/or carbonation reactions whose products do contribute to the crack sealing and to the recovery of the resistance vs. water capillary sorption.

The specimens that were re-cracked to determine the persistence of the healing capacity upon repeated cycles of cracking and healing showed healing performance comparable to that of continuously healed not-recracked specimens. As shown in Fig. 1f, if we compare re-cracked 50% recycled samples with continuously healed samples, the difference in a total reduction in water uptake is 13%. However, two successive measurements of a 50% re-cracked specimen on the third and sixth month revealed a decrease in water absorption of 16% and 5%, respectively, indicating that cracks are healing even after re-cracking. Additionally, 100% re-cracked recycled samples exhibit

(c)

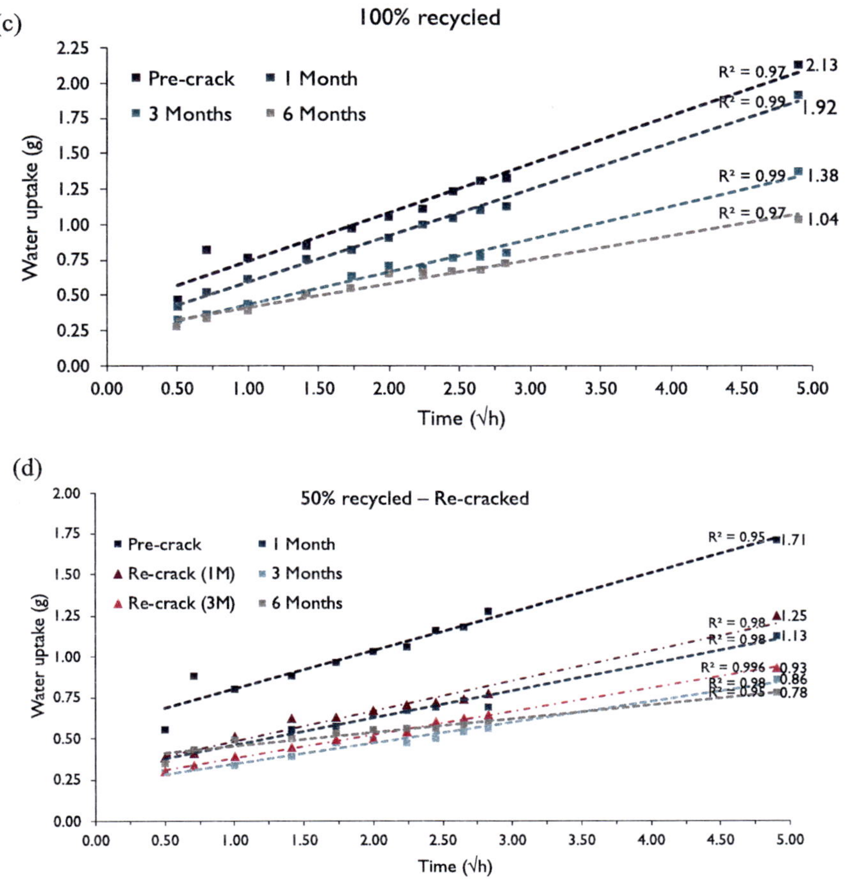

(d)

Fig. 2. (*continued*)

comparable findings to 50% re-cracked samples. This confirms that a repeatability of the healing capacity is present, likely due to the significant amount of unhydrated binder both available as such in the mix design as well as remaining attached to the recycled aggregate particles in the "old" interface transition zone. Such a healing capacity allows to recover the damage induced upon re-cracking thus guaranteeing an appreciable persistence of the overall healing capacity.

As a complement to the investigation reported above, before each sorptivity test, microscopy pictures were captured; the images were then stitched together and evaluated for crack width and area using Photoshop. Figure 3 illustrates a portion of the healing progress of a reference specimen crack, which has an average crack width of 22 μm. The second picture from the left, 65% of the crack surface area healed after one month, 95% after three months, and 100% after six months of healing in wet/dry conditions.

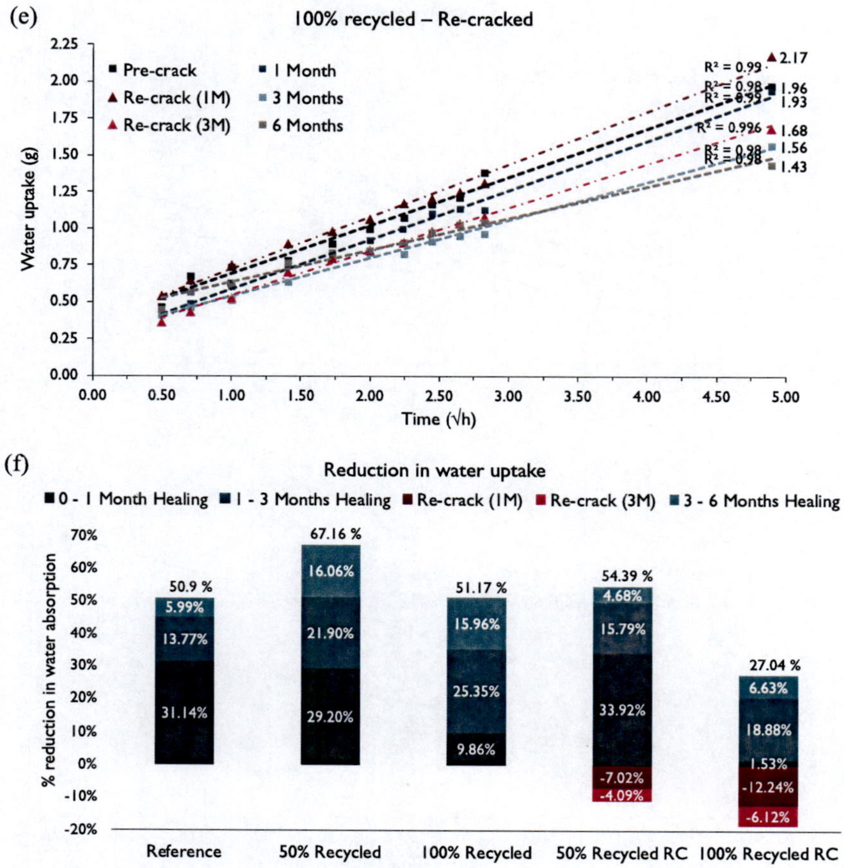

Fig. 2. (*continued*)

Correlation between the recovery of durability performance quantified as above and surface crack healing, as obtainable from image processing, is ongoing.

4 Conclusions

In this study, autogenous self-healing of Recycled-UHPFRC was evaluated by means of sorptivity test. A wet and dry healing condition was maintained throughout the experiment. The following conclusions can be made from the results shown above:

- The effective healing duration of reference specimens was three months, after which healing efficacy degraded to less than 5% each month. In six months of recovery, there was a 51% decrease in capillary water absorption.
- 50% and 100% Recycled-UHPFRCs showed an effective healing period of six months, during the last three months also substantial recovery was demonstrated.

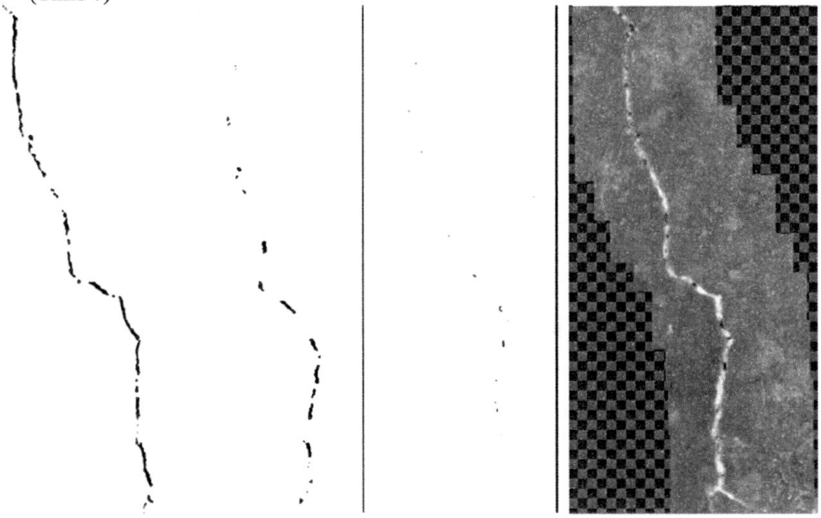

Fig. 3. Crack healing evolution of reference specimen with an average crack width of 22 μm.

- During the first month of recovery, the healing ability of reference concrete and recycled concrete were almost identical.
- Recycled specimens showed longer and better healing performance than reference specimens. Particularly 50% recycled specimens showed 16% higher healing in 6 months than the reference. Recycled specimens healing was higher from the second month of healing onwards.
- Upon repetition of cracking and healing, healing capacity is likely persisting in recycled UHPFRC specimens. Such a healing capacity is able to recover the damage induced upon re-cracking.

The results shown in this study could contribute to pave the way to further enhance the environmental sustainability of a relatively new category of cement based construction materials, such as UHPFRC, providing, in a cradle to cradle approach an added value to their full recyclability not only in terms of retention of mechanical performance, but also through maintenance and even enhancement of the overall durability.

Acknowledgements. The Recycled UHPFRC specimens employed in this investigation were produced as part of WP4-Task 4.4 activity in the framework of the ReSHEALience project (Rethinking coastal defence and green-Energy Service infrastructures through enHancEd durAbiLity high-performance fiber reinforced cement based materials), funded by the European Union in the framework of the Horizon 2020 Research and Innovation Programme, under GA

760824. The activity of the Task, to which the second, third and last author collaborated, terminated in April 2020. The self-healing investigation reported in this paper was performed, in the period October 2020-May 2021, as part of the PhD program of the first author, in the framework of the SMARTINCS project (Self-healing, Multifunctional, Advanced Repair Technologies IN Cementitious Systems). This project has received funding from the European Union's Horizon 2020 research and innovation programme under the Marie Skłodowska-Curie grant agreement No 860006.

References

1. Jin, R., Yuan, H., Chen, Q.: Science mapping approach to assisting the review of construction and demolition waste management research published between 2009 and 2018. Resour. Conserv. Recycling **140**, pp. 175–188 (2019). Elsevier B.V., https://doi.org/10.1016/j.resconrec.2018.09.029
2. López Ruiz, L.A., Roca Ramón, X., Gassó Domingo, S.: The circular economy in the construction and demolition waste sector – a review and an integrative model approach. J. Cleaner Product. **248**, 119238 (2020). Elsevier Ltd, https://doi.org/10.1016/j.jclepro.2019.119238
3. Waste statistics - Statistics Explained: https://ec.europa.eu/eurostat/statistics-explained/index.php?title=Waste_statistics#Total_waste_generation. Accessed 25 June 2021
4. U. Environmental Protection Agency, M. Conservation, R. Branch, R. Conservation, S. Division, and E. Office of Resource Conservation: Advancing Sustainable Materials Management: 2014 Fact Sheet Assessing Trends in Material Generation, Recycling, Composting, Combustion with Energy Recovery and Landfilling in the United States (2014)
5. Industry Agenda Shaping the Future of Construction A Breakthrough in Mindset and Technology Prepared in collaboration with The Boston Consulting Group (2016)
6. Zheng, L., et al.: Characterizing the generation and flows of construction and demolition waste in China. Constr. Build. Mater. **136**, 405–413 (2017). https://doi.org/10.1016/j.conbuildmat.2017.01.055
7. Borg, R.P., Cuenca, E., Garofalo, R., Schillani, F., Nasner, M.L., Ferrara, L.: Performance assessment of ultra-high durability concrete produced from recycled ultra-high durability concrete. Front. Built Environ. **7**, 78 (2021). https://doi.org/10.3389/fbuil.2021.648220
8. Borg, R.P.: The Mechanical and Durability Performance of Ultra HighDurability Concrete. 2019. SBE 19 Malta, Conference Proceedings. [Online]. https://sbe19malta.org/eBook.pdf. Accessed 22 June 2021

Environmental Impact of Calcium Sulfoaluminate Cement Manufacturing: An Indian Case Study

Atul Sharma, Anusha S. Basavaraj(✉), Piyush Chaunsali, and Ravindra Gettu

Department of Civil Engineering, IIT Madras, Chennai 600036, India
anushabasavaraj@gmail.com

Abstract. Ordinary portland cement (OPC) manufacturing contributes to about 5–8% of the CO_2 emissions globally and its consumption is expected to increase. Finding alternative to OPC is a global issue in the recent times. Calcium sulfoaluminate (CSA)-based binder is gaining attention due to the lower requirement of limestone and temperature requirement during its production. There are several lab-scale studies where CSA has been produced using different alternative raw materials. In this study, earlier research works were reviewed to understand the different types of raw materials used to produce CSA. The data for cement production was collected from a cement plant in India to conduct a life cycle assessment (LCA). Two hypothetical cases of CSA production were assumed in the same cement plant, and environmental impacts were calculated in terms of CO_2 emissions and energy consumed. It was found that with the change in alternative raw materials, the impact could vary from 460 to 590 kg CO_2 per tonne of CSA. Generally, 30–45% of CO_2 emissions reduction was observed when compared to Portland cement, whereas the substantial change in energy consumption was not evident. Transportation distance of raw materials seemed to be critical with respect to energy consumed.

Keywords: Calcium sulfoaluminate cement · Environmental impact · CO_2 emissions · LCA

1 Introduction

Demand for cement and concrete has risen because of increased population and urbanization. Concrete demand is currently estimated to be about 30 billion tonnes a year and projections have increased in the near future [1]. Ordinary Portland cement (OPC) is commonly used for concrete production and every year, more than 4 billion tons of Portland cement is manufactured worldwide [2]. After China, India is the world's second-largest cement producer, with cement production capacity projected to exceed 550 MT by the year 2025 [3]. In the financial year 2019, India produced around 329 MT of cement [3]. Around 65% of this demand was from the housing and real estate sector, while the rest from public infrastructure and industrial development [3]. The Portland cement manufacturing contributes significantly to greenhouse gases, which accounts

J. I. Escalante-Garcia et al. (Eds.): RW 2021, RILEM Bookseries 40, pp. 353–362, 2023.
https://doi.org/10.1007/978-3-031-21735-7_39

for 8% of the global anthropogenic CO_2 emissions [4]. Also, cement manufacturing is responsible for the 10% of the global energy consumption. Any small reductions in terms of energy consumption and CO_2 emissions can have large impact globally. To reduce the global carbon footprint, there are several alternatives available, among which using supplementary cementitious materials (SCMs) or alternative binders are most effective strategies. Modifying the portland cement phase assemblage to reduce the limestone requirement is another way of reducing the direct emissions from cement manufacturing [5]. In this regard, using calcium sulfoaluminate (CSA) cement offers an excellent opportunity to lower carbon footprint of concrete. Theoretical heat requirement in the manufacturing process of CSA cement is lower than that of OPC [6]. The primary phases of CSA cement clinker are ye'elimite ($4CaO.3Al_2O_3.SO_3$), belite ($2CaO.SiO_2$), and ferrite ($4CaO.Al_2O_3.Fe_2O_3$). All of the phases in CSA cement are formed at a stable temperature of about 1250 °C, which is about 200 °C lower than the temperature used in the manufacturing of OPC [5]. The lower temperature for formation not only reduces the energy requirement but also CO_2 emissions from fuels. The resulting clinker is more friable (due to high porosity), which reduces the energy required for grinding [7, 8]. The other advantage is with respect to the type of raw materials used for CSA cement manufacturing [9]. Industrial byproducts such as fluidized bed ash and flue gas desulfurization sludge with high sulfate content which are not suitable for OPC production can be used along with fly ash, slag, etc., for CSA cement manufacturing [10]. The low-cost alternative such as anodization mud and aluminum slag can be used in place of high-cost bauxite to supplement aluminum oxide in calcium sulfoaluminate cement.

The reduction of environmental impacts due to CSA cement manufacturing mainly depends on the amount of limestone used and energy required for clinkerization. In this paper, the life cycle assessment (LCA) methodology was used to calculate the environmental impacts of CSA cement manufacturing. Two lab scale CSA production studies were considered from the literature and hypothetically assumed to be produced in a cement plant in India. The industry-level data of OPC production was used to suit CSA production and environmental impacts were quantified in terms of energy consumed and CO_2 emissions per tonne of CSA clinker and cement.

2 CSA Cement Production

CSA cement production is very similar to that of OPC production. The limestone is extracted and transported to the cement plant using an appropriate mode of transportation, later crushed to a size of less than 50mm. The raw meal, a mixture of limestone and other raw materials, is finely ground and later heated to about 1250–1300 °C in a rotary kiln [11]. A sequence of reactions take place as the temperature increases in the length of the kiln. The clinker produced in a kiln has a diameter of 10–20 mm and is shaped like a ball. The clinker is then rapidly cooled before being ground to a fine powder in a ball or a tube mill. During the grinding process, gypsum is added depending upon the required composition. The significant difference between the CSA cement and OPC production is the lower amount of limestone and kiln temperature in case of CSA cement manufacturing.

3 Life Cycle Assessment (LCA)

LCA in compliance with ISO standards ISO 14040 (2006) and ISO 14044 (2006), was used as a tool to calculate the environmental impacts. There are four main steps in estimating environmental impacts: (1) identify the study's goal and scope; (2) compile the life cycle inventory; (3) perform a life cycle impact assessment of the product; and (4) evaluate and analyze the data [12, 13].

3.1 Goal and Scope of LCA

This study focused on estimating the environmental impacts of CSA cement manufacturing. To understand the influence of inclusion or exclusion of processes, two system boundaries were considered i.e. (1) ground-to-gate and (2) gate-to-gate. Mining of raw materials, transportation of raw materials, transportation of fuel and alternative fuels, calcination of raw meal, and other upstream processes were all part of the ground-to-gate system boundary. However, mining or extraction of fuels and raw materials along with transportation processes were excluded from the gate-to-gate system boundary.

3.2 Life Cycle Inventory Analysis

Inventory analysis requires data collections for performing LCA. The first phase in developing the inventories was to draw up process maps based on processes involved in production [14]. The process maps can help in understanding the inputs and outputs specific to each process in cement production. Many inventories exist for cement, such as the popular ecoinvent databases, can be accessed through the SimaPro software, which can be used to supplement the data collected from the cement plant. As the industrial production data was not available for CSA manufacturing in India, the data of OPC production, collected from a cement plant was used to suit the CSA production. The data related to OPC production in an Indian cement plant such as raw materials, alternative raw materials, type and proportion of fuels used was taken from the Gettu et al. 2018 [15]. Limestone, granulated slag, flue dust, laterite and red mud are used for OPC manufacturing with coal, pet coke and pharmaceutical wastes as a fuel for the clinkerization. The same type of fuels used for OPC production were considered with change in amount of fuel as per temperature requirement for CSA clinker production. The data for CSA cement production in terms of raw materials used per tonne of CSA cement/clinker was taken from two different studies [16, 17] from literature mentioned as case studies in the below section:

Case Studies: Two lab-scale case studies of CSA cement production were considered from two different literature studies: Chen et al. 2011 [16] and Ren et al. 2017 [17]. In one of the case studies limestone, bauxite, flue gas desulfurization (FGD) gypsum, and Class C fly ash were used as raw materials to prepare CSA clinker (CSA-1). In the other case study, limestone was replaced with decarbonized material called carbide slag. The raw materials for CSA clinker production (CSA-2) in this case study were carbide slag, coal gangue, FGD gypsum, and aluminum slag. The chemical composition of each raw materials used in this study is given in Tables 1 and 2. The maximum kiln temperature for both case studies is 1250 °C, so the amount of fuel used is same and there is no difference in emissions from fuel burning. The phase composition of the clinker produced in the two case studies is given in Table 3.

Table 1. Oxide composition of raw materials (CSA-1) [16]

Oxides (wt. %)							
	CaO	SO_3	SiO_2	Al_2O_3	Fe_2O_3	MgO	LOI at 950 °C
Limestone	**49**	**0.8**	**6**	**1.7**	**0.6**	**0.9**	**40.4**
Bauxite	0.5	–	6.2	85.9	2.2	0.2	0.5
FGD gypsum	37.9	52	2.8	0.2	0.2	0.2	6.6
Class C fly ash	27.1	1.8	35.6	18.5	6.6	6.1	0.3

Table 2. Oxide composition of raw materials (CSA-2) [17]

Oxides (wt. %)							
	CaO	SO_3	SiO_2	Al_2O_3	Fe_2O_3	MgO	LOI at 950 °C
Coal gangue	0.9	1.3	52.1	21.8	4.5	4.7	10.3
FGD gypsum	37.9	47	6.8	–	0.6	–	6.6
Aluminum slag	1.7	1.6	3.8	72.7	0.7	6.2	4.3
Carbide slag	69.9	–	2.6	1.9	0.3	–	25.2

Table 3. Mineral phase composition of CSA-1 and CSA-2

Case study	Mineral phases (wt.%)				
	Ye'elimite	Belite	Ferrite	Calcium sulfate	Free lime
CSA-1	42.4	40.5	8.3	3.8	0
CSA-2	67.5	26.2	4.7	1.3	0

3.3 Life Cycle Impact Assessment

This step of LCA aims to assess the possible environmental impacts of the elementary flows determined in the LCI. The impact indicators used are energy consumption and carbon emissions. The CO_2 emissions is a governing factor in the global warming potential and the energy demand, and significantly influences the economic and social impacts. In the present study, the characterization or conversion factors needed for the impact assessment (i.e., to obtain the impacts from the inventory) were taken from different sources which are relevant to Indian context. The characterization factors were based on the priority as stated in Gettu et al. 2018 and 2019 [15, 18].

3.4 Interpretation

The primary purpose of this study was to calculate the environmental impacts due to CSA manufacturing along with its comparison to blended cement manufacturing such as PSC (Portland slag cement), PPC (Portland pozzolana cement) produced in the same cement plant.

4 Environmental Impacts of CSA Cement Manufacturing

The amounts of raw materials used per tonne of CSA clinker and corresponding transportation distances are given in Table 4 for CSA-1 clinker manufacturing. The total emissions and energy demand per tonne of CSA-1 clinker were estimated to be 595 kg CO_2 eq. And 4917 MJ, respectively, considering the ground-to-gate system boundary. However, in gate-to-gate system boundary, it was estimated to be 503 kg CO_2 and 3188 MJ of energy demand per tonne of CSA-1 clinker. Based on the different inputs, the impacts calculated show that the CO_2 emissions are dominated by the calcination of limestone and burning of fossil fuels (accounting for 37% and 43% of the emissions, respectively), wherein the latter contributes to 70% of the energy consumed (Fig. 1).

Table 4. Input data for CSA 1

Input required for CSA clinker (1 tonne)		
Material	Amount	Source (and distance from plant
Limestone	786.2 kg	Within proximity (7km)
Bauxite	140.7 kg	Bellary, Karnataka (222 km)
FGD Sludge	106.6 kg	Udupi, Karnataka (639 km)
Class C Fly Ash	200.7 kg	Vijayawada, AP (325 km)
Coal (USA)	28.6 kg	USA (21650km)
Pet Coke	43.6 kg	South Africa (10630 km)
Pharmaceutical solid waste	2.7 kg	Hyderabad (300 km)
Carbon Black	2.7 kg	Hyderabad (300 km)
Pharmaceutical liquid waste	1.8 kg	Hyderabad (300 km)
Electricity drawn from the grid		
For homogenizing raw materials	1.8 kWh	
Preparation of the raw meal	21.3 kWh	
For the milling of coal	3.6 kWh	
For kiln operation during clinkerization	25.6 kWh	

In CSA-1 cement production at the plant, the clinker was finely ground with about 20% of FGD gypsum. The FGD gypsum was transported from Udupi, Karnataka

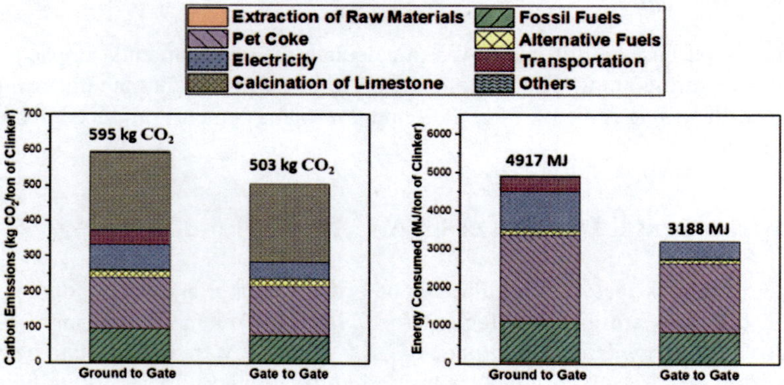

Fig. 1. CO_2 emissions and energy consumption per ton of CSA-1 clinker

(639 km) by truck. To conclude, CSA-1 cement consumed 4632 MJ of energy and produced 521 kg of CO_2 per tonne of cement in the ground-to-gate system boundary. In the gate-to-gate system boundary, only 2746 MJ of energy was consumed, and 429 kg of CO_2 was produced per tonne of CSA cement, because of the excluded processes.

Table 5. Input data for CSA 2

Input required for CSA clinker (1 tonne)		
Material	Amount	Source (and distance from plant)
Carbide Slag	368.3 kg	China (9347 km)
Aluminum Slag	334.8 kg	Bhadravati, Karnataka (465 km)
FGD Gypsum	340.9 kg	Udupi, Karnataka (639 km)
Coal Gangue	80.6 kg	Singareni, Telangana (522 km)
Coal (USA)	28.6 kg	USA (21650km)
Pet Coke	53.6 kg	South Africa (10630 km)
Pharmaceutical solid waste	2.7 kg	Hyderabad (300 km)
Carbon Black	2.7 kg	Hyderabad (300 km)
Pharmaceutical liquid waste	1.8 kg	Hyderabad (300 km)
Electricity drawn from the grid		
For homogenizing raw materials	1.8 kWh	
Preparation of the raw meal	21.3 kWh	
For the milling of coal	3.6 kWh	
For kiln operation during clinkerization	25.6 kWh	

For the CSA-2 clinker manufacturing based on the data given in Table 5, the total emissions and energy demand per tonne of CSA clinker were 467 kg CO_2 and 5484 MJ, respectively, in ground-to-gate system boundary. However, in the gate-to-gate system boundary, only 283 kg CO_2 and 3187 MJ of energy demand per ton of CSA clinker were obtained. Due to the complete replacement of limestone with carbide slag, a decarbonized material, there were no direct emissions from limestone during clinkerization. Instead, carbon emissions were dominated by the transportation of raw materials to cement plants and the burning of fossil fuels (accounting for 30% and 55% of the emissions, respectively). The latter contributed to 63% of the energy consumed.

For the CSA-2 cement production at the plant, the clinker was finely ground without gypsum. Therefore, 1 tonne of clinker was used for producing 1 tonne of CSA cement. To conclude, CSA-2 cement consumed 5966 MJ of energy and produced 500 kg of CO_2 per tonne of CSA cement in the ground-to-gate system boundary. In the gate-to-gate system boundary, only 3384 MJ of energy was consumed, and 304 kg of CO_2 was produced per ton of CSA cement.

The variation in the environmental impacts associated with CSA-1 and CSA-2 manufacturing was due to the type of raw material used and availability of raw materials within the close proximity of the cement plant. As carbide slag was considered as an imported commodity in CSA-2, energy demand and carbon emissions were higher for transportation process (Fig. 2).

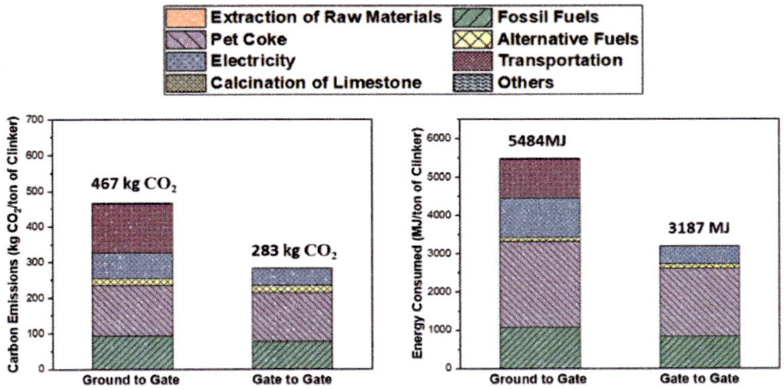

Fig. 2. CO_2 emissions and energy consumption per tonne of CSA-2 clinker

Similar trends in terms of carbon emissions and energy consumption were observed by Hanien et al. 2019 [19] and Ren et al. 2010 [20]. Conventional raw materials such as limestone, bauxite, and gypsum were used by Hanien et al. 2019 [19], which resulted in more than 600 kg CO_2 emissions. However, decarbonized material such as carbide slag was used in CSA manufacturing by Ren et al. 2010 [20] and the carbon emissions were reduced to 410 kg CO_2 per tonne of clinker.

5 Comparison with OPC and Blended Cement

The results of two CSA cement manufacturing case studies were compared with OPC and other blended cements such as PSC and PPC. These impacts of the blended cements produced at the same cement plant were calculated by Gettu et al., 2018 [15] for ground-to-gate system boundary. PSC consists of 53% OPC and 47% GGBS, whereas PPC consists of 65% clinker, 30% class F fly ash, and 5% gypsum, and OPC consists of 96% clinker and 4% gypsum. CSA-1 cement consists of 80% CSA clinker and 20% FGD gypsum, whereas CSA-2 cement consists of 100% CSA clinker. PSC, PPC, and OPC produced 550, 690, 930 kg CO_2 per tonne of cement as shown in Table 6. There is reduction of 44–46% in CO_2 emissions for CSA cement compared to OPC clinker as evident from Fig. 3. On the other hand, the carbon footprint of CSA cement is similar to that of blended cement. The use of decarbonized materials and a lower quantity of limestone per tonne of clinker reduces the carbon footprint for CSA cement. Also, the clinker ratio is higher for CSA cement when compared to blended cement, due to which there is no significant reduction in carbon footprint from clinker to cement. In terms of energy consumption, similar energy requirements for CSA cement and OPC manufacturing is observed. The primary factor contributing to the variation in energy requirement of CSA cement is the transportation of alternative raw materials. If raw materials are available within the proximity of the cement plant, energy requirements can become lower than blended cements. The lower energy requirement for blended cement is due to the meager clinker ratio. Clinker ratio for PSC and PPC were 0.53 and 0.65, respectively, whereas, for CSA cement, it was in the range of 0.80 and 1 for CSA-1 and CSA-2, respectively.

Table 6. Impact results of CSA cements considered in this study compared with other cements

Impact	CSA 1	CSA 2	PSC	PPC	OPC
Carbon footprint (kg CO_2/ton of cement)	521	500	690	550	930
Energy demand (MJ/ton of cement)	4632	5966	3840	4690	5945

6 Conclusions

This study focused on life cycle assessment of calcium sulfoaluminate cement manufacturing. The industrial level data required was taken from a cement plant which produces OPC. The environmental impacts of two lab scale studies from the literature were calculated in terms of carbon emissions and energy demand. It was found that the energy requirement for manufacturing per tonne of CSA clinker was similar to that of OPC clinker. The energy demand for CSA cement production was found to be 4917 MJ and 5484 MJ per ton of clinker for CSA-1 and CSA-2 respectively. More than 70% of the energy was required for clinkerization. The other major process, responsible for the energy demand is transportation of alternative raw materials. It was found that, with

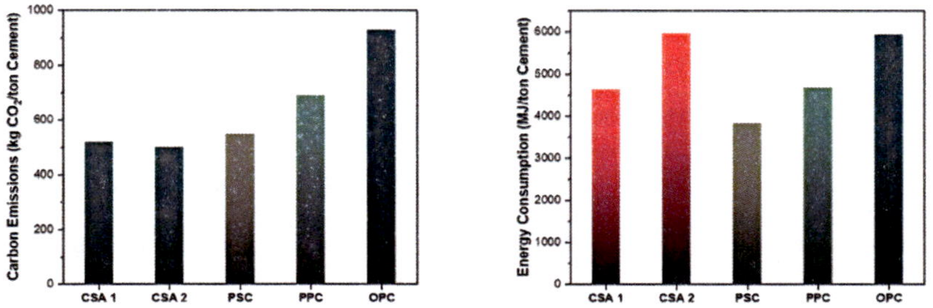

Fig. 3. Comparison of carbon emissions and energy demand of CSA cements with other commonly used cements

increase in the transportation distance of raw materials from the cement plant, the energy consumption also increased. When system boundary was changed from ground-to-gate to gate-to-gate, there was a reduction of 41% and 43% in energy consumption in CSA-1 and CSA-2 cement production. In terms of carbon emissions, it was observed that 595 and 467 kg CO_2 was emitted to produce 1 tonne of CSA clinker in CSA-1 and CSA-2, respectively. The direct emissions from limestone and combustion of fuel in the kiln was the main source of carbon emissions. In CSA-1 direct emissions from burning raw materials accounted for around 37% of total emissions. Use of decarbonized raw materials in place of limestone can help reduce carbon footprint of CSA cement manufacturing. The transportation of raw materials and alternative raw materials to cement plants influenced carbon emissions as well. For the case studies considered in this study, CSA cement manufacturing resulted in 35–45% reduction in CO_2 emissions and similar energy consumption as compared to OPC manufacturing.

Acknowledgements. The first and the second author would like to acknowledge the financial support by MHRD, Government of India. All authors gratefully acknowledge the Research Initiative grant on *Low Carbon and Lean Construction Technologies* by Indian Institute of Technology Madras.

References

1. Monteiro, P., Miller, S., Horvath, A.: Towards sustainable concrete. Nat. Mater. **16**, 698–699 (2017)
2. U.S. Geological Survey: Mineral Commodity Summaries 2015 Mineral Commodity Summaries 2015. US Geological Survey, 196 (2015)
3. IBEF Homepage, https://www.ibef.org/industry/cement-india.aspx. Accessed 10 June 2021
4. Olivier, J.G.J., Janssens-Maenhout, G., Muntean, M., Peters, J.A.H.W.: Trends in global CO_2 emissions: 2016 Report, 528 PBL NEAA and ECJRC IES (2016)
5. Mehta, P.K.: Investigation on energy-saving cements. World Cement Technol. **11**, 167–177 (1980)
6. Hanein, T., Galan, I, Elhoweris, A., Khare, S., Skalamprinos, S., Jen, G., Whittaker, M., Imbabi, M.S., Glasser, Bannerman, M.N.: Production of belite calcium sulfoaluminate cement using sulfur as fuel and as a source of clinker sulfur trioxide: pilot kiln trial. Adv. Cement Res., **28**(10) (2016)

7. Glasser, F.P., Zhang, L.: High-performance cement matrices based on calcium sulfoaluminate-belite compositions. Cem. Concr. Res. **31**(12), 1881–1886 (2001)
8. Shenbagam, V.K., Cepuritis, R., Chaunsali, P.: Influence of exposure conditions on expansion characteristics of lime-rich calcium sulfoaluminate-belite blended cement. Cement Concr. Compos. **118**, 103932 (2021)
9. Phair, J.W.: Green chemistry for sustainable cement production and use. Green Chem. **8**(9), 763–780 (2006)
10. Chatterjee, A.K.: Re-examining the potential of calcium sulfoaluminate cements from the perspective of versatility, durability and GHG emission. Indian Concrete J. **84**(11), 7 (2010)
11. Chaunsali, P., Vaishnav, K.: Calcium-sulfoaluminate-belite cements: opportunities and challenges. Indian Concrete. J. **94**, 18–25 (2020)
12. ISO 14040: Environmental management—life cycle assessment—principles and framework. International Organization for Standardization, Geneva (2006)
13. ISO 14044: Environmental management—life cycle assessment—requirements and guidelines. International Organization for Standardization, Geneva (2006)
14. Gettu, R., Prakasan, S., Patel, A., Rathi, V., Nagarath, K., Palaniappan, S., Maity, S.: Process mapping and preliminary assessment of energy consumption in Indian cement plants. In: Expanding Boundaries: Systems thinking in the Built Environment. Proceedings of Sustainable Built Environment (SBE) Regional Conference Zurich, pp. 430–435 (2016)
15. Gettu, R., Pillai, R.G., Santhanam, M., Basavaraj, A.S., Rathnarajan, S., Dhanya, B.S.: Sustainability-based decision support framework for choosing concrete mixture proportions. Mater. Struct. **51**(6), 1–16 (2018). https://doi.org/10.1617/s11527-018-1291-z
16. Chen, I.A., Juenger, M.C.G.: Synthesis and hydration of calcium sulfoaluminate-belite cements with varied phase compositions. J. Mater. Sci. **46**(8), 2568–2577 (2011)
17. Ren, C., Wang, W., Li, G.: Preparation of high-performance cementitious materials from industrial solid waste. Constr. Build. Mater. **152**, 39–47 (2017)
18. Gettu, R., et al.: Influence of supplementary cementitious materials on the sustainability parameters of cements and concretes in the Indian context. Mater. Struct. **52**(1), 1–11 (2019). https://doi.org/10.1617/s11527-019-1321-5
19. Hanein, T., Galvez-Martos, J.L., Bannerman, M.N.: Carbon footprint of calcium sulfoaluminate clinker production. J. Clean. Prod. **172**, 2278–2287 (2018)
20. Ren, C., Wang, W., Mao, Y., Yuan, X., Song, Z., Sun, J., Zhao, X.: Comparative life cycle assessment of sulfoaluminate clinker production derived from industrial solid wastes and conventional raw materials. J. Cleaner Prod. **167**, 1314–1324 (2017)

Sustainability Assessment of Concrete Pavements with Recycled Concrete Aggregate

J. Jayasuriya, Anusha S. Basavaraj(✉), Surender Singh, and Ravindra Gettu

Indian Institute of Technology Madras, Chennai, India
anushabasavaraj@gmail.com

Abstract. In the past decade, increasing attention has been given to paving the roads with concrete mixtures due to their numerous advantages over bituminous roads. Further benefits could be expected by incorporating recycled-concrete aggregates from demolition waste (DRCA) in concrete pavements. In the present study, life cycle assessment (LCA) is used as a tool to illustrate the calculation of the environmental impacts of concrete pavements made with DRCA and comparison with those of conventional concrete pavements, for both low- and high-volume applications. It was found that, for mix designs taken from the literature, the replacement of natural coarse aggregates with DRCA for high-volume roads could result in an increase in the environmental impacts (CO_2 emission and embodied energy) whereas the impacts could be similar for low-volume roads.

Keywords: Recycled concrete aggregates · C&D waste · Life cycle assessment · Rigid pavement · Sustainable pavements · Low volume roads

1 Introduction

India's road network is almost 5.9 million kilometres comprised of both paved and unpaved surfaces, making it the 2nd largest road network in the world [1]. The majority of the paved network is currently paved with bituminous material. However, more and more roads in India are expected to be constructed with concrete due to the longer service life, and low maintenance and rehabilitation needs. Subsequently, there would be an increase in the demand for the production of concrete, and in turn the demand for cement and aggregates will also increase. The environmental impacts associated with the production of concrete are high because of the use of portland cement. Although aggregates occupy about 70% of concrete volume, it only accounts for 2–3% of the emissions [2, 3]. However, it is estimated that the demand for coarse aggregates in the construction industry is more than 2 billion tonnes (as per data for 2020) [4]. The construction and demolition (C&D) waste aggregates are an alternative source for natural aggregates. It is reported that in India, around 716 million tonnes of C&D waste is being generated annually and the amount is expected to increase to 27 billion tonnes by 2041 [4].

Succinctly, the high demand for the aggregates for new infrastructure or pavement construction can be addressed by using widely available C&D waste as the source of

J. I. Escalante-Garcia et al. (Eds.): RW 2021, RILEM Bookseries 40, pp. 363–371, 2023.
https://doi.org/10.1007/978-3-031-21735-7_40

aggregates. The C&D waste has to be recycled and crushed to the required size to be used as aggregate, which is commonly called recycled concrete aggregates (RCA). However, using RCA in concrete could compromise the mechanical properties of the concrete, leading to higher pavement thickness than that in the case of pavements with pristine natural aggregates (NA), which will result in higher environmental impacts.

In this study, the consequence of using RCA obtained from demolished concrete, denoted as DRCA, in concrete pavements is evaluated. The change in thickness and corresponding environmental impacts in terms of CO_2 emissions and energy demand are calculated for the rigid pavements with and without DRCA. For the representation of the environmental impacts, the present work considers the life cycle assessment methodology.

2 Review of Demolished Recycled Concrete Aggregates (DRCA)

The RCA obtained from C& D waste can either be from construction or demolition waste. In this study, only the latter waste (i.e., DRCA) is considered, as produced by crushing concrete obtained from demolishing infrastructure at the end of its service life.

The primary factor that determines the quality of recycled aggregates is the amount of adhered cement mortar [5] which could influence the specific gravity, water absorption, abrasion resistance and durability. It was found that the adhered mortar in coarse DRCA could be up to 30% of the total aggregate volume [6]. Consequently, the specific gravity of DRCA-concrete may vary from 2.3 to 2.4, which is slightly less than that of conventional concrete [7]. Also, the additional mortar that inherently clings to aggregate increases the porosity, leading to higher water absorption [8]. Similarly, the abrasion resistance of DRCA was found to be lesser compared to pristine aggregates [6]. Tests have shown that high DRCA content could influence the mechanical properties of hardened concrete [9, 10]; the compressive and flexural strength could decrease by 30% and 10%, respectively [11]. However, the Poisson's ratio of concrete containing DRCA was found to be similar to that of concretes with NA [12]. Hansen (1986) [13] showed that the coefficient of thermal expansion of concrete with DRCA was 30% higher compared to concrete with 100% NA.

3 Illustration of the Study of Impacts of DRCA Use in Pavements

The use of DRCA for pavements, in addition to the sub-base and sub-grade [14], could be suitable, as safety or strength is not the only criteria for the design or application. Further, the environmental impact of recycled concrete aggregates in terms of the consumption of non-renewable resources is much lower compared to NA [15, 16]. Also, the use of supplementary cementitious materials (SCMs), such as GGBS, fly ash and silica fume, in concrete could further reduce the total impact of concrete while providing better performance [17]. Therefore, it is interesting to study the consequences of using DRCA on the design and performance of rigid pavement along with its implications on the impact. For the present study, the concrete mix design and mechanical parameters of the concrete with and without DRCA were taken from Yehia et al. [3]. Thereafter, the thicknesses required for the wearing slab of the pavement with these mixtures were determined, followed by an assessment of the environmental impacts associated with 1 km stretch of the pavements (which is more appropriate than considering a unit volume or mass of the material since the thickness depends on the properties of the concrete).

The concrete mix designs taken from Yehia et al. [3] (see Table 1 for properties), have the mix proportions given in Table 2; only the coarser fraction of the aggregates was replaced by DRCA. Four lane and two lane roads are considered for design purposes with high and low volume traffic, respectively. The lane and the shoulder widths considered were 3.5m and 1.5m, respectively, and the spacing between the transverse joints was assumed to be 4.5m and 2.5m for high- and low-volume pavements, respectively. The pavement was designed for three high volume traffic conditions having 1000, 6000, and 10000 commercial vehicles per day (CVPD) and two shoulder conditions, i.e., concrete and earthen shoulders. The low-volume pavements were designed for traffic volumes of 49, 149, and 449 CVPD considering earthen shoulder conditions.

Table 1. Mechanical properties of the hardened concrete considered [3].

Factors used for pavement design	NA concrete	DRCA concrete
Compressive strength (MPa)	72.6	66.6
Flexural strength (MPa)	6.59	5.96
Modulus of elasticity (GPa)	28.7	25.4
Poisson's ratio	0.15	0.15
Coefficient of thermal expansion	10×10^{-6}/°C	13×10^{-6}/°C

Table 2. Mix proportions of concrete considered in the study [3]

Raw materials needed for 1 m^3 of concrete, in kg	NA concrete	DRCA concrete
Cement	252	252
GGBS	132	132
Silica fume	57	57
Fine aggregates (dune)	429	429
Fine aggregates (crushed)	429	429
NA (coarse)	881.1	0
DRCA (coarse)	0	788
Water	184.9	207.3

The safe thickness for the pavement quality concrete (PQC) slab for low and high traffic volumes pavements (with two different shoulder conditions) was determined following IRC 58: 2015 [18] and IRC SP 62: 2014 specifications [19]. Traffic data and pavement data were taken from the examples given in IRC 58 and IRC SP 62. The design life period was assumed to be 30 years and 20 years for high- and low-volume pavements, respectively.

The safe thicknesses calculated for the pavements containing DRCA and NA with different traffic volumes and shoulder conditions are given in Fig. 1. It can be observed that replacing NA with DRCA in concrete resulted in higher pavement thickness in both high- and low-volume pavements, which is due to the lower mechanical properties of DRCA-concrete, as seen in Table 1. The thickness of pavements with earthen shoulders was found to be higher than that of concrete shoulders. Pavement thickness increased with an increase in traffic volume, as expected. For high-volume roads, the thickness of the DRCA-concrete was higher by about 15% compared to the thickness of pavement with NA. However, the increase was only about 6% for low-volume roads with DRCA.

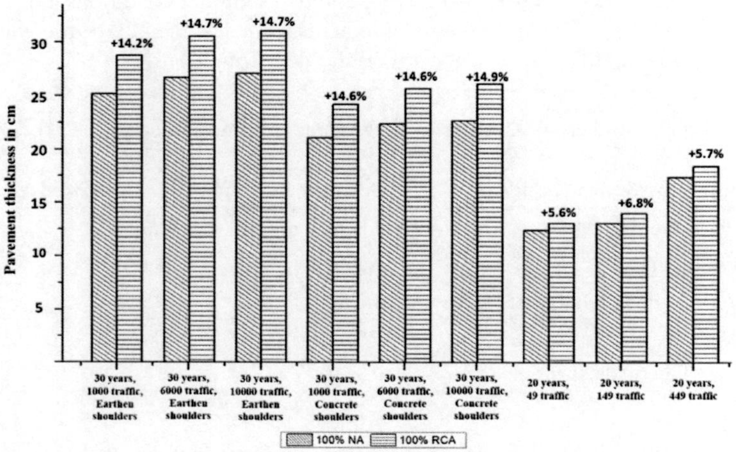

Fig. 1. Pavement thicknesses for various traffic volumes

4 Life Cycle Assessment (LCA)

LCA was used as a tool to quantify the environmental performance of concrete pavements made with NA and DRCA. The four steps used to conduct the life cycle assessment [20, 21] include the goal and scope definition, inventory analysis, impact assessment and interpretation of results. It is assumed that the mixes are made in an readymixed concrete (RMC) plant located in the Indian Institute of Technology Madras, Chennai, India (IITM) where the pavement is taken to be constructed. The natural fine aggregates (FA) were considered to be transported from Villupuram, India (192 km). The natural coarse aggregates (CA) was transported from Kancheepuram, India (75 km). Cement and GGBS are transported from Nandyal, India (400 km). Freight trucks with a capacity of 16 tonnes are assumed to be the transportation mode for all the materials. Silica fume is assumed to be transported from China through the sea route and transported from Chennai port to the RMC plant. The DRCA is assumed to be transported from Perungudi (9.5 km), where a recycling plant for C&D waste is planned to be built in Chennai. The demolished waste is considered to be stored in the Perungudi plant after collecting it from the collection centres or demolition sites. Note that the gross assumption that the mix designs of Yehia et al. [3] could be replicated in Chennai poses limitations but serves the purpose of illustration.

4.1 Goal and Scope

The goal of the LCA study was to calculate the environmental impact of concrete made of DRCA and NA. The different functional units used were a tonne of aggregates (NA and DRCA), 1 m^3 of concrete and 1 km stretch of pavement. The system considered was ground-to-gate. All the processes such as the extraction of raw materials and fuels, production of cement and aggregates (NA and DRCA), and their corresponding transportation along with SCMs, mixing of concrete, which includes all the processes till the exit gate of the ready-mix concrete plant (RMC), were considered, as shown in Fig. 2. The system used explicitly for DRCA production is also featured in Fig. 2, which includes transportation of C&D waste from the demolition site, recycling and transporting it to the RMC plant. The secondary treatment to obtain GBBS was also considered along with its transportation from the cement plant.

4.2 Inventory Analysis

The input data required for the life cycle inventory process of DRCA production was obtained from a recycling plant in Surat, Gujarat, India. It is assumed that the plant to be established in Perungudi would be similar. The input materials per tonne of DRCA was calculated as a part of the inventory analysis. The electricity needed to produce 1 tonne of DRCA was 4.5 kWh. Apart from that, 0.7 kg of diesel, 0.026 kg of lubricating oil, 28.61 kg of water was used to produce a tonne of DRCA. The transportation of materials was found to be 19.4 tonne-km. The input data for 1 m^3 of DRCA-concrete and 1 m^3 of NA-concrete includes all the materials specified in Table 2 along with the default data (from the ecoinvent database). The input data from the database considered for mixing of 1 m^3 of concrete included 4.14 kWh of electricity, 0.0119 kg of lubricating oil, 0.0238 kg of steel and 0.00713 kg of synthetic rubber.

4.3 Impact Assessment

Impact of aggregate production: The conversion factors were obtained for specific inputs consolidated in the inventory analysis. The impact due to 1 tonne of NA production was directly obtained from the ecoinvent database. Inspite of the production process being similar for DRCA and NA, the impact due to DRCA production was higher compared to NA production. This additional impact of DRCA can be attributed to the transportation of waste from the demolition site. The impacts for 1 tonne of NA was an embodied energy of 100 MJ and CO_2 eq. Emissions of 7 kg whereas it was about 200 MJ and 12 kg, respectively, per tonne of DRCA as shown in Fig. 3. The transportation process and electricity used in crushing dominate both the energy demand and CO_2 emissions in DRCA production.

Impact of concrete production: Using the impacts of DRCAs, NA and other materials such as cement, water and SCMs, along with the electricity consumed during concrete

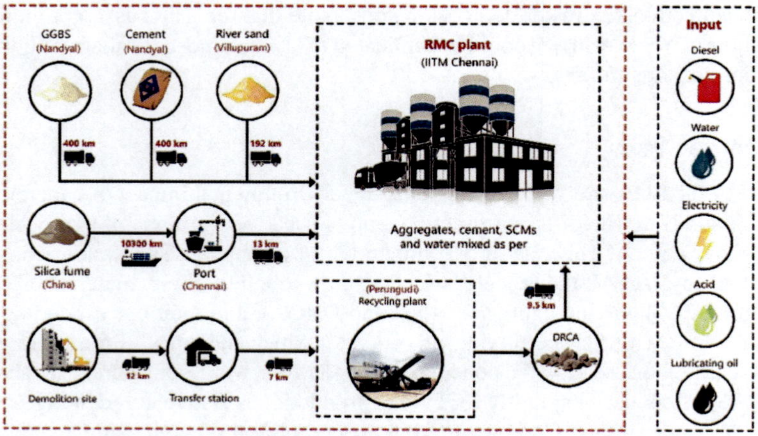

Fig. 2. System considered for concrete production with assumed locations

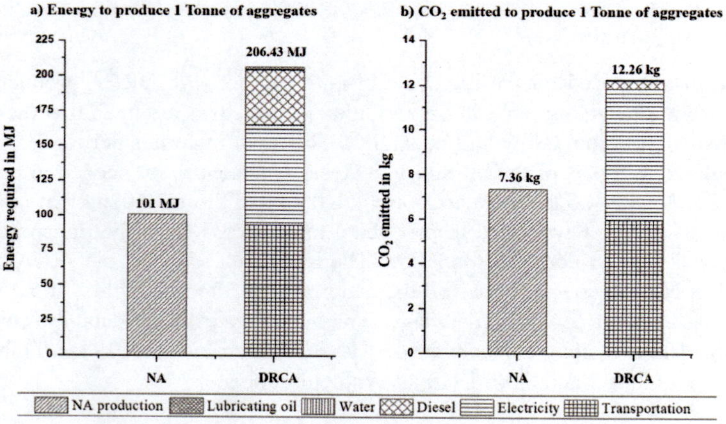

Fig. 3. Impact due to 1 tonne of aggregate production

mixing, the total impact due to the production of 1 m^3 of concrete was calculated for both the mixes. It can be observed that cement and transportation are the major contributors of impact in the concrete, with concrete with DRCA having lower impact compared to that with NA as shown in Fig. 4. Replacing coarse NA with DRCA results in practically the same energy consumption and CO_2 emission for concrete.

Impacts of 1 km stretch of pavement: The environmental impacts due to one km stretch of pavement with DRCA and NA were calculated using the safe thicknesses of the pavement and the impact due to 1 m^3 of concrete, and are shown for traffic volumes of 6000 and 449 CVPD in Fig. 5. The other traffic conditions (49, 149, 1000 and 10000) showed a similar change in the impact. In all the cases, the increase in pavement

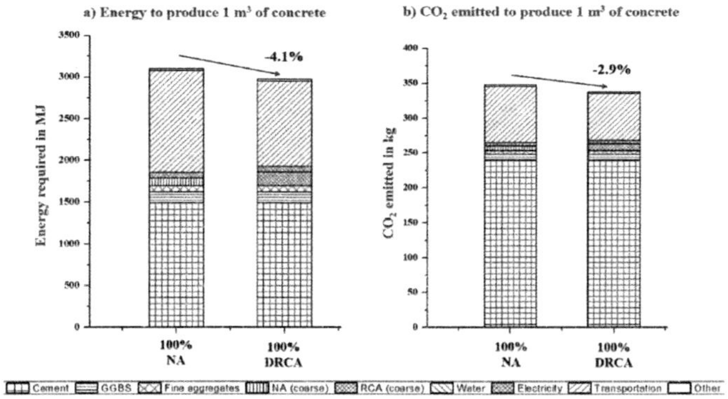

Fig. 4. Impact due to 1 m^3 of concrete production with and without DRCA

thickness contributes to the higher impact of the DRCA-pavement. The increase in impact was found to be less for low-volume roads. For high-volume roads, complete replacement of NA with DRCA resulted in about 10% increase in impacts. In the case of low volume roads, the impact is practically the same. Despite the lower thickness, pavements with concrete shoulders had a higher impact compared to the pavements with earthen shoulders, as expected.

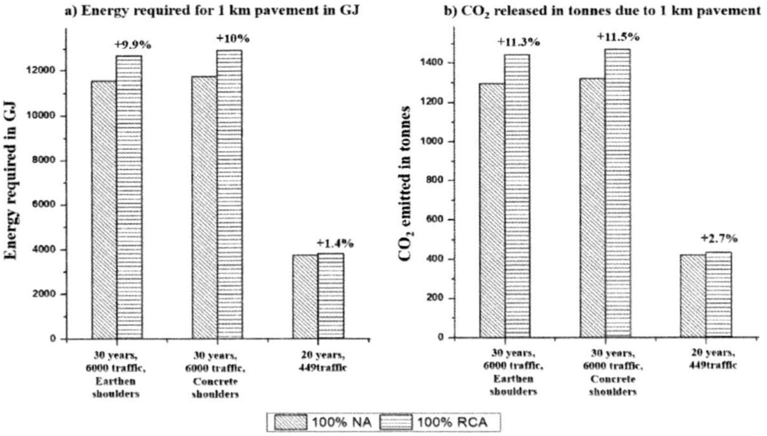

Fig. 5. Impacts due to 1 km stretch of pavements in terms of CO_2 emissions and energy consumed

5 Summary and Conclusions

The consequence of using DRCA in concrete pavements has been investigated, in terms of pavement thickness, embodied energy and carbon emissions, using mix designs from the literature. The use of DRCA in concrete pavement could increases the safe thickness required for the designed service life when the performance of the concrete with recycled aggregates is lower than that of concrete with pristine aggregates. The energy demand and carbon emissions for concrete made of DRCA could be similar to those of concrete made of NA, and governed by the transportation distances. However, other impacts due to landfilling, extraction of non-renewable resources, and resource depletion could be avoided due to the use of DRCA. The increase in the calculated impacts due to DRCA was found to be smaller in low-volume roads. This study shows that there is promise in using DRCA along with SCMs to replace NA completely for the production of concrete paving mixes.

Acknowledgements. The first and second authors would like to acknowledge the financial support by MHRD, Government of India. Funding from the Institute of Eminence Research Initiative grant on Technologies for Low Carbon & Lean Construction (TLC2) (Project SB20210809CEMHRD008100) by the Indian Institute of Technology Madras is gratefully appreciated.

References

1. MoRTH-Ministry of Road Transport & Highways. Specifications for Road and Bridge Works (2013)
2. Smith, S.H., Durham, S.A.: A cradle to gate LCA framework for emissions and energy reduction in concrete pavement mixture design. Int. J. Sustain. Built Environ. **5**(1), 23–33 (2016)
3. Yehia, S., Abdelfatah, A., Mansour, D.: Effect of aggregate type and specimen configuration on concrete compressive strength. Curr. Comput. Aided Drug Des. **10**(7), 625 (2020)
4. Banerjee, A., Paterok, K., Saluja, M.S., Arora, R., Satpathy, I., Malik, J.K., Rathi, V.: Material consumption patterns in India: a baseline study of the automotive and construction sectors. GIZ, New Delhi (2016)
5. Prajapati, R., Gettu, R., Singh, S.: Thermomechanical beneficiation of recycled concrete aggregates (RCA). Constr. Build. Mater. **310**, 125200 (2021)
6. Verian, K.P., Whiting, N.M., Olek, J., Jain, J., Snyder, M.B.: Using recycled concrete as aggregate in concrete pavements to reduce materials cost (No. FHWA/IN/JTRP-2013/18). Purdue University. Joint Transportation Research Program (2013)
7. Vieira, T., Alves, A., De Brito, J., Correia, J.R., Silva, R.V.: Durability-related performance of concrete containing fine recycled aggregates from crushed bricks and sanitary ware. Mater. Des. **90**, 767–776 (2016)
8. Olorunsogo, F.T., Padayachee, N.: Performance of recycled aggregate concrete monitored by durability indexes. Cem. Concr. Res. **32**(2), 179–185 (2002)
9. Gómez-Soberón, J.M.: Porosity of recycled concrete with substitution of recycled concrete aggregate: an experimental study. Cem. Concr. Res. **32**(8), 1301–1311 (2002)
10. Xiao, J., Li, J., Zhang, C.: Mechanical properties of recycled aggregate concrete under uniaxial loading. Cem. Concr. Res. **35**(6), 1187–1194 (2005)
11. Verian, K.P., Ashraf, W., Cao, Y.: Properties of recycled concrete aggregate and their influence in new concrete production. Resour. Conserv. Recycl. **133**, 30–49 (2018)
12. Li, X.: Recycling and reuse of waste concrete in China–Part I material behavior of recycled aggregate concrete. Resour. Conserv. Recycl. **53**, 36–44 (2008)

13. Hansen, T.C.: Recycled aggregates and recycled aggregate concrete second state-of-the-art report developments 1945–1985. Mater. Struct. **19**, 201–246 (1986)
14. Park, T.: Application of construction and building debris as base and subbase materials in rigid pavement. J. Transp. Eng. **129**(5), 558–563 (2003)
15. Hossain, M.U., Poon, C.S., Lo, I.M., Cheng, J.C.: Comparative environmental evaluation of aggregate production from recycled waste materials and virgin sources by LCA. Resour. Conserv. Recycl. **109**, 67–77 (2016)
16. Coelho, A., De Brito, J.: Economic viability analysis of a construction and demolition waste recycling plant in Portugal–part I: location, materials, technology and economic analysis. J. Clean. Prod. **39**, 338–352 (2013)
17. Gettu, R., et al.: Influence of supplementary cementitious materials on the sustainability parameters of cements and concretes in the Indian context. Mater. Struct. **52**(1), 1–11 (2019). https://doi.org/10.1617/s11527-019-1321-5
18. IRC: 58. Guidelines for the design of plain jointed rigid pavements for highways. Indian Road Congress, New Delhi (2015)
19. IRC:SP:62. Guidelines for the design and construction of cement concrete pavement for low volume roads. Indian Road Congress, New Delhi (2014)
20. Finkbeiner, M., Inaba, A., Tan, R., Christiansen, K., Klüppel, H.J.: The new international standards for life cycle assessment: ISO 14040 and ISO 14044. Int. J. Life Cycle Assess. **11**(2), 80–85 (2006)
21. Basavaraj, A.S., Gettu, R.: Life cycle assessment as a tool in sustainability assessment of concrete systems:why and how? Indian Concr. J. **6**(4), 1–20 (2022)

Characterization and Economic Viability of Local Sugarcane Bagasse Ash as Partial Replacement of Portland Cement

Ennes do Rio Abreu[1](✉), Eliana Cristina Barreto Monteiro[2], and Pedro Castro Borges[3]

[1] Universidade Católica de Pernambuco - UNICAP, Recife, Brasil
abreuennes@gmail.com
[2] Universidade Católica de Pernambuco/ Universidade de Pernambuco, Recife, Brasil
[3] Centro de Investigación y de Estudios Avanzados - CINVESTAV, Mérida, México

Abstract. Portland cement is one of the most consumed materials in civil construction, however, there is a large emission of CO2 in cement production, therefore the search for alternative materials that reduce environmental impacts and production costs arises. As Brazil is one of the main world producers in the sugar and alcohol complex, sugarcane represents a great importance of agribusiness in the country's economy. Bagasse is a sugarcane by-product which, during burning in boilers, generates sugarcane bagasse ash (CBCA). The objective of this research is to evaluate the mechanical properties and the influence of partial replacement of local CBCA in proportions of 10%, 15%, 20% and 0% as a reference. For durability analyses, specimens were made and analyzed by electrochemical techniques for corrosion assessment, through the test of corrosion potential (Ecorr), corrosion intensity (Icorr), resistivity of concrete and resistance to linear polarization. The mineralogical composition was evaluated using XRD analysis. Regarding the economic feasibility, bibliographic surveys were carried out for a better basis and findings of the research. The results indicated that the specimens produced with CBCA reached more electronegative corrosion potential values over time, providing a more favorable condition for corrosion. Concluding that the physicochemical characteristics of the local ash depend on the granulometry, degree of crystallinity of the bagasse fibers, burning conditions and that the ash has great availability and can replace cement in low contents with significant advantages.

Keywords: Sugarcane bagasse ash · Durability · Corrosion · Sustainability

1 Introduction

The National Energy Balance–BEN (2018), of the Ministry of Mines and Energy (MME), demonstrates that the Brazilian energy supply is based essentially on three sources: fossil fuels (oil and derivatives), sugarcane biomass and hydroelectricity. In this way, biomass has become an important alternative raw material for energy production. There are 574 biomass generating plants in Brazil totaling 15,404 MW and 406 thermal biomass plants in the sugar-energy sector 11,747 MW [13].

J. I. Escalante-Garcia et al. (Eds.): RW 2021, RILEM Bookseries 40, pp. 372–377, 2023.
https://doi.org/10.1007/978-3-031-21735-7_41

According to the COBEN 03/88 Technical Standard, for each ton of cane, approximately 0.27 tons of bagasse is generated. According to the Ministry of Agriculture, Livestock and Supply (MAPA), sugarcane production in calendar year 2017 reached 635.6 million tons. This amount was 5.2% lower than that recorded in the previous calendar year, when the crushing was 670.6 million tons. Based on the proportions adopted by COBEN, mentioned above, and considering all Brazilian production (2016 crop: 670.6 million tons) of sugarcane, if a quarter of this production turns into bagasse (181.06 million of tons) and is burned in the mills, then Brazil produced, in the 2016 harvest, approximately 45.26 million tons of sugarcane bagasse ash (CBCA). To take advantage of this harvest, the objective of this research is to evaluate the mechanical properties and the influence of partial replacement of local CBCA in proportions of 10%, 15%, 20% and 0% as a reference. Both characteristics could potentiate CBCA as a reliable source for local concrete industry.

The search for the use of natural resources and the use of by-products generated in industrial activities should mainly satisfy technical and socioeconomic needs, but without affecting the quality of life of the population or future generations. The uses of these materials from waste are advantageous due to the decrease in the stock of raw material derived from non-renewable natural resources, which are generally extracted in an indiscriminate, often irrational manner.

2 Experimental Procedures

2.1 Materials and Sample Dimensions

The cement used in the experiment was CPV ARI, standardized by NBR 16697 (ABNT 2018). The choice of this type of cement was the low percentage of additions having a minimal influence on the results. The ash used was supplied and collected from Usina São José LTDA., located in Igarassu, Metropolitan Region of Recife (RMR), about 48 km from the center of Recife, in the State of Pernambuco, in the northeast region of Brazil.

It is relevant to show that the ash had no processing, for the drying process, the CBCA was deposited on plastic sheets and duly spread, in which they remained for 7 (seven) days exposed to the sun and at night they were covered by plastic sheets for that were not wet by the possibility of rain, becoming completely dry. Subsequently, the ash was taken to the laboratory and submitted to the first sieving, in a 4.75 mm sieve, to remove leaves, stones and other contaminating materials, resulting from its storage in the open air in the decantation tanks. The ash from the remaining sugarcane bagasse was submitted to the sieving process in a sieve #200 with an opening of 0.075 mm, in order to obtain a granulometry similar to that of cement.

2.2 Concrete Composition

Test specimens 9cm long, 7cm wide and 5cm high were made in the laboratory. In order to obtain comparative parameters for the research, a trace was performed. Divided into 4 families, which differ by using CBCA with percentages 0, 10, 15 and 20% in partial replacement of cement. Therefore, the calculated standard unit mix was 1:1.5:2.5:0.55

by mass for cement, fine aggregate, coarse aggregate and water, respectively. From the reference mix, three other mixes were produced, partially replacing 10, 15 and 20% of Portland cement by sugarcane bagasse ash. The water used came from the public supply network in the city of Recife (COMPESA).

2.3 Electrochemical Measurements

The specimens, after drying in a laboratory environment for 28 days, were directly submitted to the induction procedure of chloride ion penetration with drying semi-cycles and partial immersion in 5% NaCl solution, a cycle lasting 24 h. The drying condition consisted of placing the specimens in an oven, keeping the temperature at 60°C. A Gamry Potentiostat/Galvanostat was used to obtain the electrochemical parameters. The instantaneous rate of corrosion, icorr, was established using the expression known as the Stern equation; Geary (1957), according to Eq. (1). The quantities measured during the test were corrosion intensity (icorr), corrosion potential (Ecorr) and ohmic resistance.

$$\mathrm{Icorr} = \mathrm{B/Rp} \quad (1)$$

The cell was assembled with 3 (three) electrodes, with the steel bars inside the concrete working electrode, the saturated calomel electrode (SCE) as the reference electrode and the counter electrode a rectangular stainless-steel plate, connected to the potentiostat/galvanostat, and the registration made by the software of the Gamry Echem Analyst equipment.

The equipment used to measure the corrosion potential consists of a multimeter (capable of recording potential difference measurements), a reference electrode and a high conductivity sponge. To obtain the resistivity results, the M.C. Miller resistivity measuring equipment, model 400A, was used.

2.4 X-ray Analysis

The mineralogical composition of sugarcane bagasse ash from the mill (CBCA) and Portland cement CPV ARI was determined using a D8 Advance diffractometer, Bruker equipment available at the National Laboratory of Nano and Biomaterials, located at Cinvestav–IPN in city of Mérida in Mexico, which contains a copper target tube with a wavelength (λ) of 1.54060 Å, current of 10 mA and voltage of 30 kV.

3 Results and Discussion

3.1 X-ray Analysis

The X-ray diffractograms shown (see Fig. 1) demonstrate that, due to the presence of minerals from the aggregates, it resulted in high-intensity quartz peaks, which impaired the visualization of other cement hydration phases, in addition to being inert and not reactive, in terms of pozzolanic activity. However, it cannot be stated that there is no presence of amorphous silica in the samples, as they are difficult to detect by the diffractometer, especially when they are present in small amounts. Therefore, the presence of silicon oxide (SiO2) in the form of quartz can be attributed to several factors, including: a) contamination of the bagasse that is exposed in the plant's yard; b) contamination during sugarcane cutting, as not all bagasse enters the furnace after the milling process, the excess remaining in the open air awaiting its return to burning; c) contamination due to bad weather.

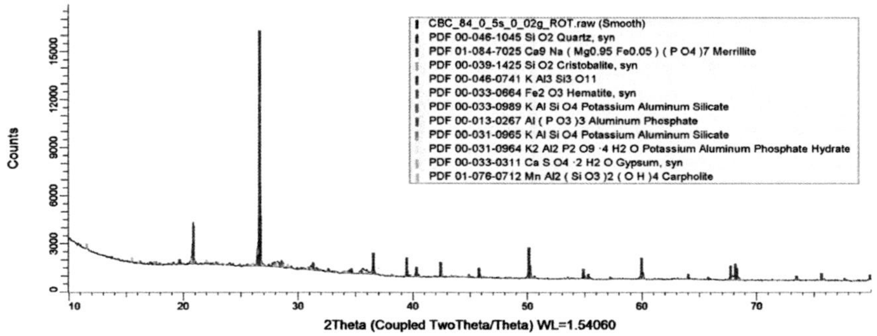

Fig. 1. Sugarcane bagasse ash diffractogram (CBCA)

3.2 Corrosion Potential (Ecorr)

Standard C 876 [4] exposes a correlation between a Cu/SO_4 reference electrode, potential ranges and the probability of corrosion occurring. The resistance to polarization corroborates the results obtained by Ecorr measurements, where the specimens show an increase in the resistance to polarization in the incorporations of 15 and 20% of sugarcane bagasse ash in partial replacement of Portland cement. Therefore, the results show that the specimens with 20% of (CBCA) in the matrix composition of the concrete present more electronegative values in the reading of the corrosion potential in the evolution of the chloride ion induction test, this being −430 mV, approximately. Concluding that the concrete produced with CBCA in partial replacement of Portland cement reached more negative values of corrosion potential over time, providing a more favorable condition for corrosion.

3.3 Corrosion Intensity (Icorr)

The technique allowed us to estimate the instantaneous corrosion velocity at each time of the test, which leads us to the total corrosion intensity through the integration of the icorr x time curve. It was possible to observe that the corrosion potential values became more electronegative when wet, confirming the trend shown in the corrosion intensity values found. With the specimens dry, only the reference samples showed an onset of corrosion in the third measurement.

Considering that to make the use of sugarcane bagasse ash viable, it must be economically viable, a survey of material costs in the region was prepared, seeking a common denominator between the application of 15% of industrial waste and the product cost. The same mix of specimens 1:1.5:2.5 (cement, sand, gravel) with a water/cement ratio of 0.55 was used. For the analysis of economic viability, the family that presented the most favorable result during the study was adopted, thus being able to arbitrate a cost reduction per $1m^3$ of concrete produced with CBCA in partial replacement of Portland cement. It was observed that the cost to produce $1m^3$ of concrete without sugarcane bagasse ash is 21.25% relative to the value of cement and the cost with sugarcane bagasse ash was 18, 60%, showing that the cost against the reference trait was $ 98,10. Thus, when analyzing the values of the compositions in the aforementioned tables, a reduction of 2.89% was found, with the proposed gray replacement. The cost reduction per m^3 of concrete using the family produced with 15% sugarcane bagasse ash in partial replacement of Portland cement was 2.89%, the use of CBCA did not significantly decrease the final value of m^3 of concrete, as it is a small amount of material, but if large-scale use is taken into account, the concrete used with CBCA would be more economically viable than concrete without the ash.

4 Conclusions

Based on the analysis presented, the following conclusions can be drawn:

From an ecological point of view, the ash from the bagasse of local sugarcane was presented as a technically viable option for the manufacture of cement or concrete, with the objective of reducing the environmental impact generated in civil construction from the inclusion of products such as the waste in question, creating sustainable initiatives. In the hardened state, the incorporation of ash with 20% of SCBA in the composition of the concrete matrix presented, in the chloride ion induction test, more electronegative values in the reading of the corrosion potential and velocity over time. We can conclude that the concrete produced with sugarcane bagasse ash in partial replacement of Portland cement reached more negative values of corrosion potential over time, providing an unfavorable condition to corrosion. It was also possible to analyze the change in the measurement of wet and dry specimens in the variation of the corrosion potential.

It can be concluded that the viability of concrete produced with CBCA in partial replacement of Portland cement, obtained a profit potential per 1 m^3 of concrete of 2.89% using the family with 15% ash incorporation, satisfactory results were obtained in some aspects, need more research regarding economic viability. Favoring the emergence of rational methods that aim to build a renewable cycle that encompasses aspects such as conscious exploitation, reuse of waste and profit.

References

1. Abreu, E.R.: Evaluation of the concrete characteristics produced with sugarcane bagasse ash in partial replacement of Portland cement and its economic viability, pp. 1–142 (2021)
2. Andrade, J.J.O., Possan, E., Dal Molin, D.C.C.: Considerations about the service life prediction of reinforced concrete structures inserted in chloride environments. J. Build. Pathol. Rehabil. **6**, 1–8 (2017)
3. Brazilian Association of Technical Standards-NBR 16697–Portland Cement–Requirements. Rio de Janeiro (2018)

4. ASTM International–ASTM C 876-Standard Test Method for Corrosion Potentials of Uncoated Reinforcing Steel in Concrete West Conshohocken, PA (2015). https://doi.org/10.1520/C0876-15
5. Berenguer, R.A.: Study of the durability and influence of sugarcane bagasse ash addition on the properties of concrete. Master's Dissertation, Federal University of Pernambuco.CTG. Graduate Program in Civil Engineering (2019)
6. Cociña, E.V., Ruiz, J.H., Frías, M., Savastano J.R.H.: Pozzolanic behaviour of a bagasse ash from the boiler of a Cuban sugar factory. Adv. Cement Res. **25**(3), 136–142 (2013)
7. Ganesan, K., Rajagopal, K., Thangavel, K.: Evaluation of bagasse ash as supplementary cementitious material. Cement Concr. Composit. **29**(6), 515–524 (2007)
8. Geiker, M.R., Polder, R.B.: Experimental support for new electro active repair method for reinforced concrete. Mater. Corros. **67**, 600–606 (2016)
9. Kigozi, J.: Investigation of sugar cane bagasse ash as a binding material for construction industry. J. Global Ecol. Environ. **2**(4), 205–208 (2015)
10. Mehta, P.K., Monteiro, P.J.M.: Concrete: microstructure, properties and materials, 2 ed. São Paulo, IBRACON, pp. 751 (2014)
11. Rerkpiboon, A., Tangchirapat, W., Jaturapitakul, C.: Strength, chloride resistance, and expansion of concretes containing ground bagasse ash. In: Construction and Building Materials, Guildford, Inglaterra, vol. 101, part. 1, pp. 983–989 (2015)
12. Sundara K.K., Venkata R.G.: Studies on production of greener concrete using agro-industrial waste. Int. J. Technol. Res. Eng. **4**(2), 219–224 (2016)
13. UNICA–Sugarcane Industry Union. São Paulo. https://www.unicadata.com.br/historico-de-producao-emoagem.php?idMn=31&tipoHistorico=2, Accessed 12 July 2020
14. Usina São José. Sustentabilidade. Igarassu, https://www.saojoseagroindustrial.com.br/sustentabilidade. Accessed 16 July 2020

Effect of Combined Action of Carbonation and Bending Load on Mortar

Elodie Piolet(✉) and Siham Kamali-Bernard

Laboratory of Civil Engineering and Mechanical Engineering (LGCGM), INSA Rennes, Rennes, France
{elodie.piolet,siham.kamali-bernard}@insa-rennes.fr

Abstract. Carbonation is a pathology that affects the majority of concrete structures as they are exposed to the ambient air and consequently to the carbon dioxide present in the air. In addition, these structures may be subject to mechanical stresses over time. The main objective of this research is therefore to investigate the effects of carbonation on the mechanical strength and conversely the effects of mechanical stress during the accelerated or natural carbonation process. Therefore, within the framework of Rilem TC 281-CCC working group 4, the coupling of accelerated carbonation and mechanical loading of Portland and slag cement based mortars has been studied. The impact of accelerated carbonation on the different properties of the mortars (porosity, masses, etc.) was also studied.

The results show a slight increase in the depth of carbonation on the mortar under tensile stress and a slight decrease in the depth of carbonation for the mortar under compressive stress.

Keywords: Carbonation · Loading · 4-point bending · Mortar · Porosity · Cement

1 Introduction

Concrete is currently the most used manufactured product in the world after water, with 190 m^3 of concrete poured every second worldwide, equivalent to 6 billion m^3 per year or one cubic meter of concrete for every person on the planet [1, 2]. However, the production of Ordinary Portland Cement (OPC) accounts for about 5–8% of all atmospheric CO_2 emissions [3]. The use of alternative cements is therefore being investigated, such as blast furnace slag cements, but key elements of the properties of these alternative materials with respect to durability and residual mechanical properties remain unknown.

There are many mechanisms that can lead to the degradation of concrete and thus to the reduction of its properties. One of the main ones is carbonation. This phenomenon concerns the majority of concrete structures as they are exposed to the ambient air and consequently to the carbon dioxide present in the air.

Indeed, the volume fraction of CO_2 is about 0.035% in the atmosphere, but depending on the location, especially in large cities, it can reach up to 1% [4].

J. I. Escalante-Garcia et al. (Eds.): RW 2021, RILEM Bookseries 40, pp. 378–387, 2023.
https://doi.org/10.1007/978-3-031-21735-7_42

Carbonation is a natural process in which CO_2 from the atmosphere diffuses in gaseous form into the pores of concrete and dissolves, forming an acid on contact with the interstitial solution of the cement paste. This phenomenon progressively modifies the chemical composition and the internal microstructure of the concrete by acting on the products resulting from the hydration of the cement. Indeed, carbon dioxide (CO_2) from the atmosphere reacts with the portlandite (calcium hydroxide ($Ca(OH)_2$)) which governs the alkalinity of the concrete's porous fluid and with the hydrated calcium silicates (C-S-H). The products of these reactions are calcium carbonate ($CaCO_3$) crystals. A higher volume of $CaCO_3$ compared to $Ca(OH)_2$ leads to a decrease in the porosity of the concrete [5]. Thus, carbonation induces a change in the chemical equilibrium between the porous solution and the hydrates leading to a decrease in the pH-value of the material (cement paste, mortar or concrete), from a value of 13 to a value below 9 [6]. The kinetics of carbonation are different depending on the cement used and in particular on the amount of portlandite. In addition, carbonation can also lead to a loss of mechanical properties [7].

The phenomenon of carbonation can be influenced by many factors such as cement composition, CO_2 concentration, relative humidity, temperature, curing conditions, concrete porosity, cracking and loading [8].

Concrete or mortar is very often subject to mechanical, compressive and/or tensile, static or cyclic loads during its use. This parameter affects the carbonation rate. To date, very little research has been carried out on the coupling of carbonation in concrete under mechanical compressive stress [9–12] and tensile stress [13–15]. These studies have shown that compressive stresses have a positive effect on the carbonation resistance of concrete [6] and tensile stresses/strain significantly increase the carbonation rate as a function of the stress/strain level [16].

Thus, the aim of the study described in this paper was to characterize the influence of static loading on the carbonation of PC mortar and GGBS mortar and vice versa. This research project is part of the Rilem TC 281-CCC on the study of carbonation of concrete with SCMs and more precisely in the working group 4, which studies the effect of the combined action of loading and carbonation. Since natural carbonation under real atmospheric conditions is a very slow process, accelerated carbonation tests with 2 and 20% CO_2 concentrations were carried out in order to obtain significant experimental results within a reasonable time. In the carbonation chambers, the relative humidity was 65 ± 5% and the temperature was 22 ± 2 °C. Mortar samples, with dimensions 7 × 2 × 28 cm, two W/C ratios (0.5 and 0.6) and two different cements were tested. After manufacturing, demoulding and curing, the samples were placed on 4-point bending frames in the carbonation chambers, one part of the samples with a load and the other without a load for comparison. After a carbonation test period, the effects of carbonation on the 4-point bending strength are studied and conversely the effects of loading during the accelerated carbonation process are observed in order to investigate their impact on the carbonation kinetics. The different results obtained will be presented below.

2 Materials and Experimental Procedures

2.1 Materials

The materials studied are mortars formulated with sand, tap water and cement.

A standardized sand CEN EN 196–1 [17] was used. It is a natural sand, siliceous in particular in its thinnest fractions. It is clean, the grains are generally isometric and rounded. Two different cements were used:

- CEM I 52.5 N CE CP2 NF cement, which was used to make reference samples. This is a Portland cement composed of 95 to 100% by mass of clinker. Its strength class 52.5 N means that at 2 days, the compressive strength of a standard mortar is greater than 20 MPa, while at 28 days it is greater than 52.5 MPa.
- CEM III/C 42.5 L-LH/SR cement which is a slag-based cement.

2.2 Test Specimens

Different series of tests were carried out. Two series of samples (series 1 and series 2) containing Portland cement were carried out in an identical manner except for the curing period: 40 days for series 1 and 28 days for series 2. A third series of tests was made using slag cement. Due to the significant sensitivity of slag cement based materials to carbonation, a lower water-to-cement ratio was chosen to try to reduce the carbonation rate. Slag mortar samples were vibrated to make them more homogeneous. The characteristics of the mixtures are as follows (Table 1).

Table 1. Characteristics of the mixtures

Series	Type of cement	w/c	Cement (kg)	Water (kg)	Sand (kg)	Placement	Demoulding time
1	Portland cement (CEM I)	0.6	0.9	0.54	2700	Self-compacting	24h
2	Portland cement (CEM I)	0.6	0.9	0.54	2700	Self-compacting	24h
3	Slag cement (CEM III)	0.5	0.9	0.45	2700	Vibrated 11s at 80Hz	48h

Different sample geometries were manufactured: 4 cm x 4 cm x 16 cm for the bending/compression tests according to NF EN 196–1 and water porosity tests and 7 cm x 2 cm x 28 cm for the carbonation tests under flexural loading (Fig. 1). The samples were demoulded after being kept in climatic chamber at RH > 97% and 20 °C in order to avoid desiccation, i.e. the too rapid evaporation of water on the surface, which could lead to cracks.

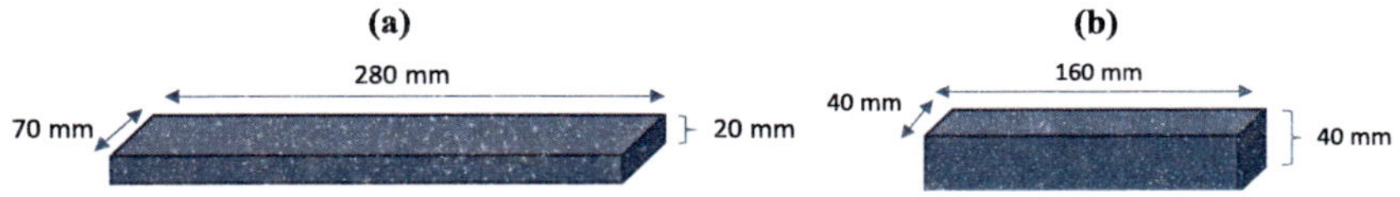

Fig. 1. **a** Geometry of sample 7 cm x 2 cm x 28 cm, **b** Geometry sample 4 cm x 4 cm x 16 cm

2.3 Pre-conditioning

After demoulding, the samples will undergo a period of pre-conditioning. The curing of the samples has an impact on their properties. The curing method used is the one prescribed by RILEM group 4, i.e. 6 days in a concentrated lime solution and 21 days (or 34 days for the first series of tests on CEM I mortars) in a humidity and temperature controlled chamber (20 °C and 65%RH) (Fig. 2). Indeed, contrary to conservation at ambient humidity, maturation under water favours hydration and thus leads to concretes that are less porous and have more carbonatable products, thus more resistant to carbonation [18].

Fig. 2. Pre-conditioning chamber with controlled humidity and temperature

2.4 Experimental Processes

At the end of the pre-conditioning period, at the age of 28 days (40 days for series 1), the various tests were carried out. Flexural and compressive characterisation tests of the mortars were performed as well as water porosity tests. In addition, mass monitoring was carried out at different stages of the process.

For the carbonation test, the chambers were set at 2% and 20% CO_2 each (Fig. 3c and d respectively), with a relative humidity of 65 ± 5% and a temperature of 22 ± 2 °C. Some samples were placed in the chamber on a 4-point bending frame with a load and others without a load for comparison. For series 1 and 2 (CEM I based specimens) the imposed load corresponds to 14.4% of the bending strength and for series 3 (CEM III based specimens) the imposed load corresponds to 25.6% of the bending strength (Fig. 3a). Samples from series 1 and 2 were carbonated for 27 days at 2% and 20% CO_2 and those from series 3 at 20% CO_2 for only 7 days because the carbonation kinetics of CEM III mortar is very fast. At the end of the carbonation cycle under loading, the specimens were broken in 4-point bending (Fig. 3b) and phenolphthalein colorimetric tests were performed on different areas of the specimens.

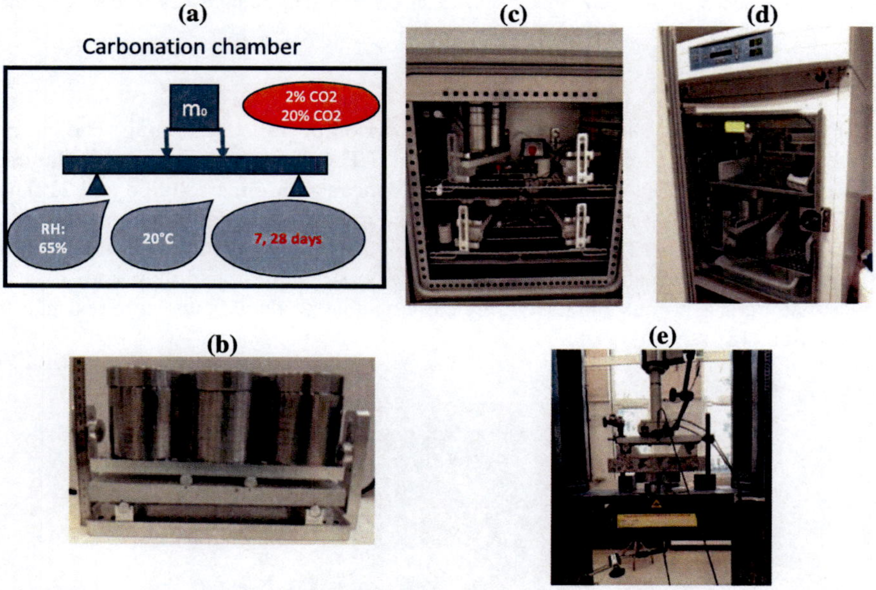

Fig. 3. **a** Scheme of the assembly and carbonation conditions, **b** 4-point bending under static load, **c** 20% carbonation chamber and **d** 2% carbonation chamber, **e** 4-point bending test.

3 Results

3.1 Mechanical and Structural Characterization of the Samples

In order to characterize the compositions of the manufactured samples, 3-point bending and compression tests were carried out in accordance with standard NF EN 196–1 on mortars measuring 4 cm x 4 cm x 16 cm at 28 days. The results are presented below (Table 2).

Water-accessible porosity is one of the main indicators of the durability of concrete and depends mainly on the cement paste of the concrete as well as the w/c ratio. Indeed, the porosity accessible to water is an important intrinsic parameter that controls the kinetics of penetration of aggressive species. The porosity tests were carried out at 28 days according to the protocol of standard NF P 18–459 of March 2010. The tests were carried out on 4 cm x 4 cm x 16 cm specimens cut into three pieces (Fig. 4a) in order to determine the differences in porosity between the specimens. For all the specimens, the results show that the top of the sample are more porous than the bottom ones.

Table 2. 3-point bending and compression results at 28 days for the 3 series.

Sample	3-point bending strength (MPa)		Compression strength (MPa)	
	Average	Standard deviation	Average	Standard deviation
Series 1 et 2 (CEM I)	6,79	0,71	51,87	1,85
Series 3 (CEM III)	4,75	0,48	46,70	1,50

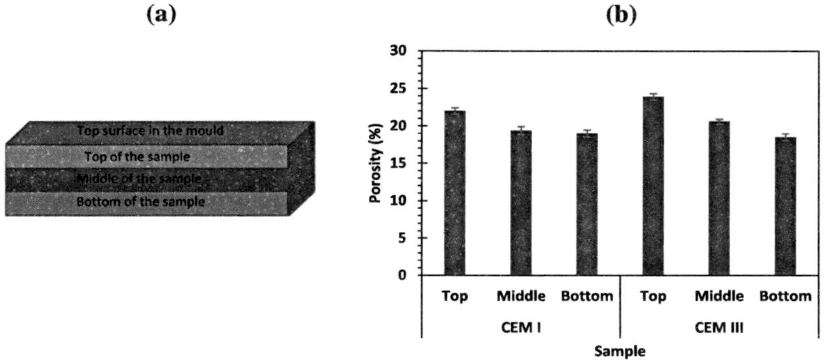

Fig. 4. **A** Scheme showing the three parts of the 4cm x 4 cm x 16 cm samples used for the porosity tests **B** porosity results of the top, middle and bottom of the CEM I and CEM III samples.

3.2 Physical Characterization of the Samples

Mass monitoring is carried out during: demoulding, change of curing, before the start of the accelerated carbonation test and after the carbonation test for each sample. The graphs below show the mass variation as a function of exchange area for series 1 and 2 with the samples carbonated at 2% (Fig. 5a) and for all series with the samples carbonated at 20% (Fig. 5b). It can be seen that samples from the same series have similar variations. However, during the 20% carbonation, series 1 and 2 regain mass while series 3 continues to lose mass. This is due to different reactions caused by the carbonation.

3.3 Carbonation/Load Coupling

The specimens are tested in 4-point bending at the end of the accelerated carbonation test to determine their 4-point flexural strength. Compared to a reference specimen, the flexural strength increases during the carbonation. However, for a specimen that are not loaded during the carbonation test, the flexural strength increases more than for a loaded specimen (Fig. 6).

In addition, the samples are also tested with phenolphthalein to determine the depth of carbonation. The results show that carbonation at 20% is significantly greater and faster than that at 2% (Fig. 7a). Furthermore, it can be observed that the loading has an impact on the carbonation front. Indeed, the top of the sample which is in compression is

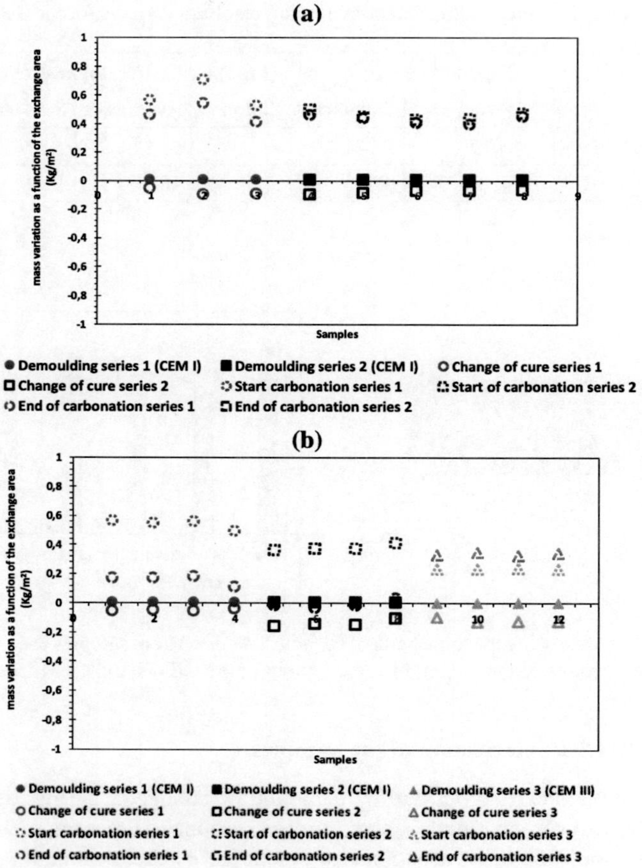

Fig. 5. Mass changes as a function of exchange area for samples carbonated at **a** 2% CO_2 and **b** 20% CO_2

less carbonated if it is loaded, the variation of the carbonation compared to an unloaded sample is −8% and −3% for the CEM I samples carbonated at 2 and 20% respectively (Fig. 7a) and −7% for the CEM III samples carbonated at 20% (Fig. 7b). In contrast, the bottom of the specimen which is in tension is more carbonated if it is loaded, the variation in carbonation compared to an unloaded specimen is 8% and 27% for the 2% and 20% carbonated CEM I specimens respectively and 12% for the 20% carbonated CEM III specimens. These results can be observed for all the three series. Furthermore, series 3, due to its composition, carbonates much faster than series 1 and 2 as shown in the photos of the samples (Table 3). In conclusion, the first results show that a moderate load has an impact on both carbonation front and flexural strength of the sample.

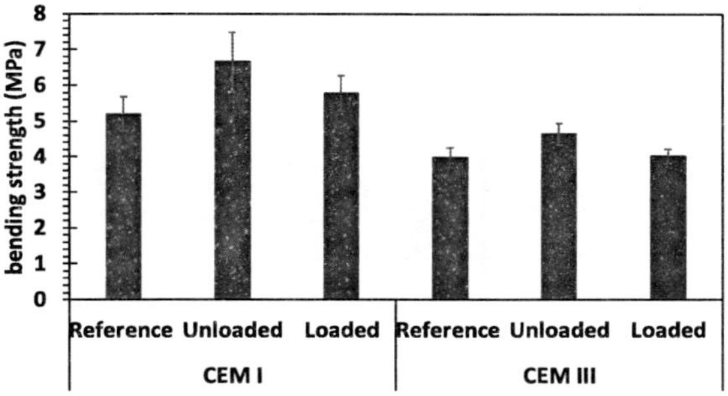

Fig. 6. 4-point bending results of carbonated specimens at 20% CO_2

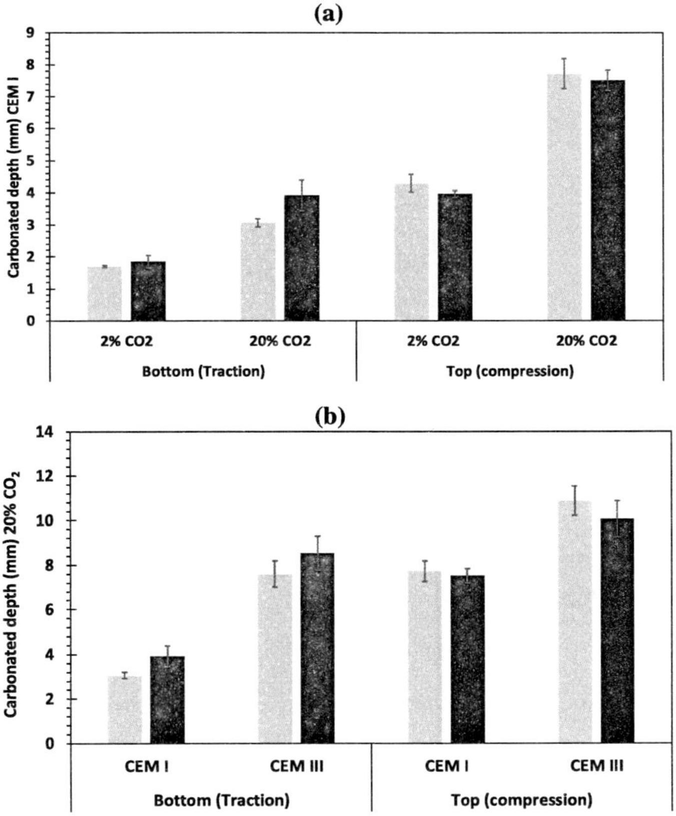

Fig. 7. Carbonated depth of the top and the bottom of the loaded and unloaded samples **a** CEM I at 2% and 20% and **b** CEM I and CEM III at 20% (legend: grey samples not loaded and black samples loaded)

Table 3. Picture of carbonated samples

	Loaded	Unloaded
CEM I 2% CO_2 28 days		
CEM I 20% CO_2 28 days		
CEM III 20% CO_2 7 days		

4 Conclusion

In conclusion, thanks to the coupling tests of carbonation and flexure loading, first results were obtained. An impact of the load can be observed on the depth of carbonation as well as on the flexural strength of the sample. The loading leads to a deformation of the sample. The flexural strength increases for a sample that has not been loaded during the carbonation test, while it decreases for a sample that has been loaded during the test.

The results also show an increase in the depth of carbonation on the mortar under tensile stress and a slight decrease in the depth of carbonation for the mortar under compressive stress. A difference was also observed on the depth of carbonation depending on the composition of the samples. Further tests are still in progress to confirm and deepen these different results.

Acknowledgements. The authors thank the SFM of INSA for the manufacture of test frames and Christian Garand for his help with the tests.

Moreover they thank the FEDER for its financial contribution to the purchase of equipment used in this work.

References

1. Planétoscope, conso Globe, site consulté le 25/08/2020. https://www.planetoscope.com/matieres-premieres/1374-production-mondiale-de-beton.html
2. Collection Technique CIM-BÉTON, « Les bétons : formulation, fabrication et mise en oeuvre » chapitre 3, Tome II., Ecole Française du Béton, pp. 96–109. www.infociments.fr/betons/composition

3. McCaslin, E.R., White, C.E.: «A parametric study of accelerated carbonation in alkali-activated slag». In: Cement and Concrete Research, p. 10 (2021)
4. Mai-Nhu, J.: «Corrosion des armatures du béton : couplage carbonatation/chlorures en présence de cycles hydriques» PhD thesis of university of Toulouse (2013)
5. Wang, X.-H., Val, D.V., Zheng, L., Jones, M.R.: «Carbonation of loaded RC elements made of different concrete types: accelerated testing and future predictions». In: Construction and Building Materials, p. 14 (2020)
6. Carevic, V., Ignjatovic, I.: «Influence of loading cracks on the carbonation resistance of RC elements». In: Construction and Building Materials, p. 12 (2019)
7. Christophe, C.: «Pathologie: La carbonatation», laboratoire de la matière, Toulouse (2006). www.betons-lemagazine.fr
8. von Greve-Dierfeld, S., et al.: Understanding the carbonation of concrete with supplementary cementitious materials: a critical review by RILEM TC 281-CCC. Mater. Struct. **53**(6), 1–34 (2020). https://doi.org/10.1617/s11527-020-01558-w
9. Wang, X.H., Val, D.V., Zheng, L., Jones, M.R.: Influence of loading and cracks on carbonation of RC elements made of different concrete types. Constr. Build. Mater. **164**, 12–28 (2018). https://doi.org/10.1016/j.conbuildmat.2017.12.142
10. Ren, Y., Huang, Q., Liu, X.L., Tong, Z.J.: A model of concrete carbonation depth under the coupling effects of load and environment. Mater. Rés. Innov. **19** (2015). https://doi.org/10.1179/1432891715Z.0000000001970
11. Tang, J., Wu, J., Zou, Z., Yue, A., Mueller, A.: Influence of axial loading and carbonation age on the carbonation resistance of recycled aggregate concrete. Constr. Construire. Mater. **173**, 707–717 (2018). https://doi.org/10.1016/j.conbuildmat.2018.03.269
12. Wan, X., Wittmann, F.H., Zhao, T.: Influence of mechanical load on service life of reinforced concrete structures under dominant influence of carbonation. Restor. Construire. Monum. **17**, 103–110 (2011). https://doi.org/10.1515/rbm-20116437
13. Castel, A., Francois, R., Arliguie, G.: Effect of loading on carbonation penetration in reinforced concrete elements. Cem. Concr. Res. **29**(4), 561–565 (1999)
14. Alahmad, S., Toumi, A., Verdier, J., Francois, R.: Effect of crack opening on carbon dioxide penetration in cracked mortar samples. Mater. Struct. **42**, 559–566 (2009)
15. Jiang, C., Gu, X., Zhang, W., Zou, W.: Modeling of carbonation in tensile zone of plain concrete beams damaged by cyclic loading. Constr. Build. Mater. **77**, 479–488 (2015)
16. Wang, X.-H.: «Influence of loading and cracks on carbonation of RC elements made of different concrete types». In: Construction and Building Materials, p. 18 (2018)
17. Norme européenne EN 196-1: Méthodes d'essais des ciments-Partie 1: Détermination des résistances mécaniques (2005)
18. Ammar, B.K.B., Mellas, M., Mezghiche, B.: «Caractérisation des bétons d'enrobage pour les bétons de préfabrication soumis à la carbonatation». Courrier du Savoir, no. 24, p. 9 (Sept 2017)

Concrete Beams with Steel Reinforcement Subjected to Fractions of Yielding Strength Under Accelerated Corrosion Tests

Rebeca Visairo-Méndez[1], Jorge Varela-Rivera[1], and Pedro Castro-Borges[2](✉)

[1] Facultad de Ingeniería, Universidad Autónoma de Yucatán, Mérida, Mexico
[2] Centro de Investigación y de, Estudios Avanzados Unidad Mérida, Mérida, Mexico
pcastro@cinvestav.mx

Abstract. The objective of this work was to verify that tensile stresses at 40% and 80% of the corresponding yielding strength are not structurally influenced by the level of corrosion under accelerated tests. One of the main concerns when having corrosion of reinforcing steel is the sustaining stress level before non-reversible corrosion damage. Eight reinforced concrete beams (15 cm × 30 cm × 350 cm) were designed by using local crushing aggregates with a water/cement ratio (w/c) of 0.62. Accelerated corrosion was applied to the tensioned longitudinal reinforcing steel bars by contaminating the concrete during the mixing process and later with wetting and drying cycles. A four-point bending load test setup was used to the posttensioned beams in pairs and simulating sustained loads. The loads were applied until reaching the tension stresses in the longitudinal reinforcing steel bars at 40% or 80% of the yielding strength of the bars. It was concluded that the average loss of the beam stiffness after flexural cracking was 76%. Few changes were observed in the corrosion potentials for those beams, then the tensile stresses at 40% and 80% of the yielding strength do not influence the corrosion potential level measured at the ages, materials and conditions tested.

Keywords: RC beams · Corrosion · Flexural behavior · Sustained load

1 Introduction

The reinforced concrete (RC) is one of the most widely used materials in the construction industry worldwide. However, the reinforcing steel embedded in it can suffer corrosion damage despite the physical and chemical protection that concrete can provide [1, 2]. Due to corrosion pathology, there are damages such as cracks and rust. This problem is due to the aggressiveness of the environment to which they are exposed, as well as an inadequate construction procedure, poor supervision during construction, the combination of all of these or an inadequate durability design [3].

Durability is the capacity of concrete to resist physical, chemical, biological and climatic actions during the service life of a structure. The life of the structure is divided into three stages or times: corrosion initiation, corrosion propagation and remaining life. The duration of the first two stages are known as the service life of the structure.

J. I. Escalante-Garcia et al. (Eds.): RW 2021, RILEM Bookseries 40, pp. 388–395, 2023.
https://doi.org/10.1007/978-3-031-21735-7_43

When the service life of a structure has ended and the remaining life period begins, it is necessary to carry out rehabilitations, repairs or reinforcements [1].

These interventions cause expenses and economic losses in the countries around the world [4]. Therefore, it is important to control the corrosion pathology and to understand it behavior under conditions very close to reality.

There are research works that study structural elements such as beams. In this works the main objective is to understand the mechanical behavior under the corrosion pathology.

Guevara et al., studied 16 beams of 20x30x200 cm dimensions using a water/cement (w/c) ratio of 0.7. The main variation factor in the specimens was the cross reinforcing steel quantity. They used the wet and dry cycles to the outdoors for the method to accelerate corrosion [5].

Wang et al., studied 12 beams of 12 × 20 × 170 cm dimensions using a w/c ratio of 0.70, 0.62, 0.55 and 0.48. The main variation factor in the specimens was the cyclic load level and the compressive strength of concrete. They used two methods to accelerate the corrosion: wet and dry cycles and application of anodic current [6].

In the mentioned works the conclusions are similar, the mechanical behavior was affected by the corrosion. However, there are some factors that must be considered such as the sustained load which represent the service load in a structural element under service. Yoon et al. studied 10 beams of 10 × 15 × 117 cm dimensions using a w/c ratio of 0.5. The main variation factor in the specimens was the sustained or previous load. They used also two methods to accelerate the corrosion: wet and dry cycles and application of anodic current [7].

Nevertheless, identifying values such as strengths, could allow the determination of a range of time in which an action can be taken to successfully repair structural elements. Therefore, an option is to study the electrochemical behavior taking as a reference the yielding stress of the steel so that, together with other contributions, results about the behavior of the RC beams can be obtained.

2 Experimental Program

2.1 Design of Beams and Materials for Construction

Eight RC beams were considered, four control beams (BC) and four beams with induced corrosion (BIC). Beam dimensions were 15 × 30 × 350 cm (width × height × length). Tension longitudinal reinforcing steel bars of beams were subjected to 40 and 80% of their corresponding yielding strength (fy) to represent beams under service load in accordance to the flexural hypothesis (Table 1).

Concrete mixture was designed using a water/cement ratio (w/c) of 0.62 (for a 250 kg/cm^2 compressive strength approximately). Design was carried out according to Standards of ACI Committee 211 [8]. Flexural design of beams was carried out using flexural theory. For the steel, the stress-strain curve was based on that proposed by Rodríguez and Botero [9]; for concrete, stress-strain curve was based on that proposed by Kent and Park [10]. Shear design of beams was based on the requirements established in the Mexico City Concrete Technical Norm [11]. Reinforcing details of beams are presented in Fig. 1.

Table 1. Stresses on tension longitudinal reinforcing steels bars and type of beams

RC beam	Stress in rebars	Type
1	0.4 fy	BIC
2	0.4 fy	BC
3	0.4 fy	BIC
4	0.4 fy	BC
5	0.8 fy	BIC
6	0.8 fy	BC
7	0.8 fy	BIC
8	0.8 fy	BC

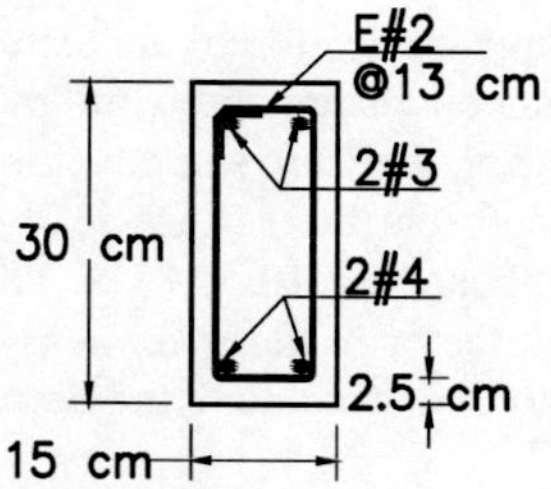

Fig. 1. Section and steel bars for the beams

The transverse steel was coated with anticorrosive enamel to guarantee that the corrosion occurred in the longitudinal steel. The concrete was elaborated in accordance with Mexican Standard NMX-C-155 [12]. In the BIC beams, the concrete was placed in two longitudinal layers of 15 cm high, the first one with concrete contaminated with chlorides (1.5% of the cement weight). Beams were cured at last 28 days after construction.

2.2 Evaluation Techniques

There were considered the contamination of fresh concrete and wetting and drying cycles as methods to accelerate the corrosion. The wetting and drying cycles began at 166 days.

Beams were tested using a four-point bending load test setup (Fig. 2). This setup was selected to avoid the interaction between bending moments and shear forces in the central part of beams. Loads were applied using a hydraulic actuator. Loads were measure using a load cell. Strain gages were attached to the tension longitudinal reinforcing steel bars to determine its unit deformation. Displacements were measured using linear potentiometers.

Beams were posttensioned to represent service loads. Posttensioned loads were applied until the tensile stress in the longitudinal steel reinforcement was equal to 40 and 80% of their corresponding yielding strength (the same amounts used for the preliminary applied load). In this system bars and steel supports were used. Beams were placed

Fig. 2. Vertical system in a preliminary applied load

one under the other. All BIC beams were under corresponding BC beams. Loads were applied after 154 days after beam construction using a hydraulic actuator and measured using a load cell. Strain gages were used to monitor corresponding stress on the steel bars. (Fig. 3).

Fig. 3. Sustained load system during the applied load

Corrosion potential was monitored using a copper - copper sulfate cell according to the standard ASTM C876–91 [13]. These measurements were taken 28 days after beam construction and during the sustained load system.

3 Results

The load–displacement curves and load–steel strain curves for the beams subjected to 40% of the yielding strength of bars are presented in Figs. 4 and 6, respectively. Loads reported correspond to the load applied at each point. Displacements correspond to those measured at the middle length of beams.

In Figs. 4 and 5, it is observed a maximum average load of 18.12 kN and 35.76 kN respectively. Also, it is observed a maximum average displacement of 4.37 and 11.20 mm respectively. In both cases, there were a vertical residual average displacement.

The load vs. displacement curves and load vs. steel strain curves for the beams subjected to 80% of the rebars yielding strength are presented in Figs. 5 and 7, respectively.

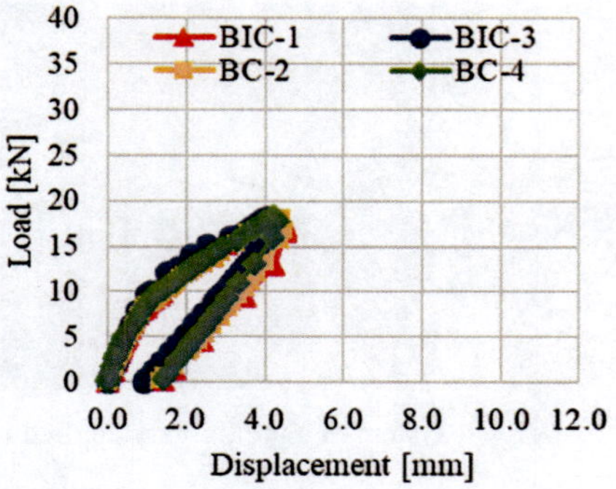

Fig. 4. Load vs. displacement of beams subjected to 40% fy

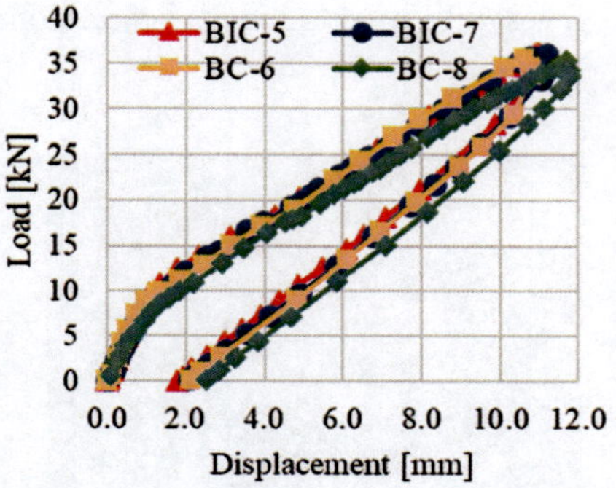

Fig. 5. Load vs. displacement for beams subjected at 80% fy

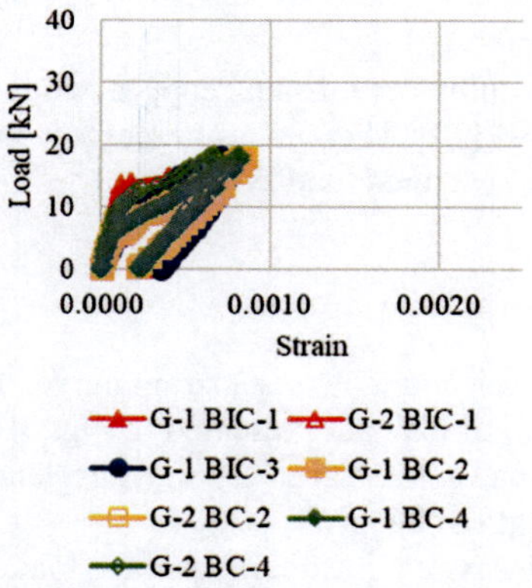

Fig. 6. Load vs. strain for beams subjected at 40% fy

In Figs. 6 and 7, it is observed a maximum average strain of 0.000797 and 0.001832 respectively. In both cases, there were a residual average strain.

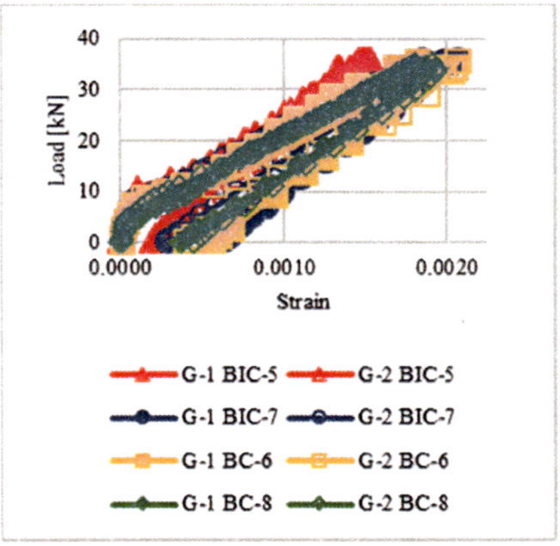

Fig. 7. Load vs. strain for beams subjected at 80% fy

The corrosion potential obtained for beams subjected to 40% and 80% of yielding strength are presented in Figs. 8 and 9, respectively. In both cases age of beams reported was up to 160 days. The other results will be available in the near future. An exhaustive analysis for both mechanical and electrochemical behavior are considered.

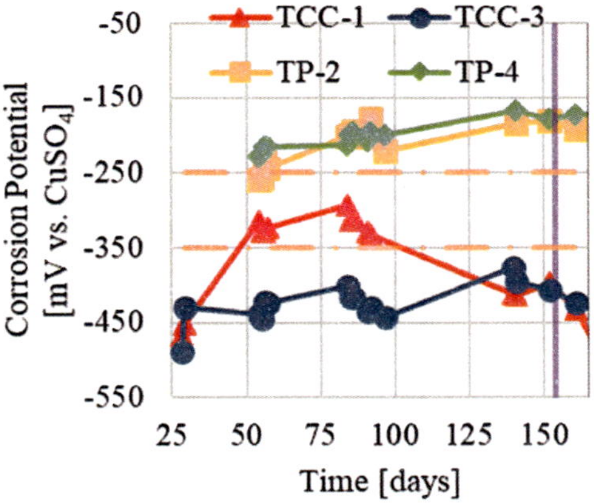

Fig. 8. Corrosion potentials for beams subjected at 40% fy

Figures 8 and 9, show that after the sustained load (represented with a vertical line at 154 days after the construction of the beams) the potential decreased to a more negative value. This behavior could indicate that the damage on the beams is greater than the damage associated without sustained loads.

4 Conclusions

In this research project the next conclusions are settled:

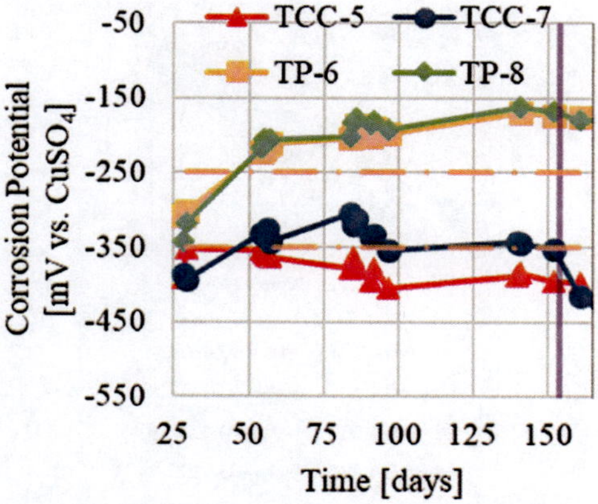

Fig. 9. Corrosion potentials for beams subjected at 80% fy

- After the preliminary test, there was residual displacements and strains. The average residual displacements were 1.28 mm for beams subjected to 40% fy and 2.21 mm for beams subjected to 80% fy. The average residual strains were 0.00025 for beams subjected to 40% fy and 0.00037 for beams subjected to 80% fy. This represents the importance of maintain the conditions of structures under service load in the different stages of any research project.
- There was a little variation of corrosion potential corresponding to 40 and 80% of yielding strength. It is recommended to induce corrosion at other level to compare results and at different ages.

Acknowledgements. The authors want to thank all the support obtained at Faculty of Engineering in the Autonomous University of Yucatán and at CINVESTAV, unit from Mérida.

The authors also remember and give a special acknowledgement for our friend Dr. Eric I. Moreno (Rest In Peace) for being a great inspiration for us.

References

1. Torres Acosta, A., Martínez Madrid, M., Muñoz Noval, A.: Capacidad remanente en vigas de hormigón que presentan corrosión localizada en el acero de refuerzo. Mater. Constr. **53**(271–172), 125–133 (2003)
2. Michel, A., Solgaard, A., Pease, B., Geiker, M., Stang, H., Olesen, J.: Experimental investigation of the relation between damage at the concrete-steel interface and initiation of reinforcement corrosion in plain and fibre reinforced concrete. Corros. Sci. 308–321 (2013)
3. Torres Acosta, A.A., Castro Borges, P.: La filosofía para obtener obras de concreto durables. IC Ingeniería Civil-Estructuras I(586), 12–15 (2018)
4. Vázquez Castillo, C.D.: Diagnóstico de la corrosión del puente de Celestún–Tesis. Universidad Autónoma de Yucatán, Mérida, Yucatán, México, Facultad de Ingeniería (1998)

5. Guevara, B., Juárez, C., Fajardo, G., Castro Borges, P.: Comportamiento a cortante en vigas de concreto reforzado con deterioro por corrosión en los estribos. Ingeniería de Construcción **24**(3), 259–270 (2009)
6. Wang, L., Li, C., Yi, J.: An experiment study on behavior of corrosion RC beams with different concrete strength. J. Coastal Res. **73**, 259–264 (2015)
7. Yoon, S., Wang, K., Weiss, W., Shah, S.P.: Interaction between loading, corrosión, and serviceability of reinforced concrete. ACI Mater. J. 637–644 (2000)
8. ACI Committee 211: Proporcionamiento de Mezclas, Concreto Normal, Pesado y Masivo. IMCYC 29–34. México, D.F. IMCYC (2004)
9. Rodríguez, M., Botero, J.: Aspectos del comportamiento sísmico de estructuras considerando las propiedades mecánicas de aceros de refuerzo producidos en México. In IX Congreso Nacional de Ingeniería Estructural, 1, pp. 302–311. Zacatecas, México (1994)
10. Kent, D., Park, R.: Flexural members with confined concrete. ASCE J. Struct. Div. **97**(7), 1969–1990 (1971)
11. FGD-2004: Normas Técnicas Complementarias para Diseño y Construcción de Estructuras de Concreto, NTC-RCDF 2004. Gaceta Oficial del Distrito Federal. Administración Pública del Distrito Federal, México, D.F. (2004)
12. NMX-C-155-ONNCCE-1987: Industria de la Construcción-Concreto Hidráulico-Especificaciones. Organismo Nacional de Normalización y Certificación de la Construcción y Edificación, S.C. (1987)
13. ASTM C876–91. Método Estándar de Ensayo para Potenciales de Media Celda para Acero de Refuerzo no Revestido en Concreto (1999)

Electrochemical Behaviour of Steel Embedded in Alkali Activated Metakaolin/Limestone Based Mortar

F. R. Vazquez-Leal[1], J. M. Mendoza-Rangel[1](✉), C. Andrade[2], P. Perez-Cortes[3], and J. I. Escalante-García[3]

[1] Facultad de Ingeniería Civil, Universidad Autónoma de Nuevo León- UANL, Ciudad Universitaria S/N, San Nicolás de los Garza C.P. 66455, N.L, México
jose.mendozarn@uanl.edu.mx
[2] International Centre for Numerical Methods in Engineering, CIMNE- UPC, Barcelona, Spain
[3] Grupo de Ingeniería Cerámica, Cinvestav Saltillo, Av. Industria Metalúrgica No. 1062, Parque Industrial, Ramos Arizpe, Coahuila, MX C.P. 25900, México

Abstract. The electrochemical behaviour of steel reinforcement was investigated as embedded in mortars based on alkali activated binders of metakaolin (MK)/limestone (LS) precursors with mass ratios of 100/0, 70/30 and 40/60, selected from previous studies of optimization and modeling of formulations, gel composition and molecular structure, as well as environmental aspects of these materials. A three electrodes arrangement and the polarization resistance technique were used to obtain the current density (I_{corr}) and the half-cell potential (E_{corr}) of the embedded steel for up to 156 days of curing under laboratory conditions. The embedded steel embedded remained in the passivation zone from the beginning of the study according to the criteria of I_{corr} used for portland cement-based materials. The incorporation of LS into the MK precursor showed a beneficial effect in the electrochemical behavior of the steel embedded, the mortar with a binder of 70/30 (30% LS) showed the best performance at 156 days of curing, with values of I_{corr} below 0.005 $\mu A/cm^2$, representing an innovative and suitable alternative to 100% metakaolin based geopolymers. On the other hand, the results showed that I_{corr} seems to be a better indicator for the condition of the steel embedded than E_{corr} in the mortars analyzed.

Keywords: Durability · Electrochemical techniques · Geopolymers · Corrosion · Passivation

1 Introduction

Alkaline activated materials (AAM) have been mainly focused on precursors of blast furnace slag (BFS), fly ash (FA), metakaolin (MK), among others, to obtain cementitious matrixes with stable reaction products and mechanical characteristics comparable to portland cement (PC)-based materials [1–3]. The use of blended precursors LS/MK with up to 60% limestone in AAM (AAM-LSMK) has been shown as an innovative and

J. I. Escalante-Garcia et al. (Eds.): RW 2021, RILEM Bookseries 40, pp. 396–404, 2023.
https://doi.org/10.1007/978-3-031-21735-7_44

sustainable alternative, with reaction products having a 3D-network molecular structure, that reached higher compressive strengths as well as lower CO_2 emissions and costs compared [4, 5] to an type CPC30 portland cement [6].

In portland cement concrete the steel is protected by a passive film formed under the chemical environment provided by the cement hydration that forms mainly $Ca(OH)_2$ (pH ~ 12.5); such film maintains the steel in passive state, acting as a barrier between the environment and the steel and protecting in against corrosion damage [7]. On the other hand, on AAM based on precursors of low CaO, such as MK, the alkalinity comes from the alkaline activator (commonly mixtures of NaOH and Na_2SiO_3); the high initial pH can decrease as the precursors dissolves and forms polymerizes into new cementitious reaction products. Although some durability aspects of AAM have been reported, studies on the electrochemical behavior of the embedded steel in AAM are not abundant and most of them are on precursors such as blast furnace slag (BFS) or metakaolin (MK) and fly ash (FA).

This investigation presents an electrochemical study on AAM based on blended precursors LS/MK of metakaolin with up to 60% limestone. The electrochemical response of embedded steel was carried to elucidate the formation of the passive film as a means to predict the service life of the reinforced concrete. The electrochemical techniques of current density (I_{corr}) and half-cell potential (E_{corr}) were used to analyze the response of the embedded steel in prismatic mortar specimens cured for up to 156 days under laboratory conditions. The usefulness of the criteria of those electrochemical parameters currently used to evaluate the state of the reinforcement in PC-based materials is elucidated for the AAM-LSMK.

2 Experimental Methodology

2.1 Materials

Mortars were prepared with a binder of AAM based on MK/LS (metakaolin/limestone) precursors; these were activated with solutions based on blends of sodium hydroxide (98% purity) flakes and waterglass (WG, 29.85%SiO_2, 9.12%Na_2O and 61.03%H_2O). The MK was obtained by firing a high purity kaolinitic clay for 4 h at 750 °C, it had an average particle size (d_{50}) of 6.19 μm (Fig. 1a) and a density of 2.54 g/cm^3; X-ray diffraction showed that it was predominantly amorphous and showed traces of quartz (SiO_2) and anatase (TiO_2) (Fig. 1b). Limestone sand was used as the fine aggregate, it had a density of 2.61 g/cm^3 and a maximum particle size of 4–75 mm; it was also used as the source for LS precursor; for that purpose, it was ball milled to an average particle size (d_{50}) of 29.2 μm (Fig. 1a) and X-ray diffraction showed mainly calcite (Fig. 1b). The chemical composition of MK and LS is shown in Table 1. AISI 1018 carbon steel corrugated rebars of 6 mm diameter were previously cleaned according to ASTM G1 and embedded into the mortars.

2.2 Mix Proportions

Three mortar compositions (Table 2) were prepared using cementitious blends selected from previous studies [5, 8]; the mass ratios MK/LS were 100/0, 70/30 and 40/60;

Table 1. Chemical composition of precursors obtained with X-ray fluorescence (expressed in wt%)

Precursor	CaO	SiO_2	Al_2O_3	TiO_2	Fe_2O_3	MgO
MK	0.14	73.49	23.11	2.23	0.51	–
LS	94.43	3.74	0.31	–	0.20	0.82

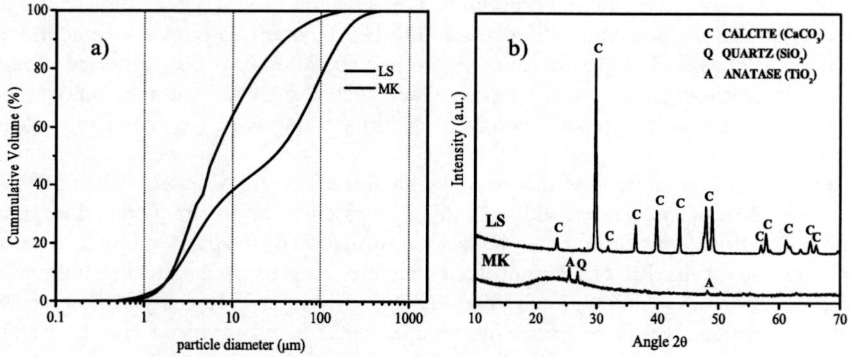

Fig. 1. Characterization of precursors a) Particle size distribution obtained by laser diffraction; b) mineralogical composition obtained with X-ray diffraction.

the oxide-molar SiO_2/Al_2O_3 ratio was kept fixed at 3.34 in all mixtures and the water/precursor ratio was adjusted to attain a fluidity of 110 ± 5 (ASTM C109). Such chosen conditions were set aiming for 28-day compressive strengths of 60–70 MPa. The sand:precursor ratio was of 2.75:1 (ASTM C109).

Table 2. Studied alkali activated MK/LS mortars and their mix proportions.

Mass				Mix proportion (g)					
Mix	MK/LS ratio	w/p	%Na_2O	MK	LS	WG	NaOH	H_2O	Sand
M1	100/0	0.62	21.46	20.02	0	21.85	3.07	0	55.06
M2	70/30	0.50	16.91	15.05	6.45	16.43	2.85	0.08	59.14
M3	40/60	0.37	10.74	9.18	13.76	10	2.06	1.92	63.08

2.3 Preparation of Mortar Specimens

The alkaline solutions were prepared 24 h in advance and left under laboratory conditions to reach thermal equilibrium for 24 h. The paste was first prepared by mixing the precursors and alkaline solutions for 30s at 60 rpm, the mixing speed was then increased to 124 rpm for 3 min and then to 255 rpm for 3 more minutes. The mortar was made by adding the fine aggregate and mixing at 60 rpm for 4 min. The mortars were poured into 2 × 5 × 8 cm prismatic molds (Fig. 2) in two layers and vibrated for 15 s to release the trapped air, these were covered with a plastic film and wet flannels to prevent shrinkage from moisture losses; the molds were left to cure at room temperature for 24 h. The freshly demolded specimens were stored wrapped with a plastic cling film at 23 °C (40–50% RH) during 28 days.

2.4 Electrochemical Techniques

The electrochemical tests on the embedded steel were conducted by using a three-electrode arrangement that consisted of a 6 mm diameter carbon steel rebar (exposed surface area: 9.425 cm^2) as the working electrode (WE), a graphite bar as the counter electrode (CE) and a saturated calomel electrode (SCE) as the reference (RE). Figure 2 shows the specimen configuration. The measurements were made with a Gamry Interface 1000 potentiostat in laboratory conditions. And the results shown in the graphs correspond to the average of 4 tests.

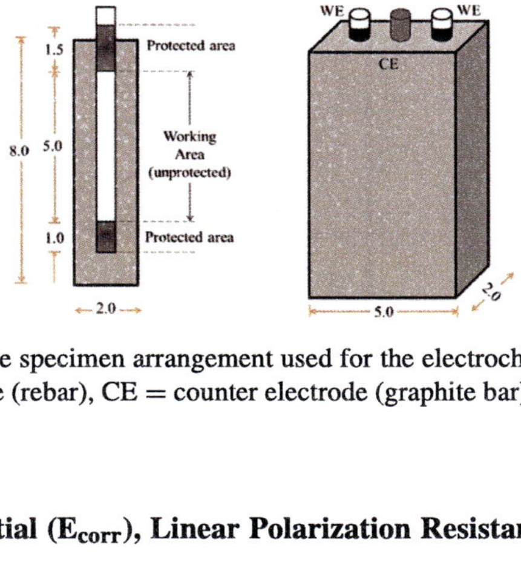

Fig. 2. Schematics of the specimen arrangement used for the electrochemical tests, units in cm. WE = working electrode (rebar), CE = counter electrode (graphite bar).

2.5 Half-Cell Potential (E_{corr}), Linear Polarization Resistance (Rp), Current Density (i_{corr})

The open circuit potential (OCP) was recorded for five minutes in the beginning of the experiment for the stabilization of the values. E_{corr} and I_{corr} were obtained with the Rp technique conducted in the range of ±20 mV, with respect to the OCP at a scan rate of 10 mV/min. The value for the constant B = 26 mV for active state of the embedded reinforcement steel proposed for Andrade et al. [9] was preliminary selected for the calculation of the corrosion rate using the Stern & Geary Eq. (1) [10].

$$I_{corr} = \frac{\beta_a \beta_c}{2.3 R_p (\beta_a + \beta_c)} = \frac{B}{R_p} \tag{1}$$

The electrochemical response of the embedded steel AAM-LSMK based mortar was measured for a curing period of up to 156 days of curing in laboratory conditions.

3 Results and Discussion

3.1 Half-Cell Potential (E_{corr})

The E_{corr} (Fig. 3) can be interpreted as the thermodynamic stability of the steel, it exhibits the electrons exchange in the metal-electrolyte interface (MEI) referenced to the saturated calomel electrode (SCE). The intervals delimited by dotted lines correspond to the criteria of ASTM C876 [11] standard for steel embedded in PC-based materials. The general behavior of the three mixtures showed incrementing values during the first 21 days; then, from 115 onwards, relatively small changes were noted and all stayed within a close interval. The curves transitioned from the "severe corrosion" to "90% of probability of corrosion" interval from 3 to 14 days; then from 115 onwards, the values stayed within the "uncertainty of corrosion" interval, showing a trend towards the passivation zone of the steel (less negative values).

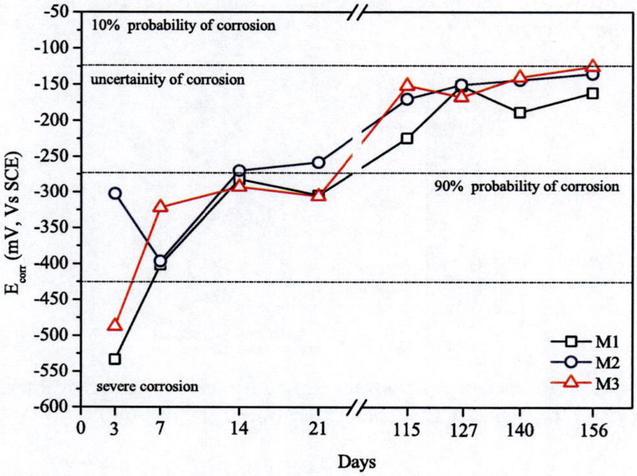

Fig. 3. E_{corr} values until 156 days of curing in laboratory conditions.

This incremental trend it can be attributed to the formation of a passive film (PF) in the metal-mortar interface; reported as a double-layer film composed by an inner layer, considered protective, composed of magnetite-type Fe oxide; and the outer, non-protective layer, composed of Fe oxy-hydroxides [12–14]. The PF prevents further oxidation of the steel and the subsequent acidification due to water hydrolysis by the steel; the oxygen availability in the aqueous phase promoted the formation of such passive film. The stabilization of the values after 115 days of curing suggests that the passive film formation process is complete and stable, with a full passivation of the rebar with no evidence of corrosion according to the electrochemical techniques.

The pH evolution in the pore solution of the AAM-LSMK is opposite to the PC-based materials, since no portlandite precipitates in the former and the alkalinity is provided by the combination of WG and NaOH leading to higher early pH than PC-based materials, which can reduce as the cementitious products solidify closing the gap over time. The alkali concentration is high at early ages, with pH values ~11.5–12.5, generating differences in the values of E_{corr} respect to the time. According to the Pourbaix diagrams, a pH > 10 maintained the embedded steel in a passive state [15]; the equilibrium of the oxidation/reduction reactions at the steel-electrolyte interface of the AAM-MKLS was maintained at higher cathodic values compared to PC-based materials. Therefore, it is necessary to particularize the scale of E_{corr} to metakaolin systems with up to 60% limestone.

3.2 Current Density (I_{corr})

Figure 4 shows I_{corr} values obtained by Eq. (1), for the embedded steel and evaluated according to the manual DURAR criteria and Alconpat recommendations [16]. The intervals delimited with dotted lines indicates the susceptibility of damage for corrosion: an $I_{corr} < 0.1\ \mu A/cm^2$ is describes a passive state. From the beginning of the measurements, the values remained in the passive state zone despite the cathodic values given by the corrosion potential (Fig. 3). The gradual decrease of the I_{corr} values with time is associated to the formation of a PF in the surface of the steel.

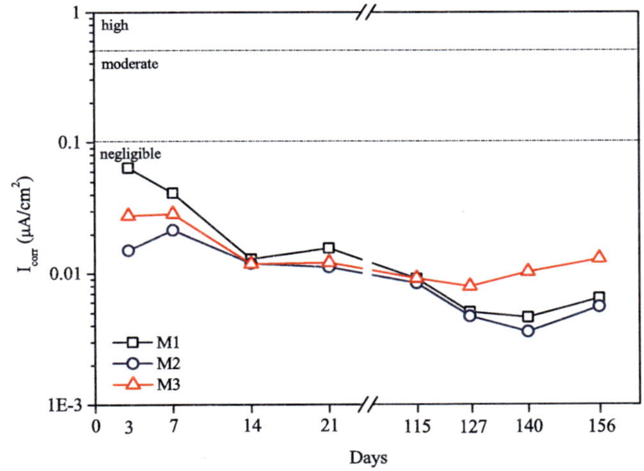

Fig. 4. Values for I_{corr} until 156 days of curing in laboratory conditions.

The high availability of OH^- in the pore solution contributed to the oxidation/reduction reactions that promoted the formation of the PF constituted by iron oxo-hydroxides, i.e. goehite (α-FeOOH), lepidocrocite (γ-FeOOH) and akagonite (β-FeOOH), covering underlying layers of possibly magnetite (Fe_3O_4), maghemite (γ-Fe_2O_3) and/or hematite (α-Fe_2O_3) as the main constituents [17, 18]. Then, the three mixes passivate the steel reinforcement from the beginning. The results indicated that the embedded steel reached the passivation following the criteria currently established for portland cement-based materials and simulated pore solutions [14].

3.3 Correlation Between Corrosion Potential (E_{corr}) and Current Density (I_{corr})

Figure 5 shows the correlation between the E_{corr} vs I_{corr} results described above; the dotted lines represent the linear fit of the 8 points points of each mortar. The behavior of E_{corr} was the inverse of I_{corr}, and there was a good lineal relation for M1 and M2 mixes in spite that the corrosion rates for the high cathodic potentials were lower than those of PC-based materials. The Tafel slopes which refer to the potential drop (in mV) in a decade (dec) of current, can be deduced as well; the results for the three mixes, i.e. ~300–350 mV/dec, were higher than those of PC based-based materials at 120 mV/dec [9], which could be associated to the existence of a reduction reaction in addition to that of O and to the high variation of E_{corr} in the alkaline environment of the mortars.

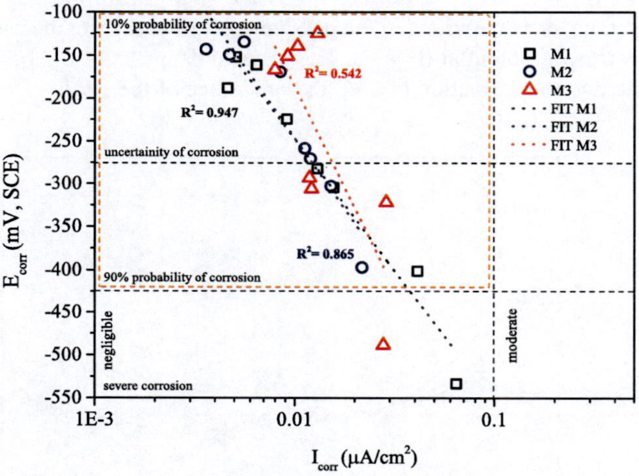

Fig. 5. Relation of E_{corr} and I_{corr} values for days of curing in laboratory conditions.

The E_{corr} is a thermodynamic parameter, while I_{corr} is a kinetics indicator of the corrosion phenomenon; hence, for the electrochemical characterization of steel in AAM-LSMK, I_{corr} seems a better indicator of the corrosion of the steel than E_{corr}.

4 Conclusions

Mortars with alkali activated precursor of metakaolin up to 60% limestone showed very similar behavior leading to an efficient passivation of steel bars, from the beginning of the study according to I_{corr}.

The E_{corr} values of steel embedded in mortars with alkaline activated binders of metakaolin with up to 60% limestone can not be evaluated with the same scale than for

PC-based materials because the alkalinity is higher and following Pourbaix's diagram (Nernst equation), the potential is more cathodic as the pH is higher.

The current density (I_{corr}) is a better indicator than the corrosion potential (E_{corr}) to establish the condition of the rebar embedded because the E_{corr} seems to be influenced by the pH value.

Acknowledgments. The authors would like to thank the National Council of Science and Technology for the Basic Science project 285976, the PhD scholarship 473068 of the student Francisco René Vázquez Leal. To the Universidad Autónoma de Nuevo León and the Centro de Investigación y de Estudios Avanzados, Unidad Saltillo.

References

1. Burciaga-Díaz, O., Magallanes-Rivera, R.X., Escalante-Garcia, J.I.: Alkali-activated slag-metakaolin pastes: Strength, structural, and microstructural characterization. J. Sustain. Cem. Mater. **2**, 111–127 (2013). https://doi.org/10.1080/21650373.2013.801799
2. Martinez-Lopez, R., Escalante-Garcia, J.I.: Alkali activated composite binders of waste silica soda lime glass and blast furnace slag: strength as a function of the composition. Constr. Build. Mater. **119**, 119–129 (2016). https://doi.org/10.1016/j.conbuildmat.2016.05.064
3. Torres-Carrasco, M., Puertas, F.: La activación alcalina de diferentes aluminosilicatos como una alternativa al Cemento Portland: cementos activados alcalinamente o geopolímeros. Rev. Ing. Constr. **32**, 5–12 (2017). https://doi.org/10.4067/s0718-50732017000200001
4. Perez-Cortes, P., Escalante-Garcia, J.I.: Gel composition and molecular structure of alkali-activated metakaolin-limestone cements. Cem. Concr. Res. **137**, 106211 (2020). https://doi.org/10.1016/j.cemconres.2020.106211
5. Perez-Cortes, P., Escalante-Garcia, J.I.: Alkali activated metakaolin with high limestone contents—statistical modeling of strength and environmental and cost analyses. Cem. Concr. Compos. **106**, 103450 (2020). https://doi.org/10.1016/j.cemconcomp.2019.103450
6. Mexicana, N.: NMX-C-414-ONNCCE-2017. In: Industria de la construcción–cementos hidráulicos–Especificaciones y métodos de prueba, NMX-C-414-ONNCCE-2017. (2017)
7. Kurdowski, W.: Cement and concrete chemistry. Springer (2014). https://doi.org/10.1007/978-94-007-7945-7
8. Perez-Cortes, P., Escalante-Garcia, J.I.: Design and optimization of alkaline binders of limestone-metakaolin—a comparison of strength, microstructure and sustainability with portland cement and geopolymers. J. Clean. Prod. **273**, 123118 (2020). https://doi.org/10.1016/j.jclepro.2020.123118
9. Andrade, C., Alonso, C.: Corrosion rate monitoring in the laboratory and on-site. Constr. Build. Mater. **10**, 315–328 (1996)
10. Stern, M., Geary, A.: Electrochemical polarization: I. A theorical analysis of the shape of polarization curves. J. Electrochem. Soc. **104**, 56–63 (1957). https://doi.org/10.1149/1.2428496
11. ASTM International.: Standard test method for corrosion potentials of uncoated reinforcing steel in concrete. ASTM C876–15. G01.14. ASTM Int. 1–8 (2015). https://doi.org/10.1520/C0876-09.2
12. Andrade, C.: Propagation of reinforcement corrosion: principles, testing and modelling. Mater. Struct. **52**(1), 1–26 (2018). https://doi.org/10.1617/s11527-018-1301-1

13. Jiang, J.Y., Wang, D., Chu, H.Y., Ma, H., Liu, Y., Gao, Y., Shi, J., Sun, W.: The passive film growth mechanism of new corrosion-resistant steel rebar in simulated concrete pore solution: Nanometer structure and electrochemical study. Materials **10** (2017). https://doi.org/10.3390/ma10040412
14. Volpi, E., Olietti, A., Stefanoni, M., Trasatti, S.P.: Electrochemical characterization of mild steel in alkaline solutions simulating concrete environment. J. Electroanal. Chem. **736**, 38–46 (2015). https://doi.org/10.1016/j.jelechem.2014.10.023
15. Floyd Brown, B., Burnett, H.C., Chase, W.T., Goodway, M., Kruger, J., Pourbaix, M. (eds.): Corrosion and metal artifacts–a dialogue between conservators and archaeologist and corrosion scientists. National Bureau of Standards (1977)
16. Terradillos, P.G., de Rincón, O.T., Bandala, E.E.M., Borges, P.C., Perdrix, A.: Recomendaciones sobre velocidad de corrosion. (2020). https://doi.org/10.21041/AlconpatInternacional/RecTec/2020-03-velocidaddecorrosion
17. Abd El Haleem, S.M., Abd El Aal, E.E., Abd El Wanees, S., Diab, A.: Environmental factors affecting the corrosion behaviour of reinforcing steel: I. The early stage of passive film formation in Ca(OH)2 solutions. Corros. Sci. **52**, 3875–3882 (2010). https://doi.org/10.1016/j.corsci.2010.07.035
18. Nieuwoudt, M.K., Comins, J.D., Cukrowski, I.: The growth of the passive film on iron in 0.05 M NaOH studied in situ by Raman micro-spectroscopy and electrochemical polarisation. Part I: Near-resonance enhancement of the Raman spectra of iron oxide and oxyhydroxide compounds. J. Raman Spectrosc. **42**, 1335–1339 (2011). https://doi.org/10.1002/jrs.2837

Additive Manufacturing in Concrete Construction

Additive Manufacturing in Architecture: 3D Printing Solutions for Vaulted Spaces

Angelo Vito Graziano(✉), Ilaria Cavaliere, Dario Costantino, Giuseppe Fallacara, and Nicola Parisi

Politecnico Di Bari, 70126 Bari, Italy
{angelovito.graziano,ilaria.cavaliere,dario.costantino, giuseppe.fallacara,nicola.parisi}@poliba.it

Abstract. Additive manufacturing (AM) will change architecture. Nowadays a trained digital designer can control both digital and physical tools for the development of a project.

The starting point of this research is Najaat, a project for an entire 3D printed village in Syria inspired by other case studies like GAIA. The project is part of a larger study about self-made architecture and materials such as raw earth, straw, rice husk and hydraulic lime which make it reusable, recyclable and affordable. The shapes chosen for Najaat are organic and easily adaptable to any environment, highlighting the total freedom that the 3D printing system allows. At the same time these geometries are also aimed at challenging the limits of 3D printing by working on the maximum projection.

This paper focuses on the possibility to print closed domed spaces using only common printing machines instead of robotic arms, in order to keep the process both economic and ecological. The printing process refers to the theory of Nubian vaults, which are structures built without supports exploiting the inclination of bricks layers. A scaled prototype of a dome has been printed through a hybrid system of layers, that helped the structure non to collapse on the top. The result still shows some imperfections due to some limitations of the machine, but it is useful to demonstrate that some typical problems of excessive overhang can be overcome through a deep study of the material itself, of the geometry and also of the printing path.

Keywords: 3D Printing · Additive manufacturing · Architecture · Digital fabrication · Raw earth

1 Introduction

Planimetric freedom has always been considered as an advantage that AM has brought into the architectural field, together with the reduction of time and material waste, a good control of the final shape, the possibility to rapidly modify the design and the possibility to use sustainable mixtures. However in some cases it might appear contradictory because it eliminates any constraints in the creative process of an architect from the planimetric point of view, but it shifts the problem to the roof of the building.

J. I. Escalante-Garcia et al. (Eds.): RW 2021, RILEM Bookseries 40, pp. 407–414, 2023.
https://doi.org/10.1007/978-3-031-21735-7_45

Usually, 3D printed monolithic buildings consist of simple extrusions covered with a traditional roof because the printing of overhanging horizontal surfaces is still challenging.

This hybrid solution (3D printed walls and wooden roof) has been used by WASP for GAIA (Fig. 1). As a matter of fact, roofing operations with the same technologies adopted for walls construction are still unresolved. It is important to highlight that the topic of roofing 3D printed manufacts is strictly related to technological limitations linked to viscous materials deposition.

We can usually distinguish two different approaches for the roofing issue: the first one consists of the use of more complex AM technologies, for example 6-axis robotic arms instead of 3-axis systems. At the same time a chemical complexification of the material to be extruded is carried out to compensate the physical deficiencies of the natural ones. However, this aspect contradicts the prospect of extreme formal freedom at a low cost. MX3D [8] technology is exemplary.

The other approach consists in simplifying the printed object to prototype with the most common technologies and the most common slicing softwares. However, this contradicts the perspective of planimentry freedom that 3D printing offers to designers.

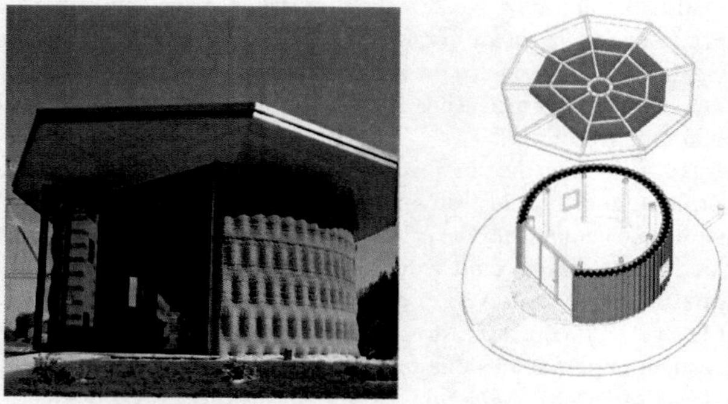

Fig. 1. GAIA, a 3D printed earth house realized by WASP in 2019.

1.1 Challenges

This research investigates the identified approach to the roofing topic and at the same time it tries to overcome the trend of constructing more complex machines, with the consequent associated costs. It also gives a contribution to a wider research that up to this point has managed to figure a printed roof on simple volumes by proposing a process which focuses on the study of geometry, machine path and rheology. This is the contribution that an architectural approach can give to the topic.

We will demonstrate that with the same geometry (dome), material (clay) and extrusion (4 mm layer thickness), through the use of a common LDM (liquid deposition modeling) technology (Delta WASP 40100), the printing path can affect the result.

2 Research

The starting point of this research has been Najaat, a project for an entirely 3D printed village in Syria.

Najaat (Fig. 2) is the result of a thesis work on architectural composition and tries to challenge the maximum printable projections keeping the planimetric freedom. Thanks to this project our research group has noticed that a traditional horizontal layering can be used to obtain the maximum overhang, managing to realize portions of vaulted spaces. Organic shapes, with both convex and concave areas have been used in order to highlight the reachable planimetric freedom, but also to create mirrored shapes that help have the maximum overhang angle.

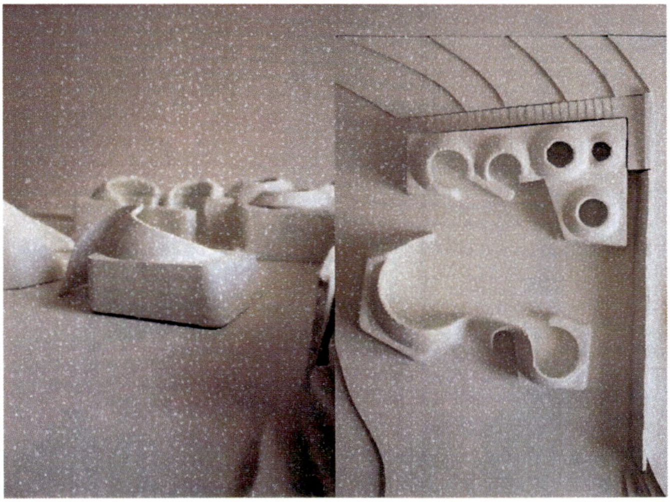

Fig. 2. A Najaat scaled model, that shows both the shapes and the 3D printing building process. The use of raw earth has been simulated through clay, with LDM technology.

The project immediately focused on the problem of the fully printed roof.

Even if the walls are tapered and geometries are conceived to support the maximum overhang, Najaat shows that an area cannot be completely covered using a horizontal layering for a monolithic architecture. This consideration led to a deeper study of the Nubian constructive system, which was used in the past to build vaulted spaces without supporting beams, through a 45° inclination of the layers of bricks. Figure 3 shows a small prototype realized at the FabLab Poliba using clay. It demonstrates that inclination of layers can be successfully applied also to 3D printing and helps to avoid the use of supports.

2.1 The Importance of the Right Orientation in 3D Printing

The way we decide to print an object can totally change the success rate of our result and that's because of the problems that supports may generate when we need them. Anyway,

Fig. 3. Nubian vault prototype made at FabLab Poliba.

there is a rule known as "THEY rule" that can be a very useful guide to avoid any hitches, telling us how to orient the parts of our geometry before the real printing process starts with the constraint of rotating the pieces only in the XZ plane. For example, the rule drives us to rotate the T letter 180° or the E letter 90° to avoid unstable projections that would usually require supports (Fig. 4).

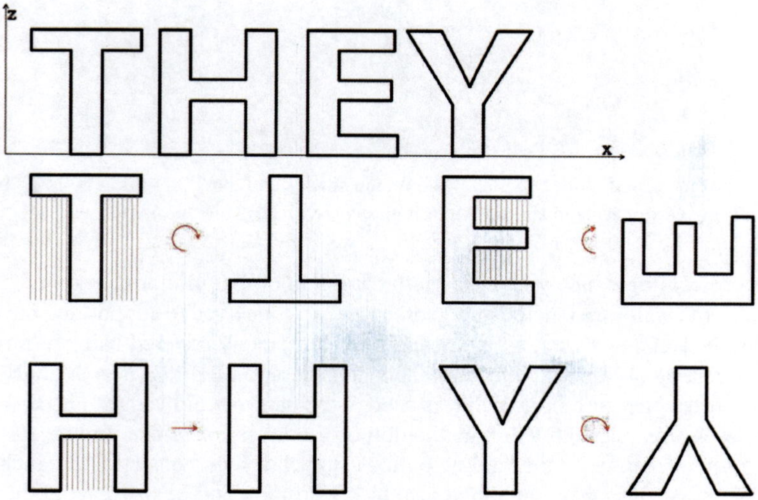

Fig. 4. Illustration of the 'THEY' rule.

3 The Right Orientation in AM for Architecture

Considerations on the Nubian vault and on the correct orientation of objects to be printed have led us to define the right orientation in the AM for architecture.

Obviously it is not possible to orient the object when it is a massive and monolithic architecture, therefore the right orientation translates into the right disposition of the extruder path.

Many tests on scaled models have been done in order to find a method that can be potentially used also for buildings in scale of 1:1 after a further study.

As we have already written, it is impossible to build a dome only through a horizontal layering, because it collapses on the top, where the tangent is horizontal (Fig. 5).

Therefore, the knowledge about the Nubian vault construction technique has been exploited to define a hybrid layering system. A dome has been cut in three parts using two planes inclined at 45°; the two pieces on the ground have been printed with a standard horizontal layering and they work as a support for the upper part, which has been printed following the inclination of the cut. Therefore it was possible to close the dome trough a layering inclined at 45° (Fig. 6).

3.1 Discussion of the Results

As the Fig. 6 shows, the prototype was successfully completed, even if it has some imperfections caused by two main reasons. First of all the nature of the material itself, which can cause cracks as it retires during the drying process. The second problem was the machine structure: as a matter of fact when the printing path isn't horizontal the nozzle tends to scrape the previous layers, compromising both the appearance and the stability of the shell. This problem could be fixed producing a more flexible nozzle or shaping it to be adapted to an inclined path without the risk of scraping.

It would be better to maintain the nozzle always tangent to the layers direction, but the result of this experiment demonstrates that some problems linked to 3D printed vaulted spaces can be easily overcome through the study of geometry and of the printing path.

At the moment the described problems are being investigated both by research groups and specialized companies, which are experimenting new strategies in the chemical field and in the technological one. As a matter of fact, advanced natural origin materials are slowly being studied and developed in order to be used as extrusion substances for 3D printing machines. Geopolymer cements are good examples to be mentioned for their high mechanical performances, durability properties and the environmental impact.

For what it may concern the technology, robotic arms are being the more and more implemented with extrusion systems, allowing more freedom in terms of printing paths even if they require more complex and expensive resources if compared to standard 3D printers.

A further development of this research is to adapt non-horizontal layering to other shapes, trying to find a general rule for the 3D printing of vaulted spaces. Some attempts have already been done (Fig. 7).

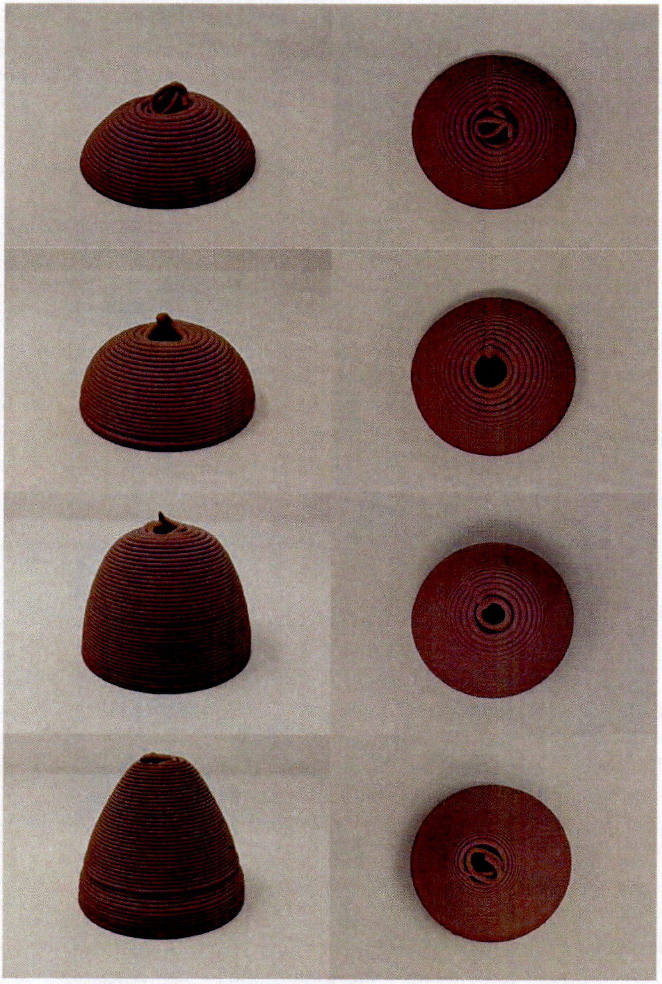

Fig. 5. Different prototypes of clay domes, realized at the FabLab Poliba using a Delta WASP 40100.

4 Conclusions

This paper shows that AM limits can be overcome adopting strategies linked to the global geometry, but also modifying the path of the nozzle and the flux of material. The current results are still imperfect and they refer only to small prototypes, but a further study can lead to an improvement of the process and also to its application to building scale prototypes.

Fig. 6. A prototype of dome printed at FabLab Poliba using a hybrid slicing.

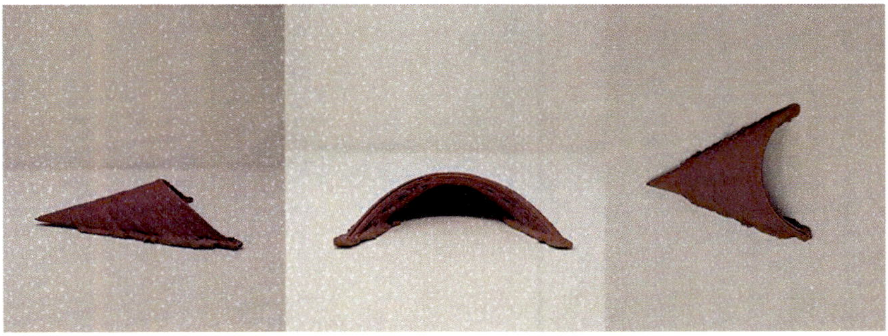

Fig. 7. A prototype of a different shape printed with an inclined layering.

References

1. Delgado, C.D., et al.: Applications of additive manufacturing in the construction industry. Autom. Constr. **89**, 110–119 (2018)
2. Ciriello, F., Graziano, A.V., Lamanna, G., Memic, H., Vukicevic, J.: Self made architecture IV, Unpublished master's thesis (2019). Politecnico di Bari, Supervisor, Parisi N
3. Craveiro, F., Duarte, J.P., Bartolo, H., Bartolod, P.J.: Additive manufacturing as an enabling technology for digital construction: a perspective on Construction 4.0. Automat. Construct. **103**, 251–267 (2019)
4. Carneau, P., Mesnil, R., Roussel, N., Baverel, O.:Additive manufacturing of cantilever—from masonry to concrete 3D printing. Automat. Construct. **116** (2020)
5. Garcia Cuevas, D.: Advanced 3D printing with Grasshopper. Clay and FDM. Printed by Amazon (2020)
6. Carbonara, F.A., Delcuratolo, F.A., Lops, M., Tina, G.: Self made architecture V, Unpublished master's thesis, (2021). Politecnico di Bari. Supervisor, Parisi N

7. WASP: Crane WASP.: Retrieved from https://www.3dwasp.com/stampante-3d-per-case-crane-wasp/, Accessed 29 June 2021
8. MX3D homepage. Retrieved from https://mx3d.com/, Accessed 29 June 2021
9. Singh, N.B., Middendorf, B.: Geopolymers as an alternative to Portland cement: AN overview. Construct. Build. Mater. **237** (2020)

Understanding the Structural Build-Up Rate of Cementitious Materials for 3D-Printing

Luiza R. M. Miranda(✉), Karel Lesage, and Geert De Schutter

Magnel-Vandepitte Laboratory, Ghent University, Ghent, Belgium
luiza.miranda@ugent.be

Abstract. The concrete 3D-printing technology is highly dependent on the material's rheology and on the deep understanding of how those properties evolve with time. This evolution occurs due to reversible structural build-up and irreversible chemical phenomena, like hydration reactions. To guarantee structural stability of the printing process, the material, as soon as it is deposited, needs to build-up an internal structure to withstand its weight and the weight of the following layers. Therefore, the static yield stress and how it increases over time at rest becomes of first interest. The concrete structural build-up happens due to both CSH bridge formation by nucleation of cement grains at their pseudo-contact points during the dormant period of hydration, and flocculation due to colloidal interactions. The currently accepted knowledge on the subject relates the yield stress growth with the structural build-up rate of the material through different models, as it also uses different measuring protocols and assumptions to its determination. The goal of this research is to summarize the literature on this topic, identifying the similarities and discrepancies in the available data to further propose a more suitable approach to evaluate this parameter. To do so, a systematic review was carried out in the Web of Science database with appropriate keywords. Then the data were organized and analyzed taking into account aspects such as the definition of structural build-up, the model used, simplifications/assumptions of the model, experimental protocol, mixture parameter and model limitations.

Keywords: Structural build-up rate · Thixotropy · 3D-printing.

1 Introduction

The concrete 3D-printing technology is highly dependent on the material's rheology and on the deep understanding of how those properties evolve with time. Cementitious materials have a complex time-dependent rheological behavior and can present shear-thinning viscosity, elastic region, static and dynamic yield stress, and structural build-up at rest [1].

The observed structural changes that the material undergoes after its first contact with water can be of both physical and chemical origins, and an evaluation of each parcel separately is of great difficulty. The term thixotropy is often used to describe the reversible changes occurring when the flow starts from a sample at rest and its

J. I. Escalante-Garcia et al. (Eds.): RW 2021, RILEM Bookseries 40, pp. 415–424, 2023.
https://doi.org/10.1007/978-3-031-21735-7_46

subsequent structural recovery when the flow stops. This can only be considered as the single phenomenon happening if one decides to neglect the irreversible chemical phenomenon due to hydration reactions that starts as soon as the binder is mixed with water. The reversible changes were reported to dominate during the first hour, as long as the static yield stress increases linearly during this period [2, 3].

The static yield stress and how it increases over time at rest is currently considered as the main rheological parameter to evaluate in the layered process [4]. The concrete structural build-up happens due to both C-S-H bridge formation by nucleation of cement grains at their pseudo-contact points during the dormant period of hydration, and flocculation due to colloidal interactions [2, 3]. The water to binder ratio, then, exerts an important effect on the structural build-up once it affects both the interparticle forces (by altering the particle distance) and the hydration process (therefore influencing the nucleation and growth of hydration products) [5].

To guarantee structural stability of the printing process, the material, as soon as it is deposited, needs to build-up an internal strength to withstand its weight and the weight of the following layers. The static yield stress is the critical stress necessary to start flow from rest and therefore is key for layer stability. It is important to notice, however, that optimization is required to find a building rate that allows structural stability and also ensures the highest bonding between the deposited fresh layers [4, 6]. The buildability of 3D printed structures is reported to be related to structural build-up, the thickness of the printed layer and the printing rate [7, 8].

The correlation between static yield stress and building time showed that, with a long enough building time, the structure is mostly supported by its yield strength [9]. The currently accepted knowledge on the subject relates the yield stress growth with the structural build-up rate of the material through different models, as it also uses different measuring protocols and assumptions to its determination. Thus, the goal of this research is to summarize the literature on this topic, identifying the similarities and discrepancies in the available data to further propose a more suitable approach to evaluate this parameter.

2 Mathematical Models

Roussel [10] presents a linear model for determination of yield stress growth with time, where A_{thix} appears for the first time as a flocculation rate. This model is developed to be simple and with parameters that can be easily measured, even if it is not that accurate in its determinations. It assumes that the increase in the yield stress is linear during the dormand period because the hydration heat release in this period is constant. This assumption makes the model suitable for resting times up to 60 min after the water contact and for mixtures with a clear dormand period, which is not always is the case for mixtures with a printing purpose. When setting accelerators are used and a rapid transition from the initial reactions to the acceleration period can be observed, there is almost no dormand period.

Experimental results showed that after the dormand period a rapid increase in the yield stress is commonly observed, making the behavior highly non-linear. Taking this observed behavior into account, Perrot et al. [6] proposed an exponential model that asymptotically tends to Roussel's model in the first hour and can predict the yield stress increase for the first 2 h after mixing. This model presents A_{thix} as a structuration rate and has a characteristic time that marks the beginning of the exponential increase of the yield stress, associated with a non-negligible solid volume fraction linear increase. This time, however, is not measured but adjusted to best fit the experimental data, which means that the parameters of the model are not linked to physical properties of the cement paste.

To overcome this limitation, Lecompte and Perrot [11] developed a model that is physically based and uses the definitions of structural build-up proposed by Roussel [10] and the hydration degree of the cement paste. The authors used Roussel's model to incorporate the reversible effect due to nucleation of the cement grains and combined it with the yield stress model "YODEL" that accounts for irreversible solid volume increase due to hydration reactions, linking this solid volume fraction increase to the hydration degree (α). In this case, the parameters related to hydration and coagulation kinetics (A_{thix}) and the hydrate volume fraction within the spherical envelope were adjusted to reach the best fitting.

Differently from those approaches, Muthukrishnan et al. [12] presented what they called a bi-linear yield strength development growth. In the first few minutes after water contact, the material presents a reversible behavior due to flocculation, that is purely due to physical interactions and can be reversed by mixing, and is characterized by the parameter R_{thix} (defined as the rate at which the material yield stress grows from dynamic to static yield stress, after shearing). Then a structuration process starts to occur, as hydration reactions begin to influence the static yield stress growth. The rate at which this structuration process occurs is determined using A_{thix}. The determination of those parameters, though, was made by an average of the peak value in static yield stress tests performed at 0.1, 0.05 and 0.01 s^{-1} shear rates for a given resting time varying from 0 to 60 min.

Pan et al. [13] state that the static yield stress increase with time can be described in two steps: one that happens quickly just after mixing until it slows down and a second where the static yield stress increases slowly with time. To fit the evolution of static yield stress with time they present a thixotropic model that considers these two steps and has five fitted parameters. The authors also considered a short-term structural build-up due to flocculation, where its approximated linear growth rate is R_{thix}; and a long-term structural build-up due to the formation of early hydrates between cement grains, where its rate is A_{thix}. The model works with a turning point (t_{perc}) in which $R_{thix} = A_{thix}$. This time represents the transition from a physical process (flocculation) to a chemical structuration, so that they further defined it as maximum operational time (MOT), that has the same physical meaning as the open time while the reversible process dominates (i.e., when the resting time is less than MOT).

Navarrete et al. [14] also used the bi-linear approach to fit the static yield stress growth results but developed a second-order polynomial regression model to predict the structural build-up as a function of supplementary cementitious materials (SCM) properties (such as particle size, chemical reactivity and surface potential) and mixture parameters (such as w/c ratio, SCM replacement and cement reactivity). They found that the structural build-up is mostly affected by water to cement ratio while it is least affected by the replacement level of SCM.

The stress decay process that a cementitious material undergoes under a constant shear rate is related to structure breakdown and therefore to thixotropy. In this context, Qian et al. [15–17] defined a thixotropic index I_{thix} as the ratio between a maximum shear stress value (τ_i) needed to start flow-onset and the equilibrium value (τ_e) that define steady-state, for a constant shear rate. The I_{thix} characterizes a relationship between static and dynamic yield stress, and the higher the parameter, the higher the thixotropy of the material. Ouyang et al. [18] state that methods like static yield stress assessment and hysteresis loop, although simple, cannot properly describe structural breakdown and structure build-up of a cementitious material. Therefore, they use the flocculation structure parameter method. This method has a strict mathematical form and defines the structure parameter (λ_0) as the ratio of the difference between the initial and equilibrium shear stress to the equilibrium shear stress. The change of this parameter can be used to describe the structure build-up at rest. Chen et al. [9] also calculated the structure parameter, which they defined as S_{thix}, and stated that high S_{thix} accounts for better thixotropy.

Ivanova and Mechtcherine [19] use both Roussel [10] and Perrot et al. [6] models to fit the experimental data, but propose two different approaches to characterize structural build-up through constant shear rate tests with a single-batch: one based on a proportionality limit and another based on the flow onset points. They determined a breaking criterion in the protocol, suitable for two types of rheometers, to avoid excessive deformation of the samples and ensure similar loading conditions. Also, the proportionality limit was introduced as the point attributed to the second maximum of the apparent viscosity. They recommend calculating A_{thix} from both approaches, being the difference between them an indicator of the capacity of the material to sustain loads before failure (in the region of plastic deformation).

Mostafa and Yahia [20] evaluated the effect of mixture parameters on the physico-chemical kinetics of structural build-up. The authors presented two indices to quantify the structural build-up, one related to the resting time needed to form a structural colloidal network (the percolation time t_{perc}); and another related to the growth of stress-bearing capacity due to chemical hydration (the rigidification rate G_{rigid}). Thus, they proposed a semi-empirical model that can predict the two aforementioned indices by considering the microstructural characteristics of cement pastes, such as inter-particle cohesion (I_C), frequency of Brownian collisions (f_C), and the nucleation rate constant (K_B). The authors found good agreement between model predictions and experimental data for the cases studied.

Zhang et al. [5] present a model that describes the relationship between structural build-up rate and inter-particle forces and nucleation and growth of hydration products. It is worth mention that Zhang et al. [5] considers A_{thix} as a structural build-up parameter and propose a new value called rate of structural build-up (r_{sbu}) that describes the growth rate of static yield stress. The authors make a differentiation from Perrot et al. [6] exponential growth model, but do not consider the same definition for the A_{thix} parameter as them. They point out that the growth rate of static yield stress does not only depend on A_{thix} and a characteristic time (t_c), but also continuously changes with time, having them found a different growth rate from 30 to 40 min than the one from 50 to 60 min.

As sometimes there is a delay during the test duration for the effectively applied shear rate to reach the intended applied shear rate, Narrela et al. [8] proposed a constant-strain method instead of the regular constant shear rate approach to evaluate structural build-up. The authors used both Roussel [10] and Perrot et al. [6] models to fit their data. The hypothesis presented is that as long as the applied strain is constant, measured structural build-up remains the same. This is true even if the shear rate varies, as long as compensation in the test duration is made according to the stiffness of the material. The authors suggest that the applied shear rate and the test duration should vary not only according to the different materials tested but also for a single material based on the changes of stiffness during time. It is worth mentioning that the authors found that Roussel's linear model overestimates the structuration rate before the characteristic time, which is especially problematic for the case of 3D-printing as this can lead to failure of the printed structure.

3 Experimental Protocol

A great part of the works that investigate the structural build-up rate of cementitious materials uses the models developed by Roussel [10] and/or Perrot et al. [6] to fit data. Then, another question appears. Not only a model that can properly describe the structural build-up rate of cementitious material, with a simple data acquisition and physically-based assumptions, is needed; but also an experimental protocol for its parameter determination would be of great interest. As there is no standard for the determination of static yield stress, which is a main parameter required in both models, the experimental protocol varies according to the researchers' preferences and expertise.

The structural build-up rate of cement-based materials can be affected by different parameters, such as properties of the constituents materials, mixture design, ambient conditions and the shear stress history of the material (i.e. pre-shear and resting time). This parameter is directly connected with the buildability of layered structures. A feasible construction rate can be determined based on the material initial static yield stress and its structural build-up rate.

The structural build-up rate of cement-based materials is most commonly accessed by the evaluation of the evolution of static yield stress with time at rest through constant shear rate (CSR) tests and small amplitude oscillatory shear (SAOS) tests. In the CSR tests, a low shear is applied at a constant rate and the minimum stress required to initiate flow from rest is recorded. This value is identified as the peak of the shear stress curve [8]. As cementitious materials exhibit both viscous and elastic properties, during SAOS, its elastic and viscous response under continuous sinusoidal shearing within a critical oscillatory strain is recorded as storage modulus (G') and loss modulus (G''), respectively. The values at the point where $G' = G''$ are reported to be the yielding point and therefore are used to determine the static yield stress [12, 15].

It is known that to accurately measure A_{thix} of a paste at rest, the tested sample should not be disturbed. According to Zhang et al. [21], it can be considered that structural build-up is only happening during the static yield stress test. That is because the applied shear rate (or stress) is so small that it causes only minimal damages to the structure of the paste. Narella et al. [8] point out that during a static test, the flow occurs at large critical strains, that according to Roussel et al. [2] are related to the network of colloidal interactions between the particles. While during the dynamic test, the flow occurs at lower critical strains, which Roussel et al. [2] related to C-S-H formation at the cement grains contact points.

What can be said, however, is that dynamic and static shear test methods evaluate different responses of the material. According to Moeini et al. [22], if the need is for the characteristic of the sheared material, the evolution of the static yield stress with time should be adopted. If the need is for the characteristic of the structuration behavior of the material, then a SAOS test is more recommended. Zhang et al. [21], on the other hand, states that static shear stress tests were used to evaluate the evolution of structural build-up in the pastes and SAOS test to characterize the instantaneous response of the structural build-up (at flowing and from flowing to standing).

SAOS is considered a non-destructive test and therefore most likely to not disturb the sample. The test requires for the sample to be sheared within the linear viscoelastic region. Yuan et al. [7] remarks on the importance of the selection of the proper strain amplitude and frequency. For cement paste a frequency of 1 Hz and a strain amplitude lower than the critical strain is appropriate. The authors also showed that the addition of superplasticizer increased the critical strain of the paste from the order of 10^{-5} to the order of 10^{-4}.

For the constant shear rate test, there are some divergences in the literature regarding the single-batch and the multi-batch approaches. For the multi-batch approach, every measure is made on a new undisturbed sample in the given age of test. According to Ivanova and Mechtcherine [19], this method has the disadvantage of not being suitable for in-situ situations besides being time, material and labor-consuming. Also, the variability between multiple samples could be larger than the variation a single-batch approach could give.

In the single-batch approach, the same sample is tested throughout all investigated ages and each measurement should stop right after the material reaches the shear stress peak value, to avoid extra deformation of the sample. Ivanova and Mechtcherine [19] state that only this approach is feasible for in-line automated control in the case of concrete 3D printing. However, this method can lead to an underestimation of A_{thix} due to the multiple disturbances of the material.

About the differences in the measurements for single and multi-batch approaches, Ivanova and Mechtcherine [19] mentioned percentage differences of A_{thix} values not greater than 9% between both approaches. Similarly, Yuan et al. [7] state small effects of the previous static yield stress measurement on the subsequent measure for the single-batch approach. Regarding pre-shearing, most of the experimental procedures adopted this step to guarantee the most dispersed state of the material's structure at the beginning of the test. Ivanova and Mechtcherine [23], however, states that pre-shearing leads to a pronounced underestimation of A_{thix}.

4 Definition of A_{thix}

According to Roussel et al. [2], two types of thixotropy can be observed in a cementitious mixture: a short term one, that is related to colloidal flocculation and it has a characteristic time of the order of a few seconds; and a long term one related to the ongoing hydrates nucleation, being long term thixotropy of practical interest. The authors also mention that the word "structuration" is often used when characterizing consequences of thixotropic behavior, as this word is broader and therefore not associated with a specific physical phenomenon.

Navarrete et al. [14] say that thixotropy is a term usually used to describe the increase of static yield stress of cementitious materials and that one of the main properties of thixotropy is to be reversible. Nonetheless, when cement-based materials are at rest, it is not trivial to separate reversible and irreversible effects that happen due to flocculation and hydration bonding (as CSH bridges), respectively. Therefore, the authors state that "structural build-up (A_{thix})", which comprises reversible and irreversible processes, is a more accurate term than thixotropy when referring to cement-based materials.

Reiter el at. [24], refers to "structuration" as the time evolution of yield stress for concrete at rest and emphasizes that definitions, as well as quantification techniques, for thixotropy, are especially ambiguous and frequently capture a range of flow rates and a particular shear rate history. Yuan et al. [7] describe that thixotropy has three characteristics: (i) its determination is often based on a viscosity decrease (although the evolution of static yield stress is more used in cement-based materials), (ii) it is a time-dependent property, and (iii) it is reversible. Therefore, the authors strictly state that the term is not applicable for cementitious materials, being structural build-up at rest more suitable. The authors of the paper at hand agree with these remarks and chose to refer to this phenomenon (i.e., the sum of reversible and irreversible effects) as structural build-up and, therefore, consider A_{thix} as a structural build-up rate, which characterizes the increase in static yield stress due to flocculation and early hydration [19]. However, the authors acknowledge that many studies affirm that thixotropy and structural build-up are correlated terminologies at early ages, as the former dominates over the latter during this period [25]. It remains necessary, though, to clearly explain the choice of nomenclature.

It is important to stress then, that a change of nomenclature to designate the same phenomena is not to be ignored. A_{thix} can be addressed as flocculation rate [2, 10], structuration rate [6], thixotropy factor [24], rate of increase of yield stress with time at rest [11], rate of thixotropic build-up [26], structural build-up rate [21], structural build-up parameter [5], long-term structural build-up rate [13], thixotropic index [22]. González-Taboada et al. [27] mention in their work A_{thix} as cement paste yield stress evolution in time, structuration rate of the paste and then cement paste structuration rate at rest. Muthukrishnan et al. [12] state that only after a re-flocculation stage (where the re-building is purely due to physical interactions that could be reversed by shearing), the hydration reaction starts to influence the growth of static yield strength with time, calling this process structuration in fresh concrete and its rate A_{thix}. The term R_{thix} also appears with different definitions, as it is defined by Muthukrishnan et al. [12] as the rate at which the yield strength grows from dynamic yield strength to static yield strength after the applied energy is removed and by Pan et al. [13] as the approximate linear growth rate of short-term structural build-up.

Those variations can be rather confusing for the reader. If the 'flocculation rate' term is used, flocculation can be considered as a part of a structuration process due to static yield stress increase at rest, but this increase also happens due to hydration reactions. If the choice is for 'structural build-up', an immediate relation can be made with a possible structural breakdown, which occurs in the case when yield stress is reached and the material starts to flow and during dynamic shear tests. 'Long-term structural build-up rate' separates the phenomena from the short-term structural build-up rate, which is the one that more directly relates to short-term thixotropy, which therefore is associated with colloidal flocculation. Once more, it is stressed how a more precise choice of terminology would allow the researches made on the topic to be properly compared.

5 Conclusions

This paper summarized the relevant literature on structural build-up rate of cementitious materials for 3D-printing technology. Different mathematical models were discussed and a tendency for the use of Roussel's linear model and Perrot et al.'s exponential model was found, although it is clear that more research is ongoing on the topic to develop more accurate models. It became also clear that an experimental protocol for the determination of the parameters would be of great interest. As there is no standard, the experimental protocol varies according to the researchers' preferences and expertise, making it difficult to properly compare the contributions.

The differences between thixotropy and structural build-up at rest were also addressed and a more suitable terminology, that comprises both reversible and irreversible parts of the structuration process, was suggested based on definitions available in the literature.

Once more, the need for a more precise choice of terminologies is stressed, as well as a detailed explanation of the physical and/or chemical effect that is being evaluated, so that the researches made on the topic can be properly correlated.

References

1. Ferron, R.P., Gregori, A., Sun, Z., Shah, S.P.: Rheological method to evaluate structural buildup in self-consolidating concrete cement pastes. ACI Mater. J. **104**(3), 242–250 (2007)
2. Roussel, N., Ovarlez, G., Garrault, S., Brumaud, C.: The origins of thixotropy of fresh cement pastes. Cem. Concr. Res. **42**, 148–157 (2012). https://doi.org/10.1016/j.cemconres.2011.09.004
3. Mostafa, A.M., Yahia, A.: New approach to assess build-up of cement-based suspensions. Cem. Concr. Res. **85**, 174–182 (2016). https://doi.org/10.1016/j.cemconres.2016.03.005
4. Perrot, A., Lecompte, T., Estelle, P., Amziane, S.: Structural build-up of rigid fiber reinforced cement-based materials. Mat. Struct. **46**, 1561–1568 (2013). https://doi.org/10.1617/s11527-012-9997-9
5. Zhang, Z., Xiao, J., Han, K., Wang, J., Hu, X.: Study on the structural build-up of cement-ground limestone pastes and its micro-mechanism. Constr. Build. Mat. **263**, 120656 (2020). https://doi.org/10.1016/j.conbuildmat.2020.120656
6. Perrot, A., Rangeard, D., Pierre, A.: Structural built-up of cement-based materials used for 3D-printing extrusion techniques. Mater. Struct. **49**(4), 1213–1220 (2015). https://doi.org/10.1617/s11527-015-0571-0
7. Yuan, Q., Zhou, D., Li, B., Huang, H., Shi, C.: Effect of mineral admixtures on the structural build-up of cement paste. Constr. Build. Mat. **160**, 117–126 (2018). https://doi.org/10.1016/j.conbuildmat.2017.11.050

8. Nerella, V.N., Beigh, M.A.B., Fataei, S., Mechtcherine, V.: Strain-based approach for measuring structural build-up of cement pastes in the context of digital construction. Cem. Concr. Res. **115**, 530–544 (2019). https://doi.org/10.1016/j.cemconres.2018.08.003
9. Chen, M., et al.: Rheological parameters and building time of 3D printing sulphoaluminate cement paste modified by retarder and diatomite. Const. Build. Mat. **234**, 117391 (2020). https://doi.org/10.1016/j.conbuildmat.2019.117391
10. Roussel, N.: A thixotropy model for fresh fluid concretes: theory, validation and applications. Cem. Concr. Res. **36**, 1797–1806 (2006). https://doi.org/10.1016/j.cemconres.2006.05.025
11. Lecompte, T., Perrot, A.: Non-linear modeling of yield stress increase due to SCC structural build-up at rest. Cem. Concr. Res. **92**, 92–97 (2017). https://doi.org/10.1016/j.cemconres.2016.11.020
12. Muthukrishnan, S., Ramakrishnan, S., Sanjayan, J.: Effect of alkali reactions on the rheology of one-part 3D printable geopolymer concrete. Cem. Concr. Comp. **116**, 103899 (2021). https://doi.org/10.1016/j.cemconcomp.2020.103899
13. Pan, T., Jiang, Y., He, H., Wang, Y., Yin, K.: Effect of structural build-up on interlayer bond strength of 3D printed cement mortars. Materials **14**(2), 236 (2021). https://doi.org/10.3390/ma14020236
14. Navarrete, I., Kurama, Y., Escalon, N., Lopez, M.: Impact of physical and physicochemical properties of supplementary cementitious materials on structural build-up of cement-based pastes. Cem. Concr. Res. **130**, 105994 (2020). https://doi.org/10.1016/j.cemconres.2020.105994
15. Qian, Y., Lesage, K., El Cheikh, K., De Schutter, G.: Effect of polycarboxylate ether superplasticizer (PCE) on dynamic yield stress, thixotropy and flocculation state of fresh cement pastes in consideration of the Critical Micelle Concentration (CMC). Cem. Concr. Res. **107**, 75–84 (2018). https://doi.org/10.1016/j.cemconres.2018.02.019
16. Qian, Y., De Schutter, G.: Enhancing thixotropy of fresh cement pastes with nanoclay in presence of polycarboxylate ether superplasticizer (PCE). Cem. Concr. Res. **111**, 15–22 (2018). https://doi.org/10.1016/j.cemconres.2018.06.013
17. Qian, Y., De Schutter, G.: Different effects of NSF and PCE superplasticizer on adsorption, dynamic yield stress and thixotropy of cement pastes. Materials **11**(5), 695 (2018). https://doi.org/10.3390/ma11050695
18. Ouyang, Y., Tan, D.J., Corr, S.P.: Shah, The thixotropic behavior of fresh cement asphalt emulsion paste. Constr. Build. Mat. **114**, 906–912 (2016). https://doi.org/10.1016/j.conbuildmat.2016.04.024
19. Ivanova, I., Mechtcherine, V.: Possibilities and challenges of constant shear rate test for evaluation of structural build-up rate of cementitious materials. Cem. Concr. Res. **130**, 105974 (2020). https://doi.org/10.1016/j.cemconres.2020.105974
20. Mostafa, A.M., Yahia, A.: Physico-chemical kinetics of structural build-up of neat cement-based suspensions. Cem Concr. Res. **97**, 11–27 (2017). https://doi.org/10.1016/j.cemconres.2017.03.003
21. Zhang, Y., Jiang, Z., Zhu, Y., Zhang, J., Ren, Q., Huang, T.: Effects of redispersible polymer powders on the structural build-up of 3D printing cement paste with and without hydroxypropyl methylcellulose. Constr. Build. Mat. **267**, 120551 (2021). https://doi.org/10.1016/j.conbuildmat.2020.120551
22. Moeini, M.A., Hosseinpoor, M., Yahia, A.: Effectiveness of the rheometric methods to evaluate the build-up of cementitious mortars used for 3D printing. Const. Build. Mat. **257**, 119551 (2020). https://doi.org/10.1016/j.conbuildmat.2020.119551
23. Ivanova, I., Mechtcherine, V.: Effects of volume fraction and surface area of aggregates on the static yield stress and structural build-up of fresh concrete. Materials **13**, 1551 (2020). https://doi.org/10.3390/ma13071551

24. Reiter, L., Wangler, T., Roussel, N., Flatt, R.J.: The role of early age structural build-up in digital fabrication with concrete. Cem Concr. Res. **112**, 86–95 (2018). https://doi.org/10.1016/j.cemconres.2018.05.011
25. Yuan, Q., Lu, X., Khayat, K.H., Feys, D., Shi, C.: Small amplitude oscillatory shear technique to evaluate structural build-up of cement paste. Mater. Struct. **50**(2), 1–12 (2016). https://doi.org/10.1617/s11527-016-0978-2
26. Reales, O.A.M., Duda, P., Silva, E.C.C.M., Paiva, M.D.M., Toledo Filho, R.D.: Nanosilica particles as structural buildup agents for 3D printing with Portland cement pastes. Constr. Build. Mat. **219**, 91–100 (2019). https://doi.org/10.1016/j.conbuildmat.2019.05.174
27. González-Taboada, I., González-Fonteboa, B., Martínez-Abella, F., Seara-Paz, S.: Thixotropy and interlayer bond strength of self-compacting recycled concrete. Const. Build. Mat. **161**, 479–488 (2018). https://doi.org/10.1016/j.conbuildmat.2017.11.157

Interlayer Bond and Porosity of 3D Printed Concrete

Manu K. Mohan(✉), A. V. Rahul, Geert De Schutter, and Kim Van Tittelboom

Magnel-Vandepitte Laboratory, Department of Structural Engineering and Building Materials, Ghent University, Ghent, Belgium
manukmohn@gmail.com, Kim.VanTittelboom@UGent.be

Abstract. Extrusion-based concrete 3D printing is being increasingly investigated for its application in the construction industry. Having a layer-wise extrusion process could impart a lack of fusion between the layers due to drying and moisture migration between the layers, resulting in higher porosity and weak interlayer bond strength. This paper presents a study to understand the effect of fresh cement paste applied in between the two layers on the porosity and bond behaviour of samples printed at a relatively high interlayer time gap of 15 min. Two different types of 3D printable mixtures were considered in this study: (i) a Portland cement-slag binary blend mixture and (ii) a calcium sulfoaluminate cement-based mixture. Two-layered wall elements were printed with a constant 15 min interlayer time interval. The interlayer bond strength was determined by pull-off tests and compared among the different mixtures with and without a Portland cement-based paste applied between the layers. Also, the porosity of the printed elements was studied with mercury intrusion porosimetry. The results indicate that the total porosity and critical pore diameter decrease with the application of the fresh cement paste. Also, the bond strength increases slightly with the application of cement paste as it reduces the total porosity of the samples.

Keywords: 3D printing · Interlayer bond · Bond strength · Porosity · CSA cement · Slag

1 Introduction

Extrusion-based concrete 3D printing (3DCP) is getting an increased attention in today's construction industry due to the economic advantages associated with it. In addition, the enhanced geometric freedom, increased construction speed, and the lack of formwork allows it to fabricate very uniquely shaped elements and structures with optimized topological designs [1–3]. In 3DCP, the elements are fabricated in a layer-wise manner with the help of a predefined computer model.

Though the research in the different aspects of extrusion-based concrete 3D printing is booming, there are several challenges associated with this technology yet to be tackled. Such challenges are the presence of high porosity at the interface and the weak interlayer bond [4, 5]. The main reasons for having higher porosity at the interface are

J. I. Escalante-Garcia et al. (Eds.): RW 2021, RILEM Bookseries 40, pp. 425–433, 2023.
https://doi.org/10.1007/978-3-031-21735-7_47

the entrapment of air voids between the two layers, the moisture migration between the layers, and the excessive drying due to the lack of formwork [4–7]. This in turn results into the anisotropic behaviour as the compressive and flexural strengths of the 3D printed concrete elements are reported to be different in different directions and generally lower compared to mould cast concrete [1, 8].

The present study aims to characterize the porosity and bond strength of 3D printed concrete made with two different types of concrete mixtures, i.e., made with a Portland cement and slag based system and a calcium sulfoaluminate cement-based system. The porosity of the 3D printed elements was characterized by mercury intrusion porosimetry tests, and the interlayer bond strength was assessed by pull-off tests.

2 Methodology

2.1 Materials and Mixtures

The materials used in this study include a CEM 52.5 N Portland cement, ground granulated blast furnace slag (GGBFS), calcium sulfoaluminate (CSA) cement, and limestone powder. The chemical composition and physical properties of the ingredients are listed in Tables 1 and 2, respectively. Fine aggregates with a maximum particle size of 2 mm were also used to make the mixtures. The particle size distributions of the binders and aggregate were determined using the laser diffraction method and sieve analysis, respectively and are shown in Fig. 1. A polycarboxylate ether-based superplasticizer, a cellulose-based viscosity modifying agent and sodium tetraborate-decahydrate were used as the chemical admixtures. The mixtures were prepared with a constant water-to-binder ratio of 0.35. The composition of the mixtures are given in Table 3. The detailed procedure of the mixture design is described in previous publications by the authors [9–11].

Table 1. Physical properties of the powder materials used

Parameter	Portland cement	GGBFS	CSA cement	Limestone powder
Blaine area (m^2/kg)	408	428	500	428
Specific gravity (–)	3.15	2.68	3.15	2.71

2.2 3D Print Experiments

3D printing experiments were carried out using a six-axis robot connected with a screw-based extrusion system. To assess the bond strength between the layers, about 1 m long two-layered wall elements were printed. Also, to improve the bond strength between the two printed layers, fresh cement paste made with the same water-to-binder ratio (see Table 3 for the composition of the cement paste) was applied manually using a brush in between the two layers. For all mixtures, the printed elements were cured until the age of 7 days at 25 °C and 95% relative humidity, after which suitable size samples were cut out to perform porosity and bond strength tests. For porosity tests two samples and bond strength tests six samples were prepared.

Table 2. Chemical composition of the powder materials

Oxide	Quantity (%)			
	Portland cement	GGBFS	CSA cement	Limestone powder
CaO	64.30	37.97	41.5	56.6
SiO_2	18.30	35.60	8.14	0.4
Al_2O_3	5.20	13.12	23.20	0.2
Fe_2O_3	4.00	0.37	1.05	0.1
$(Na_2O)_e$	0.32	0.74	0.86	0.0
MgO	1.40	7.24	3.22	0.4
SO_3	3.50	7.24	18.36	0.1

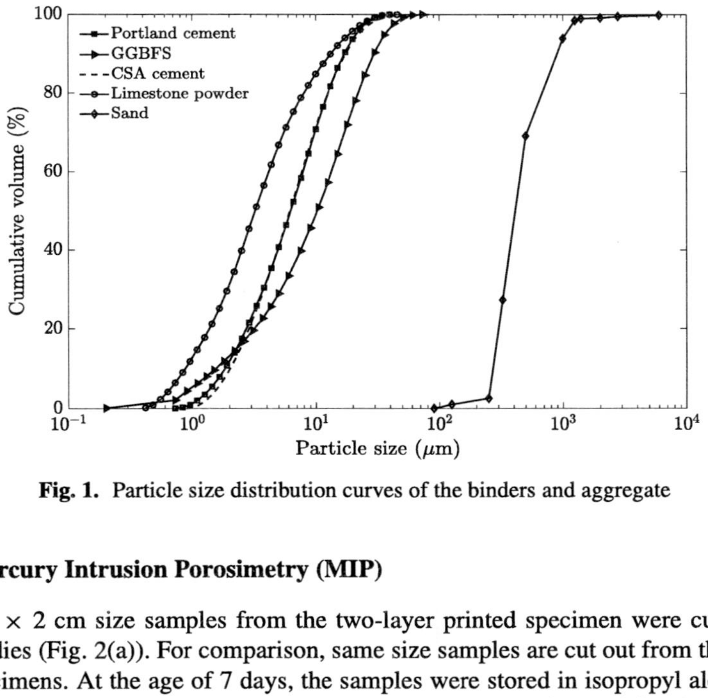

Fig. 1. Particle size distribution curves of the binders and aggregate

2.3 Mercury Intrusion Porosimetry (MIP)

About 1 × 2 cm size samples from the two-layer printed specimen were cut for the MIP studies (Fig. 2(a)). For comparison, same size samples are cut out from the mould cast specimens. At the age of 7 days, the samples were stored in isopropyl alcohol for up to 4 days to stop hydration by the solvent exchange method [12]. After four days of immersion in isopropyl alcohol, the samples were stored inside a vacuum desiccator until testing. The MIP tests were carried out by using a Pascal 140–440 series porosimeter from Thermo Scientific. Mercury was intruded into each of the samples using a pressure range varying from vacuum to a maximum pressure of 200 MPa. Using the Washburn equation, the pore sizes corresponding to each pressure step were computed by assuming the pores are cylindrical [13]. MIP tests were repeated twice to check the repeatability and a representative curve from the two tests were presented.

Table 3. Composition of the mixtures

Material	Quantity (kg/m³)		
	Mixture 1	Mixture 2	Cement paste
Portland cement	376	–	1498
GGBFS	376	–	–
CSA cement	–	561	–
Limestone powder	–	240	–
Sand	1279	1201	–
Water	263	280	524
Superplasticizer	5.27	2.80	6.0
Viscosity modifying agent	0.75	0.75	–
Borax	–	4.05	–
Total weight	2301.3	2285.6	2029
w/b	0.35	0.35	0.35
Aggregate/binder	1.7	1.5	–

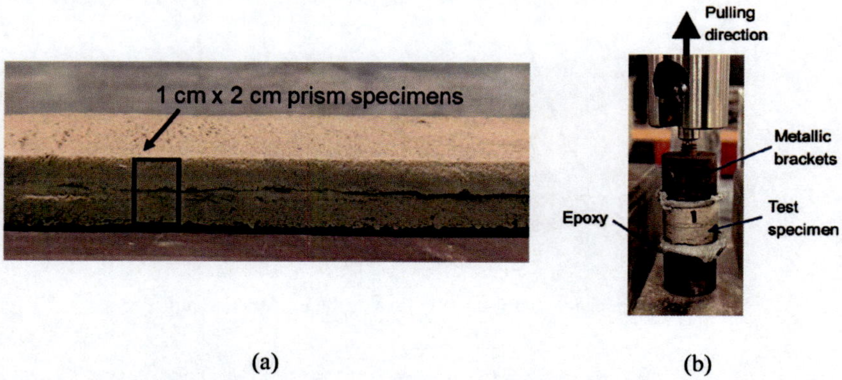

Fig. 2. 3D printed samples were sawn into (a) 1 × 2 cm prismatic specimens for MIP tests and (b) 25 mm diameter and 20 mm height cylindrical specimens for bond strength tests

2.4 Bond Strength Tests

Pull-off tests were carried out to assess the bond strength of the 3D printed elements. For that, 25 mm diameter and 20 mm length cylindrical specimens were drilled out from the two-layered printed wall elements. Same size samples were cut out from the mould cast specimens prepared as per EN 196 [14] also. The pull-off tests were performed at the age of 7 days for both the mixtures prepared at different conditions (i.e., with and without cement paste and mould cast specimens). The tests were carried out using a Macben Proceq machine in the load controlled way (Fig. 2(b)). The loading rate for the pull-off tests was 50 N/s. The drilled out cylindrical specimen was glued to the dollies using epoxy, and 24 h after application of the glue, the tests were carried out [5]. Bond strength of the mould cast samples was also determined in the same way. Pull-off tests were done with 6 number of samples and the average bond strength values were reported.

3 Results and Discussions

3.1 Porosity and Critical Pore Diameter

Figure 3(a) shows the cumulative intrusion curve of mixture 1. It can be seen that when the cement paste is present in between the two layers, the total volume of intruded mercury is significantly lower and comparable with the mould cast sample. The porosity (%) values are listed in Table 4. It can be seen that the total porosity of the mixture 1 without cement paste is about 50% higher than that of the sample with cement paste applied in between the two layers. Also, it can be seen that when cement paste is applied in between the layers, the porosity reduces to the range of mould cast samples. Figure 3(b) shows the cumulative intrusion curve of mixture 2. A similar trend can be observed in the case of mixture 2. With the application of cement paste, the porosity reduced by about 26%. However, as can be seen in Table 4, the reduction in porosity is not as significant for mixture 2 as observed in the case of mixture 1 when cement paste is applied in between the layers.

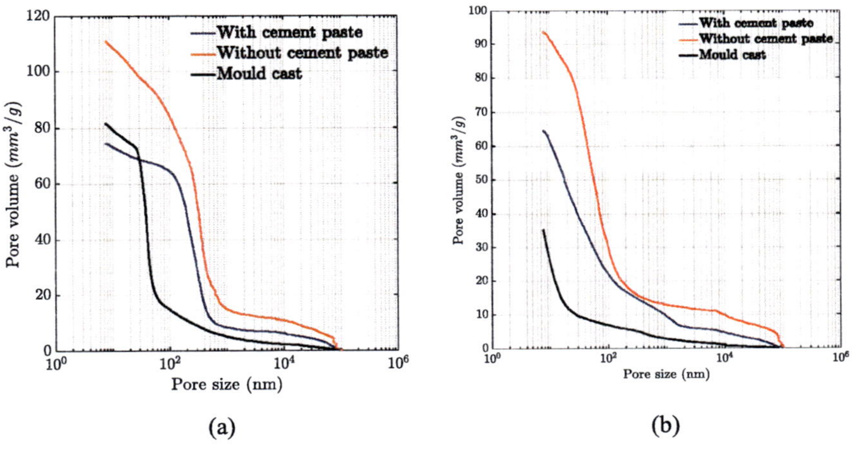

Fig. 3. Cumulative intrusion curves of (a) mixture 1 (b) mixture 2

Figure 4(a) and (b) shows the differential intrusion curve of mixture 1 and 2, respectively. The critical pore size corresponds to the maximum peak of the differential intrusion pore volume curve. The critical pore size could give an indication of the size at which maximum volume intrusion happens. In other words, it is the pore size above which no connected path could form throughout the sample and could be related to the transport of ions through the matrix [15]. The critical pore diameter values for both mixtures are listed in Table 4. For both mixtures, the sample without cement paste has the highest critical pore diameter, while the mould cast samples showed the lowest value. Further, it is seen that the critical pore sizes of mixture 2 are significantly lower than that of mixture 1. This can be attributed to higher pore segmentation in ettringite-based binder systems. In CSA cement-based systems, the formation of C-S-H gel in between the large amount of ettringite crystals formed at early ages can lead to much finer pore size distribution compared to Portland cement-based systems [16].

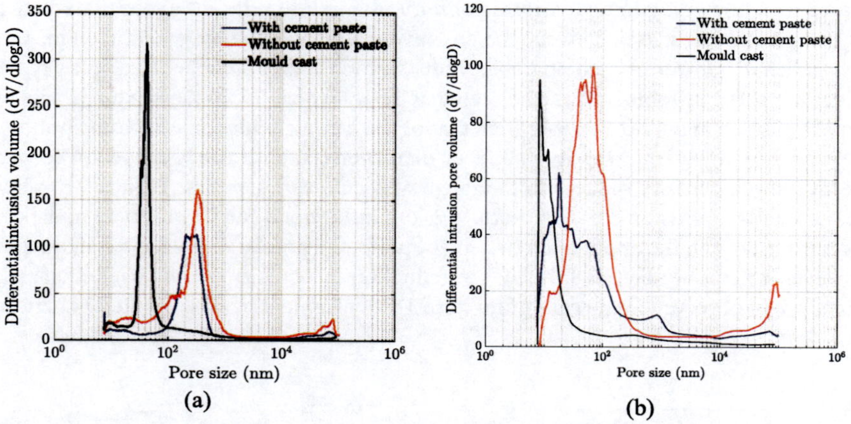

Fig. 4. Differential intrusion curves of (a) mixture 1 (b) mixture 2

Table 4. Total porosity and critical pore diameter values of different mixtures

Sample		Total porosity (%)	Critical pore diameter (nm)
Mixture 1	With cement paste	17.16	263
	Without cement past	25.48	332
	Mould cast	18.78	41
Mixture 2	With cement paste	19.22	11
	Without cement past	21.43	67
	Mould cast	8.45	8

3.2 Tensile Bond and Compressive Strength

Figure 5 shows the bond strength of the samples at the age of 7 days. The bond strength values shown in Fig. 5 are the average of 6 different samples. It can be seen that with the application of cement paste in between the layers, the bond strength of both mixtures increases by about 30%. Also, mixture 2 exhibited comparatively higher bond strength in all the different cases. Literature suggests that the bond strength of the samples depends on the porosity at the interface [4, 5, 17]. This can be due to lower porosity and the relatively finer pore size distribution for CSA-based systems as seen earlier in the MIP results.

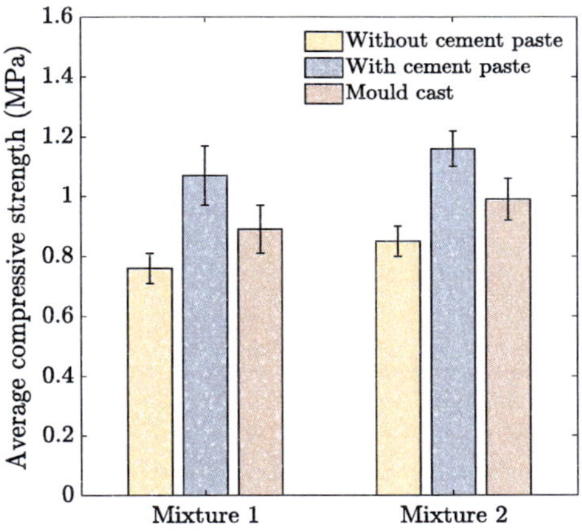

Fig. 5. Bond strength of both mixtures at 7 days

4 Conclusions

This study investigates the porosity and bond strength of 3D printable mixtures made with two different binder systems at a constant interlayer time gap of 15 min. The main conclusions drawn from this study are:

- The total porosity and critical pore sizes reduce when a fresh cement paste is applied in between the layers of the 3D printed samples observed from mercury intrusion porosimetry experiments.
- The total porosity of the samples of the Portland cement-slag mixture with cement paste applied in between is similar to the total porosity of the mould cast samples. A similar trend can also be observed in the case of CSA based printable concrete mixtures. Compared to the OPC-slag mixture, the CSA-based mixture showed lower porosity and finer pore size distribution due to the pore refinement associated with the ettringite systems.
- For both types of mixtures, the bond strength was found to be higher with the application of cement paste. Therefore, the current study points the positive effect of cement paste application at the interlayer position to improve mechanical integrity of printed samples. However, automation of the process of applying the cement paste needs further research and attention.

Acknowledgements. Authors would like to acknowledge the financial support provided by SIM (Strategic Initiative Materials in Flanders) and VLAIO (Flanders agency for innovation & entrepreneurship) towards the 3D2BGreen project. The authors also acknowledge the companies, BESIX, ResourceFull and Witteveen+Bos for being the partners of the 3D2BGreen project. The authors thankfully acknowledge the research fellows and technicians at the Magnel-Vandepitte Laboratory for their help in this study.

References

1. Mohan, M.K., Rahul, A.V., De Schutter, G., Van Tittelboom, K.: Extrusion-based concrete 3D printing from a material perspective: a state-of-the-art review. Cem. Concr. Compos. **115**, 103855 (2021). https://doi.org/10.1016/j.cemconcomp.2020.103855
2. Vantyghem, G., De Corte, W., Shakour, E., Amir, O.: 3D printing of a post-tensioned concrete girder designed by topology optimization. Autom. Constr. **112**, 103084 (2020). https://doi.org/10.1016/j.autcon.2020.103084
3. Buswell, R.A., Leal de Silva, W.R., Jones, S.Z., Dirrenberger, J.: 3D printing using concrete extrusion: a roadmap for research. Cem. Concr. Res. **112**, 37–49 (2018). https://doi.org/10.1016/j.cemconres.2018.05.006
4. Nerella, V.N., Hempel, S., Mechtcherine, V.: Effects of layer-interface properties on mechanical performance of concrete elements produced by extrusion-based 3D-printing. Constr. Build. Mater. **205**, 586–601 (2019). https://doi.org/10.1016/j.conbuildmat.2019.01.235
5. Van Der Putten, J., Deprez, M., Cnudde, V., De Schutter, G., Van Tittelboom, K.: Microstructural characterization of 3D printed cementitious materials. Materials **12**, 1–22 (2019). https://doi.org/10.3390/ma12182993
6. Van Der Putten, J., Azima, M., Van den Heede, P., Van Mullem, T., Snoeck, D., De Schutter, G., Van Tittelboom, K.: Neutron radiography to study water ingress via the interlayer of 3D printed cementitious materials. Constr. Build. Mater. Accepted (n.d.)
7. Sanjayan, J.G., Nematollahi, B., Xia, M., Marchment, T.: Effect of surface moisture on interlayer strength of 3D printed concrete. Constr. Build. Mater. **172**, 468–475 (2018). https://doi.org/10.1016/j.conbuildmat.2018.03.232
8. Rahul, A.V., Santhanam, M., Meena, H., Ghani, Z.: Mechanical characterization of 3D printable concrete. Constr. Build. Mater. 227 (2019). https://doi.org/10.1016/j.conbuildmat.2019.116710
9. Mohan, M.K., Rahul, A.V., Van Tittelboom, K., De Schutter, G.: Rheological and pumping behaviour of 3D printable cementitious materials with varying aggregate content. Cem. Concr. Res. **139** (2021). https://doi.org/10.1016/j.cemconres.2020.106258
10. Mohan, M.K., Rahul, A.V., De Schutter, G., Van Tittelboom, K.: Early age hydration, rheology and pumping characteristics of CSA cement-based 3D printable concrete. Constr. Build. Mater. **275**, 122136 (2021). https://doi.org/10.1016/J.CONBUILDMAT.2020.122136
11. Mohan, M.K., Rahul, A.V., Van Tittelboom, K., De Schutter, G.: Evaluating the influence of aggregate content on pumpability of 3D printable concrete. In: Bos, F.P., Lucas, S.S., Wolfs, R.J.M., Salet, T.A.M. (eds.) DC 2020. RB, vol. 28, pp. 333–341. Springer, Cham (2020). https://doi.org/10.1007/978-3-030-49916-7_34
12. Dhandapani, Y., Santhanam, M.: Assessment of pore structure evolution in the limestone calcined clay cementitious system and its implications for performance. Cem. Concr. Compos. **84**, 36–47 (2017). https://doi.org/10.1016/j.cemconcomp.2017.08.012
13. Diamond, S.: Mercury porosimetry. An inappropriate method for the measurement of pore size distributions in cement-based materials. Cem. Concr. Res. **30**, 1517–1525 (2000). https://doi.org/10.1016/S0008-8846(00)00370-7

14. BS:EN196–1.: Methods of testing cement–Part 1 : determination of strength. (2016)
15. Atahan, H.N., Oktar, O.N., Taşdemir, M.A.: Effects of water-cement ratio and curing time on the critical pore width of hardened cement paste. Constr. Build. Mater. **23**, 1196–1200 (2009). https://doi.org/10.1016/j.conbuildmat.2008.08.011
16. Bernardo, G., Telesca, A., Valenti, G.L.: A porosimetric study of calcium sulfoaluminate cement pastes cured at early ages. Cem. Concr. Res. **36**, 1042–1047 (2006). https://doi.org/10.1016/j.cemconres.2006.02.014
17. Chen, Y., Jansen, K., Zhang, H., Romero Rodriguez, C., Gan, Y., Çopuroğlu, O., chlangen, E.: SEffect of printing parameters on interlayer bond strength of 3D printed limestone-calcined clay-based cementitious materials: an experimental and numerical study. Constr. Build. Mater. **262**, 120094 (2020). https://doi.org/10.1016/j.conbuildmat.2020.120094

Adhesion Performance of Cement-Based Materials in the Fresh State

Yaxin Tao[1], Karel Lesage[1], Kim Van Tittelboom[1](✉), Yong Yuan[2], and Geert De Schutter[1]

[1] Ghent University, Ghent, Belgium
Kim.VanTittelboom@UGent.be
[2] Tongji University, Shanghai, China

Abstract. Cement-based materials placed onto a substrate must achieve sufficient adhesion at the interface to avoid detaching, especially for the overhead positions. Previously, more attention was paid to adhesion in the hardened state, while the adhesive properties of fresh cement-based materials were rarely studied. In this paper, a tack test was used to evaluate the adhesion of fresh cement-based materials. The influence of two aspects including the loading rate and the substrate surface roughness was investigated. Results indicated that the loading rate (varied from 0.01 to 0.08 N/s) had a limited effect on the adhesion of fresh cement-based materials. However, an increase in the substrate surface roughness resulted in a reduced adhesive strength in the fresh state.

Keywords: Adhesion · Cement-based material · Tack test · Surface roughness · Loading rate

1 Introduction

Cement-based materials, used for adhesion purposes in the context of mortar rendering [1], concrete spraying [2], and 3D concrete printing [3], must meet the requirements of wetting substrates, adhering to substrates, gaining enough strength after hardening, and remaining stable [4, 5]. Commonly, the adhesive properties of cement-based materials are measured after hardening, while the formation of contact between the substrate and the fresh cement-based material is less investigated [6, 7].

A so-called tack test provides one feasible way to measure the adhesive properties of cement-based materials in the fresh state. During a tack test, two solid surfaces, between which the fresh material is inserted, are pulled away at a fixed velocity or loading rate after a certain time period. The force versus separating displacement (or time) curves are then recorded. The tack test has been widely used to characterize the debonding properties of different types of soft materials such as pressure-sensitive adhesives [8] and smectite muds [9]. Moreover, this method has also been employed more recently to investigate the adhesive properties of cement pastes with the addition of polymers [10] and fibers [11]. However, the influence of substrate properties (e.g., the surface roughness) and the different loading rates (e.g., the thickness of sprayed concrete in a

J. I. Escalante-Garcia et al. (Eds.): RW 2021, RILEM Bookseries 40, pp. 434–441, 2023.
https://doi.org/10.1007/978-3-031-21735-7_48

specific duration) on the adhesive properties of cement-based materials has not been studied.

In this study, the influence of these two aspects (i.e., substrate surface roughness levels and loading rates) on the adhesion of cement-based materials was investigated by a tack test.

2 Experimental Program

2.1 Materials

The mixture consisted of 0–2 mm silica sand (Vandevelde NV), CEM I 52.5 cement (Holcim), deionized water, polycarboxylate ether (PCE, BASF Master Glenium 51), and hydroxypropyl methylcellulose (HPMC, Tylose MOT 60,000 YP4). More details about the materials can be found in [12]. The sand to cement ratio was 1 and the water to cement ratio was 0.35. The dosages of polycarboxylate ether and hydroxypropyl methylcellulose were both 0.1% by the mass of cement. The mixture design of the cement-based material is shown in Table 1. Around 500 g of the fresh material was prepared for each batch. The mixing procedure can be found in [3].

Table 1. Mixture design of the cement-based material (kg/m^3).

Sand	Cement	Water	PCE	HPMC
958.0	958.0	335.3	0.958	0.958

2.2 Tack Test

The tack test was performed with an Anton Paar rheometer (MCR 102) equipped with parallel plates (50 mm). A top plate with different surface roughness levels and a smooth bottom plate were used. The test was performed as follows: (1) fixing smooth bottom plate, (2) placing a plastic circular mold (height 20 mm, diameter 50 mm) on the bottom plate and fixing the mold by hand, (3) pouring the fresh material in the mold and moving away the circular mold, (4) moving down the top plate until the thickness of the fresh sample reached 10 mm, and (5) lifting the top plate with a linearly increasing load and recording the normal force and displacement. To check the reproducibility of the experimental results, five different runs were performed for each freshly prepared sample with the same batch. The testing procedure of the tack test is also shown in Fig. 1.

Four types of top plates with different roughness levels were prepared including a smooth steel plate, a sandpaper plate (sandpaper was glued to the smooth plate by using a two-face tape), a fine river sand plate (1–1.25 mm sand was glued to the smooth plate by using epoxy resin), and a coarse river sand plate (6.3–8 mm sand was glued to the smooth plate by using epoxy resin), as shown in Fig. 2.

The lower surfaces of the top plates were scanned by an automated laser scanning machine equipped with a high precision laser sensor with a resolution of 5 μm. For each

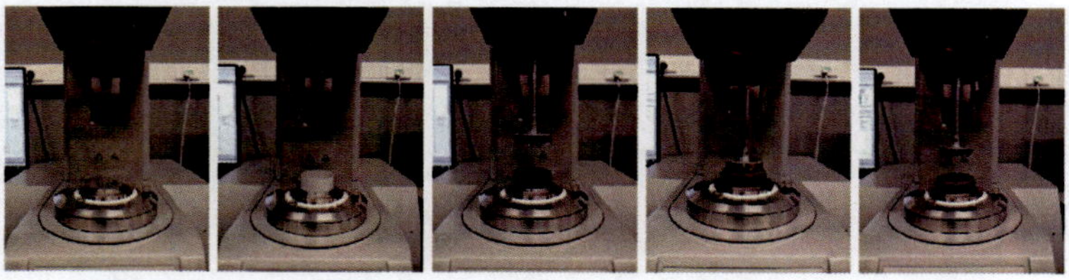

Fig. 1. Testing procedure of the tack test.

Fig. 2. Lower surfaces of top plates with a diameter of 50 mm: (P1) smooth steel, (P2) sandpaper, (P3) fine river sand, and (P4) coarse river sand.

plate, the surface was scanned four times along different diametric scanning paths and the profiles were recorded. For example, the profiles of the coarse river sand plate (P4) are shown in Fig. 3. The roughness levels of the lower surfaces were characterized by the root mean square deviation R_q (mm), which followed the standard [13]. The average values of the root mean square deviation R_q were 0.00, 0.09, 0.77, 2.36 mm for the smooth steel plate (P1), the sandpaper plate (P2), the fine river sand plate (P3), and the coarse river sand plate (P4), respectively.

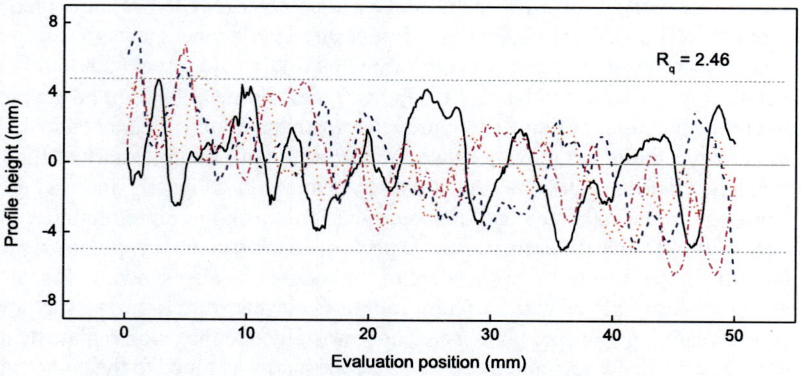

Fig. 3. Four scanned profiles of the coarse river sand plate P4.

3 Results and Discussion

3.1 The Influence of Loading Rate

The normal force evolutions with different loading rates are shown in Fig. 4. It was noticed that the normal force increased linearly. After reaching the peak force at the failure point, the normal force dropped to zero in a short time period.

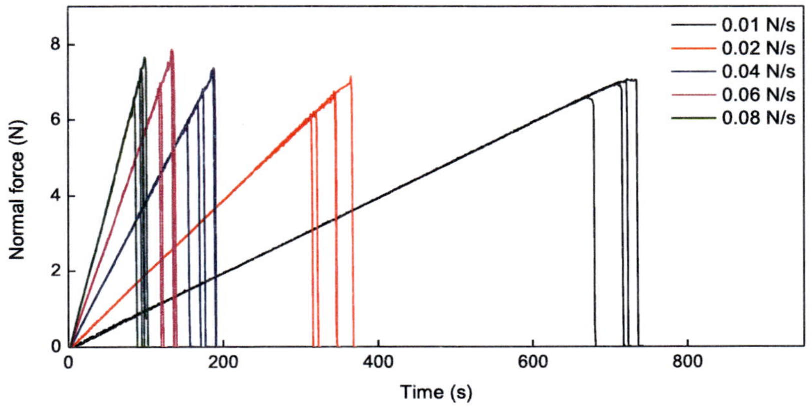

Fig. 4. Normal force evolution with different loading rates.

The average values of the peak normal force with different loading rates were 7.07, 6.61, 6.78, 7.46, and 7.19 N, respectively, as also shown in Fig. 5. A one-way ANOVA test was used to statistically analyze the influence of the loading rate on the peak normal force obtained in the tack test. A p-value less than 0.05 was considered statistically significant. According to the results, there was no significant effect of the loading rate on the peak normal force at the $p < 0.05$ level for the five conditions ($F = 2.561$, $p = 0.070$). While post hoc comparisons using the LSD test indicated that the condition with a loading rate of 0.02 N/s was significantly different when compared with the condition of 0.06 N/s loading rate ($p = 0.033$). Moreover, the condition with a loading rate of 0.04 N/s was significantly different from that of 0.06 N/s ($p = 0.010$). Kaci et al. [11] pointed out that the normal force measured in a tack test is mainly determined by the yield stress of the fresh material if the pulling velocity tended to zero. In this study, before reaching failure, the pulling velocity of the top plate remained very limited as the loading rate increased from 0.01 to 0.08 N/s, which can be further indicated by the displacement evolution curves of the top plate presented in Fig. 6. Taken together, it can be concluded that the loading rate (from 0.01 to 0.08 N/s) has a limited influence on the peak normal force within a load control mode tack test.

In addition, the displacement evolutions with different loading rates were recorded, as shown in Fig. 6. The curves with different loading rates presented a similar pattern which can be divided into three stages. As the load was applied, the fresh material showed an elastic behavior in the first stage, where the vertical deformation hardly increased.

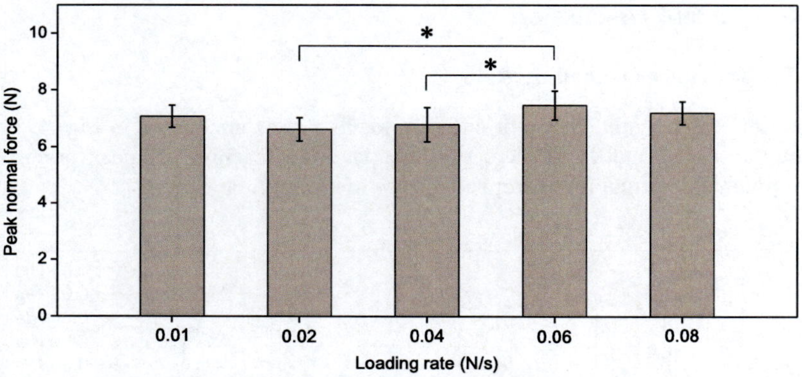

Fig. 5. Peak normal forces with different loading rates, error bars represent the standard error (n = 5), the symbol * indicates a significant difference ($p < 0.05$).

After reaching the yield point, the fresh material underwent a shear flow stage where an obvious deformation could be observed. Finally, the top plate was lifted rapidly when it was detached from the fresh sample.

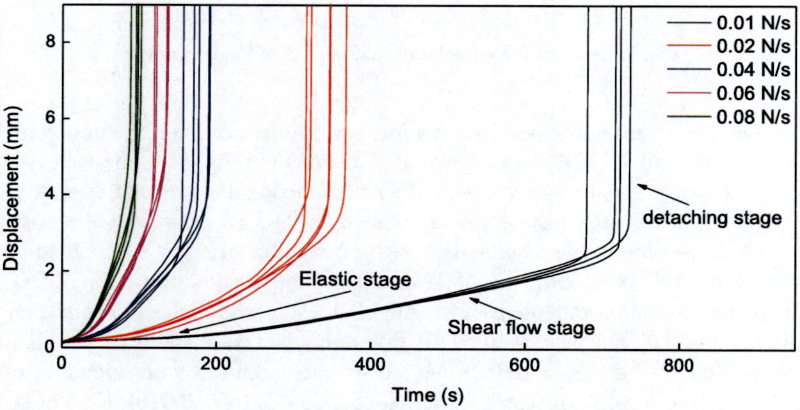

Fig. 6. Displacement evolution with different loading rates.

The displacement versus time curves at the shear flow stage and the detaching stage were linearly extrapolated, respectively. The displacement value at the intersection point of the linear extrapolations can be defined as the critical displacement, as shown in Fig. 7. A one-way ANOVA test was used to statistically analyze the influence of the loading rate on the critical displacement. Post hoc comparisons indicated that the condition with a loading rate of 0.01 N/s was significantly different when compared with other conditions, i.e., 0.02, 0.04, 0.06, and 0.08 N/s ($p = 0.001$, 0.016, 0.000, and 0.015).

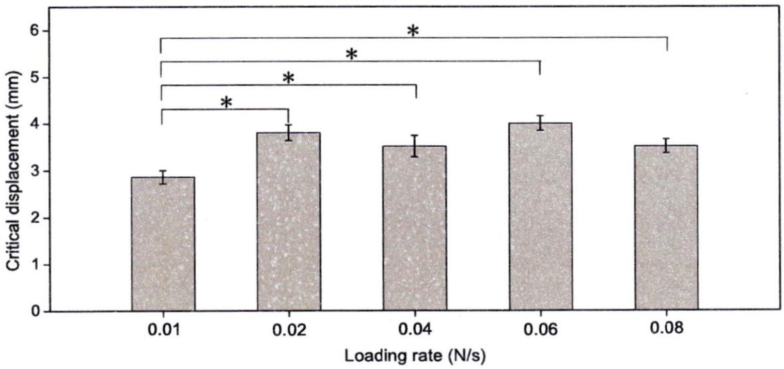

Fig. 7. Critical displacement with different loading rates, error bars represent the standard error ($n = 5$), the symbol * indicates a significant difference ($p < 0.05$).

3.2 The Influence of Substrate Surface Roughness

The normal force evolutions with the four top plates with different surface roughness levels are shown in Fig. 8. The curves present the same trend as introduced in the previous section. The average values of the peak normal force were 6.78, 3.53, 2.17, and 2.04 N for the P1, P2, P3, and P4 plates, respectively.

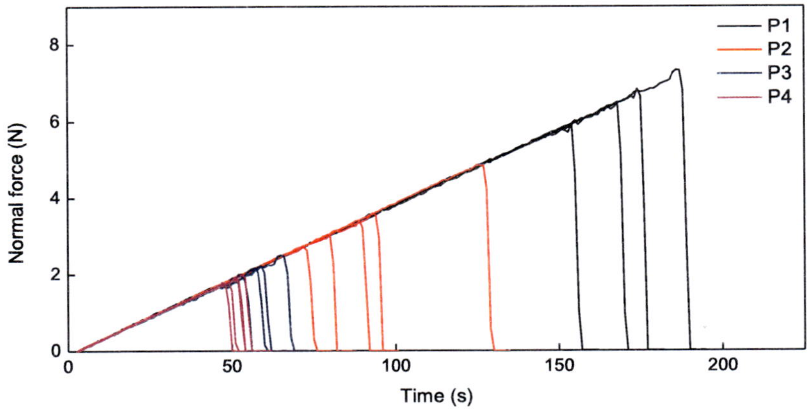

Fig. 8. Normal force evolution with different top plates.

The peak normal force versus the root mean square deviation is shown in Fig. 9. It was shown that the peak normal force reduced as the root mean square deviation increased. That is to say, the plate with a low roughness level (e.g. smooth steel plate P1) led to a better adhesion compared to that with a high roughness level. An exponential relationship was fit to describe the relationship between the peak normal force and the root mean square deviation. The fit curve indicated that the peak normal force decreased significantly and tended to reach a stable value as the roughness level of the lower surface of the top plate increased. It can be interpreted that the macro textures on the rough surface cannot be filled by the fresh cement-based material, leading to partial contact between the top plate and the fresh sample. As a result, the adhesion was limited.

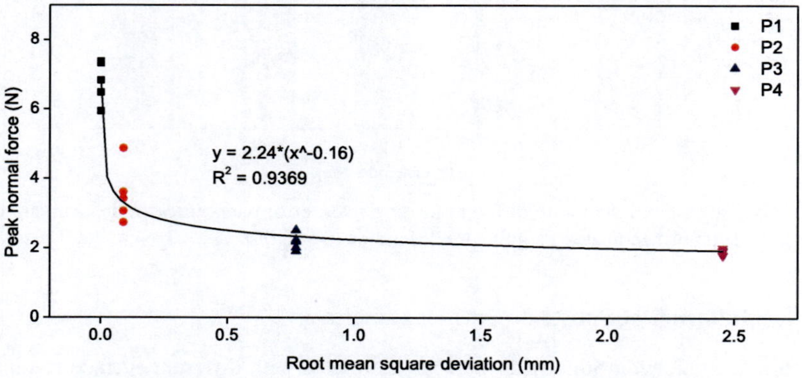

Fig. 9. Peak normal force versus root mean square deviation.

4 Conclusion

The influence of the loading rate and the substrate surface roughness on the adhesion of cement-based materials in the fresh state was investigated in this study. Based on the experimental results, the following conclusions can be drawn:

(1) The deformation of the fresh material in a tack test can be divided into three stages including an elastic stage, a shear flow stage, and a detaching stage.
(2) The loading rate (from 0.01 to 0.08 N/s) had a limited influence on the peak normal force and the adhesion of the fresh material was mainly determined by the yield stress.
(3) There was no significant difference among critical displacements when the loading rate exceeded 0.02 N/s.
(4) The peak normal force reduced as the substrate surface roughness increased, which can be attributed to the limited contact between the fresh material and the lower surface of the top plate.

References

1. Cardoso, F.A., John, V.M., Pileggi, R.G., Banfill, P.F.G.: Characterisation of rendering mortars by squeeze-flow and rotational rheometry. Cem. Concr. Res. **57**, 79–87 (2014)

2. Khooshechin, M., Tanzadeh, J.: Experimental and mechanical performance of shotcrete made with nanomaterials and fiber reinforcement. Constr. Build. Mater. **165**, 199–205 (2018)
3. Tao, Y., Vantyghem, G., Lesage, K., Yuan, Y., Corte, W.D., Tittelboom, K.V., Schutter, G.D.: Adhesion properties of printable polymer-modified concrete for rock tunnel linings. ACI Mater. J. (2020)
4. Bissonnette, B., Courard, L., Garbacz, A.: Concrete surface engineering. CRC Press (2015)
5. Zanotti, C., Banthia, N., Plizzari, G.: A study of some factors affecting bond in cementitious fiber reinforced repairs. Cem. Concr. Res. **63**, 117–126 (2014)
6. Kawashima, S., Chaouche, M., Corr, D.J., Shah, S.P.: Influence of purified attapulgite clays on the adhesive properties of cement pastes as measured by the tack test. Cem. Concr. Comps. **48**, 35–41 (2014)
7. Mohamed Abdelhaye, Y.O., Chaouche, M., Chapuis, J., Charlaix, E., Hinch, J., Roux, S., Van Damme, H.: Tackiness and cohesive failure of granular pastes: mechanistic aspects. Eur. Phys. J. E. **35**, 45 (2012)
8. Lakrout, H., Sergot, P., Creton, C.: Direct observation of cavitation and fibrillation in a probe tack experiment on model acrylic pressure-sensitive-adhesives. J. Adhes. **69**, 307–359 (1999)
9. Mohamed Abdelhaye, Y.O., Chaouche, M., Van Damme, H.: The tackiness of smectite muds. 1. The dilute regime. Appl. Clay Sci. **42**, 163–167 (2008)
10. Phan, V.-T.: Relationship between the adhesive properties and the rheological behavior of fresh mortars. École normale supérieure de Cachan-ENS Cachan (2012)
11. Kaci, A., Bouras, R., Phan, V.T., Andréani, P.A., Chaouche, M., Brossas, H.: Adhesive and rheological properties of fresh fibre-reinforced mortars. Cem. Concr. Comps. **33**, 218–224 (2011)
12. Tao, Y., Lesage, K., Tittelboom, K.V., Yuan, Y., Schutter, G.D.: Twin-pipe pumping strategy for stiffening control of 3D printable concrete: from transportation to fabrication. Cem. Concr. Res. (under review)
13. BS 1134.: Assessment of surface texture. Guidance and general information. (2010)

A Study of Early Age Shear Properties of 3D Printable Cementitious Mixes with Fiber Reinforcements

Sriram K. Kompella[1](✉), Andrea Marcucci[1], Francesco Lo Monte[1], Andrea Bassani[2], Stefano Guanziroli[2], Marinella Levi[3], and Liberato Ferrara[1]

[1] Department of Civil and Environmental Engineering, Politecnico di Milano, 20133 Milan, Italy
Kasyapasriram.kompella@polimi.it
[2] IDeCOM Erba 22036, Garbagnate Monastero, Italy
[3] Department of Chemistry, Materials and Chemical Engineering, Politecnico di Milano, 20133 Milan, Italy

Abstract. The use of Fiber Reinforcement in digitally manufactured concrete is not only one of the viable techniques in the otherwise difficult reinforcement strategy but also a segue to the use of highly durable Ultra High Performance Concretes in 3D printing. For this, the characterization of early age mechanical properties such as tensile, shear and compressive strength of printable Fiber Reinforced Concrete is very important in the design process. The printability of fresh concrete which is defined as combination of pumpability, extrudability and buildability can be characterized using the tensile strength of filament, the shear strength and the compressive strength of the freshly printed element. This study aims on defining the shear strength of 3D printable cementitious mortars at early ages focusing on the "printability window" highlighting the phase transitions in printed concrete. Highly cohesive mortar with high Sulfo-Aluminate cement content is cast with Basalt fibers in specially designed molds to understand these said properties at various times from casting namely, 30, 45, 60 and 90 min. Repeatability of the methodology has been established using 3 iterations of each experiment and crack - failure patterns have been studied closely. Further study on developing the Kupfer's failure envelopes is underway by including the compressive strength study obtained through penetration tests and tensile strength study.

Keywords: 3D-printable cementitious composites · Very early age shear behavior · Testing methods

1 Introduction

Additive manufacturing in the concrete construction industry is gaining momentum in both the industry and the research community. It addresses the shortages in skilled manual labor while conserving material. In particular, 3D printed concrete has made feasible the production of topologically optimized form-follows-force structures [1]. This allows for

J. I. Escalante-Garcia et al. (Eds.): RW 2021, RILEM Bookseries 40, pp. 442–450, 2023.
https://doi.org/10.1007/978-3-031-21735-7_49

an efficient use of resources. Quite a few recent constructions using the digital fabrication processes around the world have been reported. The complex reinforcement strategy due to the absence of formwork and questions around the durability and environmental impact of 3D printed concrete are some of the challenges holding back the scaling up of this novel technology. At present, the extrusion technology, of all the 3D printed concrete technologies, is best available in terms of technological readiness and cost effectiveness [2].

The early age strength gain is crucial in extruded concrete. The study of this strength gain is currently limited by the knowledge of printability of mixes which is in turn dependent on nebulously defined properties like buildability, and extrudability. Therefore, it is crucial to redefine the strength parameters using known material properties such as tensile strength, shear strength and compressive strength [3]. There is also a significant need to understand the strength evolution of the printable materials at very early age to understand the highly essential "open time" or "printability window". There is also a knowledge gap in terms of the relationship between the rheological properties and mechanical properties. This can be filled by a comparative study of yield strength of the material and the evolution of shear strength at very early ages. Further, a direct relationship has been established between the normal stress capacity of the cementitious mixes and their extrudability [4]. The study of the normal stresses capacities such as compressive strength and tensile strength is, therefore, very important. The establishment of the Kupfer's failure envelope and understanding the Mohr-Coulomb failure behavior by combining the study of axial and shear strengths would go a long way in establishing an otherwise non-existent design procedure for the printable concretes.

A new trend of using printed concrete formworks as a substitute to conventional steel formworks seems to accommodate the shape flexibility of the fabricated structures. This also reduces the environmental impact of the entire construction industry by reducing both material usage and transportation. These printed concrete formworks need to have a higher material toughness and stiffness. The use of fiber reinforcement in concrete is a robust solution as the structural fibers result in a higher filament strength, stiffness and resistance. This is crucial for converting the bending moment into axial forces which can be resisted by fiber reinforced concrete through tensile membrane action in thin diaphragm elements and shell structures. Fiber reinforced concrete formworks can also have a higher hoop stress resistance when the fiber orientation is aligned with the direction of printing [5, 6]. The higher cement content in 3D printed concretes causes a higher hydration rate inducing higher shrinkage cracking potential. The use of non-structural fibers such as polypropylene or basalt can resist the shrinkage cracking and also provide the necessary cohesion at very early ages.

This study at Politecnico di Milano aims to understand the evolution of shear strength of a printable cementitious mix with basalt fibers at very early and early ages, when the concrete is undergoing the phase transition.

2 Materials and Testing Methodology

2.1 Material

The binder for the cementitious mix is made up mostly of white CEM 1 Portland cement and significant quantity of Sulfo Aluminate cement (CSA) to increase the rate of hydration. White CEM 1 cement has been used simply for aesthetic and architectural purposes which include a majority of the functionality of 3D printing presently. A mix of silica sand, ground recycled glass, and quartz is used as aggregate. All the aggregates have a size less than 1 mm. A combination of bentonite and cellulose ether is used as a viscosity modifier. A polycarboxylate based water reducing admixture is also used. As mentioned earlier, basalt microfibers are used in the mix to ensure sufficient cohesion in early ages. The pre-mixed aggregates and cement are dry mixed in a pan mixer for one minute to break up the lumps. Tap water mixed with the chemical admixtures is then added to begin the hydration process and mixed for 5 min. Basalt microfibers are then added and the mixing is continued for another minute before the mix is transferred into the testing molds. Once the molds are filled, the top surface is smoothened to observe the failure patterns clearly.

2.2 Testing Setup and Methodology

A custom experimental setup has been fabricated at Politecnico di Milano [7, 8] for the purpose of testing the direct shear strength evolution of cementitious mortar mixes. Each setup has a plexiglass mold with one fixed half and one movable half. The transparency of the molds helps ensure that no voids are left behind when the cementitious mortar is poured in them.

The mold for the direct shear test is made of two rectangular elements with an opening in the center to localize the failure plane in the direction of motion. Slippage between layers is not anticipated except in the failure plane.

The movable halves of the molds are fixed to an electromechanical actuator equipped with a 600 N load cell. This ensures a displacement control with a rate of 0.1mm/s. Additionally the displacement is also recorded using LVDTs at both ends of the movable half of the molds. The load values obtained from the load cell are corrected for a friction factor and then divided by the cross-sectional area of the failure region to obtain the stress values. The frictional forces are calculated separately for each mold by placing disconnected weights equal to the weight of concrete in each half of the mold and pulling them apart at the rate of the experiment Fig. 1.

These molds are light and can easily be assembled onsite.

3 Results and Observations

The following results have been obtained from the direct shear tests done at the early ages. Figure 2 shows a plot of Shear stress (kPa) versus Displacement (mm) at 30 min from the start of hydration (contact between water and binder). The mix is clearly very fluid at 30 min and therefore no clear peak in shear stress value can be observed. There

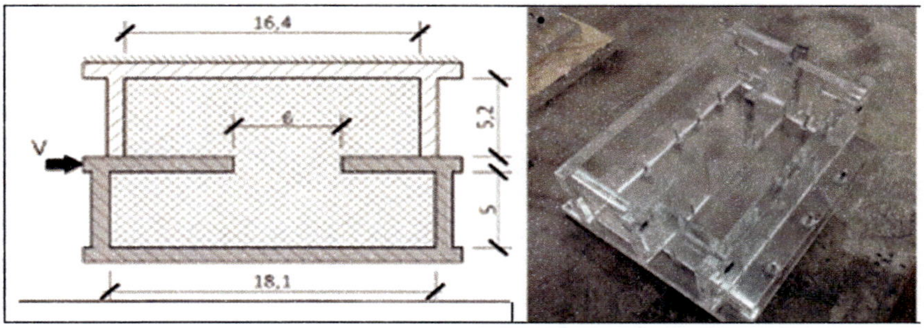

Fig. 1. Shear mold: schematic (dimension in cm) and actual image

is a repeatability in the results even at this fluid state. No specific failure point could be identified in the tests at 30 min consistent with fluid behavior.

Figure 3 shows the Shear stress (kPa) vs Displacement (mm) at 60 min from start of hydration. A change in the phase can be observed from the smooth peak and gradual increase in the stress with displacement. This is in sharp contrast to the behavior at 30 min. At 60 min, the shear stress increases continuously until a peak value and drops at failure. There is a ductility in the mix indicating that it is printable at 60 min. This difference between 30 and 60 min can also be observed in the consistency of the mix as can be seen form the of the failure pattern in Fig. 5. It is anyway worth remarking, that at 60 min, the top surface deformation is not consistent with the deformation of the rest of the specimen, initially there is a buckling like out of the plane displacement of the top surface before the ligament is sheared off completely. This could be because of higher water evaporation from the exposed surface. There is also the effect of cohesion of the mix which causes some slip along the failure plane before the formation of a shear strut.

Figure 4 shows the Shear stress (kPa) vs Displacement (mm) at 90 min from start of hydration. An even more pronounced peak observed at 90 min. The phase transition seems to be complete as can be observed from the failure pattern seen in Fig. 5. The mix at 90 min appears to be very brittle as can be seen from the failure pattern at 90 min, which clearly indicate the formation of a shear strut. The cracks began appearing at the corners of the failure plane which later propagated into the center.

From the tests at 30, 60 and 90 min, it is apparent that phase transition starts post 30 min and completes somewhere before 90 min. This is consistent with what could be categorized as the open time of the mix during which it is printable. Further investigation on the phase transition by comparing the shear stresses at smaller time intervals during this phase is underway. While at the 90 min mark, there is a consistent solid behavior, the variance in the peak value is high.

An investigation of the shear behavior at 45 min lead to some inconclusive yet interesting results (Fig. 6). Some of the specimen exhibit a solid like behavior whereas some remain fluid. This inconsistence is also observed in another study where some rheological tests are done using an ICAR rheometer. This could be because of various reasons. Around the 45 min mark, the hydration rate could be higher than at 30 or 60 min. This implies that even slight variations in the test samples could cause very large scatter.

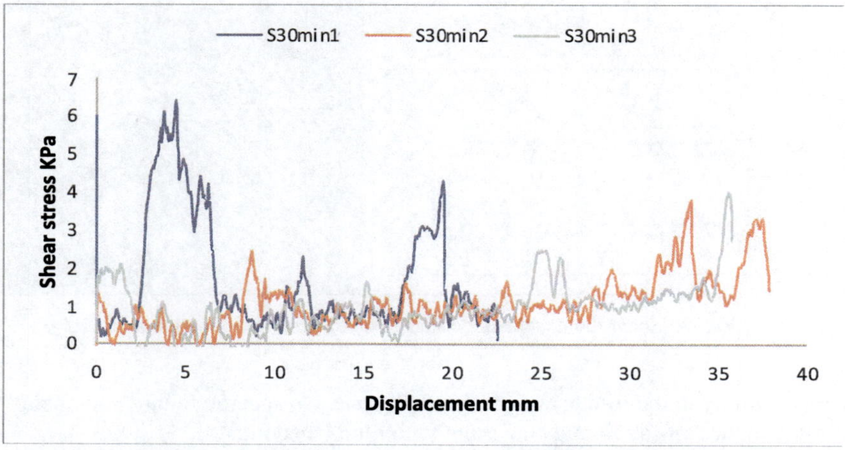

Fig. 2. Shear stress (kPa) versus Displacement (mm) at 30 min

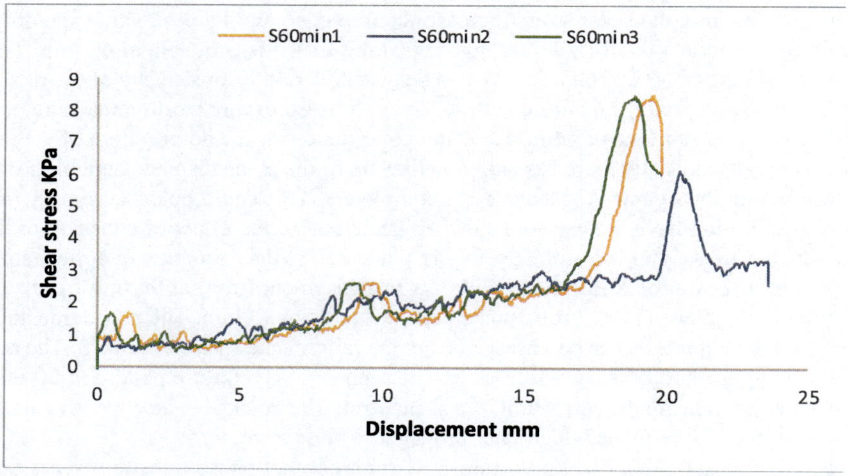

Fig. 3. Shear stress (kPa) versus Displacement (mm) at 60 min

The test seems infeasible at 120 min as the material has reached solid state. The material is seen to be very brittle at 120 min. The mix is clearly past its printability stage which is also reflected by the very high shear strength values.

The strength evolution of the mix can clearly be seen from a comparison of the shear strengths of the mix at 30, 60 and 90 min (Fig. 7). The mean shear strength at 60 min is around 7.5kPa and the same at 90 min is around 56.6kPa, an 8-fold increase in strength in 30 min indicating clear phase change.

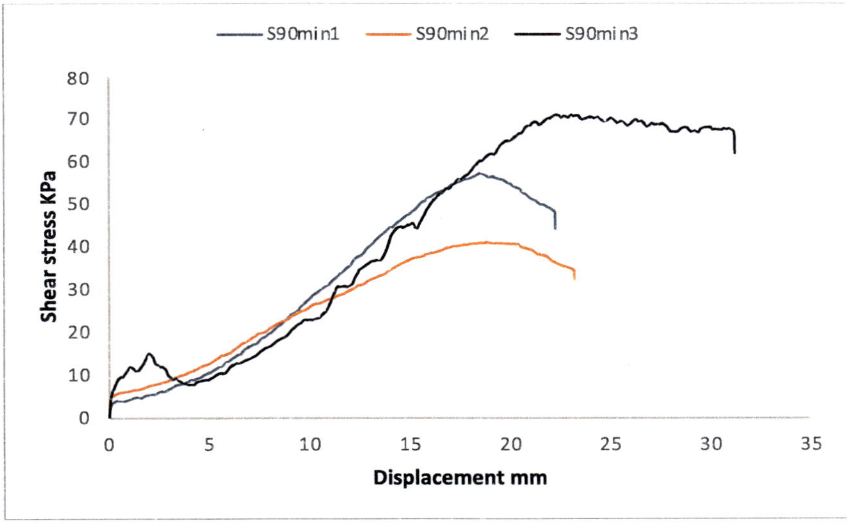

Fig. 4. Shear stress (kPa) versus Displacement (mm) at 90 min

4 Discussion and Conclusions

The proposed methodology to evaluate the direct shear stress in cementitious mortar mixes can be adopted on site to establish quantitatively the printability window and determine the phase of the mix. The basalt microfibers have no effect on the structural performance of the mix as no post-peak toughness can be observed at any time from the start of hydration. However, they add significantly to the cohesive nature of the mix. The friction between the molds seems considerable, especially in the very early ages, around 30 and 45 min from the start of hydration. This can be avoided by introducing ball-bearings or wheels on which the molds can slide on. When testing very early ages, it can be seen that due to the fluid behavior of the mix, there is a tendency to expand. This will cause an inversion in the friction direction during the progression of the test. Therefore, it is advised to keep the molds clamped when pouring the mix and to eliminate friction as much as possible.

The inhomogeneity of the ingredients in the premix could affect the strength gain vastly. The Sulfo Aluminate cement has a significantly higher rate of hydration compared to the Portland cement. It is therefore necessary to store the ingredients separately in air-tight containers for a prolonged shelf-life and consistent results.

Further study at intermediate times like at 45 min and 75 min is necessary to understand the complete behavior of the mix. This study combined with the study of uniaxial tensile strength at early ages and the study of compressive strength through penetration tests will give us a comprehensive understanding of the mechanical performance of the mix by establishing the Kupfer's failure envelope.

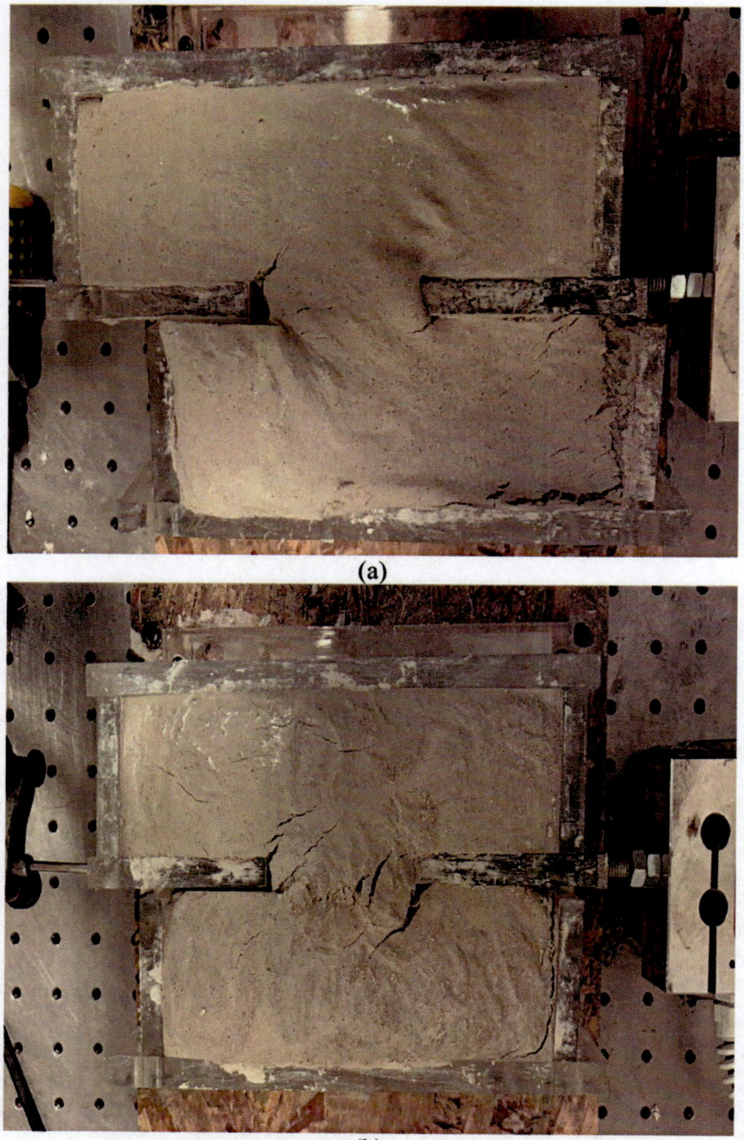

(a)

(b)

Fig. 5. Failure patterns in shear at 60 min (**a**) and 90min (**b**)

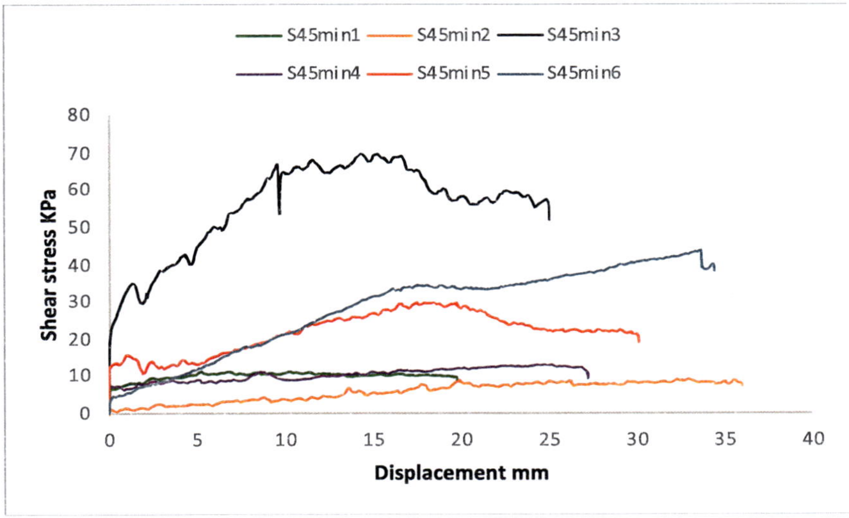

Fig. 6. Shear stress (kPa) versus Displacement (mm) at 45 min

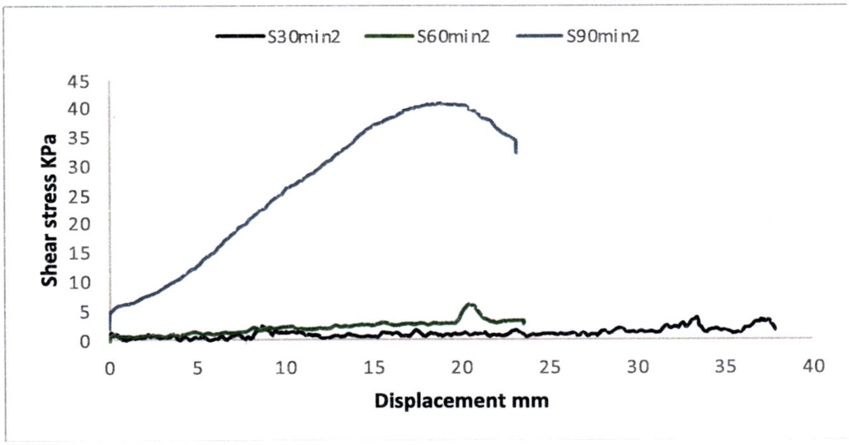

Fig. 7. Comparison of shear stress at different times

References

1. Menna, C., et al.: Opportunities and challenges for structural engineering of digitally fabricated concrete. Cem. Concr. Res. **133**, 106079 (2020)
2. Mechtcherine, V., et al.: Extrusion-based additive manufacturing with cement-based materials–production steps, processes, and their underlying physics: a review. Cem. Concr. Res. **132**, 106037 (2020)
3. Mettler, L.K., Wittel, F.K., Flatt, R.J., Herrmann, H.J.: Evolution of strength and failure of SCC during early hydration. Cem. Concr. Res. **89**, 288–296 (2016)

4. Perrot, A., Lanos, C., Estellé, P., et al.: Ram extrusion force for a frictional plastic material: model prediction and application to cement paste. Rheol Acta **45**, 457–467 (2006)
5. Arunothayan, A.R., Nematollahi, B., Ranade, R., Bong, S.H., Sanjayan, J.G., Khayat, K.H.: Fiber orientation effects on ultra-high performance concrete formed by 3D printing. Cem. Concr. Res. **143**, 106384 (2021)
6. Hambach, M., Rutzen, M., Volkmer, D.: Properties of 3D-printed fiber-reinforced Portland cement paste. In: Sanjayan, J., Nazari, A., Nematollahi, B. (eds.) 3D Concrete Printing Technology, pp. 73–113. Butterworth-Heinemann, Elsevier (2019)
7. Monte, F.L., Zago, G., Cucchi, M., Ferrara, L.: Correlation between very early age fracture performance and evolution of rheological properties of high performance fiber reinforced cementitious composites with adapted rheology. In Rheology and Processing of Construction Materials, pp. 237–245. Springer, Cham (2019)
8. Fioretti, M., et al.: Experimental investigation on the early age tensile strength of fiber reinforced mortar used in 3D concrete printing. In: Bos, F.P., Lucas, S.S., Wolfs, R.J.M., Salet, T.A.M. (eds.) DC 2020. RB, vol. 28, pp. 255–261. Springer, Cham (2020). https://doi.org/10.1007/978-3-030-49916-7_26

Influence of Crystalline Admixtures and Bacteria on the Fresh Properties of Self-healing Concrete

Harry Hermawan[1], Peter Minne[1], Enricomaria Gastaldo Brac[2], Virginie Wiktor[3], Pedro Serna Ros[4], and Elke Gruyaert[1](✉)

[1] KU Leuven, Department of Civil Engineering, Materials and Constructions, Ghent Campus, Gebroeders De Smetstraat 1, 9000 Ghent, Belgium
elke.gruyaert@kuleuven.be
[2] Penetron Italia, 10093 Collegno, Italy
[3] Cugla B.V., R&D Center, Rudonk 6b, 4824 AJ Breda, Netherlands
[4] Instituto de Ciencia y Tecnología Del Hormigón (ICITECH), Universitat Politècnica de València, Camino de Vera S/n, 46022 Valencia, Spain

Abstract. Concrete is undeniably susceptible to cracking due to its low tensile strength in spite of its ability to withstand high compressive stresses. Cracks may jeopardize the concrete core and reinforcements due to the ingress of water and aggressive substances. A promising technique to effectively seal cracks in hardened concrete is using a self-healing technology by introducing healing agents to the fresh concrete mixes. Crystalline admixtures and bacteria emerge as promising self-healing agents to promote stimulated autogenous healing and autonomous healing mechanisms, respectively, improving simultaneously the durability of the concrete. Healing agents are mostly added on top of the normal mix without changing the mix design of the concrete, while in fact, it may induce considerable effects on the fresh and hardened properties of concrete. In this study, two types of cement (CEM I 52.5N and CEM III/A 42.5N) as well as crystalline admixtures and bacterial spores, both in powder form, and PCE-based superplasticizer are utilized to evaluate the changes in consistency and setting time. The dosage of healing agents is taken between $0 - 5\%$ by weight of powders. To start, tests on paste level were executed. The results show that the addition of healing agents affects the water demand, the consistency and setting time of the pastes. The inclusion of crystalline admixtures increases the water demand, while a declined trend of setting time is observed. The gradual addition of bacteria up to 5% slightly changes the water demand, but a significant delay on the initial and final setting times was detected.

Keywords: Self-healing concrete · Crystalline admixture · Bacteria · Fresh properties · Workability

1 Introduction

During the past decades, self-healing concrete has been extensively experimented and proposed to upgrade the performance of conventional concrete with an ability to seal the

J. I. Escalante-Garcia et al. (Eds.): RW 2021, RILEM Bookseries 40, pp. 451–460, 2023.
https://doi.org/10.1007/978-3-031-21735-7_50

undesirable cracks independently. Crack occurrence in concrete is apparently inevitable and this leads to considerable maintenance cost and labor works if continuous repairs are needed. Furthermore, there is no guarantee that the repair will be efficient and will not induce more defects at other locations. Thus, a stimulated autogenous or autonomous healing of the concrete is proposed by introducing healing agents such as crystalline admixture and bacteria. The use of crystalline admixture is broadened from a porosity reducer to a promotor of self-healing effect [1]. Active chemicals present in this admixture react in the presence of water, forming precipitated crystals that increase the density of calcium-silicate-hydrate gel (CSH), and working in that way as a pore/crack blocker [2]. The implementation of bacteria as a self-healing agent in the concrete matrix emerges as one of the most promising approaches in order to extend the service life of concrete structures by sealing the concrete cracks on a biotechnology basis. Microbial precipitation of calcium carbonate inside the crack occurs due to the metabolic activation of bacteria with suitable nutrients in contact with moisture [3]. The phenomenon of intrinsic crack closure was noticeable on cementitious composite containing healing agents such as bacteria [4] and crystalline admixture [1, 5]. Nowadays, the two main goals of adding healing agent into the concrete are to activate the healing mechanism and to improve the healing efficiency, however, the evaluation of fresh properties of concrete mixes with the inclusion of healing agents is often neglected [6]. Workability is an important aspect for fresh concrete to ensure that it can be easily placed and handled in the field. Therefore, understanding the influence of adding healing agent in relation to the fresh concrete properties is of great importance. In this study, two commercial healing agents (i.e. Basilisk bacteria healing agent and Penetron Admix crystalline admixture) are introduced into the fresh paste to investigate the effects on the consistency and setting time of the paste.

2 Materials and Methods

2.1 Materials

Two types of cements (CEM I 52.5N and CEM III/A 42.5N, abbreviated as CEM I and CEM III) and demineralized water were used to make the fresh pastes. The chemical compositions of these cements are presented in Table 1. Penetron Admix crystalline admixture (CA) and Basilisk bacteria spores (BAC), both in powder form, were utilized as healing agents. The particle size distributions (PSDs) of cements, bacteria and crystalline admixtures are presented in Fig. 1. The PSDs were obtained by laser diffraction analysis. Specifically for the bacteria healing agent, laser diffraction analysis was complemented with sieve analysis because particles larger than 300 μm were present, exceeding the upper size limit of the laser diffraction apparatus. The PSD of CA is in the range of the cements, but a bit broader. The median sizes (D_{50}) of CEM I, CEM III, CA and BAC are 17.16, 11.73, 15.20 and 490.24 μm.

According to the manufacturer, the recommended dosage of healing agent is in the range of 0.8–1.0% by weight of cement for Penetron crystalline admixture and 1.0–4.0% by weight of cement for Basilisk bacteria healing agent. However, an attempt was made to use a higher dosage of healing agent to possibly increase the self-healing efficiency without lowering other properties such as workability. Healing agents, both CA and BAC, were finally added in the range of 0.8–5.0% by weight of powder. The incorporation of healing agents is mostly carried out on top of the normal mix, without any adaptation in the mix design. In this study, the healing agents are added into the cement paste by changing the powder proportion of the mixture. For instance, the powder composition of the paste CEM I + 3% BAC is composed of 97% CEM I and 3% Basilisk bacteria. Additionally, the polycarboxylate-ether (PCE) based superplasticizer (SP) of Demula Technofluid P175 supplied by Cugla B.V. was also used as a high-range water reducer (HRWR). The properties of this superplasticizer are given in Table 2.

Table 1. Chemical composition of the cements

Chemical composition (%)	CEM I 52.5N	CEM III/A 42.5N
CaO	64.35	53.58
SiO_2	21.00	27.39
Al_2O_3	4.22	6.83
Fe_2O_3	2.72	1.82
SO_3	3.29	3.02
CO_2	1.00	1.04
Cl	0.07	0.05
MgO	2.04	4.66
K_2O	0.80	0.59
Na_2O	0.24	0.27
P_2O_5	0.49	0.22

Table 2. Properties of superplasticizer (SP)

SP type	Chemical type	Density (kg/dm^3)	Solid content (%)	pH
Demula Technofluid P175	Polycarboxylate ether	1.07	30	2.0–8.0

2.2 Testing Methods

In this study, initial tests on the fresh cement paste in combination with healing agents were executed. To start, the standard consistence test and setting time test in accordance with EN 196–3:2016 were carried out. The standard consistency test is performed by use of the manual Vicat apparatus (with the use of a plunger) to determine the water demand of each modified powder mixture in comparison to pure cement without a healing agent. On the other hand, the automatic Vicat apparatus is also used to measure the initial and final setting time of the pastes. In practice, superplasticizers are widely used in

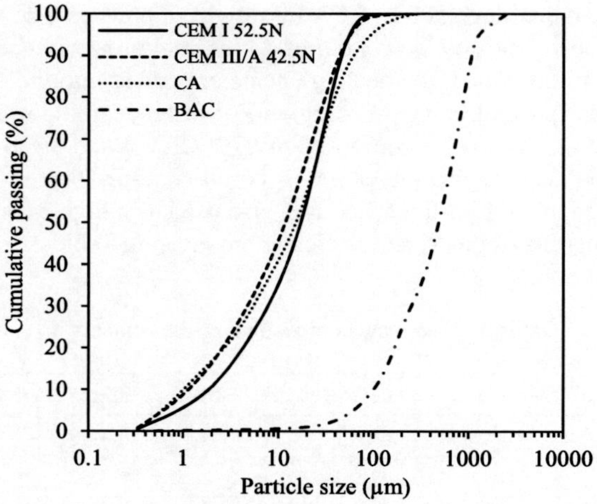

Fig. 1. Particle size distributions of cements and healing materials.

concrete mixtures to improve the workability. Consequently, an additional experiment was executed to investigate the effect of superplasticizer in relation to the reduction of water demand. The superplasticizer is added in the dosage of 0.1–2.0% by weight of powders (cement + healing agent).

3 Results and Discussions

The results on water demands of cement in combination with BAC and CA at standard consistency are presented in Fig. 2. It shows that the gradual addition of BAC slightly reduced the water demand. The combination of CEM III with 5% BAC reduces the water demand from 29.5 (pure CEM III) to 27.7% (CEM III + 5% BAC). In combination with CEM I, the reduction induced by the presence of BAC is apparently minor as the water demand reduces only from 28 to 27%. As shown in Fig. 2b, the behavior of CA is completely different on the use of different cement types which is also in contrast with the result of using BAC. The result shows that the water demand is higher in presence of CA (for all dosages from 0.8 to 5% by the weight of powder). On the use of CEM III, the amount of water needed to achieve a standard consistency raises from 29.5 to 34.5%. Although it is clear that, in combination with this type of cement, the water demand in case of 4% CA is higher than in case of 5% CA, a linear regression line was plotted (Fig. 2b). Contrarily, the result of combining CA and CEM I shows a different behavior. The tendency seems to be a parabolic regression where the gradual addition of CA up to 3% increases the water demand by 3%, while the higher dosage of CA shows a declined trend as the water demand reduces from 31 to 29.5% when the CA dosage is increased from 3 to 5%. By comparing these two results on water demand, in fact, the addition of CA showed a similar behavior on both types of cements when the dosage was limited to 3%.

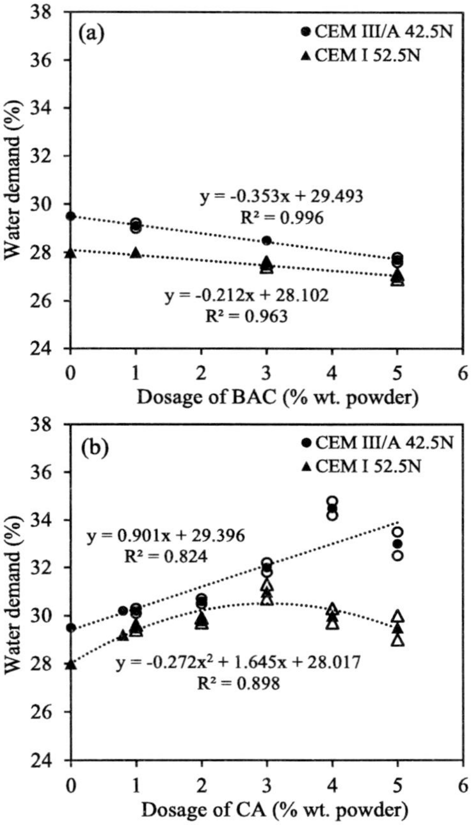

Fig. 2. Water demands of modified cement incorporated with (a) Basilisk bacteria and (b) Penetron crystalline admixture at different dosages (*note* the empty markers represent the water demand values, while the filled markers represent the mean values)

As a matter of fact, as reported by [7], in case CA is used, it is important to make a judicious selection regarding the type of cement. Cement with a high content of clinker, such as CEM I, is preferred because CA needs an appreciable amount of calcium hydroxide, which is a typical hydration product of Portland cement, to react. In the case of blended cement like CEM III, less calcium hydroxide is produced and can be even consumed.

Moreover, the setting time test was evaluated and results are shown in Fig. 3. In spite of the rather small effect by BAC on water demand, the setting time of these pastes considerably increases as a consequence of incorporating BAC into the fresh mix. Based on Fig. 3a, both initial and final setting times are delayed regardless the cement type. However, a considerable effect is identified on the mixture containing 5% BAC and cement type I. The final setting times of cement pastes (with CEM I) are 175, 221, 311 and 660 min with the addition of 0, 1, 3 and 5% BAC, respectively. It is clear that the setting time of cement paste with 5% BAC is roughly doubled in comparison with cement paste with 3% BAC. Moreover, this exceptional result is not observed in the paste with CEM III. On the use of CEM III, the addition of BAC from 1 to 5% gradually increases the setting time of the paste. The final setting times of cement pastes (with CEM III) amount to 227, 256, 333 and 430 min with the addition of 0, 1, 3 and 5% BAC, respectively. In general, the mean gap time between initial and final setting time of paste containing BAC is approximately 63 ± 9 min. Rossi et al. [8] investigated the cement hydration with the addition of Basilisk bacteria healing agent by means of an isothermal calorimetry test and they found that the addition of BAC caused a decrease of heat developed by the hydration of cement, resulting in a retardation effect. Due to the reason that BAC is composed of organic compounds, the alkaline degradation of these compounds into molecules containing the α-hydroxy carboxyl group may be responsible for the delay in the setting time of the cement pastes [8, 9].

The setting time of paste containing different CA dosage (0.8–5% by weight of powder) is presented in Fig. 3b. Similar with the consistency test results, the inclusion of CA in combination with CEM I and III shows different behavior. On the use of CEM I, the 1% CA addition increases the initial and final setting time of the paste by approximately 63 and 77 min, respectively. The higher the addition of CA the faster the setting time of the paste. Specifically the gradual addition of CA from 1 to 5% shortens the final setting time from 252 to 183 min. A significant effect in using high dosage of CA was the fast setting of the paste. By using the high CA dosage at 4 and 5% by weight of powder, the initial setting times were found to be at 68 and 22 min. From the visual observation by authors, the cement paste incorporated with a high dosage of CA was quickly stiffened after the mixing process. This may be attributed to a chemical reaction occurring when the CA came into contact with water and cement, resulting into a fast setting time. This was supported by the fact that the white powder present in CA particles which presumed to be active chemicals immediately dissolved in water. From the experiment, it was observed that when the CA came into contact with water, the mixing water turned into greenish color, indicating an immediate reaction and a hydrophilic character of CA which can react easily with water [10]. However, further investigations should be conducted to analyze these results in detail. Furthermore, it is clear that the addition of CA in combination with CEM III shows similar setting times regardless the dosage. As a matter of fact, the use of high CA dosage at 4–5% have a comparable result with the low dosage at 1% in the terms of setting time.

Additionally, the superplasticizer (SP) was added into the modified cement paste with healing agent, either BAC or CA, at the dosage between 0.1 and 2.0% by weight of powders (cement + healing agent) to measure the water reduction of the paste at normal consistency and to observe the compatibility aspect between healing agent and superplasticizer. Figure 4 shows the relationship between SP dosage and total water demand in combination with different dosage of healing agent. The BAC was fixed at the dosage up to 5%, while the dosage of CA was limited up to 3% to eliminate the unfavorable effect on the use of a higher dosage as explained in the previous results. As a note, the water demand was corrected with the water contained in the SP solution, and is further defined as total water demand.

According to Fig. 4a and b, the reduction rate of water demand is relatively similar for control paste and paste containing BAC. As shown in Fig. 4c, there is an immediate

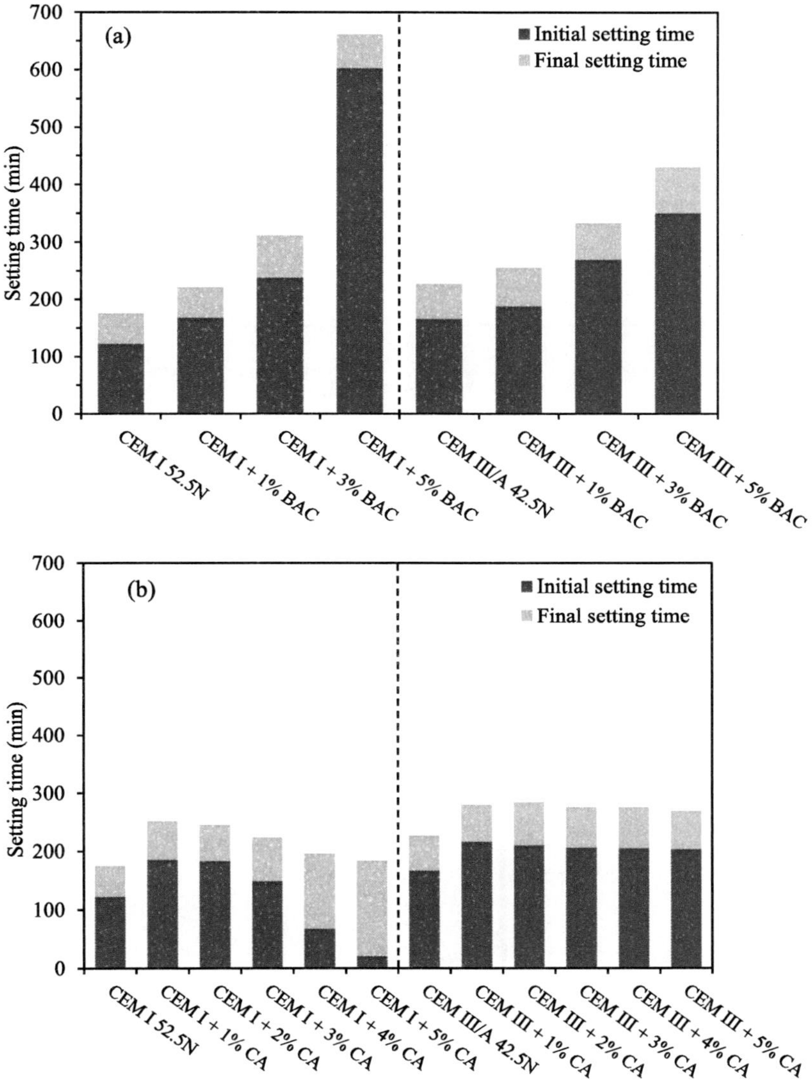

Fig. 3. Setting times of modified cement pastes incorporated with (a) Basilisk bacteria and (b) Penetron crystalline admixture at different dosages

reduction of water demand on cement pastes with CEM I containing CA at a low SP dosage. However, in general, the reduction rate of water demand with varying SP dosages of all modified pastes was quite similar with the control paste. Contrarily, this was not the case for the modified paste with CEM III and CA. As shown in Fig. 4d, the water demands of CA and CEM III are always higher than the pure cement and in addition,

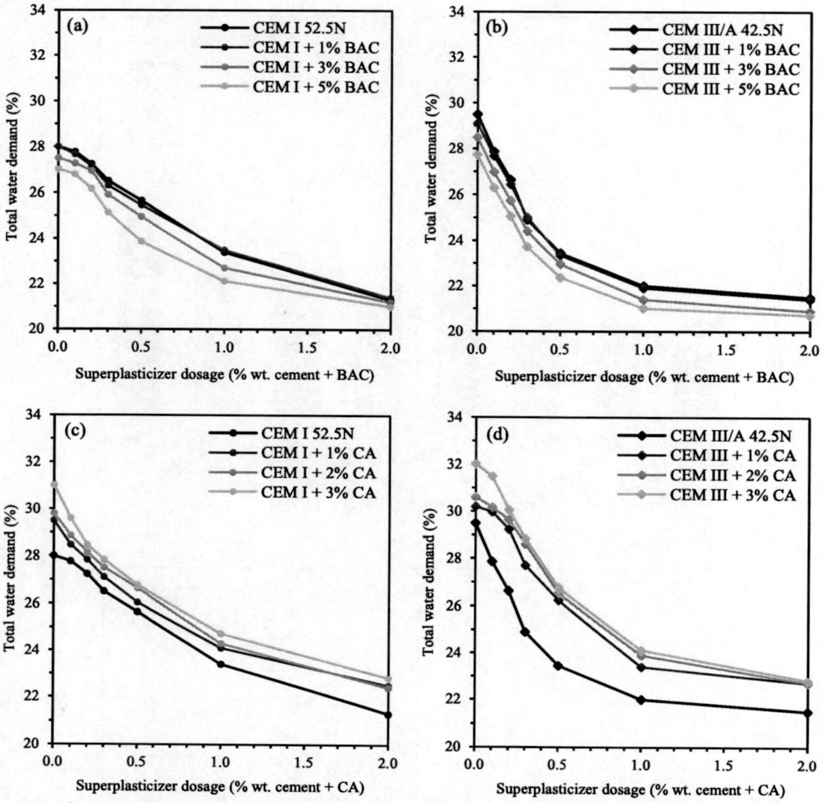

Fig. 4. Changes on total water demand of modified powder mixture due to the incorporation of different healing agents and superplasticizer in combination between (a) BAC + CEM I, (b) BAC + CEM III, (c) CA + CEM I and (d) CA + CEM III

the changes in water demand of CA and CEM III upon increasing SP dosage do not follow the behavior of water demand reduction of the pure CEM III. This behavior can be explained by the fact that the superplasticizer is adsorbed by the CA particles and the incorporation of CA considerably influences the rheological properties of the superplasticized CEM III paste by reducing the workability and increasing the viscosity of the paste [11]. Nevertheless, the incorporation of the superplasticizer decreases the water demand at all dosages of healing agents (CA or BAC), while the variations in the water demand are generally lower at a higher dosage of SP.

4 Conclusions

The effects of incorporating bacteria and crystalline admixtures in relation to the fresh properties of cement pastes were evaluated based on consistency tests and setting time tests. In addition, the reduction of water demand was investigated with the addition of a PCE-based superplasticizer. The major conclusions are presented as follows:

1. The water demand of cement with Basilisk bacteria slightly reduces as the bacteria dosage increases up to 5%, while the water demand of cement with Penetron crystalline admixture increases as the CA dosage increases up to 3%.
2. The addition of Basilisk bacteria prolongs the setting time and the higher the bacteria dosage, the longer the setting time. In contrast, the gradual addition of CA from 1 to 5% by weight of powder into the cement paste with CEM I tends to shorten the setting time. The impact of using a high CA dosage on fresh CEM I paste is rapid initial setting and loss of plasticity. However, the initial and final setting times of paste containing CA and CEM III/A are relatively similar at any CA dosage.
3. As a consequence of incorporating a PCE-based superplasticizer into fresh pastes, the water demand of cement and healing agent reduces with increasing dosage of superplasticizer. In addition, there are no incompatibility issues found between cement, crystalline admixture, bacteria and superplasticizer.

Furthermore, future works will focus on the introduction of bacteria or crystalline admixture into the fresh concrete mixture to identify the influences of healing agent on the workability aspect at concrete level.

Acknowledgement. This project has received funding from the European Union's Horizon 2020 research and innovation programme under the Marie Skłodowska-Curie grant agreement No 860006.

The authors would like to acknowledge Dr. Henk M. Jonkers for sharing Basilisk bacteria healing agent and Penetron Italia for providing Penetron crystalline admixtures.

References

1. Ferrara, L., Krelani, V., Moretti, F.: On the use of crystalline admixtures in cement based construction materials: from porosity reducers to promoters of self healing. Smart Mater Struct **25**(8), 1–17 (2016)
2. De Belie, N., Gruyaert, E., Al-Tabbaa, A., Antonaci, P., Baera, C., Bajare, D., et al.: A review of self-healing concrete for damage management of structures. Adv Mater Interfaces **5**(17), 1–28 (2018)
3. Nguyen, T.H., Ghorbel, E., Fares, H., Cousture, A.: Bacterial self-healing of concrete and durability assessment. Cem. Concr. Compos. **104** (2019)
4. Jafarnia, M.S., Khodadad, S.M., Moshtaghioun, S.M.: Use of bacteria for repairing cracks and improving properties of concrete containing limestone powder and natural zeolite. Constr. Build Mater. **242** (2020)
5. Xue, C., Li, W., Qu, F., Sun, Z., Shah, S.P.: Self-healing efficiency and crack closure of smart cementitious composite with crystalline admixture and structural polyurethane. Constr. Build Mater. **260** (2020)
6. Hermawan, H., Minne, P., Serna, P., Gruyaert, E.: Understanding the impacts of healing agents on the properties of fresh and hardened self-healing concrete: a review. Processes **9**(12) (2021)

7. RESHEALIENCE homepage, https://ec.europa.eu/research/participants/documents/downloadPublic?documentIds=080166e5be03eb21&appId=PPGMS, last Accessed 03 Sep 2022
8. Rossi, E., Vermeer, C.M., Mors, R., Kleerebezem, R., Copuroglu, O., Jonkers, H.M.: On the applicability of a precursor derived from organic waste streams for bacteria-based self-healing concrete. Front Built Environ **7**, 1–14 (2021)
9. Rodríguez, C.R., de Mendonça Filho, F.F., Mercuri, L., Gan, Y., Rossi, E., Anglani, G., et al.: Chemo-physico-mechanical properties of the interface zone between bacterial PLA self-healing capsules and cement paste. Cem. Concr. Res. **138** (2020)
10. Roig-Flores, M., Pirritano, F., Serna, P., Ferrara, L.: Effect of crystalline admixtures on the self-healing capability of early-age concrete studied by means of permeability and crack closing tests. Constr Build Mater **114**, 447–457 (2016)
11. Hermawan, H., Beltran, G.S., Wiktor, V., Serna, P., Gruyaert, E.: Effect of healing agents on the rheological properties of cement paste and compatibility with superplasticizer. MATEC Web Conf **5008**, 361 (2022)

Novel Compressive Constitutive Model for 3D Printed Concrete

Eduardo Galeote(✉) and Albert de la Fuente

Department of Civil and Environmental Engineering, Universitat Politècnica de Catalunya (UPC), Barcelona, Spain
eduardo.galeote@upc.edu

Abstract. 3D concrete printing (3DCP) is emerging as a promising technology in the construction industry. From the engineering and architectural point of view, the use of 3DCP offers a great potential in terms of enhancing construction automation, more freedom of shapes and a substantial reduction of manufacturing costs associated with production and materials. Unlike traditional poured concrete, 3DCP is a layer-to-layer based manufacturing method. Such productive method makes concrete to present a layer-based structure than can lead to an anisotropic element because of the varying properties on the different spatial directions of the material. Consequently, and given the specific manufacturing method and the internal structure of 3DCP, a new approach in terms of constitutive models for design purposes is required. The objective of this is study is to develop a novel constitutive model for compression considering the influence of the anisotropy of the material. For this, an experimental program involving the characterization of 3D printed concrete specimens was conducted. The compressive strength and the modulus of elasticity was determined to obtain the stress-strain diagram on different loading directions, taking into account the influence of the layer-based structure of 3DPC. The results revealed no significant differences between spatial directions given the great homogeneity of the 3D printed concrete in all directions. However, such approach becomes necessary so engineers and designers can fully take advantage of the material and remain on the safe side of the structural design.

Keywords: 3D printing · Compressive strength · Constitutive model · Anisotropy

1 Introduction

Additive manufacturing techniques such as 3D concrete printing (3DCP) are attracting attention in the construction industry given its potential to decreasing costs, automatize constructive procedures and widen the range of architectural and engineering designs [1]. However, 3DCP faces several challenges in terms of materials, technological developments and specifications for structural design [2]. Consequently, 3DCP technology still requires further research in order to optimize material consumption, strength performance and development of design tools.

J. I. Escalante-Garcia et al. (Eds.): RW 2021, RILEM Bookseries 40, pp. 461–468, 2023.
https://doi.org/10.1007/978-3-031-21735-7_51

3DCP hardened elements, as for conventional concrete, are required to achieve a minimum strength for its use with structural purposes. In this line, the layer-to-layer manufacturing procedure may lead 3DCP elements to present an anisotropic behaviour on the different directions of the printed element [3], which can have a significant impact on the final strength [4, 5] and affect the behaviour of a potential structure. Accordingly, this study aims on analyzing the performance of a 3DCP material as well as determine a constitutive equation for compression able to capture the anisotropic singularities.

For this, an experimental program involving 3DCP specimens manufactured through a powder-bed concrete printing technique [6] was conducted. The specimens were tested to determine the compressive strength, the modulus of elasticity and the concrete strain at failure under compression. Given the layer-based structure of the samples, the tests were performed in different directions to analyze the anisotropy of the hardened concrete. The results were used to develop a constitutive model for compression, this gathering the behaviour differences of the anisotropic structure of the specimens.

The results revealed that the anisotropic behaviour of 3DCP elements can be greatly reduced, this leading to similar strength performances in the different testing directions. For this, it is necessary to establish a thorough manufacturing procedure controlling the different parameters involved such as printing speed, water content and layer thickness. The compressive constitutive model was also developed considering the influence of the anisotropy produced by the layer-based structure. Such an approach should be taken into account to develop design equations able to reliably reproduce the characteristics and performance of the material to design safe and durable structures.

2 Experimental Program

2.1 Production and Preparation of Specimens

The material for concrete production consisted of a dry mix of prepared mortar oriented for its use in a powder-bed 3D-printer. A total number of 21 cubic specimens were manufactured, which were subsequently cored to obtain cylindrical samples (Fig. 1) with a height-diameter (h/ϕ) ratio of approximately $h/\phi = 2.0$. Due to the dimensions of the drilling core, cylindrical specimens presented a diameter of 83.5 mm and a height ranging between 144 and 187 mm given the variability of the dimensions of the cubic 3D-printed samples. The average h/ϕ ratio of the specimens is 1.97, presenting a variability of 6.9%.

2.2 Characterization Tests

As a result of the influence of the printing layers as weaknesses for potential fracture surfaces formation, the cylindrical specimens were tested applying the load both in the perpendicular and the parallel direction to the layers (Table 1). For this, three groups of data have been considered to analyze the results of compressive strength and modulus of elasticity based on the testing direction described in Table 2. The first group encompasses the results of the 21 specimens regardless of the testing direction. The second group corresponds to the specimens tested through a perpendicular direction to the layers,

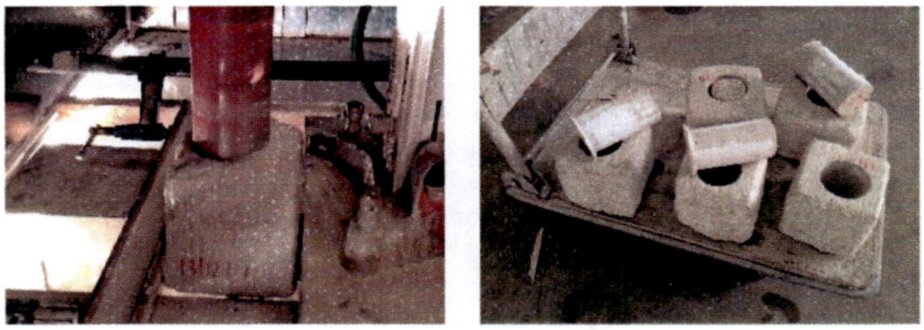

Fig. 1. Preparation and coring of specimens.

while the third group corresponds to the specimens tested through a parallel direction to the layers. A summary of the three groups of data with the corresponding number of specimens and the tests conducted is shown in Table 2.

Table 1. Direction of compressive tests.

Direction of test	Specimen	Loading representation
Perpendicular direction		
Parallel direction	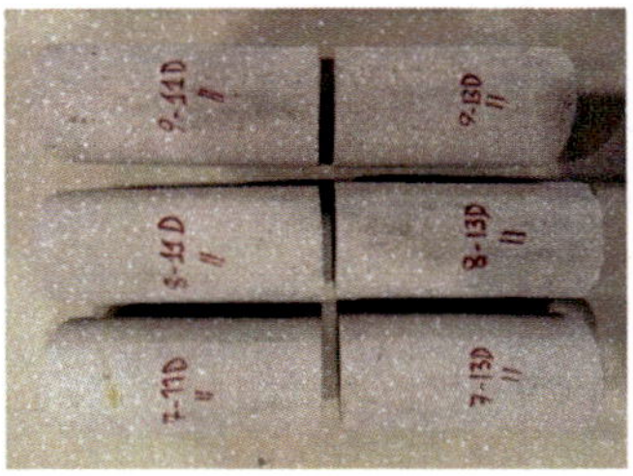	

To conduct the compressive and modulus of strength tests, all specimens were polished to obtain smooth faces and ensure full contact with the loading plates. Out of these 21 specimens, 2 were tested under simple compression to determine the load level required to conduct the test of the modulus of elasticity.

The results of the two compressive tests were used to determine the required load to conduct the test for the modulus of elasticity. The modulus of elasticity was determined using the steel frame for specimens of $\phi 100 \times 200$ mm, where three LVDTs are placed

Table 2. Description of testing direction groups.

Group of data	Description	Number of specimens	
		Compression	Elastic modulus
Group I	Perpendicular and parallel direction	21	19
Group II	Perpendicular direction	12	10
Group III	Parallel direction	9	9

at a distance of 75 mm to measure the vertical deformation of the samples. The test was conducted in three loading stages. The two first loading cycles were conducted up to a load of 65.7 kN (12 MPa) through displacement control at a constant rate of 0.5 mm/min. The third cycle was initially conducted up to a load of 150 kN at the same displacement rate and, from this point onwards, the load was increased until failure.

3 Analysis of Results

3.1 Compressive Strength

To analyze the results of the compressive tests, the Grubbs' method for outliers was conducted on the three groups of testing directions to detect any atypical values in the results. Taking the analysis of the outliers into consideration conducted on the 21 specimens tested, and after disregarding one atypical value, the average compressive strength (f_c) of the three groups as well as their respective coefficient of variation is shown in Table 3.

Table 3. Summary of compressive strength results.

Group of data	Number of data	f_c [MPa]	CV (%)
Group I (perpendicular & parallel)	20	53.4	11.5
Group II (perpendicular)	12	52.7	14.3
Group III (parallel)	8	54.4	6.3

The results reveal that there are no significant differences on the compressive strength even if testing on different directions. In this case, even loading in a direction parallel to the layers revealed a slightly better behaviour than the tests conducted when loading perpendicular to the direction of the layers. This fact highlights the high homogeneity of these specimens and its high degree of isotropy in any given direction.

3.2 Modulus of Elasticity

As in the analysis of the compressive strength, an analysis of outliers following the Grubbs' procedure has been conducted on the three loading groups to detect any atypical values of the modulus of elasticity. The average results of the modulus of elasticity, along with their coefficient of variation, are shown in Table 4 according to the loading direction.

Table 4. Summary of modulus of elasticity results.

Group of data	Number of data	E_{cm} [MPa]	CV (%)
Group I (perpendicular & parallel)	19	30275	6.0
Group II (perpendicular)	10	30406	7.2
Group III (parallel)	9	30130	4.6

The results reveal reasonably low differences on the average results of the modulus of elasticity regardless of the direction of testing, these being of the same order of magnitude when compared to traditional plain concrete. As in the compressive strength, this results indicate that the manufacturing procedure of these specimens leads to a suitable homogeneity and a high degree of isotropy in any direction.

3.3 Strain at Maximum Strength

The strain at the maximum compressive strength (ε_{cm}) is determined based on the measurements of the LVDTs at the peak strength according to the results of the modulus of elasticity. The results of ε_{cm} of the 19 specimens are presented in Table 5.

Table 5. Summary of strains at maximum strength.

Group of data	Number of data	Strain [‰]	CV (%)
Group I (perpendicular & parallel)	19	2.59	22.1
Group II (perpendicular)	10	2.53	26.9
Group III (parallel)	9	2.68	16.2

The results reveal similar strains at maximum strength regardless of the testing direction. However, specimens of Group II present a slightly lower average strain than those obtained in samples corresponding to Group III. This effect could be attributed to the loading direction, in which parallel layers may experience a greater deformation before reaching the failure maximum load.

4 Compressive Constitutive Model

4.1 Modelling of Compressive Curves

Based on the results of the compressive tests, modulus of elasticity and the strains measured using the LVDTs until failure, compressive constitutive models for compression can be determined. As in previous studies [7, 8], the model proposed by Thorenfeldt was selected since it can reproduce the behaviour under compression of non-confined concrete between 25 and 100 MPa. The equation proposed for the unconfined stress-strain relation is shown in (1), where σ_c is the theoretical compressive strength, f_{cm} is the maximum experimental compressive strength, ε_{cm} is the strain at f_{cm} and ε_c is the strain at σ_c. The parameters n and k are determined according to Eqs. (2) and (3) and depend on f_{cm}.

$$\sigma_c(\varepsilon_c) = \frac{n\left(\frac{\varepsilon_c}{\varepsilon_{cm}}\right)}{(n-1) + \left(\frac{\varepsilon_c}{\varepsilon_{cm}}\right)^{nk}} \cdot f_{cm} \tag{1}$$

$$n = 0.8 + \frac{f_{cm}}{17} \tag{2}$$

$$k = 0.67 + \frac{f_{cm}}{62} \geq 1.0 \tag{3}$$

The average compressive constitutive law of the concretes produced according to the different testing directions is shown in Fig. 2. The results are shown in relative terms and display the upper and the lower limits of the compressive model along with the average result, with the shaded area representing the scatter of the model. Notice how the scatter of the results is, in this case, smaller when the load is applied parallel to the direction of the layers. This effect takes place both at the elastic branch and, at a higher extent, at the post-failure regime.

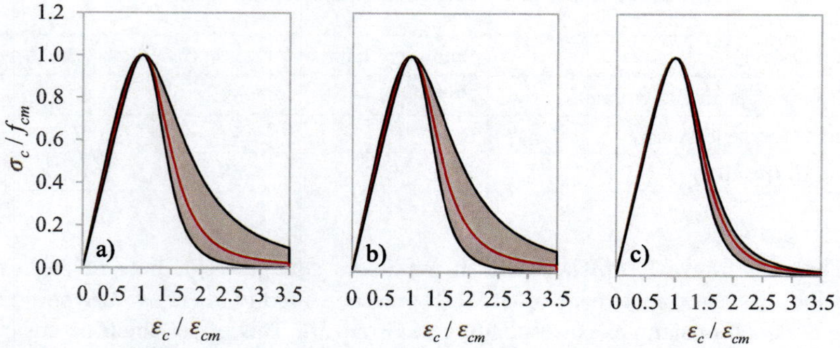

Fig. 2. Relative average compressive constitutive law according to **a** Group I, **b** Group II and **c** Group III.

4.2 Comparison of Experimental and Model Parameters

A comparison between the experimental and the theoretical results according to the Thorenfeldt model is shown in Table 6. These results show both the compressive strengths

Table 6. Numerical comparison of experimental and Thorenfeldt compressive strength results.

Group of specim	Experimental		Thorenfeldt	
	Strength [MPa]	Strain [‰]	Strength [MPa]	Strain [‰]
Group I	53.4	2.59	54.6	2.38
Group II	52.6	2.53	53.8	2.32
Group III	54.4	2.68	55.7	2.45

and the strains obtained through the experimental results and those according to the results of the Thorenfeldt approximation.

The results show there is a 2.3% average overestimation of the compressive theoretical results when compared to the experimental values. Considering the results of the strains, the values obtained at the model are in average 8.3% lower than the experimental results.

5 Conclusions

An analysis of the compressive strength and the anisotropy of 3DCP specimens due to layer-to-layer manufacturing procedure was conducted. Based on these results, a constitutive model for compression oriented to 3D printed concrete elements was determined. According to these analyses, the following conclusions were obtained:

- The specimens manufactured through a 3DCP powder-bed technique revealed average compressive strengths over 50 MPa. Such strengths meet the minimum requirements generally establishes in codes and guidelines for the use of concrete as a structural material.
- The concrete analyzed revealed a high homogeneity and a high degree of isotropy in any given direction. This is supported by the similar values of the compressive strength and the modulus of elasticity exhibited in the different testing directions.
- The compressive constitutive law can be easily obtained through the results of compressive strength and strain at maximum strength. The results revealed only small differences between the experimental values and the model developed.

Acknowledgements. The authors want to express their gratitude to AGAUR-FEDER for the financial support through the project 2019PROD00066. Likewise, the authors want to acknowledge the economic and technical support provided by ACCIONA.

References

1. Buswell, R.A., Leal de Silva, W.R., Jones, S.Z., Dirrenberger, J.: 3D printing using concrete extrusion: a roadmap for research. Cem. Concr. Res. **112**, 37–49 (2018). https://doi.org/10.1016/j.cemconres.2018.05.006

2. Zhang, J., Wang, J., Dong, S., Yu, X., Han, B.: A review of the current progress and application of 3D printed concrete. Compos. Part A Appl. Sci. Manuf. **125**, 105533 (2019). https://doi.org/10.1016/j.compositesa.2019.105533
3. Wangler, T., Roussel, N., Bos, F.P., Salet, T.A.M., Flatt, R.J.: Digital concrete: a review. Cem. Concr. Res. **123** (2019). https://doi.org/10.1016/j.cemconres.2019.105780
4. Le, T.T., et al.: Hardened properties of high-performance printing concrete. Cem. Concr. Res. **42**, 558–566 (2012). https://doi.org/10.1016/j.cemconres.2011.12.003
5. Rahul, A.V., Santhanam, M., Meena, H., Ghani, Z.: Mechanical characterization of 3D printable concrete. Constr. Build. Mater. **227**, 116710 (2019). https://doi.org/10.1016/j.conbuildmat.2019.116710
6. Lowke, D., Dini, E., Perrot, A., Weger, D., Gehlen, C., Dillenburger, B.: Particle-bed 3D printing in concrete construction – possibilities and challenges. Cem. Concr. Res. **112**, 50–65 (2018). https://doi.org/10.1016/j.cemconres.2018.05.018
7. de la Fuente, A., De Cea, A.A., Borrell, C.M.: Modelo numérico para el análisis no lineal de secciones prefabricadas construidas evolutivamente Numerical model for the non linear analysis of precast and sequentially constructed sections 57 (2008)
8. de la Fuente, A., Aguado, A., Molins, C., Armengou, J.: Numerical model for the analysis up to failure of precast concrete sections. Comput. Struct. **106–107**, 105–114 (2012). https://doi.org/10.1016/j.compstruc.2012.04.007

Effect of Nano-additives and Polymeric Viscosity Modifying Admixtures (VMA) on the Fresh and Hardened Properties of 3D Printable Concrete Mixtures

Debadri Som(✉) and Cristina Zanotti

Department of Civil Engineering, The University of British Columbia, Vancouver, BC, Canada
debadri@mail.ubc.ca

Abstract. In spite of recent advances in construction digitization and concrete 3D printing, engineering a printable concrete mixture for extrudability and buildability still presents technical challenges. Nano-additives such as nano-silica and nano attapulgite clay are gaining increasing interest for their attributes as rheological modifiers. Recent studies showed some of the potential rheological benefits of these materials, although there is a lack of comprehensive studies in this domain. Polymeric additives, e.g. a cellulose derivative (Viscosity Modifying Admixture) and polycarboxylate ether (superplasticizer), can be combined with the nanomaterials to enhance fresh properties. A comprehensive study is presented which depicted the effect of rheological modifiers on flowability and uniaxial compressive load resisting ability of fresh 3D printable concrete mixtures. A time-based flow table protocol was developed to document the change of flow diameter with time pertaining to the mixtures' thixotropic behavior. Furthermore, the effect of uniaxial compressive loading on fresh concrete ("apparent green strength") at different resting intervals from the time of mixing (i.e. at 30, 60, 90, and 120 min) was investigated. Fresh concrete specimens were subjected to compression and loaded until 40% vertical strain or complete failure, whichever occurred earlier. Based on the apparent stress-strain curves, the effect of resting time on the apparent stress, post-peak behavior and failure mode was analyzed.

Keywords: Concrete 3D printing · Flowability · Apparent green strength · Rheological modifiers · Fresh-state property · Post-peak behavior

1 Introduction

Concrete 3D printing (extrusion-based additive manufacturing) uses computer-based architectural drawings to seamlessly manufacture structures without conventional formwork. It primarily involves pumping the concrete through a pipe up to a nozzle after which the concrete is extruded layer-by-layer. A mixture that can be easily pumped, and extruded and can resist the self-weight of the current layer in addition to the weight of subsequent layers on top is referred to as a 'printable' material. ([1, 2] among others).

J. I. Escalante-Garcia et al. (Eds.): RW 2021, RILEM Bookseries 40, pp. 469–477, 2023.
https://doi.org/10.1007/978-3-031-21735-7_52

Rheometers are typically used to assess the fresh properties of 3D printable concrete (3DPC) mixtures. While they provide a quantitative assessment of yield stress and viscosity, it is often not feasible to use a rheometer on-site for the rheological quality control of 3DPC mixtures. The flow table test (as outlined in ASTM C1437 [3]) has been widely used in the concrete 3D printing sector. It follows the use of a flow table apparatus, as outlined in ASTM C230/C230M [4] to assess the fresh properties of a concrete mixture immediately after mixing. Chen et al. [5] and Moeini et al. [6] used time-based flowability assessments in order to study the effect of resting time on the flow diameter. However, unlike a rheometer, a flow table test offers limited information in terms of quantitative rheological behavior.

Since additive manufacturing follows a layer-by-layer addition of concrete layers, each layer is subjected to increasing compressive forces over time. Hence, it is critical to study the behavior of fresh concrete under a uniaxial compressive load to assess its buildability. Wolfs et al. [7] and Panda et al. [8], among others conducted tests on cylindrical specimens to enable the formation of complete shear planes during failure. They reported that the resting time after mixing influenced the failure mode of the fresh concrete specimens.

Since pumpability, extrudability and buildability have conflicting rheological requirements, it is challenging to design 3DPC mixtures satisfying them. Rheological modifiers like nano-silica, nano attapulgite clay and cellulose ether (polymeric) Viscosity Modifying Admixture have great potential in concrete 3D printing applications owing to their effects on yield stress, plastic viscosity and thixotropy. However, there exist knowledge limitations concerning the effects of these admixtures on the uniaxial compressive behavior of fresh concrete specimens and the time-dependent flowability of fresh 3DPC mixtures, which the current study tries to address.

2 Materials and Test Methods

The mixtures investigated are shown in Tables 1 and 2. The base mixture design is shown in Table 1.

Table 1. Base mixture design ([1] among others)

Sand (kg/m3)	Cement (kg/m3)	Water (kg/m3)	Superplasticizer (kg/m3)	Fly ash (kg/m3)	Silica fume (kg/m3)
1167	579	232.6	12.24	128.2	128.2

Based on initial laboratory trials and corresponding flowability results, the mixture designs were decided. In this study, a flowability value of 130–170 mm (immediately after mixing) was considered (as per suggestions by [5, 9] among others). The additives in Table 2 were added to the base mixture (in Table 1) as mass percentage of the total binder content (cement, fly ash and silica fume) to produce different mixtures.

Table 2. Summary of mixture designs used in the current study

Mixture code	Additives as mass percentage of binders
1VMA	1% VMA
1NS	1% NS
0.5NS-0.5VMA	0.5% VMA, 0.5% NS
1NC-0.5VMA	1% NC, 0.5% VMA
1.5NC-1VMA	1.5% NC, 1% VMA
1NC-0.5VMA-0.15AEA	1% NC, 0.5% VMA, 0.15% AEA
1NC-0.5VMA-0.3AEA	1% NC, 0.5% VMA, 0.30% AEA

Index: VMA[1]: Viscosity Modifying Admixture (cellulose ether), NS[2]: Nano-silica, NC[3]: Nano attapulgite clay, AEA[4]: Air Entraining Admixture.

River sand, with a maximum particle size of 4.75 mm was used as fine aggregate. General Use (GU) cement, conforming to ASTM C150 [10] was used in the current study. Class F Fly Ash (from Lafarge) and silica fume (Con-fume from Kryton International), were also used as binders. ADVA 198 and DAREX II from GCP Applied Technologies were used as a polycarboxylate-based superplasticizer and an Air Entraining Admixture (AEA) respectively. The rheological modifiers used were:

- Powdered silica nanoparticles (nano-silica) from Sigma Aldrich.
- Nano attapulgite clay from Active Minerals International (Actigel 208).
- Cellulose ether Viscosity Modifying Admixture (VMA) which is commercially available as Sika Stabilizer 4R from Sika.

Sand, cement, fly ash and silica fume were dry mixed for 2 min in a Hobart planetary mixer. Water mixed with superplasticizer, NS, NC, VMA and AEA were added in small amounts over 2 min. The mixture was finally mixed again for 3 min.

Flow table tests, pertaining to the procedure outlined in ASTM C1437 [3], were conducted at intervals of 15 min in order to study the change in fluidic behavior of fresh 3DPC concrete.

A small-scale loading frame was used to study the behavior of fresh concrete under uniaxial compressive load. Figure 1a shows a picture of the set up used. The apparatus consisted of a load cell with a capacity of 6.5 kN and an LVDT sensor. Immediately after mixing, specimens were prepared for testing in cylindrical molds (150 mm in height and 75 mm in diameter) which were detachable. The two halves of the cylindrical molds were held by a rotary clamp. The mold rested on a base plate which provided support to the specimen during sample preparation and testing. The mold-base plate set up is shown in Fig. 1b. The inner surface of the molds, the top surface of the base plates and the loading plate were coated with Teflon sheets and oiled to reduce the effect of friction. Concrete mixtures were filled in the molds in 3 layers followed by 25 hits with a tamping rod. The test was conducted at 30 min, 60 min, 90 min and 120 min after mixing, at a displacement rate of 0.5 mm/min. Based on readings from the LVDT sensor and the load cell, force-displacement curves were generated for each trial. The following formulae were used for conversion of force-displacement curves to apparent stress-vertical strain curves:

$$\text{Vertical strain} = \frac{\text{Displacement on the sensor(mm)}}{\text{Initial Height of the specimen(mm)}} \tag{1}$$

$$\text{Apparent stress} = \frac{\text{Load (N)}}{\text{Cross} - \text{sectional area of the mold}\left(\text{mm}^2\right)} \tag{2}$$

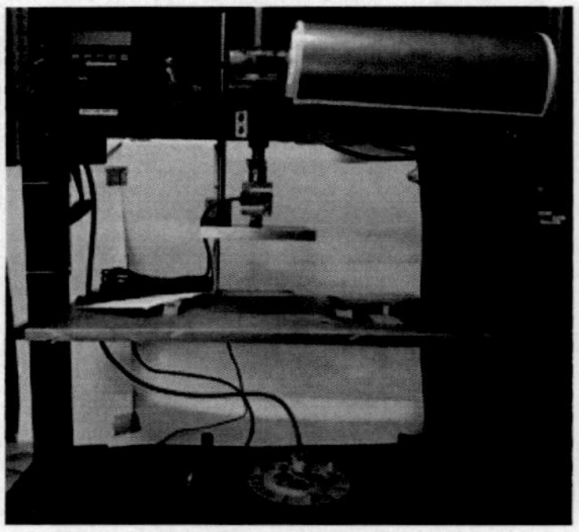

a) Experimental set up

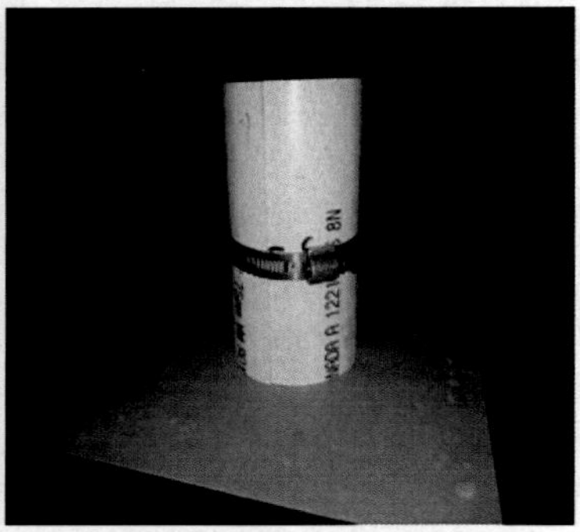

b) Mould-base plate set up

Fig. 1. Measurement of apparent green strength of 3DPC mixtures

3 Results and Discussions

3.1 Time-Based Flowability Assessments

Average flow diameters (and standard deviations) recorded immediately after mixing, 15 min and 30 min after mixing are listed in Table 3. Mixture 5 was clearly the stiffest of all the mixtures tested with the lowest flow diameter values. Considering the change in flow diameter over time, it is to be noted that for all the mixtures, the change from 0 min. to 15 min. Was higher than the change from 15 min. to 30 min. This behavior corresponded to the effects of reflocculation as explained by Kruger [11], which was further enhanced by the presence of nano-particles. Mixture 3 exhibited the highest percentage of reduction in flow diameter over the first 15 min (11.73%). It is to be noted that the combined use of nano-silica and VMA affected the thixotropic behavior more than when they were used separately as rheological modifiers (in mixtures 1 and 2).

Mixtures 4, 6 and 7 also exhibited a high percentage of reduction in flow diameter over the first 15 min (11.02%, 9.59% and 10.01% respectively). It is typically due to the build-up of lattice structures by nano clay spindles at rest, increasing the yield stress of concrete mixtures [12]. The addition of AEA to mixture 4 increased the flow diameter while reducing the rate of reduction of flow diameter over the first 15 min. Mixture 5, with a higher content of nano attapulgite clay and polymeric VMA, exhibited the lowest increase in the flow diameter during the first 15 min (6.96%). Hence, there exists an optimum dosage of nano-clay and VMA at which the thixotropic behavior is ideal for 3D printable applications, as explained by Douba et al. [13]. It is to be noted that the optimal dosages of rheological modifiers also depend on the base mixture design. For broader conclusions, further comprehensive assessments will be necessary to study the optimal dosage of rheological modifiers.

Table 3. Time-based flow diameter results for the seven mixtures

Mixture no		Mean flow diameter (Standard deviation)	
	0 min. (immediately after mixing)	15 min	30 min
1	152.33 (5.11)	139.17 (0.76)	130.83 (1.04)
2	143 (4.77)	132.83 (4.31)	128.5 (8.32)
3	160.5 (6.87)	141.67 (3.33)	134.17 (5.84)
4	148.17 (5.29)	131.83 (4.54)	125.83 (3.75)
5	134.17 (2.02)	124.83 (1.53)	119 (1.73)
6	154.67 (1.53)	139.83 (5.39)	129 (0.87)
7	164.83 (2.25)	148.33 (3.82)	(4.75)

3.2 Apparent Green Strength Test

Figure 2 shows the apparent stress-vertical strain curves for the seven mixtures tested in the current study. Each mixture was tested at 30, 60, 90 and 120 min after mixing. Three trials were conducted for each test and the average curves are plotted in Fig. 2. Mixture 1 (Fig. 2a) exhibited post-peak softening behavior starting from 90 min after mixing. It is to be noted that none of the specimens for Mixture 1 showed the formation of a shear plane. The failure was purely compressive in nature with initial barreling effect followed by yielding of the material through opening of the surface cracks (Fig. 3a). Mixture 2 (Fig. 2b.) exhibited post-peak softening behavior from 30 min after mixing. Remarkably, all the mixtures failed through the formation of a shear plane and there was slip along the surface of the plane which led to the specimen losing contact with the loading plate at 20% vertical strain (Fig. 3b). The addition of nano-silica as a rheological modifier clearly contributed to the brittle behavior of fresh concrete specimens. Mixture 3 (Fig. 2c) exhibited post-peak softening behavior from 90 min after mixing. Although 0.5% nano-silica was added to the mixture, the failure modes of the specimens for Mixture 3 exhibited controlled brittleness. The specimens were bent in order to accommodate the formation of a shear plane (Fig. 3c). Mixture 4 exhibited post-peak softening behavior starting 60 min after mixing. Although specimens at 120 min showed the formation of a shear plane, all other specimens showed a barreling effect (Fig. 3d). Mixture 5 was the stiffest mixture tested (highest apparent green strength). It exhibited a post-peak softening behavior from 60 min after mixing. Although failures at 90 min and 120 min after mixing were predominantly through the formation of a shear plane, there was no slip observed along the shear plane. For mixtures 6 and 7, the addition of 0.15% AEA and 0.30% AEA respectively to mixture 4 created a flatter post-peak region for the specimens at 90 min and 120 min. It is also to be noted that the addition of AEA reduced the apparent green strength of the mixtures.

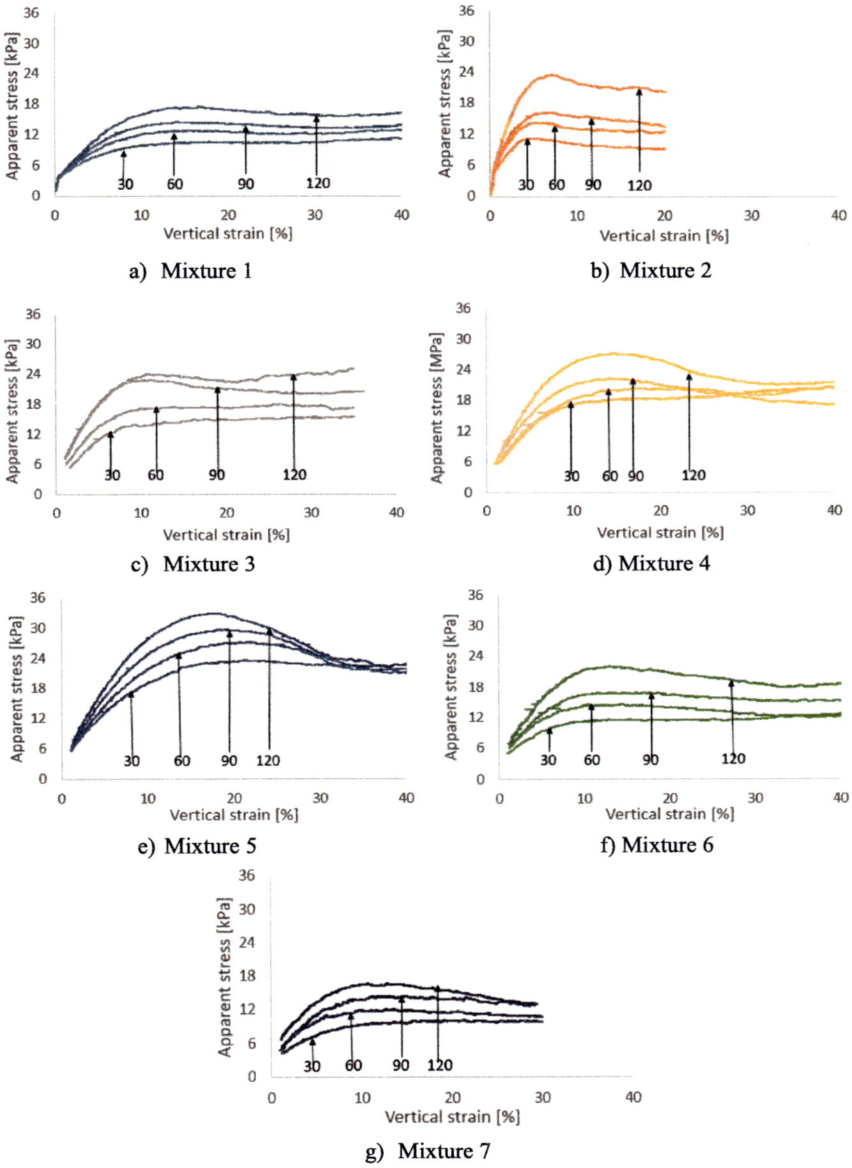

Fig. 2. Apparent stress-vertical strain curves for the seven mixtures tested

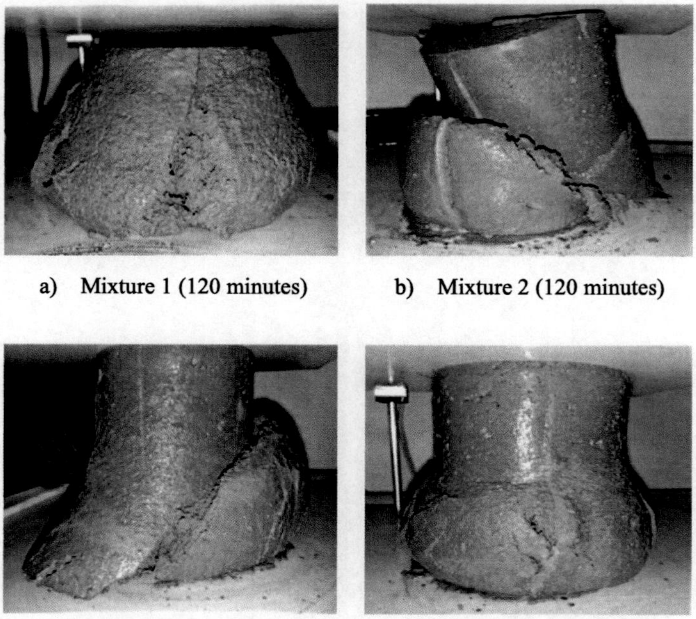

a) Mixture 1 (120 minutes) b) Mixture 2 (120 minutes)

c) Mixture 3 (120 minutes) d) Mixture 4 (90 minutes)

Fig. 3. Failure modes for the apparent green strength test

4 Conclusions

The effects of three different rheological modifiers on the fresh-state properties of 3D printable concrete mixtures were investigated. The following conclusions can be drawn from the current study:

- The time-based flowability test can be used for in-situ thixotropic assessments although it is necessary to develop correlations with static yield stress values from rheometric assessments.
- The method used to study the apparent green strength was suitable for 3D printable concrete mixtures, but only up to a flow diameter value of 170 mm. For further fluidic 3D printable mixtures, the method used in the current study proposes significant challenges with specimen demolding, causing thixotropic breakdown.
- The shape of the apparent stress-vertical strain curves and the post-peak softening behavior are dependent on two factors: resting time after mixing and type and dosage of rheological modifiers. These two factors are not mutually exclusive.
- Further assessments are necessary to optimize the dosage of nano-silica, nano attapulgite clay (NC) and polymeric Viscosity Modifying Admixture (VMA) through Design of Experiments (DoE).

- The presence of AEA increased the flow diameter and reduced the apparent green strength but further assessments are necessary in terms of their effects on the yield stress of concrete mixtures.

References

1. Le, T.T., Austin, S.A., Lim, S., Buswell, R.A., Gibb, A.G., Thorpe, T.: Mix design and fresh properties for high-performance printing concrete. Mater. Struct. **45**(8), 1221–1232 (2012)
2. Buswell, R.A., De Silva, W.L., Jones, S.Z., Dirrenberger, J.: 3D printing using concrete extrusion: a roadmap for research. Cem. Concr. Res. **112**, 37–49 (2018)
3. ASTM International. C1437–20 Standard Test Method for Flow of Hydraulic Cement Mortar. West Conshohocken, PA; ASTM International (2020). https://doi.org/10.1520/C1437-20
4. ASTM International. C230/C230M-21 Standard Specification for Flow Table for Use in Tests of Hydraulic Cement. West Conshohocken, PA; ASTM International (2021). https://doi.org/10.1520/C0230_C0230M-21
5. Chen, Y., et al.: Improving printability of limestone-calcined clay-based cementitious materials by using viscosity-modifying admixture. Cem. Concr. Res. **132**, 106040 (2020)
6. Moeini, M.A., Hosseinpoor, M., Yahia, A.: Use of the chemical and mineral admixtures to tailor the rheology and the green strength of 3D printing cementitious mixtures. In: RILEM International Conference on Concrete and Digital Fabrication, pp. 73–82. Springer, Cham (2020)
7. Wolfs, R.J., Bos, F.P., Salet, T.A.M.: Early age mechanical behaviour of 3D printed concrete: numerical modelling and experimental testing. Cem. Concr. Res. **106**, 103–116 (2018)
8. Panda, B., Lim, J.H., Tan, M.J.: Mechanical properties and deformation behaviour of early age concrete in the context of digital construction. Compos. B Eng. **165**, 563–571 (2019)
9. Tay, Y.W.: Large scale 3D concrete printing process and materials properties (2020)
10. ASTM International. C150/C150M-20 Standard Specification for Portland Cement. West Conshohocken, PA; ASTM International (2020). https://doi.org/10.1520/C0150_C0150M-20
11. Kruger, P.J.: Rheo-mechanics modelling of 3D concrete printing constructability (Doctoral dissertation, Stellenbosch: Stellenbosch University) (2019)
12. Kawashima, S., Chaouche, M., Corr, D.J., Shah, S.P.: Rate of thixotropic rebuilding of cement pastes modified with highly purified attapulgite clays. Cem. Concr. Res. **53**, 112–118 (2013)
13. Douba, A., Chan, C., Berrios, S., Kawashima, S.: Synthesis of hybridized rheological modifiers for 3D concrete printing. In: RILEM International Conference on Concrete and Digital Fabrication, pp. 32–41. Springer, Cham (2020)

Structural Performance and Design

Shrinkage-Cracking Prevention in Large-Scale Concrete Structures by Means of Superabsorbent Polymers (SAPs)

José Roberto Tenório Filho[1,2](✉), Didier Snoeck[1], and Nele De Belie[1]

[1] Department of Structural Engineering and Building Materials, Faculty of Engineering and Architecture, Magnel-Vandepitte Laboratory, Ghent University, Tech Lane Ghent Science Park, Campus A, Technologiepark Zwijnaarde 60, B-9052 Ghent, Belgium
roberto.tenorio@ugent.be

[2] SIM vzw, Technologiepark Zwijnaarde 48, B-9052 Ghent, Belgium

Abstract. SAPs have been widely studied over the past two decades as new admixtures for cementitious materials. Their ability to absorb and retain water from the cementitious mixture allowed them to be successfully used as internal curing agents preventing self-desiccation and reducing autogenous shrinkage. A lot of research has been developed and important advances have been achieved, but there is still a lack of studies reporting on the use of SAPs in large-scale structures. This paper presents results of a large-scale testing campaign in Belgium using a commercial SAP. Three reinforced concrete walls (dimensions 14 m × 2.75 m × 0.80 m) were built and monitored with regards to shrinkage and cracking. Two walls without SAPs and water-to-cement ratio of 0.44 and 0.52 were used as references. The SAP-wall was built with 0.38% (over the cement mass) of a commercial SAP, total water-to-cement ratio of 0.52 and effective water-to-cement ratio of 0.44. A laboratory campaign was performed simultaneously with the on-site one to fully characterize the concrete used in the walls. The addition of SAPs promoted a reduction of up to 75% of shrinkage strain in the first 7 days and up to 60% after 28 days. No cracks have developed in the SAP wall, while for the reference walls the first cracks were noticed 5 days after casting. The addition of SAPs caused a 17% reduction in compressive strength at 28 days compared to the reference with less water and 22% increase compared to the reference with more water.

Keywords: Superabsorbent polymers · Large-scale testing · Shrinkage · Cracking · Concrete walls

1 Introduction

Superabsorbent polymers (SAPs) have been investigated over the last two decades as new admixtures for cementitious materials and very positive results have been reported for a wide range of applications. SAPs are made of a natural and/or synthetic water-insoluble 3D network of polymeric chains cross-linked by chemical or physical bonding. They

J. I. Escalante-Garcia et al. (Eds.): RW 2021, RILEM Bookseries 40, pp. 481–488, 2023.
https://doi.org/10.1007/978-3-031-21735-7_53

possess the ability to take up a significant amount of liquids from the environment (in amounts up to 1500 times their own weight) [1]. Due to their ability to absorb and retain water from the cementitious mixture, they have been reported to be successfully used as internal curing agents preventing self-desiccation and reducing autogenous shrinkage [2–14].

Although SAPs have been widely recognized as a potential new admixture for the industry of cementitious materials, their use has been so far restricted to laboratory and small-scale research, with only a few large-scale applications being reported in literature. Van Tittelboom et al. [15] used SAPs to improve the autogenous crack healing of real-scale concrete beams (150 mm × 250 mm × 3000 mm). Mechtcherine and Dudziak [16] described the construction of a pavilion designed as a thin-walled structure with no conventional reinforcements and with very slim columns built in 2006 for the FIFA World Cup. De Meyst et al. [17] used SAPs to reduce autogenous shrinkage in small ultra-high performance concrete walls. In China, SAPs have been used in real life applications: in [18] SAPs have been successfully used as internal curing agent in a railway slab; in [19] the authors report the use of SAPs in the construction of the China Zun tower, where a reduction of 46% in shrinkage was achieved.

In this paper, a commercially available SAP is studied. It was used for the construction of large-scale reinforced concrete walls with the goal of building real life structures that would possess the features of shrinkage reduction by means of internal curing. Three walls (with a total of 96 m^3 of concrete), were built and monitored in real time. A laboratory testing campaign was also carried out to characterize the concrete mixtures and provide scientific input for the interpretation of results obtained with the monitoring of the walls.

2 Materials and Methods

All walls were produced with concrete having a consistency class S4. The concrete mixtures were produced with cement type CEM III-B 42.5N—LH/SR (CBR, Belgium); a polycarboxylate superplasticizer (Tixo, 25% conc., BASF, Belgium); a modified polycarboxylate superplasticizer (Sika-Viscoflow 26, SIKA, Belgium); sea sand 0/4 (absorption of 0.4% in mass); and limestone 2/20 (absorption of 0.5% in mass). The concrete for the wall REF2 had a lower dosage of superplasticizers in comparison to the other concrete mixtures given the higher effective water-to-cement ratio. The SAP used in the research (SAP1), made by SNF Floerger (France), is a cross-linked acrylate copolymer produced through bulk polymerization and has a mean particle size (d_{50}) of 360 μm. SAP1 can absorb up to 21 g of mixing water per gram of SAP in concrete in the first 10 min after contact with water [20]. The mixture design of the concrete mixtures produced for each wall is given in Table 1.

The walls were demolded two days after casting, and immediately prepared for the monitoring of the shrinkage on the surface. The walls were divided into two regions (over the length), where the second half of each wall received an external curing agent. To monitor the shrinkage, measuring points were glued on the surface of the walls, 0.5 from the bottom. On each side of the wall (cured and not cured), five pairs of measuring points with a measuring range of 200 mm were placed over a length of 5 m. Each

Table 1. Composition of the studied concrete mixtures, values in kg/m^3.

Mixture	Cement	Sand	Limestone	SP1	SP2	w/c_tot	w/c_eff
REF1	360	736	1116	2.42	1.56	0.44	0.44
REF2	360	695	1054	0.95	0	0.52	0.52
SAP	360	702	1078	2.42	1.56	0.52	0.44

pair of measuring points was placed at 1 m from the previous one. The monitoring was then performed using a demountable stain gage (DEMEC). For each wall, three concrete trucks were used (two batches of 10 m^3 and one batch of 10 m^3 was produced). For each truck, cubic specimens (side dimension of 100 mm) were cast to evaluate the development of compressive strength over time under laboratory conditions. The cubes were tested at the ages of 7, 28 and 56 days. Three specimens per age per concrete truck were used. They were cured under laboratory conditions (20 °C and RH > 95%) until the testing age. Together with the cubes, prismatic specimens (dimensions 100 mm × 100 mm × 400 mm) were cast for monitoring of shrinkage under laboratory conditions. Six specimens were produced per truck. The specimens were cured for 24 h in a room with controlled atmosphere of 20 °C and RH > 95% and then transferred to another room with controlled atmosphere 20 °C and 60% RH where the shrinkage was measured with a demountable strain gage (DEMEC). Half of the specimens were covered with aluminum tape to prevent drying and simulate the proper conditions for measuring of autogenous shrinkage. The rest of the specimens was left to dry under the environmental conditions of the room to assess the total shrinkage. Together with the monitoring of shrinkage, a visual inspection was carried out to identify the formation of cracks in the walls.

3 Results and Discussion

The monitoring of shrinkage under laboratory conditions showed a complete mitigation of autogenous shrinkage in the concrete mixture with SAPs and additional water for the monitoring period of 28 days (Fig. 1). In contrast, even though the addition of only water without the SAPs (REF2) promoted a significant reduction in the autogenous shrinkage strain at early ages in comparison to REF1 (around 60% at 7 days), a strong shrinking trend was observed for REF2 over the whole testing period. At the age of 28 days the average autogenous shrinkage strain of REF2 was only 10% lower than in REF1.

Considering the total shrinkage, all concrete mixtures performed in a very similar way and no beneficial effect of internal curing was noticed. REF2, which had the higher effective water-to-cement ratio presented a shrinkage strain 10% higher than the other mixtures at the age of 28 days. For both REF1 and SAP1 mixtures, at around 14 days, there is a noticeable change in the rate of shrinkage, which is less observed for REF2. That could indicate that from that moment on, drying is less severe for the first two mixtures which have a lower effective water-to-cement ratio. The fact that no positive effect of SAPs was observed in the total shrinkage is related to effects of drying. Upon exposure to the dry environment, the water stored in the SAPs might be lost to evaporation before it can be used for internal curing. That can also be affected by the dimensions of the testing specimens, as it will be discussed next.

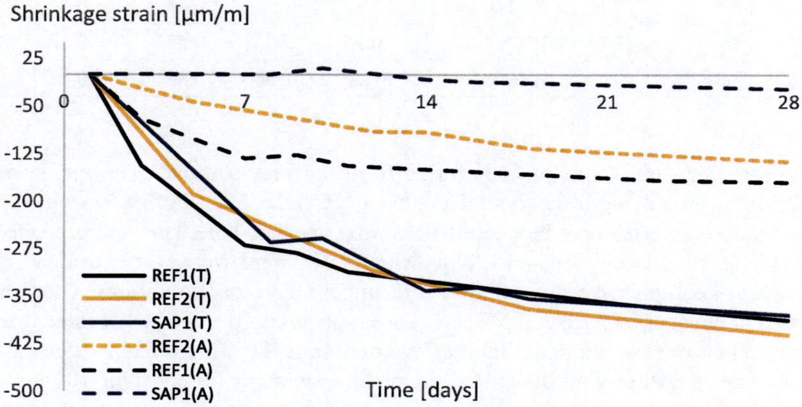

Fig. 1. Shrinkage strain of the concrete mixtures under laboratory conditions. The designation (T) refers to total shrinkage and (A) to autogenous shrinkage.

The shrinkage measured on the surface of the walls is presented in Fig. 2. During the first 28 days of monitoring, the levels of strain for the three concrete mixtures is between the ones observed for the autogenous and total shrinkage of the respective concrete mixtures tested under laboratory conditions. Even though the walls were completely exposed to external environmental conditions and the shrinkage was measured on the surface, it is possible that the bulk dimensions of the walls (in comparison to the small laboratory specimens) slowed down the drying rate and allowed for internal curing to take place (especially in the wall containing SAPs). Demolding the walls only after 48 h instead of 24 h, might also have affected the internal curing in a positive way.

The use of SAPs with additional water was also proved to be more efficient than only adding extra water to the mixture. In fact, the total strain on the SAP wall was lower than in the reference wall during the whole monitoring period. In comparison to REF1, the following reductions were found for SAP1 and REF2, respectively: 58% and 30% at 7 days, 37% and 25% at 28 days, 39% and 7% at 60 days, 38% and 9% at 100 days and 22% and 4% after 150 days. The use of SAPs and additional water to promote internal curing reduced the levels of shrinkage strain considerably for far longer than by only adding water (case of REF2). That is due to the kinetics of water release of the SAPs, where the water is released to the mixture upon self-desiccation of the cement paste.

The use of the external curing agent was proven not effective to reduce the total shrinkage strain. As it can be observed in Fig. 2, the values of shrinkage are very similar for the exposed and cured surface of all mixtures (dashed lines and full lines in the curves). For the SAP-wall, at 120 and 150 days, the shrinkage strain of the cured surface is considerably lower than the one on the exposed surface. However, an unexpected

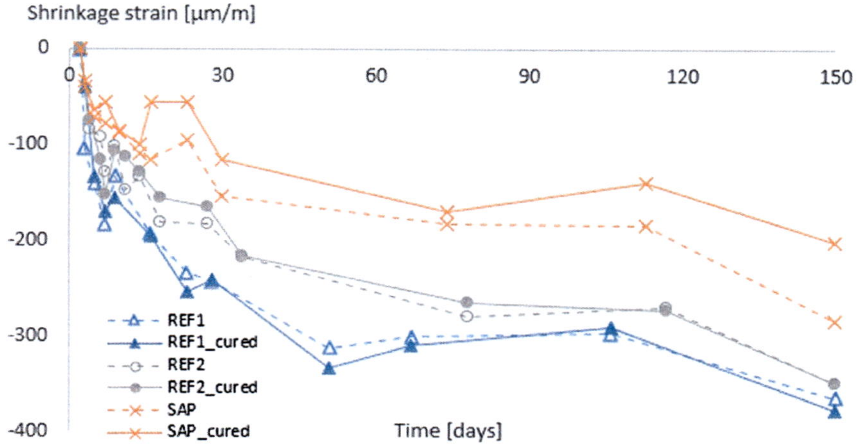

Fig. 2. Shrinkage strain on the surface of the walls. The lines represent the average of five pairs of measuring points. The full line represents the surface where the external curing agent was applied.

expansion is noticed in the curves at these moments, which might be associated with the variations inherent to the manual measurement.

The average compressive strength of the concrete mixtures at 28 days is shown in Fig. 3. Regarding the effects of the addition of SAPs, different conclusion can be drawn depending on the reference mixture used. In comparison to REF1, the addition of only extra water in REF2 promoted a general reduction in strength on all tested ages: 36%, 32% and 29% at 7, 28 and 56 days, respectively. The addition of extra water together with SAPs also promoted a general reduction in comparison to REF1, but to a lower extent than REF2: 20%, 17% and 16% at 7, 28 and 56 days respectively.

The reduction of strength in the two cases might have different causes. In the first, it is related to the higher water-to-cement ratio of the mixture, while in the second it can be associated with the formation of macro-pores in the cement matrix caused by the water release of the SAPs. However, the water released by the SAPs is mostly used for internal curing and might also be used for further hydration of unhydrated cement particles in the system. On top of that, some of the newly hydrated products might also fill in (even if partially) the air voids left by the SAPs, which contributes to a less severe reduction in strength. Additionally, if a comparison is made between the SAP mixture and REF2 (both produced with a total water-to-cement ratio of 0.52) the SAP mixture presented a compressive strength that was around 20% higher than REF2 for all tested ages.

In terms of cracking, the SAP-wall remained crack-free for the whole monitoring period of 150 days. In contrast, cracks going through the wall were first identified after 5 days in REF1 and 9 days in REF2. At the age of 28 days the cracks in REF1 had an average width of 100 µm and a maximum length of 190 cm from the bottom to the top of the wall.

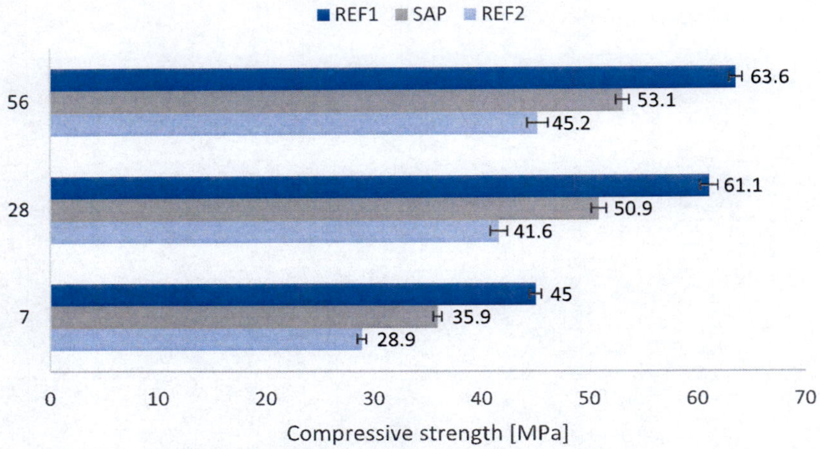

Fig. 3. Compressive strength after 7, 28 and 56 days. Each bar represents the average of nine cubes. The error bars represent the standard error of the average.

4 Conclusions

The use of SAPs and additional water in concrete mixtures was found to be effective for promotion of internal curing and complete mitigation of autogenous shrinkage in specimens tested under laboratory conditions. The effects of SAPs on the properties of cementitious materials have been investigated and well described in literature. However, those findings were still to be extended and further investigated outside of laboratory environment. In this paper, that was investigated and a contribution to the state-of-the-art of the topic is made.

Using SAPs in large scale structures under realistic conditions might bring benefits beyond the ones found in laboratory. In terms of mitigation of shrinkage, it was found that under laboratory conditions for small specimens, the total shrinkage of specimens with and without SAPs develops to the same extent and no benefits from internal curing can be observed. However, on large elements, such as the walls studied here, a bulk-effect was observed, where the bigger volume of concrete might reduce the effects of severe drying and allow internal curing to take place, highlighting the positive effects of the use of SAPs for internal curing and prevention of shrinkage cracking.

In terms of strength, the addition of SAPs and extra water promoted a limited reduction in compressive strength in comparison to a reference mixture without SAPs and extra water and even an increase (of around 20%) when compared to a reference where the same amount of extra water was used. This highlights the benefits of using SAPs beyond the mitigation of shrinkage.

Acknowledgements. The work has been financed by SIM program SHE (Engineered Self-Healing Materials) within the ICON project iSAP (Innovative SuperAbsorbent Polymers for crack mitigation and increased service life of concrete structures). As a Postdoctoral Research Fellow of the Research Foundation-Flanders (FWO-Vlaanderen), D. Snoeck would like to thank the foundation for the financial support (12J3620N). The authors also thank Tommy De Ghein, Nathan Lampens and Peter Van den bussche for the support with the laboratory activities and Artes Depret and InterBeton for the support with the activities on site.

References

1. Mechtcherine, V., Reinhardt, H.W.: Application of super absorbent polymers (SAP) in concrete construction. In: State-of-the-Art Report Prepared by Technical Committee 225-SAP, . p. 165. RILEM (2012)
2. Jensen, O.M.: Use of superabsorbent polymers in construction materials. Microstructure Related Durability of Cementitious Composites, vols. 1 and 2, 61, pp. 757–764 (2008)
3. Snoeck, D., Self-healing and microstructure of cementitious materials with microfibres and superabsorbent polymers. In: Faculty of Architecture and Engineering. Ghent University, Ghent, Belgium (2015)
4. Snoeck, D., Jensen, O.M., De Belie, N.: The influence of superabsorbent polymers on the autogenous shrinkage properties of cement pastes with supplementary cementitious materials. Cem. Concr. Res. **74**, 59–67 (2015)
5. Snoeck, D., De Belie, N.: Effect of superabsorbent polymers, superplasticizer and additional water on the setting of cementitious materials. Int. Journal 3R's. **5**(3), 721–729 (2015)
6. Snoeck, D., Pel, L., De Belie, N.: The water kinetics of superabsorbent polymers during cement hydration and internal curing visualized and studied by NMR. Sci. Rep. (2017)
7. Tenório Filho, J.R., Snoeck, D., De Belie, N.: The effect of superabsorbent polymers on the cracking behavior due to autogenous shrinkage of cement-based materials. In: 60th Brazilian Concrete Conference. Brazilian Concrete Institute, Foz do Iguacu, Brazil (2018)
8. De Meyst, L., et al.: Parameter study of superabsorbent polymers (SAPs) for use in durable concrete structures. Materials, **12**(9) (2019)
9. Snoeck, D., Pel, L., De Belie, N.: Superabsorbent polymers to mitigate plastic drying shrinkage in a cement paste as studied by NMR. Cement Concr. Compos. **93**, 54–62 (2018)
10. Geiker, M.R., Bentz, D.P., Jensen, O.M.: Mitigating autogenous shrinkage by internal curing. High-Perform. Struct. Light. Concr. **218**, 143–154 (2004)
11. Craeye, B., Geirnaert, M., De Schutter, G.: Super absorbing polymers as an internal curing agent for mitigation of early-age cracking of high-performance concrete bridge decks. Constr. Build. Mater. **25**(1), 1–13 (2011)
12. Piérard, J., Pollet, V., Cauberg, N.: Mitigating autogenous shrinkage in HPC by internal curing using superabsorbent polymers. In: International RILEM Conference on Volume Changes of Hardening Concrete: Testing and Mitigation. RILEM Publications SARL, Lyngby, Denmark (2006)
13. Jiang, C.H., et al.: Autogenous shrinkage of high performance concrete containing mineral admixtures under different curing temperatures. Constr. Build. Mater. **61**, 260–269 (2014)
14. Wu, L.M., et al.: Autogenous shrinkage of high performance concrete: A review. Constr. Build. Mater. **149**, 62–75 (2017)
15. Van Tittelboom, K., et al.: Comparison of different approaches for self-healing concrete in a large-scale lab test. Constr. Build. Mater. **107**, 125–137 (2016)
16. Dudziak L, M.V.: Mitigation of volume changes of ultra-high performance concrete (UHPC) by using super absorbent polymers. In: Ultra High Perform Concr Second Int Symp Ultra High Perform Concr. (2008)

17. De Meyst, L., et al.: The use of superabsorbent polymers in high performance concrete to mitigate autogenous shrinkage in a large-scale demonstrator. Sustainability **12**(11), 4741 (2020)
18. Zhu, C., X. L., Xie, Y.: Influence of SAP on the performance of concrete and its application in Chinese railway construction. In: Application of Superabsorbent Polymers and Other New Admixtures in Concrete Construction. RILEM Publications (2014)
19. Liu, J., C. Y., Shu, X., Ran, Q., Yang, Y.: Recent advance of chemical admixtures in concrete. In: 15th International Congress on the Chemistry of Cement. Prague (2019)
20. Tenório, J.R., et al.: Assessment of the potential of superabsorbent polymers as internal curing agents in concrete by means of optical fiber sensors. Construction and Building Materials, **238** (2020)

Effect of Artificial Lightweight Aggregates on Interfacial Transition Zone in Concrete

D. González-Betancur[1,3](✉), Ary. A. Hoyos-Montilla[1,3], J. I. Tobón[2,3], and B. Garcia[2,3]

[1] Construction Department, Universidad Nacional de Colombia, Medellin, Colombia
dgonzalezbe@unal.edu.co
[2] Materials and Minerals Department, Universidad Nacional de Colombia, Medellin, Colombia
[3] Cement and Building Materials (CEMATCO) Research Group, Universidad Nacional de Colombia, Medellin, Colombia

Abstract. Lightweight concrete has a history of more than two-thousand years and its technical development is still proceeding. Despite the diversity of lightweight aggregates available, some mechanical problems have limited their use as structural elements in the construction industry. In general, the mechanical properties of lightweight concretes are lower than normal weight concretes, being affected by factors such as the stiffness of the aggregate particles, the cement content and density. For this reason, significant efforts have been devoted to optimizing their dosage and improving their mechanical properties, to make them competent against conventional concretes. This tendency has led to the study of microstructure, particularly the mechanisms responsible for the interfacial transition zone (ITZ) formation, related with absorption and water release. This investigation focuses on analyze the characteristics of morphology of expanded clay and polymeric wastes lightweight aggregates and conclude that the impact of the ITZ on mechanical properties is of considerable significance, leading to modifying its mechanical behavior.

Keywords: Concrete · Interfacial transition zone · Alkali-activated cement · Lightweight aggregate

1 Introduction

The construction of building and structures relies heavily on the Portland cement products. Reinforced concrete is usually used in structural members in construction industry. The size and weight of reinforced concrete is a problem of great concern [1]. Accordingly, lightweight concrete (LWAC) has been developed. They are commonly used to reduce the density of concrete and to enhance its thermal and acoustic insulation, additionally. It can be produced in a wide range of unit weights and suitable strengths [2]. These materials are particularly adaptable to seismic design, due to the significant reduction of the inertial forces in the elements [3]. This material made with natural or artificial lightweight aggregates are available in many parts of the world and can be used in a great variety of applications including internal and external walls, mezzanines and floors [4].

J. I. Escalante-Garcia et al. (Eds.): RW 2021, RILEM Bookseries 40, pp. 489–498, 2023.
https://doi.org/10.1007/978-3-031-21735-7_54

Lightweight aggregates (LWA) are extracting natural or producing industrially from various materials including clay, scoria, incinerator bottom ash, fly ash and even polymers [2, 5–8]. Artificial lightweight aggregates can contribute to preservation of the environment since they recycle by-products and reduce the consumption of the natural aggregates. Despite the diversity of lightweight aggregates available, some mechanical problems have limited their use as structural elements in the construction industry [9]. In general, the mechanical properties of lightweight concretes are lower than normal weight concretes, being affected by factors such as the stiffness of the aggregate particles, the water to cement ratio (a/c), the cement content, curing age and density [8–10]. For this reason, significant efforts have been devoted to optimizing their dosage and improving their mechanical properties, to make them competent against conventional concretes. This tendency have led to the study of microstructure [11–14], particularly the mechanisms responsible for the interfacial transition zone (ITZ) formation, related with absorption and water release. Because of its nature LWA has higher absorption and subsequent release of water, increasing the hydration degree of the cementitious material around this aggregates, making the adjacent paste to develop a structure with greater porosity [4, 11].

The ITZ is a layer formed between an aggregate and the matrix of cement paste, involves an approximately 30–50 μm zone around each aggregate, within which the porosity increases as the aggregate interface is approached. Its composed of a double layer of calcium hydroxide crystals ($Ca(OH)_2$) oriented to the aggregate's side and hydrated cement paste. Crystal enrichment and orientation of $Ca(OH)_2$ is the main reason for the weak interface [4, 14, 15]. The ITZ properties are affected by the characteristics of the aggregate and the cementitious matrix, as well.

This article focuses on the study of the incidence of LWA morphology in the formation of the microstructure and the thickness of the ITZ, and its influence on the compressive strength of the LWAC using different LWA and an alkaline activated cementitious matrix.

2 Materials and Methods

2.1 Methodology

The physical properties of aggregates, such as the water absorption and density were obtained in accordance with the NTC 237 Colombian standard (ASTM C128:2015 modified adoption). The granulometry were obtained by the procedure established in the NTC 77 (ASTM C136M:2014 modified adoption).

The alkali activated concrete mixture design was performed in compliance with previous works of the cement and construction materials (CEMATCO) research group [16], using as precursor fly ash (FA) and calcium hydroxide (CH). As activators sodium hydroxide (SH) and sodium silicate (SS) were used. In addition, two types of LWA, expanded Clays (E) and polystyrene wastes (P) were used. The reference sample, contains normal weight aggregate (C). The w/b ratio remained constant at 0.40 to achieve as response variable the concrete density and afterwards the compressive strength at 3 and 7 days.

The encoding used to name the samples corresponds to the initial letter according to the type of aggregate used: Normal Weight Aggregate (C), Expanded Clays (E) and Recycle Polystyrene (P). Once the concrete specimens were fabricated, they were placed in an oven at 80 °C for 24 h [16]. They were then placed in a humidity chamber at room temperature until the day of failure. After reaching the age of curing, the compressive strength test to the corresponding samples was conducted in compliance with NTC 673 (ASTM C39:2005 identical adoption), at ages 3 and 28 days of curing.

The hydration process of the samples was stopped, drying them in an oven at 60 °C until reaching a constant mass and afterwards they were taken for cutting, sub-samples of 24.6 cm^2 were obtained for ITZ morphological analysis. Micrographs were taken in a Nikon stereoscope reference Eclipse LV100, objectives 0.75× up to 11.5×.

2.2 Materials

In the development of this study a normal weight conventional aggregate (C) and two LWA, expanded clays (E) and recycle polystyrene (P) were used. Figure 1 shows the particle size distribution of the aggregates and the values corresponding to the physical properties of aggregates used in the manufacture of alkali activated concrete are shown in Table 2. From Fig. 1, all types of aggregates used are within the gradation requirements for coarse aggregate according to NTC 4045 (ASTM C330M:2017 modified adoption), classifying the concrete as a microconcrete.

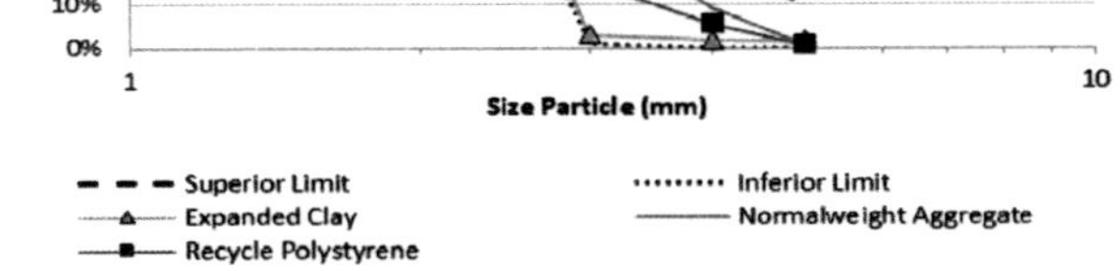

Fig. 1. Particle size distribution of the aggregates.

The aggregates physical properties results show a higher value of water absorption for the expanded clays. This behavior is due to the amount of fractured faces found in the aggregate, considering that inside they are significantly porous, the water absorption increases [12]. On the other side, the aggregate type P shows the lowest water absorption due to their hydrophobic characteristics [17, 18] (Table 1).

The coal fly ash used fits well the chemical requirements of NTC 3493 (ASTM C618:2019 modified adoption) typical of an ash type F. The chemical composition of coal fly ash (FA) and calcium hydroxide (CH) used in this study, measured in an X-ray fluorescence (XRF) equipment with a PHILIPS PW-1004 spectrometer, is shown in

Table 1. Physical properties of aggregates.

Aggregate type	Density SSS (g/cm^3)	Water absorption (%)
C	2.8	0.7
E	1.11	17
P	0.012	0.1

Table 2. For this work, no modification was made to the size of the sample particles in order to evaluate their behavior as it leaves the supplier. Additionally, the percentage of reactive silica of the coal fly ash was determined following the procedure described in the standard UNE 80-225-93, obtaining a value of the 42.20 mass%. The vitreous phase content was also found by an attack with 1% HF [19], and the value obtained was 65.71%. Values indicate that the coal fly ash can be activated to produce an alkaline cement material of good quality [20]. The calcium hydroxide (CH) used was an industrial type, its purity level was 95.6%. The density was measured instrumentally with the flask of Le Chatelier and is shown in Table 2.

The alkaline activator was sodium hydroxide (SH), sodium silicate (SS) at concentrations 8M and 4M [19–22] respectively were used, and also deionized water was used in the preparation of the all mixtures.

Table 2. Chemical composition of coal fly ash (FA) and calcium hydroxide (CH) determined by X-ray fluorescence (mass % as oxides)

Component	SiO_2 (%)	Al_2O_3 (%)	CaO (%)	Fe_2O_3 (%)	MgO (%)	Na_2O	K_2O	TiO_2	LOI[a]	Density (g/cm^3)
FA	55.26	26.95	1.00	5.66	1.78	0.44%	0.75%	2.28%	4.37%	2.160
CH	0.19	0.08	99.14	0.10	0.50	–	–	–	25.6	2.143

[a] LOI lost by ignition at 950 °C

3 Results and Discussion

In Figs. 2, 3, 4 are shown the ITZ around the different aggregates in the concretes evaluated. According to Fang et al. [23], the microstructure evolution of ITZ in an alkali activated matrix, can be primarily divided into three stages: first, at very early curing ages (<12 h), the formation of reaction products via quick alkaline activation has happened; second, at early curing ages (12 h–7 d), the modification of microstructure via the redistribution of reaction products and third, at later curing ages (7 d–28 d), reorganization of microstructure via due to the redistribution of reaction products. In Fig. 2, with the aggregate type C, ITZ is easily identified as an aura of lighter color than that of the cementitious matrix and with irregular thickness [12].

Between binder paste matrix and LWA type E (Fig. 4), a thin and denser ITZ is formed surrounding LWA [24–26]. Moreover, the density of C–S–H [12, 27] and the uniformity [14] are also improved. In addition, previous research have been found some hydration products in the pores of LWA [12]. The improvement of ITZ would be closely related to a puzzolanic powder that envelops the aggregate particles type E [4, 12, 14] which also undergoes alkaline activation when interacting with the activator present in the alkali activated paste. Water released by Expanded clays that promotes the hydration of ITZ in this case could also reduce its porosity [28]. Water release also increases chemically activity of LWA [29], this mean to provide a condition of reaction between LWA and CH. In consequence the amount of CH is reduced [27] and denser hydration products are formed in ITZ [30]. This phenomenon, added to the evolution of ITZ in alkaline activated matrix, makes this ITZ non visible in the equipment used.

With aggregate type P (Fig. 5) no mechanical interlacing is observed between the paste and the polymer. Consequently, a zone of weakness is produced between the paste and the aggregate surface [18, 31]. This phenomenon could be clearer if we consider the fact that adhesion with the cementitious matrix is very important in the manufacturing process of composite materials such as concrete. The Polymer aggregate and organic additives must interact with the components of Portland cement when they encounter with water. EPS particles and, in general, recycled polymer aggregates, do not adhere well to dry cement paste because they are hydrophobic and their surface is electrically charged [32–34].

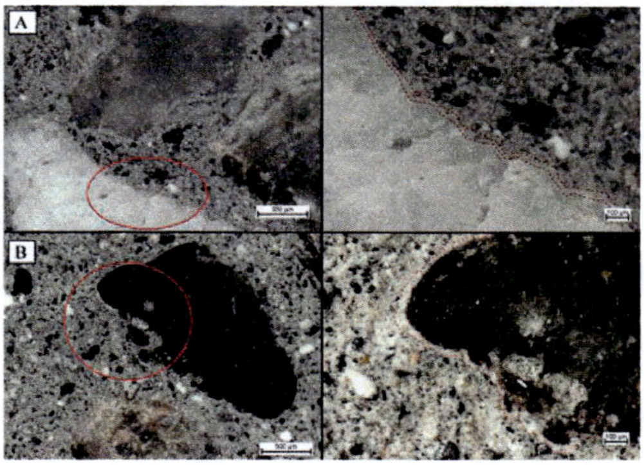

Fig. 2. ITZ micrograph in aggregate type C. 6X left, 9X right at 3 days (A) and 7 days (B).

Figures 5, 6, 7 shows the failure plans found in concretes evaluated. According to the type of aggregate that samples contained shows a typical failure plane for both 3 and 7 days of curing. In the Fig. 5, the failure is transmitted through the ITZ, separating the aggregate from the paste. The brittleness of type E aggregates is evidenced in Fig. 6 where for both 3 and 7 days, the crack goes through the aggregate. However, the ITZ densification in an alkali activated matrix described above, could delay the appearance of cracks in the aggregate, thus equalizing the compressive strengths of concretes with type E and C aggregates. In the case of aggregate type P, a significant decrease in compressive strength is observed due to the lack of adhesion of the aggregate with the cement paste. It has been reported that [18, 35] when EPS beads is added to concrete, the stresses would

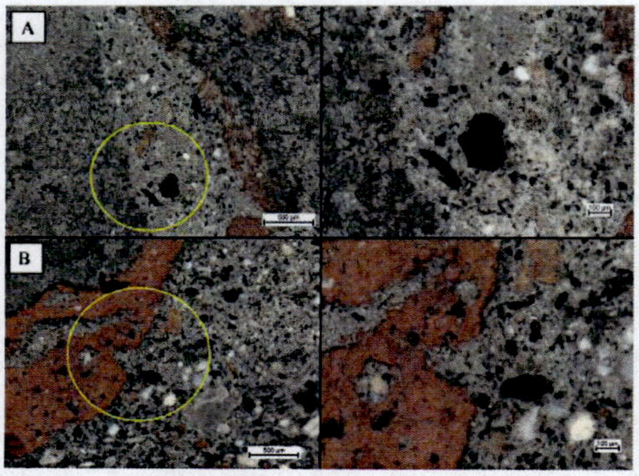

Fig. 3. ITZ micrograph in aggregate type E. 6X left, 9X right at 3 days (A) and 7 days (B).

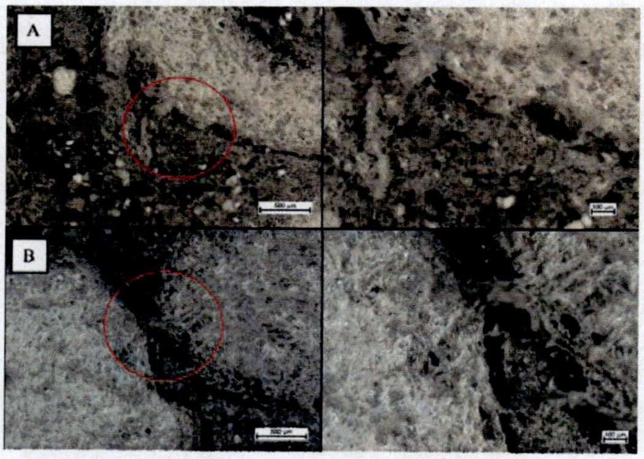

Fig. 4. ITZ micrograph in aggregate type P. 6X left, 9X right at 3 days (A) and 7 days (B).

be redistributed in the microstructure of the concrete and the damaging effect of stress concentration by the presence of EPS beads becomes more pronounced (Table 3).

Table 3. Compressive strength and density of concrete.

Sample	Density	s[a]	3 days compressive strength average (MPa)	s[a]	7 days compressive strength average (MPa)	s[a]
C	2079.29	3.7	5.97	0.31	8.59	0.35
E	1967.06	23	5.84	0.23	8.75	0.35
P	892.13	4.3	1.575	0.01	2.93	0.11

[a] Standard deviation

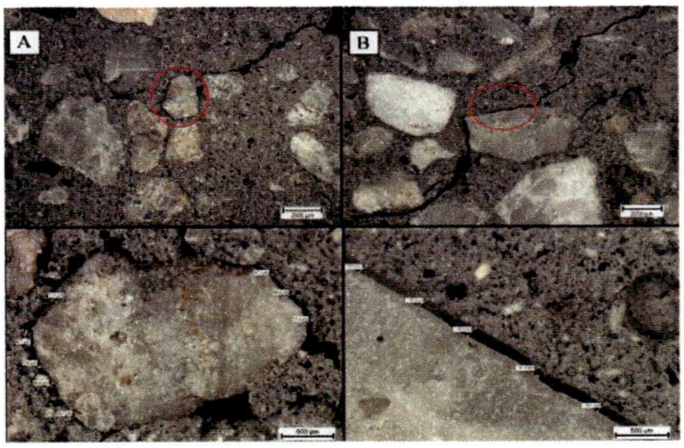

Fig. 5. Failure representation in aggregate type C, after testing compressive strength at 3 days (A) and 7 days (B).

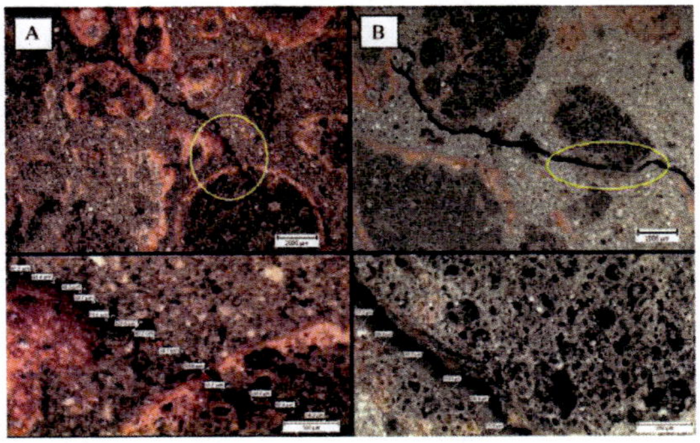

Fig. 6. Failure representation in aggregate type E, after testing compressive strength at 3 days (A) and 7 days (B).

4 Conclusions

According to the results presented above, the following conclusions are proposed:

The impact of the ITZ on mechanical properties is of considerable significance, leading to modifying its mechanical behavior.

Expanded clays aggregates promotes the ITZ densification. However, the lower strength of LWA finally results in cracking through LWA itself.

In the samples with type E aggregate, although the failure propagates through the LWA, the compressive strength obtained is similar to the type C aggregate due to the densification of the ITZ.

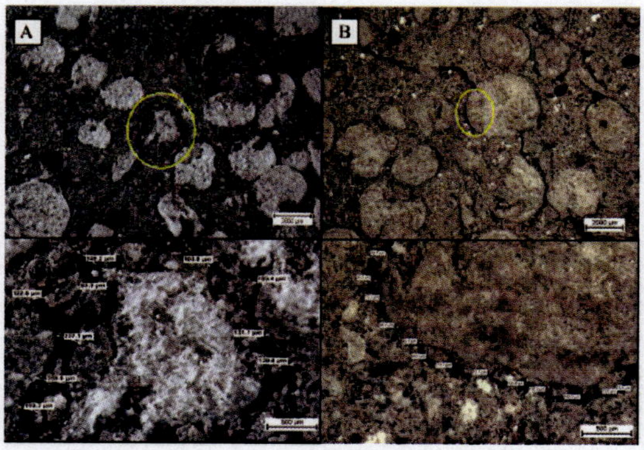

Fig. 7. Failure representation in aggregate type P, after testing compressive strength at 3 days (A) and 7 days (B).

Acknowledgments. This work was financially supported by the Universidad Nacional de Colombia, Orinoquia campus through the project: "Fortalecimiento de las capacidades científicas a través de la fabricación de cementos no convencionales para la mejora de pavimentos de vías terciarias de la región de los llanos, departamento de Arauca" (75817) and the call for young researchers and innovators in the framework of economic reactivation 2021 through the Project: efecto de la sustitución de cemento y agregados en un concreto convencional por cemento activado alcalinamente (AAC) y agregados livianos (LWA) (Hermes code: 53925).

References

1. Nadim Hassoun, M., Al-Manaseer, A.: Structural Concrete: Theory and Design, 7th ed. John Wiley & Sons, Hoboken, New Jersey (2020)
2. Sengul, O., Azizi, S., Karaosmanoglu, F., Tasdemir, M.A.: Effect of expanded perlite on the mechanical properties and thermal conductivity of lightweight concrete. Energy Build. **43**(2–3), 671–676 (2011). https://doi.org/10.1016/j.enbuild.2010.11.008
3. ACI Committee 213: Guide for structural lightweight-aggregate concrete—(ACI 213R-03). ACI Man. Concr. Pract. (2013)
4. Chandra, S., Berntsson, L.: Lightweight aggregate concrete: Science, Technology, and Applications. New York (2002)
5. Kiliç, A., Atiş, C.D., Yaşar, E., Özcan, F.: High-strength lightweight concrete made with scoria aggregate containing mineral admixtures. Cem. Concr. Res. **33**(10), 1595–1599 (2003). https://doi.org/10.1016/S0008-8846(03)00131-5
6. M.-H., Z., Gjorv, O.E.: Characteristics of lightweight aggregates for high-strength concrete. ACI Mater. J. **88**(2). DOI: https://doi.org/10.14359/1924
7. Brooks, A.L., Zhou, H., Hanna, D.: Comparative study of the mechanical and thermal properties of lightweight cementitious composites. Constr. Build. Mater. **159**, 316–328 (2018). https://doi.org/10.1016/j.conbuildmat.2017.10.102
8. Babu, D.S., Ganesh Babu, K., Wee, T.H.: Properties of lightweight expanded polystyrene aggregate concretes containing fly ash. Cem. Concr. Res. **35**(6), 1218–1223 (2005). DOI: https://doi.org/10.1016/j.cemconres.2004.11.015

9. Rossignolo, J.A., Agnesini, M.V.C., Morais, J.A.: Properties of high-performance LWAC for precast structures with Brazilian lightweight aggregates. Cem. Concr. Compos. **25**(1), 77–82 (2003). https://doi.org/10.1016/S0958-9465(01)00046-4
10. Balendran, R.V., Zhou, F.P., Nadeem, A., Leung, A.Y.T.: Influence of steel fibres on strength and ductility of normal and lightweight high strength concrete. Build. Environ. **37**(12), 1361–1367 (2002). https://doi.org/10.1016/S0360-1323(01)00109-3
11. Polat, R., Demirboğa, R., Karakoç, M.B., Türkmen, I.: The influence of lightweight aggregate on the physico-mechanical properties of concrete exposed to freeze-thaw cycles. Cold Reg. Sci. Technol. **60**(1), 51–56 (2010). https://doi.org/10.1016/j.coldregions.2009.08.010
12. Vargas, P., Restrepo-Baena, O., Tobón, J.I.: Microstructural analysis of interfacial transition zone (ITZ) and its impact on the compressive strength of lightweight concretes. Constr. Build. Mater. **137**, 381–389 (2017). https://doi.org/10.1016/j.conbuildmat.2017.01.101
13. Sáez del Bosque, I.F., Zhu, W., Howind, T., Matías, A., Sánchez de Rojas, M.I., Medina, C.: Properties of interfacial transition zones (ITZs) in concrete containing recycled mixed aggregate. Cem. Concr. Compos. **81**, 25–34 (2017). https://doi.org/10.1016/j.cemconcomp.2017.04.011
14. Zhang, M.-H., Gjørv, O.E.: Microstructure of the interfacial zone between lightweight aggregate and cement paste. Cem. Concr. Res. **20**(4), 610–618 (1990). https://doi.org/10.1016/0008-8846(90)90103-5
15. Monteiro, P.J.M., Maso, J.C., Ollivier, J.P.: The aggregate-mortar interface. Cem. Concr. Res. **15**(6), 953–958 (1985). https://doi.org/10.1016/0008-8846(85)90084-5
16. Rúa Patiño, R.F.: Evaluación de la adherencia entre un concreto activado alcalinamente con base en ceniza volante de carbón y el acero de refuerzo grado 60. Universidad Nacional De Colombia (2020)
17. Miled, K., Sab, K., Le Roy, R.: Particle size effect on EPS lightweight concrete compressive strength: Experimental investigation and modelling. Mech. Mater. **39**(3), 222–240 (2007). https://doi.org/10.1016/j.mechmat.2006.05.008
18. Bakhshi, M., Shahbeyk, S.: Experimental and microstructural study of the compressive strength of concrete samples containing low volumes of expanded polystyrene beads. Struct. Concr. **20**(4), 1379–1390 (2019). https://doi.org/10.1002/suco.201800317
19. Hoyos-Montilla, A.A., Puertas, F., Tobón, J.I.: Microcalorimetric study of the effect of calcium hydroxide and temperature on the alkaline activation of coal fly ash. J. Therm. Anal. Calorim. **131**(3), 2395–2410 (2017). https://doi.org/10.1007/s10973-017-6715-4
20. Fernández-Jiménez, A., Palomo, A.: Characterisation of fly ashes. Potential reactivity as alkaline cements. Fuel **82**(18), 2259–2265 (2003). https://doi.org/10.1016/S0016-2361(03)00194-7
21. Zhang, P., Wang, K., Li, Q., Wang, J., Ling, Y.: Fabrication and engineering properties of concretes based on geopolymers/alkali-activated binders—A review. J. Clean. Prod. **258**, 120896 (2020). https://doi.org/10.1016/j.jclepro.2020.120896
22. Hoyos-Montilla, A.A., Puertas, F., Tobón, J.I.: Study of the reaction stages of alkali-activated cementitious materials using microcalorimetry. Adv. Cem. Res. **33**(1), 1–13 (2019). https://doi.org/10.1680/jadcr.19.00025
23. Fang, G., Zhang, M.: The evolution of interfacial transition zone in alkali-activated fly ash-slag concrete. Cem. Concr. Res. **129**(2019), 105963 (2020). doi:https://doi.org/10.1016/j.cemconres.2019.105963
24. Liu, J., Shi, C., Farzadnia, N., Ma, X.: Effects of pretreated fine lightweight aggregate on shrinkage and pore structure of ultra-high strength concrete. Constr. Build. Mater. **204**, 276–287 (2019). https://doi.org/10.1016/j.conbuildmat.2019.01.205
25. Liu, X., Chia, K.S., Zhang, M.H.: Water absorption, permeability, and resistance to chloride-ion penetration of lightweight aggregate concrete. Constr. Build. Mater. **25**(1), 335–343 (2011). https://doi.org/10.1016/j.conbuildmat.2010.06.020

26. Elsharief, A., Cohen, M.D., Olek, J.: Influence of lightweight aggregate on the microstructure and durability of mortar. Cem. Concr. Res. **35**(7), 1368–1376 (2005). https://doi.org/10.1016/j.cemconres.2004.07.011
27. Agostini, F., Davy, C.A., Skoczylas, F., Dubois, T.: Effect of microstructure and curing conditions upon the performance of a mortar added with treated sediment aggregates (TSA). Cem. Concr. Res. **40**(11), 1609–1619 (2010). https://doi.org/10.1016/j.cemconres.2010.07.003
28. Ma, X., Liu, J., Shi, C.: A review on the use of LWA as an internal curing agent of high performance cement-based materials. Constr. Build. Mater. **218**, 385–393 (2019). https://doi.org/10.1016/j.conbuildmat.2019.05.126
29. Nie, S., Hu, S., Wang, F., Hu, C., Li, X., Zhu, Y.: Pozzolanic reaction of lightweight fine aggregate and its influence on the hydration of cement. Constr. Build. Mater. **153**, 165–173 (2017). https://doi.org/10.1016/j.conbuildmat.2017.07.111
30. Nie, S., Zhang, W., Hu, S., Liu, Z., Wang, F.: Improving the fluid transport properties of heat-cured concrete by internal curing. Constr. Build. Mater. **168**, 522–531 (2018). https://doi.org/10.1016/j.conbuildmat.2018.02.068
31. Angelin, A.F., Cecche Lintz, R.C., Osório, W.R., Gachet, L.A.: Evaluation of efficiency factor of a self-compacting lightweight concrete with rubber and expanded clay contents. Constr. Build. Mater. **257**, 119573 (2020). doi:https://doi.org/10.1016/j.conbuildmat.2020.119573
32. Fathi, M., Yousefipour, A., Hematpoury Farokhy, E.: Mechanical and physical properties of expanded polystyrene structural concretes containing Micro-silica and Nano-silica. Constr. Build. Mater. **136**, 590–597 (2017). doi:https://doi.org/10.1016/j.conbuildmat.2017.01.040
33. Maaroufi, M., Abahri, K., El Hachem, C., Belarbi, R.: Characterization of EPS lightweight concrete microstructure by X-ray tomography with consideration of thermal variations. Constr. Build. Mater. **178**, 339–348 (2018). https://doi.org/10.1016/j.conbuildmat.2018.05.142
34. Bouvard, D., et al.: Characterization and simulation of microstructure and properties of EPS lightweight concrete. Cem. Concr. Res. **37**(12), 1666–1673 (2007). https://doi.org/10.1016/j.cemconres.2007.08.028
35. Aslani, F., Deghani, A., Asif, Z.: Development of lightweight rubberized geopolymer concrete by using polystyrene and recycled crumb-rubber aggregates. J. Mater. Civ. Eng. **32**(2), 1–16 (2020). https://doi.org/10.1061/(ASCE)MT.1943-5533.0003008

Durability Design and Quality Control/Assurance During Construction of Concrete Structures

Andrés Antonio Torres-Acosta(✉)

Tecnologico de Monterrey, School of Engineering and Science, Department of Sustainable Technologies and Civil, Queretaro Campus, Santiago de Querétaro, Queretaro, Mexico
atorresa@tec.mx

Abstract. The procedures performed for durability quality control/quality assurance (DQC/QA) of concrete during the construction of a bridge exposed to a tropical marine environment is presented. The methodology includes the use of one durability index to estimate the transport properties of aggressive agents in concrete, called saturated electrical resistivity (ρ_S). A second index commonly used internationally to determine the concrete's typical QC/QA evaluating its compressive strength (fc) was also monitored. Both DQC/QA and QC/QA were made using standardized 10 × 20cm cylinders fabricated during the 2-year construction of the example bridge. The results obtained demonstrated that the DQC/QA for concrete should not be based solely on mechanical strength tests (fc), but also on physical tests that determine the ease (or difficulty) that the newly manufactured concrete could present to the transport of aggressive ions within. The statistical analysis performed showed that there was a greater variability in the results of some with ρ_S, using standardized cylinders, as compared to low variability obtained from fc QC/QA data obtained with same cylinders.

Keywords: Durability design · Electrical resistivity · Quality control/assurance · Rapid chloride permeability · Self-compacting concrete

1 Introduction

Ever since the industrial use of concrete started at the beginning of the last century, the most important test that has been carried out to this day, for its quality control/quality assurance (QC/QA) is the test known as the uniaxial compressive strength (fc). In any concrete quality specification, the fc is included (commonly determined at different concrete ages), and the most important being the 28-days strength, which is named worldwide as the concrete's f'c. The procedure followed for concrete's QC/QA manufacture in a typical construction site, is based on keeping a record of fc at different ages. This fc monitoring is performed by manufacturing concrete cylinders (or cubes) throughout the construction period and see possible fc values fluctuations, and thus make the necessary mixture proportion corrections to return to the project fc value.

J. I. Escalante-Garcia et al. (Eds.): RW 2021, RILEM Bookseries 40, pp. 499–509, 2023.
https://doi.org/10.1007/978-3-031-21735-7_55

However, it is increasingly common for construction professionals to be more interested in durability issues, rather than just the mechanical performance of concrete. This is largely since it is more common to encounter relatively new concrete structures (< 10 years in service) that show some damage and require corrective maintenance early in its service life. This interest in the manufacture of more durable concretes has changed paradigms to ensure and control the quality of these concretes based on durability performance instead of fc only. Thus, the need to use some durability indices that may help determine the internal transport properties of aggressive agents that concretes may have, and with these indices check if the fabricated concrete will be durable enough when are place in service.

Durability QC/QA (DQC/QA) for concrete structures initiates during material's selection. There are international regulations that recommend the material selection for a concrete mixture based on environment exposure [1]. The first stage for durability DQC/QA is the design of the concrete mixture based on durability performance. This methodology starts with the physical/chemical evaluation of the raw materials (cement, water, aggregates, additions, etc.) to be used for the manufacture of the durable concrete.

After reviewing that the raw materials for the manufacture of durable concrete meet the requirements, described in the durability regulations, the definition of the concrete design mix must be obtained before the concrete element manufacturing activities begins. Concrete mix design is based on minimum durability performance for hardened concrete according to the specifications defined in the executive project based on national or international standard recommendations [1].

This stage of designing the appropriate mixture must be carried out before the construction work begins, and this evaluation could be extended to a period greater than 90 days: performance tests of certain durability indices such as saturated electrical resistivity (ρ_S) and rapid chloride permeability (RCP) regularly recommend the need to be carried between 56 and 90 days after fabrication [2–4]. For this reason, it is recommended that the owner, external supervision, and construction company give this activity the needed time to prepare the concrete design mix prior to the start of industrial concrete manufacturing on site. Once the design mix has been met for durability performance (i.e. meets durability index values such as ρ_S and RCP), the start of concrete fabrication on site is approved and the works may be initiated [1–3].

During DQC/QA for durable concrete, close monitoring must be given to the process of dosing, manufacturing, transport, placement, consolidation, curing, and concrete cover compliance of the elements to be manufactured.

2 Research Significance

This new methodology to apply DQC/QA using durability indices during the mixture design and fabrication on site, has increasingly attracted the attention of the construction industry, but little information is available in the literature to demonstrate if typical durability indices might be used to follow a DQC/QA methodology with reliable results. That is why this investigation presents the experiences obtained in the construction of a durable bridge (located in a tropical marine environment) and the use a durability index (ρ_S) and one mechanical strength index (fc) to obtain durable concrete manufacture [1–4].

3 Experimental Procedure

3.1 Durable Concrete Mix Design

The first step before construction started, was to characterize al the concrete's components that would be used in the concrete manufacture: aggregates, cement, water, chemical and mineral admixture.

Tests were then performed to define the most appropriate concrete mixture that would meet the requested performance, which in this project were defined as saturated electrical resistivity (ρ_S). The requested performance values for both indices, defined in the executive project, was $\rho_S > 50$ kΩ cm [2, 3]. Information of the concrete mixture characteristics was presented in a previous investigation [4] and summarized in Table 1. With this value, laboratory tests were performed with mixtures manufactured at the same site, to simulate the manufacturing procedures directly in place. Once the cylinders were manufactured, they were sent to an external laboratory that could guarantee the results for a period of >90 days, since the values of ρ_S index, were specified at 90 days. More details of the results of this evaluation can be found in previous investigations [2, 3].

Table 1. Concrete mixture characteristics [4].

Material	Properties
Portland cement	Clinker content >95%; Blaine >4,000 cm^2/g; cementitious content >450 kg/m^3
Water to cement (w/c) ratio	<0.35
Coarse aggregate	MAS: 19 mm; specific gravity >2.4; Los Angeles wearing machine <40%
Fine aggregate	Specific gravity >2.4
Water	Cl^- content <90 ppm
Silica fume	>5%, but <10%
Electrical wet resistivity	>40 kΩ cm
Rapid chloride permeability	<800 Coulombs

3.2 QC/QA for Compressive Strength (fc)

For on-site concrete's QC/QA, fc tests were carried out using cylinders (10 × 20 cm) manufactured during almost two years (lasting time of construction of the bridge's second stage period). The concrete for the manufacture of these standardized cylinders was obtained directly from the concrete truck mixer during the production of AASHTO Type IV beams (AT4), Box Girder beams (BG), and cast in-place Over Pile Caps (OPiC) structural elements from the same bridge during second stage construction period (superstructure construction stage). fc results were obtained at 1, 3, 7, 14, and 28 days of age. No long-time fc performance testing was considered in this strength QC/QA procedure, because the interest was focused on early-age fc for strand release, demolding, and transporting the prefabricated elements to the curing area.

3.3 DQC/QA for ρ_S and RCP Indices

For DQC/QA, same 10 × 20 cm concrete cylinders fabricated for strength QC/QA were used to evaluate the durability index ρ_S. ρ_S measurements were made at various ages of the cylinders ranging from 1 to 400 days. The procedures are found in previous investigations [2–4] and are detailed as follows.

ρ_S has been widely used over the last twenty years to characterize concrete quality and durability [2, 4–10]. Determining ρ_S involves a simple technique that quickly measures the concrete's saturated electrical resistivity value of a concrete specimen, and it is cost effective given the low cost of the required equipment and ease to perform [5, 6].

The cylinder is removed from the saturating pond or chamber (hence the name saturated electrical resistivity, ρ_S), wet sponges placed on either end, and metal plates (normally a corrosion-resistant metal such as stainless steel or bronze) placed on either end touching the sponges. Voltage is then applied through the two plates and the resulting ionic current measured. Equipment voltage (E) and current (I) values are used to calculate concrete's electrical resistance (Re) between the plates from Ohms Law: Re = V/I. The resulting value is multiplied by a geometric constant: specimen area (A) and divided by the specimen length (L). Measurement of each specimen requires at most from one to two minutes, making it rather time efficient.

After gathering all test data from both QC/QA testing (strength and durability), empirical correlations were made between the three indices: RCP vs ρ_S and fc vs ρ_S. Finally, an analysis was performed to obtain typical statistical parameters (mean, μ; variance, σ; and variation coefficient, $V = \sigma/\mu$) to compare the reproducibility and easiness for each of the QC/QA procedures used (strength and durability) during the construction process of the bridge example presented.

4 Results and Discussion

4.1 Concrete's Compressive Strength (fc)

QC/QA methodology during concrete fabrication in a construction job is performed regularly by monitoring fc in time. Typical fc control chart, obtained with the standardized cylinders (10 × 20 cm dimension) in this investigation, is shown in Fig. 1 for AT4 beams. Similar performance was observed form control charts of the other structural element's strength QC/QA (BG beams and OPiC elements). The test dates for these control charts were 1, 3, 7, 14 and 28 days. These dates could not be extended to 90 days, such as the monitoring of the durability indices ρ_S and RCP performed with same standardized cylinders, so the information obtained in this work does not adequately characterize the final fc of the concrete, and only helped with strength QC/QA at the typical age of 28 days used in any construction work where durability is not an evaluation parameter.

As can be seen in Fig. 1 control charts, values increase by increasing the test age, and a decrease in the Δfc (fc difference between two consecutive ages) appeared between 14 and 28 days, indicative that fc may started reaching a plateau value.

Reviewing the fc values obtained for concretes used for AT4 beams (Fig. 1) showed a gradual decrease at the age of 1 day at the beginning of the implementation of these strength QC/QA's on site, obtaining values of 350–370 kg/cm^2 initially (samples 1 to 50) and after a time, it decreased to values of 300–350 kg/cm^2 (samples 51 to 78). This reduction in fc was corrected from sample 79 and remained very close to 400 kg/cm^2 on average at 1 day of age. The fc values for the other ages did not show significant changes.

In the case of the BG beams (control chart not shown), this decrease was also observed very early in the fc data at the age of 1 day (between samples 1 and 40), which was also corrected. At older ages the concrete did not show important differences in the data for ages >14 days. For OPiC fc values (control chart not shown), no significant differences were observed, so it was the structural elements that had a more uniform concrete.

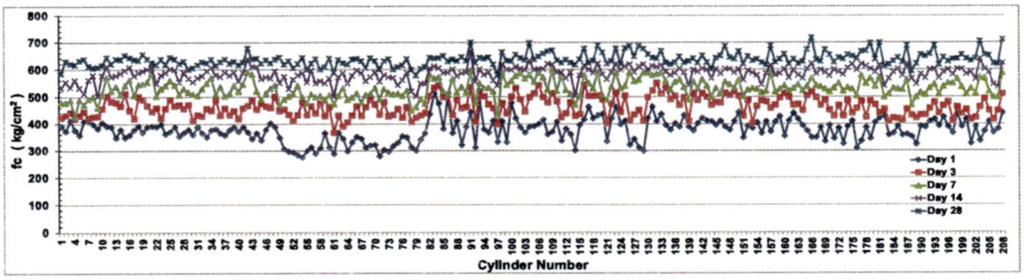

Fig. 1. fc control chart obtained with the standardized cylinders from concretes used to manufacture AT4 beams at the ages of 1, 3, 7, 14 and 28 days.

4.2 Saturated Electrical Resistivity (ρ_S)

Typical ρ_S results obtained with standardized cylinders (10 × 20cm in dimension) are shown in Fig. 2 as a function of time, for the structural elements AT4 beams. Similar performance was observed form the other structural element's DQC/QA (BG beams and OPiC elements). From this figure it is clear to observe that the ρ_S varies with time, until reaching constant values from ages >90 days. It should be noted that these results were obtained from the materials quality control laboratory, which was on-site. Inside this laboratory there was no temperature control of the environment, which may cause the observed ρ_S measurement variation at the time of the measurement for ages >90 days. [9, 10].

Figure 3 shows a composite plot of ρ_S versus time (Mixture trials #1, #2, and #3) and temperature versus time (discontinuous blue line) in the laboratory versus time for three concrete mix trials fabricated during the stage of concrete mix design, as reported from a previous investigation [10]. As observed from this figure, ρ_S is highly dependent of the temperature during measurement, especially for concretes with higher ρ_S values: AD2 concrete with $\rho_S < 40$ kΩ cm did not show broad changes, as compared to AD3 concretes (AD3-A and AD3-B) with $\rho_S >$ 50–70 kΩ cm. In the previous reference, ρ_S changes as a function of the inverse of the laboratory temperature (in K) were approximated using an Arrhenius equation. [10] Slopes for the three mixtures in Fig. 3 were obtained using

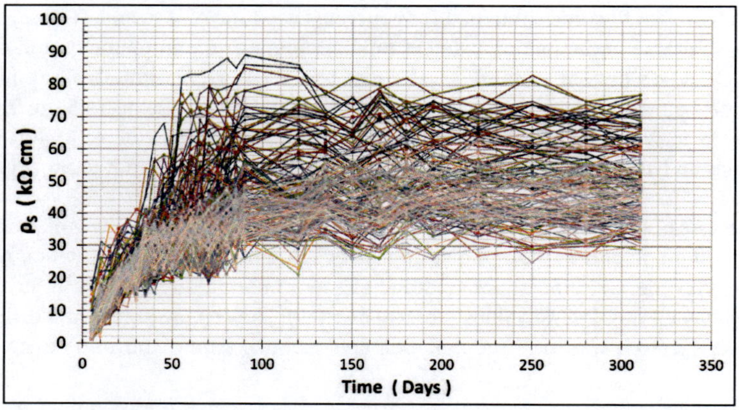

Fig. 2. ρ_S versus time for the standardized cylinders made with the self-consolidating concrete used for the construction of AT4 beams.

this approximation, giving as follows: AD3-A = 2,429.4; AD3-B = 2,498.7; and AD2 = 2,750.5 [10]. With these slopes the activation energy of the three concrete mixtures in Fig. 3 were obtained substituting theses values into the Arrhenius equation: AD3-A = −20.2 kJ/mol K; AD3-B = −20.8 kJ/mol K; and AD2 = −22.9 kJ/mol K [10]. As observed from the mixtures in Fig. 3, the higher is the ρ_S, the less negative is the activation energy, corroborating the difficulty of ion movement in such higher resistivity concretes.

With these results, the importance of the temperature effect on ρ_S measurements was confirmed, thus the measurement procedure in the on-site laboratory was modified, in such a way that air conditioning was recommended in the on-site laboratory or, as a more economical alternative, taking all measurements at very early hours, before dawn, and monitoring laboratory temperature to perform the corrections from the activation energies obtained and register ρ_S at 21 °C.

Important ρ_S variations were also observed in Fig. 2 for AT4 beams after reaching the constant value range (ages > 90 days): 25 kΩ cm < ρ_S < 80 kΩ cm. Similar performance was observed for concrete used in the fabrication of BG beams (28 kΩ cm < ρ_S < 70 kΩ cm) and for OPiC elements (30 kΩ cm < ρ_S < 65 kΩ cm). This may suggest that the ρ_S index may not be an adequate index for DQC/QA of concrete. Placing the dotted red line of Fig. 2 as the project ρ_S value (50 kΩ cm), is clear to observe that it divides the values of AT4 concretes by almost half, so it can be considered that half of the values met the design value.

4.3 fc vs ρ_S

Using the ρ_S and fc results obtained from the concrete cylinders in this project, a composite graph shown in Fig. 4 was generated (now fc is converted to MPa), where the fc–ρ_S relationship is shown as a function of the test age. Three different symbols were defined, one for each type of specimen used: AT4 beam = ○; BG beam = ◊; OPiC element = □.

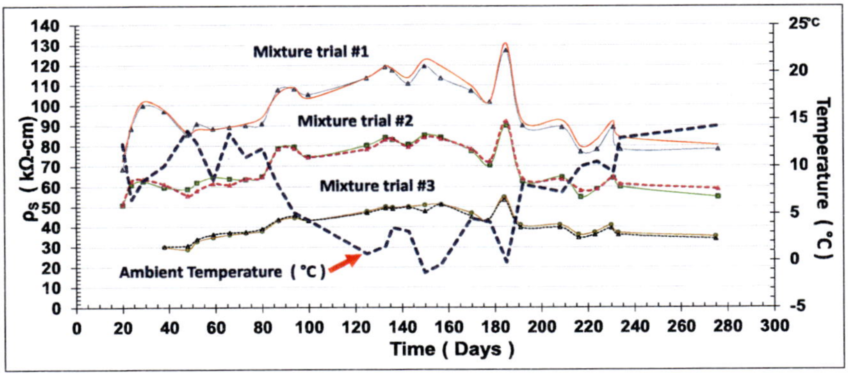

Fig. 3. ρ_S and ambient temperature versus time for the standardized cylinders manufactured with the self-consolidating concrete tested.

In this correlation it is observed that at a younger age, ρ_S values are less variable than fc values, and at older ages this behavior is reversed. The correlation equations between these two durability indices showed a correlation coefficient 0.6286, which corresponds to a fairly good one for a span of almost two years of continuous concrete manufacturing.

A fairly acceptable empirical correlation is obtained (ρ_S (kΩ cm) = 0.8566 fc (MPa)—35.026) with a correlation coefficient $R^2 = 0.6286$. This result would imply that ρ_S index could be calculated with enough approximation knowing the fc of the concrete or vice versa at <28-day age, obtaining the approximate value of the fc knowing the value of the ρ_S index for this self-compacting concrete evaluated.

4.4 Statistical Analysis of the Obtained Results

Once the data was collected from both QC/QA data (strength and durability), a statistical analysis was performed to compare the two test methods (fc, and ρ_S) to define which may be used for future applications. Data from fc and ρ_S were used as groups divided by structural element (AT4, BG, and OPiC) and age.

Data collected from every structural element (AT4, BG, OPiC) and age, was analyzed to obtain the best distribution function that fits the data, using a statistical software. The distribution function helped to determine the statistical parameters: mean (μ), standard deviation (σ), and variation coefficient ($V = \sigma/\mu$) for each of the collected data per structural element and age for fc and ρ_S. For RCP data, the statistical parameters were obtained just for structural element type, since not enough data was collected during construction due to testing costs.

As observed from Figs. 5 and 6, the three statistical parameters μ, σ, and V for fc (@28 days) varied: 60 MPa < μ < 65 MPa; 2 MPa < σ < 4 MPa; and 0.04 < V < 0.07. Same statistical parameter for ρ_S (@ >90 days) were: 48 kΩ cm < μ < 54 kΩ cm; 6 kΩ cm < σ < 10 kΩ cm; 0.12 < V < 0.20. Even though higher V estimates were obtained from ρ_S index as compared to fc index, in average the 50 kΩ cm requested in the executive project was achieved. This variations on ρ_S could be minimized if the

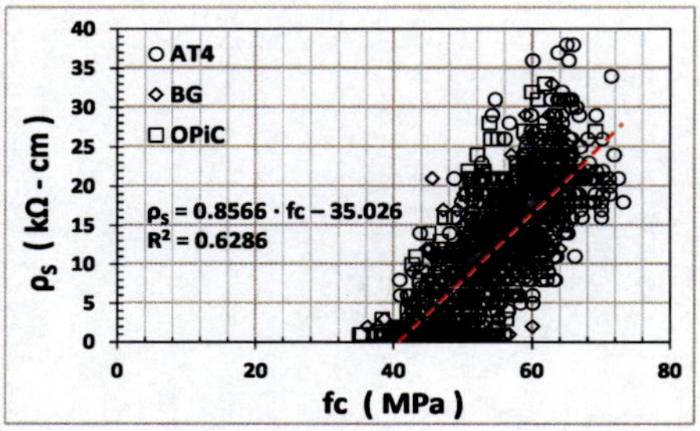

Fig. 4. Correlation between fc and ρ_S for the results obtained with all concrete cylinders tested (test age $\leq$ 28 days). Symbol difference corresponds to structural element type: AT4 beams = ○; BG beams = ◇; OPiC elements = □.

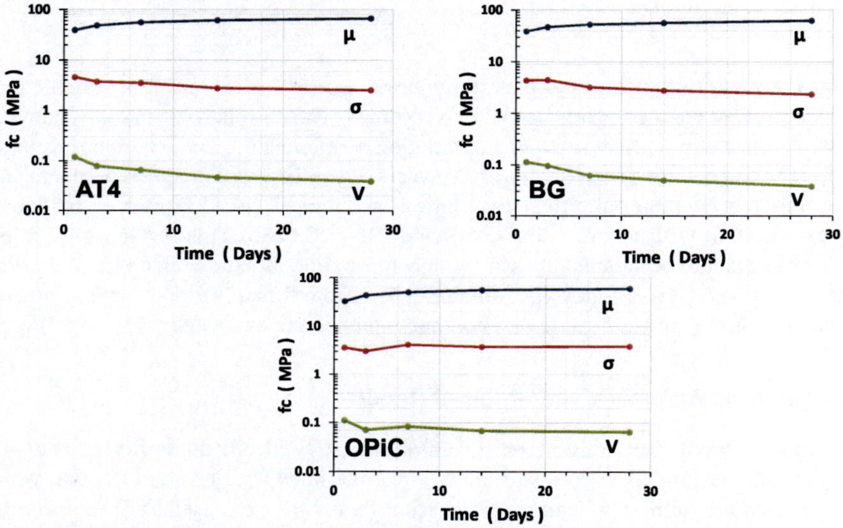

Fig. 5. Statistical parameters estimates (μ, σ, and V) for fc data vs time with the self-consolidating concrete evaluated.

testing temperature is controlled to 21 ± 2 °C in the field laboratory, where the tests are performed.

Figures 5 and 6 shows the statistical parameters μ, σ, and V for fc and ρ_S, respectively, as a function of testing age. For RCP data, the statistical parameters were as follows: AT4 beams: μ = 804.0 Coul; σ = 290.5 Coul; V = 0.36; BG beams: μ = 586.4 Coul; σ = 189.0 Coul; V = 0.32; OPiC elements: μ = 612.8 Coul; σ = 114.8 Coul; V = 0.19.

Finally, the statistical parameters obtained for RCP test data (and listed above) were quite disperse as compared to the variations of the other indices used (ρ_S and fc): 0.19 < V < 0.36. Still, the μ values were between 586 Coul—804 Coul and below the project's performance specification RCP < 1,000 Coul (only 9 out of 88 test data were above this threshold value).

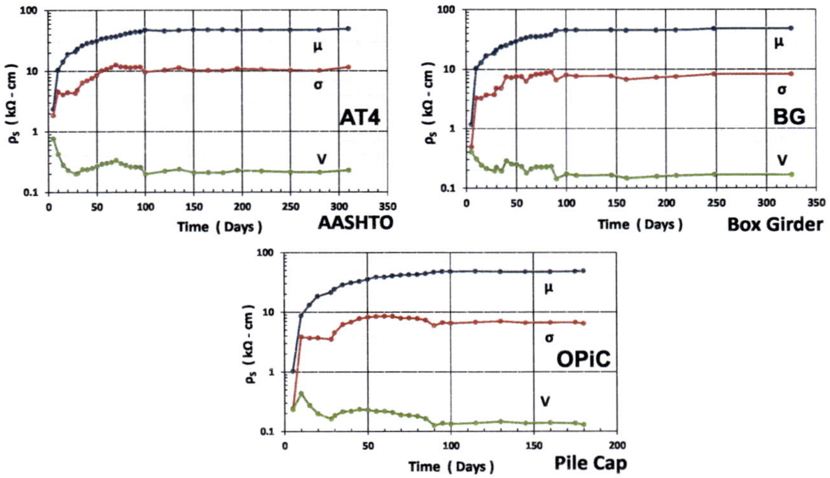

Fig. 6. Statistical parameters estimates (μ, σ, and V) for ρ_S data vs time with the self-consolidating concrete evaluated.

5 Conclusions

Based on the strength QC/QA and DQC/QA results obtained during the construction of a bridge located in a tropical marine area, the following conclusions were obtained:

1. Concrete mixture durability performance should be conducted prior to starting of concrete manufacturing activities on the job site. For this, it was shown that saturated electrical resistivity (ρ_S) tests could be adequate, due to its procedure easiness and because it is a non-destructive test that can be performed with the same specimen as many times as necessary. Special emphasis should be considered on the laboratory ambient temperature, because ρ_S is highly temperature dependent.
2. The next step for durable concrete achievement is to perform a durability quality control/quality assurance (DQC/QA) program during concrete industrial manufacturing throughout the job lasting. For this purpose, the use of the ρ_S durability index tests is also proposed, and compressive strength (fc) tests. These two tests can be performed using standardized 10 × 20 cm concrete cylinders.
3. QC/QA and DQC/QA proposed methodology using standardized 10 × 20 cm cylinders for fc, and ρ_S index was adequate, observing a greater variability for ρ_S results at ages >90 days and fc greater variability was at early ages (<28 days).
4. Care must be taken with the laboratory temperature for DQC/QA determining ρ_S index, since differences of up to 30% of the results obtained in a specimen could

be observed if there is a difference of only 10 °C between measurements. Results obtained from this investigation recommend control ambient temperature range in the testing laboratory of 21 ± 2 °C.

5. Statistical analysis performed with all data collected in this investigation for fc, ρ_S, and RCP showed low, medium, and high variability, respectively. The fc variability obtained was typical from a concrete QC/QA mechanical strength program during construction, and it was due to the fact that the laboratory in charge of this QC/QA mechanical testing had several years performing such tests. The ρ_S variability might be due to lack of experience of the field-testing laboratory in charge of the DQC/QA testing, starting with no temperature control in the laboratory.

Acknowledgements. The author thanks the participation of various institutions for the completion of this investigation: to the Tecnologico de Monterrey, School of Engineering, Queretaro Campus, for the preparation and presentation of this publication; to the port administration that owns this viaduct (API-Progreso) and who provided the resource for its construction and supervision; to the company that built this bridge (COCONAL SA de CV); and to the company that supervised it as external support to the port administration (Grupo TRIADA SA de CV). The support logistics of the Mexican Transportation Institute (IMT), a decentralized department of the Mexican Government's Secretariat of Communications and Transportation (SCT), as the former affiliation institution of the author during data collection is also appreciated. Finally, the author thanks for the research assistants, graduate and undergraduate students, who helped to collect all the data presented in this investigation.

References

1. NMX-C-530-ONNCCE: Construction Industry-Durability-General Standard for Reinforced Concrete Structures—Criteria and Specifications. Organismo Nacional de Normalización y Certificación de la Construcción y Edificación, S.C. (ONNCCE), Mexico City, Mexico (2018) (in Spanish)
2. Torres Acosta, A.A., Moreno Valdés, A., Lomelí González, M.G., Martínez Madrid, M., Díaz Cruz, L.A.: Durability of self-compacting concretes with low water-to-cement ratio and microsilica additions, Technical Publication No. 512. Instituto Mexicano del Transporte, Sanfandila, Queretaro, Mexico (2018) (in Spanish)
3. Torres-Acosta, A.A., Patrón-Solares, A., Alcantara-Lagunes, P. Durability health monitoring during construction of concrete structures in marine environment. Structuc Contr and Health Monit. **28**(3), e2674 (2021). doi:https://doi.org/10.1002/stc.2674
4. Torres-Acosta, A.A., Castro-Borges, P.: Durability control through design and health monitoring during construction. Structuc Contr and Health Monit. **28**(2), e2670 (2021). doi:https://doi.org/10.1002/stc.2670
5. Andrade, C.: Calculation of chloride diffusion coefficients in concrete from ionic migration measurements. Cement and Concrete Res. **23**, 724–742 (1993)
6. Sagüés, A.A., Kranc, S.C., Presuel-Moreno, F., Rey, D., Torres-Acosta, A.A., Yao, L.: Corrosion Forecasting for 75-year Durability Design of Reinforced Concrete, Final Report to Florida Department of Transportation, Contract Grant and Report No. BA502. Tallahassee, FL (2001)

7. Spragg, R., Bu, Y., Snyder, K., Bentz, D., Weiss, J.: Electrical testing of cement-based materials: role of testing techniques, sample conditioning, and accelerated curing. Publication FHWA/IN/JTRP-2013/28, Joint Transportation Research Program, Indiana Department of Transportation, and Purdue University, West Lafayette, Indiana (2013)
8. Torres-Acosta, A.A., Presuel-Moreno, F., Andrade, C.: Electrical resistivity as durability index for concrete structures. ACI Mater. J. **116**(6), 245–253 (2019)
9. Torres Acosta, A.A., Valdés, M., Rangel Reséndiz, J.D., Lomelí González, M.G., Martínez Madrid, M.: Relation between electrical resistivity, temperature during testing, aggregate area, and porosity in concrete cores, Technical Publication No. 547. Instituto Mexicano del Transporte, Sanfandila, Queretaro, Mexico (2019b) (in Spanish)
10. Torres Acosta, A.A., Moreno Valdés, A., Rodríguez Mendo, G., Lomelí González, M.G., Martínez Madrid, M.: Temperature effect on wet electrical resistivity in concrete cylinders, Technical Publication No. 543. Instituto Mexicano del Transporte, Sanfandila, Queretaro, Mexico (2019c) (in Spanish)

Flexural Behavior of Reinforced Concrete Beams Made with Recycled Concrete Aggregates

Marianna S. Sodré[1](✉), Maíra S. R. da Costa[1], Caroline S. Rangel[1], Mayara Amario[1,2], Marco Pepe[3,4], Enzo Martinelli[3,4], and Romildo D. Toledo Filho[1]

[1] COPPE/Federal University of Rio de Janeiro, Rio de Janeiro, Brazil
marianna.sodre@coc.ufrj.br
[2] POLI/Federal University of Rio de Janeiro, Rio de Janeiro, Brazil
[3] Department of Civil Engineering, University of Salerno, Salerno, Italy
[4] TESIS srl, Fisciano, SA, Italy

Abstract. The reuse of Construction and Demolition Waste (CDW) to produce new materials is an effective method for reducing the negative impacts from the inadequate waste disposition, in addition to bringing economic and environmental benefits to the industry. The use of Recycled Concrete Aggregate (RCA) in concrete is a technique widely disseminated, however, using this material for structural applications is a target of distrust among professionals in the construction sector. Therefore, the purpose of this study is to analyze the influence of using RCA to replace 100% of course natural coarse aggregates with size fraction 9.5–19 mm in the flexural behavior of reinforced normal-strength concrete beams (35 MPa). In addition to a reference mixture produced with only natural aggregates, a Recycled Aggregate Concrete (RAC) mixture was designed, according to the well-known Compressible Packing Model (CPM), using RCA from demolition waste. The mechanical behavior of the concrete mixtures was characterized by compressive and splitting tensile tests. A pair of flexural critical beams of each concrete mixture was tested under a supported four-point loading condition at the age of 28 days. All tested beams had 240 cm length and rectangular cross section of 20×30 cm^2. The experimental cracking, yielding and ultimate load of the beams were measured. Besides that, the load-strain behavior for concrete was obtained, as well as the load-deflection behavior. The results showed that the flexural behavior of RAC beams was not negatively affected by the presence of RCAs.

Keywords: Recycled concrete aggregate (RCA) · Recycled aggregate concrete (RAC) · Reinforced concrete beams · Normal-strength concrete · Flexural behavior

1 Introduction

The civil construction industry is responsible for an important segment of the world economy. However, its activities not only generate a huge amount of waste, but it also contributes to the degradation of the environment due to the generation and irregular disposal of the waste.

J. I. Escalante-Garcia et al. (Eds.): RW 2021, RILEM Bookseries 40, pp. 510–519, 2023.
https://doi.org/10.1007/978-3-031-21735-7_56

Thereby, considering the potential of the recycled aggregates in terms of environmental conservation, this material became a topic of high relevance for the scientific community [1]. However, based on the fact that several authors point out the high porosity as the main characteristic of recycled aggregates [2–4] and due to the Brazilian technical standard on recycled aggregate [5] that considers only its non-structural use, it is uncommon to use recycled aggregates in concretes higher than C15 grade in that country.

Arezoumandi et al. [6] produced recycled concrete beams of normal strength (35 MPa) using the Equivalent Mortar Volume (EMV) design method, with 100% volume replacement of natural coarse aggregates by RCA. The authors pointed out that, in terms of cracking, the behavior of the RAC beams was similar to the conventional beams, except for the smaller spacing between the cracks in RAC beams. In addition, the RAC beams presented a lower cracking moment, similar yielding, and higher ultimate deflection when compared to the conventional concrete beams.

Pradhan et al. [7] used the Particle Packing Method (PPM) as a mix-design along with Two Stage Mixing Approach (TSMA). Fourteen reinforced concrete beams were studied and the RAC was prepared by completely replacing natural aggregates by RCA. According to the authors, the cracking load was observed to be lower for RCA beams, which is attributed to the lower tensile strength of RAC. At the same time, very marginal difference in yield load and ultimate load was witnessed for RAC beams. The specimens, however, showed a greater number of cracks and higher crack growth rate.

Rangel et al. [8] replaced coarse aggregate by natural ones in concretes of C25 and C65 grades. Through a study of mechanical and structural properties, it was possible to evaluate the feasibility of using RCA at different levels for the production of structural concrete as long as special attention was given to the production of the aggregates and to the mixing design process.

Still considering the mix-design, Amario et al. [9] mention that a specific mix-design method should be used instead of just replacing the natural aggregates by recycled ones in volume. The authors demonstrated the applicability of the Compressible Packing Model (CPM) for concrete mixtures containing recycled aggregates and, after evaluating C25, C45 and C65 grades, they highlighted that with a rational use of RCA there is no loss in terms of mechanical performance and durability.

Since those recent studies demonstrated that it is possible to improve the characteristics of concrete with recycled aggregates through the use of specific mixing design methods like those demonstrated by Amario et al. [9] for example, it can be said that there is a gap regarding structural properties and behavior of reinforced recycled concrete. Thus, this work is dedicated to the analysis of the bending behavior of reinforced concrete beams produced with RCA.

2 Materials and Methods

2.1 Materials

The concrete mixtures were prepared using Portland cement (CP-V-ARI), superplasticizer additive "Glenium 51 BASF", clean river sand, natural coarse aggregate "Nat_B0" (grading zone 4.75 mm to 9.5 mm), natural coarse aggregates "Nat_B1" (grading zone 9.5 mm to 19 mm), recycled coarse aggregates originated from demolition waste (RCA_D_B1) (grading zone 9.5 mm to 19 mm).

Several characterization tests were carried out for the materials mentioned above. The superplasticizer used has specific gravity of 1087 kg/m^3 and solid content of 30%. The properties of the cement used are presented in Table 1.

Table 1. Characterization of Portland cement CP-V-ARI

Test	Properties
Specific gravity	3067 kg/m^3
Marsh cone flow—saturation point	0.4%
Experimental packing density	0.63
Compressive strength [9]—Age of 1, 3, 7 and 28 days	14.7, 23.6, 30.6 and 32.3 MPa

The aggregate type D was originated from a concrete demolition waste provided by a recycling plant located in São Paulo, Brazil. The particle sizes of all the aggregates are in accordance with NBR 7211 [10]. The total absorption and specific gravity were determined following NBR 16917 [11] for coarse aggregates and NBR 16916 [12] for fine aggregates. The characterization of the aggregates is presented in Table 2.

Table 2. Characterization of natural and recycled aggregates

Aggregate	Specific gravity (kg/m^3)	Total absorption (%)
Nat_B0	2618	1.7
Nat_B1	2645	1.1
RCA_D_B1	2287	5.8
Sand	2574	0.4

Circular section CA-50 ribbed steel bars of 12.5 diameters were used as longitudinal reinforcement of the concrete beams. The mechanical properties of the steel bars obtained from tensile test [13] are presented in Table 3.

Table 3. Mechanical properties of longitudinal reinforcing steel

Diameter (mm)	Yield stress (MPa)	Ultimate stress (MPa)	Elastic modulus (GPa)	Elongation (%)
12.5	527.2	684.2	174.3	12.8

2.2 Concrete Mix-Design

All the mixtures were designed by the Compressible Packing Model (CPM) [14], with the software *BetonLab Pro 3*, considering the individual properties of each material and the maximum packing density of each mixture. An individual mix design was performed for each mixture, instead of just replacing the natural aggregates by the recycled ones in mass or in volume. Two concrete mixtures of normal strength class (35 MPa) were designed: a reference one with only natural aggregates; and a recycled concrete mixture with 100% of RCA_D_B1. The mix-design is presented in Table 4.

Table 4. Concrete mix-design

Materials (kg/m^3)	Nat_B1	RCA_D_B1	Nat_B0	Sand	Cement	Total water	SP
C35-NAT	452	0	457	868	325	212	1.86
C35-D-B1	0	384	453	862	341	216	1.95

2.3 Experimental Program

The beams were designed based on *Norme Tecniche per le Construzioni* [15], which is based on Eurocode 2 [16]. Three 12.5 mm diameter bars were used as the bottom longitudinal reinforcement and two bars of the same diameter were used as the top longitudinal reinforcement. The transverse reinforcement had diameter of 8 mm spaced 160 mm apart, except in the middle of the beam, where the spacing was 320 mm.

The concrete was produced as follows: first, all the aggregates were added and mixed for one minute; second, half of the total water was added and mixed for one minute; then, the cement was added and again mixed for one minute; finally, the second half of the total water was added as well as the superplasticizer and all the materials were then mixed for 6 min. Then, the casting of beams of 20 × 30 × 240 cm^3 was performed. In addition, cylindrical specimens of 7.5 cm diameter and 15 cm height were produced for mechanical tests.

The slump test according to NBR 16889 [17] was carried out. The compressive strength test was performed according to NBR 5739 [18] at rate of 0.1 mm/min and tensile strength test according to NBR 7222 [19] at rate of 0.3 mm/min. The 4-point bending test was performed according to NBR 12142 [20] at 28 days of age. In this test, simply supported beams are loaded with two forces concentrated in the thirds of the span. The distance between the two load application points was 70 cm. The supports of the reinforced concrete beams were positioned 15 cm from the end of the beam, forming a free span of 210 cm. The set-up is illustrated in Fig. 1.

A linear variable differential transformer (LVDT) was used to measure the deflection at the beam center. Concrete strain-gages were attached to the bottom and to the top of the beam, at midspan. The load, deflection and strain for each gage were continuously measured using a data acquisition system until the beam reached failure. Figure 2

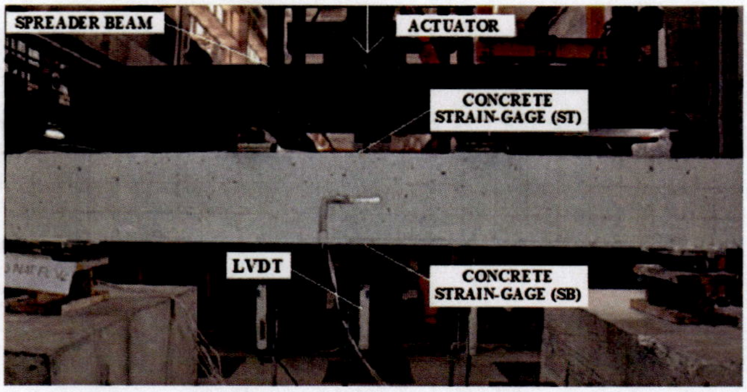

Fig. 1. Set-up of 4-point bending test

shows the details of tested beams, including the loading pattern, reinforcements, and instrumentations.

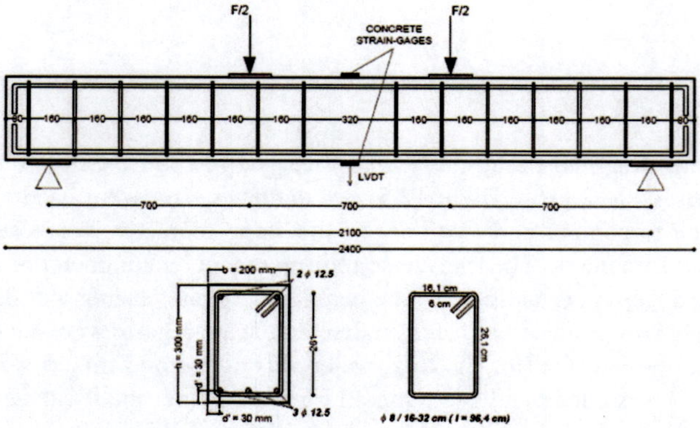

Fig. 2. Details of tested beams (not indicated units in mm)

3 Results and Analysis

3.1 Mechanical behavior of RAC

The concrete mixtures presented slump values of 130 mm ± 20 mm. Table 5 presents the means values and the coefficient of variations of 28-day compressive strength ($f_{c,28}$), strain at maximum stress ($\varepsilon_{c,28}$), elastic modulus ($E_{c,28}$), 28-day splitting tensile strength ($f_{t,28}$) and the relation between 28-day splitting tensile strength and 28-day compressive strength ($f_{t,28}/f_{c,28}$).

The compressive strength results obtained for both natural concrete and RAC were close to the target strength of 35 MPa. For the natural concrete, a variation of 4.6%

Table 5. Mechanical properties of concrete mixtures

Mixture	$f_{c,28}$ (MPa)	$\varepsilon_{c,28}$ ($\mu\varepsilon$)	$E_{c,28}$ (GPa)	$f_{t,28}$ (MPa)	$f_{t,28}/f_{c,28}$ (%)
C35-NAT	33.4 (±4.1%)	2972.8 (±7.0%)	22.4 (±5.3%)	2.2 (±10.0%)	6.5
C35-D-B1	32.4 (±4.0%)	3041.1 (±10.2%)	23.3 (±4.8%)	2.5 (±11.0%)	7.8

was observed compared to the predicted value of 35 MPa, while for the RAC mixture the variation was 7.4%. The reduction might have occurred due to the curing process performed with damp cloths over the specimens beside the beams, instead of wet chamber storing. Considering the small variation between the results, both in comparison with 35 MPa and between the different mixtures, it is possible to infer the efficiency of the scientific mix-design that was used, which is in agreement with the results obtained by Rangel et al. [8] and Amario et al. [9].

The strain at maximum stress results were also similar for the two mixtures, the RAC strain was only 2.3% higher than the natural concrete. Therefore, it can be concluded that there was no influence of the presence of recycled aggregates in this property. Likewise, the elastic modulus did not undergo major changes in mixtures with recycled aggregates.

The results of 28-day splitting tensile strength demonstrated that the RAC presented a better tensile strength (2.5 MPa) than natural concrete (2.2 MPa). On one hand, the relationship between splitting tensile strength and compressive strength was lower than expected, taking into account equations developed for conventional concrete. On the other hand, based on the equation proposed by Duan & Poon [21]—that is suitable for concrete with RCA—it can be stated that the values obtained in this study are in accordance with what is expected.

3.2 Flexural Behavior of RAC Beams

All specimens failed in a flexural ductile manner: in an early stage of loading, the initial flexural cracks appeared at midspan. As the load was increased, the flexural cracks propagated from the flexural region toward the supports, the formed cracks grew vertically, toward the compression zone, and some inclined flexure-shear cracks were observed. The tensile reinforcement yielded, and the beams deformed quickly. Finally, the failure occurred due to the crushing of concrete in the compression zone.

Figure 3 shows the crack pattern of the beams at failure. The number of cracks for RAC and conventional beams was similar. The RAC beams had 19 and 21 cracks for C35-D-B1-V1, and C35-D-B1-V2, respectively. The conventional beams had 20 and 22 cracks for C35-NAT-V1, and C35-NAT-V2, respectively. As mentioned before, flexure-shear cracks were observed in all tested beams. Therefore, the presence of recycled aggregate did not affect the number of cracks and the formation of flexure-shear cracks in RAC beams compared to conventional beams, similarly to that obtained by Arezoumandi et al. [6]. Table 6 presents the summary of flexural test results: cracking load (P_{cr}), yielding load (P_y), ultimate load (P_u), yielding deflection (δ_y) and ultimate deflection (δ_u).

The cracking load results were similar for both natural concrete and RAC, which is in agreement with the results obtained from the splitting tensile test since the presence of

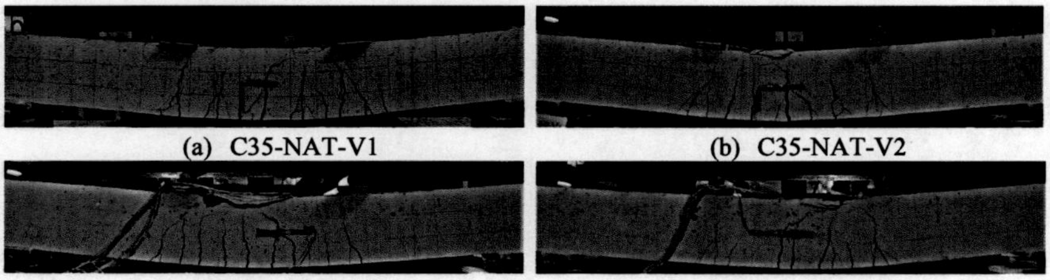

(a) C35-NAT-V1 (b) C35-NAT-V2

(c) C35-D-B1-V1 (d) C35-D-B1-V2

Fig. 3. Tested beams at failure

Table 6. Flexural test results

Beam identifier	P_{cr} (kN)	P_y (kN)	P_u (kN)	δ_y (mm)	δ_u (mm)
C35-NAT-V1	13.80	118.90	153.25	7.50	81.96
C35-NAT-V2	13.60	132.70	163.35	7.89	80.46
C35-D-B1-V1	12.80	128.90	160.35	8.87	78,78
C35-D-B1-V2	14.40	130.85	157.20	7.89	67.69

recycled aggregates had no negative impact on this property. These results are different from those obtained by Arezoumandi et al. [6] and Pradhan et al. [7]. The yielding load also presented similarities for all tested beams, except for C35-NAT-V1, whose result was significantly lower than the others. Excluding the results of C35-NAT-V1 for yielding load, it is possible to affirm the non-influence of the use of recycled aggregates in this property as obtained by Arezoumandi et al. [6] and Pradhan et al. [7]. The non-influence of the use of recycled aggregates in the ultimate load was also observed due to the similarity between the results obtained for both, RAC and conventional beams, which is in agreement with the results obtained by Pradhan et al. [7].

Figure 4 shows the load-midspan deflection behavior for each tested beam. As indicated in Table 6 as well, the beams had a similar behavior both in the linear part of the graph and in the non-linear part, except for C35-NAT-V1 yielding load and C35-D-B1-V2 ultimate deflection. However, since the ultimate deflection obtained was below others, it can be said that the influence of the recycled aggregate was not negative, differently from the results obtained by Arezoumandi et al. [6].

Figure 5 shows the load-concrete strain behavior for the strain-gages attached to the bottom (SB) and to the top (ST) of the beam. The beams presented almost the same behavior in the compression zone, which was expected due to the compressive strength test results. The concrete strains on the tension side were similar until de cracking load. After this point, under the same load, the concrete strain of RAC beams on the tension side tended to be lower than natural concrete beams.

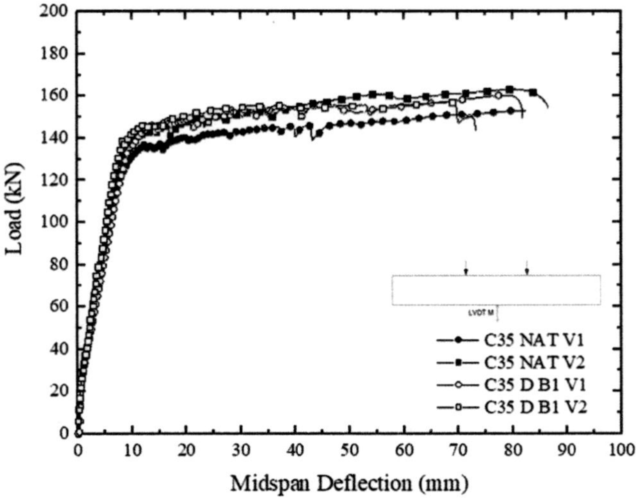

Fig. 4. Load-midspan deflection behavior for tested specimens

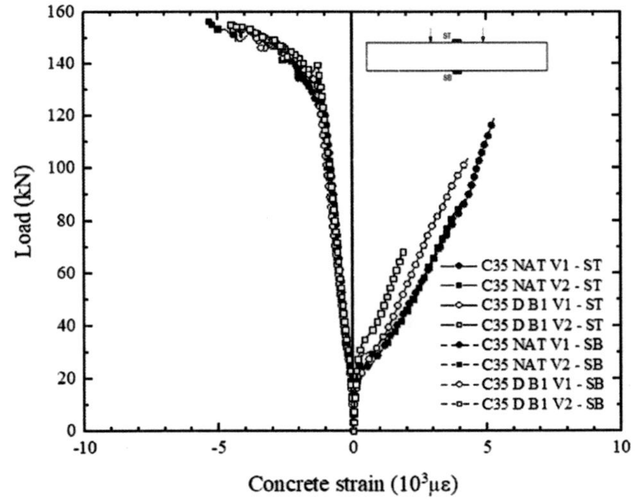

Fig. 5. Load-concrete strain behavior for tested specimens

4 Conclusion

This study reported the results of some experimental tests carried out on reinforced concrete beams of normal strength including RCA in replacement of the coarse natural aggregates. The following comments can be remarked:

- The use of recycled concrete aggregate did not affect the workability and the mechanical properties of the concrete mixtures.
- The presence of recycled aggregate did not affect the number of cracks and the formation of flexure-shear cracks in RAC beams compared to conventional beams.
- The flexural behavior of RAC beams was not negatively affected by the presence of recycled aggregates. Similar values were obtained for cracking, yielding and ultimate load for both natural concrete beams and RAC beams. The ultimate deflection, however, was lower for RAC beams than for the conventional ones.
- The load-concrete strain behavior was very similar in the compression zone for all tested beams. On the tension side, the concrete-strain behavior was similar for all beams until the cracking load. After this point, under the same load, the concrete strain of RAC beams tended to be lower than natural concrete beams.

Finally, the results of this study reinforce that, when using an adequate scientific mix-design, the presence of RCA did not negatively affect the mechanical and flexural behavior of recycled aggregate concrete in comparison to natural concrete.

References

1. Thomas, C., De Brito, J., Cimentada, A., et al.: Huda Macro—and micro-properties of multi-recycled aggregate concrete. J. Clean. Prod. **245**, 1–19. Elsevier (2020)
2. Carasek, H., Girardi, A.C.C., Araújo, R.C. et al.: Estudo e avaliação de agregados reciclados de resíduo de construção e demolição para argamassas de assentamento e de revestimento. In: Revista Cerâmica, vol. 64, pp. 288–300 (2018)
3. Sidorova, A., Vazquez-Ramonich, E., Bizinotto, B. et al.: Study of recycled aggregates nature's influence on the aggregate-cement past interface and ITZ. In: Construction and Building Materials, vol. 68, pp. 677–684 (2014)
4. Angulo, S.C., Figueiredo, A.D.: Concreto com agregados reciclados. In: Concreto: Ciência e Tecnologia, 2 ed., cap.47. Instituto Brasileiro do Concreto – IBRACON, São Paulo (2011)
5. Associação Brasileira de Normas Técnicas: NBR 15116—Recycled aggregate of solid residue of building constructions—Requirements and methodologies. Rio de Janeiro (2004)
6. Arezoumandi, M. et al.: An experimental study on flexural strength of reinforced concrete beams with 100% recycled concrete aggregate. In: Engineering Structures, vol. 88, pp. 154–162 (2015)
7. Pradhan, S., Kumar, S., Barai, S.V.: Performance of reinforced recycled aggregate concrete beams in flexure: experimental and critical comparative analysis. Mater. Struct. **51**(3), 1–17 (2018). https://doi.org/10.1617/s11527-018-1185-0
8. Rangel, C.S. et al. Flexural behavior of RC beams made from normal and high-strength recycled aggregate concrete. In: 4th Workshop on the New Boundaries of Structural Concrete, Anacapri (2016)
9. Amario, M. et al.: Optimization of normal and high strength recycled aggregate concrete mixtures by using packing model. In: Cement and Concrete Composites, vol. 84, pp. 83–92 (2017)
10. Associação Brasileira de Normas Técnicas: NBR 7211—Aggregates for concrete—Specification. Rio de Janeiro (2009)
11. Associação Brasileira de Normas Técnicas: NBR 16917—Coarse aggregate—Determination of density and water absorption. Rio de Janeiro (2021)

12. Associação Brasileira de Normas Técnicas: NBR 16916—Fine aggregate—Determination of density and water absorption. Rio de Janeiro (2021)
13. Associação Brasileira de Normas Técnicas: NBR ISO 6892-1—Metallic materials—Tensile testing Part 1: Method of test at room temperature. Rio de Janeiro (2013)
14. De Larrard, F.: Concrete Mixture Proportioning: A Scientific Approach. CRC Press, Series: Modern Concrete Technology (1999)
15. Decreto del Ministero delle Infrastrutture e dei Trasporti del 17 Gennaio 2008—Norme Tecniche per le Costruzioni—Supplemento Ordinario n. 8 della Gazzetta Ufficiale della Repubblica Italiana del 20 Febbraio 2018 (In italian)
16. EUROCODE 2: Design of concrete structures. Part 1.1: General Rules and Rules for Buildings (2004)
17. Associação Brasileira de Normas Técnicas: NBR 16889—Fresh Concrete—Slump Test. Rio de Janeiro (2020)
18. Associação Brasileira de Normas Técnicas: NBR 5739—Concrete—Concrete Test of Cylindrical Specimens. Rio de Janeiro (2018)
19. Associação Brasileira de Normas Técnicas: NBR 7222—Concrete and mortar—Determination of the Tension Strength by Diametral Compression of Cylindrical Test Specimens. Rio de Janeiro (2011)
20. Associação Brasileira de Normas Técnicas: NBR 12142—Concrete—Determination of Tension Strength in Flexure of Prismatic Specimens. Rio de Janeiro (2010)
21. Duan, Z.H., Poon, C.S.: Properties of recycled aggregate concrete made with recycled aggregates with different amounts of old adhered mortars. In: Materials & Design, vol. 58, pp. 19–29 (2014)

Analysis of Design Constitutive Model for Macro-synthetic Fibre Reinforced Concrete Through Inverse Analysis

Eduardo Galeote[1](✉), Alejandro Nogales[1,2], and Albert de la Fuente[1]

[1] Department of Civil and Environmental Engineering, Universitat Politècnica de Catalunya (UPC), Jordi Girona 1-3, 08034 Barcelona, Spain
eduardo.galeote@upc.edu
[2] Smart Engineering Ltd., UPC Spin-Off, Jordi Girona 1-3, 08034 Barcelona, Spain

Abstract. The increasing use of fibre reinforced concrete (FRC) as a structural material requires a constant development of new tools for design purposes. In this line, different codes and guidelines include constitutive models for FRC in their specifications. However, these constitutive models are usually based on the behaviour of steel fibre reinforced concrete (SFRC) and present several limitations on representing the full behaviour of macro-synthetic fibre reinforced concrete (MSFRC). With the aim of analysing the suitability of the constitutive models when applied on macro-synthetic fibres, an experimental program to determine the performance of MSFRC using different fibre contents and concrete mixes was performed. The study of the constitutive models was conducted through an inverse analysis using a two-dimensional analytical model to verify and compare the FRC constitutive equations of the MC2010 with the experimental results. The constitutive model of MC2010 was selected given that it is able to capture both the softening and hardening behaviour of FRC through a three-point constitutive equation. The results show that the original constitutive model of the MC2010 tends to overestimate the performance of PFRC. Moreover, it could be necessary to include an additional point on the post-cracking branch of the constitutive model to represent more accurately the full residual strength of PFRC. The results also show that the ultimate strain of the constitutive model should be extended to fully capture the performance of PFRC, especially for large deformations. This study suggests that it should be necessary to redefine the constitutive models through a new approach distinguishing between FRC blended with steel or macro-synthetic fibres.

Keywords: Fibre reinforced concrete · Inverse analysis · Macro-synthetic fibres · Constitutive model

1 Introduction

The use of synthetic polypropylene fibres for fibre reinforced concrete (FRC) has been attracting the attention of designers and engineers during the last years. In this line, previous studies [1, 2] have analysed the use of macro-synthetic fibres in FRC for structural

J. I. Escalante-Garcia et al. (Eds.): RW 2021, RILEM Bookseries 40, pp. 520–529, 2023.
https://doi.org/10.1007/978-3-031-21735-7_57

purposes. This has also been motivated by the presence of design constitutive equations for FRC in different codes and guidelines [3], which has led to an increasing use of fibres as reinforcement for structural elements. However, most of these constitutive models are based on studies conducted on concretes reinforced with steel fibres, this leading to certain uncertainty in case of using fibres made of other materials.

For this reason, it is necessary to develop modelling strategies that take into consideration the singularities of synthetic fibres so these can be used for designing safe and sustainable structures. Based on these principles, the present study aims at evaluating the suitability of these constitutive equations for its application on synthetic fibres. Accordingly, the constitutive equation for FRC of the Model Code 2010 [4] was selected given that is based on a performance-based approach and is able to identify a softening or hardening post-cracking behavior.

To analyse the suitability of the constitutive equation of the MC2010 for its use with polypropylene fibres, an experimental program including three different types of concretes with three polypropylene fibre contents each was conducted. Numerical simulations based on an inverse analysis were also conducted to analytically determine the post-cracking behavior of polypropylene FRC. The experimental results of flexural three-point bending tests have been compared to analytical Strength-CMOD curves obtained through inverse analyses to assess the accuracy of the constitutive model.

The results showed the necessity of modifying the parameters of the constitutive equation of the MC2010 for its use with macro-synthetic fibres. Through an iterative procedure, these parameters were modified until the analytical curves fit the experimental results, also revealing that the incorporation of an additional point for the definition of the constitutive equation would lead to improved analytical results. This study shows that constitutive models should reflect the singularities of each type of reinforcement given their different behavior and properties.

2 Experimental Program

An experimental program based on nine mixes of polypropylene fibre reinforced concrete (PFRC) was conducted. The concrete mixes (Table 1) had strength classes of C30/37 (with contents of fibres 2.5, 3.5 and 5.0 kg/m^3), C40/50 (with fibres in contents of 5.0, 7.5 and 10.0 kg/m^3) and C50/60 (fibres in contents of 5.0, 7.5 and 10.0 kg/m^3). For each concrete class, the content of superplasticizer was modified for workability purposes. The characteristics of the polypropylene fibres used as reinforcement are shown in Table 2.

The flexural strength was determined by means of three-point bending tests (3PBT) conducted on casted specimens of 150 × 150 × 600 mm [5]. The average flexural strength-CMOD curves for each concrete mix, gathered by concrete class, and for the different fibre contents are presented in Fig. 1.

As expected, the results exhibit that flexural strength f_{LOP} slightly increased with concrete strength class, while increasing the content of fibres also increased the residual strengths. However, it should be highlighted that, for concrete classes C40/50 and C50/60, increasing the content of fibres from 7.5 to 10 kg/m^3 did not significantly improve the residual strength.

Table 1. Mix compositions.

Material	C30/37	C40/50	C50/60
CEM II/A-L 42,5R	309	400	500
0–2 Sand	205	154	–
0–4 Sand	873	918	945
10–20 Gravel	757	760	750
Polypropylene fibre	2.5—3.5—5.0	5.0—7.5—10.0	5.0—7.5—10.0
Water	165	175	185
Plasticizer (% o.c.w)	0.50	0.84	0.80
Superplasticizer (% o.c.w)	0.83—1.03—1.05	0.81—0.98—1.05	0.62—0.68—0.82

Table 2. Characteristics of the polypropylene fibres.

Property	Characteristics
Material	Polypropylene
Fibre shape	Continuous embossing
Length [mm]	48
Tensile strength [MPa]	640
Elastic modulus [GPa]	12

3 Inverse Analysis

An inverse analysis based on an analysis of evolutionary sections (AES) [6] with a multi-layer approach was conducted to simulate the Strength-CMOD curves to analytically obtain the experimental results of the beams tested under three-point bending tests. For this, it is assumed that (i) sections remain plane after loading or imposed strains, (ii) there is strain compatibility, in which there is a perfect bond between concrete and fibres, (iii) shear stress strains are negligible and not considered and (iv) the internal forces are applied on the symmetrical axis of the section.

To conduct the analytical model, the constitutive law for FRC of the MC2010 was taken as reference. This constitutive model (Fig. 2) is able to represent both the hardening and softening behavior of FRC, and is based on a stress–strain (σ–ε) relationship, in which stresses σ are determined through residual strengths (f_{R1} and f_{R3}) obtained and strains (ε_{SLS} and ε_{ULS}) from crack openings of 0.5 and 2.5 mm, with all parameters obtained from three-point bending tests. The characteristic length, which is the parameter that relates the crack opening and the strain ($w = \varepsilon \cdot l_{cs}$), was defined to be equal to the depth of the beam ($l_{cs} = h$), despite several authors proposing different values for this parameter [8].

The inverse analysis is based on a discretization of the element into layers along its height. Considering that tensile stresses are located at the bottom and compression

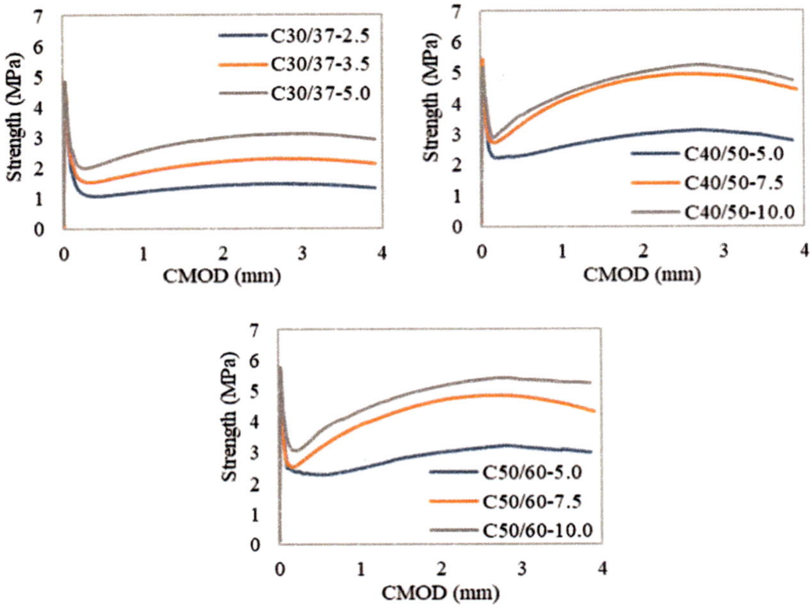

Fig. 1. Average Strength-CMOD curves for PFRC.

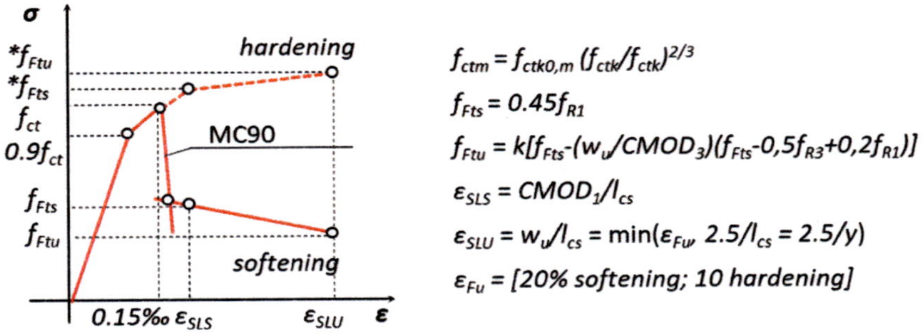

Fig. 2. Average strength-CMOD curves for PFRC.

stresses are placed at the top part of the section, a tensile strain ε_{bot} value is imposed for the initial stage, while a compressive strain at the top (ε_{top}) is iterated so the curvature χ and the strain at any layer may be calculated until the equilibrium condition between the internal and the external forces and momentums are satisfied.

Each time the iterative process to determine ε_{top} is completed, ε_{bot} is progressively increased until the iterated result of ε_{top} achieves the ultimate compressive strain ε_{cu}. While increasing the value ε_{bot}, so does the crack width at the section. When the strain of the concrete at the top layer of the section achieves the ultimate strain ($\varepsilon_c = \varepsilon_{cu}$), the area under compression collapses by crushing. At this stage, the iterative process stops and the Strength-CMOD curve for all the iterated results are determined.

4 Analysis of Results

The parameters of the constitutive model of the MC2010 were calibrated through an iterative process so the analytical results fit the experimental curves, obtaining two modified constitutive models based on 3- and 4-point equations. A comparison between the constitutive equation of the MC2010 and the 3- and 4-point adjusted equations are shown from Figs. 3, 4, 5. The results also show the Strength-CMOD curves of the experimental results and the analytical curves obtained through the MC2010 and the 3- and 4-point calibrated equations.

In line with previous studies [3, 9], the results indicate that the constitutive model of the MC2010 overestimates the residual strength for crack openings below 1.5 mm. Such overestimation might be attributed to the use of polypropylene fibres, given that the proposed constitutive equation of the MC2010 is mainly based on data and studies made of FRC with steel fibres.

The results also indicate that the residual strength of FRC with polypropylene fibres exhibit a sudden load drop after cracking, although after fibre activation the strength increases reaching in all cases a maximum residual strength at CMODs between 2.5 and 3 mm. This means that polypropylene fibres can still provide residual strength beyond such crack openings. However, the constitutive equation of the MC2010 is defined to reach a maximum strain of 20‰, which makes this constitutive law able to represent the residual strength in a Strength-CMOD curve only up to crack openings of 2.5 mm for standard specimens 150 × 150 × 600 mm considering $l_{cs} = 125$ mm. This makes such constitutive law unable to represent larger crack openings, lacking information about the residual strength behaviour beyond 2.5 mm.

The calibration of the parameters of the proposed 3- and 4-point constitutive equations leads to a greater similarity of the analytical Strength-CMOD curves with the experimental ones. For both 3- and 4-point equations, the ultimate strain was increased to 40‰, this entailing that the Strength-CMOD curve can be analytically modelled up to a crack opening of 5 mm.

The calibration of the 3-point equation with respect to the one of the MC2010 mainly consisted in increasing the strain ε_{SLS} from 4‰ to 15‰, so the maximum residual strength could be reached at a wider crack opening, as exhibited in the experimental results. Stresses f_{Fts} were also calibrated by increasing their value between 25% and 43%, being greater the increment for greater contents of fibres. Considering these modifications, the trilinear curve presents greater accuracy for contents of fibres below 5 kg/m^3 (C30/37-2.5, C30/37-3.5, C30/37-5.0, C40/50-5.0 and C60/70-5.0), whereas for contents over 5 kg/m^3 (C40/50-7.5, C40/50-10.0, C60/70-7.5 and C60/70-10.0) the analytical results underestimate the experimental values especially up to 2.5 mm. The results obtained reveal that this underestimation may reach a maximum 17% at a crack opening of 1 mm in C40/50-7.5.

With the aim of reaching to a solution for such underestimation, an additional point was included at the constitutive equation, this leading to a 4-point approach. Adjusting

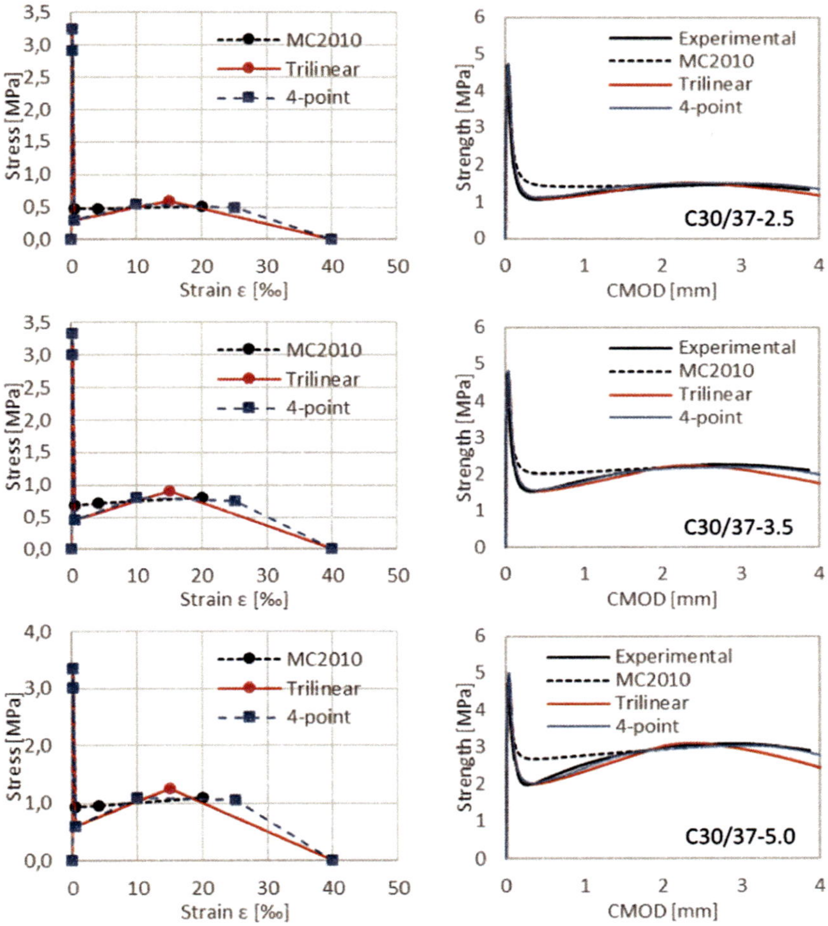

Fig. 3. Strength-CMOD curves for C30/37-2.5, C30/37-3.5 and C30/37-5.0.

the equation with a 4-point alternative leads to obtain an improved approximation of the analytical results to the experimental values, thus mitigating the differences exhibited by the 3-point constitutive calibrated equation. This 4-point proposed constitutive equation establishes its new intermediate points at strains 10‰ and 25‰. If compared to the 3-point constitutive equation, the 4-point approach provides a greater accuracy especially at the range between 1 mm and 2.5 mm of the Strength-CMOD curve. Despite this, the 4-point approach still presents a slight underestimation of the experimental results for contents of fibres over 5 kg/m^3 (C40/50-7.5, C40/50-10.0, C60/70-7.5 and C60/70-10.0), this reaching a maximum value of 11% although only up to a crack opening of 1.5 mm.

A comparison of the parameters $f_{LOP}, f_{R1}, f_{R2}, f_{R3}$ and f_{R4} between the Strength-CMOD curves of the experimental results and those obtained from the inverse analyses of the MC2010 and the trilinear and the 4-point approach are shown in Fig. 6. The results

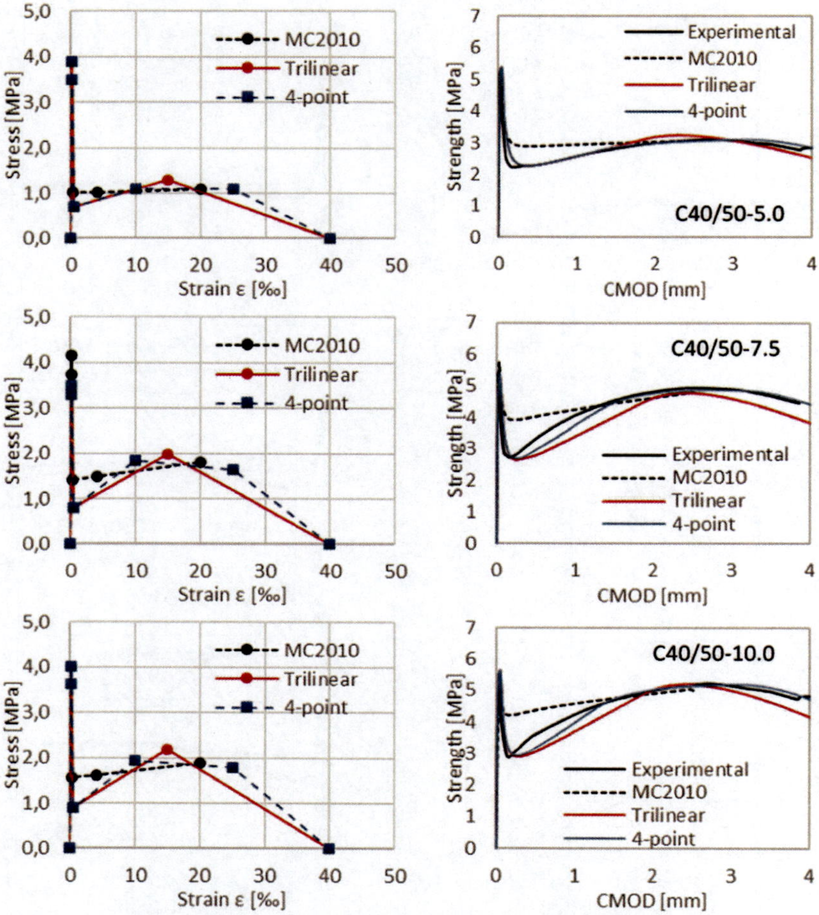

Fig. 4. Strength-CMOD curves for C40/50-5.0, C40/50-7.5 and C40/50-10.0.

exhibit that the greatest coefficient of determination (R^2) corresponds to the correlation between the experimental results and the 4-point approach, this showing a value of 0.98 and highlighting how the 4-point model presents the most accurate approximation among the analyses conducted.

5 Conclusions

An inverse analysis based on the constitutive equation for FRC of the MC2010 was conducted to determine its suitability when using polypropylene fibres as reinforcement. Based on the results of an experimental program conducted on concretes of different strength classes and contents of polypropylene fibres, the following conclusions can be drawn:

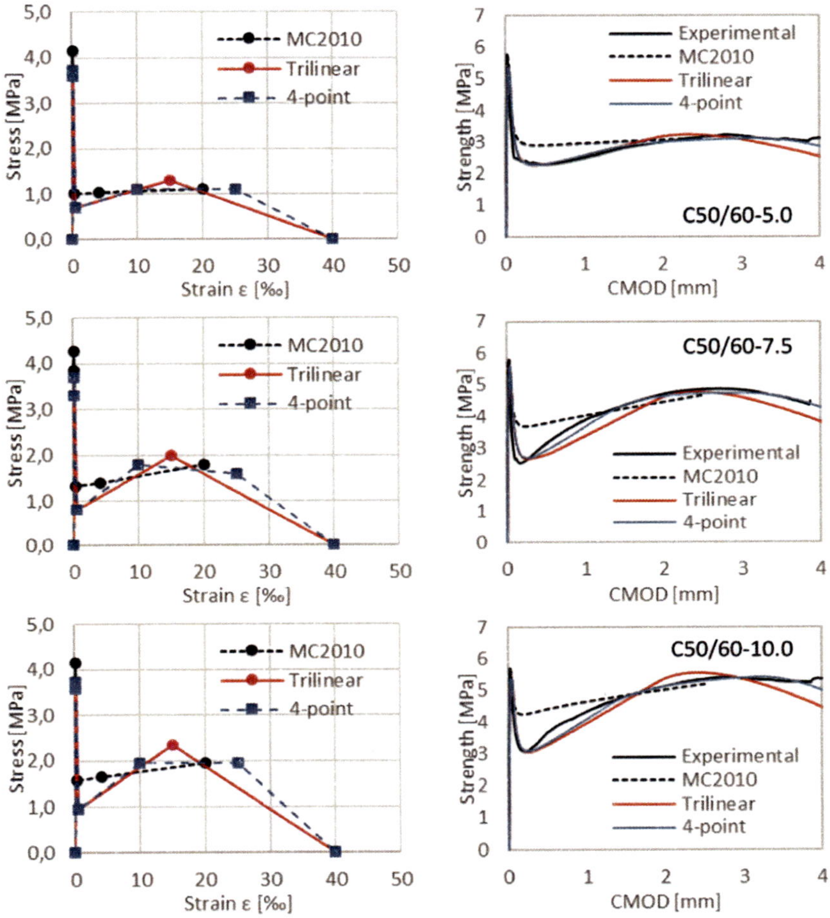

Fig. 5. Strength-CMOD curves for C50/60-5.0, C50/60-7.5 and C50/60-10.0.

- The tri-linear constitutive equation proposed into the MC2010 for simulating the tensile response of FRC tends to slightly overestimate the performance at small crack openings (SLS).
- The constitutive model is able to reproduce the Strength-CMOD curves only up to 2.5 mm, this leading to the necessity of extending the ultimate strain ε_{ULS} up to 40‰.
- The calibrated 3-point constitutive model is able to represent accurately the post-cracking results when fibres are used in contents below 5 kg/m^3. For contents of 7.5 and 10.0 kg/m^3, the analytical curves underestimate the experimental results at crack openings under 2.5 mm.
- The parameters obtained through the 4-point equation present, in average, a coefficient of determination 0.98 between the experimental results and the values obtained

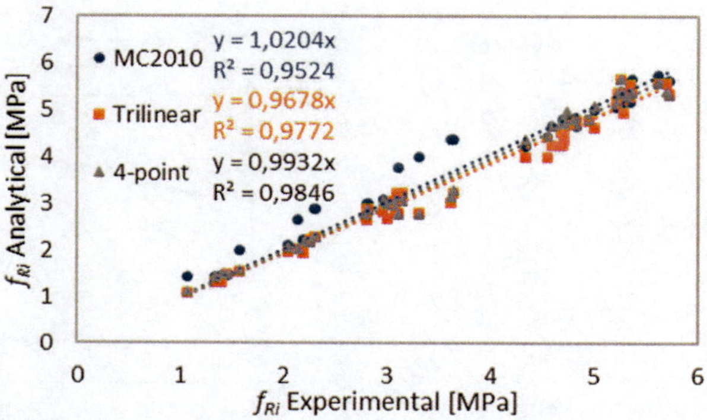

Fig. 6. Correlation between experimental and analytical parameters.

using the inverse analysis. The 4-point approach also underestimates the experimental results, although at a less extent than the 3-point equation.

Acknowledgements. The authors express their gratitude to the Spanish Ministry of Economy, Industry and Competitiveness for the financial support received under the scope of the project CREEF (PID2019-108978RB-C32). The authors wish to acknowledge the support of Elasto Plastic Concrete (EPC) during the experimental program and research. The second author also thanks the Spanish Ministry of Science, Innovation and University for providing support through the PhD Industrial Fellowship (DI-17-09390) in collaboration with Smart Engineering Ltd. (UPC's Spin-Off).

References

1. Pujadas, P., Blanco, A., Cavalaro, S., Aguado, A.: Plastic fibres as the only reinforcement for flat suspended slabs: Experimental investigation and numerical simulation. Constr. Build Mater. **57**, 92–104 (2014). https://doi.org/10.1016/j.conbuildmat.2014.01.082
2. Pujadas, P., Blanco, A., Cavalaro, S., de la Fuente, A., Aguado, A.: Fibre distribution in macro-plastic fibre reinforced concrete slab-panels. Constr. Build Mater. **64**, 496–503 (2014). https://doi.org/10.1016/j.conbuildmat.2014.04.067
3. Blanco, A., Pujadas, P., de la Fuente, A., Cavalaro, S., Aguado, A.: Application of constitutive models in European codes to RC–FRC. Constr. Build Mater. **40**, 246–259 (2013). https://doi.org/10.1016/j.conbuildmat.2012.09.096
4. International Federation for Structural Concrete: fib Model Code for Concrete Structures 2010 (2010)
5. EN 14651:2007: Test method for metallic fibre concrete. Measuring the Flexural Tensile Strength (Limit of Proportionality (LOP), Residual) (2007)
6. de la Fuente, A., Aguado, A., Molins, C., Armengou, J.: Numerical model for the analysis up to failure of precast concrete sections. Comput. Struct. **106–107**, 105–114 (2012). https://doi.org/10.1016/j.compstruc.2012.04.007
7. Galeote, E., Blanco, A., de la Fuente, A.: Design-oriented approach to determine FRC constitutive law parameters considering the size effect. Compos. Struct. **239**, 112036 (2020). https://doi.org/10.1016/j.compstruct.2020.112036

8. de Montaignac, R., Massicotte, B., Charron, J.-P.: Design of SFRC structural elements: flexural behaviour prediction. Mater. Struct. **45**, 623–636 (2012). https://doi.org/10.1617/s11527-011-9785-y
9. Blanco, A., Cavalaro, S., de la Fuente, A., Grünewald, S., Blom, C.B.M., Walraven, J.C.: Application of FRC constitutive models to modelling of slabs. Mater. Struct. **48**(9), 2943–2959 (2014). https://doi.org/10.1617/s11527-014-0369-5

Mechanical Properties of Fiber Reinforced Concrete at Low Temperatures

Stanislav Aidarov[1,2(✉)], Nikola Tošić[2], Igor Reynvart[1,2], and Albert de la Fuente[2]

[1] Smart Engineering Ltd, UPC Spin-Off, Jordi Girona 1-3, 08034 Barcelona, Spain
stanislav.aidarov@upc.edu, stanislav.aydarov@smartengineeringbcn.com

[2] Civil and Environmental Engineering Department, Universitat Politècnica de Catalunya (UPC), Jordi Girona 1-3, 08034 Barcelona, Spain

Abstract. The existing literature proves that the mechanical performance of fiber reinforced concrete (FRC) is sensitive to elevated temperatures, this being dependent on the type and amount of fibers. However, the majority of studies carried out thus far have focused on the assessment of the material behavior at elevated temperatures, disregarding the mechanical performance of FRCs at low temperatures. Taking this into account, an experimental program was carried out in order to evaluate the mechanical properties of FRC subjected to temperatures ranging from 20 °C to –30 °C; the attention was directed to the post-cracking tensile strength, i.e. the property which is of paramount importance for FRC. Additionally, different concrete mixes were analyzed, varying both the type and content of fibers. The obtained results allow proving that the residual post-cracking tensile strength of the analyzed FRC mixes increased with the decrease of temperature in all presented cases. This positive outcome might serve as a reference (which should be complemented by further investigations) to properly design elements subjected to low temperatures during transient situations and/or service life.

Keywords: Steel fiber reinforced concrete · Polymeric fiber reinforced concrete · Experimental campaign · Mechanical properties · Residual tensile strength · Low temperatures

1 Introduction

The partial or even total substitution of reinforcing bars by fibers is known to provide an enhanced structural response in terms of cracking control [1, 2], redistribution capacity [3], impact [4], and fatigue resistance [5]. These improvements, along with the acceptance of fiber reinforced concrete (FRC) as a structural material by several codes and guidelines [6–9], have boosted the use of FRC in tunnel linings [10], sewerage pipelines [11], column-supported flat slabs [12, 13], and ground-supported slabs for industrial applications [14, 15] with a recognition of beneficial outcomes in terms of sustainability.

J. I. Escalante-Garcia et al. (Eds.): RW 2021, RILEM Bookseries 40, pp. 530–538, 2023.
https://doi.org/10.1007/978-3-031-21735-7_58

Moreover, numerous studies revealed that the incorporation of fibers in the concrete mix positively affects the behavior of the concrete under temperature variations and gradients. However, practically all research programs were focused on the analysis of FRC structural responses at high temperatures [16–18], disregarding the material performance at low temperatures, which can be substantially enhanced. This is especially true considering that in certain cases FRC structures are to be subjected to low temperatures during transient situations (e.g., storage, handling and transportation of precast tunnel segments in cold regions) and in service (e.g., industrial floors for cold-storage warehouses).

Drawing an analogy with plain or traditionally reinforced (with steel bars) concrete [19, 20], the increment of compressive and tensile strength can be expected also in the case of FRC. A similar tendency is expectable for the post-cracking flexural capacity, which is a design governing parameter, however, to authors' best knowledge there is scarce research respect to that [21–23].

With this in mind, an extensive experimental program was carried out in order to characterize the compressive strength along with pre- and post-cracking flexural behavior of FRC at different temperatures (from 20 °C to –30 °C). Additionally, both steel and synthetic fibers were analyzed with varying contents. The mix composition faced no modifications apart from the fiber type/content variation and implemented additives—taking into account that the introduction of new variables (e.g., water/cement ratio) could have a significant influence on the experimental outcome.

The results of this experimental program are thought to enlarge the existing databases on FRC mechanical performance and to allow practitioners considering the mechanical properties of the FRC when designing for cold climates or conditions where low temperatures are expected.

2 Experimental Program

2.1 Concrete Mix

The selected mix (Table 1) corresponded to the normal strength concrete and S3 consistency class, i.e. the measured slump was in the range between 90 and 170 mm [24]. Based on this concrete composition, three different types of FRC were produced: macro polypropylene fiber reinforced concrete with a fiber content of 4 kg/m^3 (volume fraction 0.5%, PPFRC-4) and 8 kg/m^3 (volume fraction 0.9%, PPFRC-8), along with the concrete reinforced by 30 kg/m^3 of macro steel fibers (volume fraction 0.4%, SFRC-30); the detailed information in respect to the implemented fibers is gathered in the Table 2.

PPFRC-4 was chosen to reproduce the concrete oriented to industrial floors for which often a minimum reinforcement is required to control cracking due to thermal-hydrometric (temperature and shrinkage gradients) phenomena. In turn, PPFRC-8 and SFRC-30 were meant to reproduce medium–heavy duty pavements which, apart from the above-described indirect loads, are to be subjected to considerable load magnitudes.

2.2 Test Configuration and Procedure

The behavior of the selected FRC mixes in terms of compressive and flexural pre- and post-cracking strengths was analyzed at 20 °C (reference temperature), 0 °C, –10 °C,

Table 1. Composition of studied concrete mixes

Materials	PPFRC-4	PPFRC-8	SFRC-30
CEM II-A/L 42.5R (kg/m^3)	310	310	310
Coarse aggregate 10/20 (kg/m^3)	690	680	690
Coarse aggregate 4/10 (kg/m^3)	127	125	129
Fine aggregate 0/4 (kg/m^3)	1025	1025	1025
Water-cement ratio	0.58	0.58	0.58
Additives (% on cement content)	1.2	1.5	1.2
Synthetic fibers (kg/m^3)	4	8	–
Steel fibers (kg/m^3)	–	–	30

Table 2. Fiber properties

Property	Synthetic fiber	Steel fiber
Material	Transparent polypropylene	Steel
Shape	Embossed monofilament	Hooked-end
Diameter (mm)	0.85	0.75
Length (mm)	48	50
Aspect ratio	56.5	67
Number of fibers per kg	41200	5700
Tensile strength (MPa)	>400	1150

and –30 °C. Compressive strength was assessed in accordance with EN 12390–3 [25] by testing six Ø100 × 200 mm specimens for each temperature magnitude, resulting in 72 tested samples. The same procedure was applied to evaluate the pre- and post-cracking tensile strengths of the materials—as a result, also 72 prismatic notched beams (150 × 150 × 600 mm) were analyzed pursuant to EN 14651 [26].

After the previously listed specimens were cast and demolded (in 24 h), those were cured in a temperature (20 °C) and humidity (95%) controlled chamber during 28 days. Posteriorly, three quarters of all specimens were introduced to the laboratory freezer and each quarter was cooled down to 0 °C, –10 °C, and –30 °C, respectively.

Importantly, the prismatic specimens were equipped with a thermocouple (installed at the center of gravity of the midspan section) in order to monitor the temperature evolution and guarantee the established magnitude of the latter (Fig. 1a). Once the required temperature was reached, the specimens were placed to the IBERTEST press with 3000 kN of load capacity and INSTRON8505 (Fig. 2) with a load cell of 100 kN in order to estimate the compressive and flexural behavior, respectively.

Fig. 1. a) Freezing procedure and temperature monitoring; b) Three point bending test at –30 °C.

3 Experimental Results and Discussions

3.1 Compressive Strength

Firstly, the compressive strength of the FRCs at 28 days ($f_{cm,28}$) was estimated at the reference temperature (20 °C)—the mean values were 37.8 MPa (CV = 3.7%), 39.9 MPa (CV = 2.2%), and 39.0 MPa (CV = 6%) for PPFRC-4, PPFRC-8, and SFRC-30, respectively. Furthermore, the studied mechanical property was assessed at 0 °C. In fact, this temperature magnitude is a threshold value for water to start freezing and, therefore, the considerable increment of the compressive strength was not expected. The testing proved this assumption—minor difference was detected in the response of the material: $f_{cm,28}$ increased by 6.4%, 15.3%, and 0.5% in case of PPFRC-4, PPFRC-8, and SFRC-30, respectively (Fig. 2).

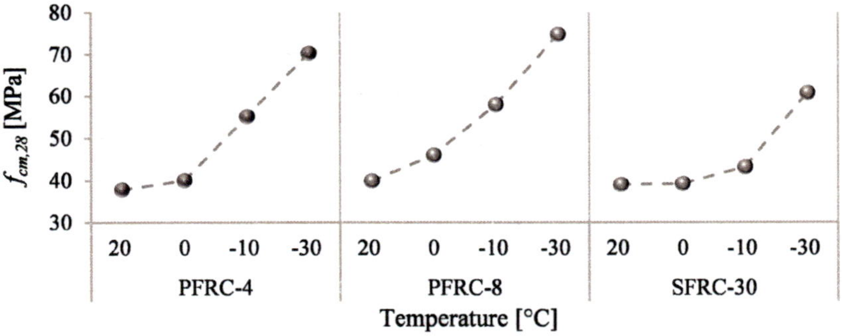

Fig. 2. Mean values of $f_{cm,28}$ of studied FRCs at different temperatures

However, once the temperature surpassed this threshold value and the capillary pores were filled with a solid material (due to water freezing), the compressive strength was increased significantly: the corresponding increment of 46.5%, 45.1%, and 10.8% was appreciated at –10 °C for PPFRC-4, PPFRC-8, and SFRC-30, whereas this enhancement was 86.7%, 87.2%, and 55.6% at –30 °C (all values are presented in respect to the measured compressive strength at ambient temperature).

Figure 2 demonstrates the described tendency for all cases—the increment of the compressive strength caused by temperature decreasing. However, the rate of these increments should be discussed in more detail: previous studies have demonstrated the absence of any considerable influence of the fiber type/content on compressive strength of the concrete as this parameter mainly depends on the material matrix, i.e. amount of the cement paste and granular skeleton [27, 28].

The tests at ambient temperature proved this statement—the mean values of three concrete mixes were in the range between 37.8 MPa and 39.9 MPa despite the different fiber type/content of the analyzed materials. Moreover, the rate of compressive strength increment was similar in case of PPFRC-4 and PPFRC-8. Nevertheless, the same parameter differed significantly in case of SFRC-30. The reason might be related with the indirect influence of fibers on the concrete matrix (e.g., entrained air content) which has a stronger effect on the compressive strength of FRC at low temperatures—however, this phenomenon should be analyzed in further studies.

3.2 Flexural Strength

The procedure was repeated to evaluate the flexural strength of the FRCs. Firstly, the prismatic notched beams were tested at the ambient temperature with following study of the flexural behavior at lower temperatures. The main attention was paid to the following parameters: the limit of proportionality (f_{LOP}) and residual tensile strengths which corresponded to the crack mouth open displacement of 0.5 and 2.5 mm, named f_{R1} and f_{R3}. The magnitudes of these values have an essential influence on the design procedure—f_{LOP} represents the behavior of the material before cracking, whereas the residual tensile strengths f_{R1} and f_{R3} are required to properly simulate the structural response at serviceability and ultimate conditions, respectively.

The limit of proportionality (f_{LOP}) is mainly dependent on the concrete matrix and, therefore, the tendency observed during the evaluation of the compressive behavior of FRCs (Fig. 2) was also expected in the pre-cracking flexural response of the prismatic notched beams. Figure 3 demonstrated that the similar tendency was detected during the flexural tests—the increment of 15.8%, 47.0%, and 28.4% was appreciated at 0 °C for PPFRC-4, PPFRC-8, and SFRC-30, respectively. The obtained magnitude of f_{LOP} demonstrated a further increment of 67.7%, 66.5%, and 65.9% for the above listed FRCs at –10 °C. The reduction of the temperature up to –30 °C provided values of similar magnitudes (in comparison with those achieved at –10 °C): total increment of 68.9%, 73.2%, and 74.1% was detected.

Finally, the influence of the temperature variation on the residual strengths (f_{R1},f_{R3}) was evaluated (Fig. 4). This parameter, apart from the properties of the concrete matrix, depends on the number of factors, such as: mechanical properties of the implemented fibers, fiber geometry (affecting mainly the anchorage capacity), fiber distribution and orientation within the critical section. Therefore, the assessment of post-cracking flexural behavior at different temperatures is a complex task which demands a consideration of previously named factors.

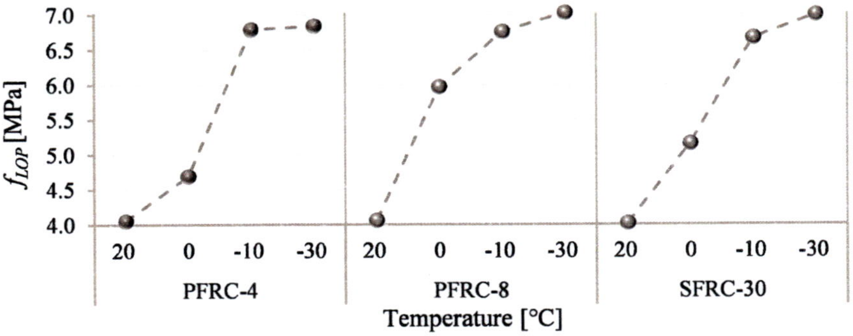

Fig. 3. Mean values of f_{LOP} of studied FRCs at different temperatures

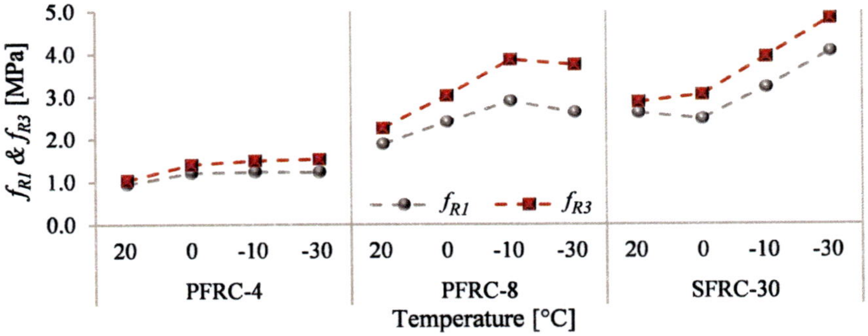

Fig. 4. Mean values of f_{R1}, f_{R3} of studied FRCs at different temperatures

Primarily, the positive effect of low temperatures on the matrix strength should be taken into account. This phenomenon, in turn, leads to the increment of the required energy to produce fiber pull-out and, therefore, tends to enhance the post-cracking flexural behavior. However, both concrete and fibers become more brittle at low temperatures—although, the achieved results (which should be complemented by further investigations) demonstrated that the studied range of temperatures had no negative effect (detectable) on the overall structural performance of the tested prismatic beams.

Additionally, the requirement in certain magnitude of residual tensile strength can be pointed out: the relatively moderate fiber content in case of PPFRC-4 did not show the continuous increment of post-cracking strength, i.e. equivalent results were obtained at –10 °C and –30 °C. In contrast, PPFRC-8 and SFRC-30 presented a considerable enhancement of f_{R1} and f_{R3} at low temperatures—remarkably, the increment of residual tensile strengths was similar in percentage terms for both concrete mixes at –30 °C. Table 3 presents more detailed information regarding the magnitudes of f_{R1} and f_{R3} variation at different temperatures (in respect to the obtained outcome at 20 °C).

Based on the results presented in Table 3, the greater increment rate of f_{R3} should be highlighted comparing the latter with the influence of low temperatures on f_{R1}—this

Table 3. Increment of residual tensile strengths (in %) at different temperatures

Temperature [°C]	PPFRC-4 [f_{R1} / f_{R3}]	PPFRC-8 [f_{R1} / f_{R3}]	SFRC-30 [f_{R1} / f_{R3}]
20	–	–	–
0	27.7 / 35.9	27.7 / 33.6	–5.4 / 5.9
−10	30.9 / 44.7	53.7 / 71.2	23.0 / 36.9
−30	29.8 / 47.6	39.4 / 65.9	55.2 / 68.3

phenomenon is also attributable to the increase of bond capacity in the matrix-fiber interaction which is usually the governing failure mechanism (fiber debonding).

4 Conclusions

In this scientific contribution, an extensive experimental program was described with an analysis of the temperature effect on the mechanical properties of FRC. The characterized elements (72 cylindrical specimens and 72 prismatic notched beams) were subjected to temperatures from 20 °C to –30 °C and compressive strength along with the pre- and post-cracking flexural behavior were assessed. The following conclusions may be derived from the obtained results:

- The compressive and tensile strength of studied FRCs significantly increased at lower temperatures due to water freezing in the capillary pores: $f_{cm,28}$ and f_{LOP} increased by 87% and 74% (maximum detected increments) at –30 °C, respectively.
- Low temperatures led to the increment of the required energy to produce fiber pull-out and, therefore, the post-cracking flexural behavior was enhanced: f_{R1} and f_{R3} increased up to 54% and 71%, respectively, for temperatures below 0 °C.
- The greater increment rate of f_{R3} at low temperatures (in comparison with the obtained values of f_{R1}) was observed for all studied cases—the factor of paramount importance for design procedures at ultimate conditions.

The outcome of the presented research program reveals the enhanced performance of FRC in terms of compressive and flexural strengths—therefore, these phenomena should be taken into account during the design procedures of the elements that are to be subjected to low temperatures during transient or in-service conditions.

Acknowledgements. This research has been possible owe to the economic funds provided by the SAES project (BIA2016–78742-C2–1-R) of Spanish Ministry of Economy, Industry and Competitiveness. The first author, personally, thanks the Department of Enterprise and Education of Catalan Government for providing support through the PhD Industrial Fellowship (2018 DI 77) in collaboration with Smart Engineering Ltd. (UPC's Spin-Off). The authors also want to extend their gratitude to MBCC Group for providing materials and supporting this research.

References

1. Groli, G., Caldentey, A.P.: Improving cracking behaviour with recycled steel fibres targeting specific applications-analysis according to fib Model Code 2010. Struct. Concr. **18**, 29–39 (2017). https://doi.org/10.1002/suco.201500170
2. Pujadas, P., Blanco, A., de la Fuente, A., Aguado, A.: Cracking behavior of FRC slabs with traditional reinforcement. Mater. Struct. Constr. **45**, 707–725 (2012). https://doi.org/10.1617/s11527-011-9791-0
3. Fall, D., Shu, J., Rempling, R., Lundgren, K., Zandi, K.: Two-way slabs: Experimental investigation of load redistributions in steel fibre reinforced concrete (2014). https://doi.org/10.1016/j.engstruct.2014.08.033
4. Nia, A.A., Hedayatian, M., Nili, M., Sabet, V.A.: An experimental and numerical study on how steel and polypropylene fibers affect the impact resistance in fiber-reinforced concrete. Int. J. Impact. Eng. **46**, 62–73 (2012). https://doi.org/10.1016/j.ijimpeng.2012.01.009
5. Carlesso, D.M., de la Fuente, A., Cavalaro, S.H.P.: Fatigue of cracked high performance fiber reinforced concrete subjected to bending. Constr. Build. Mater. **220**, 444–455 (2019). https://doi.org/10.1016/j.conbuildmat.2019.06.038
6. fib. fib Model Code for Concrete Structures (2010)
7. RILEM TC 162-TDF.: Recommendations of RILEM TC 162-TDF: Test and design methods for steel fibre reinforced concrete: bending test. Mater. Struct. **35**, 579–82 (2002). https://doi.org/10.1617/14007
8. Ministerio de Fomento. Instrucción de Hormigón Estructural (EHE-08) (2008). https://doi.org/10.1017/CBO9781107415324.004
9. Italian National Research Council CNR: CNR-DT 204 /2006 Guide for the Design and Construction of Fiber-Reinforced Concrete Structures. Consiglio Nazionale Delle Ricerche, Rome (2006)
10. de la Fuente, A., Pujadas, P., Blanco, A., Aguado, A.: Experiences in Barcelona with the use of fibres in segmental linings. Tunn. Undergr. Sp. Technol. **27**, 60–71 (2012). https://doi.org/10.1016/j.tust.2011.07.001
11. de la Fuente, A., Escariz, R.C., de Figueiredo, A.D., Aguado, A.: Design of macro-synthetic fibre reinforced concrete pipes. Constr. Build. Mater. **43**, 523–532 (2013). https://doi.org/10.1016/j.conbuildmat.2013.02.036
12. Aidarov, S., Mena, F., de la Fuente, A.: Structural response of a fibre reinforced concrete pile-supported flat slab: Full-scale test. Eng. Struct. **239** (2021). https://doi.org/10.1016/J.ENGSTRUCT.2021.112292
13. Aidarov, S., Nadaždi, A., Pugach, E., Tošić, N., de la Fuente, A.: Cost-oriented analysis of fibre reinforced concrete column-supported flat slabs construction. J. Build. Eng. **51**, 104205 (2022). https://doi.org/10.1016/j.jobe.2022.104205
14. Meda, A., Plizzari, G.A., Riva, P.: Fracture behavior of SFRC slabs on grade. Mater. Struct. Constr. **37**, 405–411 (2004). https://doi.org/10.1617/14093
15. Meda, A., Plizzari, G.A.: New design approach for steel fiber-reinforced concrete slabs-on-ground based on fracture mechanics. ACI Struct. J. **101**, 298–303 (2004). https://doi.org/10.14359/13089
16. Khaliq, W., Kodur, V.: Thermal and mechanical properties of fiber reinforced high performance self-consolidating concrete at elevated temperatures. Cem. Concr. Res. **41**, 1112–1122 (2011). https://doi.org/10.1016/j.cemconres.2011.06.012
17. Lourenço, L.A.P., Barros, J.A.O., Alves, J.G.A.: Fiber Reinforced Concrete of Enhanced Fire Resistance for Tunnel Segments, pp. 39–71. ACI Spec Publ, Am Concr Institute (2011)
18. Zhang, P., Kang, L., Wang, J., Guo, J., Hu, S., Ling, Y.: Mechanical properties and explosive spalling behavior of steel-fiber-reinforced concrete exposed to high temperature-a review. Appl. Sci. **10** (2020). https://doi.org/10.3390/app10072324

19. Lee, G.C., Shih, S., Chang, K.C.: Mechanical properties of concrete at low temperature. J. Cold Reg. Eng. **2**, 13–24 (1988)
20. Xie, J., Yan, J.B.: Experimental studies and analysis on compressive strength of normal-weight concrete at low temperatures. Struct. Concr. **19**, 1235–1244 (2018). https://doi.org/10.1002/suco.201700009
21. Aidarov, S., Nogales, A., Reynvart, I., Tošić, N., de la Fuente, A.: Effects of low temperatures on flexural strength of macro-synthetic fiber reinforced concrete: Experimental and numerical investigation. Materials (Basel) **15** (3) (2022). https://doi.org/10.3390/ma15031153
22. Caballero-Jorna, M., Roig-Flores, M., Serna, P.: Short-term effects of moderate temperatures on the mechanical properties of steel and macrosynthetic fiber reinforced concretes. In: RILEM-fib X International Symposium Fibre Reinforced Concrete, BEFIB2021. Springer, Valencia (2021)
23. Richardson, A., Ovington, R.: Temperature related steel and synthetic fibre concrete performance. Constr. Build. Mater. **153**, 616–621 (2017). https://doi.org/10.1016/j.conbuildmat.2017.07.101
24. CEN. EN 12350-2:2019.: Testing fresh concrete. Slump test (2019)
25. CEN. EN 12390-3:2019.: Testing hardened concrete. Compressive strength of test specimens (2019)
26. CEN. EN 14651.: Test method for metallic fibre concrete. Measuring the flexural tensile strength (limit of proportionality (LOP), residual) (2007)
27. Mena, F., Aidarov, S., de la Fuente, A.: Hormigones autocompactantes reforzados con fibras para aplicaciones con alta responsabilidad estructural. Campaña experimental en laboratorio. III Congr. Consult. Estructuras, pp. 1–10 (2019)
28. König, G., Kützing, L.: Modelling the increase of ductility of HPC under compressive forces—a fracture mechanical approach. In: Third International RILEM Workshop High Performance Fiber Reinforced Cementitious Composites, pp. 251–60. RILEM Publications SARL, Mainz (1999)

Finite Element Analysis Characterization of Macro Synthetic Fibre Reinforced Concrete Constitutive Equation

Alejandro Nogales[1,2](✉), Eduardo Galeote[1,2], and Albert de la Fuente[1]

[1] Civil an Environmental Engineering Department, Universtitat Politècnica de Catalunya (UPC), Carrer Jordi Girona 1-3, 08034 Barcelona, Spain
alejandro.nogales@upc.edu

[2] Smart Engineering Ltd, UPC Spin-Off, Jordi Girona 1-3, 08034 Barcelona, Spain

Abstract. Over the last years, the use of fibre reinforced concrete (FRC) has increased for structural purposes. For the structural design of FRC elements, there was a need of a model that developed the behaviour of the post-cracking response of FRC. In this sense, national and international guidelines have included models to characterise the flexural behaviour of FRC (fib Model Code, EHE-08). These models gather the performance of FRC for serviceability limit state (SLS) and ultimate limit state (ULS) for either steel or macro synthetic polypropylene fibre reinforced concrete (SFRC and MSFRC, respectively). In this regard, the codes and guidelines do not distinguish between FRC comprised of steel or synthetic fibres and establish the FRC ultimate strain in 2.5%. This limitation represents the behaviour of SFRC but limits the full potential of MSFRC for large deformations. Owing to the aspects aforementioned, an extensive experimental programme has been carried out at the Universitat Politècnica de Catalunya (UPC) to characterise the behaviour of MSFRC. This research contribution is focused on an inverse analysis to derive the MSFRC constitutive equations by means of a non-linear finite element simulation. The main goal of this study is to compare the experimental results with those obtained through the simulation using the constitutive equations of the fib MC-2010. The results show a generalised underestimation of MSFRC at ultimate strain and the necessity of adjusting the constitutive equations for SFRC and MSFRC.

Keywords: Macro synthetic fibre reinforced concrete · Non-linear analysis · Constitutive equation

1 Introduction

The addition of fibres into concrete, commonly known as fibre reinforced concrete (FRC), for structural purposes has experienced a huge growth in recent years. Fibres are either used for totally or partially replacing the traditional rebar reinforcement. In the FRC industry, there are different types of fibres such as organic, metallic or synthetic (i.e. glass fibre, steel fibres, polypropylene fibres, respectively) that are used to cover a

J. I. Escalante-Garcia et al. (Eds.): RW 2021, RILEM Bookseries 40, pp. 539–548, 2023.
https://doi.org/10.1007/978-3-031-21735-7_59

wide range of structural typologies: industrial flooring [1, 2], precast concrete segments for tunnel linings [3], elevated flat slabs [4, 5], sewer pipes [6, 7].

Due to the increasing use of this material, codes and guidelines [8–10] have been published as a demand for a reliable design tool for engineers and practitioners. These codes and guidelines developed material models that are able to reproduce the behaviour of FRC. The models consist of constitutive equations, in terms of uniaxial stress-strain (σ-ε) curves (or stress-crack width), that take into account the post-cracking residual strength of FRC given by the pull-out mechanism effect induced by fibres.

In the case of *fib* Model Code—2010 (MC-2010) [8], the post-cracking strength of the material is obtained by carrying out the beam flexural strength test (EN:14651, European Commitee for Standardization, 2005) from which the residual strength for crack mouth opening displacement (CMOD) 0.5 and 2.5 mm are extracted (f_{R1} and f_{R3}, respectively). This results in a bilinear post-cracking constitutive equation where the two main points are f_{Fts} and f_{Ftu}, which correspond to serviceability and ultimate residual strength, being the ultimate crack width (w_u) equal to 2.5. Figure 1 depicts the schematic representation of the stress-crack width curve for FRC according to MC-2010, the full curve is obtained as the combination of the post-cracking response of plain concrete (where f_{ctm} and G_F stand for mean tensile concrete strength and fracture energy, respectively) and the fibre contribution through the pull-out mechanism. Being the first point $\sigma_1 = f_{ctm}$ and $w_1 = 0$ mm, the second point σ_2 and w_2 (the intersection between the two curves) and the third one $\sigma_3 = f_{ctm}$ and $w_3 = 2.5$ mm.

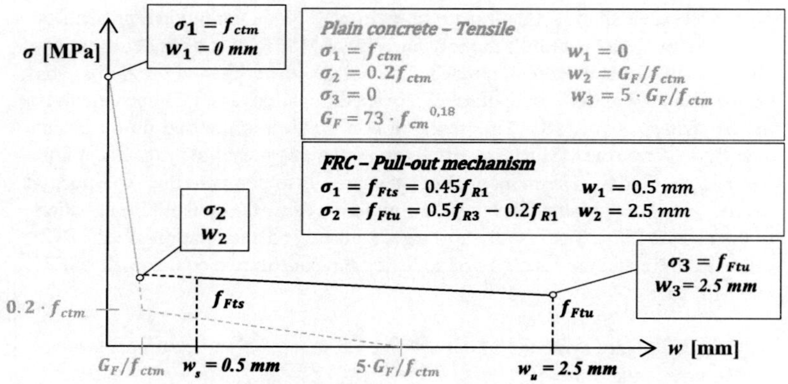

Fig. 1. MC-2010 FRC post-cracking curve

However, the aforementioned guidelines do not distinguish between different types of fibres and w_u considered in terms of ductility or durability may not be adjusted for all type of fibres. This is especially important in beam flexural strength test of macro synthetic polypropylene fibre reinforced concrete (MSFRC) where, after the sudden drop due to cracking, a hardening response is observed up to failure and the peak of the residual strength is reached beyond 2.5 mm (w_u).

In view of this, in order to assess the behaviour of MSFRC, a wide experimental programme considering a broad range of compressive concrete strength along with a representative fibre dosage used for actual applications was performed at the Universitat Politècnica de Catalunya. The concrete mixes tested in this experimental research had strength classes of C30/37 (with fibre amount of 2.5, 3.5 and 5.0 kg/m^3), C40/50 (with 5.0, 7.5 and 10.0 kg/m^3) and C50/60 (with 5.0, 7.5 and 10 kg/m^3).

The results of this experimental research were used for deriving the constitutive equations (MC-2010) and were compared with the ones obtained performing back analysis (BA) by means of a FE software. The aim of this conference proceeding is twofold: (1) compare the equations derived using the MC-2010 equations with the ones obtained by the inverse analysis and (2) propose changes in the equation in order to take into account the full potential of the MSFRC concrete. The outcome of this research work proved that the current MC-2010 constitutive equation do not take into account the full potential of MSFRC.

2 Experimental Results

In the experimental campaign, residual flexural strength and compressive tests were carried out. The mechanical and geometrical properties of macro synthetic fibres made of polypropylene (PPMSF) for structural purposes used are gathered in Table 1. In order to have sufficient representativeness of the MSFRC behaviour, nine beam tests were carried out per each concrete mix (a total amount of 81). The average curves of the beam flexural strength test of each concrete mix are depicted in Fig. 2. In some cases the tests were carried out beyond the standard CMOD for research purposes, to see any possible fibre failure. As can be seen from Fig. 2, the maximum post-cracking strength registered is beyond 2.5 mm for every individual mix, which is w_u for the constitutive equation of the MC-2010. These values are gathered in Table 2.

Table 1. Characteristics of PPMSF fibre

Material	Anchorage	Length	Young's modulus	Tensile strength	Number fibres/kg
Virgin polypropylene	Continuous embossing	48 mm	12 GPa	640 MPa	59500

Table 2. CMOD at maximum post-cracking strength

	C30/37	C40/50	C50/60
2.5 kg/m^3	2.76 mm	–	–
3.5 kg/m^3	2.69 mm	–	–
5.0 kg/m^3	2.91 mm	2.70 mm	2.80 mm
7.5 kg/m^3		2.51 mm	2.69 mm
10.0 kg/m^3	–	2.71 mm	2.68 mm

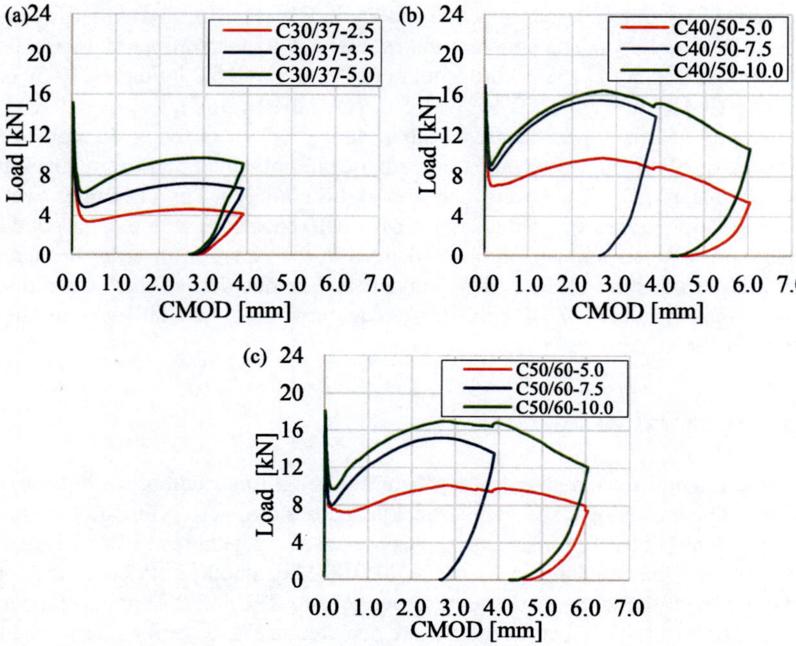

Fig. 2. Average curves of beam flexural strength test for (a) C30/37 (b) C40/50 and (c) C50/60.

In view of these results, it is highly unlikely that the σ–w constitutive curve is able to capture properly the late post-cracking behaviour of the MSFRC. In the following sections, a FE model is presented to carry out a back analysis (BA) and derive the constitutive curve suitable to reproduce the behaviour of MSFRC.

3 FE Analysis

In order to perform the BA, a non-linear 2D plain strain model was created in the FE software ABAQUS. The Concrete Damage Plasticity (CDP) model available in ABAQUS [12] was selected. This software presents a versatile tool to model a broad range of phenomena of structural concrete behaviour. The model assumes that the main two failure mechanisms for concrete are tensile cracking and compressive crushing. To model the concrete behaviour, the input data required are uniaxial σ–ε curves for compression and tension. In this study, to overcome mesh dependence due to different mesh size, the σ–w tensile curve was used instead of σ–ε.

In Fig. 3 are presented the geometry, boundary conditions and loading for the beam flexural strength test configuration. The boundary conditions were imposed so that in the vertical axis $u_y = 0$ in both supports and $u_x = 0$ in one of those. The load was applied by displacement control in order to guarantee proper convergence in case of flexural-softening response is detected. The mesh comprised of 1361 nodes and 2560 triangular linear elements (CPE3) with a mesh size of 10 mm, refined in the mid-section with 5 mm size elements. The post-cracking behaviour of FRC was captured by means of ABAQUS Explicit Dynamic algorithm (quasi-static analysis).

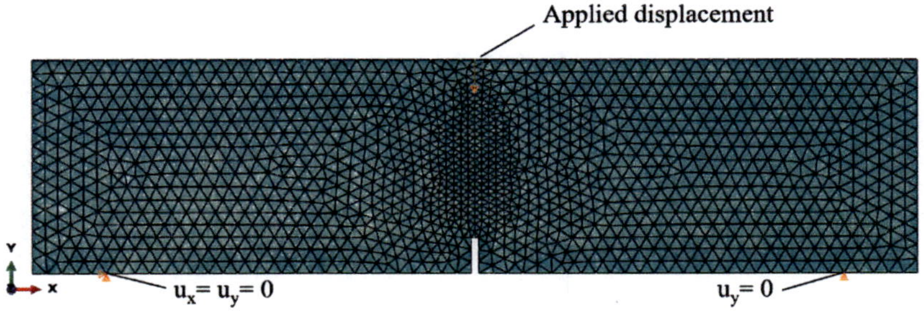

Fig. 3. 2D FEM model adopted: mesh and boundary conditions considered

4 Results

To obtain the constitutive relationships, a generalised method of back analysis was used [13] considering an iterative trial and error to fit the experimental curves with the model results. The MC-2010 constitutive equations were obtained using the results presented in Fig. 2.

From Figs. 4, 5, 6, 7, 8, 9, 10, 11 to 12 are depicted the Load—CMOD curves for the experimental results and three FE model simulations (1) using the MC-2010 constitutive curves, (2) using the constitutive curve obtained by BA and (3) an obtained simplified trilinear curve (based on the back analysis results). Additionally, the figures with stress-crack width present the constitutive curves used for the FE simulations.

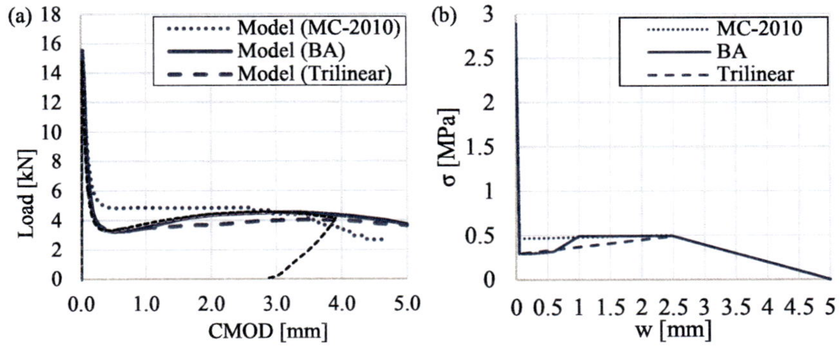

Fig. 4. C30/37 2.5 kg/m^3 **a** Load—CMOD curves **b** Constitutive equations.

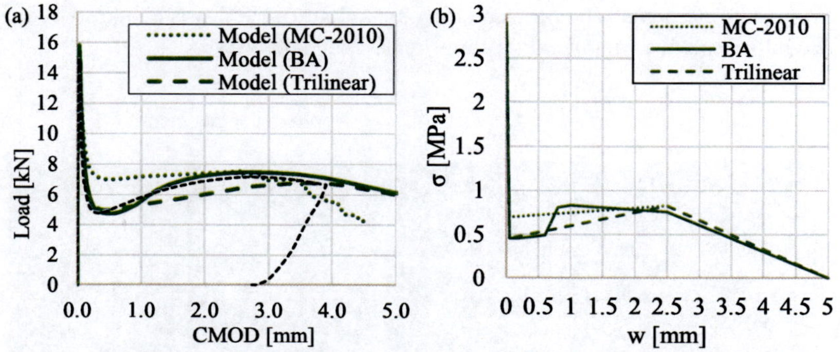

Fig. 5. C30/37 3.5 kg/m^3 **a** Load—CMOD curves **b** Constitutive equations.

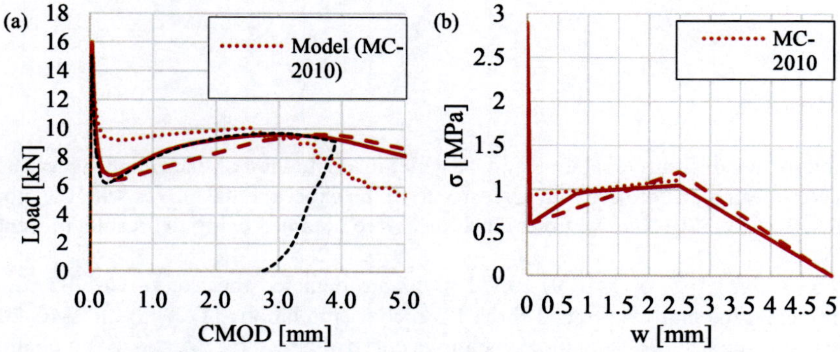

Fig. 6. C30/37 5.0 kg/m^3 **a** Load—CMOD curves **b** Constitutive equations.

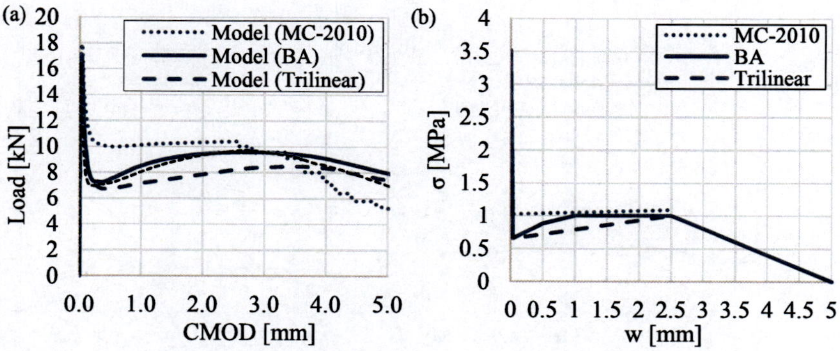

Fig. 7. C40/50 5.0 kg/m^3 **a** Load—CMOD curves **b** Constitutive equations.

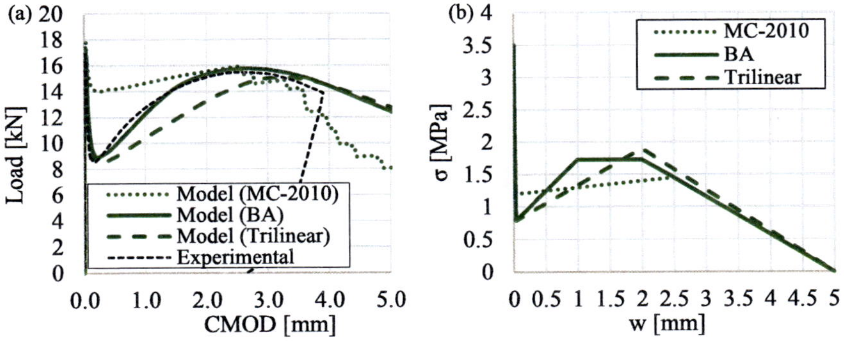

Fig. 8. C40/50 7.5 kg/m^3 **a** Load—CMOD curves **b** Constitutive equations.

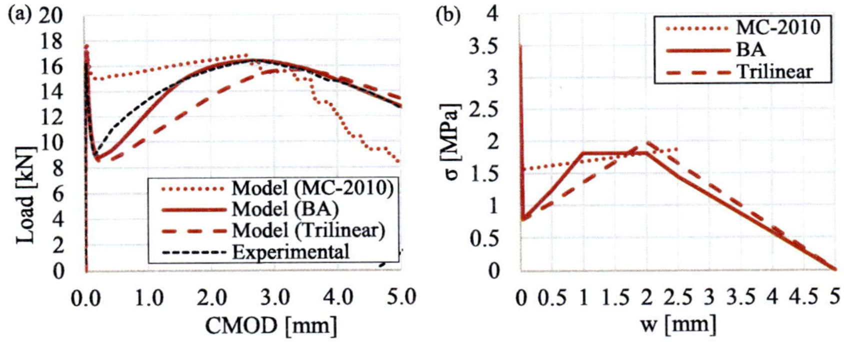

Fig. 9. C40/50 10.0 kg/m^3 **a** Load—CMOD curves **b** Constitutive equations.

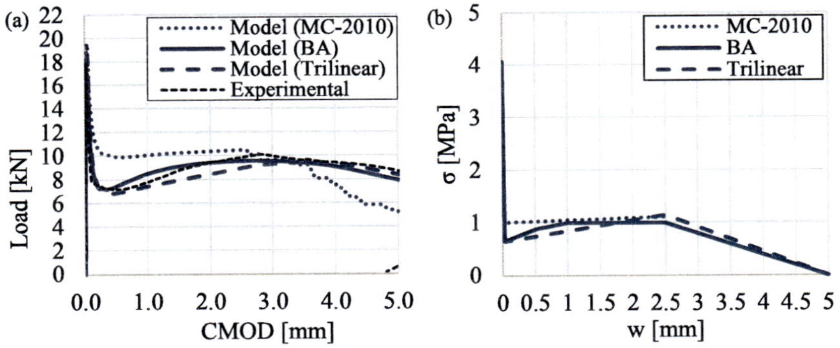

Fig. 10. C50/60 5.0 kg/m^3 **a** Load—CMOD curves **b** Constitutive equations.

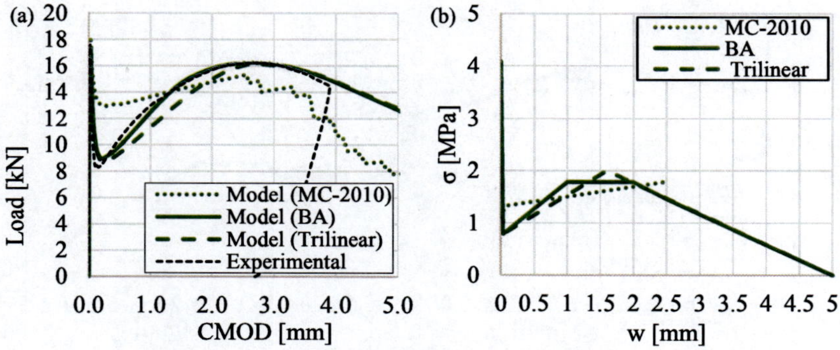

Fig. 11. C50/60 7.5 kg/m^3 **a** Load—CMOD curves **b** Constitutive equations.

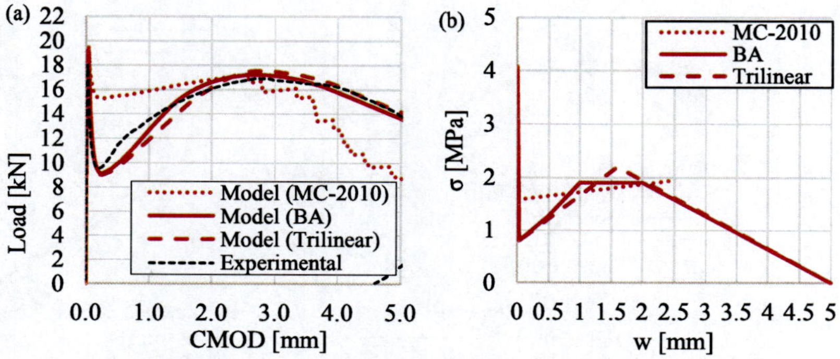

Fig. 12. C50/60 10.0 kg/m^3 **a** Load—CMOD curves **b** Constitutive equations.

The results show how the MC-2010 constitutive equation overestimates the mechanical performance for CMOD < 1.0 mm, i.e., the loss of strength is larger in the Experimental test than in the Model (MC-2010). This is of great importance, suggesting that the MC-2010 curve tends to show higher strength for serviceability limit state (f_{Fts}). Although the MC-2010 bilinear curve captures well the maximum post-cracking strength, it is reached at smaller CMOD as compared to the experimental results. Further, when the maximum post-cracking strength for MSFRC is expected (CMOD > 2.5 mm), the Load-CMOD curve starts decreasing, which does not correspond with the behaviour observed in the actual tests. This is even more evident in tests with high hardening behaviour (e.g. C40/50–10 and C50/60–10). In view of these results, two main issues can be highlighted with the MC-2010 approach: overestimates the residual strength for serviceability limit state (SLS) and it is not representative of the full potential of MSFRC reached at larger CMOD.

The constitutive curve obtained by means of BA fits well the behaviour of the actual test. In this regard, seven points (up to nine in the case of C30/37–3.5) were necessary to match the curve. The loss of strength and the mechanical performance for CMOD

< 1.0 mm was well captured, which highlights the necessity of decreasing the value of σ_2 for the MC-2010 constitutive equation. For high values of CMOD (>2.5 mm), an additional point was required to capture well the behaviour ($\sigma = 0$ MPa, $w = 5$ mm), decreasing the residual strength with a smooth slope down to zero.

Although the BA curve worked perfectly, in terms of structural design it is not a quick process due to tedious trial and error process. In view of this, a trilinear curve (four points) was proposed, so the MSFRC behaviour was better reproduced than using the MC-2010 constitutive curve. To this end, two changes in the MC-2010 are proposed to improve the results. First, a reduction in σ_2 was necessary so the loss of strength after cracking was similar to the experimental test. Moreover, the f_{Fts} was closer to the values of the experimental test being on the safe side for serviceability requirements. The second change regards the late post-cracking response (CMOD > 2.5 mm), which was sorted out adding an additional point with zero strength and $w = 5.0$ mm. In cases where a huge hardening behaviour was observed (i.e. C40/50–7.5, C40/50–10, C50/60–7.5 and C50/60–10), the third point of the trilinear equation was set for $w < 2.5$ mm to better fit the experimental behaviour.

The results using the trilinear constitutive curve showed a good agreement with the Experimental curve and a huge improvement compared to using the MC-2010 curves. However, implementing the trilinear curve there is a region (1.0 mm < CMOD < 3.0 mm) in which the Model does not fit the experimental curve (being this easily solved by setting an additional point in the σ–w curve) although the proposed trilinear approach is on the safe side for design purposes. The results evidenced that the trilinear curve proposed (based on the BA) considerably improved the results compared to adopting the MC-2010 curve.

5 Conclusions

Numerical simulations of beam flexural post-cracking strength tests have been carried out by means of a non-linear FE model. Based on the outcomes of an experimental research work, the model was used for assessing the MC-2010 FRC constitutive curve with macro synthetic fibres made of polypropylene. Moreover, a BA was performed and a trilinear curved was proposed to better fit the experimental behaviour observed.

Based on the results presented in this research contribution, the following conclusions can be drawn:

- The MC-2010 constitutive curve does not fully reproduce the potential of MSFRC. The maximum post-cracking strength for MSFRC appears for CMOD > 2.5 mm, and this is not captured for the MC-2010 constitutive curve since it is limited to $w_{\mathrm{u}} =$ 2.5 mm.
- The MC-2010 constitutive curve slightly overestimates the performance of MSFRC for CMOD < 1.0 mm as well as the loss of strength after cracking.
- Two modifications in the tri-linear constitutive curve induced a huge improvement in order to better represent the post-cracking behaviour of MSFRC: a reduction in σ_2, so the loss of strength after cracking reproduces the drop in strength in the actual behaviour, and progressively reducing the strength of the constitutive equation beyond $w_{\mathrm{u}} = 2.5$ mm up to $w = 5.0$ mm and $\sigma = 0$ MPa.

- The non-corrosive behaviour of the PP fibres justifies an opening of the crack width limitations of MC-2010. The constitutive model developed and presented here accurately represents the flexural material behaviour of MSFRC used in this experimental research. Its deviation from the MC-2010 model enables a significantly better exploitation of the favourable material response, especially at larger CMOD.

Acknowledgements. The first author acknowledges the Spanish Ministry of Science, Innovation and University for providing support through the PhD Industrial Fellowship (DI-17–09390) in collaboration with Smart Engineering Ltd. (UPC's Spin-Off). Funding for the experimental research work in this study was provided by BarChip Inc.

References

1. Meda, A., Plizzari, G.A., Riva, P.: Fracture behavior of SFRC slabs on grade. Mater. Struct. Constr. **37**, 405–411 (2004). https://doi.org/10.1617/14093
2. Meda, A., Plizzari, G.: New design approach for steel fiber-reinforced concrete slabs-on-ground based on fracture mechanics. ACI Struct. J. 298–303 (2004)
3. Nogales, A., de la Fuente, A.: Crack width design approach for fibre reinforced concrete tunnel segments for TBM thrust loads. Tunn. Undergr. Sp. Technol. **98** (2020). https://doi.org/10.1016/j.tust.2020.103342
4. Nogales, A., de la Fuente, A.: Numerical-aided flexural-based design of fibre reinforced concrete column-supported flat slabs. Eng. Struct. **232**, 1–24 (2021). https://doi.org/10.1016/j.engstruct.2020.111745
5. Aidarov, S., Mena, F., de la Fuente, A.: Structural response of a fibre reinforced concrete pile-supported flat slab: full-scale test. Eng. Struct. **239**, 112292 (2021). https://doi.org/10.1016/j.engstruct.2021.112292
6. de la Fuente, A., Escariz, R.C., de Figueiredo, A.D., Molins, C., Aguado, A.: A new design method for steel fibre reinforced concrete pipes. Constr. Build. Mater. **30**, 547–555 (2012). https://doi.org/10.1016/j.conbuildmat.2011.12.015
7. de la Fuente, A., Escariz, R.C., de Figueiredo, A.D., Aguado, A.: Design of macro-synthetic fibre reinforced concrete pipes. Constr. Build. Mater. **43**, 523–532 (2013). https://doi.org/10.1016/j.conbuildmat.2013.02.036
8. International Federation for Structural Concrete (fib).: fib-Model Code for Concrete Structures 2010. Lausanne (2010). https://doi.org/10.1002/9783433604090
9. EN1992-1-1.: Eurocode 2: Design of concrete structures—Part 1–1: General rules, rules for buildings, bridges and civil engineering structures. CEN, Brussels (2019)
10. EHE-08.: Instrucción de Hormigón Estructrural (EHE-08). Ministerio de Fomento, Madrid (2008)
11. European Commitee for Standardization. Precast concrete products test method for metallic fibre concrete-Measuring the flexural tensile strength. Br. Stand. Inst. (2005) https://doi.org/10.1002/9780580610523
12. Dassault Systèmes Simulia.: Abaqus CAE User's Manual (6.12). Providence: Dassault Systèmes (2012)
13. Roelfstra, P.E., Wittmann, F.H.: Numerical Method to Link Strain Softening with Failure of Concrete, pp. 163–175. Fract Toughness Fract Energy, Elsevier (1986)

A Discussion on the Reliability of prEN1992-1-1:2021 Shear Strength Provisions for Fibre Reinforced Concrete Members Without Shear Reinforcement

Nikola Tošić[1](✉), Jesús Miguel Bairán[1], Miguel Fernández Ruiz[2], and Albert de la Fuente[1]

[1] Department of Civil and Environmental Engineering, Universitat Politècnica de Catalunya, Barcelona, Spain
nikola.tosic@upc.edu

[2] Universidad Politécnica de Madrid, École Polytechnique Fédérale de Lausanne, Lausanne, Switzerland

Abstract. The Eurocode 2 for the design of concrete structures (EN1992-1-1:2004) is undergoing a revision that will lead to the publication of the second generation of this code to be used across all CEN member countries. Therefore, the impact of the code will reach hundreds of millions of people. Importantly, the Eurocode 2-revision will incorporate many significant changes, amongst which is the introduction of design provisions for fibre reinforced concrete (FRC). In this regard, one of the most important aspects for ultimate limit state (ULS) design will be the consideration of the shear strength of FRC members without shear reinforcement. Such a failure mode is associated with a number of uncertainties, as well as with a potentially-brittle failure response. However, so far, the design models proposed in the Eurocode 2-revision have not been accompanied by a robust reliability-based calibration of the FRC partial factor γ_{SF}. Within this context, this paper presents the results of an investigation on the safety format and the partial factor γ_{SF} for FRC members without shear reinforcement currently provisioned in the draft for the new Eurocode 2 (prEN1992-1-1:2021). Firstly, a database of experimental results on FRC beams is used to determine the model error. On that basis, a probabilistic analysis is performed using the First Order Reliability Method (FORM) to determine adequate values of γ_{SF} for varying target reliability indices. The results of the study show how γ_{SF} values need to updated in order to reach reliability indices typically considered for ULS design.

Keywords: Fibre reinforced concrete · Reliability · Partial factor · Beam · Database · Eurocode 2

1 Introduction

Fibre reinforced concrete (FRC) is a structural material that is increasingly acknowledged as an efficient and potentially more sustainable alternative to conventional reinforced

J. I. Escalante-Garcia et al. (Eds.): RW 2021, RILEM Bookseries 40, pp. 549–558, 2023.
https://doi.org/10.1007/978-3-031-21735-7_60

concrete (RC) [1–3]. A clear advantage of FRC—from the perspective of reducing construction time and material use—is the reduction of steel reinforcement in RC members, and particularly that of shear reinforcement. However, in order for designers to be able to use FRC in such a way, reliable ultimate limit state (ULS) design models are required [4]. This is instrumental, since shear resistance models, especially for members without shear reinforcement, are associated with a large number of uncertainties and shear failures of such members tend to be brittle with potentially catastrophic consequences. Therefore, it has been a topic attracting research interest for decades [5], producing a wide variety of empirical, semi-empirical and mechanical models.

In Europe, the European Committee for Standardization (CEN) has the mandate of producing structural design codes, i.e. the Eurocodes. For RC structures, the current Eurocode 2 [6] (EN-1992-1-1:2004, EC2 in the following) dates back to 2004 (with corrigenda in 2008 and 2010). Since then, numerous advances have occurred, both in the material and in the structural fields, leading to the preparation of a new generation of Eurocodes approximately 10 years ago. Within this new generation, the Eurocode 2-revision [7] (EC2-rev in the following, whose current draft is prEN1992-1-1:2021) will contain a wide range of provisions for new types of elements, concrete types and reinforcement. In particular, an informative annex is planned for steel fibre reinforced concrete (SFRC), with potential expansion in the future covering macro-synthetic fibre reinforced concretes (MSFRC) as well.

Since the CEN Enquiry phase is approaching, this study was conceived with the aim of providing a reliability-based assessment of the EC2-rev shear resistance model for FRC members without shear reinforcement. Also, it is aimed at calibrating the FRC partial factor γ_{SF} required for achieving the required code-prescribed probabilities of failure P_f according to different consequence classes. To achieve this goal, first, the model error was calculated using a database of experimental results. Then, a parametric study was carried out using the First Order Reliability Method (FORM) considering different probability distributions and parameters of input variables. Based on the results, the partial factor γ_{SF} was calibrated based on the target reliability index and failure probability.

2 Description of the Design Model and Assessment of the Model Error

2.1 prEN1992-1-1:2021 Model for the Shear Strength of FRC Members Without Shear Reinforcement

The prEN1992–1-1:2021 provisions in its annex L that, for SFRC members not requiring design shear reinforcement and with longitudinal bars in the tensile zone, the design value of the shear strength should be taken as:

$$\tau_{Rd,cF} = \eta \cdot \frac{0,6}{\gamma_C} \cdot \left(100 \cdot \rho_l \cdot f_{ck} \cdot \frac{d_{dg}}{d}\right)^{\frac{1}{3}} + f_{Ftud} \geq \eta \cdot \frac{11}{\gamma_V} \cdot \sqrt{\frac{f_{ck}}{f_{yd}} \cdot \frac{d_{dg}}{d}} + f_{Ftud} \quad (1)$$

$$\eta = \max\left\{\frac{1}{1 + 0,43 f_{Ftuk}^{2,85}}; 0,4\right\} \quad (2)$$

where:

γ_C partial factor for plain concrete (1.50 for persistent and transient design situations)
ρ_l longitudinal reinforcement ratio
f_{ck} characteristic concrete cylinder compressive strength
d_{dg} size parameter describing the crack and failure zone roughness
d cross-section effective depth
f_{Ftud} design ultimate residual tensile strength of SFRC (detailed below)
γ_V partial factor for shear and punching resistance without shear reinforcement (1.40 for persistent and transient design situations)
f_{yd} design yield strength of reinforcement
f_{Ftuk} characteristic ultimate residual tensile strength of SFRC (detailed below).

The size parameter d_{dg} is defined as

$$d_{dg} = \begin{cases} 16\,mm + D_{lower} \leq 40\,mm & for\,f_{ck} \leq 60\,MPa \\ 16\,mm + D_{lower}\left(\frac{60}{f_{ck}}\right)^4 \leq 40\,mm\,for\,f_{ck} > 60\,MP \end{cases} \quad (3)$$

where D_{lower} is the smallest value of the upper sieve size D in an aggregate for the coarsest fraction of aggregates in the concrete permitted by the specification of concrete EN 206 [8]. It should be noted here that this definition of D_{lower} is adequate for the design stage when the precise aggregate distribution is not known. However, from a physical standpoint, it seems more appropriate to use the maximum aggregate size D_{max} (when known) in Eq. (3).

As for the FRC residual strengths, the characteristic ultimate residual strength is determined as

$$f_{Ftuk} = \kappa_O \cdot 0.37 \cdot f_{R,3k} \quad (4)$$

$$f_{Ftud} = f_{Ftuk}/\gamma_{SF} \quad (5)$$

where:

κ_O factor accounting for fibre orientation (taken as 0.5)
$f_{R,3k}$ characteristic residual tensile strength corresponding to a crack mouth opening displacement (CMOD) of 2.5 mm in the notched beam three-point bending test according to EN 14651 [9]
γ_{SF} partial factor for FRC (1.50 for persistent and transient design situations).

In fact, Eq. (1) is an expansion of the prEN1992–1-1:2021 formulation for RC members:

$$\tau_{Rd,c} = \frac{0,66}{\gamma_V} \cdot \left(100 \cdot \rho_l \cdot f_{ck} \cdot \frac{d_{dg}}{d}\right)^{\frac{1}{3}} \geq \frac{11}{\gamma_V} \cdot \sqrt{\frac{f_{ck}}{f_{yd}} \cdot \frac{d_{dg}}{d}} \quad (6)$$

A difference can be observed between the left hand-sides of Eqs. (1) and (6), in the ratios $0.6/\gamma_C$ and $0.66/\gamma_V$. From Eqs. (1) and (6) it can be seen that the FRC contribution to the shear resistance mechanism is expressed through a positive effect of the design residual resistance f_{Ftud} and a negative effect of the factor η that accounts for higher crack widths that can be achieved when using FRC (penalizing the shear-transfer actions of concrete).

2.2 Experimental Database for Model Error Assessment

In order to perform a reliability-based calibration of the FRC partial factor, the model error δ of Eq. (1) should first be determined. The model error is defined as the ratio between actual behaviour/shear strength measured in experiments and shear strength predicted by models (this is a safe assumption as it neglects the geometrical and material deviations in the actual tests). For the present case, this will be performed for a model equation identical to that for design purposes, but where partial factors taken as 1.0 and considering average values for the material properties in the model equation.

Therefore, for this step, a large and representative database of experimental results is required. For this purpose, the database collected by Lantsoght, freely available online [10, 11] was considered. This database comprises 488 results on SFRC beams with longitudinal reinforcement and without shear reinforcement, sourced from 65 individual studies. The range of parameters of the original database is shown in Table 1 under the "Original database" column. All reported results were from simply supported beams tested in three- or four-point bending; the majority had rectangular cross-section, but some presented also flanges (as a first estimate, all types of cross-sections were considered). Importantly, residual strength was not reported in the database, because not all studies reported these values, for a number of reasons. However, a large number of fibre properties is reported, as well as a large variety of steel fibre types [11].

Since the range of parameters was very wide, three filtering criteria were imposed:

(a) Concrete classes between C12 and C120 were considered (mean compressive strengths between 20 and 128 MPa);
(b) Only beams with a longitudinal reinforcement ratio smaller than 4% were considered;
(c) Only beams with a clear shear span-to-effective depth ratio larger than 2.0 were considered.

These criteria were selected so as to limit the cases to the range of typical structural concrete compressive strengths, reinforcement ratios, as well as to eliminate cases where a direct load transfer to the support was occurring due to a short shear span-to-effective depth ratio. In total, using the three criteria, the number of specimens was reduced to 332. Approximately 90% of the results were on beams with $f_{cm} < 70$ MPa, effective depth between 100 and 500 mm and a longitudinal reinforcement ratio between 1.0% and 3.5%. The fibre volume fraction of 329 out of the 332 beams was below 2.0% (160 kg/m^3 for steel fibres) and for 296 beams it was below 1.5% (120 kg/m^3 for steel fibres).

Table 1. Range of parameter values in the SFRC database compiled by Lantsoght [10].

	Original database $n = 488$		Filtered database $n = 332$	
Parameter	Min	Max	Min	Max
b_w (mm)	50	610	50	610
h (mm)	100	1,220	100	1,220
d (mm)	85	1,118	85	1,118
l_{span} (mm)	204	7,823	459	7,823
a/d (–)	0.46	6.00	2.22	6.00
a_v/d (–)	0.20	5.95	2.00	5.95
ρ_l (%)	0.37%	5.72%	0.37%	3.70%
f_y (MPa)	276	900	276	610
f_{cm} (MPa)	9.8	215.0	20.2	111.5
V_f (%)	0.2%	4.5%	0.2%	4.5%
λ (–)	25	191	25	191
f_{uf} (MPa)	260	4,913	260	4,913

l_{span}—clear span of the beam; a/d—shear span-to-effective depth ratio measured from left side of loading plate to left side of support; a_v/d—clear shear span-to-effective depth ratio measured from face of loading plate to face of support; f_y—yield strength of steel reinforcement; V_f—fibre volume fraction; λ—fibre aspect ratio (ratio of fibre length to diameter); f_{uf}—tensile strength of fibres

2.3 Model Error Calculation

Since the residual tensile strengths for SFRC in the database of experimental results were not reported, they had to be estimated. For this purpose, regressions presented in Fig. 1a (for $f_{R,3}$) and 1b (for $f_{R,1}$) were developed using a statistical analysis of experimental results using the EN 14651 standard test [9], as reported by Venkateshwaran et al. [12], Tiberti et al. [13], Galeote et al. [14], and other experimental programs conducted at the Structures and Materials Technology Laboratory (LATEM) of the Polytechnic University of Catalonia (UPC).

The database includes a large variety of concrete mixes, with a range of compressive strengths of 15–117 MPa, volume fraction of fibres 0.33–2.52%, fibre aspect ratios 35–110, fibre tensile strength 1000 to 3000 MPa and fibre modulus of elasticity (E_f) 190000–210000 MPa. Figure 1 shows a good fit between the proposed linear regressions and the observed data.

The model error δ was estimated as $\delta = V_{experimental}/V_{model}$. The model shear strength V_{model} was calculated as $\tau_{model} \cdot b_w \cdot z$, where τ_{model} is the shear stress calculated using Eq. (1), b_w is the width of the cross-section and z is the internal lever arm that is taken as $0.9 \cdot d$. All partial factors were taken as 1.0, reported experimentally measured average values were used and residual strengths predicted by the regressions shown in Fig. 1.

Based on the 332 results, the mean model error was found to be 1.461, with a standard deviation of 0.394 and a coefficient of variation (CoV) of 26.9%.

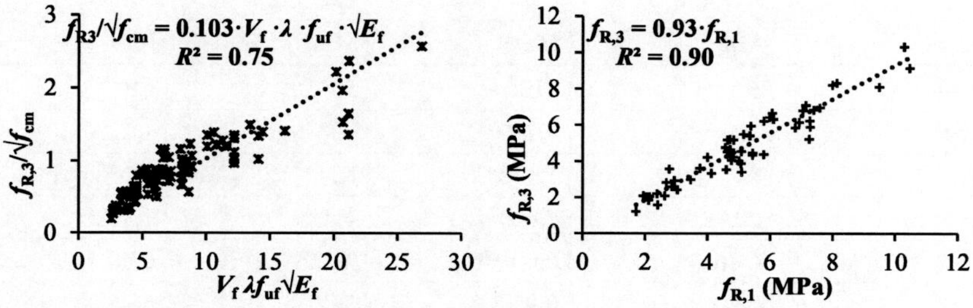

Fig. 1. Correlations used to assess $f_{R,3}$ (left) and $f_{R,1}$ (right) of the SFRC.

3 Calibration of the FRC Partial Factor γ_{SF}

3.1 Design Set and Reliability-Based Analysis

To assess the reliability of the prEN1992–1-1:2021, a set of design cases was defined, Table 2, following the methodology by Bairán and Casas [15]. A range of thicknesses typical for buildings and bridge-deck slabs, beams, footings, and mat foundations was selected, corresponding to the range of 200–1000 mm. The cross-sectional width was considered constant and equal to 300 mm, since the shear strength is linearly dependent on it. The effective depth was determined as $d = h - d_s = h - 50$ mm. The variables in Table 2 produce 420 combinations of geometry, longitudinal reinforcement, concrete class and aggregate size (expressed via d_{dg}). Different values of d_{dg} were considered in order to investigate the effect of aggregate size. The values considered for d_{dg} were the minimum and maximum values permitted (16 and 40 mm, respectively), as well as an intermediate value in order to observe the dependency on d_{dg}.

The process consisted of generating a design set of load shear stresses (Fig. 2) between 0.53 and 2.72 MPa using the currently proposed values of the resistance factors $\gamma_c = \gamma_{SF} = 1.50$ and $\gamma_V = 1.40$. FRC residual flexural capacities were limited to $f_{R,3k,min}$ and $f_{R,3k,max}$ of 3 and 10 MPa, respectively. The range was further divided in quarters, so that five design loads were obtained for each case. Considering the 5 design loads, a total of 2100 design cases (420 × 5) were generated.

The reliability of each design case was assessed using the reliability index β, related to the probability of failure (P_f) by $\beta = -\Phi^{-1}(P_f)$, where Φ is the cumulative standard normal distribution. FORM [16, 17] was used to estimate β.

A design failure was identified when a zero or negative value was found in the limit state function $G = V_R - V_S = \delta \cdot V_{R,model} - V_S$, where $V_{R,model}$ is the shear resistance predicted by the model and V_S is the shear load. Considering G a function of random variables, the probability of failure was computed as the probability of obtaining a negative value of G:

$$P_f = P(G < 0) = P(\delta \cdot V_R - V_S < 0) \tag{7}$$

Table 2. Range of variables in the design set.

Parameter	Values of parameters in the design set					
b (mm)	300					
h (mm)	200	400	600	800	1000		
ρl (–)	0.002	0.005	0.010	0.015	0.020	0.025	0.030
f_{ck} (MPa)	30	50	70	90			
d_{dg} (mm)	16	24	40				

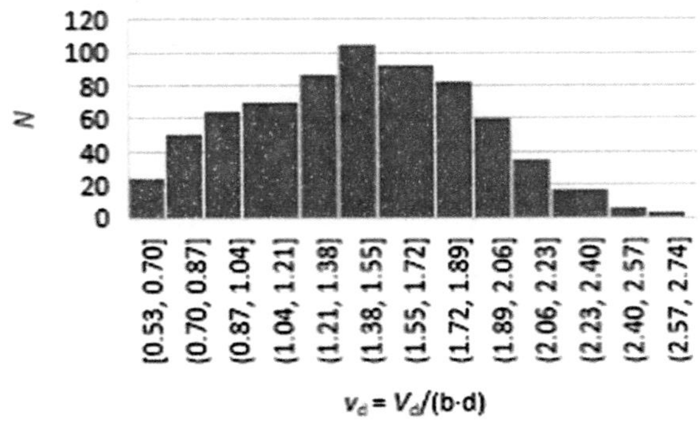

Fig. 2. Histogram of design load shear stresses (in MPa).

To calculate $V_{R,model}$, the prEN1992-1-1:2021 model was used without the safety factors and using the observed average values of the materials and geometry variables. The set of random variables and the corresponding distribution functions used are summarized in Table 3. The model error was selected as lognormally distributed according to the recommendations of the Joint Committee on Structural Safety (JCSS) and taking into account that it is a variable that only takes on positive values [18].

Table 3. Definition and distribution of random variables.

Variable	Description	Statistical model	Mean value (μ)	CoV
δ	Model error	Lognormal	1.461	0.269
f_c	Compression strength	Lognormal	f_{ck}+8 MPa	0.050–0.128
f_{Ftu}	Residual strength at w_u	Lognormal	1.412·f_{Ftuk}	0.2
Δb	Geometrical error in section width	Normal	0.003· b $\leq$3 mm	$\frac{4+0.006 \cdot b \leq 10mm}{\mu_b}$
Δd	Geometrical error in effective depth	Normal	10 mm	1

3.2 Calibration of the Partial Factor γ_{SF} for the prEN1992-1-1:2021 Model

To establish the relationship between γ_{SF} and β, the required f_{Ftu} has to be designed for each element of the design set, for different values of γ_{SF}. For this purpose, $V_{Rd} = V_{Sd}$ is imposed and the reliability index is computed for each case. The design shear load V_{Sd} was assumed to be deterministic; therefore, the computed reliability index refers to the probability of reaching a shear strength (V_R) smaller than the design resistance (V_{Rd}): $P(V \leq V_{Rd}) = \Phi(\beta_R)$ where β_R is the resistance reliability index, equal to $\beta_R = \alpha_R \cdot \beta$. α_R is the resistance sensitivity coefficient. For usual conditions this coefficient may be taken as 0.8 [19].

The reliability indexes associated to the prEN1992–1-1:2021 model for the shear strength of FRC members without shear reinforcement (Eq. (1)) were assessed for a range of safety factors γ_{SF} varying between 1.30 and 2.00, Fig. 3.

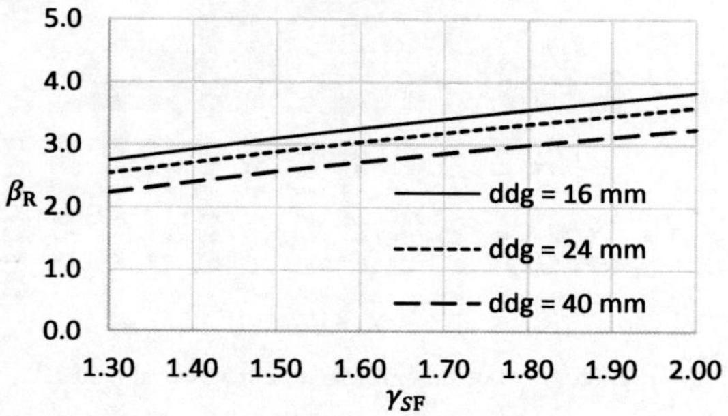

Fig. 3. Variation of the resistance reliability index with respect to γ_{SF}.

As seen from Fig. 2, the reliability index increases with the value of γ_{SF}, and decreases with increasing values of the parameter d_{dg}. The obtained values of β_R for $\gamma_{SF} = 1.50$ are 3.10, 2.89 and 2.57 for d_{dg} equal to 16, 24 and 40 mm, respectively.

As a reference, a target reliability index for ULS verifications for a period of 50 years and medium consequences of failure is $\beta_{target} = 3.8$. Then, the target resistance reliability index $\beta_{R,target} = 0.8 \cdot 3.8 = 3.04$. Therefore, it can be seen that the target reliability index is achieved (and even is slightly safe) when $d_{dg} = 16$ mm, i.e. when D_{lower} is neglected (as in the case of lightweight aggregate concrete structures). At the same time, increasing values of d_{dg} lead to a reduction of the achieved reliability index and, potentially, to a need for the increase in the FRC safety factor γ_{SF}: to achieve exactly $\beta_{R,target} = 3.04$, γ_{SF} should be 1.46, 1.59 and 1.83 for d_{dg} equal to 16, 24 and 40 mm, respectively.

4 Conclusions

This paper presents a reliability-based calibration of the FRC partial factor γ_{SF} for the shear design of FRC members without shear reinforcement according to the EC2-rev (prEN1992–1-1:2021) model. For this purpose, a database of experimental results was used for assessing the model error, after which a reliability analysis was carried out to calibrate the FRC partial factor. Based on the obtained results, the following conclusions can be drawn:

- Based on a probability analysis and a parametric study of 2100 individual cases, it was determined that the β_R values associated with $\gamma_{SF} = 1.50$ are 3.10, 2.89 and 2.57 for d_{dg} equal to 16, 24 and 40 mm, respectively. Hence, in the two cases with d_{dg} larger than 16 mm, the resistance reliability index results less than the code target for medium consequences of failure.
- In order to achieve the target resistance reliability index for ULS verifications for a period of 50 years and medium consequences of failure (0.8·3.8 = 3.04), γ_{SF} values of 1.46, 1.59 and 1.83 are needed for d_{dg} equal to 16, 24 and 40 mm, respectively. However, for practical purposes, a uniform value of γ_{SF} is desirable. This may be attained through modifications on the safety factor format or on the model, which will the subject of future research.

The results of this study can serve as a starting point for a more in-depth reliability-based analysis of the shear resistance model for FRC members without shear reinforcement to be provisioned in the future Eurocode 2. Future studies should cover a wider range of parameter values (in particular the fibre type and properties) and contain a deeper analysis of the influence of the d_{dg} parameter, so that recommendations for potential code modifications could be made.

Acknowledgements. This study has received funding from the European Union's Horizon 2020 research and innovation programme under the Marie Sklodowska-Curie grant agreement No 836270. This support is gratefully acknowledged. The authors also wish to express their acknowledgement to the Ministry of Economy, Industry and Competitiveness of Spain for the financial support received under the scope of the projects PID2019-108978RB-C32. Any opinions, findings, conclusions, and/or recommendations in the paper are those of the authors and do not necessarily represent the views of the individuals or organizations acknowledged.

References

1. de La Fuente, A., Casanovas-Rubio, M.D.M., Pons, O., Armengou, J.: Sustainability of column-supported RC slabs: Fiber reinforcement as an alternative. J. Constr. Eng. Manag. **145**, 1–12 (2019). https://doi.org/10.1061/(ASCE)CO.1943-7862.0001667
2. Winkler, A.; Edvardsen, C.; Kasper, T.: Examples of bridge, tunnel lining and foundation design with Steel-fibre-reinforced concrete. In: Proceedings of the American Concrete Institute, pp. 451–460. ACI Special Publication (2014)
3. Parra-Montesinos, G.J.; Wight, J.K.; Kopczynski, C.; Lequesne, R.D.; Setkit, M.; Conforti, A.; Ferzli, J. Earthquake-resistant fibre-reinforced concrete coupling beams without diagonal bars. In: Proceedings of the American Concrete Institute, pp. 461–470. ACI Special Publication (2014)
4. Cugat, V., Cavalaro, S.H.P., Bairán, J.M., de la Fuente, A.: Safety format for the flexural design of tunnel fibre reinforced concrete precast segmental linings. Tunn. Undergr. Sp. Technol. **103**, 103500 (2020). https://doi.org/10.1016/j.tust.2020.103500

5. Balász, G.: A historical review of shear. In: Proceedings of the Shear and Punching Shear in RC and FRC Elements, pp. 1–14. International Federation for Structural Concrete (fib), Salò (2010)
6. EN 1992-1-1 Eurocode 2: Design of Concrete Structures—Part 1–1: General Rules and Rules for Buildings. CEN, Brussels (2004). ISBN 978-0-580-62664-7
7. prEN1992-1-1 Eurocode 2: Design of Concrete Structures—Part 1–1: General Rules, Rules for Buildings, Bridges and Civil Engineering Structures. CEN, Brussels (2021)
8. EN 206: Concrete Specification, Performance, Production and Conformity. CEN, Brussels (2013)
9. EN 14651: Test method for metallic fibred concrete—Measuring the flexural tensile strength (limit of proportionality (LOP), residual). Br. Stand. Inst. (2005). 9780580610523
10. Lantsoght, E.: Database of Experiments on SFRC Beams Without Stirrups Failing in Shear (Version 1.0) . Zenodo (2019)
11. Lantsoght, E.O.L.: Database of shear experiments on steel fiber reinforced concrete beams without stirrups. Materials (Basel). **12**, 917 (2019). https://doi.org/10.3390/ma12060917
12. Venkateshwaran, A., Tan, K.H., Li, Y.: Residual flexural strengths of steel fiber reinforced concrete with multiple hooked-end fibers. Struct. Concr. **19**, 352–365 (2018). https://doi.org/10.1002/suco.201700030
13. Tiberti, G., Germano, F., Mudadu, A., Plizzari, G.A.: An overview of the flexural post-cracking behavior of steel fiber reinforced concrete. Struct. Concr. **19**, 695–718 (2018). https://doi.org/10.1002/suco.201700068
14. Galeote, E., Blanco, A., Cavalaro, S.H.P., de la Fuente, A.: Correlation between the Barcelona test and the bending test in fibre reinforced concrete. Constr. Build. Mater. **152**, 529–538 (2017). https://doi.org/10.1016/j.conbuildmat.2017.07.028
15. Bairán, J.M., Casas, J.R.: Safety factor calibration for a new model of shear strength of reinforced concrete building beams and slabs. Eng. Struct. **172**, 293–303 (2018). https://doi.org/10.1016/j.engstruct.2018.06.033
16. Melchers, R.E.: Structural Reliability Analysis and Prediction. Ellis Horwood (1987)
17. Madsen, H.O., Krenk, S., Lind, N.C.: Methods of Structural Safety. Dover, New York (2006)
18. JCSS: Probabilistic Model Code (2001). ISBN 978-3-909386-79-6
19. EN 1990: Eurocode-Basis of Structural Design, p. 2002. CEN, Brussels (1990)

Analytical Determination of Flexural Resistance of Rein-forced Concrete Beams with Corrosion

N. Vega[1(✉)], J. Moreno[1], P. Castro Borges[2], and J. Varela[1]

[1] Facultad de Ingeniería, Universidad Autónoma de Yucatán, Mérida, México
a13114865@alumnos.uady.mx

[2] Centro de Investigación y de Estudios Avanzados, del Instituto Politécnico Nacional, Unidad Mérida, Mérida, México

Abstract. Corrosion of steel reinforcement is one of the main problems associated with deterioration of reinforced concrete structures. The dimensions of the cross-section of concrete and steel reinforcement is reduced due to corrosion. Because of this, structural strength of the concrete elements is also reduced. Therefore, the behavior of the structure can be committed. In this work, analytical and experimental strengths of reinforced concrete beams affected with corrosion are compared. The experimental strengths of 11 beams found in the literature were considered. The analytical flexural strength was determined by using the fundamental hypotheses of the flexural theory. The constitutive model of steel reinforcement was based on the model proposed by Rodríguez and Botero and the experimental results found in the literature. The relationship between analytical strength and experimental flexural strength varied between 0.97 and 1.12. Based on the results obtained, it was found that the strength of reinforced concrete beams affected with corrosion is predicted accurately using flexural theory. However, it is necessary to consider the reductions in the geometry of the beam, the reduction in the section in the longitudinal and transverse steel reinforcement; and corresponding changes in their constitutive models.

Keywords: Corrosion · Reinforced concrete · Analytical models · Flexure

1 Introduction

Reinforced concrete (RC) is one of the most used structural systems for building around the world. The global annual production of concrete is estimated approximately in 11,000 million cubic meters [1], constituted mainly by columns and beams. One of the factors associated with the reduction of the useful life of RC beams is the corrosion in the steel reinforcement (SR). This phenomenon generates changes in the structure of steel, which generates modifications in its physical and mechanical properties. Reductions in the cross section of the SR can be associated with corrosion. Because of that, concentration of tensile stress of steel bars are produced in the most affected area, reducing its strength and ductility. Corrosion also generates an increase in volume associated with the transformation of steel to oxide. This produces tensile stresses in the concrete that surrounds it. The increment in tensile stress of concrete generates new cracks and eventually the

J. I. Escalante-Garcia et al. (Eds.): RW 2021, RILEM Bookseries 40, pp. 559–566, 2023.
https://doi.org/10.1007/978-3-031-21735-7_61

detachment of the concrete cover [2]. The presence of cracks favors the intrusion of aggressive agents that cause corrosion.

One of the main factors in the corrosion process in RC structures is the type of environment where they were constructed. For structures located in marine areas or near coasts, the main aggressive agents will be chlorides from the sea (NaCl). On the other hand, structures located in areas with a high content of pollution from industry or large concentrations of populations, will be affected mainly by sulfates (SO_2) and carbonates (CO_2) [3]. In 2016 the estimated cost associated with repairs and maintenance of structural elements was about 3.4 trillion dollars per year [4].

From a literature review it was found that numerous studies focused on the evaluation of the behavior of RC elements and reinforced steel bars with corrosion exist. The main variables studied were: the level of corrosion in the RS [5–13], the structural behavior of RC beams with corrosion [5–7, 9, 10, 12], the effect of corrosion on the mechanical behavior of SR [8, 11] and the models used to determine the strength of beams with corrosion [5, 10, 12].

It has been observed that corrosion of steel reinforcement produces losses in the strength, ductility, and stiffness of RC beams [9, 10, 12]. As the level of corrosion increases, the loss of the cross section of SR also increases. Therefore, tensile strength in the SR increases and produce premature failure of the steel reinforcement bars. In RC beams with high levels of corrosion, failure associated to bending behavior can be changed by shear failure [5, 10]. On the other hand, in studies focused on the behavior of steel bars with corrosion, it has been observed that the corrosion in those bars can be uniform or localized (pitting). Greater losses of the cross section of bars are generated by localized corrosion. Pitting of steel bars induce their premature failure [8, 11, 13]. Different models have been used to determine bending and shear strengths of reinforced concrete beams with corrosion. The flexural strength of RC beams with corrosion has been determined by using flexural theory [5, 12] or the Finite Element Method (FEM) [10, 12]. For both cases, the flexural strength of beams has been well predicted. However, the changes in the mechanical properties of the materials are neglected.

In order to improve the assessment of reinforced concrete structures affected by corrosion, the residual strength of those structural elements is needed. Therefore, it is necessary to develop analytical models to determine the behavior of reinforced concrete elements with corrosion. The objective of this work is to propose analytical models to determine the flexural strength of RC beams affected by corrosion. Analytical strengths are compared with corresponding experimental ones. Experimental results of 11 RC beams of the literature review were considered [5, 6].

2 Methodology

2.1 Experimental Data Obtained from the Literature

From the literature review, numerous studies focus on the behavior of RC beams with corrosion were found. However, only for 11 beams, information of geometric and mechanical properties of the concrete and the steel reinforcement before and after the corrosion is provided [5, 6]. Wenjun et al. (2013), provide information of geometric and material properties before and after the corrosion process. Rodríguez et al. [15], only provide information of geometric and material properties before corrosion. However, the steel used in their study was similar to that used by Wenjun et al. (2013). The SR models were modified with these experimental results. Table 1 shows the geometry, the lower (lc) and upper (uc) concrete cover, the longitudinal steel reinforcement for tension (LT), the longitudinal steel reinforcement for compression (LC), the transversal steel reinforcement (AV) and the spacing between stirrups (s). The B2CL2–1, T-124, T-125, T-126 and T131 beams present spalling of the concrete cover. Table 2 shows the mechanical properties of the concrete and SR, respectively, and the failure type of the RC beams considered. Table 3 shows the corresponding corrosion level (CL) for each beam. Corrosion level is based on the percentage of loss area in the steel reinforcement. Figure 1 shows the experimental set-up used on both studies, where P is the total load applied on the beams.

Table 1. Characteristics of the RC beams.

Beam	b (cm)	h (cm)	L (cm)	lc (cm)	uc (cm)	LT (cm^2)	LCS (cm^2)	Av (cm^2)	s (cm)
B2CL2–1*	13	26	95	1	0	1.67	0.63	0.29	10
B2CL2–2	15	28	95	1	1	1.76	0.63	0.29	10
T-111	15	20	220	1	1	1.43	0.63	0.32	10
T-112	15	20	220	1	1	1.43	0.63	0.32	10
T-113	15	20	220	1	1	0.89	0.49	0.11	10
T-114	15	20	220	1	1	0.92	0.47	0.05	10
T-115	15	20	220	1	1	0.98	0.55	0.05	10
T-116	15	20	220	1	1	0.60	0.50	0.05	10
T-121	15	20	220	1	1	3.66	0.80	0.32	10
T-122	15	20	220	1	1	3.66	0.80	0.32	10
T-126*	13	18	220	1	0	2.81	0.57	0.07	10

2.2 Analytical Models to Determinate Flexural Strength of RC Beams

To determine the flexural strength of RC beams, the flexural theory was used. This theory is based on three fundamental hypotheses: kinematics, constitutive and equilibrium. For concrete and steel reinforcement, the constitutive models of Kent and Park [14] "Eqs. (1)–(5)" and Rodríguez and Botero [15] "Eq. (6)", respectively, were considered. Table 4 shows the mechanical properties of bars proposed by Rodríguez and Botero. For the beams with corrosion, the model of Rodríguez and Botero was modified with the results obtained by Wenjun (Table 2). Additionally, the reduction of the concrete cross-section of beams that presented spalling of concrete cover was considered (Table 1).

$$fc = \begin{cases} fc = f'c\left(\left(\frac{2\varepsilon_c}{\varepsilon_o}\right) - \left(\frac{\varepsilon_c}{\varepsilon_o}\right)^2\right) & \varepsilon_c \leq 0.002 \\ fc = f'c(1 - Z(\varepsilon_c - \varepsilon_o)) & \varepsilon_o \leq \varepsilon_c \leq \varepsilon_{cu} \end{cases} \tag{1}$$

Table 2. Material properties of the concrete and steel reinforcement

Beam	f′c (kg/cm^2)	fy (kg/cm^2)	εy	fu (kg/cm^2)	εu	F. mode
B2CL2–1*	610	6,420	0.003	7,747	0.031	Flexure
B2CL2–2	610	6,370	0.003	7,850	0.033	Flexure
T-111	510	5,860	0.003	6,035	0.070	Flexure
T-112	510	5,860	0.003	6,035	0.070	Flexure
T-113	346	6,370	0.003	7,850	0.033	Flexure
T-114	346	6,370	0.003	7,850	0.033	Flexure
T-115	346	6,370	0.003	7,850	0.033	Flexure
T-116	346	6,370	0.003	7,850	0.033	Flexure
T-121	510	5,860	0.003	6,035	0.070	Flexure
T-122	510	5,860	0.003	6,035	0.070	Flexure
T-126*	490	6,370	0.003	7,850	0.033	Flexure

* Beams that presented spalling of concrete cover

Table 3. Corrosion levels in the longitudinal steel reinforcement for tension, for compression and the transversal steel reinforcement.

Beam	LT-CL (%)	LCS-CL (%)	Av-CL (%)
B2CL2-1	30.6	ND	64.0
B2CL2-2	34.2	ND	67.0
T-111	0.0	0.0	0.0
T-112	0.0	0.0	0.0
T-113	48.8	23.2	64.7
T-114	41.5	25.1	83.3
T-115	37.9	12.7	83.3
T-116	62.0	21.0	80.0
T-121	0.0	0.0	0.0
T-122	0.0	0.0	0.0
T-126	38.0	13.0	76.6

ND – Not determinied

$$Z = \frac{1}{2(\varepsilon_{50u} + \varepsilon_{50h} + \varepsilon_o)} \tag{2}$$

$$\varepsilon_{50u} = \frac{3 + \varepsilon_o f'c}{f'c - 1000} \tag{3}$$

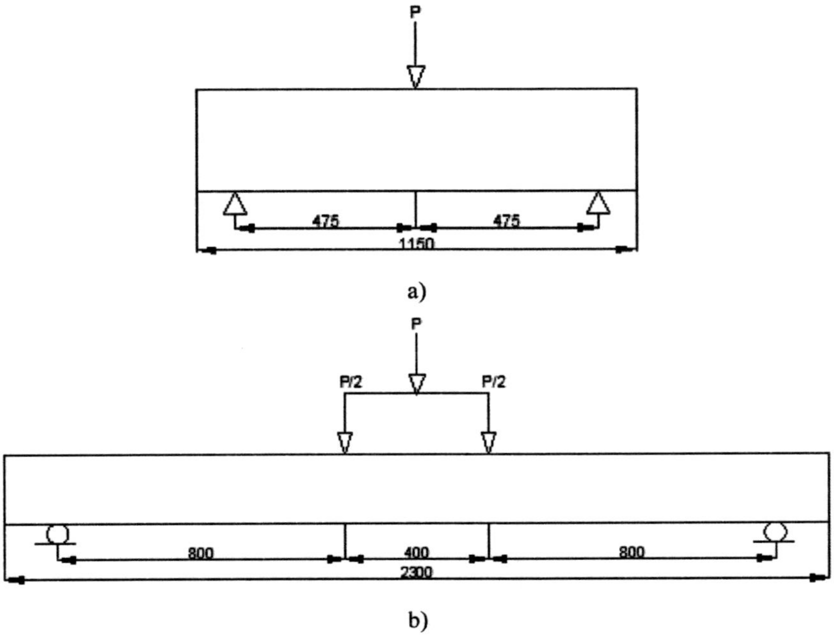

Fig. 1. Experimental set-up used on beams studied by a) Wenjun et al. (2013) and b) Rodriguez et al. [15]. Dimensions in mm.

$$\varepsilon_{50h} = \frac{3}{4}\rho_s\sqrt{\frac{b_s}{s}} \tag{4}$$

$$\rho_s = \frac{l_b A_s}{s A_c} \tag{5}$$

$$fs = \begin{cases} fs = E_s\varepsilon_s & \varepsilon_s \le \varepsilon_y \\ fs = fy & \varepsilon_y < \varepsilon_s \le \varepsilon_{sh} \\ fs = fsu + (fy - fsu)\left(\frac{\varepsilon_{su}-\varepsilon_s}{\varepsilon_{su}-\varepsilon_{sh}}\right)^p & \varepsilon_s > \varepsilon_{sh} \end{cases} \tag{6}$$

where:

fc is the compressive stress of the confined concrete.
f′c is the specified compressive strength of concrete.
ε_c is the unit strain of the concrete.
ε_0 is the unit strain corresponding to the maximum compressive strength of the concrete.
ε_{cu} is the unit strain associated with the crushing of the concrete.
E_s is the modulus of elasticity of steel.
fs is the tensile stress of the steel.

fsu is the tensile stress at breakdown of the steel.
fy is the yield stress of the steel.
ε_{su} is the unit strain corresponding to the tensile stress at breakdown of the steel.
ε_s is the unit strain of the steel.
ε_{sh} is the unit strain associated with the beginning of the de-formation hardening zone of the steel.
p it is a value assigned according to the diameter of the steel and its yield strength.

Table 4. Mechanical properties of bars proposed by Rodríguez and Botero.

Diameter	fy (kg/cm^2)	εsh	fsu (kg/cm^2)	εsu	P
<#5	4,609	0.006	7,436	0.1420	3.362
≥#5	4,577	0.0088	7,491	0.1171	3.474

3 Results and Discusion

Table 5 shows the experimental yielding strength (Py-e), the corresponding analytical results determined with the model proposed in this work (Py-a) and the ratio Py-e/Py-a. Based on the results obtained, it is observed that the relationship between the analytical and experimental yielding strength was 1.06. Therefore, the yielding strength is well predicted with the proposed analytical model. However, it is necessary to know the characteristics of SR with corrosion.

Table 5. Comparison of the yield strengths of experimental and analytical RC beams (strength in KN).

Beam	Py-e	Py-a	Py-a/Py-e
B2CL2-1	110	116.60	1.06
B2CL2-2	88	93.60	1.06

Table 6 shows the ultimate experimental strength of the beams (Pu-e), the ultimate analytical bending strength (Pu-f), the ratio Pu-f/Pu-e and the type of failure of the RC beams. It is observed that, the ratio between the analytical and experimental ultimate strength varies from 0.97 to 1.12. Therefore, the ultimate flexural strength of the studied beams considered is well predicted with the proposed model.

To determine the residual strength of the reinforced concrete beams affected by corrosion, reductions in cross-section beam (cracking and detachment of the concrete cover) must be considered. Reduction in the cross-section of the longitudinal and transverse steel reinforcement, and changes in their corresponding constitutive models are also included. Analytical models to determine shear strength of beams are also needed to define the failure mode. Finally, more studies are required on RC beams and the constitutive models of SR affected by corrosion.

Table 6. Comparison of the ultimate strength of experimental and analytical RC beams (strength in kN).

Beam	Pu-e	Pu-f	Pu-f/Pu-e	F. mode
B2CL2–1	148.00	148.94	1.01	Flexure
B2CL2–2	123.00	123.30	1.00	Flexure
T-111	20.68	20.96	1.01	Flexure
T-112	21.51	20.96	0.97	Flexure
T-113	13.84	15.56	1.12	Flexure
T-114	14.38	15.14	1.05	Flexure
T-115	15.89	17.27	1.09	Flexure
T-116	11.78	11.68	0.99	Flexure
T-121	49.45	50.96	1.03	Flexure
T-122	52.47	50.96	0.97	Flexure
T-126	39.73	40.14	1.01	Flexure

4 Conclusions

In this work, analytical and experimental strengths of beams affected with corrosion are compared. The experimental strengths of 11 beams found in the literature were considered. Based on the results obtained, the following conclusions are presented:

- The experimental flexural strength of CR of the studied beams was well predicted with the proposed model. Reductions in cross-section beam (cracking and detachment of the concrete cover) must be considered. Reduction in the cross-section of the longitudinal and transverse steel reinforcement, and changes in their corresponding constitutive models should also be included.
- Analytical models to determine shear strength of beams are needed to define the failure mode.
- It is recommended to obtain more experimental results of RC beams with corrosion, and with behavior dominated by bending and shear to validate the use of the models proposed in this work.

Acknowledgments. The first author thanks CONACYT for the scholarship granted to carry out his Doctoral studies at the Facultad de Ingeniería de la Universidad Autónoma de Yucatán.

References

1. Ceballos, M.: El concreto material fundamental para la infraestructura. Centro de Innovación Tecnológica para la Construcción 24–25 (2016)
2. Yuan, Y., Ji, Y., Surendra, P.: Comparison of two accelerated corrosion techniques for concrete structures. ACI Struct. J. 344–347 (2007)
3. Chávez, E.: El concreto material fundamental para la infraestructura. Centro de Innovación Tecnológica para la Construcción, Tesis de Doctorado. Universidad Autónoma de Campeche (In spanish)
4. NACE.: International Measures for the Prevention, Application and Economics of Corrosion Technology. NACE International (2016)
5. Rodriguez, M., Botero, J.: Evaluación del comportamiento de barras de acero de refuerzo sometidas a cargas monotonimcas y cíclicas reversibles incluyendo pandel. Revista de Ingeniería Sísmica 9–27 (1997) (In spanish)
6. Zhu, W., Francois, R., Corronelli, D., Cleland, D., Effect of Corrosion of reinforcement on the mechanical behaviour of highly corroded RC beams. Eng. Struct. 544–554 (2013)
7. Murai, R., Oyado, M., Yasojima, A., Kanakubo, T.: Compression failure of RC beams deteriorated by corrosion. In: The 11th Canadian Conference on Earthquake Engineering. Canadian Association for Earthquake Engineering (2015)
8. Kanakubo, T., Oyado, M.: Tensile characteristics evaluation of corroded reinforcing bars extracted from actual structures. J. Civ. Eng. Arch. 1439–1451 (2015)
9. Dasar, A., Hamada, H., Sagawa, Y., Yamamoto, D.: Deterioration progress and performance reduction of 40-year-old reinforced concrete beams in natural corrosion environments. Constr. Build. Mater. 690–704 (2017)
10. Bossio, A., Imperatore, S., Kiormarsi, M.: Ultimate flexural capacity for reinforced concrete elements damaged by corrosion. Buildings **160** (2019)
11. Fernandez, I., Berrocal, G.: Mechanical properties of 30-year-old naturally corroded steel reinforcing bars. Int. J. Concr. Struct. Mater. (2019)
12. Song, L., Fan, Z., Hou, J.: Experimental and analytical investigation of the fatigue flexural behaviour of corroded reinforced concrete beams. International J. Concr. Struct. Mater. (2019)
13. Chen, E., Berrocal, G., Fernandez, I., Lofgren, I., Lundgren, K.: Assessment of the mechanical behaviour of reinforcement bars with localised pitting corrosion by Digital Image Correlation. Eng. Struct. (2020)
14. Scott, B., Park, R., Priestley, M.: Stress-strain behaviour of concrete by overlapping hoops at low and high strain rates. ACI 13–27 (1982)
15. Rodriguez, J., Ortega, L., Casal, J., Load carrying capacity of concrete structures with corroded reinforcement. Constr. Build. Mater. 239–248 (1997)

Analysis of Shear Panel Elements Using Improved Fixed Strut Angle Model Based on Plane-Stress Element

Nikesh Thammishetti[1(✉)], Shanmugam Suriya Prakash[1], Trevor D. Hrynyk[2], Javad Hashemi[3], and Riadh Al-Mahaidi[3]

[1] Department of Civil Engineering, IIT Hyderabad, Telangana 502285, India
ce16resch11005@iith.ac.in
[2] Department of Civil and Environmental Engineering, University of Waterloo, Waterloo, Canada
[3] Swinburne University of Technology, Melbourne VIC – 3122, Australia

Abstract. The behaviour of reinforced concrete (RC) member can be evaluated by employing a reasonably accurate constitutive relationship and element type in the finite element (FE) framework. Important components like columns, beams and shear wall in an RC structure can be analysed in the FE framework by using a plane-stress element. In general, RC members are subjected to the combination of axial, shear, flexure, and torsional loading. Hence, it is necessary to use constitutive models which can consider the effects of combined loading. This work focuses on implementing an improved fixed strut angle model in finite element framework to analyse shear panels under combined shear and axial loads. The proposed procedure is based on secant stiffness based iterative formulations and employs constitutive relationships which consider the concrete softening and tension stiffening. The material models and the equilibrium conditions based on the fixed strut angle model were implemented into MATLAB based FE framework. The reliability of the proposed analytical model is checked against the experimental results available in the literature. Moreover, it is essential to use efficient material constitutive relationships which can be effectively implemented in a finite element framework without compromising the accuracy of the predictions. A comprehensive analytical study is carried, wherein predictions were obtained using different combinations of material models. Comparison of the results shows that the predictions of the proposed model agree well with the test results.

Keywords: Constitutive relationships · Nonlinear finite element · Plane-stress element · Shear panel · Smeared crack model

1 Introduction

The behaviour of a reinforced concrete (RC) structure can be estimated using various methods of analysis. The analysis models include simple lumped plasticity or plastic hinge, distributed nonlinearity models, finite element models and fiber models [1]. The three-dimensional finite element analysis is the most sophisticated method which

J. I. Escalante-Garcia et al. (Eds.): RW 2021, RILEM Bookseries 40, pp. 567–576, 2023.
https://doi.org/10.1007/978-3-031-21735-7_62

can consider the effect of bi-directional loading. An RC membrane element is a basic component of most of the RC structural members, and the response of members can be accurately predicted by incorporating the behaviour of an RC membrane element in finite element framework [2]. Few simplified membrane models based on the smeared crack concept were developed in the past, ignoring the contribution of dowel action and/or aggregate interlock [3, 4]. The smeared crack concept considers the cracked concrete to be a new orthotropic continuous material, and the models can be broadly classified into the rotating angle and fixed angle models [5]. The fixed strut angle model (FSAM) was proposed [6, 7], which assumes the rotation of the principal planes before the cracking of concrete and crack direction is assumed to be locked after the first crack as opposed to rotation of cracks throughout the loading assumed in modified compression field theory (MCFT) [3]. The FSAM can model crack kinematics in a better sense. To overcome the limitations of the existing FSAM model, the model is further improved to include shear transfer models, which were developed based on experimental validation [8, 9].

2 Research Significance

The present study aims to implement the improved fixed strut angle model (IFSAM) in the finite element framework. The implemented formulation is validated by comparing the analytical predictions with the experimental responses of shear panels. The IFSAM based plane stress element can accurately predict the behaviour of the shear panels subjected to different loading conditions.

3 Plane Stress Element—Finite Element Formulation

The IFSAM based plane stress element is formulated based on the implementation of a four-node rectangular element. The plane stress element has eight translational degrees of freedom (DOF), with two DOFs at each of the four nodes (Fig. 1).

3.1 Stress-Strain Relationships

A brief background of the IFSAM model covering the behaviour of membrane element under monotonic loading is presented, and a detailed description of the model can be found elsewhere [9]. Consider an RC element reinforced in perpendicular directions, as shown in Fig. 2. The directions of the longitudinal and transverse steel bars are designated as the '*l*' and '*t*' axis, respectively. The amount of longitudinal and transverse steel is given be ρ_l and ρ_t, respectively. The in-plane loads acting on the panel includes normal stresses in x, y directions which are designated as σ_x and σ_y, respectively and uniform shear stress τ_{xy}. The behaviour of the membrane element under monotonic in-plane loading can be classified into two different phases: (a) uncracked phase, (b) after the formation of the first strut.

Uncracked phase. In the uncracked phase, principal stress directions are assumed to coincide the principal strain directions. Based on the loading conditions, the principal directions may rotate with the increase in load. The equilibrium and compatibility equations can be developed in the principal directions to understand the behaviour of membrane element in uncracked phase.

After the formation of the first crack. With the increase in loading, the rotating angle changes depending on the loading condition. The direction of principal compressive strain corresponding to the loading condition at which the concrete strain reaches cracking strain is the strut direction. The compatibility and equilibrium equations in the strut direction can be developed to understand the behaviour of the membrane after the formation of the first crack.

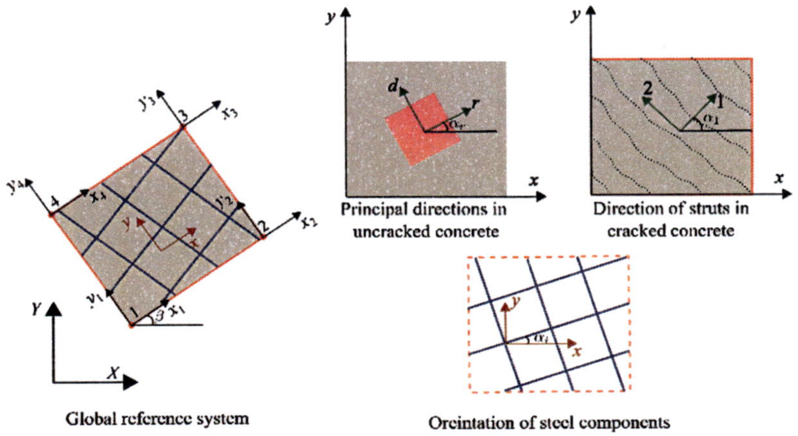

Fig. 1. Reference systems for reinforced plane stress element

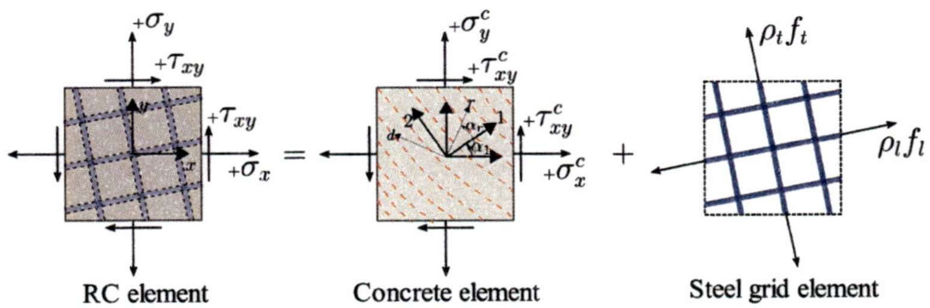

Fig. 2. RC panel subjected to in-plane shear and axial loads

3.2 Material Models

The material constitutive relationships play a significant role in the analysis. Every material model has unique features like softening, hardening. The material models used in the current study, especially the models defining concrete and aggregate interlock behaviour, use nonlinear relation between stresses and strains. It is crucial to choose a suitable material model for the type of analysis considered to predict the response accurately.

The following sub-sections provide the information related to the constitutive models used in the current work.

Compression models for concrete

Chang and Mander compression model: Chang and Mander [10] developed a material model to define the normal and high-strength concrete hysteresis behaviour under compression. Though the model can describe the pre-peak response and the post-peak stiffness degradation, it cannot consider the effect of biaxial loading. Kolozvari [9] employed a stress softening model proposed by Vecchio and Collins [11] with the Chang and Mander model, used in the current study.

Vecchio and Collins compression model: The compression model proposed by Vecchio and Collins [3] employs Hognestad parabola to define the compression response of concrete under pure compression. The model employs a stress-only softening coefficient to consider the effect of lateral tensile stresses on compressive stress. Additionally, Vecchio and Collins [11] proposed a material model to define the compressive response of high-strength concrete. The softening coefficients were modified later to consider the effect of the compressive strength of concrete [12].

Belarbi and Hsu compression model: Belarbi and Hsu [13] proposed equations for the softened stress-strain response of concrete based on a parabolic curve softened using stress and strain softening coefficients. The equations are further simplified by considering the strain and stress softening coefficients to be the same. The softening coefficient ζ is modified for high strength concrete as proposed by Zhang and Hsu [14].

Tension models for concrete

In reinforced concrete, due to the bond between the cracked concrete and the steel, the cracked concrete contributes to load resistance. The behaviour of reinforced concrete under tension will be stiffer than that of bare reinforcement. In the finite element framework, neglecting the post-cracking contribution of concrete leads to an intermittent change in stiffness estimation, promoting convergence issues. The following tension models of concrete have been implemented in the present research, considering a uniaxial cracking criterion:

Belarbi and Hsu tension model: Tamai et al. [15] have proposed a tension model based on the tension tests on concrete embedded with mild steel. Later, based on extensive testing of RC panel elements, Belarbi and Hsu [16] have proposed the tension model used in the current study.

Vecchio and Collins tension model: Vecchio and Collins [3] proposed an elastic pre-peak and post-peak decay tension model. The model considers the concrete's tensile modulus to be the same as that of the unsoftened concrete under compression.

Constitutive relationship for steel

Menegotto, Pinto and Filippou model: Menegotto and Pinto [17] developed a hysteretic model to describe the behaviour of steel under compression and tension. Filippou et al. [18] further modified the model and proposed a computationally stable material model.

The hysteretic model simplified for the monotonic case is employed in the study. The reduction in yield stress of steel in reinforced concrete is considered by incorporating the effective yield stress of steel proposed by Belarbi and Hsu [16].

Aggregate interlock mechanism

The primary assumption of the FSAM is that struts form after the cracking of concrete. The stresses developed in the strut direction constitute the resistance offered by the element. Additionally, the sliding of the concrete element results in the shear transfer through aggregate interlock. Bujadham and Maekawa [19] developed an aggregate interlock model based on the Universal Stress Transfer Model by validating experimental results and is used in the current research to model the shear transfer through aggregate interlock.

4 Plane-Stress Element Formulation

The stresses and strains in a plane stress element can be related using the material stiffness matrices of individual concrete and steel components. The stress, strain states, and the plane-stress element's sign convention are illustrated in Fig. 3.

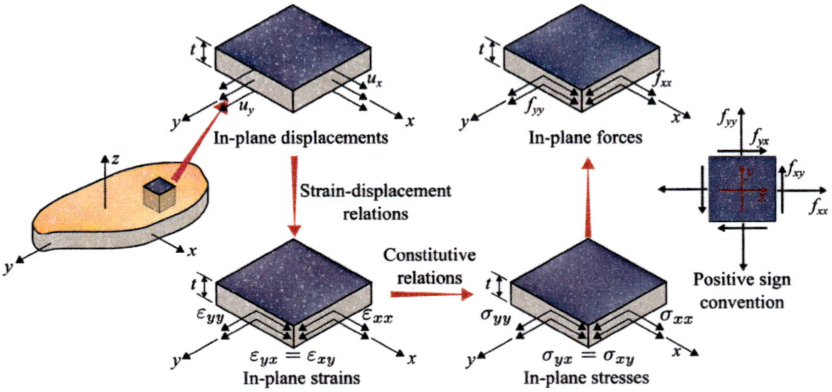

Fig. 3. Plane stress element

Material stiffness

The stress components and the strain components in the element are related based on the constitutive relationships. While representing the stress components and strain components as vectors $\{\sigma\}$ and $\{\varepsilon\}$, respectively, the material stiffness matrix $[D]$ can be used to relate both the components.

Since RC members are composite of two different materials, namely concrete and reinforcement, individual material stiffness matrices will relate the stress and strain in individual materials. The material stiffness matrices are to be defined in the local coordinate system of the individual materials. The global material stiffness is then obtained by combining the material stiffness of individual matrices, which are transformed into the global axis using appropriate transformation matrices. Considering concrete as an orthotropic material and ignoring the Poisson's ratio, the total material stiffness matrix can be evaluated as shown in Eqs. (1), (2), (3), (4) and (5).

$$[D_c]' = \begin{bmatrix} \overline{E_{c2}} & 0 & 0 \\ 0 & \overline{E_{c1}} & 0 \\ 0 & 0 & \overline{G_c} \end{bmatrix} \quad (1)$$

$$[D_s]_i' = \begin{bmatrix} \rho_i \overline{E}_{si} & 0 & 0 \\ 0 & 0 & 0 \\ 0 & 0 & 0 \end{bmatrix} \quad (2)$$

$$[D] = [\mathrm{D_c}] + \sum_{i=1}^{n} [D_s]_i \quad (3)$$

$$[D_c] = [T_c]^T [D_c]' [T_c] \quad (4)$$

$$[D_s]_i = [T_s]_i{}^T [D_s]_i' [T_s]_i \quad (5)$$

where, $\overline{E_{c2}}$, $\overline{E_{c1}}$, $\overline{G_c}$ are the secant moduli of concrete; ρ_i is the reinforcement ratio and $\overline{E}_{si}$ is the secant modulus of steel; $[T_c]$ and $[T_s]_i$ are transformation matrices for the concrete and reinforcement components, respectively.

The overall stiffness of an element in the member can be evaluated as

$$[k] = \int_v [B]^T [D][B] dv; \ [B] \text{ is the shape function matrix} \quad (6)$$

For the rectangular element, the coefficients for the elements of the stiffness matrix can be evaluated using closed-form solutions, which are available in the literature [2].

5 Solution Algorithm

A load controlled nonlinear finite element analysis of the reinforced concrete elements was implemented in MATLAB (R2018b) [20] by employing the IFSAM plane stress element. The details of the solution strategy proposed by Vecchio [2] is described in this section. Firstly, the geometrical details, loading scheme and the material properties of the element are defined. Then, the element is discretised based on the input parameters. For the first load increment, the secant stiffness of concrete, steel is assumed to be the elastic modulus of concrete and steel, respectively. In the pre-cracking regime, the shear modulus is assumed to be half of the harmonic mean of secant stiffness of concrete in the principal directions to attain convergence. Next, the individual concrete and steel material stiffness matrices are developed. The element stiffness matrices in global directions are then calculated. The element stiffness matrices are assembled to form a global structure stiffness matrix. From the loading conditions of the increment, the free joint displacements are calculated, which are subsequently used to calculate the strains in individual elements. Based on the strain condition in the global directions, the strains in principal directions or the strains in the strut directions are evaluated. The secant stiffness of the materials is re-evaluated using the strains in the principal/strut directions of each element. The new material stiffness matrices $[D]'$ are determined using the updated secant stiffness. The material stiffness matrices are checked for tolerance, if $[D]' \neq [D]$, the global stiffness matrix is evaluated using $[D]'$, else the unknown forces are evaluated.

6 Experimental Validation

The efficacy of the proposed model and the accuracy of different constitutive relationships are studied by comparing the analytical peak shear predictions with the experimental results. The test programs were chosen to consider variation in parameters like reinforcement ratio, loading criteria and compressive strength. The different programs were chosen to increase the broadness of the validation. The shear panels were modelled using the IFSAM plane stress element and the Membrane – 2000 [21]. The details of different combinations of constitutive relationships and the IDs used are presented in Table 2. The results are shown to be sensitive to the tension model. Hence, the results presented are considered through averaging the responses as VC_Tension and BH_Tension as mentioned in Table 2. The average predictions of the peak shear ratio $(\tau_{analysis}/\tau_{test})$ different tension models are presented in Fig. 4. Additionally, the peak shear ratio obtained from MCFT analysis using Membrane 2000 as τ_{MCFT}/τ_{test} is also presented in Fig. 4.

Vecchio and Collins [3] reported a test program involving thirty RC elements (PV series) subjected to different loading ratios. The details of the PV series specimens validated in the current study are presented in Table 1. The IFSAM is observed to predict the behaviour of the panels with reasonable accuracy. The predicted-to-experimental peak shear stress ratio for the PV series specimens is found to be in the range of 0.77–1.14. Additionally, both the IFSAM and MCFT underpredict the peak shear stress for the PV series specimens under combined axial and shear stresses.

Pang and Hsu [22] reported a study involving thirteen RC panel elements of 1400 × 1400 × 178 mm. Specimen of series A and series B tested under pure shear are validated in the current study (Table 1). The IFSAM is seen to overpredict the peak shear slightly, and the predicted-to-experimental peak shear ratio is in the range of 1.02–1.42.

Vecchio and Chan [23] studied the behaviour of perforated and non-perforated shear panels tested under different loading conditions. The response of four non-perforated specimens reported by Vecchio and Chan are validated, and the specimen's details are presented in Table 1. The analytical predictions accurately capture the stiffness, cracking loading for all the specimens. However, the peak shear stress is slightly higher for specimens PC1 and PC4 and is in the range of 0.98–1.14.

Vecchio et al. [12] studied the shear behaviour of high-strength concrete panels. The specimen dimensions are the same as that of the PV series. The specimens were provided with an unequal reinforcement in longitudinal and transverse directions. The analytical model is observed to over predict the peak shear for the specimen with low transverse reinforcement and subject to combined tension and or axial shear. The analytical predictions are in the range of 0.98–1.34.

Eight of the fourteen specimens cast with 100 MPa concrete, reported by Zhang and Hsu [14], subjected to pure shear, are validated as part of this study. The specimen dimensions are the same as that of the specimen tested by Pang and Hsu [22]. The model predictions slightly over-predict the peak shear and are in the range of 0.79–1.32.

The variation in the predictions of IFSAM and the MCFT is primarily due to the material models used. The average peak shear ratios from IFSAM predictions where Belabri and Hsu tension model was found to be 1.02 are in good agreement with the predictions of MCFT.

Table 1. Details of specimen validated

Series	Specimen IDs
Vecchio and Collins	PV3, PV4, PV10, PV12, PV19, PV20, PV21, PV23, PV25, PV27, PV28
Pang and Hsu	A2, A3, A4, B1, B2, B3, B4, B5, B6
Vecchio and Chan	PC1, PC1A, PC4, PC7
Vecchio et al.	PHS2, PHS3, PHS4, PHS5, PHS6, PHS7, PHS8, PHS9, PHS10, PA1, PA2
Zhang and Hsu	VA1, VA2, VA3, VA4, VB1, VB2, VB3, VB4

Table 2. Combinations of constitutive relationships validated

ID	Concrete		Steel	Agg. Interlock
	Compression	Tension		
VC_Tension	VC	VC	MPF	BM
	BH	VC	MPF	BM
	CM	VC	MPF	BM
BH_Tension	VC	BH	MPF	BM
	BH	BH	MPF	BM
	CM	BH	MPF	BM

VC—Vecchio and Collins; BH—Belarbi and Hsu; CM—Chang and Mander; MPF—Menegotto, Pinto and Filippou; BM—Bhujadham and Maekawa

7 Conclusions

Predicting the behaviour of an RC member under different combinations of loadings is challenging and depends on the constitutive relationships used. Nevertheless, over the last few decades, many promising shear panel models which can be used to predict the behaviour of RC members have been proposed. Formulations of shear panel model (IFSAM) that considers the crack direction to be locked after the formation of cracks in the members was presented. Though IFSAM captures the physical behaviour, the predictions of the mechanical behaviour depend on the material models employed. The average ratio of the predicted-to-experimental peak shear stress is in the range of 1.02 to 1.12 for all the combinations with BH and VC tension models, respectively. The predictions using the proposed model are found to be more sensitive to the tension model employed for the concrete. The average ratio of the predicted-to-experimental peak shear stress in the modelling approach where the Belarbi & Hsu tension model was used is found to be 1.02 (BH_Mean). Overall, this study has provided new insight into the influence of the material models on the predictions of IFSAM. Moreover, a simplified analysis for predicting the shear capacity of RC members can be developed using the proposed modelling approach.

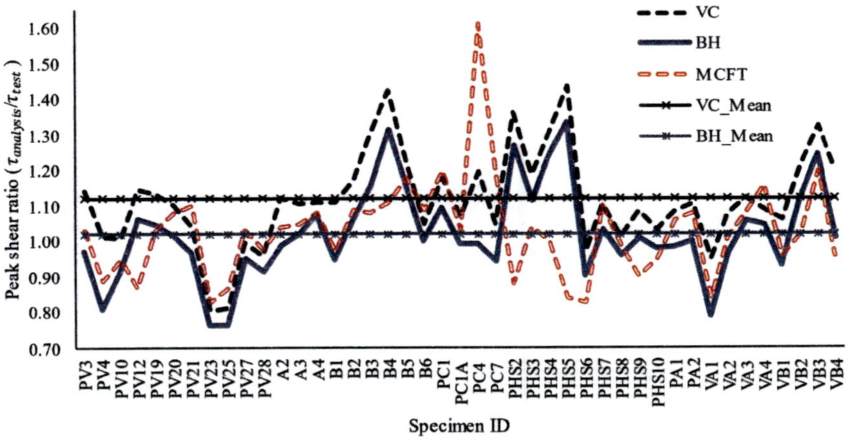

Fig. 4. Comparison of peak shear ratio

Acknowledgements. This analytical work was carried out as part of a project funded by SERB, Department of Science and Technology, India, grant No. SB/S3/CEE/0060/2013 and FAST centre of excellence for sustainable development at IIT Hyderabad. The authors also acknowledge the support provided by IITH and SUT to Mr Nikesh Thammishetti through the IITH-SUT Joint PhD program. In addition, the financial support provided by the Ministry of Education, Government of India to Mr. Nikesh Thammishetti is duly acknowledged.

References

1. Mullapudi, T.R.S. Seismic analysis of reinforced concrete structures subjected to combined axial, flexure, shear and torsional loads. PhD Thesis, University of Houston (2010)
2. Vecchio, F.J.: Reinforced concrete membrane element formulations. J Struct Eng **116**, 730–750 (1990). https://doi.org/10.1061/(ASCE)0733-9445(1990)116:3(730)
3. Vecchio, F.J., Collins, M.P.: The modified compression-field theory for reinforced concrete elements subjected to shear. ACI J. Proc. **83**, 219–31 (1986). https://doi.org/10.14359/10416

4. Hsu, T.T.C., Zhu, R.R.H.: Softened membrane model for reinforced concrete elements in shear. ACI Struct. J. **99**, 460–9 (2002). https://doi.org/10.14359/12115
5. Hsu, T.T.C., Mo, Y.L.: Unified Theory of Concrete Structures, 1st edn. Wiley, Singapore (2010)
6. Ulugtekin, D.: Analytical modeling of reinforced concrete panel elements under reversed cyclic loadings. M.S Thesis, Boğaziçi University (2010)
7. Kolozvari, K.: Analytical Modeling of Cyclic Shear—Flexure Interaction in Reinforced Concrete Structural Walls. University of California, Los Angeles (2013)
8. Thammishetti, N., Prakash, S.S., Hashemi, J., Al-Mahaidi, R.: Simplified analysis of reinforced concrete shear panel elements under static loads using modified FSAM. In: 17th World Conference On Earthquake Engineering 17WCEE, Sendai, Japan, p. 10 (2020)
9. Thammishetti, N., Prakash, S.S., Hashemi, J., Al-Mahaidi, R.: Improved fixed strut angle model for analysis of reinforced concrete panel elements under monotonic shear loads. J. Struct. Eng. **147**, 15 (2021). https://doi.org/10.1061/(ASCE)ST.1943-541X.0003015
10. Chang, G.A., Mander, J.B.: Seismic energy based fatigue damage analysis of bridge columns: Part I - Evaluation of seismic capacity. Buffalo, NY, Technical Report NCEER-94-0006 (1994)
11. Vecchio, F.J., Collins, M.P.: Compression response of cracked reinforced concrete. J Struct Eng **119**, 3590–3610 (1993). https://doi.org/10.1061/(ASCE)0733-9445(1993)119:12(3590)
12. Vecchio, F.J., Collins, M.P., Aspiotis, J.: High-strength concrete elements subjected to shear. ACI Struct. J. **91**, 423–33 (1994). https://doi.org/10.14359/4149
13. Belarbi, A., Hsu, T.T.C.: Constitutive laws of softened concrete in biaxial tension-compression. ACI Struct. J. **92**, 562–73 (1995). https://doi.org/10.14359/907
14. Zhang, L.-X., Hsu, T.T.C.: Behavior and analysis of 100 MPa concrete membrane elements. J Struct Eng **124**, 24–34 (1998)
15. Tamai, S., Shima, H., Izumo, J., Okamura, H.: Average stress-strain relationship in post yield range of steel bar in concrete. Concr Libr JSCE **11**, 117–129 (1988)
16. Belarbi, A., Hsu, T.T.C.: Constitutive laws of concrete in tension and reinforcing bars stiffened by concrete. ACI Struct. J. **91**, 465–74 (1994). https://doi.org/10.14359/4154
17. Menegotto, M., Pinto, P.E.: Method of analysis for cyclically loaded reinforced concrete plane frames including changes in geometry and non-elastic behavior of elements under combined normal force and bending. IBASE Symp. Resist. Ultim. Deform. Struct. Acted by Well Defin. Repeated Loads, Lisbon, pp. 15–22 (1973). https://doi.org/10.5169/seals-13741
18. Filippou, F.C., Popov, E.G., Bertero, V.V.: Effects of bond deterioration on hysteretic behavior of reinforced concrete joints. Berkeley, California, EERC (1983)
19. Bujadham, B., Maekawa, K.: The universal model for stress transfer across cracks in concrete. Proc. JSCE **17**, 277–287 (1992). https://doi.org/10.2208/jscej.1992.451_277
20. MATLAB. R2018b 9.5.0.944444 2018
21. Bentz, E.C.: Sectional analysis of reinforced concrete members. PhD Thesis, University of Toronto (2000)
22. Pang, X.B., Hsu, T.T.C.: Behavior of reinforced concrete membrane elements in shear. ACI Struct. J. **92**, 665–79 (1995). https://doi.org/10.14359/9661
23. Vecchio, F.J., Chan, C.C.L.: Reinforced concrete membrane elements with perforations. J Struct Eng **116**, 2344–2360 (1990). https://doi.org/10.1061/(ASCE)0733-9445(1990)116:9(2344)

Non-Portland Cements and Alkali-Activated Cementitious Materials and Eco-concrete

Feasibility Study of One-Part Alkali Activated Material with MSWI Fly Ash

Priyadharshini Perumal[1](✉) and Mirja Illikainen[2]

[1] Fibre and Particle Engineering Research Unit, Faculty of Technology, Marie Curie Individual Fellow, University of Oulu, 90570 Oulu, Finland
Priyadharshini.perumal@oulu.fi

[2] Professor, Fibre and Particle Engineering Research Unit, Faculty of Technology, University of Oulu, 90570 Oulu, Finland

Abstract. With the growing population, it is also expected that the municipal solid waste (MSW) generation would be increasing in the coming years. Management of MSW is already an identified issue and the incineration process is widely adopted as a disposal solution for non-compostable wastes. Incineration of MSW results in secondary waste residues such as almost 20% of fly ash and 80% of bottom ash. These are rich in silica and alumina content, naturally has the potential to be an aluminosilicate precursor to produce alkali-activated materials. Several issues are pointed out for the application of MSWI residues as cementing materials, either a Supplementary Cement Material (SCM) or an Alkali-Activated Material (AAM). One such issue concerning AAM application is the presence of metallic aluminium. Heavy metals from the MSWI residues are normally extracted before disposal, however metallic aluminium is not effectively removed in most of the cases. Metallic aluminium acts as a pore-forming agent by generating H_2 gas in the alkali environment, thereby reducing the strength properties. In the present study, an attempt is made to understand the influence of the metallic aluminium present in the MSWI fly ash of different size fractions, fine (0–0.5 mm) and coarse (0.5–1.0 mm). MSWI fly ash is activated with solid/liquid alkalis, such as sodium hydroxide (NaOH), sodium silicate (Na_2SiO_3) and combination of both. With the Na_2O content maintained constant across the activators, the sample activated with NaOH is the weakest and the one with sodium silicate is the strongest with small, well-distributed pore structure, irrespective of the size fraction of MSWI fly ash used. Solid activator performed equally well as that of the liquid one.

Keywords: MSWI ash · Fly ash · Metallic aluminium · Alkali activation · Porosity

1 Introduction

Waste to energy treatment is preferred for municipal wastes in places where recycling is not an option, owing to its various advantages including volume reduction and energy recovery [1]. In 2019, 502 kg of municipal waste per capita were generated in EU, of which 30% is incinerated which is double the capacity compared to 1995 [2]. However,

J. I. Escalante-Garcia et al. (Eds.): RW 2021, RILEM Bookseries 40, pp. 579–585, 2023.
https://doi.org/10.1007/978-3-031-21735-7_63

it also comes with the problem of producing other residues such as fly ash, bottom ash, and air pollution control (APC) residues [3]. Such residues were studied for their use as secondary raw materials in construction sector [4, 5]. Utilization of municipal solid waste incinerated (MSWI) fly ash was examined for applications such as supplementary cementitious materials [6–9]. Though MSWI bottom ash falls under non-hazardous category and considered for various applications, the finer fractions (MSWI fly ash) are hazardous with heavy metals and organics content [10]. Hence, stabilization of MSWI fly ash became an important and interesting point of consideration before even considering them for secondary utilization. In this context, alkali activation of this material is regarded as an promising method to immobilize heavy metals while also serving as an alternative binder material [11, 12].

In the present work, MSWI fly ash was considered for its potential to be used as precursor in one-part alkali activated materials (AAM). One-part AAM are "just add water" material like ordinary Portland cements unlike the traditional two-part alkali activated material where the aluminosilicate precursors to be added with alkali activator solution [13]. Sodium silicate and sodium hydroxide solids were mixed with MSWI fly ash acting as binder material. Compressive strength results were compared between one-part and two-part AAMs with similar Na_2O content. Influence of milling of MSWI fly ash at different time (1 and 5 min) was also focused.

2 Materials and Methodology

2.1 Materials

The MSWI fly ash used in this study was obtained from municipal waste incineration plant in Oulu, Finland. Major elements of MSWI fly ash was analyzed using X-ray fluorescence spectrometer (XRF), and the results are presented in Table 1. The fly ash has higher silica and CaO content, but much lower alumina content. There are also high amount of lead and titanium (Table 1). 3.2% of chlorides are seen in this fly ash which is also mentioned in many literature as one of the deleterious element [14]. MSWI fly ash was initially sieved through 1 and 0.5 mm sieve, which resulted in 4 and 13% residues, respectively. The residues mostly comprised of unburnt carbon (Fig. 1). Samples passing 1 mm were taken into consideration in this study, so as to utilize maximum of the material. Anhydrous sodium metasilicate (silica modulus, $SiO_2/Na_2O = 0.9$) was procured from Alfa Aesar, Germany. Liquid sodium silicate with silica modulus 3.2 and 55% water content was supplied by VMR chemicals. Reagent grade sodium hydroxide (>97%) powder was used to adjust the Na_2O content to 5% across different activators.

Table 1. Chemical composition of MSWI fly ash*.

Oxide [%]	Na_2O	MgO	Al_2O_3	SiO_2	P_2O_5	SO_3	Cl	K_2O	CaO	TiO_2	Fe_2O_3
MSWI fly ash	3.7	3.2	11.1	32.9	1.8	6.9	3.2	2.1	26.8	2.4	3.7

* 2.2% other minor oxides excluded

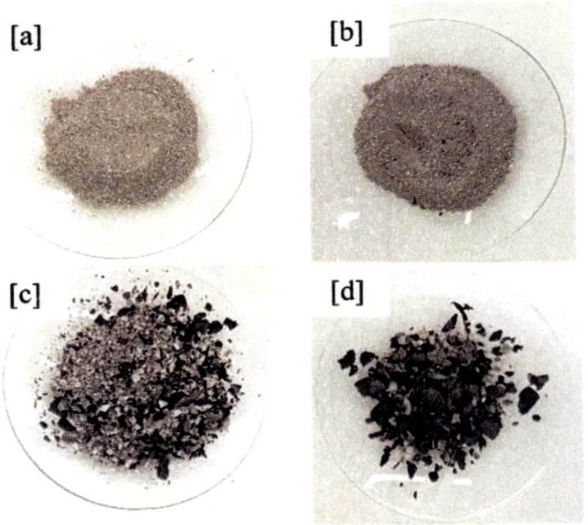

Fig. 1. MSWI fly ash passed through [a] 0.5 mm, [b] 1 mm size sieves and the residues retained in [c] 0.5 mm and [d] 1 mm size sieves.

2.2 Methodology

MSWI fly ash was dried in 100 °C for 24 h and ground in vibratory disc mill for 1 and 5 min. Disc mill Retsch RS 200 was employed at 1500 rpm in batches of 100g of MSWI fly ash. Three different activators were used, precisely, combination of solid sodium silicate and sodium hydroxide, liquid sodium silicate and liquid sodium hydroxide. Na_2O content was maintained as 5% in all the activators. Water to solids ratio was maintained constant as 0.45. To study one-part AAM, MSWI fly ash was mixed with solid sodium silicate and sodium hydroxide combination prior to introducing the water. In two-part, the alkali activator solution was mixed to the MSWI fly ash to make paste samples. Cylindrical specimens (20 × 25 mm) were cast using the one-part and two-part samples to observe their physical appearance and study the strength properties. AAM specimens were subjected to two different curing regimes; (1) 60 °C for 24 h, continued by room temperature curing, (2) room temperature curing until further testing.

3 Results and Discussion

3.1 Physical Appearance

Cylindrical specimens were demolded after 24 h and observed for physical form. Specimens cured at 60 °C for 24 h is shown in Figs. 2, 3 and 4. Unmilled MSWI fly ash appears disintegrated with solid activators and liquid NaOH, whereas, comparatively solidified with liquid sodium silicate (Fig. 2). This is also the case with room temperature cured specimens.

After milling, one-part activators performed better in MSWI fly ash pastes with solid specimens as that of two-part specimens made of liquid sodium silicate (Fig. 3).

(a) (b) (c)

Fig. 2. Unmilled MSWI fly ash activated with [a] solid Na_2SiO_3 + NaOH, [b] liquid Na_2SiO_3 and, [c] liquid NaOH at 7 days of curing age.

(a) (b) (c)

Fig. 3. Milled (1 min) MSWI fly ash activated with [a] solid Na_2SiO_3 + NaOH, [b] liquid Na_2SiO_3 and, [c] liquid NaOH at 7 days of curing age.

However, liquid NaOH does not produce stronger specimens though Na_2O content was maintained constant.

With increasing milling time to 5 min, one-part seems to be a better option as the pore size increased in two-part AAM specimens (Fig. 4). In all the three cases, Two-part specimens with liquid NaOH were very weak due to their high porosity and hence, were not considered for strength testing.

(a) (b) (c)

Fig. 4. Milled (5 min) MSWI fly ash activated with [a] solid Na_2SiO_3 + NaOH, [b] liquid Na_2SiO_3 and, [c] liquid NaOH at 7 days of curing age.

3.2 Compressive Strength

Solid activator performs equally good as that of liquid activator in the activation of MSWI fly ash as represented by compressive strength results in Figs. 5 and 6. Combination of Na_2SiO_3 and NaOH acted as solid activator making one-part AAM and liquid Na_2SiO_3 was used as activator for producing two-part AAM for the specimens used for strength testing.

It is to be noted that only two-part activation worked for unmilled MSWI fly ash samples, leading to a conclusion that one-part activation need milling as a pre-treatment. However, increasing the milling from 1 to 5 min was not favorable for alkali activation as this either does not impact (Fig. 5) or reduced the strength (Fig. 6) in case of solid activators. With the use of liquid activators, milling reduced the compressive strength, irrespective of the type of curing adopted. This can be related to the increasing amount of metallic aluminium with milling time and hence, swelling, porosity and strength reduction [15].

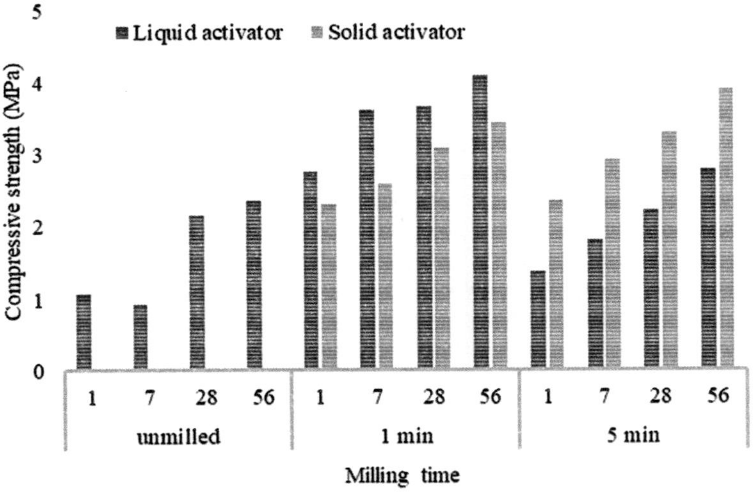

Fig. 5. Compressive strength of alkali activated MSWI fly ash pre-cured at 60 °C for 24 h and tested at different ages (1–56 days).

Type of curing shows noticeable impact on the strength development with age in one/two-part alkali activated MSWI fly ash (Figs. 5 and 6). Specimens cured at room temperature shows gradual increase in compressive strength with one-part activation (Fig. 6). It reaches a maximum strength of 4 MPa at 56 days of curing age which is similar to that of two-part AAM specimens. However, early age strength seems to be higher with two-part activation though strength improvement at later ages is not noticeable.

4 Conclusions

MSWI fly ash was studied as a source of aluminosilicate precursor for alkali activation. The possibility to use solid activators (one-part) was examined and compared with conventional two-part alkali activation method. The study concludes that milling helps in activation of MSWI fly ash in one-part method, however, it also can result in the increased release of metallic elements which can negatively affect the properties. Heat curing helps in achieving high early strength in one-part activation, however, does not contribute to the strength increase at later ages.

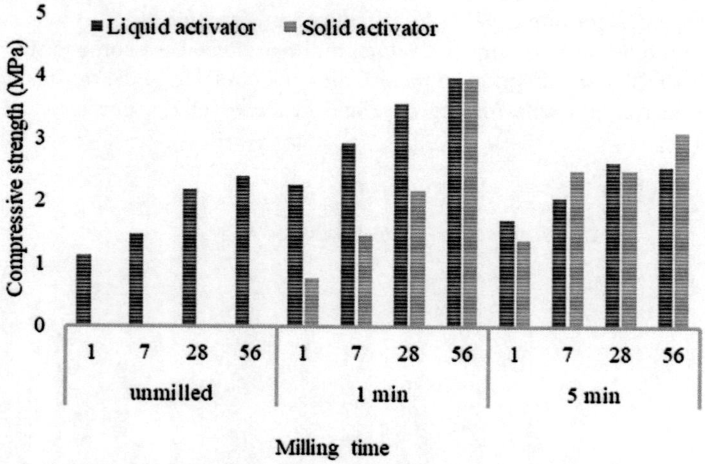

Fig. 6. Compressive strength of alkali activated MSWI fly ash cured at room temperature and tested at different ages (1–56 days).

Acknowledgement. This work was supported by the funding from the European Union's Horizon 2020 research and innovation programme under the Marie Skłodowska Curie grant agreement No [839848].

References

1. Scarlat, N., Fahl, F., Dallemand, J.-F.: Status and opportunities for energy recovery from municipal solid waste in Europe. Waste and Biomass Valorization **10**(9), 2425–2444 (2018). https://doi.org/10.1007/s12649-018-0297-7
2. Eurostat, F., Union, E., Regulation W.S.: Municipal waste statistics. Eurostat, pp. 1–6 (2021). http://ec.europa.eu/eurostat/statistics-explained/
3. Quina, M.J., Bordado, J.C., Quinta-Ferreira, R.M.: Treatment and use of air pollution control residues from MSW incineration: an overview. Waste Manag. **28**, 2097–2121 (2008). https://doi.org/10.1016/j.wasman.2007.08.030
4. Clavier, K.A., Paris, J.M., Ferraro, C.C., Townsend, T.G.: Opportunities and challenges associated with using municipal waste incineration ash as a raw ingredient in cement production—a review. Resour. Conserv. Recycl. **160**, 104888 (2020). https://doi.org/10.1016/j.resconrec.2020.104888
5. Dou, X., et al.: Review of MSWI bottom ash utilization from perspectives of collective characterization, treatment and existing application. Renew. Sustain. Energy Rev. **79**, 24–38 (2017). https://doi.org/10.1016/j.rser.2017.05.044
6. Wang, K.S., Lin, K.L., Huang, Z.Q.: Hydraulic activity of municipal solid waste incinerator fly-ash-slag-blended eco-cement. Cem. Concr. Res. **31**, 97–103 (2001). https://doi.org/10.1016/S0008-8846(00)00423-3
7. Shih, P.H., Chang, J.E., Chiang, L.C.: Replacement of raw mix in cement production by municipal solid waste incineration ash. Cem. Concr. Res. **33**, 1831–1836 (2003). https://doi.org/10.1016/S0008-8846(03)00206-0

8. Lin, K.L., Wang, K.S., Lee, T.Y., Tzeng, B.Y.: The hydration characteristics of MSWI fly ash slag present in C3S. Cem. Concr. Res. **33**, 957–964 (2003). https://doi.org/10.1016/S0008-8846(02)01002-5
9. Kan, L., Zhang, L., Shi, H.: Hydration kinetics of municipal solid wastes incineration (MSWI) fly ash-cement. J. Wuhan Univ. Technol. Mater. Sci. Ed. **34**(3), 596–603 (2019). https://doi.org/10.1007/s11595-019-2093-z
10. Joseph, A.M., Snellings, R., Van den Heede, P., Matthys, S., De Belie, N.: The use of municipal solid waste incineration ash in various building materials: a Belgian point of view. Materials (Basel) **11**, (2018). https://doi.org/10.3390/ma11010141
11. Tian, X., Rao, F., León-Patiño, C.A., Song, S.: Co-disposal of MSWI fly ash and spent caustic through alkaline-activation consolidation. Cem. Concr. Compos. **116**, (2021). https://doi.org/10.1016/j.cemconcomp.2020.103888
12. Tian, X., Rao, F., León-Patiño, C.A., Song, S.: Co-disposal of MSWI fly ash and spent caustic through alkaline-activation: immobilization of heavy metals and organics. Cem. Concr. Compos. **114**, (2020). https://doi.org/10.1016/j.cemconcomp.2020.103824
13. Luukkonen, T., Abdollahnejad, Z., Yliniemi, J., Kinnunen, P., Illikainen, M.: One-part alkali-activated materials: a review. Cem. Concr. Res. **103**, 21–34 (2018). https://doi.org/10.1016/j.cemconres.2017.10.001
14. Alam, Q., Lazaro, A., Schollbach, K., Brouwers, H.J.H.: Chemical speciation, distribution and leaching behavior of chlorides from municipal solid waste incineration bottom ash. Chemosphere **241**, 124985 (2020). https://doi.org/10.1016/j.chemosphere.2019.124985
15. Aubert, J.E., Husson, B., Vaquier, A.: Metallic aluminum in MSWI fly ash: quantification and influence on the properties of cement-based products. Waste Manag. **24**, 589–596 (2004). https://doi.org/10.1016/j.wasman.2004.01.005

Effect of the Limestone Content on the Durability of Alkali-Activated Limestone-Metakaolin Subjected to Acidic and Sulfate Environments

Pedro Perez-Cortes(✉) and J. Ivan Escalante-Garcia

Centro de Investigación y de Estudios Avanzados del IPN, Unidad Saltillo, Ramos Arizpe, Coahuila, México
Pedro.perez@cinvestav.edu.mx

Abstract. New alkali-activated cements based on calcined kaolinitic clays and limestone are a promising sustainable alternative to portland cement due to their improved mechanical properties and reduced environmental impact. Studies on durability are needed in order to gain knowledge for the scientific community and buy-in from potential stakeholders. In this investigation, alkali-activated limestone-metakaolin pastes were formulated with 20 and 60% limestone, activated with sodium silicate solutions, cured for 28 days and subjected to chemical attack by solutions of (a) 0.1 N HCl and (b) 5% $MgSO_4$ for 7, 28 and 90 days. Changes in strength, mass, microstructure and reaction products were studied. The samples exposed to HCl underwent a degradation of the external microstructure due to a dissolution of calcium carbonate and depolymerization of the N-A-S-H/N-(C)-A-S-H gel, which led to porosity, cracks and loss of weight. After the exposure to $MgSO_4$, all samples underwent a mass increase and the formation of quintinite was detected; formulation with $Na_2O/Al_2O_3 = 1.08$ underwent microcracking. Despite de microstructural and mass changes, samples with low Na_2O/Al_2O_3 of 0.60 preserved their strength after the period studied in both environments, regardless of the limestone content. This considerable chemical durability against acid and sulfate attack could extend the applications of alkali-activated limestone-metakaolin cements making them part of the future toolkit of durable binder technologies.

Keywords: Geopolymers · Durability · Acid and sulfate attack · Alkaline cements · Non-portland cements

1 Introduction

The long-term chemical resistance of building structures is of major importance for sustainable economies. Alkali Activated Cements (AACs), especially those systems with low-Ca content, are interesting alternatives as construction and repair materials due to their promising high resistance to acid [1] and sulfate [2] attack; nonetheless, contrasting results have also been reported [3–5]. The chemical composition has been

J. I. Escalante-Garcia et al. (Eds.): RW 2021, RILEM Bookseries 40, pp. 586–595, 2023.
https://doi.org/10.1007/978-3-031-21735-7_64

found as one of the parameter that influence the chemical resistance; the CaO content in the AACs appears to be important as many of the Ca-containing reaction products dissolve under acidic media [2] and/or are susceptible to form gypsum and/or ettringite under sulfate environments, thereby causing expansion and damage [5].

On the other hand, recent studies have shown that AACs of limestone and metakaolin (AAC-LSMK) are of particular interest due to the abundance of both precursors and the substantial advantages of these binders compared to the conventional portland cement; these include higher mechanical properties, lower CO_2 emissions, lower energy demand and lower cost of production [6–8]. Moreover, it has been reported that in AAC-LSMK binders, Ca^{2+} released from LS is mostly taken up by the N-A-S-H structure through an ion-exchange mechanism, resulting in a Ca-containing N-A-S-H (N-(C)-A-S-H) with a 3D network structure [9]; these types of reaction products are reported as thermodynamically stables and durable [10], which could extend the applications of AAC-LSMK to aggressive environments. Indeed, a recent study [11] showed a good thermomechanical performance of AAC-LSMK pastes under temperatures below the decarbonation of calcite.

As an advancement to the previous works, this study explores the chemical resistance of the AAC-LSMK exposed to acidic and sulfate environments.

2 Materials and Methods

A limestone powder (LS) and a high purity ground metakaolin (MK) with the features indicated in Table 1 and Fig. 1, and described in detail elsewhere [6] were activated with blends of industrial-grade waterglass (WG, 29.85% SiO_2, 9.12% Na_2O and 61.03% H_2O) and NaOH flakes of 98% purity. Three AAC-LSMK pastes were synthesized with the mix proportions listed in Table 2. The sample names describe %LS and oxide-molar Na_2O/Al_2O_3 ratio, so systems 1 and 2 were formulated with 20 and 60% LS, respectively, with $Na_2O/Al_2O_3 = 0.60$ while system 3 was formulated with 60% LS and $Na_2O/Al_2O_3 = 1.08$; in this way, a comparison among systems 1 and 2, and 2 and 3 allows to analyze the influence of the %LS and Na_2O/Al_2O_3 ratio, respectively. In all cases, the oxide-molar SiO_2/Al_2O_3 ratio was fixed in 3.17 since previous works [6, 7] indicated that such value led to AAC-LSMK binders of high strength and low environmental impact. The water/precursor ratio (w/p) was adjusted for each formulation to attain a fluidity of 110 ± 5 (ASTM C109). The synthesis procedure followed the protocol described elsewhere [6] and involves mixing of precursors and alkaline solutions for 5 min, pouring of the fresh paste into cubic stainless-steel molds of 25 mm per side, and curing at 60 °C for 24 h. After that, the samples were de-molded and stored at 20 °C (75–85% relative humidity) for 27 days, and then immersed in water for 48–72 h until reaching a constant weight. Subsequently, one set of samples was immersed in 0.1 N HCl solutions and another in 5% $MgSO_4$ solutions for up to 90 days, renewing the solutions after 7, 28 and 60 days.

The resistance of AAC-LSMK paste samples to the acidic and sulfate environments was evaluated in terms of the changes in the weight and compressive strength after 7, 28 and 90 days of exposure, in which the values were compared to those before the exposure. Furthermore, a characterization by scanning electron microscopy (SEM, Philips XL-30 ESEM) and X-ray diffraction (XRD, X'pert 3040, Philips) was carried out on the 90 day exposed samples in each environment; the sample preparation followed the procedure described elsewhere [7]. SEM observations in polished samples were taken in backscattered electron images with high vacuum mode using accelerating voltage of 25 keV, a spot size of 4–5, and a working distance of 7 mm. The XRD analysis was carried out on powders (particle size < 100 μm) in a range of 10–60° (2θ) with a step of 0.03° every 2 s using Cu Kα radiation at 40 kV and 30 mA.

Table 1. Chemical composition and other features of the LS and MK precursors

	Chemical composition (wt%)					LOI[a]	D_{50}[b], μm	Surface area[c], m^2/g
	SiO_2	Al_2O_3	CaO	TiO_2	Others			
MK	55.1	41.2	0.2	2.0	1.0	0.5	9.22	11.38
LS	2.2	1.5	49.8	–	1.2	45.3	11.30	1.90

[a] LOI—loss of ignition, 1 h at 900 °C
[b] Laser diffraction method
[c] Brunauer Emmett Teller (BET) method

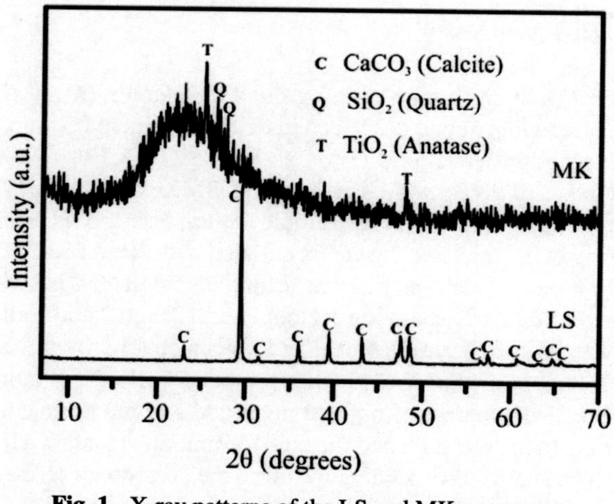

Fig. 1. X-ray patterns of the LS and MK precursors

3 Results and Discussion

Figure 2 shows the average of the weight changes (as a percentage) of the AAC-LSMK pastes after 7, 28 and 90 of exposure to (a) 0.1 N HCl solutions and (b) 5% $MgSO_4$ solutions. Under the acidic environment (Fig. 2a), the samples underwent weight fluctuations after 7 and 28 days and a remarkable weight loss after 90 days. The latter, which was more substantial for the formulation with 60% LS and Na_2O/Al_2O_3 = 0.60 (60–0.60)

Table 2. AAC-LSMK pastes and their mixture proportions

Sample name	Formulation		Components in kg/m^3					%Na_2O[a]	w/p
	%LS	Na_2O/Al_2O_3	MK	LS	WG	NaOH	H_2O		
20–0.60	20	0.60	817.4	204.4	779.6	68.8	29.9	12.0	0.51
60–0.60	60	0.60	516.0	774.0	492.1	43.4	154.3	6.0	0.36
60–1.08	60	1.08	506.6	759.9	482.7	124.8	145.9	10.9	0.37

[a] wt% relative to the total mass of the precursors

and attributed to a partial dissolution of the cementitious matrix as further discussed below. On the other hand, the weight fluctuations in the first 28 days of exposure can be related to an extended absorption of the acid solution into the specimens, which also probably induced scattering in the measures; more research may be done in the future to elucidate this hypothesis.

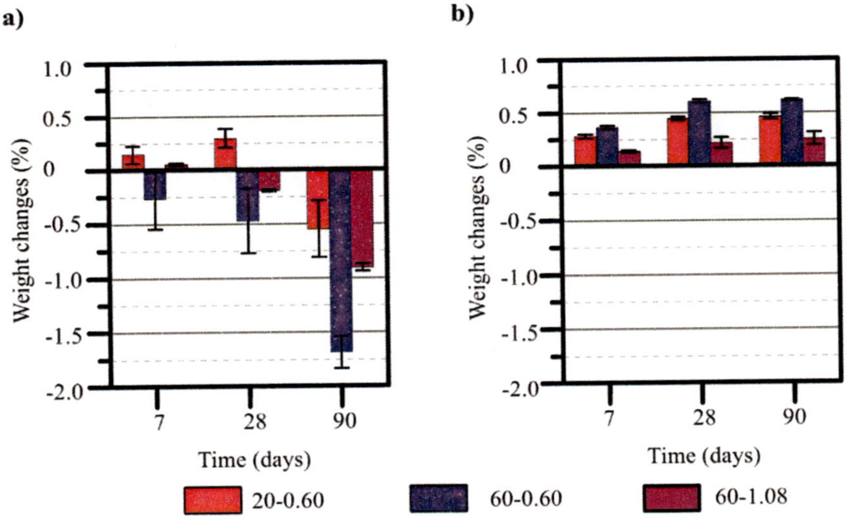

Fig. 2. Weight changes of the AAC-LSMK pastes after the exposure to (a) 0.1 N HCl solutions and (b) 5% $MgSO_4$ solutions

In contrast, the immersion in the sulfate solution (Fig. 2b) induced progressive weight gains in all samples with the exposure time. This latter was more notable for the formulation with 60% LS and Na_2O/Al_2O_3 = 0.60. The mass increase can be related to an extended absorption of the sulfate solution into the specimens, in which the ions of Mg^{2+} and $SO_4{}^{2-}$ from sulfate solution probably reacted with the Si and Al species from MK and/or Ca from LS forming new reaction products.

Figure 3 shows the compressive strength changes of the AAC-LSMK pastes after 0, 7, 28 and 90 days of immersion in the (a) 0.1 N HCl solutions and (b) 5% $MgSO_4$ solutions. No clear correlation was found between weight changes and strength in the evaluated period. Regardless of the %LS, the formulations with $Na_2O/Al_2O_3 = 0.60$ (20–0.60 and 60–0.60) did not show significant strength changes after the exposure to both environments, while formulation 60–1.08 underwent reductions and fluctuations in the strength after the exposure to the acidic and sulfate environments, respectively. As previously reported [9], low $Na_2O/Al_2O_3 = 0.60$ led to a lower dissolution of the MK, and thus cementitious matrices with higher porosity than $Na_2O/Al_2O_3 = 1.08$; so, it is probable that samples with higher porosity allowed the dissolution/formation of reaction products in the pores without affecting strength.

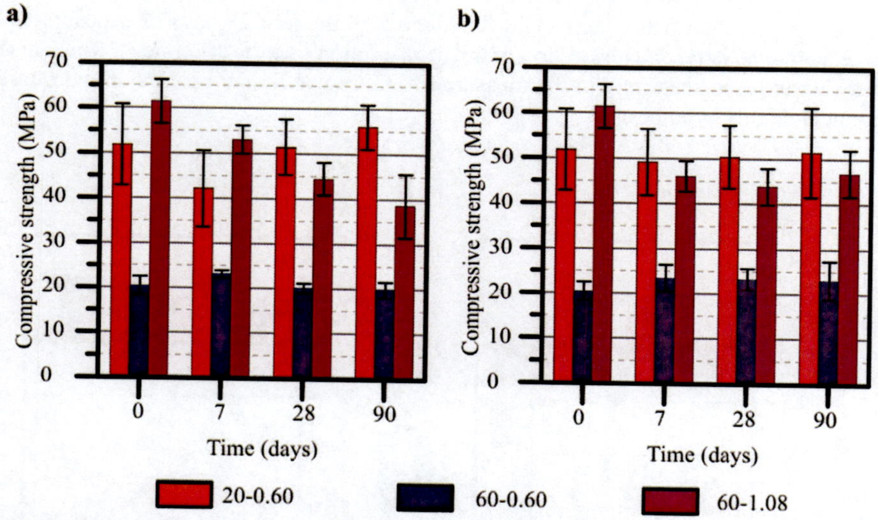

Fig. 3. Compressive strength of the AAC-LSMK pastes after the exposure to (a) 0.1 N HCl solutions and (b) 5% $MgSO_4$ solutions

Figure 4 shows representative micrographs of the samples after immersion for 90 days in the solutions of 0.1 N HCl and 5% $MgSO_4$. Under HCl, all samples showed deterioration of the cementitious matrix due to the highly corrosive action of the acid, leading to dissolution, porosity, cracking and loss of weight. Formulation 60–0.60 presented a higher depth of the corroded zone, which may be related to a higher porosity (as suggested by its lower %Na_2O, Table 2, that promoted lower initial strength) which could have favored a higher permeability of the acid solution, as well as to the high %LS, that would favor a higher dissolution of calcite. Furthermore, at the end of the corroded zone within the cementitious matrix, a change of tonality was noted, that can be related to precipitation of silica gel as a result of the dissolution of the N-A-S-H/N-(C)-A-S-H gels. A previous report [1] indicated that the precipitation of silica gel inhibited further degradation of the AAC mortars exposed to sulfuric acid. In the current study, the precipitated silica gel layer may have performed as a transport barrier for the acid attack, limiting the disintegration and contributing to the mechanical strength. This assertion could at least partially explain the strength preservation in those formulations with $Na_2O/Al_2O_3 = 0.60$, which involved a greater amount of soluble silica in the starting compositions (Table 2), thus more silica gel could have precipitated. On the other hand, strength losses of 60–1.08 can be related to its more visible microcracking, which could also account for the scattering of the strength values, and can be an indication of migration of alkali cations from the specimens into the acid solutions [12].

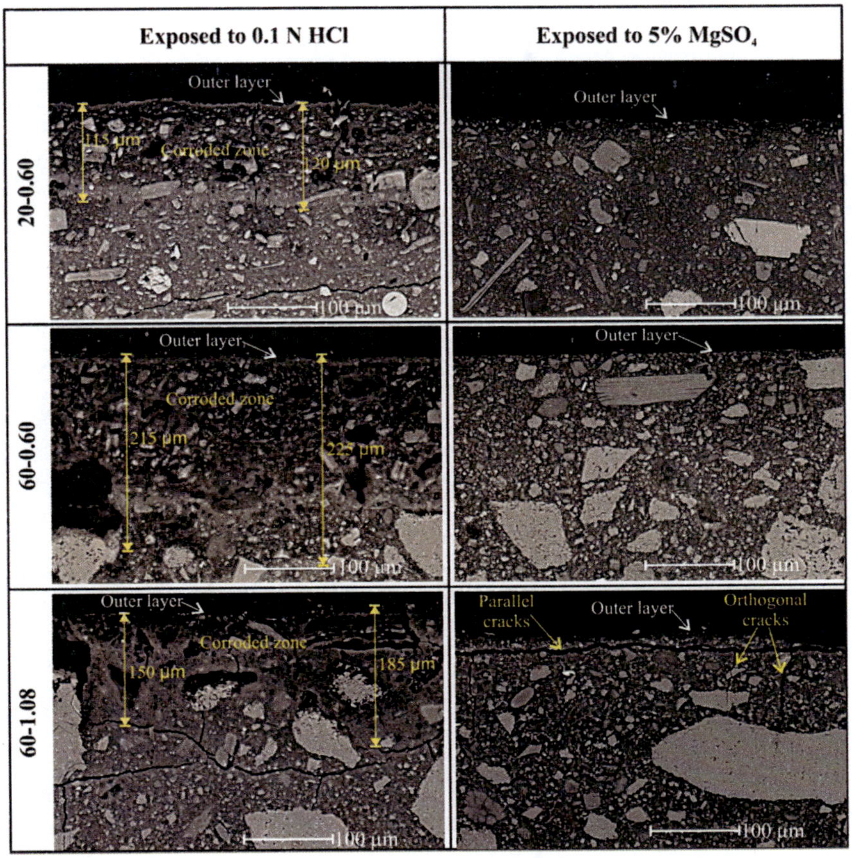

Fig. 4. SEM backscattered electron images of the AAC-LSMK pastes after the exposure to the 0.1 N HCl and 5% $MgSO_4$ solutions for 90 days.

Under the sulfate environment, both formulations with low Na_2O/Al_2O_3 of 0.60 did not undergo visible microstructural damage after the 90 days of exposure. However, parallel and orthogonal cracks to the outer layer appeared in the formulation with $Na_2O/Al_2O_3 = 1.08$. Orthogonal cracks have been reported for AAC of fly ash under sulfate solutions [4] and were attributed to the migration of alkali cations from the specimen into the sulfate solution. It has also been reported for zeolites and aluminosilicate gels, that Na^+ and K^+ can be exchanged if the material is placed in solutions of salts containing replacing cations [13]. So, the migration of alkalis could occur in all the studied systems; however, it promoted the formation of cracks only in formulation 60–1.08, probably due to its denser and less permeable unexposed microstructure [9], which may have cracked as a result of the cations exchange.

Figure 5 shows the XRD diffractograms of pastes before and after 90 days of exposure to the 0.1 HCl solutions; in the latter, patterns from the internal matrix within the intact zone and from the outer layer encompassing the corroded zone were collected. The XRD patterns of the corroded zone indicated that the acid attack promoted the dissolution of the calcite ($CaCO_3$, PDF# 5-586) and transformation of the cementitious reaction products into a more amorphous gel, related with the precipitated silica gel, in which quartz (SiO_2, PDF# 85-798) and anatase (TiO_2, PDF# 21-1272) from the MK were preserved. On the other hand, the XRD patterns from the internal matrix were similar to those of the unexposed samples, indicating that the mineralogical composition of the internal matrix was not affected by the acid attack, which is consistent with that observed in the SEM images.

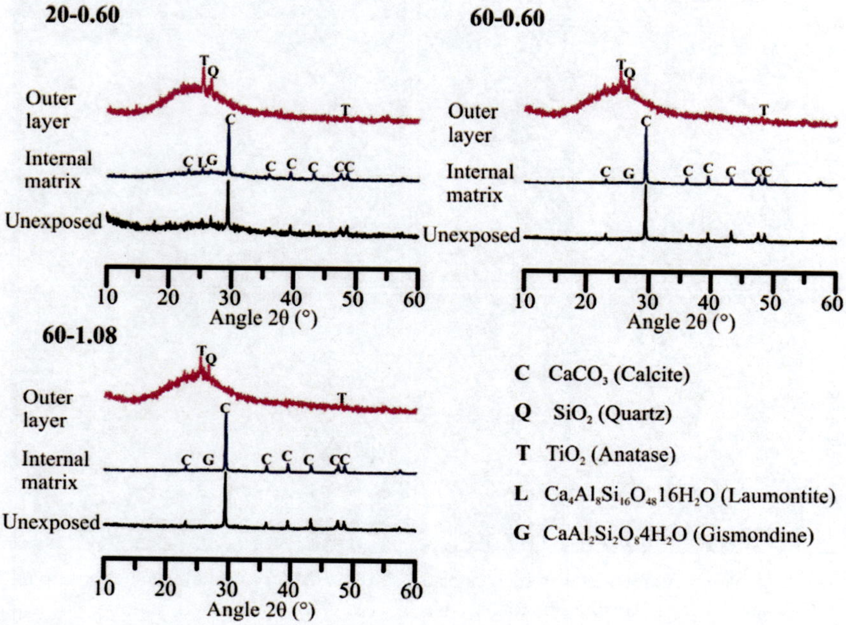

Fig. 5. X-ray diffractograms of AAC-LSMK pastes before and after their exposure to the 0.1 N HCl solutions for 90 days, in which patterns from the internal and outer layer were collected.

The XRD patterns of the pastes before and after their immersion in the sulfate solution for 90 days are shown in Fig. 6. It is noteworthy that in all cases, the mineralogical composition of the reaction products (i.e. the amorphous N-A-S-H/N-(C)-A-S-H gels, and the crystalline phases of gismondine ($CaAl_2Si_2O_8{\cdot}4H_2O$, PDF# 20-452) and laumontite ($Ca_4Al_8Si_{16}O_{48}{\cdot}16H_2O$, PDF# 47-1785) [9]) did not undergo substantial changes. Nonetheless, new low-intensity reflections at 23.54, 35.7, 39.6 and 47.2° 2θ were identified, which overlapped in some cases with those of the calcite phase increasing their intensity. Such peaks were associated with the quintinite ($Mg_4Al_2(OH)_{12}CO_3(H_2O)_3$, PDF# 00-051-1525), which is a carbonate layered double hydroxide (LDH) mineral, member of the hydrotalcite supergroup [14, 15]. Quintinite could have precipitated as a result of the reactions among the $[Al(OH)_4]^-$ species (Al released from MK), Mg^{2+} (from $MgSO_4$ solution) and $CO_3{}^{2-}$ (from the partial dissolution of LS) under the saturated conditions. Zhu et al. [16] reported that the Mg^{2+}, released during the hydrolysis of MgO, combined with $[Al(OH)_4]^-$ ions released from $NaOH/Na_2CO_3$-activated MK, led to the precipitation of Mg-Al LDH and Mg-Al-CO_3 LDH (quintinite), in which the alkali concentration played an important role. The formation of quintinite has also been reported in AACs synthesized from Mg-rich precursors such GBFS [17, 18] and mineral wools [19]; in the current study, this phase probably contributed to the mass gain after the exposure to the magnesium sulfate solution.

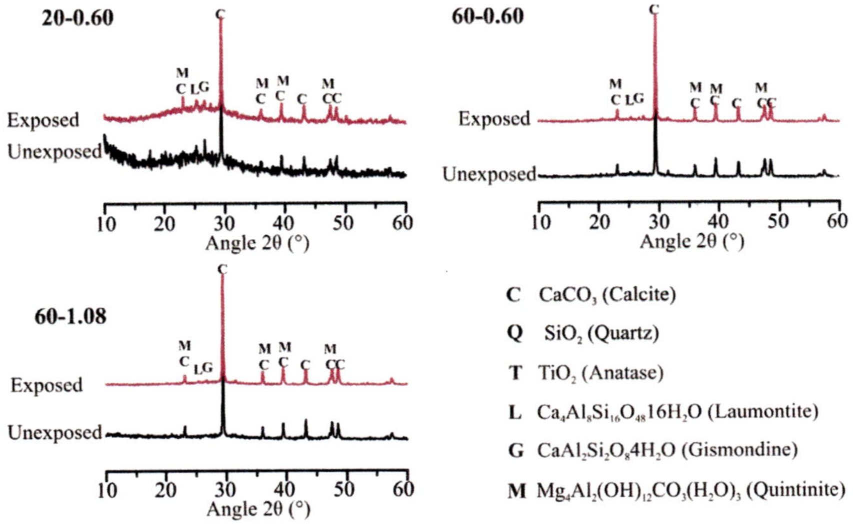

Fig. 6. X-ray diffractograms of AAC-LSMK pastes before and after their exposure to the 5% $MgSO_4$ solutions for 90 days.

4 Conclusions

This study explored the chemical resistance of the AAC-LSMK binders to acidic and sulfate environments, showing promising results that can increase the interest from the scientific community and buy-in from potential stakeholders. The main findings are summarized as follows.

The exposure to 0.1 N HCl for 90 days promoted the degradation of the samples at depths of up to 225 μm, in which calcite and the N-A-S-H/N-(C)-A-S-H gel dissolved, leading to porosity, cracks and the formation of amorphous phases related to precipitated silica gel. The depth of the damage increased with the limestone content; however, the compressive strength was preserved in formulations with $Na_2O/Al_2O_3 = 0.60$, regardless of the limestone content. The precipitated silica gel is thought to have contributed to the preservation of the mechanical strength, while performing as a transport barrier to acid attack, limiting the disintegration of the samples.

The exposure to 5% $MgSO_4$ for 90 days increased the mass of the samples, probably due to a densification of the cementitious matrix and the formation of quintinite, in which the crystalline and amorphous reaction products were preserved. Moreover, the microstructure of formulations with $Na_2O/Al_2O_3 = 0.60$ did not undergo visible changes, leading to a preservation of the compressive strength, regardless of the limestone content. On the other hand, the formulation with $Na_2O/Al_2O_3 = 1.08$ underwent microcracking, probably due to the migration of alkalis, leading to strength fluctuations.

Further research can be conducted to confirm and better understand the evidenced and suggested mechanisms of this study.

References

1. Sturm, P., Gluth, G.J.G., Jäger, C., Brouwers, H.J.H., Kühne, H.C.: Sulfuric acid resistance of one-part alkali-activated mortars. Cem. Concr. Res. **109**, 54–63 (2018)
2. Siddique, S., Jang, J.G.: Acid and sulfate resistance of seawater based alkali activated fly ash: a sustainable and durable approach. Constr. Build. Mater. **281**, 122601 (2021)
3. Bakharev, T.: Resistance of geopolymer materials to acid attack. Cem. Concr. Res. **35**(4), 658–670 (2005)
4. Bakharev, T.: Durability of geopolymer materials in sodium and magnesium sulfate solutions. Cem. Concr. Res. **35**(6), 1233–1246 (2005)
5. Ismail, I., Bernal, S.A., Provis, J.L., Hamdan, S., van Deventer, J.S.J.: Microstructural changes in alkali activated fly ash/slag geopolymers with sulfate exposure. Mater. Struct. **46**(3), 361–373 (2013)
6. Perez-Cortes, P., Escalante-Garcia, J.I.: Alkali activated metakaolin with high limestone contents—statistical modeling of strength and environmental and cost analyses. Cem. Concr. Compos. **106**, 103450 (2020)
7. Perez-Cortes, P., Escalante-Garcia, J.I.: Design and optimization of alkaline binders of limestone-metakaolin—a comparison of strength, microstructure and sustainability with portland cement and geopolymers. J. Clean. Prod. **273**, 123118 (2020)
8. Perez Cortes, P., Escalante García, J.I.: Alkaline blends of metakaolin-limestone: analysis by experimental design. In: Gemrich, J. (ed.) 15th International Congress on the Chemistry of Cement, ICCC 2019, pp. 1–11. Research Institute of Binding Materials Prague, Prague, Czech Republic (2019)
9. Perez-Cortes, P., Escalante-Garcia, J.I.: Gel composition and molecular structure of alkali-activated metakaolin-limestone cements. Cem. Concr. Res. **137**, 106211 (2020)
10. Provis, J.L., Duxson, P., Lukey, G.C., van Deventer, J.S.J.: Statistical thermodynamic model for Si/Al ordering in amorphous aluminosilicates. Chem. Mater. **17**(11), 2976–2986 (2005)

11. Perez-Cortes, P., Cabrera-Luna, K., Escalante-Garcia, J.I.: Alkali-activated limestone/metakaolin cements exposed to high temperatures: structural changes. Cem. Concr. Compos. **122**, 104147 (2021)
12. Aiken, T.A., Kwasny, J., Sha, W., Soutsos, M.N.: Effect of slag content and activator dosage on the resistance of fly ash geopolymer binders to sulfuric acid attack. Cem. Concr. Res. **111**, 23–40 (2018)
13. Breck, D.W.: Zeolite Molecular Sieves: Structure, Chemistry And Use. Krieger (1984)
14. Theiss, F., López, A., Frost, R.L., Scholz, R.: Spectroscopic characterisation of the LDH mineral quintinite $Mg_4Al_2(OH)_{12}CO_3{\cdot}3H_2O$. Spectrochim. Acta Part A Mol. Biomol. Spectrosc. **150**, 758–764 (2015)
15. Mills, S.J., Christy, A.G., Génin, J.-M., Kameda, T., Colombo, F.: Nomenclature of the hydrotalcite supergroup: natural layered double hydroxides. Mineral. Mag. **76**(5), 1289–1336 (2012)
16. Zhu, J., Zeng, B., Mo, L., Jin, F., Deng, M., Zhang, Q.: One-pot synthesis of MgAl layered double hydroxide (LDH) using MgO and metakaolin (MK) as precursors. Appl. Clay Sci. **206**, 106070 (2021)
17. Wang, S.-D., Scrivener, K.L.: Hydration products of alkali activated slag cement. Cem. Concr. Res. **25**(3), 561–571 (1995)
18. Bernal, S.A., San Nicolas, R., Myers, R.J., Mejía de Gutiérrez, R., Puertas, F., van Deventer, J.S.J., Provis, J.L.: MgO content of slag controls phase evolution and structural changes induced by accelerated carbonation in alkali-activated binders. Cem. Concr. Res. **57**, 33–43 (2014)
19. Yliniemi, J., Walkley, B., Provis, J.L., Kinnunen, P., Illikainen, M.: Nanostructural evolution of alkali-activated mineral wools. Cem. Concr. Compos. **106**, 103472 (2020)

Water Transport Properties of Hybrid Binder Concrete Containing Activated Copper Slag and Recycled Concrete Aggregate

Yury Villagrán-Zaccardi[1(✉)], Pithchai Sivakumar[1,2,3], and Nele De Belie[1]

[1] Magnel-Vandepitte Laboratory for Structural Engineering and Building Materials, Ghent University, Technologiepark Zwijnaarde 60, BE-9052 Ghent, Belgium
{yury.villagranzaccardi,pithchaipandian.sivakumar, nele.debelie}@ugent.be

[2] KU Leuven, Department of Civil Engineering, Materials and Constructions, Ghent Technology Campus, Gebroeders De Smetstraat 1, BE-9000 Ghent, Belgium

[3] SIM vzw, Technologiepark 48, BE-9052 Zwijnaarde, Belgium

Abstract. Advanced design of eco-friendly concrete is very urgent in view of the vital fight against climate change. An interesting eco-efficient strategy is the combination of (1) alternative low carbon hybrid binders with a low clinker factor and the inclusion of alkali activators, (2) substituting natural aggregate by recycled aggregate and (3) securing satisfactory durability performance. Transport properties of concrete are reliable indexes concerning the ingress rate of aggressive agents though the pore structure. The present study analyses the water transport properties of concrete made with hybrid binders and recycled concrete aggregate. Copper slag (70 wt% binder) and copper slag + ground granulated blast-furnace slag (GGBFS) (50 and 20 wt% binder, respectively) were used as precursors with a tailored activator based on sodium sulfate and Portland cement. Coarse recycled aggregate (category Rcu 90 according to EN 12620) was used as 50% of total content of coarse aggregate (particle size range 4/32 mm). The examined properties include capillary absorption rate, water penetration under pressure, and accessible porosity by vacuum saturation at curing ages of 28 and 90 days. Concrete performance was dominated by the type of binder, whereas the impact of the recycled aggregate on the transport properties was in second place. A limited content of GGBFS in the binder system proved efficient to improve concrete performance. Overall, these preliminary results reflect promising capabilities of the combined strategy of using low carbon hybrid binders and recycled aggregates in concrete.

Keywords: Low carbon binders · Recycled aggregate · Transport properties · Capillarity · Water penetration

1 Introduction

There is an urgent need to produce eco-efficient concrete mixes. The concrete industry has one of the highest CO_2 footprints, and it is also highly intensive regarding the consumption of non-renewable resources. The use of supplementary cementitious materials

J. I. Escalante-Garcia et al. (Eds.): RW 2021, RILEM Bookseries 40, pp. 596–604, 2023.
https://doi.org/10.1007/978-3-031-21735-7_65

plays a vital role in reducing the clinker factor in cements. However, consumption of cement is so large that new sources of supplementary cementitious materials become necessary. Certain policies against usage of coal power plants are already reducing the availability of fly ash in Europe, and due to the incorporation of circular economy within the major steel industries, there is a decrease in the production of ground granulated blast-furnace slag. Finding new sources of alternative SCMs that are sufficiently reactive so as to replace significant amounts of cement in the concrete is not an easy task. In some regions, the availability of natural pozzolans solves this issue, and in other regions the use of calcined-clays is being promoted with these aims. Similarly, some regions have a sensible production of other potential by-products such as copper slags and the valorization of these materials as an SCM product can help in the endeavor.

Copper slags are by-product results from pyro-metallurgical processes involving oxidation and reduction of Copper (Cu) ore or scrap to synthesize Cu metal [1]. Their composition depends on the initial Cu source and the processing. Still, their overall composition comprising mainly of SiO_2, FeO, and CaO, make them very convenient as supplementary cementitious material [2]. Copper slags have been already proven effective as supplementary cementitious material [3–6] in contents up to 20 wt% of total binder. When higher replacement ratios are aimed, the reaction of copper slag is however rather low [7], offering its contribution only at later ages. Therefore, it is necessary to search for alternative means for activating the copper slag to further reduce the clinker factor of concretes.

In case of possessing only pozzolanic characteristics, mainly in the system of non-ferrous slag like copper slag, extra activation sources particularly in form of alkalis are necessary to promote the copper slag's participation in the formation of microstructure. However, usage of high amount of alkalis can also serve as a bottle-neck in some concrete applications. An interesting strategy to reduce the content of alkalis can be usage of latent hydraulic residues such as GGBFS together with alkalis. These binders can possess only limited alkalis (according to the regulations) and cement with maximum incorporation of ferrous (GGBFS) and non-ferrous (Cu slag) slags.

Finally, another environmental concern for concrete is its circularity. Promoting the recycling of waste concrete into high value applications is key for developing a circular economy. Then, the consumption of natural rocks is reduced and landfilling is prevented. In structural concrete, however, the attached mortar in recycled aggregate particles can play a detrimental role [8]. This phase increases the water absorption and porosity of the recycled aggregates. Durability performance of recycled aggregate concretes can be limited. Still, some researchers have presented acceptable results for water transport properties when the recycled aggregate was embedded in a compact matrix that isolates their pore structure [9].

This paper presents the results of a combined design for concrete with hybrid binders based on Portland cement and activated copper slag and coarse recycled aggregate. A mix containing only copper slag is contrasted with a ternary mix containing both copper slag and ground granulated blast furnace slag. The analysis focuses on the water transport properties of both concretes and the contribution of the ground granulated blast-furnace slag to the development of microstructure. Capillary absorption rate and water penetration under pressure are analyzed as main descriptors of the pore connectivity in the concretes, and indexes of the potential durability performance of these concretes.

2 Materials and Methods

The compositions of the concrete mixes are presented in Table 1. In addition to the coarse recycled aggregate (incorporated as 50 vol% of the coarse aggregate), four natural aggregates, two fine and two coarse, were used to optimize the skeleton of the mixes. The main difference between both mixes was the binder type. One mix (Mix 1) contained 50 wt% of copper slag, 20 wt% of ground granulated blast-furnace slag, and 30 wt% of Portland cement (CEM I 52.5 R). The other mix contained 70 wt% of copper slag and 30 wt% of Portland cement (Mix 2). Both mixes were admixed with a Na_2SO_4-based activator specially designed to be compatible with the copper slag. To maintain the workability level in the mixes, a superplasticizer was added to the mixes.

Table 2 presents properties of the concretes in the fresh and hardened state. Slump values were maintained in the range of 12 ± 2 cm, and the measured entrained air was almost the same for both mixes. The properties in the hardened state showed an increased porosity and a reduced compressive strength for Mix 2 compared to Mix 1.

Table 1. Composition of concrete mixes.

Proportions (kg/m^3)	Mix 1	Mix 2
Sand 0/8	707	707
Fine sand	126	126
Gravel 4/14	142	142
Gravel 4/32	364	364
Recycled aggregate 4/32	455	455
CEM I 52.5 R	122	125
Water	180	187
Copper slag	203	291
GGBFS	81	–
Activator	19	19
Superplasticizer	1.2	1.1

For the capillary water uptake, four samples per concrete and testing age were tested. The samples were cut between 3 and 8 cm from the base of standard cylinders of 10 cm in diameter and 20 cm in height, which were laterally waterproofed with adhesive aluminum tape. The preconditioning started immediately after the curing of 28 days was finished, and consisted of saturating the samples under water for 72 h and drying them afterwards in an oven at 40 °C until the variation in weight was below 0.1 wt% in a 24 h period (approximately 10 days). Then samples were put in contact with water on their bottom

Table 2. Properties of concretes.

Properties	Mix 1	Mix 2
Slump (cm)	11.5/12.5	13.5/11.5
Entrained air (%)	2.3	2.2
Vacuum water absorption 28 days	7.80 ± 0.19	8.28 ± 0.12
Density ssd 28 days	2.38 ± 0.006	2.37 ± 0.004
Open porosity (%)	17.18 ± 0.35	18.15 ± 0.22
Compressive strength 28 days (MPa)	38.9 ± 1.0	33.4 ± 0.5

face, which laid 3 ± 1 mm below the water level. The experiment was done in a closed container to prevent evaporation, maintained in an environment at 20 °C.

For the water penetration under pressure, three specimens per concrete and testing age were tested. Standard cylinders of 15 cm in diameter by 30 cm in height were used. The cylinders were cut at half their height, and both halves were tested at their bottom faces, i.e., the bottom half was tested on its face against the mould (identified as bottom values) and the top half was tested on it cut face (identified as middle values). The purpose of these multiple measurements was to identify both the potential effect of the testing face and the potential effect of bleeding on the water penetration under pressure. All specimens were sealed with aluminum tape on their sides to prevent any leakage in case concrete was very permeable. The testing surface was a circle of 10 cm diameter, sealing the most external 2.5 cm edge. Samples were dried in the laboratory environment (60% RH and 20 °C) for 14 days and then tested by exposing the testing face to a first step of water under a pressure of 1 bar for 24 h and a second step of water under a pressure of 5 bar for 72 h. Samples were split after completing the exposure period and the profile of water penetration under pressure was measured on both sides of the split sample.

3 Results and Discussion

Figure 1 presents the results of the capillary water uptake after 28 and 90 days of wet curing. Water uptake is plotted against the fourth root of time as recommended in [10] for improved linearity. A more significant reduction of the capillary water transport with curing time is observed for the concrete containing GGBFS, whereas almost no evolution is observed for Mix 2.

Figure 2 presents the comparison of the sorptivity coefficients computed from the linear regressions to the first stage of the transport process (0–20 $s^{0.25}$). The previously explained evolution in the reduction of the sorptivity value is more clearly noticed for the concrete containing GGBFS, in comparison with Mix 2, which remains almost the same.

The values for the sorptivity index at 90 days obtained for Mix 1 (w/b = 0.44) and Mix 2 (w/b = 0.43) are comparable to values previously reported for conventional concrete (i.e. with natural aggregates and Portland cement as only binder) with w/b =

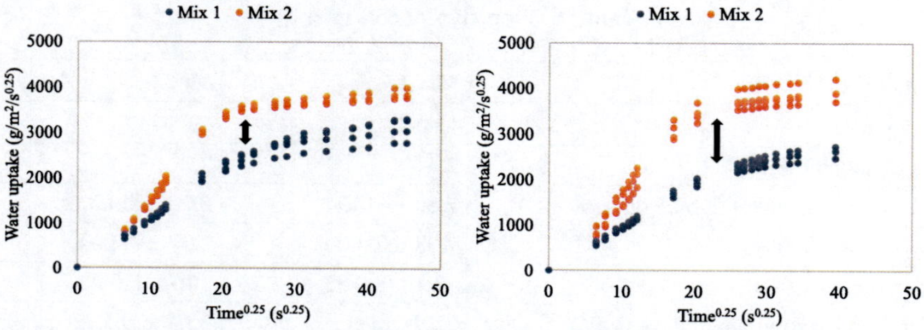

Fig. 1. Capillary water uptake of concretes after 28 days (left) and 90 days (right) of curing.

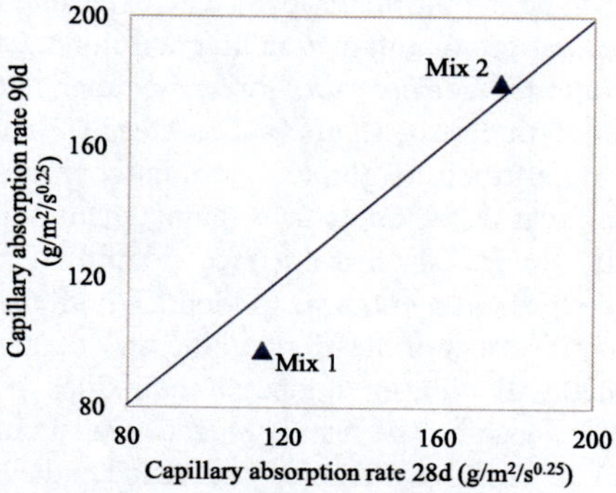

Fig. 2. Sorptivity coefficients for 28 and 90 days.

0.50 and 0.60, respectively. If this is translated into k-values for the sorptivity index, the combination of GGBFS and CS resulted in a value of 0.84 and the CS alone resulted in a factor of 0.6. The sorptivity index is addressed as a good indicator of carbonation resistance when only the physical factor is to be taken into account (i.e. no portlandite consumption due to pozzolanic action involved). This is based on both transport processes being controlled by diffusion. The obtained k-values are therefore comparable to k-values reported for carbonation of concretes blended with 20–40 wt% GGBFS (0.80) and with >60 wt% GGBFS (0.60) [11]. Considering only the physical effect of the binder system in the microstructure (assuming a neglectable consumption of portlandite by the GGBFS and CS), the results are therefore very promising as the hybrid system with alkaline activation brings results for concretes with 70 wt% CS relative to the binder that might be compared to concretes with 70 wt% GGBFS in terms of the k-factor. The results are even more valuable when considering that the concretes contained 50 vol% of coarse recycled aggregate with a higher porosity than natural aggregate.

Figure 3 presents the permeameter used to test the water penetration under pressure and an example of the water penetration profile obtained for one sample from each concrete.

Figure 4 presents the results of the water penetration under pressure. The middle face was showing a deeper water penetration than the bottom face for most of the series, except for Mix 2 at 28 days. In any case, the variability of results was high in all cases and it does not allow to establish a sound differentiation between the middle and the bottom faces. Again, increased transport values were noticed for Mix 2 in comparison with Mix 1, attributed to the contribution of the GGBFS to development of microstructure. Differently to what was noticed for the sorptivity index, there is a reduction in the transport of water under pressure with curing time for both types of concretes. In relative terms, the reduction from 28 to 90 days is similar, maintaining the relationship between both concretes at a similar value at both ages. In this case the evolution of the microstructure cannot be attributed only to the reaction of the GGBFS, but it is mostly a result of the whole binder system. The trend for this reduction is noticeable, although the significant variability of the profiles make it difficult to establish a reliable order of reduction (which is in the range of 10–30% reduction in the penetration value).

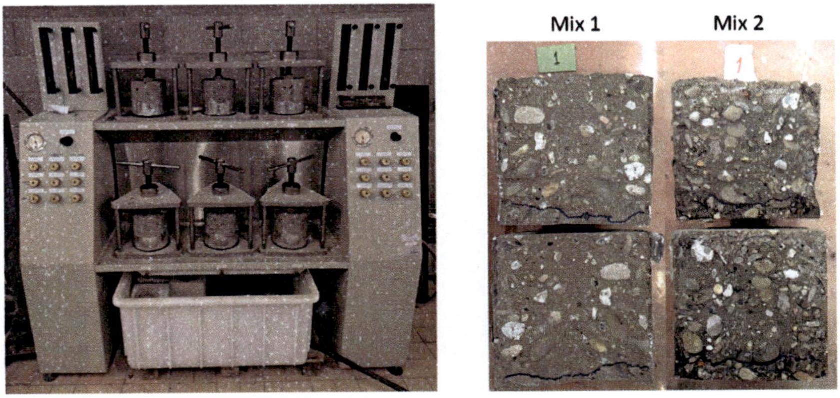

Fig. 3. Water permeameter and example of test results with the water penetration profile.

Figure 5 presents the values for Mix 1 relative to the values for Mix 2 at the ages of 28 and 90 day. The relationship maintains the same at both ages except for the sorptivity coefficient. The activation of the GGBFS provides a denser initial structure, and from 28 to 90 days the evolution of the porosity is mostly governed by the whole binder system, with not much influence of the GGBFS. It seems that the inclusion of a limited amount of GGBFS is effective in providing additional aluminate phases that react fast and reduce the potential influence of curing at later ages. The sorptivity index, however, seems to be more sensitive to additional contribution of the GGBFS. This is an indication of the GGBFS contributing to reduce the capillary pores in the concrete, while the overall porosity remains relatively unaffected. Such effect is similar to what can be expected in binary binder systems containing GGBFS and PC as the only cementitious constituents [12].

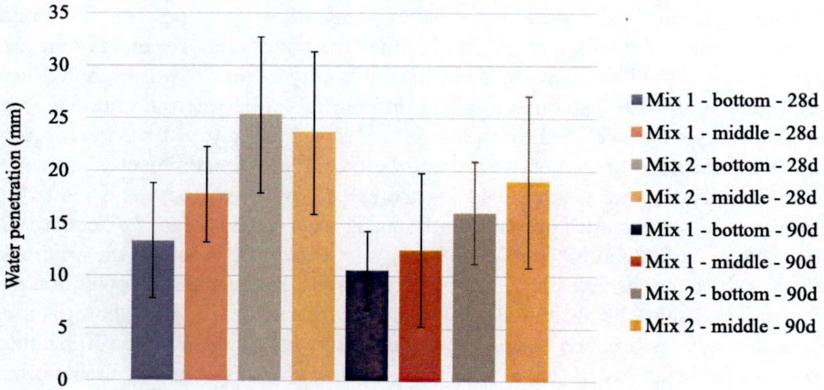

Fig. 4. Results of water penetration under pressure at 28 and 90 days.

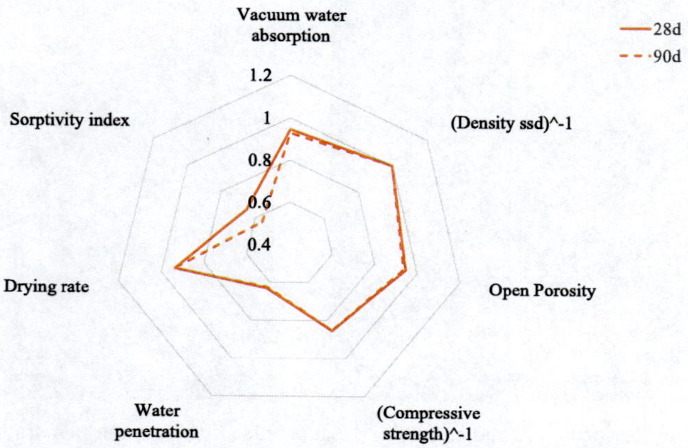

Fig. 5. Relative values of Mix 1/Mix 2 picturing the consistency of results of porous and water transport properties.

4 Conclusions

The assessment of water transport properties of concretes with hybrid binders based on copper slag and ground granulated blast furnace slag and coarse recycled aggregate allow the following conclusions:

- The incorporation of 20 wt% of GGBFS in the system CS + PC + activator improves the development of microstructure, especially concerning the refinement of capillary porosity between 28 and 90 days.
- The overall porosity of concretes with the hybrid binders and recycled aggregates was reduced from 28 to 90 days, independently of the system containing CS or CS

+ GGBFS. However, the results were more advantageous when 20 wt% GGBFS was included in the binder system.
- The predicted durability performance of the concretes is promising when considering the k-values obtained for the binder system, which are comparable to concretes containing high volumes of GGBFS and no activator.

Acknowledgments. This research was funded by SIM MARES program (DUSC project), grant number HBC.2017.0607; and Research Foundation-Flanders (FWO-Vlaanderen), grant MSCA-SoE 12ZZD21N LV (Y. Villagrán-Zaccardi). The authors would like to thank the industrial partners Resourceful, Metallo Belgium, Van Pelt and Sika for providing the activator, Cu slag, aggregates and superplasticizer.

References

1. Schlesinger, M., King, M., Sole, K.C., Davenport, W.G.I: Extractive Metallurgy of Copper, 5th edn. Elsevier, Amsterdam (2011)
2. Piatak, N.M., Parsons, M.B., Seal, R.R.: Characteristics and environmental aspects of slag: a review. Appl. Geochem. **57**, 236–266 (2015)
3. Moura, W.A., Gonçalves, J.P., Lima, M.B.L.: Copper slag waste as a supplementary cementing material to concrete. J. Mater. Sci. **42**(7), 2226–2230 (2007). https://doi.org/10.1007/s10853-006-0997-4
4. Edwin, R.S., De Schepper, M., Gruyaert, E., De Belie, N.: Effect of secondary copper slag as cementitious material in ultra-high performance mortar'. Constr. Build. Mater. **119**, 31–44 (2016)
5. Edwin, R.S., Gruyaert, E., De Belie, N.: Influence of intensive vacuum mixing and heat treatment on compressive strength and microstructure of reactive powder concrete incorporating secondary copper slag as supplementary cementitious material. Constr. Build. Mater. **155**, 400–412 (2017)
6. Sivakumar, P.P., Gruyaert E., De Belie, N., Matthys, S.: Reactivity of modified ferro silicate slag as sustainable alternative binder. In: Ganjian, et al. (eds.) Fifth International Conference on Sustainable Construction Materials and Technologies (SCMT5), vol. 3, pp. 39–49. Kingston University, London (2019)
7. Hallet, V., De Belie, N., Pontikes, Y.: The impact of slag fineness on the reactivity of blended cements with high-volume non-ferrous metallurgy slag. Constr. Build. Mater. **257**, 119400 (2020)
8. Sánchez de Juan, M., Alaejos Gutiérrez, P.: Study on the influence of attached mortar content on the properties of recycled concrete aggregate. Constr. Build. Mater. **23**(2), 872–877 (2009)
9. Zega, C.J., Santillán, L.R., Sosa, M.E., Villagrán Zaccardi, Y.A.: Durable performance of recycled aggregate concrete in aggressive environments. J. Mater. Civ. Eng. **32**(7), 03120002 (2020)
10. Villagrán Zaccardi, Y.A., Alderete, N.M., De Belie, N.: Improved model for capillary absorption in cementitious materials: progress over the fourth root of time. Cem. Concr. Res. **100**, 153–165 (2017)
11. Nobis, C., Vollpracht A.: K-value for carbonation of concretes with supplementary cementitious materials. In: Sanjayan, J., et al. (eds.) 27th Biennial National Conference of the Concrete Institute of Australia in conjunction with the 69th RILEM Week, Concrete Institute of Australia and RILEM, pp. 931–938. Concrete Institute of Australia, Melbourne (2015)

12. Zaccardi, Y.A.V., Di Maio, Á.A., Romagnoli, R.: The effect of slag and limestone filler on resistivity, sorptivity, and permeability of concrete with low paste content. MRS Online Proc. Libr. **1488**(1), 127–133 (2013). https://doi.org/10.1557/opl.2012.1552

Early Age Properties of Paste and Mortar Made with Hybrid Binders Based on Portland Cement, GGBFS and Sodium Sulfate

J. M. Etcheverry(✉), P. Van den Heede, Y. A. Villagran-Zaccardi, and N. De Belie

Magnel-Vandepitte Laboratory for Structural Engineering and Building Materials, Ghent University, Technologiepark-Zwijnaarde 60, 9052 Ghent, Belgium
{juanmanuel.etcheverry,nele.debelie}@ugent.be

Abstract. Nowadays, Ground Granulated Blast-Furnace Slag (GGBFS) is increasingly used together with Portland cement to design low carbon footprint binder systems. Such systems with a low clinker factor are generally connected with a slow strength development. Alkaline activation in hybrid binders can provide a suitable solution to cope with this issue. Among the possible activators, sodium sulfate offers the advantage of its low-carbon footprint and good performance in the presence of Portland cement; however, in addition to strength development, other early age properties are also affected and compliance with requirements in the field must be verified. In the present study, hybrid systems based on Portland cement, GGBFS and sodium sulfate were analysed. Sodium sulfate was used in doses between 0 and 10 wt% relative to GGBFS. The GGBFS content amounted to 70 wt% of the binder and water-to-binder ratio was 0.45. Accelerated setting was detected due to the addition of sodium sulfate using the Vicat test and ultrasonic pulse velocity analysis. Increases in early age strength were observed for all the analysed doses of activator. Such increases are attributed to a dual effect: the acceleration of cement hydration and the enhanced dissolution of the slag. On this basis, possible changes in the reaction mechanism in presence of sodium sulfate have been discussed. Upon a comprehensive analysis of results, an optimal content of 5–10% of sodium sulfate in the mixes is suggested.

Keywords: Sodium sulfate · Ground granulated blast furnace slag · Hybrid binders · Setting times · Ultrasonic pulse velocity

1 Introduction

Ground granulated blast furnace slag (GGBFS) has latent hydraulicity and a reduced carbon footprint associated with its production, which makes it an interesting alternative for low carbon binders. In presence of water, GGBFS has a limited degree of hydration due to a low-calcium superficial layer which delays its reaction [1]. In blended cements the GGBFS is activated by the Portland cement but at high GGBFS content a poor strength development is observed [2]. Alkaline activation of SCMs has shown to be

J. I. Escalante-Garcia et al. (Eds.): RW 2021, RILEM Bookseries 40, pp. 605–613, 2023.
https://doi.org/10.1007/978-3-031-21735-7_66

suitable to produce ultra-low carbon binders. Nevertheless, at ambient temperature the strength development is also slow [3].

Therefore, hybrid binders in which the reactivity of GGBFS is potentiated by both a small content of Portland cement and an alkali activator comes up as an intermediate alternative. For these systems, improvements in early strength as well as reductions in workability and setting times have been reported [4, 5]. Among several options for activators [6–9], sodium sulfate shows some advantages for its relatively low-carbon footprint and safety. However, little information regarding fresh properties of these hybrid systems containing sodium sulfate is available in the literature. The aim of this study is to understand how the dose of sodium sulfate impacts on 70% GGBFS–30% CEM I hybrid systems. The compressive strength of mortars was used to quantify the performance of the system, while ultrasonic pulse velocity and isothermal calorimetry were used to monitor the hydration and hardening processes. Furthermore, life cycle assessment (LCA) compliant carbon footprint calculations considering the compressive strength of mortars were executed to determine the global warming potential (GWP) of the proposed alkali activated hybrid binders.

2 Materials

The chemical composition and physical properties of the CEM I 52.5 N and GGBFS are given in Table 1. The sodium sulfate was technical grade with a purity >99%. Standard sand (EN 196-1) and tap water were used for the production of mortars.

3 Methods

3.1 Water Demand, Initial and Final Setting Times

Pastes with 70% of GGBFS and 30% of CEM I 52.5 N were made. The sodium sulfate was added dry into the mixes, in doses between 0 and 10 wt% relative to GGBFS. Water was added as needed to reach a penetration of 36 ± 2 mm of the Vicat plunger (after four minutes of contact between water and binder). The mixing procedure was 90 s at low speed (140 rpm), a pause of 30 s and another 90 s at low speed. Having determined the standard consistency, the paste was poured in the mould, and the penetrations of the 1 mm^2 section Vicat needle were recorded every 5 min until the needle penetrated only 0.5 mm in 3 different positions (final setting time). The initial setting time was the one corresponding to a needle penetration of 34 ± 3 mm (NBN EN 196-3).

3.2 Ultrasonic Pulse Velocity

Ultrasonic pulse velocity (UPV) through standard consistency pastes was determined using the FreshCon device (University of Stuttgart), and SmartPick software for data acquisition and processing. The pastes were cast and compacted in two layers and covered with polypropylene film throughout the testing time of 23 h. The test was carried out at 20 °C in a 60% relative humidity (RH) environment. Longitudinal (P) waves were measured 10 min after binder-water contact. More information about the setup, software and P-waves onset time picker is found in [10–12].

Table 1. Chemical composition and physical properties of CEM I and GGBFS.

Chemical composition		CEM I	GGBFS
CaO	[% m/m]	64.30	40.88
SiO_2	[% m/m]	18.30	33.63
MgO	[% m/m]	1.40	7.17
Al_2O_3	[% m/m]	5.20	11.16
Fe_2O_3	[% m/m]	4.00	0.34
Mn_2O_3	[% m/m]	–	0.29
Cl^-	[% m/m]	0.06	0.01
S^{2-}	[% m/m]	–	0.79
SO_3	[% m/m]	3.50	0.07
Na_2O	[% m/m]	0.32	0.13
K_2O	[% m/m]	–	0.47
Na_2O equivalent	[% m/m]	–	0.44
CaO + MgO + SiO_2	[% m/m]	84.00	81.68
(CaO + MgO)/SiO_2	–	3.59	1.43
Physical properties			
Blaine specific surface area	[m^2/kg]	408	429–484
Particle size distribution (d10/50/90)	[um]	2.35/10.82/29.44	1.33/7.63/26.77
Density	[kg/m^3]	3160	2890

3.3 Isothermal Calorimetry

Materials were first preconditioned at 20 °C to avoid temperature differences during the measurements, then they were mixed manually inside the vials ensuring a proper homogenization and placed in the isothermal calorimeter (TAM Air) 2 min after water-binder contact. The amount of paste was 14 g in all cases. The cumulative heat released was recorded for 7 days.

3.4 Flexural and Compressive Strength

Mortars were made using a binder ratio GGBFS/CEM I of 70/30, water-to-binder ratio of 0.45, and normalized sand with a binder/sand ratio of 1/3. Sodium sulfate was added dry into the mixes, in doses between 0 and 10 wt% relative to GGBFS.

GGBFS, CEM I, Na_2SO_4 and water were mixed for 30 s at low speed (140 rpm), sand was added within 30 s, then speed was increased (285 rpm) for 30 s. A pause of 1 min was adopted and then mixing was completed with 90 s at high speed. Flow was determined using the flow table (NBN EN 1015 3).

Prismatic mortar bars of 40 × 40 × 160 mm were cast (NBN EN 1015-11), demolded after 24 h and kept in humid chamber at 95% RH and 20 °C until the envisaged ages of testing.

The flexural strength of mortar was determined at 2, 7 and 28 days. Compressive strength tests were performed on the resulting prism halves. The tests were according to NBN EN 1015-11.

3.5 Carbon Footprint Calculations

Carbon footprint calculations for 1 m^3 of mortar with 0, 5 and 10% of sodium sulfate were done in SimaPro 9.1.1.1 equipped with the Ecoinvent 3.6 database. The impact method used was CML-IA for which only the impact indicator related to global warming was considered: the global warming potential (GWP) expressed in kg CO_2 equivalent.

The characteristic compressive strengths were calculated in accordance with NBN EN 1990. The impacts of basic treatment were assigned to the GGBFS. This includes granulation, drying, grinding and storage [13]. The obtained GWP values for 1 m^3 of mortar were also normalized once to the characteristic compressive strength at various ages in order to find an optimum sodium sulfate content to be used from both a mechanical and environmental viewpoint.

4 Results

4.1 Water Demand, Initial and Final Setting Times

Table 2 shows the proportions of pastes including the water necessary to reach standard consistency. The initial and final setting times were obtained as the average of two Vicat measurements. As minor differences in water demand are observed for the pastes, the effect of sodium sulfate on water demand seems to be minimal in this regard.

When it comes to setting times, significant decreases were observed with the addition of sodium sulfate, reducing the initial setting time by approximately 30% and the final setting time by approximately 35%. The reductions are for both the initial and final setting times and the time elapsed between them. However, no significant differences occurred between SS5 and SS10, implying a similar effect of the sodium sulfate within this dose range.

4.2 Ultrasonic Pulse Velocity

The UPV curves (Fig. 1) suggest a similar trend to the one observed for the Vicat tests: acceleration due to the addition of sodium sulfate but slight differences depending on the sodium sulfate dosing range (5–10%). In the reference sample (SS0), a small dormant period is noticed, followed by a rapid increase in the wave velocity. After 5 h, an attenuation to progressively reach 2200 m/s at 23 h is seen. With the addition of sodium sulfate, this dormant period disappears and the velocity started to rise immediately after the filling of the moulds. An increase in the slope of the curves with the sodium sulfate in comparison to the reference could also be noticed. Moreover, similar final velocities at 23 h of 3000 m/s for SS5 and 3100 m/s for SS10 are shown.

Table 2. Mix proportions and setting times

Materials [g]	SS0	SS5	SS10
Water	155	149	149
CEM I	150	150	150
GGBFS	350	350	350
Na_2SO_4	0	17.5	35
Initial setting time [min]	180	130	120
Final setting time [min]	390	250	260

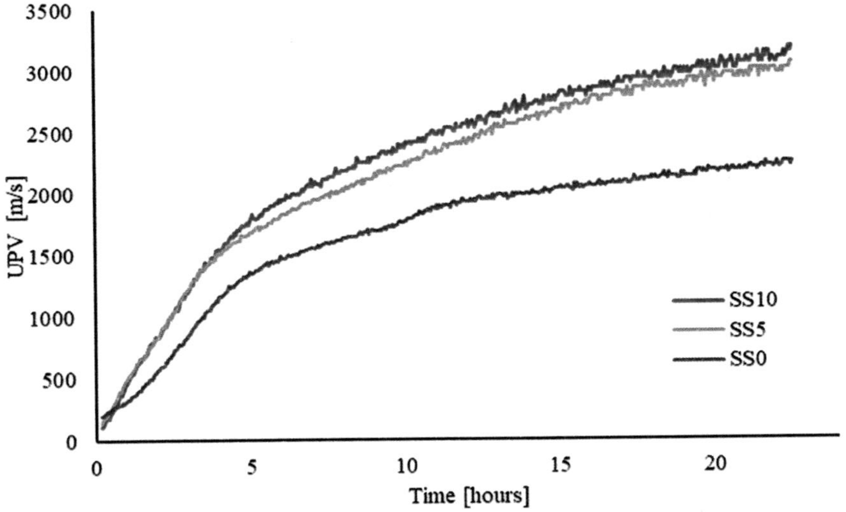

Fig. 1. Ultrasonic pulse velocity for pastes with 10, 5 and 0% of sodium sulfate versus time after water-binder contact.

4.3 Isothermal Calorimetry

Figure 2 shows the result of isothermal calorimetry, which yielded a higher cumulative heat release for the samples containing sodium sulfate. The initial slope increased with increasing sodium sulfate dose. A break in the slope is observed for the reference sample at 20 h after initial water-binder contact while for the pastes containing sodium sulfate the trend is maintained until approximately 30 h. At this point, the paste with 10% of sodium sulfate starts deviating from the 5%-sample. The heat evolution becomes almost parallel again at around 50 h. The total heat released after 2 days is 100 [J/g binder] for the reference sample, 175 [J/g binder] for SS5 and 210 [J/g binder] for SS10.

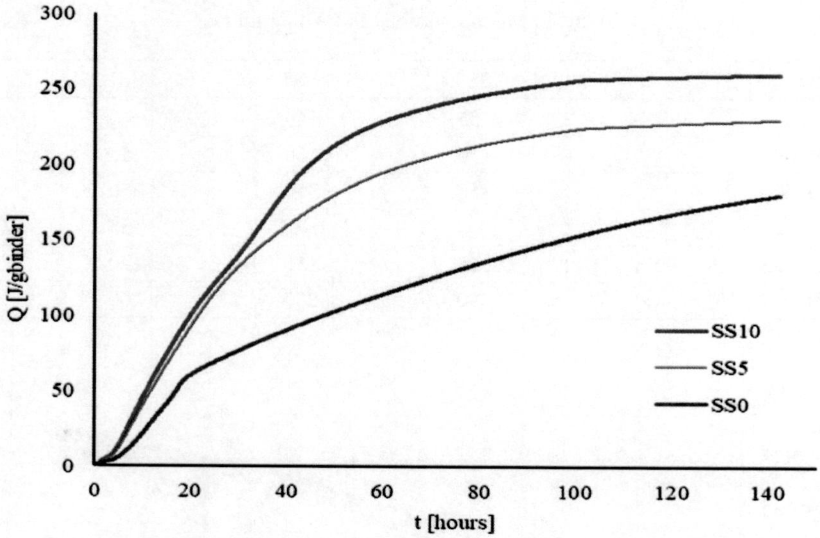

Fig. 2. Cumulative heat release curves from isothermal calorimetry.

4.4 Flexural and Compressive Strength

Table 3 shows the flexural and compressive strength and flow values of the mortars. A significant improvement in the compressive strength at 2 days with the addition of sodium sulfate is noted, being about 2.5–3 times the value of the reference. By increasing the doses of activator from 5 to 10% the compressive strength at 2 days gains an additional 30%. The absolute difference in compressive strength between the mortars at 2 days was maintained at 7 and 28 days. The relative difference reduces to the extent that all mixes gain similar strength after 2 days in absolute terms. Compressive strengths at 28 days are from 35 to 45% higher with the addition of 5 and 10% sodium sulfate.

Table 3. Flow values, flexural and compressive strength (average ± standard deviation).

	SS0			SS5			SS10		
Flow [cm]	19.0			19.0			20.5		
Age	2 days	7 days	28 days	2 days	7 days	28 days	2 days	7 days	28 days
Flexural strength [N/mm^2]	2.5 ± 0.49	4.9 ± 0.09	5.4 ± 0.01	4.8 ± 0.11	4.8 ± 0.11	9.6 ± 0.28	6.5 ± 0.12	10.8 ±0.05	11.8 ± 0.43
Compressive strength [N/mm^2]	8.5 ± 0.41	29.2 ± 2.21	43.4 ± 2.24	21.6 ± 0.12	47.5 ± 1.23	58.2 ± 1.18	27.7 ± 2.05	49.2 ± 3.16	62.3 ± 2.15

4.5 Carbon Footprint Calculations

Figure 3 shows the results of the carbon footprint calculations for 1 m^3 of mortar (Fig. 3a) and for 1 m^3 of mortar per unit of characteristic compressive strength at 2, 7 and 28 days (Fig. 3b–d). When not accounting for the mechanical performance of the mixtures, the carbon footprint obviously increases with the impact of the added amount of sodium sulfate (Fig. 3a). If the strength at 2 days is considered, the addition of sodium sulfate results in a carbon footprint of half the one observed for the reference (Fig. 3b). When considering the strength at later ages the carbon footprints associated with the different mixtures start to become slightly more similar (Fig. 3c and d).

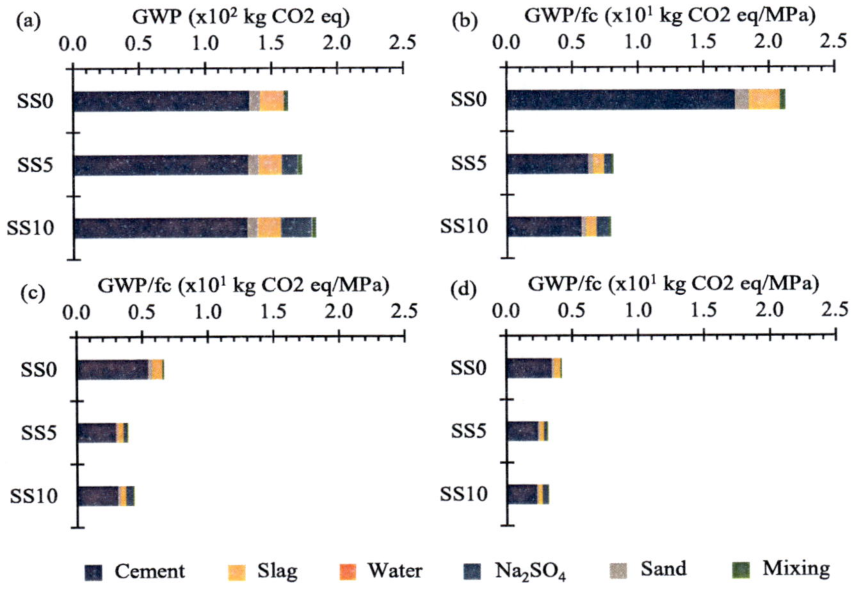

Fig. 3. (a) GWP of 1 m^3 of mortar; (b) GWP of 1 m^3 mortar per unit of 2 day characteristic compressive strength fc; (c) GWP of 1 m^3 mortar per unit of 7 day characteristic compressive strength fc; (d) GWP of 1 m^3 mortar per unit of 28 day characteristic compressive strength fc.

5 Discussion

Both the setting times measured in the Vicat test and the ultrasonic pulse velocity analysis show an acceleration due to the incorporation of sodium sulfate in the mixtures. However, doubling the dose of sodium sulfate has almost no effect on the setting. A larger effect for the ultrasonic pulse velocity result (Fig. 1) in comparison with the Vicat test result is noted. After 4 h, the ultrasonic pulse velocity curve for the sample with a 10% sodium sulfate content separates from the sample with 5%, but at around 5 h, the curves become parallel again and the trend is maintained until 23 h. The cumulative heat release curves for SS10 and SS5 (Fig. 2) also start to separate in the same time interval as the UPV, and after 7 h they become parallel. Both the calorimetry and UPV show sensitivity to the presumed activation of the GGBFS achieved with 10% sodium sulfate in comparison with 5% sodium sulfate. This suggests some very early influence of the sodium sulfate much earlier than 30 h, when dissolution of GGBFS is generally reported and also observed with the hump for SS10 in Fig. 2 at this age.

A double action of sodium sulfate can be then described; on the one hand, at very early age there is the acceleration of alite hydration [14] that commands the setting times. Then, the different sodium sulfate contents seem to have little influence. On the other hand, there is the slag dissolution occurring [15] from 30 to 50 h. It is at this stage that the sodium sulfate dosage becomes relevant. This is important because in some systems with small content of sodium sulfate it is possible to obtain increase in strength at early age, but if the content is not enough this is almost completely due to acceleration of alite hydration and not due to proper activation of the slag.

When it comes to early age strength, the cumulative heat values after 2 and 7 days seem to correlate well with the resistance levels reached. It is also observed that the action of sodium sulfate seems to have an influence mainly within the first 48 h. Afterwards, the strength development is similar for all the samples studied. Contrary to what has been reported in [14], no decrease in the 28 day compressive strength was observed. This could be attributed to the nature of the slag or to the higher doses of sodium sulfate used. However, further studies are necessary to fully describe the reaction mechanics in the mentioned systems and to elucidate the interaction between sodium sulfate, GGBFS and Portland cement. Moreover, the influence of intermediate doses to those carried out in the present study can provide additional accuracy.

Finally, the analysis of the carbon footprint considering the characteristic compressive strength at various ages seems to support the use of sodium sulfate in the range between 5 and 10%. More detailed studies are needed, considering not only other environmental parameters but also the possible microstructural changes generated by the addition of such a higher dose of activator.

6 Conclusions

The addition of sodium sulfate reduces setting of 70% GGBFS + 30% CEM I, while differences in the dosages between 5 and 10 wt% relative to GGBFS have almost no influence on setting. Still, increasing the sodium sulfate to 10% results in higher heat release and compressive strength.

The alkaline activator shows significant impact before 2 days. Later, strength development is independent of the presence and amount of sodium sulfate. Finally, based strictly on resistance parameters and carbon footprint calculations, doses between 5 and 10% of sodium sulfate are suggested to optimise hybrid mixtures containing GGBFS and Portland cement.

Acknowledgements. This research was funded by Research Foundation-Flanders (FWO-Vlaanderen) through research grants G062720N (J. M. Etcheverry and P. Van den Heede) and MSCA-SoE 12ZZD21N LV (Y. A. Villagrán-Zaccardi).

References

1. De Belie, N., Soutsos, M., Gruyaert, E.: Properties of Fresh and Hardened Concrete Containing Supplementary Cementitious Materials: State-of-the-Art Report of the RILEM Technical Committee 238-SCM, Working Group 4 (2018)
2. Labarca, I.K., Foley, R.D.: Wisconsin Highway Research Program Effects of Ground Granulated Blast Furnace Slag in Portland Cement Concrete (PCC)—Expanded Study (2007)
3. Provis, J.L., van Deventer, J.S.J.: Alkali Activated Materials-State-of-the-Art Report, RILEM TC 224-AAM. Springer, Dordrecht (2013). https://doi.org/10.1007/978-94-007-7672-2
4. Singh, N., Sarita, R., Singh, N.B.: Effect of sodium sulphate on the hydration of granulated blast furnace slag blended Portland cement. Indian J. Eng. Mater. Sci. **8**, 110–113 (2001)
5. Acevedo-Martinez, E., Gomez-Zamorano, L.Y., Escalante-Garcia, J.I.: Portland cement-blast furnace slag mortars activated using waterglass—part 1: effect of slag replacement and alkali concentration. Constr. Build. Mater. **37**, 462–469 (2012). https://doi.org/10.1016/j.conbuildmat.2012.07.041
6. Shi, C., Roy, D., Krivenko, P.: Alkali-Activated Cements and Concretes, 1st edn. CRC Press (2003). https://doi.org/10.1201/9781482266900
7. Rashad, A.M.: An exploratory study on sodium sulfate activated slag modified with Portland cement. Mater. Struct. **48**(12), 4085–4095 (2014). https://doi.org/10.1617/s11527-014-0468-3
8. Angulo-Ramírez, D.E., Mejía de Gutiérrez, R., Puertas, F.: Alkali-activated Portland blast-furnace slag cement: mechanical properties and hydration. Constr. Build. Mater. **140**, 119–128 (2017). https://doi.org/10.1016/j.conbuildmat.2017.02.092
9. Mota, B., Matschei, T., Scrivener, K.: Impact of sodium gluconate on white cement-slag systems with Na_2SO_4. Cem. Concr. Res. **122**, 59–71 (2019). https://doi.org/10.1016/j.cemconres.2019.04.008
10. Reinhardt, H.W., Grosse, C.U.: Continuous monitoring of setting and hardening of mortar and concrete. Constr. Build. Mater. **18**, 145–154 (2004). https://doi.org/10.1016/j.conbuildmat.2003.10.002
11. Manual, U.: FreshCon Duo. 0–34 (2012)
12. Krüger, M., Grosse, C.U., Lehmann, F.: Automated shear-wave techniques to investigate the setting and hardening of concrete in through-transmission. In: Güneş, O., Akkaya, Y. (eds.) Nondestructive Testing of Materials and Structures. pp. 431–436. Springer Netherlands, Dordrecht (2013)
13. Chen, C., Habert, G., Bouzidi, Y., Jullien, A., Ventura, A.: LCA allocation procedure used as an incitative method for waste recycling: an application to mineral additions in concrete. Resour. Conserv. Recycl. **54**, 1231–1240 (2010). https://doi.org/10.1016/j.resconrec.2010.04.001
14. Fu, J., Jones, A.M., Bligh, M.W., Holt, C., Keyte, L.M., Moghaddam, F., Foster, S.J., Waite, T.D.: Mechanisms of enhancement in early hydration by sodium sulfate in a slag-cement blend—insights from pore solution chemistry. Cem. Concr. Res. **135** (2020). https://doi.org/10.1016/j.cemconres.2020.106110
15. Rashad, A.M., Bai, Y., Basheer, P.A.M., Milestone, N.B., Collier, N.C.: Hydration and properties of sodium sulfate activated slag. Cem. Concr. Compos. **37**, 20–29 (2013). https://doi.org/10.1016/j.cemconcomp.2012.12.010

Review and Experimental Investigation of Retarder for Alkali-Activated Cement

Klaus Holschemacher[1(✉)] and Biruk Hailu Tekle[1,2]

[1] Structural Concrete Institute, Leipzig University of Applied Sciences (HTWK Leipzig), 04277 Leipzig, Germany
klaus.holschemacher@htwk-leipzig.de
[2] Institute of Innovation, Science and Sustainability, Federation University, Ballarat, VIC, Australia

Abstract. Alkali-activated cement (AAC) cured at ambient temperature conditions have a wider application area compared to heat cured AAC. High calcium precursor materials such as ground granulated blast furnace slag (GGBS) are commonly used to achieve ambient curing behavior. However, the GGBS results in a short setting time. Hence setting adjustment is critical in such AAC systems. This paper reviews state-of-the-art in the area of retarders for AAC systems. The most promising prospective retarders such as zinc salts, borax, sucrose, and phosphates are investigated. The retarder's effect is dependent on the precursor materials and alkaline activators used. Consequently, in the review, these are identified for each retarder discussed. Some of the retarders were then tested in AAC with a blended precursor system containing fly ash (FA) and GGBS activated with sodium hydroxide and sodium silicate. The results showed that each borax percentage, with respect to the total solid binder, increases the setting time by about 50% of the mix without borax. Sucrose, sodium acetate, acetic, and phosphoric acids have no significant effect on the investigated AAC's setting time.

Keywords: Setting retarder · Alkali-activated cement · Geopolymer · Setting time · Borax · Alkali activators · Precursors

1 Introduction

Alkali activated cements (AACs) are promising binders to replace ordinary Portland cement (OPC). The environmental issues associated with the latter binder have initiated research interest in binders with less environmental impact. AAC uses mainly byproduct materials such as fly ash (FA) and ground granulated blast furnace slag (GGBS) and has comparable or better performance to OPC [1]. AAC can be either heat or ambient cured. Heat cured AAC can be used in the precast industries; however, ambient cured AAC is preferred for a wider application. When targeting ambient cured AAC, GGBS is commonly used [2–5]. The high calcium oxide content and fineness of GGBS improve the binder system's reactivity and enhance reaction product formation at ambient temperature. However, the use of GGBS reduces the setting time of the concrete [6–8]. In

J. I. Escalante-Garcia et al. (Eds.): RW 2021, RILEM Bookseries 40, pp. 614–625, 2023.
https://doi.org/10.1007/978-3-031-21735-7_67

the case of OPC, various retarders can be used to extend the setting time. However, the research in the area of retarders for AAC is still at its early stages.

There are some researches on improving the setting time of AAC by adding different chemicals. However, most of these researches use different types AAC in terms of precursor materials and activator solutions. Due to this, a retarder recommended for a certain AAC may not work for another. This calls for a critical review of the literature, hence this study. The issue of finding a proper retarder was faced when the authors were developing an appropriate fine-grained AAC mixture for use in textile-reinforced concrete [9–11]. To check the suitability of the different reviewed retarders, some of the retarders were tested on the AAC system. The results were incorporated in the final part of this study.

2 Review of Literature

2.1 General

Different possible retarders ranging from commercial OPC retarders to chemicals such as borax, chelators, and acids have been investigated for AAC [12–14]. These studies show that retarders' performance depends on the type of activator and the type of binder. AACs are materials under development, standardization of these binders is yet to be done; consequently, there are various types of AACs. Thus it is essential to identify the type of AAC used when discussing a retarder. Table 1 gives a summary of different retarders with the specific activator and precursor material type.

To understand how some of the retarders work, it is vital to have a basic idea of the reaction mechanism of AAC. AAC can be divided into low calcium and high calcium systems. This division can be made at a calcium content of 10 wt% [15]. The formation of reaction products is based on complex polymerization reactions, which can be divided into three main phases, dissolution, reorganization, and hardening. Dissolution involves dissolving the alumina and silica rich precursor material by the highly alkaline solution such as sodium silicate and sodium hydroxide. The dissolved silicon and aluminum ions in the solution form aluminosilicates by combining with oxygen. By incorporating sodium, potassium, or calcium ions, aluminosilicates form monomers, dimers, and oligomers. This will further go through rearrangement, reorganization, and condensation while releasing water, leading to the formation of a large three-dimensionally cross-linked polymeric network called sodium aluminosilicate hydrate (N-A-S-H) [15]. In higher calcium systems, the main reaction products are calcium silicate hydrate phases containing aluminum (C-A-S-H).

2.2 Nano-ZnO and Zinc Salts

Garg and White [16] investigated nano-ZnO as a retarder for alkali-activated GGBS and alkali-activated metakaolin (MK). Sodium silicate solution was used for GGBS and a sodium silicate and sodium hydroxide solution for MK. Their results showed that nano-ZnO is very effective in retarding slag's alkali-activation reaction but has a negligible impact on metakaolin's alkali-activation. Wang et al. [17], on the other hand, reported that substitution of metakaolin by ZnO significantly hinders reaction and prolongs setting time in sodium/potassium silicate activated AAC with a more significant effect in the case of sodium silicate. The preferential formation of the Zn modified phase, which inhibits the nucleation of N/K-A-S-H, is suggested as the retardation mechanism. Cong et al. [18] reported the effectiveness of zinc nitrate at a 4% replacement of the GGBS. They also found a hybrid retarder containing zinc nitrate and sodium gluconate effective. A model for ZnO retardation, as explained by Cong et al. [18], is given by:

$$ZnO + H_2O + 2OH^- \rightarrow Zn(OH)_4^{2-}$$

$$2\left(Zn(OH)_4^{2-}\right) + Ca^{2+} + 2H_2O \rightarrow Ca(Zn(OH)_3)_2 \cdot 2H_2O + 2OH^-$$

As the zinc ion dissolves in the highly alkaline solution and reacts with the calcium ions from the GGBS, CZ $\left(Ca(Zn(OH)_3)_2 \cdot 2H_2O\right)$ is formed. This hinders the formation of the main reaction products such as C-A-S-H [16]. This reaction continues until there are no more zinc ions in the pore solution. Once all the zinc ions have dissolved and formed CZ, the pore solution's free calcium ions start forming the main reaction products. The main reaction products are more stable than the CZ. The formation of this new phase (CZ) adds a step in the polymerization reaction, which results in a higher setting time.

2.3 Borax

Bong et al. [19] reported that borax significantly increases the setting time of one part fly ash-slag blended AAC with anhydrous sodium metasilicate powder as an activator. The initial setting time of this binder without retarder was 169 min, and with 1 wt% borax to the precursor, the setting time increased to 342 min. Other researchers have also found borax effective for different types of AAC systems [13, 14, 26, 27]. Wortmann [13] recommended a combination of borax and sodium acetate or acetic acid for GGBS or FA based AACs. They also found that neither acetic acid nor sodium acetate alone can provide significant retardation. They observed that borax with other organic acids typically used as retarders for OPC, such as citric acid and tartaric acid, did not work.

For a one-part AAC made from FA, quick lime, sodium hydroxide, and manganese oxide at 75:15:6:4 ratios, Wu et al. [14] investigated borax, citric acid, and phosphoric acid. All three improved the initial setting time, as can be observed in Fig. 1. Mixes with retarders resulted in a lower pH value than those without, suggesting that the high pH of mixes without admixture could be the reason for the accelerated setting time. The admixtures could thus render set-retarding effects by lowering the pH of the solution. Incorporating the borate compound into the tetrahedral network of reaction products, which lowers the reaction rate, was suggested as a possible explanation for setting time improvement. Oderji et al. [26] observed that the setting time increased almost linearly as the borax content increased from 2 to 8% of the precursor's weight. They suggested that the increase in setting time could be due to borax's reaction with calcium ions forming a calcium-based borate layer in the alkaline solution retarding the reaction. In addition to increasing setting time, borax has also been reported to improve AAC concrete's strength [26, 28].

Table 1. Review of retarders for AAC

Binder	Activator solutions	Retarder	Dosage (by wt unless specified)	Effect on setting time	Ref.
GGBS	Na_2SiO_3	Nano-ZnO	0.5 and 1% of precursor	Effective	[16]
MK	NaOH and Na_2SiO_3	Nano-ZnO	0.5 and 1% of precursor	Negligible	[16]
GGBS	NaOH and Na_2SiO_3	ZnO and sodium gluconate	4% of GGBS	Effective	[18]
MK	Na_2SiO_3 or $K_2Si_2O_5$	ZnO	0.2/Na_2O or K_2O molar ratio	Effective	[17]
GGBS-FA	Na_2SiO_3	Borax anhydrous	1% of precursor	Effective	[19]
GGBS and FA	Na_2CO_3 and Na_2SiO_3	Borax + acetic acid/sodium acetate	0.5–1.5% borax and 0.3% acetic acid/sodium acetate	Effective	[13]
GGBS	Na_2CO_3 and Na_2SiO_3	Borax	–	Not sufficient	[13]
FA-OPC	Silica fume and NaOH	Sucrose	3%	Effective	[20]
GGBS-FA	Na_2SiO_3	Sucrose	1% of precursor	Effective	[19]
FA	NaOH and Na_2SiO_3	Sucrose	1.5 and 2.5% of FA	Effective	[21]
High CaO FA	NaOH and Na_2SiO_3	Sucrose	1 and 2% of FA	No effect on initial, effective for final	[22]
GGBS	Na_2SiO_3	Phosphoric acid	0.87 M (molarity) activator	Effective	[23]
FA, quick lime, MgO	NaOH	Citric acid, phosphoric acid, borax	2, 2, and 5%, respectively of one part cement	Effective	[14]

(*continued*)

Table 1. (*continued*)

Binder	Activator solutions	Retarder	Dosage (by wt unless specified)	Effect on setting time	Ref.
GGBS	Na_2SiO_3	Sodium phosphate	Max. 1% of P_2O_5 in sodium phosphate (with respect to GGBS)	Effective	[24]
High CaO FA	NaOH and Na_2SiO_3	Sodium hexa-metaphosphate	1.50 and 2.25% of binder	Slight improvement	[25]
GGBS	Na_2CO_3 and Na_2SiO_3	Acetic acid/sodium acetate	–	Negligible	[13]
FA	NaOH and Na_2SiO_3	Citric acid	1.5 and 2.5% of FA	Accelerating effect	[21]
FA-GGBS	$K_2Si_2O_5$	Chelators	4% of precursor	Effective	[12]
High CaO FA, FA-GGBS,	NaOH and Na_2SiO_3	Barium chloride dihydrate	0.5–1.0% of binder	effective	[25]
MK-GGBS	NaOH, KOH, Na_2SiO_3	Barium chloride dihydrate	1 and 2% of binder	Effective	[25]
High CaO FA	NaOH and Na_2SiO_3	$CaCl_2$	1 and 2% of FA	Accelerating effect	[22]
High CaO FA	NaOH and Na_2SiO_3	$CaSO_4$	1 and 2% of FA	Minor effect	[22]
High CaO FA	NaOH and Na_2SiO_3	Na_2SO_4	1 and 2% of FA	Effective for initial, Minor on final	[22]

2.4 Sucrose

Bong et al. [19] reported that sucrose increases the setting time of one part fly ash-slag blended AAC with anhydrous sodium metasilicate powder as the activator. In their investigation of sucrose, borax, and a commercial retarder (MasterSet RT 122), Bong et al. [19] observed that sucrose is the most effective retarder. Sucrose was also reported to be effective for FA-OPC blended AAC activated by sodium hydroxide [20]. A similar sucrose effect was also reported for FA based AAC with sodium silicate and sodium hydroxide activators [21]. Rattanasak et al. [22], in their study on high CaO FA with sodium hydroxide and sodium silicate, found that sucrose improves the final setting time, however, with no effect on the initial setting time.

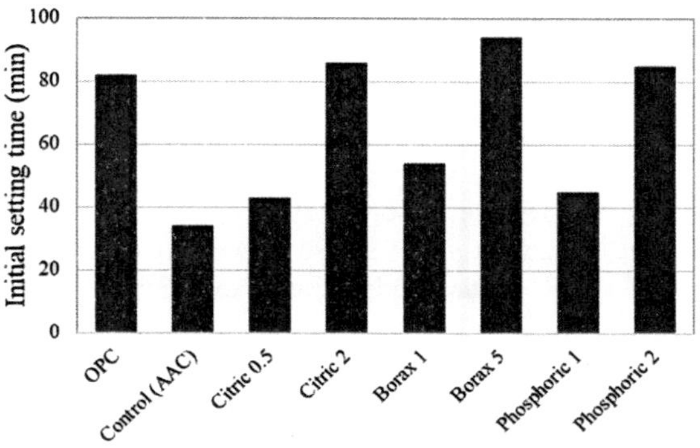

Fig. 1. Initial setting time for different retarders (wt% of the one part binder) (data from [14])

Sucrose seals off the fly ash (high-calcium) particles from the alkaline solution causing retardation of setting time [22]. It also reacts with calcium, resulting in increased viscosity, which increases the setting time [20]. The adsorption of the precipitated materials onto the surface of FA particles retard polymerization resulting in a diffusion-controlled mechanism. In OPC, sucrose inhibits the growth of reaction products such as calcium silicate hydrates leading to prolonged setting time [29]. Sucrose may perform the same in AAC by hindering the formation of C-A-S-H products. Rattanasak et al. [22] also performed scanning electron microscopy on the AAC. The mix without sucrose showed a regular-shape reaction product of C-S-H and Al-Si gel, while the mix with sucrose showed an agglomeration of small particles or units of gel and covered the FA surface.

2.5 Phosphates and Phosphoric Acid

Gong and Yang [30] reported that sodium phosphate decreases the heat evolution and retard alkali-activated red mug slag settings. The XRD pattern of the mixture with the sodium phosphate showed new characteristic peaks not found in the mixture without the phosphate. They suggested that the new phase could be attributed to the retarding mechanism of sodium phosphate. Wu et al. [14], in their study on one-part AAC, found that phosphoric acid can effectively retard the setting time (Fig. 1). Chang [23] also used phosphoric acid for sodium silicate activated slag systems. The acid was added as part of the alkaline solution. The results showed that the retardation effect is highly sensitive to the phosphoric acid concentration in the alkaline solution. When the concentration exceeds 0.84 M (molarity), the setting time was highly prolonged; at 0.87 M, the initial setting time exceeded 6 h. Kalina et al. [24] studied the sodium phosphate addition in alkali-activated slag up to 5 wt% of P_2O_5 in the sodium phosphate (with respect to the GGBS). Their results showed 1 wt% is an optimal dosage to improve the setting time. The sodium phosphate was also found to reduce the workability and improve the 28 days compressive strength.

2.6 Other Retarders

In addition to the chemicals discussed above, other chemicals have also been tested for different AACs. As shown in Fig. 1, the addition of citric acid at 2 wt% delayed the AAC setting, making it comparable with OPC [14]. Citric acid has also been investigated by Kusbiantoro et al. [21]. However, in this case, the citric acid was found to have an accelerating effect rather than retardation. In the case of Wu et al. [14], a one-part AAC with FA, quick lime and MgO binder, and sodium hydroxide activator was used, while in the case of Kusbiantoro et al. [21] FA based AAC with sodium hydroxide and sodium silicate activators were used. It is not clear if the difference came from the presence of quick lime and MgO or the sodium silicate solution; however, it is clear that the type of AAC plays a significant role in the retarder's efficiency. Wortmann [10] reported the inefficiency of combined borax and citric acid. In contrast, borax with acetic acid and sodium acetate worked well, showing the citric acid's inefficiency in GGBS based AACs.

Tartaric acid was investigated by Sun et al. [31]. The results indicated that the acid has a significant effect on the rate of reaction of sodium hydroxide activated GGBS mixes while having little effect on the reaction product. The addition of tartaric acid slightly reduced the pH of the alkaline solution. Deventer et al. [32] also recommended tartaric acid as a retarder. On the other hand, similar to citric acid, Wortmann [10] reported the inefficiency of a combination of borax with tartaric acid.

Gong and Yang [25] recommended barium chloride as a retarder for different AAC systems containing high calcium reactive aluminosilicate materials. For high calcium FA activated with sodium silicate and sodium hydroxide, 1 wt% barium chloride (to binder) increased the initial and final setting time from 29 and 45 min to 391 and 450 min, respectively. From isothermal calorimetry, they were able to observe that the heat of reaction significantly reduced with the retarder addition. With 3% barium chloride, FA-GGBS based AAC with the same activator did not show any sign of setting even after 7.4 h. Owing to such significant retardation, they recommended the use of this retarder as an emergency set brake for fresh AAC concrete.

Sasaki et al. [12] investigated the effect of three calcium ion chelators, 3-hydroxy-2,20-iminodisuccinic acid tetrasodium (HIDS), tetrasodium ethylenediaminetetraacetate (EDTA-4Na), and tetrasodium L-glutamic acid diacetate (GLDA) as retarder for FA-GGBS based AACs. Chelators are chemicals which bond with metal ion and produce a chelate compound. All the chelators improved the setting time, but the degree of prolongation depended on the type of chelator. HIDS was the most effective. The chelators bind to a metal ion by coordination with a plurality of coordination sites to form a complex and collect the metal ion. Therefore, it is thought that structure formation can be delayed.

3 Experimental Study

3.1 Materials

FA and GGBS were used as the precursor materials. The chemical compositions of these precursor materials are summarized in Table 2. The activator solution used is a mixture of sodium silicate and sodium hydroxide. The sodium silicate solution includes 26.82%

silicate, 8.2% sodium oxide, and 64.98% water. The sodium hydroxide is a 50% by weight solution. Furthermore, a fine aggregate with a maximum aggregate size of 2 mm was used.

Table 2. Chemical composition of FA and GGBS [10]

Composition	FA (%)	GGBS (%)
SiO_2	49.79	34.48
Al_2O_3	26.71	11.48
Fe_2O_3	8.57	–
MgO	2.47	7.08
CaO	4.34	42.43
K_2O	3.36	0.66
Na_2O	1.28	0.56
SO_3	1.49	2.17
TiO_2	1.23	1.14

3.2 Specimen Preparation and Test Methods

The sodium silicate and sodium hydroxide solutions were mixed in the required proportion at least 24 h before mixing. AAC specimens were prepared by first mixing the dry materials in a mixer for two minutes. Afterward, the prepared alkaline solution was mixed with the additional water, added slowly to the dry mixture, and mixed for four minutes.

The setting times were evaluated using the Vicat needle apparatus. The initial setting time is determined as the elapsed time between mixing and the time at which the penetration depth is 37 mm. To avoid uncontrolled temperature and humidity effects, the setting time test was conducted in the environmental control room (20 °C temperature, 65% humidity).

Different chemicals from the literature review were tested to check their suitability for the AAC mixture under study. The investigated chemicals were borax, sodium acetate, acetic acid, citric acid, phosphoric acid, sucrose, and zinc oxide. A hybrid retarder containing borax and sodium acetate, acetic acid, citric acid, or phosphoric acid was also investigated. The effect of increasing water content was also investigated by doubling the water content. The sand, water, FA, GGBS, sodium silicate and hydroxide solutions in the control mix are 1120, 63, 450, 300, 214 and 60 kg/m^3, respectively.

3.3 Results and Discussion

Table 3 summarizes all the initial setting time test results. The retarder chemicals were added as a percentage of the total solid binder. The total solid binder is the precursor materials (GGBS and FA) plus solids from the alkaline activator. It has been found that most of the investigated chemicals except borax have little or no effect on the present mixture.

Table 3. Initial setting time tests

Mix	Setting time (minutes)	Setting time ratio to reference mix	Mix	Setting time (minutes)	Setting time ratio to reference mix
C (reference)	37	1.0	CB2	74	2.0
CR1	36	1.0	CB3	93	2.5
CS2	34	0.9	CB0.5SA0.5	42	1.1
CSA1	37	1.0	CB1SA0.3	53	1.4
CAA1	37	1.0	CB2SA1	81	2.2
CZ1	40	1.1	CB1AA0.3	48	1.3
CZ5	51	1.4	CB1CA0.3	43	1.2
CB0.5	43	1.2	CB1PA0.3	45	1.2
CB1	55	1.5	Double water	37	1.0

S: Sucrose, R: MasterSet R 433, SA: Sodium Acetate, AA: Acetic Acid, CA: Citric Acid, PA: Phosphoric Acid, B: Borax, Z: ZnO, Naming example: CB1SA0.3—control mixture with 1 wt% B and 0.3% SA

As shown in Table 3, conventional OPC retarder MasterSet R 433 did not affect AAC's setting time. Sucrose, sodium acetate, and acetic acid also did not affect the setting time. ZnO at 1 wt% had a minor effect, while at 5 wt%, it increased the setting time by about 40%. This high dosage requirement makes the ZnO ineffective for the current AAC system. On the other hand, each percentage of borax increased the setting time by about 50%, i.e., 2 wt% borax doubled the setting time of AAC. An almost linear increase in setting time was observed with a borax increase from 0.5 to 3 wt%. A combination of borax and sodium acetate also gave good results, but the addition of sodium acetate has little effect on the setting time. Hence, borax is the preferred retarder for the current AAC system.

It was also observed that the method of adding borax was important. Two mixes with 3 wt% borax were prepared. In one of the mixes, borax was added with the precursor material. In the second one, borax was added to the water and alkaline solution and mixed before the main mixing process. When added to the solution, borax improved the initial setting time by 150%. However, when added to the solid only by 65%. Oderji et al. [25] suggested that borax reacts with calcium ions forming a calcium-based borate layer. As it has to compete with the silicates and aluminates for the calcium ions, the dissociation of borax in the solution before contacting the binder may increase its reaction capacity. Therefore, it is important to add the borax to the solution to facilitate dissociation.

Retarders should not have a significant negative influence on the mechanical performance of the concrete. In order to investigate the influence of borax in this regard, the compressive strength of the control and the CB2 mixture were tested. A 7 day compressive strength of 54.6 and 55.7 MPa was obtained for the control and CB2 mixtures, respectively. This shows that the borax at 2 wt % has no negative effect on the compressive strength. It instead showed a minor improvement. This has also been observed in previous studies [26, 28].

An increase in water increases setting time of OPC [33]. However, as can be observed in the current study, doubling the water content did not improve the setting time. This could be due to the different role water plays in OPC (main participant in reaction) and AAC [11].

4 Conclusion

Based on the literature review and the experimental investigation of this study, the following conclusions are made:

- The most effective retarder for alkali activated cements is strongly dependent on the precursor and type of alkaline solution.
- For the FA-GGBS based AAC with sodium silicate and sodium hydroxide activators, borax can improve the setting time. Each percentage of borax can increase the setting time by about 50% of the mix without borax.
- The method of addition of the retarder can impact the retarding efficiency. In the case of borax, it works better when mixed with water or the alkaline solution instead of adding to the solid binders.

References

1. Pacheco-Torgal, F., Castro-Gomes, J., Jalali, S.: Alkali-activated binders: a review. Constr. Build. Mater. **22**(7), 1305–1314 (2008)
2. Thirugnanasambandam, S., Antony, J.C.: Ambient Cured Geopolymer Concrete Products. In: Das, B.B., Neithalath, N. (eds.) Sustainable Construction and Building Materials, pp. 811–828. Springer Singapore, Singapore (2019)
3. Zannerni, G.M., Fattah, K.P., Al-Tamimi, A.K.: Ambient-cured geopolymer concrete with single alkali activator. Sustain. Mater. Technol. **23**, e00131 (2020)
4. Deb, P.S., Nath, P., Sarker, P.K.: Properties of fly ash and slag blended geopolymer concrete cured at ambient temperature. New Dev. Struct. Eng. Constr. 571–576 (2013)
5. Deb, P.S., Nath, P., Sarker, P.K.: The effects of ground granulated blast-furnace slag blending with fly ash and activator content on the workability and strength properties of geopolymer concrete cured at ambient temperature. Mater. Des. (1980–2015) **62**, 32–39 (2014)
6. Fang, G., Ho, W.K., Tu, W., Zhang, M.: Workability and mechanical properties of alkali-activated fly ash-slag concrete cured at ambient temperature. Constr. Build. Mater. **172**, 476–487 (2018)
7. Nath, P., Sarker, P.K.: Effect of GGBFS on setting, workability and early strength properties of fly ash geopolymer concrete cured in ambient condition. Constr. Build. Mater. **66**, 163–171 (2014)
8. Lee, N.K., Lee, H.K.: Setting and mechanical properties of alkali-activated fly ash/slag concrete manufactured at room temperature. Constr. Build. Mater. **47**, 1201–1209 (2013)

9. Tekle, B.H., Hertwig, L., Holschemacher, K.: Rheology of alkali-activated blended binder mixtures. Materials (Basel) **14**(18), (2021)
10. Tekle, B.H., Hertwig, L., Holschemacher, K.: Setting time and strength monitoring of alkali-activated cement mixtures by ultrasonic testing. Materials (Basel) **14**(8), 1889 (2021)
11. Tekle, B.H., Holschemacher, K.: Alkali activated cement mixture at ambient curing: strength, workability, and setting time. Struct. Concr. (2021)
12. Sasaki, K., Kurumisawa, K., Ibayashi, K.: Effect of retarders on flow and strength development of alkali-activated fly ash/blast furnace slag composite. Constr. Build. Mater. **216**, 337–346 (2019)
13. Wortmann, K.: Retarder for Alkali Activated Binder: European Patent Office (EP 3296278 A1) (2018)
14. Wu, W., Matalkah, F., Soroushian, P., Almalkawi, A., Wang, X., Xu, L.: Influence of admixtures on rheological properties and heat of hydration of alkali aluminosilicate cement. Adv. Cem. Res. **29**(9), 397–403 (2017)
15. Herrmann, A., Koenig, A., Dehn, F.: Structural concrete based on alkali-activated binders: terminology, reaction mechanisms, mix designs and performance. Struct. Concr. **19**(3), 918–929 (2018)
16. Garg, N., White, C.E.: Mechanism of zinc oxide retardation in alkali-activated materials: an in situ X-ray pair distribution function investigation. J. Mater. Chem. A **5**(23), 11794–11804 (2017)
17. Wang, L., Geddes, D.A., Walkley, B., Provis, J.L., Mechtcherine, V., Tsang, D.C.W.: The role of zinc in metakaolin-based geopolymers. Cem. Concr. Res. **136**, 106194 (2020)
18. Cong, X., Zhou, W., Geng, X., Elchalakani, M.: Low field NMR relaxation as a probe to study the effect of activators and retarders on the alkali-activated GGBFS setting process. Cement Concr. Compos. **104**, 103399 (2019)
19. Bong, S.H., Nematollahi, B., Nazari, A., Xia, M., Sanjayan, J.: Efficiency of different superplasticizers and retarders on properties of 'One-Part' fly ash-slag blended geopolymers with different activators. Materials (Basel) **12**(20), (2019)
20. Assi, L.N., Deaver, E.E., Ziehl, P.: Using sucrose for improvement of initial and final setting times of silica fume-based activating solution of fly ash geopolymer concrete. Constr. Build. Mater. **191**, 47–55 (2018)
21. Kusbiantoro, A., Ibrahim, M.S., Muthusamy, K., Alias, A.: Development of sucrose and citric acid as the natural based admixture for fly ash based geopolymer. Procedia Environ. Sci. **17**, 596–602 (2013)
22. Rattanasak, U., Pankhet, K., Chindaprasirt, P.: Effect of chemical admixtures on properties of high-calcium fly ash geopolymer. Int. J. Miner. Metall. Mater. **18**(3), 364–369 (2011). https://doi.org/10.1007/s12613-011-0448-3
23. Chang, J.J.: A study on the setting characteristics of sodium silicate-activated slag pastes. Cem. Concr. Res. **33**(7), 1005–1011 (2003)
24. Kalina, L., Bílek, V., Novotný, R., Mončeková, M., Másilko, J., Koplík, J.: Effect of Na_3PO_4 on the hydration process of alkali-activated blast furnace slag. Materials (Basel) **9**(5), (2016)
25. Gong, W., Xu, H., Lutze, I., Pegg, I.L.: Control of time of setting of geopolymer compositions containing high-Ca reactive aluminosilicate materials. US patent (2019/0084882) (2019)
26. Oderji, S.Y., Chen, B., Shakya, C., Ahmad, M.R., Shah, S.F.A.: Influence of superplasticizers and retarders on the workability and strength of one-part alkali-activated fly ash/slag binders cured at room temperature. Constr. Build. Mater. **229**, 116891 (2019)
27. Antoni, Wijaya, S.W., Satria, J., Sugiarto, A., Hardjito, D.: The use of borax in deterring flash setting of high calcium fly ash based geopolymer. In: MSF, vol. 857, pp. 416–20 (2016)
28. Revathi, T., Jeyalakshmi, R.: XPS, 29Si, 27Al, 11B MAS-NMR, ATR-IR and FESEM characterization of geopolymer based on borax modified water glass activated fly ash-GGBS blend. Mater. Res. Express **6**(8), 85337 (2019)

29. Thomas, N.L., Birchall, J.D.: The retarding action of sugars on cement hydration. Cem. Concr. Res. **13**(6), 830–842 (1983)
30. Gong, C., Yang, N.: Effect of phosphate on the hydration of alkali-activated red mud–slag cementitious material. Cem. Concr. Res. **30**(7), 1013–1016 (2000)
31. Sun, C., Sun, J., Wang, D.: Effect of tartaric acid on the early hydration of NaOH-activated slag paste. J. Therm. Anal. Calorim. **144**(1), 41–50 (2020). https://doi.org/10.1007/s10973-020-09466-5
32. van Deventer, J.S.J., Feng, D., Duxson, P.: Dry mix cement composition, methods and system involving same. US Patent (7691198B2)
33. Stefanou, G.D., Larsinos, C.: Influence of mixing water on the setting time of concrete. Int. J. Cem. Compos. Light. Concr. (1981)

Check for updates

Sustainable Cements: Hybrid Alkaline Cements Overview

A. Fernández-Jiménez(✉), I. Garcia-Lodeiro, O. Maltseva, and A. Palomo

Instituto de Ciencias de la Construcción Eduardo Torroja–CSIC, Serrano Galvache s/n, 28033 Madrid, C.P, Spain
anafj@ietcc.csic.es

Abstract. The ordinary Portland cement (PC) production contributes 2–3% of global energy use and 8% to 10% of world-wide carbon dioxide emissions (CO_2). According to estimates in 2050, the CO_2 emissions associated with the manufacture of PC, could increase between 240–260% compared to the 1990 emissions. Faced with this need, the Portland cement industry has been considering several options. One of them is to replace Portland clinker to the greatest extent possible, with low-carbon *Supplementary Cementitious Materials* (SCMs) (whether added with the cement or directly into the concrete mix, "Blended cement"). The type and quantity of mineral additions is regulated by the standards of each country, although these values do not usually exceed 35%. Higher values are not allowed because the early mechanical strength decreases. However, applying alkaline activation technology can reach replacement levels of up to 60% or 70%, in the so-called "Hybrid Alkaline Cements" (HAC). The SCMs are basically natural pozzolans, such as blast furnace slag (BFS) or fly ash (FA). To increase the reactivity of mineral additions, especially at early ages, an alkaline activator is added into the mixture, normally salts of moderate alkalinity. This work shows a review of the theoretical knowledge of the hydration of hybrid cements, as well as their mechanical performance in mortars, concrete, and in industrial-scale applications.

Keywords: Alkaline activation · Hybrid alkaline cements · Sustainability

1 Introduction

The recent climate summit held at UN headquarters (September 2019, NYC) called for a three- to five-fold increase in the effort to lower CO_2 emissions and constrain global warming to a 1.5 °C temperature rise at most, for unless the pace of the present measures is stepped up, by the end of this century the mean planet-wide temperature will be at least 3 °C higher than in pre-industrial times. The Paris Agreement (2015) must be implemented and its provisions enhanced if the world is to be carbon neutral by 2050. Although many countries and companies have committed to taking measures along those lines, just how carbon neutrality can be achieved is still poorly defined. The international community must turn its words into action. The science can help, in the certainty that with climate disruption '*We are in a battle for our lives. But it is a battle we can win'* [1].

J. I. Escalante-Garcia et al. (Eds.): RW 2021, RILEM Bookseries 40, pp. 626–639, 2023.
https://doi.org/10.1007/978-3-031-21735-7_68

There are essentially three approaches to solve the global warming problem: geo-engineering (direct intervention in the climate); adaptation (confronting the effects of change); and mitigation (greenhouse gas abatement). The emissions generated by the construction industry, portland cement (PC) manufacture in particular, is characterized by a high carbon footprint approximately el 7% to 8% of total CO_2 emissions worldwide [2–4].

However, construction is a key industry in many economies, given the huge demand for housing and infrastructure, especially in emerging countries such as India, Indonesia and many African nations. According to Chatham House (a British think tank) [5], if the cement industry was a country, it would be the world's third largest CO_2 emitter after China and the USA. The cement industry emits more CO_2 than aviation (2.5% of the total) and it is not far from agriculture (12%), the number one emitter. Lowering the emissions associated with cement manufacture is a daunting challenge, for worldwide output is growing at a lively pace and every tonne (t) produced is estimated to generate from 0.8 t to 0.9 t of CO_2. If the industry fails in reducing its emissions substantially, the planet's ability to reach the global climate goals laid down in the Paris Agreement may well be jeopardized.

Around 50% of the CO_2 generated by portland cement manufacture is attributable to the decarbonation of the raw materials used to make clinker, 40% to the fuel used in the kiln and the remaining 10%, is assigned to the electric power consumption and transport. The cement industry is aware of those statistics. Its majors are members of the World Business Council for Sustainable Development (WBCSD, http://www.wbcsdcement.org) and participate in the Cement Sustainability Initiative (CSI). The most significant measures adopted to date by those organisations include enhancing kiln efficiency; using alternative fuels; reducing the clinker factor via replacement with Supplementary Cementitious Materials (SCMs); and capturing and storing CO_2. The degree of success of such initiatives often depends as much on social pressure as on technological and economic factors.

The industry has made considerable progress in connection with the first two measures: the greater energy efficiency of new plants and the replacement of carbon fuels with waste materials have lowered the mean CO_2 emissions per tonne of cement by 18% in recent years [6]. Those lines of action have been exploited nearly as far as possible, however, with further potential for progress virtually nil. Carbon capture and storage, in turn, which has its defenders and detractors [7–9], is an expensive technology that needs fitting portland cement plants with additional facilities to capture and store CO_2. Moreover, the long-term behaviour of carbon stores is uncertain and the risk of leakage inherent.

The authors of the current paper consider that the cement industry should double down in its efforts to explore ways to produce new, less polluting binders, as cement, with very low clinker portland content or may be without clinker portland. That approach, on which this and other research groups are working, consists in developing blended cement (BC) [10, 11], Hybrid alkaline cement (HAC) [12–15] or alkaline cement (AC) [16–20]. The primary characteristics of the previous cements are described briefly below (see Fig. 1).

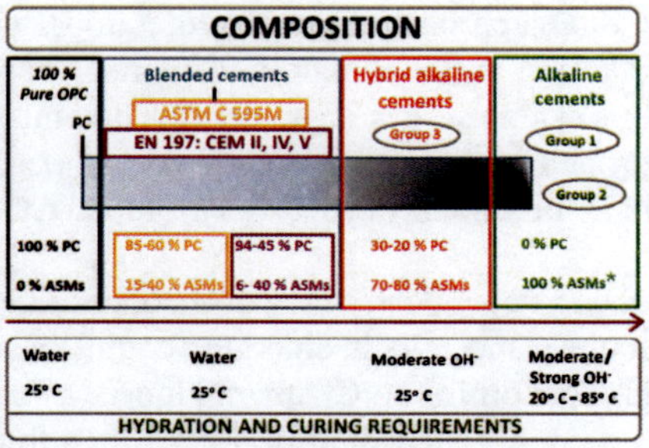

Fig. 1. Types of cement bearing Supplementary Cementitious Materials (SCMs) [Adapted from Ref. [21])

Blended or pozzolanic cements (BC) [10, 11]: as a general rule, these cements contain 60% to 70% PC clinker and 40% to 30% SCMs. The technology involved is in widespread use in today's cement plants. The type of SCMs that are deemed applicable and their dosage are regulated (see EN-197–1, Type II, III, IV and V cements). The main product of the hydration reaction is a C-S-H-like cementitious gel (CaO-SiO_2-H_2O) that may contain some aluminium in the bridging tetrahedral position of the C-S-H structure, although its structure is essentially linear (see Fig. 2(a)).

SCMs were originally deployed to lower costs and to improve certain technological properties of cement or concrete (they may enhance late-age cement strength, reduce the heat of hydration, improve the durability and so on [11]). CO_2 emissions abatement, a knock-on effect given fairly short shrift up until a few years ago, has become a key factor in the wake of the rising cost of CO_2 emissions rights and social pressure. Scrivener et al. [22] recently proved that low quality, thermally treated clays, have high potential for using in limestone calcined clay cement (LC3, limestone-bearing ternary) cements. Low purity clays don't apt for other applications (ceramic, gravel) would be usable in these ternary blends.

Alkaline activated binders (AAB) [16–20] or alkaline cements (AC) are characterised by the absence of clinker and portland cement. They are obtained from the reaction between aluminosilicate precursors (with glassy or amorphous structures) and an alkaline medium (normally NaOH or waterglass solutions). Depending on the chemical composition of the precursor (CaO-SiO_2-Al_2O_3 system), two types of binders can be distinguished. In the first type, the alkaline activation of materials with a calcium content of 30% to 40% (such as blast furnace slag [16]) generates as the main reaction product a C-A-S-H (CaO-Al_2O_3-SiO_2-H_2O)-like cementitious gel. Although this compound is similar to the gel produced during portland cement hydration, the aluminium in its structure favours bonding between linear chains, giving rise to a two-dimensional structure (Fig. 2(b)). The second type of materials [16–18] is the result of the alkaline activation of low calcium (<10%) aluminosilicates, such as fly ash or dehydroxylated kaolinite clays. Here the main reaction product is a three-dimensional, N-A-S-H (Na_2O-Al_2O_3-SiO_2-H_2O)-like cementitious gel (Fig. 2(c)).

The hybrid alkaline cements (HAC) [12–15] are the result of combining the existing knowledge of the basics of both portland and alkaline cement chemistry. In these cements,

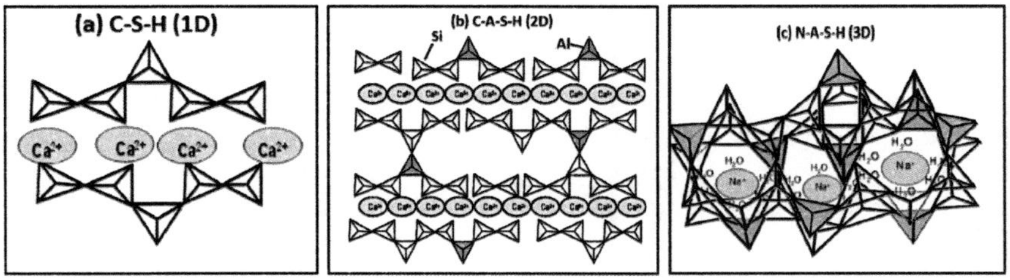

Fig. 2. Simplified gel structure: (a) C-S-H gel (linear, 1D); (b) C-A-S-H gel (cross-linked 2D); (c) N-A-S-H gel (3D). Nomenclature C = CaO, A = Na_2O, S = SiO_2, A = Al_2O_3

the SCM content is raised to 70% to 80% wt., while the clinker accounts for around 30% to 20% wt.. Particularly during early reaction stages, SCM reactivity is boosted by adding a chemical activator to the blend, normally an alkaline salt such as Na_2SO_4 or Na_2CO_3 [21]. HAC hydration generates a mix of cementitious gels, C-(A)- S-H and (C)-N-A-S-H [21]. Hybrid alcaline cement (HAC) manufacture (Fig. 1) was the outcome of the project participants' knowledge of both conventional (PC) and alkaline cements. Much progress has been made over the last 10 years [12–15], in the understanding and development of these hybrid cements, which have now even been tested on an industrial scale.

In these cements, the technological properties depends largely on the chemical composition of the precursors and the type and concentration of the alkaline products used as activator.

2 The Role of Activator

The vast majority of hybrids cement hydration studies have been carried out with activators (mainly NaOH or NaOH + Waterglass mixtures) that make the medium highly alkaline [12, 21]. The high alkalinity of these solutions (with pH values are higher than 13) favors the rapid dissolution of the precursor. However, the manufacture of NaOH and Na_2SiO_3 are quite energy intensive processes and produces high CO_2 emissions [23, 24]. For this reason, in the manufacture of alkaline cements hybrids, salts of moderate alkalinity, such as Na_2SO_4, Na_2CO_3 and K_2CO_3, K_2SO_4, are usually used [13–15, 25, 26].

The reaction of these salts with $Ca(OH)_2$ coming from the early Portland cement clinker hydration, can raise the pH in the medium due to its synergistic reaction [21, 23] see Eq. 1

$$xCa(OH)_2 + Ax\,By \leftrightarrow CaxBy(s) + 2xA(OH)(ac) \quad (1)$$

where A is an alkaline cation, normally Na^+ or K^+, and B is the anion in the respective inorganic salt.

The way in which the activator is incorporated into the cementing system, as a solution (previously dissolved in the mixing water), or in its solid form in the dry mix *(for "Just add water"),* may has an influence in the mechanical strength development, in the nature of the reaction products formed and in the technological application. If the activator is added in a solid state (pre-ground with the precursor), process control is incumbent upon the cement producers, whereas if it is dissolved in the mixing water (like concrete admixtures), control might be the responsibility of cement users (understood to be ready mix or precast concrete plants).

Fernandez-Jiménez *et al.* in 2019 [27], studied how the procedure used to add the chemical activator (in solid or liquid form) affected the development of resistance of the cement (50% PC + 50% Fly ash) or even the nature of the reaction products formed. Two experimental procedures were deployed: a) the chemical activator was dissolved in the mixing water; b) the chemical activator was ground, as a solid, into the fly ash.

The materials used to prepare the hybrid cement were **HBL** = 50% CEM 42.5R + 50% FA with a liquid activator and **HBS** = 50% CEM 42.5R + 50% FA (fly ash previously ground with the chemical activator). In both cases, same concentration and type of activator were used (5 wt.% of Na_2SO_4 respect to the precursor, fly ash). In both cases, the fly ash particles have the same granulometry (100% under 45 μm and 80% under 10 μm).

Figure 3 shows the compressive strength development of both mortars (HBL and HBS). At 2 and 7 days, higher mechanical strengths were obtained when the activator was added as a solid (pre-ground with the fly ash). Setting time data are given in Fig. 3(b). Adding the activator as a solution (dissolved in water) accelerated the chemical reactions, shortened the initial and end setting times. Those results were consistent with the authors' original premise: when the activator was added as a solid (HBS), is necessary its dissolution before reacting.

Figure 3(c) shows the XRD patterns for 2 and 28 days hydrated HBL and HBS pastes. Diffraction lines are attributed to the original phases (from clinker; C_3S, C_2S, C_3A, C_4AF, and from the fly ash, quartz and mullite). The main reaction product was an amorphous gel or gels (actually a mixture of cementitious gels; C-A-S-H, (C-N)-A-S-H), appeared on the diffractograms as a hump, stretching from 25° to 35° (2θ) [21–23, 27]. The fly ash reacted well with the $Ca(OH)_2$ or with the alkalis to generate more cementitious gels, which incorporate Al and Na into its structure.

Two secondary crystalline phases, ettringite and portlandite, were detected in all pastes, along with an increase in the intensity of the signals associated with calcite (CEM 42.5R contains calcite). At 2 days, HBL paste showed some low intensity lines attributed to the presence of U-phase ($3CaO,Al_2O_3.CaSO_4.0.5Na_2SO_4.15H_2O$), in addition to the lines corresponding to ettringite. U-Phase is considered, by several authors, the result of ettringite conversion in highly alkaline media [27]. Authors point out that the U-phase was formed during the early stages of hydration (as if it were a primary ettringite). Its presence suggests that pH value was high during the first few days of hydration.

However, none U-phase signal was observed on the diffractogram for the 2 days HBS pastes (same precursor, but different procedure for adding the chemical activator). This data are indicative of a kinetic effect: in HBS systems, sodium sulphate was added in its solid form, and therefore, Ca^{2+} ions, needs to be dissolved before reacting in the cement. The reactions were in all likelihood the same as in HBL system, but probably with a slight time lag.

This work showed that Hybrid alkaline binders could feasibly be manufactured with liquid or solid activators. Nonetheless, setting time regulation and early age mechanical

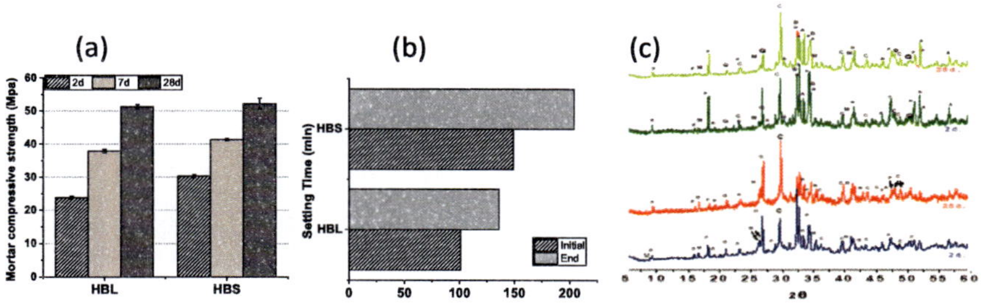

Fig. 3. (a) compressive strengths in prismatic mortar specimens (40x40x160 mm) according to European standard EN 196–1 (sand/binder ratio, S/B: 3/1; ***HBL*** = 50% CEM 42.5R + 50% FA with a liquid activator and ***HBS*** = 50% CEM 42.5R + 50% FA (fly ash previously ground with the chemical activator; water/binder ratio, W/B: 0.5). (b) Setting times in pastes (water/binder = 0.35; (c) X-ray diffraction (XRD) of pastes after frozen at 2 days and 28 days (Data adapted from reference 27). (*Legend: alite (a* = C_3S*), belite (B* = C_2S*), tricalcium aluminate (A* = C_3A*), quartz (Q* = SiO_2*), mullite (M* = $3Al_2O_3.2SiO_2$*), portlandite (P* = $Ca(OH)_2$*), calcite (C* = $CaCO_3$*), Ettringite (E); U-Phase (U).*

strength are enhanced and microstructure is more homogeneous when the solid form is used. All of this makes that, the use of solid activators in these cements, is a more real option from the point of view of its technological application. These results have been a growing interest in 'just add water' hybrid alkaline cement formulations, in which moderate alkaline solid salts are used. If it is possible, used salts from natural raw.

3 The Role of Precursors

The precursors used in these more environmentally sustainable binders HAC are basically the same used in BC and AAB, aluminosilicates materials conventionally named as SCMs; the ones most widely used are fly ash, blast furnace slag, thermally treated clays and limestone (as limestone is practically inert, with an effect that is more physical than chemical).

In hybrid cements, the chemical composition may vary substantially depending on the type of SCM used in their manufacture. It is possible to prepare binary or ternary binders. Table 1 shows the strength values of hybrid alkaline cements when are water-hydrated in the absence or presence of a solid-state alkaline activator (4% of C (Na_2CO_3) or S (Na_2SO_4) activator), for different type of precursors [28].

As Table 1 shows, when the binder consists of a blend of high % wt. of precursor and low clinker content, and is hydrated with water in the absence of activators, the early strengths are lower than when an alkaline salt is used. The presence of alkaline activators (C and S) increases the development of compressive strengths in all binder. In binders without activator, the clinker (20% of the material) reacts first, and the precursor subsequently reacts with the $Ca(OH)_2$ formed during clinker hydration. Due to the low clinker and the high aluminosilicate precursor content, the rate of reaction is a slow process, and therefore the low early strength values obtained. However, the presence of an alkaline activator induces a rise in the pH values of the pore solution, favouring

Table 1. Compressive strengths in hybrid cement pastes (Data adapted from reference [28])

Binder				Activator	Water/ binder	Compressive strengths		
KPC[1]	FA[2]	BFS[3]	MK[4]			2d	28d	90d
20	80	–	–	water	0.4	4.25	21.59	25.17
				C	0.38	20.45	30.63	33.65
				S	0.38	25,4	42,3	51,18
20	–	80	–	Water	0.26	9.46	30.02	–
				C	0.26	26.73	45.27	–
				S	0.26	32.50	48.4	–
20	60	20	–	water	0.36	4.39	33.70	44,27
				C	0.35	23.72	31.85	43.86
				S	0.35	30,12	42.43	53.45
20	–	40	20	C	0,41	14.53	23,17	–
				S	0,41	16,11	26,10	–

[1] KPC = Clinker of Portland Cement; [2] FA = Fly Ash; [3] BFS Blast Furnace Slag; [4] MK Metakaolin

the aluminosilicate precursor dissolution. The higher pH values intensify the reactivity and with it, the higher early strength values [21]. Since slag requires less aggressive activation conditions than the fly ash, mechanical strengths were slightly higher in the hybrid slag based binders than in the fly ash-bearing hybrid cements.

The multicomponent hybrid cements (ternary blends) studied also developed good cementitious properties. Mechanical strengths exceeded 20 MPa in the 2-day systems containing variable proportions of fly ash and slag and 20% clinker, and rose to 32.5 MPa in the 28-day material (the requirement for a type CEM III commercial portland cements is 32.5 MPa). In all cases, the highest strength values were obtained with activator S. These findings confirmed that the activators used enhanced precursor reactivity, with the formation of greater amounts of the cementitious gels that affords these materials their mechanical strengths.

Figure 4 shows micrographs (along with the respective EDX) of 28-days 20KPC + 80FA + S and 20KPC + 80Slag + S hydrated systems. System 20KPC + 80FA + S showed a compact matrix in which some of the fly ash particles had been attacked and others (covered of the reaction products) had not. The paste observed contained a number of gels, not morphologically distinguishable, but with different chemical compositions: a C-A-S-H gel similar to those usually identified in ordinary cement hydration; and a (N,C)-A-S-H gel, similar to the one generated in the alkaline activation of fly ash.

System 20KPC + 80Slag + S also exhibited a compact matrix, however, the main reaction product was a C-A-S-H gel with C/S ratios of around 1, gels with a slightly lower aluminium contents than the one analysed in the C-A-S-H gel identified in the previous system. The evidence indicates that the chemical composition of the gels was largely conditioned by the chemical composition of the starting materials. Gels with a higher calcium content, were observed in systems with a precursor rich in calcium, such as the slag, while silicon-rich gels were the major reaction products in systems with higher silicon content, such as 20KPC + 80FA + S.

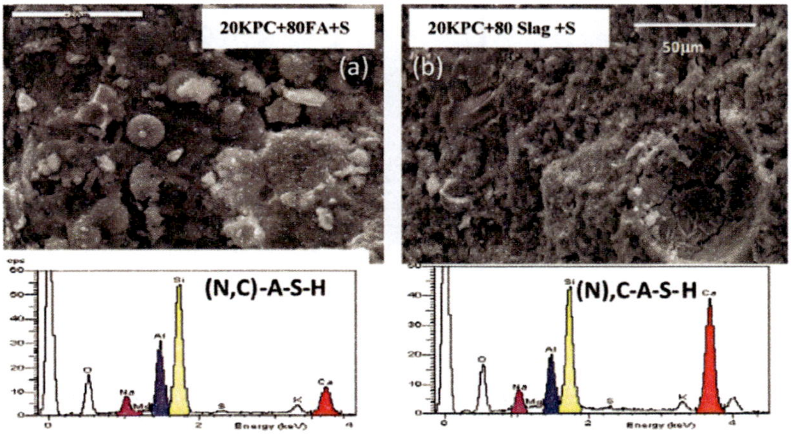

Fig. 4. SEM micrographs for (a) system 20KPC + 80FA + S and (b) system 20KPC + 80Slag + S, showing the formation of cementitious gels (C-(A)-S-H, C-A-S-H and (N,C)-A-S-H) in and EDX analyses [symbols: FA = fly ash; FA at. = fly ash under attack; CK = PC clinker; e = ettringite].

In summary, there is evidence that essentially two types of gels are formed in this kind of systems: i) as a result of the clinker reaction, a C-S-H gel that takes aluminium in its structure, a C-(A)-S-H gel; and ii) another gel with a composition that may vary with the precursor. For systems with a high silica and alumina content (i.e., BFA hybrids), a N-A-S-H gel that takes Ca up in its structure; (N,C)-A-S-H is identified; while for Ca-high systems (i.e., BSLAG hybrids) a C-A-S-H-like gel is the main reaction product generated. With time, these gels may be expected to interact and evolve into a new single (and more thermodynamically stable) gel that could be denominated (N),C-A-S-H [21]. These findings further confirm that the activators used by those authors intensify the reaction of the precursors, enhancing the formation of a larger amount of cementitious gel or gels to which hybrid alkaline cements owe their good mechanical performance.

In 2016, Garcia-Lodeiro et al., based on NMR, as well as XRD, FTIR, BSEM and TEM data, proposed a nanostructural model for gel formation in hybrid alkaline cement. The various stages of that model are illustrated in Fig. 5.

3.1 Addition of Limestone Dust

Another material that is also often incorporated into portland cement to reduce the cost of cement production, reduces the CO_2 emissions, and to improve some technological properties is finely ground limestone [22]. In PC, adding between 5% to 10% of finely ground limestone (LC) can improve the early strength development in mortars and

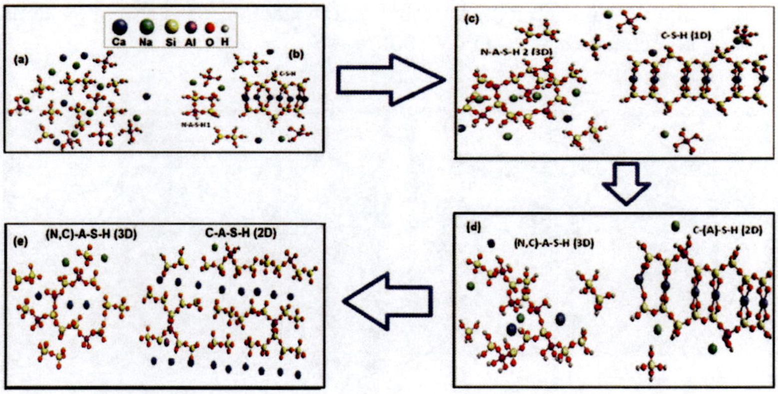

Fig. 5. Nano-structural mechanism for gel formation in hybrid alkaline cements. (*Adapt form reference* [21]).

concretes [acting as a filler, densifying the matrix; furnishing nucleation sites for the C-S-H gel; reaction with clinker aluminate phases to form hemi- or monocarboaluminate [22, 29, 30]. In blended portland cements, limestone may also reduce setting time, induce higher hydrate volume, compressive strength and hemi- and monocarbonate formation and greater ettringite stabilisation relative to the non-limestone-bearing cements [31, 32].

In Hybrid alkaline cement, the addition of finely ground limestone can also have beneficial effects [33]. Figure 6 shows the results obtained for an alkaline hybrid cement 30% PC + 65%FA + 5% S (PMs), the same blended without activator as reference (PM = 30%PC + 70%FA), and with limestone (PMLs = 30%PC + 55FA + 10LC + 5S), in all case the binders were hydrated with water (water/binder (w/c) ratio of 0.27).

The heat flow, obtained by isothermal conduction calorimetric, is shown in Fig. 6(a). In the absence of the chemical activator (PM) the fly ash (PM and PMs) induced an obvious delay in reaction kinetics. Those data corroborate the importance of the chemical activator in binders with a very high fly ash content. However, heat flow is slightly modified by the addition of 10% limestone (PMs and PMLs).

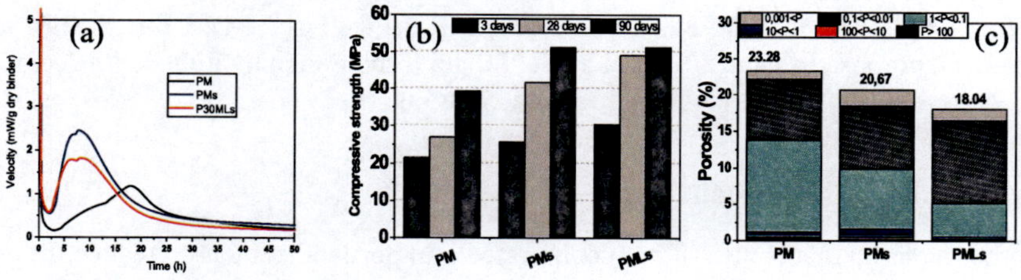

Fig. 6. Effect of limestone in Hybrid alkaline cement (PM = 30%PC + 70%FA; PMs = 30% PC + 65%FA + 5% S; PMLs = 30%PC + 55FA + 10LC + 5S); (a) Heat flow by conduction calorimetric; (b) Compressive strength; (c) Porosity (*Adapt form reference* [33]).

The presence of limestone has a benefit in the early development of compressive strengths (see Fig. 6(b)). When 10% of limestone was added (to replace the ash) the early age strengths rose significantly, from 27.73 MPa to 30.32 MPa at 3 d, and from 41.93 MPa to 49.06 MPa after 28 d. After 90 d, however, strength values tended to converge (51.25 MPa in PMs and 51.30 MPa in PMLs). In the presence of limestone, PMLs porosity and pore size distribution (see Fig. 6(c)) were much like those found in the reference (PM). These results were similar to observed in traditional portland cements [31–33]. That behaviour was partly attributable to the alteration of the porosity and pore size distribution induced by the inclusion of limestone that translated into greater PMLs uniformity. In summary, limestone can, and in fact does act as a filler in alkaline hybrid cements, densifying the matrix.

Additional characterizations studies (ATD/TG, X-ray diffraction and BSEM/EDX) on pastes shown that the presence of 10% limestone in hybrid cements, does not apparently give rise to different hydration products. In presence and absence of limestone, the main reaction product formed in alkaline hybrid binder is a mix of cementitious gels (C-A-S-H and (N,C)-A-S-H). The data also appear to show, however, that the calcium content in those gels is higher in the presence of finely ground limestone.

4 Some Pre-industrial Applications

The use of cements with very high SCMs content is a fact nowadays; cements such as those described in this work are being commercialized under different names. Although the final composition of the cementitious mixture, and the type and amount of activator used, are considered "trade secret" by the cement maker. The same happens with the dosage of concretes with a low content of Portland Clinker.

Figure 7 shows a series of images of concrete panels (2m x 2m x 10 cm) made with a concrete (300 kg/m^3 of cementitious materials (Mc); 40% was a CEM I 52.5 R and the rest 60% a mixture of fly ash, limestone and alkaline activator, all finely ground). Different types of commercial additives were also used to accelerate the setting and development of initial strengths. The requirements of the panels are adequate resistance for lifting (($>$ 10 MPa at 24h and $>$ 28Mpa at 28 days), good workability and fluidity of concrete for filling the mould, where the concrete is vibrated.

In the industrial test carried out, some samples were taken as reference in cylindrical mould (15 x 30cm) to determine the compressive strength at 1, 7, and 28 days according to the EN 12390–2 standard. The results obtained were 11.5, 23.3 and 28.7 MPa respectively. An important aspect to have in mind in the elaboration of these concretes is the need to carry out a good control of the amount of water, and of the ambient temperature at the time of mixing.

Figure 8, shows another series of photos of the application on site, specifically on a pavement. Pavements must be opened to circulation as soon as possible (normally after 7 days), therefore it is required to reach a flexural strength $>$ 2.8 MPa after 7 days and $>$ 3.5 MPa after 28 days. The resistance values are determined in prismatic specimens (15 x 15x 60 cm) according to the standards (UNE-EN 12390–1, UNE-EN 12390–2 (manufactured and cured); UNE-EN 12390–5 (procedure of test). Determination of the compressive strength of specimens, according to standard UNE-EN 12390–3: 2009, at the age of 7, 28 and 90 days.

Fig. 7. Photographs of the panel making process; (a) Abrams cone; (b) filling the panels; (c) specimens to determine compressive strength; (d) lifting off the panel; (e) panel at 24 h; (f) panel at 28 days.

The flexural strength obtained in specimens cured on site were 3.2 MPa and 3.6 MPa at 7 and 28 days respectively, and to compression 18 and 24 MPa. No segregation was observed. The developed concrete met the minimum requirements of reaching 7 days (resistance $>$ 2.8 MPa) so that construction traffic can circulate through it. However, the values obtained at 28 days are considered too fair. At this point, it indicates, that the mixing procedure carried out in this specific test was perhaps not the most appropriate, since the ashes were manually incorporated into the concrete mixer truck where the cement and water were already present. Therefore, based on the results obtained, a definitive judgment cannot be made on the concrete developed without the execution of a greater number of tests, with large-scale means, that ensure the total reproducibility of the manufacturing process.

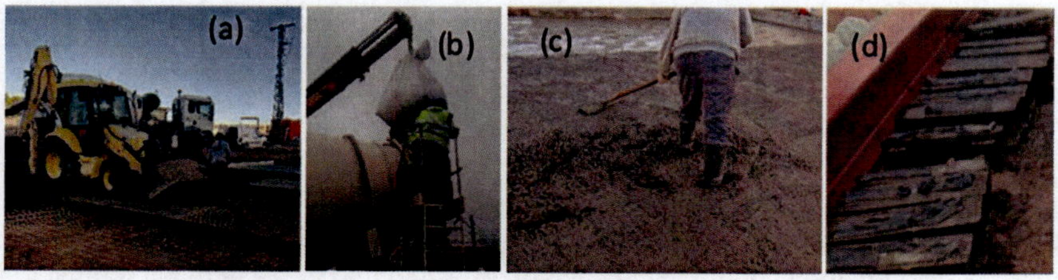

Fig. 8. Photographs of the pavement making process; (a) previous conditioning of the pavement; (b) incorporation of the ashes + activator in the concrete mixer truck; (c) pavement; (d) specimens to determine the flexural strength on site.

5 Conclusions

- The alkali activation of blends of PC, very high SCMs and alkaline activator generates the so called Hybrid alkaline Cements o Hybrid Binder, a feasible, cheap and environmentally-friendly alternative way to the use of traditional PC.
- The manufacture of hybrid alkaline cements or concretes is a very versatile technology where different precursors and activators can be used depending on the local availability of materials (kilometer zero).
- It is possible to produce cements or concretes with low clinker or PC content, with good technological properties (early age mechanical development, with 2 day values of over 10–15 MPa, and suitable setting times).
- The main reaction products (C-S-H and N-A-S-H gel), evolve with time to (N)-C-A-S-H and (N,C)-A-S-H respectively, as function of the alkalinity conditions and the percentage of SiO_2, Al_2O_3 and CaO in the starting blend. The use of one activator or another has no direct effect on the type of gels that precipitate, although it does condition the formation of secondary reaction products.

Acknowledgements. The authors wish to thank the Uropean Union for the project LIFE19 ENV IT 000165-LIFE FREEDOM, and the Spanish Ministry of the Science and Innovation and funds FEDER, under research projects PID2019-111464RB-I00.

References

1. Article/01 JUL, Guterres: Climate Action Is a "Battle for our Lives". https://unfccc.int/news/guterres-climate-action-is-a-battle-for-our-lives. Accessed 1 de octubre de 2019
2. Habert, G., Miller, S.A., John, V.M., Provis, J.L., Favier, A., Horvath, A., Scrivener, K.L.: Environmental impacts and decarbonization strategies in the cement and concrete industries, Nat. Rev. Earth Environ. (2020)
3. Schneider, M.: The cement industry on the way to a low-carbon future. Cem. Concr. Res. **124**, 105792 (2019)
4. Van Ruijven, B.J., Van Vuuren, D.P., Boskaljon, W., Neelis, M.L., Saygin, D., Patel, M.K.: Long-term model based projections of energy use and CO_2 emissions from the global steel and cement industries. Recourses Conserv. Recycling **112**, 15–16 (2016)
5. Chatham House Report, J. Lehne and F. Preston Energy, Environment and Resources Department I June 2018 Making Concrete Change Innovation in Low-carbon Cement and Concrete https://www.chathamhouse.org/sites/default/files/publications/2018-06-13-making-concrete-change-cement-lehne-preston-final.pdf. Accessed 5 Oct 2019
6. Getting the Numbers Right Database: Total Mineral Components to Produce Portland Cement Grey and White World Data With 19% Coverage in 2016, CSI Global Cement Data base on CO_2 and Energy Information, GNR. http://www.wbcsdcement.org/GNR-2016. Accessed 2 Oct 2019. 8 de 20
7. The Energy Transitions Commission estimates that at current costs CCS would increase the costs of cement production by over 50 per cent. Energy Transitions Commission, Better Energy, Greater Prosperity (2017)
8. de Lena, E., Spinelli, M., Romano, M.C.: CO_2 capture in cement plants by "Tail-End" calcium looping process. Energy Proc. **148**, 186–193 (2018)

9. Jamali, A., Fleiger, K., Ruppert, J., Hoenig, V., Anantharaman, R.: CO_2 Capture From Cement, Production, Optimised Operation of an Oxyfuel Cement Plant (2018)
10. Escalante-Garcia, J.I., Sharp, J.H.: The chemical composition and microstructure of hydration products in blended cements. Cem. Concr. Com. **26**, 967–976 (2004)
11. Skibsted, J., Snellings, R.: Reactivity of supplementary cementitious materials (SCMs) in cement blends Cem. Concr. Res. **124**, 105799 (2019)
12. García-Lodeiro, I., Fernández-Jiménez, A., Palomo, A.: Variation in hybrid cements over time: Alkaline activation of fly ash-portland cement blends. Cem. Concr. Res. **52** 112–122 (2013). https://doi.org/10.1016/j.cemconres.2013.03.022
13. Alahrache, S., Winnefeld, F., Champenois, J.-B., Hesselbarth, F., Lothenbach, B.: Chemical activation of hybrid binders based on siliceous fly ash and Portland cement. Cem. Concr. Compos. **66**, 10–23 (2016). https://doi.org/10.1016/j.cemconcomp.2015.11.003
14. Qu, B., Martin, A., Pastor, J.Y., Palomo, A., Fernandez-Jimenez, A.: Characterisation of pre-industrial hybrid cement and effect of pre-curing temperature. Cem. Concr. Comp. **73**, 281–288 (2016)
15. Askarian, M., Tao, Z., Adam, G., Samali, B.: Mechanical properties of ambient cured one-part hybrid OPC-geopolymer concrete. Constr. Build. Mater. **186**, 330–337 (2018). https://doi.org/10.1016/j.conbuildmat.2018.07.160
16. Fernandez-Jimenez, A., Puertas, F., Sobrados, I., Sanz, J.: Structure of calcium silicate hydrated formed in alkali-activated slag: influence of the type of alkaline activator. J. Am. Ceram. Soc. **86**(8), 1389–1394 (2003)
17. Palomo, A., Krivenko, P., Garcia-Lodeiro, I., Kavalerova, E., Maltseva, O., Fernández-Jiménez, A.: A review on alkaline activation: new analytical perspectives. Mater. Constr. **64**(315), e022 (2014)
18. Myers, R.J., Bernal, S.A., Provis, J.L.: Phase diagrams for alkali-activated slag binders. Cem. Concr. Res. **95**, 30–38 (2017)
19. Walkley, B., Page, S.J., Rees, G.J., Provis, J.L., Hanna, J.V.: Nanostructure of CaO-(Na_2O)-Al_2O_3-SiO_2-H_2O gels revealed by multinuclear solid-state magic angle spinning and multiple quantum magic angle spinning nuclear magnetic resonance spectroscopy. J. Phys. Chem. C **124**(2), 1681–1694 (2020)
20. Zhuang, X.Y., Chen, L., Komarneni, S., Zhou, C.H., Tong, D.S., Yang, H.M., et al.: Fly ash-based geopolymer: Clean production, properties and applications. J. Cleaner Prod. **125**, 253–267 (2016)
21. Garcia-Lodeiro I., Donatello, S., Fernandez-Jimenez, A., Palomo, A.: Hydration of hybrid alkaline cement containing a very large proportion of fly ash: A descriptive model. Materials **9**(7), 605 (2016)
22. Scrivener, K., Martirena, F., Bishnoi, S., Maity, S.: Calcined clay limestone cements (LC3). Cem. Concr. Res. **114**, 49–56 (2018)
23. Justnes, H., Østnor, T.A., Designing Alternative Binders Utilizing Synergic Reactions. In: Proceedings of the NTCC 2014, 5th International Conference on Non-Traditional Cement and Concrete, Brno, Czech Republic, pp. 1–4 (2014)
24. Habert, G., Ouellet-Plamondon, C.: Recent update on the environmental impact of geopolymers. RILEM Techn. Lett. **10**, 17–23 (2016). https://doi.org/10.21809/rilemtechlett.2016.6
25. Xue, L., Zhang, Z., Wang, H.: Hydration mechanisms and durability of hybrid alkaline cements (HACs): A review. Constr. Build. Mater. **266**, 121039 (2021)
26. du Toit, G., Kearsley, E.P., Mc Donald, J.M., Kruger, R.A., van der Merwe, E.M.: Chemical and mechanical activation of hybrid fly ash cement. Adv. Cem. Res. **30**(9), 399–412 (2018). https://doi.org/10.1680/jadcr.17.00156
27. Fernández-Jiménez, A., Garcia-Lodeiro, I., Maltseva, O., Palomo, A.: Hydration mechanisms of hybrid cements as a function of the way of addition of chemicals. J. Am. Ceram. Soc. **102**, 427–436 (2019). https://doi.org/10.1111/jace.15939

28. García-Lodeiro, I., Fernández-Jiménez, A., Palomo, A..: Low environmental impact hybrid cements: Reducing clinker content. J. ALCONPAT **5**, 1–16 (2015). https://doi.org/10.21041/ra.v5i1.73
29. Knop, Y., Peled, A.: Packing density modeling of blended cement with limestone having different particle sizes. Constr. Build. Mater. **102**, 44–50 (2016)
30. Adu-Amankwah, S., Black L.: Influence of component fineness on hydration and strength development in ternary slag-limestone cements. RILEM Techn. Lett. **4**, 81–88 (2019)
31. Zajac, M., Rossberg, A., Le Saout, G., Lothenbach, B.: Influence of limestone and anhydrite on the hydration of Portland cements. Cem. Concr. Compos. **46**, 99–108 (2014)
32. Bonavetti, V., Donza, H., Rahhal, V., Irassar, E.: Influence of initial curing on the properties of concrete containing limestone blended cement. Cem. Concr. Res. **30**, 703–708 (2000)
33. Millán-Corrales, G., González-López , J.R., Palomo, A., Fernandez-Jiménez, A.: Replacing fly ash with limestone dust in hybrid cements. Constr. Build. Mater. **243**, 118169 (2020)

One Part Blast-Furnace Slag Cements Activated with a Blend of MgO-Na_2SO_4

I. E. Betancourt-Castillo[1], J. A. Díaz-Guillén[1], J. I. Escalante-García[2], and O. Burciaga-Díaz[1,1](✉)

[1] División de Estudios de Posgrado, Tecnológico Nacional de México/I.T de Saltillo, Blvd. Venustiano Carranza # 2400, Col. Tecnológico, 25280 Saltillo, Coahuila, Mexico
oswaldo.bd@saltillo.tecnm.mx

[2] Ceramics Engineering/Sustainability of Natural Resources and Energy, Cinvestav Unidad Saltillo, Av. Industria Metalúrgica No. 1062, Parque Industrial, Ramos Arizpe, 25900 Saltillo, Coahuila, México

Abstract. In order to reduce the use of sodium silicate solutions and hydroxides for the alkaline activation of materials, the combination of alkaline activators has a positive effect to produce alternative cements with environmental advantages in relation to Portland cement-based materials. This research report results of compressive strength and reaction products of blast-furnace slag cements activated with a mixture of MgO-Na_2SO_4 with a ratio 1:1 added in 4, 6 and 8% relative to the slag weight. Pastes of alkali activated slag were prepared and cured at 20 °C and 60 °C-24h-20 °C up to 90 days. Samples at 20 °C did not show a strength gain during the first 3 days of curing, but after 7 days the strength increased noticeably of up 53 MPa at 90 days. It also was observed that curing at 60 °C favored a gain of compressive strength at early ages between 20-30MPa with a marginal increase at 90 days. The use of the mix MgO-Na_2SO_4 was favorable for the chemical activation of the slag without detrimental effects on the stability of the formulated samples. The microstructures consisted of C-S-H, ettringite, hydrotalcite and some traces of $CaCO_3$ from the raw material.

Keywords: Alkali activated slags · Blast furnace slag · Blends of MgO-Na_2SO_4 · Mechanical strength

1 Introduction

Portland cement-based concrete (PC) is the most widely used material for the construction [1], however it generates a great environmental impact due to the CO_2 emissions regarded with its manufacture process. For this situation, many worldwide initiatives are currently conducted to investigate alternative cements of lower environmental impact based on raw materials of regional abundance, as silicoaluminate clays or diverse industrial wastes as blast furnace slag (BFS). The latter has demonstrated a great potential as cementitious precursor in traditional PC binders and in alkali activated cements (AAC).

AAC are obtained by the interaction of BFS or other amorphous materials with alkaline solutions of sodium silicates and hydroxides, nonetheless, diverse alternative

J. I. Escalante-Garcia et al. (Eds.): RW 2021, RILEM Bookseries 40, pp. 640–647, 2023.
https://doi.org/10.1007/978-3-031-21735-7_69

soft alkalis as Na_2CO_3, Na_2SO_4, $Ca(OH)_2$ and CaO can be used and their effectiveness could be enhanced if some of them are used in combined form as proposed in this investigation [2–6].

An interesting additive recently explored, is the MgO which results in the densification of microstructures by the formation of hydrotalcite that fills pores and improves the mechanical properties to a reasonable degree when is combined with different soft and strong alkalis [7].

For the study, diverse cementitious pastes of BFS activated with blends of MgO-Na_2SO_4 were analyzed. The reaction products were further characterized by thermal analysis and scanning electron microscopy. The results showed that the use of Na_2SO_4 in combination with MgO was effective to obtain AAC pastes with high compressive strength and dense microstructures.

2 Experimental Procedures

2.1 Materials

BFS was used as a solid precursor with a mean particle size of (d50) = 21.2 μm, the chemical composition obtained by X-ray fluorescence (XRF) is shown in Table 1. The surface area of the BFS was of 5000 cm^2/g. The material was mainly amorphous with some traces of crystalline phases of akermanite and gehelenite. To activate the slag, two agents with 1:1 ratio of reactive MgO and Na_2SO_4 were used and supplied by Jalmek ® with chemical composition shown in Table 1. Purified water was also used to prepare pastes.

Table 1. Chemical composition by XRF (mass %) and Blaine fineness of the solid GBFS used

Wt. %	CaO	SiO2	Al2O3	MgO	SO_3	TiO_2	K_2O	Fe_2O_3	MnO	Na_2O
BFS	42.0	32.48	9.71	7.94	3.29	1.59	0.99	0.70	0.67	0.44
MgO	0.1	–	–	97.0	0.02	–	0.05	0.01	–	0.5

2.2 Sample Preparation

Six pastes were elaborated using the combination of MgO:Na_2SO_4 with a constant ratio 1:1 added at concentrations of 4, 6 and 8% by mass relative to the slag content. The water to binder ratio was 0.33 to ensure good workability and fluidity, Table 2 shows the details of the formulations studied. The pastes where formulated with 100% BFS incorporating the mixture of MgO:Na_2SO_4 (in 1:1 proportion) at 4, 6 and 8% relative to the mass of the BFS.

The BFS+ alkaline activators where dry mixed and then added to water and mixed for 3 min. The pastes were cast into cubic molds of 2.5 cm per side and vibrated on a shaking table (Controls Model 55-C0159/LZ) for 30 s to eliminate air bubbles. The samples were covered with plastic film to prevent moisture losses. The pastes were cured at 20 and 60 °C for 24 h with subsequent curing at 20 °C for 90 days.

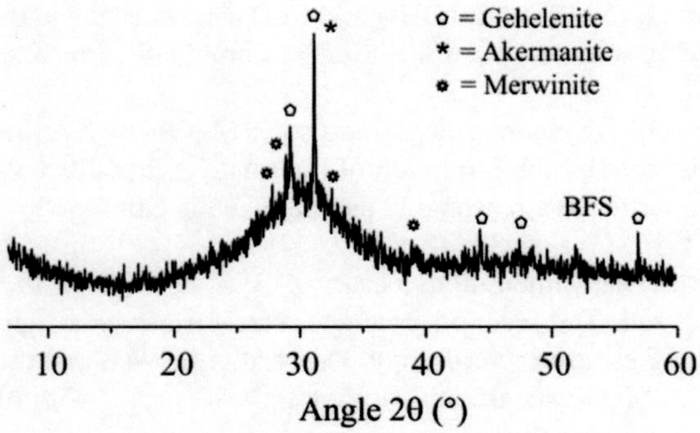

Fig. 1. XDR pattern of the BFS used

Table 2. Mix compositions of the pastes

Binder	BFS (Wt. %)	$MgO:Na_2SO_4$ (Wt. %)	Temperature	Water/binder
1	100	4	20	0.33
2		4	60 (24 h)-20	
3		6	20	
4		6	60 (24 h)-20	
5		8	20	
6		8	60 (24 h)-20	

2.3 Characterization

For compressive strength tests, four cubes of each paste were evaluated at 1, 3, 7, 14, 28 and 90 days using an automatic hydraulic machine (Controls model 50-C7024). Fragments of the fractured samples were immersed in acetone for 2 days and then dried in a vacuum oven (VWR 1430 M) for 48 h at 40 °C, with the intention of interrupting the advance of the reactions. Some samples were pulverized in a planetary mill (PM 400/2; Restch), to be characterized by thermal and differential analysis using a simultaneous device of instruments TGA/DSC SDT Q-600-TA up to 950 °C with a heating rate of 10 °C/min with Al_2O_3 crucibles. For scanning electron microscopy (SEM) analysis, dry fragments were mounted on epoxy resin and polished with silicon carbide sand paper and diamond pastes down to 0.25 μm. The samples were covered with gold to make the samples conductive under the microscope (JEOL JSM-66110LV) accessorized with an Oxford instrument X-Max detector.

3 Results and Discussion

3.1 Compressive Strength (CS)

Figure 2 shows the CS development of $MgO:Na_2SO_4$ activated pastes at 90 days of curing. Samples cured at 20 °C (Fig. 1(a)) did not show CS at 1 and 3 days of curing. After 7 days the formulations showed a slight increase of strength up to 5 MPa for samples with 4 and 8% $MgO:Na_2SO_4$. After 28 days, all pastes developed between 30–40 MPa. The lower dosage of activator, resulted in the highest compressive strength of 55 MPa at 90 days of curing while those with 6 and 8% $MgO:Na_2SO_4$ showed 45 MPa. The values of strength were higher than those previously reported for BFS activated individually with MgO that reached ~32 MPa suggesting that the combined use of MgO and Na_2SO_4 is more beneficial for the formation of cementitious phases [8–11].

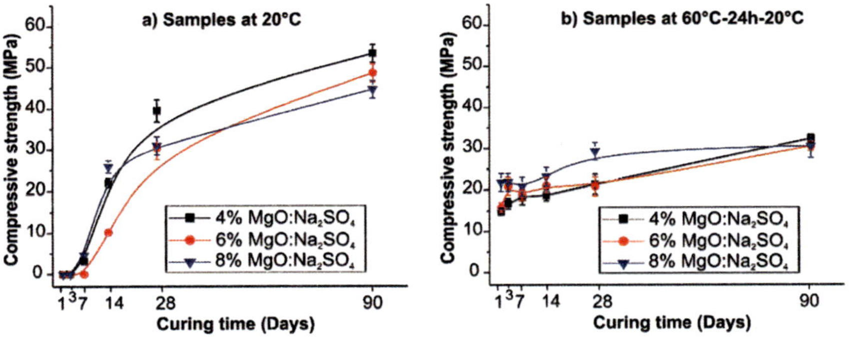

Fig. 2. Compressive strength of binders cured at 20 °C and 60 °C 24 h-20 °C and activated with $MgO\text{-}Na_2SO_4$ at concentrations of 4%, 6% and 8%.

On the other hand, a clear effect of the curing temperature was observed in Fig. 2(b) where the temperature at 60 °C per 24 h, favored a rapid gain of CS for all the pastes at early ages of curing. In this case, the temperature accelerated the kinetics of the dissolution of the BFS leading to the precipitation of reaction products with high strength. At 1 day, the binders with 8% $MgO:Na_2SO_4$ showed mechanical resistance of ~ 25 MPa. Over time, the samples showed a marginal increase of CS that stabilized in 30 MPa at 90 days regardless the dosage of activator used. This behavior can be related with the loss of moisture that occurs at high temperature at early ages limiting the further advance of the reactions [12]. From the results, the curing at 20 °C was even more feasible because under these conditions a slower dissolution of the reactive species occurred, leading to a more homogeneous condensation of mechanically strong reaction products in the matrix, due to the water retention within the microstructures as the reported elsewhere [12, 13].

3.2 Thermogravimetric Analysis

Results of thermal analysis for BFS pastes with 4 and 8% $MgO:Na_2SO_4$ cured at 20 °C and 60 °C-24 h-20 °C at 90 days are respectively shown in Fig. 3. Table 3 includes the weight losses corresponding to the decomposition of the different reaction products observed.

Table 3. Weight loss of the main phases in pastes with 4 and 8% $MgO:Na_2SO_4$ cured at 20 °C and 60 °C-24h-20 °C.

Temperature (°C) (phase)	Weight loss (%)			
	$MgO:Na_2SO_4$–4%		$MgO:Na_2SO_4$–8%	
	20 °C	60 °C 24h–20 °C	20 °C	60 °C 24h–20 °C
40–200 (C-S-H)	8.5	5.9	8.0	8.0
250–400 (Hydrotalcite)	2.5	1.8	2.4	2.4
550–700 (Calcite)	2.2	2.1	1.9	1.9

The first endothermic peak in the derivative weight loss curves at 200 °C is representative of the dehydration of C-S-H gel and ettringite phases according to [6, 9, 11]; this is also consistent with the evaporation of free water absorbed into the porous structure of the materials [14]. For samples containing 4% $MgO:Na_2SO_4$, in Figure (a) the endothermic signal intensity of C-S-H decreases as the curing temperature increased. This suggests a higher formation of C-S-H in samples at 20 °C, which agrees with the CS results previously discussed and with the higher retention of water in the samples necessary for the advance of the chemical reactions and precipitation of products.

For both dosages of activator (4 and 8% $MgO:Na_2SO_4$) an additional weight loss was observed between 250 and 400 °C by decomposition of hydrotalcite-like phases, which is a reaction product frequently reported in alkali activated BFS cements due to the presence of MgO [14, 15]. The progressive weight loss after 600 °C was essentially stimulated by the decarbonation of $CaCO_3$ present in raw materials or in products formed during the hydration reactions. For both concentrations of $MgO:Na_2SO_4$ the intensity of the endothermic signal was similar, indicating that the formation of this phase did not progress by increasing the % of alkaline activating agent [10, 14].

3.3 Scanning Electron Microscopy

Figure 4 shows the microstructures of pastes activated with 4% $MgO:Na_2SO_4$ at 20 °C and 60 °C-24h-20 °C after 90 days of curing. According to Figure (a) the curing at 20 °C favored the consolidation of a dense and compact microstructure with fully and partially reacted BFS particles of light grey tone and irregular morphology homogeneously distributed in the matrix of reaction products of darker grey tone. According to the energy dispersive spectroscopy (EDS) results, the formed products showed high concentration of Si, Ca, Mg, Al and Na, proposing the formation of the C-(A)-S-H type gel intermixed with hydrotalcite phases, some traces of S where also present and possibly incorporated in ettringite, although such phase cannot be seen in the microstructure [11, 16].

As the curing temperature increased to 60 °C-24h (Fig. 4(b)) the microstructure seemed to be less dense with a higher presence of anhydrous BFS grains dispersed in the

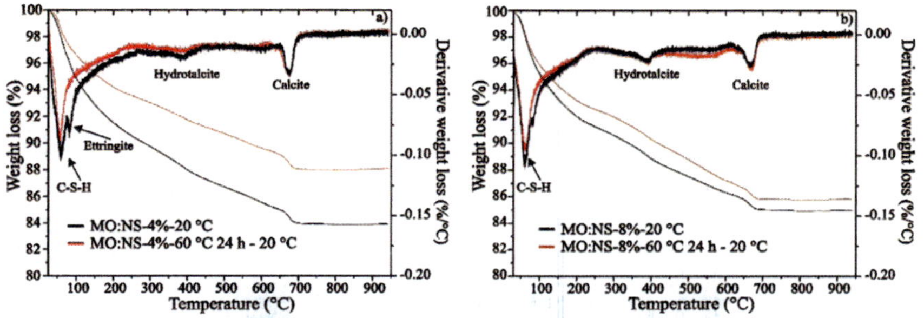

Fig. 3. Thermogravimetric analysis (TG) of binders with 4%(a) and 8% (b) MgO:Na_2SO_4 cured at 20 °C and 60 °C-24h-20 °C at 90 days.

matrix of reaction products justifying the lower values of compressive strength developed. The EDS results did not show significant differences in the chemical composition of the microstructure as the matrix also consisted of Mg, Si, Ca and Al, indicating the formation of C-(A)-S-H and hydrotalcite. Finally, evidence about the presence of unreacted MgO or Na_2SO_4 was not observed, suggesting their complete integration in the reaction products.

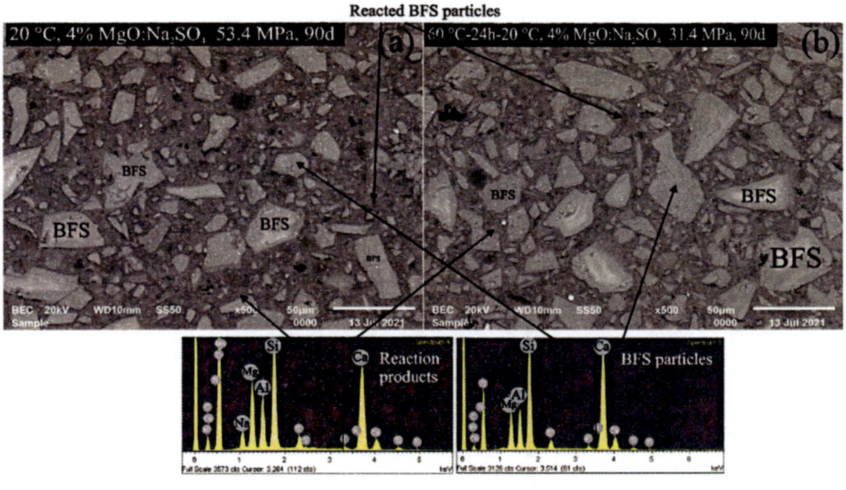

Fig. 4. SEM micrographs of binders with 4% (a) and 8% (b) MgO:NaOH cured at 20 °C and 60 °C-24 h-20 °C at 90 days.

4 Conclusions

- The mixture of MgO-Na_2SO_4 is efficient as alkaline activator of blast furnace slag.

- Under the conditions assessed, the addition of 4% $MgO\text{-}Na_2SO_4$ resulted in higher compressive strength than samples with 6 and 8% of activator
- The pastes cured at 20 °C developed of up to 53 MPa at 90 days of curing, while those exposed at 60 °C–24h-20 °C had a rapid strength gain at early ages, with a limited gain over time.
- The curing of samples at 20 °C benefits the dissolution of reactive species and condensation of dense microstructures with higher strength, while the curing at 60 °C–24h-20 °C, accelerated the reactions at early ages with a limited further gain of strength.
- The reaction products consisted of C-S-H type-gels, ettringite due to the sodium sulfate, and hydrotalcite-like phases.
- The handling of the blended $MgO:Na_2SO_4$ activators is more friendly than those containing strong alkalis as NaOH or sodium silicates. The use of powdered activators is suitable to obtain pastes just by adding water to the precursors (BFS+activator).

Acknowledgements. The Technological National of Mexico (TecNM) and the National Council of Science and Technology (Conacyt) are acknowledged for the financial support of the project 5609.15-P and for the scholarship (711629) to I. Betancourt-Castillo, respectively.

References

1. Provis, J.L., Bernal, S.A.: Geopolymers and related alkali-activated materials. Annu. Rev. Mater. Res. **44**, 3.1–3.29 (2014)
2. Kovtun, M., Kearsley, E.P., Shekhovtsova, J.: Chemical acceleration of a neutral granulated blast-furnace slag activated by sodium carbonate. Cem Concr. Res. **72**, 1–9 (2015)
3. Y. Jeong, H. Park, Y. Jun, J.H. Jeong, J. Eun Oh. Influence of slag characteristics on strength development and reaction products in CaO-activated slag system. Cem. Concr. Compos. 72 (2016), 155–167
4. Burciaga-Díaz, O., Magallanes-Rivera, R.X., Escalante-García, J.I.: Alkali-activated slag-metakaolin pastes: strenght, structural and microstructural characterization. J. Sustain. Cement-Based Mater. **2**, 111–127 (2013)
5. Shi, C., Krivenko, P.V., Roy, D.: Alkali-Activated Cements and Concretes: Taylor & Francis group (2006)
6. Kim, M.S., Jun, Y., Lee, C., Oh, J.E.: Use of CaO as an activator for producing Price competitive non-cement structural binder using ground granulated blast furnace slag. Cem. Concr. Res. **54**, 208–214 (2013)
7. He, J., Zheng, W., Bai, W., Hu, T., He, J., Song, X.: Effect of reactive MgO on hydration and properties of alkali-activated slag pastes with different activators. Constr. Build. Mater. **271**, 121608 (2021)
8. Yi, Y., Liska, M., Al-Tabbaa, A.: Properties and microstructure of GGBS-magnesia pastes. Adv. Cem. Res. **26**(2), 114–122 (2014)
9. Abdalqader, A.F., Jin, F., Al-Tabbaa, A.: Characterization of reactive magnesia and sodium carbonate-activated fly ash/slag paste blends. Constr. Build. Mater. **93**, 506–513 (2015)
10. Burciaga-Díaz, O.: Irma Betancourt-Castillo, Characterization of novel blast-furnace slag cement pastes and mortars activated with a reactive mixture of MgO-NaOH. Cem. Concr. Res. **105**, 54–63 (2018)

11. Tan, H., et al.: Compressive strength and hydration process of wet-grinded granulated blast-furnace slag activated by sodium sulfate and sodium carbonate. Cem. Concr. Compos. **97**, 387–398 (2019)
12. Burciaga-Díaz, O., Gómez-Zamorano, L.Y., Escalante-García, J.I.: Influence of the long term curing temperature on the hydration of alkaline binders of blast furnace slag-metakaolin. Constr. Build. Mater. **113**, 917–926 (2016)
13. Barnett, S.J., Soutsos, M.N., Millard, S.G., Bungey, J.H.: Strength development of mortars containing ground granulated blast-furnace slag: Effect of curing temperature and determination of apparent activation energies. Cem. Concr. Res. **36**, 434–440 (2006)
14. Lee, N.K., Koh, K.T., Kim, M.O., An, G.H., Ryu, G.S.: Physicochemical changes caused by reactive MgO in alkali-activated fly ash/slag blends under accelerated carbonation. Ceram. Int. **43**, 12490–12496 (2017)
15. Gu, K., Jin, F., Al-Tabbaa, A.: Bin Shi, Activation of ground granulated blast furnace slag by using calcined dolomite. Constr. Build. Mater. **68**, 252–258 (2014)
16. Gu, K., Jin, F., Al-Tabbaa, A., Shi, B., Liu, J.: Mechanical and hydration properties of ground granulated blast-furnace slag pastes activated with MgO-CaO mixtures. Constr. Build. Mater. **69**, 101–108 (2014)

Blended Limestone-Portland Cement Binders with Low Amounts of 2 Powdered Sodium Silicates

José Luis Santana-Carrillo[1(✉)], Oswaldo Burciaga-Díaz[2], and J. Iván Escalante-García[1]

[1] Center for Research and Advanced Studies, National Polytechnic Institute, Campus Saltillo, Mexico
luis.santana@cinvestav.edu.mx
[2] Tecnological Institute of Mexico, Tecnological Institute of Saltillo, Saltillo, Mexico

Abstract. The effect of the addition of 2 sodium silicates of modulus = 3.2, one commercial powdered sodium silicate (CSS) and one prepared on the base of waste glass-based (WGSS), was evaluated on limestone-Portland cement binders. An experimental design of three components mixture was used to formulate mortars with water/binder = 0.46, 2–20 wt.% CSS and 10–70% of Portland cement (PC); while for WGSS was used in a mortar with 40%PC, water/binder = 0.25. The aggregate/binder was kept at 2.75. The silicates were compared among similar mortar formulations in terms of compressive strength, flowability, microstructure and CO_2 emissions. The CSS had a positive effect on the strength of mortars made with binders with 10 and 40%PC, but not in those with 70%PC. The flowability of all mortars was of about 130 mm and the use of low amounts of alkaline sodium silicates did not affect the performance of the superplasticizer (added at 0.05% wt. Relative to the mass of the binder). The performance of mortars prepared with WGSS was better than those with CSS; the former reached 40 MPa at 28 days, which was similar to reference 100%CP mortar. The irregular particles of WGSS had a better interaction with the cementitious matrix compared to the spherical particles of the CSS. On the other hand, emissions of the mortar with WGSS was 30% and 40% lower than mortars with CSS and the reference with PC, respectively. Overall, the use of the WGSS in mortars with low water/binder and 40%PC showed similar performance and lower environmental impact compared to the use of 100%PC mortars.

Keywords: Limestone · Portland cement · Sodium silicate · Waste glass

1 Introduction

The use of replacement materials in PC production is an easy way to reduce the high environmental impact of the latter, which is about of 5–7% of all anthropogenic CO_2 [1]. Limestone is the main raw material to produce PC and also the most used replacement material [2]. Although the regulations allow about 40% of limestone in PC, some authors

J. I. Escalante-Garcia et al. (Eds.): RW 2021, RILEM Bookseries 40, pp. 648–655, 2023.
https://doi.org/10.1007/978-3-031-21735-7_70

pointed that more than 50% of limestone can be used without loss of strength, the key is maintain low water/binder ratios [3].

On the other hand, it has been shown that limestone is active in alkaline media, such as sodium silicates, and it has been recently used in binders such as alkali activated cements [4, 5]. Limestone has also been used as a precursor activated with high alkaline solutions of sodium silicate (SiO_2/Na_2O = 1–1.5) [5]; better results were obtained if limestone are blended with low amounts of PC (≈30%) and using low alkaline sodium silicates (SiO_2/Na_2O = 3.2) [6].

The problem of using commercial sodium silicates is its environmental impact (1140–1760 kg CO_2 eq/ton [7]); the use of alternative sodium silicates based on wastes such as: sugar cane bagasse [8], rice husk ash [9] and waste glass [10] is thus interesting and it is being explored on binders.

In this work two alkaline sodium silicates (CSS and WGSS) were evaluated in mortars using binders of limestone-PC. First, the effect of CSS on strength of limestone-PC mortars with low, medium, and high PC contents was evaluated; then the performances of CSS and WGSS were compared in terms of strength, flowability and microstructure of the limestone-PC binders.

2 Experimental

2.1 Effect of CSS on Portland-Limestone Binders

A mixture design of three components (showed in Table 1) was used to formulate 9 mortars (w/b = 0.46 and aggregate/binder = 2.75). The binders were composed of ground limestone (Blaine surface area 660 m^2/kg, 95% $CaCO_3$), PC (CPC 40 as per NMX-414-2017 [11]) and powdered sodium silicate (type G, Na_2O = 19.2, SiO_2 = 62.8). The aggregate was limestone sand (finesses modulus = 2.59). Details of the formulations are shown in Table 2. Cubic specimens of 2.5 cm per side were cast into polypropylene molds, cured at 20 °C and tested by compressive strength at the age of 28 days.

Table 1. Lower and upper limit of binder components.

Component	Lower limit	Upper limit
Limestone (LS)	10	88
Portland Cement (PC)	10	70
Commercial Sodium Silicate (CSS)	2	20

2.2 Comparison Between CSS and WGSS

The composition of Formulation 9 (40%PC, 49%LS and 11%CSS) of Table 2 was taken as a basis and two additional mortars were prepared: one using a w/b = 0.25 named

Table 2. Details of mortars formulations, as well as their 28 days compressive strength and emissions.

ID MIX	LS (wt.%)	PC (wt.%)	CSS (wt.%)	Strength (MPa)	Emissions (kg CO_2-eq/m^3)	Emissions/Strength ((kg CO_2-eq/m^3)/MPa)
1	88	10	2	1.26	82.36	65.29
2	10	70	20	30.19	474.47	15.72
3	28	70	2	30.05	344.83	11.47
4	70	10	20	5.52	231.26	41.88
5	79	10	11	4.42	159.86	36.13
6	19	70	11	31.76	392.68	12.36
7	58	40	2	11.64	212.83	18.28
8	40	40	20	17.92	352.22	19.66
9	49	40	11	16.55	285.39	17.25
Reference	0	100	0	44.9	463.79	10.33

Water/binder W/b = 0.46: Aggregate:Binder of 2.75:1

9-0.25-CSS (further details explained below) and the second also with a w/b = 0.25 but replacing the CSS with a sodium silicate synthesized in the laboratory (WGSS) using waste glass and NaOH (details below) named 9-0.25-WGSS. Both additional mortars had 11% sodium silicate and 0.5%wt. of a superplasticizer (SP) relative to mass of binder.

The synthesis of WGSS was based on the report by Vinai et al. [10] and consisted of the following: 50 g of ground waste glass, 5.46 g of NaOH (this proportions result in $SiO_2/Na_2O = 3.2$, i.e. the same as the CSS) and 20 g of deionized water were mixed and then processed at 200 °C for 2 h; the product was finally ground in a planetary ball mill for 10 min at 300 rpm.

The flowability and evolution of the compressive strength for up to 28 days were evaluated in both mortars. Additionally, pastes were prepared and analyzed by scanning electron microscopy at age of 28 days.

2.3 CO_2 Emissions

Data of Table 3 were used to calculate CO_2 per m^3 of all mortars.

Calculation of CO_2 Emissions. The starting materials (WG, NaOH and water) and necessary Processes (See Table 4) were considered to calculate WGSS emissions (Table 5).

Table 3. Density and emissions of materials.

Constituent	Unit	Density (g/cm3	Emissions (kg CO_2-eq/unit)
PC	ton	3.0	819.8 [12]
LS		2.71	27.8 [12]
CSS		0.8	1760 [7]
WGSS		2.43	273.48*
WG		2.47	0
Water		1.0	0.155 [7]
NaOH		2.1	2240 [7]
Sand		2.71	6.8 [12]
Electricity	kW/h		0.665 [12]

* Details of calculation in the next subsection

Table 4. Necessary processes to synthesize the WGSS.

Process	Energy (kWh/ton)
Crushing of materials	0.90 [13]
Grinding of hard materials (i.e. clinker, waste glass)	46.30 [13]
Grinding of soft materials (i.e. limestone, sodium silicate	19.20 [13]
Heat treatment (up to 300 °C for 2h)	20.00*
Mixing of materials	0.50**

* = Using large cabinet oven of 700 l capacity, Vinai et al. [10] estimated a consuption of 6–10 kW.
** = Using a concrete pan mixer with power of 7.5 kW and mixing capacity of 17–21 m^3/h.

3 Results and Discussions

3.1 Effect of CSS on Portland-Limestone Binders

Figure 1 shows the contour plot of compressive strength of mortars in the experimental region, it can be noted that strength was favored by higher %PC, which was as expected, since PC is the component that contributed most of the strength.

The effect of the CSS was not significant when the %PC was high (70%) and its effect became more noticeable as the %PC decreased. This behavior could be also related to the effective ratio of water/PC (w/PC) ratios (although the w/b was the same in all mortars, the w/PC was different). In mortars with 70%PC, most of the water was consumed by the hydration of PC, therefore, the lack of water would negatively affect the CSS proper dissolution and its involvement in the formation of cementitious reaction products. On the other hand, in mortars with low amounts of PC (i.e. higher w/PC), the CSS had more favorable conditions to dissolve and react with portlandite to form additional calcium

Table 5. Calculation to WGSS emissions.

		Amount	Emissions (kg CO_2-eq)
Materials (ton)	Waste glass	0.902	0.0
	NaOH	0.098	219.52
	Water	0.400	0.06
Processes (kWh)	WG crushing	0.812	0.530
	WG grinding	41.763	27.35
	Mixing of materials	0.505	0.33
	Heat treatment	20.0	13.10
	WGSS grinding	19.2	12.58
Total			273.48

silicate hydrates, which can be noted by the increase of the strength. Additionally, for low %PC there was higher %LS, promoting also the reactions between CSS and LS.

It was noteworthy that for %PC ≥ 40%, there were no differences between using 11 or 20% of CSS; so there is no point to use amounts of sodium silicate above 11% in this type of binders, which has environmental effects.

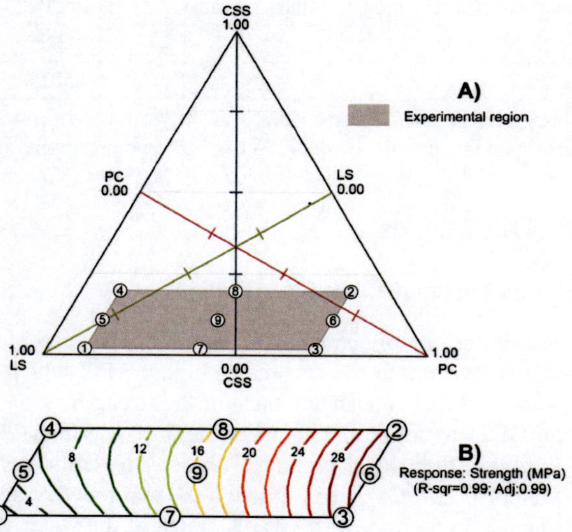

Fig. 1. A) Experimental region in trilinear diagram. B) Contour plot of 28-days compressive strength using a quadratic model.

Table 2 shows the calculated emissions of the mortars; most of those with CSS had lower emissions than the mortar of reference; however, when compressive strength is taken into account, none of the mortars had fewer emissions than the reference. The above resulted from the high emissions associated to the CSS (shown in Table 3) and the relatively low compressive strength achieved after using a normal w/b = 0.46. Therefore, a logical approach to increase the strength and reduce associated CO_2 emissions was to reduce the w/b ratio to increase compressive strength and the use of a less contaminant sodium silicate; this is discussed in the next section.

3.2 Comparison Between CSS and WGSS in Mortars with Low W/b

Figure 2 shows flowability, compressive strength and emissions of the mortars named 9–0.25-CSS, 9–0.25-WGSS, 9 and the mortar of reference. The flowability of the two mortars with w/b = 0.25 was greater than mortar 9 and reference; therefore, the SP had a good performance and it was not affected by the alkalinity of the WGSS, which contrasted to the commonly reports of SP exposed to environments of high alkalinity [14, 15].

With the approach taken, the compressive strength was also enhanced (Fig. 2 A)), both mortars with w/b = 0.25 had similar values; also, from 7 days onwards reached strengths more than 100% above that of mortar 9, and close to that of the reference mortar. The improvements of strength and the use a sodium silicate of a lower environmental impact reduced the overall emissions and the rate of emissions/MPa as shown in Fig. 2 B); mortar with WGSS had 30 and 40% less emissions than mortar with CSS and reference, respectively.

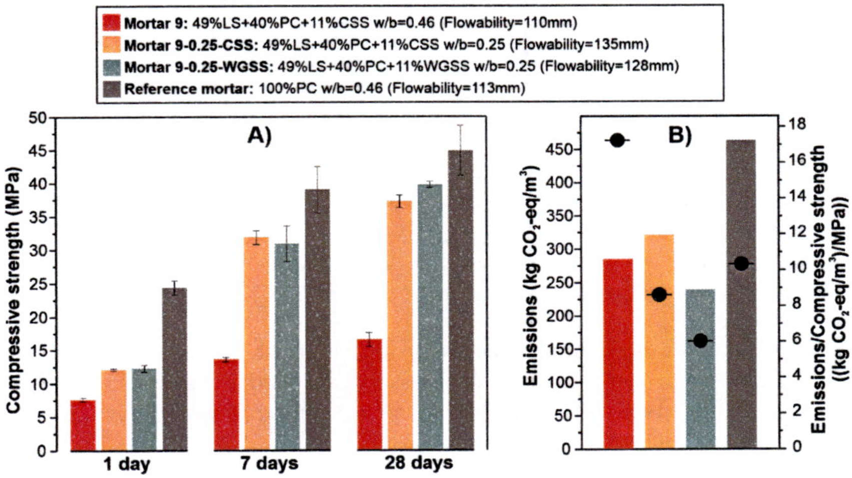

Fig. 2. Mortar 9, mortars based on formulation 9 and mortar reference. A) Evolution of compressive strength. B) Emissions (bars) and emissions/28 days compressive strength (symbols).

Figure 3 shows micrographs of the two pastes 9–0.25-CSS and 9–0.25-WGSS with w/b = 0.25; the main difference was the morphology of the two sodium silicates: spherical, for CSS; irregular and angular for WGSS. Although, both formulations had similar strength, WGSS particles seemed to have better interaction than those of CSS with the reaction products (RP).

On the other hand, a gap was noted around of CSS particles, which could have originated from a higher density of the reaction products (precipitated after the interaction of CSS with PC or LS) comparted to the density of the unreacted CSS (which is relatively low, see Table 3). Therefore, it is possible that the gap between matrix and reacted CSS might become larger at latter ages, with negative effects on the mechanical properties; which would be point in favor to using WGSS instead of CSS.

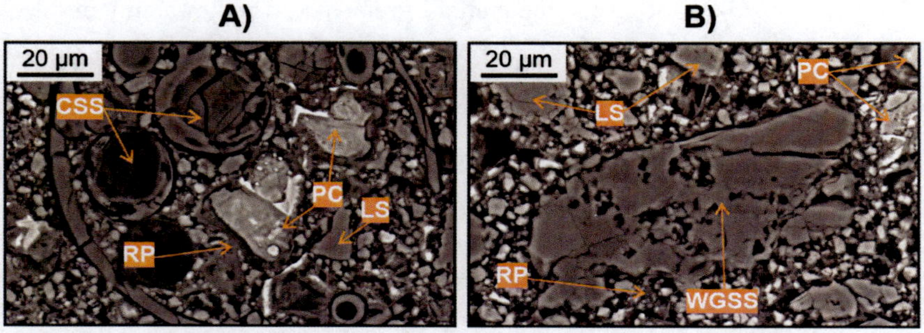

Fig. 3. A) Micrography of paste with CSS. B) Micrography of paste with WGSS.

4 Conclusions

- The positive effect of using commercial sodium silicate on the compressive strength of mortars is greater as the amount of Portland cement on the binder decreases.
- Using a commercial or a laboratory synthesized sodium silicate (CSS or WGSS) results in similar strengths and flowability on limestone-PC formulations; however, using the latter reduces the CO_2 emissions results in 30% compared to the former, and in 40% compared to the reference mortar with 100%PC.
- Microstructurally, the use of the synthetic WGSS seems advantages compared to the commercial sodium silicate in terms of a better integration/interaction with the cementitious matrix.
- Further research on these binders is necessary to: optimize the WGSS synthesis (waste-glass/alkalis, temperature and thermal processing time, particle size, etc.); evaluate the durability in mortars and concretes; and study the nature and composition of reaction products to propose structural models.

Acknowledgements. The financial support of project 154 Thanks are due to the Grant 154 SEP-Cinvestav 2018 and to the Mexican National Council of Science and Technology for the scholarship to J.L. Santana-Carrillo.

References

1. Andrew, R.M.: Global CO_2 enissions from cement production. Earth Syst. Sci. Data **10**, 195–217 (2018)
2. Ramezanianpour, A.A.: Cement Replacement Materials: Properties, Durability, Sustainability. Springer, Heidelberg (2014)
3. Palm, S., Proske, T., Rezvani, M., Hainer, S., Müller, C., Graubner, C.-A.: Cements with a high limestone content–Mechanical properties, durability and ecological characteristics of the concrete. Constr. Build. Mater. **119**, 308–318 (2016)
4. Firdous, R., Hirsch, T., Klimm, D., Lothenbach, B., Stephan, D.: Reaction of calcium carbonate minerals in sodium silicate solution and its role in alkali-activated systems. Miner. Eng. **165**, 106849 (2021)
5. Ortega-Zavala, D.E., Santana-Carrillo, J.L., Burciaga-Díaz, O., Escalante-Garcia, J.I.: An initial study on alkali activated limestone binders. Cem. Concr. Res. **120**, 267–278 (2019)
6. Santana-Carrillo, J.L., Ortega-Zavala, D.E., Burciaga-Díaz, O., Escalante-Garcia, J.I.: Modified blended limestone-Portland cement binders: Evaluation of 4 different sodium silicates. Cem. Concr. Compos. **118**, 103935 (2021)
7. Ouellet-Plamondon, C., Habert, G.: Life cycle assessment (LCA) of alkali-activated cements and concretes. In: Pacheco-Torgal, F., Labrincha, J.A., Leonelli, C., Palomo, A., Chindaprasirt, P. (eds.) Handbook of Alkali-Activated Cements, Mortars and Concretes, pp. 663–686. Woodhead Publishing, Oxford (2015)
8. Sahiron, N., Rahmat, N., Hamzah, F.: Characterization of sodium silicate derived from sugarcane bagasse ash. Malays. J. Anal. Sci. **21**, 512–517 (2017)
9. Kamseu, E., Beleuk à Moungam, L.M., Cannio, M., Billong, N., Chaysuwan, D., Melo, U.C., Leonelli, C.: Substitution of sodium silicate with rice husk ash-NaOH solution in metakaolin based geopolymer cement concerning reduction in global warming. J. Clean. Prod. **142**, 3050–3060 (2017)
10. Vinai, R., Soutsos, M.: Production of sodium silicate powder from waste glass cullet for alkali activation of alternative binders. Cem. Concr. Res. **116**, 45–56 (2019)
11. Industry of Construction Hydraulic Cements-Specifications and Test Methods. ONNCCE, Mexico City (2017)
12. Proske, T., Hainer, S., Rezvani, M., Graubner, C.A.: Eco-friendly concretes with reduced water and cement content: Mix design principles and experimental tests. In: Nazari, A., Sanjayan, J.G. (eds.) Handbook of Low Carbon Concrete, pp. 63–87. Butterworth-Heinemann, Oxford (2017)
13. Worrell, E., Price, L., Martin, N., Hendriks, C., Meida, L.O.: Carbon dioxide emissions from the global cement industry. Annu. Rev. Energy Env. **26**(1), 303–329 (2001)
14. Palacios, M., Puertas, F.: Stability of superplasticizer and shrinkage-reducing admixtures in high basic media. Mater. Constr. **54**, 65–86 (2004)
15. Nematollahi, B., Sanjayan, J.: Effect of different superplasticizers and activator combinations on workability and strength of fly ash based geopolymer. Mater. Des. **57**, 667–672 (2014)

Hydraulic Pastes of Alkali-Activated Waste Glass and Limestone Cement Using in Situ Caustification

L. E. Menchaca-Ballinas(✉) and J. I. Escalante-García

Ceramics Engineering, Cinvestav IPN Unidad, Saltillo, Mexico
luis.menchaca@cinvestav.edu.mx

Abstract. One-part waste glass (WG) and limestone (LS) alkali-activated cements were investigated in pastes using powdered activators $CaO+Na_2CO_3$. 9 formulations of pastes were selected from a full factorial experimental design with %LS, $\%Na_2O$-Na_2CO_3 and %CaO as factors. The activators promoted an in situ caustification that dissolved the WG amorphous structure leading to the formation of cementitious reaction products. The pastes developed 1-year strengths up to 35 MPa and were stable underwater. The characterization by SEM, EDS and ^{29}Si-NMR indicated that while using a Na_2CO_3:CaO ratio close to 1:1, the main reaction products were polymerized C-S-H, silica gel and modified silica gel, intimately intermixed and crosslinked through Q^3 bonds. Such coexistence of reaction products was better in terms of strength and underwater stability than those coexisting in pastes of WG+LS systems activated with solutions of only NaOH reported in literature. Moreover, curing underwater promoted the formation of shorter silicon-oxygen chains, suggesting more formation of C-S-H; nonetheless, the considerable presence of Q^3 and Q^4 signals along with a 1-year compressive strength of 32 MPa in pastes cured underwater evidenced that most of the silica gel was stable. The results indicated that in situ caustification is a promising one-part activation for SiO_2-rich precursors such as waste glass to develop hydraulic alkali activated cements with low cost and CO_2 emissions.

Keywords: Alkali-activated cement · Waste glass · Silica gel · Polymerized C-S-H · C-(N)-S-H · SiO_2-rich precursor

1 Introduction

The production of the main component of concrete, the Portland cement (PC) is highly contaminant since it emits between 0.65 to 0.9 ton of CO_2 per ton of PC, this industry contributes about 10% of the CO_2 emissions and uses 12–15% of the industrial energy [1–5]. Therefore, the development of more eco-friendly and sustainable cements is encouraged.

On the other hand, it is estimated that 130 Mt of glass are produced worldwide per year, from which 100.5 Mt are being discarded as waste; only 27% of the latter is recycled [6, 7]. The recycling by the glass industry is sensitive to chemical variations

J. I. Escalante-Garcia et al. (Eds.): RW 2021, RILEM Bookseries 40, pp. 656–666, 2023.
https://doi.org/10.1007/978-3-031-21735-7_71

and would require sorting of the waste glass (WG) by color, making this route complex. One promising way of revalorizing large volumes of this WG is through its use in construction materials. WG has been studied as supplementary cementitious material in PC composites, as aggregates and as a precursor for alkali activated cements (AAC) [6, 8–17].

Studies on WG as main precursor in AAC (referred from now on as WG-AAC) are scarce [6]. The Al_2O_3 contents in WG are low (commonly < 2 wt.%) to form a N-A-S-H (Na_2O-Al_2O_3-SiO_2-H_2O) gel [36,37], while the CaO contents (<15 wt. %) are not enough to form abundant calcium silicate hydrates (C-S-H) such as those formed in Portland cements or at least, their formation would be negligible [16, 18]. Therefore, the WG-AAC main reaction product is mainly a silica gel low in CaO and Al_2O_3, that is soluble in water [16, 19–21], resulting in non-hydraulic cements, unless additional CaO or Al_2O_3 source are used, such as metakaolin [22], ground furnace slag [23] or limestone (LS) [15, 16].

Furthermore, the alkaline activation of fly ash by Na_2CO_3 and $Ca(OH)_2$ has been reported to enhance the microstructural and strength development of the binders. These activators reacts through a caustification reaction (see Eq. (1)) that induces high alkalinity for the dissolution of the precursors, while providing calcium towards the formation of C-S-H [24–26].

$$Ca(OH)_2 + Na_2CO_3 = CaCO_3 + 2NaOH \quad (1)$$

Recently [27], a similar activation system using powders of CaO and solutions of NaOH-Na_2CO_3 resulted in hydraulic WG-AAC with strength up to 28 MPa. Considering the above, this work presents a novel 1-part activation for SiO_2-rich precursors (i.e. WG) that uses an "in situ caustification" activation system which consists of the mixing of WG + LS + CaO + Na_2O_3 powders.

2 Experimental Procedure

2.1 Raw Materials

The waste silica soda lime glass (WG) was obtained from a local glass production plant; it was washed, dried, and ball milled to a Blaine specific surface area of 600 ± 20 m^2/kg. The alkaline activators used were CaO (Piracal®, 95% Purity) and Na_2CO_3 (Alcali®, 98% purity). The chemical composition of WG by X-ray fluorescence was 69.7% SiO_2, 13.8% CaO, 12.9% Na_2O, 1.4% Al_2O_3, and 0.5% K_2O, while the limestone (LS) was composed by 56.76% CaO and 1% SiO_2, with 41% LOI due to loss of CO_2.

2.2 Experimental Procedure

A total of 9 formulations (Table 1) with 85%WG and 15%LS were selected from a larger full factorial experimental design of 18 formulations. The pastes were prepared by mixing the dry powders of WG, LS, CaO and Na_2CO_3 for 5 min; water was added to a water/binder (w/b) ratio in the range of 0.40–0.50 in order to obtain an adequate workability. Most of the pastes showed a false set, which was solved by intense mixing using an electric mortar mixer during 5–10 min, depending on the formulation. The false setting has been related to the $CaCO_3$ formation [26] in the AAC activated by caustification. The pastes were cast in cubic plastic molds of 2.55 ± 0.05 cm per side; vibration was applied to eliminate entrapped air and the molds were covered with a plastic film to prevent water evaporation. After 24 h of curing at 60 °C, the specimens were demolded and randomly grouped for curing in air (dry) and under water.

2.3 Characterization

The mean compressive strength of 4 paste specimens was reported for 1 to 365 days of curing. Crushed cubes of pastes were submerged in methanol for 72 h to stop hydration reactions, then dried in a vacuum oven at 35 °C for 72h. Pieces of selected pastes were ground manually in an agate mortar for further characterization by ^{29}Si-NMR (Bruker Avance-II 300 spectrometer); the spectra were acquired with a 59.9 MHz frequency, using a CP-MAS 4 mm probe with a rotation speed of 6.5 kHz and a relaxation delay of 7 s. As with other Si-NMR study [28], due to the wide variations and nature of Si-O species along with a highly amorphous structure of the reaction products, the acquisition of the spectra took from 12 to 72 h. The chemical shift is reported relative to tetramethylsilane $[Si(CH_3)_4]$ and the spectra deconvolutions were carried out using the freeware DMFIT with Gauss-Lorentz curve in the range of -60 to -130 ppm, attaining the best fit for every spectrum analyzed [29].

Pieces of pastes were mounted in epoxy resin, then ground, polished, and carbon coated for scanning electron microscopy (SEM, Philips XL-30 ESEM), operated at 15–25 keV; the chemical compositions of the reaction products were characterized by EDS operated at a working distance of 6–8 mm and 20 keV.

Additionally two indexes were also used as reported previously [16, 27] to elucidate the hydraulic performance of the WG-AAC: the underwater evolution index (*u.e.i.*, Eq. (2)) that compares the strength of a formulation at a given day 'X' after dry and underwater curing and the underwater stability index (*u.i.*, Eq. (3)) that compares the strength of a formulation at any given day 'X' relative to its highest strength developed from 1 day and up that 'X' day.

$$u.e.i. = \frac{under\ water\ compressive\ strength\ at\ X\ days\ of\ curing}{dry\ compressive\ strength\ at\ X\ days\ of\ curing} \tag{2}$$

$$u.i = \frac{under\ \ water\ compressive\ strength\ \ at\ X\ days\ of\ curing}{highest\ under\ water\ compressive\ strength\ \ from\ 1\ day\ and\ up\ to\ X\ days\ of\ curing} \tag{3}$$

3 Results and Discussion

The compressive strength evolution is shown in Fig. 1. The 1-day strengths were of 6–15 MPa and after 1 year the pastes reached 22–65 MPa; all showed trends towards a higher strength. The above has been related to progress of the reactions and to the molecular structure evolution of silica gel that is called aging, in which it greatly polymerizes after

Table 1. Studied formulations prepared in pastes; compositions expressed in %wt. The underwater evolution index (u.e.i.) is also shown for discussion; the u.i. of all pastes was 1. *g of activator per 100 g of precursor (WG + LS).

Trial	$\%Na_2O/Na_2CO_3$*	%CaO/CaO*	u.e.i
F1	6%/9.5 g	%10/10 g	0.9
F2	6%/9.5 g	%15/15 g	1.1
F3	6%/9.5 g	%20/20 g	0.95
F4	9%/14.3 g	%10/10 g	0.84
F5	9%/14.3 g	%15/15 g	0.86
F6	9%/14.3 g	%20/20 g	0.73
F7	12%/19 g	%10/10 g	0.41
F8	12%/19 g	%15/15 g	1.04
F9	12%19 g	%20/20 g	0.88

28 days of curing [27, 30, 31]. For example, paste F7, (activated with the lowest %CaO and the highest $\%Na_2O$, developed the highest 1-year strength, indicating the formation of silica gel as main reaction product.

Eight out of nine pastes showed good hydraulic behavior, as shown by compressive strength evolution in Fig. 2 and the u.e.i. values (see Table 1) that were within the range of 0.73–1.1; the u.i. of all pastes was of 1, indicating no strength losses underwater. Paste F7 (12% Na_2O and 10% CaO) had an u.e.i. = 0.41, showing that it developed 59% less compressive strength underwater than in dry curing; however, its evolution underwater did not show any losses from day 1 to 365 (u.i. = 1), so the hydraulic behavior decreased, but the paste was still stable underwater.

Figure 3 shows the microstructure of the paste F2, which showed a heterogeneous distribution of unreacted LS and WG particles among the matrix. Some empty spaces were also noticed which appear as WG particles taken off during the polishing of the specimen, suggesting a poor interaction between unreacted WG particles and the cementitious matrix. Some WG particles showed reaction rims, indicating a reaction via a solid-state mechanism. Table 3 shows the Ca/Si ratios of the reaction products of pastes F2 and F7 of around 0.33.

Figure 4 shows Ca-rich clusters, which have been reported in WG-AAC activated with solutions of $NaOH$-Na_2CO_3 and CaO powder [27]; however, the in situ caustification greatly influenced the microstructure of such clusters, which probably varied depending on how much the $Ca(OH)_2$ reacted with Na_2CO_3 to form $CaCO_3$ (Eq. (1)). The Ca-rich cluster of Fig. 4a, probably composed mainly by $Ca(OH)_2$, had an agglomeration of rosette-like porous structures; its interaction with the Na_2CO_3 might have hindered the growth of the rosette-like structures, resulting in a partially reacted cluster with lesser apparent porosity (Fig. 4b). Such interaction can reach a point of total reaction as shown in Fig. 4d, changing its porous microstructure into a very dense structure,

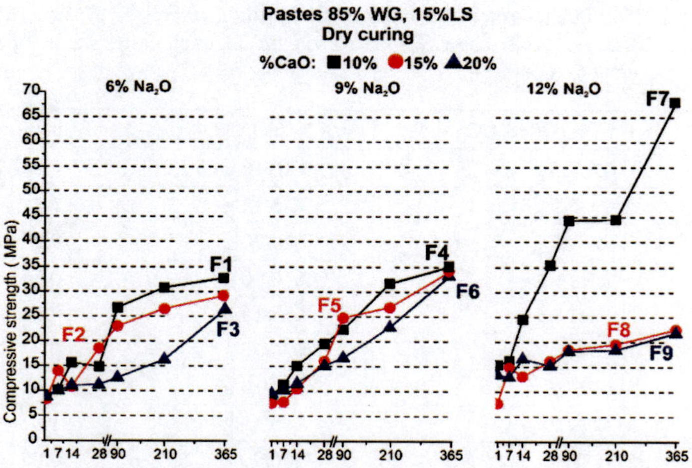

Fig. 1. Compressive strength evolution of pastes composed by 85% WG, 15% LS from 1 to 365 days of curing in dry conditions.

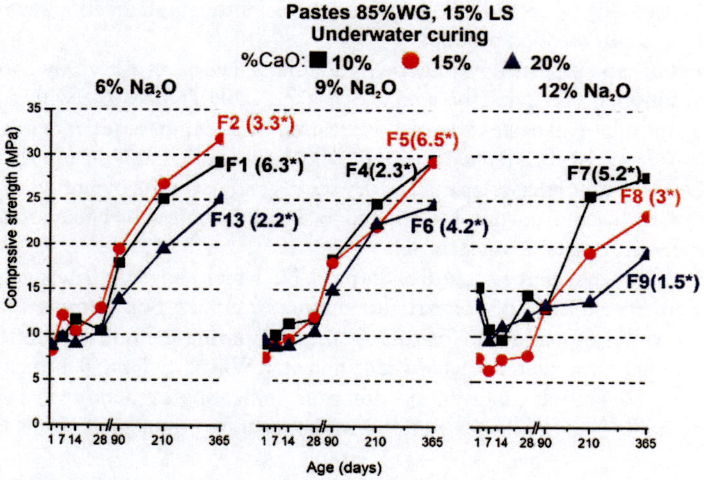

Fig. 2. Compressive strength evolution of pastes composed by 85% WG, 15% LS from 1 to 365 days of curing in underwater conditions. *Standard deviation

which would be composed probably by $CaCO_3$ or some double calcium/sodium carbonate such Gaylussite; the latter has been reported in other LS-WG composites [15]. Is noteworthy that paste F7 showed significantly more totally reacted Ca-rich clusters, which would be related to the high %Na_2O that resulted in a Na_2CO_3:CaO ratio 2:1 (see Table 1), noticeably increasing the chances of $Ca(OH)_2$-Na_2CO_3 interaction.

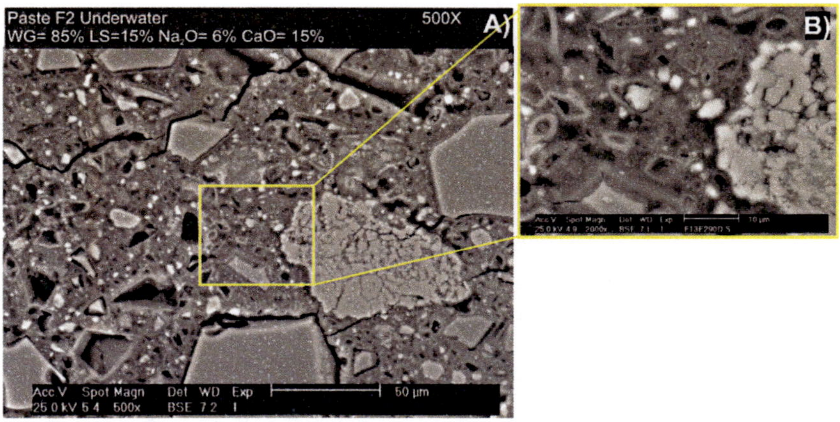

Fig. 3. Microstructure of paste F2: A) at 500x, and B) 2000x.

Figure 5 shows the ^{29}Si NMR spectra of selected pastes and unreacted WG. Table 2 shows the Q^n signal contributions to the area of the spectra after the deconvolution of the unreacted WG, pastes F2 and F7 spectra. The notation Q^n ($0 \leq n \leq 4$) represents a SiO_4 tetrahedron connected to 'n' SiO_4 tetrahedra and the Q^n ranges were selected according to Davidovits [32].

The 3D framework of the unreacted WG was corroborated by the predominance of signals Q^4 (−103.75 to −117 ppm) and Q^3 (−94.5 to −101 ppm). The paste F7 (12% Na_2O, 10% CaO) showed a predominance of Q^2 (−85 to −90 ppm) and Q^3 (−95 to −101.3 ppm), indicating the formation of long branched silon-oxygen chains [16, 31, 33]; if the lower hydraulic behavior and higher compressive strength (relative to the other pastes) are considered as well, the above seems to support the idea of the formation of silica gel as the main reaction product in paste F7. A recent discussion on the LS-silica gel interaction,[16] evidenced that LS can chemically participate, providing of sufficient Ca to modify the molecular structure of silica gel, hindering its polymerization. It is known that if the Na content is high, significant amount of -Si-O-Na bonds can form, which are sites prone to hydrolyzation [34]. Thus, in spite of using 12% Na_2O, the silica gel interaction with the $CaCO_3$ from the LS and that formed from Eq. (1), was enough to make the paste stable underwater (u.i. = 1) after 1 year, probably through the formation of a Ca-modified silica gel.

On the other hand, the 28-day spectrum of Paste F2 (6% Na_2O,15% CaO) was similar to that of 90-day of F7; nonetheless, the 90-day F2 spectra in dry and underwater conditions changed, increasing the Q^1 and Q^2 signals (see Table 2). The paste F2 was activated with a Na_2CO_3:CaO ratio of ~ 1:2 (see Table 1), meaning that part of the CaO-$Ca(OH)_2$ would have not interacted with Na_2CO_3, making it possible for some pozzolanic reactions to form C-S-H. However, the Q^2/Q^1 ratios of 3.9–8.5 suggest the formation of a polymerized C-S-H intermixed with silica gel [16, 27] but the latter predominates in the molecular structure as both dry and underwater spectra had high Q^3 contributions of 30–37%.

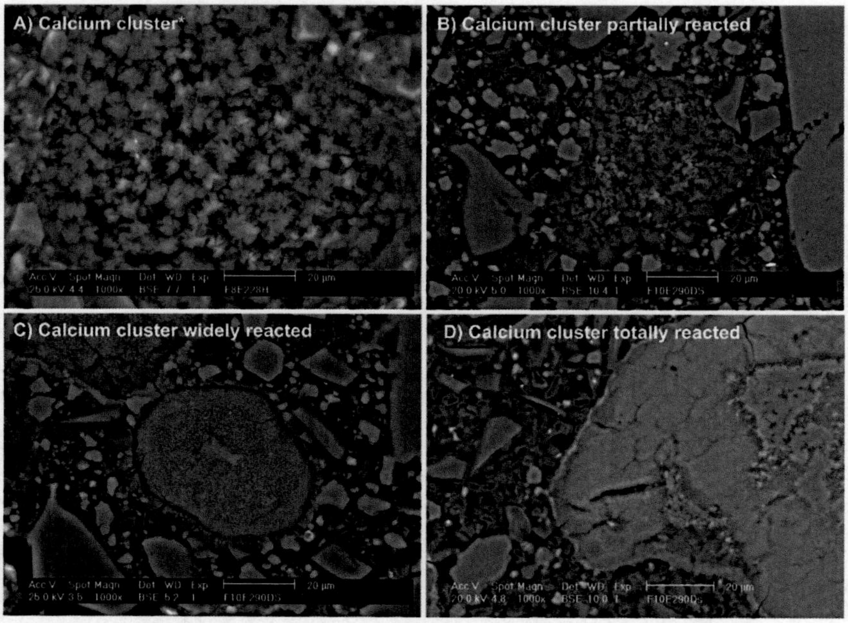

Fig. 4. Microstructure of Ca-rich clusters: A) cluster without/low reaction with Na_2CO_3 (*adapted from Menchaca-Ballinas et al. [27]), B) cluster partially reacted, C) widely reacted and D) totally reacted with Na_2CO_3; micrographs B, C and D are from F8 paste.

Table 2. Q^n signal contributions to the ^{29}Si NMR MAS spectra and Q^2/Q^1 ratios of unreacted WG and the selected pastes.

Paste	Q^1 integral (%)	Q^2 integral (%)	Q^3 integral (%)	Q^4 integral (%)	Q^{poly} integral (%)
F7 90 days dry, *12% Na_2O, 10% CaO*	2.5	32.8	36.9	13.6	14.3
F2 90 days dry, *6% Na_2O, 15% CaO*	6.55	45.2	33.9	14.3	0
F2 90 days underwater	11.9	46.9	30.4	10.8	0
F2 28 days	4.75	40.44	23.6	18.9	12.35

The comparison of the dry and underwater 90-day spectra of F2 showed that the underwater curing spectrum had sharper Q^1 and Q^2 signals, as well as a smaller Q^2/Q^1

ratio (see Table 3). This would corroborate that underwater curing benefits the activation of WG with CaO and Na_2CO_3 by means of water saturation of the paste matrix, enhancing the ions transport and interactions, forming more C-S-H-like products [27]. Furthermore, the paste 90-day spectrum of F2 cured underwater was dominated by Q^2 and Q^3, evidencing the presence of branched linear modified silica gel. This would point to that the modification of silica gel by calcium preserved most of the silica gel branched polymerized structure, while the presence of sharp Q^1 and Q^2 peaks would indicate the presence of polymerized C-S-H, probably crosslinked with modified silica gel through Q^3 bonds.

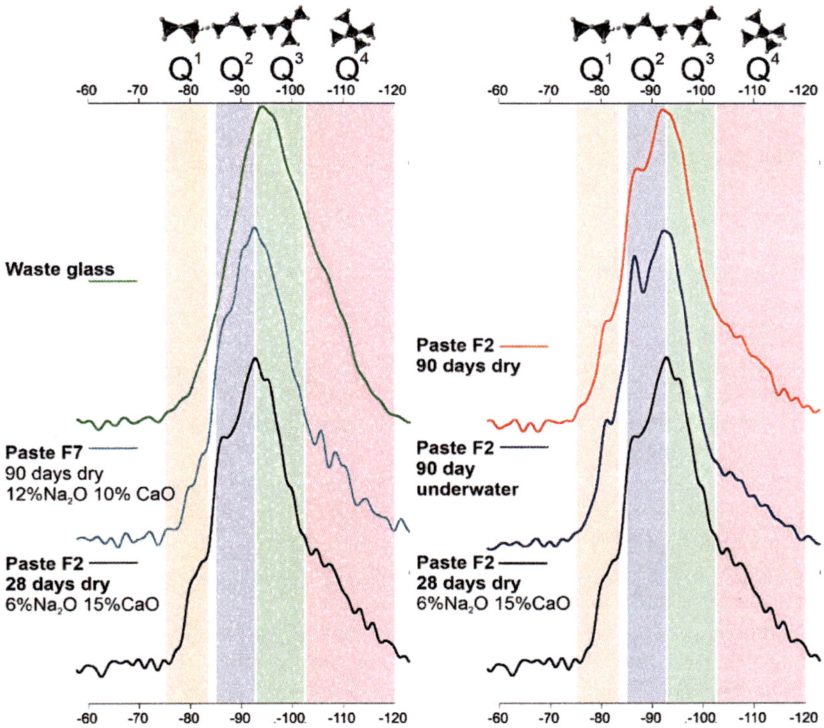

Fig. 5. ^{29}Si-NMR spectra of unreacted WG, pastes F2 (at 28 and 90 days) and F7.

Table 3 shows a summary of 1-year strength, u.e.i., Ca/Si and Q^2/Q^1 ratios data of the selected pastes. The Ca/Si ratios of F2 and F7 pastes were fairly similar, although their molecular structures varied significantly. For a Na_2CO_3:CaO of ~2:1 (F7) more silica gel was formed, resulting in highly polymerized overall structure with a Q^2/Q^1 ratio > 13, with underperformance underwater (u.e.i. = 0.41). Conversely, a Na_2CO_3:CaO ratio of ~ 1:2, encouraged the formation of polymerized C-S-H, silica gel and Ca-modified silica gel, which might be crosslinked, benefiting the underwater performance of the paste (u.e.i. = 1.1), which are better results that using CaO + NaOH as activators for WG-LS systems reported elsewhere [16].

This point to the success of the in situ caustification activation for the stabilization of silica gel underwater by means of in its interaction with polymerized C-S-H and the $CaCO_3$ from LS and Eq. (1), resulting in a hydraulic WG-AAC.

Table 3. Summary of compressive strength, u.e.i., Ca/Si and Q^2/Q^1 ratios of selected pastes.

Paste	Compressive strength (MPa) 1 year	u.e.i 1 year	Ca/Si ratio	Q^2/Q^1 ratio
F7 90 days dry *12% Na_2O, 10% CaO*	68	0.41	0.33	13.18
F2 90 days dry *6% Na_2O, 15% CaO*	29.1	1.1	0.34	6.90
F2 90 days underwater	32	1.1	0.33	3.93
F2 28 days	18.5 (28 days)	1.1	–	8.51

4 Conclusions

The study of a novel 1-part activation of waste glass alkali activated cement by the use of an in situ caustification was successfully implemented. The highlights of the study were of the following:

- The pastes developed 1-year compressive strengths of up to 68 MPa.
- The underwater stability index indicated that all the formulations were stable underwater. Paste F2 (85% waste glass, 15% limestone, 6%Na_2O/15%CaO) developed 31 MPa underwater at 1 year.
- The underwater evolution index indicated that paste F7 activated with 12% Na_2O and 10% CaO had the lower hydraulic behavior, developing only 41% of the compressive strength developed in dry conditions. However, it did not lose any strength underwater after 1 year.
- The in situ caustification activation resulted in the formation of $CaCO_3$ and polymerized C-S-H when the Na_2CO_3:CaO ratio is less than 1.
- The in situ caustification promoted the underwater stability of silica gel.

Acknowledgements. Thanks are due to the Mexican National Council of Science and Technology (Conacyt) for the scholarship to Menchaca-Ballinas and the financial support of project 154. Thanks are due to the Grant 154 SEP-Cinvestav 2018.

References

1. Summerbell, D.L., Barlow, C.Y., Cullen, J.M.: Potential reduction of carbon emissions by performance improvement: A cement industry case study. J. Clean. Prod. **135**, 1327–1339 (2016). https://doi.org/10.1016/J.JCLEPRO.2016.06.155
2. Van Vliet, K., et al.: Set in stone? A perspective on the concrete sustainability challenge. MRS Bull. **37**(4), 395–402 (2012). https://doi.org/10.1557/mrs.2012.55
3. Pour-Ghaz, M.: Sustainable infrastructure materials: Challenges and opportunities. Int. J. Appl. Ceram. Technol. **10**, 584592 (2013). https://doi.org/10.1111/ijac.12083
4. Ali, M.B., Saidur, R., Hossain, M.S.: A review on emission analysis in cement industries. Renew. Sustain. Energy Rev. **15**, 2252–2261 (2011). https://doi.org/10.1016/j.rser.2011.02.014
5. Hendrik G. van Oss: Cement-mineral commodity summaries. U.S. Geol. Surv. 38–39 (2015)
6. Siddika, A., Hajimohammadi, A., Mamun, M.A. Al, Alyousef, R., Ferdous, W.: Waste glass in cement and geopolymer concretes: A review on durability and challenges. Polymers (Basel) **13**, 2071 (2021). https://doi.org/10.3390/polym13132071
7. Kaza, S., Yao, L., Bhada-Tata, P., Woerden, F. Van: What a Waste 2.0. (2018)
8. Jani, Y., Hogland, W.: Waste glass in the production of cement and concrete—A review. J. Environ. Chem. Eng. **2**, 1767–1775 (2014). https://doi.org/10.1016/j.jece.2014.03.016
9. Johnston, C.D.: Waste glass as coarse aggregate for concrete. J. Test. Eval. **2**, 344–350 (1974). https://doi.org/10.1520/JTE10117J
10. Bignozzi, M.C., Saccani, A., Barbieri, L., Lancellotti, I.: Glass waste as supplementary cementing materials: The effects of glass chemical composition. Cem. Concr. Compos. **55**, 45–52 (2015). https://doi.org/10.1016/j.cemconcomp.2014.07.020
11. Ismail, Z.Z., AL-Hashmi, E.A.: Recycling of waste glass as a partial replacement for fine aggregate in concrete. Waste Manag. **29**, 655–659 (2009). https://doi.org/10.1016/j.wasman.2008.08.012
12. Shayan, A., Xu, A.: Performance of glass powder as a pozzolanic material in concrete: A field trial on concrete slabs. Cem. Concr. Res. **36**, 457–468 (2006). https://doi.org/10.1016/j.cemconres.2005.12.012
13. Sobolev, K., Türker, P., Soboleva, S., Iscioglu, G.: Utilization of waste glass in ECO-cement: Strength properties and microstructural observations. Waste Manag. **27**, 971–976 (2007). https://doi.org/10.1016/j.wasman.2006.07.014
14. Matos, A.M., Sousa-Coutinho, J.: Durability of mortar using waste glass powder as cement replacement. Constr. Build. Mater. **36**, 205–215 (2012). https://doi.org/10.1016/j.conbuildmat.2012.04.027
15. Avila-López, U., Almanza-Robles, J.M., Escalante-García, J.I.: Investigation of novel waste glass and limestone binders using statistical methods. Constr. Build. Mater. **82** 296303 (2015) https://doi.org/10.1016/j.conbuildmat.2015.02.085
16. Menchaca-Ballinas, L.E., Escalante-García, J.I.: Limestone as aggregate and precursor in binders of waste glass activated by CaO and NaOH. Constr. Build. Mater. **262** (2020). https://doi.org/10.1016/j.conbuildmat.2020.120013
17. Menchaca-Ballinas, L.E., Escalante-Garcia, J.I.: Low CO_2 emission cements of waste glass activated by CaO and NaOH. J. Clean. Prod. **239**, 117992 (2019). https://doi.org/10.1016/j.jclepro.2019.117992
18. Marchand, B., Lanier, S., Davy, C.A., Albert-Mercier, C., Tricot, G.: Are calcium silicate hydrates (C-S-H) present in alkali-activated glass cullet cement? Mater. Lett. **219**, 104–108 (2018). https://doi.org/10.1016/j.matlet.2018.02.046
19. Redden, R., Neithalath, N.: Microstructure, strength, and moisture stability of alkali activated glass powder-based binders. Cem. Concr. Compos. **45**, 46–56 (2014). https://doi.org/10.1016/j.cemconcomp.2013.09.011

20. Torres-Carrasco, M., Puertas, F.: Waste glass as a precursor in alkaline activation: Chemical process and hydration products. Constr. Build. Mater. **139**, 342–354 (2017). https://doi.org/10.1016/j.conbuildmat.2017.02.071
21. Dimas, D., Giannopoulou, I., Panias, D.: Polymerization in sodium silicate solutions: A fundamental process in geopolymerization technology. J. Mater. Sci. **44**(14), 3719–3730 (2009). https://doi.org/10.1007/s10853-009-3497-5
22. Burciaga-Díaz, O., Diaz-Guillen, A., Durón-Sifuentes, M., Escalante García, J.I.: Effect of waste glass incorporation on the properties of geopolymers formulated with low purity metakaolin. Cem. Concr. Compos. **107**, 103492 (2019)
23. Martinez-Lopez, R., Ivan Escalante-Garcia, J.: Alkali activated composite binders of waste silica soda lime glass and blast furnace slag: Strength as a function of the composition. Constr. Build. Mater. **119**, 119–129 (2016). https://doi.org/10.1016/j.conbuildmat.2016.05.064
24. Jeon, D., Jun, Y., Jeong, Y., Oh, J.E.: Microstructural and strength improvements through the use of Na_2CO_3in a cementless $Ca(OH)_2$-activated Class F fly ash system. Cem. Concr. Res. **67**, 215–225 (2015). https://doi.org/10.1016/j.cemconres.2014.10.001
25. Suh, J. Il, Jeon, D., Yoon, S., Oh, J.E., Park, H.G.: Development of strong lightweight cementitious matrix for lightweight concrete simply by increasing a water-to-binder ratio in $Ca(OH)_2$-Na_2CO_3-activated fly ash system. Constr. Build. Mater. **152**, 444–455 (2017). https://doi.org/10.1016/j.conbuildmat.2017.07.011
26. Jeon, D., Yum, W.S., Song, H., Sim, S., Oh, J.E.: The temperature-dependent action of sugar in the retardation and strength improvement of $Ca(OH)_2$-Na_2CO_3-activated fly ash systems through calcium complexation. Constr. Build. Mater. **190**, 918–928 (2018). https://doi.org/10.1016/j.conbuildmat.2018.09.164
27. Menchaca-Ballinas, L.E., Escalante-García, J.I.: Waste glass as a precursor in sustainable hydraulic cements activated with CaO-NaOH-Na_2CO_3. Waste Biomass Valorization (2021)
28. Beaudoin, J., Raki, L., Alizadeh, R.: A 29Si MAS NMR study of modified C-S–H nanostructures. Cem. Concr. Compos. **31**, 585–590 (2009). https://doi.org/10.1016/j.cemconcomp.2008.11.004
29. Massiot, D., Fayon, F., Carpon, M., Le Calvé, S., Alonso, B., Durand, J.O., Bujoli, B., Gan, Z. Hoatzon, G.: DMFIT, http://nmr.cemhti.cnrs-orleans.fr/dmfit/ (2002)
30. Scherer, G.W.: Aging and drying of gels. J. Non. Cryst. Solids. **100**, 77–92 (1988)
31. Scherer, G.W.: Sol-Gel-Glass I: Gelation and gel structure. J. Non. Cryst. Solids. **70**, 301–322 (1985)
32. Davidovits, J.: Geopolymer Chemistry and Applications. Geopolymer Institute (2008)
33. Scherer, G.W.: Structure and properties of gels. Cem. Concr. Res. **29**, 1149–1157 (1999). https://doi.org/10.1016/S0008-8846(99)00003-4
34. Visser, J.H.M.: Fundamentals of alkali-silica gel formation and swelling: Condensation under influence of dissolved salts. Cem. Concr. Res. **105**, 18–30 (2018). https://doi.org/10.1016/j.cemconres.2017.11.006

RILEM TC 281-CCC Working Group 6: Carbonation of Alkali Activated Concrete—Preliminary Results of a Literature Survey and Data Analysis

Gregor J. G. Gluth[1](✉), Xinyuan Ke[2], Anya Vollpracht[3], Susan A. Bernal[4], Özlem Cizer[5], Martin Cyr[6], Katja Dombrowski-Daube[7], Dan Geddes[8], Cyrill Grengg[9], Cassandre Le Galliard[8], Marija Nedeljkovic[10], John L. Provis[8], Zhenguo Shi[11], Luca Valentini[12], and Brant Walkley[13]

[1] Division 7.4 Technology of Construction Materials, Bundesanstalt für Materialforschung und -prüfung (BAM), Berlin, Germany
gregor.gluth@bam.de
[2] Department of Architecture and Civil Engineering, The University of Bath, Bath, UK
[3] Institute of Building Materials Research, RWTH Aachen University, Aachen, Germany
[4] School of Civil Engineering, The University of Leeds, Leeds, UK
[5] Department of Civil Engineering, KU Leuven, Louvain, Belgium
[6] Laboratoire Matériaux et Durabilité des Constructions (LMDC), Université de Toulouse, Toulouse, France
[7] Institut für Bergbau und Spezialtiefbau, Technische Universität Bergakademie Freiberg, Freiberg, Germany
[8] Department of Materials Science and Engineering, The University of Sheffield, Sheffield, UK
[9] Institute of Applied Geosciences, Graz University of Technology, Graz, Austria
[10] Faculty of Civil Engineering and Geosciences, Delft University of Technology, Delft, The Netherlands
[11] Laboratory for Concrete & Construction Chemistry, Swiss Federal Laboratories for Materials Science and Technology (Empa), Dübendorf, Switzerland
[12] Department of Geosciences, University of Padua, Padua, Italy
[13] Department of Chemical and Biological Engineering, The University of Sheffield, Sheffield, UK

Abstract. The current understanding of the carbonation of alkali-activated concretes is hampered *inter alia* by the wide range of binder chemistries used. To overcome some of the limitations of individual studies and to identify general correlations between their mix design parameters and carbonation resistance, the RILEM TC 281-CCC working group 6 compiled carbonation data for alkali-activated concretes and mortars from the literature. For comparison purposes, data for blended Portland cement-based concretes with a high percentage of SCMs ($\geq$ 66% of the binder) were also included in the database. A preliminary analysis of the database indicates that *w*/CaO ratio and *w*/b ratio exert an influence on the carbonation resistance of alkali-activated concretes but, contrary to what has been reported for concretes based on (blended) Portland cements, these are not good indicators of their carbonation resistance when considered individually. A better

J. I. Escalante-Garcia et al. (Eds.): RW 2021, RILEM Bookseries 40, pp. 667–676, 2023.
https://doi.org/10.1007/978-3-031-21735-7_72

indicator of the carbonation resistance of alkali-activated concretes under conditions approximating natural carbonation appears to be their $w/(CaO + Na_2O + K_2O)$ ratio. Furthermore, the analysis points to significant shortcomings of tests at elevated CO_2 concentrations for low-Ca alkali-activated concretes, indicating that even at a concentration of 1% CO_2, the outcomes may lead to inaccurate predictions of the carbonation coefficient under natural exposure conditions.

Keywords: Alkali-activated materials · Carbonation · Natural exposure · Accelerated testing

1 Introduction

The carbonation of cementitious materials is an important degradation phenomenon, in many cases being the limiting factor for the service life of reinforced concrete structures. For concretes based on conventional cements (Portland and blended Portland cements), the $w/CaO_{reactive}$ ratio has been found to be the major parameter determining their carbonation resistance [1, 2]; related to this ratio are concepts that involve *w/b* ratio and clinker content [3] as well as the k-value concept of EN 206 [4]. However, for concretes and mortars based on alkali-activated binders and other alternative cements these approaches seem to be not applicable without restrictions [5, 6]. Relatedly, the factors and mechanisms determining the carbonation resistance of alkali-activated concretes are not yet fully understood, partly due to the wide range of binder chemistries and the fact that individual carbonation studies are generally limited to only a few mix designs, curing conditions and exposure conditions [7].

To contribute to these issues, the RILEM TC 281-CCC working group 6 compiled carbonation data for a wide range of alkali-activated concretes and mortars from the literature. This has been done with the aim of resolving the shortcomings of individual studies and to identify general correlations between mix design parameters of alkali-activated concretes and their carbonation resistance. For comparison purposes, data for blended Portland cement-based concretes with a high percentage of supplementary cementitious materials (SCMs) in the binder were included in the database. In this paper, the database, the assumptions made in the evaluation of the data, and preliminary results and conclusions are presented.

2 Survey Scope and Data Processing

2.1 Data Inventory

The scientific literature was searched for studies that report carbonation depth (d_c)-*versus*-carbonation duration (t) data for alkali-activated concretes or mortars, or for concretes or mortars based on CEM III/B (according to EN 197–1; $\geq 66\%$ blast furnace slag in the cement) or blended cements with at least 70% of other or mixed SCMs in the binder. Only data obtained at controlled relative humidity (conditions approximating indoor or "sheltered" natural carbonation and accelerated carbonation) were included in the database, while results for unsheltered natural exposure were not considered, as the degree of saturation of the pore systems of the exposed concretes is not controlled in the latter experiments. Carbonation data of pastes were not considered, as the potential effects of aggregates on the carbonation coefficient would not be observed. Studies with incomplete or internally inconsistent information about the mix design or the test conditions were also excluded from the database.

The mix designs of the concretes or mortars, the chemical compositions of the binders (solid precursors and activators), the physical properties of the constituents as well as curing, preconditioning and carbonation conditions were included in the database together with the respective carbonation depth data. In most of the studies, the carbonation depths were determined by spraying a phenolphthalein solution (usually 1%) on fracture surfaces; in one case alizarin yellow R was used as indicator, and in another case thin sections were examined.

The current database contains data for 74 concretes/mortars based on blended cements, *i.e.* binders with no activator other than Portland clinker (BCC); 125 concretes/mortars based on alkali-activated ground granulated blast furnace slag (GGBS-based AAC); 12 concretes/mortars based on alkali-activated fly ash or metakaolin (FA- or MK-based AAC); and 18 concretes/mortars based on other alkali-activated binders (other AAC). GGBS-based AAC are defined here as concretes/mortars with $\geq$ 50% GGBS in the solid fraction of the binder, *i.e.* excluding Na_2O, SiO_2 etc. dissolved in the liquid activator. FA- or MK-based AAC are defined as concretes/mortars with >50% fly ash or metakaolin, respectively, in the solid fraction of the binder; only one of the twelve materials was solely metakaolin-based. The other AAC include materials with binders based on municipal solid waste incineration bottom ash, silico-manganese slag or natural pozzolans; two hybrid cements were assigned to this class as well.

Most of the AAC were activated with sodium silicate solutions (of various SiO_2/Na_2O ratios), while fewer were activated with NaOH, Na_2CO_3, or a mixture of sodium silicate and Na_2CO_3. In some cases, portlandite [$Ca(OH)_2$] or gypsum [$CaSO_4{\cdot}2H_2O$] was employed as an additional activator,

2.2 Data Evaluation

When only one d_c-*versus*-t value was given for a specific material and carbonation conditions, the carbonation coefficient was calculated as $k = d_c/\sqrt{t}$, *i.e.* $d_c = 0$ mm at $t = 0$ d was implicitly assumed. When multiple d_c-*versus*-t values were available, the carbonation coefficient was obtained by first fitting a linear function to all available data points plotted as $d_c(\sqrt{t})$ up to $t = 1$ year. If that fit yielded a positive intercept with the ordinate, the slope of the function was used as k for the evaluation. If the fit yielded a negative intercept with the ordinate (*i.e.* a negative carbonation depth at $t =$ 0 d, which would be unphysical), the fit was repeated with the linear function forced through the origin of the coordinates, and the value thus obtained was used. It is noted that the assumption of $d_c = 0$ mm at $t = 0$ d is not necessarily true, particularly for materials based on blended cements or alkali-activated binders [8, 9]. However, in the absence of data for the start of the exposure to CO_2, this assumption was made.

The data of the round robin testing programme of RILEM TC 247-DTA [8] was treated as if it was obtained in a single laboratory, using the average of the carbonation depths obtained in the participating institutes. However, the d_c obtained in lab G in that testing programme were not included in the present analysis, because that laboratory had employed a curing regime that differed from the curing in the other laboratories. The coefficient of variation (COV) of the d_c in the reduced dataset was in the range 21.2–42.1% for carbonation under conditions approximating natural exposure, and 1.0–45.4% for accelerated carbonation. To translate these uncertainties into the uncertainty of the resulting k, it was assumed that the COV of the latter was 40%, *i.e.* the standard deviation of the carbonation coefficient was estimated to be $0.40 \times k$.

In the following, k_{nat} will be used as the symbol for carbonation coefficients obtained under conditions approximating natural carbonation (sheltered), and k_{acc} will be used as the symbol for carbonation coefficients obtained under accelerated conditions. If the coefficients refer to a specified CO_2 concentration or concentrations, this will be noted in brackets (e.g., $k_{acc}[1\%]$).

From the mix designs of the concretes/mortars and the chemical compositions of their constituents, several parameters that may exert an influence on the carbonation resistance were computed. In cases where these parameters are water-to-oxide(s) mass ratios [*w*/oxide(s)], the value used for the oxide(s) is the total amount of that oxide(s) in the binder paste, including the contribution of the solid fraction of the activator. For example, w/Na_2O is the ratio between water and the total amount of Na_2O in the paste, *i.e.* the sum of the Na_2O contributed by the activator and the Na_2O present in the solid fraction of the binder. No attempt was made to discriminate "reactive" and "non-reactive" oxides in the paste, as this is usually not possible with high accuracy, if at all. The sum $Na_2O + K_2O$ is written as $(Na,K)_2O$ to reflect the fact that Na^+ and K^+ play a similar role in these binders and can be treated equivalently for the present purpose. However, all AAC in the current database were produced with Na-dominated activator solutions.

3 Preliminary Results and Discussion

3.1 Carbonation Under Conditions Approximating Natural Exposure

Figure 1 shows that the carbonation resistance of alkali-activated concretes and mortars cannot be predicted based on their *w*/CaO ratio alone. The FA-based AAC, and particularly the MK-based AAC, deviate considerably from a linear relationship that can be discerned for the combined BCC and GGBS-based AAC. Within the group of BCC, a good correlation between k_{nat} and *w*/CaO was observed, while a correlation was less evident in the GGBS-based AAC (Fig. 2). Compared to k_{nat} *versus* *w*/CaO, a slightly more reasonable correlation was obtained when plotting k_{nat} *versus* *w*/*b* of the GGBS-based AAC (Fig. 3B).

The observation that FA- and MK-based AAC do not follow the same trend as the BCC as regards *w*/CaO can be attributed to the fact that in binders with low CaO content, carbonation is determined to a large extent by the amount of alkali ions present in the pore solution [5, 10]. This is supported by a plot of k_{nat} *versus* $w/[CaO + (Na,K)_2O]$ (Fig. 4). In this plot, a correlation between the two parameters is evident for all concretes combined, though the scatter of the data is considerable. In particular, there is significant deviation between the MK-based and FA-based AAC with comparatively high k_{nat}, indicating that additional parameters substantially influence the carbonation resistance of these concretes.

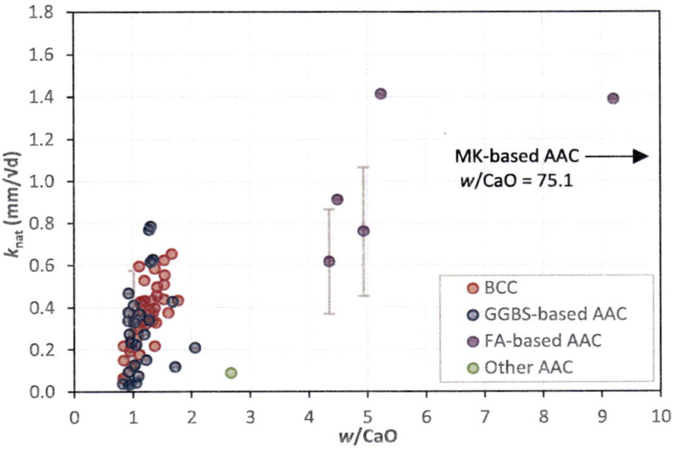

Fig. 1. Carbonation coefficients obtained under conditions approximating natural carbonation *versus w*/CaO ratio. Error bars represent the estimated standard deviations of the results obtained by RILEM TC 247-DTA.

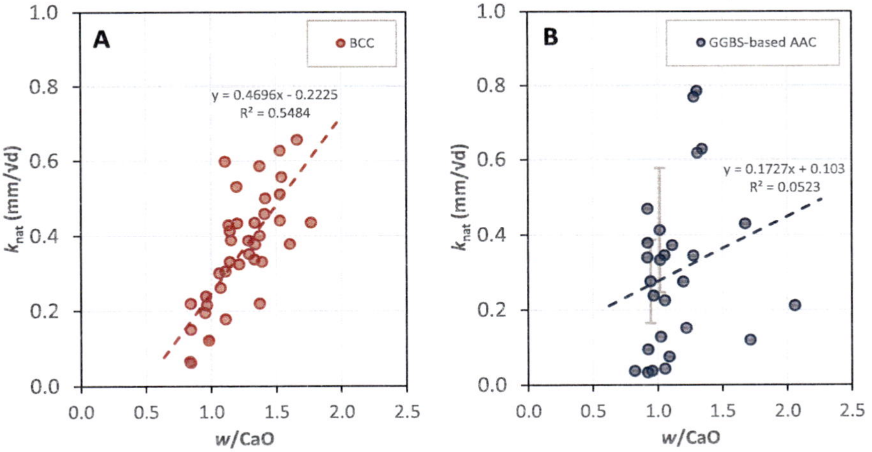

Fig. 2. Carbonation coefficients obtained under conditions approximating natural carbonation *versus w*/CaO ratio (**A**, BCC; **B**, GGBS-based AAC). Error bars represent the estimated standard deviations of the results obtained by RILEM TC 247-DTA.

3.2 Accelerated Carbonation

Accelerated carbonation testing involves exposing concrete or mortar samples to CO_2 concentrations (c_{CO2}) higher than the natural CO_2 concentration present in air (~0.04% CO_2). Assuming that the square root-of-time law holds for estimating the depth of carbonation, the relationship between the carbonation coefficients obtained with accelerated

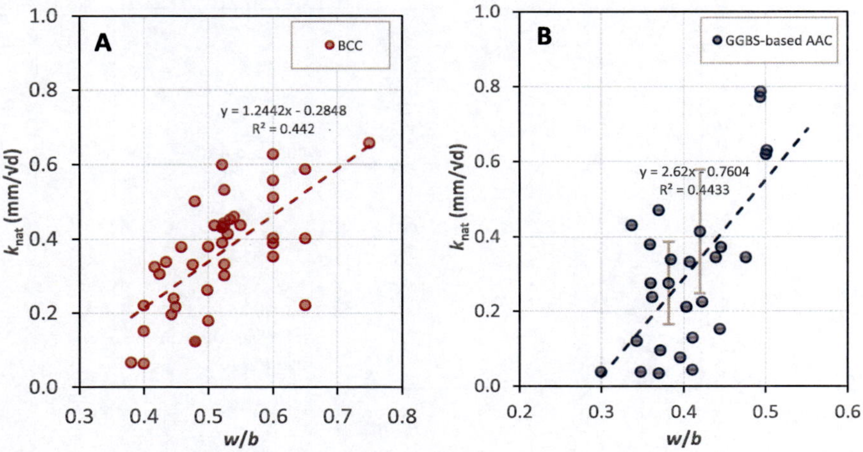

Fig. 3. Carbonation coefficients obtained under conditions approximating natural carbonation *versus* w/b ratio (**A**, BCC; **B**, GGBS-based AAC). Error bars represent the estimated standard deviations of the results obtained by RILEM TC 247-DTA.

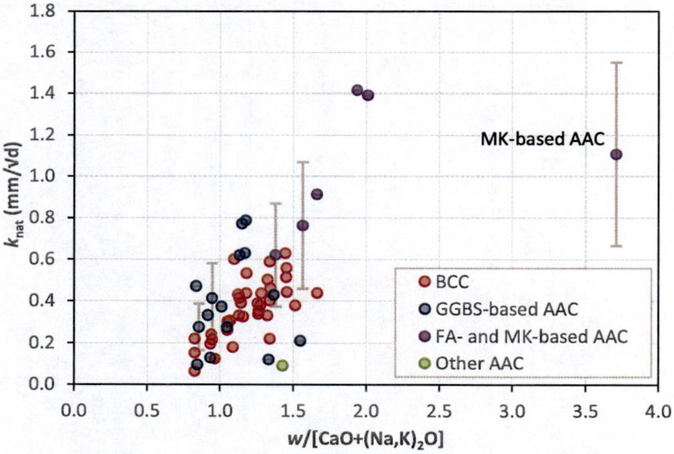

Fig. 4. Carbonation coefficients obtained under conditions approximating natural carbonation *versus* $w/[CaO + (Na,K)_2O]$ ratio. Error bars represent the estimated standard deviations of the results obtained by RILEM TC 247-DTA.

carbonation and natural carbonation can be derived as [11, 12]:

$$k_{acc}/k_{nat} = \sqrt{}(c_{CO2,acc}/c_{CO2,nat}) \quad (1)$$

For example, inserting 1% for $c_{CO2,acc}$ and 0.04% for $c_{CO2,nat}$ yields $k_{acc}[1\%]/k_{nat} = \sqrt{(1/0.04)} = 5$. There are some experimental data supporting at least the approximate validity of Eq. (1) for BCC [6, 12], though the deviations between computed and measured carbonation coefficients can be significant [1, 4].

Figure 5 shows that for BCC and GGBS-based AAC the relationship between k_{acc} and k_{nat} appears to generally conform to Eq. (1), while FA- and MK-based AAC yield considerably lower k_{acc} than what would be expected from Eq. (1). The finding that the behaviour of the FA- and MK-based AAC deviates significantly from the expected relationship between k_{acc} and k_{nat} has important implications. Notably, it indicates that the outcomes of accelerated carbonation testing of low-Ca alkali-activated materials (even at $c_{CO2} = 1\%$) lead to inaccurate predictions of their carbonation resistance under natural exposure. However, the number of available studies of these materials that include accelerated and natural carbonation testing results for the same materials is currently very limited; thus, additional data is required to verify the above conclusion. It is also noted that the testing programme of RILEM TC 247-DTA [8] gave similar ranking of a diverse set of AAC in natural and accelerated carbonation testing, indicating the usefulness of accelerated testing at least for this purpose.

The present analysis further indicates that even for GGBS-based AAC (*i.e.* high-Ca alkali-activated materials) and—in line with a previous analysis of RILEM TC 281-CCC [4]—for BCC the informative value of accelerated carbonation testing is limited. For example, no relationship between k_{acc} and *w*/CaO is apparent for CO_2 concentrations of 1% and 3–5% (Fig. 6). When k_{acc} is compared with *w*/[CaO + $(Na,K)_2O$] (Fig. 7), a correlation may possibly exist for $c_{CO2} = 1\%$ for the BCC, the GGBS—and the FA-based AAC, but for $c_{CO2} = 3–5\%$ no relationship can be discerned. Consideration of the MgO content of the binders did not lead to a significant improvement, *i.e.* did not enable the observation of any clear relationship. It is also possible that this relates to deviations from the assumptions that underpin the use of the square root-of-time relationship to obtain a characteristic coefficient to describe carbonation kinetics for materials that differ significantly from plain Portland cement; this merits further investigation.

4 Concluding Remarks

The reported data for carbonation resistance of alkali-activated concretes and mortars are very unevenly distributed: While a comparatively high number of studies concerning GGBS-based AAC could be found, only a small number of studies measuring the carbonation performance of FA– and MK-based AAC appear to exist. Slightly more data concerning "other" AAC, *i.e.* materials with binders based on municipal solid waste incineration bottom ash, non-ferrous slag or natural pozzolans, are available, but the majority of these were obtained under highly accelerated carbonation conditions, e.g. with $c_{CO2} \geq 10\%$; thus, these results might not be directly comparable with those obtained under lower CO_2 concentrations.

An equally important issue is the comparison of results of accelerated carbonation conditions with results of conditions approximating natural exposure: While several studies have investigated this issue for Portland cement concretes and BCC, there is only limited data available in this regard for GGBS-based AAC, and only two major studies [8, 13] were found for FA–and MK-based AAC.

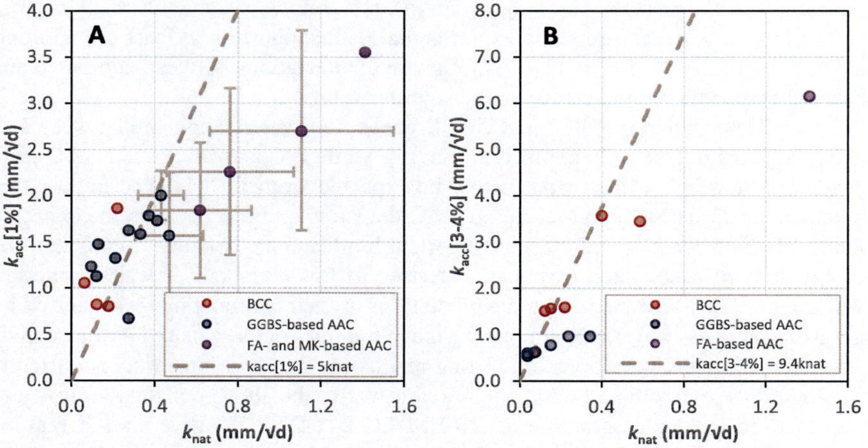

Fig. 5. Carbonation coefficients obtained under accelerated conditions (**A**, $c_{CO2} = 1\%$; **B**, $c_{CO2} = 3$–4%) *versus* the carbonation coefficients obtained under conditions approximating natural exposure. Error bars represent the estimated standard deviations of the results obtained by RILEM TC 247-DTA.

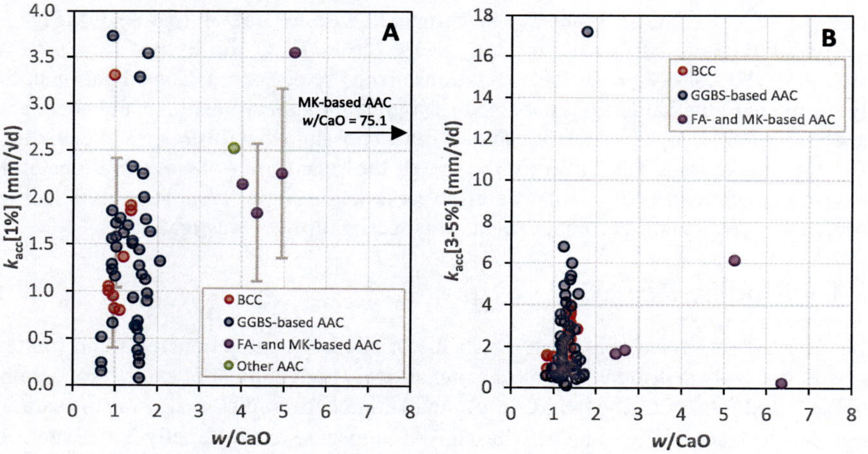

Fig. 6. Carbonation coefficients obtained under accelerated conditions (**A**, $c_{CO2} = 1\%$; **B**, $c_{CO2} = 3$–5%) *versus w*/CaO ratio. Error bars represent the estimated standard deviations of the results obtained by RILEM TC 247-DTA.

Thus, to better understand the carbonation of alkali-activated materials as a class of cements, *i.e.* over their complete range of chemistries, and to reliably apply accelerated carbonation testing methods to these materials, it will be necessary to obtain more data on the carbonation resistance of low-Ca alkali-activated concretes and mortars with a focus on the comparison between accelerated conditions and conditions approximating natural

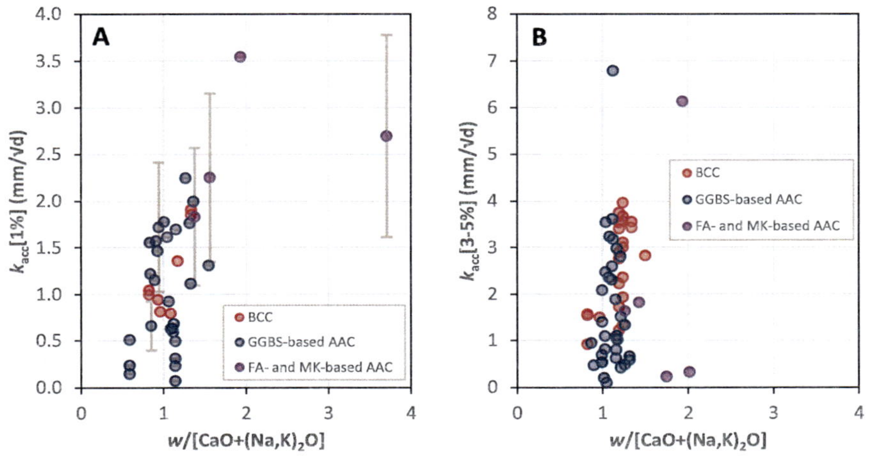

Fig. 7. Carbonation coefficients obtained under accelerated conditions (**A**, $c_{CO2} = 1\%$; **B**, $c_{CO2} = 3\text{–}5\%$) *versus* w/[CaO + (Na,K)$_2$O] ratio. Error bars represent the estimated standard deviations of the results obtained by RILEM TC 247-DTA.

exposure. Though not touched by the present analysis, it appears also necessary to focus on the effects of different relative humidities in the laboratory and under unsheltered natural conditions (*cf*. Ref. [1]).

As regards the present analysis, with the limitations as described, two findings are particularly significant:

(1) The carbonation coefficient under conditions approximating natural exposure appears to be related to the binder w/[CaO + (Na,K)$_2$O] ratio, but not the w/CaO ratio, for GGBS- and FA-based AAC. However, the scatter of the data is considerable; thus, the former ratio may serve at best as a rough indicator of the carbonation resistance.

(2) For FA- and MK-based AAC (low-Ca alkali-activated materials), accelerated carbonation testing leads to considerably lower carbonation coefficients than what would be expected from the square root-of-time law; thus, the application of that law to predict the carbonation resistance of these concretes under natural conditions may be inaccurate. Due to the limitations of the present database, however, additional work is required to verify this conclusion.

Acknowledgements. This article was prepared within the framework of RILEM TC 281-CCC. The article has been reviewed and approved by all members of the TC. Lia Weiler (RWTH Aachen University) is thanked very much for assistance in extending and reviewing the database. Participation of S. A. Bernal in this study was sponsored by the Engineering and Physical Sciences Research Council (EPSRC) through the ECF EP/R001642/1.

References

1. Leemann, A., Nygaard, P., Kaufmann, J., Loser, R.: Relation between carbonation resistance, mix design and exposure of mortar and concrete. Cem. Concr. Compos. **62**, 33–43 (2015)
2. Leemann, A., Moro, F.: Carbonation of concrete: The role of CO_2 concentration, relative humidity and CO_2 buffer capacity. Mater. Struct. **50**, 30 (2017)
3. von Greve-Dierfeld, S., Gehlen, C.: Performance-based durability design, carbonation part 2—classification of concrete. Struct. Concr. **17**, 523–532 (2016)
4. von Greve-Dierfeld, S., et al.: Understanding the carbonation of concrete with supplementary cementitious materials: a critical review by RILEM TC 281-CCC. Mater. Struct. **53**(6), Article number: 136 (2020). https://doi.org/10.1617/s11527-020-01558-w
5. Bernal, S.A., Provis, J.L., Brice, D.G., Kilcullen, A., Duxson, P., van Deventer, J.S.J.: Accelerated carbonation testing of alkali-activated binders significantly underestimates service life: The role of pore solution chemistry. Cem. Concr. Res. **42**, 1317–1326 (2012)
6. Leemann, A., Pahlke, H., Loser, R., Winnefeld, F.: Carbonation resistance of mortar produced with alternative cements. Mater. Struct. **51**(5), Article number: 114 (2018). https://doi.org/10.1617/s11527-018-1239-3
7. Bernal, S.A., Provis, J.L.: Durability of alkali-activated materials: Progress and perspectives. J. Am. Ceram. Soc. **97**, 997–1008 (2014)
8. Gluth, G.J.G., et al.: RILEM TC 247-DTA round robin test: Carbonation and chloride penetration testing of alkali-activated concretes. Mater. Struct. **53**(1), Article number: 21 (2020). https://doi.org/10.1617/s11527-020-1449-3
9. Herterich, J., Black, L., Richardson, I.: Microstructure and phase assemblage of low-clinker cements during early stages of carbonation. In: 14th International Congress on the Chemistry of Cement (2015)
10. Pouhet, R., Cyr, M.: Carbonation in the pore solution of metakaolin-based geopolymer. Cem. Concr. Res. **88**, 227–235 (2016)
11. Papadakis, V.G., Vayenas, C.G., Fardis, M.N.: Fundamental modelling and experimental investigation of concrete carbonation. ACI Mater. J. **88**, 363–373 (1991)
12. Sisomphon, K., Franke, L.: Carbonation rates of concretes containing high volume of pozzolanic materials. Cem. Concr. Res. **37**, 1647–1653 (2007)
13. Khan, M.S.H., Castel, A., Noushini, A.: Carbonation of a low-calcium fly ash geopolymer concrete. Mag. Concr. Res. **69**, 24–34 (2017)

Alkali Binder Based on Waste Glass and Limestone

U. Avila-López[1](✉) and J. I. Escalante-García[2]

[1] Universidad Autónoma de Coahuila-Facultad de Ingeniería, Blvd. Fundadores Km. 13, Ciudad Universitaria, 25354 Coahuila, Arteaga, México

[2] Centro de Investigación y de Estudios Avanzados del IPN, Unidad Saltillo, Av. Industria Metalúrgica, 1062, Ramos Arizpe Coahuila, México

Abstract. Investigation was carried out about to characterize the reaction products and strength of pastes of alkaline binders based on urban waste glass and limestone activated at 20, 40 and 60 °C with 6, 9 and 12% Na_2O using NaOH, Na_2CO_3 and a mixture of these. The waste glass and limestone were ground to pass through the #50 (150μm) and #325 (45μm) sieves. The investigation was developed in two stages. First, the Taguchi Method, which is a fractional experimental design, was used to determine the conditions that promote the greatest compression strength. Second, pastes were prepared using a factorial experimental design to be able to evaluate and compare the effect of activators on compressive strength and microstructure. The results showed that the use of Na_2CO_3 promotes the highest compressive strength (71 MPa), but the structural stability decreases after 14 days of curing at temperatures greater than 40 °C. The use of the Na_2CO_3/NaOH mixture was considered optimal, according to Taguchi statistical method, the compressive strength at 28 days of curing was of 64.8 MPa ± 15 MPa, maintaining structural stability in late curing ages. The pastes prepared with NaOH developed the least strength at early ages, but in later stages it can promote compressive strength similar or greater than those developed by the pastes prepared with the other activators used (59 MPa, 28 days). The characterization carried out by XRD, SEM and TGA-DSC indicates that the reaction products formed were pirssonite, gaylussite, recrystallized calcium carbonate and C-S-H intimately intermingle with silica gel.

Keywords: Waste glass · Limestone · Taguchi method · Alkaline activation · Sixth keyword

1 Introduction

The different productive activities carried out by societies inevitably generate a series of solid, liquid and gaseous wastes that can have negative effects on the environment and human health. In México (2017) 52% of the municipal solid waste (MSW) was organic and the remainder was inorganic waste, such as paper and cardboard (14%), plastic (11%), glass (6%), metal (4%), textile (1%) and other types of waste (12%) [1]. Municipal solid waste glass is an inert material that does not harm nature. However, it

J. I. Escalante-Garcia et al. (Eds.): RW 2021, RILEM Bookseries 40, pp. 677–686, 2023.
https://doi.org/10.1007/978-3-031-21735-7_73

occupies a larger volume that other types of waste in landfills, and its recycling by the glass industry is made hard by various factors. Nevertheless, vitreous MSW can become an area opportunity, allowing its reuse in the form of useful cements for construction industry [2]. Common glass is essentially made of silica and is susceptible to activation by alkaline compounds, and depending on the availability of calcium, it can form silica gel or C-S-H [3–5]. To compensate for the CaO deficiency in the glass and promote the formation of C-S-H, it is necessary to combine the glass with a source rich in calcium, such as limestone. Limestone is abundant in earth´s crust and represents an interesting option as a raw material in alternative binders. Based on the above, this work presents results of the study of the effects of alkali activation on the reaction products, microstructure and mechanical properties of alkali binders based on waste glass and limestone.

2 Experimental Procedure

2.1 Raw Materials

Waste limestone (WLS) from local quarry and ground commercially available limestone (CLS) were used on the alkaline binders; these were milled to pass the sieves #50 (S50) and #325 (S325). The urban waste glass (UWG) was obtained from local urban wastes. The main chemically components of raw materials are given in Table 1. The alkaline activators were NaOH, Na_2CO_3 and Na_2CO_3/NaOH solutions.

Table 1. Chemical composition (%wt) of waste glass and limestone by X-ray florescence.

Compound	SiO_2	CaO	$CaCO_3$	Na_2O	Al_2O_3
Glass (UWG)	70.34	12.76		13.58	1.52
Limestone (CLS)	1.44		97.29	0.132	0.318
Waste limestone (WLS)	1.16		97.61		0.359

2.2 First Stage: Taguchi Method

The Taguchi method allow optimize the experimental work, by reducing the numbers of trials. This method uses orthogonal arrays (OA) with factors at different levels randomly combined. The selection of inners factors, indicated in Table 2, was reported in a previous contribution [6]; additionally, the study of three outer factors was proposed. The inner factors were studied with an inner OA L_{15} (15 trials), and outer factors were studied with an outer OA L_4. The evaluation criteria of bigger is better was applied to analyze the compressive strength at 28 days of curing.

Table 2. Factors and levels for the experimental design. PA: Chemical preactivation process

Inner factors		Level 1	Level 2	Level 3	Level 4	Level 5
A	CaO/SiO_2 ratio	0	0.5	1	1.5	2
B	% with PA*	0	30	40		
C	PA order	Limestone	Glass	Mixture		
D	Alkaline activator	NaOH	Na_2CO_3/NaOH	Na_2CO_3		
E	% Na_2O	6	9	12		
F	Water/Binder ratio (W/B)	0.3	0.4	0.5		
G	Curing temperature	20 °C	40 °C	60 °C		
Outer factors						
H	Limestone (LS)	S50	S325			
I	Urban waste glass	S325	S325			
J	Limestone type	CLS	WLS			

* PA (Preactivation): means that a given fraction of the raw material (% with PA) was activated by simple mixing with the whole activating solution; then the rest of the raw materials were added.

2.3 Second Stage: Factorial Experimental Design

The Taguchi method randomly proposes the values for each factor to be studied, therefore, comparing the effect of the activating agent between different trials is hard. From the trials proposed by the Taguchi method, those that developed the greatest compressive strength at 28 days of curing were selected for each alkaline activator. For each selected formulation, each alkali activator was used to evaluate the effect of the activators on the mechanical properties at the same conditions, the details are shown in Table 3. For comparative purposes, Portland cement pastes were prepared for each curing temperature used.

2.4 Sample Preparation

The pastes were prepared using a blender with planetary movement for 5 min. On the trials with preactivation, a given fraction (%PA) of the raw powder materials was mixed for 5 min with the whole of the activating solution; the rest of material were added later and mixed by 3 more min. The pastes were cast into plastic molds and vibrated. After 24h of curing at the specified experimental conditions by inner and outer arrays, the specimens were demolded and dried store at 20, 40 or 60 °C in plastic containers in isothermal chambers.

Table 3. Formulation of selected trials to analyze the effect of alkali activator.

Formulation	CaO/SiO_2	%PAQ	Or PAQ	$\%Na_2O$	A/C	Temperature	Alkali activator
Formulation trial 2	0.5	0	LS	12	0.3	40 °C	Na_2CO_3
							Na_2CO_3/NaOH
							NaOH
Formulation trial 12	0.5	40	UWG	9	0.3	40 °C	Na_2CO_3
							Na_2CO_3/NaOH
							NaOH
Formulation trial 4	1.5	40	UWG	9	0.3	60 °C	Na_2CO_3
							Na_2CO_3/NaOH
							NaOH
Portland cement	CPC 30R				0.4	40 °C	
	CPC 30R				0.4	60 °C	

2.5 Characterization

The 28 days compressive strength results were fed to the calculations of the Taguchi methodology. Fragmented pieces of crushed cubes were dried in vacuum oven at 35 °C, which were grounded in planetary mill to pass a sieve of 150 μm. The powders were characterized by X-ray diffraction (XRD, X'Pert 3040, Phillips) using Cu Kα radiation at 40 kV and 30 mA; as well as by thermal analysis (TG-DSC, STA PT 1600, Linseis) heating up to 1100 °C at 10 °C/min. Fragments of selected samples were mounted in epoxy resin, ground using SiC sand paper and the polished using diamond pastes from 3 to 1.25μm; these were further coated with carbon for observation under the scanning electron microscopy (SEM, XL-30 ESEM, Philips).

3 Results

3.1 Compressive Strength

The Taguchi methodology uses the signal to noise ratio (S/N), which is a performance characteristic. There are three categories of performance characteristics: larger is better, smaller is better and nominal is better. In this research, the quality characteristic of bigger is better was applied using Eq. (1) [7] to maximize the strength.

$$(S/N) = -10\log_{10}\left[\frac{1}{n}\sum_{j=r}^{n}\frac{1}{Y_{ij}^{2}}\right] \tag{1}$$

where: *S/N* are performance statistics, *n* is the number of repetitions for an experimental combination, and Y_i is the result of the *i*th experiment.

The levels with the highest S/N ratio value for each factor were chosen as the optimal conditions that maximize the development of compressive strength at 28 days. The optimal conditions were: CaO/SiO_2 ratio = 0.5, % with PA = 40, PA order = UWG, Alkaline activator = $Na_2CO_3/NaOH$, $\%Na_2O = 9$, W/B = 0.3 and curing temperature = 40 °C. The complete statistical analysis of the compressive strength using the Taguchi methodology will be reported in a future contribution.

The average compressive results from each trail are given in Table 4. The compressive strength results showed that only three trials exceeded 20 MPa, which are trial 2 (36 MPa), trial 12 (40 MPa) and trial 14 (24 MPa), in which a different alkali activator was used. However, because the formulation of each of these is different, it was not possible to define which alkaline activators promotes the greatest development of compressive strength. For this reason, the formulations of these trials were selected to prepare the pastes again but using the three alkaline activators proposed in this study.

Table 4. Average compressive strength results.

Inner array						Outer array					
						J	LS type	CLS	WLS	WLS	CLS
						I	UWG	S50	S325	S50	S325
						H	LS	S50	S50	S325	S325
						No.	Repetition	1	2	3	4
Trial no.	CaO/SiO_2	% PA	Order PA	Alkaline activator	% Na_2O	W/B	°T	*Average compressive strength (MPa)*			
1	0	0	LS	NaOH	6	0.3	20	2.1			
2	*0.5*	*0*	*LS*	*Na_2CO_3*	*12*	*0.4*	*40*	*36.1*			
3	1	0	LS	NaOH	12	0.5	60	13.3			
4	1.5	0	LS-UWG	Na_2CO_3	6	0.3	60	9.5			
5	2	0	LS-UWG	NaOH	6	0.5	20	0.6			
6	0	30	LS-UWG	Na_2CO_3	9	0.5	40	19			
7	0.5	30	LS-UWG	$Na_2CO_3/NaOH$	12	0.3	20	5			
8	1	30	LS-UWG	NaOH	12	0.4	60	19.3			
9	1.5	30	LS	Na_2CO_3	6	0.4	60	3.6			
10	2	30	UWG	Na_2CO_3	12	0.4	20	0.2			
11	0	40	UWG	$Na_2CO_3/NaOH$	6	0.5	40	15.7			
12	*0.5*	*40*	*UWG*	*$Na_2CO_3/NaOH$*	*9*	*0.3*	*40*	*40*			
13	1	40	UWG	$Na_2CO_3/NaOH$	9	0.5	40	4.3			
14	*1.5*	*40*	*UWG*	*NaOH*	*9*	*0.3*	*60*	*24.1*			
15	2	40	LS	$Na_2CO_3/NaOH$	9	0.4	20	1.1			

Figure 1 shows the mechanical behavior of the pastes prepared with the formulation of the selected trials of the first stage. The mechanical behavior of pastes prepared with Portland cement were included as a reference. It was observed that most of the pastes-based waste glass-limestone developed a higher compressive strength than Portland cement pastes at 28 and 90 days of curing. The pastes activated with NaOH showed the best mechanical performance after 14 days of curing (strength higher than 40 MPa). The pastes activated with Na_2CO_3 and Na_2CO_3/NaOH showed a good initial mechanical development; however, after 14 days of curing they showed a decrease in compressive strength. The paste that showed the greatest decrease in compressive strength was that activated with Na_2CO_3 (decrease of 38.5 MPa), from the formulation of trial 14. The results suggest that the presence of Na_2CO_3 in the alkaline activator can be harmful to the structural stability of cementitious pastes based on waste glass-limestone.

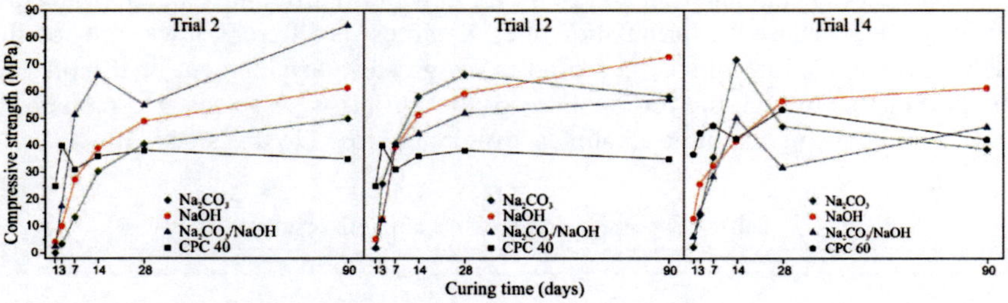

Fig. 1. Compressive strength vs time for pastes to each formulation of second stage.

3.2 X Ray Diffraction (XRD)

The XRD results for the first and second stages for each alkali activator were similar, for this reason, it was not considered necessary to include the results of each stage. Figure 2 shows the diffraction patterns of paste activated with Na_2CO_3 from 1 to 28 days of curing of the trial 14. This cementitious system was selected because it presented the greatest decrease in compressive strength after 14 days of curing, which suggested that in this cementitious system the possible cause of the mechanical deterioration presented by various systems could be more easily identified. Data from limestone and glass are given as reference.

The diffractograms of pastes showed that the gaylussite ($Na_2Ca(CO_3)_2 \cdot 5H_2O$) and pirssonite ($Na_2Ca(CO_3)_2 \cdot 2H_2O$) phases formed in the first 24 h of curing [8], and later to that age there was no significant increase in the formation of these phases; which suggest that these phases cannot be the cause of the deterioration of the mechanical properties. On the other hand, the main reflection of the calcite at 29.9° 2θ showed a decrease in its intensity from 1 to 14 days; however, at later age this reflection showed an increase in intensity. This suggest that after 14 days of curing there was a recarbonation reaction of the Ca ions, which would be giving rise to the formation of calcite as a reaction product. Therefore, it is assumed that calcium recarbonation could be a cause of the decrease in mechanical properties.

3.3 Scanning Electron Microscopy (SEM)

Figure 3 shows the microstructure of the pastes of formulation activated with NaOH and Na_2CO_3. The microstructure of paste activated with Na_2CO_3/NaOH was as the

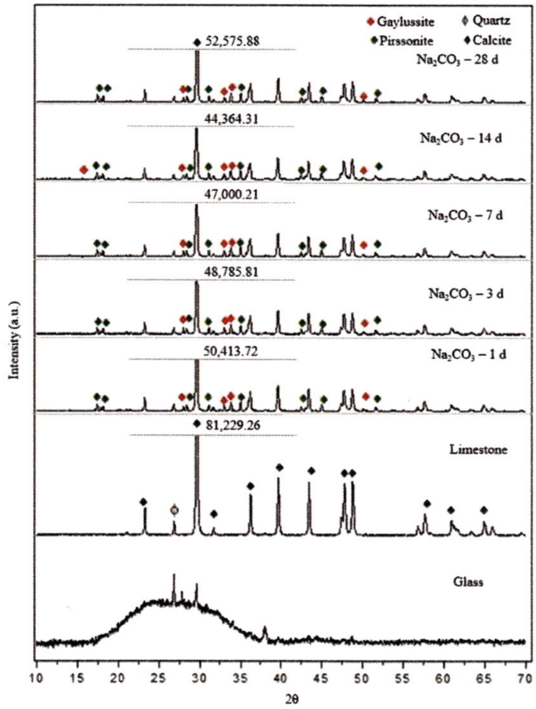

Fig. 2. XRD patterns of pastes with formulation of trial 14 activated with Na_2CO_3.

microstructure of paste activated with Na_2CO_3. The microstructure at 28 days of paste activated with NaOH, Fig. 3, showed a heterogenous morphology with low microporosity and unreacted particles of limestone (LS) and waste glass (UWG); most of the UWG particles showed a rim of reaction products.

Figure 3b and c shows the microstructure of the paste activated with Na_2CO_3 at 14 and 28 days of age. From 14 day of curing two main zones were noted. EDS analysis indicates that the zone labelled 1 is rich in Ca, suggesting the presence of pirssonite and gaylussite finely intermixed. The zone labelled 2 was rich in Si, suggesting that it is composed mainly of silica gel finely intermixed with C-S-H. At 28 days the unreacted particles were reduced, mainly the UWG particles; the zone 2 densified respect to 14 days conforming the bulk of the microstructure, indicating the progressive reaction of the UWG and the diffusion and precipitation of reaction products. The areas of zone 1 decreased at 28 days, which suggest that if Ca participated in subsequent reactions, it was to form C-S-H and to recrystallize in $CaCO_3$, according to the results obtained by XRD.

Fig. 3. Microstructures of the pastes of formulation 14 activated with NaOH at 28 curing day **a** and Na_2CO_3 at 14 **b** and 28 **c** curing days.

3.4 Thermal Analysis

The results of DTA of the paste activated with Na_2CO_3 of trial 14 at 14 and 28 curing days are present in Fig. 4, thermograms of limestone and waste glass are given as reference. The UWG showed an exothermic peak at around 700 °C (Tg), associated to the glass transition temperature (Tg) [9]; the LS showed an endothermic peak at about 750 °C, which correspond to the decomposition of $CaCO_3$ into CaO and CO_2. The thermograms of paste showed similar features at 28 and 90 days. The thermograms of paste showed peaks about 75–95 °C, which were associated to the decomposition of C-S-H [10, 11], suggesting its formation. Additionally, peaks around 125 °C and 175 °C from the dehydration of gaylussite and pirssonite, respectively. In the 28-day thermogram, the gaylussite peak was not observed because of the overlapping with C-S-H peak. The peaks at about 675 °C and 700 °C were associated with the Tg and the decomposition and melting of the carbonates. Thermal analysis confirmed the C-S-H formation in the prepared pastes.

4 Conclusions

The use of Na_2CO_3 as alkaline activator promotes compressive strength of 40–70 MPa at 7 and 14 curing days. However, at later ages the use of this alkaline activator compromises the structural stability of the pastes, and the compressive strength decreases due to the possible recrystallization of calcium ions in calcite. The use of NaOH as an activator

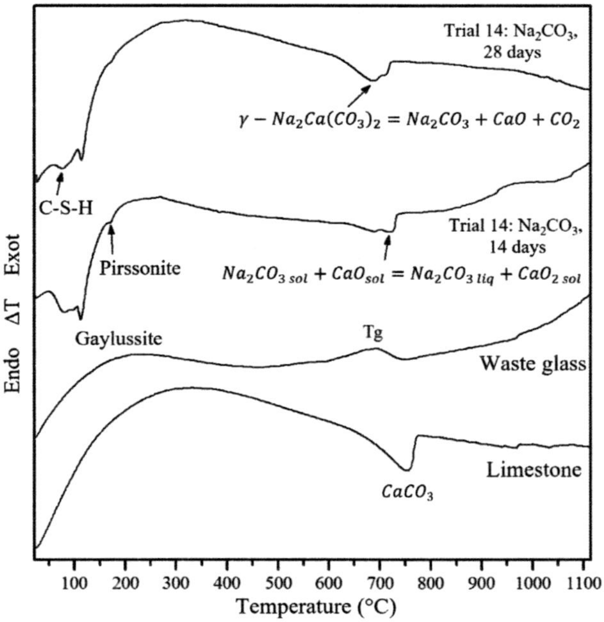

Fig. 4. Differential thermal analysis of trial 14 activated with Na_2CO_3.

promotes values of compressive strength of 40 MPa at 14 days, and the pastes prepared with this activator did not show any negative effect on late ages.

The main reaction products were C-S-H and silica gel type, and as crystalline phases were gaylussite ($Na_2Ca(CO_3)_2 \cdot 5H_2O$) and pirssonite ($Na_2Ca(CO_3)_2 \cdot 2H_2O$).

The compressive strength of most pastes prepared based waste glass-limestone was higher than PC, making this binder a promising options as alternative cement.

References

1. Semarnat.: Informe de la situación del medio ambiente en México, edición 2018, Semarnat, Mexico (2019)
2. Puertas M.F., Torres C.J.J., Varga F.C., Torres M.: Procedimiento para la fabricación de cementos alcalinos a partir de residuos vítreos urbanos industriales. OMPI, Patente WO 2012/172138 A1 (2012)
3. Bunker, B.C., Arnold, G.W., Beauchamp, E.K.: Mechanisms of alkali leaching in mixed Na-K glasses. J. Non-Cryst. Solids **58**, 295–322 (1983)
4. Escalante-García, J.I., Palacios-Villanueva, V.M., Gorokhovsky, A.V., Mendoza-Suarez, G., Fuentes, A.F.: Characteristics of a NaOH activated blast furnace slag blended with a fine particle silica waste. J. Am. Ceram. Soc. **85**(7), 1788–1792 (2002)
5. Vega-Cordero E, Gómez-Zamorano LY, Escalante-García JI.: Synthesis of metakaolin-based geopolymers and the effect of the incorporation of geothermal silica waste. In: XIII International Congress on the Chemistry of Cement, p. 7. Madrid (2011)

6. Avila-López, U., Almanza-Robles, J.M., Escalante-García, J.I.: Investigation of novel waste glass and limestone binders using statistical method. Constr. Build. Mater. **82**, 296–303 (2015)
7. Taguchi, G., Chowdhury, S., Wu, Y.: Taguchi´s Quality Engineering Handbook, 1st edn. Wiley, New Jersey (2004)
8. Heinz, K.H.: Crystal Grown in Gels, 1st edn. Dover Publications, New York (1970)
9. Ghasemzadeh, M., Nemati, A., Nozad, A., Hamnabard, Z., Baghshahi, S.: Crystallization kinetics of glass ceramics by differential thermal analysis. Ceram Silik **55**(2), 188–194 (2011)
10. Ben, H.M., Le, S.G., Winnefeld, F., Lothenbach, B.: Influence of activator type on hydration kinetics, hydrate assemblage and microstructural development of alkali activated blast-furnace slags. Cem. Concr. Res. **41**, 301–310 (2011)
11. Wang, S.D., Scrievener, L.K.: Hydration products of alkali activated slag cement. Cem. Concr. Res. **25**, 561–571 (1995)

Cultural Heritage

Materials Characterization and Compressive Strength of Compressed Earth Blocks Reinforced with Oat Straw

Sulpicio Sánchez Tizapa[1](✉), Gerardo Altamirano de la Cruz[2], Alfredo Cuevas Sandoval[1], Adelfo Morales Lozano[1], Ángel Moreno Dimas[1], and Luis Alberto Patiño[1]

[1] Universidad Autónoma de Guerrero, 39087 Chilpancingo, Gro, Mexico
sstizapa@uagro.mx

[2] Universidad Autónoma de Nuevo León, 66450 Monterrey, NL, Mexico

Abstract. This paper describes the material characterization, compressive strength, and chemical composition of two compressed earth blocks (CEBs) stabilized with oat straw. The properties of the oat straw and soils used for the improvement of the CEBs were evaluated. The compressive strength was assessed by randomly choosing and testing 26 pieces. In addition, the chemical composition was obtained. As a result, the unit density of the oat straw is within the top limit of the experimental referenced values; however, the absorption records the highest value, i.e., one outside of the experimental referenced range. The soil properties are within the ranges defined by international standards. The compressive strength is adequate for both pieces (2.9 MPa for CEB A and 4.7 MPa for CEB B), but the initial absorption is high. Silica and calcium oxide comprise 90% of the chemical composition of the two pieces. Considering the high absorption of these CEBs, their colocation is not recommendable in the lower parts of walls. This is to avoid standing water in the foundation soil. The highest compressive strength of CEB B is related to the plastic index and chemical composition.

Keywords: Compression · Soil · Oat straw · Earth block

1 Introduction

Earth-building materials have been used for centuries worldwide in housing construction. Overall, 30% of the world population lives in these types of structures, with the percentage rising to 50% in developing countries [1]. In Mexico, adobe houses represent 5.36% of total housing [2]. The advantages of these types of houses over others are their accessibility, utilization of recycling, thermal-acoustic insulation, fire resistance, and low environmental cost.

Elaboration of the compressed earth blocks (CEBs) began in the 19th century, with the use of mechanical presses to pressure raw materials to improve their mechanical behaviors [3, 4]. Adding stabilized lime or reinforcing soils with natural fibers are two common processes for improving the properties of CEBs.

J. I. Escalante-Garcia et al. (Eds.): RW 2021, RILEM Bookseries 40, pp. 689–698, 2023.
https://doi.org/10.1007/978-3-031-21735-7_74

A high compressive strength guarantees an excellent structural behavior in adobes or CEBs. For reference, artisanal adobes reach an average compressive strength value of 1.5 MPa in Guerrero state (Mexico) [5], and reinforced fiber CEBs reach within interval from 1.7 MPa to 8.5 MPa [1, 6]. The applied pressure for improving the properties of the CEBs is between 2.0 MPa and 10 MPa [7, 8].

The goal of this experimental research was to evaluate the mechanical-physical properties and chemical composition of two CEBs, as well as the physical properties of their components (oat straw and soil). A part of the reported work is described in [9].

2 Methodology

2.1 Experimental Program and Standards

The experimental research was developed in three phases. In the first phase, the soil and oat straw were tested; in the second phase, two types of CEBs were studied; and in the third phase, the discussion and conclusions were presented.

The choices of the materials, dosages, and production were out of the control of this study, because the pieces were provided by the only local producer of CEBs. According to the raw materials, i.e., reinforced soil with natural fibers without lime, the New Mexico standard applied for the compressive strength tests of the CEBs [10], and the Mexican standard [11, 12] defines the absorption and unit weight tests.

The soil tests were performed according to the Mexican standard [13–16]. For the fiber properties (absorption and unit density), the Rilem recommendations were used [17]. Finally, the values of the fiber and soil parameters were compared with those in international standards and references [1, 6–8, 18, 19, 22, 23].

2.2 Production of Compressed Earth Blocks (CEBs)

The local manufacturer used empirical dosages in the production (Table 1). For CEB A, the mix was a combination of 30% reddish sandy clay and 70% brown clayey sand, whereas for CEB B, the soil was yellowish sandy clay. The reinforcing natural fiber was oat straw with lengths between 30 and 50 mm. The production process included: a) extracting the soils from the bank, b) selecting a 3/8″ (9.52 mm) mesh passing material, c) for CEB A, providing the combination of 30% reddish sandy clay and 70% brown clayey sand, d) adding the water and oat straw fiber and mixing the materials, e) improving the CEB properties by mold filling and pressure application, f) drying the CEBs in the shade for two weeks, and g) storing the CEBs.

To improve the compressive strength, a compressor applied a pressure of 2.50 MPa to the wet mixture for ten seconds during CEB production. The equipment produced one piece per cycle. The average values in weight of the water-soil and oat straw-soil ratios were 0.09 and 0.007, respectively. Figures 1 and 2 present images of the oat straw, equipment, and a manufactured CEB.

Table 1 Dosage by weight (KN) [9].

Type of compressed earth block (CEB)	Reddish sandy clay	Brown clayey sand	Yellowish sandy clay	Oat straw	Water
CEB A	41.7	92.3		1.0	11.8
CEB B			135.0	1.0	11.8

Fig. 1. Oat straw [9].

Fig. 2. a) Press and compressor system, b) Compressed earth block (CEB) [9].

2.3 Tests in Oat Straw

Absorption and unit density. The reference [17] defines a process for evaluating the water absorption and unit density of hemp shiv. The authors used this recommendation, as there is no corresponding standard for oat straw. Nine samples were evaluated for each test.

2.4 Tests of Soils

Granulometry. The three soils used in the CEB production were analyzed to obtain the particle size distribution. In all cases, altered samples were used, and the fine material was removed by washing. The meshes used were 9.5 mm, 6.3 mm, 4.75 mm, 2.0 mm, 0.850 mm, 0.425 mm, 0.250 mm, 0.150 mm, and 0.075 mm [13]. The test was performed with a mechanical sieve shaker, followed by recording the weights retained in each mesh. According to the specifications [14], gravel was defined as the material retained in mesh # 4, sand was the material passing through mesh # 4 and retained in mesh # 200, and fine material was the material passing through mesh # 200. The maximum aggregate size did not exceed 25.4 mm.

Limits of Atterberg. To calculate the liquid limit of the soil (w_L), a sample was placed in a standardized Casagrande cup and grooved, and the cup was rotated a defined number of strokes. Then, w_L was the water content of the sample for which the groove closed, corresponding to 25 strokes. Thus, the soil shear strength was 2.45 kPa. The plastic limit (w_P) was defined as the water content when a cylindrical soil sample rolled over a glass reached a diameter of 3 mm and cracks appeared, prevented further movement. The plastic index (*PI*) was the difference between the liquid and plastic limits [15]. The soil classification was conducted according to the Unified soil classification system [14].

Unit density (Dss). For calculating the unit density, a sample of soil and distilled water was placed in a long neck flat bottom flask, where the weight plus the water weight at the test temperature was known. The flask was filled to the gauge mark without the presence of air and weighed. Then, the density of the solids could be evaluated [16].

2.5 Tests of CEBs

Physical properties and compressive strength of CEBs. A sample size of 26 pieces (13 per each CEB type) was randomly selected from a production batch of 1,000 pieces (500 blocks per each type) to determine the unit weight, compressive strength, average dimensions, and water content [10, 11]. According to the standard, the pieces must be tested in the same position as they would be in the built walls (Fig. 3). Accordingly, sand and a leather fragment were placed on both sides to guarantee the uniform distribution of the load. The hydraulic press was an ELE model 3600657 with an 800-KN capacity; it had a load application system for ensuring a perpendicular application of the force, with a speed of 0.06 MPa/s. The maximum relationship force to the transverse surface was the compressive strength.

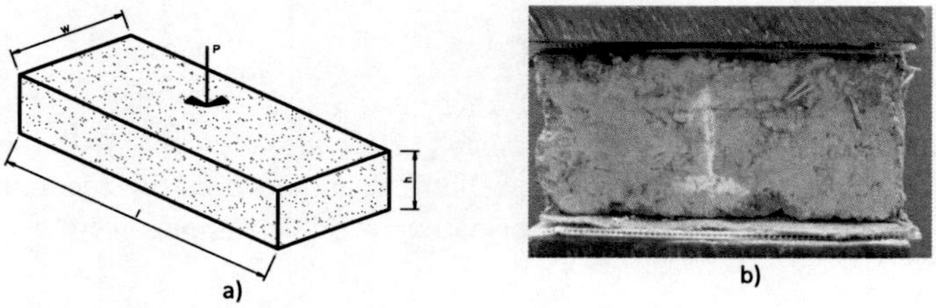

Fig. 3. **a** Specimen size and load application, **b** setup of compression test.

Initial absorption. According to the standard, the initial absorption is the amount of water absorbed by the capillary effect in a piece supported on two metallic profiles inside a container with water added for 10 min. The water does not exceed 5 mm from the bottom surface of the piece. The calculus with the initial and final weights provides the value of this parameter [12].

Mineralogical analysis. X-ray diffraction is a high-tech analytical method that allows for the quick and easy identification of the crystalline phases of materials. For this purpose, a sample is pulverized and placed on a sample holder. The powder generates a high number of small crystals, and each crystal reflects the X-rays if it is properly oriented to the beam of rays. The result is a diffractogram graph with the angle of diffraction in the abscissa and intensity in the ordinates (in photons per second). A database comparison allows for the identification of the minerals in the sample [20]. For this case and to provide another reference, a sample of Portland concrete CPC-30R was also analyzed [24].

3 Results

3.1 Oat Straw and Soils

The average values of the unit density and absorption of the oat straw are 1.55 kg/m^3 and 513.5%, respectively. In the same order, the coefficients of variation are 6.2% and 12.5%, respectively. The distribution of the particle size is shown in Fig. 4, as well as the percentages of gravel, sand, and fine soil (12.18%, 41.4%, and 46.38% for CEB A, respectively; 22.71%, 34.69%, and 42.6% for CEB B, respectively). The classifications of the soils according to the Atterberg limits ($\boldsymbol{w_L}$, $\boldsymbol{w_P}$, $\boldsymbol{PI}$) are shown in Fig. 5. As can be seen, the yellowish sandy clay (CEB B) and reddish sandy clay (RSC) are low plasticity clays (CL), whereas the brown clayey sand (BCS) is classified as a clayey sand of low plasticity (SC). The soil combination of CEB A is a clay of low plasticity (CL). Table 2 shows the values of the Atterberg limits and unit density ($\boldsymbol{Ds}$).

3.2 CEBs

The average dimensions of the length, width, and height are 403.3 × 201.7 × 96.9 mm and 403.2 × 201.0 × 99.5 mm for CEB A and CEB B, respectively. In the same order, the unit weight registered values are 19.3 KN/m^3 and 18.6 KN/m^3 with coefficients of variation (CVs) of 8.0% and 4.5%, respectively. The water contents for CEB A and CEB B are 3.4% and 2.3%, respectively. The length/width ratio is two for both pieces. The average compressive strengths (f_c) are 2.9 MPa and 4.7 MPa for CEB A and CEB B, with CVs of 5.0% and 36.7%, respectively. The standardized compressive strengths are 1.55 MPa and 2.50 MPa, respectively, based on using a value of the CV greater than or equal to 0.35 [21]. This was done in consideration of two factors; a) the artisanal production process lacked quality control, and b) the tested lot was less than 30 pieces. Figure 6 shows the final states after the compression tests of the two specimens.

Regarding the initial absorption, the CEBs presented detachment of the material at the end of the test; therefore, the data were not reliable. As another approach, the chemical composition showed high amounts of silica (SiO_2) and calcium oxide (CaO) and low amounts of alumina (Al_2O_3), magnesium oxide (MgO), and iron oxide (Fe_2O_3), as shown in Fig. 7. The chemical composition of the local cement was also evaluated.

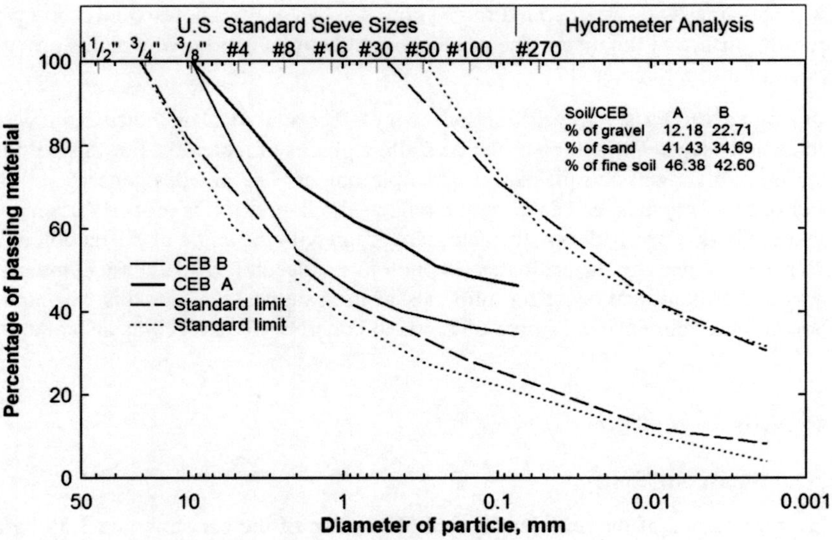

Fig. 4. Particle size curves.

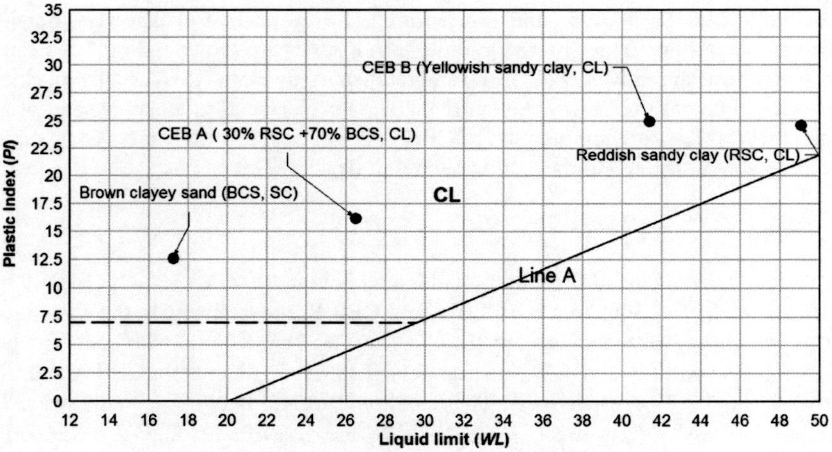

Fig. 5. Classification of soils.

4 Discussions

The unit density is 1.55 kg/m^3, i.e., inside of the experimental referenced range (0.56–1.67); in contrast, the absorption has the highest value (513%), with the experimental values ranging from 54% to 349% [1, 7, 8, 22, 23]. This situation could generate deficient behaviors in the CEBs, especially in moist environments, causing a rapid decrease in the compressive strength. In contrast, the granulometric distribution of the soils is within the

Table 2 Classification and unit density of soils.

Soil	w_L, %	w_P, %	*PI*, %	D_s kg/m^3	Classification
Reddish sandy clay	49.1	24.6	24.5	2.70	CL
Brown clayey sand	17.2	4.6	12.6	2.74	SC
CEB A (30% of reddish sandy clay + 70% brown clayey sand	26.5	10.3	16.2	2.73	CL
CEB B (Yellowish sandy clay)	41.4	16.4	25.0	2.67	CL

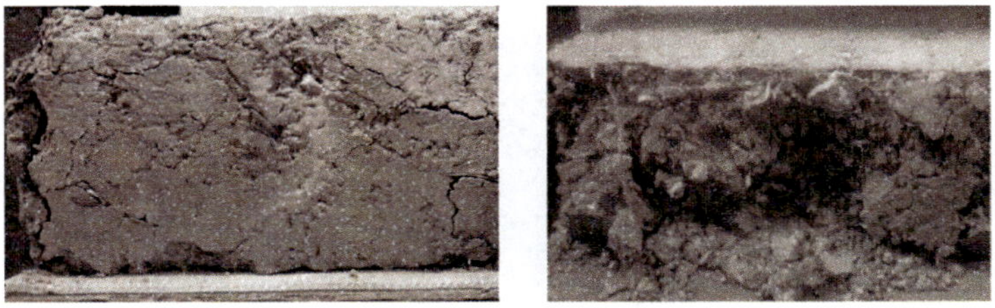

Fig. 6. Final condition after the compression tests.

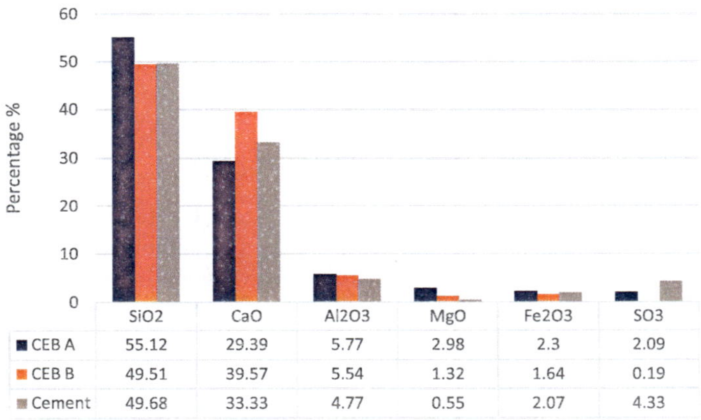

Fig. 7. Chemical compositions of the CEBs.

limits defined by the European standards, i.e., between 9.52 and 0.008 mm in diameter (Fig. 4) [18].

Despite the different specimen sizes and setup of the compression test, by considering the experimental referenced values [1, 5–8, 19], the compressive strength of CEB B is found to be the higher value (4.7 MPa).

Figure 8 shows the relationship among the compressive strength and w_L-***PI*** values [1, 7, 8, 19]; the highest value corresponds to CEB B. In fact, a directly proportional relationship exists between these variables. As can be seen, both CEBs are in the range defined by the European standards (continuous and dotted line) [18]. CEB B is at the top of the normative enclosure with high values of the w_L and PI, but without reaching the limit between low- and high-compressibility clays ($w_L = 50\%$). The numbers in the parentheses are defined as follows: the size of the specimen (P, piece, HP, half piece), pressure applied during the piece fabrication (U if the pressure is unknown), and compressive strength, respectively; the applied pressure and compressive strength are in MPa.

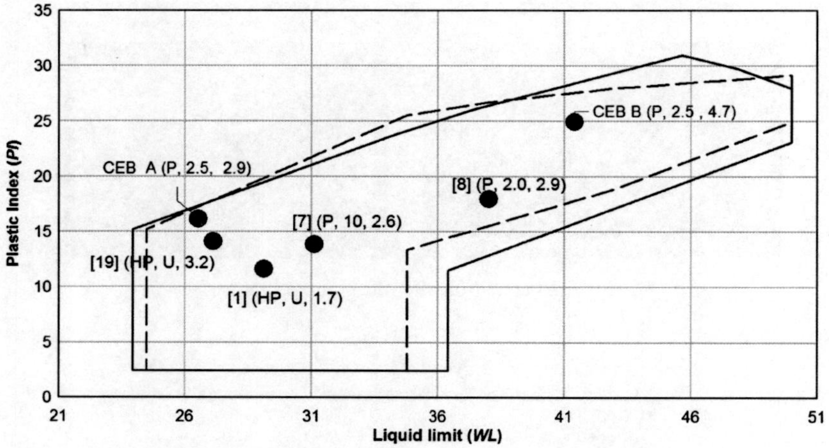

Fig. 8. Liquid limit vs. plastic index relationship and compressive strength [1, 7, 8, 18, 19].

Regarding the chemical compositions of the two pieces, CEB A has 5.6% more silica, and CEB B has 10% more calcium oxide; the differences diminish for the remaining compounds (Al_2O_3, MgO, and Fe_2O_3). The compressive strength of CEB B is 1.62 times that of CEB A. This might be related to the high percentage of calcium oxide (39.5%), i.e., higher by 6% with respect to that existing in the local cement. When using this as a reference, CEB B has a more similar composition to the reference than CEB A.

5 Conclusions

Different techniques are widely used for building housing with various earth materials (e.g., adobe, rammed earth, bahareque). In Mexico, the most frequently used material is adobe. Another lesser-known system uses CEBs, i.e., pieces stabilized with lime or natural fibers to reach a higher compressive strength.

As the CEBs analyzed in this study did not include lime, the Mexican Standard could not be used as a reference for the compression tests. For this reason, the New Mexico and European standards (or similar experimental work results) were used to evaluate and compare the physical-mechanical properties.

The oat straw has a density of 1.55 kg/m^3 and absorption of 513%; the first value is inside of the referenced interval, in contrast to the second one (54–349%). Thus, it is recommendable to consider the flattening of pieces to reduce the humidity generated by the environment.

The values of w_L, w_P, and ***PI*** are inside of the recommended intervals. The soils used are low-compressibility clays (CL), and their granulometric distribution is adequate. For CEB A and CEB B, the average measured compressive strengths are 2.9 MPa and 4.7 MPa, respectively, meeting the New Mexico standard. According to the literature, the last value is the second-highest value recorded. However, this piece has a high CV (36.7%), which suggests a plan to improve the quality control. For CEB B, a directly proportional relationship is detected between the ***PI*** and compressive strength; its chemical composition shows the highest percentage of calcium oxide relative to the other materials (CEB A and cement).

The initial absorption was high, and it was not possible to evaluate it owing to the disintegration of the material during the test. According to the New Mexico Standard, owing to the lack of lime as a material stabilizer and high absorption of the oat straw, these pieces should only be used to build walls 10 cm above the floor level, i.e., to avoid failures caused by environmental moisture.

References

1. Araya-Letelier, G., Concha-Riedel, J., Antico, F.C., Valdés, C., Cáceres, G.: Influence of natural dosage and length on adobe mixes damage-mechanical behavior. Constr. Build. Mater. **174**, 645–655 (2018)
2. Instituto Nacional de Estadística y Geografía Homepage. http://www.beta.inegi.org.mx/programas/intercensal/2015/. Accesed 01 Feb 2021
3. Boubekeur, S., Houben, H.: Blocs de terre comprimée. Centre pour le développement industriel-CRA Terre-EAG, Bruxelles (1998)
4. Rigassi, V.: Compressed earth blocks: manual of production. Deutsches Zentrum für Entwicklungstechnologien, Eschborn (1985)
5. Arroyo, R., Sánchez, S., Catalán, P.: Caracterización experimental de las propiedades mecánicas de la mampostería de adobe del sur de México. Ingeniería **17**(3), 167–177 (2013)
6. Montes, J.: Estudio del efecto de la fibra del bagazo de Agave Angustifolia en la resistencia a flexión y compresión del adobe compactado. Tesis de maestría, Oaxaca (2009)
7. Danso, H., Martinson, D.B., Ali, M., Williams, J.B.: Physical, mechanical and durability properties of soil building blocks reinforced with natural fibres. Constr. Build. Mater. **101**, 797–809 (2015)
8. Millogo, Y., Morel, J.C., Aubert, J.E., Ghavami, K.: Experimental analysis of Pressed Adobe Blocks reinforced with Hibiscus cannabinus fibers. Constr. Build. Mater. **52**, 71–78 (2014)
9. Patiño, L. A., Moreno, A.: Evaluación de las propiedades físicas y mecánicas de piezas de adobe compactado en el estado de Guerrero. Tesis de Licenciatura, Chilpancingo (2019)
10. New Mexico State records centers and archives Homepage. http://164.64.110.134/nmac/T14C007. Accesed 12 Dec 2020
11. Organismo Nacional de Normalización y Certificación de la Construcción y Edificación.: NMX-C-038-ONNCCE-2013. Industria de la construcción mampostería-determinación de las dimensiones de bloques, tabiques o ladrillos y tabicones-método de ensayo. ONNCCE, S. C., México, D. F (2013)

12. Organismo Nacional de Normalización y Certificación de la Construcción y Edificación.: NMX-C-037-ONNCCE-2013. Industria de la construcción mampostería-determinación de la absorción total y la absorción inicial de agua en bloques, tabiques o ladrillos-método de ensayo. ONNCCE, S. C., México, D. F (2013)
13. Secretaría de Comunicaciones y Transportes.: M MMP 1 06/03. Métodos de muestreo y pruebas de materiales. Suelos y materiales para terracería. Granulometría de materiales compactables para terracerías (2003)
14. Secretaría de Comunicaciones y Transportes.: M MMP 1 02/03. Métodos de muestreo y pruebas de materiales. Suelos y materiales para terracería. Clasificación de rocas y suelos (2003)
15. Secretaría de Comunicaciones y Transportes.: M MMP 1 07/07. Métodos de muestreo y pruebas de materiales. Suelos y materiales para terracería. Límites de consistencia (2007)
16. Secretaría de Comunicaciones y Transportes.: M MMP 1 05/03. Métodos de muestreo y pruebas de materiales. Suelos y materiales para terracería. Densidades relativas y absorción (2003)
17. Amziane, S., Collet, F., Lawrence, M., Magniont, C., Picandet, V., Sonebi, M.: Recommendation of the RILEM TC 236-BBM: characterization testing of hemp shiv to determine the initial water content, water absorption, dry density, particle size distribution and thermal conductivity. Mater. Struct. **50**(3), 4–11 (2017)
18. Jiménez, N., Cañas, I.: The selection of the soils for unestablished earth buildings: A normative review. Constr. Build. Mater. **21**, 237–251 (2007)
19. Jové-Sandoval, F., Barbero-Barrera, M.M., Flores Medina, N.: Assessment of the mechanical performance of three varieties of pine needles as natural reinforcement of adobe. Constr. Build. Mater. **187**, 205–213 (2018)
20. Melgarejo, J., Proenza, J., Gali, S., Lovet, X.: Técnicas de caracterización mineral y su aplicación en exploración minera. Boletín de la Sociedad Geológica Mexicana **62**(1), 1–23 (2010)
21. Gaceta oficial CDMX.: Normas técnicas complementarias para diseño y construcción de estructuras de mampostería. CDMX (2017)
22. Pickering, K.L., Aruan, M.G., Le, T.M.: A review of recent developments in natural fibre composites and their mechanical performance. Compos. Part A **83**, 98–112 (2016)
23. Moreno, L.E., Trujillo, E.E., Osorio, L.R.: Estudio de las características de haces de fibra de Guadua Angustifolia. Scientia et Technica **34**, 613–617 (2007)
24. Maciel, N., et al.: Comparative analysis of curaua fiber density using geometric characterization and pycnometry technique. In: Characterization of Minerals Metals, and Materials, pp. 11–19. Springer, Cham (2017)

How Different is the Deteriorating Mechanism of Fired Clay Bricks Due to NaCl Salt Compared to the Highly Damaging Na_2SO_4?

Swathy Manohar[1](✉) and Manu Santhanam[2]

[1] Department of Civil Engineering, IIT Bombay, Mumbai, India
swathym@iitb.ac.in
[2] Department of Civil Engineering, IIT Madras, Chennai, India

Abstract. Deterioration due to salt crystallization or physical salt attack is one of the crucial issues faced by the world's architectural heritage. Sodium sulphate is known for its highly damaging nature and hence adopted universally as the deteriorating agent by various laboratory salt degradation tests. There exists a general disagreement in the research community regarding the damaging nature of NaCl to masonry structures, perhaps due to a lack of sufficient information regarding the mechanism of decay. The current paper analyses the degree of deterioration and mechanisms in various brick types in the presence of NaCl and compares that with the mechanism of Na_2SO_4 damages. Accelerated salt weathering tests were conducted as per the RILEM V 1.b test procedure followed by the micro-analytical characterization of samples post-degradation. The difficulty in attaining higher supersaturations was observed in the case of NaCl, which resulted in a much slower rate of degradation. The role of microstructure of the material in altering the damage mechanism is elucidated using data collected with Mercury Intrusion Porosimetry and Scanning Electron Microscopy. Also, a question on choosing the right protective treatment for the samples on various exposures- sodium sulphate and sodium chloride is addressed.

Keywords: Salt crystallization · Sodium chloride · Masonry structures

1 Introduction

Sodium sulphate is considered as the most deleterious salt causing deterioration in masonry materials, predominantly due to the crystallization pressure of the salt. This pressure is a function of supersaturation of the salt solution (exposure condition) and pore geometry (size and shape of pores) [1–4, 6–8]. In general, salt crystallization occurs due to the generation of high supersaturation, which occurs due to evaporation of water or due to the phenomenon called dissolution-mediated phase transformation, or can even be due to a sudden change in atmospheric conditions [4, 5, 11].

Sodium sulphate has been prescribed to be used as the deteriorating agent by various laboratory salt degradation tests. This is due to its highly damaging nature Rodriguez-Navarro and Doehne [10]. Even though at a given supersaturation, the crystallization

J. I. Escalante-Garcia et al. (Eds.): RW 2021, RILEM Bookseries 40, pp. 699–706, 2023.
https://doi.org/10.1007/978-3-031-21735-7_75

pressure due to sodium chloride is much higher than that due to sodium sulphate, the damage due to sodium sulphate is very high as compared to sodium chloride [10, 12]. This can be attributed to the temperature-dependent solubility of sodium sulphate, and the dissolution mediated phase transformations undergone by the salt. Even though sodium chloride is considered as an omnipresent salt, and there exist a disagreement in the literature regarding the damage mechanisms and degeneration of masonry with the exposure of sodium chloride. Besides the nature of salt, the physicomechanical and microstructural properties of the masonry units (say, bricks) also play a role in determining the degree of damage. In the current study, a comparative analysis is demonstrated to understand the differences in the degree of deterioration and mechanisms in various brick types in the presence of NaCl and of Na_2SO_4 salt solutions.

2 Materials and Methods

Fresh commercial brick samples of two different types were used in the study, named C1 and C2, C1 with a porosity of 49% and compressive strength of 8 MPa; and C2 with porosity 31% and compressive strength 12 MPa. Two types of protective treatments were used on the bricks to observe the alterations in the salt weathering mechanisms in the presence of treatments. The treatments used were water repellents, first was a silicone-based treatment (denoted as SC throughout the paper), and second is an acrylic siloxane based treatment (AS).

The uncoated and coated brick specimens (4 cm cubes) were subjected to accelerated salt crystallization as per RILEM V1 b [9], separately with 10% Na_2SO_4 anhydrite solution and 15.6% NaCl solution for 112 cycles. Each cycle constituted immersion in salt solution for 2 h, followed by drying at 60 °C for 19 h and subsequent cooling at room temperature for 3 h. The mass changes and visible damages occurred to the specimens were recorded at the end of every cycle. Ultrasonic pulse velocity was measured using 54 kHz transducers as per the standard ASTM C597-2016 [2] on the bricks after each weathering cycle for a period of 15 days. Mercury intrusion porosimetry (MIP) was used to study the porosity and pore structure alteration with weathering for various specimens.

3 Results and Discussions

3.1 Mass Changes

The mass changes for C1 and C2 bricks with uncoated and coated variants when exposed to sodium sulphate weathering are presented in Fig. 1a and b respectively (marked as C1-S and C2-S; S for sulphate weathering). Figure 1c and d shows the same kind of specimens when exposed to sodium chloride weathering, marked as C1-C and C2-C; C stands for chloride weathering. SC denotes the specimens coated with silicone based water repellent and AS denotes the ones with acrylic siloxane based water repellent.

In case of sodium sulphate exposure, uncoated C1 had a very low but progressive mass loss, whereas C1-AS and C1-SC specimens did not show any significant mass differences throughout the long-term test. For C2, the uncoated sample showed progressive mass loss but at a very fast rate. While coated with the acrylic-siloxane based coating (C2-AS), the mass loss was resisted remarkably throughout the cycles. Meanwhile, C2-SC specimens had a sudden mass loss without any warning at around 15 cycles, which was much lower weathering resistance than that of C2 left uncoated. The uncoated samples of C1 and C2 exhibited slow mass increase up to the 60th cycle, although both showed visible signs of damage to different extents. The mass gain due to salt accumulation dominated the mass loss due to damage of specimens. C1-SC suffered no damage, but C2-SC underwent sudden mass loss as in the case of sodium sulphate exposure. The mass of these specimens remained constant throughout the test for both the brick types coated with acrylic siloxane coating (C1-AS and C2-AS).

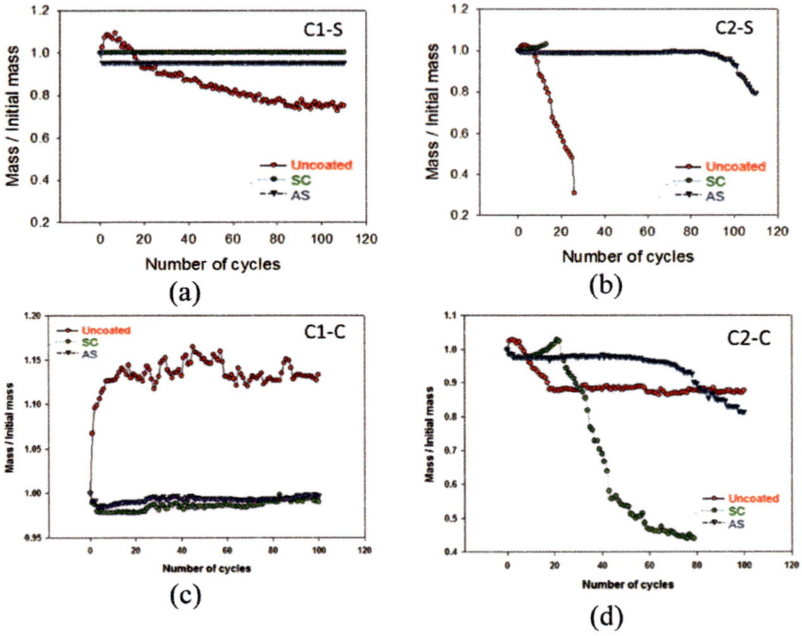

Fig. 1. Mass changes with salt crystallization cycles **(a)** Specimen C1 exposed to sulphate salt, **(b)** Specimen C2 exposed to sulphate salt, **(c)** Specimen C1 exposed to chloride salt, **(d)** Specimen C2 exposed to chloride salt.

3.2 Visual Damages

From the sodium sulphate weathering test, it was observed that the commonly available bricks fired at low temperatures (C1) and the wire-cut bricks fired in a modern kiln (C2) exhibited the two extreme cases of damage pattern. Figure 2 shows the damages incurred to the variants of C1 and C2 specimens, respectively, after 30 cycles of sodium sulphate weathering tests. Signs of deterioration like roughening of surface and deposition of powder around the samples were visible even before the beginning of the mass loss phase. For C1, the deterioration pattern was uniform, without any spalling or flaking. Whereas, the mass loss and deterioration were found to be the maximum in C2 where heterogeneous and deep damage was observed after the 6th cycle itself. The specimens coated with the silicone-based coating (SC) exhibited distinguishable results. Except for the commonly available non-wire-cut bricks fired at a relatively lower temperature (C1), the samples coated with the silicone-based coating (SC) underwent severe damage in the very early stages.

Figure 3 shows the images of each type of specimen before and after 90 days of salt weathering with NaCl exposure. The damage pattern in the uncoated C1 was homogeneous with granular disintegration. On the other hand, the damage pattern observed in the uncoated sample of C2 was heterogeneous with more visible signs of damage like powdery surface and rounding of edges. Another vital feature exhibited by the uncoated samples was massive efflorescence, that was uniformly distributed over the surface of C1 and non-uniformly distributed over C2. The C1 samples coated with the silicone-based coating (C1-SC) suffered little or no damage up to the 60th cycle, while the wire-cut brick samples coated with the silicone-based coating (C2-SC) underwent sudden damage with cracking, like their behaviour in sodium sulphate weathering test. This was followed by a period of rapid mass loss. Both the samples coated with the acrylic-based coating (AS) performed very well in the sodium chloride weathering test.

(a) Damage state of specimens C1, C1-SC and C1-AS respectively after 30 weathering cycles.

(b) Damage state of specimens C2, C2-SC and C2-AS, respectively after 30 weathering cycles.

Fig. 2. Images of specimens deteriorated with sodium sulphate exposure.

(a) Damage state of specimens C1, C1-SC and C1-AS respectively after 90 weathering cycles.

(b) Damage state of specimens C2, C2-SC and C2-AS, respectively after 90 weathering cycles.

Fig. 3. Images of specimens deteriorated with sodium chloride exposure.

3.3 Ultrasonic Pulse Velocity Test

Figure 4 shows the UPV readings for C1 and C2 specimens (both uncoated and coated variants) taken during sodium sulphate and sodium chloride exposure, respectively. After 15 days of weathering, some of the specimens lost most of their surfaces, and some other specimens disintegrated into several pieces; hence the UPV readings were disrupted. It shows that the velocities become stable with uniform salt depositions inside the pores. However, this trend was seen applicable only before the onset of larger cracks. Hence, UPV represents the degradation only once larger cracks are formed, and not with the formation of microcracks and fissures, which would get filled with salts. The trend is similar with both NaCl and Na_2SO_4.

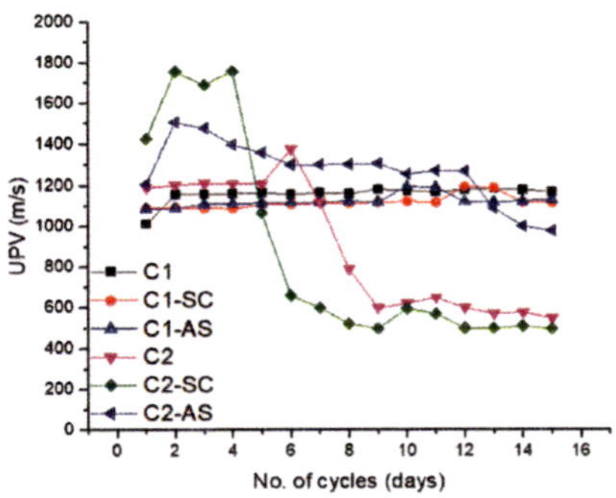

(a)

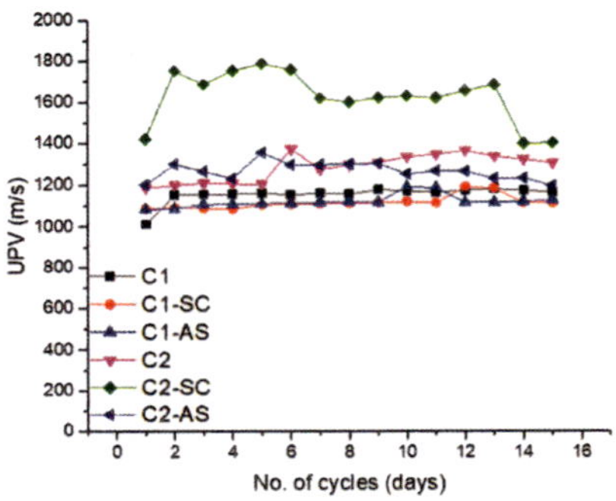

(b)

Fig. 4. Variation of UPV in bricks being exposed to **(a)** sulphate weathering cycles **(b)** chloride weathering cycles.

3.4 Porosity

There are serious limitations with respect to consideration of mass change as the only way to keep track of the damage. Mass change is not observed in those cases when damage occurs as cracking without any material loss. Such damage can be identified by studying the pore structure, in the current case, MIP was made use of. Figures 5 and 6 compares the differential intrusion curves of unweathered and weathered samples-coated and uncoated. C1 S indicates the brick C1 after sulphate weathering, C1-SC S denotes the brick C1 which was treated with the silicone-based coating (SC) subjected to sulphate weathering, and C1-AS S refers to brick C1 treated with acrylic-siloxane based coating (AS) subjected to sulphate weathering. Similarly, C1 cl indicates the brick C1 after chloride weathering. The nomenclature pattern is similar to the other specimens. If the differential intrusion graphs of salt weathered samples are subtracted from the differential intrusion graphs of fresh samples in each case (from Figs. 5 and 6), the remaining portion denotes the pores filled with the salt, which is very high in case of C2 and almost negligible in case of C1. This supports the observation from Benavente et al. [3]. Also, the critical pore sizes were almost unaffected between unweathered and weathered samples. A minimal reduction of critical pore diameters for C2 was observed, which indicates that a large share of the most common sized pores was slightly filled by the crystals of sodium sulphate.

In case of sodium chloride weathering, for C1, as expected from the similarities in the visual damage patterns between sulphate weathering and chloride weathering, the curves were very identical with each other. The curves for both the coated specimens (C1-SC cl and C1-AS cl) curves were overlapping with that of unweathered C1 specimen. This indicates both the coatings could prevent any action of weathering on the material and preserve the material as such even after 100 days of severe exposure with no difference among the type of coating. Unlike the case of sulphate weathered specimens, uniform salt deposition occurred in uncoated C1 specimens when subjected to weathering cycles, which shows a difference in the 'C1-cl' curve. In the case of samples from brick C2, critical pore size of the curve for C2-AS cl is overlapping with that of uncoated C2. This observation again indicates the higher efficiency of AS coating compared to that of SC. The weathered samples, which were uncoated and those coated with silicone-based treatment (C2-SC cl) exhibited overlapping differential curves.

4 Conclusions

Accelerated weathering tests were conducted on brick specimens of various pore structure by exposing coated and uncoated specimens to sodium sulphate and sodium chloride salt solutions separately with alternate dry-wet cycles as per RILEM V 1.b test procedure. The study investigates the differences in the mechanisms of deterioration with exposure of sodium sulphate salt and sodium chloride salt on bricks, and the role of the microstructure of the bricks in altering the damage mechanism.

Sodium sulphate was considered for the apparent reason that it is the most deleterious salt among the common exposures for a masonry structure. And, sodium chloride was also studied in view of its wide availability in natural environment. The performance of the treatments towards salt weathering resistance was found contradictory to that of its

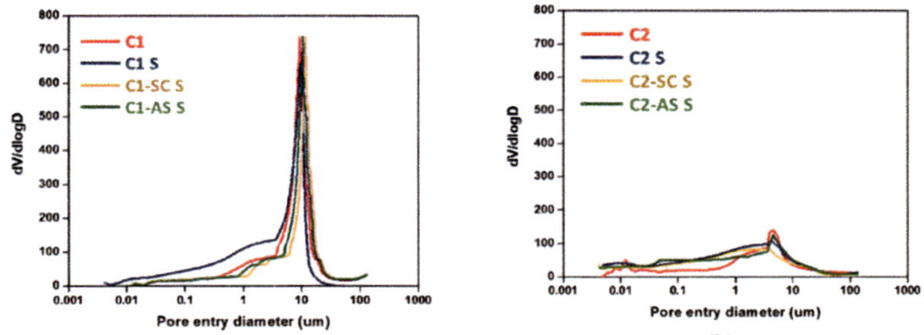

(a) (b)

Fig. 5. Comparison of differential intrusion curves for unweathered and weathered specimens with coatings with Na_2SO_4 **(a)** for C1, **(b)** for C2.

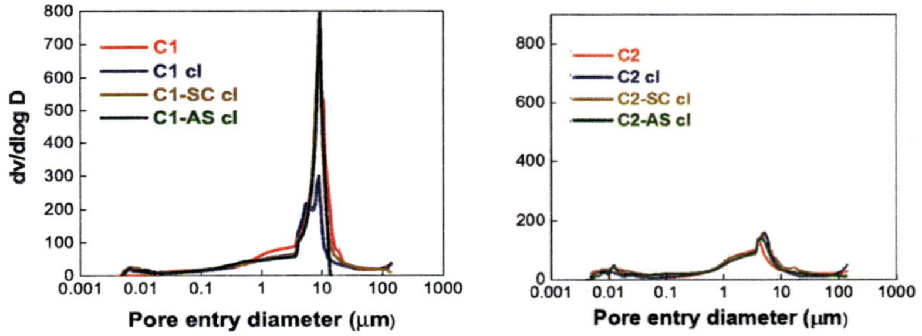

(a) (b)

Fig. 6. Comparison of differential intrusion curves for unweathered and weathered specimens with coatings with NaCl **(a)** for C1, **(b)** for C2.

water-repellence. Acrylic-siloxane based treatment was found to be providing a high salt weathering resistance to the substrate material, due to its property of cracking with salt exposure. On the other hand, the silicone-based coating blocked many pores, causing the specimens to be in a sealed condition. Thus, crystallization stress was created and retained within the material leading to further damages. It could be concluded from the study that there exists a relationship between the properties of water-repellent treatment chosen and the pore structure of the substrate (brick) material, to provide the required salt weathering resistance. For a material (brick in this case) which is more prone to salt crystallization damage (due to a large amount of smaller pores), the treatment provided should not be one which blocks the pores.

The study demonstrates that development of crystallization pressure and subsequent crack formation predominates in case of sodium sulphate exposure, whereas salt deposition in the pores is predominant for sodium chloride exposure. This is because NaCl salt does not face phase transitions, but gets precipitated and deposited on evaporation. In the case of C2, the exposure of sodium chloride solution resulted in the deposition of NaCl crystals in the larger portion of micropores which are abundant in the C2 system. Smaller ionic size and easy mobility of NaCl aids its deposition in micropores. Mass depositions were found predominating for NaCl because of its low crystallization pressures, resulting mostly in decreased porosities and fewer damages. However, in the long run, the damaging pattern was similar for NaCl and Na_2SO_4 among the range of sample types used indicating the similar choice of restoration strategies for effective conservation.

References

1. Alves, C., Figueiredo, C., Maurício, A.: A critical discussion of salt weathering laboratory tests for assessment of petrological features susceptibility. Proc. Earth Planet. Sci. **17**, 324–327 (2017)
2. ASTM International: ASTM C597 Standard Test Method for Pulse Velocity Through Concrete. ASTM International (2016)
3. Benavente, D., García del Cura, M.A., García-Guinea, J., Sánchez-Moral, S., Ordóñez, S.: Role of pore structure in salt crystallisation in unsaturated porous stone. J. Cryst. Growth **260**(3–4), 532–544 (2004)
4. Flatt, R.J., Caruso, F., Sanchez, A.M.A., Scherer, G.W.: Chemo-mechanics of salt damage in stone. Nat. Commun. Nature Publishing Group **5**, 1–5 (2014)
5. Flatt, R.J.: Salt damage in porous materials: How high supersaturations are generated. J. Cryst. Growth **242**(3–4), 435–454 (2002)
6. Manohar, S., Santhanam, M.: Correlation between mineralogical properties and weathering resistance using characterization case studies in historic Indian bricks. Int. J. Archit. Heritage 1–14 (2020)
7. Manohar, S., Santhanam, M., Chockalingam, N.: Performance and microstructure of bricks with protective coatings subjected to salt weathering. Constr. Build. Mater. **226**, 94–105 (2019)
8. Manohar, S., Chockalingam, N., Santhanam, M.: Experimental comparison between salt weathering testing procedures on different types of bricks. J. Mater. Civil Eng. ASCE (2021). ISSN 0899-1561
9. RILEM Recommended tests to measure the deterioration of stone and to assess the effectiveness of treatment methods, Test V.1a—crystallization test by total immersion (for untreated stone); Test V.1b—crystallization test by total immersion (for treated stone). Mater. Struct. **13**(75), 175–253 (1980)
10. Rodriguez-Navarro, C., Doehne, E.: Salt weathering: influence of evaporation rate, supersaturation and crystallization pattern. Earth Surf. Proc. Land. **24**(3), 191–209 (1999)
11. Scherer, G.W.: Stress from crystallization of salt. Cem. Concr. Res. **34**(9), 1613–1624 (2004)
12. Steiger, M.: Crystal growth in porous materials - I: The crystallization pressure of large crystals. J. Cryst. Growth **282**(3–4), 455–469 (2005)

Effect of Soil Characteristics on the Physico-mechanical Properties of Non-stabilized Compressed Earth Blocks

M. A. Kyriakides[1,2](✉), R. Panagiotou[2], R. Illampas[2], and I. Ioannou[2]

[1] Department of Engineering, University of Nicosia, 2417 Nicosia, Cyprus
kyriakides.m@unic.ac.cy

[2] Department of Civil and Environmental Engineering, University of Cyprus, 1678 Nicosia, Cyprus

Abstract. In an attempt to investigate the effect of soil characteristics on the physico-mechanical behavior of non-stabilized Compressed Earth Blocks (CEBs) produced in Cyprus, four types of locally sourced soils, with different mineralogical/granular composition and plasticity characteristics, have been examined and used for CEB production. A series of tests, including particle-size analysis, Atterberg limits determination, shrinkage and compaction measurements, and X-ray diffraction (XRD) analyses have been carried out to determine the characteristics of the raw materials selected. In addition, compression and three-point bending tests, capillary absorption measurements, thermal conductivity analyses, and scanning electron microscopy (SEM) analyses have been conducted to assess the physico-mechanical properties of the CEBs produced. The results indicate that, despite the different granular composition of the soils used, all CEBs produced demonstrate adequate mechanical properties in terms of compressive (over 5 MPa) and flexural (over 0.5 MPa) strength. The X-ray diffraction (XRD) analyses confirm that the soils investigated are mainly composed of carbonates and silicates at different ratios, in line with their geological origin. The thermal conductivity of all specimens tested ranges between 0.60–0.85 W/mK, while capillary absorption measurements have shown higher sorptivity in the specimens with extended micro-cracks and voids.

Keywords: Compressed earth block · Compression test · Flexural test · Capillary absorption · Thermal conductivity · Mineralogical composition

1 Introduction

Compressed Earth Blocks (CEBs) are building materials produced primarily with soil mixed at low moisture content and compressed under controlled pressure, without any firing. The end-products can be stabilized or non-stabilized, i.e., produced with or without the use of a small quantity of cement, lime or other stabilizer ($<12\%$ by weight, [1]), for enhancing their mechanical and durability properties. Non-stabilized CEBs are currently attracting increasing interest among construction industry professionals, due to

J. I. Escalante-Garcia et al. (Eds.): RW 2021, RILEM Bookseries 40, pp. 707–717, 2023.
https://doi.org/10.1007/978-3-031-21735-7_76

their lower production cost (compared to traditional fired clay bricks) and excellent recyclability potential [2]. However, the lack of widely accepted standards for soil selection and for the mechanical characterization of CEBs, as well as of relevant design procedures, limits the use of this environmentally friendly building material in contemporary architecture. This study is part of an ongoing research that focuses on the design, production, and characterization of a sustainable and eco-friendly prototype CEB masonry system that will be fabricated using raw materials originating from Cyprus.

2 Materials

2.1 Raw Material Characterization

Four different soil samples sourced from Alambra, Delikipos, Tseri and Potamia areas (Fig. 1), hereafter referred to as A-soil, D-soil, T-soil and L-soil, respectively, were used in the present study. The latter is, in fact, a fine aggregate waste from a quarry operating in the aforementioned area. The first three areas of the island were selected because of their rich tradition in earthen architecture.

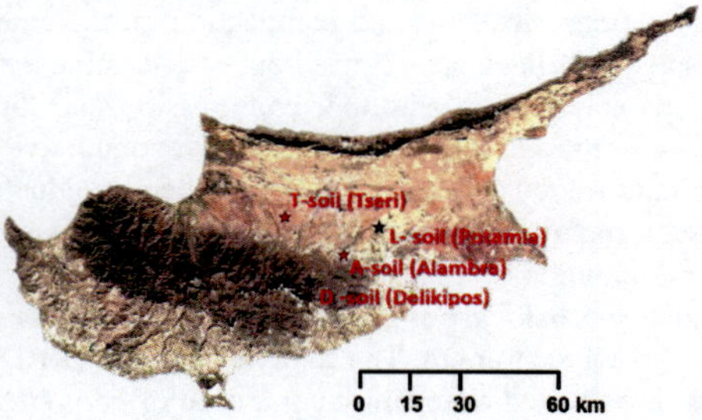

Fig. 1. Map of Cyprus with sampling locations.

The granular composition of each soil sample was determined according to ASTM D422 [3] and is presented in Fig. 2, together with the limits suggested by the literature for CEB production [1].

Table 1 provides the other essential characteristics of the soils examined in this study, namely the specific gravity, the Atterberg limits and linear shrinkage, the optimum moisture content and the corresponding maximum dry density, that were determined according to ASTM D854 [4], BS 1377–2 [5], and BS 1377–4 [6], respectively.

In a review of the soil selection criteria for non-stabilized earthen materials by Delgado and Guerrero [7], the recommended values for the liquid limit and plasticity index suited to the production of CEBs are 25–50% and 2–33%, respectively. While only one sample, namely D-soil, satisfies the requirements regarding the granular composition, three out of the four soil samples tested satisfy the liquid limit and plasticity index recommendations specified by the literature [8]. One specific sample, namely L-soil, is strikingly different in that it comprises of much more silt, less sand and no gravel, and

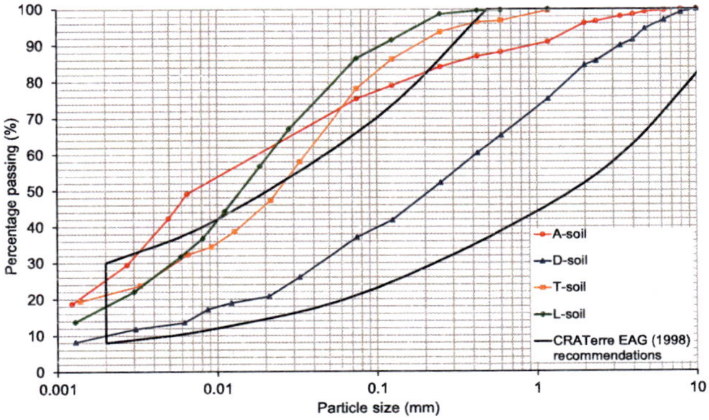

Fig. 2. Granular composition of the four soils used to produce CEBs.

Table 1. Characteristics of soils used in the present study. LL = Liquid Limit, PL = Plastic Limit and PI = Plastic Index.

Soil	Specific gravity	Atterberg limits			Linear shrinkage	Compaction test	
		LL	PL	PI		Maximum dry density	Optimum moisture content
		(%)	(%)	(%)	(%)	(g/cm^3)	(%)
A	2.56	36	21	15	7.89	1.75	17.7
D	2.68	44	28	16	9.06	1.68	19.5
T	2.56	41	24	17	6.67	1.66	19.0
L	2.66	29	20	9	4.34	1.85	15.5

has lower Atterberg limits, linear shrinkage and optimum moisture content, compared to the other three samples hereby tested.

Table 2 provides the results from the X-ray diffraction analyses conducted on the four soils using a scanning range of 2–70° at a rate of 1°/min.

The mineralogical composition of A-soil is in line with its geological origin. This soil probably originated following the erosion of carbonate rocks. Thus, it has a high concentration of carbonates, while it has substantially less silicates, followed by little chlorite and traces of other clay minerals. D-soil seems to be the result of the weathering of ophiolitic rocks found on the nearby mountains of Troodos. Silicates are found in abundance in this soil, while chlorite, carbonates and oxides are also identified. In addition, D-soil has traces of other clay minerals. T-soil has a mineralogy that consists of both silicate and carbonate minerals. Less chlorite is found in this soil, together with some other clays and gypsum. Finally, L-soil probably originated following the erosion of carbonate rocks. Thus, it has a high concentration of carbonates, substantially less silicates, followed by little chlorite and traces of other clays.

Table 2. Mineralogical composition of soils used in the present study.

Mineral (%)	Soil			
	A	D	T	L
Calcite	81	4	30	80
Quartz	4	6	14	9
Dolomite	–	4	Traces	5
Feldspars	7	–	38	4
Chlorite	8	16	12	Traces
Rhodochrosite	–	–	–	Traces
Gypsum	–	–	–	–
Plagioclase	–	34	–	–
Kaolinite	–	–	5	–
Actinolite	–	Traces	–	–
Talc	–	4	–	–
Enstatite	–	12	–	–
Goethite	–	8	–	–
Sodalite	–	6	–	–

2.2 Production of CEBs

A commercial mobile press (AECT Impact 2001A) for the on-site production of 300 mm × 150 mm × 100 mm (length x width x height) CEBs was used to produce the blocks examined in this study. Each of the four soils characterized above was first mixed with approximately 10% water by volume—as per the recommendation of practitioners—in a planetary mixer, until uniform consistency was achieved. The mixtures were then placed in the mobile press and a pressure of about 15 MPa was applied, as per the mobile press manufacturer's recommendations. After the fabrication of the four different batches of CEBs, the samples were kept in the laboratory under ambient conditions (relative humidity 55% ± 10% and temperature 25 °C ± 5 °C) for at least three months before testing, to achieve air-dried conditions.

3 Test Methods for the Characterization of CEBs

3.1 Compression Tests

Compression tests were conducted on eight full-size CEB units from each production batch as per ASTM C67-03a [9]. Four 20 mm-range Linear Variable Differential Transducers (LVDTs 1–4 shown in Fig. 3a) were symmetrically attached to the test setup to monitor the relative displacement of the loading platen's corners. This enabled the uniaxial strain deformation of each specimen to be recorded and the secant elastic modulus to be calculated. Loading was imposed at a displacement-controlled rate of 1 mm/min. Two 10-mm LVDTs (i.e., LVDTs 5 and 6 in Fig. 3a) were attached on either side of the CEB for applying the displacement-controlled procedure.

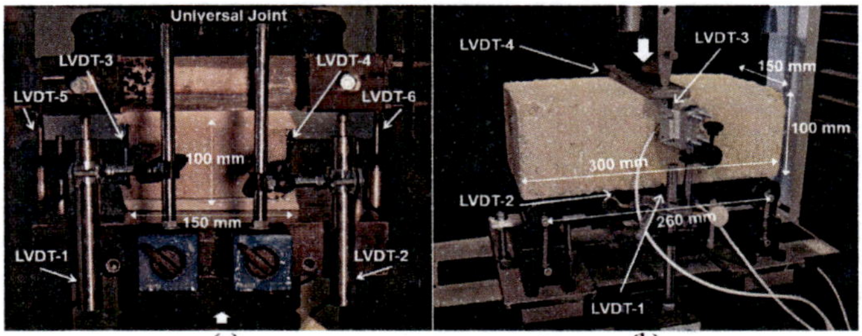

(a) (b)

Fig. 3. (a) Compression, and (b) three-point bending test setups.

3.2 Three-Point Bending Tests

Three-point bending tests were conducted on eight full-size CEB units from each production batch as per ASTM C67-03a [9]. Each specimen was simply supported on rollers with 260 mm spacing between them (Fig. 3b). The load was applied at mid-span through another roller and a steel bearing plate placed between the roller and the upper surface of the specimen. The loading rate was set at 0.2 mm/min. Four 10 mm-range LVDTs (i.e., LVDTs 1–4 shown in Fig. 3b) were used to monitor the mid-span displacement; two of these were measuring the deflection of the bottom face of the specimen (LVDTs 1 and 2) and the other two (LVDTs 3 and 4) the movement of the upper face.

3.3 Scanning Electron Microscopy

Scanning electron microscopy (SEM) analyses were conducted on gold sputtered samples from all CEB batches. These samples were obtained from the surface of the CEBs that would have been exposed to the external environment in a structure, to provide a better inside on the potential effect of the micro-structure on the durability performance of the blocks, and were analyzed using secondary electron imaging mode.

3.4 Capillary Absorption Tests

Capillary absorption tests were conducted according to EN 1925 [10] on six 70-mm cubic specimens per CEB batch that were carefully dry cut using a masonry saw from full-sized blocks to avoid possible disturbance. The tests were carried out using a non-reactive liquid, namely acetone (CH_3COCH_3), to overcome problems leading to anomalous results arising with the use of water (i.e., non-linear relationship between the volume of water absorbed per unit surface area of absorption and the square root of time) [11]. Due to the high volatility of acetone, throughout the tests the specimens were kept in a sealed container under controlled temperature (ca. 20 °C), thus minimizing possible evaporation of the absorbing liquid.

3.5 Thermal Conductivity Tests

Direct dynamic heat transfer measurements were conducted using an ISOMET analyzer equipped with surface heating probes, to evaluate the thermal conductivity (λ_b), the volume heat capacity (C_p), and the thermal diffusivity *(a)* of the sampled materials. Eight 100-mm cubic specimens per CEB batch were used in these measurements. The test specimens were dry cut with a masonry saw from full blocks. The path of heat flow during the heat transfer experiments was set parallel to the direction along which the CEBs would have transferred heat though convection in a structure (i.e., along the thickness of the masonry wall).

4 Results and Discussion

4.1 Compression Response of CEBs

Figure 4a shows the characteristic compression failure of the CEBs hereby tested. All blocks experienced a similar failure pattern, namely near-vertical surface cracks. This failure pattern is attributed to the lateral restraint imposed on the specimens by the loading platens; this was further enhanced by the low height-to-width ratio of the specimens.

Fig. 4. Typical failure mode of CEBs in (a) compression, and (b) three-point bending.

The compression stress-strain response of all CEBs tested is shown in Fig. 5a.

The average compressive strength and secant elastic modulus of all CEBs are provided in Table 3, along with the average moisture content determined for each batch. Specimens manufactured with A-, D- and T-soils exhibited similar average compressive strength (>9 MPa), even though only D-soil satisfies the grading requirements reported in the literature (Fig. 2). In contrast, CEBs made of L-soil exhibited much lower average compressive strength (ca. 5 MPa), despite the lower moisture content recorded. All compressive strengths obtained, however, were higher than the lower bounds recommended by various international standards and guidelines, e.g., the NZS 4298 [12] (1.5 MPa), and the German Association for Building with Earth guidelines (2007) [13] (2–4 MPa).

In terms of secant elastic modulus, CEBs made of T-soil exhibited the highest average stiffness, i.e., 694 MPa, albeit also recording the highest coefficient of variation. Specimens made of L- and D- soils showed the lowest moduli of elasticity (ca. 400–450 MPa).

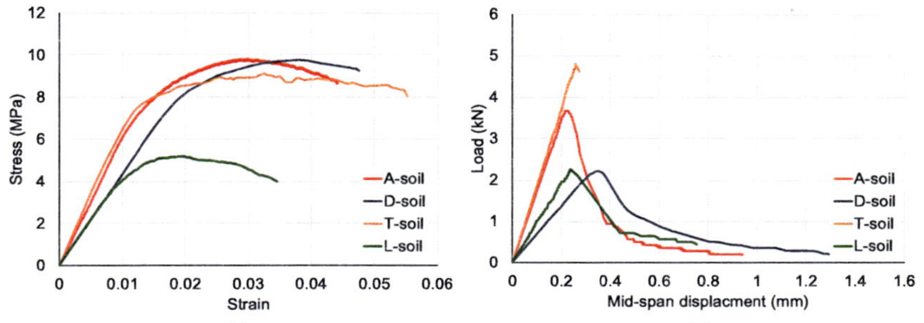

Fig. 5. (a) Characteristic compression stress-strain response, and (b) characteristic load vs bottom-face mid-span displacement of CEB units.

Table 3. Average values and coefficient of variation (COV) for the compressive strength and secant elastic modulus of CEBs.

Soil	Compressive strength MPa (COV)	Secant elastic modulus MPa (COV)	Moisture content % (COV)
A	9.79 (5%)	587 (15%)	2.97 (3%)
D	9.77 (3%)	408 (9%)	7.25 (1%)
T	9.18 (4%)	694 (23%)	4.88 (2%)
L	5.31 (7%)	449 (12%)	1.46 (1%)

4.2 Three-Point Bending Response of CEBs

The characteristic failure mode of the CEBs under three-point bending is shown in Fig. 4b. All specimens made of A-, D- and T-soil and most specimens made of L-soil experienced brittle failure caused by a midspan flexural crack at the bottom face.

Figure 5b shows the characteristic load-midspan (bottom face) displacement response of all specimens under three-point bending. All specimens demonstrated a linear behavior almost until the corresponding peak load sustained. At peak, a flexural crack formed, causing the load to drop rapidly. The most brittle behavior was demonstrated by specimens made of T-soil.

The average flexural strength and average bending stiffness of all CEBs tested under three-point bending are provided in Table 4. The flexural strength, f_{fl}, has been estimated from Eq. (1),

$$f_{fl} = 3P_{max}l/\left(2bh^2\right) \tag{1}$$

where P_{max} is the maximum applied load, l is the distance between the supports, b and h are the width and height, respectively, of the cross-section of the specimen at midspan (i.e., at plane of failure). The bending stiffness, k_{fl}, is defined as the slope of the linear branch of the load-displacement response of CEBs under three-point bending.

CEBs made of T-soil exhibited the highest flexural strength, followed by those made of A- and D-soil. These three groups of specimens also had the highest compressive strengths (Table 3). The lowest flexural strength was exhibited by CEBs made of L-soil, although these specimens also recorded the lowest moisture content. All flexural strengths obtained were higher than the lower bounds recommended by various international normative documents, e.g., the NZS 4298 [12] (0.25 MPa). A similar trend to the flexural strength was observed for the bending stiffnesses.

Table 4. Average values and coefficient of variation (COV) for the flexural strength, f_{fl}, and stiffness, k_{fl}, of CEBs.

Soil	Flexural strength, f_{fl} MPa (COV)	Stiffness, k_{fl} kN/mm (COV)	Moisture content % (COV)
A	0.93 (13%)	15.52 (24%)	2.94 (6%)
D	0.72 (6%)	6.80 (21%)	7.02 (2%)
T	1.21 (7%)	18.90 (15%)	5.41 (4%)
L	0.50 (16%)	9.34 (18%)	1.52 (2%)

4.3 Scanning Electron Microscopy Analyses

Figure 6 provides characteristic magnified images obtained from the scanning electron microscopy (SEM) analyses conducted on samples of CEBs from all batches. The images confirm the heterogenous nature of all samples; denser microstructure, smaller micro-cracks and voids <20 μm are observed in the specimens made of A-soil and T-soil (Figs. 6a and c, respectively), compared to those made of D-soil and L-soil (Figs. 6b and d, respectively). While CEBs fabricated with A-soil did not have large areas with significant amounts of phyllosilicates, those produced with D-soil presented areas consisting mainly of phyllosilicates. This is also confirmed by the results of the mineralogical analysis of D-soil. It is also noted that all specimens examined exhibited horizontal cracks, perpendicular to the direction of compression of the CEBs during fabrication, due to the lower—recommended by practitioners - amount of moisture (10% by weight) used in their mix design, compared to the corresponding optimum moisture content (Table 1).

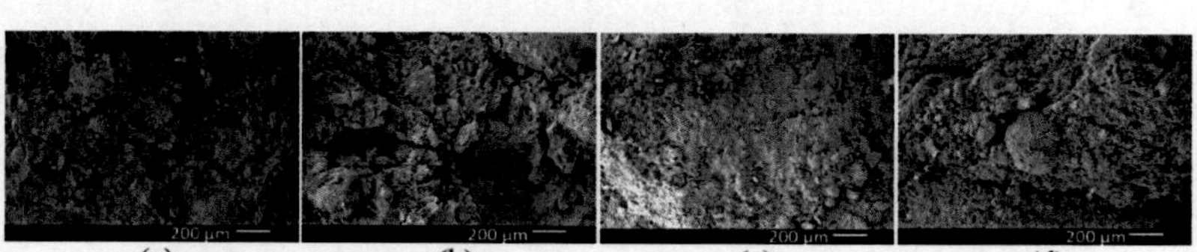

(a) (b) (c) (d)

Fig. 6. Scanning electron microscopy (SEM) images of CEB samples made of a) A-soil, b) D-soil, c) T-soil, and d) L-soil.

4.4 Capillary Absorption Response

Figure 7 shows characteristic capillary absorption vs square root of time relationships for the four groups of specimens tested, namely specimens made of A-, D-, T- and L-soils. The resulting capillary absorption coefficients are provided in Table 5.

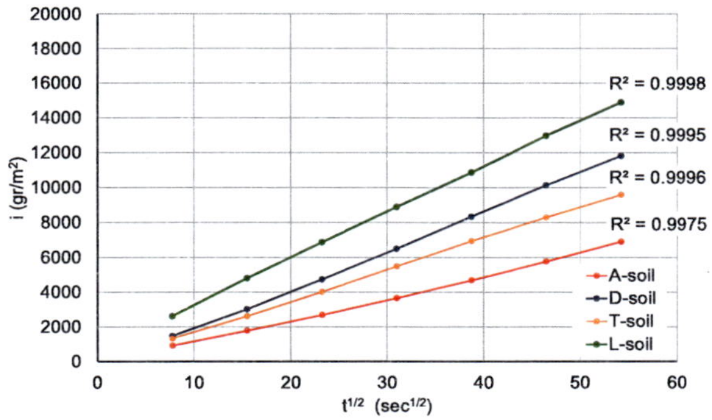

Fig. 7. Characteristic response of liquid mass absorbed due to capillary phenomena per unit surface area relative to the square root of time, for A-, D-, T-, and L-soil specimens.

The lowest capillary absorption coefficient, $c_{i,b}$, was obtained with the specimens made of A-soil, followed by those made of T-soil. The highest values were obtained from the specimens made of D-soil and L-soil. These values agree with the observations of the SEM analyses. CEBs made of A- and T-soils had denser micro-structure, compared to those made of D- and L-soils, thus leading to lower values of $c_{i,b}$. Although CEBs made of T-soil had denser micro-structure, compared to those made of A-soil, the presence of voids seems to have led to a higher $c_{i,b}$.

4.5 Thermal Response of CEBs

All specimens have a thermal conductivity, λ_b, between 0.63 and 0.85 W/mK. The values of λ_b obtained are typical for CEBs, according to the literature. For example, Zhang et al. [14] demonstrated that the thermal conductivity of compressed earth bricks increased linearly from 0.52 W/mK to 0.93 W/mK with increasing bulk density.

5 Conclusions

Raw material characterization tests, followed by tests on CEBs, have been used to investigate the effect of soil characteristics on the physico-mechanical properties of non-stabilized CEBs. While most soils hereby studied do not satisfy granular composition requirements specified by guidelines and recommendations found in the literature, all soils satisfy the requirements with regard to the liquid limit and plasticity index. Compression and flexural tests have shown that granular composition alone cannot fully provide an insight on the mechanical properties of non-stabilized CEBs. CEBs made

Table 5. Average values and coefficient of variation (COV) for capillary absorption coefficients, direct dynamic heat transfer measurements and air-dry densities of specimens made of A-, D-, T-, and L-soil.

Soil	Capillary absorption coefficient, $c_{i,b}$	Thermal conductivity, λ_b	Volume heat capacity, C_p	Thermal diffusivity, α	Gross air-dry density, ρ_b
	$g/m^2\ s^{1/2}$ (COV)	W/mK (COV)	J/m^3K (COV)	m^2/s (COV)	g/cm^3 (COV)
A	144 (14%)	0.853 (0.9%)	1527917 (0.7%)	5.57E-07 (1.0%)	1.93 (0.5%)
D	230 (7%)	0.705 (4.0%)	1515417 (0.5%)	4.65E-07 (3.8%)	1.96 (2.0%)
T	182 (11%)	0.700 (3.1%)	1490833 (2.2%)	4.70E-07 (1.0%)	1.88 (1.6%)
L	263 (9%)	0.629 (3.2%)	1458750 (0.9%)	4.31E-07 (2.2%)	1.80 (0.1%)

of soils with a more uniform particle-size distribution and a small percentage of gravel demonstrated superior mechanical properties, over CEBs manufactured with soils with larger percentages of silt and clay and no sand or gravel. Furthermore, soils with high concentration of carbonates and silicates demonstrated adequate mechanical properties in terms of compressive (over 5 MPa) and flexural (over 0.5 MPa) strength, despite their different granular composition. The determination of the mineralogical composition of the soil is therefore recommended for better understanding the mechanical properties of CEBs. SEM analyses and sorptivity tests have shown that CEBs made of A- and T-soils exhibited lower capillary absorption than CEBs made of D- and L- soils, due to their denser micro-structure. Thermal tests have shown that all specimens had similar thermal conductivities to those reported in the literature, with the thermal conductivities of CEBs made of D-, T- and L- soils being lower to those made of A-soil.

Acknowledgements. The authors would like to acknowledge financial support by the European Regional Development Fund and the Republic of Cyprus, through the Cyprus Research and Innovation Foundation (Project ENTERPRISES/0618/0007), and the University of Cyprus. They would also like to thank Mr. M. Tapakoudis from Gigantas Antaios Touvlopiio Ltd for his help with the production of CEBs.

References

1. CRATerre-EAG: Compressed earth blocks: Standards–Technology series No.11. Brussels, CDI (1998)
2. Malkanthi, S.N., Balthazaar, N., Perera, A.A.D.A.J.: Lime stabilization for compressed stabilized earth blocks with reduced clay and silt. Case Stud. Constr. Mater. **12**, e00326 (2020)

3. ASTM D422–63: Standard test method for particle-size analysis of soils, ASTM International, West Conshohocken, PA (2002)
4. ASTM D854: Standard Test Methods for Specific Gravity of Soil Solids by Water Pycnometer, ASTM International, West Conshohocken, PA (2014)
5. BS 1377–2: Methods of test for soils for civil engineering purposes. Classification tests. British Standards Institution (1990)
6. BS 1377–4: Methods of test for soils for civil engineering purposes. Compaction-related tests. British Standards Institution (1990)
7. Delgado Jiménez, M.C., Guerrero, I.C.: The selection of soils for unstabilised earth building: A normative review. Constr. Build. Mater. **21**(2), 237–251 (2007)
8. Rigassi, V.: CRATerre-EAG: Compressed Earth Blocks, Vol. I: Manual of Production. Braunschweig, Allemagne: Friedrich Vieweg & Sohn, (1995)
9. ASTM C67–03a: Standard Test Methods for Sampling and Testing Brick and Structural Clay Tile, ASTM International, West Conshohocken, PA (2003)
10. EN 1925: Natural stone test methods. Determination of water absorption coefficient by capillarity, CEN, Brussels (1999)
11. Hall C., Hoff W. D.: Water transport in brick, stone and concrete, 2nd edition, Taylor and Francis, London (2012)
12. NZS 4298: Materials and workmanship for earth buildings, New Zealand Standards, New Zealand (2020)
13. The German Association for Building with Earth, Dachverband Lehm e.V.: Rules governing building with earth, 3rd revised edition (2007)
14. Zhang, L., Jelle, B.P., Wang, Y., Gustavsen, A.: Hygrothermal properties of compressed earthen bricks. Constr. Build. Mater. **162**, 576–583 (2018)

Salt Deterioration of Heritage Structures—Correlating the Insights from Field and Lab Studies

Swathy Manohar(✉), V. A. Anupama, and Manu Santhanam

Department of Civil Engineering, Indian Institute of Technology Bombay, Mumbai, India
swathym@iitb.ac.in

Abstract. Salt deterioration is unequivocally reported as the most crucial cause of the degeneration of coastal heritage structures. However, the deterioration patterns differ greatly for various sources and conditions, which are difficult to simulate exactly in laboratories. There are mixed effects and mechanisms of various salts getting crystallized in the same structure depending upon the differences in the microclimate. Insights developing from laboratory studies, even though they cannot replicate similar weathering, can be suggestive of the decay variations observed in the field conditions. In the current study, brick samples of the same origin but distinctive damage patterns collected from a heritage fort near the seashore were analyzed for physical and microanalytical properties. Combined and separate effects of sodium chloride and sodium sulphate were observed in the historic bricks and were then simulated in studies on fresh bricks in the laboratory. The extent and nature of damage in the field and lab samples were assessed with the help of characterization techniques such as XRD, SEM and MIP. Phase transitions of sodium sulphate were analyzed in the accelerated weathering studies conducted in the lab and correlated with the varying exposure conditions of the field. The study contributed to providing a clear understanding of the different decay mechanisms and damage patterns observed at the site in relation to the standard salt deterioration tests conducted in labs.

Keywords: Salt crystallization · Heritage conservation · Bricks · Characterization

1 Introduction

The crystallization of soluble salts within the pores has been found to cause the greatest damage for porous masonry units such as bricks or stones [3, 8, 11, 12]. Extensive research is ongoing in developing and modifying standard laboratory test procedures for accelerated salt weathering tests and in understanding the mechanisms of the crystallization of various salts [5, 7, 10]. However, the possible bias in the responses from lab studies is accepted, considering the complications in the nature of salts and environmental conditions in real scenarios. The tests are being suggested and developed by the RILEM TC 271 ASC considering the acceleration of the degradation process without alteration of

J. I. Escalante-Garcia et al. (Eds.): RW 2021, RILEM Bookseries 40, pp. 718–728, 2023.
https://doi.org/10.1007/978-3-031-21735-7_77

damage mechanism [9]. The existing recommendations of test procedures by RILEM [14] and European standard [2] have their limitations and differences, which suggests a possible difference in the resulting mechanisms, too [13]. Also, in laboratories, it is more convenient to control and change the variables in an accelerated weathering study. In real cases of salt damage in masonry structures, the variables will be changing with respect to space and time, and hence it would be inappropriate to interpret the mechanism without caution.

The current paper considers a historical brick masonry structure of 400 years old from Southern India, severely affected by salts – but with various degrees and kind of damage at various locations. A physical, chemical, mechanical and microstructural analysis of the samples collected from the site from different locations were analyzed to correlate the causes for the various damage patterns and the mechanisms of action. Subsequently, fresh brick samples of different properties were subjected to accelerated salt crystallization tests in the laboratory with different salts, to relate the findings from the field. This paper puts forward a comparative analytical study of the properties and performance of salt-affected bricks in natural as well as laboratory condition.

2 Materials and Methods

Historic brick samples collected from a deteriorated fort were used for analyzing the field effects of salt weathering, and fresh commercial bricks were used to understand the salt weathering mechanisms by accelerated tests at the laboratory. Historic brick samples T1 and T2 were collected from the fort of Tranquebar, T1 samples were collected from the external walls of the fort, from a lower level. These bricks were in an exposed condition but not in direct contact with the sea. Samples T2 were collected from the part of a wall that was partially in contact with the sea; hence, they were wet bricks, subjected to erosive forces as well. C1 and C2 are the two types of fresh bricks used in this study for accelerated weathering studies, where C1 is a high porosity brick, and C2 is a low porosity brick.

Physicomechanical properties were determined for the brick samples, such as water absorption, bulk density and compressive strength. The mineralogical characterization was performed using XRD with a Bruker Discover D8 power diffractometer and EDS spectra from Scanning Electron Microscopy (SEM) with an EMcrafts machine. XRD patterns were obtained by step scanning from 5° to 90° 2θ with a count of 0.5 s/step. The diffractograms were studied with X'PERT HighScore Plus software. Accelerated salt crystallization studies were conducted on C1 and C2 bricks (on cubical specimens of 40 mm side) by immersion dry-wet cycling with sodium chloride and sodium sulphate salts according to RILEM V 1.b, (1980). Specimens were exposed to 10% sodium sulphate anhydrite solution for 2 h, followed by drying at 60 °C for 19 h and subsequent cooling at room temperature for 3 h. Mass loss and damages were monitored for each cycle for a period of 100 cycles to demonstrate the difference in the action of sodium sulphate and sodium chloride on the same brick specimens under similar atmospheric conditions.

3 Field Study

3.1 Background

Structures along the coastal stretch experience severe damage due to the encroachment of the sea, resulting in salt weathering. The Tranquebar Fort, located at the shores of the Bay of Bengal in Tamil Nadu, is a historic brick masonry structure exposed to extreme salt attack. This fort has manifestations of deteriorations at several locations exposed to NaCl. Thus, understanding the characteristics and response of clay bricks used in constructing this fort can explain the mechanisms of damage due to NaCl crystallization. Understanding the effect of salt crystallization in altering the pore structure of the material is critical in the conservation and restoration of coastal masonry structures. The scientific knowledge on the crystallization mechanism is inevitable to ensure a compatible and durable repair.

Based on the extent of contact with the sea, two sets of samples were collected from the fort, T1 from relatively dry walls and T2 from wet walls. The representative samples of full bricks were carefully removed using a chisel and hammer. Splashing waves on the T2 bricks exerted erosive forces, which resulted in softening and progressive erosion from the sides. Heavy efflorescence and blistering were prominent in T1 bricks. Even though efflorescence was not observed in T2 bricks, salt crystallization beneath their surface caused sub-florescence. Figure 1 shows the locations from where the samples were collected.

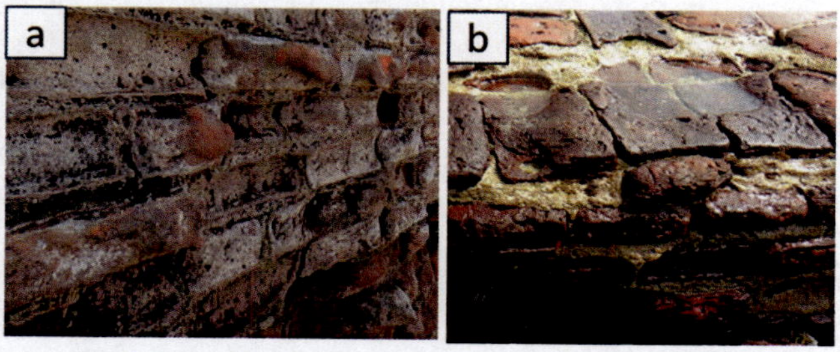

Fig. 1. **(a)** Brick wall from where sample T1 is collected **(b)** T2 sample bricks which are partially immersed in seawater

3.2 Physicomechanical Properties

The moisture content of T2 bricks, which were partially or fully immersed in seawater, was 30%, whereas that of T1 bricks from the relatively dry wall was 7%. The water absorption of T2 was lesser (11.5%) compared to that of T1 samples (15.5%). These values indicate that the T2 bricks which were subjected to more tidal forces and salt have lesser open porosity than T1 samples. The physicomechanical properties of the brick samples are consolidated in Table 1. The higher compressive strength of T2 bricks (10.1 MPa) can be correlated to the lesser porosity.

Table 1. Physical properties of the bricks

Sample	Moisture content (%)	Compressive strength (MPa)	Water absorption (%)
T1	7	4.2	15.5
T2	30	10.1	11.5

3.3 Mineralogy

X-ray diffraction technique was carried out for mineralogical characterization of brick samples. The analysis of XRD patterns indicates that quartz and K-feldspar were predominant in both T1 and T2 bricks. Halite and thenardite were also present in both sets of samples. T2 bricks, which were in direct contact with the seawater, contained halite at an increased amount compared to T1 samples. The blistering observed in the T1 samples can be explained by the presence of gypsum observed from the diffractograms. The analysis of XRD peaks of T1 samples also reinforced the hypothesis about the volumetric expansion and subsequent cracking by the phase conversion from thenardite to mirabilite. The major peaks obtained from the XRD analysis are shown in Fig. 2. The absence of high-temperature phases like diopside and gehlenite implies that the brick samples were fired at low temperature. The porous structure resulted from the low-temperature firing makes the bricks weak and vulnerable to deterioration in a deleterious environment.

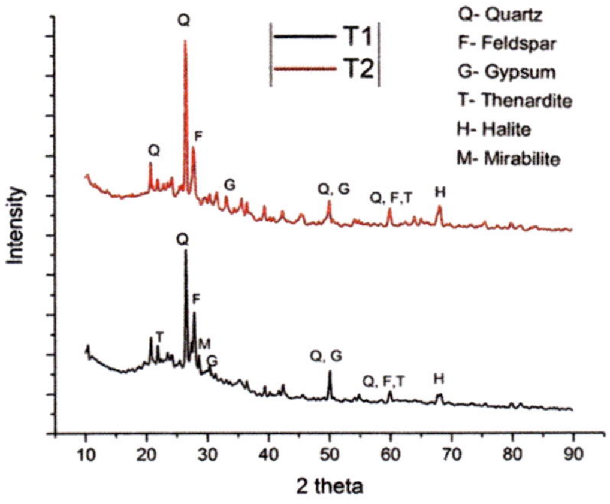

Fig. 2. Diffractograms showing the peaks of major mineral phases

3.4 Pore Structure

The growth of chloride crystals in the pores of the bricks can increase porosity by the initiation of microcracks and superficial mass loss during the initial stages. At later stages, the crystal growth densifies the microstructure and reduces the porosity [16].

Mercury Intrusion Porosimetry was used to analyze the pore structure and measure the porosity of the brick samples. The lesser porosity of T2 samples can be attributed to the halite crystallization, which fills the pores without exerting much crystallization pressure and volumetric expansion, unlike the crystallization of sulphate salts. NaCl molecules are highly soluble and mobile due to their small ionic size. Hence, they can penetrate smaller pores easily and crystallize, resulting in a compact microstructure. The critical pore diameter of T2 bricks is obtained from MIP is 1.01 μm.

The higher porosity of T1 bricks can be related to the crystallization of sulphate salts accompanied by the conversion from thenardite to mirabilite, causing volumetric expansion, stress development and generation of microcracks. The effect of sulphate crystallization is predominant in the pores of diameter greater than 0.5 μm, whereas halite crystallization happened in pores smaller than even 0.5 μm. Hence the relative percentage of pores lesser than 0.5 μm is more for T1 bricks compared to that in T2 bricks, as shown in Table 2. Table 3 shows the pore parameters obtained from the MIP test. The critical pore diameter of T1 bricks is obtained from MIP is 6.27 μm, which is about six times more than the critical pore diameter of T2 bricks. The higher critical pore diameter of T1 samples confirms that T1 bricks are subjected to a greater extent of damage than T2 bricks. Figure 3 shows the Porosimetry curves of both sets of samples.

Table 2. Relative proportions (as % volume) of various pore sizes

Pore diameter ranges (μm)	Relative volume percentage	
	T1	T2
100 –10	19.69	22.64
10 –1	43.28	39.93
1–0.3	7.43	17.81
0.3–0.2	1.83	3.46
0.2–0.1	2.62	4.34
0.1– 0.05	3.30	3.38
0.05–0.01	10.87	6.31

Figure 4(a) and (b) are the representative micrographs of T1 and T2 bricks, respectively. The porosity analysis of these SEM images using the MATLAB code developed by Caglar Arslan estimated the porosity of images through the K-Means Clustering algorithm [1]. This unsupervised classification algorithm segments the area of our interest from the background. Hence, the pores seen as black regions in the SEM micrographs are segmented from the solid products. The image is divided into four discrete regions by assigning the cluster number as four. Randomly chosen cluster centroids are further improved by iteration. Depending on the distance between data points and the centroids, reassignment of data points to nearby clusters is performed. The iteration is continued till the data points are not reassigned or till the assigned number of iterations is reached. Prior to the porosity calculation, the Image Serializer function serializes the pixels of the

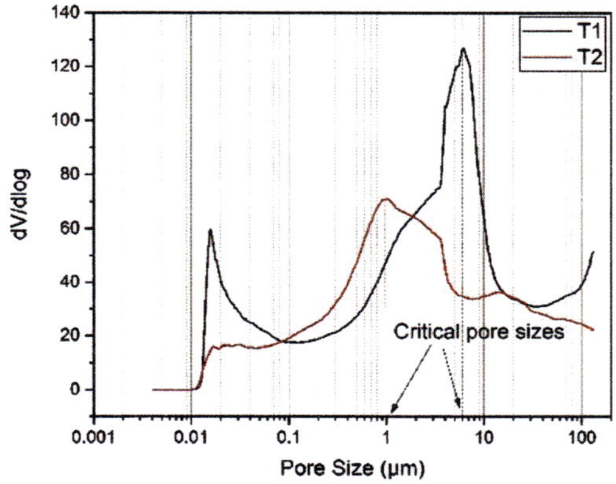

Fig. 3. Differential intrusion curves showing the critical pore diameter for the bricks

Table 3. Density and pore parameters obtained from MIP study

Properties	T1	T2
Bulk density (g/cm^3)	1.89	2.03
Porosity by Hg intrusion (%)	35.29	27.88
Threshold pore diameter (μm)	8.01	2.5
Critical pore diameter (μm)	6.27	1.01

image. Through image analysis with the SEM images as shown in Fig. 4(a and b), T1and T2 samples showed 29.39% and 27.64% porosity, respectively. The binarised images generated from the SEM micrographs of T1 and T2 samples are shown in Fig. 4(c) and (d), respectively. From the analysis of about 3 micrographs corresponding to each type of brick sample, T1 samples showed higher porosity than T2 samples.

On comparing the porosities obtained from Mercury Intrusion Porosimetry and SEM image analysis, it can be seen that the porosities are essentially identical in both cases for the sample T2, which was exposed to sodium chloride weathering predominantly. In contrast, the porosity obtained from image analysis is much lower (29.39%) case of T1 brick compared to porosity from MIP (35.29%). This trend of T1 is general and stems from the limitation of SEM to capture smaller pores, which can be captured in a MIP test even up to 3 nm. The mercury intrusion under pressure can fill even the nanopores in the sample. Thus, the reduction of porosity of the T1 sample in image analysis indicates the presence of nanopores, with size below the resolution of SEM. It is possible that this expected general reduction in porosity from image analysis was not observed in the case of T2, in accordance with the understanding of the mechanism of action of sodium chloride salt. With the low ionic size and high mobility of the ions, sodium chloride can precipitate and fill pores of micro ranges—which would have contributed to the porosity value in MIP otherwise [11]. Hence, the results of image analysis and MIP studies are

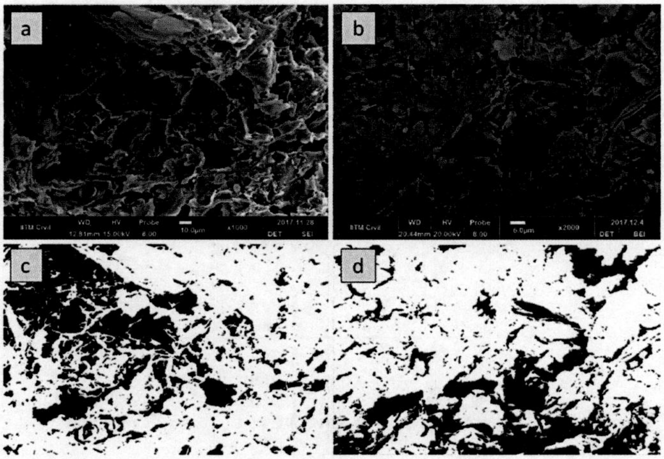

Fig. 4. (a, b) Representative micrographs of the T1 and T2 bricks, respectively (c, d) Binarised images from the SEM micrographs of T1 and T2 bricks, respectively.

corroborating to each other. Therefore, the results suggest that porosity analysis using SEM images coupled with K-means cluster algorithm ensure the calculation of open and close pores. However, SEM resolution limits their accuracy in determining finer micro and nanopores. Consequently, the best-suited technique has to be chosen depending on the range of pores and pore structure likely to be present in the sample [4].

4 Laboratory Weathering Tests

The performance of bricks with various microstructures on exposing to the salt solution was studied by subjecting the specimens to accelerated salt weathering tests as per RILEM Test V 1.a, exposing to sodium sulphate (10% sodium sulphate anhydrate) and sodium chloride (15.6%) solutions separately. Mass changes were documented for the samples after each day (each cycle) for 100 cycles, and Fig. 5 shows the mass change ratios calculated for the specimens. C1-Cl and C2-Cl are C1 and C2 specimens exposed to sodium chloride solution, and C1-S and C2-S are those exposed to sodium sulphate solution, respectively. For sodium sulphate exposure, all the specimens had a mass gain period, followed by a period of mass loss. The mass gain was due to the continuous deposition of salt crystals in the pores of the material, and mass loss initiates on the onset of damages due to crystallization pressure. However, in the case of sodium chloride exposure, for specimens C1-Cl and C2-Cl, the mass remains unchanged after the initial turbulence because the halite crystals get deposited inside the pores, and they do not cause crystallization pressure or damage. The observations were similar throughout the continuous accelerated exposure of 100 cycles. These results are in line with the observations on the salt-affected bricks from the field, where NaCl affected samples were of low porosity and high strength because of salt deposition, whereas Na_2SO_4 affected samples were highly disintegrated.

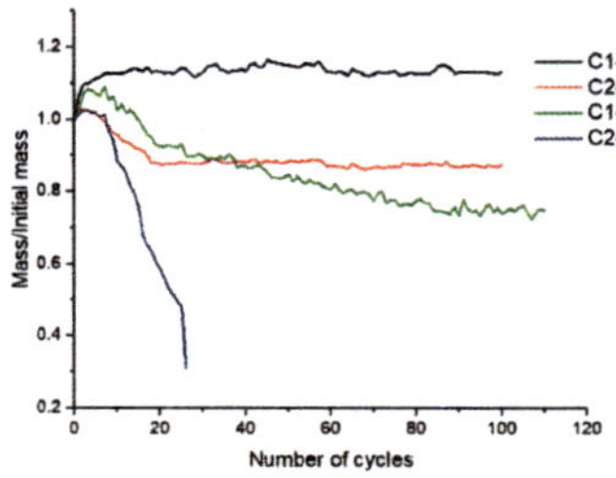

Fig. 5. Mass changes for the samples during accelerated salt weathering

5 Correlating the Characteristics and Performance Parameters

The higher compressive strength of T2 bricks (10.1 MPa) can be correlated to the lesser porosity, attributed to the presence of excess NaCl. The NaCl molecules being highly soluble, penetrate the pores efficiently and get crystallized on evaporation which densifies the microstructure. Though expansion, cracking, and subsequent damage are also reported in the literature because of salt crystallization, T2 bricks did not show cracks but got softened and eroded progressively. T1 bricks from the dry walls had both chlorides and sulphates present, as observed from the X-ray diffractograms and showed higher porosity, lower strength, and excessive deterioration than T2 bricks. The phase change of sodium sulphate from thenardite (Na_2SO_4) to mirabilite ($Na_2SO_4.10H_2O$) is accompanied by volumetric expansion and consequent intergranular cracks. The higher solubility of sulphates can result in high supersaturation producing higher crystallization pressure. The expansion caused by hydration of the anhydrous phase of sodium sulphate and stress caused by crystallization weaken the system and is a crucial damage mechanism in porous materials [7]. The presence of sulphates is hence critical to the durability of a masonry structure.

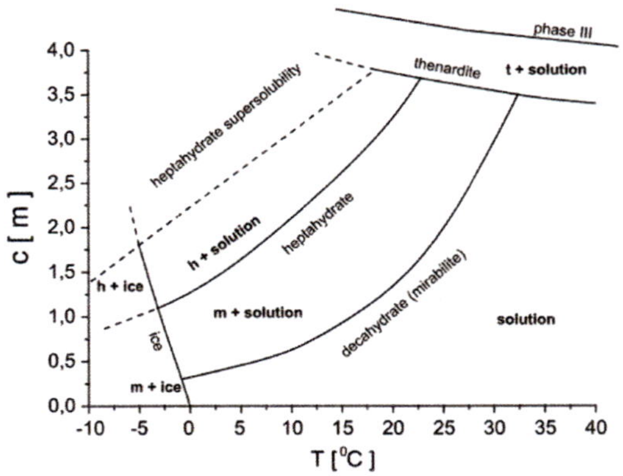

Fig. 6. Phase diagram of sodium sulphate showing the equilibrium solubilities of various phases (h, t, m for heptahydrate, thenardite and mirabilite, respectively)

The sample T1 was collected from the coastal region, where the temperatures range between 25 °C in the night and 40 °C in the day most of the time. As it can be seen from the phase diagram given in Fig. 6, thenardite is the only stable phase at temperatures greater than 32 °C. So, thenardite crystallizes directly from the solution at these temperatures [6]. On the other hand, mirabilite is stable at temperatures below 32 °C. However, the crystalline phase of mirabilite crystallizing directly from the solution is stable with respect to dehydration only at relative humidity greater than 70%. Tranquebar being a coastal region, the relative humidity increases to more than 75%. This facilitates direct precipitation of mirabilite and also the dissolution-mediated phase transformation of thenardite to mirabilite. Furthermore, the mechanism of crystallization in natural systems is heterogeneous, nucleating from imperfections and dust (multiple phases), rather than the homogeneous crystallization from pure solution (two-phase) as suggested by the phase diagram [15]. Heterogeneous crystallization results in non-equilibrium crystals of thenardite along with equilibrium crystals of mirabilite. Therefore, the reasons for the presence of both the mirabilite and thenardite simultaneously in T1 bricks can be summarised as,

- Surrounding temperature varying across a wide range (both less than and greater than the phase transition point of 32 °C).
- High relative humidity (stabilizing crystalline mirabilite).
- Dissolution-mediated phase transformations.
- Heterogeneous (multiple-phase) crystallization.

6 Conclusions

The current study focused on analyzing two historic salt-affected brick samples from the field, T1 and T2, of distinct damage patterns, followed by linking the interpretations to the results of an accelerated salt crystallization test conducted in a controlled environment on undamaged brick samples—C1 and C2. Correlating the insights from the field and the lab studies helped reinstate the effects of NaCl and Na_2SO_4, as arrived at from the analysis of the field samples. Even though the potential caution on the complexities of multiple salts and environmental conditions cannot be neglected, the artificial laboratory studies confirmed the observation made in line with the mechanism of action of salts in the field samples. T1 bricks, which were exposed to the coastal environment without direct contact with seawater, and T2 bricks, which were partially or fully immersed in seawater, were tested for physicomechanical properties, water absorption, bulk density and compressive strength; mineralogy and pore structure. X-Ray Diffraction was performed, and EDS analysis from SEM was conducted for mineralogical characterization. Results of the porosity analysis using MIP was compared with the porosity values obtained by image analysis with MATLAB. The major conclusions from the study are the following.

- Despite an increase in moisture content of about 23% for T2 bricks, they showed a higher compressive strength of 10.1 MPa, whereas T1 samples of lower moisture content exhibited lower compressive strength of 4.2 MPa. The increased strength of T2 samples can be correlated to their lower porosity compared to T1 samples, as obtained from MIP and possible sodium chloride salt accumulation in the pores.

- The mineralogical characterization using XRD and EDS analysis confirmed the presence of halite (NaCl) in the T2 samples, which were partially or fully immersed in seawater. Blistering of T1 samples can be correlated to the presence of gypsum. The phase conversion from thenardite to mirabilite and the volumetric conversion were evident from the XRD peaks of T1 samples. The absence of diopside and gehlenite indicated that both the brick samples were fired at low temperature.
- The results of Mercury Intrusion Porosimetry (MIP), where T1 samples showed a higher porosity (35.29%) than T2 samples (27.88%), suggested the possibility of halite crystallization in the pores of T2 bricks resulting in porosity reduction. It also indicated sulphate crystallization in T1 samples accompanied by volumetric expansion, causing microcracks. The notable damage undergone by the T1 samples were reflected in the critical pore diameter (6.27 μm, which is about six times more than that of T2 samples (1.01 μm).
- Image analysis of SEM micrographs using MATLAB indicates the same porosity trend for the samples as obtained from MIP. The porosity of T1 bricks (29.39%) was lower than the value obtained from MIP, whereas T2 bricks showed almost comparative porosity (27.64%). This indicates the presence of micro and nanopores of size lesser than the resolution of SEM in the T1 bricks. Such smaller pores were not of the considerable amount in T2 samples. This might be attributed to the halite crystallization on the pores of T2 bricks.
- Accelerated salt weathering tests in the laboratory with NaCl and Na_2SO_4 salts on bricks with different microstructure confirmed the difference in the action of salts as observed in the field samples. Sodium sulphate exposure in the lab resulted in initial mass gain due to the deposition of salt crystals in the pores and subsequent mass loss by the deterioration due to crystallization pressure. Sodium chloride exposure retains a constant mass of the samples after the initial dissolution-precipitation phase. This might be due to the penetration of small NaCl molecules and the deposition of halite crystals on the pores without exerting crystallization pressure on the pore walls. Also, the condition for the coexistence of the two phases of sodium sulphate -thenardite and mirabilite in the field samples T1 could be explained from the phase diagram of sodium chloride, and the varying temperature and humidity conditions in the particular site.

References

1. Caglar Arslan.: Porosity Calculator for SEM Images, https://www.mathworks.com/matlabcentral/fileexchange/66331-porosity-calculator-for-sem-images, MATLAB Central File Exchange. Accessed 13 May 2021
2. CEN (European Committee for Standardization).: Natural stone test methods: Determination of resistance to salt crystallization. EN 12370:1999. CEN, Brussels, Belgium (1999)
3. Charola, A.E.: Salts in the deterioration of porous materials: An overview. J. Am. Inst. Conserv. **39**, 327–343 (2000)
4. Divya Rani, S., Rahul, A.V., Santhanam, M.: A multi-analytical approach for pore structure assessment in historic lime mortars. Constr. Build. Mater. **272**, 121905 (2021)
5. Doehne, E.: Salt weathering: A selective review, salt weathering: A selective review. Geol. Soc. Spec. Pub. **205**(1), 51–64 (2002)

6. Flatt, R.J.: Salt damage in porous materials: How high supersaturations are generated. J. Cryst. Growth **242**(3–4), 435–454 (2002)
7. Flatt, R.J., Caruso, F., Sanchez, A.M.A., Scherer, G.W.: Chemo-mechanics of salt damage in stone. Nat. Commun., Nature Publishing Group **5**, 1–5 (2014)
8. Goudie, A.S., Viles, H.A.: Salt Weathering Hazards. Wiley, Chichester (1997)
9. Lubelli, B., et al.: Towards a more effective and reliable salt crystallization test for porous building materials: state of the art. Mater. Struct. **51**(2), 1–21 (2018). https://doi.org/10.1617/s11527-018-1180-5
10. Lubelli, B., Hees, R.P.J. van, Nijland, T.G.: Salt crystallization damage: How realistic are existing ageing tests? In: AMS '14 Proceedings of the International Conference on Ageing of Materials and Structures, Vol. 2(1), pp. 103–111 (2014)
11. Manohar, S., Santhanam, M.: Correlation between mineralogical properties and weathering resistance using characterization case studies in historic Indian bricks. Int. J. Archit. Heritage 1–14 (2020)
12. Manohar, S., Santhanam, M., Chockalingam, N.: Performance and microstructure of bricks with protective coatings subjected to salt weathering. Constr. Build. Mater. **226**, 94–105 (2019)
13. Manohar, S., Chockalingam, N., Santhanam, M.: Experimental comparison between salt weathering testing procedures on different types of bricks. J. Mater. Civil Eng. ASCE (2021). ISSN 0899-1561
14. RILEM: Recommended tests to measure the deterioration of stone and to assess the effectiveness of treatment methods, Test V.1a—crystallization test by total immersion (for untreated stone); Test V.1b—crystallization test by total immersion (for treated stone); Test V.2— crystallization test by partial immersion. Mater. Struct. **13**(75), 175–253 (1980)
15. Rodriguez-Navarro, C., Doehne, E., Sebastian, E.: How does sodium sulfate crystallize? Implications for the decay and testing of building materials. Cem. Concr. Res. **30**(10), 1527–1534 (2000)
16. Scherer, G.W.: Stress from crystallization of salt. Cem. Concr. Res. **34**(9), 1613–1624 (2004)

Durability of Heritage Masonry Structures: Review on the Substrate-Mortar Interaction

V. A. Anupama(✉) and Manu Santhanam

Department of Civil Engineering, IIT Madras, Chennai, India
vaanupama9@gmail.com

Abstract. Much of the built heritage stock comprises masonry units of stone or brick, adhered by lime mortar. The properties of the materials used in the construction play a pivotal role in the strength and long-term durability of the masonry structures. The breathability of lime mortar, the porosity of substrates, and the properties of the substrate-mortar interface are critical to the extended serviceability of the structure. The current study focuses on a review of the substrate-mortar interface properties developed over short-term and long-term durations. The fundamental mechanisms of adherence and material transport across the interface were studied in detail. Analysis of physico-hygro-thermal influences on masonry structures was found to give crucial insights on the porosity and pore size distribution of the unit, mortar and the interface. The mortar properties such as water content and composition were found to have a significant impact on the bond strength development. The microstructural investigations using characterization techniques confirmed that the network of products formed at the interface by hydration and carbonation also governed the interface porosity. The study on the weathering characteristics that modify the transport parameters is critical for understanding the fundamental principles behind moisture attack and salt crystallization in ancient structures, typically on coastal monuments. The study provided a comprehensive insight into the transport mechanisms and the critical parameters for consideration during the design of mortars for restoration.

Keywords: Masonry structures · Repair mortar · Salt crystallization · Heritage masonry

1 Introduction

Masonry structures are a heterogeneous assemblage of substrates (brick, stone or concrete) and mortar joints, the latter providing integrity to the system [1, 2]. The properties of each component and their geometry influence the characteristics of the masonry. The predominant discontinuity in the masonry system can be attributed to the interface between the substrate and the mortar, which carries most of the material variation in a masonry building [3]. Several engineering properties demand detailed investigation for assessing current structural performance and the design of repair mortars for masonry structures. Examination of properties focusing on the interaction between the components is critical to ensure a durable repair.

J. I. Escalante-Garcia et al. (Eds.): RW 2021, RILEM Bookseries 40, pp. 729–739, 2023.
https://doi.org/10.1007/978-3-031-21735-7_78

1.1 Important Engineering Properties of Masonry

Properties such as compressive strength, tensile strength, stiffness, bond strength, porosity, pore size distribution and water transport mechanism are significant in ensuring strength and durability of structures and structural components.

Adhesion is critical to ensure load transmission and prevent detachment [4]. The compressive strength of the brickwork mainly depends on adhesion, mortar strength, unit strength, relative ratio between them, the orientation of the unit in relation to the load application direction and the thickness of the mortar joint [5–8]. A uniaxial compressive load on the masonry prism induces triaxial compressive stresses on the mortar whereas, the brick unit experiences biaxial tension and uniaxial compression, as presented in Fig. 1 [10]. The failure manifests as cracks parallel to the loading direction on the brick units when the biaxial tensile stress exceeds the tensile strength of the unit, where failure is by crushing [9].

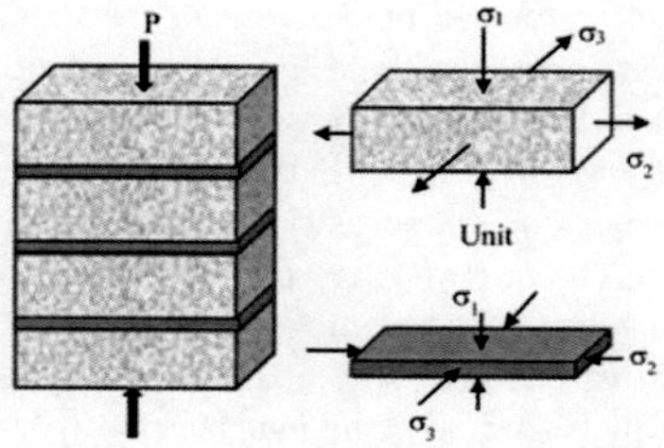

Fig. 1. Schematic of the masonry composite behaviour under uniaxial compression [10]

If the bricks possess high tensile strength, the cracks in flexure tend to pass through the mortar joints in a stepped fashion, whereas the cracks split the bricks with low tensile strength [11]. Costigan and Pavia [12] state that the mechanics of mortar and masonry are similar, though it demands further investigation. Highly hydraulic mortars are elastic, and so is the masonry constructed using them. The strength parameters are also influenced by the moisture transport mechanism, which depends on porosity, pore size distribution, and contact configuration of the building materials [12, 13].

1.2 Factors Affecting the Performance of Masonry Structures

Properties of the individual components, bond, and environmental conditions play a significant role in maintaining the strength and durability of masonry structures. Lime-based mortars retain more water than cement mortars and were used for bedding, plastering and incised works. An optimum rate of initial water transport is essential to acquire adherence and complete hydration of the mortar [14]. Masonry mortar comparatively has lesser strength than the unit. Consequentially, the structure in the Flemish bond by virtue of the higher volume of mortar shows a lesser compressive strength than the English bond style [15]. As shown in Fig. 2, Flemish bond has alternate headers and

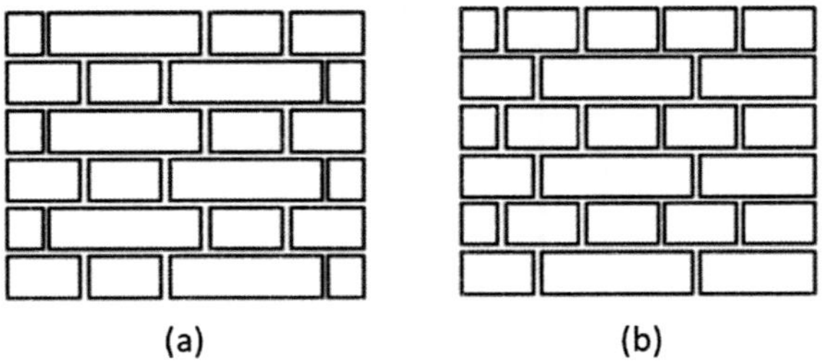

Fig. 2. **a** Flemish bond, **b** English bond [15]

stretchers in the same course whereas English bond has alternate courses of headers and stretchers.

Extreme environments such as heavy precipitation, freeze-thaw and saline water limit the durability of buildings. Masonry structures at locations of heavy rainfall can be subjected to moisture infiltration through walls and rising dampness by capillary action, which might carry soluble salts [2]. Literature suggests that mortar shall allow minimum moisture infiltration and maximum evaporation under conditions that minimize salt crystallization, considering the wide range of factors affecting salt ingress [14].

1.3 Significance of Research

Until the early 19th century, masonry mortar was predominantly composed of non-hydraulic lime, later partially replaced by hydraulic lime and Portland cement [4]. Lime-based binders are flexible to accommodate minute building movements without cracking [2]. Moreover, the higher porosity of the former contributes to the thermal insulation of the building. The sacrificial nature of lime mortars is attributed to the breathability property, which allows the free evaporation of moisture without damaging the substrate unit. However, the practical challenges of using lime-based binders include longer setting time, higher shrinkage, slower stiffness gain, low strength potential and higher porosity [16, 17]. Though cement mortar is used widely for restoration, it is a poor alternative to the historic lime mortars in repairing heritage masonry structures. Aesthetic incompatibility, low plasticity, excessive brittleness and early stiffness gain are a few obstacles in using cement-based mortars [2, 14]. In addition to their thermal expansion coefficient, low deformability damages stone structures [18].

This review examines the vital engineering properties to be considered during repair mortar design to ensure the durability of masonry structures.

2 Selection of Mortars for Masonry Repair

Mortars are cohesive mixes of binder, water and fine aggregates. The physico-chemical and thermal compatibility of the mortar and unit that can be examined through their microstructural analysis is expected to reduce discontinuity in the masonry [4]. Thus, knowledge about the parent characteristics aid in designing repair mortars with the target properties. Strength in compression and tension, stiffness, porosity and transport mechanisms are the major factors governing compatibility [1].

2.1 Adherence

The bond between substrates and mortar is maintained by the cohesion and friction between the two surfaces [9]. Chemical composition and microstructure of the individual components and their interface influence the bond development [20]. To ensure failure through mortar or interface instead of unit failure, adhesive strength should not be higher than the tensile strength of the unit [19].

The bond strength is established by wetting and the hydraulic drag of fines towards the substrate [21]. The higher water content in the mortar creates channels to facilitate the flow of water and mortar fines (sand fines and binder) into the porous substrate by virtue of capillary suction driven by the difference in suction potential between the substrate and the mortar [7]. The transport of fines stops within seconds if the substrate has a high Initial Rate of Absorption (IRA), whereas it can extend for a few minutes if the substrate has a low IRA [6]. Consequently, mechanical bond forms between the two materials. The formation of cement hydration products such as portlandite and CSH within the brick surface pores and on the brick surface strengthens the interface. This network formation is the primary bonding mechanism between brick and mortar, aided by prewetting [5, 7]. The mortar's volume of coarse pores (pores above the critical pore diameter of the substrate) influences the transport of fines and strengthens the bond between these elements. Insufficient solids at the interface can result in adhesive failure. However, excess solids at the interface can result in less-paste porous zones next to the interface having poor cohesive strength [7].

The higher initial moisture content of the brick can result in a weaker bond because of less suction pressure, resulting in less transport of moisture and fines to the interface. Although, dry brick absorbs excess water from the mortar leaving insufficient water for cement hydration, causing drying shrinkage. Hence, both these conditions are to be avoided [22]. For a brick with moderate suction, lime addition reduces bond strength because of high water retention property, leading to fewer hydrates at the interface. However, lime addition improves bond strength when used along with high absorption brick because lime retains water within the mortar providing sufficient water for cement hydration [23]. Using pure cement mortar in such a scenario shall be avoided as water drawn towards the substrate leaves insufficient water within the mortar for complete hydration. Excess lime addition increases water retentivity and can result in brick pores devoid of water which ultimately causes weaker bond. Hence, optimum lime content in the mortar composition must be determined [23]. Like lime addition, silica fume addition also increases the water retention capacity of mortar because of its high surface area. Polypropylene fibres can control cracks and block the pores, thereby increasing water retention and reducing permeability. Hamdy et al. (2019) suggest adding 2% of lime, silica fume or polypropylene fibres to obtain acceptable results. An increase in the percentage of additives shows a decrease in bond strength due to increased water retention and reduced permeability [23]. Hence, mortars of higher water retentivity are preferred along with bricks of higher suction. The influence of the Ca/Si ratio in mortar and fineness of the sand used in mortar in the bond development between the unit and mortar is contradictory in literature. Marusin [20] suggests that as the Ca/Si ratio increases, adherence also increases.

2.2 Compressive Strength

Mortar strength is influenced by the type of binder, degree of fineness, quality of ITZ (the zone between the aggregate and the paste) and presence of additives. The addition of pozzolans improves the hydraulic reactivity and thereby enhances the mechanical properties of the mortar at the hardened state [1, 4]. However, mortar strength must be lesser than unit strength to facilitate pore-collapse mechanism before unit breakage [24]. As the challenge in repairing or replacing the substrates is more than the challenge in repairing the mortars, interface failure is preferred over substrate failure. The mechanical strength of the interface is designed to be lower than the substrate strength. Hence interface is the weakest link in a masonry structure [19, 20]. Furthermore, the strength of the interface influences the post crack behaviour of the masonry with low bond strength. Masonry structures that experience mortar deterioration due to weathering or mechanical actions exhibit failure through the interface and are manifested as debonding [25].

For cement mortars, a higher percentage of strength and stiffness develops within the 28th day, which is contributed by the hydration of cement. In contrast, lime carbonates at a later stage, causing strength and stiffness gain after 28 days. As the grade and lime content increase, flexural and compressive strength of the mortar increase over age, initially by hydration followed by carbonation. Carbonation at a later stage and higher specific surface area requiring more water to attain the same workability might be the reasons for the reduced early strength of lime mortars [26]. The compressive strength variation among mortars can thus be represented as: cement mortar > hydraulic lime > air lime[27]. Cohesion increases and strength decreases with a higher binder:aggregate ratio. However, even when the binder:aggregate ratio is too low, mechanical strength decreases due to poor binding and cohesion [28].

2.3 Tensile Strength

Air lime mortars show higher tensile strength than natural hydraulic lime mortars. Santos et al. (2020) suggest that the higher tensile strength can be attributed to the higher water retentivity and larger coarse porosity of air lime mortars. Hence, air lime mortars require more water to attain the same flow compared to hydraulic lime mortars. The excess water carries fines while penetrating unit pores and establishes an efficient bond with the substrate [29]. Microcracking precedes the tensile failure in a unit-mortar interface. They initiate at the weaker stress-concentrated zones and ultimately merge to one large crack before failure in tension [3].

2.4 Stiffness

Masonry mortar is expected to be deformable enough to accommodate substrate movements and dissipate the stresses by relaxation. Hence, modulus of elasticity lower than that of masonry is preferred for the mortar [4]. As the hydraulic strength of the mortar increases, they become more elastic and brittle. HL mortars are rigid mixes having long elastic regions with less or no plastic regions, whereas CL90 mortar has a lesser elastic region with an extended plastic region [4, 12]. Nezerka et al. (2014) suggest that the brick-fragment addition to lime mortar increases water absorption, promoting stiffness gain. In lime-metakaolin mixes, the formation of hydration products, such as calcium silicate hydrates (C-S-H), C_4H_{13}, C_3AH_6 and C_2ASH_8, contributes to the early stiffness compared to the pure lime mixes [28].

2.5 Porosity

Porosity and pore size distribution is crucial for the long-term durability of the masonry as it governs moisture transport and the space available for salt crystallization [4]. Paste, ITZ, and entrapped air/cracks are the three sources of porosity in a masonry mortar [25]. Based on the size range, IUPAC classifies pores into macropores (>0.05 μm), mesopores (0.002–0.05 μm), and micropores (<0.002 μm) [30]. Macropores influence the mortar strength, mesopores influence the transport properties, and micropores indicate the sorption properties[31]. Generally, historic lime mortars show pores of size ranging between 0.1 μm and 4 μm. Non-hydraulic lime mortars exhibit higher porosity (about 50–60%) than hydraulic lime and cement mortars.

The volumetric ratio between the binder and the aggregate, granulometry of aggregates, water to binder ratio, water content in the mortar, grade of compaction and binder type influence porosity. A higher binder to aggregate ratio results in higher paste content, resulting in higher porosity [5, 24]. Mortar porosity generally reduces with time as portlandite coverts to calcium carbonate because this conversion causes an 11.8% volume increase [32]. However, if ceramic dust and fragments are added and cured under laboratory conditions, porosity increased due to the fissures and fracture porosity formed in the mortar. This increase in porosity is due to self-healing, and the tabular $CaCO_3$ formation, which increases stress on the surrounding matrix leading to crack initiation [33]. The carbonation rate of mortars with limestone aggregate is higher because of the compositional similarity, which acts as the nucleation sites for newly formed calcite and CSH. Hence, a more compact structure is formed with a lesser mean pore diameter [34]. The addition of PCE increases porosity by dispersing the CH particles. With the addition of cellulose, water retentivity increases, which reduces rapid evaporation and thereby decreased porosity. Crushed clay brick addition to lime mortars reduces pore radii by the formation of hydration products at the lime-fragment interface. The hydration products which contribute to the early stiffness of lime-metakaolin mixes are also responsible for porosity reduction; as the percentage of metakaolin in a lime mortar increases, porosity decreases.

Extraction of water by the units from the fresh mortar applied between them reduces the water to binder ratio, which further decreases the mortar porosity [35, 36]. Subsequently, mortar penetrates the substrate pores and alters the interface, as a result of which the pore size distribution moves towards a finer range at the interface. The penetration of CSH crystals into the smaller pores is critical in establishing a stronger bond. Hence, greater the percentage of pore volume on the brick surface represented by pores with sizes below 1μm, better bond strength is observed when used along with hydraulic lime mortars [34].

2.6 Transport Mechanisms

The contact between mortar and units involving interpenetration of their porous structure, namely, hydraulic continuity, is responsible for the transport mechanism observed in masonry structures [37]. Lime mortar facilitates liquid water flow and vapour transport, mainly affected by capillary conductivity (K) and capillary suction (∇p_c). As ∇p_c increases and K decreases with a reduction in pore size, a balance between both the parameters is essential to facilitate moisture transport. In addition, the finer pores in the fine mortars result in higher porosity, increasing moisture flow and vapour transport [4].

Capillary transport in multi-layered masonry systems differs from the transport through homogeneous systems [24, 38]. Variations in the hygric properties are attributed to the heterogeneity in the system involving porous unit and mortar. Additionally, intra-component abnormalities due to the difference in manufacturing processes and conditions also contribute to the disparity. The hygric interaction between the porous substrate and the mortar results in interface resistance, which retard the moisture transfer rate as the moisture reaches the interface [39]. Interface resistance increases due to transport of mortar fines to the substrate, shrinkage microcracks and difference in the pore systems of unit and the mortar [18, 24, 27]. The interface resistance also is governed by the moisture extraction while curing, which is nil for mould-cured mortar, low for wet-cured mortar and comparatively higher for dry-cured mortar [40]. The mortar application prevents the free flow of gases from the top face of the unit and changes the capillary absorption coefficient of the unit. The air spaces between the unit and mortar influence transport by contributing a high capillary coefficient to the interface and retarding the moisture absorption rate [41]. The interface resistance develops only during the hydration or carbonation of the binder, and hence fresh mortar does not exhibit discontinuity of pore structure [14].

3 Influence of Mortars in Salt Crystallization

Lime mortar is characterized by lower mechanical resistance and higher porosity, enhancing the susceptibility to long-term deterioration, the major being salt crystallization [42]. Heritage masonry structures along the coastal line experience the threat of salt attack. The salt crystal, whose growth is restrained by the pore walls, exerts tensile stresses on the surroundings. Cracks initiate, and the structure undergoes deterioration when the crystallization stresses exceed the tensile strength of the material [43]. The stresses due to crystallization are significant in smaller pores compared to the pressure developed in bigger pores. Hence, the strength and porosity of the mortar play a crucial role in the susceptibility to salt crystallization. The higher absorption values of mortars result in water build-up, causing an increased probability for salt crystallization [44]. Apart from the material parameters, one major factor in the crystallization process is moisture availability. The ions are introduced in the building pores and transported into the different parts through moisture. In multilayer renders, the inner coats experience a higher probability of moisture infiltration and inefficient drying than the external mortar layer, resulting in crypto-florescence [4]. If the evaporation rate is low, the water escapes from near the surface, resulting in needle-like salt structures, known as efflorescence. Whereas, if the evaporation rate is high, the evaporation occurs within the structure, resulting in salt crystals forming inside the walls and the pores, known as Subflorescence [45]. Phase conversion of sodium sulphate from thenardite (Na_2SO_4) to mirabilite ($Na_2SO_4.10H_2O$) occurs during favourable temperature and humidity conditions. The transition from the anhydrite form to the decahydrate form involves expansion and causes subsequent intergranular cracks Hence, the pore networks within the system play a significant role in ion mobility and the extent of deterioration [45].

4 Research Gaps to Be Addressed

Although the deterioration of heritage masonry structures is a global scenario causing substantial economic damage, a scientific approach to analyzing the parent material properties and formulating repair material compositions are limited. As the properties of the individual components change at the interface and are non-additive, the substrate-mortar interface needs to be studied at the microstructural level. Though the unit properties and mortar characteristics are examined in the literature, the substrate-mortar interface requires additional focus. The interface modifies the porosity and pore size distribution and plays a pivotal role in moisture transport through multi-layered masonry systems. As it influences adhesion between the masonry components as well, the impact of the interface in ensuring the durability of structures is significant. Several mechanisms of adhesion are addressed in the literature. However, they are limited to the physical bond development process though unidentified XRD peaks are observed in a study [5]. Hence, the studies need to be extended to the prospects of chemical interaction between the components. The role played by different polymorphs of calcium carbonate in providing strength and interface resistance can be examined. The majority of the studies focus on adhesion, considering brick as the substrate and hence the mechanism of hydraulic drag of fines is addressed only within that context. A broader exploration inclusive of brick and stone substrates is inevitable to validate and extend the existing theories on bond formation. Analysis of transport mechanism on stone masonry structures is essential for a better understanding. However, compatibility can be ensured only by extending the studies to accommodate the weathering characteristics of the substrate, mortar, and the assemblage. Research focusing on mortar compositions and their response to different long-term exposure conditions is hence the need of the hour.

5 Conclusions

The study focused on reviewing various properties of the masonry, substrate and mortar, which are significant in designing repair mortars for masonry structures. The fresh and hardened properties of the mortar, engineering properties of the unit, and interface characteristics influence the durability of a repair system. Formulation of repair mortars should consider bond strength, compressive strength, stiffness, porosity and pore size distributions. Adhesion between masonry components involves hydraulic drag of fines to the unit pores and surface, modifying the interface. Compressive strength and porosity follow an opposite trend in masonry mortar with lime addition; higher the lime content, higher the porosity, and lower the compressive strength. Stiffness of the mortar influences the mechanics of the masonry. As stiffness increases with the hydraulic strength of the mortar, air lime mortars are plastic, and hydraulic lime mortars are elastic. The hygric interaction between the unit and the mortar develops interface resistance which modifies the transport mechanism. Thus the capillary transport in a heterogenous masonry system differs from transport in a homogenous system. Salt crystallization can cause severe damage to the masonry structure, especially in the low-strength systems with small pores. The network of pores in the mortar and substrate plays a vital role in the susceptibility to salt crystallization.

References

1. Ioannou, I., Philokyprou, M., Ilia, A.: Design and production of repair mortars for historic masonry taking into account the characteristics of old mortars 1051–1058 (2010)
2. Hamiane, M., Aitouakli, S., Boumghar, L., Zekagh, W., Saidi, M.: Design of jointing mortar compatible with the masonry of the Villa Maheddine of the "Ottoman Era" 1023–1033 (2010)
3. Shadlou, M., Ahmadi, E., Kashani, M.M.: Micromechanical modelling of mortar joints and brick-mortar interfaces in masonry structures: a review of recent developments (2019)
4. Nogueira, R., Ferreira Pinto, A.P., Gomes, A.: Design and behavior of traditional lime-based plasters and renders. Review and critical appraisal of strengths and weaknesses. Cem. Concr. Compos. **89**, 192–204 (2018)
5. Chase, G.W.: Characterization of the interface between brick and mortar (1983)
6. NPTEL__Civil Engineering—NOC_Design of Masonry Structures (n.d.)
7. Page, A., Lawrence, S.J.: The development of mortar/unit bond (2014)
8. Fortes, E.S., Parsekian, G.A., Camacho, J.S., Fonseca, F.S.: Compressive strength of masonry constructed with high strength concrete blocks. Rev. IBRACON Estruturas e Mater. **10**, 1273–1319 (2017). https://doi.org/10.1590/s1983-41952017000600008
9. Binda, L., Tedeschi, C., Cardani, G.: The role of mortar joints in the mechanical behaviour of masonry structures. Work. Repair Mortars Hist. Mason. Pro **067**, 35–46 (2009)
10. Zahra, T., Dhanasekar, M.: A generalized damage model for masonry under compression. Int. J. Damage Mech. **25**, 629–660 (2016)
11. Zhou, Z., Walker, P.: Strength characteristics of hydraulic lime mortared brickwork, 139–146 (2008)
12. Costigan, A., Pavía, S.: Influence of mechanical properties of lime mortar on the strength of masonry, 457–466 (2010)
13. Delgado, J.M.P.Q., Azevedo, A.C., Guimarães, A.S.: Influence of hydraulic contact interface on drying process of masonry walls. Dry. Technol. **38**, 1121–1137 (2020)
14. Hendrickx, R., Roels, S., Van Balen, K.: Water transport between mortar and brick: the influence of material parameters. RILEM Bookseries. **7**, 329–341 (2013)
15. Taguchi, T., Cuadra, C.: Influence of bond types on brick masonry strength. 15th World Congr. Adv. Struct. Eng. Mech. (2015)
16. Papayianni, I., Papadimitriou, D.: Lime based mortars. Relationships Between Compos. Parameters Mech. Strength **1**, 1316–1324 (2019)
17. Nežerka, V., Němeček, J., Slížková, Z., Tesárek, P.: Investigation of crushed brick-matrix interface in lime-based ancient mortar by microscopy and nanoindentation. Cem. Concr. Compos. **55**, 122–128 (2015)
18. Isebaert, A., Van Parys, L., Cnudde, V.: Composition and compatibility requirements of mineral repair mortars for stone—a review. **59**, 39–50 (2014)
19. Stefanis, N., Theoulakis, P.: Design of conservation mortars for the restoration of piraeus stone at the monuments of the acropolis of Athens, 1181–1188 (2010)

20. Marusin, S.L.: Investigation of shale brick interface with cement-lime and polymer-modified mortars. J. Am. Ceram. Soc. **73**, 2301–2308 (1990)
21. Betânia, E., Costa, C., Santos, M., França, D., Luiz, F., Bergossi, N.: Squeeze flow of mortars on brick substrate and its relation with bond strength, 265 (2020)
22. Cao, H.T.: National Bui lding Technology Centre, 194–204 (n.d.)
23. Hamdy, G.A., El-hariri, M.O.R., Farag, M.F.: Use of additives in mortar to enhance the compression and bond strength of masonry exposed to different environmental conditions. J. Build. Eng. **25**, 100765 (2019)
24. Verhelst, F., Kjaer, E., Jaeger, W., Middendorf, B., Van Balen, K., Walker, P.: Sustainable masonry : the importance of mortar quality (2012)
25. Sarhosis, V., Garrity, S.W., Sheng, Y.: Influence of brick-mortar interface on the mechanical behaviour of low bond strength masonry brickwork lintels. Eng. Struct. **88**, 1–11 (2015)
26. Ramesh, M., Azenha, M., Louren, P.B.: Mechanical characterization of lime cement mortars : E-modulus and fracture energy, 119–126 (2019)
27. Torres, I., Silveira, D., Flores-colen, I., Pinto, R., Matias, G.: Influence of the substrate on the mechanical characteristics of the applied mortars **2**, 1306–1315 (2019)
28. Nežerka, V., Slížková, Z., Tesárek, P., Plachý, T., Frankeová, D., Petráňová, V.: Comprehensive study on mechanical properties of lime-based pastes with additions of metakaolin and brick dust. Cem. Concr. Res. **64**, 17–29 (2014)
29. Lopes dos Santos, A.R., do R. da Silva Veiga, M., dos Santos Silva, A.M., Caliço Lopes de Brito, J.M.: Tensile bond strength of lime-based mortars: The role of the microstructure on their performance assessed by a new non-standard test method. J. Build. Eng. **29**, 101136 (2020)
30. Divya Rani, S., Rahul, A.V., Santhanam, M.: A multi-analytical approach for pore structure assessment in historic lime mortars. Constr. Build. Mater. **272**, 121905 (2021)
31. Maria, S.: Methods for porosity measurement in lime-based mortars. Constr. Build. Mater. **24**, 2572–2578 (2010)
32. Cazalla, O., Rodriguez-Navarro, C., Sebastian, E., Cultrone, G., De la Torre, M.J.: Aging of lime putty: effects on traditional lime mortar carbonation. J. Am. Ceram. Soc. **83**, 1070–1076 (2000)
33. Ergenç, D., Gómez-Villalba, L.S., Fort, R.: Crystal development during carbonation of lime-based mortars in different environmental conditions. Mater. Charact. **142**, 276–288 (2018)
34. Lawrence, M., Walker, P., Zhou, Z.: Influence of interfacial material pore structure on the strength of the brick/lime mortar bond. RILEM Bookseries. **7**, 373–381 (2013)
35. Derluyn, H., Janssen, H., Carmeliet, J.: Influence of the nature of interfaces on the capillary transport in layered materials. Constr. Build. Mater. **25**, 3685–3693 (2011)
36. Rani, S.D., Deb, M., Santhanam, M., Gettu, R.: Evaluation of repair mortar materials for old monuments in Southern India. In: Proceedings of 4th International Conference on Concrete Repair, Rehabiliation Retrofit ICCRRR 2015, pp. 138–139 (2016)
37. Guimarães, A.S., Delgado, J.M.P.Q., De Freitas, V.P.: Influence of mortar joints on the moisture transfer in layered materials, Defect Diffus. Forum **365**, 160–165 (2015)
38. Szemerey-kiss, B., Balázs Szemerey-Kiss, Á.T., Török, Á.: Bull. Eng. Geol. Environ. (2016)
39. J.M.P.Q.D.V.P.D.F.A.S. Guimarães: Water movement in building walls : interfaces influence on the moisture flux. Heat Mass Transf. **52**, 2415–2422 (2016)
40. Janssen, H., Derluyn, H., Carmeliet, J.: Moisture transfer through mortar joints: a sharp-front analysis. Cem. Concr. Res. **42**, 1105–1112 (2012)
41. Guimarães, A.S., Delgado, J.M.P.Q., Azevedo, A.C., De Freitas, V.P.: Interface influence on moisture transport in buildings **162**, 480–488 (2018)
42. Thomson, M., Lindqvist, J.-E., Elsen, J., Groot, C.: Porosity of historic mortars. In: 13th International Brick Block Masonry Conference (2004) CD

43. Flatt, R.J.: Salt damage in porous materials: how high supersaturations are generated. J. Cryst. Growth. **242**, 435–454 (2002)
44. Oan, A.F., Badawy, R.: Deterioration of masonry and concrete structures due to salt weathering : state of art **10** (2019)
45. Nogueira, R., Pinto, A.P.F., Gomes, A.: Artificial ageing by salt crystallization: test protocol and salt distribution patterns in lime-based rendering mortars **45**, 180–192 (2020)

Mexico's Caminos Reales (Royal Roads) Bridges: Location Based on Georeferencing Techniques

Andrés A. Torres-Acosta[1](✉) and Joel Bustamante-Altamirano[2]

[1] Tecnologico de Monterrey, School of Engineering and Science, Campus Querétaro, Epigmenio González 500, 76130 San Pablo, Santiago de Querétaro, Qro., México
atorresa@tec.mx

[2] SATELSA SA de CV Paseo de La Reforma 184, Lomas del Marqués, 76146 Santiago de Querétaro, Qro., México

Abstract. After the establishment of the Viceroyalty of New Spain in the 16th century, interest in precious metals and mining influenced the emerging economic activities and the communications network of New Spain. All communication infrastructures were built along two mutually perpendicular axes. The first axis ran east to west (connecting Veracruz–Mexico City–Acapulco), and the other axis ran north to south (linking the northern part of the Viceroyalty with New Mexico-Zacatecas-Mexico City–Oaxaca–Guatemala). As part of the road infrastructures needed to get a continuous flow of goods, bridges were very important for crossing rivers, cliffs, and mountains. Fabrication of these structural elements was based mainly on Roman old-school methods. The geometry, length, and materials used were varied, depending on local labor expertise and the technical improvements at the time. Vestiges of viceregal bridges are scattered across the 2 million km^2 of present-day Mexico. Most of them are still in use, but other old Royal Roads were forgotten when they went out of use. This investigation deals with the use of methodology based on information technology to create a web-based management system for viceregal bridges, which is now available for use by the three levels of government in Mexico (Municipality, State, and Federal) and by research institutions.

Keywords: Architectural heritage · Camino real de Tierra Adentro · Georeferencing techniques · Historical bridges · Information technology · Royal road

1 Introduction

1.1 Royal Roads

Mexican pre-Columbian societies used mainly footpaths for trade and to move between different places. Their means of transportation was by foot without the use of wheels

J. I. Escalante-Garcia et al. (Eds.): RW 2021, RILEM Bookseries 40, pp. 740–750, 2023.
https://doi.org/10.1007/978-3-031-21735-7_79

or pack animals. The *tameme* Indians were in charge of transporting any items or merchandize from one place to the other on foot; they could carry loads weighing ~23 kg on their backs and went as far as Guatemala.

Central Mexico initially was conquered by a small group of Spaniards, thus beginning an extensive period of conquest that continued across the land. The Conquistadors opened bridle paths made of dirt in order to reach different indigenous populations on horseback to subdue them for the Spanish crown. In these early stages, the invaders were content with merely broadening the existing pre-Columbian roads for use by horses, until then unknown in the New World.

During the Viceroyalty (16th–19th centuries), interest in different products substantially affected the network of exchange and drove the conquerors to open new roads to enable travel to regions with difficult access. During this period, mining (mainly for silver) clearly was the most prominent activity. Newly accessed regions were turned into colonies, and road networks were shaped according to the locations of the *real de minas* (mining *reales* or centers), with the purpose of exporting the products from the mines. Paralleling these developments, other activities (agriculture, ranching, and some manufacturing needed for daily life and for trade) surfaced as they were indispensable for the continued exploitation at the mine sites, mostly by providing goods for consumption by the workers and raw materials for the mining activities. As new seams were discovered or as mining centers were consolidated, population centers emerged and more roads with their corresponding branches were opened around them [1].

Starting in the 16th century, when the conquerors settled in different regions of New Spain such as the New Kingdom of Galicia, New Biscay, New Santander, and the New Kingdom of León, new routes for the transport of merchandize were opened. Communication routes during the viceregal period were built along two perpendicular axes. The first one, the east–west axis, connected Orient to Occident, linking Veracruz port–Mexico–Acapulco/Navidad ports; at the time it was considered a colossal bridge between the Atlantic and Pacific Oceans, known back then as the North and South Seas, respectively. The other axis, running north–south, united the northern Viceroyalty with México City–Oaxaca–Guatemala. Once opened, these axes of communication were called *Caminos Reales* or Royal Roads; four such main channels were built:

Royal Road of México–Veracruz or "Viceroys Royal Road,"

Royal Road of México–Acapulco or "Royal Road of the Orient,"

Royal Road of México–Santa Fe or "Interior Land Royal Road," and

Royal Road of México–Guatemala or "Royal Road to the Salazones" [2].

Military campaigns, missionaries, merchants, stockmen, and expeditionary forces dispersed from these roads carrying seeds and culinary traditions as well as creeds, customs, rituals, ideas, and knowledge that came from different places. Accordingly, they thereby established in New Spain and its provinces a complex economic and cultural system well beyond the mining centers from which were extracted precious metals that for a long time supported the economy of the old continent. The complex new growth included cities and villages; baserris, haciendas, and ranches; presidiums (named for military fortifications); hospitals; and convents.

Before conquest, the means of travel were the same people that walked the roads (*tamemes* and armies). Then, after the conquerors arrived, the use of draught animals (such as horses, mules, and oxen) resulted in radical modification of the system of roads and their construction. The clash between the pre-Columbian and European cultures, the shifts in economic activities, and the introduction of different means of transportation resulted in changes to the road network that went beyond the creation of new roads compared to what existed prior to the conquest. Such modernization of cross-land connections not only modified the configuration and extent of the roads network but also the outfitting of the Royal Roads, which corresponded to purposes served (e.g., transport of minerals, raw materials, foodstuffs, textiles, and many other products; transit of people for cultural and religious purposes).

The Royal Roads are of considerable extent, stretching across a notable variety of physical settings and correspondingly different weather. For instance, the 2600-km-long *Camino Real de Tierra Adentro* (or Interior Land Royal Road) is the longest of the four and crosses 11 Mexican States and one USA State). Conflicts with the indigenous population persisted throughout the Viceroyalty in different regions surrounding the roads. Therefore, the outfitting of these roads needed to provide safety both for the travelers and their loads, shelter from the elements and for rest, and points where provisions could be found. The lodging structures so that the travelers could rest or spend the night, change mules, and get provisions consisted of *ventas*, *mesones*, and inns; outfitting on the road itself consisted of abreuvoirs, watch posts, post houses, fortifications, checkpoints, and bridges [3].

1.2 Bridges

The first bridges used by humans are thought to have been natural, subsequently imitated by humans in their processes of settlement and movement across the land—such as rock in appropriate shapes, dimensions, and placements or fallen tree trunks and protruding tree roots. In Mexico, a notable example of such natural bridges is Puente de Dios (God's Bridge), which is a rock span ~80 m above the Amajac River in the State of Hidalgo [4].

The pre-Columbian footpaths were traveled by people and therefore were gradually flattened and shaped by the passage of the walkers; it did not matter if the routes were not straight or had pronounced slopes. The bridges along these footpaths were very simple; to solve immediate obstacles, they were made of materials at hand and commonly incorporated natural rock formations. However, there is evidence of more sophisticated structures, such as the *calzadas* or roads found in the city of Tenochtitlan, the center of the Aztec Empire. They were outfitted with bridges that allowed the connection between different urban sectors while maintaining the hierarchic urban order. Some collapsible bridges served as protection for the city, and other structures served as simple bridges for expeditions on foot and by horse. Bridges played an important role in the conquest of the city and the exploration of unknown, commonly treacherous, territories [5].

The establishment of the Viceroyalty of New Spain and other kingdoms within the Spanish Empire (such as the New Kingdom of Galicia, New Biscay, the New Kingdom of León, Santa Fe of New Mexico, and New Santander) affected many aspects of life in the territory, such as religion, art, cuisine, economic activities, technology, and road networks. The arrival of horses, pack mules, and carts came to substitute for the *tamemes* during the 16th century, and both roads and bridges had to be widened, reinforced, and made to have subtler slopes so that the pack animals could go through easily, thus ensuring the safety of the travelers and their loads [6].

In modern times, transformation of roads and highways, including their outfitting, continues to update the network in the country. During these changes, new roads are frequently modernized, re-align, and commonly the existing roads' number of lanes are increased, and/or materials initially used for construction are modified or replaced [7]. Communities and road-network infrastructures (such as bridges) emerge into the public spotlight or vanish or lose prominence as the ever-shifting needs of the population demand a constant updating of the roads. Changes in road infrastructures can be used as an analytical element to trace the socioeconomic and technical changes within a society; the bridges are especially significant because they exhibit the evolution and refinement of systems of construction and reflect the cunning of human beings to solve obstacles present in a physical setting. Also, bridges commonly display the prevailing aesthetic elements of a society within a specific time frame through the architecture, craftsmanship, construction materials, and ornamentation [8].

The Royal Roads reflect the connectivity of the viceregal society given the cultural and economic impact they had on the general society of New Spain through the centuries of its existence. The bridges, being part of these roads' infrastructure, played an important role in the flow of merchandize and people and in the sharing or influence of customs, creeds, and knowledge throughout the territory of New Spain.

The information that viceregal bridges can provide to diverse fields of knowledge and disciplines is one of the main reasons for performing research on them. Another reason for such research is to contribute to the effort to preserve these constructions and their legacy, thereby highlighting their cultural and historical value through the dissemination of research results.

2 Problem Statement

In Mexico, property of historical value is regulated by the National Institute of Anthropology and History (INAH, by its acronym in Spanish) within the Federal government of Mexico (https://www.inah.gob.mx). This institute owns information regarding all the historical properties in the country, and such information is in an inventory generated by each of the 31 INAH offices (one for each of the States in Mexico). Bridges built during the viceregal period (16th–19th centuries) are some of the structures included in the INAH's inventory.

The inventories include data fields of information such as location, sketches or architectural outlines, photographs (as many as a dozen), and some historical details. Unfortunately, some of the included properties lack specific geographic coordinates, and their historical details are relegated to an observations field, which may be sparse.

The inventories generated by each of the 31 States are kept in each of their INAH's Regional Centers as well as in the General Coordination of Historical Monuments of the INAH, in Mexico City (https://www.monumentoshistoricos.inah.gob.mx/). The organizations in charge of harboring the assets of cultural interest have diverse typologies and capacities, which leaves them in the position of having to manage them in an effective way with very limited resources. The information they have is diversified, decentralized, and commonly paper dependent. Only very rarely is the information digital, in standardized formats, and safeguarded by one entity.

While an investigation regarding the viceregal bridges in the Royal Roads could prove very useful for any potential initiatives of salvaging and/or repurposing these historic bridges as well as to generate related research projects, it is difficult to gather and process all the information for subsequent analysis, as the data contained in the inventories are incomplete, obsolete, incorrect, and/or sparse in many cases. Also, the changes that the road networks undergo through the years usually result in communities' letting stretches of roads and structures fall into oblivion with common outcomes being disrepair or complete loss, thus leaving no material evidence of their prior existence.

We perceived the central need to build an updated inventory of these bridges using the information provided in the 31 inventories (paper and digital resources) along with spatial research and georeferencing to complete or reconstruct existing information and generate new, key information regarding these structures. Information and communications technologies (ICT) will be used to release the compiled research results to the general public, research institutions, and government agencies and thus be able to safeguard and broadcast the heritage that these historical bridges have to offer.

3 Research Project

This research project started in 2011 as an independent initiative with the objective of creating a digital compilation or database of the bridges constructed from the viceregal period (16th–18th centuries) up to the present time and determine which bridges are located on the route of the Royal Road of the Interior Land. The project started with the reconnaissance of few bridges, most of them in the State of Querétaro: Bridge of Calesa, Bridge of Juárez, Bridge of Sauz (Pedro Escobedo), Bridge of History (San Juan del Río); also included were two other bridges outside of this State: Bridge of Atongo (Tepeji del Río, Hidalgo) and Bridge of Tres Guerras (Celaya, Guanajuato). For this reconnaissance, we conducted trips over the corresponding stretches in the Royal Road, obtained photographic evidence, and collected or reconfirmed such information as geographic coordinates, constructions materials, and physical state of the structure.

During these trips, we found some bridges for which we had no previous knowledge or whose current state was deteriorating to the point of abandoned use. For this reason, support from the INAH was requested in order to obtain an updated catalog of the bridges in Mexico to get background information on their condition and location. The INAH provided the information in Portable Document Format, including fact sheets on ~850 of registered bridges.

This information was transferred into an Excel spreadsheet to be able to better manage and work it. All the relevant information was transcribed into a document that comprises 635 registries of bridges was created.

While the data transcription was underway, a situation was made clear as a problem: the scarcity of information or precise detail that would allow correct location of bridges. However, this awareness supported a new direction in our research. In terms of location of a structure, basic data such as State, city, and street were available, but geographic coordinates were not included. Although some of the fact sheets included handmade sketches showing the spatial location of the structure and/or a poorly detailed sketch of the structure, without the coordinates, these sketches were insufficient to pinpoint precise location. The fact sheets also gave information regarding the physical condition of the structure and whether it can be accessed and/or on what type of property it is located.

Once the Excel spreadsheet was populated with data for the 635 bridge registries, the next step was to determine precise bridge locations (mostly missing) in order to start the creation of the catalog. Because the information was collected at the national level and given that the project is an independent initiative, we did not have enough resources to perform a direct (fieldwork) search and georeferencing of most of the registered bridges, as we had done in the exercises performed for six bridges in 2011 at the beginning of the project. We also decided that only structures cataloged as belonging to the viceregal period (16th–18th centuries) would be identified and worked on.

A direct reconnaissance would not be possible in many cases as many of the structures have limited access for various reasons: the structure itself may be covered from view by excess weeds or other natural overgrowths, there may be no passable roads in the vicinity, and/or there may be blockages of different types such as barbed wire or cattle fencing). There are cases for which no vestiges of the structures were to be found. In other cases, structures were on private land or on restricted government property and were inaccessible to the general public. Therefore, under these various circumstances, it was impossible to get evidence of a structure's existence or current conditions.

For the cases in which direct georeferencing was not possible, information technologies were used to locate each of the bridges. For software applications, we used Environmental Systems Research Institute's (ESRI's) commercial geographic information systems software (ArcGIS) and made extensive use of Google Earth's satellite maps with georeferencing and information on roads and highways. A one-by-one search was performed for each of the structures by using the available information regarding State, municipality, and street contained in the Excel file and making use of any relevant information extracted from the images and sketches included in the INAH fact sheets. Information gathered from the Internet, books, and other documents were collected and reviewed to complement the information gleaned from the fact sheets.

3.1 Remote Location and Georeferencing of Bridges

Figures 1, 2 and 3 shows the viceregal bridges located form INAH's bridge catalogue, and separated by centuries depending on when the bridge was approximately constructed: Fig. 1—16th century; Fig. 2—17th century; and Fig. 3—18th century. A classification system was setup according to the difficulty in locating each bridge and its distance from starting points based on information provided in the fact sheets or derived from satellite maps by using tools included in Google Earth. The color-coded classes presented in Figs. 1, 2 and 3 were based on what we determined to be the range of distance in which the structure could be found, measured from the chosen starting point:

GREEN indicates bridges that can be found within a 100 m radius from the starting point, which could be the location suggested in the fact sheets of the catalog or the location pinpointed on satellite maps. These bridges may or may not be visible on the satellite map, but, by using Google Earth tools such as Street View, it was possible to establish a location for the bridge or where it was constructed. Some of these bridges now may be completely hidden from sight by modern structures surrounding or covering them.

ORANGE indicates bridges that could not be identified on satellite maps upon initial investigation, nor could significant information on them be found in other sources, although the normally available municipality, community, and street specifying their location could be used as the starting point. We determined that these bridges could be located within a 101–500 m radius from such a starting point.

RED includes all the remaining bridges whose location could not be properly established and for which we found only minimal data (such as community, neighborhood, and street names that were not always correctly captured) and some sketches. Without precise locations starting points, we decided to base them on the center of a community or neighborhood or near mentioned streets.

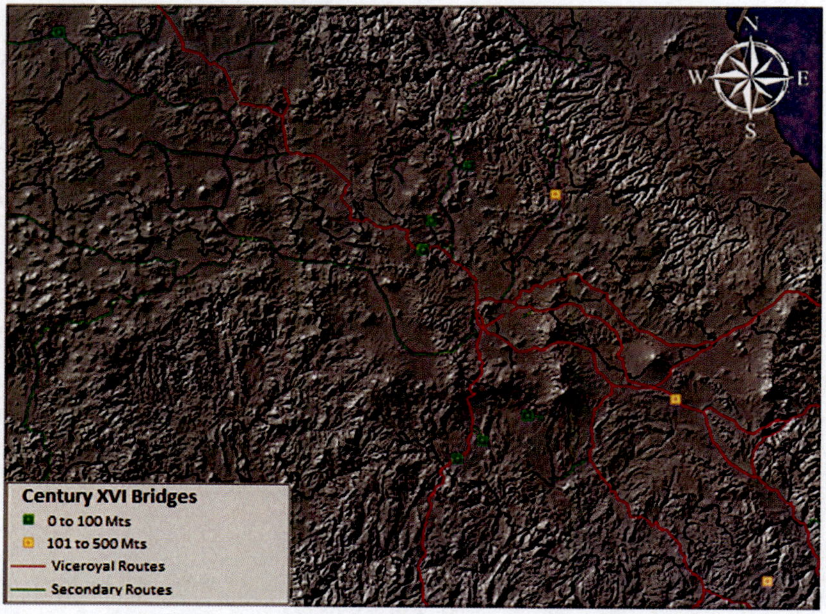

Fig. 1. Fifteen identified bridges built during 16th century. Red lines indicate Royal Roads, and green lines, transverse roads. Bridges that could be identified with varying accuracy are marked by squares and are colored green/orange/red according to our classification system.

It is visually apparent that many of the bridges shown in green (Figs. 1 and 2) can be found near modern roads and highways. Some of these bridges are still in use and in other cases, modern bridges or other structures were built near the old bridges for their replacement, making their historic preservation possible. A notable case is the Bridge of La Laja or Bridge of Tres Guerras at the entrance to Celaya, Guanajuato.

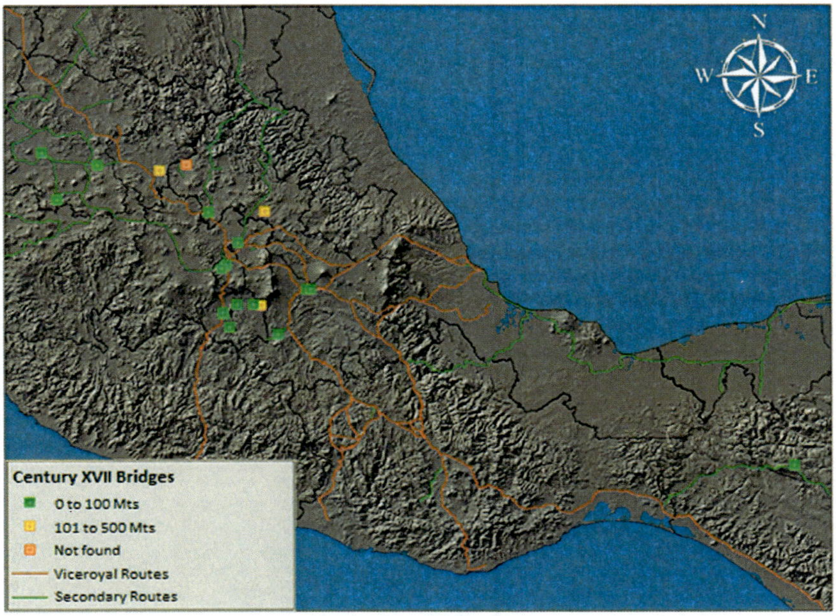

Fig. 2. Twenty-three identified bridges built during 17th century. Red lines indicate Royal Roads, and green lines, transverse roads. Bridges that could be identified with varying accuracy are marked by squares and are colored green/orange/red according to our classification system.

One of the objectives of this project has been to share with the general public the catalog of viceregal bridges that exist or existed in the country. Each registry in the catalog is meant to include information regarding its construction, materials used, present condition, photographs, and possible geographic coordinates. For this objective, we sought a digital platform with a graphical user interface and capable of managing geographic information, which led us to ESRI's ArcGIS. However, this is commercial software that requires a high-cost license per computer for individual use and/or a special license for online access, and without a certain degree of previous knowledge of the software, it has a steep learning curve. These issues forced us to replan our research approach.

3.2 Platform Generation

After the graphic user interface was developed with the data fields to be used and definitions for each layer (corresponding to administrative divisions, terrain, Federal highways, etc.), the information contained in the Excel file was converted into tabular format used by database software to subsequently be linked to the section where queries may be run and results obtained. Initial information can be filtered regarding the century of the bridge's construction and location by State, after which the structure of interest can be selected from a list of search results. Selecting the desired structure displays a complementary screen that provides access to all the information gathered from the fact sheets (such as the type of structure and related general observations, commentary made by the local population, the bridge's condition, the government or private entity to which it belongs, and any other obtained detail) and to any available photographic archive of the structure.

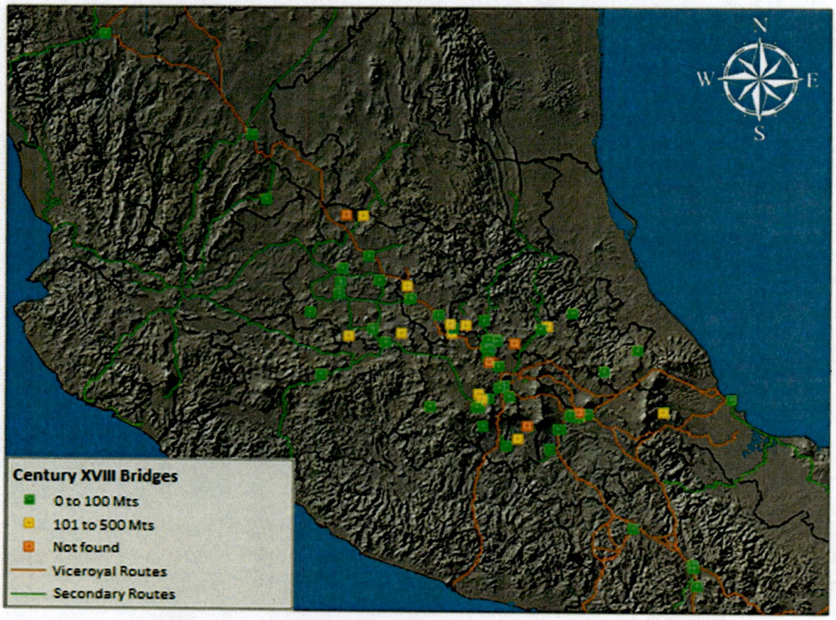

Fig. 3. Eighty-one identified bridges built during 18th century. Red lines indicate Royal Roads, and green lines, transverse roads. Bridges that could be identified with varying accuracy are marked by squares and are colored green/orange/red according to our classification system.

Technology such as Apache Web Server was used to host all the information gathered, web pages, images, 3D models, etc. MySQL was used as the engine to manage the information in dBase; PHP was used for dynamic handling and structure of the site, while JQuery and JS were used for the dynamic handling from the user's perspective. By using all these technologies, it was not necessary to be limited to any specific operating system such as Linux, Windows, or Mac (Figs. 4 and 5).

The web-based application-development phase of the project is ongoing to provide the user with the most information possible. The platform is intended to be an interactive space where the public can share information online either through a blog or a forum (yet been determined).

The site can be accessed through the link http://sgpa.desacomp.com, and comments or suggestions can be submitted via the email address joel.bustamante@satelsa.com.mx.

4 Conclusions

Providing the obtained information as well as a means of interaction with the public will not only serve to complement the information presented but also to link the public

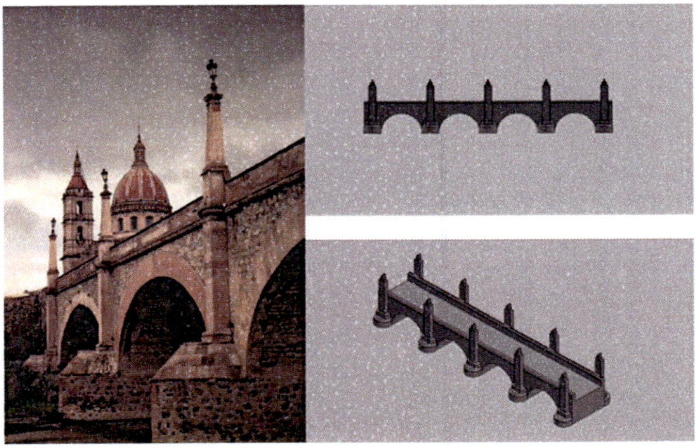

Fig. 4. Bridge of Lagos. Built in 16th century in city of Lagos de Moreno in Jalisco. [Top right: elevation projection; bottom right: isometric projection]

Fig. 5. Puente Grande (Big Bridge) built in 18th century in city of Acámbaro in Guanajuato. [Bottom: elevation projection]

to the historic bridges, considering their artistic, social, and cultural importance. If the government institutions and other organizations involved with historic and cultural heritage and the public are aware of the heritage still standing in the country and its value, perhaps we will see an invigorated interest in the repurpose and preservation of these structures.

Acknowledgments. This investigation is part of the investigation activities of the Multidisciplinary Research Collaboration Agreement from the Grupo Camino Real de Tierra Adentro

Queretaro: Rosa Estela Reyes, Virginia Guzmán Monroy, García, Cynthia Leija Sánchez, Eva Lilia Acosta Garnica, Centro INAH Qro; Frida Sarete Orozco Loeza, José Ignacio Urquiola Permisán, Marja González Juárez, Gaspar Real Cabello, Universidad Autónoma de Querétaro. The authors greatly acknowledge the funding provided by SATELSA SA de CV for the project since 2009. Authors acknowledge the Tecnologico de Monterrey, School of Engineering, Queretaro Campus, for the preparation and presentation of this publication. The opinions and findings in this paper are those of the author and not necessarily those of the funding agencies.

References

1. Tutino, J.: Making a new world. Founding Capitalism in the Bajío and Spanish North America, pp. 159–208. Duke University Press (2011)
2. Escalante, P.: Los Caminos del México Antiguo. In: Cramaussel, C. (Ed.): Rutas de la Nueva España, pp. 28–36. El Colegio de Michoacán (2006)
3. Cramaussel, C.: El Camino Real de Tierra Adentro. De México a Santa Fe. In: Cramaussel, C. (Ed.): Rutas de la Nueva España, pp. 301–303, 314–323. El Colegio de Michoacán (2006)
4. Ortiz, J., Hernández, J., González, J., Esteva, P.: Puentes de México, Arte e Historia, pp. 7–10. Grupo Tolteca (1988)
5. Ortiz, J., Hernández, J., González, J., Esteva, P.: Puentes de México, Arte e Historia, pp. 30–39. Grupo Tolteca (1988)
6. Ortiz, J., Hernández, J., González, J., Esteva, P. Puentes de México, Arte e Historia, pp. 69–73. Grupo Tolteca (1988)
7. González de Cosio, F.: Historia de la Obras Públicas en México, vol. I, pp. 483–484. Secretaría de Obras Públicas (1971)
8. Avilés, M., Brambila, R.: La puerta del camino a tierra adentro. En busca de evidencias materiales. In: Cramaussel, C. (Ed.), Rutas de la Nueva España, pp. 330–337. El Colegio de Michoacán (2006)

Towards Green Materials for Cultural Heritage Conservation: Sustainability Evaluation of Products for Stone Consolidation

Alessandro Dal Pozzo, Giulia Masi, Alessandro Tugnoli, and Enrico Sassoni(✉)

University of Bologna, 40131 Bologna, Italy
enrico.sassoni2@unibo.it

Abstract. Formulation of effective stone consolidants for the restoration of weathered carbonate substrates is an active area of research in conservation science. In addition to consolidation performance, sustainability is increasingly taken into account in the selection of the most suitable conservation intervention. Here, a comparative life cycle assessment of three consolidants (a novel product based on a diammonium hydrogen phosphate solution and two commercial alternatives: nanolimes and acrylic resin) was carried out to determine the environmental burdens that might arise from their production, their application processes, and the supply chain of precursors and auxiliary materials. The impacts of the three consolidants were compared considering the respective amounts required to treat 1 m^2 of two alternative substrates (limestone and lime mortar), as determined by application until apparent refusal. The results of the study highlight the lower environmental footprint of the phosphate-based treatment compared to the commercial solutions on several impact categories. The analysis allows to point out, for each product, the components that contribute the most to the environmental impact, namely solvents in the case of the commercial consolidants and production of DAP itself in the case of the phosphate treatment.

Keywords: LCA · Consolidants · Mortar · Ammonium phosphate · Nanolime · Acrylic resin

1 Introduction

Many historic buildings are made of porous carbonate substrates (e.g. limestone and air lime-based mortars), which undergo severe deterioration processes when exposed outdoors. Salt crystallization cycles and freezing-thawing cycles are among the major causes of deterioration, as they induce severe stress inside the pores [1]. When the tensile stress exerted by the growing crystals exceeds the tensile strength of the substrate, damage occurs, in the form of cracking, powdering and material loss. Notably, climate change predicted in the near future is expected to worsen these weathering processes in some areas [2, 3].

J. I. Escalante-Garcia et al. (Eds.): RW 2021, RILEM Bookseries 40, pp. 751–760, 2023.
https://doi.org/10.1007/978-3-031-21735-7_80

To extend the life of historic buildings, consolidants are often applied to restore the cohesion of weathered substrates. Consolidants are liquid products that, by different mechanisms, lead to the formation of a new solid inside the treated substrate, thus increasing its resistance to further deterioration processes. Because commercial consolidants often exhibit limitations when applied onto carbonate substrates, research on innovative products is still ongoing. A promising new approach is using aqueous solutions of diammonium hydrogen phosphate (DAP) to form calcium phosphates inside the treated substrates, from the reaction between calcium ions from the substrate and phosphate ions from the consolidant [4]. The technical performance of DAP solutions has been thoroughly investigated, with very encouraging results [5], which has led to an increasing number of applications onto real buildings and monuments [6–8]. Nonetheless, the environmental sustainability of this novel approach has not been specifically investigated yet.

Sustainability of innovative materials and techniques is a parameter of paramount importance in all areas of technology. Within the building sector, the environmental sustainability of materials and techniques for the construction of new buildings has received attention since many years [9], while focus has been put also on historic buildings more recently [10]. With regard to stone and mortar conservation [11], life cycle assessment (LCA) has been carried out on materials and techniques for cleaning [11, 12], self-cleaning [13] and consolidation [14, 15]. Among stone consolidants, the products that have been evaluated include calcium acetoacetate, calcium ethoxide and calcium isopropoxide for carbonate substrates [14], as well as a TEOS-based consolidant for silicate substrates [15].

The present study was aimed at performing a life cycle assessment (LCA) of DAP solutions, in comparison with alternative consolidants frequently used on carbonate substrates, namely nanolimes and acrylic resin. Nanolimes are suspensions of $Ca(OH)_2$ nanoparticles, which react with atmospheric CO_2 to form new calcium carbonate inside the treated substrate, with consolidating ability [16]. Acrylic consolidants are solutions of acrylic resin dissolved in a suitable solvent (e.g. acetone); after application onto the substrate, the solvent evaporates and acrylic resin solidifies inside the substrate, thus strengthening it [17]. In this study, two kinds of porous substrates were considered, namely porous limestone and lime-based mortar, which were traditionally used in historic buildings for both structural and decorative purposes. The comparison among the consolidants was performed considering the amount of product (in L/m^2) necessary to treat the two reference substrates until apparent refusal, as recommended in the product's technical data sheets. To determine the amount of product corresponding to apparent refusal, experimental tests were performed on limestone and mortar samples, treated with the three commercial consolidants. The resulting amounts were used as reference flows for the setup of a "cradle-to-gate" LCA analysis, aimed at quantifying the environmental burdens related to the production, transportation and application of the alternative consolidants and identifying the "hotspots" of environmental impacts along their life cycles.

2 Materials and Methods

2.1 Consolidating Treatments

The following commercial products, all kindly supplied by C.T.S. s.r.l. Italy, were used:

- diammonium hydrogen phosphate (DAP, $(NH_4)_2HPO_4$), used to prepare an aqueous solution containing 1M DAP + 1 mM $CaCl_2$ ($CaCl_2$ supplied by Sigma Aldrich). This treatment (labelled "AmmPhos") was selected based on previous results that pointed out its effectiveness on marble, limestone and mortar [5, 18, 19].
- Nanorestore®, a suspension of calcium hydroxide nanoparticles in ethanol, with 5 g/L concentration (treatment labelled "n-$Ca(OH)_2$");
- Paraloid B72®, dissolved in acetone to prepare a solution with 5 wt% concentration (treatment labelled "Res").

2.2 Substrates

Two types of substrate were considered:

- porous limestone (labelled "limestone"): Lecce stone, widely used in Baroque architecture in Southern Italy and severely affected by salt weathering, was selected. Lecce stone is an organogenic limestone, with open porosity ~26%. Specimens with 5 x 5 x 2 cm^3 size were cut from a single slab.
- lime mortar (labelled "mortar"): samples (3 x 3 x 2 cm^3 in size) were prepared using slaked lime and calcareous sand (binder-to-aggregate ratio of 1:2, water-to-binder ratio of 1:1 by volume). The samples were left to cure for 4 months at RH = 90 ± 2% and T = 21 ± 2 °C and then for additional 6 months in laboratory conditions. The resulting mortar had open porosity ~21%.

2.3 Consolidant Application

Limestone and mortar samples were treated with each consolidant by brushing on a 5 x 5 and 3 x 3 cm^2 face, respectively, until apparent refusal, intended as the condition when the specimen surface remained wet for more than 1 min after a single brush stroke [20]. The amount of consolidant corresponding to the condition of apparent refusal was determined by weighing the specimens before and after consolidant application.

2.4 Goal and Scope of the LCA Study

Aim of the LCA study was to identify the environmental impacts arising from the relevant life cycle stages of the three alternative consolidants introduced in Sect. 80.1. The functional unit (FU) of the study was set as the treatment of 1 m^2 of substrate, either limestone or mortar. Considering that restorers typically apply consolidant products until apparent refusal, as recommended by the product manufacturers, the reference flow that satisfies the FU for each consolidant was determined by the tests described in Sect. 80.3.

The study adopted a cradle-to-gate approach, including in the system boundaries (Fig. 1) the production of the consolidant material (synthesis of the consolidant, production of its precursors, production of its packaging), the relevant transportation stages, and the application onsite. The end-of-life stage was discarded, as consolidants once applied become part of the cultural heritage structure to be preserved [15].

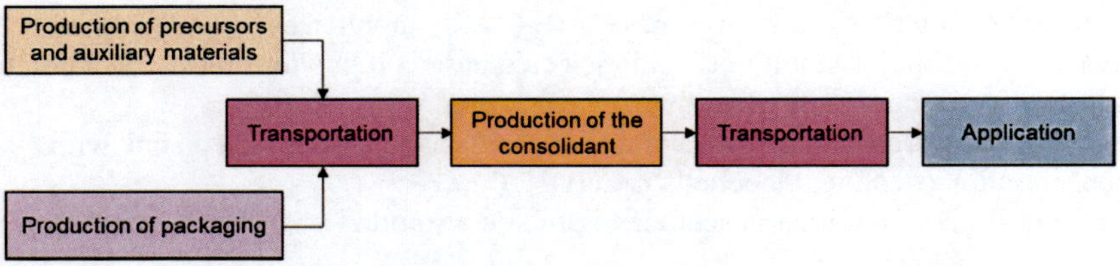

Fig. 1. Life cycle stages of the consolidants included in the system boundaries.

2.5 Life Cycle Inventory

Considering the generic system boundaries depicted in Fig. 1, the specific life cycles of the three alternative consolidants were modelled as follows.

For the "AmmPhos" treatment, DAP was considered to be synthesized from ammonia (25% w/w in water) and phosphoric acid (40% w/w in water) in a batch reactor (volume: 5 m^3). The reaction is exothermic and proceeds to completion under moderate stirring. Simple assumptions for process sizing were followed to calculate the energy demand of the batch process [21]: power consumption for stirring was estimated considering an axial flow impeller at a rotational speed of 50 rpm [22], while a cooling duty was estimated in terms of water consumption to keep the batch at a temperature lower than 50 °C to prevent DAP decomposition.

For the "n-$Ca(OH)_2$" treatment, the synthesis of lime nanoparticles was supposed to happen by aqueous precipitation [23]. 0.4 M $CaCl_2$ and 0.8M NaOH solutions in DI water are mixed at 90 °C in a 5 m^3 batch reactor under vigorous stirring (100 rpm). Power consumption for heating duty and stirring was considered. Precipitated $Ca(OH)_2$ nanoparticles, net of an estimated 5% loss due to solubility in water, are then dispersed in anhydrous ethanol to reach the desired concentration. An extra 30% consumption of ethanol was assumed to be needed to rinse the nanoparticles from water.

For the "Res" treatment, it was considered that beads of acrylic resin Paraloid B72® are directly dissolved in the adequate amount of acetone at the site of application.

Inventory data for the production processes of the involved chemicals (ammonia, phosphoric acid, calcium chloride, sodium hydroxide, ethanol, acrylic resin, acetone), were retrieved from the Ecoinvent Environmental Footprint database.

5 L HDPE plastic buckets (0.25 g tare weight) were assumed as conventional packing for all the three consolidants, while transportation distances were set to 500 km for the transportation of precursor chemicals and to 300 km for the transportation of the consolidant product to the site of application. Transportation was assumed to take place by road on <7.5 t metric ton lorries (mix Euro 0–5).

Application on the substrate was supposed to happen by brushing. After application, it was assumed that volatile compounds in the consolidants evaporate. For the "n-$Ca(OH)_2$" and "Res" systems, the entire mass of ethanol and acetone, respectively, is released into atmosphere. For the "AmmPhos" system, the application of DAP may induce the formation of hydroxyapatite, as favorable phase, by reaction with the substrate [4] and the process entails the release of carbon dioxide, as follows:

$$CaCO_3 + (NH_4)_2HPO_4 = Ca_{10}(PO_4)_6(OH)_2 + (NH_4)_2CO_3 + CO_2 + H_2O \quad (1)$$

The direct emissions of CO_2 were accounted for, while for ammonium carbonate, $(NH_4)_2CO_3$, any further decomposition pathway that might result in the air emission of CO_2 and NH_3 was not considered, since the excess product at the end of application is typically rinsed with water [12].

2.6 Impact Assessment

A set of six midpoint impact indicators, namely abiotic depletion potential of fossil fuels (ADP), global warming potential (GWP), acidification potential (AP), photochemical ozone creation potential (POCP), human toxicity potential (HTP) and eutrophication potential (EP), were considered for the evaluation of environmental burdens. The characterization factors for these impact categories were retrieved from the most recent update of the well-established CML-IA database [24].

3 Results and Discussion

3.1 Determination of Product Consumption

The amount of product consumed to treat the two substrates until apparent refusal is reported in Table 1. In the case of the "AmmPhos" treatment, the consumption on limestone was a little higher than on mortar (6.9 and 6.4 L/m^2, respectively), in agreement with the respective values of open porosity (26% and 21%, respectively). The consumption of n-$Ca(OH)_2$, quite similar between the two substrates, was generally higher than that of "AmmPhos" (9.0–9.3 L/m^2, compared to 6.4–6.9 L/m^2), which can be explained considering the higher volatility of the solvent in the "n-$Ca(OH)_2$" treatment (ethanol), compared to the "AmmPhos" treatment (water). The amount of product consumed for the "Res" treatment was actually quite different between the two substrates (4.3 L/m^2 for limestone and 8.0 L/m^2 for mortar) and opposite to what would be expected simply based on the substrate's open porosity (higher for limestone). This can be explained considering that, besides the volume of open pores, also their size strongly influences the rate of product uptake and hence the product consumption until refusal, which was defined based on the time of absorption into the substrate (cf. Sect. 80.3).

3.2 Comparison of the Life Cycle Impacts of the Alternative Consolidants

The environmental burdens related to the use of the three alternative consolidants were determined via LCA analysis for the two case studies of application to a limestone and a mortar substrate, based on the data in Table 1.

Table 1. Product consumption (L/m^2) corresponding to apparent refusal (values are averages for duplicate specimens).

Consolidant	Limestone	Mortar
AmmPhos	6.9	6.4
n-$Ca(OH)_2$	9.0	9.3
Res	4.3	8.0

The results obtained for the six impact categories considered in the analysis (cf. Sect. 80.6) are reported in Table 2 and are shown in Fig. 2 as indicators normalized to the highest score in each impact category.

Table 2. Environmental impacts in the selected impact categories.

	AmmPhos	n-$Ca(OH)_2$	Res
Case: *limestone*			
ADP (MJ)	0.57×10^2	4.40×10^2	2.15×10^2
GWP (kg CO_2 eq.)	2.90×10^0	13.4×10^0	8.14×10^0
AP (kg SO_2 eq.)	1.61×10^{-2}	2.45×10^{-2}	2.19×10^{-2}
POCP (kg ethylene eq.)	0.01×10^{-1}	28.5×10^{-1}	3.19×10^{-1}
HTP (kg 1,4-DCB eq.)	7.29×10^{-3}	19.8×10^{-3}	9.68×10^{-3}
EP (kg $PO_4{}^{3-}$ eq.)	7.87×10^{-4}	8.48×10^{-4}	9.93×10^{-4}
Case: *mortar*			
ADP (MJ)	0.53×10^2	4.55×10^2	3.99×10^2
GWP (kg CO_2 eq.)	2.69×10^0	13.9×10^0	15.1×10^0
AP (kg SO_2 eq.)	1.50×10^{-2}	2.53×10^{-2}	4.07×10^{-2}
POCP (kg ethylene eq.)	0.01×10^{-1}	29.4×10^{-1}	5.92×10^{-1}
HTP (kg 1,4-DCB eq.)	6.76×10^{-3}	20.5×10^{-3}	18.0×10^{-3}
EP (kg $PO_4{}^{3-}$ eq.)	7.30×10^{-4}	10.3×10^{-4}	15.7×10^{-4}

For the limestone substrate, the "n-$Ca(OH)_2$" treatment, which is associated with the highest amount of product consumption (cf. Sect. 80.1), exhibits the highest impacts across all the considered categories. For the mortar substrate, where the required consumption of acrylic resin is higher, the "Res" treatment shows the highest impacts in the GWP, AP and EP indicators, while the "n-$Ca(OH)_2$" alternative retains the highest scores in the remaining categories.

Significantly, for both substrates, the "AmmPhos" treatment exhibits lower burdens in all the categories. In the limestone case, despite requiring a higher specific consumption of consolidant (cf. Table 1), the reduction of impact of the "AmmPhos" system compared to the "Res" system is higher than 20% for all the categories except eutrophication. In the mortar case, where the "AmmPhos" system is also associated to the lowest consumption of consolidant, the reduction of impact compared to the closest alternative is higher than 50% in all the categories. Notably, the environmental advantage of the "AmmPhos" solution is the highest in the POCP indicator (−99.6% impact compared to the closest alternative), thanks to the absence of organic solvents in the formulation of the consolidant, as discussed in Sect. 80.3.

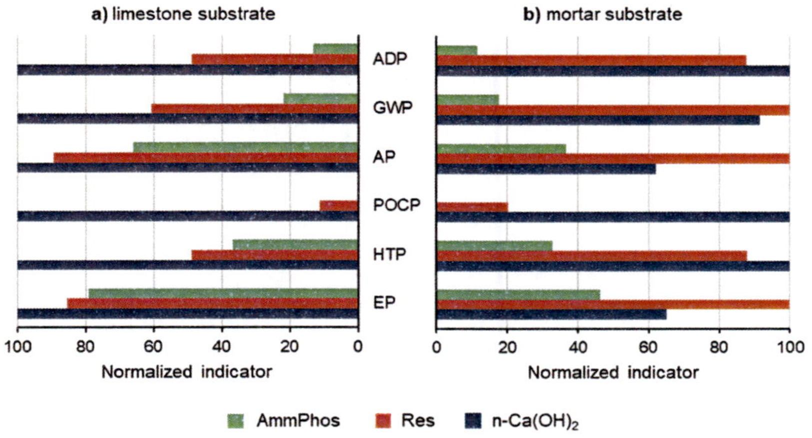

Fig. 2. Impacts of the three consolidants for the two case studies of application: **a** limestone, **b** mortar. Results are shown according to internal normalization.

3.3 Contribution Analysis

A contribution analysis was carried out to better understand the differences between the three alternative consolidants and to identify the environmental "hotspots" in their life cycles. Figure 3 details the contribution of the different life cycle stages to the overall impacts of the three systems under study.

For the "Res" system, the main contributor to the impact in 5 categories is the production of acetone, the solvent which constitutes 95% by mass of the consolidant product. However, the manufacturing of the active component, the acrylic resin Paraloid B72®, which is a more energy-intensive process, claims a relevant share of the impact (9%/24%) in the ADP, GWP and AP indicators. Packaging and transportation combined contribute at most for the 19% in the EP category. The POCP category, instead, is dominated (99.5%) by the impacts arising from the evaporation of acetone after the application of the consolidant, which dwarf the release of VOCs and NO_x in the other life cycle stages.

A similar picture can be traced for the "n-$Ca(OH)_2$" system. The production of the organic solvent, which in this case is ethanol and is 99.3% by mass of the final product, is the main contributor to the impact in 5 categories, while the evaporation of the solvent itself after application is responsible for almost the entirety of the burdens in the POCP indicator. The ozone formation potential of ethanol is four times higher than that of acetone, hence the significantly higher impact in terms of POCP observed in Fig. 2 for the "n-Ca(OH)$_2$" system. The synthesis of the lime nanoparticles and the production processes of the precursors ($CaCl_2$ and NaOH) together contribute for less than 10% of the total impact in all the categories.

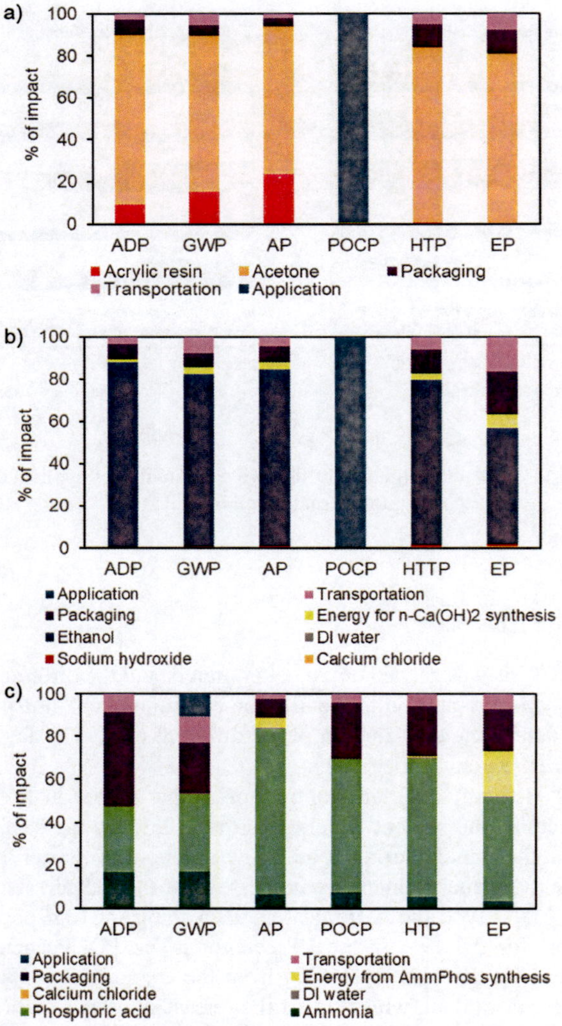

Fig. 3. Contribution analysis to the overall environmental impact for the three consolidants: **a** "Res", **b** "n-Ca(OH)$_2$", **c** "AmmPhos".

Lastly, the “AmmPhos” system benefits from the use of water as solvent, which contributes for less than 0.4% of the impact in all the categories. The main responsible for environmental burdens is thus phosphoric acid, whose contribution ranges from 31% for ADP to 78% for AP. Given the relatively low impact of the constituents of the consolidant product, in this system the share of impact ascribed to the packaging (plastic bucket) increases as high as 27% for POCP and 44% for ADP.

4 Conclusions

The novel phosphate-based treatment was found to exhibit a lower environmental impact than the commercial alternatives, mainly thanks to the absence of an organic solvent, which is the driver of impacts for nanolimes and resin in all the six impact categories considered, because of its energy-intensive production process and its volatilization after application. As for the phosphate-based treatment, here the analysis of the application process was limited to the quantification of CO_2 release, but further investigation on the ultimate fate of the ammonium carbonate released by reaction (1) could offer a more complete picture on the related environmental impacts.

Methodologically speaking, the present study takes a step towards a more accurate comparison of the environmental profiles of alternative consolidant materials by considering that product consumption during application might vary significantly for each consolidant and depending on the treated substrate, whereas previous studies compared different consolidants considering the same consumption [14, 15]. The approach adopted to differentiate consumption for the three consolidants was to measure the absorption of consolidant by the substrate until apparent refusal, reproducing the typical mode of application used during restoration. The definition of an even more accurate functional unit for the comparative analysis would require considering the actual consolidating effectiveness achieved by the alternative consolidants, by identifying a suitable metric for its determination.

References

1. Scherer, G.W.: Crystallization in pores. Cem. Concr. Res. **29**, 1347–1358 (1999)
2. Grossi, C.M., Brimblecombe, P., Harris, I.: Predicting long term freeze–thaw risks on Europe built heritage and archaeological sites in a changing climate. Sci. Total Environ. **377**, 273–281 (2007)
3. Hall, C., Hamilton, A., Hoff, W.D., Viles, H.A., Eklund, J.A.: Moisture dynamics in walls: response to micro-environment and climate change. Proc. R. Soc. A **467**, 194–211 (2011)
4. Sassoni, E., Naidu, S., Scherer, G.W.: The use of hydroxyapatite as a new inorganic consolidant for damaged carbonate stones. J. Cult. Herit. **12**, 346–355 (2011)
5. Sassoni, E.: Hydroxyapatite and other calcium phosphates for the conservation of cultural heritage: a review. Materials **11**, 557 (2018)
6. Ma, X., Balonis, M., Pasco, H., Toumazou, M., Counts, D., Kakoulli, I.: Evaluation of hydroxyapatite effects for the consolidation of a Hellenistic-Roman rock-cut chamber tomb at Athienou-Malloura in Cyprus. Construct. Build. Mater. **150**, 333–344 (2017)
7. Sassoni, E., Ugolotti, G., Pagani, M.: Nanolime, nanosilica or ammonium phosphate? Laboratory and field study on consolidation of a byzantine marble sarcophagus. Constr. Build. Mater. **262**, 120784 (2020)

8. Sassoni, E., et al.: Phosphate treatments for conservation of stone: 3-year field study in the royal palace of Versailles (France). Mater. Struct. **54**, 140 (2021)
9. Ortiz, O., Castells, F., Sonnemann, G.: Sustainability in the construction industry: a review of recent developments based on LCA. Constr. Build. Mater. **23**, 28–39 (2009)
10. Vilches, A., Garcia-Martinez, A., Sanchez-Montanes, B.: Life cycle assessment (LCA) of building refurbishment: a literature review. Energy Build. **135**, 286–301 (2017)
11. Balliana, E., Ricci, G., Pesce, C., Zendri, E.: Assessing the value of green conservation for Cultural heritage: positive and critical aspects of already available methodologies. Int. J. Conser. Sci. **7**, 185–202 (2016)
12. Franzoni, E., Volpi, L., Bonoli, A., Spinelli, R., Gabrielli, R.: The environmental impact of cleaning materials and technologies in heritage buildings conservation. Energy Build. **165**, 92–105 (2018)
13. Ferrari, A.M., Pini, M., Neri, P., Bondioli, F.: Nano-TiO_2 coatings for limestone: which sustainability for cultural heritage? Coatings **5**, 232–245 (2015)
14. Turk, J., et al.: Environmental performance of three innovative calcium carbonate-based consolidants used in the field of built cultural heritage. Int. J. Life Cycle Asses. **22**(9), 1329–1338 (2017). https://doi.org/10.1007/s11367-017-1260-8
15. Mauko Pranji, A., et al.: Life cycle assessment of novel consolidants and a photocatalytic suspension for the conservation of the immovable cultural heritage. J. Clean. Prod. **181**, e293–e308 (2018)
16. Rodriguez-Navarro, C., Ruiz-Agudo, E.: Nanolimes: from synthesis to application. Pure Appl. Chem. **90**, 523–550 (2018)
17. Amoroso, G.G., Fassina, V.: Stone Decay and Conservation. Elsevier, New York (1983)
18. Sassoni, E., Graziani, G., Franzoni, E., Scherer, G.W.: Calcium phosphate coatings for marble conservation: influence of ethanol and isopropanol addition to the precipitation medium on the coating microstructure and performance. Corros. Sci. **136**, 255–267 (2018)
19. Masi, G., Sassoni, E.: Air lime mortar consolidation by nanolimes and ammonium phosphate: compatibility, effectiveness and durability. Constr. Build. Mater. **299**, 123999 (2021)
20. Ferreira Pinto, A.P., Delgado Rodrigues, J.: Stone consolidation: the role of treatment procedures. J. Cult. Herit. **9**, 38–53 (2008)
21. Righi, S., Baioli, F., Dal Pozzo, A., Tugnoli, A.: Integrating life cycle inventory and process design techniques for the early estimate of energy and material consumption data. Energies **11**, 970 (2018)
22. Piccinno, F., Hischier, R., Seeger, S., Som, C.: From laboratory to industrial scale: a scale-up framework for chemical processes in life cycle assessment studies. J. Cleaner Prod. **135**, 1085–1097 (2016)
23. Ambrosi, M., Dei, L., Giorgi, R., Neto, C., Baglioni, P.: Colloidal particles of $Ca(OH)_2$: properties and applications to restoration of frescoes. Langmuir **17**, 4251–4255 (2001)
24. Van Oers, L.: CML-IA database, characterisation and normalisation factors for midpoint impact category indicators. Version 4.5 (2015)

Limestone Consolidation: How Much Product is Enough?

Giulia Masi and Enrico Sassoni(✉)

Department of Civil, Chemical, Environmental and Materials Engineering, University of Bologna, Via Terracini 28, 40131 Bologna, Italy
enrico.sassoni2@unibo.it

Abstract. This study was aimed at evaluating the performance of four different consolidants (ammonium phosphate, nanolimes, ethyl silicate, acrylic resin) when applied onto a porous limestone (Lecce stone) by different number of brush strokes, with the aim of identifying the optimal product consumption for each consolidant. The results of the study indicate that, in the case of nanolimes and acrylic resin, insufficient mechanical consolidation and resistance to accelerated ageing is achieved even increasing the number of applications, likely because of a scarce penetration depth. In the case of ethyl silicate, increasing the product consumption leads to increasing mechanical effectiveness, which however is linked to significant color change and alteration in water absorption, as well as scarce resistance to freezing-thawing cycles. Ammonium phosphate was able to induce significant consolidation even for the lowest product consumption. By increasing the number of brush strokes, an increase in mechanical effectiveness was registered. For the highest number of applications, the mechanical improvement caused by ammonium phosphate was slightly lower than that caused by ethyl silicate, but ammonium phosphate had the important advantage of not causing any compatibility issue and resisting to freezing-thawing cycles better than all the other alternative consolidants.

Keywords: Consolidation · Carbonate stone · Ammonium phosphate · Freezing-thawing · Tensile strength · Aesthetical compatibility

1 Introduction

Consolidation is a common practice during restoration of porous substrates in historic buildings [1–5]. Consolidation of carbonate substrates (marble, porous limestone, calcareous sandstone and lime-based mortars) is a challenging task, because consolidants have to fulfil all the requirements expected in Cultural Heritage, i.e. effectiveness, compatibility, durability and reversibility [1, 2].

In the market, many products for consolidation of carbonate stones are already available, including organic treatments (such as acrylic and epoxy resins) and inorganic materials (such as nanolimes and ethyl silicates) [1, 3, 6]. In addition, a new inorganic treatment based on aqueous solutions of diammonium hydrogen phosphate ($(NH_4)_2HPO_4$,

J. I. Escalante-Garcia et al. (Eds.): RW 2021, RILEM Bookseries 40, pp. 761–770, 2023.
https://doi.org/10.1007/978-3-031-21735-7_81

DAP) has been developed in the last ten years [7]. The concept of this consolidation treatment is to form new calcium phosphates that act as a binder in the cracks of the deteriorated substrates [7, 8]. Various formulations, involving different concentrations of DAP and different inorganic or organic additions, have been developed, depending on the substrate that needs to be treated [8–10]. The calcium ions necessary to form new calcium phosphates can be supplied either from the substrate [7] or from the exterior, by adding calcium ions in the aqueous solutions [11] or by pre-treating the substrate with a calcium source [12–14]. The DAP treatment generally exhibited a satisfactory consolidation performance, also ensuring physical, microstructural and aesthetical compatibility [8, 15]. Finally, considering durability, promising results have been reported on porous stones, in terms of resistance to freezing-thawing cycles [15–17] and salt crystallization [13, 16, 17]. Thanks to all these promising features, DAP has recently been introduced into the market of consolidants.

When selecting the consolidating product to be applied on a degraded substrate, the product consumption is a fundamental piece of information, because the higher the amount of product applied, the higher the expected consolidating ability (at least to a certain extent), but also the higher the time needed for the application and the higher the cost. Products' technical data sheets usually suggest application until apparent refusal but, depending on the properties of the substrate (e.g., porosity and pore size distribution) and of the consolidant (e.g., viscosity and volatility of the solvent), this condition may correspond to very different amounts of product (usually expressed as L/m^2), with implications on the economic and the environmental sustainability of the restoration campaign. In addition, even if the same amount of product is applied, different consolidants may exhibit significantly different consolidating abilities.

To investigate this open problem, the present study aims at assessing the performance (in terms of consolidating effectiveness, compatibility and durability) of four consolidants (DAP, nanolimes, ethyl silicate and acrylic resin) applied onto a porous limestone (Lecce stone). Samples were treated by an increasing number of brush strokes and the resulting changes in dynamic elastic modulus, tensile and compressive strength, color and water absorption were evaluated, together with the durability to freezing-thawing cycles, with the aim of correlating the performance of each treatment with the amount of product consumed and hence deriving indications about the optimal product to be used in each case.

2 Materials and Methods

2.1 Materials and Consolidation Treatments

The porous limestone selected for this study was Lecce stone, widely used in the Cultural Heritage in Southern Italy (Salento, Puglia). Lecce stone is an organogenic limestone, characterized by open porosity of about 26%. Cylindrical samples (5 cm diameter, 2 cm height) were obtained from a single slab. After cutting, all the samples were rinsed with water and dried at 40 °C until constant mass. Before consolidation, all the samples were artificially pre-deteriorated by heating at 250 °C for 3 h [7].

The samples were treated with the four consolidants by applying 10, 20 or 30 brush strokes (except for acrylic resin, applied by 10, 15 or 20 strokes) over one circular face, waiting for the product to be absorbed between subsequent strokes. In total, 5 conditions were considered as follows:

1. *REF*, untreated reference.
2. *AmmPhos*, aqueous solution of 3 M DAP (DAP supplied by C.T.S. s.r.l., Italy). In previous studies [8, 18–21], this concentration exhibited a significant consolidating performance, thanks to the high availability of phosphate ions to form new calcium phosphates. At the end of the brushing application, the samples were wrapped in a plastic film for 24 h. After removing the film, the samples were rinsed with water and dried at room temperature. The treated surface was then covered with a sheet of Japanese paper to avoid sticking and then a poultice of cellulose pulp and a saturated solution of calcium hydroxide (1:4 weight ratio) was applied. This would prevent unreacted DAP from remaining in the limestone. Then, samples were again wrapped in a plastic film for 24 h to facilitate calcium ions penetration into the limestone [21]. Finally, samples were unwrapped and left to dry still covered with the poultice.
3. *n-Ca(OH)*$_2$, nanolimes (Nanorestore Plus® Ethanol 5 by C.T.S. s.r.l., Italy), applied as received. This product consists of a dispersion of calcium hydroxide nanoparticles in ethanol (5 g/L concentration). A sheet of Japanese paper was preliminarily applied onto the sample surface to prevent whitening, as suggested by the technical data sheet. At the end of the application, a poultice of cellulose pulp imbibed with deionized water (1:4 weight ratio) was applied over the Japanese paper, to favor carbonation.
4. *TEOS*, ethyl silicate (ESTEL 1000 by C.T.S. s.r.l., Italy), applied as received. This commercial product contains 75 wt% ethyl silicate with also 1% dibutyltindilaurate as catalyst and 25 wt% white spirit.
5. *RES*, acrylic resin (PARALOID B-72® by C.T.S. s.r.l., Italy). This commercial product consists of 100% ethyl-methacrylate copolymer. A solution of 5 wt% resin in acetone was prepared. To accelerate the solubility of the copolymer in acetone, the acetone was heated up to 50 °C during solution preparation. Immediately at the end of the consolidant application, excessive resin on the sample surface was removed by using acetone applied by brush, to avoid a glitter effect.

All the specimens were cured in laboratory conditions for 4 weeks before testing. This is the curing time recommended for n-Ca(OH)$_2$ and TEOS (while AmmPhos and RES treatments need a shorter curing time).

2.2 Characterization

The consolidants performance was assessed considering effectiveness, compatibility, and durability, relating the results to the product consumption. At least 3 replicates were used for each test.

Effectiveness. The product consumption was evaluated by weighing beakers with the various consolidants, before and after application by brushing. The treatment effectiveness was assessed in terms of variation in dynamic elastic modulus (E_d), tensile (σ_t) and compressive ($\sigma_{c,DPT}$) strengths. E_d was measured on each sample, before and after treatment, according to the formula $E_d = \rho \times UPV^2$, where ρ is the geometric density and UPV is the ultrasonic pulse velocity, measured using a PUNDIT instrument with 55 kHz transducers. The tensile strength of untreated and treated samples was determined by Brazilian splitting test, loading the cylindrical samples along two opposite lines on the lateral surface. The two halves resulting from the Brazilian test were finally subjected to double punch test (DPT), to determine the sample's compressive strength. The DPT consists in loading samples using two circular steel plates (2 cm diameter) and then calculating the compressive strength ($\sigma_{c,DPT}$) as the ratio of the failure load to the circular plate area.

Compatibility. The evaluation of the aesthetic and physical compatibility was performed measuring the color change and the difference in water sorptivity after consolidation. In particular, color change ΔE^* was assessed using a NH310 colorimeter. The CIE Lab color parameters (L^* = black-white, a^* = green-red, b^* = blue-yellow) of untreated and treated samples were measured on three samples for each condition, taking the measurements on three different spots for each sample. Then, the average CIE Lab color parameters for each condition were used to calculate the color difference between the treated and the untreated conditions, according to the formula $\Delta E^* = (\Delta L^{*2} + \Delta a^{*2} + \Delta b^{*2})^{1/2}$. Water sorptivity was measured in accordance with the European Standard EN 15801 [22]. The behavior of untreated and treated samples was evaluated by allowing water to penetrate the samples by capillarity through the treated face, until saturation that occurred after 72 h.

Durability. Untreated samples and samples treated with the highest number of brush strokes for each consolidant were subjected to accelerated durability tests by freezing-thawing cycles. The cycles were carried out by partly modifying the European Standard EN 12371 [23]. Samples were preliminarily saturated with deionized water for 3 days. Then, they were subjected to freezing at -20 ± 2 °C for 2 h, followed by thawing in deionized water at $+20 \pm 2$ °C for additional 2 h. The effects of accelerated ageing were then assessed at the end of 70 freezing-thawing cycles by measuring the weight variation and E_d, following the same procedures described above.

3 Results and Discussion

3.1 Effectiveness

Increasing the number of applications caused an increase in product consumption, as reported in Table 1. In the case of AmmPhos and RES, the consumption of the maximum number of applications (30 and 20, respectively) was comparable, while the consumption of n-$Ca(OH)_2$ and TEOS were respectively higher and lower than that of AmmPhos. This can be explained considering the different volatility of the solvents of the products (water in AmmPhos, ethanol in $Ca(OH)_2$ and white spirit in TEOS).

After curing, the various consolidants formed new phases that would lead to an increase in the limestone mechanical performance (i.e., increase in dynamic elastic modulus, tensile and compressive strengths). However, the various consolidants exhibited different behavior, as reported in Table 1. The untretaed limestone was characterized by $E_d = 9.9(\pm 0.4)$ GPa, $\sigma_t = 2.1(\pm 0.1)$ MPa and $\sigma_{c,DPT} = 14.5(\pm 2.1)$ MPa.

Limestone consolidation with the n-$Ca(OH)_2$ and RES treatments caused a modest improvement in the mechanical performance. This behavior was observed, in terms

Table 1. Product consumption, variation in dynamic elastic modulus (ΔE_d), tensile ($\Delta \sigma_t$) and compressive ($\Delta \sigma_{c,DPT}$) strengths (values are averages for 3 samples)

Consolidant	Brush strokes	Consumption (L/m^2)	ΔE_d (%)	$\Delta \sigma_t$ (%)	$\Delta \sigma_{c,DPT}$ (%)
AmmPhos	10	1.3	22	23	21
	20	2.6	27	25	28
	30	3.9	28	26	33
n-Ca(OH)$_2$	10	1.6	5	0	2.0
	20	3.2	11	0	0
	30	4.8	11	0	0
TEOS	10	0.9	12	2	4
	20	1.9	19	11	13
	30	2.8	33	43	36
RES	10	1.9	6	0	3.0
	15	2.9	8	2	5.0
	20	3.8	11	0	0

of E_d, σ_t and $\sigma_{c,DPT}$, even when the number of brush strokes was increased. For both these treatments, an increase in E_d by +11% was registered for the maximum number of applications, but no improvement in tensile and compressive strength was recorded at all. This may be ascribed to a scarce penetration of these consolidants, so that some increase in E_d was measured (as the ultrasonic pulse travels through both the consolidated and non-consolidated parts) but no substantial increase in σ_t and σ_c was possible (as σ_t and σ_c are strongly influenced by the bulk properties of the material). On the contrary, both the AmmPhos and TEOS treatments showed a significant consolidation effect, thanks to their ability to penetrate in depth in the substrate, the mechanical benefit increasing for increasing number of applications. When 10 brush strokes were applied, the AmmPhos treatment already exhibited a significant consolidating ability ($\Delta \sigma_t$ = +23%, $\Delta \sigma_{c,DPT}$ = +21%), while only a modest improvement was recorded for TEOS ($\Delta \sigma_t$ = +2%, $\Delta \sigma_{c,DPT}$ = +4%). However, when the maximum number of brush strokes was applied, the TEOS treatment induced slightly higher improvements compared to AmmPhos ($\Delta \sigma_t$ = +43%, $\Delta \sigma_{c,DPT}$ = +36% for TEOS, $\Delta \sigma_t$ = +26%, $\Delta \sigma_{c,DPT}$ = +33% for AmmPhos). The good performance of TEOS on Lecce stone can be attributed to the presence of small amounts of naturally contained quartz, as confirmed by FT-IR.

3.2 Compatibility

Results about aesthetic and physical compatibility of the investigated consolidants are reported in Table 2.

The new phases formed after application of the various consolidants induced some color variation. Both AmmPhos and n-Ca(OH)$_2$ caused color changes below the visibility limit (ΔE^* = 2.3 [24]), except for 30 brush strokes of AmmPhos that caused a color

Table 2. Color difference (ΔE^*), variation in CIE Lab color parameters and water absorption after 72 (values are averages for 3 samples).

Consolidant	Brush strokes	ΔE^*	ΔL^*	Δa^*	Δb^*	Water absorption (wt%)
AmmPhos	10	0.65	0.15	−0.24	−0.58	15.6
	20	0.90	0.45	−0.32	−0.71	16.2
	30	2.52	1.94	−0.65	−1.48	15.2
n−$Ca(OH)_2$	10	0.57	0.39	0.32	0.25	16.3
	20	0.72	0.59	−0.34	−0.22	17.3
	30	0.73	0.46	0.17	0.55	16.9
TEOS	10	5.66	−4.12	2.49	2.97	13.2
	20	6.39	−5.19	2.22	3.01	7.9
	30	7.20	−6.11	2.13	3.17	9.5
RES	10	9.34	−7.17	3.40	4.92	2.6
	15	8.16	−6.80	2.49	3.74	4.7
	20	9.06	−7.15	2.43	5.00	2.5

variation of 2.5, slightly higher than the previously mentioned limit but still below the maximum acceptability threshold usually adopted in Cultural Heritage field ($\Delta E^* = 5.0$ [25]). On the contrary, both TEOS and RES caused a strong color variation, both above the acceptability threshold even for the lowest number of brush strokes. These results suggest not negligible issues of aesthetic compatibility for both TEOS and RES when applied onto this stone.

About the physical properties, the untreated limestone samples showed water absorption after 72 h of 15.9(±0.4) wt%. Both the AmmPhos and $Ca(OH)_2$ treatments caused modest variations in water transport properties, independently of the number of brush strokes, while application of both TEOS and RES induced a strong variation in water absorption (Table 2). For these latter consolidants, a significant reduction in water sorption rate was recorded, giving rise to a marked hydrophobic behavior (Fig. 1). In the case of a porous limestone treated with TEOS, a similar temporary hydrophobic behavior had been registered in previous studies [15, 26]. In the cited studies [15, 26], after one week of contact with water the final water absorption was similar to the untreated reference, suggesting that the temporary hydrophobicity was mainly due to residual ethoxy groups remaining until TEOS hydrolysis was completed [15, 26]. Differently, the RES treatment exhibited a hydrophobic behavior even after prolonged contact with water. All things considered, a limited alteration in water absorption (like in the case of AmmPhos and n-$Ca(OH)_2$) can be considered as a positive feature, while consolidants that induce a strong alteration in water transport properties (like TEOS and RES) may give rise to compatibility issue [27].

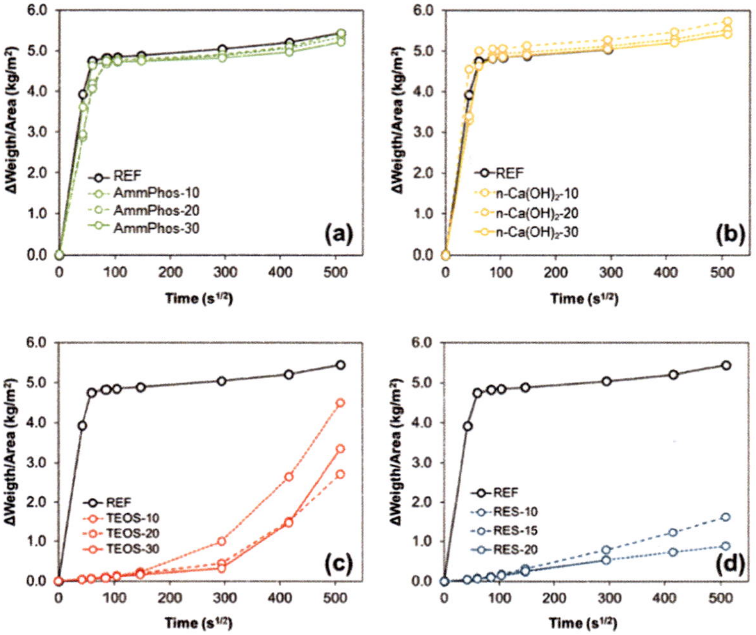

Fig. 1. Water sorptivity of untreated samples and samples treated by different brush strokes with **a** AmmPhos, **b** n-$Ca(OH)_2$, **c** TEOS and **d** RES.

3.3 Durability

Accelerated ageing by freezing-thawing cycles had a dramatic effect on untreated samples, which exhibited significant material loss after 70 freezing-thawing cycles, as depicted in Fig. 2. Among the various consolidants, RES showed a dramatic deterioration, showing a mass loss slightly higher than the untreated samples, as well as a similar decrease in E_d. To a lower extent, also TEOS and n-$Ca(OH)_2$ showed visible degradation after 70 freezing-thawing cycles. Thus, only the application of AmmPhos ensured a promising resistance to this type of accelerated ageing, considering that after 70 freezing-thawing cycles the limestone samples treated with AmmPhos did not show any weight loss and the mechanical performance was still superior compared to the reference samples before ageing (Fig. 2). To further investigate the frost resistance of AmmPhos applied on limestone, freezing-thawing cycles are still ongoing.

All things considered, while consolidation with TEOS by 30 brush strokes caused slightly higher improvements compared to AmmPhos right after application (Table 1), both compatibility and durability tests clearly highlighted some issues regarding TEOS application, which caused a significant variation in color and water transport properties and lower resistance to freezing-thawing cycles. Thus, the performance of the various consolidants may be regarded as follows: UT $\approx$ n-$Ca(OH)_2$ $\approx$ RES $<$ TEOS $<$ AmmPhos.

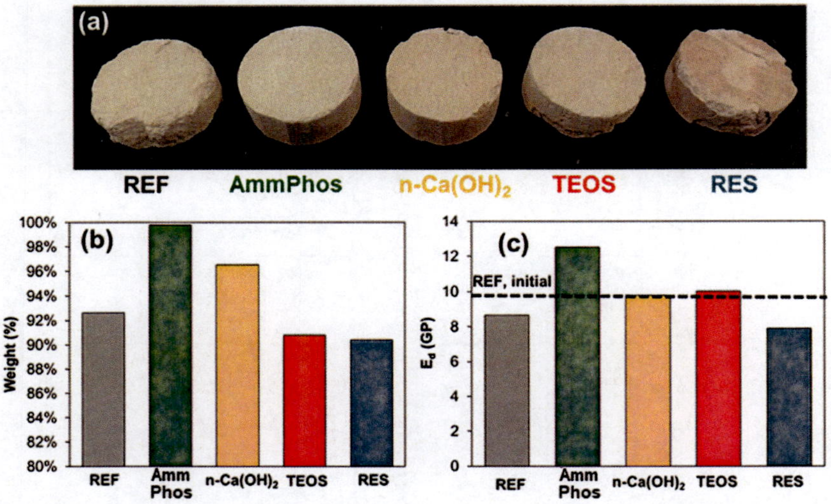

Fig. 2. **a** Appearance, **b** weight change and **c** dynamic elastic modulus of untreated and treated specimens after 70 freezing-thawing cycles.

4 Conclusions

This study investigated the optimal amount of four different consolidants applied onto a porous limestone, with the aim of maximizing the effectiveness, compatibility and durability to freezing-thawing cycles. The following conclusions can be drawn:

- The application of acrylic resin (RES) and nanolimes (n-$Ca(OH)_2$) caused negligible consolidation, independently of the number of brush strokes, as highlighted by the unsuccessful improvement in mechanical properties. RES treatment was also responsible for some aesthetic and physical compatibility issues, in terms of strong color alteration and hydrophobic behavior. Finally, both these products also showed durability issues already after 70 freezing-thawing cycles, as highlighted by weight loss and decrease in dynamic elastic modulus.
- Increasing the number of applications of ethyl silicate (TEOS) caused an improvement in the consolidation effectiveness, as the mechanical properties of the substrate progressively increased for increasing product consumption. After 30 brush strokes (consumption of 2.8 L/m^2), the highest consolidation effect was reached. However, some aesthetical compatibility issues ($\Delta E^* > 5$) and insufficient performance in terms of frost resistance should be highlighted.
- Ammonium phosphate, already after 10 brush strokes (consumption of 1.3 L/m^2), caused significant consolidation of the substrate ($\Delta\sigma_t = +23\%$, $\Delta\sigma_{c,DPT} = +21\%$). Increasing the number of applications up to 30 strokes (3.9 L/m^2) led to an additional improvement in effectiveness. Even though the level of consolidation reached by ammonium phosphate after 30 brush strokes was slightly lower than that reached by TEOS, ammonium phosphate had the advantage of not causing any compatibility issues and resulting the most resistant treatment to freezing-thawing cycles.

Acknowledgments. Greta Ugolotti (University of Bologna) is gratefully acknowledged for collaboration to the durability tests. Dr. Leonardo Borgioli (C.T.S. s.r.l., Italy) is gratefully acknowledged for providing the consolidants.

References

1. Amoroso, G.G., Fassina, V.: Stone Decay and Conservation. Elsevier, New York (1983)
2. Lazzarini, L., Laurenzi Tabasso, M.: Il restauro della pietra. CEDAM, Padua (1986)
3. Zendri, E., Biscontin, G., Nardini, I., Riato, S.: Characterization and reactivity of silicatic consolidants. Constr. Build. Mater. **21**, 1098–1106 (2007)
4. Toniolo, L., Paradisi, A., Goidanich, S., Pennati, G.: Mechanical behaviour of lime based mortars after surface consolidation. Constr. Build. Mater. **25**, 1553–1559 (2011)
5. Vacchiano, C.D., Incarnato, L., Scarfato, P., Acierno, D.: Conservation of tuff-stone with polymeric resins. Constr. Build. Mater. **22**, 855–865 (2008)
6. Ferreira Pinto, A.P., Delgado, R.J.: Stone consolidation: the role of treatment procedures. J. Cult. Herit. **9**, 38–53 (2008)
7. Sassoni, E., Naidu, S., Scherer, G.W.: The use of hydroxyapatite as a new inorganic consolidant for damaged carbonate stones. J. Cult. Herit. **12**, 346–355 (2011). https://doi.org/10.1016/j.culher.2011.02.005
8. Sassoni, E.: Hydroxyapatite and other calcium phosphates for the conservation of cultural heritage: a review. Materials **11**, 557 (2018). https://doi.org/10.3390/ma11040557
9. Graziani, G., Sassoni, E., Scherer, G.W., Franzoni, E.: Resistance to simulated rain of hydroxyapatite- and calcium oxalate-based coatings for protection of marble against corrosion. Corr. Sci. **127**, 168–174 (2017). https://doi.org/10.1016/j.corsci.2017.08.020
10. Sassoni, E., Graziani, G., Franzoni, E., Scherer, G.W.: Calcium phosphate coatings for marble conservation: influence of ethanol and isopropanol addition to the precipitation medium on the coating microstructure and performance. Corr. Sci **136**, 255–267 (2018). https://doi.org/10.1016/j.corsci.2018.03.019
11. Naidu, S., Scherer, G.W.: Nucleation, growth and evolution of calcium phosphate films on calcite. J. Colloid Interface Sci. **435**, 128–137 (2014). https://doi.org/10.1016/j.jcis.2014.08.018
12. Yang, F., et al.: Biomimic conservation of weathered calcareous stones by apatite. New J. Chem. **35**, 887–892 (2011). https://doi.org/10.1039/c0nj00783h
13. Weththimuni, M.L., Licchelli, M., Malagodi, M., Rovella, N., La Russa, M.: Consolidation of bio-calcarenite stone by treatment based on diammonium hydrogen phosphate and calcium hydroxide nanoparticles. Measurement **127**, 396–405 (2018). https://doi.org/10.1016/j.measurement.2018.06.007
14. Pesce, C., Moretto, L.M., Orsega, O., Pesce, G., Corradi, M., Weber, J.: Effectiveness and compatibility of a novel sustainable method for stone consolidation based on di-ammonium phosphate and calcium-based nanomaterials. Materials **12**, 3025 (2019). https://doi.org/10.3390/ma12183025
15. Sassoni, E., Graziani, G., Franzoni, E.: An innovative phosphate-based consolidant for limestone. Part 1: Effectiveness and compatibility in comparison with ethyl silicate. Constr. Build. Mater. **102**, 918–930 (2016). https://doi.org/10.1016/j.conbuildmat.2015.04.026
16. Sassoni, E., Graziani, G., Franzoni, E.: An innovative phosphate-based consolidant for limestone. Part 2: Durability in comparison with ethyl silicate. Construct. Build. Mater. **102**, 931–942 (2016). https://doi.org/10.1016/j.conbuildmat.2015.10.202

17. Shekofteh, A., Molina, E., Rueda-Quero, L., Arizzi, A.; Cultrone, G.: The efficiency of nanolime and dibasic ammonium phosphate in the consolidation of beige limestone from the Pasargadae World Heritage Site. Archaeol. Anthrop. Sci. (2019)
18. Masi, G., Sassoni, E.: Air lime mortar consolidation by nanolimes and ammonium phosphate: compatibility, effectiveness and durability. Constr. Build. Mater. **299**, 123999 (2021). https://doi.org/10.1016/j.conbuildmat.2021.12399
19. Defus, A., et al.: Di-ammonium hydrogen phosphate for the consolidation of lime-based historic mortars—preliminary research. J. Cult. Herit. (2021). https://doi.org/10.1016/j.culher.2021.01.005
20. Sassoni, E., Andreotti, S., Scherer, G.W., Franzoni, E., Siegesmund, S.: Bowing of marble slabs: can the phenomenon be arrested and prevented by inorganic treatments? Environ. Earth Sci. **77**(10), 1–18 (2018). https://doi.org/10.1007/s12665-018-7547-7
21. Franzoni, E., Sassoni, E., Graziani, G.: Brushing, poultice or immersion? The role of the application technique on the performance of a novel hydroxyapatite-based consolidating treatment for limestone. J. Cult. Herit. **16**, 173–184 (2015). https://doi.org/10.1016/j.culher.2014.05.009
22. European Standard EN 15801, Conservation of cultural property—Test methods—Determination of water absorption by capillarity (2010)
23. European Standard EN 12371, Natural stone test methods - Determination of frost resistance (2010)
24. Sharma, G.: Color Fundamentals for Digital Imaging. Digital Color Imagining Handbook; CRC Press, Boca Raton, FL, USA (2003)
25. Rodrigues, J.D., Grossi, A.: Indicators and ratings for the compatibility assessment of conservation actions. J. Cult. Herit. **8**, 32–43 (2007). https://doi.org/10.1016/j.culher.2006.04.007
26. Franzoni, E., Graziani, G., Sassoni, E.: TEOS-based treatments for stone consolidation: acceleration of hydrolysis–condensation reactions by poulticing. J. Sol-Gel. Sci. Technol. **74**(2), 398–405 (2015). https://doi.org/10.1007/s10971-014-3610-3
27. Scherer, G.W., Wheeler, G.S.: Silicate consolidants for stone. Key Eng. Mater. **391**, 1–25 (2009). https://doi.org/10.4028/0-87849-365-4.1

Effects of Steam-Slaking on the Characteristics of Lime from Three Different UK Manufacturers

Cecilia Pesce[1](✉), Martha C. Godina[1], Alison Henry[2], and Giovanni L. Pesce[1]

[1] Department of Architecture and Built Environment, Northumbria University, Newcastle Upon Tyne NE1 8ST, UK
cecilia.pesce@northumbria.ac.uk

[2] Historic England, The Engine House, Fire Fly Avenue, Swindon SN2 2EH, UK

Abstract. The practice of producing lime mortars by slaking the quicklime directly in the sand rather than water is known as hot-mixing and, according to numerous historic accounts, was much more common in the past centuries than is generally appreciated by the majority of practitioners. During the last decade, a renewed interest has been dedicated toward the study of the hot-mixed mortar technology, as these mortars are regarded by most craftsmen of superior quality with respect to putty-based mortars in terms of workability, durability and other physical and mechanical properties. In such systems, the slaking of the quicklime is carried out by the moisture surrounding the sand grains and a small amount of added water. The steam developed by such initial slaking promotes further slaking which is supposed to be crucial in determining the characteristics of the lime, and, consequently, of the mortars. However, there are very few in-depth investigations regarding the role of steam on the characteristics of slaked lime. In this study, we have investigated the effects of steam slaking on the characteristics of lime produced by slaking the quicklime from three different UK manufacturers: Lhoist UK, Singleton Birch and Tarmac. Microstructural and mineralogical characteristics of the steam slaked limes are compared with those of the same limes slaked in excess of water and of the related oxides. Analyses were performed using X-ray fluorescence, scanning electron microscopy, and crystallite size and platelet abundance measured by X-ray diffraction data. The results provide useful information on the relations between the characteristics of the limestone used to produce the oxides and the effects of different slaking methods on the characteristics of the slaked lime. This can aid professional conservators to choose the right material and slaking method for the production of compatible mortars in conservation works.

Keywords: Lime · Steam slaking · Water slaking · Calcium oxide · Microstructure

1 Introduction

The hot-mixed mortar technology has seen a renewed interest over the last decade [1]. This type of mortar production entails mixing quicklime and sand together with a small amount of water. All water added to the mix is either consumed by hydration of the

J. I. Escalante-Garcia et al. (Eds.): RW 2021, RILEM Bookseries 40, pp. 771–780, 2023.
https://doi.org/10.1007/978-3-031-21735-7_82

quicklime or evaporates as steam. This creates a 'dry' mix. Once slaking is substantially complete, more water is added to produce a workable mortar. This procedure substantially differs from the procedure based on the use of lime putty, where the quicklime is first slaked in excess of water and the paste produced is only subsequently (sometimes even several years after) mixed with the aggregate to produce the mortar.

Hot-mixed mortars are of great interest to conservators and restoration practitioners, since the use of this technology was extremely common in the past centuries for the production of construction mortars and for plaster and render base coats. Hot-mixed mortars were cheaper and more practical to produce on site [2] and, according to some empirical record, outperform putty mortars in terms of stickiness, water retentivity, and workability. Increased mechanical properties and faster carbonation rate (with respect to putty mortars) have also been reported [3, 4].

In a recent work [5], we investigated some aspects of this technology that could explain the enhanced performance of the mortars. We started from the fact that, across the numerous reported variants of the procedure for hot-mixing, a common element is the fact the quicklime is initially slaked in a limited amount of water, before adding any water. This led us to infer that in such circumstance the quicklime is partially slaked by the moisture, first, and partially by the steam formed by the heat released as a result of the hydration reaction. Results of our study show that: (1) steam-slaked lime exhibits a different microstructure from the same lime slaked in an excess of water; (2) a mortar based on steam-slaked lime has a significantly higher water retention and lower water demand (i.e. it requires less water to reach an appropriate consistency) compared to a putty mortar. However, this study is limited to just one type of quicklime (i.e. produced by a single UK manufacturer).

It is generally known that limes sourced from different producers can have different properties due to differences in the geological properties of the limestone (across different quarries or even within the same quarry), calcination conditions, or other manufacturing processes [6]. In this work, we investigate the effects of the steam-slaking on the characteristics of slaked lime produced using three commercial quicklimes, provided by three UK manufacturers. The microstructure and mineralogy of both, the raw materials and the relative hydroxides obtained through water-slaking and steam-slaking are analysed and compared with those of the related oxides.

2 Methodology

2.1 Raw Materials

Commercial information on the three quicklimes used in the study are reported in Table 1 together with some information on the geology of the limestones used for their production. All quicklimes are classified as high calcium building limes according to the BS EN 459-1 [7]. Nominal granulometry varied from less than 0.180 mm for the Microlime (Singleton Birch) to less than 6 mm for the Calbux Fine 6 (Tarmac). Chemical composition of the quicklimes as determined byd the X-ray fluorescence analysis is reported in Table 2 (details of the analysis in Sect. 82.3).

Table 1 highlights the fact that the limestones used by Tarmac and Lhoist share the same geological characteristics since they are extracted from quarries that are just 4 miles apart, located in the Bee Low limestone formation, in the Peak District, (Derbyshire). Limestones from the Bee Low formation are grey coloured, fine-to medium-grained calcarenites, mainly biosparites (i.e. the cement is composed of sparry calcite: clean, coarse-grained calcite crystals) and bioclasts (i.e. shell fragments and skeletons of microorganisms) [8]. The quicklime from Singleton Birch, instead, is produced with limestone extracted from the Melton Ross quarry (North Lincolnshire), in the Welton Chalk geological formation. Limestones originating in this formation are described as white, thickly bedded chalk with common flint (siliceous) nodules [9].

Table 1. Characteristics of the quicklime from each selected producer.

Producer	Product	Lime type	Nominal granulometry	Source	Geology
Tarmac	Calbux fine 6	CL90	≤6 mm	Tunstead quarry, Buxton (Derbyshire) [10, 11]	Bee low limestone formation (Carboniferous), highly pure calcarenite [12]
Lhoist UK	Milled quicklime	CL90	<2 mm	Brierlow quarry, Buxton (Derbyshire) [13]	
Singleton Birch	Microlime 90	CL90	≤0.18 mm	Melton Ross quarry, Melton Ross (North Lincolnshire) [14]	Welton chalk formation (cretaceous), biocalcarenite [12]

2.2 Slaking Process

The quicklimes tested in the project were used as received soon after delivery to the laboratory and slaked using two methods: with water and with steam, as described below.

Water slaking. A sample of 333 g of each quicklime, in thermal equilibrium with the laboratory temperature (20 °C), was placed in a metal container, where 1 L of degassed de-ionised water (also at 20 °C) was subsequently added, so that the $CaO:H_2O$ mass ratio was 1:3. The lime was, then, thoroughly mixed to promote hydration and, then, left to rest in water at 20 °C for 1 h. At the end of the slaking process, three samples of 5 g each were collected from each batch, dried for 6 h in a vacuum oven, and stored in a desiccator for analysis.

Steam slaking. A sample of 100 g of each quicklime was sieved to obtain particles with diameter 0.5 mm. Each sample was placed in an inert PTFE container. The containers were placed in a perforated steamer resting on a pan filled with de-gassed water. The steamer was, then, placed in an oven at 90 °C for 8 h to allow the steam generated to slake the quicklime. The temperature inside the oven was regularly monitored using a thermocouple and adjusted throughout the test to maintain the water temperature at about 90 °C. At the end of the test, the temperature inside the oven was gradually reduced to room temperature. Subsequently, three samples of 5 g each were collected from each batch and dried in a desiccator for at least 24 h prior to analysis.

2.3 Analytical Techniques

The chemical composition of the quicklimes was determined by quantitative X-ray fluorescence (XRF), performed using a Spectro Xepos benchtop XRF analyser. The 'Geochemistry traces' method was used as internal calibration, and each analysis consisted of 4 passes: (i) atmosphere air; voltage 45.1 keV; (ii) atmosphere air; voltage 60.1 keV; (iii) atmosphere helium; voltage 22.5keV; (iv) atmosphere helium, Voltage 22.5keV. Resolution was 131.5eV at 5.89keV. Pressed pellets with a mixture of CaO and an inert wax binder (Fluxana Cereox wax, $C_{38}H_{76}O_2N_2$) were produced for analysis to improve consistency of the results.

The micro-morphological characterization of the oxides and hydroxides (both steam- and water-slaked) was carried out using a Tescan Mira 3 scanning electron microscope (SEM) in high vacuum mode, voltage at the gun 10 kV. Prior to analysis, the samples were coated with a 5 nm thick platinum layer.

Mineralogical characteristics were investigated using a Rigaku SmartLab X-ray diffractometer (Cu-K_α radiation, parallel beam geometry, 2-ϑ range 10–90°, step 0.05°, scan speed 1.5°/min, 40 kV, 50 mA). Data evaluation (i.e. phase identification and crystallite size) was carried out using the Rigaku SmartLab Studio II software. Crystallite size was calculated using the Halder-Wagner equation [15].

3 Results and Discussion

3.1 Quicklime Composition

The results of the XRF analysis for all quicklimes are shown in Table 2.

Results show that all three quicklimes have a high calcium content (purity > 98%). However, Tarmac's has a slightly reduced CaO content compared to the other products, and shows small traces of magnesium, silicon, and aluminum (these elements in Lhoist and Singleton Birch quicklimes are below level of quantification). Singleton Birch's quicklime shows a slightly higher Fe_2O_3 content compared to the other samples, although still a very small amount.

3.2 SEM Analysis

SEM micrographs of the quicklime, water-slaked lime and steam-slaked lime from each producer are shown in Fig. 1.

The quicklimes show the typical micromorphology of calcined calcium carbonate, i.e. agglomerates made of a network of interconnected round sub-particles [16]. The microstructure of Lhoist and Tarmac quicklimes are similar, with diameter of the

Table 2. Chemical composition of the quicklimes determined by XRF. Values marked with a (*) are below level of quantification.

Phase	Tarmac	Lhoist	Singleton Birch
CaO	98.8	99.9	99.7
MgO	0.530	<0.017*	<0.017*
SiO_2	0.361	<0.0011*	<0.0011*
Al_2O_3	0.083	<0.0038*	<0.0038*
Fe_2O_3	0.075	0.057	0.124
S	0.050	<0.001*	<0.001*
Cl	0.015	<0.003*	<0.003*
P_2O_5	0.009	<0.0007*	<0.0007*

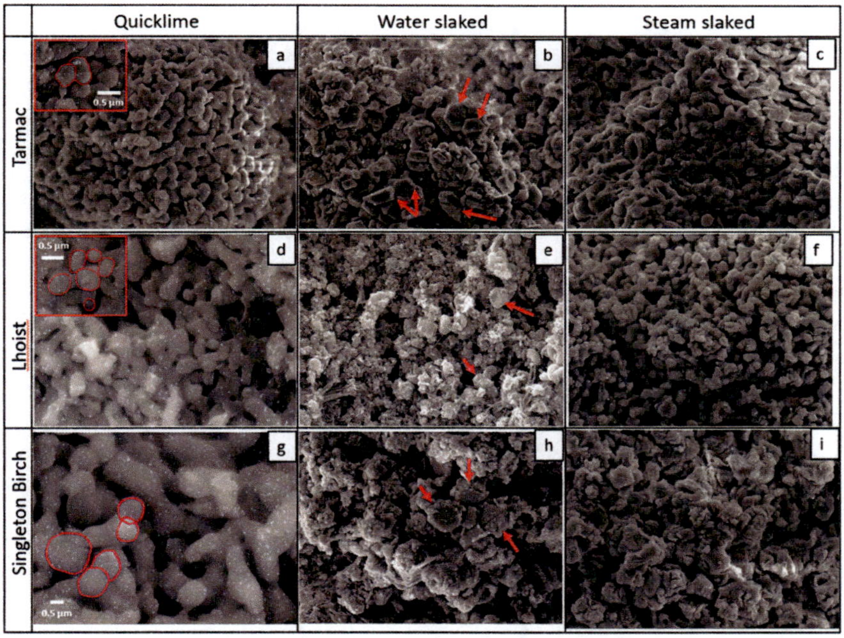

Fig. 1. SEM micrographs of quicklime, water-slaked hydroxide, and steam-slaked hydroxide of each selected producer. View field of all micrographs is 10 µm. The features measured in the quicklime are highlighted in insets in (**a**, **d**, **g**). In (**b**, **e**, **h**) well-developed hexagonal portlandite crystals are highlighted by the arrows.

sub-particles in the range 0.4–0.7 µm (Fig. 2a, d), whereas Singleton Birch quicklime displays a significantly coarser microstructure, with diameter of the features in the range 0.8–1.5 µm (Fig. 2g). It is worth noting that the textural features of the quicklime do not necessarily reflect the granulometry of the product. Indeed, Tarmac and Lhoist quicklimes are made of particles with diameter up to 6 and 2 mm, respectively. However, under the SEM, they show a finer microstructure than the Singleton Birch quicklime which is made of finer granules (up to 0.18 mm) but shows a coarser microstructure under the microscope. Such textural difference between quicklimes is likely to be due to the different microstructure of the parent limestones.

The microstructure of the water-slaked limes are characterised by the presence of several well-developed equi-axed/short hexagonal portlandite crystals (with width up to 0.7 μm for Lhoist and 1.5 for Tarmac and Singleton Birch) embedded in a matrix of nanometric, granular-shaped crystals. The slightly bigger crystals observed in the Tarmac and Singleton Birch quicklime in comparison with Lhoist may be linked with a higher reactivity of the Lhoist quicklime, as an increased reaction kinetics often promotes the formation of smaller crystals [17].

The steam-slaked limes display a distinctly different microstructure compared to the water-slaked limes. The microstructure of the steam-slaked samples resembles that of the parent quicklime, i.e. made of agglomerates of interconnected, rounded sub-particles stacked together (the full conversion of the oxide into hydroxide is confirmed by XRD analysis; see Sect. 82.3). This type of microstructure has been previously observed by the authors in steam-slaked Tarmac lime [5, 18]. At higher magnification, it is possible to observe that the particles display the hexagonal geometry typical of portlandite (Fig. 2). The observed microstructure is probably related to the fact that, when lime is slaked by steam, the hydration reaction occurs on a gas/solid interface, resulting in piles of thin platelets of portlandite stacked along the *c* axis [5, 19, 20]. This type of crystal growth results in mechanical stress in the newly-formed portlandite crystals [5], which exhibit nano-cracks on their surface (Fig. 2b). The size of the sub-particles slightly varies across producers. Similarly to what observed in the water-slaked limes, Tarmac and Singleton Birch samples show slightly bigger crystals (up to 0.7 μm) than Lhoist (up to 0.4).

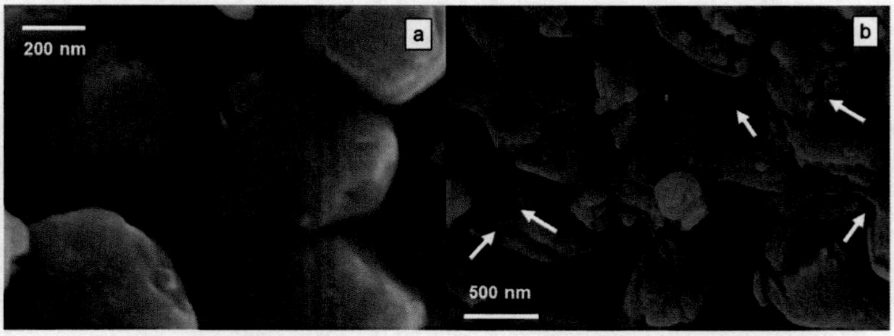

Fig. 2. SEM micrographs of steam-slaked portlandite crystals (Tarmac) at high magnification. **b** Arrows highlight nano-cracks due to stress developed during steam slaking.

3.3 XRD Analysis

The quantitative analysis of the samples revealed that in the quicklimes less than 5% of the mineral was portlandite, likely due to partial hydration upon contact with moist air during material handling at the factory or inside the laboratory. Traces of calcite (<1%) were found in the hydrated limes, both water- and steam-slaked (scans of the latter limes shown in Fig. 3), likely due to partial carbonation upon contact with atmospheric moisture and CO_2, together with traces of aluminum, due to the signal from the sample holder, and zincite, used as internal standard. Apart from these minerals, the analyses showed that all quicklimes are predominantly made of the mineral CaO (i.e. lime), and the hydrated limes are predominantly made of $Ca(OH)_2$ (i.e. portlandite).

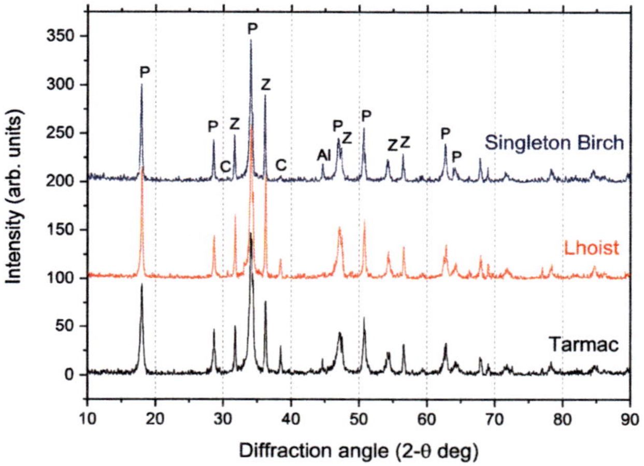

Fig. 3. XRD scans of steam-slaked limes (keys: P = portlandite; C = calcite; Z = zincite; Al = aluminum).

In our previous works [5, 18], we showed that the XRD profiles of steam-slaked lime systematically display peaks with a remarkable broadening, when comparing with the same lime slaked in water. The peak broadening is inversely correlated to the crystallite size by the Scherrer's equation, meaning that steam slaked lime has smaller portlandite crystallites. The reduction in crystallite size of steam-slaked lime is likely due to the internal stress developed when portlandite crystals nucleate and grow on the CaO surface and not in solution, as in water-slaked lime [5].

The crystallite size measurements as obtained from the XRD data on the quicklimes and slaked limes are shown in Fig. 4. The graph shows that the quicklimes have a remarkably higher crystallite size (60–70 nm) compared to the hydrated phases. Indeed, it has been previously reported that calcium oxide is a mineral phase of high crystallinity [21–23] and in particular, Brunauer et al. [23] note that there is no significant difference in crystallite size between soft-burned and hard-burned quicklimes. Our results are in agreement with the literature and do not suggest substantial difference in crystallite size across the three manufacturers. Conversely, all the hydrated limes show a small crystallite size (all below 40 nm), suggesting that the slaking results in a hydrated product of lower crystallinity, possibly due to the highly exothermic nature and very fast rate of the hydration reaction. Figure 4 also shows that for the Lhoist and Tarmac samples, the steam slaking leads to a lime with lower crystallite size compared with lime produced through water slaking. This is likely the consequence of the internal mechanical stress developed during steam-slaking [5], as also suggested by the presence of nano-cracks, visible on the surface of crystals by SEM analysis (Fig. 2b). Remarkably, Singleton Birch lime seems to behave differently, with the crystallite size of the steam-slaked lime only slightly smaller than that of the water-slaked lime. As already observed in the SEM images, the Singleton Birch quicklime displays a different microstructure compared to Lhoist and Tarmac, and such difference is likely to be related to the different characteristics of the geological formation of the parent limestone. It is also worth reporting, to this regard, that during the water slaking it was empirically observed that, whereas Tarmac and Lhoist underwent a very fast, violent hydration reaction which brough the water to a boil in a few seconds from the moment when the water was added, the sample from Singleton Birch took a few minutes before the material started to visibly break, and the overall slaking rate seemed slower than that of the other products. It is likely that the coarser microstructure of the quicklime accounts for its different behaviour on slaking, due to a reduced specific surface area available for reaction. Specific surface area of CaO has been reported to be correlated with hydration rate [23]. A microscopic examination of the limestones used to produce the quicklimes through petrographic analysis and SEM would provide further information on this matter.

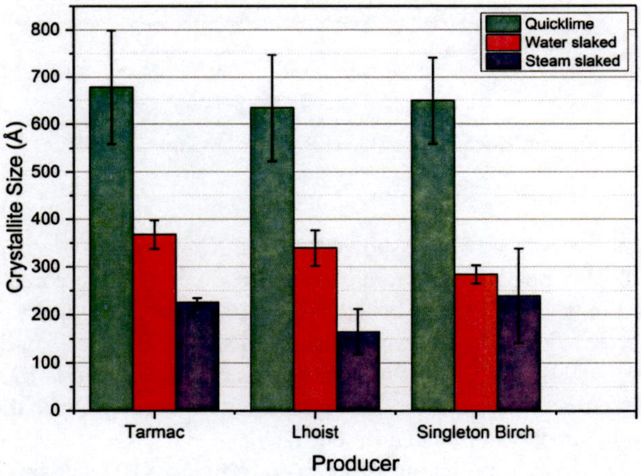

Fig. 4. Crystallite size of quicklime, water-slaked hydroxide, and steam-slaked hydroxide from each producer, as calculated by XRD.

4 Conclusions

The following conclusions can be drawn from this study:

- The hydrated phases (both water- and steam-slaked) have a much lower crystallinity than the quicklimes, for all three producers;

- The quicklime produced by Singleton Birch has a coarser microstructure in comparison with Tarmac and Lhoist which accounts for a different behaviour on slaking: XRD analyses highlighted that steam-slaked limes from Lhoist and Tarmac have a significantly smaller crystallite size compared to the water-slaked samples. Conversely, Singleton Birch steam-slaked lime shows a crystallite size comparable to that of the water-slaked lime.
- The microstructural differences between Singleton Birch and Tarmac/Lhoist are ascribable to differences in mineralogical characteristics of the parent limestones.
- The reduced crystallinity of steam-slaked lime from Tarmac and Lhoist in comparison with Singleton Birch can be linked to a higher reactivity towards CO_2 and enhanced properties of mortars based on this type of lime.

References

1. Henry, A.: Hot-mixed mortars: the new lime revival. Context (Inst. Hist. Build. Conserv.) **54**, 30–33 (2018)
2. Copsey, N.: Hot Mixed Lime and Traditional Mortars. The Crowood Press, Marlborough (2019)
3. Válek, J., Matas, T.: Experimental study of hot mixed mortars in comparison with lime putty and hydrate mortars. RILEM Bookseries **7**(September), 269–281 (2013). https://doi.org/10.1007/978-94-007-4635-0_21
4. Moropoulou, A., Tsiourva, T., Bisbikou, K., Biscontin, G., Bakolas, A., Zendri, E.: Hot lime technology imparting high strength to historic mortars. Constr. Build. Mater. **10**(2), 151–159 (1996)
5. Pesce, C., Godina, M.C., Henry, A., Pesce, G.: Towards a better understanding of hot-mixed mortars for the conservation of historic buildings: the role of water temperature and steam during lime slaking. Herit. Sci. **9**(1), 1–18 (2021). https://doi.org/10.1186/s40494-021-00546-9
6. Ontiveros-Ortega, E., Ruiz-Agudo, E.M., Ontiveros-Ortega, A.: Thermal decomposition of the CaO in traditional lime kilns. Applications in cultural heritage conservation. Constr. Build. Mater. **190**, 349–362 (2018). https://doi.org/10.1016/j.conbuildmat.2018.09.059
7. British Standards Institution: Building lime. Definitions, specifications and conformity criteria. BS EN 459-12015 (2015)
8. British Geological Survey: Bee Low Limestone Formation. https://webapps.bgs.ac.uk/lexicon/lexicon.cfm?pub=BLL. Accessed 25 Jun 2021
9. British Geological Survey: Welton Chalk Formation. https://webapps.bgs.ac.uk/lexicon/lexicon.cfm?pub=WCK. Accessed 25 Jun 2021
10. Tarmac: Cement and Lime Tunstead. https://tunstead.tarmac.com/cement-and-lime/. Accessed 24 Jun 2021
11. Tarmac: Environment Tunstead. https://tunstead.tarmac.com/environment/. Accessed 24 Jun 2021
12. British Geological Survey: Geology of Britain viewer. https://www.bgs.ac.uk/map-viewers/geology-of-britain-viewer/. Accessed 24 Jun 2021
13. Derbyshire County Council and Derby City Council: Derbyshire and Derby Minerals Local Plan (2017). https://www.derbyshire.gov.uk/site-elements/documents/pdf/environment/planning/planning-policy/minerals-local-plan/industrial-limestone-background-paper.pdf. Accessed 24 Jun 2021

14. Singleton Birch: Contact us. https://www.singletonbirch.co.uk/home/contact-us/
15. Halder, N.C., Wagner, C.N.J.: Separation of particle size and lattice strain in integral breadth measurements. Acta Crystallogr. **20**(2), 312–313 (1966). https://doi.org/10.1107/s0365110x66000628
16. Witoon, T.: Characterization of calcium oxide derived from waste eggshell and its application as CO_2 sorbent. Ceram. Int. **37**(8), 3291–3298 (2011). https://doi.org/10.1016/j.ceramint.2011.05.125
17. Ginebra, M.P., Driessens, F.C.M., Planell, J.A.: Effect of the particle size on the micro and nanostructural features of a calcium phosphate cement: a kinetic analysis. Biomaterials **25**(17), 3453–3462 (2004). https://doi.org/10.1016/j.biomaterials.2003.10.049
18. Pesce, C., Pesce, G.: Effects of steam slaking on the characteristics of portlandite crystals. In: Proceedings of 39th Cement Concrete Science Conference, pp. 58–61 (2019)
19. Ramachandran, V.S., Sereda, P.J., Feldman, R.F.: Mechanism of hydration of calcium oxide. Nature **201**, 288–289 (1964). 1475-2875-9-181[pii]\n. https://doi.org/10.1186/1475-2875-9-181
20. Beruto, D., Barco, L., Belleri, G., Searcy, A.W.: Vapor-phase hydration of submicrometer CaO particles. J. Am. Ceram. Soc. **64**(2), 74–80 (1981). https://doi.org/10.1111/j.1151-2916.1981.tb09579.x
21. Fischer, H.C.: Calcination of calcite: II, size and growth rate of calcium oxide crystallites. J. Am. Ceram. Soc. **38**(8), 284–288 (1955). https://doi.org/10.1111/j.1151-2916.1955.tb14946.x
22. Blanton, T.N., Barnes, C.L.: Quantitative analysis of calcium oxide desiccant conversion to calcium hydroxide using X-ray diffraction. Adv. X-ray Anal. **48**, 45–51 (2005). https://doi.org/10.1154/1.1779815
23. Brunauer, S., Kantro, D.L., Weise, C.H.: The surface energies of calcium oxide and calcium hydroxide. Can. J. Chem. **34**, 729–742 (1956)

A New Research Strategy in Studying and Improving the Bond Between Concrete and Repair Mortar

Mohammad Ali Yazdi[1], Elke Gruyaert[2], Kim Van Tittelboom[1], and Nele De Belie[1](✉)

[1] Magnel-Vandepitte Laboratory for Concrete Research, Ghent University, Tech Lane Ghent Science Park, Campus A, Technologiepark Zwijnaarde 60, 9052 Ghent, Belgium
Nele.DeBelie@UGent.be

[2] KU Leuven, Department of Civil Engineering, Ghent Technology Campus, Gebroeders De Smetstraat 1, 9000 Ghent, Belgium

Abstract. The required depth of concrete removal should dictate the selection of concrete removal technique and removal energy in repair systems. This study presents a new research approach in the investigation of the impact of concrete removal techniques on the substrate, and as a result, the bond with repair mortar. On the other hand, it introduces a remedy to improve the bond strength. To this end, crushed limestone concrete (CC) was made and the substrate surface was removed until a certain depth (2–3 cm) by using two commonly used techniques: jack-hammering (7 kg), and hydrodemolition/water-jetting (100 MPa). The selection of removal technique and energy was according to the defined removal depth. Automated laser measurements (ALM) were employed to study the surface roughness of the substrates. Later, the bond strength results of jackhammered-CC samples were compared with those of another study, for which the same removal depth was applied to a gravel concrete (GC) substrate by using a lighter jackhammer (1.5 kg). Moreover, nano-silica was used to improve the bond strength of jack-hammer treated substrates with the repair mortar.

Keywords: Concrete removal depth · Concrete removal techniques and energy · Bond strength · Nano-silica

1 Introduction

Apart from roughness and saturation degree [1, 2] of the existing concrete, removal technique and energy [3] highly affect the quality of the bond strength in repair systems. Concrete removal techniques like jackhammering (JH) or high-pressure water jetting (WJ) are usually referred as "aggressive" [4]. In research studies, the mentioned techniques are sometimes compared with techniques such as blasting, while they belong to different classifications and so aim at different applications. As a matter of fact, JH and WJ are mainly used for removal, while blasting is used for cleaning or roughening of the surface of concrete [5]. Yazdi et al. [4, 6] for the first time expressed the importance of the selection of the removal technique based on the concrete removal classifications

J. I. Escalante-Garcia et al. (Eds.): RW 2021, RILEM Bookseries 40, pp. 781–788, 2023.
https://doi.org/10.1007/978-3-031-21735-7_83

in research studies. In the current paper, the selection of removal techniques and energy was based on the removal depth of the substrates. To this end, a certain removal depth (2–3 cm) was defined and then both removal technique and energy were selected and adopted accordingly.

JH has been reported to weaken the bond strength between concrete substrate and repair mortar by creating microcracks in the superficial surface of the existing concrete [4, 6–8]. However, not much attention has been paid to find a technique to improve the weakened bond strength caused by JH. This paper aims to introduce a remedy to this problem.

2 Materials and Methods

2.1 Sample Preparation

Two concrete slabs with dimensions of 1000 mm × 1000 mm × 200 mm and 2000 mm × 1000 mm × 200 mm were prepared with crushed limestone aggregates and a mix composition as shown in Table 1. Superplasticizer Master Glenium-51 was added to the mixture until the slump of 120 ± 20 mm was reached. These crushed stone concrete (CC) slabs were exposed to jackhammering (JH) and water jetting (WJ) removal techniques, respectively, at the age of 3 months or older. Half of the larger slab was not exposed and remained as control. JH and WJ were selected based on the removal depth of 2–3 cm. Removal energy was also adjusted according to the removal depth. The intention was to keep the removal depth similar regardless of the removal technique. Afterwards, each slab was sawn into samples of 330 mm × 330 mm × 200 mm. JH was performed using a Hitachi H45MEY(35JR), 7 kg, 1150 W, 13.6 J and WJ with a high pressure of 100 MPa was performed by the Crahay company in Belgium. Gravel concrete (GC) substrates (300 mm × 300 mm × 100 mm) were also prepared and underwent JH as explained in [4] to be compared with JH-CC substrates. For GC substrates an Air Hammer model 29062 with the weight of 1.5 kg was used. Mix composition of GC substrate is given in [4]. In this study, concrete substrates were later on covered by BASF MasterEmaco S 5400 repair mortar with a 15 ± 5 mm thickness. Colloidal nano-silica containing synthetic amorphous nano-silica with nominal particle size of 12 nm supplied by the GRACE company and, was used to treat the JH-CC samples. Information regarding the surface treatment of the substrates is given in [9].

2.2 Automated Laser Measurement (ALM) Method

Roughness of the treated substrates (JH and WJ) was measured using Automated Laser Measurements (ALM). The height of the surface peaks and valleys (Z) has been measured with a high precision laser beam (sensor ILD 1800-50 and interface optoNCDT 1800, Micro-Epsilon Mess-technik GmbH, Ortenburg, Germany; resolution = 5 microns) mounted on an in-house developed measurement table equipped with two stepping motors that control the movement in X and Y direction [10]. For each concrete substrate, the surface profile was obtained in X-direction for 3 different Y-positions, and vice versa. The sampling length and the evaluation length were 40 mm and 280 mm,

Table 1. Mix composition of CC substrate.

Materials	Amount
CEM I 52.5 N (kg/m^3)	340
Water (kg/m^3)	153
Sand (kg/m^3)	604
Crushed limestone 2/6.3 (kg/m^3)	336
Crushed limestone 6.3/20 (kg/m^3)	967
Superplasticizer Master Glenium-51	*
W/C	0.45
90 days characteristic compressive strength (f_{ck}) (N/mm^2)	68.40

*The amount of superplasticizer was adjusted to obtain 120 ± 20 mm slump

respectively, see Fig. 1. The shorter sampling length is chosen to separate the surface roughness from long wave components. Two samples (330 mm × 330 mm × 200 mm) per treatment technique underwent ALM analysis. In order to evaluate the roughness, Eqs. (1) and (2) were employed based on BS 1134 (2010). The former (R_a) is to calculate the centre-line roughness value (mm) and the latter (R_q) to measure the root-mean-square roughness value (mm).

$$R_a \approx \frac{1}{n}\sum\nolimits_{i=1}^{n}|Z_i| \tag{1}$$

$$R_q \approx \sqrt{\frac{1}{n}\sum\nolimits_{i=1}^{n}{Z_i}^2} \tag{2}$$

where n is the number of measurements, i is the measurement point number and Z is the amplitude (mm).

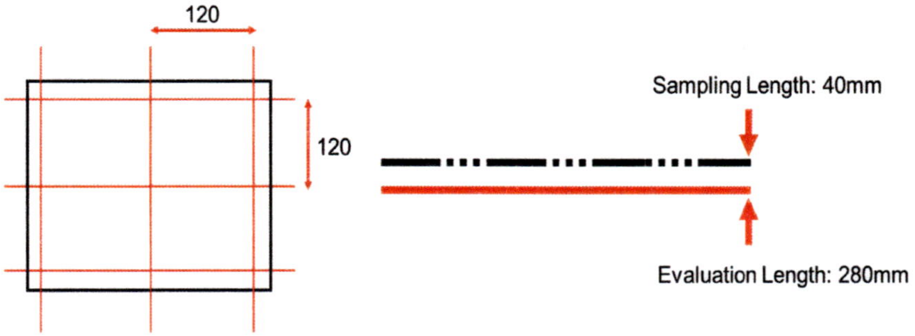

Fig. 1. ALM direction on a slab with 330 mm x 330 mm surface area (left), evaluation and sampling lengths (right).

2.3 Pull-Off Test

The surface tensile and bond strengths of samples were evaluated using the pull-off test method according to NBN EN 1542. Bond strength measurements were done 28 days after the repair mortar was applied. Cores of 50 mm diameter and height of 25 ± 5 mm, depending on the thickness of the repair mortar (15 ± 5 mm), were drilled into the concrete. For surface tensile strength, measurements were done on substrates with no overlay material. Epoxy resin was used to glue a dolly to each core. Overall, at least 3 pull-off tests were performed per concrete composition. The adhesion strength was calculated following Eq. (3):

$$f_{hs} = \frac{4F_h}{\pi D^2} \tag{3}$$

where f_{hs} is the final strength (N/mm^2), F_h the failure load (N) and D the mean diameter (mm).

3 Results

3.1 Roughness

As the roughness results in Fig. 2 show, WJ resulted in a higher surface roughness compared to JH. It is the same for both R_a, the centre-line roughness value, and R_q, the root-mean-square roughness value. A higher roughness value was also reported in [4] for WJ when it was compared to JH.

Fig. 2. R_a and R_q results of CC samples (mm). Error bars represent the standard error.

3.2 Adhesion

Surface tensile and bond strength results are given in Fig. 3. Surface tensile strength (STS) was measured to study the characteristics of the prepared surface prior to repair application. As shown, STS decreased in order of untreated, WJ and JH substrates. JH achieved the lowest STS, the obtained strength is significantly lower than that of untreated and WJ substrates. With regard to the bond strength (BS), the same trend is observed.

In fact, untreated samples have the highest BS following by WJ and JH. The low bond strength of JH is attributed to the microcracks created during the removal process [4, 7]. The presence of microcracks at the superficial layer of the existing concrete can impair the bond at the interface [4, 7]. In general, Fig. 3 confirms the applicability of STS tests for the estimation of the BS. From the results given in Fig. 3, the ineffectiveness of increased roughness to improve the BS is evident when the state of stress is tension (in contrast with shear where interlocking would play a role). WJ samples compared to untreated samples have a significantly higher roughness but lower BS and STS. This is also seen in the results reported in [4].

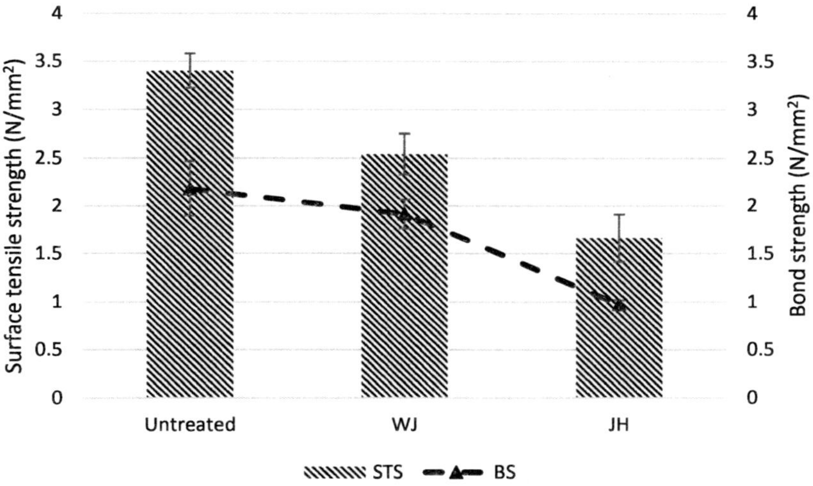

Fig. 3. Surface tensile and bond strength results of CC samples. Error bars represent the standard error (n $\geq$ 5).

Figure 4 shows the bond strength results of JH samples from this study (GC2-2 cm-1.5 kg and CC-2 cm-7 kg) and another study performed by the authors (GC1-1 cm-1.5 kg) [4]. Results shown in dots and diagonal lines belong to samples with a similar mix composition (GC1 and GC2) and the one represented with vertical lines is made of CC. Interestingly, removal depth has shown to influence the bond strength results. Although bond strength results shown for GC1 and GC2 were obtained from slabs with a similar mix composition, the difference in removal depth shows to significantly influence the substrate properties and so does the bond strength between these samples and the overlay. With regard to GC2 and CC, both achieved a similar bond strength. For these samples, the composition and the weight of the jackhammer used for the concrete removal were different. However, the removal depth was similar. The comparison of these results once again shows the importance of concrete removal depth in research studies. Taking removal depth constant regardless of removal technique seems to be important in research studies on concrete removal techniques. For example, for a technique like JH, the higher the removal depth, the higher the removal energy. Therefore, if the removal depth is not similar for each removal technique, comparison of the effects of removal techniques on the existing concrete becomes difficult. In addition, the removal depth affects the extent to which the aggregates are exposed, which will influence the roughness.

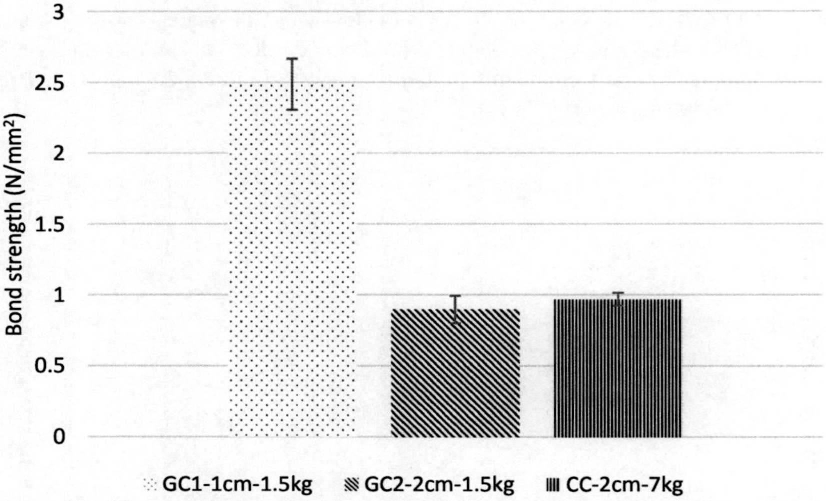

Fig. 4. Bond strengths results. 1 cm and 2 cm indicate the removal depth, and 1.5 kg and 7 kg indicate the weight of the jackhammer. Error bars represent the standard error (6 replicates for GC1, 8 for CC, and 3 for GC2).

3.3 Bond Strength Improvement of JH Concrete

To attain a robust bond between the repair material and the JH concrete, JH samples were pre-treated with nano-silica. JH results do not meet the minimum threshold value of 1.5 MPa given by EN 1504-3 for laboratory performances for structural repair. However, according to the results shown in Fig. 5, a noticeable improvement by 50% is achieved when the surface of the host concrete is treated with nano-silica, reaching the minimum value suggested by the standard.

3.4 Conclusions

This study introduces a new strategy in studying the effect of removal techniques on the existing concrete and so the bond strength with the overlay material. According to the results, in the comparison of removal techniques, removal depth is an important parameter to be taken into account. The removal energy and technique should be adjusted and selected according to a defined removal depth for a more accurate comparison of removal techniques. Moreover, nano-silica showed to significantly improve the bond strength of JH samples with the overlay material.

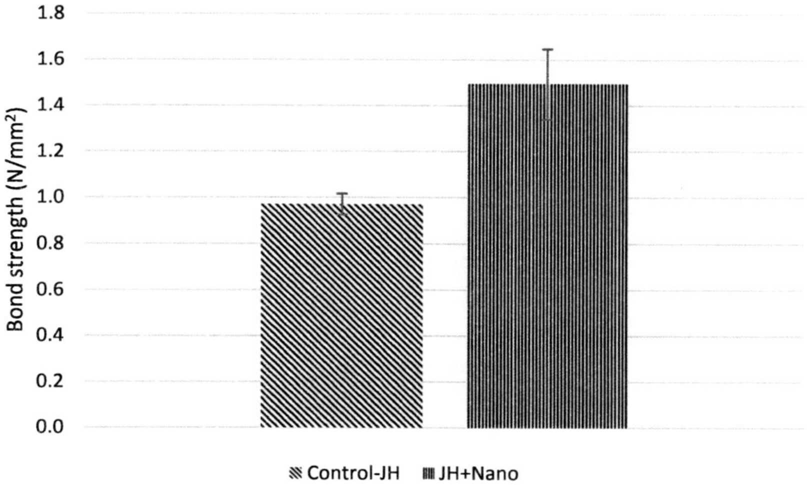

Fig. 5. Bond strengths of JH and JH treated with nano-silica. Error bars represent the standard error (n $\geq$ 5).

References

1. Perez, F., Morency, M., Bissonnette, B., Courard, L.: Correlation between the roughness of the substrate surface and the debonding risk. In: Concrete Repair, Rehabilitation and Retrofitting II: 2nd International Conference on Concrete Repair, Rehabilitation and Retrofitting, ICCRRR-2, p. 347. CRC Press, Cape Town (2008)
2. Courard, L., Bissonnette, B., Garbacz, A., Vaysburd, A. M., von Fay, K. F.: Guidelines for concrete surface preparation: 10 years research and experience. In: MATEC Web of Conferences, Vol. 199, p. 08004. EDP Sciences (2018)
3. Matthews, S.: CONREPNET: performance-based approach to the remediation of reinforced concrete structures: achieving durable repaired concrete structures. J. Build. Apprais. **3**(1), 6–20 (2007)
4. Yazdi, M.A., Dejager, E., Debraekeleer, M., Gruyaert, E., Van Tittelboom, K., De Belie, N.: Bond strength between concrete and repair mortar and its relation with concrete removal techniques and substrate composition. Constr. Build. Mater. **230**, 116900 (2020)
5. Vorster, M.C., Merrigan, J.P., Lewis, R.W., Weyers, R.E.: Techniques for concrete removal and bar cleaning on bridge rehabilitation projects No. SHRP-S-336 (1992)
6. Yazdi, M.A., Dejager, E., Debraekeleer, M., Gruyaert, E., Van Tittelboom, K., De Belie, N.: How concrete removal techniques affect the bonding between concrete and repair mortar. In: MATEC Web of Conferences, vol. 289, p. 02008. EDP Sciences (2019)
7. Courard, L., Piotrowski, T., Garbacz, A.: Near-to-surface properties affecting bond strength in concrete repair. Cement Concr. Compos. **46**, 73–80 (2014)
8. Courard, L., Bissonnette, B., Belair, N.: Effect of surface preparation techniques on the cohesion of superficial concrete: comparison of jack-hammering and water jetting. In: Concrete Repair, Rehabilitation and Retrofitting, pp. 1027–1031. Taylor & Francis Group, Cape Town (2006)
9. Yazdi, M.A., Gruyaert, E., Van Tittelboom, K., Boon, N., De Belie, N.: Treatment with nano-silica and bacteria to restore the reduced bond strength between concrete and repair mortar caused by aggressive removal techniques. Cement Concr. Compos. **120**, 104064 (2021)

10. De Belie, N., Monteny, J., Beeldens, A., Vincke, E., Van Gemert, D., Verstraete, W.: Experimental research and prediction of the effect of chemical and biogenic sulfuric acid on different types of commercially produced concrete sewer pipes. Cem. Concr. Res. **34**(12), 2223–2236 (2004)

Ammonium Phosphate for "Green" Conservation of Cultural Heritage: 10 Years of Research in the Laboratory and in the Field

Enrico Sassoni(✉)

University of Bologna, 40131 Bologna, Italy
enrico.sassoni2@unibo.it

Abstract. This paper presents a summary of results obtained in the last 10 years on the use of ammonium phosphate solutions for conservation of heritage stones and mortars. After a brief introduction about the genesis of this treatment, the most used formulations are presented. Various functionalities of the treatment are reviewed, then its compatibility and durability are discussed. Recent results on the treatment environmental sustainability are also reported. The treatment's performance in comparison with alternative consolidants (ethyl silicate, nanolimes, acrylic resin) is then evaluated and, finally, some pilot applications of ammonium phosphate onto real buildings and monuments are reviewed. All the laboratory and field studies published so far point out ammonium phosphate as one of the most promising inorganic treatments currently available for conservation of heritage materials.

Keywords: Ammonium phosphate · Hydroxyapatite · Protection · Consolidation · Bowing · Marble

1 Introduction

The use of diammonium hydrogen phosphate (DAP) for the conservation of cultural heritage was proposed in 2011 [1–3], with the aim of overcoming the limitations that commercial consolidants often exhibit when applied onto carbonate stones. The idea is to react calcium-rich substrates with aqueous solutions of DAP to form new calcium phosphates (CaP) with consolidating and protective abilities [1–4]. Ideally, hydroxyapatite (HAP, $Ca_{10}(PO_4)_6(OH)_2$), which is the CaP with lowest water solubility at pH > 4, is formed, as schematically illustrated in Fig. 1 [4].

Conversion of calcium carbonate to calcium phosphate was inspired by the so-called ammonium oxalate treatment, aimed at converting calcium carbonate to calcium oxalate to form a protective layer over the treated substrate. Compared to calcium oxalate, HAP has the advantage of having significantly lower water solubility and dissolution rate and a good match with calcite in terms of lattice parameters, which should ensure a good adhesion between the substrate and the new phase [1]. Starting from this idea, in the following years extensive laboratory testing has demonstrated the significant potential of the DAP treatment for several functions.

J. I. Escalante-Garcia et al. (Eds.): RW 2021, RILEM Bookseries 40, pp. 789–798, 2023.
https://doi.org/10.1007/978-3-031-21735-7_84

The present paper aims at reviewing the main studies dealing with the DAP treatment. First, the various formulations proposed over the years are reviewed, highlighting the rationale behind each of them. Then, the effectiveness, the compatibility and the durability of the treatment are discussed. A recent study investigating the environmental sustainability of the treatment is also reported. The performance of the DAP treatment in comparison with alternative commercial products is then evaluated. Finally, some recent studies reporting the outcome of the application of the DAP treatment to real monuments are summarized.

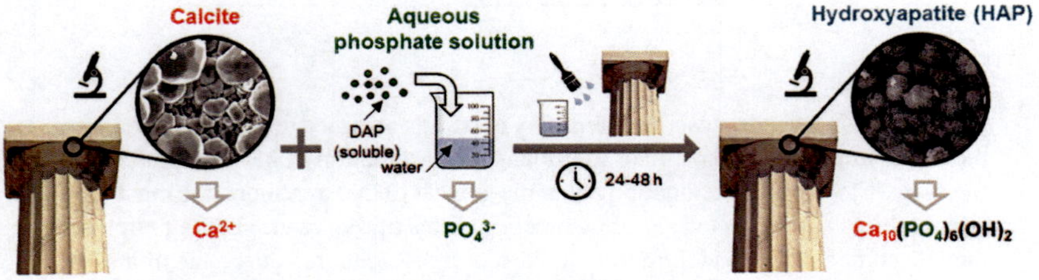

Fig. 1. Scheme illustrating *in situ* formation of HAP [4].

2 Formulations

Several different formulations of the DAP treatment have been proposed. The most relevant ones are discussed in the following. Unless otherwise specified, water is intended as the solvent and 24 h as the reference treatment duration.

1 M DAP. This was one of the first formulations to be tested [1]. The new phase formed after treatment is HAP [5], which is the ideal CaP to form being the least soluble, but the new HAP coating formed over marble was found to be incomplete [6].

5 wt% DAP or 10 wt% DAP. These concentrations, corresponding to 0.38 and 0.76 M DAP respectively, were proposed in [3, 7], following a more application-oriented approach. Good consolidating ability was reported with 5 wt% DAP, thanks to formation of HAP [3]. With 10 wt% DAP, the new consolidating phase was identified as poorly-crystalline stoichiometric HAP, with minor formation of octacalcium phosphate (OCP, $Ca_8(HPO_4)_2(PO_4)_4{\bullet}5H_2O$) and brushite ($CaHPO_4{\bullet}2H_2O$) [7].

1M DAP+1mM $CaCl_2$. With respect to the previous formulations, the addition of $CaCl_2$ as a calcium source proved to significantly accelerate CaP formation, thus also allowing more complete coverage of the treated surface [5]. Upon addition of a calcium source, the newly formed CaP coating was found to contain both HAP and OCP, which is more soluble than HAP but still less than calcite [5].

3M DAP, then limewater. This formulation was proposed following the rational that the higher the concentration of PO_4^{3-} ions, the more abundant the formation of HAP [8]. However, treating porous substrates with high DAP concentrations may lead to unreacted fractions remaining in the pores at the end of the treatment. To remove them, the application of a poultice of limewater (i.e. a saturated solution of calcium hydroxide) was recommended, with the twofold aim of providing additional calcium ions to form new CaP and transporting the unreacted fractions into the poultice during drying [8]. The new phase formed after this treatment was reported to be mainly HAP, although formation of OCP could not be completely excluded [8].

0.1M DAP+1mM $CaCl_2$ in 10 vol% ethanol or isopropanol. As an alternative route to increase the reactivity of phosphate ions in solution, without increasing the DAP concentration, the addition of alcohols to the DAP solution was proposed [9–11]. Indeed, in the presence of alcohol the formation of new CaP is strongly favored, even at DAP concentration as low as 0.1M. This has been attributed to the weakening of the hydration sphere of phosphate ions made possible by the alcohol molecules [11].

3 Effectiveness

3.1 Protective Ability

Several routes have been proposed to form continuous, pore- and crack-free coatings over marble to protect it from dissolution in rain, thanks to the low solubility of HAP.

Chemical route. CaP films can be formed by reaction of marble with DAP solutions, possibly also containing a calcium source. By using 1M DAP+1mM $CaCl_2$ solutions, a reduction in the dissolution rate of the substrate by ~40% was achieved [6]. The HAP+OCP coating formed with $CaCl_2$ addition proved to be more complete than the HAP coating formed using only 1M DAP [5], but still was porous and cracked, likely because the high DAP concentration leads to excessive film growth [11]. Better protective efficacy was achieved when crack- and pore-free coatings were obtained, by reducing the DAP concentration and by adding alcohols to the solution [9–11].

Electrochemical route. To obtain continuous coating with no bare areas, electrodeposition (i.e. the application of an electric current to a solution to improve the transport of ions and to make a compound precipitate near the surface of an electrode) was explored, taking inspiration from the biomedical field [12]. By placing a metallic grid close to the marble surface to be coated, by using a DAP solution as the electrolyte and by making the grid work as the cathode, more rapid and more continuous formation of a HAP film was obtained, compared to similar conditions with no electric potential applied [12].

Biomimetic and biogenic routes. A "biomimetic" route (involving the use of collagene, over which HAP is formed using DAP and $CaCl_2$ as precursors [13]) and a "biogenic" route (involving the use of bacteria that promote HAP precipitation [14]) have also been investigated. According to the latter approach, the substrate is inoculated with the bacterium *Pseudomonas fluorescens* and a solution containing LB broth (a medium used for the growth of bacteria), KH_2PO_4 and K_2HPO_4. Thanks to the bacterial metabolic processes, which were responsible for an increase in pH, poorly crystalline carbonated HAP was formed after 20 days, whereas brushite was found when no bacteria were used [14].

3.2 Consolidating Ability

Thanks to the formation of new CaP with binding action in intergranular fissures and inside the pores of weathered substrates, the significant consolidating ability of various DAP formulations has been assessed by several techniques (ultrasonic measurements, tensile strength, compressive strength, resistance to abrasion, resistance to material loss by peeling) on a variety of substrates, including marble [11, 15], limestone [1, 2, 16], calcarenite [17], sandstone [18], mortars based on air lime [19–21], natural hydraulic lime [19] and cement [19], gypsum stuccoes [22]. As a general trend, the higher the DAP concentration, the higher the strengthening effectiveness [11, 23, 24], because in the case of consolidation the possible presence of pores and cracks in the new CaP phases is not as detrimental as in the case of protection.

3.3 Ability to Arrest Bowing

Thin marble slabs exposed outdoors, like in monumental cemeteries or in modern buildings façades, are subject to progressive bowing, as a response to temperature fluctuations. This phenomenon can be significantly mitigated by application of the DAP treatment [23, 25], because HAP formed in intergranular fissures is able to strengthen the material without completing occluding voids. This is important because, in case all the intergranular fissures are filled with a rigid material, any new temperature variation would lead to stress at the grain boundaries and microcrack formation. By application of DAP solutions, further bowing of already-bowed slabs can be reduced, while bowing of initially planar slabs can be delayed [23].

3.4 Desulfating Ability

After prolonged exposure in urban environment, carbonate substrates are often covered with a layer of gypsum, resulting from the sulfation of calcium carbonate induced by atmospheric sulfur oxides. The DAP treatment can be used to transform soluble gypsum into insoluble CaP, thus performing a desulfating action [17, 22]. The same type of conversion can also be used for conservation of gypsum stuccoes and mortars [22]. However, given the high water solubility of gypsum and the rapid reaction with DAP solutions, measures should be taken to reduce gypsum solubility and improve the penetration depth, such as adding ethanol to the DAP solution [22]. In case particulate matter is entrapped in the gypsum layer (so-called black crusts), cleaning should be performed before the DAP treatment, to avoid embedding the particulate matter in a CaP layer that is less soluble and more resistant than gypsum.

3.5 Durable Self-cleaning Ability

The self-cleaning action of commercial products based on photocatalytic nano-TiO_2 is often limited because nano-TiO_2 is easily leached by rainfall. To prevent this, nano-TiO_2 can be embedded in HAP films that have demonstrated good resistance to dissolution in rain [26]. Different ways of combining DAP and nano-TiO_2 have been investigated, demonstrating that a significantly higher resistance to leaching by rain can be obtained when nano-TiO_2 is added directly to the DAP solution, which allows for chemical bonding between the nanoparticles and the newly formed HAP coating [26].

4 Compatibility

4.1 Aesthetical Compatibility

The DAP treatment generally causes modest color alterations. In white substrates, such as marble [4, 11] and air lime mortars [20], color changes are usually invisible to the human eye, while slightly higher changes were reported on pale stones, anyways generally below the common acceptability limit [4]. When the treatment was tested on colored substrates (e.g. pink mortars containing brick dust, brown mortars based on NHL, gray mortars based on cement), the color alterations varied as a function of the DAP concentration, resulting basically negligible for low concentrations and clearly visible for higher ones [19]. In view of the possible use of the DAP treatment for conservation of frescoes and wall paintings, the DAP effect has been evaluated also on traditional pigments, resulting generally compatible except for red lead [27].

4.2 Microstructural-Physical Compatibility

Because a very little amount of new phases is formed after the treatment, DAP is usually responsible for minor alterations in the pore system and in water transport properties of the treated substrates. Mercury intrusion porosimetry on treated stones [1] and mortars [20] has revealed modest reductions in open porosity and a general shift of the pore size distribution towards smaller pores, because newly formed CaP partially occlude bigger pores, thus apparently increasing the amount of smaller ones [4]. Because of the modest changes in the pore system and in the contact angle [16], the treated substrates generally undergo minor alterations in water sorptivity and absorption, drying rate and water vapor permeability [4, 16, 19].

4.3 Thermal Compatibility

Thanks to the limited alterations in the pore system, the DAP treatment proved to be suitably compatible also in terms of thermal properties [23, 28], causing modest changes in thermal expansion coefficient and residual strain after heating-cooling cycles, as assessed both on marble [23, 28] and limestone [16].

5 Durability

5.1 Resistance to Chemical Deterioration

The low solubility and the slow dissolution rate of HAP ensure the good resistance of DAP-treated substrates to acidic attack, as demonstrated mostly on marble [6, 9–11]. Such protection from chemical attack is provided as long as the newly formed CaP coatings are continuous and free from cracks and pores, as discussed in Sect. 84.1.

5.2 Resistance to Physical Deterioration

Experimental tests have demonstrated the good resistance of DAP-treated substrates to the main deterioration processes affecting porous materials. Wetting-drying cycles had no effects on porous limestone [29], thanks to the good stability of HAP. Freezing-thawing cycles performed on porous limestone [29, 30] and air lime mortars [20] showed that, in spite of the slight shift of the pore size distribution towards smaller pores after treatment, DAP-treated substrates were more resistant than untreated references (and also commercial alternatives). Similarly, the DAP treatment improved the resistance of porous limestone to salt crystallization cycles [16, 30, 31].

5.3 Resistance to Biological Deterioration

The possible risk connected to the permanence of some residual DAP inside the pores of the treated substrate when highly concentrated DAP solutions are used has been specifically investigated [32]. In fact, DAP is also used as a fertilizer, so it might increase the substrate's sensitivity to biological growth, in case DAP residues are present. However, no DAP residues have been found when DAP solutions with low concentration were used, while application of a limewater poultice after treatment with highly concentrated DAP solution was sufficient to remove unreacted DAP [8, 33]. Nonetheless, combination of DAP with biocides was found to be even more effective in preventing biological growth, as expected [32].

6 Sustainability

The environmental sustainability of the DAP treatment has recently been investigated by life cycle analysis (LCA) [34]. Phosphoric acid necessary for production of DAP was found to be the factor that contributes the most to the impact of the DAP treatment on several categories [34]. In comparison with nanolimes and acrylic resin, all applied until apparent refusal on porous substrates, the DAP treatment proved to have the lowest impact in all the categories [34].

7 Comparative Studies

7.1 Ethyl Silicate

Compared to ethyl silicate, a first advantage of DAP is that it requires curing for only 24 h [1] or at most 1 week when limewater is also applied [8], while ethyl silicate requires curing for at least 4 weeks [35]. In terms of effectiveness, ethyl silicate has shown a comparable and sometimes superior consolidating ability on limestones containing quarzitic fractions [16], but the increase in mechanical properties was lower than DAP on marble [15, 23]. A major difference between the two consolidants is the different impact on the transport properties of the treated substrate: while DAP leaves the treated stone hydrophilic [4], ethyl silicate causes hydrophobicity for more than 6 months [35], which may be an inconvenience for other restoration works and may be a risk if a water source is present behind the consolidated layer. Because of the temporary hydrophobicity and the more pronounced alteration in the pore system, porous limestone treated with ethyl silicate showed lower resistance to freezing-thawing cycles and salt crystallization cycles compared to DAP [29, 36].

7.2 Nanolimes

A first advantage of DAP compared to nanolimes is its shorter curing time, as nanolimes require curing for 2–4 weeks. Moreover, DAP solutions are able to penetrate deeply into the treated substrate and cause formation of new CaP for several centimeters from the treated surface [16], while nanolimes are often affected by limited penetration and/or are transported back to the treated surface during drying. As a result, the strengthening effectiveness of DAP has been found to be superior to that of nanolimes both on marble [37] and air lime mortars [20]. Accordingly, the DAP treatment was able to significantly improve the resistance of air lime mortar to freezing-thawing cycles, while nanolimes provided no significant benefit [20].

7.3 Acrylic Resin

The performance of DAP and acrylic resin has been compared only in a few studies. While in the case of vacuum impregnation acrylic resin was able to penetrate in depth in weathered marble and cause increases in cohesion and reductions in open porosity more pronounced that those of DAP [25], in the case of application by brushing DAP solutions demonstrated much higher ability than acrylic resin to penetrate in depth into porous limestone, to strengthen it and to improve its durability [36]. However, compatibility and stability over time of organic consolidants are known major issues.

8 Field Studies

An increasing number of applications of the DAP treatment to buildings and monuments is being reported, but only in a limited number of cases systematic data are available to evaluate the actual performance of the treatment.

8.1 Rock-Cut Tomb in Cyprus

A rock-cut chamber tomb in Cyprus, made of a porous marlstone, was treated with 1M DAP solution for 3 h and the resulting effects were evaluated right after treatment and after 1 year [38]. After treatment, HAP was the only new CaP phase detected, but some unreacted DAP was most likely present, as the amount of HAP was found to be increased after 1 year. If present, unreacted DAP did however not significantly promote biological growth, because some traces of bioorganisms were detected after 1 year both in treated and untreated areas, thus without a systematic trend. In some areas where gypsum was originally present, brushite was formed. Overall, the treatment effects after 1 year were regarded as highly positive [38].

8.2 Byzantine Sarcophagus in Ravenna (Italy)

A byzantine sarcophagus made of Proconnesian marble, permanently exposed outdoors in the San Vitale Basilica in Ravenna (Italy), was treated with 1 M DAP+1mM $CaCl_2$ applied by poultice for 24 h [37]. The consolidating ability was evaluated on site by ultrasounds, which confirmed an increase in cohesion after treatment, especially in the parts that initially exhibited the strongest deterioration [37]. The aesthetic compatibility of the treatment, preliminary assessed in laboratory conditions by spectrophotometry, was confirmed onsite by visual observation [37].

8.3 Baroque Sculpture in Versailles (France)

A baroque sculpture made of marble, permanently exposed outdoors in the park of the Royal Palace in Versailles (France), was treated with 1M DAP+1mM $CaCl_2$ for 24 h and its conservation state was periodically monitored for 3 years by non-destructive measurements (ultrasounds, water absorption, color change) [24]. The DAP treatment was able to cause some improvement in marble cohesion, especially in the most decayed parts of the sculpture, without dramatic alterations in aesthetic appearance and physical properties. After field exposure for 3 years, the initial benefit was sensibly diminished but still detectable [24].

9 Conclusions

In the last 10 years, ammonium phosphate has shown great potential for conservation of carbonate substrates, as it is able to provide protection, consolidation and several other functionalities, without causing compatibility issues and with encouraging results in terms of resistance to accelerated ageing. Compared to commercial alternatives, the DAP treatment offers several advantages in terms of technical performance, speed of action and environmental sustainability. In the light of these positive features, the treatment has been applied onto several monuments around Europe, again exhibiting promising results. The first long-term study on the treatment's performance in the field has highlighted that some reduction in the treatment effectiveness is experienced over time, but a benefit is still present after 3 years of field exposure.

References

1. Sassoni, E., Naidu, S., Scherer, G.W.: The use of hydroxyapatite as a new inorganic consolidant for damaged carbonate stones. J. Cult. Herit. **12**, 346–355 (2011)
2. Yang, F., et al.: Biomimic conservation of weathered calcareous stones by apatite. New J. Chem. **35**, 887–892 (2011)
3. Matteini, M., Rescic, S., Fratini, F., Botticelli, G.: Ammonium phosphates as consolidating agents for carbonatic stone materials used in architecture and cultural heritage: preliminary research. Int. J. Archit. Herit. **5**, 717–736 (2011)
4. Sassoni, E.: Hydroxyapatite and other calcium phosphates for the conservation of cultural heritage: a review. Materials **11**, 557 (2018)

5. Naidu, S., Scherer, G.W.: Nucleation, growth and evolution of calcium phosphate films on calcite. J. Colloid Interface Sci. **435**, 128–137 (2014)
6. Naidu, S., Blair, J., Scherer, G.W.: Acid-resistant coatings on marble. J. Am. Ceram. Soc. **99**, 3421–3428 (2016)
7. Possenti, E., et al.: New insight on the interaction of diammonium hydrogen phosphate conservation treatment with carbonatic substrates: a multi-analytical approach. Microchem. J. **127**, 79–86 (2016)
8. Franzoni, E., Sassoni, E., Graziani, G.: Brushing, poultice or immersion? The role of the application technique on the performance of a novel hydroxyapatite-based consolidating treatment for limestone. J. Cult. Herit. **16**, 173–184 (2015)
9. Graziani, G., Sassoni, E., Franzoni, E., Scherer, G.W.: Hydroxyapatite coatings for marble protection: Optimization of calcite covering and acid resistance. Appl. Surf. Sci. **368**, 241–257 (2016)
10. Graziani, G., Sassoni, E., Scherer, G.W., Franzoni, E.: Resistance to simulated rain of hydroxyapatite- and calcium oxalate-based coatings for protection of marble against corrosion. Corr. Sci. **127**, 168–174 (2017)
11. Sassoni, E., Graziani, G., Franzoni, E., Scherer, G.W.: Calcium phosphate coatings for marble conservation: influence of ethanol and isopropanol addition to the precipitation medium on the coating microstructure and performance. Corr. Sci **136**, 255–267 (2018)
12. Sassoni, E., Masi, G., Bignozzi, M.C., Franzoni, E.: Electrodeposition of hydroxyapatite coatings for marble protection: preliminary results. Coatings **9**, 207 (2019)
13. Liu, Q., Zhang, B.J.: Synthesis and characterization of a novel biomaterial for the conservation of historic stone buildings and sculptures. Mater. Sci. Forum **675–677**, 317–320 (2011)
14. Turner, R.J., Renshaw, J.C., Hamilton, A.: Biogenic hydroxyapatite: a new material for the preservation and restoration of the built environment. ACS Appl. Mater. Interfaces **9**, 31401–31410 (2017)
15. Sassoni, E., Graziani, G., Franzoni, E.: Repair of sugaring marble by ammonium phosphate: comparison with ethyl silicate and ammonium oxalate and pilot application to historic artifact. Mater. Des. **88**, 1145–1157 (2015)
16. Sassoni, E., Graziani, G., Franzoni, E.: An innovative phosphate-based consolidant for limestone. Part 1: Effectiveness and compatibility in comparison with ethyl silicate. Constr. Build. Mater. **102**, 918–930 (2016)
17. Molina, E., Rueda-Quero, L., Benavente, D., Burgos-Cara, A., Ruiz-Agudo, E., Cultrone, G.: Gypsum crust as a source of calcium for the consolidation of carbonate stones using a calcium phosphate-based consolidant. Constr. Build. Mater. **143**, 298–311 (2017)
18. Sassoni, E., Franzoni, E., Pigino, B., Scherer, G.W., Naidu, S.: Consolidation of calcareous and siliceous sandstones by hydroxyapatite: comparison with a TEOS-based consolidant. J. Cult. Herit. **14**, e103–e108 (2013)
19. Sassoni, E., Franzoni, E.: Lime and cement mortar consolidation by ammonium phosphate. Constr. Build. Mater. **245**, 118409 (2020)
20. Masi, G., Sassoni, E.: Air lime mortar consolidation by nanolimes and ammonium phosphate: compatibility, effectiveness and durability. Constr. Build. Mater. **299**, 123999 (2021)
21. Defus, A., Possenti, E., Sansonetti, A., Tedeschi, C., Colombo, C., Biondelli, D., Vet-tori, S., Realini, M.: Di-ammonium hydrogen phosphate for the consolidation of lime-based historic mortars—preliminary research. J. Cult. Herit. **48**, 45–53 (March-April 2021)
22. Sassoni, E., Graziani, G., Franzoni, E., Scherer, G.W.: Conversion of calcium sulfate dihydrate into calcium phosphates as a route for conservation of gypsum stuccoes and sulfated marble. Constr. Build. Mater. **170**, 290–301 (2018)
23. Sassoni, E., Andreotti, S., Scherer, G.W., Franzoni, E., Siegesmund, S.: Bowing of marble slabs: can the phenomenon be arrested and prevented by inorganic treatments? Environ. Earth Sci. **77**(10), 1–18 (2018). https://doi.org/10.1007/s12665-018-7547-7

24. Sassoni, E., et al.: Phosphate treatments for conservation of stone: 3-year field study in the royal palace of Versailles (France). Mater. Struct. **54**, 140 (2021)
25. Menningen, J., Sassoni, E., Sobott, R., Siegesmund, S.: Constraints of the durability of inorganic and organic consolidants for marble. Environ. Earth Sci. **80**(10), 1–18 (2021). https://doi.org/10.1007/s12665-021-09664-w
26. Sassoni, E., D'Amen, E., Roveri, N., Scherer, G.W., Franzoni, E.: Durable self-cleaning coatings for architectural surfaces by incorporation of TiO_2 nano-particles into hydroxyapatite films. Materials **11**, 177 (2018)
27. Ma, X., Pasco, H., Balonis, M., Kakoulli, I.: Investigation of the optical, physical, and chemical interactions between diammonium hydrogen phosphate (DAP) and pigments. Sustainability **11**, 3803 (2019)
28. Sassoni, E., Graziani, G., Ridolfi, G., Bignozzi, M.C., Franzoni, E.: Thermal behavior of Carrara marble after consolidation by ammonium phosphate, ammonium oxalate and ethyl silicate. Mater. Des. **120**, 345–353 (2017)
29. Sassoni, E., Graziani, G., Franzoni, E.: An innovative phosphate-based consolidant for limestone. Part 2: Durability in comparison with ethyl silicate. Construct. Build. Mater. **102**, 931–942 (2016)
30. Shekofteh, A., Molina, E., Rueda-Quero, L., Arizzi, A., Cultrone, G.: The efficiency of nanolime and dibasic ammonium phosphate in the consolidation of beige limestone from the Pasargadae World Heritage Site. Archaeol. Anthrop. Sci. (2019)
31. Weththimuni, M.L., Licchelli, M., Malagodi, M., Rovella, N., La Russa, M.: Consolidation of bio-calcarenite stone by treatment based on diammonium hydrogen phosphate and calcium hydroxide nanoparticles. Measurement **127**, 396–405 (2018)
32. Barriuso, B.C., Botticelli, G., Cuzman, O.A., Osticioli, I., Tiano, P., Matteini, M.: Conservation of calcareous stone monuments: screening different diammonium phosphate based formulations for countering phototrophic colonization. J. Cult. Herit. **27**, 97–106 (2017)
33. Graziani, G., Sassoni, E., Scherer, G.W., Franzoni, E.: Penetration depth and redistribution of an aqueous ammonium phosphate solution used for porous limestone consolidation by brushing and immersion. Constr. Build. Mater. **148**, 571–578 (2017)
34. Dal Pozzo, A., Masi, G., Tugnoli, A., Sassoni, E.: Towards green materials for cultural heritage conservation: sustainability evaluation of products for stone consolidation, RILEM Book series (Merida proceedings). J. I. Escalante-Garcia et al. (Eds.) RW 2021, vol. 40, pp. 1–10 (2023). https://doi.org/10.1007/978-3-031-21735-7_78
35. Scherer, G.W., Wheeler, G.S.: Silicate consolidants for stone. Key Eng. Mater. **391**, 1–25 (2009). https://doi.org/10.4028/0-87849-365-4.1
36. Masi, G., Sassoni, E.: Limestone consolidation: how much product is enough? RILEM Book series (Merida proceedings). J. I. Escalante-Garcia et al. (Eds.) RW 2021, vol. 40, pp. 1–10 (2023). https://doi.org/10.1007/978-3-031-21735-7_79
37. Sassoni, E., Ugolotti, G., Pagani, M.: Nanolime, nanosilica or ammonium phosphate? Laboratory and field study on consolidation of a byzantine marble sarcophagus. Constr. Build. Mater. **262**, 120784 (2020)
38. Ma, X., Balonis, M., Pasco, H., Toumazou, M., Counts, D., Kakoulli, I.: Evaluation of hydroxyapatite effects for the consolidation of a Hellenistic-Roman rock-cut chamber tomb at Athienou-Malloura in Cyprus. Constr. Build. Mater. **150**, 333–344 (2017)

Characterization and Physical-Mechanical Properties of Adobes from La Huacana, Mexico

Adria Sanchez-Calvillo[1(✉)], Elia M. Alonso-Guzman[1,2], Wilfrido Martinez-Molina[2], Hugo L. Chavez-Garcia[2], and Melissa Ruiz-Mendoza[2]

[1] Faculty of Architecture, Universidad Michoacana San Nicolás de Hidalgo, 58000 Morelia, MICH, Mexico
adria.sanchez.9@hotmail.com

[2] Faculty of Civil Engineering, Universidad Michoacana San Nicolás de Hidalgo, 58000 Morelia, MICH, Mexico

Abstract. Earthen architecture is one of the oldest construction solutions known, with archaeological evidences as ancient as the first human settlements. Nevertheless, in the last decades, adobe constructions have been abandoned and severely modified, been displaced by the use of industrial materials like concrete blocks and ceramic bricks. The importance of these traditional systems in the local communities is a crucial aspect, since the techniques have been transmitted for generations and are part of the inner culture. Adobe samples from heritage buildings, including complete specimens and cylinders, were collected in the town of La Huacana, in the state of Michoacan, in Mexico. Also these structures were surveyed and analyzed from the technological and architectural study of the buildings and materials. After the field work the samples were brought to the materials laboratory of the Michoacan University of Saint Nicholas of Hidalgo and were analyzed with non-destructive tests, mechanical resistance and soil mechanics trials. The experimentation allowed to calculate the mechanical resistance as well as the ultrasonic pulse velocity; the results were compared to the sieve analysis, the Atterberg limits and the USCS (Unified Soil Classification System) of the samples. The material characterization of the adobes allowed to determine the properties of these construction materials, which commonly are not deeply studied. The samples from La Huacana presented a very singular composition with great percentages of fine material, while the mechanical properties were quite satisfactory in comparison with the background researches. It is expected to perfect and regulate the characterization of earthen materials by means of research works like the one presented, achieving a complete perspective of these traditional systems.

Keywords: Materials properties · Earthen construction · Non-destructive tests · Sustainable materials · Soil mechanics

1 Introduction

Earthen constructions are ancient cultural, artistic and architectural representations of our different peoples and cultures. Many civilizations from all the inhabited continents

J. I. Escalante-Garcia et al. (Eds.): RW 2021, RILEM Bookseries 40, pp. 799–807, 2023.
https://doi.org/10.1007/978-3-031-21735-7_85

have used clays in combination with other materials to build their monumental, religious, military and residential constructions, being earth a natural and available resource since they could easily obtain it. These materials present numerous advantages and benefits, being totally sustainable as they can be reused multiple times for construction purposes or other activities [1].

It is essential to consider these materials as main priorities for sustainable design and construction, considering that currently only embraces small markets and local initiatives [2]. Furthermore, in the last decades, the earthen architecture has been neglected and abandoned, being considered as an obsolete technology, characteristic of the rural territories associated to the lower incomes and development index [3].

Mexico has an enormous cultural heritage built with earthen materials, being the vernacular adobe houses the most common and representative typologies, specially for the rural communities and the indigenous culture. Nevertheless, the adobe housing in Mexico has rapidly decreased over the last decades, being substituted by modern materials and systems which do not take under consideration the environmental surroundings and the local resources. There are few initiatives to preserve, restore, refurbish and retrofit this cultural heritage [4], however, most of the rural communities have several problems to maintain this local architecture in front of the modern materials and techniques.

Tierra Caliente is a geographical region of Mexico which includes territories from the states of Michoacan, Guerrero and State of Mexico. The nomenclature means "dry land" due to the extremely hot climate, and culturally, this place also presents uniqueness. The vernacular architecture of the Tierra Caliente region incorporates local materials which have a low environmental impact, using traditional construction techniques with thick adobe walls and simple solutions like light roof structures which allow the ventilation of the interior spaces (See Fig. 1) [5]. The dwellings are totally adapted to the local conditions, since they proportionate a great thermal performance due to the thermal mass of the walls and the ventilated arcades in the façades. These local strategies respond to the adaptation of human beings to the local environment as well as the combination of the many cultures which have inhabited this region for centuries.

Fig. 1. Adobe house in Turicato, Tierra Caliente Region.

2 Methodology

In the recent years, the research on adobe masonry from vernacular buildings has increased, with several works regarding the benefits and virtues of these construction systems [6–8]. Most of the investigations have revised the earthen construction from the perspective of the construction technology and the cultural significance of the traditional architecture; while the physico-chemical characterization as a construction material has often been overlooked [6–9]. Nevertheless, we can find interesting researches about the material characterization which have performed complete studies of the material, working with historic constructions built with adobes, emphasizing the mechanical properties of the blocks [8, 9], but also incorporating innovative techniques of the study of the cultural heritage [6, 7].

Besides the material characterization, the vernacular houses of the region were studied during the field work, identifying the study cases and cataloguing and analyzing them from the architectural, technological and structural perspectives. This methodology included the survey of constructive systems and materials, structural damages and deteriorations of the traditional adobe buildings. The representative adobe dwelling resulted in the model shown in Fig. 2, with adobe walls with regular proportions, compartmentalized in two or three dormitories, wooden diaphragms, light roof structures and a ventilated façade.

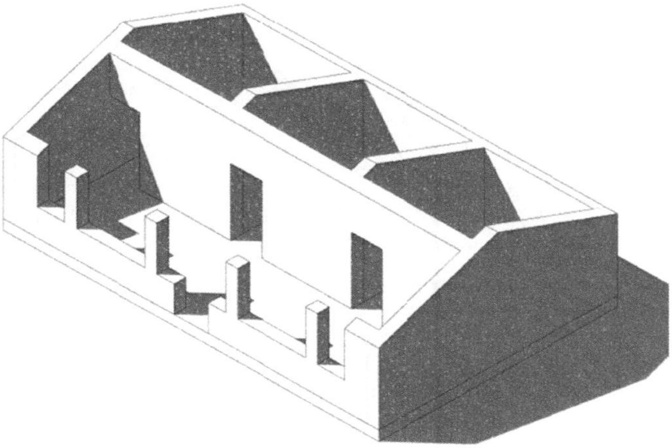

Fig. 2. 3D model of the representative vernacular adobe housing of the Tierra Caliente Region.

During the field work in the town of La Huacana, in the Tierra Caliente region, adobes from heritage buildings and "haciendas" were collected. Complete adobe pieces were obtained and brought to the laboratory, and cylindrical cores were extracted from these historical constructions. The samples were analyzed under control conditions in the materials laboratory of the Universidad Michoacana San Nicolás de Hidalgo (UMSNH), where the study was conducted.

Regarding the material, two classes of specimens were identified, one of them presenting a higher proportion of vegeral fibers, as it was later confirmed with the sieve analysis. The classification of the fibers and the petreous and natural aggregates were also different for each of the both classes.

Currently, we can find very few regulations about earthen construction, specially from the laboratory analysis and characterization, like limited guidelines for the study of the materials [10]. Therefore, the methodology followed was retaken from previous works of the group with adobes from heritage buildings [6, 11]. The methodology included non-destructive tests, mechanical resistance tests, and soil mechanics analysis.

The main non-destructive experiment was the Ultrasonic Pulse Velocity test (UPV), which was performed following the Mexican and the international standards, NMX-C-275-ONNCCE-2004 and ASTM D2845-08 [12, 13], with the equipment V-Meter MK IV from James Instruments. The cylinders were tested with the application of gel in the edges as it is recommended in the regulations (See Fig. 3).

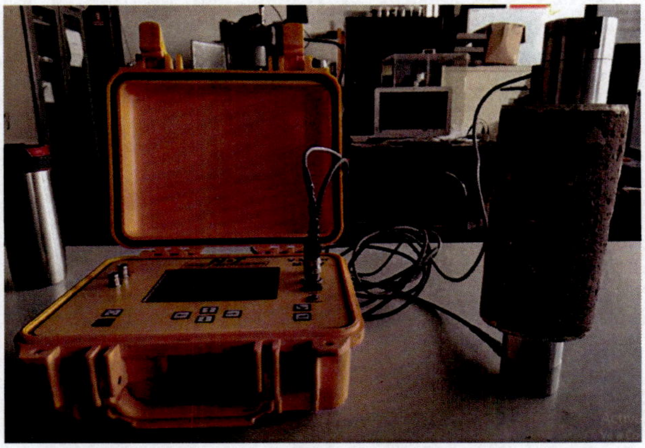

Fig. 3. Ultrasonic pulse velocity test with cylindrical sample.

Regarding the mechanical tests, the uniaxial compressive strength was calculated. The cylinders were capped with sulphur, as it is done with the concrete samples, and performed with the Tinius Olsen Universal Test Machine of the laboratory.

Last, the soil mechanics analysis included the sieve analysis, calculation of Atterberg Limits, and classification by the USCS. The material was dried in an oven before the performance of the analysis, at a maximum temperature of 50–60 °C to avoid the change in the composition of the clay minerals. Afterwards, the trials were fulfilled according to the applicable regulations (See Fig. 4) [14].

3 Results and Discussion

UPV allows to assume the internal structure of the material, as well as its porosity. The test is designed for the study of natural and artificial rocks [12, 13]; and the regulations

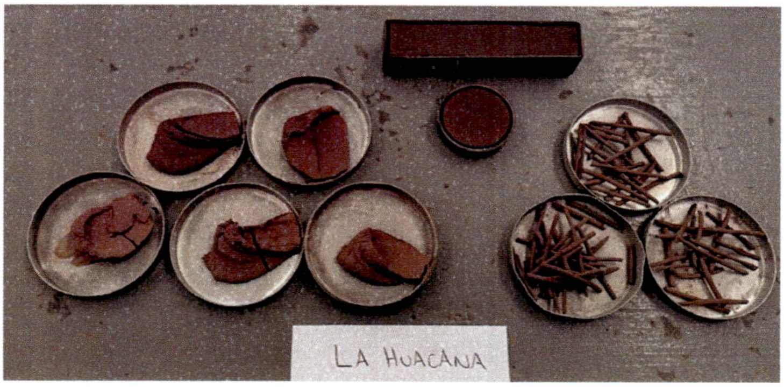

Fig. 4. Material used for the Atterberg limits.

followed during the experimentation are developed mainly for concrete and rocks, not for adobes or earthen materials. Only few background investigations have experimented with UPV and adobe blocks [11, 15], with the advantage of allowing to calculate the elastic modulus of the material. In this study case, the results were much more reliable than previous researches with heritage buildings [6], since the samples obtained in la Huacana were kept in an exceptional condition (see Table 1).

Table 1. Ultrasonic pulse velocity results of the adobes.

Sample	Diameter (m)	Height (m)	Density (kg/m^3)	UPV (m/s)
H-CI1	0.091	0.18	1780.02	1050
H-CI2	0.091	0.18	1656.26	876
H-CI3	0.090	0.18	1754.25	1270
H-CI4	0.091	0.18	1706.07	1104
Average	0.091	0.18	1724.15	1075

The tested cubes presented reasonably good compressive strength values (See Table 2). A slenderness correction factor of 0.91 was applied, according the NMX-C-083-ONNCCE-2014 standard for concrete [16]. From the results, we can find a clear differentiation between the two adobe typologies found: the first three samples (H-C2, H-C3 and H-C5), which showed a very compact and smooth aspect reached the higher compressive strength values; while the last two samples (H-C4 and H-C9), presented higher amounts of straw and their compressive strength was notably lower.

Regarding the soil mechanics analysis, the sieve analysys allowed to understand the granulometry and composition of the adobes, including the natural fibers and organic material (mainly straw) contained in the blocks. Figure 5 shows the granulometric curves

Table 2. Uniaxial compressive strength of the adobes.

Sample	Density (ton/m^3)	Load (kg)	Area (cm^2)	UCS (kg/cm^2)
H-C2	0.89	1243.80	91.69	12.34
H-C3	0.91	1285.50	91.44	12.79
H-C5	0.93	949.50	91.24	9.47
H-C4	0.89	583.20	90.26	5.88
H-C9	0.89	712.80	90.29	7.18
Average	0.90	954.96	90.98	9.53

of the material from La Huacana, the adobes presented high fine soils content, an impressive percentage taking into account the previous mentioned values of the earthen construction literature [17], which consider lower amounts of clays and silts to produce durable adobes.

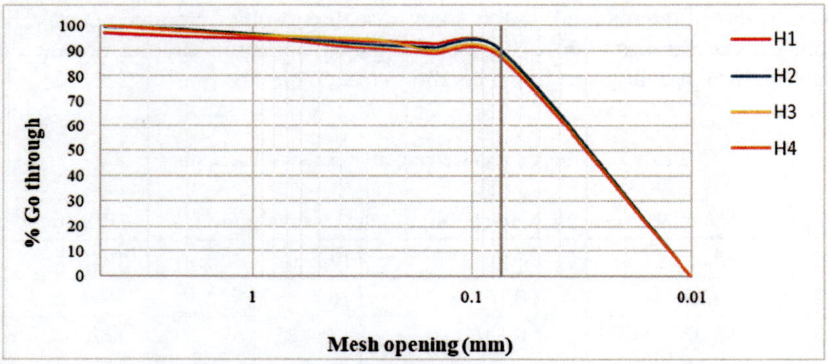

Fig. 5. Granulometric curves of the adobes from La Huacana.

Figure 6 shows the results of the plasticity index and liquid limit and the position of each one of the specimens in the Casagrande plasticity chart. All the samples were located in the range of the clays, in the threshold between high and low plasticity.

Table 3 indicates the classification of all the samples tested by the Unified Soil Classification System (USCS). The adobes are classified as organic clays/silts, except H1 which is an elastic silt (MH). Considering the great amount of fine soils and clays, this low plasticity of many of the specimens could be due to the stabilization of the adobes with other materials, like the most commonly used lime.

4 Conclusions

The research on characterization of earthen materials needs to be pursued and strengthened, as these construction systems could be a solution for the bioclimatic design and

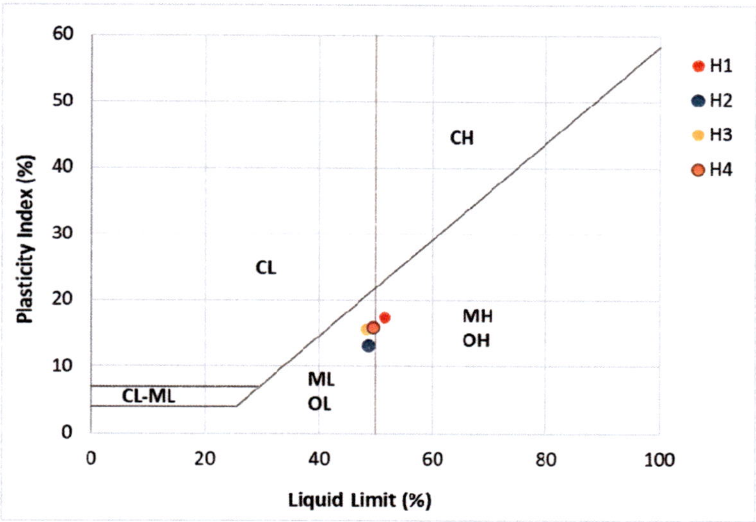

Fig. 6. USCS classification of the samples according the Casagrande plasticity chart.

Table 3. USCS classication of the specimens.

Soil	H1	H2	H3	H4
Gravel	3.06%	0.30%	0.00%	0.00%
Sand	6.72%	9.08%	11.58%	12.45%
Fine	90.23%	90.62%	88.42%	87.55%
Organic	0.26%	0.37%	0.25%	0.40%
USCS	MH/OH	ML/OL	ML/OL	ML/OL

architecture in many regions. These traditional techniques have proven its worth for centuries, but nowadays they have been neglected by modern materials which do not relate with the cultural scene and local traditions of the rural communities.

Regarding the material characterization, the adobe blocks from La Huacana, Mexico, presented a good mechanical performance in comparison with the international literature about earthen materials. The high content of fine soils employed for the production of the bricks was impressive, considering all the background research that has been published; nevertheless, this greater proportion translated into good features, as the pieces resisted over centuries and were found in good shape when they were collected during the field work.

Very few research papers have explored non-destructive tests as characterization techniques for earthen materials, therefore it is important to continue exploring towards a more complete and calibrated methodology. It is important to find correlations between different laboratory tests, with both non-destructive and mechanical resistance analysis, also implementing mineralogical and elemental studies of the materials.

The use of clay for the elaboration of blocks is a relevant alternative in certain types of constructions, like vernacular housing and sustainable architecture, since if they are structurally well designed, they can provide good thermal and acoustic comfort, in addition to complying with goals 11, 12 and 13 of the United Nations sustainable development agenda.

Acknowledgements. The authors acknowledge the Materials Laboratory "Ing. Luis Silva Ruelas" of the Faculty of Civil Engineering of the UMSNH for the equipment support, the Postgraduate Program PIDA of the Faculty of Architecture of the UMSNH, the economic support given by CONACYT, CIC-UMSNH in Morelia, Mexico; and CIMNE, in Barcelona, Spain.

References

1. Costa, C., Cerqueira, Â., Rocha, F., Velosa, A.: The sustainability of adobe construction: past to future. Int. J. Arch. Herit. **13**, 639–647 (2018). https://doi.org/10.1080/15583058.2018.1459954
2. Morel, J.-C., Charef, R., Hamard, E., Fabbri, A., Becket, C., Bao Bui, Q.: Earth as construction material in the circular economy context: practitioner perspectives on barriers to overcome. Philos. Trans. R. Soc. B: Biol. Sci. (2021). https://doi.org/10.1098/rstb.2020.0182
3. Sánchez Calvillo, A., Alonso Guzmán, E.M., López Núñez, M.: Vulnerabilidad sísmica y la pérdida de la vivienda de adobe en Jojutla, Morelos, México, tras los sismos de 2017. Vivienda Comunidades Sustentables **10**, 9–29 (2021). https://doi.org/10.32870/rvcs.v2i10.162
4. Haji Sadeghi, N., Correia, M., Oliveira, D.V., Azizi-Bondarabadi, H.: Preventive conservation of vernacular adobe heritage located in seismic-prone regions. Int. Arch. Photogramm., Remote. Sens. Spat. Inf. Sci. **XLIV-M-1**, 855–860 (2020). https://doi.org/10.5194/isprs-archives-XLIV-M-1-2020-855-2020
5. Ayala Ortega, L.A.: Arquitectura doméstica tradicional de tierra caliente, Michoacán. Carácuaro: un caso de estudio. Master Thesis, Universidad Michoacana San Nicolás de Hidalgo, Morelia, Mexico (2019)
6. Sanchez-Calvillo, A., et al.: Characterization of adobe blocks: point-load assessment as a complementary study of damaged buildings and samples. Heritage **4**, 864–888 (2021). https://doi.org/10.3390/heritage4020047
7. Silva, A., Oliveira, I., Silva, V., Mirão, J., Faria, P.: Vernacular Caramel´s adobe masonry dwellings—material characterization. Int. J. Arch. Herit., 1–18 (2020). https://doi.org/10.1080/15583058.2020.1751343
8. Vicente, E., Torrealva, D.: Mechanical properties of historical adobe in Perú. In: Joffroy, T., Guillaud, H., Sadozaï, C. (eds.) Terra Lyon 2016: Articles Selected for On-Line Publication, Lyon. CRAterre, Villefontaine, France (2018)
9. Silveira, D., Varum, H., Costa, A., Martins, T., Pereira, H., Almeida, J.: Mechanical properties of adobe bricks in ancient constructions. Constr. Build. Mater. **28**, 36–44 (2012). https://doi.org/10.1016/j.conbuildmat.2011.08.046
10. The Earthen Architecture Initiative: Guidelines for Teaching Earthen Conservation. Material Analysis—In Situ and Laboratory Material Characterization. The Getty Conservation Institute, Los Angeles, CA, USA (2011)

11. Ruiz-Mendoza, M., Sanchez-Calvillo, A., Martinez-Molina, W., Chavez-Garcia, H.L., Alonso-Guzman, E.M., Navarrete-Seras, M.A., Preciado-Villicaña, D., Ramirez-Carranza, A.: Ensayo de Velocidad de Pulso Ultrasónico y cálculo del módulo de elasticidad en adobes de Tierra Caliente. In: 15th Congreso Nacional de Ciencia, Tecnología e Innovación. Instituto de Ciencia, Tecnología e Innovación del Estado de Michoacán de Ocampo, Morelia, Mexico (2020)
12. ONNCCE: Norma mexicana NMX-C275-ONNCCE-2004. ONNCCE S.C., Mexico City, Mexico (2004)
13. ASTM International: ASTM D2845-08, Standard Test Method for Laboratory Determination of Pulse Velocities and Ultrasonic Elastic Constants of Rock (Withdrawn 2017). ASTM International, West Conshohocken, PA (2008). https://doi.org/10.1520/D2845-08
14. ASTM International: ASTM D2487—17e1, Standard Practice for Classification of Soils for Engineering Purposes (Unified Soil Classification System). ASTM International, West Conshohocken, PA, USA (2017)
15. Aguilar, R., Saucedo, C., Montesinos, M., Ramírez, E., Morales, R., Uceda, S.: Caracterización mecánica de las unidades de adobe del complejo arqueológico Huaca de la Luna mediante ensayos de ultrasonido. In: Proceedings of 15th Seminario Iberoamericano de Arquitectura y Construcción con Tierra, pp. 28–39. Universidad de Cuenca, Cuenca, Ecuador (2015)
16. ONNCCE: Norma mexicana NMX-C-083-ONNCCE-2014. ONNCCE S.C., Mexico City, Mexico (2015)
17. Martins Neves, C.M., Borges Faria, O., Rotondaro, R., Cevallos Salas, P., Hoffmann, M.V.: Selección de Suelos y Métodos de Control en la Construcción con Tierra—Prácticas de Campo, 1st edn. Rede Ibero-Americana Proterra, Lisboa, Portugal (2009)

Conservation of Cultural Heritage Building: Evaluation of $Ca[Zn(OH)_3]_2{\cdot}2H_2O$ Nanoparticles Coating Behavior Under Salt Crystallization Cycles

Montserrat Soria-Castro[1](✉), José Faisal-Sulub[1], Katia Josceline Pérez-Ostos[1], Aketzali Abigail García-Reyes[1], Javier Reyes-Trujeque[1], Patricia Quintana-Owen[2], Susana del Carmen De la Rosa-García[3], Claudia Araceli García-Solís[4], and Sergio Alberto Gómez-Cornelio[5]

[1] National Science Laboratory for Research and Conservation of Cultural Heritage-Corrosion Research Center, LANCIC-CICORR, UAC, Campeche, Campeche, México
monsoria@uacam.mx
[2] Applied Physics Department, CINVESTAV-IPN, Mérida, Yucatán, México
[3] Laboratory of Applied Microbiology, DACBioL, UJAT, Villahermosa, Tabasco, México
[4] National Institute of Anthropology and History, INAH, Mérida, Yucatán, México
[5] Polytechnic University of Center, UPC, Villahermosa, Tabasco, México

Abstract. Calcareous stones are composed of a mixture of minerals and different cementing agents. Due to their heterogeneity, they are sensitive to the action of multiple environmental factors, including temperature, relative humidity, solar radiation, microorganisms, and salt crystallization processes that are the cause of degradation, inducing changes in their physical and chemical properties. Then it is desirable to account for specific studies on new nanomaterials that improve stone characteristics to extend its performance for long periods. In Yucatán, the calcareous stone is a traditional building material used since the prehispanic period, however, its susceptibility to deterioration has created a demand to elaborate protective treatments that improves its consolidation. This work shows the performance of new nanomaterials synthesized via sol-gel, based on calcium and zinc hydroxide dihydrate, with the formula $Ca[Zn(OH)_3]_2{\cdot}2H_2O$ (CZ) and applied on three Yucatán calcareous stones Calcehtok, Chichén Itzá and, Mayapán, and further evaluated under the sodium sulphate crystallization aging test. Results indicated that after applying CZ nanoparticles (NPs) to stones, physical and mechanical properties like capillarity, colour, propagation of ultrasonic velocity, and structure improved as they showed a decrease in stone cracks, cavities, and pores. The salt crystallization tests indicate significant changes in the control properties of untreated stones when compared to those specimens coated with CZ. In addition, a better performance was observed for the stones from Calcehtok, followed by Mayapán and Chichén Itzá, respectively.

Keywords: Stone · Salt crystallization · Conservation · Nanoparticles

J. I. Escalante-Garcia et al. (Eds.): RW 2021, RILEM Bookseries 40, pp. 808–817, 2023.
https://doi.org/10.1007/978-3-031-21735-7_86

1 Introduction

The Yucatán Peninsula, formed by Campeche, Quintana Roo, and Yucatán, is geologically a platform with powerful strata of carbonate rocks [1]. From such strata, the Mayan civilization extracted and used these materials to construct its characteristic monumental buildings as temples, palaces, and other architectural elements that are currently found as vestiges in various archaeological sites [2]. Chichén Itzá and Mayapán are clear examples; since they were built entirely with limestone rock, they represent an essential economic element (tourism) and social identity in the Yucatán Peninsula [3].

Carbonate rocks, such as limestone, are easily disturbed; this process depends on intrinsic and extrinsic factors, leading to deterioration mechanisms, which produce changes in their properties, known as alteration indicators [4, 5]. An obvious example is the crystallization of salts in historical monuments built in stone [6] because they are generally exposed to very aggressive atmospheric conditions that are difficult to predict. Therefore, accelerated alteration tests are frequently used in a specialized laboratory to predict and understand these processes [7, 8]. Consequently, in this work, the effect of a saline environment on three limestone lithotypes was evaluated; two are from the archaeological sites of Chichén Itzá and Mayapán, and a commercial one, extracted from the town of Calcehtok. The test was evaluated recording the variation of weight and the visual deterioration observed; in addition, its physical-mechanical changes were analysed when coated with nanoparticles based on double zinc and calcium hydroxide dihydrate (CZ).

2 Materials and Methods

2.1 Materials and Stone Samples Characterization

The calcium zincate synthesized by the sol-gel method was characterized by X-ray diffraction and electron microscopy analyses, according to Soria Castro et al. [9]. The formation of CZ NPs can be described by the following reaction:

$$CaCl_2 \cdot 2H_2O + 2ZnO + 2H_2O + 2NaOH \rightarrow 2Zn(OH)_2Ca(OH)_2 \cdot 2H_2O + 2NaCl$$

This study used three different types of limestone from the Yucatan Peninsula, which were provided by the Institute of Anthropology and History (INAH). The stones correspond to two decontextualized rock blocks found isolated in the vicinity of buildings from the archaeological sites of Chichén Itzá (CH) and Mayapán (CO). In addition, a stone of commercial origin was selected with contrasting visual characteristics, such as porosity and compaction to the previous two, named Calcehtok (CA).

Specimens consisting in stone cubes with dimensions of 3 × 3 × 3 cm, were analysed by propagation of ultrasonic velocity (Vp), spectrophotometry and the capillarity coefficient, before and after 20 days of applying CZ NPs on the surface of the coupons, to compare the differences in both and assess the protective treatment effectiveness. Only for the X-Ray diffraction (DRX) technique, a portion of the stone samples was powdered before analysis.

Propagation of ultrasonic velocity was measured to evaluate the mechanical properties and the distribution of the nanoparticles. In this study Vp was carried as specified in UNE EN 14579 [10], using a Proceq Pundit PL2 Pulser instrument. By means of this technique, the anisotropy index (dM %) can be determined, using the velocities (V) found for Vp, which are obtained measuring the three orthogonal directions in space of the stones evaluated. The anisotropy index is calculated as proposed by [11], as shown below in Eq. (1):

$$dM\,\% = [1 - (2Vpmin/(Vpmax + Vpmean))] \times 100 \quad (1)$$

where Vp_{max}, Vp_{min}, and Vp_{mean} are respectively the maximum, minimum, and mean values [11].

The chromatic parameters on the stones surface were measured with a spectrophotometer MINOLTA CM-700d using the Lab system or Lab colour space or CIELAB (Commission Internationale de l'Eclairage CIE 1976) [12]. The standard illuminant was D50 and observer 10°; measured parameters were L*, which accounts for luminosity, a*, and b* for coordinates chromatic parameters defined in the ASTM 313-76 system (ASTM 2000) [13]. The total colour difference ΔE* [14] is obtained by the Eq. (2):

$$\Delta E* = \left[(\Delta L*)^2 + (\Delta a*)^2 + (\Delta b*)^2\right]^{1/2} \quad (2)$$

The capillarity coefficient was determined to analyse the hydric behaviour through the stone. This technique shows the amount of water absorbed, expressed by the mass increase of the stone per unit area for the square root of time ($kg/m^2\ s^{1/2}$), and the time interval of the amount of water absorbed by the sample, was automatically monitored (every 1, 3, 5, 10, 15, 30 min and 1, 3, 5, 8, 24 h), when its lower surface is in contact with the water reservoir, following the norm AENOR, UNE-EN 1925:1999 [15].

The resistance to salt crystallization test was carried out following the UNE-EN 12370 [16] standard to evaluate the effect of the salt crystallization mechanism in rocks, the influence and behaviour of CZ NPs as a protective coating. The procedure consists of immersing the specimens with and without the treatment of the CZ NPs in a saturated solution of sodium sulphate decahydrate for 2 h. Once the time has elapsed, the stone samples are placed in the oven for 16 h at 105 °C (±5 °C) to leave only the salt crystals in the samples. Subsequently, they are removed from the oven and stored in a desiccator for 2 h before repeating the cycle. Thus, a total of 15 repetitions were performed.

3 Results

The X-ray diffraction for the untreated (natural) stones with contrasting characteristics showed that CA contains only 100% calcite ($CaCO_3$), CH is formed by two minerals, 90% calcite and dolomite ($CaMg(CO_3)_2$) at 10% and CO has only 100% dolomite (see Fig. 1). The petrophysical characterization tests that were carried out for the three lithotypes are ultrasound speed transmission, anisotropy index, colorimetry, and the capillary coefficient.

Table 1 shows the average data of the three types of limestone evaluated without and with the nanomaterial treatment, using the techniques of propagation of ultrasonic velocity (Vp) and the anisotropy index (dM%).

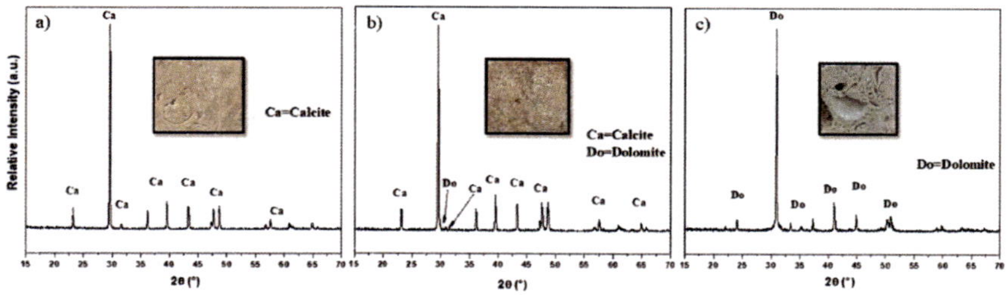

Fig. 1. X-ray diffraction patterns of the stones: **a** Calcehtok, **b** Chichén Itzá and **c** Mayapán

Table 1. Propagation of ultrasonic velocity values and anisotropy index for the three types of limestone

Stone		Vp (m/s)	dM (%)
CA	Untreated	995 ± 27	2.7
	Treated	1045 ± 137	1.1
CH	Untreated	879 ± 27	17.2
	Treated	815 ± 122	3.2
CO	Untreated	832 ± 73	24.3
	Treated	883 ± 81	21

Calcehtok (CA) without the treatment shows an ultrasound transmission speed of 995 m/s; after applying the CZ NPs, the transmission speed was 1045 m/s; the same trend was observed for Mayapán, as the speed increased from 832 to 883 m/s. In the Chichén Itzá sample, the wave transmission speed decreased from 879 to 815 m/s. The anisotropy index decreased in the three cases after treated with CZ NPs. This observation could be due to the transformation of calcium hydroxide to calcium carbonate phase, which causes a consolidating action [17]. This result is relevant, considering that CZ in its chemical formula is composed of $Ca(OH)_2$, and the stones evaluated in this investigation are limestones in nature.

The spectrophotometry results for Calcehtok, Chichén Itzá, and Mayapán without and with CZ NPs are presented in Table 2, where L* = Luminosity, a* = Red tone, b* = Yellow tone, ΔE* = Global change of color. Calcehtok showed that the luminosity increases from 82.04 to 85.64, and Chichén Itzá and Mayapán decreased. The parameter a* represents the red hue, which decreases for CH from 2.09 to 2.01 as it is a reddish stone. Previous studies by X-ray diffraction of González et al. [18] showed that the red coloration for the Ticul stone from the State of Yucatán is due to iron minerals in the form of iron hydroxide-oxide, such as goethite (FeO(OH)) and hematite (Fe_2O_3). In the case of CA, such parameter decreases from 2.12 to 1.94 since it is a stone with a cream hue; and the compound has a white coloration, which could cause this decrease. For this reason, the parameter b* increases for CH from 10.686 to 11.91 and Mayapán from 10.71 to 11.21.

Table 2. Color parameters values for three types of limestone

Stone		L*	a*	b*	ΔE
CA	Untreated	82.04 ± 2	2.12 ± 0.3	12.34 ± 2	4.60
	Treated	85.64 ± 1	1.94 ± 0.3	9.44 ± 0.8	
CH	Untreated	85.05 ± 2	2.09 ± 0.4	10.686 ± 2	2.17
	Treated	83.25 ± 1	2.01 ± 0.2	11.91 ± 1.6	
CO	Untreated	85.53 ± 0.8	1.25 ± 0.2	10.71 ± 0.7	1.96
	Treated	83.64 ± 2	1.49 ± 0.2	11.21 ± 0.7	

Figure 2 shows the trend for a* and b* coordinates of the untreated and treated stones with CZ NPs, and it is seen with the help of the chromatic circle, also we can see the parameter L* representing the luminosity from 0 to 100. However, the concentration of the nanomaterial used (5 mg/mL), is not perceptible to the human eye (ΔE $<$ 5), and therefore it could be used in cultural heritage conservation.

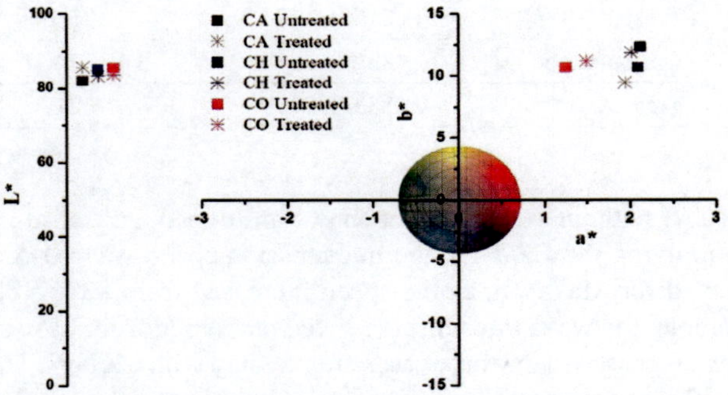

Fig. 2. Parameters of a*, b* (graph right), and L* (graph left) of limestone samples comparing with the chromatic circle, as a reference.

The mass loss at the end of the 15 salt crystallization cycles is presented in Table 3. Values for initial mass (IM), final mass (FM), and the difference (ΔM) were assigned.

The effect of CZ NPs the stones was satisfactory when compared for those that do not contain nanoparticles, since there is a more significant degradation and consequently greater loss of mass in the untreated stones. This behaviour indicates that the test specimens coated with $Ca[Zn(OH)_3]_2 \cdot 2H_2O$ delay crystallization of salts; this comparison shows the behaviour of weights values as the test progresses (see Fig. 3).

In the case of Calcehtok, the stones showed changes in their behaviour from cycle 5. These stones with and without treatment obtained a significant increase in their mass because the test tubes reached their maximum saturation, since, from that cycle, the specimen tends to have a gradual loss in its weight. Chichén Itzá behaved differently, in the case of stones without nanoparticles, there was a loss in mass at cycle 6, shedding

Table 3. Mass loss ratio of untreated and treated samples with CZ NPs

Stone		IM (g)	FM (g)	ΔM (g)
CA	Untreated	110.20 ± 4	109.83 ± 2	−0.37
	Treated	105.39 ± 3	105.42 ± 2	0.03
CH	Untreated	104.61 ± 2	97.01 ± 0.9	−7.6
	Treated	99.63 ± 1	99.26 ± 0.8	−0.37
CO	Untreated	63.03 ± 1	61.82 ± 0.8	−1.21
	Treated	60.95 ± 0.9	61.16 ± 0.7	0.24

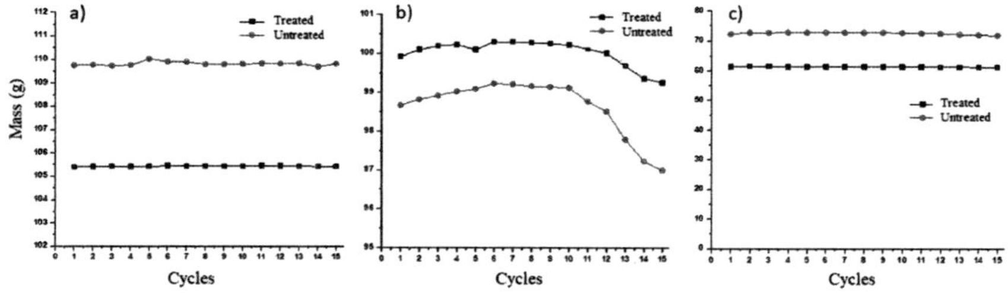

Fig. 3. Cycles of resistance to salt crystallization: **a** Calcehtok, **b** Chichén Itzá, and **c** Mayapán.

visible part of its material due to degradation, and recovering the weight loss for the absorption of the salt solution, tended to be constant until cycle 10, and after that, it lost more matter abruptly. With the CZ NPs, it was observed that the change in weights between cycles 6 and 10 remained more stable and then slightly decreased, even losing rocky material. The stone from Mayapán was the one that maintained almost imperceptible stability in its weights throughout the test. However, the detachment of the control stone was observed higher compared to the stone with $Ca[Zn(OH)_3]_2{\cdot}2H_2O$.

Table 4 shows the comparative results of the capillarity coefficient test of the limestone lithotypes with and without the CZ NPs and after the tests of resistance to salt crystallization. The average capillarity coefficient was compared before and after treatment with the CZ NPs for Calcehtok was 0.0014 and 0.00049 kg/m^2 s$^{1/2}$, and for Chichén Itzá, 0.015 and 0.015 kg/m^2 s$^{1/2}$, finally for Mayapán was 0.008 and 0.006 kg/m^2 s$^{1/2}$, respectively. These results are similar to those obtained by López-Arce et al. [19].

In the case of untreated CA, which is a more compact stone with lower porosity according to its anisotropy index (Table 1), being under the accelerated effect of salts, an increase in weight is observed, and for stones with greater porosity is observed to decrease or maintain their weight, this is due to the loss of rocky material. Some authors [8] attribute this behaviour to the growth of saline crystals inside the stone pores. In some cases, the increase in capillarity coefficient values is due to their weight; in others, the crystals lead to greater internal tension. The stone gives rise to exceed the threshold of the resistance to traction of the stone, causing its sandblasting.

Table 4. Capillary coefficient results of limestone specimens

Stone	CA kg/m^2 s$^{1/2}$	CH kg/m^2 s$^{1/2}$	CO kg/m^2 s$^{1/2}$
Untreated	0.0014	0.015	0.008
Treated	0.00049	0.015	0.006
Salt/untreated	0.0016	0.012	0.003
Salt/treated	0.0011	0.012	0.002

When applying the CZ NPs in the limestone coupons, the results show greater resistance to hydric tests and resistance to salt crystallization; these can be observed in the capillarity curves (Fig. 4).

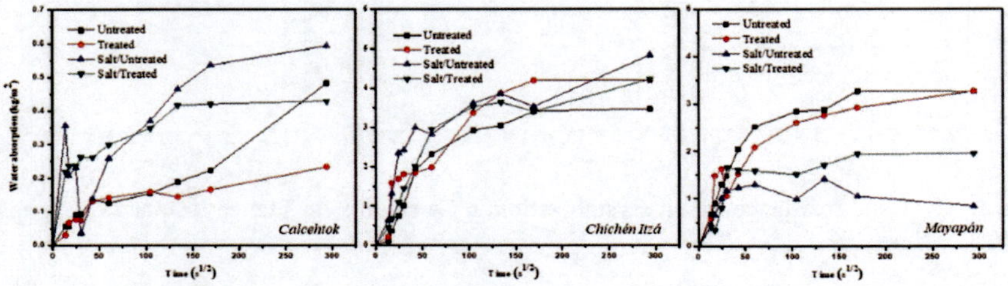

Fig. 4. Capillarity curves obtained during water absorption of limestone specimens

The results of the water absorption by capillarity of the specimens without and with CZ NPs, and after the exposure of accelerated crystallization cycles, show an improvement in the water properties of the treated stone. For Calcehtok, the values decreased when CZ NPs were applied, compared to the untreated stone, therefore the amount of water absorption is less. After the salt cycles the mass increases as a result of the presence of mineral thenardite, that corresponds to sodium sulphate (salts) found by XRD, which can be observed on the outer surface of the sample (see Fig. 5). The Chichén Itzá stone presents its maximum water absorption when the specimens were analysed after the salt test, which caused its fragility, producing erosion in the upper face in addition to its fragmentation and multiple cracks, thus, the results showed a decrease in the values of the capillary coefficient. Mayapán showed similar behaviour to that of Chichén Itzá. In both cases, initially, the water tests affected its original mass after the accelerated salt test, the maximum deterioration was observed. However, the deterioration was less for nanoparticles treated stones.

Table 4 shows the values of visual deterioration for the three limestone specimens without treatment and treated with CZ NPs after the salt test. The treated stones showed a more compact appearance, and greater resistance compared to the untreated stones, which showed a more significant loss of mass volume in the vertices and edges of the

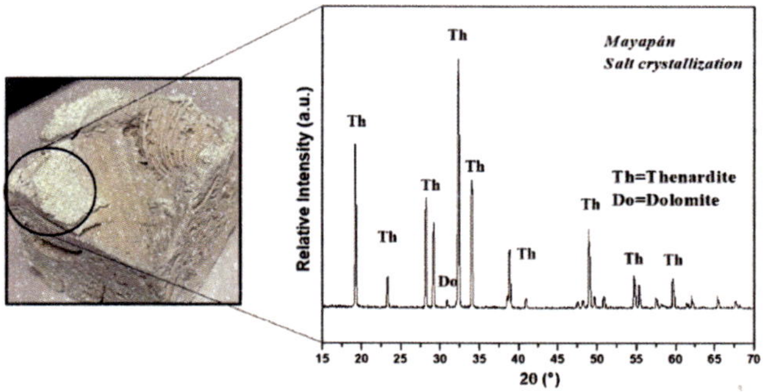

Fig. 5. X-ray diffraction pattern of the Mayapán after salt crystallization.

coupon (Fig. 6) similar to the work reported by Fort et al. [8], since the orientation of the stones for the test was according to the anisotropy measurements.

Fig. 6. Limestone specimens after the salt crystallization test (cube dimension ~ 3 cm^3)

4 Conclusions

The CZ nanoparticles used for the protection of the carbonaceous nature control stones have shown to be effective in improving the petrophysical properties of Calcehtok and Mayapán, by achieving an increase in the degree of cohesion of the stone, also an increase in the ultrasound propagation speed and a decrease in the anisotropy index. In addition, there is a decrease in the capillarity coefficient, which gives rise to an improvement against the water deterioration factors except for Chichén Itzá, which could be affected in the coefficient test of capillarity creating more cracks or micro-fissures in its interior. Therefore, the measurement values are affected. The concentration of nanomaterial used is imperceptible to the human eye, and no presence of by-products that could damage the material could be detected. Furthermore, the test of resistance to salt crystallization shows maximum values of visual erosion in limestone specimens without the treatment of CZ NPs than those in which the nanomaterial was applied, so they showed greater resistance and a more compact structure. Therefore, it is concluded that the calcium-zinc hydroxide dihydrate compound could be considered as an innovative nanomaterial to preserve and conserve historical monuments.

Acknowledgements. We are grateful for the financial support of the National Council for Science and Technology (Consejo Nacional de Ciencia y Tecnología [CONACYT]) through the projects, Estancias Posdoctorales por México, CONACYT LANCIC LN314846, Ciencia de Frontera No. 61526 and SCORE Project No. 101007531. Additional thanks to National Science Laboratory for Research and Conservation of Cultural Heritage-Corrosion Research Center (LANCIC-CICORR), UAC Campeche, for the use of their laboratorios. Thanks go also for the assistance in SEM and XRD analyses was provided by Dora A. Huerta and D. Aguilar at LANNBIO (CINVESTAV, Mérida). Also, we thank T. López for the technical assistance provided during the antimicrobial assays and S. García López and L. Díaz Flores for their support with TEM-SAED (Tabasco Science and Applied Technology Reserch Center). Thanks to INAH for providing the stones for this research.

References

1. Bautista, F., Palacio, G.: Península de Yucatán. En: Krasilnikov, P., Jiménez Nava, F.J., Reyna Trujillo, T., García Calderón, N.E. (eds.) Geografía de los suelos de México, 1st edn., pp. 355–406. Prensas de Ciencias, Ciudad de México, México (2011)
2. Videla, H.A., Guiamet, P.S., Gómez de Saravia, S.G.: Biodeterioro de materiales estructurales de sitios arqueológicos de la civilización maya. Rev Mus. Plata **1**(44), 1–11 (2003)
3. Vargas Velásquez, A.G.: Impacto socio-económico y turístico de la inscripción de la ciudad maya Palenque en la Lista del Patrimonio Mundial de la UNESCO, pp. 5–7. Universidad de Barcelona, Barcelona, España (2015)
4. Gonzales Sanchez, M., et al.: Caracterización y estudios de deterioro/conservación de materiales pétreos en monumentos históricos. En: Museu Nacional d'Art de Catalunya (eds.) Investigación en Conservación y Restauración, 1st edn., pp. 269–275. Universidad de Barcelona, Barcelona, España (2005)
5. Alonso Rodriguez, F.J.: Durabilidad de materiales rocosos, p. 1. Departamento de Geología (Petrología y Geoquímica), Oviedo, España (2013)
6. Grossi, G.M., Esbert, R.M.: Las sales solubles en el deterioro de rocas. Mater. Constr. **44**(235), 15–29 (1994)
7. Alonso Rodríguez, F.J., Ordaz, J., Valdeón, L., Esbert, R.M.: Revisión crítica del ensayo de cristalización de sales. Mater. Constr. **37**(2006), 53–60 (1987)
8. Fort, R., Fernández Revuelta, B., Varas, M.J., Álvarez de Buergo, M., Taborda Duarte, M.: Influence of anisotropy on the durability of Madrid-region Cretaceous dolostone exposed to salt crystallization processes. Mater. Constr. **58**, 289–290 (2008)
9. Soria Castro, M., De la Rosa García, S.C., Quintana, P., Gómez Cornelio, S., Sierra Fernández, A., Gómez Ortíz, N.: Broad spectrum antimicrobial activity of $Ca(Zn(OH)_3)_2{\cdot}2H_2O$ and ZnO nanoparticles synthesized by the sol-gel method. J. Sol-Gel Sci. Technol. **89**(1), 284–294 (2018)
10. AENOR: UNE-EN 14579:2005. Natural Stone Test Methods. Determination of Sound Speed Propagation. AENOR, Madrid, Spain (2005)
11. Guydader, J., Denis, A.: Propagation des ondes dans les roches anisotropes sous contrainte évaluation de la qualité des schistes ardoisiers. Bull. Eng. Geol. **33**, 49–55 (1986)

12. AENOR: UNE-EN 1925:1999. Métodos de ensayo para piedra natural, determinación del coeficiente de absorción de agua por capilaridad, pp. 1–13. Normas UNE: Minería y explosivos Asociación Española de Normalización y Certificación (1999)
13. CIELAB (Commission Internationale de l'Eclairage): Recommendations on Uniform Color Spaces, Color Difference Equations, Psychometric Color Terms. Supplement No 2 to CIE Publication No 15, Colorimetry. Bureau Central de la CIE, Paris (1976)
14. ASTM (American Society for Testing and Materials): E313-00, Standard Practice for Calculating Yellowness and Whiteness Indices from Instrumentally Measured Color Coordinates. ASTM, West Conshohocken, PA (2000)
15. Delgado Rodrigues, J., Grossi, A.: Indicators and ratings for the compatibility assessment of conservation actions. J. Cult. Herit. **8**(1), 32–43 (2007)
16. AENOR: UNE-EN 12370. Métodos de ensayo para piedra natural. Determinación de la resistencia a la cristalización de sales. Madrid, Spain (1999)
17. López Arce, P., Gómez Villalba, L.S., Pinho, L., Fernández Valle, M.E., Álvarez de Buergo, M., Fort, R.: Influence of porosity and relative humidity on consolidation of dolostone with calcium hydroxide nanoparticles: effectiveness assessment with non-destructive techniques. Mater. Charact. **61**, 168–184 (2010)
18. González Gómez, W.S., Quintana, P., May Pat, A., Aviles, F., May Crespo, J., Alvarado Gil, J.J.: Thermal effects on the physical properties of limestones from the Yucatan Peninsula. Int. J. Rock Mech. Min. Sci. **75**, 182–189 (2015)
19. López-Arce, P., Tagnit-Hammou, M., Menéndez, B., Mertz, J.-D., Kaci, A.: Durability of stone-repair mortars used in historic buildings from Paris. Mater. Struct. **49**(12), 5097–5115 (2016). https://doi.org/10.1617/s11527-016-0846-0

New Sensors for Moisture Monitoring in Historic Walls: Preliminary Results

Mattia Bassi[1] and Elisa Franzoni[1,2(✉)]

[1] Interdepartmental Centre for Industrial Research in Building and Construction (CIRI Building and Construction), University of Bologna, Bologna, Italy
elisa.franzoni@unibo.it

[2] Department of Civil, Chemical, Environmental and Materials Engineering (DICAM), University of Bologna, Bologna, Italy

Abstract. The presence of moisture in historic masonry walls represents one of the main sources of decay for cultural heritage and materials, besides causing unhealthy indoor conditions and worsening the thermal performance of the building envelope. In fact, moisture causes and/or exacerbates different deterioration mechanisms, such as salt crystallization, freeze-thaw cycles, biological growth, etc. Given the importance of moisture in materials, there are many techniques aimed at determining moisture content in structures, but the most common ones only provide qualitative results or do not allow reliable repeated measurements over time, which is a severe limitation when continuous monitoring is necessary. In this perspective, the MIMESIS project (funded by Emilia-Romagna Region, Italy) aims at developing a range of sensorized materials for the remote monitoring of moisture, temperature, pH and detachment, useful to determine the 'health state' of historic buildings. It is also expected that the sensorized materials give a warning when critical situations are reached, allowing to carry out adequate interventions before irreversible damage occurs. In this paper, a brick-based sensor suitable for measuring moisture content in masonry was developed and tested. The sensing element was originally designed for agriculture, to measure the 'soil water tension', hence an investigation about the applicability to building materials and a calibration in bricks was necessary. The results are encouraging and suggest that this new sensorized brick could be applied for the continuous monitoring of moisture in historic masonry.

Keywords: Rising damp · Historic masonry walls · Moisture measurement · Monitoring · Calibration

1 Introduction

One of the main problems of historic buildings is the presence of moisture in porous materials, such as brick, natural stone, mortar, plaster, etc. Moisture leads to deterioration of materials, unhealthy indoor conditions, and a decrease of the walls' thermal insulation. In particular, the presence of water in materials leads to damage from salt crystallization, freeze-thaw cycles, chemical and biological attack, wind erosion, etc.,

J. I. Escalante-Garcia et al. (Eds.): RW 2021, RILEM Bookseries 40, pp. 818–828, 2023.
https://doi.org/10.1007/978-3-031-21735-7_87

thus threatening historic buildings' conservation [1–5]. Hence, a fundamental step in any restoration project is to determine the moisture content of the elements affected by damp, in order to evaluate if it is necessary to take measures to reduce it. Notably, if adequate measures are not taken, interventions made on moist materials will be useless, if nor counterproductive [3]. Unfortunately, phenomena such as capillary rise from ground, rain precipitation, surface condensation or broken pipes and gutters may act together as sources of water and their complete elimination is very difficult [4]. Besides determining moisture content and distribution inside the walls, its monitoring over time is equally important, to prevent damage to cultural heritage [3]. However, measuring moisture in historic buildings is not an easy task, for many reasons. First of all, the authorities in charge of cultural heritage allow very limited destructive operations in buildings, thus excluding too invasive techniques. Secondly, many non-destructive techniques provide data that require a calibration in order to obtain the moisture content, and this is a severe problem in historic buildings, due to the limited destructive actions allowed and the heterogeneity of materials [1, 3, 6]. In the case of monitoring over a period of time, such measurement is even more challenging [4].

The most accurate and reliable method to determine moisture content in walls is commonly considered gravimetry, which consists in collecting samples and determining their moisture amount by comparing the wet mass and the dry one, the latter obtained after oven drying. This method is very simple, but exhibits several disadvantages, because it provides only local information, is time consuming, destructive and non-repeatable, and does not allow monitoring over time [7]. Moreover, the collection of the samples is a delicate operation especially in heritage buildings, as it is necessary to minimize the number and size of samples, without losing representativeness. For this reason, gravimetry is often used to calibrate non-destructive methods or during preliminary survey campaigns [1, 3, 7, 8], rather than for monitoring.

In recent years, several non-invasive probes and remote techniques were developed, suitable for buildings having historic and artistic value [1, 4, 7, 9–14]. These are generally indirect techniques, in which a parameter depending on the moisture content of the material is measured [1]. The most diffused methods are based on electrical properties, such as resistance, impedance, capacitance and dielectric constant, which are strongly influenced by the presence of moisture [5]. Both invasive and non-invasive sensors have been developed [1, 4, 10]. However, the results are affected by the characteristics of the material, state of the surface, sensor/surface contact, frequency of current used and, mainly, temperature and salt content. The calibration is complex, due to the heterogeneity of historical building materials and it could be necessary to know their absorption properties and salt content to have a more accurate measurement [4, 5, 15]. An interesting example of method based on electrical resistance is the so-called 'gypsum blocks', designed for continuous monitoring of soil moisture in agriculture [16]. It exploits a low-cost sensor made of a cylindrical block of gypsum mounted on a plastic base and provided with two electrodes. The moisture in the block goes into equilibrium with the surrounding material and the electrodes allow to measure the electrical resistance. The main advantage of this technique is that gypsum, rich in calcium and sulphate ions, is not sensitive to the presence of salts, acting as a buffer towards the effect of salts on electrical conductivity [16, 17]. Other quite diffused techniques are based on the use of microwaves, where an electromagnetic field interacts with the material and water contained in it, losing part of its energy, according to an extinction coefficient which depends on the dielectric properties of the dry material and the moisture amount [1, 6, 15, 18, 19]. Thanks to the high frequency of microwaves, the presence of salts is not an issue [2, 5, 6, 19].

Another completely non-invasive method to detect moist zones in a walls is infrared thermography, as the water evaporation from the masonry surface causes a lowering of the temperature and can be easily seen [1, 4, 5, 15]. Infrared thermography is normally used to obtain qualitative information, while quantitative techniques should be used to quantify moisture. Moreover, the measurement refers only to a surface layer of the material, influenced by atmospheric conditions [1, 15]. Other methods exploits the thermal conductivity of materials, which is not affected by salt presence [1]. The sensors are thermocouples, which can heat the materials and register temperature increase, allowing to determine the moisture content, provided that the heat capacity of the dry material is known [1]. A more recently introduced thermal method exploits optical fibres, which are chemically inert, light, remotely controllable and allow a mapping [1]. They are widely used in distributed temperature sensing (DTS) in agriculture for the detection of soil moisture over long distances [14, 20].

As shown above, many methods and sensors are available to measure moisture in porous building materials. However, most of them only provide qualitative information and a calibration on each material would be necessary, which is basically impossible in historical buildings that are made, by their nature, of heterogeneous materials and often even from different periods. The MIMESIS project (Materiali Smart Sensorizzati e Sostenibili per il Costruito Storico [Sensorized and sustainable smart materials for historic buildings]), funded by Emilia-Romagna Region, Italy, aims at developing a sensorized ceramic material to be applied in masonry walls, which is able to overcome the difficulties described above and to provide reliable values of moisture (liquid water content). The new sensor is expected to provide data remotely and in real time, giving a warning in case of 'critical' moisture conditions and preventing irreversible damages.

2 Materials and Methods

2.1 Sensorized Ceramic Material

A preliminary research was carried out on the sensors available in the market for moisture measurement. Particular attention was paid to the possibility to integrate the sensors into a ceramic material (brick) and to the ability to provide remote and real time data. Finally, a type of sensor was selected, which is currently used in agriculture ('Watermark Soil Moisture Sensor', by IRROMETER Company, Fig. 1a). It is an electrical resistance sensing device used to measure the 'soil water tension' (SWT, in cbar), which represents the tension required by the roots of the plants to extract water from soil, hence providing an indication about the need of irrigation [21, 22]. The sensor consists of a pair of corrosion-resistant electrodes encapsulated in a granular matrix (gypsum). The diameter of the sensor is 22 mm and its length is 76 mm. It is equipped with a cable for the connection to a data collection and transmission control unit ('Plug & Sense! Smart

Agriculture Pro Wi-fi' control unit, by Libelium), which allows to simultaneously connect up to three Watermark sensors and a temperature sensor (Fig. 1b). The temperature sensor is a PT-1000 probe, again used in agriculture for soil. The granular matrix constituting the Watermark sensor reaches the hygrometric equilibrium with the surrounding material in which it is fixed (generally soil) and the electrodes measure the electrical resistance of the granular matrix, finally sending the data to the collection unit. This translates electrical resistance values into SWT values through a proprietary calibration equation provided by the manufacturer. The data are sent to a cloud in which they can be seen real time, thanks to an online platform designed for the MIMESIS project. The SWT values are plotted over time for the three moisture sensors and the temperature sensor (Fig. 2). Data are available both as single values, obtained every minute, and as hourly average.

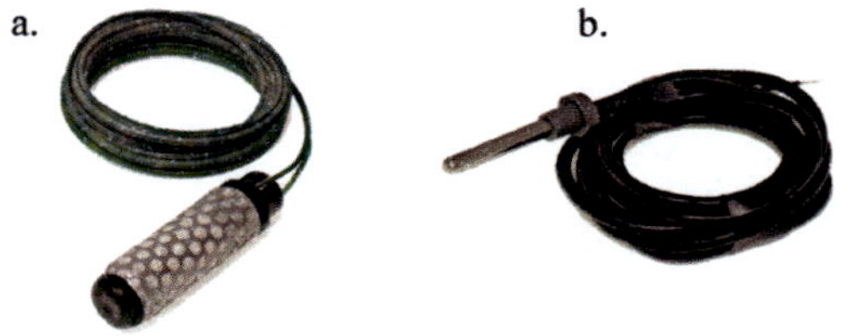

Fig. 1. **a** Sensor selected to measure moisture content. **b** Probe selected to measure temperature.

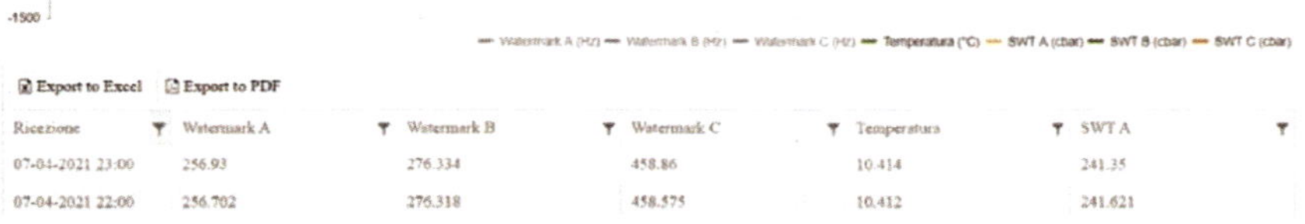

Fig. 2. Example of an online platform screen, where curves of SWT values over time (dates from March 16 to April 7, 2021) referred to the three Watermark sensors are visible (yellow, orange and dark green curves) together with temperature curve provided by the PT-1000 probe (light green curve).

Once the moisture sensor was selected, a system to integrate it inside a ceramic material was designed. The idea was to create a sort of 'plug' to be inserted in the masonry, consisting in a cylinder of brick with the sensor embedded inside, as shown in Fig. 3. The cylinder is 40 mm in diameter and 100 mm in length, with a central hole having diameter 24 mm and depth 90 mm, suitable to insert the sensor without forcing it and without leaving too much free space between sensor and hole. The cable connecting the sensor to the control unit can be hidden in the mortar joints or under the render.

Four different types of commercial solid bricks (manufactured in Italy by IBL, SanMarco, Pica, and RDB, and here labelled as 'brick 1–4', respectively) were used for the tests, to find the most suitable one for the embedding of the sensor. According to preliminary tests, all the tested bricks have similar microstructure characteristics to those currently found (on average) in historic buildings. In this way, the sensorized ceramic plug can be considered quite similar and compatible with historic masonries, although it is well known that ancient bricks exhibit an impressive heterogeneity [4].

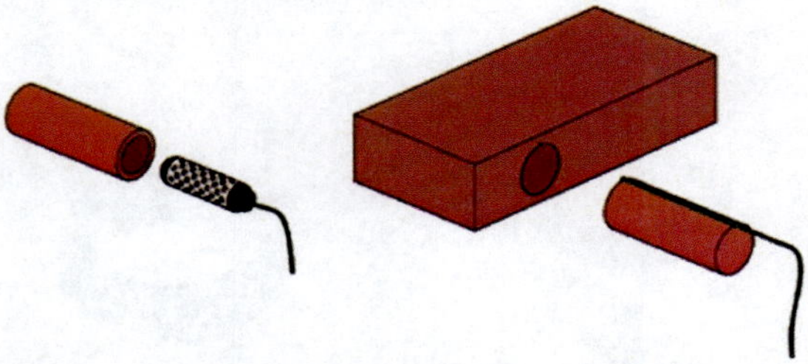

Fig. 3. Scheme of the sensorized ceramic material and its way of application onsite. The brick cylinder is drilled, and the moisture sensor is positioned inside. For the application to a real masonry, the sensorized element is turned with the base towards the outside, so the sensor is more protected from the external masonry surface, and it is also more easily hidden.

2.2 Preliminary Tests on the Sensors

Despite a careful research in the scientific literature, no examples were found on the use of the selected sensor in applications different from agriculture. Hence, preliminary tests were carried out to understand if the sensor can be used in different substrates.

At first, the bare sensors were positioned inside loose materials (Fig. 4a), namely brick powder (obtained by grinding a sample of brick 1) and different types of sand (fine, coarse, and mixed). Then, water was added and the values of SWT were measured. Another set-up was prepared to follow a complete wetting and drying cycle, somehow simulating capillary rise from the ground. In this second set-up, the sensors were inserted in open containers (Fig. 4b) filled with the same loose materials previously used and positioned on a pack of filter paper, which was kept wet by adding tap water in the container. After wetting, the set-up was let dry at laboratory conditions and the wetting and drying curves of the sensors were monitored through the online platform, also observing the different response in the selected loose materials.

Therefore, the bare sensors were positioned inside solid bricks having size 12 × 25 × 5.5 cm, by previously drilling holes having diameter equal to 24 mm and depth equal to 2/3 of length of the brick. The diameter of the hole was selected in order to be only slightly higher than that of the sensor (22 mm), for the reasons above. Different coupling materials were investigated to fill the space between the sensor and the hole. One brick was left without coupling material, with the sensor simply inserted into the hole, while the others were filled with brick powder and fine sand. The hole was then sealed with plasticine and the bricks were positioned in a basin under a water head of 5 mm, to let them absorb water by capillarity up to complete saturation (Fig. 4c). After 3 days in this condition, the bricks were let dry at laboratory conditions.

2.3 Calibration of the Sensorized Ceramic

As the sensors are calibrated by the manufacturer for the measurement of moisture in different kinds of soils, a tailored calibration in bricks was necessary, in order to obtain the moisture content value starting from the SWT. Different tests were carried out.

In the first calibration, brick 1 was used. The methodology adopted was inspired by the experimental procedure in [23], aimed at obtaining uniformly distributed partial saturation conditions in concrete samples. In this study, cylindrical brick samples (Fig. 3) were let absorb fixed amounts of water by capillarity, corresponding to a moisture percentage [(moist mass − dry mass)/dry mass] equal to 4, 6, 10, 12, 15, and 18%. The value 18% corresponds to saturation at atmospheric pressure for the selected brick. The sensor was then immediately inserted in the cylinder and the hole sealed with plasticine, finally wrapping all the set-up with plastic film and adhesive tape, to prevent any water loss by evaporation (Fig. 5). No coupling material was used between the sensor and the brick cylinder, following the results obtained in the preliminary tests (see Sect. 87.1). Once sealed, the sensorized ceramic elements were left at laboratory until moisture uniformly distributed and became constant, which was checked by monitoring the SWT (variation in 24 h $< 1\%$). The temperature probe was kept close to the samples.

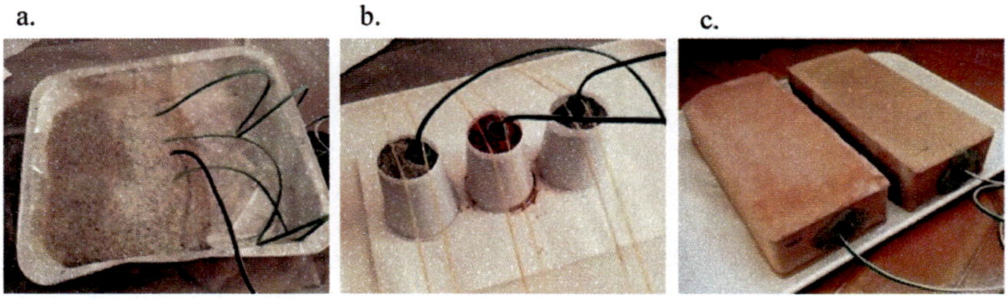

Fig. 4. **a** Experimental set-up with the sensors inserted in sand (mixed grain size), wetted with water. **b** Capillary rise set-up with various materials (sand with different granulometry and brick powder). **c** Set-up with the sensors inserted in full bricks, with and without coupling material inside the hole.

In the light of the results obtained in this first calibration test and the problems found in measuring low moisture amounts (see Sect. 87.2), new tests were carried out, following the same testing procedure described above (Fig. 5), but using: (1) different kinds of bricks for the cylinders (bricks 1–4); (2) a fixed moisture amount, namely 10%.

After this further group of measurements, a specific brick was selected as the most suitable for the manufacturing of the sensorized ceramic (brick 4) and the calibration

Fig. 5. Sensorized ceramic elements wrapped with plastic film and insulating adhesive tape during the calibration phase.

was repeated using this material, at different moisture contents (4, 6, 10, 12, 15, and 18%), putting the wrapped cylinders with sensors inside a climatic chamber at a constant temperature of 20 ± 1 °C.

3 Results and Discussion

3.1 Preliminary Tests on the Sensors

From the preliminary tests carried out using loose materials and full bricks with different coupling powders, some basic observations were derived about the functioning of the sensor in substrates different from soil. In all the materials under investigation, SWT decreases as the moisture content in the material increases, as expected, and the sensor is strongly sensitive to changes in moisture content, hence the measurement in building materials seems feasible. The tests performed inserting the sensor in solid bricks with holes showed that the presence of a coupling material between sensor and hole (fine sand or brick powder) is not necessary, as the SWT values obtained with and without coupling materials were fully comparable. Moreover, comparing the SWT values obtained inserting the sensors in the perforated solid brick and in the brick cylinders, no substantial differences were found, confirming that SWT does not depend on the size of the sample in which it is inserted.

3.2 Calibration of the Sensorized Ceramic

The results obtained from the first calibration with brick 1 are reported in Fig. 6, where a clear correlation between SWT and moisture can be observed. However, two main issues were found. Firstly, the range of measurement of SWT recommended by the manufacturer's datasheet is 0–239 cbar, therefore the measurement is likely less reliable for moisture <8%, which is a significant limit in the envisaged application, as a masonry exhibiting moisture in this range cannot be considered dry. Secondly, a certain degree of dispersion was found in the values, and this was ascribed to the slight temperature variations in laboratory at the different dates in which the calibration tests were carried out. In fact, monitoring SWT in the online platform showed it depends on temperature.

To overcome the first issue, further tests were carried our using different kinds of brick at the same moisture amount (10%). The results are shown in Fig. 7, where the SWT values are reported versus time, and the attainment of the equilibrium in the sensorized materials can be observed. Considering that the measurement reliability limit of SWT is 239 cbar, the most suitable brick resulted brick 4, i.e., the one which provided the lowest SWT value for the same moisture content.

Hence, brick 4 was selected to be used for the plug and a second calibration was carried out, this time at a constant temperature of 20 °C, in order to overcome also the second issue found in Fig. 6 (data dispersion). The results are reported in Fig. 8, where it can be observed that only for moisture approximately below 5% the value of SWT falls above the range recommended by the manufacturer (>239 cbar), which is quite acceptable for masonry walls. Moreover, the linear equation describing the relationship between SWT and moisture fits in an excellent way with the data obtained ($R^2 = 0.9672$).

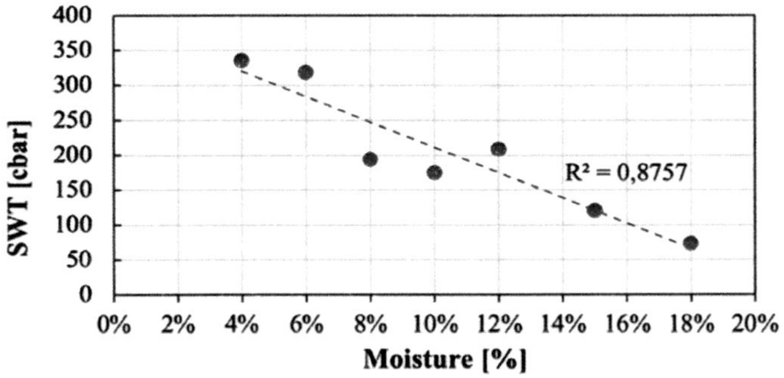

Fig. 6. Correlation between SWT and moisture amount obtained using brick 1.

Following the results obtained, a sensorized plug based of the use of brick 4 will be manufactured according to the scheme in Fig. 3 and used for further testing. The brick element of the plug will reach the equilibrium with the surrounding masonry in which it is inserted, while the sensor will always remain in contact with the same material.

4 Conclusions

The present paper represents a preliminary study aimed at the development of a sensorized ceramic plug to be inserted inside historic masonry walls for moisture monitoring purposes. The results obtained so far are very encouraging, but several aspects need further investigation, such as the role of temperature, the role of soluble salts that are often present in buildings affected by rising damp and the kind of equilibrium which establishes between the ceramic plug and the surrounding historic bricks. All these aspects are currently under investigation within the MIMESIS project, also including a trial testing in-the-field, in two historic buildings in Italy.

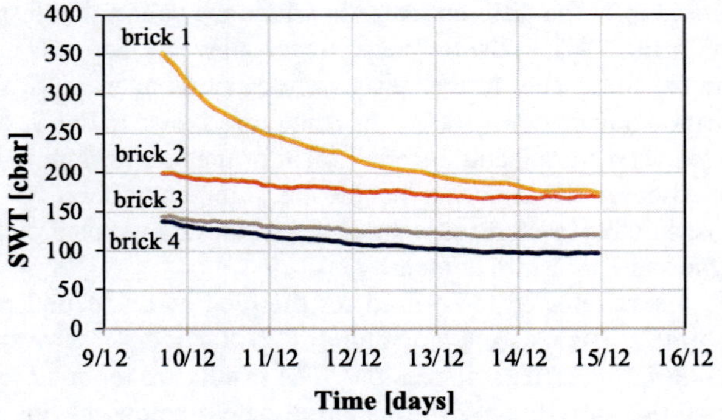

Fig. 7. SWT variation curves for the four different types of brick, at moisture content 10%. SWT is plotted versus time, thus showing the attainment of the equilibrium in the system brick + sensor.

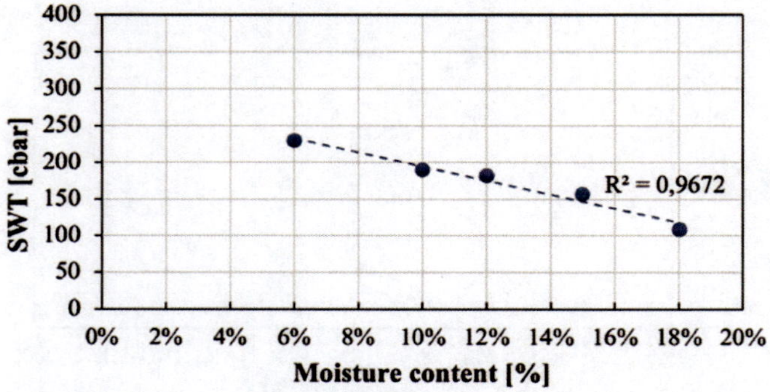

Fig. 8. SWT versus moisture content, obtained using brick 4 and performing the measurement at a constant temperature of 20 °C.

Acknowledgements. This research has received funding from Regione Emilia Romagna, Italy, through the Project 'MIMESIS: Materiali smart sensorizzati e sostenibili per il costruito storico' (funded under the Call POR-FESR Emilia Romagna 2014–2020 - Asse 1 - Ricerca e innovazione, Azione 1.2.2). Geo.Smart.Lab (by Sis-Ter SRL), Imola, Italy, is gratefully acknowledged for the support in the sensors selection phase.

References

1. Agliata, R., Greco, R., Mollo, L.: Moisture measurements in heritage masonries: a review of current methods. Mater. Eval. **76**, 1468–1477 (2018)
2. Goetzke-Pala, A., Hoła, J.: Influence of burnt clay brick salinity on moisture content evaluated by non-destructive electric methods. Arch. Civ. Mech. Eng. **16**(1), 101–111 (2015). https://doi.org/10.1016/j.acme.2015.08.001
3. Hola, A.: Measuring of the moisture content in brick walls of historical buildings-the overview of methods. In: IOP Conference Series: Materials Science and Engineering (2017)

4. Sandrolini, F., Franzoni, E.: An operative protocol for reliable measurements of moisture in porous materials of ancient buildings. Build. Environ. **41**, 1372–1380 (2006). https://doi.org/10.1016/j.buildenv.2005.05.023
5. Phillipson, M.C., et al.: Moisture measurement in building materials: an overview of current methods and new approaches. Build. Serv. Eng. Res. Technol. **28**, 303–316 (2007). https://doi.org/10.1177/0143624407084184
6. Camuffo, D., Bertolin, C.: Towards standardisation of moisture content measurement in cultural heritage materials. e Preserv. Sci. **9**, 23–35 (2012)
7. Guizzardi, M., Derome, D., Mannes, D., Vonbank, R., Carmeliet, J.: Electrical conductivity sensors for water penetration monitoring in building masonry materials. Mater. Struct. **49**(7), 2535–2547 (2015). https://doi.org/10.1617/s11527-015-0666-7
8. Zhao, J.H., Rivera, E., Mufti, A., Stephenson, D., Thomson, D.J.: Evaluation of dielectric based and other methods for moisture content measurement in building stones. J. Civ. Struct. Health Monit. **2**, 137–148 (2012). https://doi.org/10.1007/s13349-012-0024-1
9. Moropoulou, A., et al.: Structural health monitoring of the Holy Aedicule in Jerusalem. Struct. Control Health Monit. **26**, (2019). https://doi.org/10.1002/stc.2387
10. Cacciotti, R., Valach, J., Wolf, B.: Innovative and easy-to-implement moisture monitoring system for brick units. Constr. Build. Mater. **186**, 598–614 (2018). https://doi.org/10.1016/j.conbuildmat.2018.07.125
11. Walker, R., Pavía, S., Dalton, M.: Measurement of moisture content in solid brick walls using timber dowel. Mater. Struct. **49**(7), 2549–2561 (2015). https://doi.org/10.1617/s11527-015-0667-6
12. Minardo, A., Catalano, E., Zeni, L., Agliata, R., Greco, R., Mollo, L.: Measurement of moisture content in masonry materials by active distributed optical fiber sensors. In: 18th Italian National Conference on Photonic Technologies (Fotonica 2016). Institution of Engineering and Technology (2016)
13. Cetrangolo, G.P., Domenech, L.D., Moltini, G., Morquio, A.A.: Determination of moisture content in ceramic brick walls using ground penetration radar. J. Nondestr. Eval. **36**(1), 1–11 (2017). https://doi.org/10.1007/s10921-016-0390-4
14. Agliata, R., Bogaard, T.A., Greco, R., Minardo, A., Mollo, L., Steele-Dunne, S.C.: Non-invasive water content estimation in a tuff wall by DTS. Constr. Build. Mater. **197**, 821–829 (2019). https://doi.org/10.1016/j.conbuildmat.2018.11.250
15. Bison, P., et al.: Limits and advantages of different techniques for testing moisture content in masonry (2011)
16. Larsen, P.: Determination of water content in brick masonry walls using a dielectric probe. J. Arch. Conserv. **18**, 47–62 (2012). https://doi.org/10.1080/13556207.2012.10785103
17. Friis Dela, B.: Measurement of soil moisture using gypsum blocks. SBI forlag (2001)
18. Kääriäinen, H., Rudolph, M., Schaurich, D., Tulla, K., Wiggenhauser, H.: Moisture measurements in building materials with microwaves. NDT E Int. **34**, 389–394 (2001). https://doi.org/10.1016/S0963-8695(01)00005-6
19. Kruschwitz, S., Niederleithinger, E., Trela, C., Wöstmann, J.: Use of complex resistivity tomography for moisture monitoring in a flooded masonry specimen. J. Infrastruct. Syst. **18**, 2–11 (2012). https://doi.org/10.1061/(ASCE)IS.1943-555X.0000053
20. Cao, D., Shi, B., Zhu, H., Wei, G., Chen, S.-E., Yan, J.: A distributed measurement method for in-situ soil moisture content by using carbon-fiber heated cable. J. Rock Mech. Geotech. Eng. **7**, 700–707 (2015). https://doi.org/10.1016/j.jrmge.2015.08.003
21. Shock, C.C., Flock, R., Feibert, E., Shock, C.A., Pereira, A., Jensen, L.: Irrigation monitoring using soil water tension. Oregon State University (2005)
22. Shock, C.C., Wang, F.-X.: Soil water tension, a powerful tool for productivity and stewardship. In: HortScience: A Publication of the American Society for Horticultural Science, pp. 178–185 (2010)

23. Antón, C., Climent, M.A., De Vera, G., Sánchez, I., Andrade, C.: An improved procedure for obtaining and maintaining well characterized partial water saturation states on concrete samples to be used for mass transport tests. Mater. Struct. Constr. **46**, 1389–1400 (2013). https://doi.org/10.1617/s11527-012-9981-4

Non-Destructive Testing Techniques

Fresh Concrete Curing Monitoring Using Acoustic Emission

Ashwin P. S. Dias, Gerlinde Lefever, and Dimitrios G. Aggelis(✉)

Department of Mechanics of Materials and Constructions, Vrije Universiteit Brussel (VUB), Pleinlaan 2, 1050 Brussels, Belgium
dimitrios.aggelis@vub.be

Abstract. In the present paper, acoustic emission (AE) is applied to monitor internal curing of fresh concrete. Processes taking place during the fresh state of concrete prove to be very acoustically active allowing detailed monitoring of the events. These processes include, settlement, production of hydrates, shrinkage cracking and correlate well with other measured parameters like ultrasonic pulse velocity and compressive strength. It is also characteristic that internal curing owing to the addition of superabsorbent polymers (SAPs) contributes to large numbers of AE activities, exhibiting the potential of the technique to distinguish between different sources based only on passive and non-intrusive use of sensors on the external side of the mold. The heavily damping nature of fresh concrete necessitates elaborate design of the setup and optimization of the sensors' placement. Parameters like the total AE activity, amplitude, average frequency and time domain characteristics prove very useful in characterizing the process of internal curing. The possibility to control the curing based on the results of acoustic monitoring is also discussed.

Keywords: Fresh concrete · Acoustic emission · Superabsorbent polymers · Non-destructive testing

1 Introduction

The control of the quality of concrete in its fresh state is of paramount importance to ensure the desired strength and durability in its final state. Hence, monitoring and understanding the wide variety of processes that occur in fresh concrete are crucial to meet the ever-increasing demands on the quality of concrete. This has resulted in many innovations in the fields of monitoring as well as construction materials, some of which are discussed herein.

Acoustic emission (AE) based testing methods have proven to be sensitive for the detection of processes such as settlement, bubble migration [1], hydration [2] and cavitation [3] in fresh concrete. This testing method that works on the principle of detecting low-energy elastic waves, is capable to detect events with energy of the order of 10^{-18} J [4]. This high level of sensitivity is particularly important in the context of fresh concrete due to the strongly attenuative nature of the medium. Apart from the trends seen in the

J. I. Escalante-Garcia et al. (Eds.): RW 2021, RILEM Bookseries 40, pp. 831–837, 2023.
https://doi.org/10.1007/978-3-031-21735-7_88

cumulative detected bursts of energy or 'hits', the detected waveform parameters also play an important role in understanding the processes occurring in the fresh concrete. One study [5] found that trends seen in the average frequency and rise time parameters of the detected hits had a strong correlation with the water-to-cement (w/c) ratio during the first 24 h of monitoring. Another study [6] also investigated the effect of additives such as fly ash and it was observed that the total number of hits decreased because of higher fly ash contents, which resulted in reduced shrinkage cracking.

Among concrete additives, superabsorbent polymers (SAPs) have been found to be highly acoustically active [7]. These polymers can swell by absorbing up to 500 times their weight [8] when exposed to water and release it as the concrete undergoes drying, due to the increase in capillary pressure. SAPs have shown to be efficient in reducing autogenous shrinkage, which occurs due to the hydration reaction, particularly in mixes with low w/c ratios [8]. One of the key reasons for this is that the SAPs act as an internal curing agent, having the potential to effectively cure the entire volume of concrete as opposed to traditional water-based curing methods. The properties of the SAP utilized, such as the particle size distribution, chemical composition and the cross-linking degree along with the method of SAP implementation (dry mixing or pre-soaking), play a role in determining the final properties of the concrete [9]. One of the concerns regarding the application of SAPs in concrete mixes is that once the SAPs release their entrained water, they leave behind voids, which could have a negative impact on the porosity, permeability, compressive strength and in general the durability of concrete.

The objective of the present study was to monitor the process of internal curing in mortar with superabsorbent polymers and to identify methods of characterizing the activity of the SAPs through AE parameters. Additionally, the possibility to control the course of internal curing through real-time AE feedback was investigated.

2 Materials and Methodology

2.1 Acoustic Emission Setup

A suitable experimental setup is crucial in order to capture the processes occurring inside the concrete. Previous studies have identified that the type of sensors used and the method of connecting the sensors to the concrete can have an impact on the type of processes detected [10]. In this experiment five R15α sensors (150 kHz resonance frequency) were attached to the steel mold (one per side, except for the top open surface) by means of a Vaseline layer to ensure acoustic coupling. The highly transmissive mold would serve as a waveguide to help capture events within the attenuative fresh concrete.

Apart from the AE sensors, a thermocouple was also embedded in the concrete to help determine the ongoing processes based on temperature trends. Signals from the sensors were pre-amplified prior to being processed. Figure 1 shows the experimental setup used for acquisition. A pencil lead break test was carried out prior to commencement of monitoring to ensure sufficient coupling between the sensors and the mold.

2.2 Concrete Mix Design

The reference concrete mix proportions are indicated in Table 1.

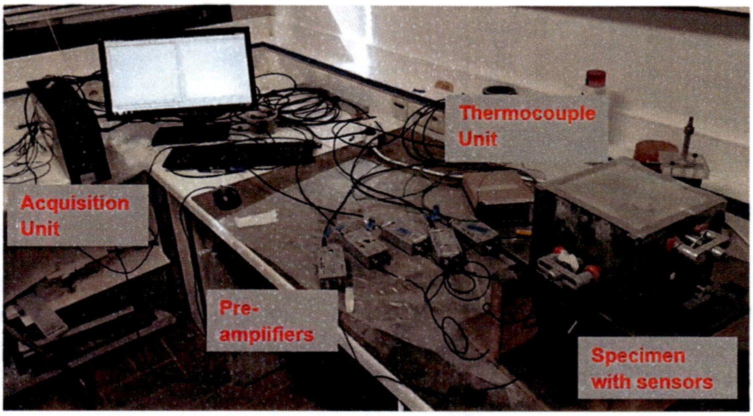

Fig. 1. Experimental setup

Table 1. Concrete mix proportions

Compound	Proportion (kg/m^3)	Ratio w.r.t. cement
Cement	384	1
Sand (0–2 mm)	488	1.27
Fine aggregate (2–4 mm)	488	1.27
Gravel (4–16 mm)	906	2.36
Water	134	0.35
Plasticizer	2.30	0.006

As the w/c ratio of the mix was relatively low, a superplasticizer - MasterGlenium® 51 (conc. 35%), was included in the mix in order to ensure sufficient workability of the concrete. In cubes with SAPs, a dosage of 0.2% by mass of cement was included in the dry mix. The SAP used in the experiment was a copolymer of acrylamide and sodium acrylate, provided by BASF. Together with the SAPs, an additional amount of water equal to 26 g per gram of SAP was added, to achieve an identical workability of the SAP and reference mixture.

The dry components were initially mixed at a low speed for 60 s prior to the addition of water and super-plasticizer. Following the addition of water another 60 s of slow mixing was followed by 60 s of a faster mixing speed. Finally, another 60 s of slow mixing was done, following which the concrete was added to the mold and vibrated in two layers until sufficiently compacted. The AE monitoring was initiated immediately after.

3 Results and Discussion

3.1 Comparison of Hits

The comparison of the total number of hits in the samples is shown in Fig. 2. The main source of activity in the reference samples occurred during the first 2 h of monitoring, which corresponds well to the activity expected due to settlement. Another smaller burst of activity can also be seen around 14 h, where activity corresponding to hydration was detected by the setup. In the REF-1 sample, 2 sources of activity are seen around 45 and 62 h, which could be due to shrinkage cracking or debonding between the concrete and the mold. The SAP activity, which occurred from around 15 to 40 h, accounts for a large percentage of the activity detected in the cubes with SAPs (SAP-1, SAP-2). After the internal curing activity due to SAPs, the rate of emissions in the SAP samples reduces to the levels of reference specimens.

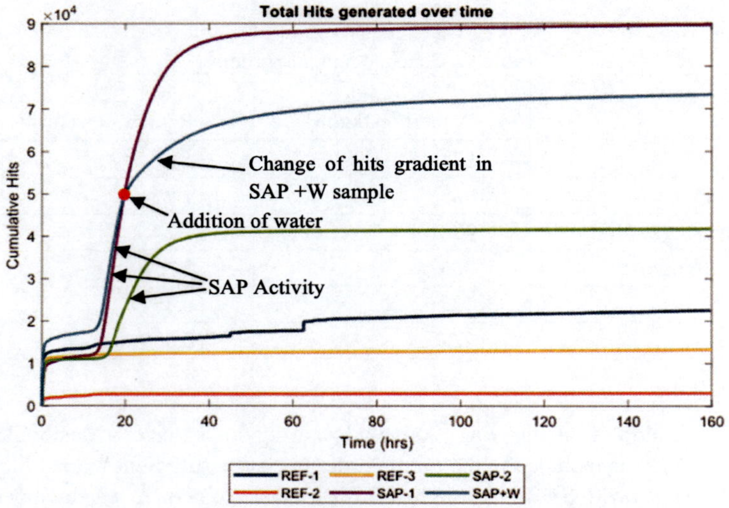

Fig. 2. Comparison of hits

The SAP + W sample was used to test the hypothesis that the internal curing could be controlled based on AE activity. Specifically, when AE was recorded with high rates, indicating that the internal curing was ongoing, a layer of water was added on top of the cube and specifically at 19 h after mixing, see SAP + W curve in Fig. 2. The addition of water appeared to slow down the rate of activity of SAPs, as can be seen from Fig. 2. It was observed that the slope of the hits curve decreased and that the SAP activity period was prolonged when compared to the other SAP samples. This is a clear indication that the internal curing of the SAPs can be modified if required through the addition of water at the appropriate time as indicated by AE activity.

Another notable observation during the study is that the samples with SAPs displayed a high variability with regards to the total number of hits detected in each sample, as well as the activity among the different sensors. This fact is indicative of the heterogeneous distribution of internal curing and deserves further study.

3.2 Comparison of AE Parameters

The behavior of basic AE parameters is discussed herein.

Amplitude. The AE parameters of the samples were analyzed in order to determine any distinguishable parametric trends with regard to the activity of internal curing. Looking at the trends seen in the 250-hit moving average for the amplitudes detected in the samples in Fig. 3, several key observations stand out. In general, the amplitude in all samples increases during the first few hours, however, in samples with SAPs a sharp, sudden drop is observed in the amplitude, corresponding to the onset of SAP activity.

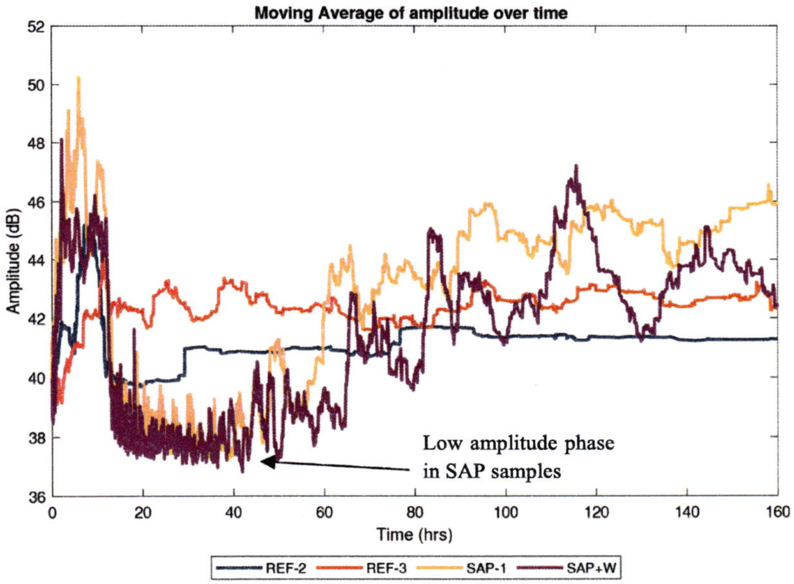

Fig. 3. Comparison of amplitude moving average

One possible reason for the drop in amplitude could be due to the SAPs introducing an additional heterogeneity in the medium, as once the SAPs release the entrained water, a void is left behind. This in turn increases the attenuation of the medium, resulting in lower detected amplitudes. However, if this was the case, a gradual decrease in amplitude would be more likely, as the water is released over time. Therefore, given the rapidity of the drop in the amplitude, it could be concluded that the process of water release from the SAPs is a relatively low-amplitude process when compared to that of settlement or hydration.

Average Frequency. Apart from the amplitude, the trends observed in the average frequency (AF) of the samples were also studied. As can be seen in Fig. 4, clear trends could be identified to characterize SAP activity. The average frequency in the samples with SAPs shows a distinct sharp increase when the SAP activity begins. Following the completion of SAP activity, the AF average starts to decrease. Thus, the AF trend of samples could provide a means of characterizing the SAP activity in concrete.

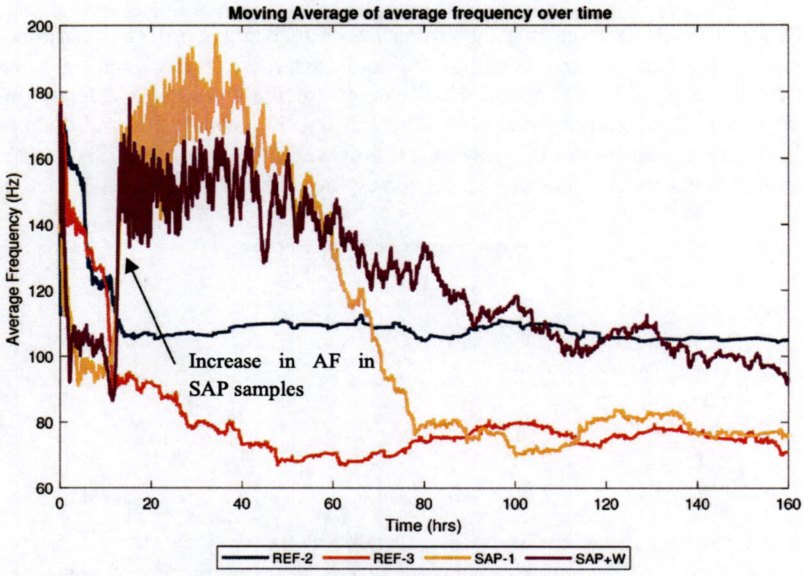

Fig. 4. Comparison of average frequency trends

4 Conclusions

Some of the key conclusions of the study are listed below.

- Internal curing owing to SAPs can be monitored by non-intrusive AE monitoring. SAPs activity contributes to a great number of hits during the hydration of the concrete.
- AE activity can be used as a controlling signal to modify the internal curing by adding water on top of the sample. This causes a slowdown in SAP activity, as monitored by a reduction of AE recording rate resulting in prolonged SAPs activity period.
- The amplitude of SAP-related activity was found to be relatively low in nature. This could partly explain the higher variability of the number of hits observed in these samples. Depending on the position of the sources, the detection of a single hit by multiple sensors is less likely, due to the high attenuation of the medium and the low amplitude of the wave.
- AE parameters such as average frequency and amplitude showed distinct trends during SAP activity. This could prove to be a useful tool in characterizing the activity of the SAPs in fresh concrete.

References

1. Dzaye, E., De Schutter, G., Aggelis, D.G.: Study on mechanical acoustic emission sources in fresh concrete. Arch. Civ. Mech. Eng. **18**, 742–754 (2018)
2. Assi, L., Soltangharaei, V., Anay, R., Ziehl, P., Matta, F.: Unsupervised and supervised pattern recognition of acoustic emission signals during early hydration of Portland cement paste. Cem. Concr. Res. **103**, 216–225 (2013)
3. Lura, P., Couch, J., Jensen, O.M., Weiss, J.: Early-age acoustic emission measurements in hydrating cement paste: evidence for cavitation during solidification due to selfdesiccation. Cem. Concr. Res. **39**(10), 861–867 (2009)
4. Aggelis, D.G., Grosse, C.U., Tomoki, S.: Acoustic emission characterization of fresh cement-based materials (chap. 1). In: Serdar, M., et al. (ed.) Advanced Techniques for Testing of Cement-Based Materials, pp. 1–22. Springer International Publishing (2020)
5. Iliopoulos, S., El Khattabi, Y., Aggelis, D.G.: Towards the establishment of a continuous nondestructive monitoring technique for fresh concrete. J. Nondestr. Eval. **35**(3), 1–11 (2016)
6. Pazdera, L., Topolar, L., Korenska, M., Smutny, J., Bilek, V.: Forecasting of concrete strength during the hardening process by means of acoustic emission method. In: 11th European Conference on Non-Destructive Testing, Prague, Czech Republic, pp. 1–8 (2014)
7. Lefever, G., Tsangouri, E., Snoeck, D., Aggelis, D.G., De Belie, N., Van Vlierberge, S., et al.: Combined use of superabsorbent polymers and nanosilica for reduction of restrained shrinkage and strength compensation in cementitious mortars. Constr. Build. Mater. **251**, 118966 (2020)
8. Jensen, O.M.: Use of superabsorbent polymers in concrete. Concr. Int. **35**(1), 48–52 (2013)
9. Woyciechowski, P.P., Kalinowski, M.: The influence of dosing method and material characteristics of superabsorbent polymers (SAP) on the effectiveness of the concrete internal curing. Materials **11**, (2018)
10. Thirumalaiselvi, A., Saptarshi, S.: Acoustic emission monitoring and classification of signals in cement composites during early-age hydration. Constr. Build. Mater. **196**, 411–427 (2019)

Crack Closure Assessment in Cementitious Mixtures Based on Ultrasound Measurements

Gerlinde Lefever[1(✉)], Nele De Belie[2], Danny Van Hemelrijck[1], Dimitrios G. Aggelis[1], and Didier Snoeck[1,2]

[1] Department of Mechanics of Materials and Constructions, Vrije Universiteit Brussel, Pleinlaan 2, 1050 Brussels, Belgium
Gerlinde.Lefever@vub.be

[2] Magnel-Vandepitte Laboratory for Structural Engineering and Building Materials, Department of Structural Engineering and Building Materials, Faculty of Engineering and Architecture, Ghent University, Tech Lane Ghent Science Park, Campus A, Technologiepark Zwijnaarde 60, 9052 Ghent, Belgium

Abstract. Cracking of cementitious materials poses severe problems for the mechanical performance and the durability of concrete structures. Self-healing cementitious materials provide a means of countering these issues, but the testing procedures to assess the self-healing efficiency are mostly impractical for in-situ evaluation. This study concerns the evaluation of the crack closure in cementitious mixtures with and without superabsorbent polymers by ultrasound measurements. Ultrasonic waves are sent through the cracks by means of pencil lead break tests, thereby offering information on the sound material as well as on the healing products formed within the cracks. The healing process is promoted by placing the specimens in wet-dry curing cycles and is monitored for a period of 14 days. Ultrasound was found to be sensitive to the closure of cracks, which was confirmed by the visual reduction in crack width opening, as investigated by microscopic analysis.

Keywords: Cement · Superabsorbent polymers · Self-healing · Ultrasound

1 Introduction

Concrete is known as a solid construction material when loaded in compression. On the contrary, concrete is highly sensitive to tensile stresses, which lead to cracking. These cracks pose problems for the mechanical performance and the durability of the structures, as deleterious substances can enter the material and harm the inner matrix and the present reinforcement. To safeguard the construction's service life, repair of cracks is of high importance. However, manual crack repair is labour-intensive and costly, especially when multiple repetitions need to be performed throughout the service life of a structure.

Self-healing concrete using superabsorbent polymers (SAPs) aims to provide a solution to these issues [1–3]. SAPs are highly absorbent materials that swell upon contact

J. I. Escalante-Garcia et al. (Eds.): RW 2021, RILEM Bookseries 40, pp. 838–845, 2023.
https://doi.org/10.1007/978-3-031-21735-7_89

with water or moisture and retain these liquids for a certain period of time. The time of retention depends on the relative humidity of the environment, which, as it decreases, drives desorption of water from the SAPs. When SAPs are present inside the cracks, they absorb water from the outer environment and thereby block the entrance through a crack (self-sealing). At a later stage, the water is released again and promotes the further hydration of unhydrated cement particles and the precipitation of calcium carbonate (self-healing).

The evaluation of the crack healing capacity is mostly performed by means of permeability tests, microscopic analysis and/or mechanical experiments. Whereas these tests have shown to be valuable on the lab scale, they are impractical for in-situ testing. A different testing procedure is thus necessary to assess the extent of crack closure inside cementitious materials, being ultrasound monitoring. Ultrasonic experiments on cementitious materials have proven to be sensitive to the elastic properties and microstructural developments of the medium [4–6]. Therefore, ultrasound can be used to follow up the closure of cracks, which facilitates the propagation of ultrasonic waves through the crack [7]. Being a non-destructive method, ultrasound allows for continuous monitoring of the process over time as well as its adoption for in-situ evaluation.

In this study, ultrasound measurements were executed for the evaluation of the crack closure over time. Pencil lead breaks, generating a broadband wave signal, were performed during curing of fractured cementitious specimens in wet-dry cycles. These healing cycles promote the autogenous healing mechanism, as water is essential for its initiation. As the emitted wave signals are sensitive to the material they travel through, crack closure or filling can be estimated by an analysis of the wave parameters after propagating across the cracks. A comparison with microscopic analysis revealed a correlation between the visual crack closure and the restoration of several wave parameters.

2 Materials and Methods

2.1 Materials

To evaluate the efficiency of superabsorbent polymers for crack healing and their effect on the ultrasonic wave parameters, two cementitious mixtures were considered, being a reference mortar and a mortar with SAPs. The cement used was CEM I 52.5 Strong from Holcim (Belgium). Small aggregate, being oven-dried river sand 0/2, was added in an amount of 2 to 1 with respect to the weight of cement. The water-to-cement ratio is 0.35 and an additional amount of superplasticizer (MasterGlenium 51 conc. 35%, BASF, Germany) was necessary to improve the workability. The amount of superplasticizer used was 0.4% by weight of cement. The SAP utilized is a copolymer of acrylamide and sodium acrylate and was provided by BASF. The dry particle size is equal to 100 $\pm$ 21.5 μm and can reach up to 257 $\pm$ 55 μm when swollen in a cementitious matrix. An amount of 0.2% SAP by weight of cement was included together with an additional amount of water of 26 g per gram of SAP, which was determined upon a flow table test. Including an additional 26 g of water per gram of SAP, the workability of the SAP and reference mortars were identical.

2.2 Self-healing Evaluation

Sample preparation and crack creation

Three mortar prisms of each mixture, measuring 30 × 30 × 360 mm, were cast with a central steel reinforcement bar of 6 mm in diameter and a length of 700 mm. The specimens were wrapped in plastic foil and cured at 20 ± 2 °C for 28 days.

After this curing period, multiple cracks were created within the mortar specimens by exerting a tensile load on the steel reinforcement. The load was increased until no new cracks occurred and an average crack width between 100 and 200 μm was observed.

To promote the closure of cracks, the fractured specimens were cured in wet-dry cycles, which consist of 23 h in dry conditions at 20 ± 2 °C and 60 ± 5% RH and 1 h submersed in water at 20 ± 2 °C, and this for a period of 14 days.

Ultrasound

To evaluate the self-healing ability, ultrasound measurements were performed by means of pencil lead break (PLB) excitation. By breaking a thin pencil lead of a mechanical pencil, a broadband wave signal is emitted and travels through the material under test. In this study, PLB tests were repeatedly executed, i.e. before cracking, right after cracking and after 3, 7 and 14 days of curing in wet-dry cycles at the end of the dry curing stage.

Two broadband pico sensors with a peak sensitivity at 450 kHz were used. These were placed on the mortar's surface with aid of a coupling agent at a constant distance of 30 mm apart (see Fig. 1). In case of the sound material, five PLBs were performed. In the cracked state, the sensors were placed at opposite sides of specific cracks, so that the generated wave travels across the crack opening.

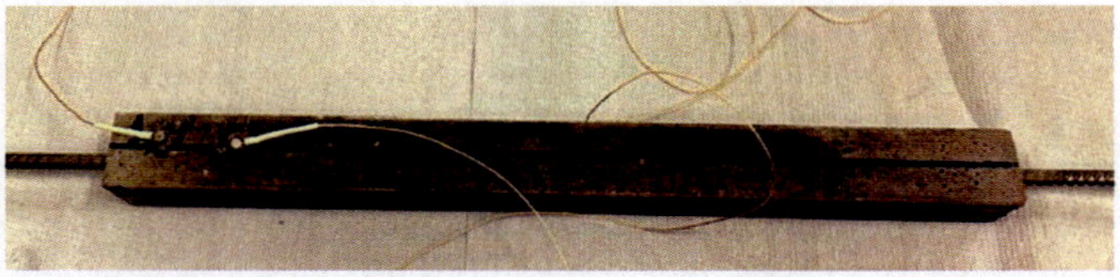

Fig. 1. Ultrasound set-up showing two pico sensors on the mortar surface.

Microscopic analysis

Crack width measurements were performed to investigate the evolution of the crack closure during wet-dry curing. An optical microscope Leica S8 APO, mounted with a DFC 295 camera was used. To measure the crack widths of the prism specimens, a central line was drawn on each of the four sides of the sample. A picture of the crack opening was then taken right underneath each line, allowing to repeat a subsequent measurement at the exact same location. Per picture, five crack width measurements were conducted, meaning that, in total, 20 measurements were available per crack. These measurements were executed right after fracture and repeated after 14 days in wet-dry cycles.

3 Results and Discussion

After 28 days of curing in plastic foil, three reference and three SAP specimens were subjected to five PLB tests for the first time. The sensors at the surface normally capture both the longitudinal and Rayleigh waves. However, in the case of cracked media, only the longitudinal wave was received, as the Rayleigh wave is not able to propagate through a crack opening in the absence of shear rigidity. The longitudinal wave velocity was calculated by dividing the distance between both sensors, being 30 mm, by the travel time of the wave between the two sensors. The travel time, indicated in Fig. 2, is the time difference between the arrival of the wave at sensor 2 and the arrival at sensor 1, taken as the first disturbance above noise level, calculated by a simple threshold crossing algorithm. Besides the wave velocity, the attenuation of the wave was calculated. The attenuation is a measure for the reduction in amplitude that occurs as the wave travels from one sensor to the other. The amplitude of the wave signal at the first signal is indicated in Fig. 2. When looking at the waveform in sensor 2, it can be seen that the highest voltage reached is much lower than the amplitude of the first sensor, showing the attenuative effect of the mortar. The attenuation (dB/mm) is calculated by Eq. (1), where A_1 (V) is the amplitude from the first sensor and A_2 (V) the amplitude from the second sensor.

$$\text{Attenuation} = -\frac{1}{30\,\text{mm}} * 20 * \log\left(\frac{A_2}{A_1}\right) \tag{1}$$

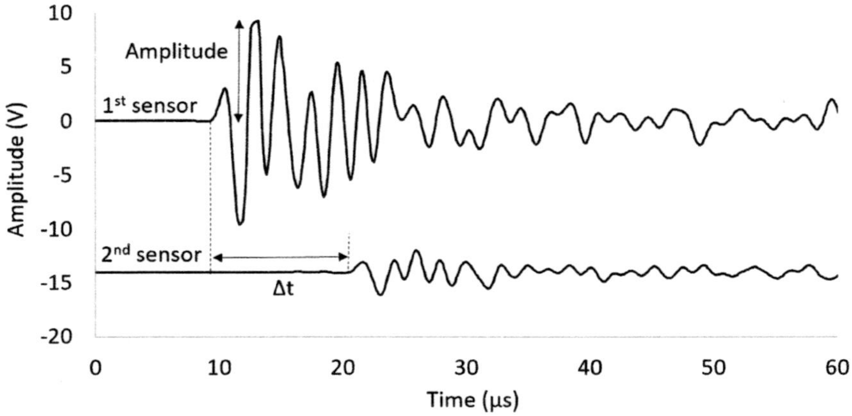

Fig. 2. Representative ultrasound results of an unfractured reference specimen. The amplitude of the wave signal and the travel time of the longitudinal wave between both sensors are indicated. The amplitude of the lower waveform is shifted downwards with respect to the original amplitude value for clarity.

After cracking of the specimens, one to two cracks per specimen were randomly chosen to be monitored by ultrasound testing. PLB repetitions were always performed right before the onset of the wet curing period to limit the effect of moisture on the measurements. Similar to the first PLB tests on sound specimens, the longitudinal velocity and attenuation were calculated for each chosen crack. Before listing the results of these parameters, the evolution of the received waveforms over time is presented in Fig. 3 for a reference specimen. As already mentioned for the uncracked specimens, the arrival to the second sensor is delayed compared to the first sensor, which is closer to the PLB source. Also, the amplitude is reduced compared to the waveform received by the first sensor. When a crack is created between the sensors, the wave arriving at the first sensor remains unchanged as it does not interact with the crack. However, the waveform received by the second sensor in the cracked state is flattened, due to the discontinuity in the wave's travel path. Afterwards, during wet-dry curing, it can be noticed that the waveforms tend to restore partially, indicating that (partial) healing of the crack takes place.

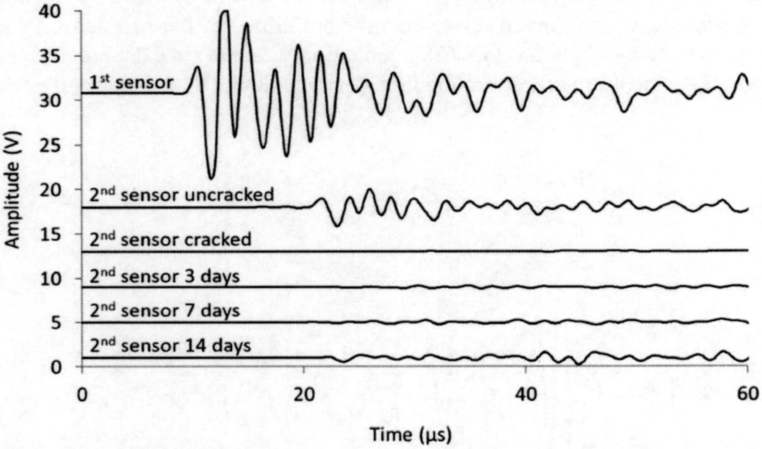

Fig. 3. Evolution of waveforms over time of a representative reference specimen. The amplitudes of the waveforms are shifted with respect to their original values.

The average longitudinal velocity and attenuation are summarized in Table 1, together with the standard deviations. The standard deviations of all experiments conducted are relatively high, which is caused by the heterogeneity of the material. Still, a qualitative comparison can be done. When looking at the results of the uncracked specimens, the longitudinal velocity of reference specimens is observed to be higher than the one of SAP mortars. This phenomenon is due to the creation of air-filled cavities inside SAP mixtures upon the release of water by the SAPs for internal curing. In agreement, the attenuation of the reference blend is lower than seen for mortar with SAPs [7]. Again, this is caused by the presence of cavities, which increase scattering of the signal and lower the wave's amplitude.

After crack creation, the longitudinal velocity decreases drastically for both reference and SAP specimens, caused by the discontinuity between the two sensors. As the curing

Table 1. Average longitudinal velocity and attenuation of reference and SAP mortar specimens over time.

		Longitudinal velocity (m/s)	Attenuation (dB/mm)
Reference	Uncracked	4647 ± 315	0.32 ± 0.10
	Cracked	2534 ± 292	0.90 ± 0.18
	3 days	2753 ± 455	0.75 ± 0.12
	7 days	3214 ± 339	0.72 ± 0.21
	14 days	3240 ± 473	0.70 ± 0.14
SAP	Uncracked	4376 ± 247	0.37 ± 0.14
	Cracked	2452 ± 235	0.99 ± 0.12
	3 days	3035 ± 412	0.90 ± 0.15
	7 days	3051 ± 289	0.84 ± 0.14
	14 days	3297 ± 332	0.74 ± 0.10

time in wet dry cycles proceeds, the longitudinal velocity increases gradually, thereby showing a partial restoration towards the initial velocity values in reference as well as SAP samples. These results suggest that a healing process took place inside the crack, which fills the gap and reduces the effect of the discontinuity. Moreover, the restoration of longitudinal velocity is higher for SAP mortars, being 845 m/s on average, than seen in the reference material (approximately 706 m/s). The SAPs thus promote the healing ability of cementitious materials, but the extent remains limited due to the small size and included amount of the SAP particles. Typically, a fivefold is necessary to observe complete healing [8].

Regarding the attenuation, upon cracking, a strong increase for all tested specimens is noticed. The difference in attenuation between reference and mortar samples can be explained partly by the presence of voids in the SAP mortar and the difference in crack size and geometry. In case of the reference mixture, as the samples are cured in wet-dry cycles, the attenuation reduces significantly within the first three days and levels out afterwards. This can be explained by the high amount of unhydrated cement particles that is present inside the cracks from the start of curing, but is consumed quickly. Concerning the SAP mortars, the restoration of the attenuation is slower at first, but continues up to 14 days of curing. In this case, the formation of healing products does not only depend on the formation of hydrates, but also on the precipitation of calcium carbonate, which is promoted by the moisture present inside the cracks. Again, when comparing the reduction attenuation between the cracked and the healed state after 14 days, it can be seen that the inclusion of SAPs lead to a higher extent of healing; the reduction is 0.20 dB/mm for the reference material compared to 0.25 dB/mm for the SAP mortars.

To visually assess whether crack closure had occurred, crack width measurements were performed right after cracking and after 14 days of curing in wet-dry cycles. Care is taken that each measurement after 14 days is performed in the exact same location as after cracking. From the obtained crack widths, the healing ratio (%) is calculated, using

Eq. 2.

$$\text{Healing ratio} = \frac{\text{Initial crack width} - \text{Final crack width}}{\text{Initial crack width}} \tag{2}$$

The healing ratio of all measured cracks plotted with respect to the original crack width can be seen in Fig. 4a, b for the reference and SAP samples, respectively. It can be observed that the initial crack widths vary significantly. This is due to the fact that the crack openings differ at different sides of the specimen, so the cracks are not exactly parallel. In case of the reference mixture, healing ratios up to 45% are observed, showing that partial filling of the cracks takes place, but with a low efficiency. The closure of cracks is caused by the continued hydration of unhydrated cement particles inside the crack. An average healing ratio of 26 ± 10% is found.

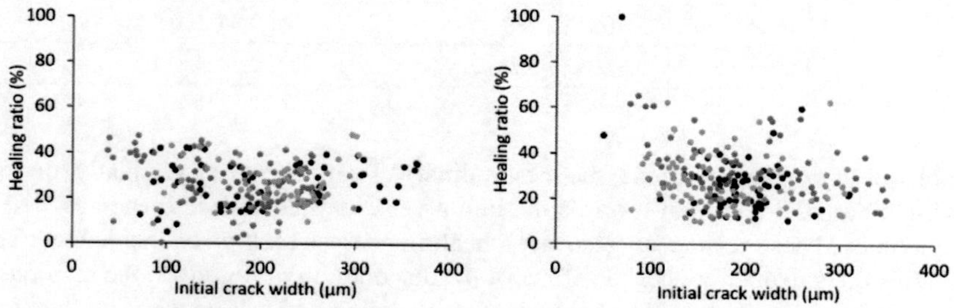

(a) Reference (b) SAP

Fig. 4. Healing ratio including all measured cracks of **a** reference mortar and **b** SAP mortar.

Similar to the reference mortar, SAP mortars show a wide range of healing ratios and initial crack widths. While there is a large overlap in values with the reference mortar, it is demonstrated that the healing ratio of SAP mixtures reaches higher values up to 65%, while the minimum healing ratio is 10%. In agreement therefore, with the moderate improvement of the ultrasound parameters, the promotion of the healing ability by SAPs can also be confirmed by the average healing ratio of 29 ± 12%, even with this low amount of SAP.

4 Conclusions

Ultrasound experiments were performed on hardened specimens in the cracked and uncracked state by means of pencil lead breaking. Before cracking, the longitudinal velocity showed a reduction when SAPs were added to the mixture, which is due to the macropore formation upon release of water by the SAPs. The existence of voids in SAP mixtures was also seen by an increase in the attenuation of the wave amplitude.

After characterizing the sound material, the specimens were cracked and ultrasound tests were performed at different moments during wet-dry curing. Immediately, after cracking, the longitudinal velocity was largely decreased compared to the sound material for reference as well as SAP mortars. However, over time, the velocity increased again and suggested that closure of the cracks took place. Similarly, the attenuation was increased after cracking, caused by the created discontinuity. As wet-dry curing cycles were performed, a partial restoration towards the original attenuation value was observed. This restoration was more pronounced for mixtures with SAPs, showing their advantage for self-healing purposes.

Using microscopic analysis, the percentage of visual crack closure was assessed by calculating the healing ratio. The obtained results confirmed the closure of cracks over time and the improved effect when SAPs were included. Therefore, it can be concluded that ultrasound monitoring is a valuable tool for studying the self-healing mechanism within cementitious materials, allowing a real time, non-destructive evaluation of the healed stages.

References

1. Snoeck, D., Van Tittelboom, K., Steuperaert, S., Dubruel, P., De Belie, N.: Self-healing cementitious materials by the combination of microfibres and superabsorbent polymers. J. Intell. Mater. Syst. Struct. **25**(1), 13–24 (2014)
2. Mignon, A., et al.: pH-responsive superabsorbent polymers: a pathway to self-healing of mortar. React. Funct. Polym. **93**, 68–76 (2015)
3. Lefever, G., Snoeck, D., Aggelis, D., De Belie, N., Van Vlierberghe, S., Van Hemelrijck, D.: Evaluation of the self-healing ability of mortar mixtures containing superabsorbent polymers and nanosilica. Materials **13**(2), 380 (2020)
4. Philippidis, T., Aggelis, D.: Experimental study of wave dispersion and attenuation in concrete. Ultrasonics **43**, 584–595 (2005)
5. Chaix, J.-F., Garnier, V., Corneloup, G.: Ultrasonic wave propagation in heterogeneous solid media: theoretical analysis and experimental validation. Ultrasonics **44**(2), 200–210 (2006)
6. Aggelis, D.G., Shiotani, T., Polyzos, D.: Characterization of surface crack depth and repair evaluation using Rayleigh waves. Cement Concr. Compos. **31**(1), 77–83 (2009)
7. Lefever, G., Snoeck, D., De Belie, N., Van Vlierberghe, S., Van Hemelrijck, D., Aggelis, D.: The contribution of elastic wave NDT to the characterization of modern cementitious media. Sensors **20**(10), 2959 (2020)
8. Snoeck, D., De Belie, N.: Autogenous healing in strain-hardening cementitious materials with and without superabsorbent polymers: an 8-year study. Front. Mater. **6**(48), 1–12 (2019)

Resistivity of Repair Materials for Concrete Repair Prior to the Application of a Cathodic Protection System

Bjorn Van Belleghem(✉), Mathias Maes, and Tim Soetens

SANACON bv, Ghent University spin-off, Nijverheidsweg 1/A, 9820 Merelbeke, Belgium
bjorn.vanbelleghem@sanacon.be

Abstract. Suitability of repair materials for local concrete repair in combination with a cathodic protection system depends on the compatibility of their electrical resistivity characteristics with the parent concrete. As data on resistivity of repair mortars is very limited, selection of a compatible product is not always possible. In this research, the electrical resistivity of 10 different structural repair mortars was tested at three different environmental conditions (temperature of 20 °C and relative humidity of 60, 80, and 100%) by means of three resistivity test methods: four-point electrode method (Wenner Probe), two-plate electrode method and method with embedded electrodes. In this way, a database is created through which the most suitable repair mortar can be selected for each concrete structure in different environments. The relative humidity (RH) of the environment has a large influence on the resistivity of the repair products: on average, at the age of 91 days, a decrease of the RH of 20% tripled the resistivity of the repair mortars. Furthermore, based on the obtained results it is expected that this effect will be even larger on the long term (e.g. after several years). The two bulk resistivity methods showed comparable results. However, the two-plate electrode method was found to be unsuitable for resistivity measurements of non-saturated mortar specimens. The surface resistivity, measured by the Wenner Probe, was found to be a factor of 4 to 5 times higher compared to the bulk resistivity. The difference can mainly be attributed to the small size of the specimens used in this research.

Keywords: Resistivity · Repair materials · Concrete repair · Cathodic protection

1 Introduction

Cathodic protection (CP) of steel in concrete, either galvanic (GCP) or through impressed current (ICCP), is a widely accepted repair technique to significantly reduce or completely stop corrosion of steel in reinforced concrete structures [1, 2]. Especially when corrosion is induced by chlorides, CP provides a durable and cost efficient repair on the long term [3–5].

Cathodic protection is mostly applied on concrete structures which already have a certain amount of visible damage due to reinforcement corrosion, mostly in the form of cracks, delamination and spalling of concrete. Therefore, before the application of a

J. I. Escalante-Garcia et al. (Eds.): RW 2021, RILEM Bookseries 40, pp. 846–855, 2023.
https://doi.org/10.1007/978-3-031-21735-7_90

CP system, local repair of delaminated concrete is required. The compatibility of repair mortars for local concrete repair with the CP system is a major concern to obtain efficient protection of the steel reinforcement over the whole structure.

The European standard on cathodic protection of steel in concrete, EN ISO 12696:2016 [6], provides certain guidelines concerning the suitability of repair materials in the sense that electrical resistivity characteristics of the repair material should be compatible with the parent concrete. Typically, this is defined as repair materials which have an electrical resistivity within the range of 0.5 to 2 times the mean resistivity of the parent concrete, measured under the same environmental conditions.

The comparison of the electrical resistivity of repair materials and parent concrete of an existing structure is difficult, since the resistivity of the parent concrete is mostly measured by means of an on-site test method, while the resistivity of repair materials is measured in laboratory conditions, mostly with a different test method. Moreover, the environment at which the resistivity is tested is different for the parent concrete and the repair materials.

In this research, the resistivity of 10 different repair mortars was tested at three different environmental conditions (temperature of 20 °C and relative humidity of 60, 80, and 100%) by means of three generally accepted test methods.

This paper discusses the influence of the used test method as well as the relative humidity of the environment on the resistivity of repair mortars.

2 Materials and Methods

2.1 Repair Mortars

Ten different structural repair mortars, commonly applied in Europe for concrete repair, were used in this study. All repair mortars comply with the requirements for a class R4 (structural repair) according to the European standard EN 1504-3 [7] and are suitable to be applied either by hand or by a spraying method.

The repair products are numbered from 1 to 10 in this paper. Repair mortars 1 to 7 are based on a hydraulic binder (denoted as "CC"), while repair mortars 8, 9 and 10 are hydraulic mortars modified by means of polymer additives (denoted as "PCC").

The main binder type of all repair mortars, except for mortar CC-4, is Portland cement (mostly sulfate resistant). Repair mortar CC-4 on the other hand contains cement replacement (not further specified) and recycled aggregates.

Repair products CC-1 to CC-7 and PCC-10 also contain polymeric fibers. When the type of fiber is specified in the technical data sheet of the product, it is always polyacrylonitrile.

2.2 Mortar Specimens

Two types of mortar specimens were prepared for every repair product to determine the resistivity:

- Mortar prisms with dimensions 200 × 60 × 60 mm

– Mortar prisms with dimensions 100 × 100 × 50 mm containing two embedded stainless steel bars (diameter 6 mm), positioned 50 mm apart

The mortar mix of each repair product was prepared by mixing a predefined amount of water, described in the technical datasheet of the respective product, with the mortar powder by means of a portable handheld paddle mixer. The mix procedure and mixing time as described in the technical datasheet for each product was followed. After mixing, the mortar was brought into the molds and compacted shortly on a vibration table. All mortar specimens were produced in triplicate.

Immediately after casting, the mortar specimens were put in an air-conditioned room at a temperature of 20 ± 2 °C and a relative humidity of at least 95% for 24 h. Subsequently, the specimens were demolded and placed under constant environmental conditions for resistivity testing (see Sect. 90.4).

2.3 Resistivity Measurement Methods

The resistivity of the repair mortar products was measured by three different methods, described in the recommendations of RILEM TC 154-EMC [8] and CUR-Aanbeveling 45 (in Dutch) [9].

A. **Four-point electrode method (FPEM) - Wenner Probe (surface resistivity)**

The first method to measure resistivity was performed by means of the so-called 'Wenner Probe' on the mortar prisms with dimensions 200 × 60 × 60 mm. The Wenner Probe consists of four equally spaced point electrodes which are pressed onto the mortar surface. A current is induced between the two outer electrodes and a potential difference is measured between the two inner electrodes. The ratio between the voltage and the current gives a resistance value, which can be converted to resistivity using a cell constant based on the electrode spacing [10].

The Wenner Probe used in this research is a standard device developed by Proceq called the Resipod meter with an electrode spacing of 50 mm. For each mortar specimen, four resistivity measurements with the Resipod were performed, one measurement on each side (200 × 60 mm) of the specimen (see Fig. 1a).

B. **Two-plate electrode method (TPEM) - bulk resistivity**

The second resistivity measurement method was also performed on the mortar prisms with dimensions 200 × 60 × 60 mm, by placing the prisms along their length direction between two stainless steel plates (Fig. 1b). Wet cloths are placed at the contact surfaces of the plates with the prism and a weight with a mass of about 2 kg was placed on top of the upper plate to ensure uniform electrical contact.

The resistance between the two plates was measured by means of an LCR meter at a frequency of 120 Hz. The resistivity was subsequently calculated from the resistance measurement by Eq. (1):

$$\rho = R \cdot A / L \tag{1}$$

where R is the measured resistance (Ω), A is the area of the prism in contact with the stainless steel plates (60 × 60 = 3600 mm^2) and L is the length of the prism (200 mm).

C. **Embedded electrodes method (EEM) - bulk resistivity**

The resistivity measurement method with embedded electrodes is described in the Dutch recommendation document regarding cathodic protection of steel in concrete structures, CUR-Aanbeveling 45 [9]. It is performed by measuring the resistance between two embedded stainless steel bars of the mortar prisms with dimensions 100 × 100 × 50 mm by means of an LCR meter (measurement frequency of 120 Hz), see Fig. 1c.

The resistivity is calculated by multiplying the measured resistance by a cell constant C which depends on the geometry of the mortar prism and embedded electrodes. The cell constant for the specimen geometry used in this research was determined experimentally by filling a mold with the same geometry as the mortar prisms with a 0.01 M KCl solution and measuring the resistance between the electrodes in the same way as for the mortar prisms, i.e. by means of an LCR meter. The conductivity of the 0.01 M KCl solution was measured to calculate its resistivity. Finally, the cell constant was calculated by dividing the resistivity of the solution by the measured resistance.

(a) (b) (c)

Fig. 1. Resistivity measurement methods: **a** four-point electrode method (FPEM), **b** two-plate electrode method (TPEM), and **c** embedded electrodes method (EEM).

2.4 Environmental Test Conditions and Measurement Times

The resistivity of the repair mortars was tested in three different environments at a constant temperature of 20 °C and a relative humidity (RH) of 60, 80, and 100% (immersed in water).

For each test environment, three replicate specimens of each type (see Sect. 90.2) were made for all repair mortars. The resistivity measurements with the three aforementioned methods were performed at 7 different ages of the specimens: at the age of 3, 7, 14, 21, 28, 56, and 91 days.

3 Results and Discussion

3.1 Resistivity in Function of Time

The resistivity of the repair mortars increased with time for all test environments. One example of the evolution of the resistivity in function of time for all tested repair mortars is given in Fig. 2 where the measurements were performed by the embedded electrodes method (EEM) at test conditions of 20 °C and 80% RH. Similar graphs can be obtained for other test conditions and for measurement with other test methods.

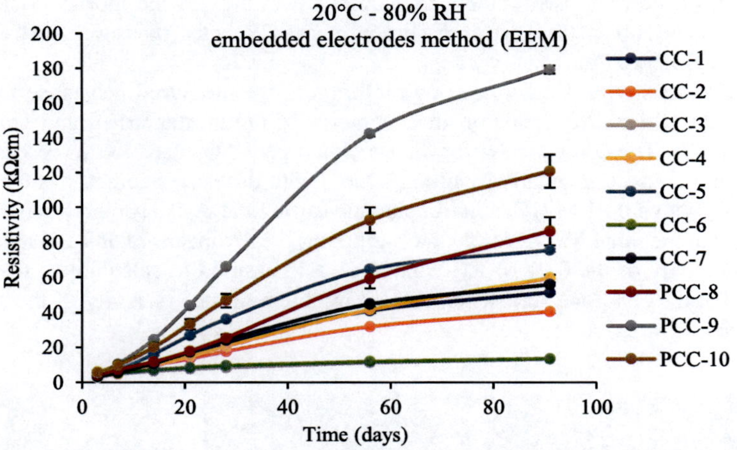

Fig. 2. Resistivity in function of time for all repair products at 20 °C and 80% RH, measured with the method with embedded electrodes (EEM).

In general, there is a large variation of resistivity between the different tested repair mortars: the resistivity of the repair mortar PCC-9, which generally has the highest resistivity in all test conditions, is up to 15 times higher than the repair mortar with the lowest resistivity (mortar CC-6).

Based on the measurements in all test conditions, there are two repair mortars (CC-3 and CC-6) which have a relatively low resistivity compared to the other products, while there is one repair mortar (PCC-9) which has a resistivity that is much higher than all other tested products.

Generally, as can be noticed in Fig. 2, the tested PCC mortars tend to have a higher resistivity compared to the CC mortars. However, this difference is certainly not always significant, for example for mortar PCC-8.

3.2 Influence of Relative Humidity

The relative humidity of the environment has a big influence on the resistivity of the repair products. Figure 3 shows an example of the development of the resistivity in time for repair mortar #1 for storage at different relative humidities.

For storage at 100% RH, the resistivity of the repair products tends to reach a stable value at the end of the testing period, i.e. the resistivity does not increase much anymore between 56 and 91 days for most products. For lower relative humidities, especially 60%, the resistivity keeps increasing significantly after 56 days.

Consequently, the difference in resistivity between the different environments increases in time. Considering all products, the average resistivity increase factor from

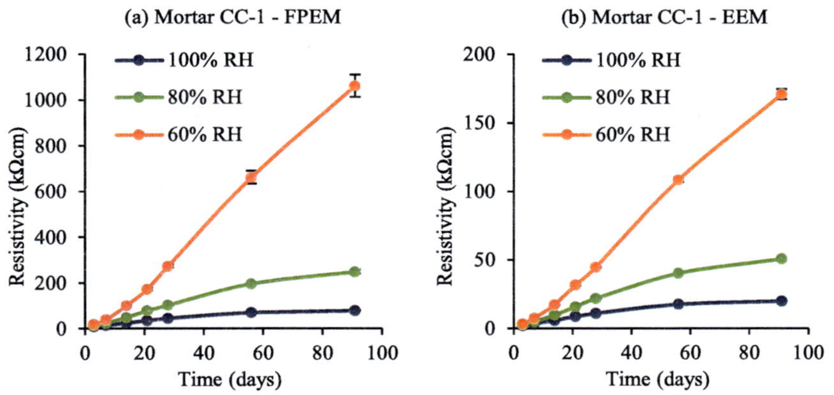

Fig. 3. Influence of the relative humidity (RH) on the development of resistivity in time, measured by **a** four-point electrode method (FPEM) and **b** embedded electrodes method (EEM).

100 to 80% RH and from 80 to 60% RH is very similar. At the age of 91 days, this average factor ranges from 2.6 to 3.3 for all testing methods. Thus, for a decrease of the relative humidity of the environment of 20%, the resistivity of a repair mortar is about 3 times higher at 91 days age. Note that as the difference increases with time, it is expected that this factor will be even higher at larger testing ages.

Figure 4 summarizes the measured resistivity of all repair products at the age of 91 days for the three test conditions (i.e. variation of environmental RH from 100 to 60%) and the three different test methods. The large influence of the environmental relative humidity on the resistivity is also clearly visible in this graph for all tested repair mortars.

3.3 Influence of Measurement Method

Next to the influence of RH, Fig. 4 also shows that there is a difference in the measured resistivity between the different test methods. The bulk resistivity measurement methods, i.e. the two-plate electrode method (TPEM) and the method with embedded electrodes (EEM), mostly show similar results while the resistivity measurements with the four-point electrode method (FPEM) are consistently much higher.

In Fig. 5, the mean ratio of the surface resistivity ($\rho_{surface}$), measured with the FPEM, and the bulk resistivity (ρ_{bulk}), measured either by the TPEM or EEM, is plotted for the different test environments. As can be seen in the figure, a mean factor between 4 and 5 is obtained for the ratio between surface and bulk resistivity based on all executed measurements. The testing age did not have an influence on this factor.

Despite of the large difference in absolute value of the resistivity between surface and bulk measurement methods, a good correlation between both methods is obtained, see Fig. 6.

The main reason for the higher measured surface resistivity by the FPEM can be linked to the relatively small size of the test specimens used in this research. As has been

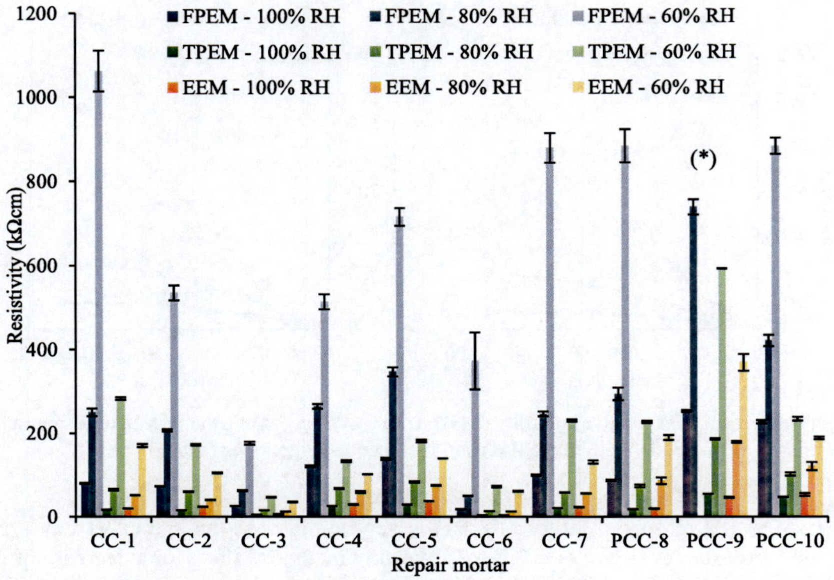

Fig. 4. Resistivity of all repair products at 91 days age. (*) The resistivity of repair mortar #9 at 60% RH could not be measured by the FPEM as it was higher than the measurement range (>2000 kΩ cm) of the test device.

established in literature [10–12], the measurement of surface resistivity by means of the FPEM (Wenner Probe) on small scale laboratory specimens will give an overestimation of the true resistivity of the material due to the fact that a semi-infinite medium is taken into account in the (automatic) calculation of the resistivity. Therefore, the measurements by the FPEM method on the specimens in this research should be considered as an apparent resistivity, not the true resistivity of the material.

Lastly, it should be noted that at a high relative humidity (80 or 100%) the results of the resistivity measurement by both bulk resistivity methods (TPEM and EEM) show very good agreement. At a relative humidity of 60% however, the agreement between both methods is worse, in the sense that the TPEM always leads to higher resistivity values (see Fig. 4). This can be attributed to the fact that the resistivity measurements by means of the TPEM were unstable at a lower RH. The measured resistance values decreased over time during execution of the measurement. An example of this is given in Fig. 7 for the measurement at 60% RH at an age of 56 days.

The effect of decreasing resistance during the measurement was more pronounced for increased age of testing. The measurement was stopped after about 5 to 10 min until a more or less stable measurement value was obtained. However, probably the measured resistance would have slowly decreased further if the measurement was continued. This effect was also noticed for the specimens stored at 80% RH, but less pronounced.

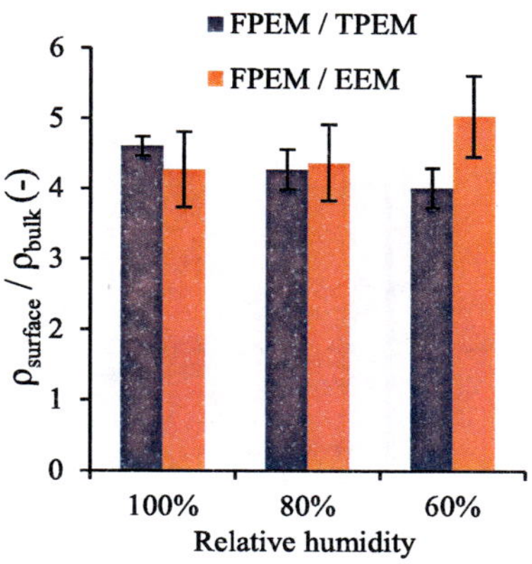

Fig. 5. Ratio of surface and bulk resistivity measurements.

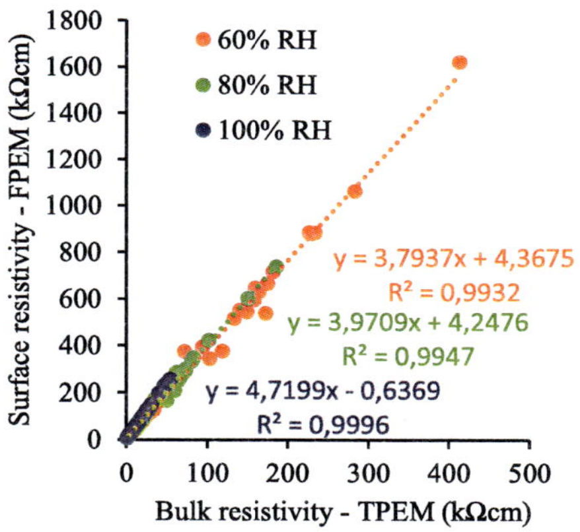

Fig. 6. Correlation between surface and bulk resistivity measurements.

Therefore, the TPEM was found to be unsuitable for measurement of resistivity of mortar/concrete specimens which are stored at low RH or, more generally, unsuitable for unsaturated mortar/concrete specimens due to the long required measurement time.

A possible way of avoiding this would be to attach electrodes by means of a conductive adhesive layer to each sample separately, instead of using the two plates with a wet cloth at the contact surface with the specimens.

4 Conclusions

Ten different structural repair mortars, commonly used for concrete repair in Europe, were tested in this research on their electrical resistivity in order to evaluate their possible use for concrete repair in combination with the application of a cathodic protection system.

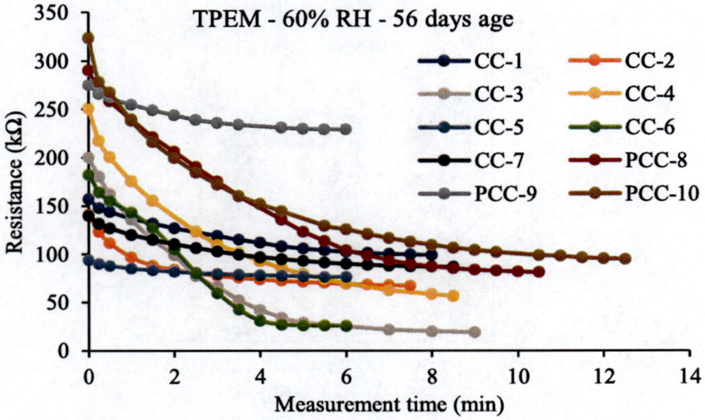

Fig. 7. Example of decreasing resistance during measurement with the two-plate electrode method (TPEM).

Repair mortar specimens were prepared and stored at three different test environments at 20 °C and varying relative humidity (RH) of 60, 80, and 100% for a period of 91 days. The resistivity of the mortar specimens was measured at regular time intervals by means of three commonly used test methods for both bulk and surface resistivity.

Two out of the ten tested repair mortars have a very low resistivity compared to the other products, while there was one product (PCC repair mortar) which has a resistivity that is much higher than all other tested products.

A clear dependency of the resistivity on the relative humidity of the environment was noticed, in the sense that the resistivity is lower for at a higher RH. At 100% RH, the resistivity nearly reached a stable value after a testing period of 91 days for most products, while at lower relative humidities the resistivity kept increasing. Especially at a RH of 60% there was still a large increase in resistivity from 56 to 91 days. In order to reach a stable resistivity at this environment, a much longer test period would be required.

Comparable results were obtained with both bulk resistivity test methods in most cases. Only in the case of a relatively low RH of the environment (unsaturated specimens), the two-plate electrode method was found to unsuitable due to unstable measurement readings.

By means of the surface resistivity method (four-point electrode method or Wenner Probe), the obtained resistivity was on average 4 to 5 times higher compared to the bulk resistivity methods. The higher measured resistivity can most probably be linked to a geometrical factor or correction factor which should be applied to account for the effect of the small sample size of the specimens used in this research. However, this shows that it is very important to mention the measurement method and specimen size, next to the age and storage conditions, when reporting resistivity measurements.

Based on the measurements performed in this research, a database is created of the resistivity of commonly used repair mortars at different environmental conditions which makes it possible to choose the most suitable repair mortar for local concrete reinstatement prior to the application of a cathodic protection system.

In further research, the influence of temperature on the resistivity of repair mortars will be investigated and modelling of the resistivity in function of time, relative humidity and temperature will be performed in order to be able to predict the resistivity of other repair products under specific environmental conditions.

References

1. Chess, P., Broomfield, J.P.: Cathodic Protection of Steel in Concrete, 1st edn. Taylor & Francis, London (1998)
2. Araujo, A., Panossian, Z., Lourenço, Z.: Cathodic protection for concrete structures. Ibracon Struct. Mater. J. **6**(2), 178–193 (2013)
3. Christodoulou, C., Glass, G.K., Webb, J., Austin, S., Goodier, C.: Assessing the long term benefits of impressed current cathodic protection. Corros. Sci. **52**(8), 2671–2679 (2010)
4. Polder, R.B., Leegwater, G., Worm, D., Courage, W.: Service life and life cycle cost modelling of cathodic protection systems for concrete structures. Cem. Concr. Compos. **47**, 69–74 (2014)
5. Polder, R.B., Peelen, W.: Cathodic protection of reinforcement in concrete – experience and development over 30 years. In: Concrete Solutions 2019 – 7th International Conference on Concrete Repair, MATEC Web Conf. **289**, 03006 (2019)
6. EN ISO 12696. Cathodic protection of steel in concrete (2016)
7. EN 1504-3. Products and systems for the protection and repair of concrete structures - definitions, requirements, quality control and evaluation of conformity - part 3: structural and non-structural repair (2006)
8. Polder, R.B., et al.: Test methods for on site measurement of resistivity of concrete. Mater. Struct. **33**, 603–611 (2000)
9. CUR-Aanbeveling 45. Kathodische bescherming van wapening in betonconstructies (1996)
10. Layssi, H., Ghods, P., Alizadeh, A.R., Salehi, M.: Electrical resistivity of concrete. Concr. Int. **37**(5), 41–46 (2015)
11. Morris, W., Moreno, E.I., Sagüés, A.A.: Practical evaluation of resistivity of concrete in test cylinders using a Wenner array probe. Cem. Concr. Res. **26**(12), 1779–1787 (1996)
12. Azarsa, P., Gupta, R.: Electrical resistivity of concrete for durability evaluation: a review. Adv. Mater. Sci. Eng. **2017**, 8453095 (2017)

Non-destructive Tests for Estimating the Tensile Strength in Concrete with Deep Learning

José A. Guzmán-Torres(✉), Carlos A. Júnez-Ferreyra, Ramiro Silva-Orozco, and Wilfrido Martínez-Molina

Civil Engineering Faculty, UMSNH, Morelia, Michoacán, México
jaguzman@umich.mx

Abstract. Currently, non-destructive tests (NDTs) are one of the most well-known procedures considered for monitoring concrete behavior because it is possible to short-cut laboratory experimentation time. There are many methods for estimating the tensile strength values with acceptable accuracy, although it is necessary to increase the accuracy for this parameter. This research proposes to use an artificial intelligence method to achieve this aim. Artificial Intelligence has been one of the most efficient approaches for solving engineering and material problems because of its effective performance and strength to achieve higher accuracy. This paper proposes a novel approach, using a Deep Learning model for predicting the tensile strength in concrete based on just standard non-destructive test measurements. The considered model for this research is a Deep Neural Network applied in a particular concrete mixture. This model allows predicting the tensile strength with acceptable accuracy and low computational cost. Also, it is an adequate alternative to estimate the resistance of tensile strength in several structures just by taking standard measurements like Electrical resistivity, Resonance Frequencies, and Ultrasonic Pulse Velocity. For this study, 250 concrete samples were analyzed. The model employs a Rectified Linear Unit function in their architecture as an activation function in all hidden layers, and the model contains 450 neurons. This model considers the NDTs measurements as input data, and the output value is the tensile strength. The model shows an excellent accuracy of over 96%, which is impressive. This approach showed satisfactory performance in tensile strength prediction for the concrete mixture analyzed.

Keywords: Concrete · Deep learning · Non-destructive tests · Tensile strength

1 Introduction

During the last years, there have been a growing number of methods for estimating different concrete properties such as compressive strength, flexural strength, and non-destructive parameters. Artificial intelligence, or *AI* for short, is a specialized approach that has come to give a hand to material's analysis field. For instance, in [1], the concrete performance was analyzed using the gradient descent algorithm to estimating compressive strength, *Cs* for short, using non-destructive parameters reaching an accuracy over 75%. Another study involves the tensile strength prediction supporting the electrical

J. I. Escalante-Garcia et al. (Eds.): RW 2021, RILEM Bookseries 40, pp. 856–866, 2023.
https://doi.org/10.1007/978-3-031-21735-7_91

resistivity (*Er*) as an input parameter using a regression tree algorithm, accomplishing a modest accuracy of about 83% [2]. These studies show the strong relationship between the non-destructive tests (NDTs) and the destructive tests. This assumption is proved and detailed in [3], where the analysis of the *Cs* and the *Er* through support vector regression algorithm was made, obtaining an average accuracy of 86%. Despite the moderate accuracy of the mentioned Machine Learning (*ML*) algorithms, these have been applied in several applications of the construction field. However, there is a specialized field that can enhance the accuracy obtained for estimating concrete features. This field is known as Deep Learning, and this approach has been used for detecting damage in high structures, as is detailed in [4], estimating the flexural strength as seen in [5], and classifying cracks on masonry surfaces as is described in [6].

This paper presents the outcomes about concrete tensile strength, *Ts* for short, and NDTs such as ultrasonic pulse velocity (UPV), *Er*, and resonance frequencies (*Rf*). In order to estimate the *Ts* parameter, a Deep Neural Network (DNN) model was developed, using the NDTs as input parameters. In theory, the NDTs provide material properties information such as homogeneity, density and porosity [7]. Further, the use of NDTs is a crucial consideration in the development of infrastructure renewal [8]. For this study, the NDTs were validated and calibrated using actual *Ts* results. Also, *Ts* tests were performed to validate the results shown in this study. The approach proposed in this work provides an advantage for predicting the *Ts* value since only NDTs values are needed. Therefore, the *Ts* value is obtained efficiently with high accuracy compared to other *ML* algorithms.

The concrete mixture used for this work was analyzed and characterized previously in a laboratory always under controlled and supervised conditions. Before the modelling process, an examination between the predictor variables was done in order to establish the correlation value among the tests.

The DNN model for this paper is composed of many layers, neurons and several connections between them. The DNN architecture contains an input layer with three neurons, 150 neurons in each hidden layer, and an output layer. Of note is that the activation function selection has a significant effect on the training process and in model performance. For the DNN presented in this research, a Rectified Linear Unit (ReLu) activation function was selected by a deep and exhaustive inspection. Further, the ReLu activation function is one of the most thriving and widely accepted in DNN regression problems and is recommended for preventing gradient vanishing problems [9].

2 Research Significance

This work proposes a DNN model with a specific architecture capable of estimating the tensile strength of concrete using just standard Non-destructive tests as input parameters. The developed model for this study is considered robust and is able to get satisfactory accuracy in the *Ts* forecasting task. Codes, data, and network corresponding to this research are available at JaGuzmanT/Split-Tensile-Strength-prediction-using-non-destructive-tests (github.com).

3 Development of Database

The dataset used for training the DNN model presented in this paper was prepared through a review of examined laboratory results. The dataset comprised NDTs and *Ts* test and their respective behavior through different tests stages. The dataset includes 285 concrete cylinders, with dimensions of 10 cm diameter by 20 cm height, of which all of them were analyzed with non-destructive equipment before the *Ts* test. The dataset has four different predictive variables, the NDTs (*Er*, *Rf*, UPV) and the variable *Ts*, defined as the target value. The dataset was mixed in order to add randomness and generate a robust model. This task was handled on the domain expertise of the authors. Also, a preliminary correlation matrix was developed to determine the correlation between the features analyzed. The Pearson's correlation parameter was chosen for this task. This coefficient estimates the linear relationship among the properties. Pearson's correlation assumes that the dataset has a normal distribution. The correlation value fluctuates between -1 and $+1$, indicating 0 as no correlation. The correlation of -1 or $+1$ indicates an exact linear relationship.

The mixture analyzed in this work was characterized in a fresh state in order to obtain a concrete slump. Further, the mixture design was elaborated following the American Concrete Institute methodology. It is a standard American method that considers the physical properties of materials. Table 1 shows the mixture properties for this study.

Table 1. Mixture properties.

Material	Type amount (g/cm^3)	
Cement		CEM II 40R CPC
Fine aggregate	Shape	Crushed volcanic
	Percent mesh 8 mm Percent mesh 16 mm Percent mesh 30 mm Percent mesh 50 mm Percent mesh 100 mm Tray	12.350 28.720 31.030 12.660 7.280 7.95
Coarse aggregate	Shape	Crushed limestone
	Maximum size	19.05 mm
	Amount (kg/m^3)	1435
Water	Amount (kg/m^3)	205
Air	Percent included air	2
Concrete w/c rate	F′c (kgf/cm^2) 0.35	350
Concrete slump	8 cm	

4 Deep Neural Network Development

The Artificial Neural Network, ANN for short, simulates the biological neuron function to imitate the human brain work. ANN is based on a set of connected units called artificial neurons [10]. In this scheme, each neuron transmits a signal to another neuron by a connection. In each union is assigned a weight, which is able to modify the strength of the signal sent [11]. The ANN is considered a supervised learning problem and can solve highly nonlinear problems with relative efficiency through a mapping function developed using multivariable regression and pattern recognition methods. An ANN comprises different crucial elements such as an input layer, hidden layers, and an output layer. The input layer receives all the predictor variables that are going to be analyzed by the model. The hidden layers contain a complex structure of several neurons that store weights in each one and are interconnected across the layers through activation functions such as, ReLu, TanH, Sigmoid, etc. The activation function determines if a particular neuron in some layer must be activated. The output layer gives the predicted value by the model. Of note is that the use of more than one hidden layer often improves the ability of the ANN to identify complex and sophisticated relations between the predictor variables and the target, improving the prediction accuracy in the model. Therefore, ANN with more than one hidden layer is known as Deep Neural Network, DNN for short [12].

4.1 Analysis Methodology

Firstly, the dataset used in this research was randomly discretized into the training and testing dataset. The distribution was 70% and 30%, respectively. Also, a new dataset as a validation set was considered to provide certainty to the model. For this study, it is assumed that the uncertainty of data is ignored. It is known that this kind of assumption is commonly accepted in the literature [13]. The DNN structure, such as number of hidden layers, number of neurons in each layer, activation functions and the learning rate, is assumed for the initial model. All of these parameters were changing during the fine-tuning hyperparameters task.

Each neuron receives the sum of the products made between the weights and the input data. Thus, it is possible to generalize the equation of a neural network as:

$$\sum_{i=0}^{n} f\left(w_{ij}a_i + b_j\right) \tag{1}$$

and, for this case, the Eq. (1) can be rewritten as:

$$k_n = \sum_{i=0}^{2} f\left(w_{ij}a_i + b_j\right) \tag{2}$$

where k_n is the outcome of the model, w_{ij} is the weight matrix, a_i corresponds to the input vector, b_j is the bias vector and f represents the activation function. Specifically, i, j, denotes the neuron number and a denotes the input variables.

The DNN training begins with certain assumed weights and biases for the inputs provided to the neurons of the input layers, which are then used to generate initial outcomes. Then, the outcomes are compared with the actual outcome of the training data, where the aim is to minimize the difference between them, i.e., reduce the residual.

Before the modelling process, the dataset was standardized. This process is a type of normalization. Normalization is a technique often applied as part of data preparation for ML. Normalizations aim to change the values of numeric columns in the dataset to a standard scale without distorting differences in the ranges of values. The model development was performed in Phyton's environment using frameworks such as TensorFlow [14] and Scikit-learn [15]. In the beginning, the values contained in the dataset were standardized to minimize both errors and computation time. The DNN comprises three hidden layers with 150 neurons in each one. ReLu activation function breaks the linearity on all hidden layers, and the final hidden layer employs a Linear activation function, and this function allows to linearize with higher accuracy the answers by the predictions. The architecture of the DNN is presented in Table 2, and the plot of the multilayer perceptron model Graph is shown in Fig. 1. Once the network has been defined, it is possible to compile it. The compilation is a crucial step for efficiency. It transforms the simple sequence of layers previously defined into a highly efficient series of matrix transforms in a format intended to be executed. The model was trained for a modest 350 training epochs with a default batch size equal to one example. The model employs an Adam optimizer with a Learning rate equal to 0.01 and a Hessian uniform (he_uniform) kernel initializer. The Adam name is taken from adaptive moment estimation, and Adam optimizer is an optimization algorithm that can be used instead of the classical stochastic gradient descent procedure to update network weights iterative based on training data [16].

Table 2. DNN description.

Layer	Output shape	Parameters #	Learning rate	Activation function	Cost function
Dense	(None, 150)	600	0.01	ReLu	MAE
Dense_1	(None, 150)	22,650	0.01	ReLu	MAE
Dense_2	(None, 150)	22,650	0.01	ReLu	MAE
Dense_3	(None, 1)	151	0.01	Linear	MAE

4.2 Evaluation Method

The model accuracy was measured making a comparison between the actual value y against the predicted ($y_{predicted}$). Prior to the modeling process, the correlation coefficient (p), known as Pearson's coefficient, was estimated in order to find the relationship between the features involved in the dataset. This coefficient estimates the linear relationship between two variables. This correlation fluctuates within -1 and $+1$, with 0 implying no correlation. Correlations of -1 or $+1$ imply an exact linear relation [17]. The metric used for evaluating the performance of the DNN was the Mean Absolute Error (MAE) and is given by Eq. (3). For estimating the model accuracy, the coefficient

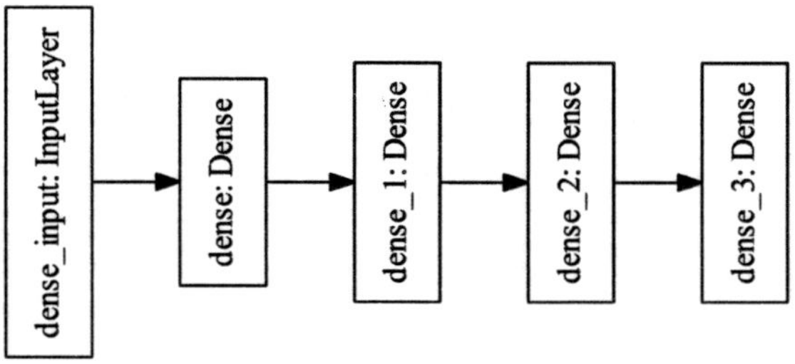

Fig. 1. Plot of the multilayer perceptron model graph.

of determination R^2 was implemented. This coefficient represents how well the values fit compared to the original values and is given by the Eq. (4)

$$MAE = \frac{1}{n}\sum_{i=1}^{n}\left|y_i - y_{predicted}\right| \tag{3}$$

$$R^2 = 1 - \frac{\sum \left(y_i - y_{predicted}\right)^2}{\sum (y_i - \bar{y})^2} \tag{4}$$

5 Performance Evaluation and Discussions

Initially, the input parameters involved in the model were handled because of the expertise of the authors. The histograms of these parameters are shown in Fig. 2, and their correlation is shown in Fig. 3. The histograms aid to show the frequency of each value in each test, i.e., the main goal in this kind of diagrams is to approximate the underlying probability density function that generated the data by binning and counting observations. For data validation, it is necessary to inspect the data. It is shown in Fig. 4, which shows in the main diagonal the kernel density estimation (KDE), which is a method for visualizing the distribution of observations in the dataset, analogous to the histogram.

For this case, KDE distribution represents the data using a continuous probability density curve in two dimensions.

It is essential to take into consideration the correlation values between predictor variables and the target. According to Fig. 3, *Er* and *Rf* have a high correlation of over 90% with the *Cs*, while the UPV has a low correlation value equal to 64% compared with the *Cs*. It is well known that correlation values above 77% suggest a linear trend and a high correlation.

During the modeling process, one of the principal challenges is to find the optimal values for the hyperparameters involved in the DNN architecture. The hyperparameters values were founded by an exhaustive inspection to provide stability and robustness to the model. All the hyperparameters are crucial, but the learning rate is fundamental to providing a correct learning process to the DNN. For the present DNN, a conservative configuration for the Adam optimizer with a learning rate of 0.01was used. The MSE loss function was optimized in order to monitor the accuracy metric, which is appropriate

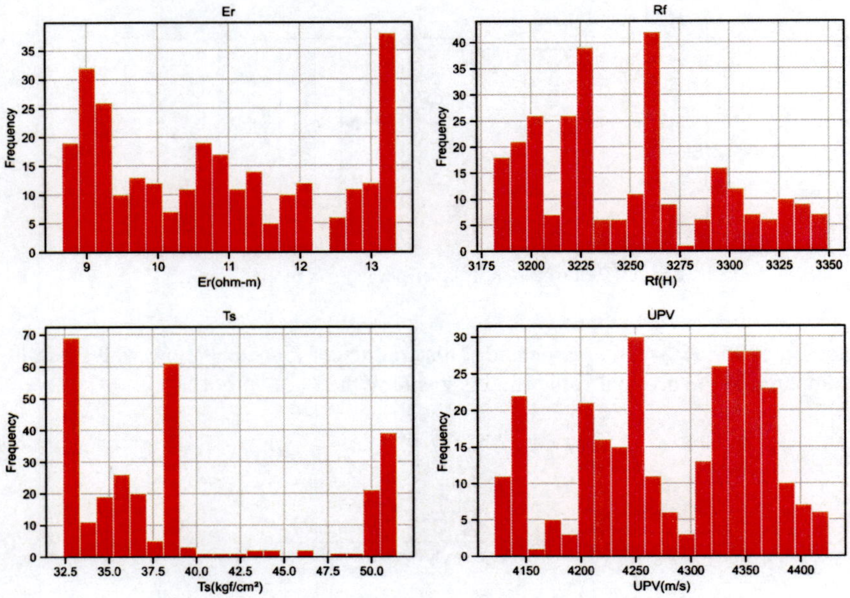

Fig. 2. Histograms of the input parameters in the dataset.

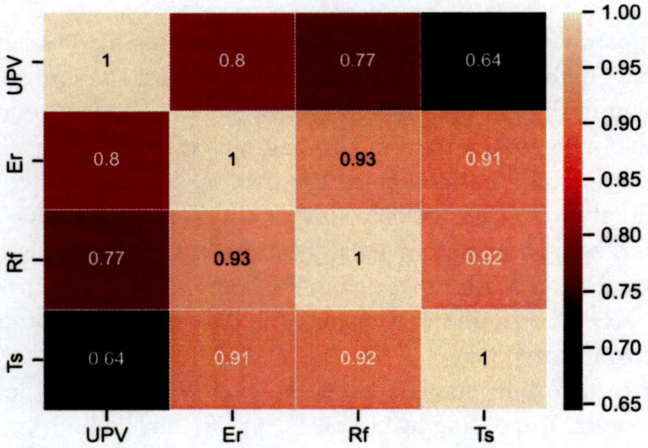

Fig. 3. Correlation matrix corresponding to the input parameters of the dataset.

given the predictor variables. After the model was defined, it was trained through a generic 350 training epochs. Therefore, the epoch value of 350 was taken as an optimal epoch value for the final DNN model. A batch size equal to 1 was chosen for the present DNN. It means that each value of the dataset was analyzed for avoiding a stochastic analysis. This task is recommended when the amount of data is not high (lower than 1000 data). In this case, it is possible to see in Fig. 5 that the model generally achieves a good fit, with train and test learning curves converging.

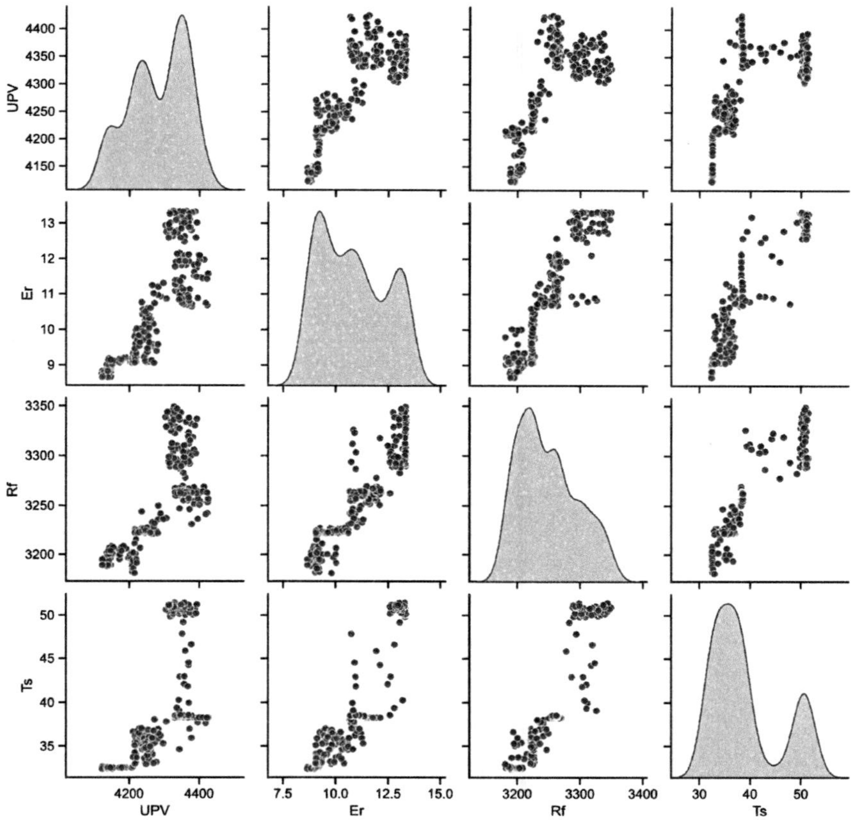

Fig. 4. Dataset distribution and KDE in the main diagonal.

To avoid the overfitting problem, in the algorithm, an early stopping method was implemented with patience equal to 3. This kind of regularization technique affords a review of the differences between the validation loss and validation of the metric used. The early stopping process aims to take an accounting of the variations between the validations aforementioned, not allowing a higher difference among them and avoiding overfitting problem in the DNN layers.

The training and test process show that the loss function and the error metric MSE decrease through the iterations number. Figure 6 shows the comparison of the predicted values for the model against the actual values in the training phase, and Fig. 7 shows the same comparative but for the test phase. The R^2 values for the validation and test datasets are 95.6% and 96.2%, respectively, which are practically the same. Thus, the selected DNN architecture with optimal hyperparameters trained on a scaled dataset can predict the concrete Ts. Finally, the validation performance set achieves an accuracy of 95.8%. Of note is that the validation set considers a concrete mixture with different properties.

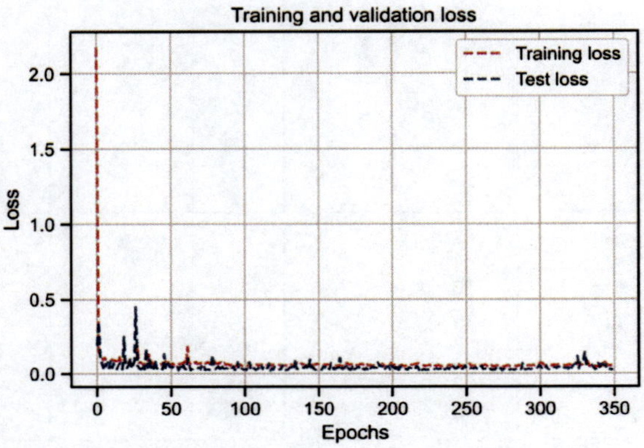

Fig. 5. Loss performance trough the number of epochs.

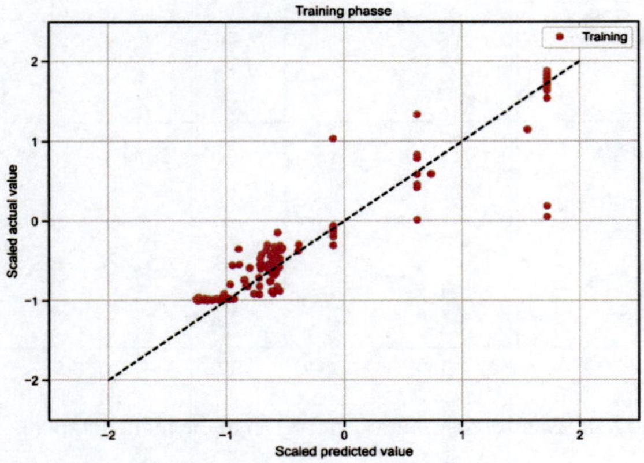

Fig. 6. Model performance in training phase.

6 Conclusions

This study presents an Artificial Intelligence approach for estimating the concrete tensile strength employing a data collection of non-destructive tests. This study proposes a novel approach based on a Deep Neural Network (DNN) model to evaluate the concrete tensile strength. The DNN model is a suitable alternative for short-cut the laboratory work time. In particular, the model proposed in this paper suggests a new, high-order neuron to account for the non-linear relationship among input activations at layers. The DNN model has three hidden layers and is validated with a dataset obtained from [18], also available in the literature. The model architecture is determined through extensive hyperparameter fine-tuning. The DNN model developed in this work can predict the concrete tensile strength with an accuracy over 95% in the test phase and achieves an accuracy of almost 96% in a validation dataset.

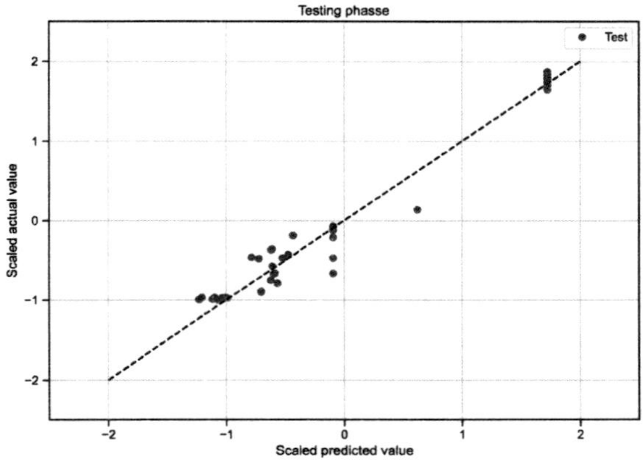

Fig. 7. Model performance in testing phase.

These results are superior compared to the results shown in [2], where the ML model proposed in that research depicts a 93.61%. Further, a similar approach that estimates a concrete feature such as the flexural strength using DNN achieves a 90% of accuracy [5], denoting that for our investigation, the hyperparameter fine-tuning process is a crucial task to improve the model performance. In addition, the results of [10] show an accuracy of 95% in the prediction of concrete tensile strength task; however, the authors considered a 90% for the training set and 10% for the testing set, of which, in some cases it is a routinary practice to increase the model accuracy, contrasting with the robustness proposed in our research; where the segmentation was 70% and 30%, training and test respectively.

Of note is that the scaling process in the input parameters significantly reduces the computational time and improves the accuracy of the model. The proposed model performs adequately, given the input parameters. However, there are some outliers in the training dataset that might be addressed. Outlier detection was not considered for this analysis in order to generate a robust and generalized model. The outlier's presence in the data might correspond to equipment calibration and variation of the moisture concrete samples. Future work will consider the arduous task of improving the model performance by adding new input features or eliminating some of the considered features in this paper. The approach used in this study can be extended to predict other concrete features or concrete properties.

References

1. Guzmán Torres, J.A., Mota, F.J.D., Guzmán, E.M.A., Molina, W.M., Ruiz, G.T.: A review of concrete performance employing a starch as addition using several regression techniques. In: Advanced Materials Research, vol. 1160, pp. 1–14. Trans Tech Publications Ltd. (2021)
2. Guzmán Torres, J.A., et al.: Prediction of the tensile strength and electrical resistivity of concrete with organic polymer and their influence on carbonation using data science and a machine learning technique. In: Key Engineering Materials, vol. 862, pp. 72–77. Trans Tech Publications Ltd. (2020)

3. Guzmán-Torres, J.A., Zalapa-Damian, A., Domínguez-Mota, F.J., Alonso-Guzmán, E.M.: Data science and machine learning technique for predicting electrical resistivity in recycled concrete with nopal as addition. In: Advanced Engineering Forum, vol. 40, pp. 43–62. Trans Tech Publications Ltd. (2021)
4. Rafiei, M.H., Adeli, H.: A novel machine learning-based algorithm to detect damage in high-rise building structures. Struct. Des. Tall Spec. Build. **26**(18), e1400 (2017)
5. Guzmán-Torres, J.A., Domínguez-Mota, F.J., Alonso-Guzmán, E.M.: Estimating the flexural strength of concrete using compressive strength as input value in a deep learning model. In: IOP Conference Series: Materials Science and Engineering, vol. 1150, no. 1, p. 012019. IOP Publishing
6. Dais, D., Bal, I.E., Smyrou, E., Sarhosis, V.: Automatic crack classification and segmentation on masonry surfaces using convolutional neural networks and transfer learning. Autom. Constr. **125**, 103606 (2021)
7. Guzmán Torres, J.A., Alonso Guzmán, E.M., Dominguez Mota, F.J., Martínez Molina, W., Tinoco, J.G., Navarrete Seras, M.A.: El comportamiento del pulso ultrasónico en un concreto de alto desempeño adicionado con un polímero orgánico comparado con un concreto sin adiciones. In: Congreso CONPAT 2019, Tuxtla Gutiérrez, Chiapas (2019)
8. Kaiser, H., Karbhari, V.M., Sikorsky, C.: Non-destructive testing techniques for FRP rehabilitated concrete. II: an assessment. Int. J. Mater. Prod. Technol. **21**(5), 385–401 (2004)
9. Ramachandran, P., Zoph, B., Le, Q.V.: Searching for activation functions (2017). arXiv preprint arXiv:1710.05941
10. Bui, D.K., Nguyen, T., Chou, J.S., Nguyen-Xuan, H., Ngo, T.D.: A modified firefly algorithm-artificial neural network expert system for predicting compressive and tensile strength of high-performance concrete. Constr. Build. Mater. **180**, 320–333 (2018)
11. Singh, T.N., Sinha, S., Singh, V.K.: Prediction of thermal conductivity of rock through physico-mechanical properties. Build. Environ. **42**(1), 146–155 (2007)
12. Schmidhuber, J.: Deep learning in neural networks: an overview. Neural Netw. **61**, 85–117 (2015)
13. Antoniades, A., et al.: Deep neural architectures for mapping scalp to intracranial EEG. Int. J. Neural Syst. **28**(08), 1850009 (2018)
14. Gulli, A., Pal, S.: Deep Learning with Keras. Packt Publ. Ltd. (2017)
15. Pedregosa, F., et al.: Scikit-learn: machine learning in Python. J. Mach. Learn. Res. **12**, 2825–2830 (2011)
16. Kingma, D.P., Ba, J.: Adam: a method for stochastic optimization. In: International Conference on Learning Representations (ICLR) (2015)
17. Bonamente, M.: Statistics and Analysis of Scientific Data. Springer (2017)
18. Del Valle Moreno, A., et al.: Solicitaciones mecánicas y estáticas a concreto hidráulico simple elaborado con agregados pétreos redondeados y adicionados con fibras deshidratadas de cactus opuntia. Publicación Técnica 448 (2015)

A Combined Electrical and Electromechanical Impedance Study of Early-Age Strength Gain in Cement Mortars

Hussameldin M. Taha[1(✉)], Richard J. Ball[1], Andrew Heath[1,2], and Kevin Paine[1]

[1] Department of Architecture and Civil Engineering, BRE Centre for Innovative Construction Materials, University of Bath, Bath, UK
H.M.Taha.Abdalgadir@bath.ac.uk

[2] Department of Architecture and Civil Engineering, Centre for Infrastructure, Geotechnical and Water Engineering Research (IGWE), University of Bath, Bath, UK

Abstract. From simple strain gauges to more sophisticated techniques such as acoustic emission and digital image correlation methods, a variety of non-destructive methods are proposed in the literature for inspecting civil structures. However, in addition to accuracy and simplicity, automation and timely data collection are the main requirements for an efficient in situ non-destructive testing technique. Electrical impedance and the electromechanical impedance tests of cementitious members, are perceived to be suitable candidates for fulfilling the simplicity, automation and accuracy requirements, hence providing an efficient yet relatively non-expensive remote monitoring system. Throughout the first 28 days of a cementitious composites age, significant gain of strength is typically developed. It is crucial during this period to characterize the in situ strength gain profile for both planning and quality control purposes. This study investigates the viability of both the electromechanical impedance technique and the electrical impedance spectroscopy technique as non-destructive strength testing methods. These were assessed against their ability to detect strength changes due to the ongoing hydration reaction in cementitious mortars conforming with BS EN 196-1:2016. Results from both techniques were compared with the compressive and flexural strengths. The electromechanical impedance was assessed in terms of the change in the electrical impedance signature response through time, which was obtained through surface attached PZT sensors. Concurrently the electrical impedance response was collected through embedded stainless-steel electrodes across the frequency range of 1–10 MHz. This study will act as a guide for selection of a suitable technique for in situ strength gain monitoring using electrical methods.

Keywords: Impedance spectroscopy · Electromechanical Impedance · Piezoelectric sensors (PZT) · Strength monitoring · Non-Destructive testing

J. I. Escalante-Garcia et al. (Eds.): RW 2021, RILEM Bookseries 40, pp. 867–876, 2023.
https://doi.org/10.1007/978-3-031-21735-7_92

1 Introduction

Cement hydration is a complex chemical reaction that occurs through dissolution and precipitation processes within the interstitial liquid phase [1, 2]. During the first 24 h, the cementitious matrix transforms from the fluid to the solid-state, showing enhancement in its mechanical properties through both stiffness and compressive strength increase. During the construction phase, it is crucial to have adequate knowledge about the development of the mechanical properties of in situ cementitious members. Obtaining timely knowledge relating to the in-place strength evolution increases the efficiency of managing the construction schedule in terms of the earliest possible timing for formwork removal. This also enhances safety against catastrophic failures due to lack of strength when loading newly constructed cementitious members.

Different non-destructive tests and methods have been proposed for monitoring the in situ mechanical property development of cementitious members. These tests typically measure a specific material property and then relate it to the strength evolution through time, or they excite the material and then measure the response [3]. However, in addition to the accuracy, the capability of remote monitoring, automation, and the simplicity of the measuring set-up, are all required when selecting a suitable non-destructive testing method for monitoring strength. This paved the way for simple testing methods such as the maturity method, which utilizes the exothermic nature of the hydration reaction to relate the in-place strength with the temperature released during the hydration reaction [4]. In this regard, the electrical measurements are suitable candidates for remote and timely data collection testing.

The application of electrical measurements to cementitious members can be performed either intrinsically, such as in the electrical impedance spectroscopy method (EIS), or extrinsically such as in the electromechanical impedance method (EMI). Both the EIS and the EIM have been extensively referred to in the literature as damage detection methods [5, 6] and for evaluating strength evolution through time [7]. However, despite the fact that both of these techniques share the same type of devices for data collection, the EIS excites the cementitious member directly by applying an alternative electrical voltage or current before obtaining the impedance response of the cementitious mix, whereas in the EMI method, the member is excited using high-frequency mechanical waves through an attached piezoelectric sensor exploiting the direct and the inverse properties of the piezoelectric materials. Therefore, for both damage detection and strength assessment purposes, the EMI and the EIS assess the investigated cementitious material differently, as the EIS parameters are more closely related to the cementitious material pore structure development [8, 9], whereas the EMI parameters are more closely related to the mechanical impedance of the investigated material [10].

This study assesses the accuracy in determining the strength development through time for mortar using both the EIS and the EMI responses of the samples. Importantly, it highlights the fundamental difference between the information that can be extracted experimentally from both methods.

2 Methodology

2.1 Samples Casting and Preparation

In this study, mortar samples were used throughout the experimental programme. The cement used was CEM I 52.5 N conforming to BS EN 197-1 [11], and standard sand confirming to BS EN 196-1 [12] was used as fine aggregate. The water/cement ratio used was 0.5, while the sand to binder ratio was 3.0. Samples were cast in 40 × 40 × 160 mm stainless steel moulds. For the electrical measurement, 40 × 40 × 40 mm mortar cubes, see Fig. 1, were used, while for the strength measurement the full sample size was used. The 40 × 40 × 40 mm cubes were cast by dividing the casting mould into two compartments by inserting a waterproofed 40 × 40 × 80 mm wooden cube in the centre of the mould. Two 2 mm diameter stainless steel electrodes were embedded in these mortar cubes on casting. The stainless-steel electrodes embedded depth was 20 mm, and the distance between the centre of the cube and the centre of each electrode was 10.5 mm.

After casting, all the samples were covered with cling film before demolding after 24 h. After demolding, one side of the 40 × 40 × 40 mm mortar cubes was prepared by rubbing with sand paper, and a Lead Zirconate Titanate (PZT) piezo ceramic disc type PIC 153 with the specifications shown in Table 1 was bonded to the surface by using an Araldite epoxy resin.

Following demolding all the samples were water cured at 20 °C until testing.

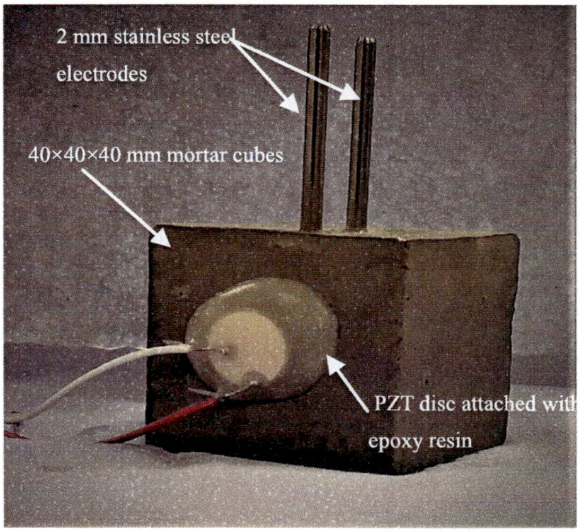

Fig. 1. 40 × 40 × 40 mm mortar cubes, with embedded stainless-steel electrodes and attached PZT sensor, for monitoring both the EIS and the EMI responses through time.

Table 1. Properties of PZT material [13]

Property	Value
Material type	PIC 153
Density, (g/cm^3)	7.60
Electric permittivity, $\varepsilon_{33}^T/\varepsilon_0$	4200
Piezoelectric strain coefficient, d_{33} (10^{-12} C/N)	600
Elastic compliance coefficient, S_{11}^E (10^{-12} m^2/N)	1960
Dielectric loss factor, tanδ (10^{-3})	30

2.2 Test and Methods

Electrical impedance test:

The electrical impedance of the samples was measured in a surface-dry condition, requiring removal from and return to the curing tank, throughout the testing period. Electrical impedance data was collected using a Newtons4th PSM 3750 frequency analyzer (Newtons4th Ltd., Leicester, UK) interfaced with an impedance analyzer. The impedance analyzer was connected to the stainless-steel electrodes embedded in the samples, through 50 Ω coaxial cables with BNC connectors. A potential of 350 mV was applied over the frequency range 1–10 MHz. Measurements were performed on three replicate samples.

Electromechanical impedance test:

The same samples and devices used for the EIS measurements were also used to collect the EMI response. In this case, the 50 Ω coaxial cables were connected to wires which were soldered to silver paint electrodes on the surface of the PZT disc. The frequency range used in this case was 15–350 kHz, and the actuating voltage was 2 V.

Flexural and compressive strength tests:

Both the flexural and compressive strengths were determined for the samples following the BS EN 196-1 [12]. After performing the flexural strength test, the halves of prisms were used to measure the compressive strength. Both the flexural and the compressive strength tests were performed at the ages of 3, 7, 14, 21 and 28 days. Three replicate samples for the flexural test were used; hence the compressive strength was measured on six halves.

3 Results and Discussion

3.1 Electrical Impedance (EIS) Response of Mortar Samples

Several electrical parameters can be readily extracted from the complex electrical impedance response of the samples over time. These include polarization related and conduction related electrical parameters. In this study, the samples' electrical resistivity and bulk frequency responses were studied and compared to the development of mechanical properties over time. Figure 2a shows the complex impedance Nyquist plots, and Fig. 1b shows both the electrical impedance real part and the phase angle responses for one representative mortar sample at the ages of 1, 3, 7, 14, 21 and 28 days. The complex plots exhibit distinctive features which can be categorized relatively to the frequency ranges in which these features appear. The real part on the Nyquist plot records large values at lower frequency ranges, and the imaginary part records the largest negative value. These two values are due to the double layer polarization in the sample/ electrodes interface, which gives rise to high electrical impedance.

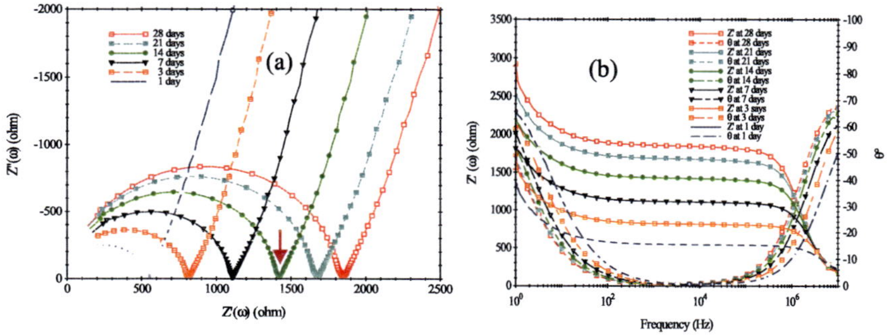

Fig. 2. **a** Complex Nyquist plot, and **b** the real impedance part and the phase angle through frequencies for mortar sample at the ages of 1, 3, 7, 14, 21 and 28 days.

As the measuring frequency increases, the electrode/sample polarization process relaxes; hence both the impedance imaginary part in Fig. 2a and the phase angle in Fig. 2b reach their minimum negative values. The high-frequency arc, representing the sample bulk response, starts to develop and forms as the measuring frequency increases, which indicates the interaction of the bound charges within the developing pore structure and the multi-frequency electric field. This interaction gives rise to several simultaneous polarization processes inside the material, which in turn relax as the measuring frequency increases. The relaxation of the bulk material polarization processes is inferred through the dispersive behavior observed on both the phase angle and the impedance real part in Fig. 2b as the measuring frequency increases.

By considering the electrical impedance behavior through time, it is observed that as the age of the sample increases, the bulk resistance of the sample increases. This is well illustrated by comparing the average bulk resistivity of the samples at the ages of 1 day and 28 days which was extracted from the cusp point between the electrode's arc and the bulk's arc as shown by the arrow in Fig. 2a on the 14 days age Nyquist plot. The resistance through this period increased from 540.2 Ω to 1848 Ω. In addition, it is also evident from Fig. 2a that the center of the high-frequency arc becomes more depressed under the real axes of the complex plot as the age of the sample increases. The increase in the sample's bulk resistance indicates the decrease in the continuous conductive paths, whereas the more depressed high-frequency arc indicates that multiple polarizations' relaxation processes take place at different frequency ranges; hence more pore size populations are developing as the age of the sample increases.

In addition to the above-mentioned effects on the complex impedance response, the bulk frequency in which the bulk resistance of the material is obtained decreased over time. This can be observed by considering the phase angle response through time in Fig. 2b, as the minimum negative phase angle values for the sample at the age of 1 day, and the age of 28 days were recorded at the bulk frequency values of 6.721 kHz and 2.984 kHz, respectively as shown in Fig. 3.

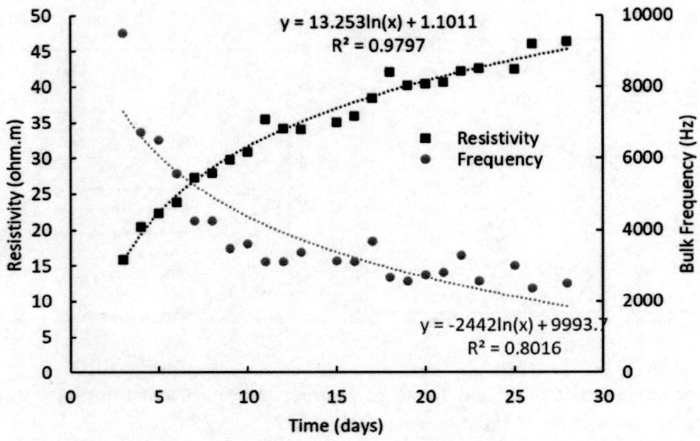

Fig. 3. Average resistivity and bulk frequency responses of the mortar samples through time.

This decrease in the bulk frequency values through time is related to the inability of the free charges in the sample to follow high frequencies alternative electrical field due to the reduction in the pore size due to the ongoing hydration process.

3.2 The Electromechanical Impedance (EMI) Response

Figure 4 shows a representative impedance signature response for a PZT sensor attached to the surface of a mortar sample. The impedance signature response generally changed through time as the age of the sample increased, as shown in Fig. 4a. A distinguishable shift in the peaks frequencies values towards higher frequency ranges was observed, as well as changes in the values of the impedance signature amplitudes. These two changes can be demonstrated by considering peaks 1 and 2 in Fig. 4b, as peak 1 and 2 at the age of 1 day took place at the frequency values of 43.168 kHz and 77.037 kHz, respectively, however by the age of 28 days, these two peaks' frequencies shifted to 49.540 kHz and 88.438 kHz.

With regard to the amplitude values for both peaks 1 and 2, they also showed a noticeable change through time, as at the age of 1 day the amplitudes of these two peaks recorded the values of 120.98 Ω and 103.76 ohms respectively, whereas at the age of 28 days the amplitudes recorded the values of 192.64 ohms and 153.46 ohms respectively.

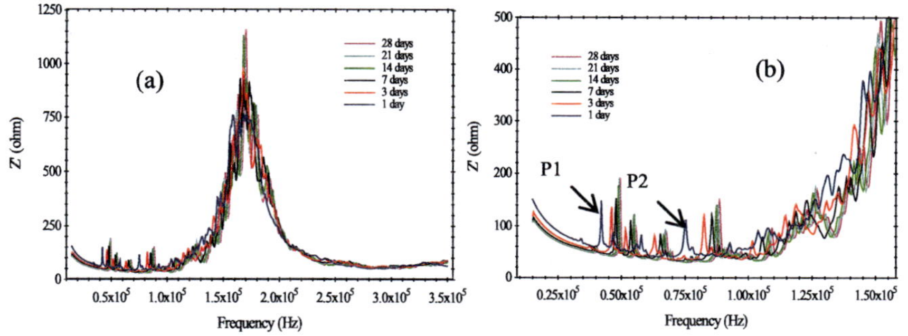

Fig. 4. A representative impedance signature plot for a mortar surface attached PZT sensor

These changes in the impedance signature peak frequencies and amplitudes indicate an increase in the coupled PZT/sample stiffness due to the ongoing hydration process, especially when considering the increase in the peaks' frequencies values.

The changes in the impedance signature through time can also be studied using the root mean square deviation, RMSD [5], as a pattern recognition metric to give a single numerical value that can quantify both the changes in the amplitude and the frequency shifts through time, as shown in Fig. 5. By using the 1 day impedance signature as a baseline, it is observed that the RMSD values increase gradually through time, which indicates that the RMSD can capture the changes in the samples due to the ongoing hydration and hardening processes. It worth mentioning that the RMSD values for samples 1, 2, 3 were assessed at the frequency ranges of 100–115 kHz, 160–175 kHz and 130–145 kHz, respectively.

3.3 Relation Between the EIS/EMI and the Mechanical Properties of the Sample

To compare the compressive strength and the flexural strength development against the parameters extracted from both the EIS and the EMI methods, Fig. 6 shows the relation between the compressive and the bending stresses with the EIS and the EMI parameters.

From Fig. 6, it can be seen that both the EIS and the EMI extracted parameters show a high correlation with both the compressive and the flexural strength of the samples. Despite the noticeable variation in the R^2 values observed for the three replicate samples, the mean R^2 recorded for the relationship between the RMSD, and the flexural, and the compressive strength was 0.90 and 0.89, respectively These values indicate a reasonable correlation between the RMSD and the mechanical properties of the samples.

On the other hand, and with regard to the EIS extracted parameters, the resistivity showed a significant correlation with the compressive and the flexural strengths, as the R^2 recoded 0.97, whereas for the bulk frequency, the R^2 recorded the value of 0.84.

These results indicate that both the EMI and the EIS extracted parameters can be used as indicators for the development of both the compressive and the flexural strength of cementitious materials at the first 28 days of the sample age.

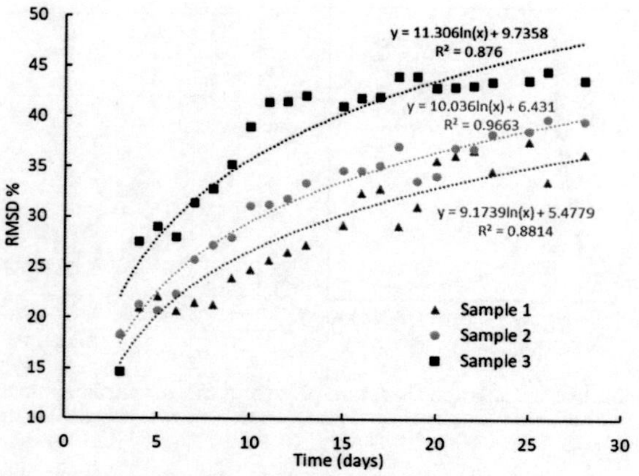

Fig. 5. Root mean square, RMSD, response through time for three replicate mortar specimens

4 Conclusion

In this study, the effect of the ongoing hydration and hardening processes on both the electrical impedance spectroscopy response and the electromechanical impedance response of mortar samples through the first 28 days of the age of the samples was studied. The impedance spectroscopy response was collected through the frequency range of 1Hz to 10 MHz through embedded stainless-steel electrodes, while the electromechanical response was collected through surface attached PZT sensors at the frequency range of 15–350 kHz.

Both the EIS and the EMI extracted parameters showed changes as the age of the samples increased. The electrical resistivity, which was extracted from the Nyquist plot, showed a continuous increase through time, indicating a decrease in the connected capillary porosity in the samples, while the bulk frequency at which the bulk resistivity was extracted showed a decreasing trend through time, indicating a decrease in the charge carriers mobility due to the reduction in the pores diameters.

With regard to the EMI parameters, it was observed that the impedance signature peaks frequencies were shifted to higher frequency ranges as the samples age increases. In addition, the impedance signature amplitude at the lower frequency range was observed to increase through time. These mentioned changes on the peaks frequencies and the amplitudes indicate an increase in the stiffness of the sample, particularly the noticeable shift in the impedance signature peak frequencies.

By comparing the relationship between the compressive and flexural strengths for the EIS and the EMI extracted parameters, it was observed that the relationships showed R^2 values >0.85. The RMSD recorded an average R^2 value in the range of 0.9, whereas the electrical resistivity of the samples showed a stronger correlation showing R^2 value of 0.97. It is noteworthy that the lowest correlation recorded was for the bulk frequency, as the R^2, in this case, recorded a value of 0.84.

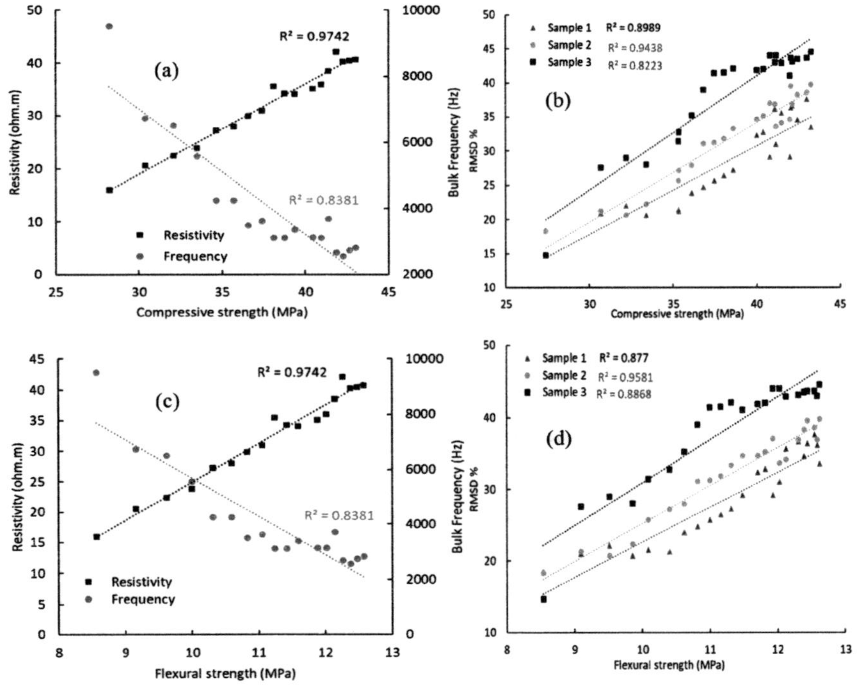

Fig. 6. **a** and **b** show the relation between the compressive strength of the mortar samples and the EIS and the EMI extracted parameters. **c** and **d** show the relation between the flexural strength and the EIS and the EMI

Acknowledgements. This research was funded by UK-EPSRC, grant number No. EP/P02081X/1, Resilient Materials 4 Life, RM4L. The authors gratefully acknowledge the technical staff within the Department of Architecture and Civil Engineering at the University of Bath for technical support and assistance in this work.

References

1. Gartner, E., et al.: Hydration of Portland cement. Struct. Perform. Cem. **13**, 978 (2002)
2. Le Chatelier, H.L., Mack, J.L.: Experimental researches on the constitution of hydraulic mortars (1905)
3. Lim, Y.Y., Smith, S.T., Soh, C.K.: Wave propagation based monitoring of concrete curing using piezoelectric materials: review and path forward. NDT E Int. **99**, 50–63 (2018)
4. Bickley, J.A.: Practical application of the maturity concept to determine in situ strength of concrete. Transp. Res. Record, 558 (1975)
5. Taha, H., Ball, R.J., Paine, K.: Sensing of damage and repair of cement mortar using electromechanical impedance. Materials **12**(23), 3925 (2019)
6. Azhari, F.: Cement-based Sensors for Structural Health Monitoring. University of British Columbia (2008)

7. Su, Y.-F., et al.: Novel methodology on direct extraction of the strength information from cementitious materials using piezo-sensor based electromechanical impedance (EMI) method. Constr. Build. Mater. **259**, 119848 (2020)
8. Taha, H., McCarter, W.J., Suryanto, B., Starrs, G.: Frequency- and time- domain dependency of electrical properties of cement-based materials during early hydration. Adv. Civil Eng. Mater. (2016)
9. McCarter, W.J., Brousseau, R.: The A.C. response of hardened cement paste. Cem. Concr. Res. **20**(6), 891–900 (1990)
10. Bhalla, S., Soh, C.K.: Structural health monitoring by piezo-impedance transducers. I: modeling. J. Aerosp. Eng. **17**(4), 154–165 (2004)
11. EN197-1:2011. In Cement: Composition, specifications and conformity criteria for common cements. London, UK, BSI (2011)
12. BSI, BS EN 196-1: 2016: Methods of testing cement. Determination of strength. London, UK, BSI (2005)
13. PI_Piezo_Technology. Piezoelectric Materials. (n.d). https://www.piceramic.com/en/products/piezoelectric-materials/#c15163. Accessed 23 May 2021

Bituminous Materials, Polymers, Timber, Bamboo, Recycling, Masonry, etc.

Using Neutron Tomography to Study the Internal Curing by Superabsorbent Polymers in Cementitious Materials

Didier Snoeck[1,2](✉), Wannes Goethals[3], Jan Hovind[4], Pavel Trtik[4], Tim Van Mullem[1], Philip Van den Heede[1], and Nele De Belie[1]

[1] Magnel-Vandepitte Laboratory, Department of Structural Engineering and Building Materials, Faculty of Engineering and Architecture, Ghent University, Tech Lane Ghent Science Park, Campus A, Technologiepark Zwijnaarde 60, B-9052 Gent, Belgium
didier.snoeck@ulb.be

[2] BATir Department, Université Libre de Bruxelles (ULB), 50 F.D. Roosevelt Avenue, 1050 Brussels, Belgium

[3] Centre for X-Ray Tomography, Department of Physics and Astronomy, Faculty of Sciences, Ghent University, Proeftuinstraat 86, N12, B-9000 Gent, Belgium

[4] Laboratory for Neutron Scattering and Imaging (LNS), Paul Scherrer Institut, Forschungsstrasse 111, 5232 Villigen PSI, Switzerland

Abstract. Superabsorbent polymers (SAPs) are novel admixtures in cementitious materials in order to have internal curing. Neutron tomography is used to investigate and to visualize the water release by SAPs with different kinetics over time. This gives insight into the time window for effective internal curing as the results are linked to shrinkage and internal relative humidity measurements. One SAP showed complete mitigation of autogenous shrinkage and even a partial expansion effect (+121 $\mu m \cdot m^{-1}$ after two days), while another type of SAP proved to be less effective as the shrinkage (-189 $\mu m \cdot m^{-1}$) was more resembling a reference system with a higher effective water-to-cement ratio. The internal relative humidity after two days of hardening showed that the effective mixture still possessed a high relative humidity at 98% while the ineffective mixture had a lower internal relative humidity at 95% after 2 days. It was still unclear whether this ineffective SAP released its water too slow or too fast. The study with neutron tomography showed the necessary kinetics to mitigate autogenous shrinkage of one SAP between final setting and 30 h. However, the ineffective SAP released its water quickly after final setting and prior to 20 h, even though water is still necessary at later time intervals to mitigate further autogenous shrinkage and to maintain the internal relative humidity.

Keywords: Internal curing · Hydrogel · Cement paste · Durability

1 Introduction

Autogenous shrinkage cracking at early age is a problem in cementitious materials, especially in systems with a low water-to-binder ratio (i.e. below 0.42 [1]), either in

J. I. Escalante-Garcia et al. (Eds.): RW 2021, RILEM Bookseries 40, pp. 879–887, 2023.
https://doi.org/10.1007/978-3-031-21735-7_93

high-strength concrete structures or in automated printed elements [2]. This is due to self-desiccation as the internal relative humidity decreases over time. By including water-filled admixtures, i.e. saturated lightweight aggregates [3, 4] or superabsorbent polymers (SAPs) [5–7], internal curing can be provided. The principal task is to maintain the internal relative humidity at high levels by the release of entrained water from the included materials, thus avoiding the increasing capillary forces due to receding water menisci upon hardening. This may counteract the occurring autogenous shrinkage and associated cracking.

SAPs have been used in concrete technology to improve several properties. They are used as admixtures to change the rheological behavior [8], for increasing the freeze-thaw resistance [9], for including self-sealing characteristics [10, 11] and to promote autogenous healing [12, 13]. The SAPs' main use is the mitigation of shrinkage and the SAP water-release kinetics upon hardening and curing need to be investigated in detail. When self-desiccation occurs, the SAPs will release their absorbed additional mixing water due to the increasing capillary forces, thereby mitigating autogenous shrinkage. However, not all SAP types show the same behavior and some SAPs may release their entrained water too slow or too fast, linked to their water-release kinetics. In previous research one type of SAP, further on being called SAP A, proved to be ideal in completely counteracting autogenous shrinkage [14] and even partially plastic shrinkage [15]. However, another type of SAP, further on being called SAP B, was less effective in mitigating autogenous shrinkage [14]. The hypothesis suggested that it was either not releasing its stored water during the necessary internal curing window or that it was releasing its water prematurely or too fast. Nuclear Magnetic Resonance (NMR) testing showed that both SAPs were releasing their water at the correct time, being the time at final setting [16]. SAP A seemed to release its water slightly from 11 h onwards and more notably from 22 h onwards till approximately 30 h. SAP B on the other hand showed some minor premature release of water followed by reabsorption and a noteworthy and quick and complete release of entrained water from 10 h till 20 h.

As it is difficult to study and visualize the water release kinetics of SAPs within the cementitious matrix without destroying the fragile microstructure and stopping the hydration, the use of non-destructive high-tech equipment is necessary. Neutron tomography measurements were performed and linked to autogenous shrinkage and internal relative humidity measurements found in literature [14, 16]. Neutron tomography is an ideal technique to study the water within a cementitious materials but the resolution may be limited. SAPs with a size of around 2 mm in swollen size were already successfully studied in terms of entrained water release [17, 18]. The time of water release and the relative remaining swollen size of the SAP could be monitored. In this study smaller SAPs are studied as they are more effective in mitigating shrinkage [5], in order to find an answer on the supporting water-release kinetics by SAPs to have internal curing. By using neutron tomography a visualization is possible.

2 Materials and Methods

2.1 Materials and Mixing Procedure

SAP A is a bulk polymerized copolymer of acrylamide and sodium acrylate with a particle size of 100.0 ± 21.5 μm (BASF, Germany), and SAP B is a bulk polymerized cross-linked potassium salt polyacrylate with a particle size of 476.6 ± 52.9 μm (BASF, Germany). They are both able to absorb around 300 g of demineralized water per g of SAP, and around 60 g of cement filtrate (10:1) solution, as determined by means of a filtration method [19]. Both SAPs swell within a minute and thus during mixing, as determined by means of a Vortex method [20].

The mixtures studied were two reference and two SAP mixtures. The water used was demineralized water in order to exclude possible additional ions or cations influencing the water absorption capacity of the SAPs [19, 21–23]. One reference mixture (**R0.300**) had a water-to-cement ratio of 0.300 and a superplasticizer (Glenium 51, conc. 35%, BASF) amount of 0.5 m% (mass-% of cement weight) and showed clear autogenous shrinkage [14]. A theoretical amount of entrained water of 0.054 [1] is necessary to counteract shrinkage for this type of cement, being CEM I 52.5 N (Holcim, Belgium). Therefore, a second reference (**R0.354**) was additionally studied having a water-to-cement ratio of 0.354 and a superplasticizer amount of 0.3 m%. This second reference also showed prominent autogenous shrinkage albeit slightly less compared to **R0.300**.

The two SAP mixtures (**SAPA** and **SAPB**) had a total water-to-cement ratio of 0.354 and an effective water-to-cement ratio of 0.300 meaning that 0.054 was entrained in the SAPs. Both mixtures had the same superplasticizer amount as compared to **R0.300**, being 0.5 m%. **SAPA** had 0.22 m% of SAP A included and **SAPB** had 0.45 m% SAP B included in its mixture design. Upon mixing, the SAP particles were first dry mixed with the cement and the total amount of mixing water with dissolved superplasticizer was added subsequently. All mixtures were designed to show the same flow value as determined by EN 12350-5. This resulted in an absorption capacity during mixing of 23 $g{\cdot}g^{-1}$ SAP A and 11 $g{\cdot}g^{-1}$ SAP B.

2.2 Neutron Tomography Measurements

Small fresh cementitious samples with a diameter of 5 mm and a height of 10 mm were studied in sealed Plexiglass cups. The different samples were stacked and positioned in the neutron radiography chamber using a rotating sample holder and an aluminum tube. The setup can be seen in Fig. 1.

The neutron beam line NEUTRA with a proton current of 1.5 mA used is located at the Swiss spallation source SINQ of the Paul Scherrer Institute (PSI) [24]. For two days, the measurements continued to visualize the water release from the SAPs as a function of time. The time necessary to complete one tomogram was around 1 h using 47 projections with 60 s exposure each per 360° rotation. After correcting all radiographs for dark current, flat field, background scattering and fluctuations in beam intensity per radiograph, the radiographs were combined to a series of 3D tomographic grey-level representations of the water level intensity per voxel. Due to the used setup, the sample size, the exposure time, the amount of projections and the algebraic reconstruction technique, a final 42 μm voxel size was obtained. Afterwards, using piecewise constant functions [25], the significant decreases in water levels per voxel were determined and studied. The first complete tomogram was used as a reference state to compare the different obtained subsequent tomograms. Bright colors were assigned to water while air-filled cavities were assigned to dark colors.

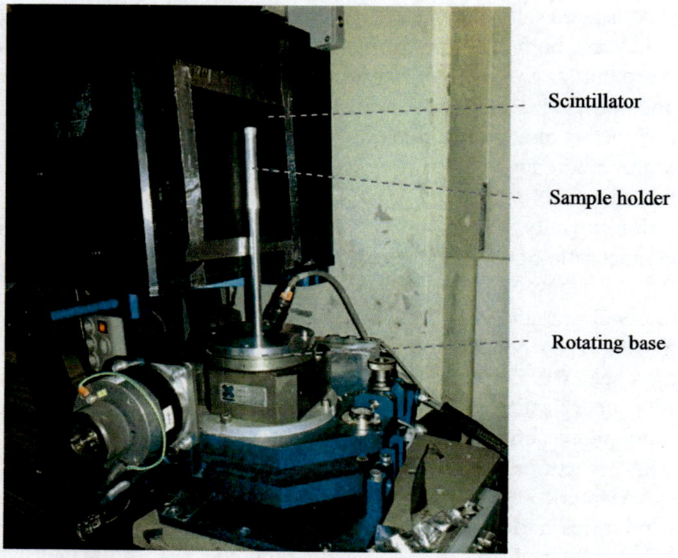

Fig. 1. Neutron tomography setup showing the aluminum sample tube and the rotating base.

2.3 Autogenous Shrinkage and Internal Relative Humidity Measurements

Previously published results [14, 16] were used to support the findings with neutron tomography. All results were compared for the first two days of hardening.

The autogenous strain was measured following ASTM C 1698-09 using a corrugated tube system with a diameter of 29 mm and automated dilatometers, regulated in a polyalkylene thermobath at 20.0 ± 0.1 °C [14].

The internal relative humidity was monitored with tubes having a diameter of 14 mm, and a hoovering and equilibrated relative humidity (RH) sensor 1 mm above the sample surface. A protective lid system was used to protect the RH probe and to maintain the internal conditions of the sample near the probe. The temperature was regulated by putting the setup in a water bath at 20 ± 1 °C [16]. The relative humidity was determined three times: around final setting, after 1 day and 2 days of hardening.

3 Results and Discussion

The reference samples did not show much variation in grey level intensities of the neutron tomograms. The **R0.354** mixture was showing slightly brighter tomograms as more mixing water was included compared to **R0.300**. Over time, no decrease in total water intensity was found, meaning that the samples were tightly sealed and not exposed to the environment. The cementitious matrix of the **SAPA** and **SAPB** samples were visually resembling grey levels of the **R0.300** while the total cumulative average grey values were more close to the **R0.354**. This is due to the differences in total and effective water-to-cement ratios.

Upon analyzing the tomographic horizontal slices in SAP mixtures, it became clear that water-filled inclusions (bright color) transformed to air-rich regions (dark color) over time. These locations, as seen in Fig. 2, correspond to the expected SAP sizes for in this case sample **SAPB**, as the swollen size of a SAP B particle is around 981 μm in size [26]. Initially, right after mixing, white saturated SAPs are seen (left), while during hardening, the SAPs are found to release the entrained water (middle) to end up as dark air-filled cavities (right).

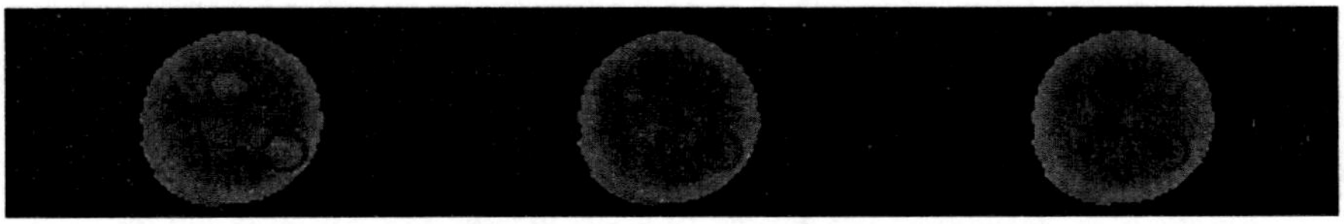

Fig. 2. Internal curing visualized with qualitative grey levels: light-colored water-filled inclusions which empty over time due to the water release by SAP B, leaving behind dark air-filled macro pores. The shown cylinder of 5.5 mm in diameter includes the Plexiglass walls.

On post-processing, the voxels which showed a decrease in water intensity from the first tomogram compared to the final tomogram were combined into a volume and the relative decrease in water was determined as a function of time. Both reference systems did not show a significant number of such voxels. However, such voxels could be found in the SAP mixtures. In order to study the water-release from the SAP voxels, the initial reference state (first tomogram) was assumed to show SAP voxels completely filled with water and the last tomogram completely filled with air (after 56 h, being the last recorded tomogram). This was confirmed upon analysing the last tomogram where only a small portion of dry SAP particle leftover was found without water-filled inclusions (see Fig. 2). In Fig. 3 the relative signal as a function of time for the first two days in the **SAPA** and **SAPB** mixtures is plotted. In this way, combining all SAP voids, the relative water signal could be studied. Even though the end tomogram was taken after 56 h, the results are plotted till 48 h as most water was already released at this stage. In this way, the results could be compared to autogenous shrinkage measurements on the same mixtures. In this graph the relative water intensity signal in the SAP voids is shown on the left vertical axis in bold while the autogenous shrinkage measurements are shown on the right vertical axis in dashed lines. The curves of the autogenous shrinkage are plotted starting from time zero. In this paper, this time zero corresponds to the final setting as determined using a Vicat needle which also corresponded to the so-called knee point in the shrinkage curves.

The water release seen in the **SAPA** mixture is starting around final setting and gradually decreases over time. Most of the water is released after 30 h. These results

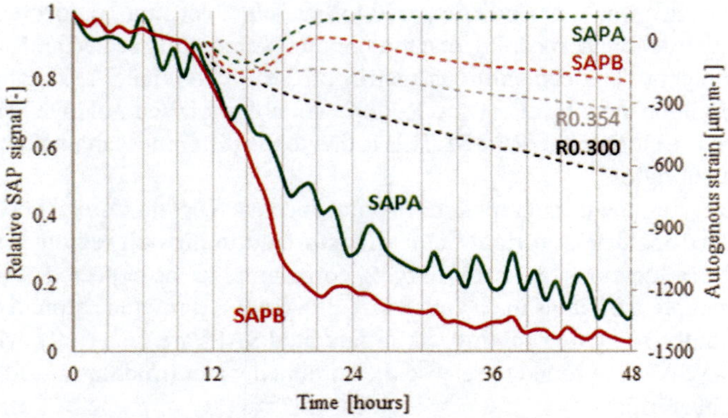

Fig. 3. Water intensity signal in SAP-containing mixtures (**SAPA** and **SAPB**), relative to the signal at 0 h and 56 h, showing the qualitative water release during internal curing. The results are compared to autogenous shrinkage results during the first 48 h of testing.

are comparable to the release as found with NMR results on the same mixture [16]: starting after final setting at 11 h and more gradually in between 20 h and 30 h. For **SAPB** a clear and quick release of stored water in between 10 h and 20 h is found, which is comparable to the NMR results. This already shows that SAP B particles seem to release their stored water prematurely as autogenous shrinkage still plays a role after 20 h when all entrained water is already released. This is translated to the autogenous shrinkage curves also shown in Fig. 3. In **SAPA**, starting from final setting, there first is a slight expansion. This is due to the water release by the SAPs to stimulate further hydration. Due to the formation of hydration products, there is an internal expansion, which is initially also the case in **SAPB**. In **SAPA** internal curing is still possible after 20 h and this results in a complete mitigation of autogenous shrinkage. On the other hand, for the **SAPB** mixture no internal curing water is available after 20 h and the autogenous shrinkage rate becomes the same as for the **R0.354** mixture having the same total water-to-cement ratio. As there was an initial mitigation, the total shrinkage value of the **SAPB** mixture is lower than for the reference systems, yet cracking may still occur. The **R0.300** mixture shows clear autogenous shrinkage. After only 2 days, the shrinkage amounted to -647 $\mu m \cdot m^{-1}$ while a higher water-to-cement ratio (**R0.354**) resulted in -326 $\mu m \cdot m^{-1}$. **SAPA** showed an expansion of $+121$ $\mu m \cdot m^{-1}$ after two days, while **SAPB** showed -189 $\mu m \cdot m^{-1}$ shrinkage.

Generally, a two-stage desorption behavior is found and also seen in literature [27]: an osmotic stage with free water release followed by a humidity stage driven by the humidity gradient. Both SAPs are slightly different in chemical nature, one being a polyacrylate neutralized by a sodium salt while the other one is neutralized by a potassium salt. The difference in cations influences the osmotic behavior in the pore liquid, leading to different kinetics of water desorption, due to the ion exchange between absorbed Ca^{2+} and the monovalent cations in the SAP [23]. Furthermore, the size of SAP A has a d_{50} of 100 μm while SAP B is on average 477 μm in size. It was already stipulated in literature [28] that the size plays a major role due to limited water diffusion around 100 μm when the capillary pore space is depercolated. In addition, the water release is more easy when using a larger particle due to more adhesion of the desaturating SAP to the cementitious pore wall. This adhesion speeds up desorption [29, 30], which is the case for the larger SAP B particles.

The internal relative humidity around final setting, after one day and two days is shown in Table 1 for the **R0.300**, **R0.354**, **SAPA** and **SAPB** mixtures, respectively. As can be seen, the **R0.300** mixture already showed a decrease in internal relative humidity after one day while the others more or less maintained their humidity at a high level. At two days, only the **SAPA** mixture kept its high internal relative humidity while all other mixtures showed a decrease in time. This shows the effective behavior of SAP A over SAP B, and supports the findings in terms of autogenous shrinkage behavior linked to the internal-curing water release by the two types of studied SAPs. SAPs can maintain the internal relative humidity, while the reference systems show an overall steady decrease in internal relative humidity. This decrease is responsible for the self-desiccation effect and linked observed shrinkage.

Table 1. Internal relative humidity around final setting, after one day and two days for the **R0.300**, **R0.354**, **SAPA** and **SAPB** mixtures, respectively.

	R0.300	R0.354	SAPA	SAPB
Final setting	97.2%	97.4%	97.6%	97.6%
24 h	95.4%	97.1%	98.5%	97.8%
48 h	89.9%	94.6%	98.0%	95.1%

4 Conclusions

Neutron radiography is a powerful non-destructive test technique to determine the water release from SAPs in a cementitious material as a function of time. One SAP, being a polyacrylate neutralized by a sodium salt and having size of 100 μm, showed complete mitigation of shrinkage and kept the internal relative humidity high due a steady internal-curing water release. Another type of SAP, being a polyacrylate neutralized with a potassium salt and having an average size of 477 μm, is not as efficient. After final setting, it releases its stored water too fast and has no sufficient water remaining at later stages when autogenous shrinkage is still taking effect. This results in a partial drop in internal relative humidity and a behavior closely linked to a reference system with the same total water-to-cement ratio and a higher initial effective water-to-cement ratio.

These results may open new pathways for the further understanding of internal curing by SAPs and the modeling of the cementitious materials in order to increase their practical application in concrete structures for mitigating autogenous shrinkage where necessary.

Acknowledgements. As a Postdoctoral Research Fellow of the Research Foundation-Flanders (FWO-Vlaanderen), D. Snoeck would like to thank the foundation for the financial support

(12J3620N). Wannes Goethals was supported by the Research Foundation—Flanders in the Junior Research Project program (3G036518). T. Van Mullem and N. De Belie were supported by a grant (21SCIP-C158977-02) from the Construction Technology Research Program funded by the Ministry of Land, Infrastructure and Transport of the Korean government.

References

1. Powers, T.C., Brownyard, T.L.: Studies of the Physical Properties of Hardened Portland Cement Paste. Portland Cement Association, Research Laboratories, Cornell (1948)
2. Van Der Putten, J., Snoeck, D., De Coensel, R., De Schutter, G., Van Tittelboom, K.: Early age shrinkage phenomena of 3D printed cementitious materials with superabsorbent polymers. J. Build. Eng. **102059**, 1–12 (2021)
3. Trtik, P., et al.: Release of internal curing water from lightweight aggregates in cement paste investigated by neutron and X-ray tomography. Nucl. Instrum. Meth. A **651**, 244–249 (2011)
4. Bentz, D.P., Weiss, W.J.: Internal Curing: A 2010 State-of-the-Art Review. In: N. Interagency (Ed.), p. 82 (2011)
5. Jensen, O.M., Hansen, P.F.: Water-entrained cement-based materials I. Principles and theoretical background. Cem. Concr. Res. **31**, 647–654 (2001)
6. Jensen, O.M., Hansen, P.F.: Water-entrained cement-based materials II. Experimental observations. Cem. Concr. Res. **32**, 973–978 (2002)
7. Mechtcherine, V., Wyrzykowski, M., Schröfl, C., Snoeck, D., Lura, P., De Belie, N., Mignon, A., Van Vlierberghe, S., Klemm, A.J., Almeida, F.C.R., Tenório Filho, J.R., Boshoff, W.P., Reinhardt, H.-W., Igarashi, S.-I.: Application of super absorbent polymers (SAP) in concrete construction-update of RILEM state-of-the-art report. Mater. Struct. **54**, 1–20 (2021)
8. Mechtcherine, V., Secrieru, E., Schröfl, C.: Effect of superabsorbent polymers (SAPs) on rheological properties of fresh cement-based mortars—Development of yield stress and plastic viscosity over time. Cem. Concr. Res. **67**, 52–65 (2015)
9. Mechtcherine, V., et al.: Effect of superabsorbent polymers (SAP) on the freeze-thaw resistance of concrete: results of a RILEM interlaboratory test. Mater. Struct. **50**, 1–19 (2017)
10. Snoeck, D., Steuperaert, S., Van Tittelboom, K., Dubruel, P., De Belie, N.: Visualization of water penetration in cementitious materials with superabsorbent polymers by means of neutron radiography. Cem. Concr. Res. **42**, 1113–1121 (2012)
11. Snoeck, D., Van den Heede, P., Van Mullem, T., De Belie, N.: Water penetration through cracks in self-healing cementitious materials with superabsorbent polymers studied by neutron radiography. Cem. Concr. Res. **113**, 86–98 (2018)
12. Snoeck, D., Pel, L., De Belie, N.: Autogenous healing in cementitious materials with superabsorbent polymers quantified by means of NMR. Sci. Rep. **10**, 1–6 (2020)
13. Snoeck, D., Van Tittelboom, K., Steuperaert, S., Dubruel, P., De Belie, N.: Self-healing cementitious materials by the combination of microfibres and superabsorbent polymers. J. Intell. Mat. Syst. Struct. **25**, 13–24 (2014)
14. Snoeck, D., Jensen, O.M., De Belie, N.: The influence of superabsorbent polymers on the autogenous shrinkage properties of cement pastes with supplementary cementitious materials. Cem. Concr. Res. **74**, 59–67 (2015)
15. Snoeck, D., Pel, L., De Belie, N.: Superabsorbent polymers to mitigate plastic drying shrinkage in a cement paste as studied by NMR. Cem. Concr. Comp. **93**, 54–62 (2018)
16. Snoeck, D., Pel, L., De Belie, N.: The water kinetics of superabsorbent polymers during cement hydration and internal curing visualized and studied by NMR. Sci. Rep. **7**, 1–14 (2017)

17. Trtik, P., Muench, B., Weiss, W.J., Herth, G., Kaestner, A., Lehmann, E., Lura, P.: Neutron tomography measurements of water release from superabsorbent polymers in cement paste. In: Brameshuber, W. (ed.) International RILEM Conference on Material Science, RILEM Publications S.A.R.L., Aachen, pp. 175–185 (2010)
18. Schröfl, C., Mechtcherine, V., Vontobel, P., Hovind, J., Lehmann, E.: Sorption kinetics of superabsorbent polymers (SAPs) in fresh Portland cement-based pastes visualized and quantified by neutron radiography and correlated to the progress of cement hydration. Cem. Concr. Res. **75**, 1–13 (2015)
19. Snoeck, D., Schröfl, C., Mechtcherine, V.: Recommendation of RILEM TC 260-RSC: testing sorption by superabsorbent polymers (SAP) prior to implementation in cement-based material. Mater. Struct. **51**, 1–7 (2018)
20. Zohuriaan-Mehr, M.J., Kabiri, K.: Superabsorbent polymer materials: a review, Iran. Polym. J. **17**, 451–477 (2008)
21. Schröfl, C., Snoeck, D., Mechtcherine, V.: A review of characterisation methods for superabsorbent polymer (SAP) samples to be used in cement-based construction materials—Report of the RILEM TC 260-RSC. Mater. Struct. **50**, 1–19 (2017)
22. Mechtcherine, V., et al.: Testing superabsorbent polymer (SAP) sorption properties prior to implementation in concrete: results of a RILEM round-robin test. Mater. Struct. **51**(1), 1–16 (2018). https://doi.org/10.1617/s11527-018-1149-4
23. Kang, S.-H., Hong, S.-G., Moon, J.: Importance of monovalent ions on water retention capacity of superabsorbent polymer in cement-based solutions. Cem. Concr. Res. **88**, 64–72 (2018)
24. Lehmann, E., Vontobel, P., Wiezel, L.: Properties of the radiography facility NEUTRA at SINQ and its use as European reference facility. NDT E Int. **16**, 191–202 (2001)
25. Goethals, W., Snoeck, D., De Belie, N., Boone, M.: Analysing dynamic neutron tomography of water absorption in cementitious materials using piecewise constant functions. In: The International Conference on Tomography of Materials and Structures, ICTMS2019, Cairns, pp. 1–2 (2019)
26. Snoeck, D., Dewanckele, J., Cnudde, V., De Belie, N.: X-ray computed microtomography to study autogenous healing of cementitious materials promoted by superabsorbent polymers. Cem. Concr. Comp. **65**, 83–93 (2016)
27. Yang, J., Wang, F., Liu, Z., Lui, Y., Hu, S.: Early-state water migration characteristics of superabsorbent polymers in cement pastes. Cem. Concr. Res. **118**, 25–37 (2019)
28. Bentz, D.P., Snyder, K.A.: Protected paste volume in concrete—extension to internal curing using saturated lightweight fine aggregate. Cem. Concr. Res. **29**, 1863–1867 (1999)
29. Farzanian, K., Ghahremaninezhad, A.: The effect of the capillary forces on the desorption of hydrogels in contact with a porous cementitious material. Mater. Struct. **50**(5), 1–15 (2017). https://doi.org/10.1617/s11527-017-1068-9
30. Farzanian, K., Ghahremaninezhad, A.: Desorption of superabsorbent hydrogels with varied chemical compositions in cementitious materials. Mater. Struct. **51**(1), 1–15 (2017). https://doi.org/10.1617/s11527-017-1128-1

Modular Lightweight Wastewater Treatment Plants Made of Textile Reinforced Concrete—Means to Reliable Wastewater Treatment in Rural Areas

Kira Heins[1(✉)], Gözdem Dittel[1], Komathi Murugan[2], Mohit Raina[3], Smitha Gopinath[4], Oliver Hentzschel[5], Sachin Paul[2], Ravindra Gettu[2], and Thomas Gries[1]

[1] Institut Fuer Textiltechnik of RWTH Aachen University, 52074 Aachen, Germany
kira.heins@ita.rwth-aachen.de

[2] Department of Civil Engineering, Indian Institute of Technology Madras, Chennai 600036, India

[3] Raina Industries Private Limited, Mumbai 400102, India

[4] CSIR-Structural Engineering Research Centre, Chennai 600113, India

[5] Betonwerk Hentzschel GmbH, 04910 Elsterwerda, Germany

Abstract. The safe supply of clean water is of such prominent importance that it is even defined by the United Nations Sustainable Development Goal 6 "Clean Water and Sanitation". In order not to lose sight of those regions, outside the conurbations, that are difficult to access, wastewater treatment plants (WTP) must adapt to new requirements. As a novel material, textile reinforced concrete (TRC) opens up new possibilities to ensure the treatment infrastructure. A solution that can be used on a large scale requires consideration of the composite material, adaptation of the structure design to the requirements of material, process and load as well as the economic consideration of manufacturing, distribution and sales. The focus here is on a modular product solution. In this way, the low weight of the TRC components in transport can be fully utilised. The operating loads are defined in accordance with conventional WTPs. The life cycle loads from production to operation are included in the structural simulation and design development. Based on this, the requirements for textile reinforcement and concrete are deduced. Both single components as well as the composite material are evaluated with mechanical testing, regarding their suitability for the present application. Six compositions of matrices and four varieties of textiles with two different types of coatings in each were examined. TRC composites were prepared and tested for the best combination of these materials. Numerical models, based on the material and composite results, considered various practical conditions, to yield the primary configuration and dimensions of the treatment unit. The details of the steps involved in arriving at the suitable material selection and the numerical simulation are discussed in this paper.

Keywords: Textile reinforced concrete · Structural simulation · Wastewater treatment plant · Textile reinforcement · Structural design development

J. I. Escalante-Garcia et al. (Eds.): RW 2021, RILEM Bookseries 40, pp. 888–896, 2023.
https://doi.org/10.1007/978-3-031-21735-7_94

1 Introduction

The availability of clean water for all population groups and in all countries is becoming increasingly important. This insight led the United Nations to include water supply in 'the 2030 Agenda for Sustainable Development' and the associated 17 Sustainable Development Goals. In the United Nations World Water Development Report of 2020 [1], the importance of water management for the environment and the population has been described. In addition to the direct burden on human health, the failure to treat wastewater has a negative impact on the environment, as well as natural water resources and, by extension, on agriculture and the food industry. Nevertheless, the rate of untreated wastewater in developing countries is 80–90% [1].

In the field of wastewater treatment plants (WTPs), centralized treatment plants are considered as the ideal solution [2]. Those wastewater management systems usually are connected to large infrastructure, which is mostly unavailable or unaffordable in rural areas. Libralato et al. [3] provided an overview over the advantages and disadvantages of centralised as well as decentralised wastewater management systems. Centralised WTPs require a large area for plant construction. In addition, alone the required supply system accounts for up to 90% of the implementation costs. Furthermore, even with continuous maintenance, complete renewal of parts or the entire plant are required after 50–60 years. The plants require a lot of energy and water for cleaning during operation. Decentralisation can be realized with smaller WTPs affiliated to individual blocks, clusters or individual households [3]. Such decentralised plants are particularly suitable for suburban and rural communities. The effort required to implement the feeder system is significantly reduced due to the proximity to the origins of the wastewater. They are ideal for reducing water source contamination and for enabling water reuse. In addition, their compact and robust technology makes them comparatively easy to maintain and operate [3].

1.1 Wastewater Management in India

In a tropical country like India, cleaning and recycling of waste water is highly recommended, to overcome the issue of water scarcity and to avoid contamination of ground and surface water. At present, around 70000 million litres per day (MLD) of sewage get generated in India, including domestic, industrial and agricultural sources. In this, the domestic sewage amounts to 35000 MLD. Out of this quantity, only 26% is estimated to be treated. Even though the metropolitan localities have centralised sewage treatment systems with well-developed sewer networks, wastewater may often be released without proper treatment in rural regions. The figures from a statistical survey conducted in 2019 show the presence of 920 WTPs in different states in India, out of which 615 are operational, 80 are non-operational, 154 are under construction, and 71 are under planning stage. The types of treatment systems vary from the most conventional septic and settling systems (deployed in rural areas), to the centralised plants with different units including clarifiers, filters, aerators, digesters etc. (operated in urban areas). In recent times, several agencies (NGOs, academia and industry as part of collaborative funded projects, etc.) have been involved in the installation of decentralised treatment units over various regions of the country, treating sewage via processes employing units such as sequencing batch reactors, moving bed biofilm reactor, membrane bioreactor etc.

The tanks of the small decentralised plants are usually precast concrete elements that are subsequently placed in pits on-site. This construction method entails thick-walled (e.g. 100 mm) heavy elements.

For about 25 years, textile reinforced concrete (TRC) has been investigated as an alternative construction material. This involves the use of reinforcement made of grid-like high-performance textiles using carbon or glass fibres and fine-grained concrete matrix. This high-performance reinforcement absorbs tensile forces acting under load in a similar way to steel reinforcement. In contrast to steel reinforcement, the materials used are not corrosive, which means that the concrete cover to protect the reinforcement against the corrosion and thus wall thickness as well as the weight of the components can be significantly reduced [4].

The methodology for the choice of materials, composite tests and structural design development involved in the development of a WTP made of TRC are explained in this paper.

2 Development of a Modular TRC Wastewater Treatment Plant

The above-mentioned needs for a decentralised wastewater treatment system for use in remote and rural regions, as well as the potential of TRC, motivate considerations for the implementation of small modular systems. In the long term, these are to be transported to the municipalities and implemented on-site. The use of TRC is expected to promote lightweight construction and to significantly facilitate the transport. To implement such a project, research is required on both, the material and the structural component, as well as on the design basis. The preliminary work and results are explained below.

2.1 Methodology

The idea of small-scale WTPs is not novel in its basic concept. Therefore, based on existing product variants, the requirements for the plant are derived and adapted to the material TRC. The design of the system has to be based on the location, which is intended to be in the region of Jhansi, India. The components of the composite, the textile reinforcement and the concrete matrix, have been characterized. The material characteristics determined are used for the numerical simulation of the designs under different load cases.

Characterisation of the textile reinforcement structure

The textile reinforcement is assessed by means of tensile tests. An initial selection is made on the basis of previous trials and existing experience in the industry. In addition, the basic material properties as well as the influence of the coating material are validated in yarn tensile tests. Subsequently, textile structures with four different binding types shown in Fig. 1—pillar, tricot, plain and counterlaid tricot—are characterised. Those textiles are produced by using 2400 tex alkali resistant (AR) glass-rovings with a centreline distance of adjacent rovings of 8 mm. The pre-selection of textile structure was supported by strip textile testing according to DIN EN ISO 13934 Part 1 [5] at a clamping length of 200 mm, testing speed of 20 mm/min and under an initial pre-stress of 10 N.

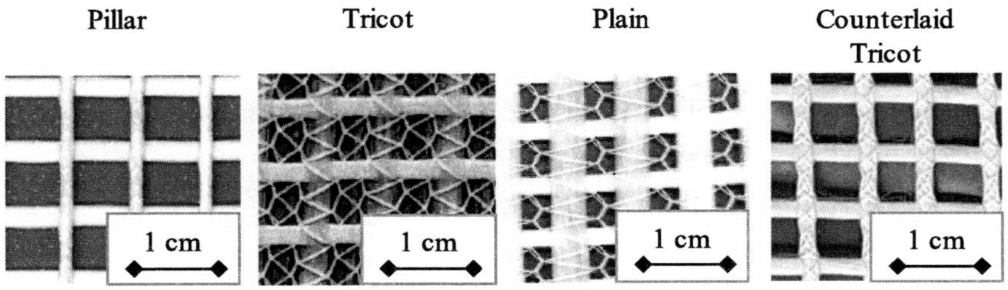

Fig. 1. Biaxial warp-knitted structures—Pillar, Tricot, Plain, Counterlaid tricot (left to right)

Additionally, single yarn material characterisation was conducted in yarn testing on 2400 tex AR-glass rovings, used in the textile reinforcement. The samples are tested with a servo-controlled tensile testing machine in accordance with DIN EN ISO 9163 [6]. A preload of 0.5 cN/tex is applied at a clamping length of 250 mm. Then the tensile test is performed at a speed of 10 mm/min. Based on the properties determined, an initial extrapolation of the textile strength is possible independently of the textile structure with variable grid openings. Consequently, initial estimates of the number of reinforcement layers required can be made at the beginning of development.

Characterisation of the concrete matrix

Concrete for TRC should satisfy special requirements regarding the fluidity, mechanical properties, and durability of the reinforcing material. The parameters for mix design depend upon the fabrication technique, application and type of textiles used. The maximum size of aggregates is generally limited to 2 mm for better penetration of concrete through grid openings. Use of mineral admixtures like fly ash, metakaolin and blast furnace slag has proven to enhance the durability of the composite by reducing the calcium hydroxide content [7–9]. Further, silica fume incorporation enhances the bond between the textile and the matrix [10, 11].

The primary ingredients of the mix are as follows:

- Binder: (a) Portland cement, conforming to [12] (b) Class F fly ash (c) silica fume, conforming to [13]
- Aggregates: (a) quartz sand, with maximum particle size limited to 1.2 mm (b) quartz powder, with particle size in the range of 20 to 100 microns
- Chemical admixtures: polycarboxylic ether based superplasticizer in liquid form
- Water: Clean and potable.

The initial binder proportion is obtained by limiting the total mineral admixture percentage to 30% of the total binder and restricting the silica fume content within 5–10%. The particle size distribution for portland cement, fly ash and quartz powder is determined by laser diffraction. For silica fume, it is taken as given in the product specifications. Sieve analysis is performed to obtain the particle size distribution of quartz sand. The final proportioning of all the dry materials is done based on particle packing method, using EMMA software. The water-to-binder ratio is fixed based on the strength and workability requirements. The optimum dosage of superplasticizer is determined through flow tests. The ingredients are mixed in a planetary mixer. Initially, dry ingredients are mixed for two minutes at a lower rpm. Next, water mixed with superplasticizer is added to the dry mix, and it is blended for four minutes. Finally, the ingredients are mixed at a higher rpm for two minutes.

The sewage consists of organic sulphides, proteins and sulphates. The primary component causing biogenic corrosion in concrete is H_2S. Under anaerobic conditions, H_2S is biologically produced in sewer, which rises above the sewage level to the crown/roof portion, interacts with oxygen and moisture to produce biogenic sulphuric acid (H_2SO_4). This acid corrodes the concrete. Hence, the rate of degradation of mass and the effectiveness of surface coatings in minimizing this corrosion are under investigation. After 28 days of mist curing, 70 mm size cube specimens were immersed in H_2SO_4 [2% concentration and pH $\sim$ 1], with and without coating. This is the simulated acidic environment in the laboratory, which represents the real field condition in an accelerated method. The selected pH mimics the worst field condition.

Five types of coating materials available in Indian market were examined. They were two-component mineral powder, coal tar epoxy (two sources), silicate gel solution and calcium aluminate cement. The mass changes are recorded every week for the first one month and then once in every two weeks thereafter. The acid was replenished during every measurement activity, to maintain the pH. The study is under progress.

Numerical simulation of treatment units

For the structural simulation of TRC WTP, finite element modelling is used. The geometrical and the material modelling of the TRC components are carried out with the finite element analysis software ABAQUS. In addition to the conventional design of a cylindrical plant, a cuboidal configuration is also simulated. The size of rectangular shaped TRC tank is set at $L \times B \times H$ of $3 \times 2 \times 2.7$ m^3, and the wall and base slab thickness is 25 mm. For the cylindrical tank, the diameter $\times$ depth is 3×2.5 m; with height of the frustum as 0.5 m; and the wall thickness is 25 mm uniform throughout. All the walls and base slab are considered to be reinforced with three layers of AR glass textile of 2400 tex. Both cylindrical and rectangular TRC WTPs are simulated under different load conditions, such as bending, pressure and their combined effect and checked for the structural suitability. Five load cases, as described in Table 1, are considered. The TRC WTPs were assumed to be resting on a rigid base for all the cases.

2.2 Results and Discussions

The size of the plant is defined on the basis of the local conditions in the region of Jhansi, India. Due to the predominant clustering of households, the daily operating volume of wastewater is estimated at a value of 2.5 m^3. In order to simplify the production process, each chamber shall be housed in individual tanks. Furthermore, after discussions with the population, the need for an underground facility becomes clear, which might allow further usage of the area above. One of the main challenges in the design is the

Table 1. Load cases and combinations

Combination	Load case
Load combination 1	Tank empty with earth pressure
Load combination 2	Tank empty with earth pressure and surcharge
Load combination 3	Tank empty with uplift pressure
Load combination 4	Tank full with no pressure
Load combination 5	Handling stresses

compromise between economic efficiency and technical optimum. An example of this is the correlation between the grid opening and the maximum grain size of the aggregates in the concrete mix. The smaller the grid opening, the smaller the number of layers and consequently the wall thickness of the tank. However, small grid openings also require comparatively small aggregate diameters in the concrete mix. The cost of these fine-grained mixtures increases with decreasing grain diameter.

The textile reinforcement

The reinforcement will be realised with biaxial warp knitted textile, which allows an open grid structure with a straight roving course. The initial characterisation of the textile structures yields the maximum tensile strength shown in Fig. 2. It can be seen that pillar and counterlaid tricot in uncoated status show higher average maximum tensile strength than plain and tricot. Pillar reaches 4.6 kN, counterlaid tricot, 4.5 kN, whereas plain amounts to a maximum tensile strength of 3.7 kN and conventional tricot to 3.7 kN. This is caused by the course of the stitch thread. In the case of pillar and counterlaid tricot, the individual filaments of the roving are pressed tightly together by the knitting process. This increases the friction between them compared to the loose bond in tricot and plain. Load transfer from outer to inner individual filaments is improved and stress distribution across the cross-section is enhanced.

With regard to the effect of the coating, it is clear that finishing with both epoxy resin and styrene-butadiene-rubber (SBR) lead to an increase in the maximum tensile load. An exception here is the finishing of pillar with SBR, which, however, can be justified by the high scatter of the values. The increase in tensile strength is generally explained by the creation of a good bond between the individual filaments of the roving. In this way, the force transmission between the individual filaments is improved and made more uniform. It is clear that epoxy resin leads to a greater increase in strength, as it is more stiff after curing than the SBR.

Additionally, the single yarn characterisation yields a tensile strength of 1391 N for one epoxy-coated yarn, with a standard deviation of 230 N. SBR coated rovings achieve an average maximum tensile strength of 1056 N and a standard deviation of 86 N. With a maximum occurring force of 25 kN/m and a cross-sectional area of the rovings of 0.96 mm, the required number of textile layers was calculated taking into account safety factors based on, for example, the actual effective utilisation of the yarn strength. Table 2 shows an overview of the development of the required number of layers as a function

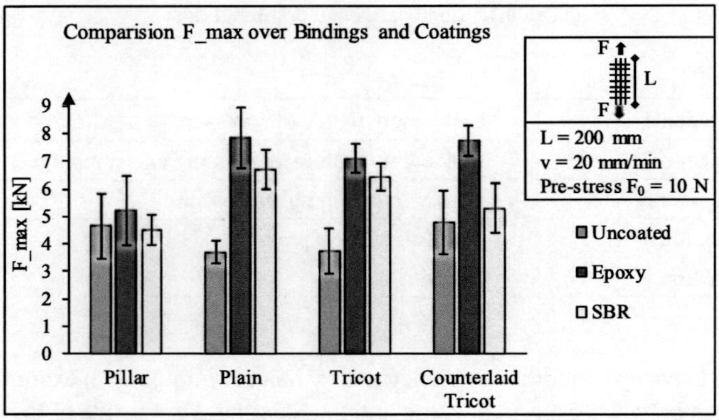

Fig. 2. Maximum tensile strength F_max of biaxial warp knitted textiles in form of pillar, plain, tricot and counterlaid tricot with different coatings.

of the grid opening and the coating material. As expected, with smaller grid openings and thus increasing reinforcement cross-sectional area per meter, the number of layers required becomes smaller. Likewise, the higher strength of the epoxy resin coated textiles described above reduces the number of layers compared to the SBR coated textiles.

Table 2. Calculation of numbers of reinforcement layers

Textile roving centreline distance (mm)	Coating	Required number of layers
8	Epoxy	1
	SBR	2
16	Epoxy	2
	SBR	3
25	Epoxy	4
	SBR	4

For the stipulated loads and their combinations on this treatment unit, the major effects would be bending and hoop tension. The layers of textile reinforcement are placed at appropriate direction to take the loads. This system does not necessitate placement of the layers of reinforcement in differential or shifted angles, as may be done in other composite systems, as the unidirectional or bidirectional loads and their effects are taken care of appropriately by a single orientation of textile reinforcement at any particular location.

The concrete matrix

Compression tests on 70.6 mm size cubes were conducted in compliance with IS: 4031 (Part 6) - 1988 [14]. The mix proportions by weight are as follows:

OPC 53: Fly ash: Silica fume: Quartz sand: Quartz powder = 1: 0.36: 0.07: 1.02: 0.61; Superplasticizer = 0.15% of binder content; Water-to-binder ratio = 0.38. The cast specimens were subjected to mist curing and tested at 28 days. The achieved compressive strength of the cubes was 58 MPa. Also, the flow diameter was found to be 320 mm, assessed in compliance with IS: 5512 - 1983 [15]. This mix was primarily found to satisfy the requirements and hence used to prepare the TRC composite panels along with counterlaid tricot textiles. Further trials are in progress to arrive at a robust mix with modified binders and varied aggregates, to achieve improved mechanical and durability performances.

The numerical simulation

From the FE analysis of both, rectangular and circular WTPs, the following conclusions were drawn. The loading condition of TRC WTP being empty with earth pressure and surcharge (Load combination 2) is found to be critical for both cylindrical and rectangular configurations. The maximum principal stress under tension was around 2.2 MPa for cylindrical WTP whereas the value is 4.4 MPa for rectangular case. The maximum tensile stress of TRC considered for the analysis was 4.3 MPa; indicating that the rectangular WTP needs redesigning with respect to the investigations carried out. The principal compressive stress is well below the compressive strength of the TRC in both cylindrical and rectangular WTPs. The displacement experienced is found to be higher for rectangular WTP than for cylindrical WTP. For the critical load combination, the maximum displacement experienced by rectangular WTP was around 29 mm whereas for cylindrical case the value is 0.15 mm.

3 Summary and Conclusions

The overall objective of the research presented in this paper was the preliminary determination of suitable materials for small WTPs made from TRC with regards to both, textile reinforcement and concrete matrix, as well as to understand the numerical simulation of the WTP considering different load cases.

It was found that while the measurements of tensile strength suggest the selection of epoxy-coating, the textile selection has to be based on the mechanical properties and the requirements of the production and handling. Based on this, the further development will be conducted under the utilisation of a biaxial warp knitted textile with a binding type of counterlaid tricot, which is produced from 2400 tex AR-glass rovings with a centreline distance of 16 mm. The coating will be chosen depending on the final structural design (rectangular and circular) and the resulting requirements form the manufacturing perspective. The high-strength matrix was chosen based on the mechanical and durability characterization. Further investigations on the modifications of matrix for better performance and reduced production cost are under progress. Based on the numerical simulation of both cylindrical and rectangular geometries, the final geometry to be considered is arrived as cylindrical, due to the structural performance advantage it offers.

Acknowledgement. The authors would like to thank the BMBF (Federal Ministry of Education and Research) Germany, the DLR Projektträger, Germany and the DST (Department of Science and Technology, Government of India) for funding and the coordinating the research project "CleanWater" within the Indo-German Science & Technology Centre (IGSTC). The mechanical testing at IIT Madras was carried out at the Laboratory for Mechanical Performance of Civil Engineering Materials, partially funded by the FIST Grant SR/FST/ETII-054/2012.

References

1. UN Water: The United Nations World Water Development Report 2020—Water and Climate Change, United Nations Educational, Scientific and Cultural Organization (2020)
2. Schellenberg, T., Subramanian, V., Ganeshan, G., Tompkins, D., Pradeep, R.: Wastewater discharge standards in the evolving context of urban sustainability–The case of India; policy and practice reviews. Front. Environ. Sci. **8**, Article 30, April (2020)
3. Libralato, G., Ghirardini, A.V., Avezzù, F.: To centralise or to decentralise: an overview of the most recent trends in wastewater treatment management. J. Environ. Manag. **94**(1), 61–68 (2012)
4. Peled, A., Bentur, A., Mobasher, B.: Textile reinforced concrete. CRC Press, Boca Ranton, Florida, USA (2017)
5. DIN EN ISO 13934-1: Textiles—Tensile properties of fabrics—Part 1: Determination of maximum force and elongation at maximum force using the strip method (ISO 13934-1:2013); German version EN ISO 13934-1:2013
6. DIN EN ISO 9163 Textile glass—Rovings—Manufacture of test specimens and determination of tensile strength of impregnated rovings (ISO 9163:2005); German version EN ISO 9163:2005
7. Marikunte, S., Aldea, C., Shah, S.P.: Durability of glass fiber reinforced cement composites: effect of silica fume and metakaolin. Adv. Cem. Based Mater. **5**(3–4), 100–108 (1997)
8. Butler, M., Mechtcherine, V., Hempel, S.: Experimental investigations on the durability of fibre-matrix interfaces in textile-reinforced concrete. Cem. Concr. Compos. **31**(4), 221–231 (2009)
9. Kong, K., Mesticou, Z., Michel, M., Larbi, A.S., Junes, A.: Comparative characterization of the durability behaviour of textile-reinforced concrete (TRC) under tension and bending. Compos. Struct. **179**(1), 107–123 (2017)
10. Scheffler, C., et al.: Interphase modification of alkali-resistant glass fibres and carbon fibres for textile reinforced concrete I: fibre properties and durability. Compos. Sci. Technol. **69**(3–4), 531–538 (2009)
11. Mechtcherine, V., Schneider, K., Brameshuber, W.: Textile Fibre Composites in Civil Engineering. Elsevier (2016)
12. IS 12269. Indian Standard ordinary portland cement, 53 grade—specification. Bureau of Indian Standards (2013)
13. ASTM C1240-15. Standard specification for silica fume used in cementitious mixtures. ASTM International (2015)
14. IS : 4031 (Part 6) - 1988, Indian Standard methods of physical tests for hydraulic cement—determination of compressive strength of hydraulic cement other than masonry cement. Bureau of Indian Standards, New Delhi
15. IS : 5512 - 1983, Indian Standard specification for flow table for use in tests of hydraulic cements and pozzolanic materials, Bureau of Indian Standards, New Delhi

Influence of Textile Orientation on the Quasi-static and Repeated Loading Behavior of Textile Reinforced Cementitious (TRC) Composites

M. El Kadi(✉), M. Ahmad, and T. Tysmans

Department Mechanics of Materials and Constructions, Vrije Universiteit Brussel (VUB), Pleinlaan 2, 1050 Brussels, Belgium
michael.el.kadi@vub.be

Abstract. Textile Reinforced Cementitious (TRC) composites combine the compressive strength of cement and the tensile resistance of fiber reinforcements. These materials have become increasingly popular in recent years; due to their desirable properties such as high tuneability, slenderness and freedom of form they provide smart solutions for progressive building designs. Currently, a lack of research exists that tackles complex and repeated loading conditions of these promising materials. The objective of this research campaign was threefold. Firstly, the influence of the angle of unalignment between the textile reinforcement and the loading direction was investigated in quasi-static loading conditions (0°, 22.5° and 45° were considered), focusing on the TRC strength, post-cracking stiffness and crack evolution. Secondly, these parameters were investigated after repeated loading conditions. For this purpose, an experimental campaign was performed where the specimens were repeatedly loaded during $N = 25$ cycles between a residual stress in the pre-cracked stage and 0.5 times the failure load obtained from the quasi-static tests. Thirdly, the residual properties of the beams tested in repeated loading were characterized by means of a quasi-static test until failure.

A total of 18 (500 mm × 150 mm × 15 mm) TRC samples were tested in four-point bending (three for each test series and orientation). The samples were reinforced by means of a single textile reinforcement grid, the longitudinal fiber volume fraction (V_f) was 0.74%. From the experimental results, a clear influence of the angle of unalignment on the quasi-static and repeated loading behavior of the TRCs was observed.

Keywords: Repeated loading · Off-axis loading · Quasi-static loading · Textile Reinforced Cement (TRC) composites

1 Introduction

Textile Reinforced Cement (TRC) composites offer a high-potential alternative to traditional building elements by combining a cementitious matrix with textile reinforcement [1–3]. The matrix warrants the compressive resistance whereas the textile reinforcement

J. I. Escalante-Garcia et al. (Eds.): RW 2021, RILEM Bookseries 40, pp. 897–905, 2023.
https://doi.org/10.1007/978-3-031-21735-7_95

provides the tensile capacity of the elements. Due to their high-performance aspects, TRCs can be found in many industrial applications, like firewalls [4], large-size façade elements [1, 5], noise and water barriers [1], gypsum board joint finishing systems [1], pedestrian bridges [1, 6], barrel vaults [7] and grid structures [7]. Exploratory research projects and pilot studies have also been performed worldwide, incorporating TRC in prefabricated slabs and wall elements [8], stay-in-place formwork elements for floors and beams [9] and seismic retrofitting systems [10]. TRCs can also efficiently be tuned and optimized for the reinforcement of structures subjected to complex loading conditions such as thin shell structures that exploit both the slenderness of the material and the form freedom of the fiber reinforcement. However, to accurately design these structures, the state-of-the-art knowledge on TRC behavior is not sufficiently advanced: while the loading of TRCs aligned with the textile orientation has already been extensively studied and described in literature [11–13], one of the main barriers is the very limited insight on the material's behavior under loading unaligned with the fiber direction.

Only a very limited number of unaligned tests have been performed on TRC [7, 14] and as a result, there is little knowledge available on the mechanical behavior of TRC subjected to these loading conditions. Therefore, the first objective of this study was to investigate the influence of the loading orientation with respect to the fiber orientation in TRC composites. For this purpose, three different fiber orientations were considered in a quasi-static four-point bending experimental campaign: (i) a reference orientation with fibers aligned with the loading direction (0°) and two orientations with angles of respectively (ii) 22.5° and (iii) 45° between the loading direction and the fiber alignment. The second objective of this research was to investigate the influence of these unalignment on the early decay of mechanical properties of the TRCs. For this purpose, the same configurations were considered in a repeated loading experimental campaign, up to 25 loading cycles. Lastly, the residual mechanical properties after the repeated loading experiments were assessed by means of quasi-static tests up to failure. All displacements and crack monitoring happened by means of the Digital Image Correlation (DIC) technique [15].

Section 2 gives an overview of the selected materials for the TRC composites as well as the employed loading and monitoring methods. Section 3 gives an overview of the experimental results and discussion and lastly, the main conclusions are summarized in Sect. 4.

2 Materials and Methods

2.1 Matrix Material

A commercial grout [16] was selected as matrix for its low viscosity before hardening and compensated shrinkage properties. The mixture consists of Ordinary Portland Cement, siliceous sand and admixtures and the relatively small aggregate size allows for optimal flow through the textile meshes. Table 1 summarizes the properties of the mortar.

Table 1. Properties of the cementitious matrix, based on [16]

Aggr. Size (mm)	Compr. Strength (MPa)	Modulus of rupture (MPa)	Density (kg/m^3)	Young's modulus (GPa)	W/C ratio (–)
0–1.6	70	12	2010	9	0.15

2.2 Textile Reinforcement

The selected textile reinforcement was a commercially available, bi-directional, SBR-coated AR-Glass textile [17] with two textile yarns per mesh side (see Fig. 1). The properties of the textile are given in Table 2.

Fig. 1. Planar AR-Glass textile reinforcement.

Table 2. Properties of the textile reinforcement, obtained from [17].

Material	Grid size (mm × mm)	Textile density coated (g/m^2)	Textile density uncoated (g/m^2)	Yarn stiffness (GPa)	Yarn strength (MPa)
AR-Glass 2400 tex	17.5 × 17.5	653	568	67	526

2.3 TRC Geometry and Manufacturing Process

The considered specimen geometry was 500 mm × 150 mm × 15 mm (length × width × thickness). All configurations had a bottom mortar cover of 2 mm, one textile reinforcement layer of 1.25 mm and a top mortar cover of 11.75 mm. The unidirectional fiber volume fraction was 0.74%. For each of the configurations (0°, 22.5° and 45°), a total of six specimens were manufactured (three for the quasi-static loading campaign and three for the repeated loading campaign + the residual quasi-static campaign). The six specimens were manufactured in one mold system where one textile reinforcement layer was continuously clamped throughout the multiple specimens (see Fig. 2). After application of demolding oil, the pre-mixed mortar was poured over the textiles until the molds were filled. The molds were then sealed-off and left to harden at ambient conditions for 28 days. After demolding, the extruding textile reinforcement was cut-off and a speckle pattern was applied on the specimen's sides for DIC monitoring.

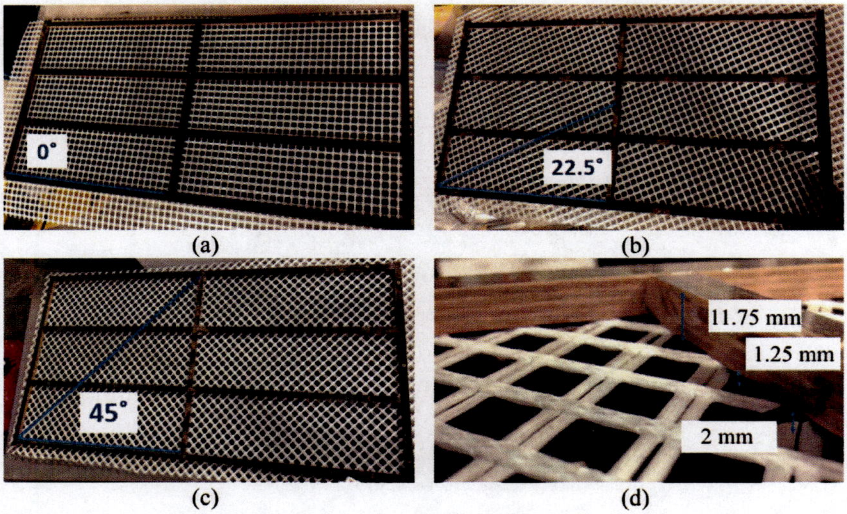

Fig. 2. Image of molds with textile reinforcement layer clamped at **a** 0°, **b** 22.5° and **c** 45° as well as close-up of textile placement within the mold.

2.4 Experimental Setup and Loading Conditions

The four-point bending experiments were performed using an Instron 5885H electromechanical test bench at a rate of 2 mm/min. Two DIC cameras monitored the displacements of the specimens' sides, the setup is shown in Fig. 3. The distance between the supports and loading pins were respectively 350 mm and 100 mm.

The quasi-static experiments were continued until specimen failure. The repeated loading experiments were loaded from an unloaded state to 50% of the failure load (for the 0° and 22.5° configurations) or 35% of the failure displacement (for the 45° configuration) and then unloaded until a loading state in the pre-cracked zone of the samples (250 N f or all configurations), during 25 cycles. This testing procedure was defined by the authors as most optimal to identify the effects of low-cycle repeated loading on TRC composites. The exact lower and upper load limits of the repeated loading campaign are given in Table 3.

Fig. 3. Experimental four-point bending test setup.

Table 3. Lower and upper load limits of the repeated loading campaign.

	0°	22.5°	45°
Lower load limit (N)	250	250	250
Upper load limit (N)	1250	675	–
Upper displacement limit (mm)	–	–	8.4

3 Experimental Results and Discussions

3.1 Quasi-static Experimental Results (Uncracked)

This section shows the results of the quasi-static experimental campaign performed on the three TRC configurations (0°, 22.5° and 45°). Figure 4 shows an overview of the quasi-static four-point bending test results for all three configurations. Table 4 shows the values of the post-cracking stiffness and failure loads of all configurations. Figure 5 shows the DIC images of a representative specimen's side and lastly, Table 5 gives an overview of the number of cracks, crack spacing and crack width for all configurations at a constant displacement value of 20 mm.

As can be seen in Fig. 4 and Table 4, a clear decrease in post-cracking stiffness and failure stress was observed for increasing angle of unalignment between the textile reinforcement and the loading direction. Compared to the reference 0° configuration, the 22.5° and 45° showed a decrease in post-cracking stiffness of respectively 20% and 28%. For the failure load, this decrease was respectively 45% and 60%. For both the post-cracking stiffness and the failure load, the value of decrease of the 22.5° was located closer to the 45° than the 0°. Even though only three data points were available, these results hinted that the decrease in mechanical properties might not be linear with increasing angle of unalignment.

At the level of the crack analysis, Fig. 5 and Table 5 show that the higher the unalignment between the textile reinforcement, the lower the amount of cracks and consequently,

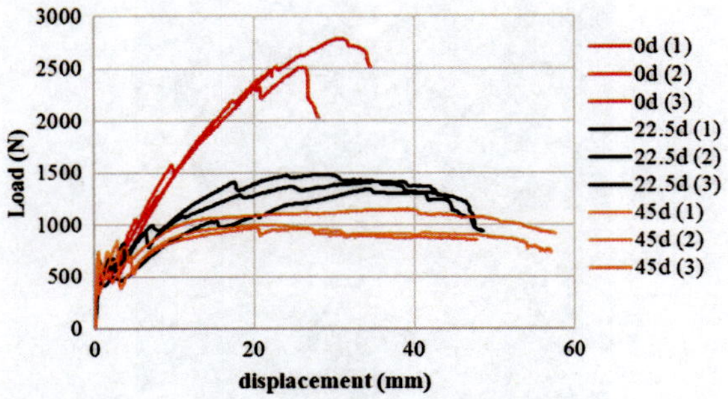

Fig. 4. Quasi-static test results for all configurations.

Table 4. Post-cracking stiffness and failure load for the quasi-static test.

	0°	22.5°	45°
Post-cracking stiffness (N/mm)	127.73 ± 6.87	102.11 ± 10.81	91.61 ± 7.72
Failure load (N)	2590.40 ± 135.59	1415.51 ± 58.94	1039.84 ± 77.51

the higher the individual crack openings. A notable observation was that in the configurations with textile unalignment (22.5° and 45°), the same number of cracks could not be reached as in the aligned (0°) configuration. Instead, the limited number of cracks that formed, opened further until specimen failure.

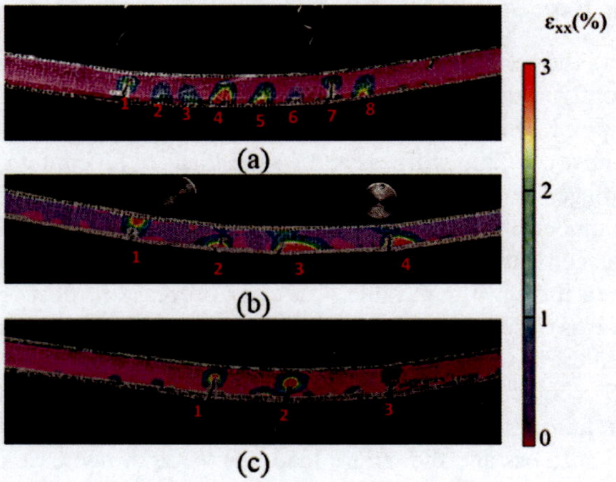

Fig. 5. Longitudinal strain field as obtained from DIC analysis for the **a** 0°, **b** 22.5° and **c** 45° configurations at a constant displacement of 20 mm.

Table 5. Average number of cracks, crack spacing and crack widths for all configurations at a constant displacement of 20 mm.

	0°	22.5°	45°
Nr. of cracks	7.33 ± 0.94	3.67 ± 0.47	2.67 ± 0.47
Crack spacing (mm)	14.92 ± 6.37	30.8 ± 2.57	45.54 ± 9.73
Crack width (mm)	0.37 ± 0.07	0.74 ± 0.16	1.08 ± 0.22

3.2 Repeated Loading Experimental Results

Figure 6a–c show the curves of the repeated loading (25 cycles) experimental campaign for a representative sample of each configuration (respectively 0°, 22.5° and 45°). Compared to the first loading cycle, the three configurations showed an immediate decay in post-cracking stiffness of 10% in the first reloading cycle. In the subsequent cycles, the post-cracking stiffness decreased almost linearly up to a total decay of 20% compared to the initial post-cracking stiffness of the respective configurations. This decaying trend was observed both in the loading and the unloading parts of the consecutive cycles.

These results showed that a notable decay in post-cracking stiffness was already observed in very early stages of repeated loading. Furthermore, the rate of decay did not seem to depend on the angle of unalignment.

3.3 Residual Quasi-static Experimental Results

After the 25 cycles of repeated loading, the samples were tested again from an unloaded state until failure. Figure 6d shows the residual quasi-static experimental curves for all configurations. Table 6 shows the values of the post-cracking stiffness and failure loads of all configurations. The residual quasi-static stiffness of the specimens that were previously tested in repeated loading (Table 6) was consistently higher (38%, 54% and 11%) than the post-cracking stiffness of the uncracked specimens (Table 4). These results showed that even though a decay of the stiffness was observed during the repeated loading cycles, the residual stiffness of the TRC (starting from an unloaded state) remained higher than the post-cracking stiffness of an uncracked specimen. In other words, this means that the 25 loading cycles (between the given load limits) did not fully damage the bond between matrix and textiles or thus that the composite action was not fully lost. On the other hand, the failure load of the damaged specimens was consistently lower than the reference samples (7%, 9% and 14%).

4 Conclusions

This research presented the experimental results of three different TRC configurations in four-point bending. The angle of alignment between the textile reinforcement and the specimens' longitudinal direction was varied between 0°, 22.5° and 45°.

In a first instance, quasi-static tests until failure showed a clear reduction in post-cracking stiffness and failure load with increasing angle of unalignment. At a constant

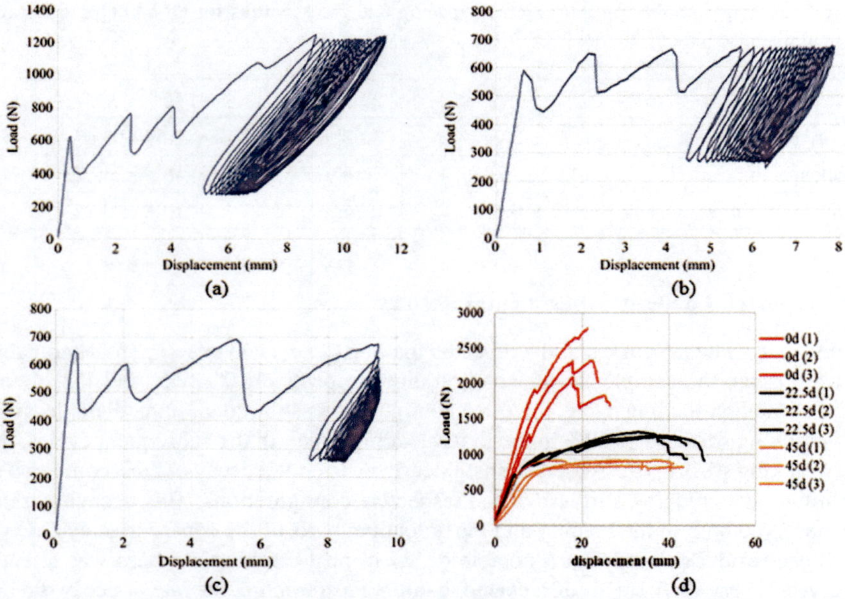

Fig. 6. Repeated loading test results for the **a** 0°, **b** 22.5° and **c** 45° configurations and **d** residual quasi-static test results for all configurations.

Table 6. Post-cracking stiffness and failure load for the residual quasi-static test.

	0°	22.5°	45°
Post-cracking stiffness (N/mm)	176.56 ± 11.72	157.60 ± 6.18	101.78 ± 7.64
Failure load (N)	2403.39 ± 189.14	1281.44 ± 55.12	892.44 ± 62.38

displacement value, less but larger cracks were observed for the configurations with higher angle of unalignment.

Secondly, the effect of 25 repeated loading cycles on the mechanical behavior of the TRC was investigated. The experimental results showed that for all the configurations, the first reloading cycle exhibited a decay of approximately 10% in post-cracking stiffness, which further decayed up to 20% compared to the initial post-cracking stiffness.

Lastly, the specimens that underwent the repeated loading cycles were tested from an unloaded state up to failure. The residual stiffness of these samples was (up to 54%) higher than the post-cracking stiffness of the quasi-static, uncracked specimens. This showed that between the given loading stages, the specimens did not entirely lose their composite action during the 25 stages of repeated loading.

Further research should investigate the evolution of these parameters after a larger amount of repeated loading cycles. Additionally, the decrease in post-cracking stiffness with the angle of unalignment as well as the decay with increasing cycles should be investigated in wider specimens, where the unaligned fibers are embedded over larger distances within the samples.

References

1. Brameshuber, W.: Textile Reinforced Concrete—State-of-the-Art Report of RILEM TC 201-TRC, 1st edn. RILEM, Aachen (2006)
2. Bentur, A., Mindess, S.: Fibre reinforced cementitious composites. Mod. Concr. Technol. Ser. (2007)
3. Naaman, A.E.: Textile reinforced cement composites: competitive status and research directions ferrocement. In: International RILEM Conference on Material Science, vol. I, pp. 3–22 (2010)
4. Hiel, C.: Improving the reliability of the electric grid infrastructure: case study of firewalls manufactured with fiber reinforced inorganics (FRI). In: Proceedings of the American Society for Composites, p. 14 (2009)
5. Engberts, E.: Large-size façade elements of textile reinforced concrete. In: Proceedings of International RILEM Symposium Textile Reinforced Concrete, pp. 309–318 (2006)
6. Helbig, T., Rempel, S., Unterer, K., Kulas, C., Hegger, J.: Fuß- und Radwegbrücke aus Carbonbeton in Albstadt-Ebingen. Die weltweit erste ausschließlich carbonfaserbewehrte Betonbrücke. Beton- und Stahlbetonbau **111**, 676–685 (2016)
7. Hegger, J., Voss, S.: Investigations on the bearing behaviour and application potential of textile reinforced concrete. Eng. Struct. **30**, 2050–2056 (2008)
8. Holschemacher, K., Klug, Y., Kobisch, S., Weiβe, D., Zscheile, H.: Textile reinforced slabs and prefabricated double walls. In: Proceedings of international RILEM symposium Textile Reinforced Concrete, pp. 319–330 (2006)
9. Papantoniou, I.C., Papanicolaou, C.G.: Textile Reinforced Concrete (TRC) for precast Stay-in-Place formwork elements. Tailor Made Concr. Struct. **8**, 475–481 (2008)
10. Papanicolaou, C.G., Triantafillou, T.C., Karlos, K., Papanthanasiou, M.: Seismic retrofitting of unreinforced masonry structures with TRM. In: Proceedings of International RILEM Symposium Textile Reinforced Concrete, pp. 341–350 (2006)
11. Peled, A., Haik, R., Sasi, E.A.: Influence of three-dimensional (3D) fabric orientation on flexural properties of cement-based composites. **80** (2017)
12. Peled, A., Bentur, A., Mobasher, B.: Textile Reinforced Concrete (2017)
13. El Kadi, M., Tysmans, T., Verbruggen, S., Vervloet, J., De Munck, M., Wastiels, J., Van Hemelrijck, D.: Experimental study and benchmarking of 3D textile reinforced cement composites. Cem. Concr. Compos. **104** (June 2018), 103352 (2019). https://doi.org/10.1016/j.cemconcomp.2019.103352
14. Bruckermann, O., Hegger, J., Sherif, A.G.: Modeling of textile reinforced concrete with inclined textile orientation. ACI Mater. J. **104**(5), 511–519 (2007)
15. Sutton, M.A., Orteu, J.-J., Schreier, H.W.: Image Correlation for Shape, Motion and Deformation Measurements. Springer Science + Business Media (2009)
16. Sika. Sikagrout 217, Note de produit. https://fra.sika.com/fr/solutions-produits/marche-construction/voirie/produits-systemes/scellement-calage-grandes-dimensions.html (2016)
17. Fraas, V.: Technical Datasheet SitGrid200. V. Fraas (2017). www.solutions-in-textile.com. Accessed 01 May 2017

Check for updates

Compressive Behaviour of Alkali Stabilized Quarry Sludge Blocks

Nicolas Zapata(✉), Andres Restrepo, Yhan Arias, and Juan Ochoa

Universidad Nacional de Colombia, Medellín, Colombia
nizapatape@unal.edu.co

Abstract. There is a global interest in alternative building elements that accomplish an eco-friendly fabrication process and offer economical solutions. Raw quarry sludge was a residue with no acceptable compressive strength nor well-durability performance; hence, stabilization through physical and chemical additions was performed. Alkali stabilized quarry sludge (ASQS) can contribute to the reuse of industrial mining wastes and even use them as main materials on construction prefabricated elements, aligned with the circular economy principles. This study presents the experimental validation of a previously mechanical simulation of a block with solid geometry, using ASQS as the main material. Validation was carried out through a compressive strength test. Besides, block characterization is effectuated with different standardized tests such as capillary absorption, Abrasion, and Efflorescences. Thus, it could be determined that quality block standards are not accomplished due to issues in compaction as traditional techniques are employed. This way, further research, and development must be applied to the masonry element fabrication process, seeking the achievement of a new sustainable material and an economical prefabricated element with adequate mechanical performance.

Keywords: Waste · Construction · Reutilization · Stabilization

1 Introduction

Currently, the construction sector remains traditionalist in some implementations. For instance, cement-based materials, concrete is the second most used material in the world, just after water [1]. These, in turn, result in demand for raw materials exploitation such as gravel, sand, limestone, and clay, i.a. Extracting 60 billion tons of minerals worldwide annually and growing at very high rates [2, 3]. Besides, the output products include in their manufacturing process stages with high energy consumption, becoming responsible for most greenhouse gas emissions. Flower and Sanjayan [4] reports 0.29 t CO_2-e/m^3 of a typical 25 MPa concrete, where cement production represents 82%. Research on alternative cement and replacements has been continuously made, which has demonstrated good performance in concrete admixtures [5] and can reduce concrete CO_2 emissions by around 15% [4]. Although efforts have been made in this matter, other proposals are based on more radical options that reduce cement [6] or even neglect

J. I. Escalante-Garcia et al. (Eds.): RW 2021, RILEM Bookseries 40, pp. 906–915, 2023.
https://doi.org/10.1007/978-3-031-21735-7_96

its use by implementing industrial residues. This way, some contributions are made to shift from a linear economy conception to a new paradigm known as circular economy, mainly applying the 3R's principle, recycling, reducing, and reusing [7]. Aligned with this paradigm, a process called alkaline activation and its subgroup geopolymerization, where a variety of wastes rich in silica and alumina, such as clay, fly ash, silica fume, slag, i.a, called precursors are reacted under alkaline conditions, using an alkaline solution called activator, ending up in a hardened material with alkali-aluminosilicate phases [8, 9].

Quarry Sludge (QS), the studied residue, is a raw materials exploitation by-product, specifically generated from the production of aggregates used in concrete. QS's nature is clayey. This type of soil can be physically stabilized by adding CaO charges, seeking C-A-S-H and C-S-H gel production [10]. Besides, it is chemically composed mostly of silica and alumina. Then, is also suitable for geopolymerization. The authors have previously developed through a response surface experimental design an improvement process for QS adding Na(OH) and $Ca(OH)_2$, giving, as a result, Alkali Stabilized Quarry Sludge (ASQS) with better mechanical behavior and durability characteristics [11]. This process is defined as alkaline stabilization as alkali-aluminosilicate phases remain unproven. Albeit, mechanical properties improvement are evinced.

Initially, ASQS seems to be appropriate for being implemented as a building material, nevertheless, its low workability discards in-situ use, but there is still the alternative of the precast element, as its mechanical behavior and water exposition durability were upgraded. An application niche is the concrete block and fired-clay bricks industry, as masonry elements have been present since the ancient Egyptian civilization [12] and it is a fundamental element in current building projects. In addition, aside from the energy consumption of cement production as it is used in concrete block fabrication, clay bricks also result in negative environmental impacts. A brick kiln emits about 70–282 g of carbon dioxide per kilogram of brick fired [13].

Last but not least, since less polluting materials are by the same token cheaper as reused residues, it raises the possibility of contributing to social issues such as some cities' housing deficit. In Colombia, 36.6% of housing presents a habitational deficit [14]. Economical gaps do not allow access to appropriate building materials, leading the poorest families into the development of weak structures without dimensional stability nor enough strength that certainly will not bear seismic activities, putting at risk their integrity. Undoubtedly the employment of residual stabilized materials, such as ASQS, in prefabricated elements production also represents a contribution to this matter.

2 Materials and Methods

2.1 Raw Materials

QS was collected from "Arenas del Sur" quarry located in Amaga, a west Colombian mining town known for supplying building materials to nearby cities. This quarry employs a standard washing process to remove fines presence in aggregates as ASTM-C117 establishes [15], separating the sludge in a sedimentation tank, where samples were taken. Sodium Hydroxide flakes Na(OH) with 98% purity and commercial Lime powder

38.47%—$Ca(OH)_2$, (5%—CaO) and (56% $CaCO_3$ and minor elements). Were acquired for the alkali stabilization process.

2.2 Characterization and Mix Design

QS Particle size distribution presenting D_{10}: 6 μm; D_{30}: 17 μm; D_{60}: 34 μm. Chemical and physical properties are shown in Table 1, it is classified as silty clay (Low plasticity) according to ASTM D2487 [16]. As QS is extracted from a tank it is found saturated with 100% natural water content (ASTM D2216-19) [17]. Specific Gravity (ASTM D5550-14)[18] was measured with Ultrapyc 5000 Anton Paar's nitrogen gas pycnometer. And X-ray fluorescence was used to determine the chemical composition. Finally, qualitative determination of crystalline phases was analyzed by X-Ray Diffraction CuKα1 ($\lambda = 1.54059$ Å) in a PANalytical X´Pert MPD PRO between 5° and 60° with a step size of 0.02° and counting time of 56 s. (NaOH-9.36M) was used in solution and added to a mixture of QS(powder dry) 92%-mass and commercial Lime 8%-mass. This dosage was previously evaluated by Restrepo et al. in 2020 [11]. Called Alkali Stabilized Query Sludge ASQS in this research. When alkali stabilization is done, results of ASQS present a 23% environmental water content. Plastic Limit (ASTM D4318-17) [19] reduction to the point it was not possible to test it and a 2% Specific Gravity decrease.

Table 1. QS and ASQS properties.

Properties	QS value	ASQS values
Physical		
Natural water content	100%	23%
Liquid limit	30.2%	41.08%
Plastic Limit	22.4%	–
Plasticity index	7.6%	–
Specific gravity	2.66	2.61
Chemical		
SiO_2	53.04	47.4
Al_2O	16.16	14.6
Fe_2O_3	7.24	7.76
K_2O	4.19	3.51
CaO	1.05	5.13
MgO	2.18	2.38
Na_2O	3.20	7.86
Others	0.3	0.3
LOI	12.64	11.06

The high content of silica and alumina (78% mass) for QS, with a SiO_2/Al_2O_3 modulus of 1.49 is related to the formation of conglomerating phases under alkaline activation conditions, according to [20], although the activation conditions are associated with low microstructure crystallinity of the alumina and silica precursors. For the QS, the diffractogram (Fig. 1A) shows typical clay minerals such as montmorillonite, illite, muscovite, and albite. Quartz and orthoclase are more related to the nature of the sandy exploitation. Therefore, it could be classified QS as a material with high crystallinity, however, under alkaline activation conditions some peaks disappear, in addition, when stabilized ASQS (Fig. 1B) sodalite and zeolite phases are formed, which suggests microstructural changes.

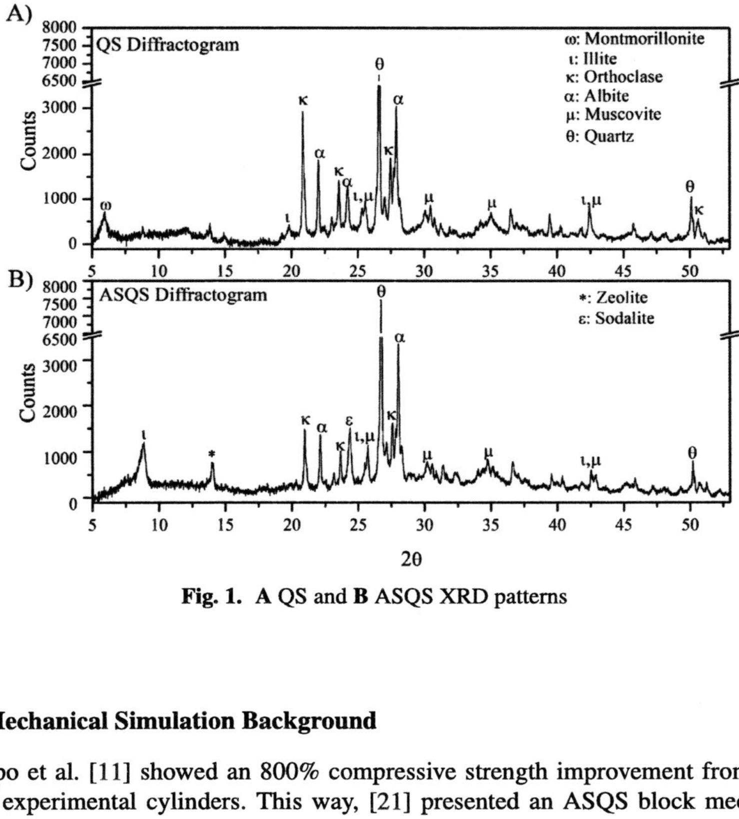

Fig. 1. **A** QS and **B** ASQS XRD patterns

2.3 Mechanical Simulation Background

Restrepo et al. [11] showed an 800% compressive strength improvement from QS to ASQS experimental cylinders. This way, [21] presented an ASQS block mechanical simulation using the Finite Element Method (FEM), employing those ASQS cylinder experimental properties obtained, where it was concluded that solid block elastic stress is around 0.5 times QS material compressive bearing capacity. Stating that an 8 MPa ASQS compressive strength bears only 4 MPa stress on the element. ASQS block experimental mechanical performance is sought to be validated on this simulation basis, focusing the attention on this material-block compressive relation.

2.4 Block Fabrication and Tests

The experimental diagram is presented in Fig. 2. Based on [11], before QS stabilization, waste processing was done. Following, solid blocks were fabricated using a traditional mechanical compactor machine called CINVA-RAM [22] with specific dimensions of 30 cm length, 15 cm width, and 10 cm height. Mixing was done employing a concrete mixer. Blocks were packed with plastic film to ensure hydration in the admixture. Besides, curing was maintained for 28 days at 40 °C. Subsequently, samples were environment dried for 24 h and finally, standard tests were carried out to validate the initial simulations and characterize the element.

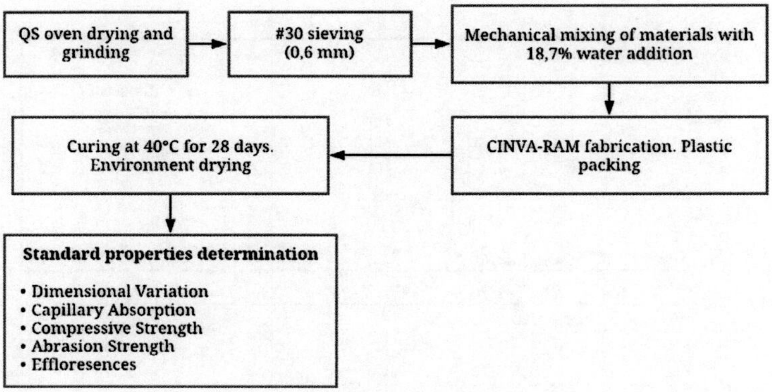

Fig. 2. General experimental diagram

Assessment of the suitability of ASQS blocks was rectified through several tests proposed to characterize according to Colombian standard NTC 5324 [23] for soil-cement blocks classification, as it is the most similar material, such as dry and wet compressive strength, capillary absorption, Abrasion, and Efflorescence. Albeit, due to difficulties in material workability, wet and dry compressive strength could not be done. So, it was chosen to simply assess unconfined compressive strength in three orthogonal directions.

3 Results and Discussion

3.1 Physical Characterization

Colombian Standard classifies soil-cement blocks according to their physical characterization, durability, and mechanical behavior. In the first section can be found tests such as Capillary Absorption and Lateral Abrasion. Additionally, apparent density was included in Fig. 3 where each bar represents the average and thin vertical lines the standard deviation. The first one measures weight increment caused by block material absorption due to its porosity and capillary water property, this test was carried out for five specimens, maintaining a 5 mm water level on the bottom for ten minutes and measuring mass gain expressed in percentage. Elements are classified as weakly capillary when weight increase is less than 20% or little capillary when is more than 20%. Thus, ASQS specimens are labeled in the first category as having an average of 19.3.

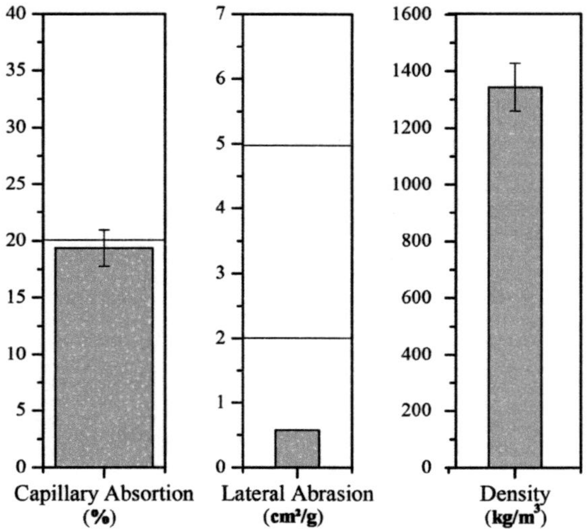

Fig. 3. Physical characterization outcomes

Abrasion test consists of a 2 kg weight steel brush that traverses 60 times across the block's visible face. It is expressed in terms of brushed surface per weight lost. Classifying this behavior on BSC20, BSC40 BSC 60 according to 2, 5, and 7 abrasion coefficients respectively. Three specimens were assessed resulting in a 0.5 coefficient, neglecting its adequate performance. Representing 2.6% of a soil-cement block abrasion test. It was observed that best abrasion performance blocks also had more density, indicating a possible cause of low results, in support of capillary absorption, denser blocks achieve more weight increase explained by smaller voids propitiate water rising [24]. Then, it was noted that 1340 kg/m^3 average apparent density represents a 4% decrease concerning the initial ASQS cylinder.

3.2 Compressive Strength Test

Wet and Dry compressive strength and classification proposed in [23] standard states cutting the element in two. This action could not be done as ASQS material plasticity does not allow a clean cut. Alternatively, Unconfined Compressive Strength, applying a load rate of 30 kN/min, was measured at different orientations. Just as [25] demonstrated that one-dimensional compaction generates anisotropic fabric, bearing this in mind it is sought to establish this strength direction relation. Figure 4 shows three different loading directions and the average result on each bar with its respective standard variation.

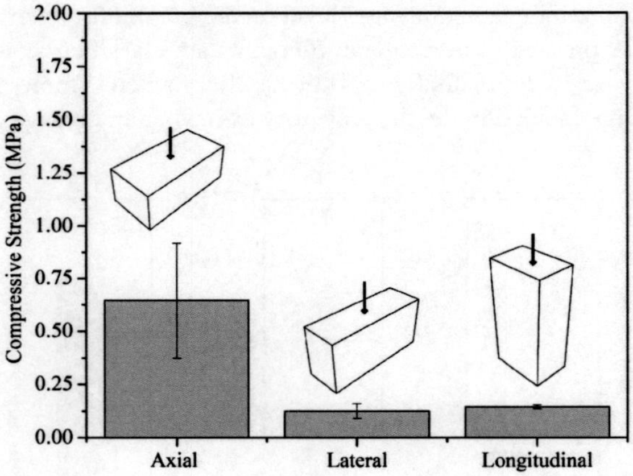

Fig. 4. Compressive strength on different directions

Table 2. Compaction parameters of ASQS cylinder and block.

Properties	Applied Load (kgf)	Area (cm^2)	Compaction pressure		
			kgf/cm^2	MPa	psi
ASQS cylinder	4000	19.635	203.7	20.3	2944
ASQS CINVA RAM block	1060	450	2.3	0.23	33.5

Initial ASQS cylinder compressive strength by [11] presented a 7 MPa bearing capacity. Furthermore, [21] mechanical simulation relation establishes that for a 7 MPa material result, ASQS block will only bear 3.5 MPa. Nevertheless, experimental results showed quite the opposite, an axial load resulted in 0.65 MPa. 0.13 and 0.14 for lateral and Longitudinal respectively, emerging on an orthotropic relation of around 0.2 times axial compressive strength equal to a Lateral and a Longitudinal Strength. Then, further research was done related to fabrication conditions, this was justified by density decrease from cylinder to block, even corrected density due to acquired humidity (Fig. 5). Ends up on a 15% decrease, this is a consequence of fabrication parameters difference (Table 2), as in cylinder production 4 tonf load is applied for compaction, on contrast for block only 1060 kgf, representing around just 25% of initial conditions impeding total density reach. This way can be proposed with the aim of confirming [25] statements. Density influences block mechanical response. Also, Fig. 5. Allows appreciating Compressive Strength growing tendency as density increments.

3.3 Efflorescences

Altough the efflorescences test is a common sintered brick test, considering the Na(OH) addition for Alkali Stabilization, it was proposed for ASQS block. This test was referenced on NTC 4017 [26], it consists of a visual inspection comparing a witness sample

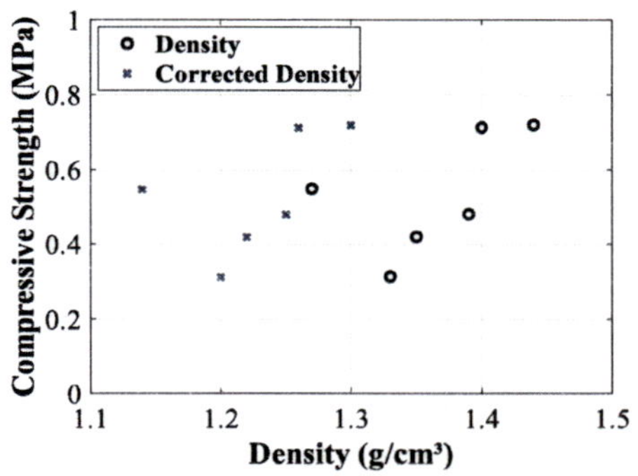

Fig. 5. Density-Strength relation

with the tested ones, determining three levels; No efflorescent, Negligible Efflorescent, and Efflorescent according to the qualitative characterization of soluble salts on the specimen's surface. Salts outward flux is promoted by letting samples bottom submerged 25 mm for seven days. Four blocks were tested and one block served as a witness sample.

As can be observed in Fig. 6B–E specimens showed crusting, patching with an abundance presence of salts, according to this they must be classified as Efflorescent. While C specimen was only affected on the edge with little salt presence, being classified as Negligible Efflorescent.

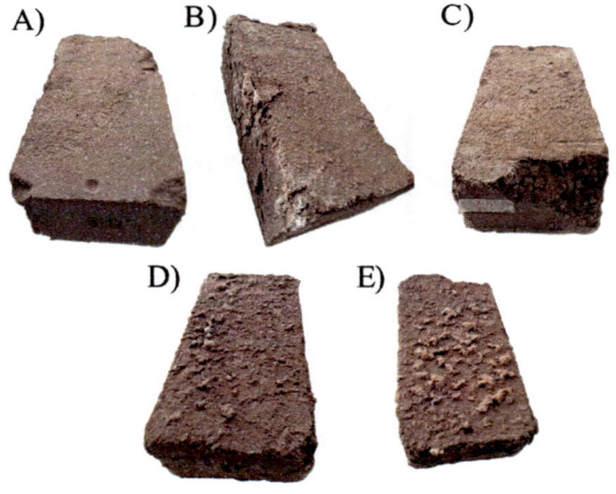

Fig. 6. **A** Efflorescence witness specimen. **B–E** Efflorescence tested specimens

4 Conclusions

This Study proposed ASQS implementation on building elements such as solid blocks employing traditional fabrication techniques. To assess its suitability, a Compressive

Strength test was carried out comparing outcomes with previous mechanical simulations. Furthermore, Physical and durability characterization was also done. Following contributions on this matter are made:

Clay and traces of silica in the QS composition did not allow efficient alkaline activation of the waste materials. However, the decrease in plasticity was high enough to consolidate a building material. It is required to increase the content of silica and alumina reactive or amorphous to achieve the formation of binder materials.

ASQS cylinder mechanical outcomes and ASQS block mechanical simulation prediction are not fulfilled. Caused by lack of cylinder conditions compaction replica on fabrication process, representing 1% load applied, which directly affects block density. Nevertheless, according to material-block relation, material strength must increase by 67% to classify ASQS blocks for non-structural use.

ASQS block's first iteration could not achieve a suitable performance based on [22] national standards. Even so, it was possible to understand difficulties in the transition from a material to a building element. Especially how density affects Capillary absorption due to pores size. Along with Abrasion as voids presence promoted brush penetration.

Efflorescences allowed evaluation of salts arising on block surfaces, this is connected with excessive amounts of Na(OH) in the stabilization process, as only a fraction reacts inside ASQS admixture. This way, a more optimal stabilization mix must be defined.

ASQS block general performance must be improved by researching and figuring out a better fabrication process including higher compaction loads, and by understanding microstructural admixture behavior to define a new stabilization protocol.

5 Future Work

This research projection is to get an optimal design of ASQS masonry units. This design will be driven by improving stabilization and fabrication processes through material comprehension. The main goal is to come up with an applicable prototype.

Funding. Thanks are given to "Aprovechamiento de lodos de flotación como subproducto de la industria minera antioqueña para el desarrollo de un prototipo de prefabricado constructivo que pueda integrarse en el sector de la construcción" HERMES 50012 Project, Construction Laboratory, Arenas del Sur and Universidad Nacional de Colombia for research funding.

References

1. Gagg, C.R.: Cement and concrete as an engineering material: an historic appraisal and case study analysis. Eng. Fail. Anal. **40**, 114–140 (2014)
2. Krausmann, F., Erb, K.-H., Gingrich, S., Lauk, C., Haberl, H.: Global patterns of socioeconomic biomass flows in the year 2000: a comprehensive assessment of supply, consumption and constraints. Ecol. Econ. **65**(3), 471–487 (2008)
3. Krausmann, F., Gingrich, S., Eisenmenger, N., Erb, K.-H., Haberl, H., Fischer-Kowalski, M.: Growth in global materials use. GDP and population during the 20th century. Ecol. Econ. **68**, 2696–2705 (2009)
4. Flower, D.J.M., Sanjayan, J.G.: Greenhouse gas emissions due to concrete manufacture. Int. J. Life Cycle Assess **12**, 282 (2007)

5. Xu, G., Shi, X.: Characteristics and applications of fly ash as a sustainable construction material: a state-of-the-art review. Resour. Conserv. Recycl. **136**, 95–109 (2018)
6. Meyer, C.: The greening of the concrete industry. Cem. Concr. Compos. **31**(8), 601–605 (2009)
7. Prieto-Sandoval, V., Jaca, C., Ormazabal, M.: Towards a consensus on the circular economy. J. Clean. Prod. **179** (2018)
8. Singh, N.B., Middendorf, B.: Geopolymers as an alternative to Portland cement: an overview. Constr. Build. Mater. **237**, 117455 (2020)
9. Provis, J.L.: Alkali-activated materials. Cem. Concr. Res. **114**, 40–48 (2018)
10. Shubbar, A.A., Sadique, M., Kot, P., Atherton, W.: Future of clay-based construction materials—A review. Constr. Build. Mater. **210**, 172–187 (2019)
11. Restrepo, A., Zapata, N., Arias, Y.P., Ochoa, J.C.: Poster: Residual quarry sludge stabilization to prefabricate sustainable building elements. In: TMS 2020 149th Annual Meeting and Exhibition, San Diego, CA, United States (2020)
12. Hewlett, P.C.: Lea's Chemistry of Cement and Concrete. Elsevier Ltd. (1998)
13. Maithel, S., Kumar, S., Lalchandani, D., Kanungo, S.: Factsheets About Brick Kilns in South and South-East Asia: Fixed Chimney Bull's Trench Kiln (2014)
14. DANE.: Boletín Técnico: Déficit Habitacional CNPV 2018, p. 20 (2020). www.dane.gov.co
15. ASTM C117-17, Standard Test Method for Materials Finer than 75-μm (No. 200) Sieve in Mineral Aggregates by Washing, ASTM International, West Conshohocken, PA (2017). www.astm.org
16. ASTM D2487-17e1, Standard Practice for Classification of Soils for Engineering Purposes (Unified Soil Classification System), ASTM International, West Conshohocken, PA (2017). www.astm.org
17. ASTM D2216-19, Standard Test Methods for Laboratory Determination of Water (Moisture) Content of Soil and Rock by Mass, ASTM International, West Conshohocken, PA (2019). www.astm.org
18. ASTM D5550-14, Standard Test Method for Specific Gravity of Soil Solids by Gas Pycnometer, ASTM International, West Conshohocken, PA (2014). www.astm.org
19. ASTM D4318-17e1, Standard Test Methods for Liquid Limit, Plastic Limit, and Plasticity Index of Soils, ASTM International, West Conshohocken, PA (2017). www.astm.org
20. Hoyos-Montilla, A.A., Arias-Jaramillo, Y.P., Tobón, J.I.: Evaluation of cements obtained by alkali-activated coal ash with NaOH cured at low temperatures. Mater. Constr. **68**(332), e170 (2018). https://doi.org/10.3989/mc.2018.10117
21. Zapata, N., Restrepo, A., Marin, S., Arias, Y.P., Torrenegra, J., Ochoa, J.C.: Mechanical analysis of a sustainable building prefabricated element based on a stabilized residual quarry sludge. In: Proceedings of International Structural Engineering and Construction (2021)
22. C. Universidad Nacional, «Centro Inter-Americano de Vivienda», AUNC, pp. 213–223, ene (1954)
23. NTC 5324. Soil Cement Blocks for walls and divisions. Definitions. Specifications. Test methods. Conditions of delivery. ICONTEC (2004)
24. Holtz, R., Kovacs, W.: An Introduction to Geotechnical Engineering. Prentice Hall (1981)
25. Bruno, A.W., Gallipoli, D., Perlot, C., Mendes, J.: Mechanical behaviour of hypercompacted earth for building construction. Mater. Struct. **50**(2), 1–15 (2017). https://doi.org/10.1617/s11527-017-1027-5
26. NTC 4017. Standard Test Methods for Sampling and Testing Brick and Structural Clay Tile. ICONTEC (2018)

Properties of Concrete with Thermo-mechanically Beneficiated Fine Recycled Aggregates

Rohit Prajapati(✉), Stefie J. Stephen, Ravindra Gettu, and Surender Singh

Indian Institute of Technology Madras, Chennai, India
rohitprajapati1992@gmail.com

Abstract. There is a worldwide shortage of fine aggregates/sand, and recycled concrete aggregates could be an alternate source to meet the demand. However, the main problem associated with the fine recycled concrete aggregate (FRCA) is the high amount of low-density porous adhered cement paste. This, along with the presence of finer particles (<300 μm), could increase the absorption capacity of FRCA significantly. Considering this in the present study, an attempt has been made to use thermo-mechanical beneficiation, where waste concrete was heated at a temperature of 400–500 °C followed by cooling at ambient temperature, after which it is processed further with the help of a ball mill. After this, the finer particles (<300 μm) are separated to give FRCA that has been used in concrete mixtures of 40 MPa design compressive strength. The mechanical properties were found to be comparable to the reference mix with natural sand. Also, the results from durability tests show that the use of beneficiated FRCA would not significantly reduce the durability of concrete, which could be a major concern.

Keywords: Fine recycled concrete aggregates (FRCA) · Concrete recycling · Thermo-mechanical beneficiation · C&D waste · Concrete

1 Introduction

The emphasis on sustainability in the construction industry has driven the need for ways to use environmentally friendly and low emission products. The use of alternative raw materials such as C&D waste recycled products is becoming more and more popular. Such recycling of C&D waste could solve two major problems—the disposal of waste material and alternative supply of aggregates. Further, it also promotes the circular economy. With the rapid increase in construction activities especially in developing countries like India, the C&D waste generation has increased many folds. The main products of C&D waste recycling are coarse and fine aggregates. The use of coarse aggregates is more common, and it has been extensively explored by researchers and field engineers, whereas the use of fine recycled aggregates is still restricted due to higher variability and inferior properties compared to natural sand; presence of adhered cement paste and fine particles generated during the crushing process is the main reason for these aspects.

J. I. Escalante-Garcia et al. (Eds.): RW 2021, RILEM Bookseries 40, pp. 916–922, 2023.
https://doi.org/10.1007/978-3-031-21735-7_97

In the current scenario when the mining of river sand is banned in most places, the demand for fine aggregate is mostly fulfilled by crushed/manufactured sand, whereas other alternatives such as crushed C&D waste and slag are also finding their way to fit in. Additionally, there are many beneficiation methods available that focus on improving the quality of fine recycled concrete aggregates (FRCA) by removing the adhered cement paste (ACP) using different techniques. In chemical beneficiation, strong acids, such as hydrochloric acid (HCL), are used for dissolving the ACP [1]. The ACP can also be removed using mechanical action, through the crushing of the C&D waste with a series of jaw and vertical shaft impact (VSI) crushers [2]. The microbial treatment has been found to improve the FRCA properties by precipitating a layer of carbonate, which could reduce porosity, and subsequently, the water absorption [3]. Thermal beneficiation is done at high temperatures (350–500 °C) that weakens the cement paste by dehydrating the hydrated cement products. Additionally, due to the differential thermal expansion, the ACP detaches from aggregate. Further, this loose ACP could be separated using the mechanical action [4, 5].

This paper presents the properties of thermo-mechanically beneficiated FRCA and the evaluation of its performance in concrete. The details of treatment parameters are discussed in the next section.

2 Experimental Programme

2.1 Materials

The waste concrete was obtained from a 15-year-old reinforced concrete building; the demolition site is shown in the Fig. 1a. The concrete was free from other materials, such as brick, wood, plastic, ceramics and the reinforcement bars that were already recovered by the demolition contractor. The big concrete chunks were then crushed into 40–50 mm size pieces (Fig. 1b).

a b

Fig. 1. Building demolition site and the crushed waste concrete

The crushed concrete was taken to the laboratory and heated in an electric furnace at 450–500 °C for about 1 h. After heating, it was allowed to cool at room temperature for 24

h. Further, the material is subjected to mechanical abrasion using a ball mill containing steel balls. The thermo-mechanical process is shown schematically in Fig. 2. Thereafter, the bigger particles (>4.75 mm) were separated to be used as coarse aggregate. The aggregate fraction with grain sizes ranging from 4.75 mm to 300 μm was separated by sieving, and is referred as FRCA in this paper. Further applications for the concrete fines, such as filler material, raw meal for clinker manufacturing, and partial replacement of cement are being explored [4, 6, 7].

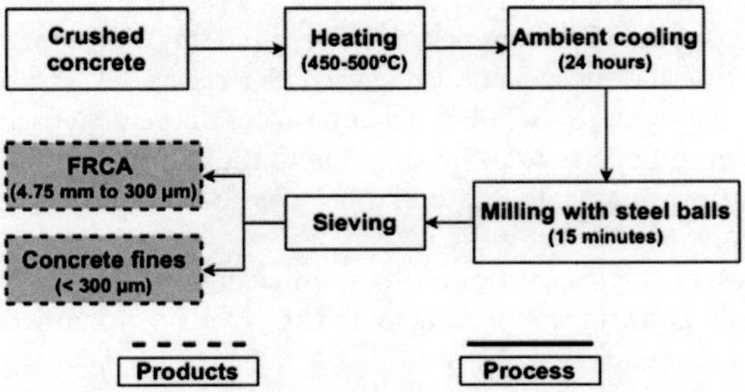

Fig. 2. Thermo-mechanical beneficiation process

2.2 Concrete Mixing Procedure

Table 1 shows the properties of the aggregates used in this research study. The beneficiated FRCA is seen to have better properties compared to the untreated aggregates, which could have water absorption as high as 10–15% [8, 9].

Table 1. Properties of aggregates

Aggregate type	Size	Water absorption
Coarse granite aggregate	10 mm	0.39
	20 mm	0.32
River sand	<4.75 mm	0.67
FRCA	4.75–300 μm	5.91

Since the water absorption of beneficiated FRCA is not significantly high and concrete fines were already removed, the normal mixing approach was followed, without any pre-soaking of the FRCA. Figure 3 shows the picture of FRCA, it can be seen that most of the sand particles are clean after the milling operation, though there is still some hydrated cement paste (HCP) present, which could explain the high amount of water absorption of FRCA compared to river sand. It should also be noted that small granite particles, i.e., fragments of the coarse aggregate from the milling process, are also present in the FRCA.

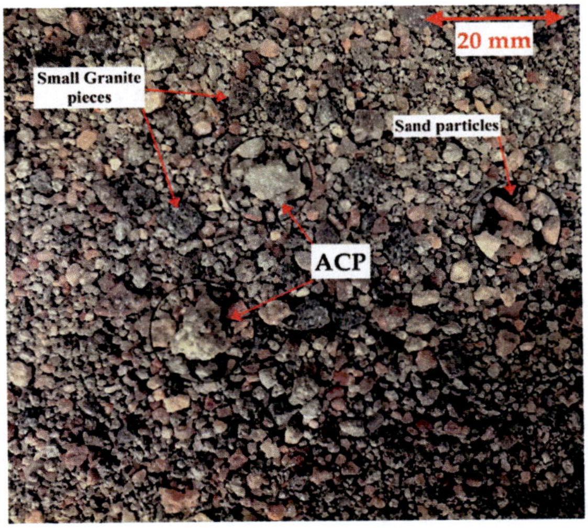

Fig. 3. Beneficiated fine recycled concrete aggregate (FRCA)

The mix proportion of the concrete used in this research work is shown in Table 2. To assess the suitability of FRCA in concrete, the reference mix was prepared using portland pozzolana cement (PPC), river sand and granite aggregate for a characteristic strength of 40 MPa. Mixes with river sand and FRCA were prepared to assess the full potential of replacing natural sand with FRCA. PCE based superplasticizer was added in all mixes to obtain a slump of 80–120 mm.

Table 2. Concrete constituents for the mixes (kg/m^3)

Cement	Fine aggregate	Coarse aggregate (10 mm)	Coarse aggregate (20 mm)	Water	Super plasticizer
380	767	380	681	171	0.67

2.3 Testing Methodology

The two concrete mixes were prepared for assessing the mechanical and durability properties. The compressive strength test was done using cubes of 150 mm size, and the four-point flexural test was done with beams of size 100 mm × 100 mm × 500 mm. For the durability tests, cylindrical specimens were prepared by extracting concrete cores that are epoxy coated over the sides. The specimen of 70 mm diameter and 30 mm height were used for the oxygen permeability index (OPI) test, and for the chloride migration test, 100 mm diameter and 50 mm height specimens were used. For the OPI test, specimens were kept at 50 °C for 7 days for preconditioning and placed in permeability cell under pressure. The chloride migration specimens were kept in a desiccator under vacuum conditions and the pores were saturated with calcium hydroxide solution, and then electrical potential is applied axially across the specimen to facilitate the migration of chloride ions.

3 Results and Discussion

The samples were cured in a mist room at 25 °C until testing. The cube compressive and flexural strength tests were done at the age of 28 days. The durability tests were performed on specimens stored in ambient conditions for 90 days.

3.1 Mechanical Properties

The mechanical properties of the concretes are shown in Fig. 4. The compressive strength (Fig. 4a) of fine recycled aggregate concrete (FRAC) is about 10% lower than the reference natural aggregate concrete (NAC); this reduction appears lower than that obtained with untreated FRCA, which could yield upto 30% reduction in strength [10]. In terms of the average flexure strength (Fig. 4b), it could be seen that the complete replacement of river sand with beneficiated FRCA does not show significant variation.

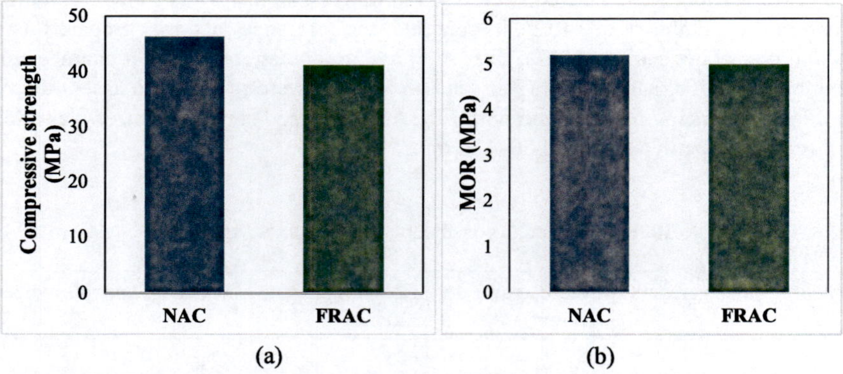

Fig. 4. Mechanical properties of FRAC: **a** compressive and **b** flexural strengths

3.2 Durability Properties

Figure 5 shows the durability parameters obtained for FRAC: the oxygen permeability index (OPI) and the chloride migration coefficient. The OPI test measures the resistance against gas permeation through concrete which could be correlated with the carbonation of concrete. The non-steady state chloride migration coefficient was measured as per NT Build 492. It estimates the rate of chloride ions ingress through the concrete. The OPI of FRAC (Fig. 5a) is lower than that of the reference concrete (NAC), though it still qualifies as "good" concrete (i.e., OPI > 9.5) [11]. Interestingly, the chloride migration coefficient (Fig. 5b) indicates a slight improvement over the reference concrete.

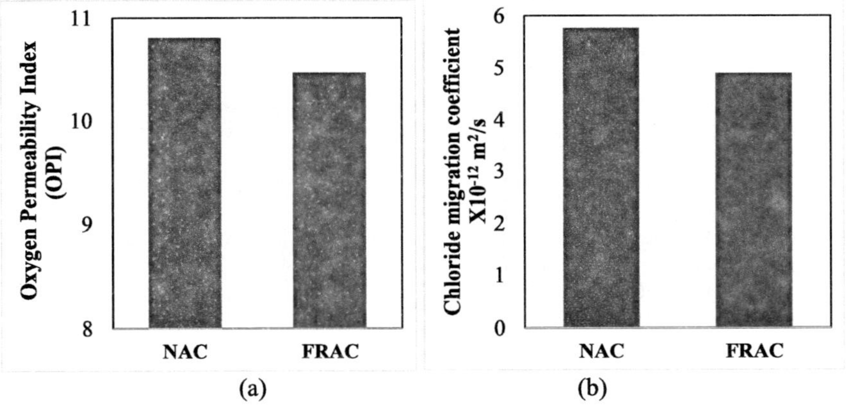

Fig. 5. Durability properties of FRAC: **a** Oxygen permeability index and **b** Chloride migration coefficient

4 Conclusions

The beneficiation of fine recycled concrete aggregates (FRCA) has been shown to have a positive impact on the water absorption properties, that could be relatively high in the case of untreated crushed FRCA. The removal of the concrete fines (<300 μm) plays a major role in reducing the quantity of hydrated cement paste in the fines, and consequently the water demand. Subsequently, the mechanical and durability parameters of fine recycled aggregate concrete (FRAC) were found to be satisfactory even at 100% replacement of natural sand with FRCA. These preliminary results show the feasibility of completely replacing river sand with thermo-mechanically beneficiated FRCA in concrete for structural applications.

Acknowledgements. The authors wish to express their gratitude for partial financial support from UKIERI (UK-India Education and Research Initiative) and DST (Department of Science & Technology), Government of India [Project DST/INT/UK/P-172/2017]; and the Institute of Eminence Research Initiative on Technologies for Low Carbon Lean Construction [Project SB20210809CEMHRD008100]. The authors also acknowledge the use of facilities in Mechanical Performance of Civil Engineering Materials Laboratory, IIT Madras, established through the FIST Grant SR/FST/ETII-054/2012, and the Lafarge-IIT Madras Laboratory for Durability and Long-Term Performance of Concrete.

References

1. Santha Kumar, G., Minocha, A.K.: Studies on thermo-chemical treatment of recycled concrete fine aggregates for use in concrete. J. Mater. Cycles Waste Manag. **20**(1), 469–480 (2017). https://doi.org/10.1007/s10163-017-0604-6
2. Ulsen, C., Kahn, H., Hawlitschek, G., Masini, E.A., Angulo, S.C., John, V.M.: Production of recycled sand from construction and demolition waste. Constr. Build. Mater. **40**, 1168–1173 (2013)

3. Feng, Z., Zhao, Y., Zeng, W., Lu, Z., Shah, S.P.: Using microbial carbonate precipitation to improve the properties of recycled fine aggregate and mortar. Constr. Build. Mater. **230**, 116949 (2020)
4. Shima, H., Tateyashiki, H., Matsuhashi, R., Yoshida, Y.: An advanced concrete recycling technology and its applicability assessment by the input-output analysis. J. Adv. Concr. Technol. **3**(1), 53–67 (2005)
5. Ahn, J.W., Kim, H.S., Han, G.C.: Recovery of aggregates from waste concrete by heating and grinding. Geosyst. Eng. **4**(4), 123–129 (2001)
6. Florea, M.V.A., Ning, Z., Brouwers, H.J.H.: Activation of liberated concrete fines and their application in mortars. Constr. Build. Mater. **50**, 1–12 (2014)
7. Shui, Z., Xuan, D., Chen, W., Yu, R., Zhang, R.: Cementitious characteristics of hydrated cement paste subjected to various dehydration temperatures. Constr. Build. Mater. **23**(1), 531–537 (2009)
8. Sosa, M.E., Carrizo, L.E., Zega, C.J., Villagrán Zaccardi, Y.A.: Water absorption of fine recycled aggregates: effective determination by a method based on electrical conductivity. Mater. Struct. **51**(5), 1–13 (2018). https://doi.org/10.1617/s11527-018-1248-2
9. Evangelista, L., De Brito, J.: Concrete with fine recycled aggregates: a review. Eur. J. Environ. Civ. Eng. **18**(2), 129–172 (2014)
10. Khatib, J.M.: Properties of concrete incorporating fine recycled aggregate. Cem. Concr. Res. **35**(4), 763–769 (2005)
11. Dhanya, B.S., Santhanam, M., Gettu, R., Pillai, R.G.: Performance evaluation of concretes having different supplementary cementitious material dosages belonging to different strength ranges. Constr. Build. Mater. **187**, 984–995 (2018)

Bond-Behavior of Bamboo Strips and Bamboo/Wood Bio-Concretes

Nathalia Andrade da Silva(✉), Amanda Lorena Dantas de Aguiar, M.'hamed Yassin Rajiv da Gloria, and Romildo Dias Toledo Filho

Universidade Federal do Rio de Janeiro, Rio de Janeiro, Brasil
nathalia.silva@coc.ufrj.br

Abstract. Seeking the development of light structural elements with less environmental impact, the use of bamboo strips as bio-concretes reinforcement can provide an interesting solution in civil construction. Therefore, in order to use the bamboo strip with bio-concrete, it is important to understand the bond-behavior between both materials. This paper aims to analyze the bond-behavior of coated and uncoated bamboo strips with bamboo bio-concretes (BBC) and wood bio-concretes (WBC). The coating was made with a mixture of epoxy resin and sand on the lateral surface of the strips, and its effect on the adherence with bio-concretes was analyzed through the pull-out test. The bio-concretes were produced with a volumetric fraction of 40% of bio-aggregates and characterized through uniaxial compression and splitting tests. The results showed that, after 28 days of curing, the bio-concretes presented compressive strength of 6 MPa, and a tensile strength ranging from 1.14 to 1.23 MPa for both bio-concretes. As expected, the coated strips presented a bond higher with both bio-concretes than the uncoated. Comparing the bio-concretes, the force required to pull the coated bamboo strip from the BBC was greater than that of the WBC.

Keywords: Bamboo strips · Wood bio-concrete · Bamboo bio-concrete · Bonding

1 Introduction

The use of plant materials in civil construction emerged as an alternative due to factors such as global warming, energy savings, and life cycle problems [1]. The United Nations Program [2] documented that buildings are consuming about 40% of global energy, 25% of global water, and 40% of global resources, one of the largest energy consumers in the world. The search for innovative materials, such as plant-based cementitious composites, is an alternative that has been increasingly valued and one of the great advantages is that they are materials that allow the sequestration and storage of CO_2, which is the main greenhouse gas [3].

Most plant materials come from the reuse of plants and residues from industry and agriculture for the production of bio-sustainable insulating materials and bio-aggregates in concrete [4]. Bamboo and wood, for example, can be used as reinforcement elements

J. I. Escalante-Garcia et al. (Eds.): RW 2021, RILEM Bookseries 40, pp. 923–933, 2023.
https://doi.org/10.1007/978-3-031-21735-7_98

in civil construction or bio-aggregates for bio-concrete production. Bio-concrete is a concrete where mineral aggregates, such as sand and gravel, can be totally or partially replaced by bio-aggregates [5]. Da Glória [6] highlighted that bio-concrete can be used in civil construction as core in sandwich panels as they are lightweight and thermally efficient materials.

The use of bamboo (strip or culm) as a reinforcement element of concrete has been studied by researchers in recent decades since bamboo has high tensile strength [7]. However, for the efficient of such a system, adherence between the bamboo and concrete is necessary. As bamboo is a hygroscopic material when in contact with a water environment, dimensional variations occur before the concrete cure, and those variations promote internal local stress-inducing cracks [8]. Performing treatments on bamboo contribute to increasing its bond strength with concrete. One of the treatments that have been studied by researchers is the coating with a combination of epoxy resins and sand [7–9].

This work aims to analyze the adhesion between untreated and treated bamboo strips in wood bio-concrete and bamboo bio-concrete.

2 Materials and Methods

2.1 Production and Treatment of Bamboo Strips

The bamboo species used to make the strips was *Dendrocalamus Giganteus*, ranging in age from 3 to 7 years. The culm of bamboo was treated with boric acid. The bamboo culms were divided in the entire diameter and strips of the rectangular section were obtained from the region closest to the stalk wall, where the highest concentration of fibers is found (Fig. 1). The dimensions of the strips used were 6 × 12 × 200 mm (thickness × width × length).

Fig. 1. Preparation of bamboo strips

Two types of strips were used in this work: coated and uncoated. The coating was realized according to Silva [9]. A layer of approximately 1.5 mm of epoxy resin (Sikadur 32®) was applied on a part of the strip surface and after 5 min layer of sand was applied to the epoxy layer and the whole dried for 24 h (Fig. 2).

2.2 Bio-Aggregates

The bio-aggregates used in this research came from the state of Rio de Janeiro, Brazil.

Fig. 2. Coated bamboo strips

The bamboo aggregates (Fig. 3a) were obtained from a mix of the species *Dendrocalamus asper*, *Phyllostachys aurea,* and *Phyllostachys edulis.* As these wastes came in the shape of culms, the first stage of processing was the culms cutting and grinding, followed by sieving in order to obtain particles of nominal diameter lower than 4.75 mm. Next, the bamboo bio-aggregates were treated in water at 80 °C to remove the extractives using a water-to-bamboo ratio of 10. After the treatment, the bamboo bio-aggregates were air-dried for 48 h and homogenized through an "elongated pile" method. The apparent density [10] and the moisture content [11] were 580 kg/m^3 and 10.18%, respectively.

The wood waste (Fig. 3b), from the species *Hymenolobium petraeum, Cedrela fissilis, Erisma uncinatum warm* and *Manilkara Salzmann,* were obtained in the shape of shavings. After sieving (1.18 mm), the wood shavings were immersed twice in a 0.73% calcium hydroxide solution, with a water-to-wood ratio of 10, as realized by Aguiar et al. [12]. Each immersion lasted 1 h and was followed by air-drying and homogenization. The wood bio-aggregates presented 530 kg/m^3 of apparent density and 12.50% of moisture content.

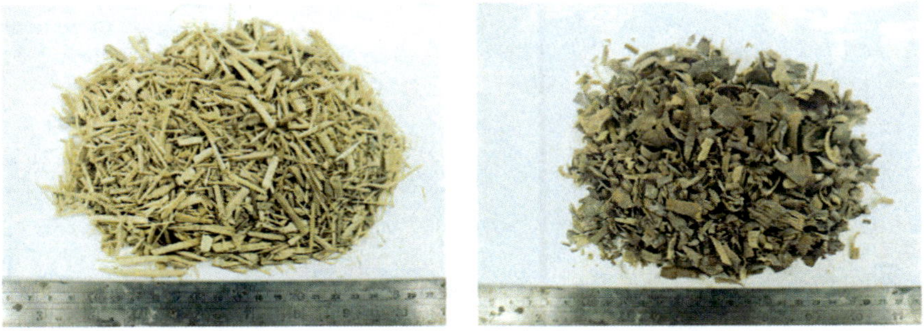

(a) (b)

Fig. 3. Bio-aggregates: **a** Bamboo; **b** Wood shavings

2.3 Bio-Concretes Production

Two bio-concretes were produced: bamboo bio-concrete (BBC) and wood bio-concrete (WBC). The matrix used was composed of Brazilian Portland Cement (CPV-ARI), partially replaced in mass by 30% Metakaolin and 30% Fly Ash. It is important to highlight that the addition of pozzolans contributes to the reduction of cement consumption. The volumetric fraction of bio-aggregates was 40% and the water-to-cement ratio 0.35. Calcium Chloride was used as the setting accelerator and fixed at 3% of the mass of the cementitious material. To guarantee good workability of the mixtures, additional water based on the bio-aggregates absorption was used. The water absorption coefficient of bamboo and wood shavings were 82 and 80%, respectively. Table 1 shows the proportion of materials used to produce 1 m^3 of BBC and WBC.

Table 1. Mixtures composition (kg/m^3)

Materials	BBC	WBC
Total water	489.41	501.17
Cement	305.97	305.97
Fly ash	229.47	229.47
Metakaolin	229.47	229.47
$CaCl_2$	27.04	27.04
Bio-aggregates	212.00	232.00

The mixture was prepared in a 20 L mixer and the mixing process was performed as follows: mixing of cementitious materials and dry bio-aggregates for 1 min, progressive addition of total water, previously mixed with calcium chloride, for 1 min, and mixing until homogenization, until the total time of 4 min.

Fresh bio-concrete was introduced into lubricated molds, distributed in three layers, and densified on a vibrating table (68Hz) for 30 s. After molding, the samples were placed in a chamber with a relative humidity of 55% and a temperature of 21 °C.

2.4 Uniaxial Compression Test

The uniaxial compression tests were performed using the Universal Testing Machine model Whykehan Ferrance 250 kN at a speed rate of 0.3 mm/min. For each mixture, four cylindrical specimens of diameter 50 mm and a height 100 mm were tested at the age of 28 days. The parameters analyzed in this test were the compressive strength and the modulus of elasticity according to NBR 5739 [13] and NBR 8522 [14], respectively. A setup coupled to the center of the specimens was used to measure the axial strain with two coupled LVDTs.

2.5 Splitting Test

The splitting test was performed using the Shimadzu AG-X 100 kN testing machine, with a speed of 0.1 mm/min. The dimensions of the samples were 50 × 50 mm^2 (diameter × height). Three pairs of notches were made on the samples. The first pair of notches (Fig. 4a) in V-shape with an angle of 90° was inducted during the bio-concrete production, through the used mold internal shape. The second and third pairs of notches (Fig. 4b) were made through a saw with a depth of 3 mm. The purpose of the notches was to induce a single straight crack.

The load was applied in the middle of a set of steel rollers and two aluminum plates positioned in the V-notch (Fig. 4c). To control the crack opening, two clip-gauges were positioned between two steel plates glued at the center of the opposites faces of the specimen (Fig. 4d).

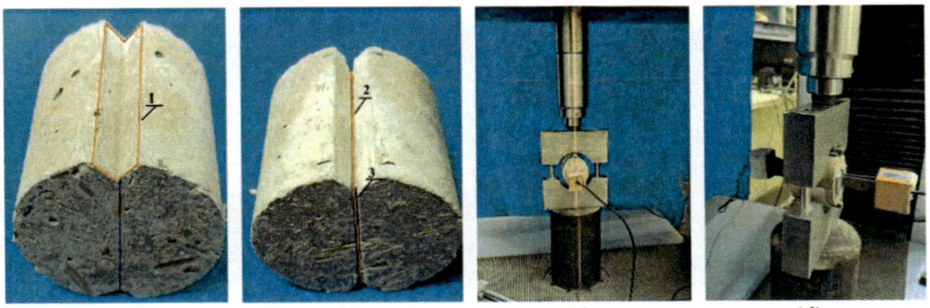

(a) (b) (c) (d)

Fig. 4. **a** The First notch made in the sample during molding; **b** Notches made with a saw; **c** Splitting test set-up; **d** Positioning of clip gauges during the test

2.6 Pull-Out Test

The test was performed on five samples of each bio-concrete, in a 100 kN Shimadzu AG-X machine, at a speed of 1 mm/min. Before molding, the bamboo strips were glued to the bottom of the mold to ensure the embedding length (100 mm) (Fig. 5a). The fresh bio-concrete was placed in three layers until filling the molds of 75 × 100 mm^2 (diameter x height) (Fig. 5b, c). After 24 h, the samples were demolded and tested within 28 days of cure. In the setting of the test, the bio-concrete was fixed between two steel plates, as shown in Fig. 5d, while the strip was pulling off through a claw.

3 Results and Discussions

3.1 Bulk Density and Compressive Strength

The WBC and BBC produced presented consistency indexes [15] of 200 and 245 mm, respectively.

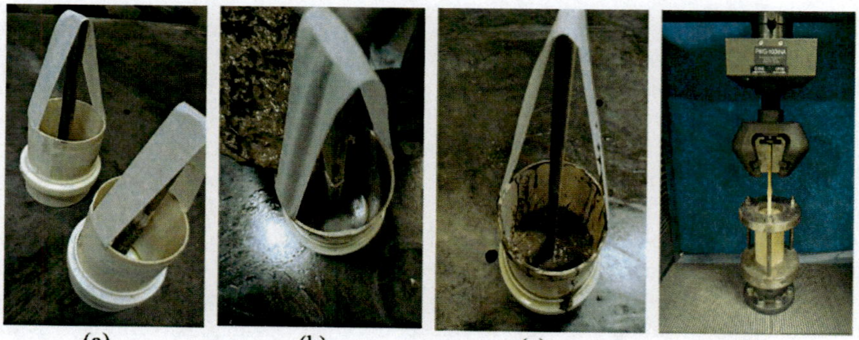

Fig. 5. **a** Bamboo strips glued in the molds; **b** Molding; **c** The First layer of concrete in the mold; **d** Pull-out test set-up

Table 2. Bulk density, compressive strength, and Young modulus (variation coefficients in % in brackets)

Bio-concrete	Bulk density (kg/m^3)	Compressive strength (MPa)	Young modulus (GPa)
BBC	1088.90 (0.60)	6.31 (3.29)	1.92 (6.57)
WBC	1051.65 (1.46)	6.26 (5.81)	1.51 (3.72)

The bulk density, compressive strength, and modulus of elasticity values of bio-concretes are shown in Table 2.

Regarding the bulk density, the values found for BBC and WBC do not have a significant difference. According to the Rilem functional classification (1978), the two bio-concretes can be considered lightweight since they present a density of less than 1800 kg/m^3.

Analyzing the stress-strain curves of the bio-concretes (Fig. 6), it is possible to notice that both BBC and WBC presented a linear behavior, up to 50% of the peak strength. After this linear phase, the curves showed a non-linear region until reaching the stress peak value. In the post-peak, there is a gradual decrease in tension and an increase in deformation.

Comparing the two bio-concretes, a similarity is observed in the modulus of elasticity showing values in the same trend. Aguiar et al. [16] found a modulus of elasticity of 1.66 GPa for the same bio-concrete mixture with 40% wood shavings. Andreola et al. [17] in their study produced a BBC with the same volumetric fraction as this study and the same proportion of pozzolans, but with a water-cement ratio of 0.32 and found a modulus of elasticity of 1.48 GPa for 28 days.

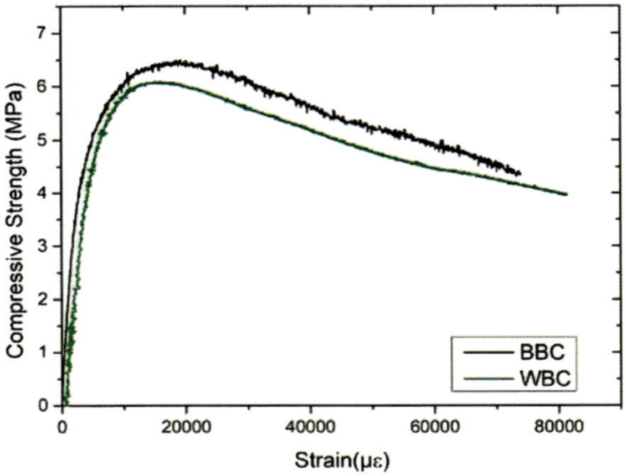

Fig. 6. Stress-strain curves for bamboo and wood bio-concretes after 28 days of curing

3.2 Splitting Tests

Figure 7 shows the stress × CMOD (Crack Mouth Opening Displacement) curves of the samples tested for the splitting test. BBC presented an average value of tensile strength of 1.28 MPa with a coefficient of variation of 9.78% while the WBC reached 1.14 MPa and 8.80% of the coefficient of variation. The bio-concretes showed a linear behavior up to approximately 90% of the peak value, as well as found by da Gloria [18].

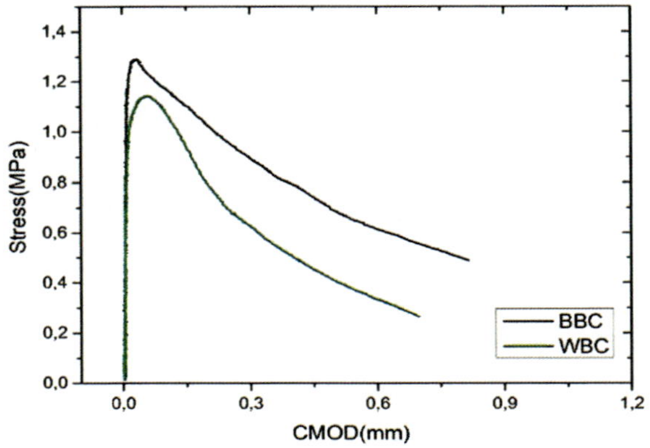

Fig. 7. Stress-CMOD curves

The peak stress of BBC and WBC was reached for CMOD of 0.02 and 0.05, respectively. At the end of the test, a complete rupture in two parts was observed on the WBC samples, while bamboo particles were still acting like a bridge between the two parts of the samples after the crack propagation (Fig. 8).

Fig. 8. BBC after the crack propagation

This behavior of the BBC can be explained by the elongated shape and tensile strength of the bamboo particles, which remind the vegetable fibers (short sisal fibers) used for composites reinforcement. It is worth noting that few bamboo particles broke during the test, as observed with the wood shavings. Figure 9a illustrates the fracture region and the points where the bamboo has been completely broken and the particles that were pulled out. For the WBC, the bio-aggregate fractured surface presents breakage points, pull-off, and pores (Fig. 9b).

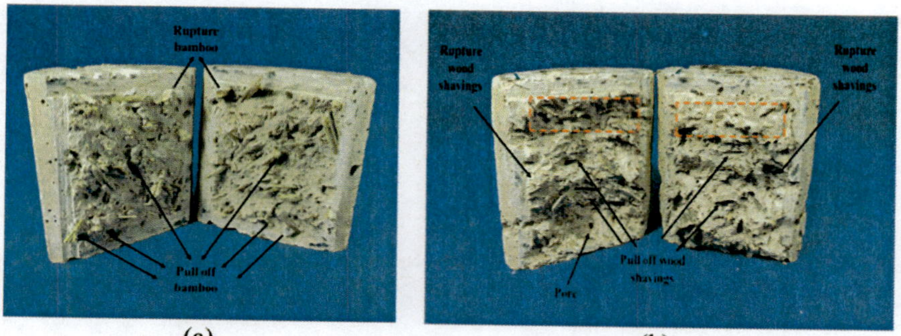

Fig. 9. Fractured surfaces after splitting tests **a** BBC; **b** WBC

3.3 Pull Out

Table 3 and Fig. 10 show the results of the bond behavior of uncoated (REF) and coated bamboo strips (SIK).

The force required to pull out the coated strip (SIK) from the BBC and WBC was 2.4 and 4.11 times higher than the force used to pull the uncoated strip (REF), respectively. For untreated strips, the pullout strength was greater for BBC, but with treated strips,

Table 3. Pull out force and bond strength and variation coefficient in % in brackets

Bio-concrete	Treatment	Pull out force (N)	Bond strength (MPa)
BBC	REF	2104.64 (1.31)	0.54 (11.99)
BBC	SIK	5107.18 (1.30)	1.49 (12.08)
WBC	REF	1212.54 (12.20)	0.33 (12.20)
WBC	SIK	4992.39 (5.53)	1.43 (6.24)

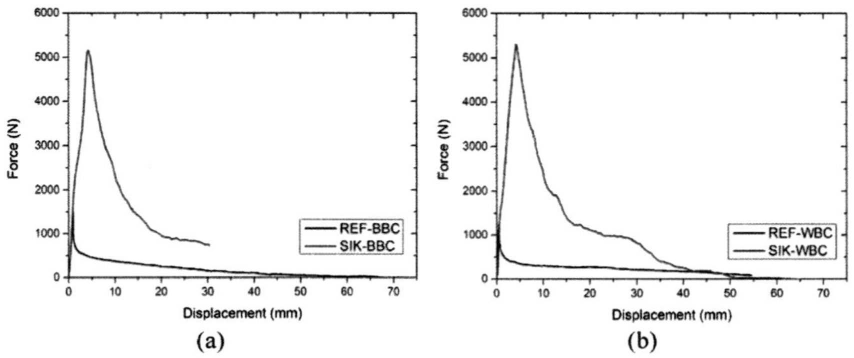

Fig. 10. Typical pull-out force-displacement curve: **a** BBC; **b** WBC

this difference decreased. These results indicated that for both bio-concretes, the coating increased significantly the bond with the bamboo strips.

In Fig. 11 it can be seen that after the peak value, an abrupt displacement followed by a fragile post-peak behavior, for both REF and SIK curves.

After the pullout test, it is noted that the REF strips do not have bio-concrete adhered to their surface (Fig. 11a) and the hole created in the bio-concrete does not show damage. In Fig. 11b with the SIK strips, bio-concrete was retained on its surface and small fractures in the sample orifice, due to friction caused by the treatment, were observed.

Observing the behavior of the two mixtures of bio-concrete, the results show that the use of wood and bamboo bio-aggregates did not interfere with the adherence of bamboo to the bio-concretes produced. A significant adhesion between bamboo and bio-concrete resulted from the combination of sand and Sikadur, which created a roughness on the strip surface. Javadian et al. [8] and da Silva et al. [9] observed that the sand coating with epoxy resin creates a roughness on the strip surface, increasing the mechanical interlock between the bio-concretes and the bamboo strips.

4 Conclusions

Based on the results, the following main points can be highlighted:

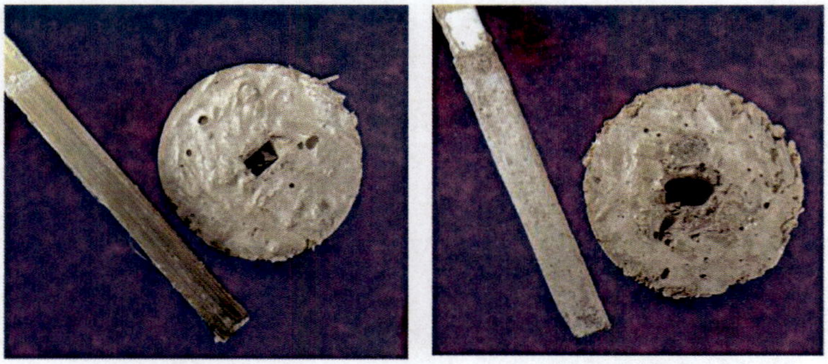

(a) (b)

Fig. 11. Samples after testing **a** REF; **b** SIK

- Bio-concretes had similar mechanical properties, both in compression and traction, regardless of the type of bio-aggregate used.
- In the splitting modified test, it is noteworthy that the shape of the bio-aggregate interferes in the fracture of bio-concrete.
- On the bond behavior of bamboo strips with bio-concrete matrices, the type of bio-aggregate does not significantly interfere in the adherence.
- The bamboo strips treatment with sand and epoxy resin is important to promote adhesion between the bio-concrete matrix and the bamboo.

References

1. Amziane, S., Sonebi, M.: Overview of biobased building material made with plant aggregate. RILEM Tech. Lett. **1**, 31–38 (2016)
2. PROGRAMA DAS NAÇÕES UNIDAS PARA O MEIO AMBIENTE. Ambiente pra o desenvolvimento (2016)
3. Caldas, L.R., Da Gloria, M.H.Y.R., Pittau, F., Andreola, V.M., Habert, G., Toledo Filho, R.D.: Environmental impact assessment of wood bio-concretes: evaluation of the influence of different supplementary cementitious materials. Constr. Build. Mater., 121146 (2020)
4. Sáez-pérez, M.P.; Brümmer, M., Durán-Suárez, J.A.: A review of the factors affecting the properties and performance of hemp aggregate concretes. J. Build. Eng. 101–323 (2020)
5. Amziane, S., Arnaud, L., Challamel, N.: Bio-aggregate-based building materials. Appl. Hemp Concr., Ediia **1**, 1–11 (2013)
6. da Gloria, M.H.Y.R., Andreola, V.M., dos Santos, D.O.J., Pepe, M., Toledo Filho, R.D.: A comprehensive approach for designing workable bio-based cementitious composites. J. Build. Eng. 101696 (2020)
7. Mali, P.R., Datta, D.: Experimental evaluation of bamboo reinforced concrete beams. J. Build. Eng. **28**, 101071 (2020)
8. Javadian, A., Wielopolski, M., Smith, I.F., Hebel, D.E.: Bond-behavior study of newly developed bamboo-composite reinforcement in concrete. Constr. Build. Mater. **122**, 110–117 (2016)

9. da Silva, N.A., Gomes, B.M.C., Andreola, V.M., da Gloria, M.H.Y.R., Toledo Filho, R.D.: Analysis of bond-behavior of bamboo strip and bamboo bio-concrete. In : Non-Conventional Materials and Technologies (NOCMAT), Nairobi Kenya (2019)
10. ABNT NBR NM 52. Agregado miúdo - Determinação da massa específica e massa específica aparente (2019)
11. ABNT NBR 9939. Agregado graúdo – Determinação do teor de umidade total – Método de ensaio (2011)
12. Aguiar, A.L.D, et al.: Influência do tratamento da serragem de madeira na resistência à compressão do bio-concreto de madeira. Congresso de Construção Civil, Brasília-DF, Brasil (2020)
13. ABNT NBR 5739. Concreto – Ensaio de compressão de corpos de prova cilíndricos (2018)
14. ABNT NBR 8522. Concreto – Determinação dos módulos estáticos de elasticidade e de deformação à compressão (2017)
15. ABNT NBR 13276. Argamassa para assentamento de paredes e revestimento de paredes e tetos-Determinação do teor de água para obtenção do índice de consistência padrão (2016)
16. Aguiar, A.L.D., et al.: Influence of high temperatures on the mechanical properties of wood bio-concretes. In: 4th International Conference on Bio-Based Building Materials, Barcelona-Spain (2021)
17. Andreola, V.M., et al.: Partial replacement of cement by combination of fly ash and metakaolin in bamboo bioconcretes. Acad. J. Civil Eng. **37**(2), 102–106 (2019)
18. da Gloria, M.H.Y.R.: Comportamento experimental e numérico de bio-concretos de madeira e de painéis sandúiches com faces em laminados reforçados com fibras longas de sisal. (PhD thesis). Universidade Federal do Rio de Janeiro, Rio de Janeiro, Brasil (2020)

Controlled Microencapsulation of Rejuvenators by Jet Vibration to Promote Self-healing in Bituminous Materials

José L. Concha(✉), Erik Alpizar Reyes, and Jose Norambuena-Contreras

Department of Civil and Environmental Engineering, LabMAT, University of Bío-Bío, Concepción, Chile
jlconcha@ubiobio.cl

Abstract. This study aimed to synthesise enhanced biocapsules containing oil as an additive to promote self-healing in bituminous materials. The capsules consisted of a biopolymeric matrix of sodium alginate as encapsulating material and sunflower cooking oil as a non-hazardous rejuvenating agent. Oil was incorporated in five concentrations by weight of alginate, 1:1, 1:3, 1:5, 1:7, and 1:9, respectively. Thus, five oil-in-water (O/W) emulsions were prepared to synthesise biopolymeric capsules via ionic gelation using an encapsulator device that works by jet vibration technique. The physical properties, stability and droplet size distribution were measured in the O/W emulsions using microscopy techniques and laboratory tests. The morphological and the physical-mechanical properties of the biocapsules were also evaluated. The main results showed that the O/W emulsions increased their viscosity and physical stability when increased the oil content, reducing their creaming rate over time. The synthesised biocapsules presented spherical morphology and high encapsulation efficiency. The size of the produced biocapsules was mainly dependent on the selected nozzle size, the flow rate and viscosity of the extruded O/W emulsions, and the vibrational frequency applied and controlled on the encapsulator device. A reduction in the encapsulated rejuvenator content was quantified in the biocapsules, resulting in a lower volume and higher density and compressive strength values. Finally, the capsules effectively resisted the load associated to on-site compaction in a way that they can be potentially used as an effective additive for asphalt self-healing purposes.

Keywords: O/W emulsion · Microencapsulation · Biopolymeric capsules · Compressive strength · Asphalt self-healing

1 Introduction

A major problem affecting the asphalt pavements performance during their service life is bitumen oxidative ageing [1] promoted by environmental aspects like temperature, weather, and sunlight [2]. As a solution, researchers are nowadays carrying out intense research to promote the autonomic healing of bituminous materials using microencapsulated rejuvenating agents [1]. In this context, encapsulated rejuvenators are described as

J. I. Escalante-Garcia et al. (Eds.): RW 2021, RILEM Bookseries 40, pp. 934–941, 2023.
https://doi.org/10.1007/978-3-031-21735-7_99

spherical particles or capsules added to the asphalt mixture as an additive to recover the original properties of aged bitumen [3]. The principle behind this approach is that when micro-cracks happen next to the capsules, they will break and release the rejuvenator, which will dissolve bitumen and improve its flow capacity. The rejuvenator then mixes the bitumen to seal the crack, preventing further propagation [4].

Overall, rejuvenating agents are encapsulated in petroleum-based polymers (e.g. melamine-formaldehyde and urea-formaldehyde) through chemical and physicochemical methods (i.e. in-situ polymerisation and two-step coacervation processes) to produce single and core-shell microcapsules [5]. The main drawback of this type of capsule remains in the toxicity of the materials used in the manufacturing process [6]. As an alternative to these methods, physical-mechanical encapsulation methods based on ionic gelation has been explored. In this technique, bio-based polymers such as alginate and rejuvenating agents have been recently proposed to develop polynuclear bio-capsules as a sustainable approach to extend the lifetime of roads [5]. In previous works led by Zhang et al.[7] and Xu et al. [8] hypothesise that polynuclear microstructure provides a more controlled liberation of the healing agent once activated the capsule, allowing multiple healings on aged bitumen over time and good thermal and mechanical stability.

Most of the encapsulation methods used to synthesise polynuclear capsules via ionic gelation principle are based on the dropping funnel or microfluidic pressure pump techniques [9]. However, these methods do not allow appropriate control of the parameters needed for the encapsulation of the rejuvenating agent resulting in capsules with irregular shapes. The main objective of this study was to synthesise polynuclear microcapsules using a controlled microencapsulation method for asphalt self-healing purposes. The effect of its control parameters was evaluated on the morphological, physical, and mechanical properties of the microcapsules. With this aim, five different capsules were fabricated based on stable components and characterised by laboratory tests.

2 Materials and Method

2.1 Materials

Biopolymeric capsules containing Sunflower oil (SO) were synthesised in this study. The polymeric structure of the capsules consisted of low-viscosity biopolymer sodium alginate powder (viscosity $\leq$300 mPa·s, density 1020 kg/m^3 in a 2%w/w solution), provided by Buchi and calcium-chloride dihydrate ($CaCl_2 \cdot 2H_2O$) with 77% purity, provided by Winkler. SO presented density 850 kg/m^3, viscosity 70 mPa·s @ 20 C, and pH 5.3–5.5 @ 25 C. This oil was used for encapsulation since its proven capacity as rejuvenating agent for aged bituminous materials [1, 4, 8].

2.2 Synthesis and Characterisation of the O/W Emulsions

In this study, five different O/W emulsions were prepared for encapsulation purposes. First, a sodium alginate solution in a concentration of 2% w/w into deionised water was prepared by continuous agitation in a magnetic stirrer (SCI550-S) at 250 rpm for 24 h. After this period, the SO was incorporated dropwise to the alginate solution at five different biopolymer:oil mass (B/O) ratios; 1:1, 1:3, 1:5, 1:7 and 1:9, respectively. During this process, the biopolymer solution was mechanically agitated at 1200 rpm for 40 min (Fig. 1a). The physical stability of the synthesised emulsions was characterised by their creaming index (CI), as proposed by McClements [10], indicating the phase separation extent of the SO from the alginate solution over time. CI was reported as the average of three measurements at 0h, 1h, 2h, 3h, 6h, 12h, 24h, 26h, 28h, and 30h response. Moreover, for each of the abovementioned periods, the oil droplet size distribution in the O/W emulsions was determined by fluorescence microscopy (ICOE IV 5100FL) and the software ImageJ® (Fiji, version 1.53c).

2.3 Synthesis and Characterisation of the Biopolymeric Microcapsules

Polynuclear capsules were synthesised based on the ionic gelation principle by electrostatic interactions between sodium alginate and the Ca^{2+} ions from the $CaCl_2{\cdot}2H_2O$. An encapsulator device (Buchi, model B-390) was used to produce capsules by vibrating jet technique. The method consisted of pumping each O/W emulsion by applying air pressure (A_p) to feed a vibrating nozzle system composed of a vibrating unit with control of frequency (V_f) and mounted on a nozzle with a specific size (N_s). A_p, V_f, and N_s were settled to 450–550 mbar, 350 Hz, and 750 μm, respectively. With these parameters, the outgoing emulsion with a laminar flow from the nozzle was broken up by vibration, then separated into droplets and collected in a hardening solution of 5% w/w $CaCl_2{\cdot}2H_2O$ in deionised water, continuously agitated at 250 rpm using a magnetic stirrer (see Fig. 1b). This process resulted in the formation of polynuclear microcapsules.

To prevent the droplets from hitting each other in flight and as they entered the hardening solution, an electrostatic charge of 1500 V was applied on the surface of the outgoing emulsion droplets from the nozzle. The freshly prepared microcapsules were filtered from the hardening solution and rinsed with 500 mL of deionised water later dried in an oven at 30 C for 24 h (see Fig. 1c. Finally, the capsules were stored in a freezer at −10 C, preventing the oxidation of the encapsulated oil.

The morphology of each type of microcapsule was statistically characterised by the size distribution and the sphericity factor (SF) of 100 capsules using optical microscopy (Leica EZ4) and the image processing software ImageJ® (Fiji, version 1.53c). SF was determined as proposed by Voo et al. [11]. Furthermore, the physical properties of the capsules were characterised by their encapsulation efficiency (*EE*) of the SO, determined according to the method proposed by Guadarrama-Lezama et al. [12] and their bulk density according to Alpizar-Reyes et al. [13]. Finally, the mechanical properties of the microcapsules were evaluated by compressive test as follows. First, each type of capsule was dried in an oven at 30 C for 2 h. Once reaching room temperature (20 C), the capsules were compressed using an Universal Testing Machine (Zwick/Roell, model Z005), with a load cell of 5kN and a load speed of 0.2 mm/min, as proposed by Norambuena-Contreras et al. [9]. The representative compressive strength of each type of capsule was determined based on its yield compressive strength, as the average of 5 measurements.

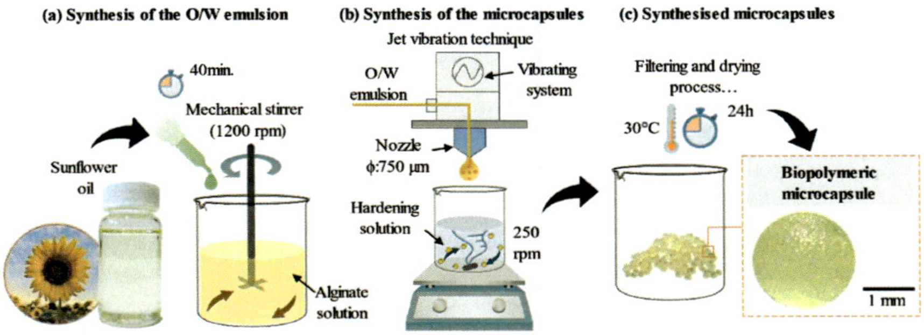

Fig. 1. Representation of the **a** synthesis of the O/W emulsions; **b** controlled microencapsulation by jet vibration technique; and **c** synthesised microcapsules.

3 Result and Discussion

3.1 Evaluation of the Physical Stability of the O/W Emulsions

To ensure the encapsulation of the SO, the O/W should be physically stable during the synthesis of the capsules. In this context, Fig. 2a presents the average results of the CI values for the different emulsions with time. Overall, it should be noticed an increase in the CI with time, identifying three distinctive tendencies. A first one (G1) for the 1:1 emulsion, a second one (G2) for the 1:3 and 1:5 emulsions, and a third one (G3) for the 1:7 and 1:9 emulsions. For G1, the emulsion rapidly creamed in two different stages; an almost linear increase of CI from 0h to 6h, reaching a value up to 90%, and a steady phase, from 6 to 30 h, where the CI was maintained at this value. For G2, the 1:3 and 1:5 emulsions presented a similar tendency to G1 during the first 3 h, but with lower CI values; 67.27% and 66.56%, respectively. From the 3h-6h range, a transition phase was noticed, reducing the rate at which the emulsions were creamed compared to G1. Then, from the 6h-30 h range, a steady phase was reached for the emulsions with CI values closer to 76.5% and 76.0%, respectively. For G3, the 1:7 and 1:9 emulsions creamed at a rate lower than G1 and G2. From 0h to 3h, a first stage was identified with CI values increasing up to 22.63% and 22.75%, respectively. Then, from the 3h-12h range, a transition stage was identified where the 1:7 and 1:9 emulsions reached average CI values of 59.62% and 53.50%, respectively. Both emulsions presented a steady phase from 12h to 30h, with average CI values around 61% and 55%. The above results evidenced that emulsions were affected by the gravitational separation of their components, becoming unstable over time. As an example of the creaming phenomenon occurring in the emulsions, Fig. 2b shows representative images for each type of emulsion at 1h and 30 h, revealing the separation process over time.

Identifying the previous tendencies can be attributed to the following factors affecting the physical stability of the emulsions. (1) A first factor is the increment of the viscosity of the emulsions with the incorporation of SO. In this context, O/W emulsions with B/O ratios 1:1, 1:3, 1:5, 1:7 and 1:9 presented viscosity values of 188.9 mPa·s, 201.8 mPa·s, 203.1 mPa·s, 222.3 mPa·s, and 262.1 mPa·s, respectively. Thus, the increase in the viscosity with the incorporation of SO retarded the oil droplets ascension to the top

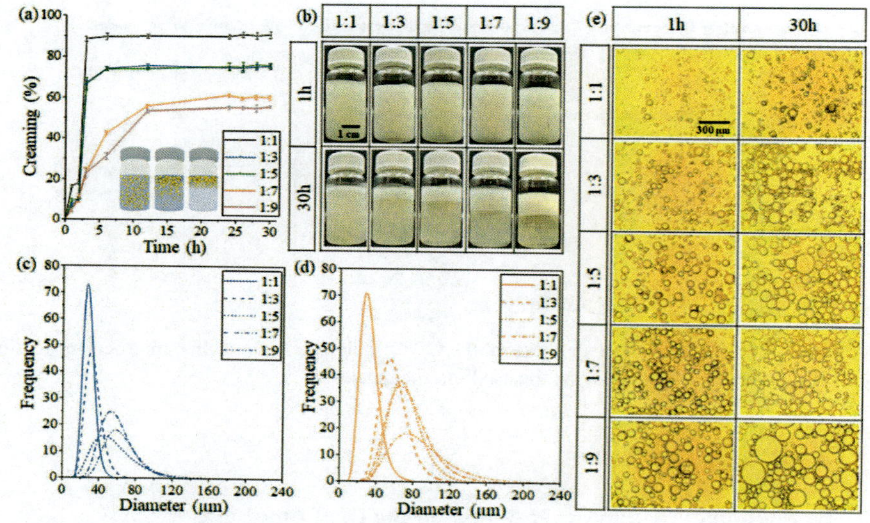

Fig. 2. **a** Average results of creaming index evolution over time for the emulsion with different B/O ratios; **b** Representation of the creaming phenomenon at 1h and 30 h occurring in the O/W emulsion with different B/O ratios; **c–d)** Oil droplet size distribution for the different em O/W emulsion fitted to a log-normal distribution; **e** Fluorescence microscopy images at 1h and 30 h occurring in the O/W emulsion with different B/O ratios.

of the emulsion, and so, reducing their extent of separation from the alginate solution. (2) A second factor is the increase in the SO droplets size and concentration with its incorporation to the emulsion, blocking the ascension of the oil droplets to the top of the emulsion, and so retarding the phase separation between the SO and the alginate solution. To prove this, Fig. 2c and d shows the oil droplet size distributions of the different emulsions at 0h and 30h, respectively. Overall, an increase in the size of the oil droplets is observed with the incorporation of SO into the emulsion. On the other hand, Fig. 2e shows representatives fluorescence microscopy images for the O/W emulsions, evidencing an increase in the number of droplets blocking from each other with the increment of the SO. Based on the previous analysis, it can be concluded that the incorporation of SO to the O/W emulsions reduced their CI values, improving their physical stability over time.

3.2 Effect of the Oil Content on the Morphological and Physical-Mechanical Properties of the Biocapsules

Figure 3a–e shows representative optical microscopy images for each synthesised capsule. Overall, capsules were synthesised with spherical morphology and average SF values varying from 0.053 for capsules with B/O ratio 1:1 to 0.036 for capsules with B/O ratio 1:9. This Figure also shows that the capsules increased their size with the incorporation of the SO. This result is supported by Fig. 3f, showing box-plot representations of the diameter for each type of capsule adjusted to a log-normal distribution. The representative size values for the capsules with B/O ratios 1:1, 1:3, 1:5, 1:7 and 1:9

were 1.145 mm (SD = 0.126 mm), 1.426 mm (SD = 0.127 mm), 1.476 mm (SD = 0.166mm), 1.572 mm (SD = 0.147 mm), and 1.734 mm (SD = 0.169 mm), respectively. The variation in size of the capsules were mainly attributed to their respective emulsion viscosity values, previously reported. Thus, with the incorporation of SO, the emulsion viscosity was increased, retarding the realease of the emulsion droplets from the nozzle.

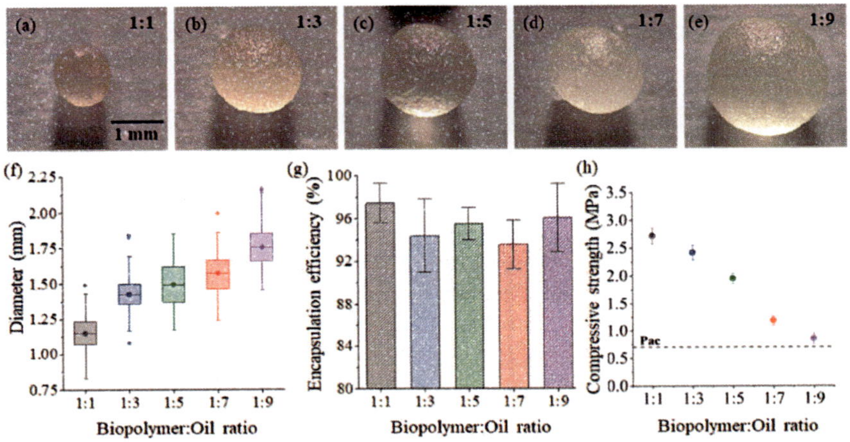

Fig. 3. **a–e** Optical microscopy images for the polynuclear capsules synthesised with alginate:oil ratios 1:1, 1:3, 1:5, 1:7, and 1:9, respectively; **f** Box-plot representation of the diameter of the capsules with different B/O ratios; **g** Average results of encapsulation efficiency for the capsules with different B/O ratios; and **h** Average results of yield compressive strength for the capsules with different B/O ratios.

A consequence of the increase of the size of the capsules with the incorporation of SO was a reduction of their apparent density, with average values of 1.115 g/cm^3 (SD = 0.0191 g/cm^3), 0.987 g/cm^3 (SD = 0.0148 g/cm^3), 0.971g/cm^3 (SD = 0.0165 g/cm^3), 0.901 g/cm^3 (SD = 0.0139 g/cm^3), and 0.875 g/cm^3 (SD = 0.0161 g/cm^3) for the capsules with B/O ratio 1:1, 1:3, 1:5, 1:7, and 1:9, respectively. These results meant that the increase in the volume of the capsules was more significant than the increment of their mass with the incorporation of SO. As a result, the proportion of encapsulated SO was quite similar, as shown in Fig. 3g. Overall, from this Figure, it can be concluded that the encapsulation process by vibrating jet technique was highly efficient, with EE values over 90%, meaning that almost all the SO was encapsulated.

Furthermore, Fig. 3h shows the average results of compressive tests carried out on the capsules. This Figure shows a reduction in the compressive strength of the capsules with the increase of the SO content. Thus, capsules with B/O ratios 1:1, 1:3, 1:5, 1:7 and 1:9 presented average values of 2.71 MPa (SD = 0.14 MPa), 2.42 MPa (SD = 0.13 MPa), 1.96 MPa (SD = 0.09 MPa), 1.18 MPa (SD = 0.09 MPa), and 0.85 MPa (SD = 0.09 MPa), respectively. Besides, a marginal decrease of the compressive force between the capsules with 1:5, 1:7 and 1:9 B/O ratios could be attributed to their similar average size, as previously reported. For an effective healing process of the capsules, they must be able to resist the compaction pressure during the on-site construction of the road. Delgadillo et al. [14] have established roller compact pressures (Pac) up to 0.7 MPa. Based on this value, all the capsules presented values over Pac indicating that capsules can potentially resist the on-site compaction activity. Nonetheless, once the microcracks propagates on the bituminous matrix, the capsules must be able to opportunely release the encapsulated rejuvenator. In consequence, capsule designs with compressive strength values near to Pac could facilitate its opportune activation, i.e., capsules with B/O ratios of 1:7 and 1:9.

4 Conclusions

The following conclusions can be drawn from this study: (1) the creaming rate of O/W emulsions was reduced with the increase of the oil content, reducing their physical instability over time. (2) The increase of the viscosity and the oil droplet size in the emulsions increased their physical stability. (3) The encapsulation method used resulted in enhanced capsules with spherical morphology and high encapsulation efficiency, over 90%. (4) the increase of the encapsulated amount of oil in the enhanced capsules results in increased size and reduced bulk density and compressive strength. Finally, (5) the compressive strength of the capsules was enough to resist the compaction load pressure in a way that they can be potentially added as an effective additive for asphalt self-healing purposes.

Acknowledgments. The authors thank the funding given by the National Research and Development Agency (ANID) through the FONDECYT Regular research project 1190027. The first author wishes also to thank the financial support given by the University of Bío-Bío for his internal PhD scholarship granted.

References

1. Gonzalez-Torre, I., Norambuena-Contreras, J.: Recent advances on self-healing of bituminous materials by the action of encapsulated rejuvenators. Constr. Build. Mater. **258**, 119568 (2020)
2. Mirwald, J., Werkovits, S., Camargo, I., Maschauer, D., Hofko, B., Grothe, H.: Investigating bitumen long-term-ageing in the laboratory by spectroscopic analysis of the SARA fractions. Constr. Build. Mater. **258**, 119577 (2020)
3. Xu, S., García, A., Su, J., Liu, Q., Tabaković, A., Schlangen, E.: Self-healing asphalt review: from idea to practice. Adv. Mater. Interfaces **5**, 1–21 (2018)
4. Al-Mansoori, T., Norambuena-Contreras, J., Micaelo, R., Garcia, A.: Self-healing of asphalt mastic by the action of polymeric capsules containing rejuvenators. Constr. Build. Mater. **161**, 330–339 (2018)
5. Norambuena-Contreras, J., Arteaga-Pérez, L-E., Concha, J.L., Irene, G.-T.: Pyrolytic oil from waste tyres as a promising encapsulated rejuvenator for the extrinsic self-healing of bituminous materials. Road Mater. Pavement Des. **22**(1), S117–S133 (2021)
6. Ruiz-Riancho, N., Garcia, A., Grossegger, D., Saadoon, T., Hudson-Griffiths, R.: Properties of Ca-alginate capsules to maximise asphalt self-healing properties. Constr. Build. Mater. 284 (2021)

7. Zhang, L., et al.: Synthesis and characterisation of multi-cavity Ca-alginate capsules used for self-healing in asphalt mixtures. Constr. Build. Mater. **211**, 298–307 (2019)
8. Xu, S., Tabaković, A., Liu, X., Palin, D., Schlangen, E.: Optimisation of the calcium alginate capsules for self-healing asphalt. Appl. Sci. 9 (2019)
9. Norambuena-Contreras, J., Arteaga-Perez, LE., Guadarrama-Lezama, AY., Briones, R., Vivanco, JF., Gonzalez-Torre, I.: Microencapsulated bio-based rejuvenators for the self-healing of bituminous materials. Materials **13**(6),1446 (2020)
10. McClements, D.J.: Critical review of techniques and methodologies for characterisation of emulsion stability. Crit. Rev. Food Sci. Nutr. **47**, 611–649 (2007)
11. Voo, W.P., Ooi, C.W., Islam, A., Tey, B.T., Chan, E.S.: Calcium alginate hydrogel beads with high stiffness and extended dissolution behaviour. Eur. Polymer J. **75**, 343–353 (2016)
12. Guadarrama-Lezama, A.Y., Dorantes-Alvarez, L., Jaramillo-Flores, M.E., Pérez-Alonso, C., Niranjan, K., Gutiérrez-López, G.F., Alamilla-Beltrán, L.: Preparation and characterization of non-aqueous extracts from chilli (capsicum annuum L.) and their microencapsulates obtained by spray-drying. J. Food Eng. **112**, 29–37 (2012)
13. Alpizar-Reyes, E., Varela-Guerrero, V., Cruz-Olivares, J., Carrillo-Navas, H., Alvarez-Ramirez, J., Pérez-Alonso, C.: Microencapsulation of sesame seed oil by tamarind seed mucilage. Int. J. Biol. Macromol. **145**, 207–215 (2020)
14. Delgadillo, R., Bahia, H.U.: Effects of temperature and pressure on hot mixed asphalt compaction: field and laboratory study. J. Mater. Civ. Eng. **20**(6), 440–448 (2008)

A Novel Spore-Based System with Rejuvenator Controlled Release for the Self-healing of Bituminous Materials

Erik Alpizar-Reyes(✉), José L. Concha, and José Norambuena-Contreras

Department of Civil and Environmental Engineering, LabMAT, University of Bío-Bío, Concepción, Chile
ealpizar@ubiobio.cl

Abstract. Self-healing bituminous materials by adaptive encapsulated rejuvenators system is a hot topic within road materials. For the first time, this work explores the natural spores for the encapsulation of rejuvenators as a novel controlled release system to promote the self-healing capability in aged bituminous materials. Treated spores were obtained from natural spores by chemical processes. To synthesize spore-based microcapsules, firstly, spores were defatted, and then the protein and cellulosic materials were removed. Spores were processed, and sunflower oil as a rejuvenator was encapsulated inside spore microcavities by encapsulating passive and vacuum loading technologies. Physicochemical and microstructural properties of the natural spore formulations were evaluated. Changes in the composition of the spores among the process were studied by FTIR spectroscopy. The microstructure analysis showed that the spores after treatment were empty, showing the complete removal of all contents inside, keeping intact microstructure, and providing large cavities for molecular loading. Furthermore, sunflower oil encapsulation efficiency into the spore cavities was considerably higher with vacuum loading than by the passive loading process. The controlled release mechanism and the healing efficiency of encapsulated rejuvenators in the spore capsules were also quantified. It was proved that the proposed approach was able to control the release of the rejuvenator with an attractive biomaterial possessing a very strong structure. The main results provide the basis for further exploration into the encapsulation of rejuvenators in natural plant spores as a promising novel controlled release spore-based system for the self-healing of bituminous materials.

Keywords: Asphalt self-healing · Microencapsulated rejuvenators · Spore-based system · Controlled release

1 Introduction

Viscoelastic bituminous materials are widely used for asphalt pavement construction in the world. However, ageing is a physicochemical phenomenon that promotes the oxidation of bitumen and consequently its damage by cracking [1–3]. This cracking continues damaging the bituminous materials resulting in eventual structural damage. As a result,

J. I. Escalante-Garcia et al. (Eds.): RW 2021, RILEM Bookseries 40, pp. 942–952, 2023.
https://doi.org/10.1007/978-3-031-21735-7_100

recently, researchers developed a new field of research on the self-healing of bituminous materials. This emerging field aims to develop solutions to promote autonomous crack-healing on bituminous materials [3]. One of the most attractive alternatives to encourage the autonomic healing of cracked bituminous materials is the development of solutions at micro-scale levels with the capacity to release compounds known as rejuvenating agents. These compounds must be able to lubricate and reconstitute the chemical integrity of the aged bitumen [4]. Furthermore, healing agents demand to be retained into multi-scale capsules to be released when the crack appears. With this in mind, microcapsules with different shapes, layers of wall materials, numbers, and positions of cores can be synthesized for this purpose. Spore shells (i.e. sporopollenin exine) are highly efficient and robust bio-based microcarriers that protect their genetic materials and cellulose from the exposition to the environment and mechanical damages [5]. Moreover, spore shells can be chemically emptied [6] to synthesize a new kind of spore-based system to act as carriers ready to confer protection and controlled release of rejuvenators when the damage appears. No other naturally occurring polymer used in microencapsulation has such extensive protective properties armoury [5]. So, this work is a preliminary exploration to the use of natural spores for the encapsulation of rejuvenators as a novel controlled release system for the self-healing of bituminous materials.

2 Materials and Experimental Methods

2.1 Materials

Spore-based microcarriers with a controlled release of rejuvenating agent (cooking oil) to promote the extrinsic self-healing in bituminous materials were synthesized in this work. Natural spores from Sporopollenin (*Lycopodium clavatum*) were purchased from Merck (Darmstadt, Alemania) with a Carbon content of 64.8%, Hydrogen content of 9.4%, and Nitrogen content of 1.0%. Cocking oil (cold-pressed sunflower oil, ®Aceite Natura) with density @20 °C 0.85 g/cm^3, viscosity @20 °C 70 cP and pH @25 °C ~ 5.4 acquired in a local supermarket (Concepción-Chile), was used as an asphalt rejuvenator. Aged bitumen with density 1.04 g/cm^3 and penetration grade 50/70 (penetration $60 \times 10 - 1$ mm @ 25 °C and a softening point 50 °C) was provided by Laboratory LEMCO from The Federico Santa María Technical University (Valparaiso-Chile). The chemical reactants (Petroleum ether, phosphoric acid, acetone, hydrochloric acid, ethanol, and n-hexane) were provided by Merck (Darmstadt-Alemania). Ultra-pure water (TOC <30 ppb) was used in all experiments.

2.2 Chemical Conditioning of the Spores to Produce Spore-based Microcarriers

To obtain spore-based microcarriers, natural spores were previously conditioned by a process consists of three stages of chemical treatments (as shown in Fig. 1). This procedure comprises of a preliminary investigation of the method reported by Alpizar-Reyes et al., 2022 [6].

In the first stage (see Fig. 1b), 15 g of natural spores were defatted by Soxhlet extraction with petroleum ether at four intervals of 8 h per treatment. Then, they were

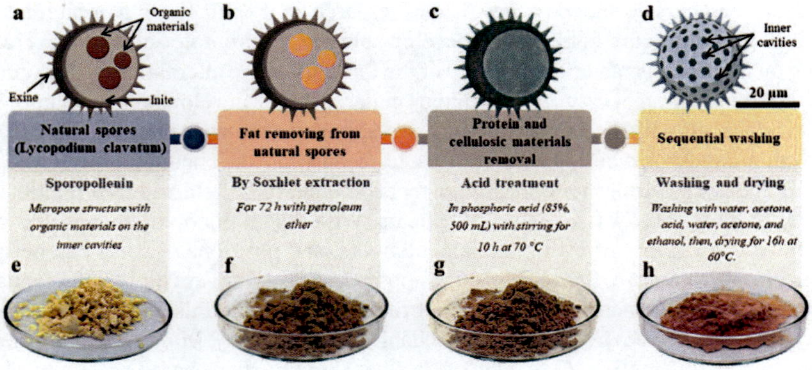

Fig. 1. Chemical extraction of Sporopollenin Exine Capsules from natural spores. **a** A schematization of the main components of the natural spores; **b** Defatting process from natural spores (Stage 1); **c** Organic materials remotion from natural spores (Stage 2); **d** Treated empty Sporopollenin Exine Capsules (Stage 3); **e–f** Images showing the appearance change between the chemical processing of natural spores to turn into empty shells (treated natural spores).

dried at 60 °C for 24h in a convection oven. The second stage consisted of the elimination of organic materials, mainly pollen and cellulose. Defatted spores were dispersed in hot phosphoric acid (85%, 500 mL), stirring at 70 °C for 10 h. Finally, to neutralize the pH and ensure the emptiness of the SEC´s, on the third stage (see Fig. 1d), the SEC´s were collected by filtration and sequentially washed at room temperature (~20 °C) as follows: (a) 5 times with 250 mL of ultrapure water; (b) 2 times with 250 mL of acetone; (c) Once with 500 mL of hydrochloric acid 2M; (d) 5 times with 250 mL of ultrapure water; (e) 2 times with 250 mL of acetone; (e) and, finally once using 250 mL of ethanol and the empty SEC´s were dried for 16h at 60 °C and keep for further use [6].

2.3 Experimental Characterization of the Spores

To evaluate the changes that occurred on the spores due to the chemical processing, natural and treated spores were characterized by microstructure and Fourier-Transform Infrared spectroscopy (FT-IR) analysis. The microstructure was examined by morphology, size, and shape of the spores. Morphology of the natural and treated spores was investigated by Scanning Electron Microscopy (SEM) using a JSM-6010PLUS/LA (JEOL, Tokyo, Japan) by applying an acceleration voltage of 5 keV. Thus, SEM images were used to determine the size and the sphericity factor distribution of 100 spores randomly selected and processed by the image processing software ImageJ®, version 1.52p, Fiji distribution (Maryland-USA). The sphericity factor (*SF*) was calculated as follows, Eq. (1):

$$SF = \frac{D_{max} - D_{per}}{D_{max} + D_{per}} \tag{1}$$

where D_{max} is the maximum diameter passing through the centroid of the spore (μm) and D_{per} is the diameter perpendicular to D_{max} passing through the centroid (μm).

Finally, infrared spectra were recorded on a Thermo Scientific Nicolet iS20 FTIR Spectrometer (Massachusetts-US) with a DTGS detector and KBr windows and supported with OMNIC Specta Software. Each spectrum was recorded over 20 scans in the range from 4000 to 400 cm^{-1}, with a resolution of 2 cm^{-1}.

2.4 Obtention of the Spore-based System of Microcapsules

Once the treated spores were emptied, spore-based microcarriers of cooking oil were produced by loading the cooking oil as a healing agent into the treated spores. Thus, two different microencapsulation methods were used: ***(1) Encapsulation by passive loading**:* 500 mg of treated spores were suspended in 2500 mg of cooking oil, keeping a relationship between treated spores and cooking oil equal to 1:5 mass-to-mass. The suspension was mixed by vortexing for 60 min in a Vortex V1 plus (Hamburg-Germany) and then was homogenized for 120 min at 950 rpm with a mechanical stirrer, Scilogex OS20-Pro (Connecticut-USA). ***(2) Encapsulation by vacuum**:* 500 mg of treated spores were suspended in 2500 mg of cooking oil and then subjected to vacuum (120 mbar pressure) for one hour with a vacuum pump, Boeco R-300 (Hamburg-Germany) into an ultrasonic bath, MRC DC-120H (Cape Town-South Africa) to avoid the clumping formation. Vacuum encapsulation was carried out at room temperature.

2.5 Characterization of the Spore Microcapsules

The surface oil, payload, and encapsulation efficiency of the spore microcapsules loaded by passive and vacuum encapsulation were quantified based on the method described by Alpizar-Reyes et al. [7]. These properties were determined by extracting the cooking oil on the surface with n-hexane dispersions, the total and internal oil with a Soxhlet extraction system for 8 h at room temperature. The controlled release of the cooking oil out of the spore microcapsules was analysed over time as follows. First, 1 g of sample of microcapsule was washed for 60s with n-hexane to remove the oil on the surface. Then, the sample was dispersed at room temperature in 20 mL of n-hexane with continuous string for a testing time of 1 min. Later, the spores were filtered from the solution. Finally, to measure the weight of oil released, the n-hexane was evaporated overnight from the solution in a convection oven at 40 °C. The same procedure was repeated by varying only the testing time to 5, 10, 15, 20, 25, 30, 45, and 60 min.

2.6 In-Vitro Self-healing Efficiency of the Spore Microcapsules

The in-vitro self-healing efficiency of the cooking oil released by the spore-based system was quantified in aged bitumen samples by fluorescence microscopy following the proposed method by Norambuena-Contreras et al. [2]. Thin-film bitumen samples with dimensions of 20 × 20 × 0.5 mm were built-in microscope slides. Then, hot bitumen test samples were conditioned for 2h until to reach room temperature. A 100 μm-width microcrack was made in the centre of the bitumen sample using a metallic cutting element. A sample of ~50 μL was dropped on the cracked region of the thin-film of the aged bitumen to induce the in-vitro healing phenomena. By the effect of cooking oil diffusion over time, the crack closure of the bitumen sample was periodically recorded using an inverted fluorescence microscope, ICOE IV 5100 FL (Ningbo-China) with phase contrast and a magnification of 40×. Crack-width in the images was measured every ten minutes until reach crack closure by image processing software ImageJ®, version 1.52p, Fiji distribution (Maryland-USA). Finally, the self-healing efficiency of the BO rejuvenating over time was quantified as follows, see Eq. (2):

$$\mathrm{SHE_{ti}} = \left(\frac{\mathrm{ICW_0} - \mathrm{PCW_{ti}}}{\mathrm{ICW_0}}\right) \times 100 \tag{2}$$

where $\mathrm{SHE_{ti}}$ is the healing efficiency at a time ti measured in %, $\mathrm{PCW_{ti}}$ is the average partial crack-width at a time ti measured in μm, and $\mathrm{ICW_0}$ is the average initial crack-width measured in μm. Finally, four fractions, namely Saturates, Aromatics, Resins and Asphaltenes (SARA), were obtained from pyrolytic oil according to the standard ASTM D4124-09.

3 Results and Discussion

3.1 Extraction and Characterization of the Spore-based Microcarriers from Natural Spores

In this study, a spore-based microcarrier system containing a healing agent has been proposed to rejuvenate aged bitumen. To synthesize the spores-based system, an approach to modify natural spores by chemical treatments was implemented. The chemical extraction process (see Fig. 1) was selected considering the variety and nature of the different species of spore species. In this case, this process was adequate for *Lycopodium Clavatum* spores. This method comprises three stages involving moderate acid heating conditions and various washing to remove the organic materials of the spores and produce "empty shells" [8] able to be loaded by cooking oil to act as a healing agent of aged bitumen. Figure 1e–f shows the appearance change between the chemical processing of natural spores to turn into empty shells. The colour changes are associated with the remotion of different specific components of the spores, such as fatty acids with a percentage in weight of 2.73 ± 0.52% (Fig. 1f); genetic material, cellulosic content and protein remotion of 28.73 ± 2.06% (Fig. 1g); and a total mass loss on the chemical treatment of 31.06 ± 2.56% (Fig. 1h).

Additionally, the microstructure of the natural and chemically treated spores are observed in Fig. 2. In Fig. 2a–f it can be seen the morphology of the spores before and after chemical treatments. Thus, Fig. 2a shows natural pollens with a structure containing the cellular material, and Fig. 2c shows an intact outer wall mainly composed of exine, which can behave like a "shell" protecting the organic cellulosic materials contained among the spores.

Figure 2b,e–f shows that the treated spores kept intact their structural integrity after chemical treatments without any damage by denaturation. Also, the cellulosic materials, including fats and proteins (Fig. 2b), were removed from the spores showing in Fig. 2e–f the presence of large empty cavities with pores traduced in available sites which allow us to fill the inner structure of the spores with cooking oil to act as a healing agent [8]. In Fig. 2g we can appreciate the mean diameter of the particle of the spores with an average

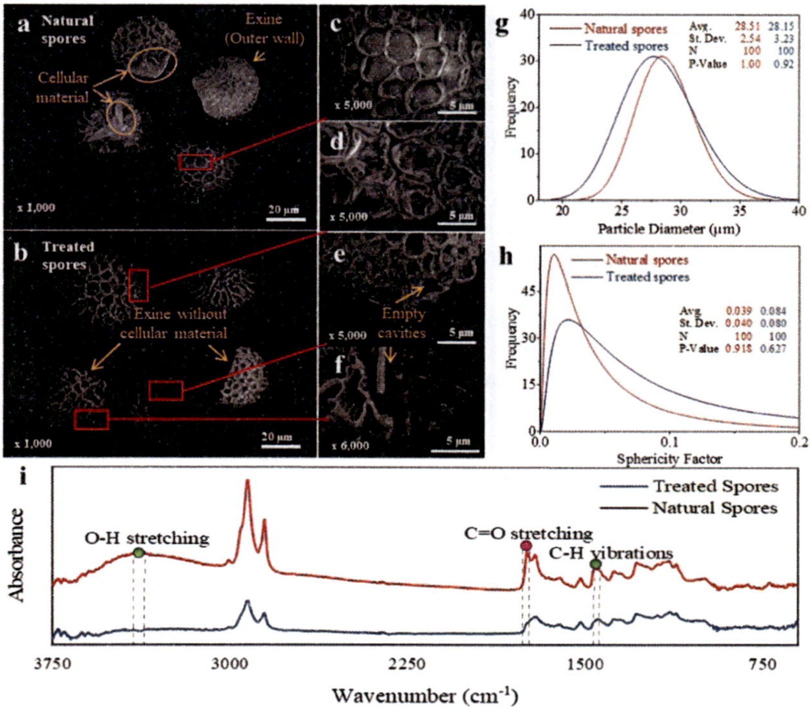

Fig. 2. Spore-based microcarriers. **a** SEM image of natural spores; **b** SEM image of treated spores; **c** SEM image of the spore surface; **d** SEM detail image of the treated spore surface; **e–f** SEM image detail of the empty cavities of treated spores; **g** Frequency histogram of the size of the spores fitted to a log-normal distribution; **h** Frequency histogram of the porosity of the spores fitted to a log-normal distribution; **i** Normalized infrared spectra of the spores.

of 28.51 µm for natural and 28.15 µm for treated spores. The uniformity of the size of the pollens can be appreciated in Fig. 2g before and after the chemical shedding of spore shells. In addition to the greatness of the uniformity in diameter of the spores, the sphericity factor was measured to determine whenever the particles tend to be spherical (values nearby 0) or irregular shapes (values nearby 1). The shape of spores can be considered non completely spheric due to the characteristic distribution of the external shell with average values of 0.051 a.u. for natural, and 0.095 a.u. for treated spores with a sphericity factor > 0 (Ideal sphere $= 0$). Finally, according to Fig. 2, the uniformity on the microstructure observed for the spores is incredibly consistent. These features allow us to exploit this biobased material to encapsulate cooking oil as a self-healing agent of aged bitumen.

Finally, to ensure that the chemical treatment does not affect the spores shell's chemical structure, natural and treated spores were characterized by FT-IR. Figure 2i shows the FT-IR spectra for natural and treated spores. Both spectra presented remarkable similarities, where three peaks that are observed in the natural spores no longer appear in the spectrum of the treated spores. The first peak that only appears on the natural spores corresponds to a strong broad between 3500 to 3000 cm^{-1}. This broad has been extensively attributed to O-H stretching [9]. The second peak is observed at 1740 cm^{-1} [10], this strong peak is attributed to C = O stretching, and the third peak was located at 1460 cm^{-1}, with a medium stretching due to C-H bending vibrations [9]. Strong broad observed at 3500–3000 cm^{-1}, and the peak at 1460 cm^{-1} are the finger paint of the fatty acids [11], and the strong peak at 1740 cm^{-1} is a characteristic peak of the cellulose [11]. With this in mind, we can ensure that the cellulose was completely removed from natural spores. The peaks observed for both spectra has been previously reported for exine spores, the saturated aliphatic components of the spores are located at the strong characteristic peaks at 2914 and 2856 cm^{-1} and which are attributed to -CH_2 asymmetric and symmetric stretching vibrations, respectively [6], and -CH2 scissoring at 1179 and 1445 cm^{-1} [10, 12]. The amide I band at 1656 cm^{-1} and amide II bands at 1528 cm^{-1} demonstrating C = O stretching and C-N stretching, were two prominent characterization peaks [13]. The two absorption peaks located at 1708 cm^{-1} and 1114 cm^{-1} belong to C = O and C-O stretching vibration of the ester [6]. The peaks at 1380, 1360, and 1288 cm^{-1} are typical of the -C-C- skeletal and the -C-H interactions on the structure of the organic materials [10].

3.2 Encapsulation of Cooking Oil into the Spore-based System of Microcapsules and the Controlled Release

In this section, two different encapsulation technologies (i.e. passive and vacuum encapsulation) were tested to determine the most appropriate technology to design a spore-based system of microcapsules of rejuvenator for ageing bitumen. Figure 3a summarizes the surface oil, payload, and encapsulation efficiency for spore microcapsules. The main results show that the surface oil was significantly higher for passive encapsulation (17.42%, SD = 2.11) than for vacuum encapsulation (9.21%, SD = 1.94). In addition, Fig. 3a shows the payload percentages of both encapsulation technologies tested. The payload is a parameter that indicates the adequate amount of oil that the empty shells can retain on the surface and into the shell. In short, the payload of the vacuum microencapsulation (70.28%, SD = 4.11) presented higher values than passive encapsulation (62.12%, SD = 3.65). Hence, with the vacuum encapsulation, a higher quantity of cooking oil was loaded inside the empty cavities of the spores offering the biggest control on the release of the cooking oil into aged bitumen when needed.

Finally, Fig. 3a shows the average result of the encapsulation efficiency for both technologies. From this Figure, it can be concluded that passive encapsulation (59.10%, SD = 1.73) presented lower encapsulation efficiency compared with vacuum encapsulation (86.14%, SD = 2.19). This result means that vacuum encapsulation technology can retain a higher amount of oil into the spore shells as vacuum encapsulation is the technology with the best performance for encapsulating cooking oil into the empty shell of the treated spores. The explanation for this phenomenon is evident from applying an external force as a vacuum facilitates the forced diffusion of the cooking oil inside the spore shells and facilitates encapsulation. Meanwhile, with the passive loading, the oil diffuses inside the spore shell only by a natural diffusion [14, 15]. To determine which is the best encapsulation technology is also necessary to study the controlled cooking

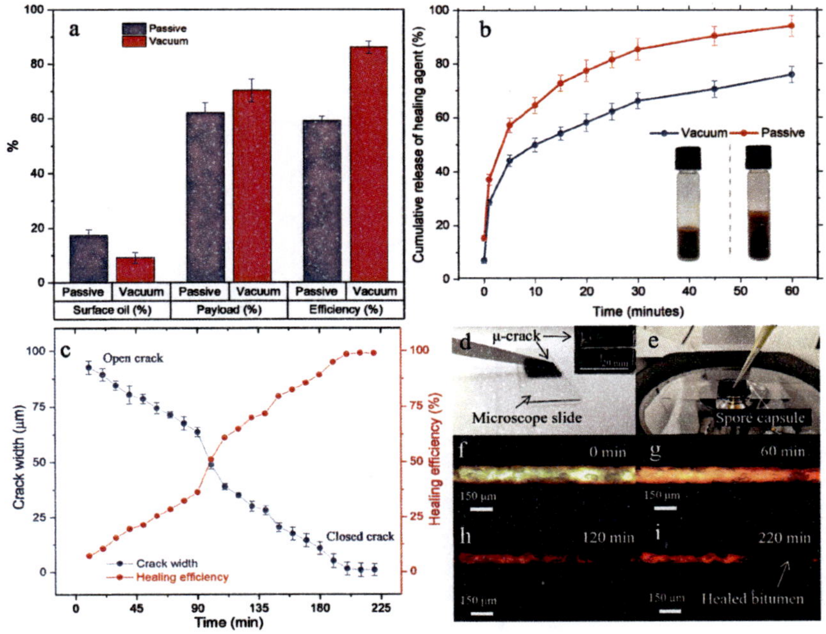

Fig. 3. Characterization and self-healing efficiency of the spore-based microcapsules over time. **a** Surface oil, payload and encapsulation efficiency of passive and vacuum encapsulation technologies; **b** Release profiles of cooking oil out of the loaded spores over time; **c** Results of healing efficiency and μ-crack over time; **d–e** Procedure to determine the healing efficiency of spore-capsules on a film of aged bitumen test samples; **f–i** Fluorescence microscopy showing the closure crack phenomena by the action of the oil release from the spore-capsules over time.

oil release for each encapsulation technology (see Fig. 3b). The cooking oil release out of the spore microcapsules is faster with passive encapsulation than vacuum encapsulation. This result can be related to the payload and encapsulation efficiency because, on passively loaded spores, there is more space on the cavities of the shell that allow a faster release outside of the shell. In summary, the ideal encapsulation technology was vacuum encapsulation, leading to the optimal encapsulation efficiency and a better-controlled release for the microencapsulation of cooking oil with the potential to be used for asphalt self-healing applications.

3.3 Self-healing Efficiency of the Spore Microcapsules

The self-healing capability results of the spore microcapsules loaded with cooking oil are shown in Fig. 3d–i. For instance, in Fig. 3f–i is possible to observe the autonomic healing phenomena induced by the cooking oil release outside the spore capsules to the micro-crack on the aged bitumen. The crack closure followed a slight linear trend on the closure time over time, and the complete crack healing was successfully achieved at ~200 min (Fig. 3c). Resins of the cooking oil promoted the self-healing of the aged bitumen, this is because resins are the main component of the maltenes, crack healing is induced by the action of the maltenes, when a maltene diffuses through aged bitumen, it will deposit the maltene-enriched blend on the films of aged binder, restoring the proper balance among asphaltenes/maltenes within the aged binder, Fig. 4 [1].

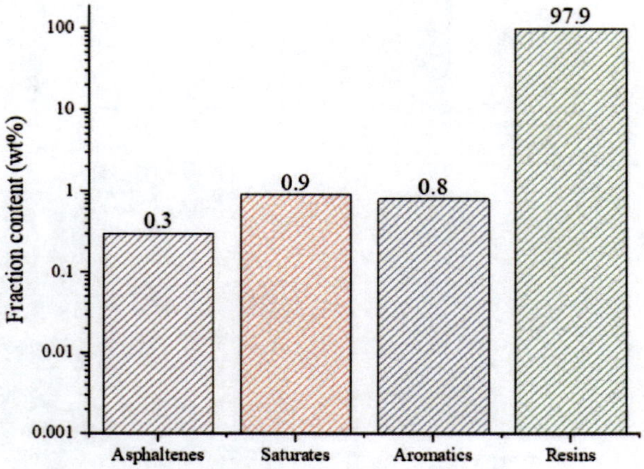

Fig. 4. SARA fractions of the cocking oil analysed.

It was reported [1, 2] that self-healing takes more than 120 min to be reached in a sample of aged bitumen. Thus, it is necessary to design microcarriers with the ability to ensure a continued release of the healing agent for longer times. Therefore, to offer capsules with a higher degree of control on releasing the healing agent into the aged bituminous materials, it is necessary to explore and external coating technologies to be applied on the microcapsules developed in this work. The self-healing capability results of spore-based microcarriers of cooking oil in bitumen are a promising technology with a future perspective to improve the control over the release of the healing agents to develop more bio-based sustainable solutions to replace synthetic materials.

4 Conclusions

This work studied for the first time the use of natural spores for the encapsulation of rejuvenators as a novel controlled release system to promote self-healing in aged bitumen. Based on the results, the following conclusions have been obtained:

The chemical treatment proposed in this study successfully removed the organic materials inside the shells of the natural spores, obtaining a spore-based system with treated spores that work as a shell able to encapsulate cooking oil to act as a healing agent of aged bitumen.

The microstructure and FTIR results of the treated spores proved that the treated spores were uniform in size and sphericity and kept intact their structural integrity with the presence of large empty cavities with a consistent uniformity on the microstructure.

Encapsulation by vacuum technology led to the optimal encapsulation efficiency and a better-controlled release for the microencapsulation of cooking oil with excellent potential for asphalt self-healing applications.

Self-healing tests confirmed that the spore-based microcarriers are a spore-based system that can promote the self-healing of the artificial microcracks on aged bitumen, achieving a self-healing efficiency near 100% at 200 min study.

Acknowledgments. The authors thank the funding given by the National Research and Development Agency (ANID) through the Research Projects FONDECYT Postdoctoral 3200227 and FONDECYT Regular 1190027. The second author wishes to thank the financial support given by the University of Bío-Bío for his internal PhD scholarship granted. Finally, the authors also appreciate the testing support (FT-IR and SEM) given by Dr. Luis E. Arteaga-Perez (LPTC-UBB) and Dr. Manuel Melendrez (DIMAT-UDEC).

References

1. Norambuena-Contreras, J.L., Arteaga-Pérez, E., Concha, J.L., Gonzalez-Torre, I.: Pyrolytic oil from waste tyres as a promising encapsulated rejuvenator for the extrinsic self-healing of bituminous materials. Road Mater. Pavement Des. **22**, 117–133 (2021)
2. Norambuena-Contreras J., Arteaga-Perez, L.E., Guadarrama-Lezama, A.Y., Briones, R., Vivanco, J.F., Gonzalez-Torre, I.: Microencapsulated bio-based rejuvenators for the self-healing of bituminous materials. Materials (Basel) **13**(6), 1446 (2020)
3. Gonzalez-Torre, I., Norambuena-Contreras, J.: Recent advances on self-healing of bituminous materials by the action of encapsulated rejuvenators. Constr. Build. Mater. **258**, 119568 (2020)
4. Karlsson, R., Isacsson, U.: Investigations on bitumen rejuvenator diffusion and structural stability. Asph. Paving Technol.: Assoc. Asph. Paving Technol.-Proc. Tech. Sess. **72**, 463–501 (2013)
5. Mackenzie, G., Beckett, S., Atkin, S., Diego-Taboada, A.: Pollen and spore shells—nature's microcapsules. Microencapsul. Food Ind. 283–297 (2014)
6. Alpizar-Reyes, E., Concha, F.J.L., Martín-Martínez, F., Norambuena-Contreras, J.: Biobased spore microcapsules for asphalt self-healing. ACS Appl. Mater. Interfaces **14**, 27, 31296–31311(2022)
7. Alpizar-Reyes, E., Varela-Guerrero, V., Cruz-Olivares, J., Carrillo-Navas, H., Alvarez-Ramirez, J., Pérez-Alonso, C.: Microencapsulation of sesame seed oil by tamarind seed mucilage. Int. J. Biol. Macromol. **45**, 207–215 (2020)
8. Mundargi, R.C., Potroz, M.G., Park, J.H., Seo, J., Lee, J.H., Cho, N.J.: Extraction of sporopollenin exine capsules from sunflower pollen grains. RSC Adv. **6**(20), 16533–16539 (2016)
9. Sherazi, S.T.H., Talpur, M.Y., Mahesar, S.A., Kandhro, A.A., Arain, S.: Main fatty acid classes in vegetable oils by SB-ATR-Fourier transform infrared (FTIR) spectroscopy. Talanta **80**(2), 600–606 (2009)
10. Hegedüs, K.: Facile isolation and analysis of sporopollenin exine from bee pollen. Sci. Rep. **11**(1), 1–16 (2021)
11. Tomak, E.D., Topaloglu, E., Gumuskaya, E., Yildiz, U.C., Ay, N.: An FT-IR study of the changes in chemical composition of bamboo degraded by brown-rot fungi. Int. Biodeterior. Biodegrad. **85**, 131–138 (2013)
12. Deng, Z.: Carboxymethylpachymaran-zein coated plant microcapsules-based β-galactosidase encapsulation system for long-term effective delivery. Food Res. Int. **128**, 108867 (2020)

13. Wu, D., Wang, X., Wang, S., Li, B., Liang, H.: Nanoparticle encapsulation strategy: Leveraging plant exine capsules used as secondary capping for oral delivery. J. Agric. Food Chem. **67**(29), 8168–8176 (2019)
14. Anthony, T.C., Fahrenbruch, A.L., Bube, R.H.: Growth of CdTe films by close-spaced vapor transport. J. Vac. Sci. Technol. A **2**, 1296 (1984)
15. Firon, N., Nepi, M., Pacini, E.: Water status and associated processes mark critical stages in pollen development and functioning. Ann. Bot. **109**(7), 1201–1213 (2012)

Permanent Deformation Characteristics of Hot Recycled Blends of Three Different RAP Sources

Ankit Sharma(✉), Dheeraj Adwani, G. D. Ransinchung R.N., and Praveen Kumar

Indian Institute of Technology Roorkee, Roorkee, Uttarakhand, India
asharma1@ce.iitr.ac.in

Abstract. As the road networks are expanding worldwide, recycling pavements, both flexible and rigid pavements, is an important area of research nowadays. To achieve the sustainability objectives in the road construction industry, extensive research focused on reclaimed asphalt pavement (RAP) is required. Rheological parameters like Superpave stiffness parameter (G*/sinδ), true high PG temperature (PGH), and Zero Shear Viscosity (ZSV) have been widely used to study the permanent deformation characteristics of asphalt binder. The experimental campaign includes three RAP sources (RAP-D, RAP-UP, and RAP-UK) and one Recycling Agent, RA (soft binder grade VG-10). The proportion of RAP binder in total binder was 0%, 15%, 25%, 40%, 65%, 80%, and 100% (RAP binder) by weight of total binder. It was found that for highly aged RAP source having PGH of 154 °C, the linear blending chart concept is not applicable. Similarly, the tiered approach of RAP incorporation was found to over-predict the minimum RAP content when the RAP binder is highly aged. For the three RAP sources, it was observed that a log-linear relationship can improvise the conventional log-log relationship between PGH and ZSV. The same was verified with the statistical analysis and using a scatter plot of matrices (SPLOM). The effect of RAP content on mixing and compaction temperature of the hot recycled mix was analyzed based on the equi-viscous approach.

Keywords: Reclaimed asphalt pavement · Recycling agent · Soft grades · Hot mix asphalt · Recycled blends · Rejuvenators

1 Introduction

In the past decade, Zero Shear Viscosity (ZSV) has been widely used as a rutting criterion for bituminous binders, especially in Europe. Researchers have widely used this parameter to evaluate the high-temperature permanent deformation characteristics of asphalt binder [1, 2]. The ZSV is the theoretically calculated viscosity during shear deformation when the shear rate is approaching to zero [3, 4]. The absolute viscosity at 60 °C is a conventional parameter used to estimate the stiffness of a binder. However, the conventional apparatus (Cannon Manning Viscometer) can measure viscosity up to a certain stiffness level of the binder. Hence, sometimes it is beyond capacity to measure the absolute viscosity of a stiff RAP recovered binder. In the present study,

J. I. Escalante-Garcia et al. (Eds.): RW 2021, RILEM Bookseries 40, pp. 953–962, 2023.
https://doi.org/10.1007/978-3-031-21735-7_101

it was found that even "400R" and "800R" capillary viscosity tubes were not able to lift the highly aged RAP binder under the specified vacuum pressure of 300 mm Hg, and the maximum time limit of test (1000 s), conforming to ASTM D2171 [5]. In such cases, extrapolation of complex viscosity (at 60 °C) from frequency sweep data to the zero frequency provides ZSV, which can be considered as a replacement to the absolute viscosity at 60 °C [3]. The Superpave stiffness parameter G*/sinδ represents the irrecoverable deformation of binder during load application, and simulates the rutting resistance [6, 7]. In this study, two rutting criteria viz. G*/sinδ and ZSV are compared to establish the relationship between the two parameters. As per the current recycled asphalt mix design practices, true high PG temperature (PGH), G*/sinδ, and ZSV are widely used to finalize the minimum RAP content in a mix which can provide minimum stiffness to the recycled binder to resist permanent deformations. Further, the mixing and compaction temperature of asphalt mixes also differ as the RAP source changes. It results in different fuel requirements for asphalt mix production and greenhouse gas emission levels of bituminous plants. In the present study, an effort is made to develop a relationship of ZSV with PGH. The variation in mixing and compaction temperature is also analyzed using the equi-viscous approach.

2 Objective

To develop a relationship between ZSV and PGH, and to analyze the variation in mixing and compaction temperature using the equi-viscous approach.

3 Materials and Methods

3.1 Materials

In the present study, three different RAPs were collected from two states and one union territory of India viz. Uttarakhand (UK), Uttar Pradesh (UP), and Delhi (D). The age of RAP-UK, RAP-UP, and RAP-D were 20, 10, and 6 years, respectively. The three RAP materials were milled from the wearing course and base course of the pavement, having a dense-graded mix. The target binder (TB) grade was VG-40, whereas the Recycling Agent (RA) was a soft binder grade, VG-10. Both virgin binders are part of the Indian binder specification, Indian Standard (IS): 73 (2013) [8].

RAP Binder Recovery Process

To recover the RAP binder, firstly, RAP material was submerged in toluene. The resulting solution was extracted from RAP material using centrifuge extraction method conforming to ASTM D2172 [9]. The solution was filtered through a 20μ filter paper. The filtrate was filled in centrifuge bottles and further subjected to a high-speed centrifuge process at 3600 RPM for 30 min [10]. The centrifuged filtrate was decanted in a flask and was taken for RAP binder recovery using a rotary evaporator conforming ASTM D5404 [11]. The experimental results found that the "20μ paper vacuum filtration process" and the "high speed centrifugation process" can remove up to 0.23% and 0.94% fines by weight of RAP material (average values for the three RAP sources), respectively. The characteristic properties of binders used in the study are shown in Table 1.

Table 1. Characterization of virgin and RAP binders.

S.No	Property/ Binder	RA	TB	RAP recovered binder		
		VG-10	VG-40	RAP-D	RAP-UP	RAP-UK
1	High PG Temperature, PGH (°C)	64	76	124	130	154
2	Softening Point, S.P. (°C)	43	59	84	86	92
3	Absolute Viscosity at 60 °C (P)	1126	4473	NA	NA	NA
4	Age of binder (Years)	NA	NA	6	10	20

3.2 Methodology

The RAP recovered binder was blended with RA. The proportion of RAP binder in total binder was 0%, 15%, 25%, 40%, 65%, 80% and 100% by weight of total binder. The experimental campaign included a total of nineteen binders viz. Fifteen RAP-RA combinations, three RAP binders, and one RA. The method of blending was in conformation with ASTM D4887 [12]. It was observed that for stiff RAP binder e.g. RAP-UK recovered binder, oven heating with intermittent stirring is more suitable for blending than the hot plate method. The target is to produce a homogenous blend at the end of the blending process.

For rheological evaluation of binders, original binder grading test (at 10 rad/s and 12% strain level) and frequency sweep test at 60 °C (at 0.1–100 rad/s, 1% strain level) were performed on all considered binders. An Anton Paar rheometer (Physica 101) was used for the rheological measurements. The ZSV values were calculated using the Carreau-Yasuda model applied on the frequency sweep data. To calculate the ZSV value, the regression analysis was performed with the help of the Rheology App. Rotational viscosity of binders was measured at 135 °C and 150 °C using a Brookfield viscometer, which was further used to estimate the mixing and compaction temperatures employing equi-viscous approach. The results presented in this study are the averaged values of three replicates.

4 Results and Analysis

4.1 Variation of G*/sinδ with RAP Content at 64 °C

The Superpave stiffness parameter G*/sinδ is used as a measure of the rutting resistance of the asphalt binder. According to the Indian road recycling specification IRC 120 [13], the minimum RAP content in a Hot Recycled Mix Asphalt (HRMA) can be determined using a G*/sinδ blending chart. The blending chart assumes a linear relationship between G*/sinδ and the RAP content. In this study, the blending chart was prepared using logarithmic of G*/sinδ and RAP content as shown in Fig. 1. However, the variation between log G*/sinδ and RAP content was not found to be linear. The ordinary linear regression analysis of the test data showed the Pearson's correlation coefficient (r) of 0.99, 0.99, and 0.96 for RAP-D, RAP-UP, and RAP-UK, respectively. It can be inferred that the assumption of linear variation between RAP content and G*/sinδ depends on the stiffness of binary components of the recycled binder viz. RAP binder and RA. The minimum RAP content in the recycled binder representing an equivalent G*/sinδ of TB was found to be 14.6%, 12.9%, and 11.3% for RAP-D, RAP-UP, and RAP-UK, respectively. The trend of log G*/sinδ vs. RAP content was found to be similar for RAP-UP and RAP-D. It is because of only one grade difference between PGH of RAP-UP (PGH = 130 °C) and RAP-D (PGH = 124 °C) binders. The PG difference between TB (PG 76–16) and RA (PG 64–22) is two grades and one grade for high and low PG temperatures, respectively. According to Manual Series-2 (MS-2) guidelines [14], for the RAP content between 15 to 25%, the PG of RA should be one grade lower (high as well as low PG) than TB. In the present study, even though the PG of RA is one grade lower than TB, the minimum RAP content was found to be less than 15%. This can be attributed to the fact that the stiffness difference between the two components is not adequate to provide the desirable RAP content. It should be noted that the tier criterion of MS-2 (2014) [14] does not apply for such highly stiff RAP binders. Thus, it can be concluded that the minimum RAP content is dependent on the stiffness gap between the RAP binder and the RA.

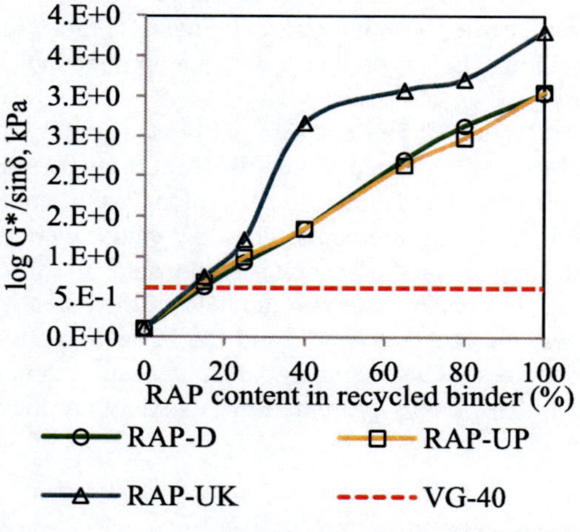

Fig. 1. Variation of G*/sinδ at 64 °C with varying RAP content of RAP sources.

4.2 Variation of ZSV with RAP Content at 60 °C

Figure 2 indicates the variation of logarithmic of ZSV with RAP content. The ZSV of RAP binders had very high magnitudes compared to the ZSVs of blended recycled binders (hereafter designated as recycled binders). Hence for graphical representation, ZSVs of pure components viz. RAP binder and RA were excluded. A sharp increase in ZSV was observed for RAP content of more than 40%. The ZSVs of RAP-UK recycled binders were relatively higher, which agrees with the G*/sinδ results.

The G*/sinδ is calculated at a constant frequency of 10 rad/s and 12% strain, whereas ZSV is an extrapolated theoretical viscosity, indicating the viscosity at zero frequency corresponding to 1% strain. The stiffness values of RAP-UP recycled binders were found to be lower than the stiffness of RAP-D recycled binders at zero frequency/stagnant vehicle condition, whereas an opposite trend was observed at 10 rad/s (simulating vehicle moving at around 80 kmph). This dissimilarity in trend can be explained by difference in viscoelastic behaviour of the recycled binders. Also, tt may be due to better compatibility and thus, better mixing between RAP-UP binder and RA molecules, which resulted in comparatively lower ZSVs of RAP-UP recycled binder. Further, the difference between ZSVs of RAP-D and RAP-UP recycled binders increased with the increase in RAP content. From the ZSV criteria, the allowable RAP content for RAP-D, RAP-UP, and RAP-UK was established as 11.4%, 16.7%, and 7.8%, respectively.

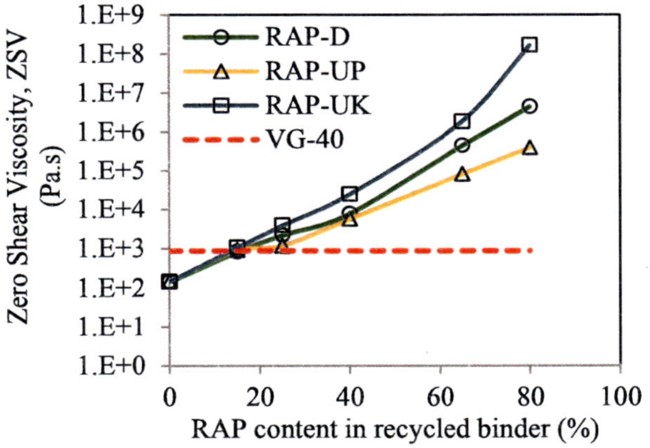

Fig. 2. Variation of ZSV at 60 °C with varying RAP content of RAP sources.

4.3 Relationship Between ZSV and G*/sinδ at 60 °C

The relationship between ZSV and G*/sinδ at 60 °C for various recycled binders is shown in Fig. 3a–d.

In the above figures, the data points from left to right (except VG-40) represent the increase in RAP percent as 0% (VG-10), 15%, 25%, 40%, 65%, 80% and 100% (RAP binder). Both parameters are widely used to characterise the rutting resistance of binders. From the 45° line, it was observed that at higher RAP content (>40%), ln ZSV has more magnitude than ln G*/sinδ. The relationship between ln G*/sinδ and ln ZSV can be best described as linear with r-values for RAP-D, RAP-UP, and RAP-UK as 0.99, 0.97, and 0.98, respectively. The relationship is perfectly linear up to 25% RAP content. The curves of *ln G*/sinδ* vs. *ln ZSV* intersect with 45° line at 59%, 64%, and 45% for RAP-D, RAP-UP, and RAP-UK, respectively. Results indicate that with an increase in stiffness of recycled binder, the increase in ZSV is more than the relative increase in G*/sinδ at the same temperature. The allowable RAP content based on the G*/sinδ criterion is 3.2% and 3.5% higher for RAP-D and RAP-UK respectively as compared to the ZSV

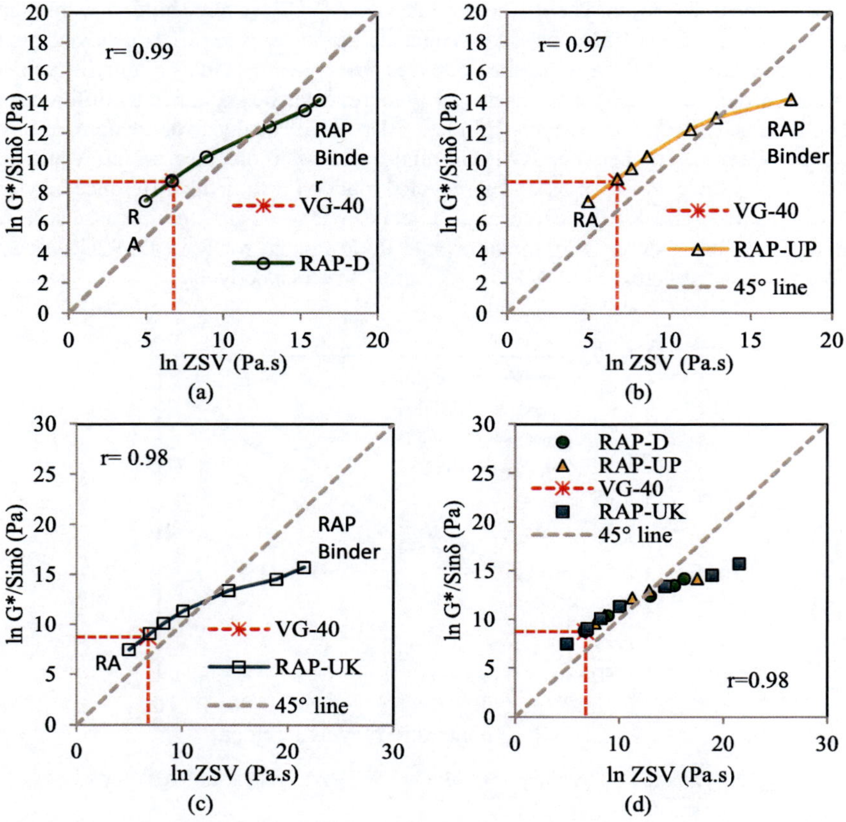

Fig. 3. G*/sinδ v/s ZSV at 60 °C for **a)** RAP-D, **b)** RAP-UP, **c)** RAP-UK, **d)** combine G*/sinδ for all RAP sources.

criteria. Whereas for RAP-UP, the ZSV criteria resulted in 3.8% higher RAP content than the G*/sinδ criteria.

4.4 Modelling of PGH Versus ZSV

Rowe et al. [15], in their study have established a relationship between log (ZSV) and log (PGH). The modelled relationship between ZSV and PGH is as indicated by Eq. (1)

$$\log(ZSV) = k - n * log(PGH) \tag{1}$$

where, k and n are the regressions coefficients. The model shown in Eq. (1) was fitted on the ZSV and PGH data and regression coefficients “k” and “n” were calculated, and are shown below in Table 2.

It was observed that the “sum of the squared estimate of error” can be reduced and a better model fitting can be obtained by considering the log-linear model instead of the

Table 2. Regression coefficients for the log(ZSV)-log(PGH) model.

Regression coefficients/ RAP source	RAP-D	RAP-UP	RAP-UK
K	−32.45	−33.99	−39.87
N	−18.71	−19.38	−22.36

existing log-log model. After shifting to the log-lin model, the percent change in error and correlation between modelled and measured ZSV are shown in Table 3.

Table 3. Change in parameters by shifting on log-lin model from the log-log model.

RAP source	RAP-D	RAP-UP	RAP-UK
% decrease in "Sum of squared estimate of error"	67.18	46.46	55.19
log-log model correlation coefficient	0.99	0.97	0.98
log-lin model correlation coefficient	1	0.98	0.99

As indicated in Table 3, on shifting to the log-lin model from the log-log model, the error has decreased by 67.18%, 46.46%, and 55.19%, respectively, for RAP-D, RAP-UP, and RAP-UK. A scatter plot of matrices (SPLOM) was drawn using the "R programming language" [16]. The SPLOM plot showed a perfect ellipse on shifting from the "log-log" to the "log-lin" model, indicating a better fit. Overall, it can be inferred that for all the RAP sources, the association between log (ZSV) and PGH is better compared to log (ZSV) and log (PGH). The new log-lin model can be written as shown in Eq. (2):

$$\log(ZSV) = K - N * (\text{PGH}) \quad (2)$$

where, K and N are the regression coefficients for the log-lin model. The coefficients for the proposed model are as shown below in Table 4.

Table 4. Regression coefficients for log-lin Model for ZSV and PGH.

RAP source/regression coefficients	RAP-D	RAP-UP	RAP-UK
K	−3.4604	−3.8049	−4.0898
N	−0.0826	−0.0844	−0.0872

The conventional log-log model (Eq. 1) has been extensively used for the polymer modified binders. In the present study, the accuracy of the log-lin model (Eq. 2) has been examined for all three different type of recycled binders (three RAP sources) and was found to be suitable for RAPs used in the study. Further validation is suggested to verify this relationship for different recycling agents and RAP sources.

4.5 Mixing and Compaction Temperatures of Recycled Binders

The blending efficiency of the recycled binder is a function of mixing temperature and mixing time. The effect of RA on the mixing and compaction temperatures of the recycled mix can be seen from Fig. 4, as shown below. The mixing and compaction temperatures have been calculated based on the eqi-viscous approach for all the recycled blends. The results show consistency in mixing and compaction temperatures of the blends. As shown in Fig. 4, with the increase in RAP content up to 40%, mixing and compaction temperatures increase, whereas after 40% RAP content, the relative increase in mixing and compaction temperatures reduces. For 15% and 25% RAP-D recycled blends, the mixing and compaction temperatures are equivalent to VG-40. The allowable RAP percent based on workability criterion is found to be 27%, 17% and 15% for RAP-D, RAP-UP, and RAP-UK, respectively. With increase in RAP percent, the susceptibility of change in compaction temperature is found to be more than mixing temperature. The slope of percent increase in mixing and compaction temperature with unit increase in RAP percent with respect to TB's mixing and compaction temperatures is found to be maximum for RAP-D (0.04% & 0.12%), followed by RAP-UP (0.04% & 0.11%) and RAP-UK (0.02% & 0.09%). It can be inferred from these results that the stiffness level of RAP-UP is nearly equivalent to that of RAP-D. It can also be observed that the difference between mixing and compaction temperatures decreases with increase in RAP percent. It indicates that with an increase in stiffness of the recycled blends, the temperature susceptibility decreases.

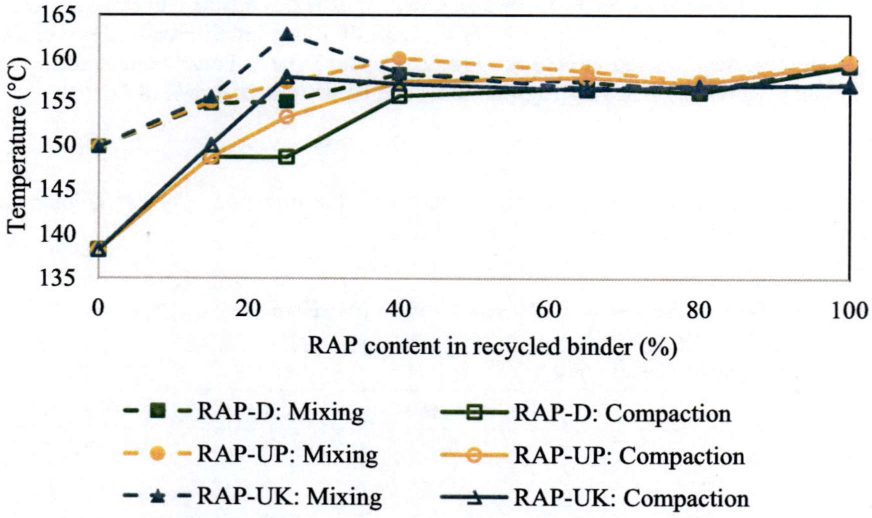

Fig. 4. Variation of mixing and compaction temperature for RAP-D, RAP-UP, and RAP-UK.

5 Conclusions and Future Scope

- For highly aged RAP binders, the applicability of PG based pure mixing rule was found to be dubious. It was observed that, for RAP binder having PGH higher than 124 °C, the allowable RAP content of 15–25% was overestimated even with a RA having PG one grade lower than TB's PG.
- It was observed that rutting parameters viz. ZSV and G*/sinδ have excellent correlation. However, at higher RAP content (more than 40%), the increase in ZSV with RAP content is comparatively more than the corresponding increase in G*/sinδ.
- It was established that a log-lin model is better than a log-log model for relating parameters PGH and ZSV. For all the three RAP sources, an improvement was observed in the modelled values with the proposed log-lin model.
- The results of mixing and compaction temperatures indicate that with an increase in RAP content, corresponding mixing and compaction temperatures also increase. However, at RAP content more than 40%, the relative difference between mixing and compaction temperatures becomes relatively less. It indicates that at higher RAP contents, lower thermal energy is required to change the viscosity of binder.
- It is recommended to measure the PG of recycled binder (in addition to mixing rule), and its thermal susceptibility should also be examined to control the thermal cracking.

References

1. Sybilski, D.: Zero-shear viscosity of bituminous binder and its relation to bituminous mixture's rutting resistance. Transp. Res. Rec. **1535**, 15–21 (1996). https://doi.org/10.1177/0361198196153500103
2. Biro, S., Gandhi, T., Amirkhanian, S.: Determination of zero shear viscosity of warm asphalt binders. Constr. Build. Mater. **23**, 2080–2086 (2009). https://doi.org/10.1016/j.conbuildmat.2008.08.015
3. Heneash, U.: Effect of the repeated recycling on hot mix asphalt properties (2013). http://eprints.nottingham.ac.uk/13283/
4. Sharma, A., Rongmei Naga, G.R., Kumar, P., Raha, S.: Development of an empirical relationship between non-recoverable creep compliance & zero shear viscosity for wide-ranging stiffness of asphalt binders. Constr. Build. Mater. **326**, 126764 (2022). https://doi.org/10.1016/j.conbuildmat.2022.126764
5. ASTM D2171: Standard Test Method for Viscosity of Asphalts by Vacuum Capillary Viscometer (2018). astm.org/Standards/D2171.htm
6. Pradhan, S.K., Sahoo, U.C.: Performance assessment of aged binder rejuvenated with Polanga oil. J. Traffic Transp. Eng. (English Ed.) **6**, 608–620 (2019). https://doi.org/10.1016/J.JTTE.2018.06.004
7. Sharma, A., Naga, G.R.R., Kumar, P., Raha, S.: Rheological characterization of recycled asphalt binders and correlating the zero shear viscosity to the superpave rutting parameter. J. Mater. Civ. Eng. **34**, 4022218 (2022). https://doi.org/10.1061/(ASCE)MT.1943-5533.0004358
8. IS 73: Paving Bitumen- Specification (2013)
9. ASTM D2172: Standard Test Methods for Quantitative Extraction of Asphalt Binder from Asphalt Mixtures (2017). https://compass.astm.org/download/D2172D2172M.480.pdf

10. McDaniel, R.S., Soleymani, H., Anderson, R.M., Turner, P., Peterson, R.: Recommended Use of Reclaimed Asphalt Pavement in the Superpave Mix Design Method, Washington, DC (2000)
11. ASTM D5404: Standard Practice for Recovery of Asphalt from Solution Using the Rotary Evaporator (2012). https://www.astm.org/Standards/D5404
12. ASTM D4887: Standard Practice for Preparation of Viscosity Blends for Hot Recycled Asphalt Materials BT - Standard Practice for Preparation of Viscosity Blends for Hot Recycled Asphalt Materials (2016). https://compass.astm.org/download/D4887D4887M.24527.pdf
13. IRC: 120: Recommended practice for recycling of bituminous pavements (2015)
14. Asphalt Institute: MS-2 Asphalt Mix Design Methods (2014) https://bookstore.asphaltinstitute.org/catalog/book/ms-2-asphalt-mix-design-methods
15. Rowe, G.M., Rowe, G.M., D'Angelo, J.A., Sharrock, M.J.: Use of zero shear viscosity as a specification parameter for the high temperature binder specification. In: 2nd International Symposium on Binder Rheology and Pavement Performance. p. 25. San Antonio, Texas (2002)
16. R Core Team: R: a language and environment for statistical computing (2020). https://www.r-project.org/

Influence of Treated Mixed Recycled Aggregates in Concrete Properties

Miren Etxeberria(✉) and Carla Vintimilla

Universitat Politecnica de Catalunya. Barcelona TECH, 08034 Barcelona, Spain
miren.etxeberria@upc.edu

Abstract. Extensive construction and demolition waste (CDW) is generated during the construction process. That waste is currently deposited in landfills. It is necessary to produce and employ mixed recycled aggregates (MRA) in low strength (up to 30 MPa) concrete elements. This study aimed to describe the properties (physical, mechanical and durability) of concrete made with treated (employing water glass, WG5, and silane admixture, Si6) MRA aggregate compared to untreated recycled concrete. Two types of MRA aggregates (with different amounts of ceramic particles) were employed. The concretes used 100% coarse and fine recycled aggregates with 300 kg of cement and an effective water/cement ratio of 0.50. These concrete mixes were suitable for use as 25 MPa concrete. The concrete produced with MRA aggregates with the WG5 surface treatment achieved the desired strength for reinforced concrete of 30–35 MPa. However, all the concretes obtained a low value of the elastic modulus. Although all the concretes manufactured with 100% recycled aggregate met the EHE requirements (Spanish Standard of Structural Concrete, EHE) regarding the depth of water penetration under pressure and relatively low carbonation coefficient, the drying shrinkage was high. However, the surface treatments of aggregate reduced recycled concrete shrinkage considerably.

Keywords: Recycled aggregate concrete · Treated mixed aggregates · Silane · Sodium-silicate · Durability

1 Introduction

The generation of CDW is one of the primary economic and environmental concerns of the European Union because it represents one of the heaviest and most voluminous waste streams [1]. Based on a review of policies for the use of CDW, a Deloitte report [2] concluded that Spain achieved a recycling rate of only 30% of CDW by 2017, well below the limit established for 2020 of 70% [3]. In addition, the Spanish Government published the State Waste Framework Plan (PEMAR) 2016–2022 [4], which establishes new goals for increasing the quality of recycled materials.

The applicability of MRA was the focus of this study due to their abundance and low current use. They are composed of concrete, bricks, tiles and ceramic materials, which do not contain dangerous substances [5]. The physical properties of MRA aggregate range

J. I. Escalante-Garcia et al. (Eds.): RW 2021, RILEM Bookseries 40, pp. 963–972, 2023.
https://doi.org/10.1007/978-3-031-21735-7_102

between 1,800–2,400 kg/m^3 density and 6–14% absorption. In addition, with lower mechanical performance and higher wear rates than natural aggregates, they are highly dependent on the ceramic materials and mortar content [5, 6]. The study of MRA for use as concrete aggregates has expanded considerably in the last decade. Still, it has only focused on the use of the coarse fraction for MRA non-structural concretes [6, 7], or low degree of requirement applications [8]. It was described that the concretes produced with MRA achieved lower mechanical and increased creep, which can be 55% higher than that of conventional concrete (HC). Furthermore, the use of MRA aggregates increases the permeability, chloride penetration and shrinkage of the concrete [6, 9, 10].

However, the properties of recycled aggregate concretes could improve with the application of surface treatments in recycled aggregates. The sodium silicate treatment of recycled concrete aggregate (RCA) improves the concrete properties more than those of concretes produced with untreated RCA [11, 12]. The modulus of elasticity and drying shrinkage of recycled concrete could be increased and reduced when the concretes are produced with treated RCA using calcium silicate or carbonation process [12, 13]. Ding et al. [14] obtained a chloride penetration reduction of up to 50%, almost reaching the values of conventional concrete when the ceramic recycled particles (CRA) were treated with silane, reducing the porosity of ceramic particles.

In this study, CDW produced by Hercal Digger company was cleaned and sized in the authors' university to have two types of MRA for concrete fabrication. After the surface treatment of recycled aggregates, the improved properties of concrete were analysed when they were produced with 100% of treated fine and coarse aggregates compared to the properties of concrete made with 100% untreated recycled concrete.

2 Materials

High strength and rapid hardening cement CEM I 52.5 R was employed in all concrete production. A superplasticiser based on modified polycarboxylate was also used.

2.1 Recycled Aggregates

The fine (F) and coarse (G) fractions of two types of MRA aggregates (M and MC, with different constituents) were used for concrete fabrication. The coarse fractions, GM and GMC, were characterised by 4/20 mm fraction size. In contrast, the constituents (determined following EN 933–11) of the GM aggregate were 56% concrete (MRc), 9% raw aggregates (MRu), 34% ceramic (MRb), and less than 1% other material (asphalt, etc.). The constituents of the GMC aggregates included 44% MRc, 6% MRu and 50% MRb. The fine fraction of both types of aggregates (FM and FMC), characterised by 0/4 mm fraction size, with 9% for the FM and 2% for the FMC of particles finer than 63 μm. In addition, the total soluble chloride and sulphate amount in both types of MRA were lower than the 0.05% and 0.8% limits, respectively, as defined by the Spanish Concrete Structural Specification (EHE). Table 1 summarises the physical properties of M and MC aggregates. In contrast, the GM fraction had a lower absorption capacity and higher density than the GMC aggregates. The FM fraction had higher absorption than the FMC fraction. The ceramic particles typically increase the absorption capacity [5], decreasing the particle density. This fact was more evident in coarse fraction particles because a high amount of mortar was accumulated in the fine fractions. As a result, the absorption also increased. In addition, the flakiness index of the GM and GMC was lower than the maximum limit of 35 defined by the EHE.

Table 1. Properties of mixed recycled aggregates

		Untreated recycled aggregates			
		GM (4/20)	GMC (4/20)	FM (0/4)	FMC (0/4)
Dry density, ρ_d, (kg/dm^3)		2.17	2.03	2.08	2.07
Absorption, WA (%)	24 h	8.2	10.8	13.8	12.1
	1 h	6.9	9.0	12.8	11.7

2.2 Treated Recycled Aggregates

The coarse and fine recycled aggregates were subjected to surface treatment using two methods: (a) immersion in a solution (I) and (b) spraying (S) the material with a solution. Two solutions were employed: (1) a solution with 5% water-glass, and (2) a solution with 4% or 6% silane. The original water glass liquid used to solve with 5% sodium silicate had a SiO2/Na2O ratio of 3.35 with 8% Na2O and 26.4% SiO2. The original silane liquid (hydrophobic component) was a low viscous water-based chemical admixture with a modified silane dispersion of 20%. The used silane solution was diluted in water to achieve a 4% or 6% silane concentration.

Table 2. Properties of treated mixed recycled aggregates

M-Type	Treatments employed	ρ_d (kg/dm^3)	WA (%) 24h	MC-Type	Treatments employed	ρ_d (kg/dm^3)	WA (%) 24h
GM (4/20)	I-WG5	2.16	7.54	GMC (4/20)	I-WG5	2.01	9.72
	I-Si4	2.11	6.20		I-Si4	1.99	7.22
	S-WG5	2.20	7.8		S-WG5	2.1	9.83
	S-Si4	2.16	6.81		S-Si4	2.03	9.37
FM (0/4)	I-WG5	2.13	9.33	FMC (0/4)	I-WG5	2.11	10.03
	I-Si4	2.1	6.13		I-Si4	2.03	4.64
	S-WG5	2.18	11.2		S-WG5	2.1	10.6

The results for dry density and absorption capacity of the treated aggregates are summarised in Table 2. We concluded that the treatments reduced the absorption capacity of all the aggregates. However, they were more effective in fine fractions than in coarse fractions. In addition, the silane solution was more effective in MC fractions than the M fraction. The chemical composition and mineralogical structure of the treated aggregates did not modify for untreated aggregates.

The treatment by immersion (I-WG5 and I-Si4) process was carried out by submerging the dry particles for one hour in the solution and then drying them in an oven at 100°C. For the treatment by spray (S-WG5 and S-Si4), the aggregates were dried before the spray treatment. For assessing the sprayed surface durability, the aggregates were submitted to two actions: (a) they were introduced into a plastic container with distilled water and shaken for 24 h, and (b) the aggregates were exposed to the outside atmospheric environment for three months. After those two steps, the density and absorption of the aggregates were determined, and it was concluded that the obtained values were similar to those of the initially sprayed aggregates. Thus, the spray process was qualified as durable.

Although the spray method was not as effective as immersion (see Table 2), it was much easier to apply in higher-scale production than immersion. In addition, due to the high cost of the silane solution and the high effective absorption of fine recycled aggregates, we considered the possibility of processing the treatment of fine aggregates by adding the solution directly into the mixer during the concrete production process. In that case, the concrete was produced in two stages: stage (1) the fine aggregates were mixed with silane solution, and stage (2) the standard concrete production process was conducted. 6% concentration of silane (Si6) was used.

After the analysis of the results, two treatments were carried out. On the one hand, the fine and coarse aggregate were pre-treated with a WG5 using the spray method (S). On the other hand, the fine recycled aggregates were treated in situ (IS) in the concrete production process with a Si6 solution.

3 Mix Proportions and Test Procedure

3.1 Mix Proportions

All the concretes were produced using 300 kg of cement, 100% fine and coarse recycled aggregates and a water/cement ratio of 0.50. Table 4 summarises the mix proportions of all the concretes produced. The C-MC and C-M concrete were fabricated with untreated MC and M recycled aggregates, respectively. The C-M-WG5 and C-MC-WG5 concrete were made with fine and coarse aggregates treated with the WG5 sprayed solution. In addition, the C-M-Si6 and C-MC-Si6 concretes were produced using the silane solution but only applied in fine aggregates. As mentioned above, the concrete was produced in two stages. While the coarse recycled aggregates were always used under high humidity conditions to reduce their water absorption capacity, the fine aggregates were used under low humidity or dry conditions. Because the fine aggregates absorb 100% of their absorption capacity in the first 10–20 min, it was easy to achieve saturation in the concrete production process.

All the concrete achieved a similar slump. However, the treated recycled concrete required more superplasticiser. In addition, a higher amount was needed in the concrete produced with MC recycled aggregates, probably due to a higher amount of ceramic material.

Table 3. Concrete Mix proportion. Weight (in kg) of each component for 1 m^3 of concrete

	Cem	Total water	Mixing water	Treat-water (IS)	Fine AR	Coarse AR	Effec w/c	Adm. (%)	Slump (cm)
C-MC	300	333			734	813	0.5	2.1	12
C-MC-WG5	300	325			790	853	0.5	3.2	12
C-MC-Si6	300		172	96.6	734	813	0.5	2.7	14
C-M	300	316			786	845	0.5	1.6	15
C-M-WG5	300	296			801	881	0.5	2	12
C-M-Si6	300		164	95.5	786	845	0.5	2.4	12

3.2 Test Procedure

The mechanical properties of compressive strength (UNE-EN 12390–3:2020) at 7 and 28 days, splitting tensile strength (UNE-EN 12390–6:2010) and modulus elasticity (UNE-EN 12390–13:2014) at 28 days were determined in all the concretes produced. In addition, the physical properties (UNE-EN 12390–7:2020), including total porosity using mercury intrusion porosimetry (MIP), was determined at 28 days. Cube specimens of 100 mm were used for compressive strength and physical properties. Cylindrical samples 100 mm in diameter and 20 m long were used to test splitting tensile strength and modulus of elasticity in the concrete. Three specimens were used for each test to calculate the average value and its standard deviation.

Sorptivity (ISO 15148:2002) and water penetration under pressure (UNE-EN 12390–9:2009) were determined using three cylindrical specimens. Three prismatic samples were used for the drying shrinkage (ASTM C596) test. In addition, the carbonation coefficient (GB T50082–2009) was determined for all concrete types. The specimens were exposed for seven days in a CO_2 chamber at 20% CO_2, 20 °C and 60% RH. The concrete samples were subjected to pre-conditioning at 60% and 20 °C for 15 days following 28 days of curing.

4 Results

4.1 Mechanical Properties

Table 4 summarises the mechanical properties of the produced concretes. C-MC and C-M concrete manufactured with 100% untreated MRA obtained a compressive strength (fcm) of 32–34 MPa (for cube specimens) at 28 days.

The concretes made with FM and GM aggregates achieved slightly lower strength than those made with FMC and GMC aggregates. The FM likely had more fine material (<63 μm) than the FMC, with some impurities present. Therefore, the quality of the aggregate was a fundamental factor to achieve high-quality concrete. However, the C-M-Si6 and C-M-WG5 concretes reached 35 MPa and 44 MPa of compressive strength, respectively, suitable for the manufacture of reinforced concrete of 25 MPa and 30 MPa (fck, cylindrical). Finally, the C-MC-Si6 and C-MC-WG5 concrete reached 40 MPa and 51 MPa compressive strength, respectively, optimal for standard 30–35 MPa reinforced concrete.

Table 4. Mechanical and physical properties of produced concretes

	Mechanical Properties (MPa)				Physical properties			
	f_{cm}		f_{ctm}	E_{cm}	ρ_d(kg/dm^3)	WAbs (%)	P-MIP (%)	S(mm/min$^{0.5}$)
	7d	28 d						
C-MC	29 (0.6)	35 (0.2)	2.96	18829	1.88	10.9	22.6	0.466
C-MC-WG5	43 (0.3)	52 (1.3)	3.14	21571	2.01	8.0	18.3	0.114
C-MC-Si6	37 (0.2)	41 (0.4)	3.19	21250	1.97	8.9	19.8	0.247
C-M	26 (0.1)	32 (0.3)	2.84	19766	1.98	10	19.8	0.277
C-M-WG5	36 (0.3)	44 (0.6)	2.82	20617	2.02	8.6	16.8	0.172
C-M-Si6	30 (0.2)	35 (0.6)	2.78	21210	2.02	9.6	18.4	0.139

* between the brackets (), the standard deviation of each average value

Based on splitting tensile strength (f_{ctm}), the concrete produced with GMC and FMC also achieved a higher splitting tensile strength than those made with M aggregates. The C-MC-Si6 concrete achieved the highest splitting tensile strength.

The concrete produced with 100% of fine and coarse recycled aggregates obtained a low modulus of elasticity due to the high porosity of those aggregates. The use of recycled aggregates increased the modulus of elasticity up to 15% compared to concrete produced with 100% untreated aggregates. The C-MC-WG5 achieved the highest value of 21,571 MPa.

4.2 Physical Properties

Due to the low density and high water absorption capacity of recycled aggregates, the concrete produced with 100% of MRA achieved low dry density (ρd) and high absorption (WA) capacity (see Table 3). With a higher number of ceramic particles, the C-MC concrete obtained a lower density and higher absorption capacity than C-M concrete. The density did not increase with the use of treated recycled aggregates. However, the treated aggregates reduced the concrete absorption capacity slightly. The C-M-WG5 and C-MC-WG5 concrete achieved the lowest absorption capacity. In addition, the concrete produced with treated MC achieved the highest reduction.

Figure 2 shows that the C-MC concrete achieved a higher cumulative intrusion than that of the C-M concrete. However, the concrete produced with treated MC aggregates reached a lower cumulative intrusion value, consequently lower porosity. In addition, the WG5 treatment was the most effective. The C-M-WG5 and C-MC-WG5 concrete achieved a 15% and 19% porosity (P-MIP) reduction, respectively, compared to the concrete produced with corresponding untreated recycled aggregates (see Table 4). Therefore, while the treatment of FM and GM aggregates influenced the reduction of the volume of the smaller pores, the treatment of FMC and GMC reduced the volume of all pore sizes. Figure 1 shows the pore size distribution of the concrete.

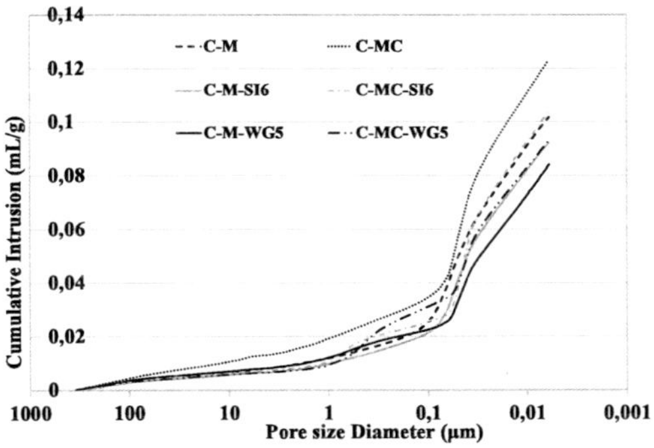

Fig. 1. Pore size distribution of produced concretes

4.3 Durability Properties

Sorptivity and Water Penetration Under Pressure

Table 4 shows that the WG5 treatment effectively reduced the capillary absorption capacity (sorptivity, S) of the concrete manufactured with 100% MRA. In addition, the C-M-WG5 and C-MC-WG5 concrete reached suitable properties for structural concrete. We estimated that concrete fabricated with a w-c ratio of 0.4 to 0.6 reaches a suction of 0.09-0.17mm / $min^{0.5}$. Although, in general, the values obtained by the produced concretes were higher than the described values, it is detected that the treatment process reduced the sorptivity considerably.

According to the EHE, the maximum and medium water penetration depth of concrete specimens under pressure must be 50 mm and 30 mm, respectively. All the produced concretes achieved those two requirements (see Fig. 2). In addition, the C-MC-WG5 achieved the lowest value.

Drying Shrinkage

The concretes with 100% porous fines, and coarse MRA were produced using a high volume of water. In addition, as expected, the shrinkage value was higher in concrete made with a higher amount of water (see Table 3). Figure 3 shows that the C-MC concrete achieved the highest shrinkage values. However, treated aggregates reduced the shrinkage values: the C-MC-WG5 and C-MC-Si6 concrete achieved a 13% of shrinkage value reduction compared to the C-MC value. In addition, the concrete produced with treated M aggregates, C-M-Si6 and C-M-WG5 were the most adequate, achieving a reduction of 21% compared to C-M concrete. The obtained values were acceptable according to the values described by ACI [15] as typical drying shrinkage values (-200 to -800 $\mu\varepsilon$).

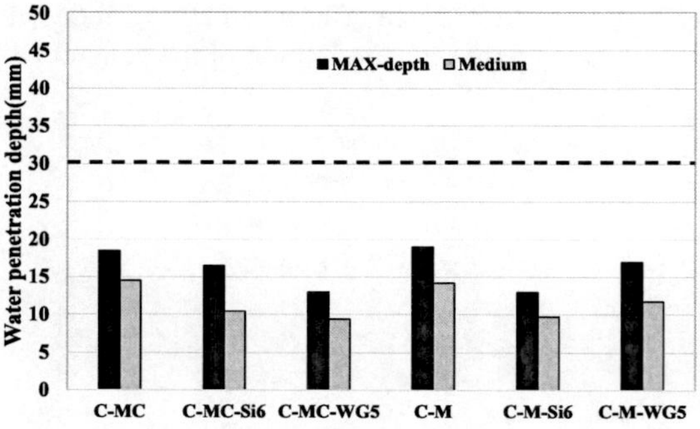

Fig. 2. The water penetration depth under pressure

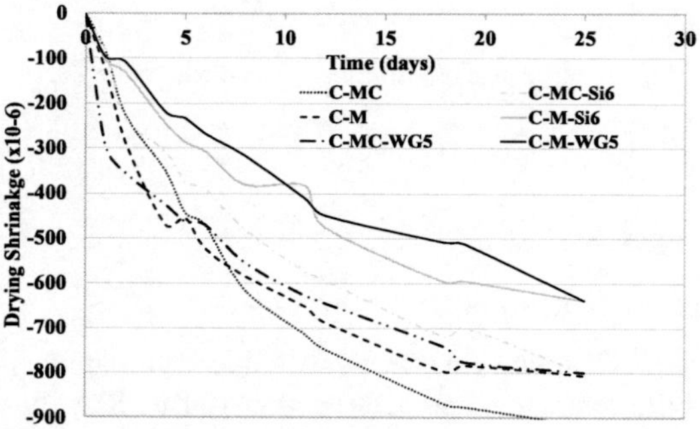

Fig. 3. The drying shrinkage values of concretes produced

Carbonation Resistance

Figure 4 shows the graph, carbonation depth (mm) versus time(days$^{0.5}$), in which the experimentally obtained accelerated carbonation coefficient (K_{acc}, mm/day$^{0.5}$) after the carbonation process at 20% CO_2 for seven days, is described. The ratio of the accelerated (K_{acc}) and natural (K_{nat}) carbonation coefficient can be expressed as

$$K_{acc}/K_{nat} = (\text{Ø})_{acc})^{0.5}/(\text{Ø})_{nat})^{0.5} \quad (1)$$

where, $Ø_{acc}$ and $Ø_{nat}$ are the CO_2 concentrations, the $Ø_{acc}$ was 200000ppm, and $Ø_{nat}$ was 400ppm in Barcelona. The obtained values of K_{nat} were: C-M of 2.66 mm/year$^{0.5}$; C-M-SI6 of 2.37 mm/year$^{0.5}$; C-M-WG5 of 2.19 mm/year$^{0.5}$; C-MC of 2.46 mm/year$^{0.5}$; C-MC-SI6 of 2.03 mm/year$^{0.5}$; and C-MC-WG5 of 2.09 mm/year0.^{5}. According to the values obtained of K_{nat}, the speed of carbonation was low enough to achieve adequate durability. The treatment of the aggregates reduced the carbonation coefficient up to 18% compared to the untreated aggregates. However, these values should be validated with tests at lower CO_2 concentrations and at the natural carbonation rate.

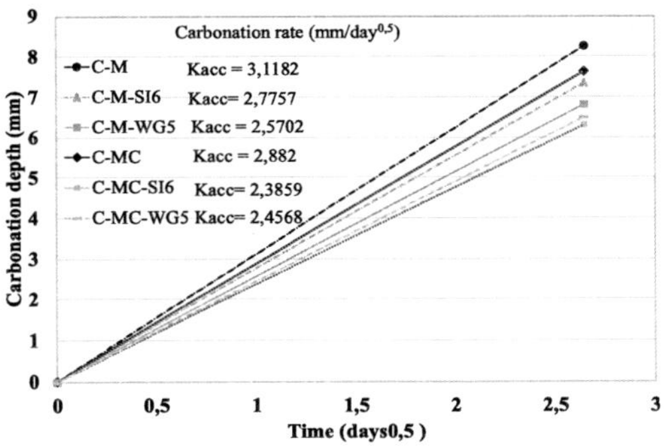

Fig. 4. Carbonation coefficient in accelerated test

5 Conclusions

The following conclusions were obtained:

- The absorption capacity of the two MRA aggregates was high, which could be reduced efficiently with WG5 and Si4 treatments.
- The presence in fine aggregates of material finer than 63 μm influenced the quality of the concrete. Consequently, the concrete produced with M recycled aggregates (C-M, C-M-WG5 and C-M-Si6) achieved lower compressive and splitting tensile strength than their corresponding concrete (C-MC, C-MC-WG5 and C-MC-Si6).
- The modulus elasticity of C-M was slightly higher than C-MC. The higher number of ceramic particles reduced the stiffness. In addition, the C-MC concretes achieved a higher absorption value. However, the concretes produced with treated MC aggregates reached a higher modulus of elasticity and lower absorption capacity than concretes produced M aggregates. Hence, the WG5 treatment the most effective.
- The C-MC-WG5 achieved the lowest sorptivity and water penetration values. Both treatments reduced up to 50% of the capillary absorption capacity of the C-M and C-MC.
- The shrinkage value and carbonation coefficient were reduced considerably with the use of treated aggregates. The C-M-WG5 and C-MC-Si6 achieved the most significant improvements.

The concrete produced with recycled aggregates treated with the WG5 spray process showed the most significant improvement in achieving satisfactory hardened properties. The in situ treatment with the Si6 solution was adequate when the recycled aggregates were composed of a high number of ceramic particles. Every treated recycled concrete could be used for 30 MPa strength (fck, cylindrical) concrete. In addition, a concrete of 40 MPa would be employed when the WG5 was applied.

Acknowledgement. The authors wish to thank Hercal Diggers company for their interest and support. The authors also want to acknowledge the financial support of The Ministry of Science and Innovation CDTI Project.

References

1. Euroslag, "Euroslag, 2012. Statistics.": http://www.euroslag.com/%0Aproducts/statistics/2012/
2. Deloitte: Study on Resource Efficient Use of Mixed Wastes, Improving management of construction and demolition waste – Final Report (2017).
3. Directive 2008/98/Ec of the European Parliament abd of the Council of 10 November 2008 on waste. Official Journal of the European Union (2008)
4. D. E. (Environmet M. Agricultura and M. Ambiente), "National framework plan for waste management, plan estatal marco de gestión de residuos (PEMAR)" (2016)
5. Silva, R.V., de Brito, J., Dhir, R.K.: Properties and composition of recycled aggregates from construction and demolition waste suitable for concrete production. Constr. Build. Mater. **65**, 201–217 (2014)
6. Etxeberria, M., Gonzalez-Corominas, A., Valero, I.: Application of low grade recycled aggregates for nonstructural concrete production in the city of Barcelona. In: Third International Conference on Sustainable Construction Materials and Technologies (2013)
7. Beltrán, M.G., Barbudo, A., Agrela, F., Galvín, A.P., Jiménez, J.R.: Effect of cement addition on the properties of recycled concretes to reach control concretes strengths. J. Clean. Prod. **79**, 124–133 (2014)
8. Martínez, I., Etxeberria, M., Pavón, E., Díaz, N.: A comparative analysis of the properties of recycled and natural aggregate in masonry mortars. Constr. Build. Mater. **49**, 384–392 (2013)
9. Hoffmann, C., Schubert, S., Leemann, A., Motavalli, M.: Recycled concrete and mixed rubble as aggregates: influence of variations in composition on the concrete properties and their use as structural material. Constr. Build. Mater. **35**, 701–709 (2012)
10. Lopez-Uceda, A., Ayuso, J., López, M., Jimenez, J.R., Agrela, F., Sierra, M.J.: Properties of non-structural concrete made with mixed recycled aggregates and low cement content. Materials (Basel) **9**(74), 1–19 (2016)
11. Gao, H., Wu, G., Yuan, X.: Experimental research on mechanical properties of recycled aggregate concrete with different modification methods. In: no. Ic3me, pp. 1957–1962 (2015)
12. Galindo, A., Etxeberria, M.: Influence of treated recycled concrete aggregate on concrete properties. In: International Union of Laboratories and Experts in Construction Materials, Systems and Structures (RILEM) (2017)
13. Xuan, D., Zhan, B., Poon, C.S.: Assessment of mechanical properties of concrete incorporating carbonated recycled concrete aggregates. Cem. Concr. Compos. **65**, 67–74 (2016)
14. Ding, X., Qi, J., Fang, W., Chen, M., Chen, Z.: Improvement on properties of recycled concrete with coarse ceramic vase aggregates using KH-550 surface treating technology. Eur. J. Environ. Civ. Eng. 1–16 (2018, 2017)
15. ACI Committee 209: "Report on Factors Affecting Shrinkage and Creep of Hardened Concrete," Farmington Hills (2005)

Check for updates

Use of Corncob Granules as Sand Replacement in the Production of Cement Mortars

Taghried Abdel-Magid(✉), Pete Walker, Kevin Paine, and Stephen Allen

Centre for Innovative Construction Materials, Dept. Architecture & Civil Engineering, University of Bath, Bath, UK

Abstract. Low-cost housing is urgently needed in low- and middle-income countries. Concrete is often too costly and raw material production, including aggregate extraction, is known to be harmful to the environment. In this paper the use of an agricultural waste, corncob, as a potential lightweight replacement of sand in cement materials has been studied. Corncob granules were added to the mix in percentages of 5%, 10%, 15% and 20% by volume. The corncob granules were used in their natural state as well as following coating with either cement slurry or gum Arabic. The effect of sand replacement was determined through flexural and compressive testing of materials, while the chemical composition was evaluated using Thermogravimetric Analysis (TGA) and X-Ray Diffraction (XRD). The porosity of the mixes was also determined using a mercury inclusion porosimeter (MIP). The results showed that while the early age strength (7 days) was extremely low, the late age's strengths improved drastically between 10% up to 3500% in 28 days.

Keywords: Corncob · Maize · Cement · Low-cost housing · Strength

1 Introduction

The use of agricultural waste in construction has been a rich area of study in recent decades. Natural materials have been used for millennia for construction in rural areas of developing countries, and although suitable for the specific weather, are not particularly durable. While concrete is the most widely used construction material, its harm to the environment and contribution to greenhouse gas emissions from cement production is increasingly recognized. Replacing components of traditional concrete mix with alternatives, with lower environmental impacts, provides an opportunity to reduce these impacts. Replacing aggregates, often in short supply, can, if successful, serve in reducing the need to mine aggregates, as well as decrease weight, which result in saving materials for foundation and reducing transport and handling costs. Construction industry alone consumes 40–50 billion tonnes per year of sand [1, 2] which, as it needs to obtain specific properties, is extracted from rivers and crushed rocks to be used as fine aggregate in concrete and cement constructions. This activity is indeed harmful for the environment in many ways, including altering the river bed, lowering the water table, degrading the natural habitat and disrupting the local ecological balance [2–4]. The high demand on sand

J. I. Escalante-Garcia et al. (Eds.): RW 2021, RILEM Bookseries 40, pp. 973–982, 2023.
https://doi.org/10.1007/978-3-031-21735-7_103

resulted in a corresponding increase in its cost, hence a rise in the price of construction. Studies to find alternatives to natural sand evaluated the use of agricultural wastes, such as groundnut shell, rice husk, bagasse and coconut shell, as replacement of sand in lower strength application such as masonry blocks, have found mixes using replacements can meet the minimum performance requirements of the appropriate standards [3–6].

Corncob, which is considered an agricultural waste once the kernels are removed, is obtained from corn (maize). Corn is one of the most important cereal crops worldwide, with approximately 1150 million tonnes being produced in 2019 [7]. The cobs are utilised by industry as a bedding material for poultry and reptiles, and as an abrasion material. Its ash has been studied as partial replacement of cement and as a soil stabiliser due to its potential pozzolanic properties [8–10]. Furthermore, corncob granules have been studied by a number of researchers in construction industry due to its availability, biodegradability and high porosity, which result in improving thermal and sound insulation traits, as well as producing low-weight construction materials [11, 12]. Studies on the use of corncob granules in load-bearing construction materials are limited and their effect on the material properties is yet to be established.

This research paper is a part of an ongoing research study that focuses on the possible use of agricultural wastes available in the country of Sudan for construction. The authors appreciate the limitations of the natural materials concerning strength and consistency of properties; hence, the focus is to utilise these materials as mortar between bricks or in low-cost construction materials, which require lower strength. For this intent, the aim of the study is to understand the effect the corncob granules have on masonry construction materials for low-cost housing. Objectives are: (1) testing the viability of corncob granules-cement matrix as a building material; (2) testing the viability of replacing sand in cement mortars with corncob granules; and (3) understanding the effect certain coating agents have on the addition of the corncob granules.

2 Materials

2.1 Cement and Sand

The cement used in this study was CEM I 42.5N conforming to BS EN 197-1, and the sand was a CEN standard sand conforming to BS EN 196-1. Properties of the sand are given in Table 1.

Table 1. Physical properties of sand and corncob granules.

ID	Density (kg/m^3)	Moisture content (%)	WA (%)
CEN Sand	1500	<0.2%	
CC	587 ± 60	6.6	170
CCem	543 ± 47	5.9	135
CGum	504 ± 44	6.4	199

2.2 Corncob Granules

The corncob granules used in this study were sourced from the company 'Oberflächentechnik Seelmann e.K.' with a grain size ranging between 200 and 400 μm. Granules were used in three forms: (i) as-is (referred to herein as CC), (ii) coated with cement slurry (CCem) and (iii) coated with Gum Arabic (CGum). CCem was prepared by submerging the granules in a 1:1 water:cement slurry. CGum was prepared by submerging the granules in a 1:4 water:gum-Arabic mixture. Both CCem and CGum were left to absorb the water for 12 h then dried at 80°C in an oven for 3 h. All three types of granules are shown in Fig. 1.

Fig. 1. Corncob granules used in the study: pure (top), covered with gum (left), covered with cement slurry (right).

3 Methods

3.1 Corncob Granules Characterisation Tests

Density. The bulk densities used to calculate the volume replacement in the specimen mixes was measured by layering the granules in a container, each layer was compacted by letting the container fall 10 times from a height of around 5 cm. The mass of the granules was measured after the container was filled and the density was calculated by dividing that mass by the volume. The test was repeated for a minimum of 20 times, then average value was recorded. Table 1 reports the density results of the granules.

Moisture Content. The granules moisture content (m.c.) was tested according to BS EN 1377-2 [13] by drying them in a 105°C oven until a constant mass was achieved. The results are given in Table 1.

Water absorption. Water absorption (WA) and WA rate were used to understand the effect the coating agents had on the absorption properties of the granules. WA was tested by submerging the granules in water for 24 h, WA rate was tested by submerging the granules for different durations. 24 h WA values are shown in Table 1. WA rate is shown in Fig. 2.

Particle size distribution. Granules' particle size distribution was determined by sieving the specimens through a set of standard sieves in accordance with BS EN 1377–2.

Pore size distribution. Mercury intrusion porosimetry (MIP) was performed on the granules to define their pore size distribution.

3.2 Cement Mortar

Mortar specimens were prepared in accordance with BS EN 196–1 [14]. Two control mixes were prepared by using 450g of cement, 1350g of sand and mixing water of 225ml. 1% plasticizer by cement mass was added to the first set to maintain the workability of the mix that includes CC. The corncob was added as replacement to sand volume. All mixes in this study are detailed in Table 2. Mixes herein are referred to using the format X-%, where "X" stands for the corncob used (CC, CCem and CGum), as defined in 1.2, and "%" refers to the percentage of replacement of sand with corncob, e.g., CC-5 is the mix prepared by replacing 5% of sand's volume with pure corncob.

Table 2. Mortar mix details.

Replacement % by volume	0	5	10	15	20
CEN Sand (g)	1350	1282	1215	1148	1080
CC (g)*	0	26.4	52.8	79.2	105.6
CCem (g)	0	24.4	48.9	73.3	97.7
CGum (g)	0	22.7	45.4	68.0	90.7

* This set is referred to as Set-1, all mixes of Set-1 contain 1% plasticizer

Flexural and Compressive Strength Tests. Tests were conducted following BS EN 196-1. After three days, mortar prisms of 40 mm × 40 mm × 160 mm were cured in a water bath of 20°C until test day. Tests were performed at 7, 28 and 90 days of curing. Three prisms were tested for each mix for flexural strength, and the six halves were subsequently tested for compressive strength. In some mixes, due to the water absorption of the corncob granules, hydration retarding resulted in insufficient strength gain after one day, and so some specimens were damaged during demoulding process. Thus, specimens were subsequently left in moulds for three days before demoulding.

Thermal Stability. Thermal decomposition was analysed with a thermogravimetric technique (TG) using a High Temperature Platform 400 STA 449 F1 Jupiter in a nitrogen atmosphere. TG was carried out on powdered mortars of mass 92 ± 1 mg at a rate of 10°C/min starting from 30 °C up to 950 °C.

Crystalline Phase and Mineralogic Composition. X-ray diffraction (XRD) was used to determine the crystalline phases of mortar specimens and to qualitatively analyse their mineralogic composition. The range used was 2Theta of 2 to 76 with a step of 0.015 and 60 s/step on a STOE STADI P (Cu radiation) Double setup in transmission mode.

4 Results and Discussion

4.1 Corncob Granules Versus Sand

From Table 1, the moisture content of the granules used was considerably higher than the standard sand used for control mixes. This was expected to increase the water/cement ratio in the mix, but it was found to be less significant when put in line with the high water absorption of the granules. The higher water absorption was managed at the mixing stage by adding plasticiser to the mix containing plain corncob (CC). The water absorption rate shown in Fig. 2 suggested that, for the mixing stage, the other two types of granules, CCem and CGum, did not need a plasticiser to manage workability. Nonetheless, the noticeable increase in water absorption after the first hour (Fig. 2) no doubt affected the cement hydration process, which was evident when the mortar specimens had to be left to set for three days before demoulding.

The density of the granules, as can be seen from Table 1, was approximately one third of that of standard sand. This resulted in specimens with reduced density (between 2 and 10% lower than control), which is beneficial from a construction perspective; lowering the deadload of the building, ease the handling and reducing the cost.

The results of MIP test did not show a clear threshold diameter, which is the diameter after which the volume of the mercury intruded into the specimen increases significantly. This result suggests that water can intrude the corncob easily and quickly, as proved with Table 1 and Fig. 2, which may adversely affect the brick's shrinkage and swelling capacity.

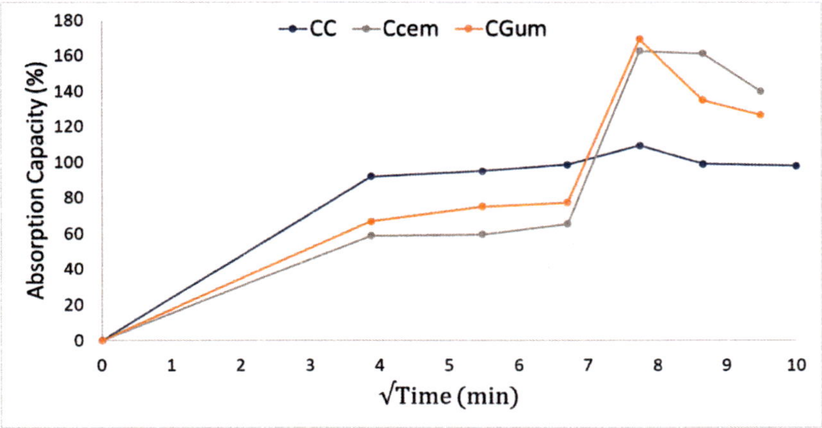

Fig. 2. Water absorption rate of corncob.

4.2 Mortars Flexural Strength

The average values of flexural strength are illustrated in Fig. 3 and detailed in Table 3. Set 1 (the control mix of which is labelled 'Control + P'), is shown in black and Set 2 (the control mix of which labelled 'Control') in grey. For all replacement percentages the flexural strength decreased compared to the control mix. For both sets, having only 5% of corncob replacing sand led to the highest strengths, while replacing 15% and 20% reported significantly low strength, with the exception of CCem, which had a strength of 3.3MPa in 28 days and 4.1 MPa at 90 days.

The bond between units and the mortar was not studied here. More tests should be made to confirm the viability of these units in a wall setting. Further tests are also required to confirm that the high water absorption does not adversely affect the unit's shrinkage.

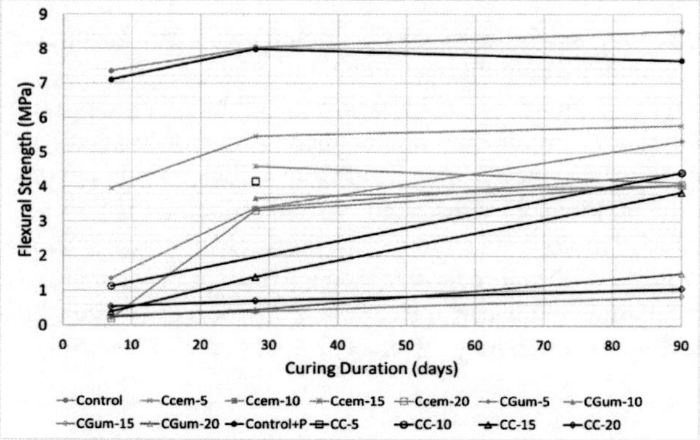

Fig. 3. Flexural strength for different mixes versus curing duration

4.3 Compressive Strength

The average values of compressive strength are illustrated in Fig. 4 and detailed in Table 3. Set 1 (controlled by 'Control + P'), is shown in black and Set 2 (controlled by 'Control') in grey. It is, again evident that all specimens have a lower compressive strength compared to the control mixes. The hydration-retarding effect of corncob granules was clear as well, where the early strength at 7 days was between 48% and 99% lower than the control. Different standards provide minimum requirements for the compressive strength of masonry, depending on their intended use and the mortar utilised. All CCem specimens had late-stage strengths (28 and 90 days) that is higher than what is commonly used for low-rise residential housings, while only 5% and 10% replacement with CC and CGum met the minimum values for 28 and 90 days.

All CGum specimens obtained a yellowish colour when tested, this was attributed to the natural dye of the corncobs when soaked in water. CCem specimens with lower percentages of replacement did not show that significant change in colour. The effect of this natural dye on the durability of the brick needs to be examined.

The hydration retarding effect was clear when the percentage of increase in strength was calculated as illustrated in Fig. 5. The change was measured in two levels: the

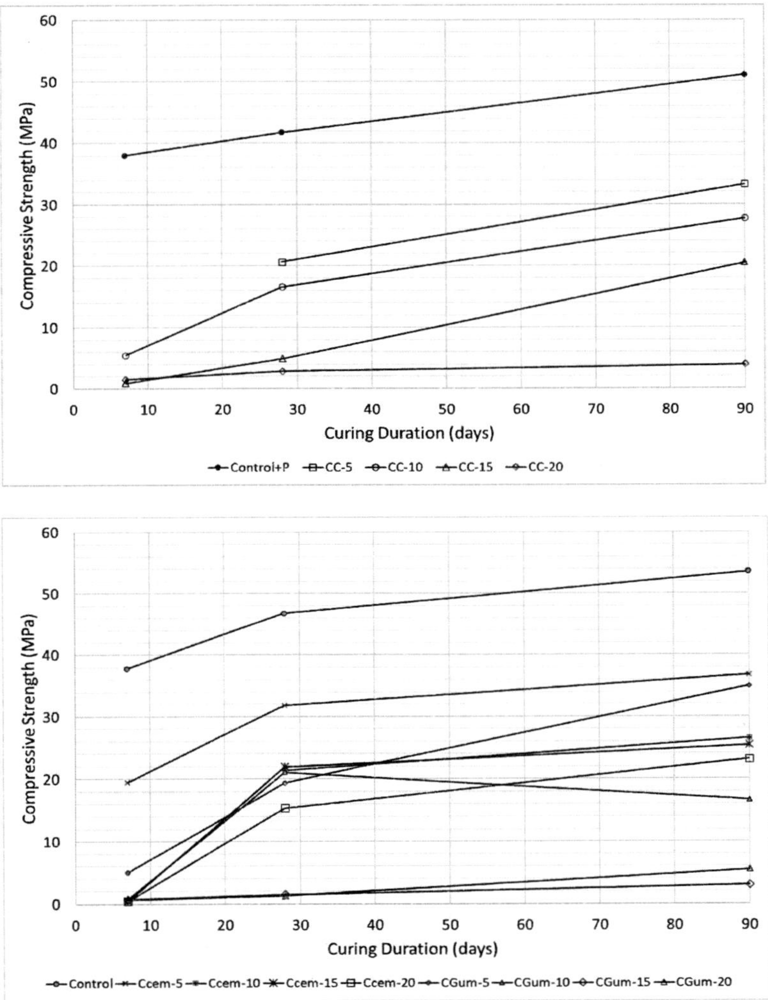

Fig. 4. Compressive strength for different mixes versus curing duration: CC (top) CCem and CGum (bottom)

duration between 7 and 28 days, and the duration between 28 and 90 days according to "Eq. (3)", where i and j represent the later and the earlier age, respectively.

$(fc_i - fc_j)/fc_j \times 100\%$(3)

The increase in compressive strength was extreme (up to 4800%) for CCem specimens in the first duration. In the late duration though, the CGum specimens showed huge increase from the 28-day strength, this was attributed to sugar in the gum retarding hydration.

Table 3. Compressive and flexural strength details.

Days	Control + P	CC-			Control			CCem-				CGum-		
	0	5	10	15	20	0	5	10	15	20	5	10	15	20
Compressive strength (MPa)														
7	38.0	N.A	5.5	0.9	1.5	37.8	19.5	N.A	0.4	0.4	5.0	0.7	0.7	0.7
28	41.7	20.6	16.5	4.9	2.8	46.8	31.8	21.3	21.9	15.4	19.4	21.0	1.4	1.2
90	51.1	33.3	27.7	20.4	3.9	53.5	36.8	26.5	25.3	23.1	34.9	16.7	3.0	5.4
Flexural strength (MPa)														
7	7.1	N.A	1.1	0.4	0.6	7.4	4.0	N.A	N.A	0.2	1.3	N.A	0.3	0.2
28	8.0	4.2		1.4	0.7	8.0	5.5	3.4	4.6	3.3	3.4	3.7	0.4	0.4
90	7.7	N.A	4.4	3.9	1.1	8.5	5.8	4.4	4.1	4.1	5.3	4.1	0.8	1.5

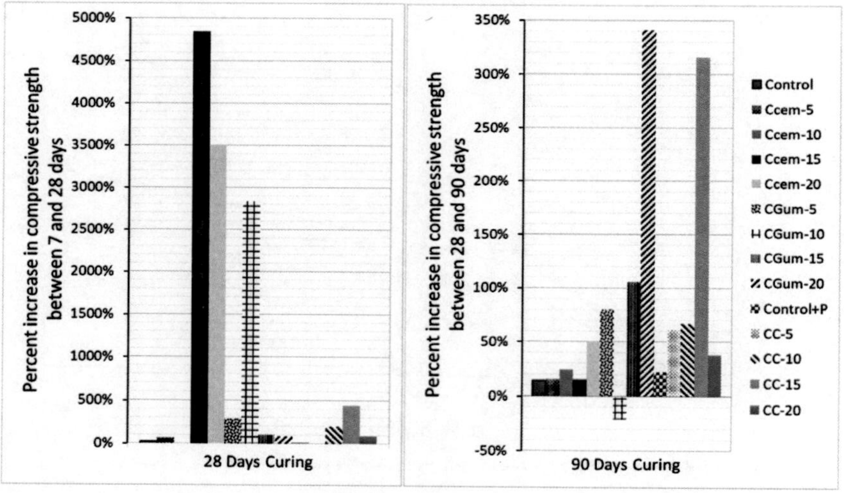

Fig. 5. Percent of change in compressive strength between 7th and 28th day of curing (left), and between 28th and 90th day of curing (right)

4.4 Mineralogical Composition

TG and XRD results were compared to define the mineralogical composition of the mortars. In TG graphs, a peak corresponding to Portlandite decomposition (at 450–480 °C) was present in all CCem specimens and at 5%, 10% of CGum specimens. A further peak at around 720 °C, corresponds to the decomposition of calcite (calcium carbonate).

The presence of Portlandite was also observed in XRD results of mortars cured for 28 days, as shown in Fig. 6. The major phase expected to be present was the quartz. Other phases, such as calcite and calcium silicate hydrates were also present in all the mixes with different intensities. The difference in intensities can be attributed to the different size and preparation procedure of the samples. A phase corresponding to calcium magnesium aluminium oxide silicate was present in the CC specimens exclusively. This indicated that the pre-treatment applied on the granules altered their reactivity in the alkaline environment which resulted in a more strong final matrix. All corncob specimens showed a small hump at the start of their XRD pattern. This was credited to the amorphousness of corncob granules being an organic matter. Further TG tests are planned in air atmosphere to ensure the stability of the organic matter and confirm their reactivity so as to further understand their durability.

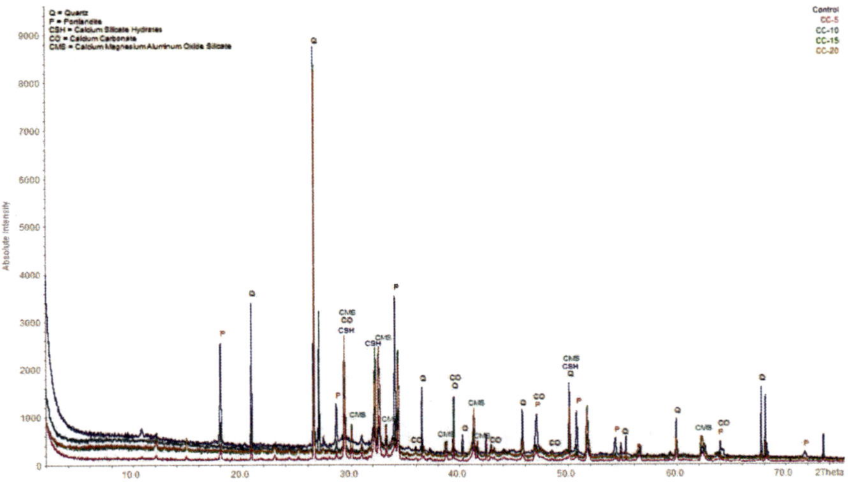

Fig. 6. XRD patterns for Set-1 specimens cured for 28 days.

5 Conclusions

The following conclusions and recommendations can be drawn of this research study:

1. Corncob granules can be potentially used as sand replacement in cement mortars and bricks for low-cost housing.
2. Further testing for the granules' durability needs to take place, such as swelling and shrinkage tests.
3. Pre-treating the granules with cement slurry can positively alter their water absorption and chemical reactivity.
4. Pre-treating the granules can provide mortars with strength better than the ones prepared with pure granules. Strength, compared to pure granules, can be improved by up to 400% when coated with cement slurry and by up to 70% when coated with gum Arabic.

5. Early-stage rate of strength gain of specimens containing corncob might hinder the construction process.
6. Gum Arabic resulted in altering the absorption rate in a way similar to the cement slurry, nonetheless, it resulted in retarding the hydration of cement and hence reducing the strength of mortars by up to 100% in all curing days.

References

1. United Nations Environment Programme: "Sand and sustainability: finding new solutions for environmental governance of global sand resources." (2019)
2. Cousins, S.: Shifting sand: why we're running out of aggregate. Constr. Res. Innov. **10**(3), 69–71 (2019). https://doi.org/10.1080/20450249.2019.1656448
3. Sathiparan, N., De Zoysa, H.T.S.M.: The effects of using agricultural waste as partial substitute for sand in cement blocks. J. Build. Eng. **19**, 216–227 (2018). https://doi.org/10.1016/j.jobe.2018.04.023
4. Sada, B., Amartey, Y., Bako, S.: An investigation into the use of groundnut shell as fine aggregate replacement. Niger. J. Technol. **32**(1), 54–60 (2013)
5. de Almeida, A.C., da Silva, M.A.L., de Abreu, Q.C., da Silva, A.L., Martins, S.P.R., Pereira, C.D.S.S.: Evaluation of partial sand replacement by coffee husks in concrete production. J. Environ. Sci. Eng. B **8**(4), 129–133 (2019). https://doi.org/10.17265/2162-5263/2019.04.001.
6. Sales, A., Lima, S.A.: Use of Brazilian sugarcane bagasse ash in concrete as sand replacement. Waste Manag. **30**, 1114–1122 (2010). https://doi.org/10.1016/j.wasman.2010.01.026
7. FAO, "FAOSTAT-Crops-Maize" (2021). http://www.fao.org/faostat/en/#data/QC/visualize. Accessed 12 Jun 2021
8. Olafusi, O.S., Kupolati, W.K., Sadiku, E.R., Snyman, J., Ndambuki, J.M.: Characterization of corncob ash (CCA) as a pozzolanic material. Int. J. Civ. Eng. Technol. **9**(12), 1016–1024 (2018)
9. Akinwumi, I.I., Aidomojie, O.I.: Effect of Corncob ash on the geotechnical properties of Lateritic soil stabilized with Portland cement. Int. J. Geomatics Geosci. **5**(3), 375–392 (2015)
10. Aprianti, E., Shafigh, P., Bahri, S., Farahani, J.N.: Supplementary cementitious materials origin from agricultural wastes – a review. Constr. Build. Mater. **74**, 176–187 (2015). https://doi.org/10.1016/J.CONBUILDMAT.2014.10.010
11. Pinto, J., et al.: Corn's cob as a potential ecological thermal insulation material. Energy Build. **43**(8), 1985–1990 (2011). https://doi.org/10.1016/j.enbuild.2011.04.004
12. Njeumen Nkayem, D.E., Mbey, J.A., Kenne Diffo, B.B., Njopwouo, D.: Preliminary study on the use of corn cob as pore forming agent in lightweight clay bricks: physical and mechanical features. J. Build. Eng. **5**, 254–259 (2016). https://doi.org/10.1016/j.jobe.2016.01.006
13. British Standards Institution BSI, "BS 1377-2:1990 Methods of test for Soils for civil engineering purposes — Part 2: Classification tests," no. 1 (1990)
14. BSI, "BS EN 196-1:2016 Methods of Testing Cement Part 1: Determination of Strength (2016). https://doi.org/10.1590/S1413-294X2011000200001

Properties of High-Performance Concrete with Coarse Recycled Concrete Aggregate for Precast Industry

Xiaoguang Chen[1], Hanne Vanoutrive[2], Elke Gruyaert[2], and Jiabin Li[1](✉)

[1] Department of Civil Engineering, KU Leuven, B-8200 Bruges, Belgium
jiabin.li@kuleuven.be

[2] Department of Civil Engineering, KU Leuven, B-9000 Ghent, Belgium

Abstract. Nowadays, industrially produced high-quality coarse recycled concrete aggregate (RCA) complying with NEN B 15-001 Type A + are available in Belgium. The Flemish precast concrete industry is interested in incorporating such RCA in concrete products to enhance competitive advantages. This paper provides a feasibility study of using the commercial RCA to develop high-performance concrete (HPC) for structural precast concrete elements. The coarse natural aggregate was partially or fully replaced by RCA with a similar particle size distribution, and the replacement percentages were 0%, 30%, 50% and 100% in volume. The physical and mechanical properties and durability of concrete were tested. The results indicated that the incorporation of RCA had different effects on the various properties of concrete. It is feasible to produce HPC with RCA when concrete is appropriately designed and produced.

Keywords: High-performance concrete · Recycled concrete aggregate · Early-age strength · Carbonation resistance · Freeze-thaw resistance with de-icing salts

1 Introduction

High-performance concrete (HPC) is defined as concrete meeting special combinations of performance and uniformity requirements that cannot always be achieved routinely using conventional constituents and normal mixing, placing, and curing practices [1]. These requirements for HPC typically involve qualified mechanical properties, high workability, high durability and high volume stability [2]. Carefully selecting, controlling, and proportioning all of the ingredients will achieve effective production of high-strength concrete [3], but this will not, by itself, ensure durable concrete. Appropriate placement practices and workmanship are essential to the production of durable concrete [4].

Recycled concrete aggregate (RCA) is defined as aggregate resulting from the processing of inorganic material previously used in construction, comprising a minimum of 90% crushed concrete [5]. Compared with natural aggregates, RCA generally has

J. I. Escalante-Garcia et al. (Eds.): RW 2021, RILEM Bookseries 40, pp. 983–993, 2023.
https://doi.org/10.1007/978-3-031-21735-7_104

lower purity, higher porosity and lower resistance to fragmentation, and therefore, it is questionable whether RCA can be used to develop HPC.

A few experimental investigations [6–12] have shown some feasibilities of producing HPC with RCA, and the following general conclusions have been drawn: (a) the mix design of concrete with RCA is very similar to natural aggregate concrete (NAC) [6]; (b) properties of concrete with RCA may be significantly improved by using mineral additions and chemical admixtures, similarly to cases of NAC [6, 11]; (c) the quality of RCA has significant influence on the mechanical properties of concrete [6–10], and concrete containing RCA derived from high-strength concrete (approximately 80–100 MPa) may achieve performance similar to NAC [8]; (d) the replacement percentage of natural aggregate significantly affects concrete properties [7, 10–12]; (e) an extra amount of water is required to compensate for the high water absorption of RCA [6]; (f) utilizing RCA in concrete has satisfactory workability, but the concrete with RCA will have slump-loss higher than NAC due to the absorption process of RCA [8]; (g) the compressive strength at an early age (24 h) is more dependent on the free water content during mixing rather than the use of RCA [7]. In addition to those findings, knowledge about the effect of high-quality RCA on structural precast concrete elements is rather limited. In other words, it is not clear whether the existing conclusions still apply to HPC with a strength class greater than C50/60.

This study aims to evaluate the feasibility of incorporating commercial high-quality coarse RCA in HPC for precast industry. Taking the findings above into account, the following approaches were adopted, such as using conventional HPC mix design, adding superplasticizer and compensation water, considering appropriate replacement percentage, and following proper mixing, placing and curing practices. For the HPC, structural precast concrete elements commonly require high early compressive strength for hosting, reaching 40 MPa in cubic specimens at the age of 14 h for example. When concrete is exposed to significant attack by freeze-thaw cycles whilst wet, such as precast concrete pavements and bridge decks in cold climates, high freeze-thaw resistance with de-icing salts (scaling) is required. In this work, a total of eight HPC mixes were produced and tested.

2 Experimental Program

2.1 Materials

Portland cement, CEM I 52.5 R conforming to NBN EN 197-1 [13], was used to obtain high early strength. The specific surface of the cement was 5270 cm^2/g. Superplasticizer, Sika ViscoCrete-4035M, was used to obtain a high workability. The used RCA was derived from demolished concrete pavement, produced using a jaw crusher followed by an impact crusher at a local recycling plant, and conformed to the Type A + category according to the Belgian standard NBN B 15-001 [14]. Figure 1 shows a photo of the RCA and Fig. 2 presents the particle size distributions of RCA and limestone. Table 1 summarizes the basic properties of the aggregates.

Fig. 1. Photo of 6.3/14 mm recycled concrete aggregate

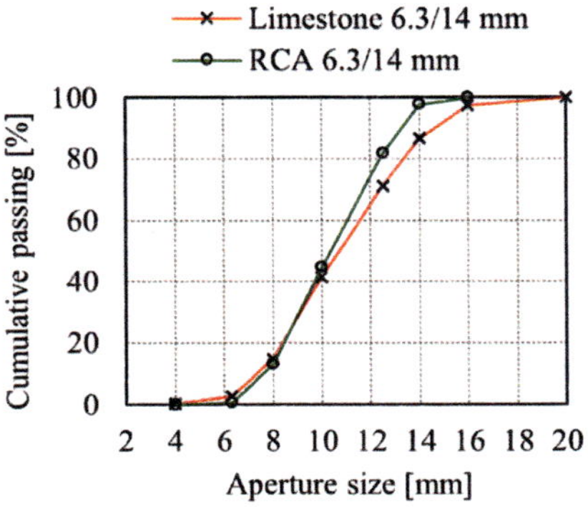

Fig. 2. Particle size distributions of recycled concrete aggregate (RCA) and limestone

2.2 Concrete Mixes and Specimen Preparations

According to ACI 211.4R [3] and NBN B 15-001 [14], eight concrete mixes were designed, as shown in Table 2. The moisture content of RCA was between 2.2% and 3.2%. This value is a function of the aggregate storage locations and the weather conditions on site [22]. A two-stage mixing approach [23] was adopted in the production of concrete. Cylindrical specimens Ø150 mm × 300 mm were used to determine the density, ultrasonic pulse velocity, compressive strength, elastic modulus and surface electrical resistivity. Prismatic specimens 100 mm × 100 mm × 500 mm were used to determine the carbonation resistance. Cylindrical specimens Ø113 mm × 50 mm were used to determine the freeze-thaw resistance with de-icing salts. The cylindrical specimens were cured in water at a temperature of (20 ± 2) °C until 28-day age. The prismatic specimens were cured in the climate room with the temperature of (20 ± 2) °C and the relative humidity of (80 ± 10) % until 7-day age, and then dried under indoor ambient conditions until 28-day age. Cubic specimens were cured in water until 7-day age and then dried under ambient conditions. At 21-day age, cylindrical specimens Ø113 mm

Table 1. Physical and mechanical properties of the aggregates

Property	Sand	Limestone	Limestone	RCA	Standard
d/*D* [mm]	0/4	2/6.3	6.3/14	6.3/14	NBN EN 933-1 [15]
FM	2.3	–	–	–	NBN EN 12620 [16]
FI [%]	–	–	14	10	NBN EN 933-3 [17]
Rc [%]	–	–	–	Rc_{90}	NBN EN 933-11 [18]
M_{DE} [%]	–	–	16	17	NBN EN 1097-1 [19]
LA [%]	–	–	17	24	NBN EN 1097-2 [20]
ρ_a [kg/m^3]	2600	2710	2680	2660	NBN EN 1097-6 [21]
ρ_{rd} [kg/m^3]	2580	2680	2630	2380	
ρ_{ssd} [kg/m^3]	2590	2690	2650	2470	
WA_{24} [%]	0.4	0.5	0.8	4.8	

d/*D*: Particle size; *FM*: Fineness modulus; *FI*: Flakiness index; *Rc*: Mass content of crushed concrete and crushed mortar in coarse recycled aggregates; M_{DE}: Micro-Deval coefficient; *LA*: Los Angeles coefficient; ρ_a: Apparent particle density; ρ_{rd}: Oven-dried particle density; ρ_{ssd}: Saturated and surface-dried particle density; WA_{24}: Water absorption

× 50 mm were drilled and sawed out of these cubic specimens, and then dried under ambient conditions. At 28-day age, the test surfaces of the cylindrical specimens were saturated with water. At 31-day age, the water was replaced by 3% NaCl solution, and the prepared specimens were placed into a freeze-thaw chamber.

2.3 Test

The consistency of the freshly mixed concrete was determined according to NBN EN 12350-2 [24]. The air content of freshly mixed concrete was determined according to NBN EN 12350-7 [25]. The density of hardened concrete was determined according to NBN EN 12390-7 [26]. The ultrasonic pulse velocity of concrete was determined according to NBN EN 12504-4 [27]. The compressive strength was determined according to NBN EN 12390-3 [28]. The secant modulus of elasticity was determined according to NBN EN 12390–13 [29]. The surface electrical resistivity was determined using Proceq Resipod. The resistance to carbonation was determined according to NBN EN 13295 [30]. The freeze-thaw resistance with de-icing salts was determined according to NBN B 15-100 [31], but the saw cut surface, instead of the side, was used as the test surface according to CEN/TS 12390-9 [32].

3 Experimental Results and Discussion

3.1 Physical Properties of Fresh and Hardened Concrete

For the slump test, according to visual inspections, all concrete cones remained substantially intact and symmetrical, and thus these slump tests were valid. The results are

Table 2. Mix proportions and design parameters

Component [kg/m³]	H40-0	H40-30	H40-50	H40-100	H30-0	H30-30	H30-50	H30-100
Cement	424	424	424	424	492	492	492	492
Water	170	170	170	170	148	148	148	148
Compensation *w*	10	18	23	37	10	18	23	37
Sand 0/4	758	758	758	758	758	758	758	758
Limestone 2/6.3	191	191	191	191	191	191	191	191
Limestone 6.3/14	763	534	381	0	763	534	381	0
RCA 6.3/14	0	207	345	691	0	207	345	691
Superplasticizer	1.9	1.9	1.9	1.9	4.4	4.4	4.4	4.4
w/c_{eff}	0.40	0.40	0.40	0.40	0.30	0.30	0.30	0.30
Paste [m³/m³]	0.30	0.30	0.30	0.30	0.30	0.30	0.30	0.30
Agg. [m³/m³]	0.66	0.66	0.66	0.66	0.66	0.66	0.66	0.66
RCA [m³/m³]	0	0.09	0.15	0.29	0	0.09	0.15	0.29

shown in Table 3. For the H40 series mixes, the concrete with RCA achieved similar or even higher slump than NAC. As reviewed by Verian et al. [33], due to the high water absorption, rough surface and irregular particle shape of RCA, the replacement of natural aggregate by RCA usually decreased the workability of concrete. However, in this study, a certain amount of water was already used to compensate the high water absorption of RCA. For the H30 series mixes, the slump varied largely, and the variation in fines content was considered one of the decisive parameters.

The measured air contents of fresh concrete are shown in Table 3. No clear relation between the RCA content and the air content was observed. According to ACI 318 [34], for the concrete considered in this study, the minimum entrained air content is required to be 4% to 6% to resist damage from freeze-thaw cycles.

The results of density and ultrasonic pulse velocity of hardened concrete are shown in Table 3. As expected, an increase in RCA content reduced the density and ultrasonic pulse velocity. The ultrasonic pulse velocity was not less than 4.50 km/s [7], indicating excellent performance for normal weight concrete (density of 2400 kg/m³).

3.2 Mechanical Properties of Concrete

The results of the compressive strength tests are presented in Table 4. For both series, the concrete with RCA obtained lower compressive strength than NAC, and the maximum strength loss was 8.0% when comparing H30-50 with H30-0, but H40-100 reached a similar compressive strength to H40-0. González-Fonteboa et al. [35] developed a database and reported that in most cases the use of coarse recycled aggregates decreased

Table 3. Physical properties of freshly mixed and hardened concrete at 28-day age

Notation	Slump class	Air content	No. of samples	Density		Ultrasonic pulse velocity	
				μ	σ	μ	σ
		[%]		[kg/m^3]		[km/s]	
H40-0	S4	3.9	9	2380	10	4.97	0.05
H40-30	S4	3.5	12	2360	10	4.89	0.03
H40-50	S5	3.7	12	2340	10	4.85	0.02
H40-100	S4	4.3	12	2330	10	4.50	0.03
H30-0	S4 [a]	4.2	12	2420	40	5.00	0.04
H30-30	S3 [b]	2.9	6	2400	10	4.99	0.02
H30-50	S3	3.2	12	2390	10	4.94	0.02
H30-100	S5	3.0	12	2370	10	4.85	0.01

[a] The second batch of concrete obtained S1 slump class, showing a low workability. After 15s of vibration, the fresh concrete liquefied and the top surface leveled, showing that the concrete was well compacted, so the produced specimens were retained for tests

[b] The first batch of concrete obtained no slump, and the fresh concrete did not liquefy after compaction, so this batch of concrete was rejected

the compressive strength, and in a few cases increased the strength. In the current study, the variation in the water absorption of RCA might cause a change in the real *w/c* ratio of concrete, resulting in a scatter in concrete strength.

Figure 3 presents the rate of compressive strength developed at 18 h to that at 28 days. According to EN 1992-1-1 [36], when CEM 52.5 R is used, the compressive strength of concrete at the age of 18-h is estimated to be 36% of the concrete strength at the age of 28 days, which was lower than the values in the current study. Gonzalez et al. [7] reported a similar case study on precast concrete in which cement dosage was only 380 kg/m^3, no mineral additions were used, and the slump was S1. Their concrete achieved about 55% ~ 75% of the 28-day strength at 1-day age.

The results of elastic modulus are shown in Table 4. For both series mixes, the concrete with RCA obtained lower elastic modulus than NAC, and the maximum decline was 7.8% when comparing H40-100 with H40-0, but H40-30 and H30-50 obtained similar values to NAC. Gonzalez et al. [7] reported a clear decline in the elastic modulus of concrete with RCA, and they commended that elastic modulus seemed to be more sensible than strength to the RCA content. Verian et al. [33] commented that the variation in the elastic modulus of concrete appeared to be lower compared to the compressive strength. The current study obtained a similar tendency but with a certain variation. In addition, the H40 series mixes showed relatively high values of elastic modulus. Figure 4 presents the test data for each test specimen, and estimations according to ACI 318 [34], CEB-FIP [37] and EN 1992-1-1 [36].

Table 4. Compressive strength and elastic modulus of concrete

Notation	$f_{cm,cyl-18h}$			$f_{cm,cyl-28d}$			$E_{c,s-28d}$		
	n	μ	Σ	n	μ	σ	n	μ	σ
		[MPa]			[MPa]			[GPa]	
H40-0	3	35.4	0.6	9	71.3	1.5	3	43.4	1.3
H40-30	6	31.8	2.6	12	66.7	2.7	6	43.9	1.6
H40-50	6	32.7	1.1	12	65.9	2.6	6	40.7	0.8
H40-100	6	36.1	2.3	12	70.2	4.6	3	40.0	0.9
H30-0	3	54.5	1.0	12	91.5	2.6	6	44.5	1.9
H30-30	3	55.3	0.3	6	85.6	2.9	3	42.9	0.8
H30-50	3	52.4	0.5	12	84.2	2.4	6	44.8	1.4
H30-100	3	54.1	0.8	12	85.4	6.8	6	43.0	1.7

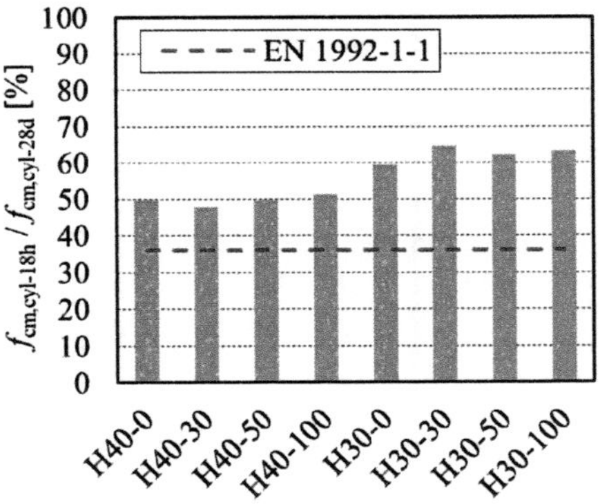

Fig. 3. Ratio of the compressive strength developed at 18 h to the compressive strength at 28 days

3.3 Durability of Concrete

The results of surface electrical resistivity are shown in Table 5. The higher the surface electrical resistivity, the lower the likelihood of corrosion. The H30 series mixes achieved higher resistivity than the H40 series mixes. It was likely due to the fact that a lower *w/c* ratio or a lower amount of mixing water resulted in less porosity of hardened concrete. This was consistent with the change in concrete density. For both series mixes, an increase in RCA content decreased the resistivity, and the maximum reduction was 7.6% when comparing H30-100 with H30-0. Langford et al. [38] proposed the following four corrosion risk levels: very high (0 ~ 5 kΩcm), high (5 ~ 10 kΩcm), moderate (10 ~ 20 kΩcm) and low (larger than 20 kΩcm). Gonzalez et al. [7] reported resistivity of about 10 ~ 33 kΩcm for HPC with or without RCA. Therefore, all the concrete mixes considered in the current study showed a low or moderate risk to corrosion.

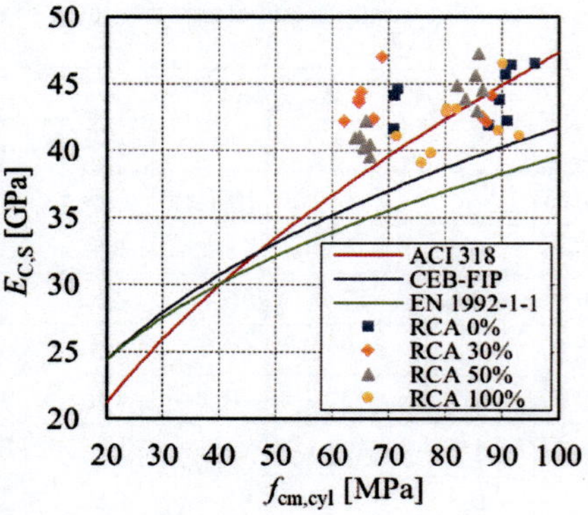

Fig. 4. Relationship between the secant elastic modulus and the cylindrical compressive strength

The results of carbonation depth are shown in Table 5. Both series mixes did not exhibit a clear carbonation front. Mixes of the H40 series had fade edges, whereas mixes of the H30 series had negligible edges, so the latter achieved higher resistance than the former. It was likely due to the fact that a lower *w/c* ratio or a higher amount of ordinary Portland cement resulted in less porosity and higher CO_2 buffer capacity in concrete. Sáez del Bosque et al. [39] adopted derivative thermogravimetry (DTG) to analyze the carbonation profile, and then compared with the carbonation depth determined by the phenolphthalein method. Both methods showed that the degree of carbonation in recycled aggregate concrete was higher than that in NAC. In the current study, the concrete with RCA showed comparable resistance to carbonation as NAC. A continuation of this test is necessary to draw further conclusions.

Table 5. Surface electrical resistivity, carbonation depth and salt scaling of concrete

Notation	ρ_s			$d_{k\text{-}56d}$		$d_{k\text{-}91d}$	s_{28}			s_{56}	
	n	μ	Σ	n	μ	μ	n	μ	Σ	μ	σ
		[kΩcm]			[mm]	[mm]		[kg/m^2]		[kg/m^2]	
H40-0	9	14.0	0.3	2	1.9	2.2	12	0.06 [a]	0.06	0.20 [a]	0.21
H40-30	12	13.4	0.3	2	2.0	2.1	6	0.18	0.13	0.26	0.16
H40-50	12	13.1	0.3	2	2.0	2.2	6	0.27	0.24	0.47	0.45
H40-100	6	12.6	0.7	2	2.0	2.3	6	0.07 [a]	0.10	0.12 [a]	0.13
H30-0	12	22.3	1.6	2	0	1.2	6	0.01	0.02	0.03	0.02
H30-30	6	22.1	0.3	2	0	0.6	6	0.06	0.04	0.09	0.04
H30-50	12	21.3	0.7	2	0	1.0	6	0.06	0.06	0.10	0.08
H30-100	12	19.4	1.3	2	0	0.9	6	0.15	0.04	0.23	0.04

[a] Test surfaces were formed surfaces

The results of freeze-thaw test with de-icing salt are shown in Table 5. The H30 series mixes obtained lower mass loss than the H40 series mixes. It was likely due to the fact that a lower *w/c* ratio or a lower amount of mixing water resulted in higher strength and less capillary pores in the hardened cement paste. For both series, the concrete with RCA obtained higher mass loss than NAC, except for H40-100 which obtained lower mass loss. Based on visual inspection, it was observed that the scaling was strongly related to the RCA exposed on the saw cut surface. In particular, the attached mortar and crushed mortar particles in the RCA were prone to crumbling and peeling off under freeze-thaw attack with deicing salts. In addition, all concrete mixes exhibited limited scaling. NBN EN 13877-2 [40] specifies three freeze-thaw resistance categories (FT0, FT1 and FT2) for concrete pavements cast in situ and recommends the tests in accordance with CEN/TS 12390-9 [32]. In fact, the two test methods specified in NBN B 15-100 [31] and CEN/TS 12390-9 [32] are based on the so-called slab test [41]. The main differences lie in the size of the test specimens, the test surface of specimens and the methods of evaluating the degree of freeze-thaw damage. In the current study, the salt scaling on the saw cut surface was studied, and all the concrete mixes were considered to achieve high salt scaling resistance and conform to the FT2 category.

4 Conclusions

A Belgian case study was carried out to evaluate the feasibility of using commercial high-quality coarse RCA to develop HPC for precast industry. The following conclusions were drawn:

Concrete with RCA achieved workability and air content comparable with NAC. However, fines present in the RCA may largely decrease the workability.

Concrete with RCA obtained density and ultrasonic pulse velocity lower than NAC.

Concrete with RCA showed 18-h compressive strength comparable with NAC, and 28-day compressive strength and elastic modulus lower than NAC.

Concrete with RCA obtained surface electrical resistivity lower than NAC carbonation resistance comparable with NAC, and freeze-thaw resistance with de-icing salts lower than NAC. The salt scaling mainly occurred at the attached mortar and crushed mortar particles present in the RCA.

Although the incorporation of RCA affects various properties of concrete, HPC can be developed when low water/cement ratio, superplasticizer, compensation water and two-stage mixing approach are considered.

Acknowledgement. The authors gratefully acknowledge the financial support of Lvnong Chair in Construction Waste Recycling at KU Leuven (No. VAR-LSLVNG-O2030) and China Scholarship Council (No. 201808110212). The authors sincerely thank O.B.B.C NV and Willy Naessens Group for supplying raw materials used in this study.

References

1. American Concrete Institute: ACI Concrete Terminology. ACI CT-13, Michigan (2013)

2. China Association for Engineering Construction Standardization: Technical specification for application of high performance concrete. CECS 207, Beijing (2006)
3. American Concrete Institute: Guide for Selecting Proportions for High-Strength Concrete Using Portland Cement and Other Cementitious Materials. ACI 211.4R, Michigan (2008)
4. American Concrete Institute: Guide to Durable Concrete. ACI 201.2R, Michigan (2016)
5. European Committee for Standardization: Concrete - Specification, performance, production and conformity. EN 206:2013+A1:2016, Brussels (2016)
6. Ajdukiewicz, A., Kliszczewicz, A.: Influence of recycled aggregates on mechanical properties of HS/HPC. Cement Concr. Compos. **24**, 269–279 (2002)
7. Gonzalez, A., Etxeberria, M.: Experimental analysis of properties of high performance recycled aggregate concrete. Constr. Build. Mater. **52**, 227–235 (2014)
8. Kou, S., Poon, C.: Effect of the quality of parent concrete on the properties of high performance recycled aggregate concrete. Constr. Build. Mater. **77**, 501–508 (2015)
9. Gonzalez-Corominas A., Etxeberria M., Poon C.: Influence of steam curing on the pore structures and mechanical properties of fly-ash high performance concrete prepared with recycled aggregates. Cement Concr. Compos. **71**, 77–84 (2016)
10. Gonzalez-Corominas, A., Etxeberria, M.: Effects of using recycled concrete aggregates on the shrinkage of high performance concrete. Constr. Build. Mater. **115**, 32–41 (2016)
11. Pedro, D., de Brito, J., Evangelista, L.: Evaluation of high-performance concrete with recycled aggregates: Use of densified silica fume as cement replacement. Constr. Build. Mater. **147**, 803–814 (2017)
12. Gonzalez-Corominas, A., Etxeberria, M., Fernandez, I.: Structural behavior of prestressed concrete sleepers produced with high performance recycled aggregate concrete. Mater. Struct. **50**, 1–14 (2017)
13. Bureau of Normalization: Cement -Part 1: Composition, specifications and conformity criteria for common cements. NBN EN 197–1, Brussels (2011)
14. Bureau of Normalization: Concrete – Specification, performance, production and conformity - National supplement to NBN EN 206:2013+A1:2016. NBN B 15-001, Brussels (2018)
15. Bureau of Normalization: Tests for geometrical properties of aggregates – Part 1: Determination of particle size distribution – Sieving method. NBN EN 933-1, Brussels (2012)
16. Bureau of Normalization: Aggregates for concrete. NBN EN 12620, Brussels (2013)
17. Bureau of Normalization: Tests for geometrical properties of aggregates – Part 3: Determination of particle shape – Flakiness index. NBN EN 933-3, Brussels (2012)
18. Bureau of Normalization: Tests for geometrical properties of aggregates – Part 11: Classification test for the constituents of coarse recycled aggregate. NBN EN 933-11, Brussels (2011)
19. Bureau of Normalization: Tests for mechanical and physical properties of aggregates – Part 1: Determination of the resistance to wear (micro-Deval). NBN EN 1097-1, Brussels (2011)
20. Bureau of Normalization: Tests for mechanical and physical properties of aggregates – Part 2: Methods for the determination of resistance to fragmentation. NBN EN 1097-2, Brussels (2020)
21. Bureau of Normalization: Tests for mechanical and physical properties of aggregates – Part 6: Determination of particle density and water absorption. NBN EN 1097-6, Brussels (2013)
22. Soares, D., de Brito, J., Ferreira, J., Pacheco, J.: Use of coarse recycled aggregates from precast concrete rejects: mechanical and durability performance. Constr. Build. Mater. **71**, 263–272 (2014)
23. Tam, V.W.Y., Gao, X.F., Tam, C.M.: Microstructural analysis of recycled aggregate concrete produced from two-stage mixing approach. Cem. Concr. Res. **35**, 1195–1203 (2005)
24. Bureau of Normalization: Testing fresh concrete – Part 2: Slump test. NBN EN 12350-2, Brussels (2019)

25. Bureau of Normalization: Testing fresh concrete – Part 7: Air content – Pressure methods. NBN EN 12350-7, Brussels (2019)
26. Bureau of Normalization: Testing hardened concrete – Part 7: Density of hardened concrete. NBN EN 12390-7, Brussels (2019)
27. Bureau of Normalization: Testing concrete – Part 4: Determination of ultrasonic pulse velocity. NBN EN 12504-4, Brussels (2004)
28. Bureau of Normalization: Testing hardened concrete – Part 3: Compressive strength of test specimens. NBN EN 12390-3, Brussels (2019)
29. Bureau of Normalization: Testing hardened concrete – Part 13: Determination of secant modulus of elasticity in compression. NBN EN 12390-13, Brussels (2014)
30. Bureau of Normalization: Products and systems for the protection and repair of concrete structures – Test methods – Determination of resistance to carbonation. NBN EN 13295, Brussels (2004)
31. Bureau of Normalization: Beton – Methodologie voor de evaluatie en attestering van de gebruiksgeschiktheid van cementen en van toevoegsels bestemd voor beton. NBN B 15-100, Brussels (2018)
32. Bureau of Normalization: Testing hardened concrete – Part 9: Freeze-thaw resistance with de-icing salts – Scaling. CEN/TS 12390-9, Brussels (2016)
33. Verian, K.P., Ashraf, W., Gao, Y.: Properties of recycled concrete aggregate and their influence in new concrete production. Resour. Conserv. Recycl. **133**, 30–49 (2018)
34. American Concrete Institute: Commentary on Building Code Requirements for Structural Concrete. ACI 318, Michigan (2014)
35. González-Fonteboa, B., Seara-Paz S., de Brito, J., González-Taboada, I., Martínez-Abella, F., Vasco-Silva, R.: Recycled concrete with coarse recycled aggregate. An overview and analysis. Mater. Construcción **68**(330), e151 (2018)
36. European Committee for Standardization: Eurocode 2: Design of concrete structures – Part 1–1: General rules and rules for buildings. EN 1992-1-1, Brussels (2004)
37. International Federation for Structural Concrete: fib Model Code for Concrete Structures. Switzerland (2010)
38. Langford, P., Broomfield, J.: Monitoring the corrosion of reinforcing steel. Constr. Repair **1**(2), 32–36 (1987)
39. Sáez del Bosque, I.F., Van den Heede, P., De Belie, N., Sánchez de Rojas, M.I., Medina, C.: Carbonation of concrete with construction and demolition waste based recycled aggregates and cement with recycled content. Constr. Build. Mater. **234**, 117336 (2020)
40. Bureau of Normalization: Concrete pavements – Part 2: Functional requirements for concrete pavements. NBN EN 13877-2, Brussels (2013)
41. RILEM TC 176-IDC: Slab test: Freeze/thaw resistance of concrete – Internal deterioration. Mater. Struct. **37**, 754–759 (2004)

Artificial Rocks Made from Dredged Sands of the Magdalen Islands (Canada): Preliminary Study

Amine El Mahdi Safhi[1,2,3] (✉), Patrice Rivard[3], Mahfoud Benzerzour[2], and Nor-Edine Abriak[2]

[1] Mohammed VI Polytechnic University, EMEC Program, Benguerir, Morocco
Amineelmahdi.safhi@um6p.ma
[2] Université de Lille, IMT Lille Douai, 4515 - LGCgE Douai, France
[3] Université de Sherbrooke, Département de Génie Civil, Sherbrooke, Canada

Abstract. The consumption of natural materials for construction purpose is increasing each year while the resources are limited. To reduce the environmental burden and efficiently optimize resources, the use of local and alternative resources as building materials is necessary. This solution is not always convenient, especially for the isolated places like islands, and the Magdalen Islands in Quebec are not an exception. Those islands are currently in a shortage of granular materials and must import them, which represents considerable economic and environmental costs. The general goal of this contribution is to take advantage of the local dredged sands, to be reused in civil engineering infrastructures. This study explores the feasibility of consolidated sandy sediments as an artificial rock, which may serve as mass elements for coastal protection against erosion.

Keywords: Artificial rocks · Dredged sand · Durability · Multivariable optimization · Waste management

1 Introduction

The Magdalen Islands (Fr: Iles-de-la-Madeleine), is a small archipelago of eight major islands located in the Gulf of Saint Lawrence. The erosion of their coastline is already a serious issue. Based on 4614 measurements alongside the 235 km coastline, and a data from 1963 to 2008, it was found that 81% of the coastline will be eroded by 2060 [1]. A significant acceleration in the rise in sea level was measured between 1964 and 2010, i.e., a rate of 3.5 mm/year according to tide data. The use of local dredged sand (DS) can have a significant economic interest and environmental impact [2]. Such valorization can only take place through a good knowledge of the physio-chemical, mineralogical, and environmental properties. After a detailed characterization, DS consolidation with cement was investigated to produce artificial rocks to could be used as protection against coastal erosion. Identifying an optimal mixture with a total incorporation of DS was determined based on the mechanical and physical performance. A study of the durability was performed on concrete samples to evaluate the resistance to deicer scaling and freeze-thaw cycles.

J. I. Escalante-Garcia et al. (Eds.): RW 2021, RILEM Bookseries 40, pp. 994–1003, 2023.
https://doi.org/10.1007/978-3-031-21735-7_105

2 Materials and Methods

2.1 Materials Characterizations

The DS were dredged by *Dragage IM Entreprise* from three different harbor: Grosse-Ile (GI), Pointe-Basse (PP), and Millerand (M). The loss of weight at 550 °C measured on six samples ranged from 1.1–11% ($\bar{x} = 4.4\%$). The lowest value was obtained by the DS of PP ($\bar{x} = 1.4\%$), with an apparent density of 2.66 kg/m^3. A General Use cement (GU) was used with a Blaine specific surface area (SSA) of 396 m^2/kg. Table 1 represents the chemical properties of the cement and DS. It can be seen that DS are mainly composed of SiO_2. X-ray diffraction analyses revealed the presence of quartz, clinochlore, microcline, blodite, dolomite, calcite, halite, and Gehlenite (Fig. 1).

Table 1. Physio-chemical characteristics of the used materials.

Oxide, %	Al_2O_3	CaO	Fe_2O_3	MgO	SO_3	SiO_2	TiO_2	K_2O	Na_2O	LOI
Cement	4.90	63.2	2.20	1.90	3.40	20.0	–	0.40	0.60	2.70
DS	2.20	8.90	0.70	0.80	0.20	79.9	0.06	0.60	0.80	5.00

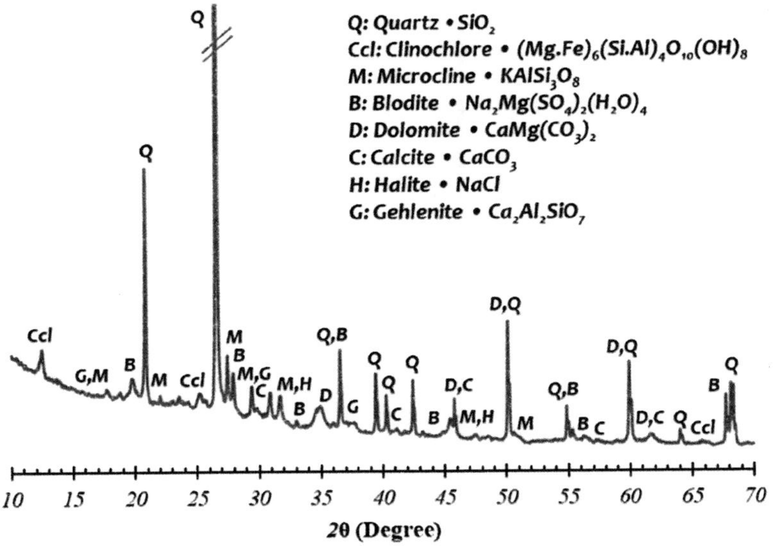

Fig. 1. XRD analyses of the three DS.

Figure 2 represents the particle size distribution of two samples of each DS. A very tight and poorly graded particle size was found. The mean of the six samples has a curvature coefficient of 1.0, a uniformity coefficient of 1.3, and a fineness modulus of 1.1, while the ASTM C33 [3] recommends a value between 2.3 and 3.1. The SEM observation shows that DS has an irregular and spherical morphology.

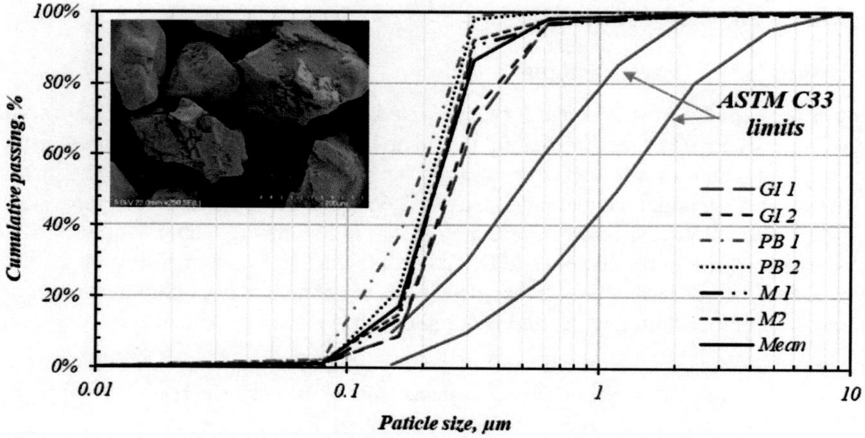

Fig. 2. Particle size distribution and a SEM observation of the DS.

2.2 Experimental Methods

For the following experimental tests, only the DS from Pointe-Basse were used. A factorial design was developed by varying three factors as illustrated in Fig. 3. Exactly 35 mixtures were generated, prepared, and cast in 50 mm^3 molds. The samples were demolded after 24 h and kept on a controlled temperature chamber at 20 °C until the age of testing. The dry density, electrical resistivity (ASTM C1760 [4]), and compressive strength (ASTM C39 [5]) were measured at 7 and 28 days of curing. Additional experimental tests were performed on the optimized mixture including capillary absorption (ASTM C642 [6]), and accessible porosity (NF P18–459 [7]).

The study of durability was performed on concrete (Table 2), where an entrained air agent was used with a dosage of 15 ml/100 kg to cement. Deicer scaling test (ASTM C672 [8]) which determine the scaled resistance of a horizontal concrete surface subjected to freeze-thaw cycles, where "calcium chloride" de-icing chemicals is added, was performed on two plates of 250 × 250 × 75 mm. The freeze-thaw resistance test (ASTM C666 [9]) was performed on two beams of 75 × 75 × 350 mm.

3 Results and Discussion

3.1 Mortar Optimization

Figures 4, 5, and 6 represent the evolution of the apparent density, the electrical resistivity, and the compressive strength, respectively, at 7 and 28 days. Increasing the incorporation rate of DS from 0 to 100%, decreases the apparent density by 300 g/m^3. This is due to their lower density compared with the reference sand. Increasing the *w/c* ratio increased the apparent density. It is inferred that the volume of water promotes the rearrangement of the particle, leading to higher wet density.

Figure 5 shows that the electrical resistivity increased between 7 and 28 days, due to the formation of hydrates that refined the pore structure. Increasing both DS content

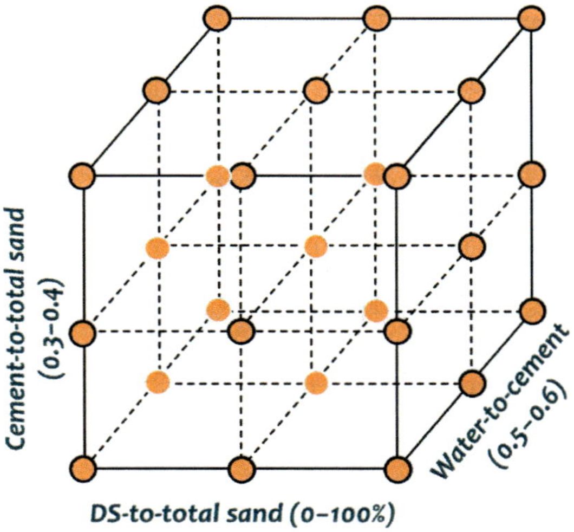

Fig. 3. Design of experiments: factorial design

Table 2. Mix design of the concrete.

Composition	Cement	*Water*	CA_1, 10–20 mm	CA_2, 5–10 mm	CA_3, 10–14 mm	DS
Mass, kg/m^3	350	210	290	587	286	875

and w/c ratio decreased the electrical resistivity. The same tendency was observed with respect to the evolution of the compressive strength (Fig. 6). Increasing DS content decreases the strength alongside with the increase of *w/c* ratio. Similar results were founded in previous studies [10].

A correlation between the founded results (Fig. 7) indicated a strong correlation between the apparent density and the compressive strength, and a moderate correlation between the latter and the electrical resistivity, and between the apparent density and the electrical resistivity.

3.2 Mortar Mix Design Optimization

In order to optimize the mix design, two constraints have been set: a minimum electrical resistivity of 15 and 20 kΩ.cm at 7 and 28 days respectively, and a minimum compressive strength of 25 and 30 MPa was targeted at 7 and 28 days, respectively. This optimization is depicted in Fig. 8; the yellow area represents the response surface. In order to maximize the incorporation of DS, the mixture that led to a total incorporation was chosen. These mix design corresponds to cement-to-total sand content of 0.4 and a *w/c* ratio of 0.6.

To ensure the reproducibility of the results, additional tests were carried out on this optimal mixture (100 Ø 200 mm). Table 3 shows that the targeted strengths were

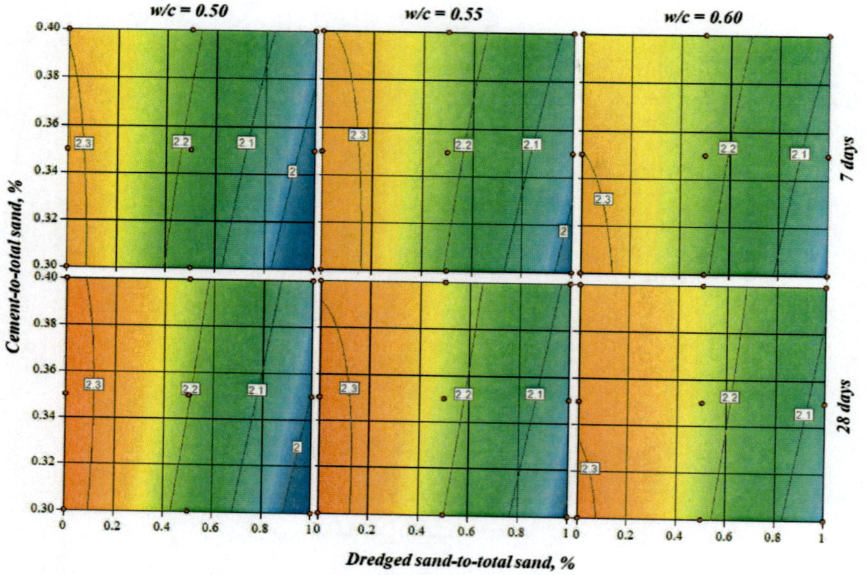

Fig. 4. Evolution of the apparent density (kg/m^3) at 7 and 28 days for the different *w/c* ratios.

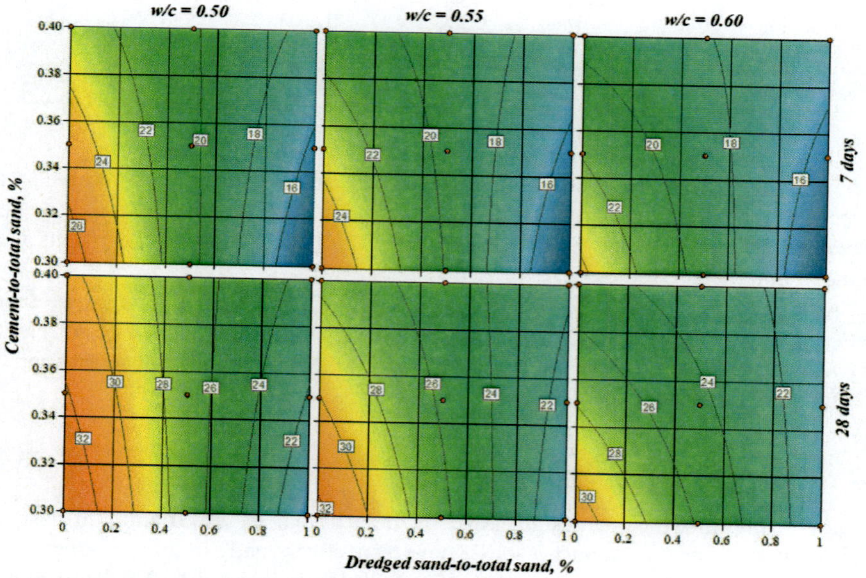

Fig. 5. Evolution of electrical resistivity (kΩ.cm) at 7 and 28 days for the different *w/c* ratios.

achieved. The electrical resistivity and the compressive strength increased with curing time, while capillary absorption and permeable pores decreased. The accessible water

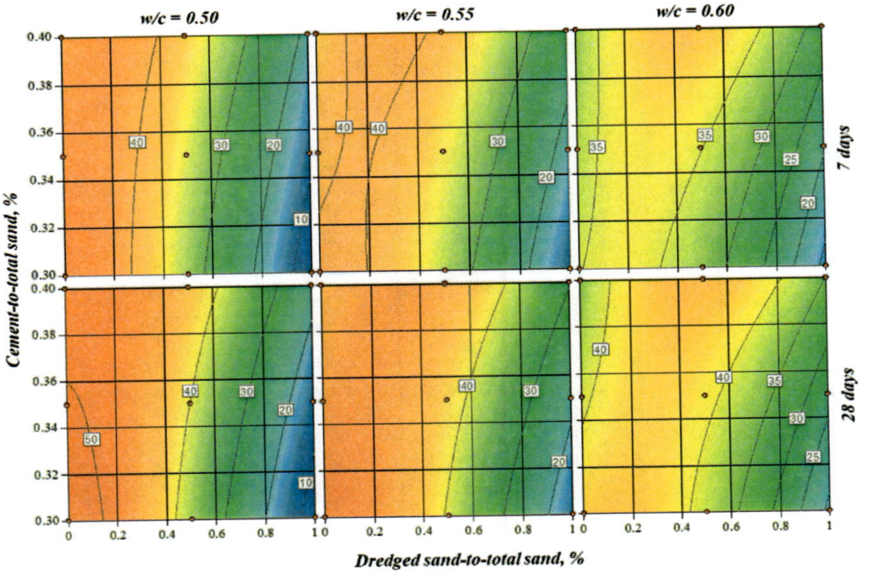

Fig. 6. Evolution of compressive strength (MPa) at 7 and 28 days for the different *w/c* ratios.

porosity for all the curing age was around 25%, this could be related to the discontinuous granulometry of DS.

SEM observations were conducted on faction of mortar at 7 days (Fig. 9) and 91 days (Fig. 10). Large quantities of ettringite are observed, which is associated with many factors, such as temperature, humidity, and variation in calcium sulfate. It can be noticed that the structure is disordered.

3.3 Durability Study on Concrete

The fresh properties of the concrete can be summarized as follows: a slump flow of 180 mm, which corresponds to S4 consistency class indicating high workability, a fresh density of 2196 kg/m^3, with an air content of 5.8%. The properties of the hardened concrete are presented in Table 4. The compressive strength was lower than expected with a maximum value of ~15 MPa at 56 days. An electrical resistivity of 32 KΩ.cm was measured at the same age. It was noted during the mixing that the solid concentration was low, which influenced the properties of the hardened concrete. This behavior can be associated with poor grain size class of the DS. It was noticed that there was a weak cohesion between the DS and the gravel in the hardened mortars. To improve the performance of concrete, a granular correction is required by adding coarser aggregates.

Results of the freeze-thaw resistance test are presented in Fig. 11. The transverse resonant frequency, the change in length, and the change in mass were measured at several times, as well as the variation in the dynamic modulus of elasticity. Until the 25th cycle, no significant mass loss was observed. At the end of the 60th cycle, a gradual increase in mass loss (7%) was then observed. This increase continues slightly until reaching 8%

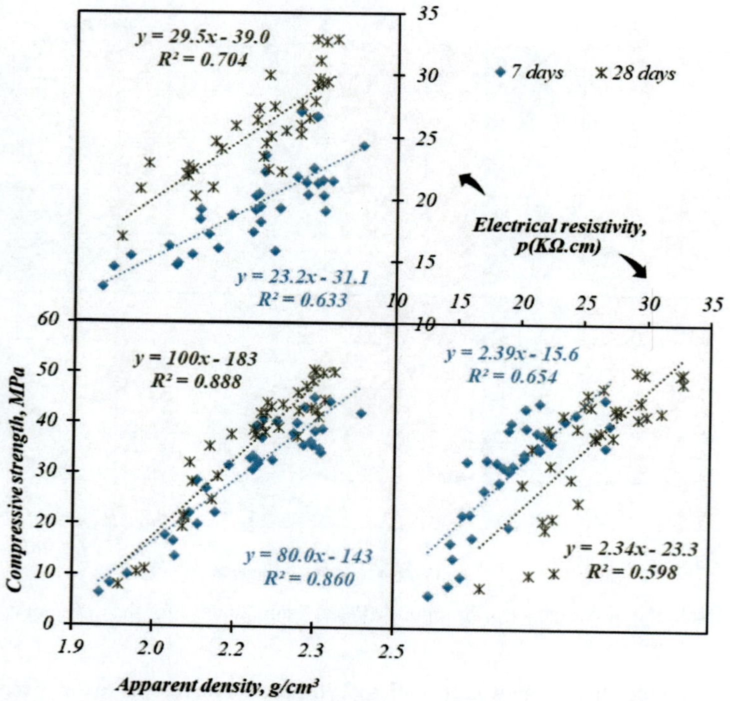

Fig. 7. Correlation between compressive strength, electrical resistivity, and apparent density.

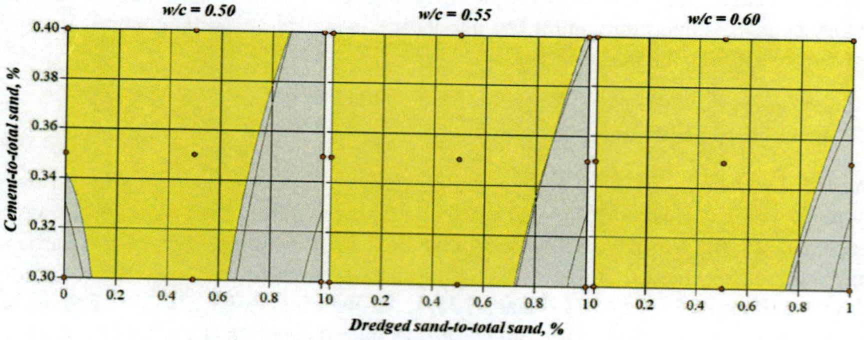

Fig. 8. Mix design optimization.

after 87 cycles, then gradually increasing until reaching 14% of the beam's mass after 164 cycles. The intensity of the degradation of the beams is relative to the intensity of the cycles and this explains the decrease in the dynamic modulus which reaches around 50% at 164 cycles. Concrete made with DS did not show good performance with respect

Table 3. Properties of the optimal mixture (Average ± SD).

Age	Density, kg/m^3	Resistivity, kΩ.cm	Compressive strength, MPa	Accessible water porosity, %	Capillary absorption, %	Permeable pores, %
7	2047 ± 4.0	14.6 ± 0.3	23.0 ± 0.5	25.7 ± 0.9	12.8 ± 0.4	23.9 ± 0.3
14	2030 ± 21	17.4 ± 0.2	23.3 ± 0.7	22.2 ± 0.9	12.7 ± 0.4	23.6 ± 0.7
28	2035 ± 14	20.4 ± 0.2	27.2 ± 0.3	25.8 ± 0.4	12.1 ± 0.8	23.2 ± 1.2
56	2064 ± 9.0	24.0 ± 0.2	29.6 ± 0.7	24.4 ± 1.1	11.9 ± 0.7	22.2 ± 1.0

Fig. 9. SEM Observations on a mortar fracture at 7 days.

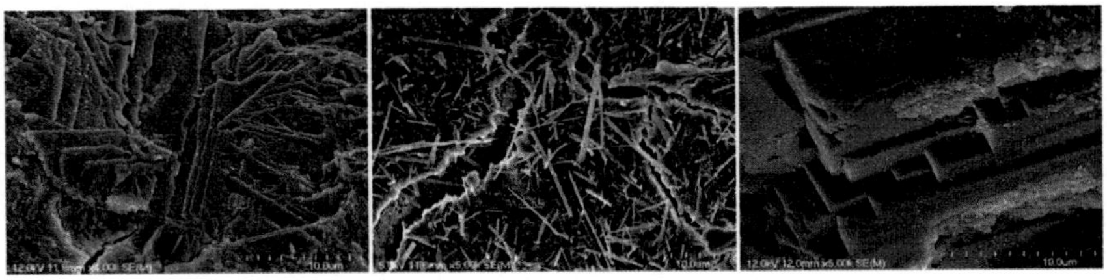

Fig. 10. SEM Observations on a mortar fracture at 91 days.

Table 4. Hardened properties of the concrete (Average ±SD).

Age	Density, kg/cm^3	Electrical resistivity, kΩ.cm	Compressive strength, MPa
7 days	2275 ± 12	22.8 ± 0.7	7.01 ± 0.1
28 days	2257 ± 18	28.1 ± 1.4	10.9 ± 0.6
56 days	2302 ± 11	32.0 ± 1.0	14.8 ± 0.1

to external climatic attack. This is mainly due to the high porosity thus higher embanked water.

Results of deicer scaling test are presented in Fig. 12. The formation of gels a few mm below the concrete surface causes internal pressures. The degradation of the plates is then observed through spalling of the concrete surface. The mass loss due to the attack

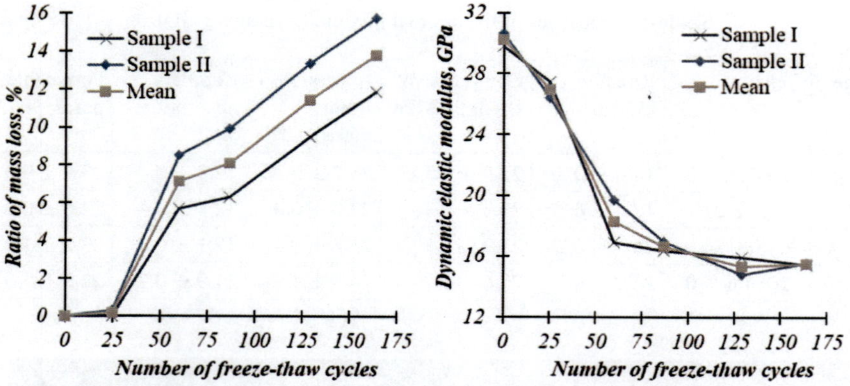

Fig. 11. Freeze-thaw cycles: mass loss and dynamic elastic modulus evolution.

of calcium chlorides on the plate surface after freeze-thaw cycles was measured. Starting from the 10th cycle, a slight to moderate scaling started to occur. A moderate scaling was noticed at the 25th cycle. The vulnerability of the damage is related to the formulation of the concrete which is influenced by the non-continuous granulometry of the DS. The test was stopped after 25 cycles because the plates were scratched as could not withstand additional freeze/thaw cycles. This concrete based on DS has a poor performance against freeze/thaw cycles.

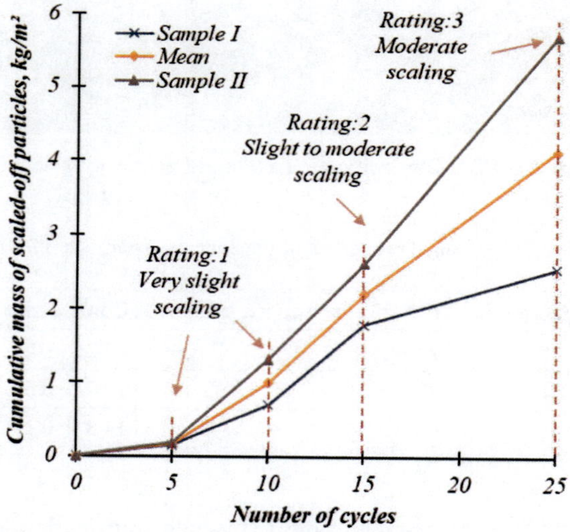

Fig. 12. Cumulative mass of scaled-off particles due to the deicer scaling.

4 Conclusion

This research is part of a recovery vision seeking to provide an alternative to the granular material shortage in the Magdalen Islands. The mineralogical and physical characterization of the DS from the Magdalen Islands show a great potential for upgrading as found in the mortar study. The poor performance of the cementitious products produced is due to the granular discontinuity of the DS. It is strongly recommended to correct the particle size of these sands by incorporating coarser materials which will improve the compactness of the matrix by reducing the porosity. Due to the low resistance to freeze/thaw cycles and scaling, the elements produced could be used in immersion where the thermal variation is of low importance. Overall, the reuse of those DS will have a great economic and environmental added value locally.

Acknowledgements. This study has been funded by the *Fonds québécois de la recherche sur la nature et les technologies* (FRQNT). The contribution of the *Centre de recherche sur les milieux insulaires et maritimes* (CERMIM), in particular Marc-Olivier Masse, is also acknowledged. The authors wish to thank *Dragage IM* for providing the dredged sediments.

References

1. Bernatchez, P., Drejza, S., Dugas, S.: Marges de sécurité en érosion côtière : évolution historique et future du littoral des îles de la Madeleine. (Rapport remis au ministère de la Sécurité publique du Québec) (p. 71). Laboratoire de dynamique et de gestion intégrée des zones côtières, Université du Québec à Rimousk (2012)
2. Safhi, A.E.M. (ed.).: Valorization of Dredged Sediments as Sustainable Construction Resources. CRC Press, Boca Raton (2022). https://doi.org/10.1201/9781003315551
3. ASTM C33, C09 Committee: Specification for concrete aggregates. ASTM International (2018). https://doi.org/10.1520/C0033_C0033M-18
4. ASTM C1760. (2012). Test Method for Bulk Electrical Conductivity of Hardened Concrete. ASTM International. https://doi.org/10.1520/C1760-12
5. ASTM C39. (2018). Test method for compressive strength of cylindrical concrete specimens. ASTM International. https://doi.org/10.1520/C0039_C0039M-18
6. ASTM C 642, C09 Committee. (2013). Test method for density, absorption, and voids in hardened concrete. ASTM International. https://doi.org/10.1520/C0642-13
7. AFNOR. (2010, March). NF P18-459: Concrete - Testing hardened concrete - Testing porosity and density
8. ASTM C672, C09 Committee: Test method for scaling resistance of concrete surfaces exposed to deicing chemicals. ASTM International (2012). https://doi.org/10.1520/C0672_C0672M-12
9. ASTM C666, C09 Committee: Test method for resistance of concrete to rapid freezing and thawing. ASTM International (2015). https://doi.org/10.1520/C0666_C0666M-15
10. el Mahdi Safhi, A., Benzerzour, M., Rivard, P., Abriak, N.E., Ennahal, I.: Development of self-compacting mortars based on treated marine sediments. J. Build. Eng. **22**, 252–261 (2019). https://doi.org/10.1016/j.jobe.2018.12.024